OSMOTIC
AND
IONIC
REGULATION

Cells and Animals

OSMOTIC AND IONIC REGULATION

Cells and Animals

Edited by
David H. Evans

CRC Press
Taylor & Francis Group
Boca Raton   London   New York

CRC Press is an imprint of the
Taylor & Francis Group, an **informa** business

CRC Press
Taylor & Francis Group
6000 Broken Sound Parkway NW, Suite 300
Boca Raton, FL 33487-2742

ISBN 13: 978-0-367-45250-6 (pbk)
ISBN 13: 978-0-8493-8030-3 (hbk)

Library of Congress Cataloging-in-Publication Data

Osmotic and ionic regulation : cells and animals / David H. Evans.
    p. cm.
  Includes bibliographical references.
  ISBN 978-0-8493-8030-3 (alk. paper)
  1. Osmoregulation. 2. Cells--Physiology. 3. Ions--Physiological transport. I. Evans, David H. (David Hudson), 1940-

QP90.6.O87 2008
572'.3--dc22   2008029540

Visit the Taylor & Francis Web site at
http://www.taylorandfrancis.com

and the CRC Press Web site at
http://www.crcpress.com

Dedication

To Bill Potts and Gwyneth Parry,
who wrote the book
that started my career.

For my family, especially Jean.

Contents

Foreword

One of the aphorisms of the pioneer French physiologist Claude Bernard might be paraphrased as: "The regulation of its internal medium frees an animal from its external environment." The subject of osmotic and ionic regulation is the salt and water content of the internal environment. Protists perform the regulation at the surface of the cell. More efficiently, metazoa devote a small part of their surfaces to the task, providing a more stable medium for their cells and, in effect, carrying a fossil ocean within. As a result, animals have been able to adapt to fresh waters (even rain water), acid and alkaline waters, saturated salt solutions, and deserts.

The comparative physiology of osmotic and ionic regulation began with measurements of the concentrations of blood or plasma of animals in a variety of conditions. This inevitably led to the study of the regulation of salt and water movements across the body surface. Before the introduction of isotopes, such movements could only be demonstrated by upsetting the equilibrium and measuring the rate of restoration, which required very precise measurements. Nevertheless, Homer Smith was able to demonstrate in this way that marine teleosts drank seawater and excreted the salt. The introduction of isotopes transformed the subject and made it possible to measure salt and water movements when the animal was in equilibrium. It was soon discovered that the fluxes of water and ions were often much greater than the simple models had indicated.

As a wider range of animals was examined, it was found that similar mechanisms had evolved many times independently, and unsuspected excretory and secretory organs were discovered—from orbital glands in birds and reptiles to epipodites in crustacea. As is often the case, although comparative studies disclosed an ever-increasing number of regulatory systems, further work showed that at the molecular level the range of transport mechanism was more limited. Pores, gates, and ion transporters were once hypotheses, but advances in molecular biology have made it possible to analyze them down to the atomic detail.

The regulation of the concentrations of ions and of water volumes is usually very complex. A single on or off control produces a fluctuating concentration. Several overlapping control mechanisms, with complex feedback loops, result in more stable regulation. Intensive studies of the mammal have discovered a surprisingly complex system, involving several interacting hormones, that regulates the volume and composition of mammalian plasma. The mammals may be unusually sophisticated animals, but we have a long way to go before we have fully analyzed the regulatory systems of other animals.

One is always astounded by the elegance and complexity of the solutions that natural selection has developed to what might seem, at first sight, to be insoluble problems. Leaf-eating insects have an almost sodium-free diet, although sodium is essential for nerve function. By concentrating the traces of sodium available around the nerves and making up the osmotic pressure of the hemolymph with potassium and organic acids caterpillars thrive. Although water cannot be actively transported, it is routinely manipulated by creating osmotic gradients. Some animals can even extract water vapor from the atmosphere in this way. Similarly, although mammals have not evolved the ability to actively transport urea, the mammalian kidney produces a concentrated salt solution by active transport and then uses this to concentrate urea. No doubt similar mechanisms remain to be discovered.

Apart from providing a suitable medium for the tissues, the composition of the internal media in marine animals may provide buoyancy in pelagic marine animals. A reduction in the concentrations of sulfate and magnesium ions in seawater results in a reduction in the density of the solution, even when it is maintained isosmotic with seawater. This was first demonstrated in jellyfish but may be used more widely in many pelagic larvae.

It is always dangerous to attempt to predict where the next advances may be made. We still know little of the function of vesicles in the bulk transport of water and ions, but new techniques in microscopy and fluorescence look promising.

It is almost 70 years since August Krogh wrote his pioneering book, *Osmotic and Ionic Regulation in Aquatic Animals*, and over 40 since Potts and Parry's *Osmotic and Ionic Regulation in Animals* appeared. It is indicative of the growth of the subject that as time passes the number of authors needed to review the subject grows exponentially. The time is ripe for a new survey of the subject, and Dr. Evans is to be congratulated on the expert crew that he has recruited.

W.T.W. Potts

Preface

I read Potts and Parry's *Osmotic and Ionic Regulation in Animals*[13] when it was published in 1964, and I was hooked. The structured and elegant presentation made clear where the study of osmoregulation was at that time and where it needed to go. It described the classic experiments of Croghan, Forster, Lockwood, Keys, Krogh, B. and K. Schmidt-Nielsen, Shaw, Smith, Ramsay, Robertson, Ussing, Wigglesworth, and others that established the basic patterns of osmoregulation in both invertebrate and vertebrate animals. I was lucky enough to spend the summer of 1965 working with Bill Potts in Ladd Prosser's training course at Woods Hole and then secured an NIH PostDoc to continue studying with Bill at the University of Lancaster from 1967 to 1968. I was also fortunate to spend 3 months in late 1968 in Villefranche-sur-mer, France, in the lab of Jean Maetz. Jean had probably the largest and most active group working in osmoregulation at that time. His life was tragically cut short by an automobile accident in 1977.

In the intervening years, various chapters and volumes have reviewed osmoregulation in cells and specific animal groups,[1–12,14–16] but no treatment of this depth and breadth has appeared since 1979. For the past 40 years I have thought about updating the original Potts and Parry, but I never found the time to put together a review of a field that was growing so rapidly. I had intended to spend the first few years of retirement reviewing the literature and writing the book, but it immediately became obvious that one person could not properly review osmoregulation in cells and animals—hence, this edited volume. The authors have been recruited from four continents and for their relative longevity in their respective areas of expertise, as well as their writing styles. In so doing, I hoped to generate a volume that was current, well-organized, and of interest to others in this field, as well as colleagues and students in comparative and general physiology.

REFERENCES

1. Gilles, R., *Mechanisms of Osmoregulation in Animals*, John Wiley & Sons, Chichester, 1979.
2. Gilles, R., *Animals and Environmental Fitness*, Pergamon Press, Oxford, 1980.
3. Greger, R., Ed., *Comparative and Environmental Physiology*. Vol. 1. *NaCl Transport in Epithelia*, Springer-Verlag, Berlin, 1988.
4. Gupta, B.L., Moreton, R.B., Oschman, J.L., and Wall, B.J., *Transport of Ions and Water in Animals*, Academic Press, London, 1977.
5. Hoar, W.S. and Randall, D.J., *Fish Physiology*, Vol. X, Academic Press, Orlando, FL, 1984.
6. House, C.R., *Water Transport in Cells and Tissues*, Edward Arnold, London, 1974.
7. Jorgensen, C.B. and Skadhauge, E., *Osmotic and Volume Regulation*, Munksgaard, Copenhagen, 1978.
8. Kirschner, L.B., Water and ions. In *Comparative Animal Physiology*, 4th ed., Prosser, C.L., Ed., Wiley-Liss, New York, 1991, pp. 13–107.
9. Lahlou, B., *Epithelial Transport in the Lower Vertebrates*, Cambridge University Press, Cambridge, U.K., 1980.
10. Malins, D.C. and Sargent, J.R., *Biochemical and Biophysical Perspectives in Marine Biology*, Academic Press, London, 1974.
11. Maloiy, G.M.O., *Comparative Physiology of Osmoregulation in Animals*, Academic Press, London, 1979.
12. Pequeux, A., Gilles, R., and Bolis, L., *Osmoregulation in Estuarine and Marine Animals*, Springer-Verlag, Berlin, 1984.

13. Potts, W. T. W. and Parry, G., *Osmotic and Ionic Regulation in Animals*, Macmillan, New York, 1964.
14. Rankin, C. and Davenport, J.A., *Animal Osmoregulation*, John Wiley & Sons, New York, 1981.
15. Schmidt-Nielsen, K., Bolis, L., and Maddrell, S.H.P., *Comparative Physiology: Water, Ions and Fluid Mechanics*, Cambridge University Press, Cambridge, U.K., 1978.
16. Wood, C.M. and Shuttleworth, T.J., Cellular and molecular approaches to fish ionic regulation. In *Fish Physiology*, Hoar, W.S., Randall, D., and Farrell, A.P., Eds., Academic Press, San Diego, CA, 1995.

The Editor

David H. Evans, PhD, is professor emeritus of Zoology at the University of Florida. Dr. Evans received his AB in Zoology from DePauw University, Indiana, in 1962 and his PhD in Biological Sciences from Stanford University, California, in 1967. He held postdoctoral positions in the Biological Sciences Department of the University of Lancaster, United Kingdom, and the Groupe de Biologie Marine du C.E.A., Villefranche-sur-mer, France, in 1967 and 1968. In 1969, he joined the Department of Biology at the University of Miami, Florida, as assistant professor and served as professor and chair from 1978 to 1981, when he became professor of Zoology at the University of Florida. He served as chair of that department from 1982 to 1985 and 2001 to 2006. He also held affiliate professorships in the departments of Fisheries and Aquatic Sciences and Physiology and Functional Genomics at the University of Florida.

Dr. Evans served as Director of the Mount Desert Island Biological Laboratory (MDIBL), Salisbury Cove, Maine, from 1983 to 1992, as well as Director of the MDIBL's Center for Membrane Toxicity Studies from 1985 to 1992. He served on a White House Office of Science and Technology Policy Acid Rain Peer Review Panel from 1982 to 1984 and on the Physiology and Behavior Panel of the National Science Foundation from 1992 to 1996. He is a member of the Society for Experimental Biology, Sigma Xi, and the American Physiological Society. He has served on the editorial boards of *The Biological Bulletin, Journal of Experimental Biology, Physiological and Biochemical Zoology, Journal of Experimental Zoology,* and *Journal of Comparative Physiology.* He is currently serving on the editorial board of the *American Journal of Physiology (Regulatory, Comparative, Integrative Physiology).*

Dr. Evans received the University of Miami Alpha Epsilon Delta Premedical Teacher of the Year Award in 1974, the University of Florida and College of Liberal Arts and Sciences Outstanding Teacher Awards in 1992, the University of Florida Teacher–Scholar of the Year Award in 1993, the Florida Blue Key Distinguished Faculty Award in 1994, the University of Florida Professorial Excellence Program Award in 1996, the University of Florida Chapter of Sigma Xi Senior Research Award in 1998, and a University of Florida Research Foundation Professorship in 2001. In 1999, he was elected a Fellow of the American Association for the Advancement of Science. In 2008, he was awarded the August Krogh Lectureship of the Comparative and Evolutionary Physiology Section of the American Physiological Society.

Dr. Evans has presented over 75 invited lectures and seminars, has published nearly 120 papers and book chapters and 160 abstracts, and has edited three editions of *The Physiology of Fishes* for CRC Press. He has been the recipient of research grants from the National Science Foundation, the National Institute of Environmental Health Sciences, and the American Heart Association. His current research interests center on the hormonal and paracrine control of fish gill perfusion and epithelial transport.

The Editor

Contributors

Richard D. Allen, PhD
Pacific Biosciences Research Center
University of Hawaii
Honolulu, Hawaii

Klaus W. Beyenbach, PhD
Department of Biomedical Sciences
Cornell University
Ithaca, New York

S. Donald Bradshaw, PhD
Emeritus Professor
Senior Honorary Research Fellow
School of Animal Biology
The University of Western Australia
Crawley, Perth, Australia

Eldon J. Braun, PhD
Department of Physiology, College of Medicine
University of Arizona
Tucson, Arizona

Guy Charmantier, PhD
Adaptation Ecophysiologique et Ontogenèse
Université Montpellier II
Montpellier, France

Mireille Charmantier-Daures, PhD
Adaptation Ecophysiologique et Ontogenèse
Université Montpellier II
Montpellier, France

Keith Choe, PhD
Anesthesiology Research Division
Vanderbilt University Medical Center
Nashville, Tennessee

James B. Claiborne, PhD
Department of Biology
Georgia Southern University
Statesboro, Georgia

William H. Dantzler, MD, PhD
Department of Physiology, College of Medicine
University of Arizona
Tucson, Arizona

David C. Dawson, PhD
Department of Physiology and Pharmacology
Oregon Health and Science University
Portland, Oregon

Lewis Deaton, PhD
Biology Department
University of Louisiana
Lafayette, Louisiana

David H. Evans, PhD
Department of Zoology
University of Florida
Gainesville, Florida, and
Mount Desert Island Biological Laboratory
Salisbury Cove, Maine

Stanley D. Hillyard, PhD
School of Dental Medicine
University of Nevada
Las Vegas, Nevada

Rolf K.H. Kinne, MD, PhD
Max Planck Institute for Molecular Physiology
Dortmund, Germany

Erik Hviid Larsen, PhD
Department of Biology
August Krogh Institute
Copenhagen, Denmark

Xuehong Liu, PhD
Department of Physiology and Pharmacology
Oregon Health and Science University
Portland, Oregon

Nadja Møbjerg, PhD
Department of Biology
August Krogh Institute
Copenhagen, Denmark

Yutaka Naitoh, PhD
Department of Neurophysiology
Kagawa School of Pharmaceutical Sciences
  and Institute of Neuroscience
Tokushima Bunri University
Kagawa, Japan, and
Institute of Biological Sciences
University of Tsukuba
Tsukuba, Japan

Peter M. Piermarini, PhD
Department of Biomedical Sciences
Cornell University
Ithaca, New York

Robert L. Preston, PhD
Department of Biological Sciences
Illinois State University
Normal, Illinois, and
Mount Desert Island Biological Laboratory
Salisbury Cove, Maine

Kevin Strange, PhD
Departments of Anesthesiology and Molecular
  Physiology and Biophysics
Vanderbilt University Medical Center
Nashville, Tennessee

Shigeyasu Tanaka, PhD
Graduate School of Science and Technology
Shizuoka University
Suruga-ku, Shizuoka, Japan

Takashi Tominaga, PhD
Department of Neurophysiology
Kagawa School of Pharmaceutical Sciences
  and Institute of Neuroscience
Tokushima Bunri University
Kagawa, Japan

David Towle, PhD
Mount Desert Island Biological Laboratory
Salisbury Cove, Maine

Mark L. Zeidel, MD
Department of Medicine
Beth Israel Deaconess Medical Center
Harvard Medical School
Boston, Massachusetts

1 Osmoregulation: Some Principles of Water and Solute Transport

David C. Dawson and Xuehong Liu

CONTENTS

I. INTRODUCTION: MAINTAINING THE NONEQUILIBRIUM COMPOSITION OF LEAKY COMPARTMENTS

Living cells have developed the ability to persist in the face of a fundamental contradiction. On the one hand, they preserve an internal composition that is an optimal milieu for metabolic processes that are essential to the maintenance of the living state and maintain the ability to

TABLE 1.1
Samples of Solute Composition of Muscle Cells and Extracellular Fluids

Species	ICF/ECF	Na⁺	K⁺	Cl⁻
Sepia officinalis (cuttlefish)	ICF	31	189	45
	ECF	465	22	591
Loligo forbesi (squid)	ICF	78	152	91
	ECF	419	21	522
Mytilus edulis (mussel)	ICF	79	152	94
	ECF	490	13	573
Carcinus maenus (green crab)	ICF	54	146	53
	ECF	468	12	524
Limulus polyphemus (horseshoe crab)	ICF	126	100	159
	ECF	445	12	514
Squalus acanthias (spiny dogfish)	ICF	18	130	13
	ECF	296	7	276
Rana esculenta (frog)	ICF	10	124	2
	ECF	109	2	78
Rattus norvegieus (rat)	ICF	16	152	5
	ECF	150	6	119

Note: ICF, intracellular fluid; ECF, extracellular fluid.

Source: Kirschner, L., Ed., *Environmental and Metabolic Animal Physiology: Comparative Animal Physiology*, Wiley-Liss, New York, 1991. With permission.

regulate that composition as a defense against external perturbations. On the other hand, the maintenance of this environment and the nature of the associated regulatory processes demand that matter be continuously shuttled in and out of the cell. In other words, cells must maintain an internal composition that is constant but is also not in equilibrium with its environment. Cellular composition is maintained in a so-called "steady state" in the face of constant in and out traffic across the cell membrane.

This chapter focuses on two significant elements of cellular composition: water and small solutes such as ions, sugars, and amino acids. The aim is to provide the foundation for a quantitative understanding of osmoregulatory phenomena that makes intuitive sense and is also in accord with physical reality. The emphasis is on basic principles that apply to all osmoregulatory phenomena regardless of the organism or its environment. As such, much of what follows is devoted to developing analytical tools and sound ways of thinking so the content will continue to be useful even as new osmoregulatory mechanisms, perhaps not even envisioned today, are discovered. Along the way, we will point out conceptual pitfalls that we hope will be useful to the researcher and also to the teacher who must deliver these concepts to beginning students.

A. NONEQUILIBRIUM CELL COMPOSITION

Even the most cursory glance at the tabulated values for the composition of cells reveals striking differences between the inside of cells and their external environment (Table 1.1). The term *external environment*, as used here, could refer to the interstitial fluid of vertebrates or invertebrates or, for some single-celled organisms, to either pond water or seawater. The intracellular concentration of potassium, for example, is typically of the order of 100 to 200 m*M*, and the extracellular concentration is of the order of 2 to 20 m*M*, so transmembrane potassium gradients are generally tenfold or greater. Sodium gradients, with some notable exceptions, typically have the opposite orientation, sodium being more concentrated outside of cells than inside. Gradients of calcium are even more impressive, the concentration of this essential divalent cation typically being of the order of 10,000-fold higher outside of cells than inside. So, how are these gradients that are so profoundly important to the activities of the cell established and maintained?

Over the years, a number of explanations have been advanced to explain the differing composition of cells and their environment, including a proposal that some substances were simply unable to cross the cell membrane (that is, they were held to be *impermeant*). Another model, advanced by Gilbert Ling[59] and developed to a considerable degree of sophistication, held that gradients such as that of potassium arise due to specific associations of the ion with intracellular, macromolecular constituents. These models envisioned the cell as a sort of gel-like, semisolid in which potassium ions would behave differently than in free solution. This point of view has had vigorous proponents,[15,16,66,67] but enormous advances in our ability to accurately determine cell composition (and the state of solutes and water in cells), as well as the development of methods that enable us to discern the properties of single proteins, often in simplified environments, have led to the demise of these notions in favor of a relatively simple pump–leak model. According to this model, cell composition is maintained by the coordinated activity of two sets of macromolecules: the pumps (energy-converting, coupled transporters) and the leaks (non-energy-converting transporters). Here we use the term *pump* to refer to any transporter that has the ability to couple a free energy source to the transport of matter where the source might be adenosine triphosphate (ATP) or the free energy inherent in a gradient of sodium or protons. The leak elements, channels, and so-called facilitated transporters confer upon the cell membrane selective leakiness or, more properly, selective permeability.

B. Selective Permeability and Energy Conversion

The basis for osmoregulation lies in the stringent control of the water and solute content of body fluid compartments. This regulation is achieved by surrounding the compartments with membranes that are specially equipped to regulate solute and water traffic by means of an array of transport proteins that confer upon them the essential features of *selective permeability* and selective *energy-converting transport*. Consider, for example, the plasma membrane surrounding a typical excitable cell. In the resting state it may exhibit significant permeability (i.e., leakiness; see below) to potassium and be nearly impermeable to sodium. Less than a millisecond later, however, the sodium permeability can increase by 100-fold. Likewise, some cell membranes are highly permeable to D-glucose and sparingly permeable to the L isomer. Water permeation can vary widely from cell to cell and in the presence of or absence of hormones such as antidiuretic hormone (ADH).[30–32] These are examples of selective permeability that is brought about by the presence in the plasma membrane of specific proteins that create a selective permeation path through the lipid bilayer. Such pathways, ion channels, water channels, nonelectrolyte channels, glucose transporters, etc., although critically important to the regulation of fluid compartment composition, cannot explain the departure of cell composition from equilibrium. This requires the addition of a second feature: the ability to carry out the electrochemical work of transport—that is, the ability to use energy from the hydrolysis of ATP or an ion gradient to drive the transport of a second species.

Historically, membrane transporters were defined operationally on the basis of their functional attributes, usually related to qualities such as selectivity or the presence or absence of the coupling of transport to some source of free energy. Thus, in the 1960s, one spoke of ion channels (e.g., sodium channels, potassium channels, and calcium channels) absent any knowledge of the molecular entities that were responsible for their functional properties and with only a superficial appreciation of the variety that might exist within a particular class. These operational definitions turned out to be very accurate, however, and were verified in the era of cloning beginning in the 1980s, during which specific functional activities were linked to specific membrane proteins. It is interesting in this regard that Hodgkin and Huxley[48,49] and Watson and Crick[83] published their seminal papers in the same 2-year period!

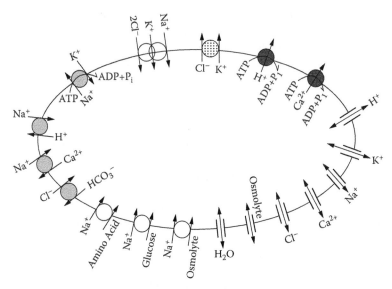

FIGURE 1.1 Depiction of the variety of transporters that might contribute to osmoregulation in cells. Shown are energy-converting transporters, such as ATPase and transporters coupled to a sodium gradient, as well as passive leak pathways for some nonelectrolytes and water.

The application of cloning technology revealed an underlying subtlety and complexity in the cellular repertoire of transporters that was only hinted at in the earlier functional definitions, so today we speak of *families* of membrane proteins, such as the family of potassium-selective channels that now numbers dozens of members. For channels and some transporters the advent of single-channel, patch clamp recording made it possible to actually determine the functional attributes of a single protein for which one knew the primary structure and could predict (albeit with some uncertainty) the topology within the membrane. The ability to identify the amino acid sequence of specific transporters has led to the expression of amounts of protein (particularly for bacterial transporters) sufficient to grow crystals and determine atomic-scale, three-dimensional structures. The availability of atomic-scale structures for membrane proteins has given rise to a renewed interest in structure–function studies aimed at determining how actual transport events are effected by local changes in protein structure.[43]

Figure 1.1 illustrates the varieties of transport proteins that we will encounter in our brief tour through osmoregulatory mechanisms. Shown are channels for water, ions, and nonelectrolytes, as well as energy-converting transporters, such as the nearly ubiquitous Na,K-ATPase and the calcium and proton ATPases. Also indicated are cotransporters for glucose or amino acids that are driven by the free energy in a sodium gradient, as well as countertransporters for protons, calcium, and other species. The sections that follow explore these transporters from the perspective of energetics and consider a few examples for which molecular mechanisms have become apparent.

C. THE PUMP–LEAK MODEL: GENERAL EXPECTATIONS

The mechanistic questions that arise universally in the context of pump–leak systems can be appreciated by considering the simple, hydraulic model shown in Figure 1.2. Depicted here are two situations. In one, a pump *fills* a leaky can and in the other a pump *empties* a leaky can. The can-filling example might represent the maintenance of intracellular potassium by a Na,K-ATPase. In this example, equilibrium would be represented by equal levels of water in the can and its external environment—in this case, a bath comprising a larger can. This hydrostatic equilibrium would represent the state of the system if the pump is turned off: no pump, no leak. If we activate the pump, water will flow into the can, raising the water level and causing an outward water leak

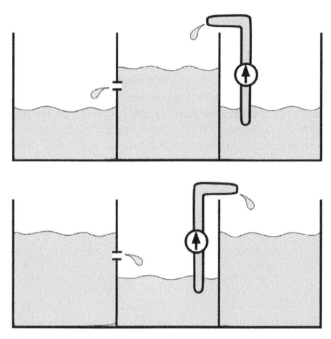

FIGURE 1.2 Two hydraulic pump–leak systems that you can build in your backyard to illustrate why maintenance of the steady state depends on the properties of *both* the pump and the leak.

into the surrounding bath. (Here, we assume, for the sake of simplicity, that the outer can is sufficiently large that emptying the inner can does not change the level of the bath.) The outward leak is slow at first, but as the water level rises in the can the hydrostatic pressure, the driving force for outward water flow, increases so the leak rate increases correspondingly.

The rate of outward water leakage is a function of two parameters: the magnitude of the driving force (hydrostatic pressure) and the water permeability of the can (the size of the hole). As the water level rises, it will eventually attain a value such that the outward leak of water just balances the inward, pump-mediated water flow. At this point, the system is said to be in a *steady state* (stationary state); that is, the water level in the can will be time independent as long as neither the pump rate nor the leak rate changes. Note that the steady-state water level depends on *both* the pump rate and the leak rate; for example, if we enlarge the hole in the can, the steady-state water level will be reduced, even though the pump rate is unchanged, because the same outward flow can be achieved at a reduced hydrostatic pressure.

This simple, mechanical analogy foreshadows the questions that we will ask about the pump–leak systems that are the molecular basis for osmoregulation. We need to identify the pumps and leaks in the cell membrane and identify and quantify the driving forces, particularly those for leak flows. To accomplish this, we will first develop some analytical tools based on thermodynamics. This line of questioning also brings us face to face with the central biological question that we do not consider in detail here—namely, that of *selectivity*. How do pumps and leaks select certain ions or nonelectrolytes over others? How is it that a leak can allow the flow of water but not ions and *vice versa*? The molecular, even atomic, basis for selectivity, or molecular recognition, is one of the frontiers in biological transport.[24,68,86]

D. Thermodynamic Tools and Equilibrium

Our thinking about the thermodynamics of solutions owes an enormous debt to the work of J. Willard Gibbs, a Yale physical chemist. In his now legendary treatise,[42] "On the Equilibrium of Heterogeneous Substances," he set down the basic principles that now permit us to describe the thermodynamics of solutions in a concise, quantitative form. From the perspective of membrane transport, the goal is to derive, for solutes and water, a measure of the driving force for transmembrane movement that is akin to the role of hydrostatic pressure in determining the rate and direction of water flow through a pipe. We usually think of this in terms of the weight of water producing hydrostatic pressure, but we could also express the driving force in terms of the difference in *potential energy* associated with the hydrostatic gradient. For a pressure difference (ΔP), the difference in potential energy could be calculated by simply multiplying the pressure by the volume of water moved.* The thermodynamics of Gibbs, developed to treat heterogeneous substances such as solutions, enables us to express concisely the potential energy of *each component* of a solution and to compare it with its potential energy in any other solution that might be, for example, on the other side of a membrane separating the two.

1. The Chemical Potential

For the sake of simplicity, we begin by considering a binary solution: a single, uncharged solute (s) dissolved in water. The chemical potential (μ_i) for either component can be written as:

$$\mu_i = \mu_i^0 + RT \ln X_i + P v_i \tag{1.1}$$

where μ_i^0 is the standard chemical potential, the necessary reference point for the potential energy that is defined in more detail below. X_i is the *mole fraction* of the ith component, be it solute or water, where the sum of the solute and solvent mole fractions ($X_s + X_w$) must equal 1.0. P is the hydrostatic pressure, and v_i is the partial molar volume of the component, the volume *per mole* of the constituent. For water this value is 18 cm^3 per mole.

Note that μ_i has the units of *free energy per mole*; it is the *partial free energy* of one component of a mixture, such that the *total* free energy, usually denoted G in honor of Gibbs, is given by the sum of these partial free energies:

$$G = \sum n_i \mu_i \tag{1.2}$$

where n_i is the number of moles of the ith component. The analysis of the partial molar free energy of components of a solution implied by Equation 1.1 has been of inestimable value in the analysis of biological transport, but this form of the relation is not very useful because the dependence of μ_i on composition is expressed in terms of mole fraction. Practical equations are derived by relating the mole fractions of solute and solvent to the more commonly used measure of composition: *molar solute concentration*. Considering first the solute we have for μ_s:

$$\mu_s = \mu_s^0 + RT \ln X_s + P v_s \tag{1.3}$$

The mole fraction of the solute (X_s) can be written as:

$$X_s = \frac{n_s}{n_s + n_w} \tag{1.4}$$

where n_s and n_w represent the number of moles of solute and solvent, respectively, in the solution. But, in dilute solutions, such as those commonly encountered in biology, $n_s \ll n_w$, so, to an acceptable approximation, X_s is given by:

---

* The units of pressure are force/area, and the units of volume can be thought of as area × height, so the product $P \times V$ will have the units of work (or energy), or force × distance.

$$X_s = \frac{n_s}{n_w} \qquad (1.5)$$

The molar concentration of s in the solution (C_s) is simply the number of moles of s divided by the total volume of the solution:

$$C_s = \frac{n_s}{n_s v_s + n_w v_w} \qquad (1.6)$$

But, $n_s \ll n_w$, so, taking into account Equation 1.5, we can express the mole fraction of a solute in dilute solution in terms of concentration as:

$$X_s = C_s v_w \qquad (1.7)$$

So, the practical expression for the chemical potential of the dilute solute can be written as:

$$\mu_s = \mu_s^0 + RT \ln(v_w C_s) + P v_s \qquad (1.8)$$

or

$$\mu_s = \mu_i^* + RT \ln C_s + P v_s \qquad (1.9)$$

where $\mu_i^* = \mu_s^0 + RT \ln v_w$; that is, the value of the standard chemical potential of the *solute* differs depending on whether composition is expressed as solute mole fraction or solute concentration.

For charged solutes, we must add an additional term to the partial molar free energy that reflects the influence of the local electrical potential (V):

$$\bar{\mu}_i = \mu_i^* + RT \ln C_s + zFV + P v_s \qquad (1.10)$$

where the addition of the overbar ($\bar{\mu}_i$), denotes the *electrochemical potential* of a charged species. The last three terms in Equation 1.10 relate the partial molar free energy of the solute to three parameters that we might reasonably expect to influence solute movement: solute concentration, electrical potential, and hydrostatic pressure. The latter is generally of less importance because the difference in hydrostatic pressure across most cell membranes is vanishingly small. The term standard chemical potential (μ_i^*) can be more of a mystery intuitively. It is best thought of as depending on how the ion (or nonelectrolyte in Equation 1.9) is solvated; for example, ions (as well as nonelectrolytes) are solvated by water inside and outside of cells, so typically no difference exists in the standard chemical potential across a biological membrane. In other words, $\Delta \mu_i^* = 0$. Differences in μ_i^* will be significant, however, if we compare the state of an ion in water with its state within a channel[24,68,86] or the state of a nonelectrolyte in water with its state after it has dissolved into a lipid membrane.[30–32]

2. Chemical Potential of Water

Recalling Equation 1.3, the chemical potential of water has three components:

$$\mu_w = \mu_w^0 + RT \ln X_w + P v_w \qquad (1.11)$$

This fundamental relationship is only rarely used in biology, however, because it is more convenient (although somewhat confusing) to express the term $RT \ln X_w$ in terms of *solute concentration* rather than the mole fraction of water. For a binary solution comprised of water and a single solute (*s*):

$$RT \ln X_w = RT \ln(1 - X_s) \tag{1.12}$$

But, for dilute solutions, $X_s \ll 1.0$, so we can approximate Equation 1.12 by:

$$RT \ln X_w = -RTX_s \tag{1.13}$$

To convert to units of *solute concentration*, we recall Equation 1.7 ($X_s = v_w C_s$) and, combining this with Equation 1.11 and Equation 1.13, we arrive at a relation for the chemical potential of water comprised of three components:

$$\mu_w = \mu_w^0 - v_w RTC_s + Pv_w \tag{1.14}$$

One component is the standard chemical potential, one relates to the hydrostatic pressure (*P*), and a third represents the effect of water concentration that is (to confuse the enemy) expressed in terms of *solute* concentration. This change in perspective is emphasized by the negative sign of the term RTC_s; as the solute concentration increases, the water concentration decreases![22]

II. WATER PERMEATION

An understanding of the movement of water across cell membranes is essential to an understanding of the physical basis for osmoregulation. The volume of fluid compartments (intracellular or extracellular) is for, practical purposes, equal to the volume of water that resides therein, and the (passive) distribution of water is determined entirely by the distribution of solutes.

A. DRIVING FORCES FOR WATER MOVEMENT

In one sense, water movements are incredibly simple; water permeation is a passive process. There is no evidence for active water transport; water movement is driven entirely by the gradient of the chemical potential of water. For biological membranes separating two aqueous solutions (for which standard chemical potential of water will be the same), the transmembrane difference in the chemical potential of water given by Equation 1.14 is:

$$\Delta\mu_w = v_w (\Delta P - RT\Delta C_s) \tag{1.15}$$

which we can express in the practical units of *pressure* by dividing by the partial molar volume of water (v_w) to yield:

$$\frac{\Delta\mu_w}{v_w} = \Delta P - RT\Delta C_s \tag{1.16}$$

The driving force for passive water transport, as it is often described, is the difference between the hydrostatic pressure gradient (ΔP) and the gradient of "osmotic pressure" ($\Delta\pi$) where $\Delta\pi = RT\Delta C_s$. This conventional usage can be confusing because π is *not* a pressure; rather, π is a so-called *colligative property* of the solution, a measure of composition. Likewise, $\Delta\pi$ is *not* a pressure difference; it is an expression of the *difference in water concentration* across the membrane. The association of $\Delta\pi$ with a pressure arises from an analysis of the equilibrium distribution of water

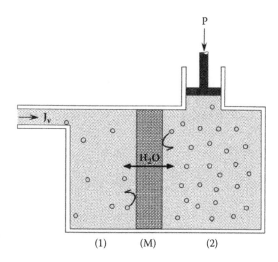

FIGURE 1.3 A membrane that is permeable to water but impermeable to solute separates two solutions, 1 and 2. The *solute* concentrations are such that $C_s(2) > C_s(1)$, so a gradient of *water* concentration will tend to drive water from side 1 to side 2. Applying pressure to the piston will counter the osmotic flow and can be adjusted to produce osmotic equilibrium.

across a membrane that is permeable to water but impermeable to solute and is configured such that a hydrostatic pressure can be applied to one side as indicated in Figure 1.3. If a single, impermeant solute (s) is present on both sides of the membrane such that $C_s(2) > C_s(1)$, then the resulting concentration gradient of water will drive water from side 1 to side 2—that is, from high water concentration to low water concentration. An equilibrium can be established by applying a pressure to the piston on side 2 such that the water flow, denoted here as the volume flow J_v (see below) is reduced to zero. In this condition, $\Delta\mu_w$ equals zero, and from Equation 1.16 we obtain the classic van't Hoff equation[22] for the equilibrium osmotic pressure:

$$\Delta P = \Delta\pi = RT\Delta C_s \qquad (1.17)$$

Here, ΔP, the difference in *hydrostatic pressure*, is numerically equal to the opposing difference in *osmotic pressure*, $\Delta\pi$. The unwary are thus led to equate values of π with some sort of solute pressure. In fact, a fair bit of nonsense has been written about presumed solute pressures that have no basis in reality,[6,45,61,63–65] but we can avoid the linguistic pitfall simply by remembering that π is *not* a pressure; it is a measure of water concentration expressed, for convenience, in terms of solute concentration.

1. An Equation of Motion for Water Flow

Our analysis of osmotic equilibrium and the identification of two forces that can drive water flow—a gradient of hydrostatic pressure and a gradient of water concentration—leads naturally to an equation of motion for water flow that is akin to Ohm's law for electrical current flow:

$$J_v^w = L_p\left(\Delta P - RT\Delta C_s\right) \qquad (1.18)$$

Here, water flow is measured as a *volume flow* (J_v^w) (e.g., cm³ per second); L_p, with a nod to Ohm's law, is denoted the *hydraulic conductivity* of the membrane, a quantitative measure of the leakiness of the membrane to water.

The ease with which we arrived at Equation 1.18 hides a subtle and perhaps questionable assumption that numerically equivalent values of ΔP and $RT\Delta C_s$ would produce identical volume flows. Is the value of L_p, in fact, the same regardless of whether the driving force is a difference in hydrostatic pressure or water concentration? It turns out that Equation 1.18 does, indeed, describe volume flow across solute-impermeable membranes, but it breaks down when the solute is permeant; however, the volume flow equation can be rescued for solute-permeable membranes by including a correction factor, the so-called *reflection coefficient* (σ), where $\sigma = 1.0$ for a solute-impermeable membrane, and $\sigma < 1.0$ for a solute-permeable membrane. For volume flow across a solute-permeable membrane, we obtain:

$$J_v^m = L_p\left(\Delta P - \sigma RT\Delta C_s\right) \tag{1.19}$$

where the volume flow is superscripted by m to denote the measured volume flow, which may be equal to the sum of that due to water and that due to permeant solute:

$$J_v^m = J_v^w + J_v^s \tag{1.20}$$

$$J_v^m = v_w J_w + v_s J_s \tag{1.21}$$

where J_w and J_s represent the *molar* flows of water and solute, respectively. Herein lies one of the mechanistic issues that can be buried in the reflection coefficient—that the *total* measured volume flow and the volume flow of *water* will differ in the case of a permeant solute (see also Section II.E, below). In the case of a membrane containing water-permeable pores, we will see that the permeability of the solute, reflected in a value of $\sigma < 1.0$, determines whether water moves through pores by bulk flow or simple diffusion or some combination thereof. The critical point here is that the interpretation of σ can be complicated because it depends on the nature of the underlying water transport mechanisms, which we consider in the next section.

B. Mechanisms of Water Transport: Diffusion vs. Bulk Flow through Pores

Water movements across biological membranes are passively driven by the chemical potential gradient of water. These passive flows fall into two categories: *diffusional water flow*, generally through the lipid portion of the cell membrane, and the *bulk flow of water* through water-conducting pores. The distinction is important mechanistically because pore-mediated water flow is orders of magnitude more efficient than simple diffusion. Early measurements of water flow[32] provided evidence for water-conducting pores in cell membranes and foreshadowed the discovery of a family of water-conducting pores, the aquaporins, that inhabit a wide variety of water-transporting membranes and can be inserted into, or retrieved from, cell membranes to effect the rapid alteration of water permeability.[30–32] We can contrast the two modes of water flow by comparing water movements across a solute-impermeable, lipid membrane with that across a membrane containing pores that conduct water but exclude solute.

C. Crossing the Lipid Membrane by Solubility–Diffusion: The Equivalence of Osmotic and Diffusional Permeability

Figure 1.4 diagrams two methods for measuring the water permeability of the lipid bilayer, represented here as a layer of oil. In Figure 1.4A, an osmotic gradient (a gradient of water concentration) is imposed by bathing the two sides with solutions containing different concentrations of an impermeant solute (s). In Figure 1.4B, there is no gradient of water concentration ($\Delta\mu_w = 0$,

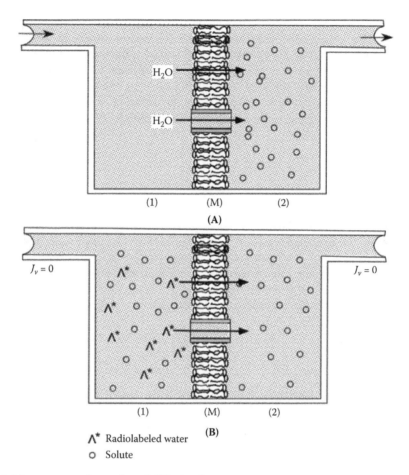

FIGURE 1.4 Measurement of osmotic and diffusional water permeability. (A) A lipid membrane containing a water-conducting pore is exposed to a water concentration gradient ($\Delta P = 0$). Water can move down the water concentration gradient from side 1 to side 2 by two routes: solubility–diffusion across the lipid bilayer and bulk flow through the pore. (B) The same membrane now separates two solutions configured such that $\Delta P = \Delta \Pi = 0$. Radioactive (tracer) water is added to side 1, and the rate of movement to side 2 is monitored.

$RT\Delta C_s = 0$), but radioactively labeled (tracer) water has been added to one side so the unidirectional flux of tracer water can be measured. In neither case is there a hydrostatic pressure gradient.

In the presence of a water concentration gradient produced by imposing a gradient of impermeant solute, water flow is generally measured as a volume flow, where:

$$J_v^m = J_v^w = L_p \left(RT\Delta C_s \right) \tag{1.22}$$

and the value of the hydraulic conductivity (L_p) is obtained. In the tracer flow experiment (conducted in the absence of net volume flow such that $\Delta C_s = 0$; see Figure 1.4B), the flow of labeled water is determined by sampling the cold side of the membrane periodically and determining the unidirectional flow of tracer:

$$J_w^* = P_d^* A_m \left(\Delta C_w^* \right) \tag{1.23}$$

where J_w^* is the flux of tracer that might be measured in such units as counts per minute (cpm) per unit time, A_m is the area of the membrane, and the value of P_d^* is obtained.

The mechanism of water transport across an oil membrane, be it measured by net volume flow or tracer flow, is identical; in either case, water crosses the lipid membrane by solubility–diffusion. Water molecules must partition into the hydrocarbon region of the bilayer on one side and diffuse across and exit into the opposite solution. The partition coefficient of water in oil (about 10^{-6}) is sufficiently small to ensure that any water molecules, labeled or unlabeled, will behave as a *dilute component* within the lipid membrane. This solubility–diffusion mechanism is captured in the familiar expression for the tracer (diffusional) water permeability (P_d^*), where:

$$P_d^* = \frac{\beta^* D_w^* \left(v_w / v_m\right)}{\delta_m} \tag{1.24}$$

and β^* is the oil-water partition coefficient for tracer water (assumed to be equal to that for bulk water), D_w^* is the diffusion coefficient for tracer water within the oil membrane (again, assumed to be equal to that for unlabeled water), δ_m is the thickness of the membrane, and v_w and v_m are the partial molar volumes of water and the oil membrane phase, respectively.

The terms in partial molar volume arise because, following Finkelstein,[32] we here define the water–membrane partition coefficient for tracer water (β^*) as the *ratio of the mole fraction* of tracer water just inside the membrane boundary, $X_w^*(M)$, to that in the bulk solution, $X_w^*(B)$; that is,

$$\beta^* = \frac{X_w^*(M)}{X_w^*(B)} \tag{1.25}$$

This differs from the more conventional definition for a partition coefficient typically utilized for uncharged solutes that is defined as the ratio of *molar concentrations* in each phase,[30–32,34] but this form facilitates the critical comparison between water permeation measured by tracer flow and that measured by applying a gradient of water concentration. As shown in Appendix 1, the *molar* water flow in the osmotic flow experiment (J_w) can be expressed as:

$$J_w = \left(\frac{D_w \beta \cdot v_w}{\delta_m v_m} \right) A_m \Delta C_s \tag{1.26}$$

where ΔC_s is the gradient of *solute* concentration, and the term in brackets is the osmotic water permeability, denoted P_f because it is determined by measuring the volume flow of water:

$$P_f = \frac{D_w \beta \cdot v_w}{\delta_m v_m} \tag{1.27}$$

Comparison of Equation 1.24 and Equation 1.27 confirms our intuition that for a lipid (oil) membrane:

$$P_f = P_d^* = P_d^w \tag{1.28}$$

The permeability coefficient for water is predicted to be identical in either an osmotic flow experiment or tracer diffusion experiment, because in either case water must cross the lipid membrane by means of simple solubility–diffusion — that is, by dissolving in the hydrocarbon phase and diffusing across to the other side. The reader will note that Equation 1.26 contains the symbolic dissonance characteristic of all expressions for osmotically driven water flow because we utilized dilute solution approximations to express the water concentration gradient in terms of *solute* concentration.

Another important expression for osmotic permeability is obtained by recalling the expression for volume flow of water when $\Delta P = 0$:

$$J_v^w = L_p \left(RT \Delta C_s \right) \tag{1.29}$$

Substituting $v_w J_w$ for J_v and recalling that:

$$J_w = P_f A_m (\Delta C_s) \tag{1.30}$$

we obtain the relation between the osmotic permeability (P_f) and the hydraulic conductivity (L_p):

$$P_f = RT \frac{L_p}{v_w A_m} \tag{1.31}$$

For the lipid membrane, P_f and L_p represent the same coefficient in different units, an equality that, in fact, holds regardless of the water permeation mechanism.

D. POROUS MEMBRANES, IMPERMEANT SOLUTE

Here, we consider water flow through pores that conduct water molecules but exclude solute—that is, osmotic flow through water-filled pores as diagrammed in Figure 1.4A. In this example, water flow is driven by an applied osmotic gradient, and the transmembrane difference in hydrostatic pressure is zero. Careful examination of this situation should result in a sort of double take. On the one hand, to an observer external to the pore, the physics are transparently clear; a transmembrane difference in water concentration will drive water flow from side 1 to side 2. On the other hand, if we peer *inside* the water-conducting pore we find, as specified, *only pure water* and not a gradient of water concentration. What, then, is the local driving force for the bulk water flow through the pore? The answer, obtained by Alex Mauro,[32,61–65] is that a gradient of impermeant solute creates, within the pore, a gradient of hydrostatic pressure despite the fact that across the membrane $\Delta P = 0$! The origin of this pressure gradient lies at the interface between the water-filled pore and the solute-containing bulk solution. At the interface, the concentration gradient of water tends to drive the escape of water from the pore with a consequent lowering of the local pressure within the pore on the solute-containing side of the membrane. Once this is realized, it is then no surprise that water flow through pores is predicted to be identical whether driven by a hydrostatic or osmotic gradient, as indicated by Equation 1.18.[20,32]

The water permeability of a porous membrane is expected to be higher than that of a simple layer of oil because within the pores water can move by what is commonly referred to as *bulk flow*. Bulk flow is more like water flow through a pipe in which water–water viscous interactions and the resulting transfer of momentum predominate. We can gain some appreciation for the water permeability of a single, right circular pore by assuming that it is sufficiently large to allow water molecules to slide past one another as described by Poisseuille's law which, for an applied pressure (ΔP), a pore of radius r, and length l, yields:

$$j_v = \left(\frac{\pi \cdot r^4}{8l\eta} \right) \Delta P \tag{1.32}$$

where j_v is the volume flow of water through a *single pore*, and η is the viscosity of water. The quantity in parentheses is the predicted *hydraulic conductivity* of a single pore. The water permeability (P_f) for such a pore is obtained from Equation 1.31, where $A_m = A_p$, the pore area. Thus, P_f for a single pore is given by:

$$P_f = \frac{RTr^2}{8l\eta\nu_w} \qquad (1.33)$$

The ratio of the osmotic water permeability to the diffusional water permeability (P_f/P_d) has often been used as an experimental test for the mechanism of water flow. The reason is seen by comparing the results for our two simple models, the lipid membrane and the porous membrane. For the former we found that P_f/P_d was predicted to be unity, reflecting the identical, solubility–diffusion mechanism for osmotic flow or tracer water flows. In the case of our simple, right circular pore, however, the water permeability contains a term in r^2, suggesting that bulk flow driven by an osmotic gradient is more efficient due to the transfer of momentum between water molecules specified in Poisseuille flow. For such a pore, tracer permeability would be determined *in the absence of osmotic flow* to eliminate the solvent drag effect of bulk flow through the pore on the flow of labeled water. In this setting P_d^w is given by $P_d^w = D_w/l$, where D_w is the diffusion coefficient for tracer water in bulk water and l is the length of the pore. Note that there is no term in the partition coefficient, because the tracer water is assumed to move from bulk water in the bathing solution to bulk pore water such that $\beta = 1$. The ratio P_f^w/P_d^w for a single, water-filled pore is, therefore:

$$\frac{P_f^w}{P_d^w} = \left(\frac{RT}{8\eta D_w \nu_w}\right) \cdot r^2 \qquad (1.34)$$

For a pore of radius 4 Å, P_f^w/P_d^w would be predicted to be about 3, similar to that estimated for the pores formed by the polyene antibiotics amphotericin-B and nystatin in lipid bilayer membranes.[33,35,38,51]

1. Unstirred Layers

In principle, the ratio P_f^w/P_d^w is an experimental test for water flow through pores. In practice, however, the determination and interpretation of P_f^w/P_d^w can be confounded by at least two phenomena: unstirred layers and single-file water flow.[32] The term *unstirred layer* refers to stagnant layers of fluid that necessarily form at the membrane–solution interfaces. These layers present a series resistance to the permeation of tracer water that becomes more important the greater the intrinsic water permeability of the membrane. If, for computational purposes, we lump together layers at two membrane–solution interfaces, we can obtain a relationship between the *actual* permeability of the lipid membrane to tracer water (P_m) and the measured water permeability (P_{meas}):

$$\frac{1}{P_{meas}} = \frac{1}{P_1} + \frac{1}{P_m} \qquad (1.35)$$

where P_1 represents the tracer permeability of the stagnant slab of water characterized by an effective thickness (δ_1) so $P_1 = D_w/\delta_1$, and:

$$P_{meas} = \frac{P_m}{1 + P_m/(D_w/\delta_1)} \qquad (1.36)$$

If the water permeability of the membrane (P_m) is sufficiently high (e.g., $P_m = P_1$), then the value of P_m is significantly *underestimated*, an effect that could produce the false impression that P_f^w/P_d^w is greater than 1.

Unstirred layers also compromise the estimation of P_f^w due to solute polarization, the accumulation and depletion of solute (caused by solvent drag) in the unstirred boundary layers. This sweeping effect tends to *decrease* the actual osmotic gradient encountered by the membrane, leading to an underestimation of P_f^w.

2. Single-File Water Pores

The case of the single-file pore deserves attention because this is the mechanism most likely to pertain to the now well-characterized water pores, the aquaporins.[4,40,53] Equation 1.34 gives the (false) impression that, as pore radius is reduced, P_f^w/P_d^w would approach unity for pores that are the size of the water molecule. This extrapolation is undermined, however, by the fact that when water molecules cannot pass one another the mechanism of water flow is altered dramatically. In the single-file pore, water flow driven by an osmotic gradient is, as intuition suggests, expected to be more diffusion like. The frictional interactions between the water molecules and the pore walls, rather than water–water viscous interaction, dominate the process. For tracer water flow, however, the mechanism differs strikingly from simple diffusion. Tracer flow through a single file pore that is occupied, on the average, by n water molecules is attenuated because, for one tracer molecule to traverse the pore, the entire pore contents must move. Due to the low molar abundance of labeled water in a tracer experiment, a water channel containing a single, labeled water molecule is most likely to contain $n - 1$ *unlabeled* water molecules as well.[32] In the case of the single-file pore, the ratio P_f^w/P_d^w is, in fact, a measure of the number of water molecules occupying the pore:

$$\frac{P_f^w}{P_d^w} = n_w \tag{1.37}$$

E. REFLECTION COEFFICIENT

The reader will recall the general expression for volume flow measured across a membrane leaky to solute:

$$J_v^m = L_p \left(\Delta P - \sigma RT \Delta C_s \right) \tag{1.38}$$

where the term representing the driving force due to a water concentration gradient ($RT\Delta C_s$) is modified by σ, the reflection coefficient, to account for the modification of volume flow in the condition of nonzero solute permeability. Although σ is often dismissed as an empirical correction factor, it is instructive to consider its mechanistic significance in terms of our two permeation models, the lipid bilayer (layer of oil) and the transmembrane, water-conducting pore.

1. σ for a Lipid Bilayer

The interpretation of σ for osmotic flow across a lipid bilayer is straightforward. In the presence of a gradient of permeant solute, the *measured* (*total*) volume flow across the membrane (J_v^m) now has two, oppositely directed components—the volume flow of water (J_v^w) and the volume flow of solute (J_v^s) such that:

$$J_v^m = J_v^w - J_v^s = \nu_w J_w - \nu_s J_s \tag{1.39}$$

In the presence of a permeable solute, the *measured total volume flow* is less than the volume flow of water by the amount of the oppositely directed volume flow of solute; however, in the presence of an osmotic gradient ($\Delta P = 0$), the volume flow of *water* is identical to that seen with an impermeant solute:

$$J_v^w = L_p RT \Delta C_s \qquad (1.40)$$

whereas the measured (total) volume flow is:

$$J_v^m = \sigma L_p RT \Delta C_s \qquad (1.41)$$

So, σ is simply the ratio of the measured (total) volume flow to the volume flow of water; that is, inserting Equation 1.39:

$$\sigma = \frac{J_v^m}{J_v^w} = 1 - \frac{J_v^s}{J_v^w} \qquad (1.42)$$

Or, recalling that the molar flows of water and solute are given by:

$$J_s = P_s A_m \Delta C_s \quad \text{and} \quad J_w = P_w A_m \Delta C_s \qquad (1.43)$$

We obtain for σ:

$$\sigma = 1 - \frac{v_s P_s}{v_w P_w} \qquad (1.44)$$

For the lipid membrane, σ is simply a measure of the solute/solvent permeability ratio. The inclusion of σ corrects our description of total volume flow for the contribution of the volume of *solute* flow. Water flow, *per se*, is unaffected.

2. σ for Water-Conducting Pores

In the case of a water-conducting pore and a permeant solute, the physical significance of σ is less straightforward, so much so in fact that there is a fair bit of confusion in the older literature regarding its physical meaning.[32] In the case of a porous membrane, the counter flow of solute will also, of course, diminish the measured total volume flow, as with the lipid membrane. A second effect, however, is much more profound. If solute can enter the pore, the gradient of hydrostatic pressure within the pore (created as a result of ΔC_s) is reduced or eliminated, depending on the relative permeability of the solute.[32] In the case of a freely permeable solute, volume flow is reduced to zero because the intrapore pressure gradient is abolished. Now J_v^w equals $-J_v^s$ because solute and water are mixing in the pore as they would in free solution. This all seems simple enough, but the failure to recognize the effects of solute permeation on intrapore pressure gradients has resulted in substantial confusion, particularly as regards the interpretation of so-called *anomalous osmosis* or the *wrong-way water flow* as described in detail by Finkelstein[32] and summarized in Dawson.[20]

III. THE THERMODYNAMICS OF SOLUTE PUMPS AND LEAKS

To dissect the coordinated symphony of pumps and leaks that act to maintain cellular solute composition, it is convenient to begin, as we did with water, by exploring the energetics of the leak pathways—that is, by identifying the driving forces for the passive leakage of ions or nonelectrolytes across the cell membrane. It turns out that we can do this with a fair level of precision by utilizing our thermodynamic tools, particularly the concept of equilibrium. We begin by defining a leak pathway as a flow of matter that is driven solely by external influences such as electrical potential or concentration gradients. A leak flow may be regarded, in the thermodynamic sense, as a purely dissipative flow or, as one often hears, purely passive. For a leak flow, no energy conversion occurs

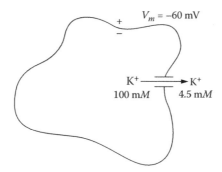

FIGURE 1.5 A hypothetical cell showing the potassium ion distribution and the membrane potential in the steady state. The direction of passive movement of K^+ is determined by the balance between V_m and E_K.

(except, of course, to heat, as in the heat effect of current flow through a resistor) nor coupling of the flow to metabolic energy either directly or indirectly. The most obvious example of a purely dissipative element is an ion channel through which ions move only when they are driven by transmembrane differences in electrical potential or ion concentration.

Ionic leak flows through a channel require a departure from equilibrium. This corresponds to the intuitive notion that a departure from equilibrium will generally set in motion events that tend to restore equilibrium. If a ball is rolled up an incline and released, it will roll down the incline of its own accord, seeking, as it were, mechanical equilibrium. In the ball example, the force that is the cause of motion is obvious; gravitational attraction drags the ball down the incline just as a flow of water is driven by a gradient of hydrostatic pressure. A description of ionic equilibrium, however, requires that we consider the net result of *two* driving forces, a difference in concentration and a difference in electrical potential.

A. THE DRIVING FORCES FOR IONIC FLOWS

The driving forces for passive ion flows are easily identified. Ions move in the electrical field produced by a difference in electrical potential across a membrane. A concentration gradient imposed across a membrane will also result in passive ion flow from high to low concentration. This much is pretty obvious, especially when only one of these forces is present. In Figure 1.5, however, we depict a nonequilibrium situation in which the passive movement of potassium through an ion channel is driven by a combination of gradients of concentration and electrical potential. Fortunately, we can treat the two forces individually; that is, we can employ the principle of superposition and treat the electrical potential difference and the concentration gradient as being additive. The situation depicted in the figure, however, may initially cause some consternation. Here we have an outwardly directed gradient of potassium (100 mM to 4.5 mM) and an electrical potential difference that is negative inside of the cell by 60 mV. The two component forces are inconveniently oriented in opposite directions, so, to determine the *net* result, we need to compare the two forces quantitatively in the same units. This turns out to be pretty simple using the electrochemical potential of potassium that we developed earlier (in Section I.D). Using Equation 1.10 and taking the difference across the membrane (inside-outside) we arrive at:

$$\Delta\overline{\mu}_K = \Delta\mu_i^* + RT \ln \frac{[K]_i}{[K]_o} + zF(V_i - V_o) + (P_i - P_o)v_s \tag{1.45}$$

where the difference in the standard chemical potential ($\Delta\mu_i^*$) is zero because its value is presumed to be the same inside and outside the cell. Similarly, the pressure difference across most cell membranes is vanishingly small and is usually neglected.

The difference in electrochemical potential ($\Delta\bar{\mu}_K$) represents the total (or net) passive driving force on potassium ions and has the units of *Joules per mole of potassium*. We can arrive at a more user-friendly version by dividing through by zF to change the units to volts or millivolts, converting to common logs ($\log 10 = 2.3\log e$), and inverting the terms in potassium concentration (note sign change!) to arrive at:

$$\Delta\bar{\mu}_K / zF = (V_i - V_o) - 2.3\left(\frac{RT}{zF}\right)\log\left(\frac{[K]_o}{[K]_i}\right) \qquad (1.46)$$

This form has the advantage of corresponding to two widely used conventions for the description of the electrochemical driving force. First, the term $V_i - V_o$ represents the most common definition of the cell membrane potential (V_m), which is the potential of the inside *with respect to* the outside. In practical terms, this means that using this convention the sign of the membrane potential will always have the sign of the *inside* of the cell. So, if we define V_m as $V_i - V_o$, then a V_m of –60 mV means that the inside of the cell is more negative than the outside by 60 mV. Second, the term $(2.3RT/zF)\log([K]_o/[K]_i)$ is the conventional definition for what is referred to as the *equilibrium potential* or *reversal potential* for potassium, denoted as E_K. This leaves us with a practical equation that is, in fact, one of the most useful in all of biology:

$$\frac{\Delta\bar{\mu}_K}{zF} = V_m - E_K \qquad (1.47)$$

where:

$$E_K = \frac{2.3RT}{zF}\log\left(\frac{[K]_o}{[K]_i}\right)$$

This says that the *total* or net driving force on the potassium ions, expressed in electrical units, can be described as the difference between two components, one the difference in electrical potential (V_m) and the other the driving force due to the concentration gradient which, as the result of our algebraic manipulations, is expressed here, like the electrical potential, in units of millivolts.

The equation for the total driving force accomplishes the essential goal of expressing the two components of the total force *in the same units* so they can be compared. With the knowledge that at 25°C the term $2.3RT/zF$ is approximately 60 mV, we can now return to our previous dilemma and immediately discern that the *outward* force due to the K+ gradient (81 mV) exceeds the *inward* force due to the membrane potential (60 mV), so the *net* force (19 mV) will drive potassium out of the cell. Returning to our earlier mechanical analogy, for potassium ions, downhill is out of the cell.

Note that we have not worried too much here about the sign of the calculated forces. That is because the sign conveys the orientation of the driving force (inward vs. outward), and, in this example, the orientation of the driving forces is intuitively obvious. We can, however, look at this in a more formal way and expand this analysis to an actual equation of motion for ions, also known as Ohm's law (see Section IV.A). Also, if we insert into Equation 1.47 the values for V_m (–60 mV) and E_K (–82 mV), we get the intuitively obvious result of 19 mV for the net outward-driving force for potassium ions.

In developing Equation 1.47, we have adopted implicitly a mechanical view of the movement of potassium. The basic notion is that something moves if it is pushed. This perspective fits quite nicely with the action of the difference in membrane potential, a measure of the electric field within the membrane. Each individual ion experiences a net force due to this electric field; that is, each ion acquires a component of average velocity because it is pushed by the electric field. In the case

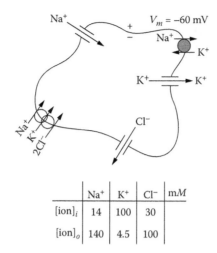

	Na$^+$	K$^+$	Cl$^-$	mM
[ion]$_i$	14	100	30	
[ion]$_o$	140	4.5	100	

FIGURE 1.6 A hypothetical cell illustrating representative, steady-state, nonequilibrium distributions of Na$^+$, K$^+$, and Cl$^-$.

of the driving force attributed to the concentration gradient (E_K), however, we find that the mechanical analogy, although it is *very useful and produces the correct results*, does not accurately represent the underlying process. According to our development, it would seem quite reasonable to say that the concentration gradient is the driving force for ion flow. The deficiency in this mechanical view is immediately apparent, however, when we ask, "What is the driving force due to the concentration gradient as experienced by an individual ion?" Clearly, there is none! Ions move down their concentration gradient by diffusion, a process that is driven by the random thermal motions of the individual ions. The direction of net diffusional flow is from high concentration to low concentration, as a result of this random motion and the fact that there are more ions on one side of the membrane than the other.[19] Despite this conceptual discontinuity, however, expressing the driving force in this thermodynamic form produces results that are in accord with experiment (see Section IV).

B. Diagnosing Active Transport

The utility of the thermodynamic approach developed in the preceding section can be demonstrated by considering the model cell diagrammed in Figure 1.6. Shown is the steady-state composition of the cell and its surroundings along with the value of the membrane potential, V_m. Is there evidence here for active transport—transport that is coupled to some source of energy other than an electrochemical potential gradient? Calculating $\Delta\bar{\mu}$ for Na$^+$, K$^+$, and Cl$^-$, we find that all of the values are nonzero. The steady-state distributions of all three ions are maintained *away from thermodynamic equilibrium*. Put another way, each ion experiences a passive, electrochemical driving force that is inward for Na$^+$, outward for K$^+$, and outward for Cl$^-$. Referring to our pump–leak model, this means that leak flows in this example must be inward for sodium, outward for potassium, and outward for chloride. The fact that the ion distributions are not changing with time demands, for each ion, an equal and opposite flow that is moving uphill—that is, against a prevailing electrochemical potential gradient. These latter flows are, by definition, active or energy requiring; they require that the uphill flow be coupled to some source of free energy.

For most cells, the gradients of sodium and potassium typically observed are largely attributable to the Na,K-ATPase that can couple the efflux of Na$^+$ and the influx of K$^+$ to the free energy of hydrolysis of ATP. A gradient of chloride such as that depicted in Figure 1.6 is generally attributable to some sort of cotransporter such as the one that carries out the coupled transport of chloride together with sodium and potassium. This thermodynamic analysis does not reveal the mechanism by which the gradients are produced, but it tells us unequivocally that the observed gradients can be maintained in the steady state *only* if some form of active transport is operating to move ions. As a practical matter, it is important to note that the gradients of Na^+ and K^+ shown in Figure 1.6 are typical, at least qualitatively, for a wide variety of animal cells that utilize the Na,K-ATPase. Chloride gradients in cells, however, are a bit of a wild card. That shown in Figure 1.6 is typical of cells that make up the epithelial cell layers that carry out transepithelial chloride secretion. In such cells, chloride enters across the basolateral membrane and up an electrochemical gradient, usually coupled to Na^+ (and K^+) via a cotransporter.[44,69] In contrast, the intracellular Cl^- concentration in some nerve cells is such that the passive, electrochemical driving force is *inward* ($[Cl]_i$ < 10 mM) so the Cl^- *efflux must be active* and requires coupling to an energy-converting mechanism such as a KCl cotransporter.[69,74] In some skeletal muscle cells, the Cl^- distribution is near equilibrium (i.e., $E_{Cl} = V_m$).

Another approach to diagnosing active transport that is often used in connection with transport across epithelial cell layers is to measure ion or nonelectrolyte flows under conditions that permit passive, electrochemical driving forces to be completely eliminated. Such experiments are based on the classic studies of sodium transport across sheets of isolated frog skin by Hans Ussing and his colleagues.[77] The sheets of skin (or intestine, etc.) are mounted so as to separate two identical solutions and are voltage clamped (see Section IV.C) so the transepithelial electrical potential difference can be maintained at 0 mV. In this condition, Ussing and Zerahn[80] observed net transport of sodium (measured using a radioactive isotope) from the pond side to the blood side of the skin, despite the fact that $\Delta\bar{\mu}_{Na}$ across the cell layer had been reduced to zero. This landmark experiment was the first unequivocal demonstration of active transport, and this test has been repeated many hundreds of times over the years as investigators have used the method to identify active transport of sodium, potassium, chloride, protons, and other substances across epithelial cell layers derived from organs from a wide variety of species. It is important to note that the conclusions related to the active nature of the transepithelial transport process are not compromised by the fact that the frog skin is a multicellular, multilayered structure—an organ, one might say—comprised of multiple cell membranes.

1. Flux Ratio Analysis

Another approach to the analysis of coupled transport is determination of the ratio of forward and backward, unidirectional rates of transport—that is, the *flux ratio*. Operationally, this means measuring (usually in separate experiments) the forward and backward, unidirectional rates for transport of, for example, sodium by determining rates of radioactive tracer flow. Ussing and colleagues[77–79] used this approach to analyze the ratio of one-way fluxes of sodium across sheets of isolated frog skin. Using the Nernst–Planck equation for electrodiffusion[77] as a starting point, they showed that if sodium crosses the frogs skin by simple diffusion, uncoupled to any energy-donating process, the flux ratio is given by:

$$\frac{(J_{Na})_{pb}}{(J_{Na})_{bp}} = \frac{[Na]_p}{[Na]_b} \exp\left(\frac{-zFV_{te}}{RT}\right) \tag{1.48}$$

where $(J_{Na})_{pb}$ and $(J_{Na})_{bp}$ are the one-way fluxes from pond side to blood side and blood side to pond side, respectively. $[Na]_p$ and $[Na]_b$ are the respective concentrations of sodium, V_{te} is the transepithelial electrical potential difference (defined such that the value of V_c has the sign of the blood side of the skin), and z, F, R, and T have their usual meanings.

Deviation of the flux ratio from this prediction indicates a departure from simple diffusion suggestive of a coupled, energy-converting mechanism. Later analysis of the flux-ratio equation emphasized its similarity to the relation between the ratio of the forward and reverse rate coefficients for a chemical reaction to the free-energy change between reactants and products.[13,14,17,18] In the case of diffusion, Equation 1.48 can be rewritten as:

$$\frac{(J_{Na})_{pb}}{(J_{Na})_{bp}} = \exp\left(\frac{\Delta\bar{\mu}_{Na}}{RT}\right)$$

where $\Delta\bar{\mu}_{Na}$ is the difference in the electrochemical potential for sodium across the skin. For transport that is not coupled to any driving force other than simple diffusion, the free-energy change inherent in the transport event reflects only the passive, electrochemical driving force. Subsequent derivations[13,14,17,18] served to emphasize this point and underline the conclusion that the validity of the equation does not depend on any assumptions about the nature of the transport processes other than the lack of any coupling of the flow, directly or indirectly, to any driving force apart from the electrochemical potential difference.

2. Single-File Diffusion

The flux-ratio equation has also been utilized to analyze the movement of ions[50] and water[32,33,38] through so-called *single-file pores*, pores that are sufficiently narrow so individual ions or water molecules cannot pass one another. This sort of behavior was first recognized by Hodgkin and Keynes,[50] who measured fluxes of radioactive potassium across the membrane of the squid giant axon. The resulting flux ratio differed from that predicted for simple diffusion, and they postulated that the K$^+$ channel pore could be occupied by more than one potassium ion at a time, a condition that leads to a coupling of the flow of radioactive K$^+$ to the flow of nonradioactive (abundant) K$^+$.[18,54] Crystal structures have confirmed in atomic detail the basis for the binding of two or more K$^+$ ions within the selectivity filter of K$^+$ channels.[23,84,85]

C. COUPLED, ENERGY-CONVERTING TRANSPORT: FEASIBILITY ANALYSIS

Earlier we recognized that the term *active transport* could in principle be used to describe a range of transport mechanisms sharing the ability to function as *energy converters*, in that they possess the ability to couple a source of free energy to the performance of electrochemical work. We also noted that the source of the donated free energy could be an energy-yielding chemical reaction such as the hydrolysis of ATP or it could be an ion gradient (sodium for example) that could be used to drive the transport of a second species. The free energy that is invested (by the Na,K-ATPase) in a transmembrane sodium gradient can be used to drive the uptake of amino acids or sugars or the extrusion of calcium or protons. The thermodynamic tools we have developed can provide some important insights into the mechanisms of these coupled transport processes. Any such analysis is based on the First Law of Thermodynamics, the conservation of energy. If we somewhat arbitrarily divide the coupled process into two parts, one energy-donating and the other energy-utilizing, then the First Law says that the work done to effect the energy-utilizing process cannot exceed the energy available from the energy-donating process. This simple inequality provides an important check on our reasoning in relation to a coupled process, particularly as regards the question of stoichiometry.

Consider, for example, a sodium/proton exchanger (or antiporter) of the sort that is found in many cell membranes. If the cell maintains an inwardly directed sodium gradient of 140:14 m*M* in the steady state, what is the maximum pH gradient that could be maintained? This simple question brings us face-to-face with two important issues pertaining what we might think of as *feasibility analysis*. The first is that we must admit that thermodynamics can never, in and of itself, tell us what the actual steady-state proton gradient will be; that would require detailed knowledge of the properties of both the energy converter (the antiporter in this case) and any leak pathways. From such details we might calculate the point at which the net proton flow would be zero, the steady state in which the pump rate and the leak rate are equal and opposite. The analysis of the energetics of the coupled process can tell us, however, *what is the best we can do* — what is the *maximum proton gradient* possible under the specified conditions if there were no leak? When we attempt this apparently simple calculation, however, we run smack into the second issue — namely, that of stoichiometry. The work available from the Na⁺ gradient, per transport cycle, is simply:

$$\left(\frac{n_{Na}}{N_A}\right)\Delta\bar{\mu}_{Na} \qquad (1.49)$$

where n_{Na} is the number of sodium ions transported per transport cycle, $\Delta\bar{\mu}_{Na}$ is the Na⁺ electrochemical potential gradient, and N_A is Avogadro's number, required because $\Delta\bar{\mu}_{Na}$ is in molar units; therefore,

$$n_{Na}\left(\frac{ions}{cycle}\right)\times\frac{1}{N_A}\left(\frac{moles}{ion}\right)\times\Delta\bar{\mu}_{Na}\left(\frac{Joules}{mole}\right)=\frac{Joules}{cycle}$$

Similarly, the *work done* per cycle to move protons is given by:

$$\left(\frac{n_{H}}{N_A}\right)\Delta\bar{\mu}_{H} \qquad (1.50)$$

and the *maximum* proton electrochemical potential gradient achievable for a given value of $\Delta\bar{\mu}_{Na}$ is:

$$(\Delta\bar{\mu}_{H})_{max}=\left(\frac{n_{Na}}{n_{H}}\right)\Delta\bar{\mu}_{Na} \qquad (1.51)$$

We have defined $\Delta\bar{\mu}_{Na}$ and $\Delta\bar{\mu}_{H}$ for convenience so they will both be positive numbers for an inward gradient. The maximum gradient of protons depends, of course, on the magnitude of the Na⁺ gradient, but it also depends on the stoichiometry, the coupling ratio n_{Na}/n_{H}. The pivotal role of stoichiometry becomes even more obvious if we rearrange Equation 1.51 to yield an expression for the maximum ratio of hydrogen ion concentrations in the condition $V_m = 0$:

$$\left(\frac{[H]_o}{[H]_i}\right)_{max}=\left(\frac{[Na]_o}{[Na]_i}\right)^{\alpha} \qquad (1.52)$$

where $\alpha = n_{Na}/n_{H}$. The stoichiometry or coupling ratio appears as an *exponent*. Whereas a 1Na:1H⁺ stoichiometry would produce, at best, a 10-fold gradient of proton concentration, a stoichiometry of 2Na:1H would predict in the same condition a maximum possible proton gradient of 100:1! The stoichiometry is a critical parameter in any estimation of the energetics of a coupled process.

Implicit in the foregoing analysis of the energetics of sodium–proton exchange is the notion that the stoichiometry is, in fact, constant. That is to say, the influx of sodium and outflow of protons is tightly coupled; there is no slip, the coupling ratio does not vary. The behavior of many coupled process indeed appears to resemble that of a chemical reaction that occurs with a fixed stoichiometry. Perhaps the best demonstration of this is the fact that, like a chemical reaction, these coupled transport processes exhibit an equilibrium and can actually be reversed or driven backward by manipulating the electrochemical driving forces. Let's assume, for example, that the stoichiometry of our sodium–proton exchange is 1Na:1H. This is not only typical for these proteins but also simplifies the analysis because for a one-to-one Na/H exchange no net charge transfer occurs during the catalytic cycle. As a result, the *energetics* of the cycle are independent of the value of V_m.* Imagine a cell membrane containing a 1Na:1H, sodium/proton exchanger and having a negligible proton leak. If an inward sodium gradient is maintained at a constant, 10:1 ratio by the Na,K-ATPase, what is the prediction for the proton gradient? It would gradually increase until it reached a value of 10:1, and then the inflow of sodium and efflux of protons would stop. The Na/H exchange cycle would be at equilibrium. There would certainly be some transport events, exchanging an outside sodium for an intracellular proton and *vice versa*, but these would occur with the same average frequencies so the net result would be zero flow. If the external pH were maintained at say, 6.5, the predicted maximum internal pH would be 7.5, 1 pH unit less acidic than the external environment.

Now, consider making the external pH somewhat more acidic, say pH = 6.0. The energetics predict that the exchange process should now reverse—protons should move down the now larger proton electrochemical potential gradient, entering the cell and causing a net efflux of sodium, against the sodium gradient.[8,9] A functional test for such an exchange would be to determine whether it can be driven backward, or, better, can the transport cycle be driven in either direction depending on the prevailing electrochemical potential gradients?

The dependence of the direction of the transport cycle on the prevailing electrochemical potential gradient mirrors the effect of concentration on the direction of a chemical reaction embodied in the law of Le Chatlier. This behavior depends on the process having a fixed stoichiometry and has important implications for the energy economy of the cell. Substantial gradients can be generated by co- or countertransport and maintained near or at equilibrium for the transporter, thereby minimizing energy expenditure. It seems appropriate to view the entire subset of transporters that utilize the sodium gradient as having been designed to utilize free energy that has been invested in the sodium gradient by the Na,K-ATPases to drive other energy-requiring tasks, such as exporting protons or calcium or importing sugars or amino acids or even neurotransmitters. In each case, the transporter, although serving as an energy converter, is also clearly a leak pathway for sodium, and therein lies the cost of the coupled export or import process. That cost is minimized, however, because the turnover will cease when the coupled transporter is at equilibrium. Consider, for example, the situation for a $3Na^+:1Ca^{2+}$ exchanger[46,47,58] that is worked out in Appendix 2. When external $[Na]_o = 140$ mM, $[Na]_i = 14$ mM, $[Ca^{2+}]_o = 1$ mM, and V_m equals –60 mV, the predicted minimum value of internal calcium is 10^{-7} mM. If $[Ca^{2+}]_i$ is of the order of this value, no net turnover of the exchanger occurs until intracellular calcium increases, perhaps due to the opening of some calcium channels, in which case the exchange cycle begins to operate until the gradient is restored.

IV. ION FLOWS

A. ION PERMEABILITY AND CONDUCTANCE

Small ions such as sodium, potassium, and chloride make up a major portion of the osmotically active solute population in cells and body fluids, so the distribution of these species is critical to osmotic homeostasis. The steady-state distribution of ions is determined by the coordinated interplay of a variety of transporters, including pumps, co- and countertransporters, and ion channels. These elements are also key working components of salt- and water-transporting epithelial cell layers such as those that make up much of the kidney and gastrointestinal tract as well as specialized organs such as fish gills and avian and elasmobranch salt glands. Ion channels are particularly important for rapid changes in ion distribution because ion channels are designed to support high throughput.

---

* This does not preclude some effect of V_m on the protein that could, in principle, alter protein conformation and, thereby, the *rate* of transport.

Any operational definition of passive ion permeability is potentially complicated by two factors: the existence of multiple parallel pathways by which any single ion can cross a membrane and the fact that ion leak flows often are generally driven by some combination of ion concentration gradients *and* the transmembrane electrical potential difference. Accordingly, the most general operational definition of the permeability of an ion is based on the measurement (real or imagined) of tracer flow across the membrane so as to define what we refer to as the tracer *rate coefficient* (λ^*). If we imagine determining the unidirectional flow of labeled potassium across a cell membrane, the tracer rate coefficient λ_K^* is given by:

$$\lambda_K^* = \frac{J_K^*}{C_K^*} \tag{1.53}$$

where J_K^* is the one-way flow of tracer and C_K^* is the concentration of tracer on the hot side of the membrane. As operationally defined, however, this rate coefficient will be less than satisfactory for analytical purposes for two reasons. First, it will reflect contributions from multiple transport pathways, and, second, buried in its value will be the influence of the transmembrane potential (V_m) as a driving force for ion flow. The first problem is usually addressed through the judicious use of various experimental conditions, in particular the use of inhibitors, to isolate specific transport pathways. The influence of the membrane potential is in general more difficult but is relatively straightforward for a pathway of major importance to osmotic regulation—ion channels, which we consider here. (It is important to note that an electrophysiological approach to the analysis of ion permeablility, which we do not consider here, is to measure the impact of ion substitutions on reversal potentials. This approach is described in detail in Dawson.[21])

Ion channel proteins allow ions to cross the normally ion-impermeable lipid bilayer by forming pores that create a polar pathway through which an ion can move by diffusion, and the movement of ions through such channels can be described by two distinct, but related parameters: permeability or conductance.[21] Here, we examine the relation between these two parameters as an example of how channel properties are assessed. For the sake of concreteness, we will consider a single channel that is highly selective for potassium. The conductance of the K^+ channel is most conventionally described using a form of Ohm's law:

$$i_K = \gamma_K \left(V_m - E_K \right) \tag{1.54}$$

In this equation of motion, i_K represents the flow of potassium through a *single* channel measured as an electrical current expected to be of the order of pico-amperes. The term in brackets is simply the value of the total electrochemical potential driving force ($\Delta\bar{\mu}_K$), expressed an electrical units ($\Delta\bar{\mu}_K/zF$), and γ_K is the single-channel conductance. The definition of γ_K is based on the determination of total charge flow measured as an electric current. As such, it represents a highly condensed summary of the physical process by which an ion leaves an aqueous solution and enters the channel, diffuses across the channel, and jumps out to rejoin the water molecules on the other side. Note that the equation assumes that the same coefficient (γ_K) describes the flow of potassium driven by *either* a gradient of electrical potential (an electric field) or a gradient of potassium concentration, despite the fact that the nature of these two driving forces is clearly quite different. In the case of the electric field, each ion experiences a driving force due to the electric field in the membrane, whereas E_K represents a virtual force or "pseudo-force" that reflects the statistical tendency of ions to move down the potassium concentration gradient.[19,37] In both cases, however, the mechanism of ion flow is the same (simple diffusion), so, as long as the driving forces are thermodynamically equivalent, the equivalence of γ_K seems justified. Note that nothing in the development of Equation 1.54 requires that γ_K be constant. In fact, it is most likely that γ_K will vary with ion concentration and voltage; that is, the $i–V$ relationship for the channel will be nonlinear.[21,36,37]

How does the single-channel conductance relate to the permeability of the channel to potassium? To explore this connection and its implications it is useful to consider the simplified case in which $V_m = 0$ and the sole driving force for potassium flow is a transmembrane concentration gradient. Ohm's law predicts that in this condition:

$$i_K = \gamma_K (E_K) \tag{1.55}$$

An alternative description can be obtained by applying Fick's law in the condition that $V_m = 0$; that is:

$$j_K = P_K ([K]_i - [K]_o) \tag{1.56}$$

where j_K is the net potassium flow through a single channel, and P_K is the permeability of the single channel to potassium which could be operationally defined in a hypothetical measurement of the movement of labeled (tracer) potassium through the channel at $V_m = 0$:

$$P_K = \left(\lambda_K^* \right)_{V_m=0} \tag{1.57}$$

(Pitfalls and complications in the interpretation of permeability are addressed in Dawson.[21]) For the single-channel potassium current, we obtain:

$$(i_K)_{V_m=0} = zF(j_K)_{V_m=0} = zFP_K ([K]_i - [K]_o) \tag{1.58}$$

We can connect these two very different-looking descriptions of single-channel current by analyzing a very simple channel model: a cylindrical pore containing potassium ions that diffuse as if in free solution. As shown by Finkelstein and Mauro,[36] we can derive an expression for the single-channel current in the condition $V_m = 0$:

$$(i_K)_{V_m=0} = \frac{(zF)^2}{RT} P_K [\Theta] E_K \tag{1.59}$$

where $[\Theta]$ is the *logarithmic mean* of the K^+ concentrations on the two sides of the membrane, that is:

$$[\Theta] = \frac{[K]_i - [K]_o}{\ln ([K]_i / [K]_o)} \tag{1.60}$$

This value can be thought of as a measure of the average concentration of K^+ within the pore. The term in brackets is the single-channel conductance (γ_K) where:

$$\gamma_K = \frac{(zF)^2}{RT} P_K [\Theta] \tag{1.61}$$

We see here that, even in the simplest channel model, conductance depends not only on ion permeability but also on the *abundance* (concentration) of the conducted ions. Conductance is proportional to ion concentration. If we plug in the expressions for Θ and E_K we end up where we started, with Equation 1.58:

$$(i_K)_{V_m=0} = zFP_K ([K]_i - [K]_o) \tag{1.62}$$

demonstrating that the two approaches produce the same result. It is instructive, however, to contrast the intuitive sense of the physical basis for ion conduction conveyed by Equation 1.58 and Equation 1.59. In the latter, the driving force for K$^+$ flow is identified in the thermodyamically correct form as E_K, which depends on the potassium concentration gradient. The rate of ion flow for any given value of E_K is determined by the conductance (γ_K) that reflects the channel permeability (P_K) but is also, itself, concentration dependent. In Equation 1.58, the thermodynamic driving force has been obscured, but the current is shown to depend simply on the product of the concentration difference and the permeability, as expected for simple diffusion. The effect of these different formulations on our view of the permeation process can be illustrated by comparing two different ways of applying a transmembrane concentration difference of 50 mM K. In both cases, we let $V_m = 0$. In one case, we impose a gradient of 55:5 mM K$^+$ and in the other 150:100 mM K$^+$. Equation 1.58 tells us immediately that for our model channel the current will be identical in either case, because the transmembrane difference in K$^+$ concentration is identical. Equation 1.59 arrives at the same endpoint, but via a different path. The electrochemical driving force for K$^+$ is clearly greater for the 55:5-mM gradient; the concentration ratio is greater. How, then, can the currents be equal? The answer is found in the conductance. The value of γ_K is larger for the 150:100-mM gradient (because the average concentration of K$^+$ is greater) by an amount just sufficient to compensate for the reduced driving force and equalize the currents.

B. Ionic Currents from Cells to Cell Layers

To fully appreciate the role of ion channels in the regulation of cellular composition, it is necessary to consider the impact of an ensemble of channels as measured by the macroscopic conductance determined, for example, in the whole-cell configuration of the patch clamp technique or in a transepithelial voltage-clamp experiment. In the case of potassium, to choose one example, it would not be unusual to find that a particular cell type was equipped with as many as three to six distinctly different K$^+$ channel types (different gene products) that are differentially regulated—that is, opened or closed under different conditions. These might include channels activated by depolarization of V_m, increases in cytosolic cAMP, Ca^{2+}, or cell swelling. For any single population of K$^+$ channels, however, we can define a *macroscopic K$^+$ current* (I_K), that would be given by:

$$I_K = g_K^i \left(V_m - E_K \right) \tag{1.63}$$

where g_K^i is the macroscopic conductance for a particular ensemble of N_K^i K$^+$ channels and is given by:

$$g_K^i = \gamma_K^i N_K^i P_o \tag{1.64}$$

Here γ_K^i is the single-channel conductance characteristics of a particular channel population, N_K^i is the number of these channels, and P_o is the probability of finding a channel in the open state, assumed here to be the same for all channels in the population. Equation 1.64 is a useful reference for thinking about channel regulation because contained within it are all of the means by which cells can regulate their macroscopic conductance. Much of conductance regulation revolves around the modulation of P_o by factors such as voltage and intracellular messengers, including phosphorylation. In addition, however, channel number can be modulated by the insertion and retrieval of channels from the cell membrane.

C. Transepithelial Ionic Flows and the Short-Circuit Current

The ability of epithelial cell layers, such as those that make up the lining of the gastrointestinal tract, kidney tubules gills, and other organs, to effect the net absorption and secretion of salt is a key element of osmotic regulation. In many cases, epithelial cell layers can be isolated and studied

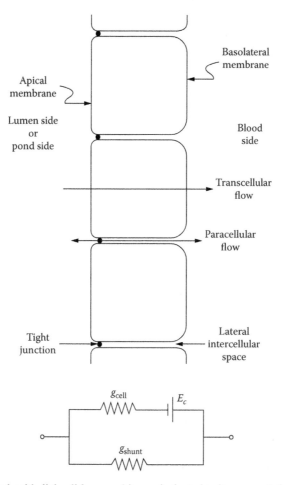

FIGURE 1.7 An idealized epithelial cell layer and its equivalent circuit representation.

under voltage-clamp conditions so the ion flows can be detected as a transepithelial electrical current. The simplest experimental configuration involves mounting a portion of the cell layer (e.g., a piece of intestine or bladder or gill epithelium) as a flat sheet separating two solutions configured such that each bath contains two electrodes: one for measuring the transepithelial electrical potential difference and another for passing electrical current across the cell layer. Chambers designed for this purpose are universally referred to as *Ussing chambers* in recognition of the ground-breaking studies by Koefoed-Johnson and Ussing[56] on sodium transport across isolated frog skin. The unique significance of this measurement is best appreciated by considering how a simple cell layer can be represented by an electrical equivalent circuit as suggested in Figure 1.7. Here we depict a homogeneous layer of cells connected by tight junctions such that, on purely morphological grounds, one can recognize a *transcellular* pathway comprised of the series arrangement of apical and basolateral membranes of the epithelial cells and a *paracellular* pathway comprised of the tight junctions and lateral intracellular spaces that lie between adjacent epithelial cells.

It is often found that, when an isolated epithelial cell layer is bathed on both sides by identical solutions such that all transepithelial ion gradients are eliminated, it is nevertheless possible to measure a transepithelial electrical potential ranging from several millivolts to over 100 mV, generally negative on lumen side. The origin of this transepithelial electrical potential difference can be illuminated by the analysis of a simple equivalent circuit representing the cell layer. The paracellular pathway (or shunt pathway) is represented in the equivalent circuit by a conductance, which accounts for the fact that, in the presence of a transepithelial driving force (voltage or ion concentration gradient), ion flows can occur through this shunt, which bypasses the cells. The transcellular pathway is represented by a transcellular conductance (g_c) and an electromotive force (emf) (E_c). The transepithelial conductance (g_c) represents the conductive properties of the series arrangement of apical and basolateral cell membranes; the emf is the most concise description of the capacity of the cell layer for electrogenic (charge-transporting) ion transport and accounts for the observation of a nonzero, transepithelial electrical potential in the absence of any transepithelial ion gradients. The latter observation constitutes unequivocal evidence for active ion transport across the cell layer.

The measured transepithelial potential (V_t) can be related to the elements of the simple equivalent circuit by treating it as a voltage divider:

$$V_{te} = \frac{g_c}{g_c + g_s} E_c \qquad (1.65)$$

It is immediately apparent that the transepithelial potential is, as expected, dependent on the electrogenic property of the cellular pathway, concisely represented by E_c. In addition, however, the value of V_t depends on the relative values of the cellular and paracellular conductances. This sort of perspective has given rise to the general expectation that a leaky epithelial cell layer $(g_s \gg g_c)$ should be characterized by a smaller value of V_{te} than a so-called *tight* epithelium $(g_s \ll g_c)$, a useful general guide.[39] The significance of the short-circuit current (I_{sc}) is readily apparent if we compute the current required to reduce V_t to zero. In this condition, current through the shunt path (g_s) must be zero due to the lack of any driving force, so the short-circuit current is given by:

$$I_{sc} = g_c E_c \qquad (1.66)$$

Equation 1.66 makes it clear that I_{sc} measures the properties of the *cellular transport path*, thereby providing a sensitive, time-resolved assay of active ion transport. Although we arrived at this perspective on the significance of I_{sc} using a particularly simple example, the range of applications for this approach is remarkably broad and has proven useful for characterizing a wide range of transporting epithelia, including frog skin, intestine, airways, gills, and others.[29] Indeed, some 60 years after its introduction by Ussing and Zerahn,[80] the short-circuit technique remains a key analytical tool in studies of epithelial transport.

V. A GLANCE AT THE MOLECULAR BASIS FOR SOLUTE AND WATER TRANSPORT

We close this chapter with a fleeting glance at the molecular basis for the solute and water transport processes that are the basis of osmoregulation. This area of inquiry is undergoing a virtual explosion as ever more crystal structures for transport proteins are obtained, particularly structures representing different transporting states of the same protein. The future of such studies clearly lies in understanding the dynamic changes in protein structure that are the basis for the wide variety of transport activities that underlie osmoregulation. Here, we aim to establish a framework for the motivated reader who can follow these developments on his or her own. One excellent Web site on membrane proteins can be found at http://portal.acm.org/citation.cfm?id=1181568.1181585 (*Bioinformatics*, 22(5), 623–625, 2006).

A. CHANNELS AND TRANSPORTERS

For the purpose of structural categorization, it is useful to group all of the proteins engaged in transmembrane solute and water traffic into two categories: channels and transporters. In both cases, the translocation event is accompanied by (or brought about by) some sort of conformational change.

A channel is distinguished by the fact that, at some point during its conformational cycle, it represents an open pathway across the membrane through which ions, for example, can move under the influence of an electrochemical potential gradient at a high rate, 10^6 per second or faster. Channel proteins cycle through conformations that are open (conducting) or closed (nonconducting), and the movement of protein segments underlying these opening and closing (gating) reactions are now beginning to be resolved.[57,60,75] A transporter can also be described as being *gated*, but in such a way that a binding site (say, for sodium ions or glucose) is accessible from only one side of the membrane at any instant. These so-called *alternating access* mechanisms can now be visualized in atomic detail.[1,7,12,43,52]

Transporter mechanisms are distinguished by a somewhat slower rate of turnover because the rate of solute transport is limited by the rate of the conformational change that must accompany the switch from the outward-facing to the inward-facing form of the protein. Within this simple difference, the open pathway of channels and alternating access of transporters, lies the basis for the unique features of both groups. The presence of an open pathway permits high rates of ion translocation through channels, and alternating access provides the basic substrate for coupling the transport event to some energy-yielding process, such as the hydrolysis of ATP. Emerging evidence suggests that for some proteins, at least, the structural basis for this critical difference between transporters and channels may be relatively minor. In the case of the CLC family of chloride channels, for example, the bacterial cousin, once thought to be an anion-conducting channel, has been shown to be a proton–chloride antiporter. The translocation pathway of channel family members differs from the transporter by a single glutamate residue.[2,3] Likewise, the product of the cystic fibrosis gene, a chloride channel known as CFTR, is a member of a large family of transporters, a heritage that may become even more apparent when the mechanism of CFTR gating is fully resolved.[41,82] The ubiquitous Na,K-ATPase is possessed of an underlying channel character that can be revealed by modifying the protein with a marine toxin that converts the transporter to a channel gated by permeant ions.[10,73]

B. Selectivity

The exquisite selectivity of transporters and channels has intrigued investigators for decades, and the basis for this discrimination is now being revealed at the atomic scale for membrane proteins. Here, we consider two examples to whet the appetite of the reader: potassium selectivity of K^+ channels and water selectivity of the aquaporins.

1. Physical Basis of Potassium Selectivity

One of the most enduring puzzles in biology is the ability of a potassium channel to select potassium over the smaller univalent cation sodium. Sodium exclusion from the pore is so effective that highly selective K^+ channels basically do not conduct Na^+. This Na^+ exclusion is a critical factor in the creation of resting membrane potentials and a variety of excitability phenomena. Early theories focused on two contributions to selectivity that might be grouped under the headings "chemistry" and "geometry." Eisenman's approach emphasized chemistry—specifically, the electrostatic interactions of permeant cations with specific, coordinating ligands, such as carbonyl oxygens, in comparison to water coordination.[27,28] Bezanilla and Armstrong,[11] on the other hand, emphasized the role of geometry. They advanced the *snug-fit hypothesis*, the notion that the geometry of the intrachannel coordination site, like the cage of the potassium-selective antibiotic valinomycin,[25,26] rendered potassium coordination energetically more favorable than that of sodium. The advent of atomic-scale structures for potassium channels provided the basis for a detailed appraisal of selectivity theories and a reconsideration of the relative roles of chemistry and geometry. The first crystal structure for the bacterial potassium channel (KcsA) revealed the ion-conducting pore and, in particular, the long sought-after *selectivity filter*.[23] The potassium selectivity filter is constructed in such a way as to orient backbone carbonyl oxygens toward the center of the pore, where they can coordinate and stabilize visiting potassium ions, replacing, in a sense, the waters of hydration that stabilize potassium ions in the aqueous solution. The geometry of the potassium coordination sites is controlled by the interaction of the corresponding amino acid side chains (oriented away from the pore) that interact with elements of the surrounding protein. The geometry of a functioning selectivity filter is not expected to be rigid but instead to vary in space and time due to thermal influence.[5]

Valiyaveetil and Mackinnon[81] recently proposed that the K channel selectivity filter exhibits a sort of conformational selection. They proposed that the binding of a potassium ion to the potassium channel selectivity filter stabilizes a specific conformation that is favorable to K^+ conduction, whereas the binding of the sodium ion does not.[81] Side-chain, protein-surround interactions may determine the allowable conformations of the selectivity filter and may be tuned in such a way as to favor a conducting conformation that requires a bound K^+ for stability.

2. Water Selectivity of Aquaporins

The founding member of the aquaporin family, discovered by Peter Agre (who shared the 2003 Nobel Prize with Rod MacKinnon), was originally named CHIP28 because it was a 28-kDa red cell membrane protein.[71,72] The subsequent discovery that expression of CHIP28 conferred water permeability on cell membranes led to its identification as a water channel.[72] Family members are widely expressed in a variety of salt- and water-transporting organs.[4,53] It was expected that a channel that conducts water molecules would be highly conductive to protons. Protons exhibit an anomalously high mobility in water because they can be passed from one water molecule to the next, an effect that requires the reorientation of water molecules along the chain. Anomalously high proton conductance is also observed in the gramicidin channel through which water molecules must move in single file, a phenomenon referred to as a *proton wire effect* that is aided by the reorientation of successive water molecules in the chain.[70] Aquaporin 7, however, was found to conduct water, but not protons! This surprising selectivity was explained by comparing experimental results with molecular dynamics simulations of the single-file translocation of water molecules through the pore.[76,87] As water molecules translocate in single file, the structure and electrostatic properties of the pore prevent the uniform orientation and reorientation necessary for the proton wire effect, thereby blocking proton conduction.

ACKNOWLEDGMENTS

We are grateful for editorial suggestions from Erik Hviid Larsen, Stanley Hillyard, and David Evans and for the many insights provided by the writings of Professor Alan Finkelstein. The writing of the chapter was supported by grants from NIDDK and the Cystic Fibrosis Foundation. The chapter would never have been completed but for the patient prodding of brother Daryl.

REFERENCES

1. Abramson, J., Smirnova, I., Kasho, V., Verner, G., Kaback, H.R., and Iwata, S., Structure and mechanism of the lactose permease of *Escherichia coli. Science*, 301(5633), 610–615, 2003.
2. Accardi, A. and Miller, C., Secondary active transport mediated by a prokaryotic homologue of ClC Cl⁻ channels. *Nature*, 427(6977), 803–807, 2004.
3. Accardi, A., Walden, M., Nguitragool, W., Jayaram, H., Williams C., and Miller, C., Separate ion pathways in a Cl⁻/H⁺ exchanger. *J. Gen. Physiol.*, 126(6), 563–570, 2005.
4. Agre, P. and Kozono, D., Aquaporin water channels: molecular mechanisms for human diseases. *FEBS Lett.*, 555(1), 72–78, 2003.
5. Allen, T.W., Andersen O.S., and Roux, B., On the importance of atomic fluctuations, protein flexibility, and solvent in ion permeation. *J. Gen. Physiol.*, 124(6), 679–690, 2004.

6. Andrews, F.C., Colligative properties of simple solutions. *Science*, 194(4265), 567–571, 1976.
7. Armstrong, N, Jasti, J., Beich-Frandsen M., and Gouaux, E., Measurement of conformational changes accompanying desensitization in an ionotropic glutamate receptor. *Cell*, 127(1), 85–97, 2006.
8. Aronson, P.S., Identifying secondary active solute transport in epithelia. *Am. J. Physiol.*, 240(1), F1–F11, 1981.
9. Aronson, P.S., Electrochemical driving force for secondary active transport: energetics and kinetics of Na^+–H^+ exchanger and Na^+–glucose cotransporter. In *Electrogenic Transport: Fundamental Principles and Physiological Implications*, Blaustein, M.P. and Lieberman, M., Eds., Raven Press, New York, 1984, pp. 49–70.
10. Artigas, P. and Gadsby, D.C., Na^+/K^+-pump ligands modulate gating of palytoxin-induced ion channels. *Proc. Natl. Acad. Sci. USA*, 100(2), 501–505, 2003.
11. Bezanilla, F. and Armstrong, C.M., Negative conductance caused by entry of sodium and cesium ions into the potassium channels of squid axons. *J. Gen. Physiol.*, 60(5), 588–608, 1972.
12. Boudker, O., Ryan, R.M., Yernool, D., Shimamoto, K., and Gouaux, E., Coupling substrate and ion binding to extracellular gate of a sodium-dependent aspartate transporter. *Nature*, 445(7126), 387–393, 2007.
13. Britton, H.G., Induced uphill and downhill transport: relationship to the Ussing criterion. *Nature*, 198, 190–191, 1963.
14. Britton, H.G., The Ussing relationship and chemical reactions: possible application to enzymatic investigations. *Nature*, 205, 1323–1324, 1965.
15. Cope, F.W. and Damadian, R., Biological ion exchanger resins. IV. Evidence for potassium association with fixed charges in muscle and brain by pulsed nuclear magnetic resonance of 39K. *Physiol. Chem. Physiol.*, 6(1), 17–30, 1974.
16. Damadian, R., Structured water or pumps? *Science*, 193(4253), 528–530, 1976.
17. Dawson, D.C., Tracer flux ratios: a phenomenological approach. *J. Membr. Biol.*, 31(4), 351–358, 1977.
18. Dawson, D.C., Thermodynamic aspects of radiotracer flow. In *Biological Transport of Radiotracers*, Colombetti, L.G., Ed., CRC Press, Boca Raton, 1982, pp. 79–95.
19. Dawson, D.C., Principles of membrane transport. In *Handbook of Physiology*. Section 6, Vol. IV. *Intestinal Absorption and Secretion*, Schultz, S.G., Field, M., and Frizzell, R.A., Eds., American Physiological Society, Bethesda, MD, 1991, pp. 1–44.
20. Dawson, D.C., Water transport: principles and perspectives. In *The Kidney: Physiology and Pathophysiology*, Seldin, D.W. and Giebisch, G., Eds., Raven Press, New York, 1992, pp. 301–316.
21. Dawson, D.C., Permeability and conductance in ion channels: a primer. In *Molecular Biology of Membrane Transport Disorders*, Andreoli, T.F., Hoffman, J.F., Fanestil, D.D., and Schultz, S.G., Eds., Plenum Press, New York, 1996, pp. 87–109.
22. Denbigh, K., *The Principles of Chemical Equilibrium*. Cambridge University Press, New York, 1971.
23. Doyle, D.A., Morais, J., Cabral, R., Pfuetzner, A., Kuo, A., Gulbis, J.M., Cohen, S.L., Chait, B.T., and MacKinnon, R., The structure of the potassium channel: molecular basis of K^+ conduction and selectivity. *Science*, 280(5360), 69–77, 1998.
24. Dutzler, R., Campbell, E.B., Cadene, M., Chait B.T., and MacKinnon, R., X-ray structure of a ClC chloride channel at 3.0 Å reveals the molecular basis of anion selectivity. *Nature*, 415(6869), 287–294, 2002.
25. Eigen, M. and Winkler, R., Alkali ion carriers: dynamics and selectivity. In *The Neurosciences Second Study Program*, Schmitt, F.D., Ed., Rockefeller Press, New York, 1970, pp. 685–696.
26. Eigen, M. and Winkler, R., Carriers and specificity in membranes. II. Characteristics of carriers. Alkali ion carriers: specificity, architecture, and mechanisms: an essay. *Neurosci. Res. Progr. Bull.*, 9(3), 330–338, 1971.
27. Eisenman, G., Some elementary factors involved in specific ion permeation. In *Proc. of the 23rd Int. Congr. Physiol. Sci.*, Vol. 4, Excerpta Medical Foundation, Amsterdam, 1965, pp. 489–506.
28. Eisenman, G. and Horn, R., Ionic selectivity revisited: the role of kinetic and equilibrium processes in ion permeation through channels. *J. Membr. Biol.*, 76(3), 197–225, 1983.
29. Evans, D.H., Piermarini P.M., and Choe, K.P., The multifunctional fish gill: dominant site of gas exchange, osmoregulation, acid–base regulation, and excretion of nitrogenous waste. *Physiol. Rev.*, 85(1), 97–177, 2005.

30. Finkelstein, A., Nature of the water permeability increase induced by antidiuretic hormone (ADH) in toad urinary bladder and related tissues. *J. Gen. Physiol.*, 68(2), 137–143, 1976.

31. Finkelstein, A., Water and nonelectrolyte permeability of lipid bilayer membranes. *J. Gen Physiol.* 68(2), 127–135, 1976.

32. Finkelstein, A., *Water Movement Through Lipid Bilayers, Pores, and Plasma Membranes: Theory and Reality*, Wiley Interscience, New York, 1987.

33. Finkelstein, A. and Andersen, O.S., The gramicidin A channel: a review of its permeability characteristics with special reference to the single-file aspect of transport. *J. Membr. Biol.*, 59(3), 155–171, 1981.

34. Finkelstein, A. and Cass A., Permeability and electrical properties of thin lipid membranes. *J. Gen. Physiol.*, 52(1), 145–173, 1968.

35. Finkelstein, A. and Holz, R., Aqueous pores created in thin lipid membranes by the polyene antibiotics nystatin and amphotericin B. *Membranes*, 2, 377–408, 1973.

36. Finkelstein, A. and Mauro, A., Equivalent circuits as related to ionic systems. *Biophys. J.*, 3, 215–237, 1963.

37. Finkelstein, A. and Mauro, A., Physical principles and formalisms of electrical excitability. In *Handbook of Physiology*. Section 1, Vol. 1. *Cellular Biology of Neurons*. American Physiological Society, Bethesda, MD, 1977, pp. 161–213.

38. Finkelstein, A. and Rosenberg, P.A., Single-file transport: Implications for ion and water movement through gramicidin A channels. In *Membrane Transport Processes*, Stevens, C.F. and Tsien, R.W., Eds., Raven Press, New York, 1979, pp. 73–88.

39. Fromter, E. and Diamond, J., Route of passive ion permeation in epithelia. *Nat. New Biol.*, 235(53), 9–13, 1972.

40. Fujiyoshi, Y., Mitsuoka, K., de Groot, B.L., Philippsen, A., Grubmuller, H., Agre P., and Engel, A., Structure and function of water channels. *Curr. Opin. Struct. Biol.*, 12(4), 509–515, 2002.

41. Gadsby, D.C., Vergani, P., and Csanady, L., The ABC protein turned chloride channel whose failure causes cystic fibrosis. *Nature*, 440(7083), 477–483, 2006.

42. Gibbs, J.W., *On the Equilibrium of Heterogeneous Substance: The Collected Works of J. Willard Gibbs*. Yale University Press, New Haven, CT, 1928.

43. Gouaux, E. and Mackinnon, R., Principles of selective ion transport in channels and pumps. *Science*, 310(5753), 1461–1465, 2005.

44. Haas, M. and Forbush, 3rd, B., The Na–K–Cl cotransporter of secretory epithelia. *Annu. Rev. Physiol.*, 62, 515–534, 2000.

45. Hammel, H.T., Colligative properties of a solution. *Science*, 192(4241), 748–756, 1976.

46. He, Z., Tong, Q., Quednau, B.D., Philipson, K.D., and Hilgemann, D.W., Cloning, expression, and characterization of the squid Na^+–Ca^{2+} exchanger (NCX-SQ1). *J. Gen. Physiol.*, 111(6), 857–873, 1998.

47. Hilgemann, D., Matsuoka, W.S., Nagel, G.A., and Collins, A., Steady-state and dynamic properties of cardiac sodium–calcium exchange: sodium-dependent inactivation. *J. Gen. Physiol.*, 100(6), 905–932, 1992.

48. Hodgkin, A.L. and Huxley, A.F., Currents carried by sodium and potassium ions through the membrane of the giant axon of *Loligo*. *J. Physiol.*, 116, 449–472, 1952.

49. Hodgkin, A.L. and Huxley, A.F., A quantitative description of membrane current and its application to conduction and excitation in nerve. *J. Physiol.*, 117, 500–544, 1952.

50. Hodgkin, A.L. and Keynes, R.D., The potassium permeability of a giant nerve fibre. *J. Physiol. (Lond.)*, 128, 61–88, 1955.

51. Holz, R. and Finkelstein, A., The water and nonelectrolyte permeability induced in thin lipid membranes by the polyene antibiotics nystatin and amphotericin B. *J. Gen. Physiol.*, 56(1), 125–45, 1970.

52. Kaback, H.R., Dunten, R., Frillingos, S., Venkatesan, P., Kwaw, I., Zhang, W., and Ermolova, N., Site-directed alkylation and the alternating access model for LacY. *Proc. Natl. Acad. Sci. USA*, 104(2), 491–494, 2007.

53. King, L.S., Kozono, D., and Agre, P., From structure to disease: the evolving tale of aquaporin biology. *Nat. Rev. Mol. Cell Biol.*, 5(9), 687–698, 2004.

54. Kirk, K.L. and Dawson, D.C., Basolateral potassium channel in turtle colon: evidence for single-file ion flow. *J. Gen. Physiol.*, 82(3), 297–329, 1983.

55. Kirschner, L., Water and ions. In *Comparative Animal Physiology*. Vol. 1. *Environmental and Metabolic Animal Physiology*, Prosser, C.L., Ed., Wiley-Liss, New York, 1991.

56. Koefoed-Johnsen, V. and Ussing, H.H., The nature of the frog skin potential. *Acta Physiol. Scand.*, 42, 298–308, 1958.

57. Lee, S., Lee, Y.A., Chen, J., and MacKinnon, R., Structure of the KvAP voltage-dependent K+ channel and its dependence on the lipid membrane. *Proc. Natl. Acad. Sci. USA*, 102(43), 15441–1546, 2005.

58. Linck, B., Qiu, Z., He, Z., Tong, Q., Hilgemann D.W., and Philipson, K.D., Functional comparison of the three isoforms of the Na^+/Ca^{2+} exchanger (NCX1, NCX2, NCX3). *Am J. Physiol.*, 274(2, Pt. 1), C415–C423, 1998.

59. Ling, G.N., *A Physical Theory of the Living State: The Association-Induction Hypothesis*, Blaisdell, New York, 1962.

60. Long, S.B., Tao, X., Campbell E.B., and MacKinnon, R., Atomic structure of a voltage-dependent K+ channel in a lipid membrane-like environment. *Nature*, 450(7168), 376–382, 2007.

61. Mauro, A., Nature of solvent transfer in osmosis. *Science*, 126(3267), 252–253, 1957.

62. Mauro, A., Some properties of ionic and nonionic semipermeable membranes. *Circulation*, 21, 845–854, 1960.

63. Mauro, A., Osmotic flow in a rigid porous membrane. *Science*, 149, 867–869, 1965.

64. Mauro, A., Forum on osmosis. III. Comments on Hammel and Scholander's solvent tension theory and its application to the phenomenon of osmotic flow. *Am. J. Physiol.*, 237(3), R110–R113, 1979.

65. Mauro, A., *The Role of Negative Pressure in Osmotic Equilibrium and Osmotic Flow*, Alfred Benzon Foundation, Copenhagen, Denmark, 1981.

66. Minkoff, L. and Damadian, R., Caloric catastrophe. *Biopolymers*, 13(2), 167–178, 1973.

67. Minkoff, L. and Damadian, R., Energy requirements of membrane pumps [letter]. *N. Engl. J. Med.*, 292(3), 162–163, 1975.

68. Morais-Cabral, J.H., Zhou, Y., and MacKinnon, R., Energetic optimization of ion conduction rate by the K+ selectivity filter. *Nature*, 414(6859), 37–42, 2001.

69. Payne, J.A. and Forbush, 3rd, B., Molecular characterization of the epithelial Na–K–Cl cotransporter isoforms. *Curr. Opin. Cell Biol.*, 7(4), 493–503, 1995.

70. Pomes, R. and Roux, B. Structure and dynamics of a proton wire: a theoretical study of H+ translocation along the single-file water chain in the gramicidin A channel. *Biophys. J.*, 71(1), 19–39, 1996.

71. Preston, G.M. and Agre, P., Isolation of the cDNA for erythrocyte integral membrane protein of 28 kilodaltons: member of an ancient channel family. *Proc. Natl. Acad. Sci. USA*, 88(24), 11110–1114, 1991.

72. Preston, G.M., Carroll, T.P., Guggino, W.B., and Agre, P., Appearance of water channels in *Xenopus* oocytes expressing red cell CHIP28 protein. *Science*, 256(5055), 385–387, 1992.

73. Reyes, N. and Gadsby, D.C., Ion permeation through the Na+,K+-ATPase. *Nature*, 443(7110), 470–474, 2006.

74. Rivera, C., Voipio, J., and Kaila, K., Two developmental switches in GABAergic signalling: the K+-Cl- cotransporter KCC2 and carbonic anhydrase CAVII. *J. Physiol.*, 562(Pt. 1), 27–36, 2005.

75. Schmidt, D., Jiang, Q.X., and MacKinnon, R., Phospholipids and the origin of cationic gating charges in voltage sensors. *Nature*, 444(7120), 775–779, 2006.

76. Tajkhorshid, E., Nollert, P., Jensen, M.O., Miercke, L.J., O'Connell, J., Stroud, R.M., and Schulten, K., Control of the selectivity of the aquaporin water channel family by global orientational tuning. *Science*, 296(5567), 525–530, 2002.

77. Ussing, H.H., The distinction by means of tracers between active transport and diffusion. *Acta Physiol. Scand.*, 19, 43–56, 1949.

78. Ussing, H.H. Some aspects of the application of tracers in permeability studies. *Adv. Enzymol.*, 13, 21–65, 1952.

79. Ussing, H.H., Interpretation of tracer fluxes. In *Membrane Transport in Biology*, Giebisch, G., Testeson, D.C., and Ussing, H.H., Springer-Verlag, Berlin, 1978, pp. 115–140.

80. Ussing, H.H. and Zerahn, K., Active transport of sodium as the source of electric current in the short-circuited isolated frog skin. *Acta Physiol. Scand.*, 23(2–3), 110–127, 1951.

81. Valiyaveetil, F.I., Leonetti, M., Muir, T.W., and Mackinnon, R., Ion selectivity in a semisynthetic K+ channel locked in the conductive conformation. *Science*, 314(5801), 1004–1007, 2006.

82. Vergani, P., Lockless, S.W., Nairn, A.C., and Gadsby, D.C., CFTR channel opening by ATP-driven tight dimerization of its nucleotide-binding domains. *Nature*, 433(7028), 876–880, 2005.

83. Watson, J.D. and Crick, F.H., Molecular structure of nucleic acids: a structure for deoxyribose nucleic acid. *Nature*, 171(4356), 737–738, 1953.
84. Zhou, M. and MacKinnon, R., A mutant KcsA K(+) channel with altered conduction properties and selectivity filter ion distribution. *J. Mol. Biol.*, 338(4), 839–846, 2004.
85. Zhou, Y. and MacKinnon, R., The occupancy of ions in the K$^+$ selectivity filter: charge balance and coupling of ion binding to a protein conformational change underlie high conduction rates. *J. Mol. Biol.*, 333(5), 965–975, 2003.
86. Zhou, Y., Morais-Cabral, J.H., Kaufman, A., and MacKinnon, R., Chemistry of ion coordination and hydration revealed by a K$^+$ channel–Fab complex at 2.0 Å resolution. *Nature*, 414(6859), 43–48, 2001.
87. Zhu, F., Tajkhorshid, E., and Schulten, K., Theory and simulation of water permeation in aquaporin-1. *Biophys. J.*, 86(1, Pt. 1), 50–7, 2004.

APPENDIX 1. OSMOTIC FLOW ACROSS A LIPID MEMBRANE

(See Finkelstein, 1987, and Dawson, 1992.) We assume here that diffusional water flow through the oil layer (lipid membrane) can be adequately described by Fick's law, which for the molar flow of water yields:

$$J_w = \left(\frac{D_w A_m}{\delta_m} \right) \Delta C_w (M) \tag{A1.1}$$

where the term $\Delta C_w(M)$ refers to the concentration of water *within the oil layer* where water is a dilute component. D_w is the diffusion coefficient of water in the oil membrane, A_m is the membrane area, and δ_m is the membrane thickness.

To relate this description to the difference in water concentration between the two bathing solutions we need to investigate the boundary condition at the interface between the layer of oil and the bulk water-containing solution. $\Delta C_w(M)$, the water concentration difference within the membrane, can be written as:

$$\Delta C_w (M) = C_w (M1) - \Delta C_w (M2) \tag{A1.2}$$

where the latter two terms refer to the concentration of water just inside the oil layer on either side. To relate these to the water concentration in the solutions bathing the membrane, we define the equilibrium partition coefficient (β_w) as:

$$\beta_w = \frac{X_w (M)}{X_w (B)} \tag{A1.3}$$

which is that ratio of the mole fraction of water just inside the membrane to that in the bulk solution, $X_w(B)$.

Recalling our dilute solution approximation, we can express $C_w(M)$ within the membrane as:

$$C_w (M) = \frac{X_w (M)}{v_m} \tag{A1.4}$$

where v_m is the partial molar volume of the oil phase in which the water is dissolved. Inserting Equation A1.3 we obtain the relation between the *concentration* of water within the lipid layer and the *mole fraction* of water in the bulk solution:

$$C_w (M) = \frac{1}{v_m} \beta_w X_w (B) \tag{A1.5}$$

But, the mole fraction of water in the bulk solution (solute-containing) can be replaced with $1 - X_s$, which with Equation A1.5 yields:

$$C_w(M) = \frac{1}{\nu_m} \beta_w \left(1 - C_s \nu_w\right) \tag{A1.6}$$

where C_s and ν_w refer to the aqueous bathing solution. Finally, for $\Delta C_w(M)$ we obtain:

$$\Delta C_w(M) = \beta_w \left(\frac{\nu_w}{\nu_m}\right) \Delta C_s \tag{A1.7}$$

and for the molar water flow:

$$J_w = \left(\frac{D_w \beta_w \nu_w}{\delta \nu_m}\right) A_m \Delta C_s \tag{A1.8}$$

where the term in the bracket represents the *osmotic water permeability* (P_f):

$$P_f = \frac{D_w \beta_w \nu_w}{\delta_m \nu_m} \tag{A1.9}$$

APPENDIX 2. ENERGETICS OF SODIUM–CALCIUM EXCHANGE

(See Dawson, 1991.) We assume a Na^+/Ca^{2+} exchange with a fixed stoichiometry of $3Na^+:1Ca^{2+}$. The equilibrium point for this process would be described by:

$$3\Delta\bar{\mu}_{Na} = \Delta\bar{\mu}_{Ca} \tag{A2.1}$$

where we define the difference so an inwardly direct gradient is a positive number.

Inserting the variables for concentration and voltage and assuming no difference in pressure or standard chemical potential, we obtain from Equation 1.45:

$$-3RT \ln \frac{[Na^+]_o}{[Na^+]_i} + 3z_{Na} FV_m = -RT \ln \frac{[Ca^{2+}]_o}{[Ca^{2+}]_i} + z_{Ca} FV_m \tag{A2.2}$$

where $V_m = V_i - V_o$. Inserting $z_{Na} = 1$ and $z_{Ca} = 2$ yields:

$$-RT \ln \frac{[Ca^{2+}]_o}{[Ca^{2+}]_i} = -3RT \ln \frac{[Na^+]_o}{[Na^+]_i} + FV_m \tag{A2.3}$$

Inserting common logs and dividing by F we obtain:

$$2.3\frac{RT}{F} \log \frac{[Ca^{2+}]_o}{[Ca^{2+}]_i} = 3\left(2.3\frac{RT}{F} \log \frac{[Na^+]_o}{[Na^+]_i}\right) - V_m \tag{A2.4}$$

So, using $2.3(RT/F) = 60$ mV and assuming that $V_m = -60$ mV, we see that the minimum intracellular concentration of calcium will be $10^{-7} M$ assuming $[Ca^{2+}]_o = 1$ mM.

2 Volume Regulation and Osmosensing in Animal Cells

Keith Choe and Kevin Strange

CONTENTS

I. INTRODUCTION

Maintenance of a constant volume in the face of extracellular and intracellular osmotic perturbations is a critical problem faced by all cells. Most cells respond to swelling or shrinkage by activating specific membrane transport or metabolic processes that serve to return cell volume to its normal resting state. These processes are essential for normal cell function and survival. This chapter provides an overview of the cellular and molecular events underlying cell volume homeostasis. Our discussion is focused on animal cell volume regulation; however, where appropriate, we discuss studies in bacteria and fungi that may provide insights into the molecular mechanisms of how animal cells detect volume changes and activate regulatory responses.

II. WATER FLOW ACROSS CELL MEMBRANES

The bulk movement of water across a semipermeable membrane is termed *osmosis*. An ideal semipermeable membrane is one that is permeable only to water. If such as membrane separates solutions with different solute concentrations—for example, 0.1-M NaCl on one side and 1-M NaCl on the other—water will move from the dilute into the concentrated NaCl solution. Water flow will continue until the NaCl concentrations in both solutions are equalized. The driving force for water flow is the concentration gradient for water. The concentration of water is higher in the 0.1-M NaCl solution compared to the 1-M NaCl solution.

Osmotic water flow across the membrane can be prevented by applying an opposing hydrostatic force. The pressure required to stop water flow is termed the *osmotic pressure*. The mathematical expression that defines osmotic pressure was derived by van't Hoff:

$$\Delta\pi = RT\Delta C_i \tag{2.1}$$

where $\Delta\pi$ is the osmotic pressure difference, R is the gas constant, T is the absolute temperature, and ΔC_i is the difference in solute concentration across the membrane.

Osmotic pressure is dependent upon the total concentration of dissolved solute particles. The terms *osmolality* and *osmolarity* indicate the total number of dissolved particles present in a kilogram of water and a liter of solution, respectively. One osmole is 1 mole of particles, which is 6.02×10^{23} individual particles. Osmolality and osmolarity are used interchangeably when referring to the relatively dilute intracellular and extracellular solutions of animals.

The above discussion of osmosis is based on the simplifying concept that water flow is occurring across a membrane permeable only to water. Real membranes are not quite so simple. All membranes have finite solute permeabilities. Although many biologically relevant solutes have permeabilities substantially lower than water and behave as if they were effectively impermeable, some solutes have permeabilities approaching that of water. These high-permeability solutes diffuse across the membrane down their concentration gradient. As they do so, the osmotic pressure driving water flow is reduced. If the movement of solute is fast enough, the concentrations of the solute on the two sides of the membrane can become equalized before significant osmotic water flow occurs.

To account for the nonideal behavior of membranes, Staverman defined the term *reflection coefficient* for solute i (σ_i) as:

$$\sigma_i = \frac{\Delta\pi_{obs}}{\Delta\pi_{th}} \tag{2.2}$$

where $\Delta\pi_{obs}$ is the observed osmotic pressure and $\Delta\pi_{th}$ is the theoretical osmotic pressure obtained from Equation 2.1. The reflection coefficient is a dimensionless term that ranges from 1 for a solute that behaves as if it were effectively impermeant (i.e., the solute is reflected by the membrane) to 0 for a solute whose permeability is similar to that of water. The *effective osmotic pressure* ($\Delta\pi_{eff}$) across a membrane generated by solute i is, therefore,

$$\Delta\pi_{eff} = \sigma_i RT\Delta C_i \tag{2.3}$$

The flow of water (J_v) across a membrane is defined as:

$$J_v = L_p(\sigma_i\Delta\pi_{th} - \Delta P) \tag{2.4}$$

where L_p is the *hydraulic conductivity coefficient* of the membrane, and ΔP is the hydrostatic pressure difference across the membrane. The hydraulic conductivity coefficient is a measure of the water permeability of the membrane. Cell membranes do not generate and maintain significant hydrostatic pressure gradients. Thus, when considering water flow into and out of animal cells, the ΔP term in Equation 2.4 is usually ignored; however, in organisms with relatively rigid cell walls, such as bacteria, plants, and yeast, significant hydrostatic pressure gradients can be generated and play important roles in driving water flow.

Water flow across most biological membranes occurs by simple diffusion of water molecules through the lipid bilayer; however, some cells possess specialized proteins that form transmembrane water-selective pores termed *aquaporins*.[104] Aquaporins dramatically increase cell membrane water permeability.

III. FUNDAMENTALS OF CELL VOLUME REGULATION

A. Mechanisms of Cell Volume Perturbation

Water is effectively in thermodynamic equilibrium across the plasma membrane of animal cells. In other words, the osmotic concentration of cytoplasmic (π_i) and extracellular (π_o) fluids are equal under steady-state conditions. Changes in intracellular or extracellular solute content generate a transmembrane osmotic gradient ($\Delta\pi$). Cell membranes are freely permeable to water, so any such gradient results in the immediate flow of water into or out of the cell until equilibrium is again achieved. Because animal cell membranes are unable to generate or sustain significant hydrostatic pressure gradients, water flow causes cell swelling or shrinkage.

Cell volume changes are usually grouped into two broad categories: *anisosmotic* and *isosmotic*. Anisosmotic volume changes are induced by alterations in extracellular osmolality. Under normal physiological conditions, most mammalian cells, with a few noteworthy exceptions (e.g., cells in the renal medulla and gastrointestinal tract), are protected from anisosmotic volume changes by the precise regulation of plasma osmolality by the kidney; however, a variety of disease states can disrupt the regulation of plasma osmolality.[127,229] In addition, many nonmammalian animals have limited abilities to regulate extracellular osmolality or are osmoconformers. The cells of these animals can be exposed to substantial osmotic stress during fluctuations in the osmolality of extracellular body fluids.

Isosmotic volume changes are brought about by alterations in intracellular solute content. All cells are threatened by possible isosmotic swelling or shrinkage. Under steady-state conditions, intracellular solute levels are held constant by a precise balance between solute influx and efflux across the plasma membrane, as well as by the metabolic production and removal of osmotically active substances. A variety of physiological and pathophysiological conditions, however, can disrupt this balance.[127,229] For example, the cell swelling that occurs in the mammalian brain following a stroke or head trauma is an example of isosmotic volume increase and is due to intracellular accumulation of NaCl and other solutes.

B. Cell Volume Regulation

Cells respond to volume perturbations by activating volume-regulatory mechanisms. The processes by which swollen and shrunken cells return to a normal volume are termed *regulatory volume decrease* (RVD) and *regulatory volume increase* (RVI), respectively (Figure 2.1A). Cell volume can only be regulated by the gain or loss of osmotically active solutes, primarily inorganic ions such as Na^+, K^+, and Cl^- or small organic molecules termed *organic osmolytes*.

Volume-regulatory electrolyte loss and gain are mediated exclusively by membrane transport processes.[112,145] In most animal cells, RVD occurs through loss of KCl via activation of separate K^+ and Cl^- channels or by activation of K–Cl cotransporters. Regulatory volume increase occurs by uptake of both KCl and NaCl. Accumulation of these salts is brought about by activation of Na^+/H^+ and Cl^-/HCO_3^- exchangers or Na–K–2Cl cotransporters. Figure 2.1B illustrates the ion transport systems commonly involved in cell volume regulation. Activation of these transport pathways is rapid and occurs within seconds to minutes after volume perturbation. Certain volume-sensitive ion transport systems play multiple roles, participating in volume regulation as well as transepithelial salt and water movement and intracellular pH control.

Organic osmolytes are found in high concentrations (tens to hundreds of millimolar) in the cytosol of all organisms from bacteria to humans.[242] These solutes play key roles in cell volume homeostasis and may also function as general cytoprotectants. In animal cells, organic osmolytes are grouped into three distinct classes: (1) polyols (e.g., sorbitol and *myo*-inositol), (2) amino acids and their derivatives (e.g., taurine, alanine, and proline), and (3) methylamines (e.g., betaine and glycerophosphorylcholine).

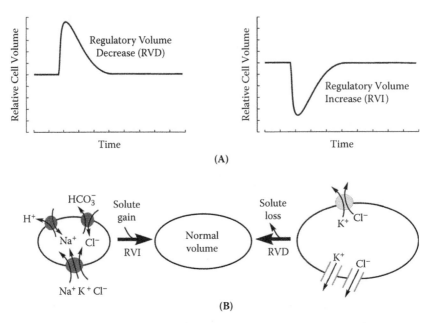

FIGURE 2.1 Cell volume regulation. (A) Cell swelling is induced by extracellular hypotonicity or by accumulation of intracellular solutes, whereas shrinkage is induced by solute loss or increases in extracellular osmolality. Cells respond to volume perturbations by activating regulatory mechanisms that mediate net solute loss or accumulation. Volume-regulatory solute loss and gain are termed *regulatory volume decrease* (RVD) and *regulatory volume increase* (RVI), respectively. The time course of RVD and RVI varies with cell type and experimental conditions. Typically, however, RVI mediated by electrolyte uptake and RVD mediated by electrolyte and organic osmolyte loss occur over a period of minutes. (B) Volume-regulatory electrolyte loss and accumulation are mediated by changes in the activity of membrane carriers and channels. Activation of these transport pathways occurs rapidly after the volume perturbation.

Organic osmolytes are *compatible* or *nonperturbing solutes*.[242] They have unique biophysical and biochemical properties that allow cells to accumulate them to high levels or to withstand large shifts in their concentration without deleterious effects on cellular structure and function. In contrast, so-called *perturbing* solutes such as electrolytes or urea can harm cells or disrupt metabolic processes when they are present at high concentrations or when large shifts in their concentrations occur; for example, elevated electrolyte levels and intracellular ionic strength can denature or precipitate cell macromolecules. Even smaller changes in cellular inorganic ion levels can alter resting membrane potential, the rates of enzymatically catalyzed reactions, and membrane solute transport that is coupled to ion gradients. Thus, although animal cells typically use inorganic ions for rapid RVI following shrinkage, they will replace these solutes by nonperturbing organic osmolytes when exposed to hypertonic conditions for prolonged periods of time.

Accumulation of organic osmolytes is mediated either by energy-dependent transport from the external medium or by changes in the rates of osmolyte synthesis and degradation.[22,23] Volume-regulatory organic osmolyte accumulation is typically a slow process relative to electrolyte uptake and requires many hours after initial activation to reach completion. Activation of organic osmolyte accumulation pathways usually requires transcription and translation of genes coding for organic osmolyte transporters and synthesis enzymes (Figure 2.2).

Loss of organic osmolytes from cells is elicited by swelling and occurs in two distinct steps. First, swelling induces a very rapid (i.e., seconds) increase in passive organic osmolyte efflux via channel-like transport pathways (Figure 2.2).[93,98] Downregulation of organic osmolyte synthesis and uptake mechanisms also contribute to the loss of these solutes from the cell. Overall, this

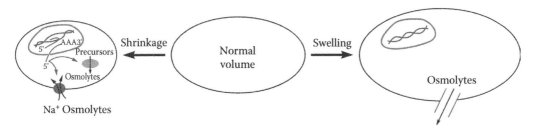

FIGURE 2.2 (See color insert following page 208.) Mechanisms of organic osmolyte accumulation and loss. Shrinkage-induced organic osmolyte accumulation in animal cells is mediated largely by increased transcription and translation (green arrows) of genes encoding Na$^+$-coupled membrane transporters or enzymes involved in organic osmolyte synthesis. During swelling, organic osmolytes are lost from animal cells largely by passive efflux through channel-like transport pathways. In addition, cell swelling inhibits the expression of genes involved in organic osmolyte accumulation.

process is slow. Cell swelling inhibits transcription of the genes coding for organic osmolyte transporters and synthesis enzymes.[23,57] As transcription decreases, mRNA levels drop and the number of functional proteins declines over a period of many hours to days.

IV. CELL VOLUME SIGNALS AND SENSORS

Volume-sensing mechanisms appear to be extremely sensitive; for example, studies by Lohr and Grantham[120] on the renal proximal tubule and Kuang et al.[105] on cultured corneal endothelial cells have demonstrated that cells can sense and respond to volume changes induced by osmotic perturbations of only 2 to 3 mOsm. Our understanding of the mechanisms by which cells sense volume perturbations and transduce those changes into regulatory responses is limited. To further complicate the picture, recent evidence suggests that cells can detect more than simple swelling or shrinkage. Cells most likely possess an array of volume detector and effector mechanisms that respond selectively to the magnitude and rate of the volume perturbation as well as the mechanism of volume change (i.e., isosmotic vs. anisosmotic).[17,55,123,133,192] Such functionally distinct sensor and effector pathways may afford the cell simultaneous control over a variety of parameters, such as intracellular pH and ionic composition, as well as volume. The signals that cells may use to detect volume perturbations fall into two broad categories: changes in mechanical force and changes in cytoplasmic composition. In the following sections, we describe possible cell volume signals and sensing mechanisms that may detect those signals.

A. CHANGES IN MECHANICAL FORCE

1. Mechanosensitive and Volume-Sensitive Channels

Swelling and shrinkage are mechanical perturbations and can impact cell architecture on global as well as microdomain levels. Changes in cellular architecture can have two broad effects that are relevant to cell volume sensing and activation of volume-regulatory mechanisms. Cell architecture can directly affect the arrangement of signaling complexes and hence the activity of signaling pathways. In addition, changes in cellular architecture can generate mechanical forces that directly alter macromolecule conformation and function.

Volume changes can alter mechanical forces on macromolecules in two ways. First, cell volume changes can deform the plasma membrane and in theory alter lipid bilayer tension, thickness, and curvature, which may be sensed by membrane-embedded proteins (see Figure 2.3A). Second, membrane proteins that are tethered to relatively immobile extracellular matrix or cytoplasmic proteins may be displaced relative to those proteins during cell swelling or shrinkage. This

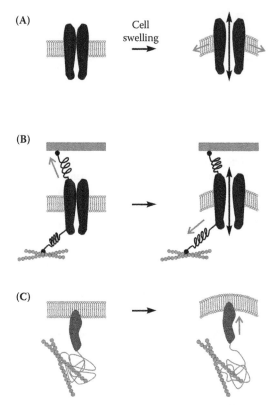

FIGURE 2.3 (See color insert following page 208.) Possible mechanisms of cellular mechanical force detection and transduction. Panels A and B illustrate mechanisms by which ion channel gating can be modulated by mechanical force; however, the function of any protein embedded in the plasma membrane may be subject to similar mechanical regulation. (A) Bilayer model of mechanosensitive channel gating; changes in mechanical force generated within the bilayer during cell swelling or shrinkage directly alter channel conformation and gating. (B) Tethered model of mechanosensitive channel gating; the channel is tethered to relatively immobile extracellular matrix and/or intracellular cytoskeletal proteins. Mechanical force is placed on the channel through tether proteins during swelling or shrinkage. (C) Mechanical-force-induced change in conformation of an intracellular macromolecule; in addition to membrane proteins, mechanical force can alter the conformation and hence function of intracellular macromolecules. The illustration shows a cytoplasmic protein (blue) tethered to a membrane-embedded protein and a cytoskeletal network. Cell swelling displaces the membrane protein relative to the cytoskeletal network, thereby stretching the cytoplasmic protein. The resulting conformational change could alter enzyme activity, expose functional domains such as phosphorylation sites, alter protein–protein interactions, etc. Green arrows in all panels indicate direction of applied mechanical force.

displacement in turn will alter mechanical forces on both membrane and extracellular/cytoplasmic proteins (Figure 2.3B).

Mechanosensitive channels[72,198] are obvious candidates for sensing swelling- or shrinkage-induced changes in mechanical force. Figure 2.3A and B illustrates two different mechanisms by which mechanical force might regulate channel gating. Membrane channels may directly sense mechanical force exerted on the protein through the lipid bilayer (Figure 2.3A). The *Escherichia coli* MscL (mechanosensitive channel large) channel exhibits this type of gating and is the most extensively characterized mechanosensitive channel.[197] Increases in turgor pressure during hypotonic stress activate MscL. The channel is poorly selective and allows molecules as large as 1 kDa to exit the cell. Solute extrusion prevents cell lysis during extreme hypotonic shock. MscL is comprised of five identical subunits. The channel exhibits mechanosensitive gating when reconstituted in artificial membranes. As the membrane is stretched, force is exerted on the protein at the lipid–protein interface which in turn alters channel conformation and gates it open.[197]

Mechanical force can also be transmitted to channels by tethering them to the cytoskeleton or extracellular matrix (Figure 2.3B). In this model, cell volume changes displace the channel with respect to relatively immobile intracellular and extracellular proteins. Mechanical force placed on the channel through the tether proteins in turn alters channel conformation and gating. The tethered model is best illustrated by studies of ion channels required for touch sensitivity in the model organism *Caenorhabditis elegans*. Nematodes sense gentle body touch by five "touch" neurons. Forward genetic analysis has identified genes required for touch neuron mechanosensory functions. Touch-insensitive mutant worms are mechanosensory abnormal, and the genes responsible for this abnormality are termed *mec*. Approximately 15 *mec* genes have been identified to date. Several of these genes are required for normal touch neuron development. At least eight *mec* genes encode proteins that have been postulated to form a mechanosensitive ion channel complex.[8,201]

Both *mec-4* and *mec-10* encode ion-channel-forming proteins that share significant homology with epithelial Na+ channels (ENaCs).[8,201] Genetic, molecular, and biochemical studies indicate that MEC-4 and MEC-10 proteins interact with cytoskeletal and extracellular matrix proteins. Deformation of the cuticle by touch is thought to displace extracellular proteins relative to the MEC-4/MEC-10 channel. This displacement in turn exerts a mechanical force that alters channel gating. Recent elegant *in vivo* electrophysiological studies have demonstrated the presence of mechanically gated cation currents in touch neurons, and null or loss-of-function mutations in various *mec* genes abolish the currents or alter their properties.[144]

It is important to stress that channels may be directly as well as indirectly regulated by mechanical force.[29] In addition to the models shown in Figure 2.3A and B, it is conceivable that mechanical stimuli may alter the conformation of one or more accessory proteins that in turn control channel gating. Alternatively, mechanical force could alter the activity of signaling pathways that regulate channel activity.

Numerous cell-volume-sensitive ion currents have been described.[90,112,231] Several channels whose activity is modulated by cell volume perturbations have also been identified at the molecular level. These include the TRP channels TRPV4,[9,64,117,195,236] TRPV2,[134] TRPM3,[68] and TRPM7;[142,143] the *Drosophila* TRPV channels IAV[67] ("inactive") and NAN[97] ("nanchung"); the ClC anion channels ClC-2[69,92] and CLH-3b;[45,176] the human homolog of the *tweety* (hTTYH1) Cl− channel;[200] the human and mouse bestrophin Cl− channels hBest1 and mBest2;[61] and the K+ channels KCNQ1, KCNQ4,[70] KCNQ5,[89] TREK-1,[160] and TASK-2.[137] Activation of a swelling-activated, outwardly rectifying anion current, $I_{Cl,swell}$, which is also known as the volume-sensitive, outwardly rectifying Cl− channel (VSOR), volume-regulated anion channel (VRAC), and volume-sensitive organic osmolyte/anion channel (VSOAC), is a ubiquitous response of animal cells to volume increase.[146,193] The ClC channel ClC-3 has been proposed to be the long-sought channel responsible for $I_{Cl,swell}$;[51] however, ClC-3 knockout mice have normal $I_{Cl,swell}$ channel activity, making this hypothesis untenable.[5,230]

The role, if any, that most of these molecularly identified channels play in cell volume regulation is not fully defined. In addition, it must be stressed that it is unclear whether they are regulated either directly or indirectly by mechanical force. The exceptions to this generalization are TRPM7,[143] TREK-1,[160] and TRPV2,[134] which have been shown to be gated by membrane stretch. IAV and NAN play roles in *Drosophila* hearing,[67,97] which clearly involves mechanical gating of ion channels.[225] The *Caenorhabditis elegans* TRPV channels OSM-9 and OCR-2 are expressed in sensory neurons and play an essential role in detecting both hypertonic environments and mechanical force.[32,208] The function of these channels in mechanosensation suggests that hypertonic environments may be detected by changes in membrane tension that likely occur when sensory neurons shrink during exposure to hypertonic conditions. TRPV4 is activated by shear stress[64] and can rescue OSM-9 loss-of-function mutants,[118] suggesting that it may detect cell volume perturbations through changes in mechanical force.

2. Other Possible Sensors of Mechanical Force

In addition to ion channels, numerous other sensors of cellular mechanical force have been identified. Any macromolecule in theory can function as a mechanosensor as long as mechanical force of sufficient magnitude to alter its structure is exerted on it (see Figure 2.3).[81] The activity of phospholipase A_2 (PLA_2), for example, is sensitive to membrane lipid-packing density.[21] When reconstituted in liposomes, PLA_2 is activated by osmotic swelling, which causes membrane stretching and a reduction in lateral lipid packing (see Figure 2.3A).[114] Changes in mechanical forces on DNA may affect gene transcription.[16,66,99] Actin polymerization is sensitive to mechanical force,[26] as well as cell volume changes.[30] Recent studies have shown that actin polymerization/depolymerization regulates gene expression by modulating the movement of transcription factors in and out of the nucleus.[220]

The Src (sarcoma) kinase substrate Cas is a cytoskeletal associated scaffolding protein that is involved in numerous signaling pathways.[42] In HEK293 cells, Cas is phosphorylated in a stretch-dependent manner. *In vitro* stretching of a Cas substrate domain protein dramatically increases phosphorylation without changing Src kinase activity, and Cas stretching and phosphorylation are detected in peripheral regions of spreading cells where traction forces are high.[180] Taken together, these results suggest that mechanical stretching of Cas exposes Src phosphorylation sites (see Figure 2.3C).

The physical interaction of cells with their neighbors and the extracellular matrix (ECM) controls numerous cellular processes. Adhesion of cells to the ECM requires transmembrane proteins termed *integrins* that interact with the cytoskeleton. An extensive body of evidence has shown that integrins play essential roles in cellular mechanotransduction.[44,60,81] Integrins also function in the regulation of osmoprotective gene expression,[131] as well as other osmotic stress-induced signaling processes.[19,185,227]

MEC-5 is an extracellular collagen that is required for mechanosensation in *Caenorhabditis elegans*.[8,201] Recent studies have demonstrated that cuticle collagens also play a role in regulating organic osmolyte accumulation in worms, possibly by detecting and transducing hypertonic stress-induced mechanical signals.[111,232]

Caveolae, membrane invaginations 60 to 80 nm in diameter, play important roles in cellular mechanotransduction.[156] Studies by Eggermont and coworkers have demonstrated that caveolin-1, a principal caveolae coat protein, controls $I_{Cl,swell}$ activity.[211,212,218] Interestingly, they have also shown that targeting of Src kinase to caveolae inhibits swelling-induced activation of $I_{Cl,swell}$. Inhibition does not require kinase function but is instead dependent on Src homology domains 2 and 3, suggesting that Src disrupts a caveolae signaling cascade that regulates the channel.[210] A simple and attractive hypothesis suggested by these studies is that cell volume changes alter the conformation of caveolae (as well as other membrane microdomains), which in turn causes rearrangement of scaffolding proteins and associated signaling components such as kinases, phosphatases, and their substrates. It is easy to envision how such a rearrangement could bring kinases and their targets into close apposition, allowing phosphorylation to occur. Alternatively, rearrangement of microdomain architecture could move kinases and substrates apart, allowing phosphatases to associate with and dephosphorylate kinase targets.

B. Changes in Cytoplasmic Composition

1. Intracellular Ionic Strength

Cell swelling or shrinkage leads to changes in intracellular water activity as well as the concentrations of intracellular solutes and macromolecules. A number of studies have identified ionic strength as a signal that activates or modulates volume-regulatory mechanisms. As noted above (Section III.B), the swelling-activated K–Cl cotransporter plays an important role in RVD in animal cells. Studies in fish[71,133] and dog[155] red cells have shown that the volume setpoint of the cotransporter is sensitive to intracellular ionic strength. Specifically, as cytoplasmic inorganic ion levels rise, less swelling is required to trigger cotransporter activation.

The $I_{Cl,swell}$ channel (see Section IV.A.1) appears to play a critical role in RVD and may be an important pathway for organic osmolyte efflux.[93,98] The molecular identity of the channel is unknown, and the precise mechanism of swelling-induced activation is unclear; however, two groups have demonstrated that cytoplasmic ionic strength is an important modulator of channel activity.[24,55,138,224] Less swelling is required to activate the channel at reduced intracellular ionic strength. Cytoplasmic ionic strength has a similar effect on swelling-induced organic osmolyte efflux from trout red cells[71,133] and C6 glioma cells.[55] Strange and coworkers[24,55] suggested that ionic strength modulates the volume setpoint of the $I_{Cl,swell}$ channel, whereas Voets et al.[224] concluded that channel activation is triggered by swelling-induced reductions in ionic strength. Similarly, Guizouarn and Motais[71] concluded that reductions in intracellular ionic strength activate the swelling-induced organic efflux pathway in trout red blood cells.

The differential effect of intracellular ionic strength on volume-regulatory electrolyte and organic osmolyte transport pathways may have important physiological implications.[55] Isosmotic swelling induced by net salt uptake increases cell inorganic ion content. In addition, cells that have undergone RVI have elevated intracellular ionic strength. When cell swelling occurs concomitantly with increased cytoplasmic inorganic ion content, it is advantageous for cells to use electrolytes selectively for RVD via activation of an electrolyte-selective transport pathway such as the K–Cl cotransporter. The loss of organic osmolytes under such conditions would mediate RVD but would also further concentrate intracellular electrolytes as cells undergo volume-regulatory water loss and concomitant shrinkage. Changes in intracellular ionic strength may therefore play an important role in coordinating the activities of various volume-regulatory transport pathways. This postulated coordinated regulation could in turn contribute to the long-term maintenance of cytoplasmic ionic composition.

In addition to regulating RVD transport pathways, intracellular ionic strength may also regulate the transcription of genes encoding organic osmolyte transporters and enzymes involved in their synthesis. This idea was first proposed by Uchida et al.,[215] who demonstrated that the activity of aldose reductase, an enzyme required for the synthesis of the organic osmolyte sorbitol, increases as a function of cellular inorganic ion content.

Kwon and coworkers have extensively characterized the mechanisms of organic osmolyte accumulation in the mammalian kidney. Increased transcription of genes encoding organic osmolyte transporters and synthesis enzymes is controlled by a *cis*-regulatory element termed *tonicity-responsive enhancer*, or TonE.[58,174.202] TonE-binding protein (TonEBP) binds to TonE and stimulates gene transcription.[130] Hypertonic shrinkage stimulates TonEBP to translocate from the cytoplasm into the cell nucleus.[37] Studies by Neuhofer et al.[136] suggest that increased intracellular ionic strength increases the activity and nuclear localization of TonEBP. Regulation of organic osmolyte efflux (see discussion above) and expression of genes involved in organic osmolyte accumulation by cytoplasmic ionic strength would provide an important feedback mechanism that allows coordinated regulation of both cell inorganic ion levels and volume.

Any macromolecule whose conformation is sensitive to physiologically relevant shifts in ionic strength could function as a cell volume sensor. In animal cells, no such sensor has been identified at the molecular level; however, recent studies in bacteria have demonstrated that cystathionine-β-synthase (CBS) domains function as ionic strength and cell volume sensors.[13,124] The CBS domain is a ubiquitous motif found in diverse proteins, including adenosine triphosphate (ATP)-binding cassette (ABC) transporters, ClC channels and transporters, transcription factors, and various enzymes.[80]

OpuA is an osmoregulatory ABC transporter that mediates the uptake of organic osmolytes in hypertonically stressed bacteria. When reconstituted in proteoliposomes, OpuA is activated by shrinkage and by elevation of luminal ionic strength. The threshold for ionic-strength-dependent activation is sensitive to the content of anionic lipids in the liposome membrane.[166,239] Poolman and coworkers[13,124] have shown that deleting the CBS domains renders OpuA transport activity

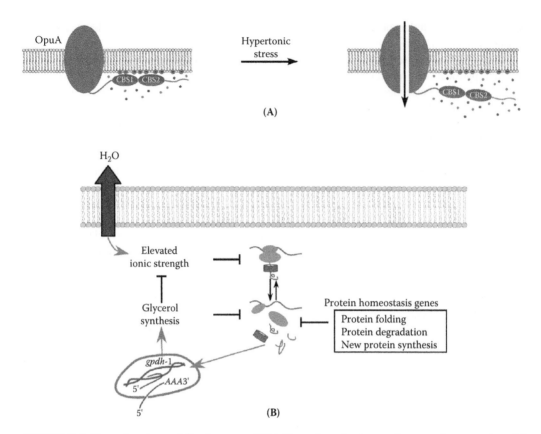

FIGURE 2.4 (See color insert following page 208.) Examples of osmoregulatory pathways activated by increases in intracellular ionic strength. (A) OpuA is a bacterial organic osmolyte transporter activated by hypertonic stress. The cationic surface (shown by red shading) of the CBS domains interacts with anionic membrane lipids (shown in green) and inactivates the transporter. Hypertonic stress and cell shrinkage increase intracellular ionic strength (small red and green circles), which in turn disrupts this electrostatic interaction, leading to increased OpuA activity. The anionic C-terminus (green) of OpuA is expected to be repelled away from anionic membrane lipids which may modulate ionic strength sensitivity by weakening the interaction between the membrane and CBS domains. (B) Model for regulation of *Caenorhabditis elegans* osmosensitive gene expression by disruption of protein homeostasis. Hypertonic-stress-induced water loss causes elevated cytoplasmic ionic strength which in turn disrupts new protein synthesis and cotranslational protein folding. Misfolded and incompletely synthesized proteins function as a signal that activates *gpdh-1* expression and glycerol synthesis. Glycerol replaces inorganic ions in the cytoplasm and functions as a chemical chaperone that aids in the refolding of misfolded proteins. Loss of function of protein homeostasis genes also causes accumulation of damaged proteins and activation of *gpdh-1* expression.

largely insensitive to intracellular ionic strength. Deletion of the 18-residue anionic C-terminus of the protein shifts ionic strength sensitivity to higher values. They propose that the cationic surface of the CBS domains interacts with anionic membrane lipids and inactivates the transporter. Increasing intracellular ionic strength disrupts this electrostatic interaction, leading to increased OpuA activity. The anionic C-terminus is expected to be repelled away from anionic membrane lipids, which presumably modulates ionic strength sensitivity by weakening the interaction between the membrane and CBS domains (see Figure 2.4A).

KdpD is a membrane-bound histidine kinase that regulates hypertonicity-induced expression of a high-affinity K^+ uptake system in *Escherichia coli*. When reconstituted in liposomes, KdpD autophosphorylation activity is increased by vesicle shrinkage and by increasing luminal ionic strength.[95] A cluster of five positively charged arginine and lysine residues is critical to the function of KdpD. Single-point mutation of these amino acids to glutamine alters kinase and phosphatase activity of the protein and its association with the membrane.[94] Anionic membrane lipids increase kinase activity.[190] Taken together, these findings suggest that, like OpuA, KdpD activity may be modulated by ionic-strength-dependent changes in the interaction of the protein with the lipid bilayer.

Rather than detecting shifts in ionic strength *per se*, cells could detect macromolecular damage induced by alterations in cytoplasmic inorganic ion levels; for example, elevation of intracellular ionic strength can disrupt protein folding and protein synthesis.[18,242] Recent studies in *Caenorhabditis elegans* suggest that ionic-strength-induced protein damage functions as a signal that increases the transcription of the gene encoding glycerol 3-phosphate dehydrogenase-1 (GPDH-1), which is required for synthesis and accumulation of the organic osmolyte glycerol during hypertonic stress.[111] Using a GFP reporter of GPDH-1 expression and a genome-wide RNA interference screen, Lamitina et al.[111] identified 122 genes that function as negative regulators of GPDH-1 expression. Loss of function of these genes causes constitutive expression of GPDH-1 and glycerol accumulation in the absence of hypertonic stress. The largest class of genes identified functions in protein homeostasis and includes genes required for RNA processing, protein synthesis, protein folding, and protein degradation. Protein homeostasis genes function to maintain levels of properly folded and functioning cellular proteins. Inhibition of these genes is expected to increase the levels of damaged cellular proteins.[141]

Interestingly, protein damage induced by numerous stressors including heat shock does not activate GPDH-1 expression.[111] Previous studies have shown that hypertonic stress but not heat or oxidative stress inhibits protein synthesis in yeast.[216] The initiation and elongation steps of protein synthesis *in vitro* are inhibited by increases in salt concentration of as little as 10 mM, and this inhibition is fully reversed by organic osmolytes.[18] Disruption of elongation would cause accumulation of incomplete and aberrantly folded polypeptides in the cytoplasm. Importantly, the majority of the protein homeostasis genes identified by Lamitina et al.[111] function in RNA processing, protein translation, and cotranslational protein folding. Inhibition of these genes is predicted to disrupt protein synthesis.

Taken together, the findings of Lamitina et al.[111] are consistent with a model in which glycerol accumulation is specifically activated by hypertonicity-induced increases in intracellular ionic strength that disrupt new protein synthesis and cotranslational folding. Increased levels of damaged or denatured proteins act as a signal that triggers increased GPDH-1 expression. Accumulation of organic osmolytes such as glycerol is expected to stabilize protein structure and decrease protein misfolding and aggregation,[6,79] which in turn would autoregulate pathway activity (see Figure 2.4B).

2. Intracellular Inorganic Ions

In addition to ionic strength, changes in the concentration of specific ions could function to signal cell volume perturbation; for example, the activity of several ion channels, cotransporters and exchangers is sensitive to intracellular Cl⁻ levels.[7,77,148,151,175,245,246] Furthermore, intracellular Cl⁻ regulates a variety of other proteins and physiological processes, including the transmembrane molecular motor prestin,[38] G-protein signaling,[76] and exocytotic secretory activity in endocrine cells.[214] Interestingly, recent studies by Delpire and coworkers have shown that the activity of PASK/SPAK (proline alanine-rich Ste20-related kinase) and the closely related kinase OSR1 (oxidative stress response 1) is regulated by physiologically relevant levels of intracellular Cl⁻.[62] PASK, OSR1, and the *Caenorhabditis elegans* homolog GCK-3 (germinal center kinase 3) play critical roles in regulating volume-sensitive Na–K–2Cl and K–Cl cotransporters[3,50,63,132,222,223] and a swelling-activated *C. elegans* ClC channel[46,177] (see Section V.A.2). GCK-3 also plays an essential role in systemic osmotic homeostasis in *C. elegans*.[28]

BetP is a bacterial Na⁺-coupled glycine betaine uptake system that is activated by hypertonic stress *in vivo* and when reconstituted in liposomes.[173] Hypertonicity-induced activation is mediated by increases in internal K⁺ concentration.[172,181] Mutagenesis studies indicate that the transporter

C-terminus is required for osmosensing.[162,183] Schiller et al.[182] suggested that shrinkage-induced increase in intracellular K^+ concentration induces a conformational change in the C-terminus of BetP. This conformational change in turn disrupts the interaction of the positively charged C-terminus with negatively charged membrane lipids, leading to transporter activation.

3. Intracellular Water Activity

Water activity affects the hydration state and thus the conformation and function of macromolecules.[170,171] It is conceivable then that changes in intracellular water activity in osmotically stressed cells may function as a signal to activate osmoregulatory effector mechanisms. A possible sensor that detects water activity is *Escherichia coli* ProP, a H^+-coupled organic osmolyte transporter that is activated by hypertonic stress.[239] When reconstituted in proteoliposomes, ProP is activated by increasing concentrations of extracellular solutes that cause vesicle shrinkage. Interestingly, ProP is also activated when medium osmolality is increased by the addition of poly(ethylene)glycols (PEGs) that permeate the vesicle and cause no measurable shrinkage.[167] Furthermore, loading proteoliposomes with PEGs of a specific size activates ProP.[36] Wood[239] has proposed that cytoplasmic and luminal PEGs and proteins compete with ProP for water of hydration. As water activity is decreased during hypertonic stress, ProP is partially dehydrated, and this dehydration in turn alters the conformation and activity of the transporter.

4. Macromolecular Crowding

Most *in vitro* studies of biochemical processes employ relatively dilute solutions of reactants and the proteins that catalyze their reaction. The cytoplasm of a real cell is considerably more complicated. Typically, 5 to 40% of the total cytoplasmic volume is occupied by macromolecules.[1,53,54] (See Medalia et al.[128] for electron tomographs that illustrate the crowded nature of the cytoplasm.) Thus, an intracellular macromolecule functions in an environment that is crowded with other macromolecules. It is now widely appreciated that macromolecular crowding has profound effects on the equilibria and kinetics of biological reactions; for example, the addition of macromolecular polymers such as polyethylene glycol, glycogen, or Ficoll to the reaction mixture stimulates the activity of T4 polynucleotide kinase[74,75] and DNA ligase several orders of magnitude.[73,163] Macromolecular crowding alters biological reactions by altering the rates of diffusion of reactants, thermodynamic activities, and association and dissociation kinetics of macromolecules and their substrates.[1,53,54]

Given the dramatic effects of crowding on biochemical processes, it stands to reason that changes in crowding change reaction rates and kinetics. Macromolecular crowding is altered any time a cell swells or shrinks. Zimmerman and Harrison[248] were the first to suggest that swelling- or shrinkage-induced changes in macromolecular crowding could provide cells with a mechanism to detect volume perturbations. Studies by Colclasure and Parker[33,34] on resealed dog red cell ghosts suggested that the activities of the swelling-activated K–Cl cotransporter and the shrinkage-activated Na^+/H^+ exchanger are regulated not by cell volume *per se* but by the concentration of intracellular proteins. Parker and Colclasure[154] suggested that macromolecular crowding regulates kinases and phosphatases that control transporter activity (see Section V.A.2). Minton et al.[129] explored this idea in detail using modeling approaches. Their results suggest that, in the crowded cytoplasmic environment, the association of a soluble regulatory kinase with an insoluble (i.e., membrane-associated) transporter is much more sensitive to cell volume changes than would be suggested by mass action alone. Association of the kinase with the transporter results in a net increase in transporter phosphorylation and concomitant change in activity.

Replacement of intracellular macromolecules with sucrose triggers cell shrinkage in perfused barnacle muscle fibers that is dependent on plasma membrane verapamil-sensitive Ca^{2+} channels and Ca^{2+} influx.[199] RVD in this cell type also requires Ca^{2+} influx via verapamil-sensitive Ca^{2+}

channels that presumably activates volume-regulatory solute efflux pathways.[11] Summers et al.[199] proposed that macromolecular crowding governs the association of an inhibitory regulator with the Ca^{2+} channels. Dilution of macromolecules with sucrose or by cell swelling decreases crowding and increases the fluid volume accessible to the inhibitor. This in turn alters the association equilibrium of the channel and putative inhibitor, leading to channel activation.

V. CELL-VOLUME-SENSITIVE SIGNALING MECHANISMS

When a signal indicating cell volume perturbation has been detected, it must be transduced into a regulatory response. Transduction may be direct as is the case for mechanosensitive channels such as MscL or the ionic-strength-sensitive OpuA transporter. Alternatively, cell volume changes may trigger signaling pathways that activate downstream effector mechanisms. In the following sections, we review well-characterized and emerging signaling proteins and pathways that are sensitive to cell volume changes.

A. OSMOTICALLY REGULATED PROTEIN KINASES

Changes in protein phosphorylation control innumerable cellular processes, and protein kinases are one of the largest protein families in eukaryotes.[125] In humans, protein kinases can be divided into as many as 209 subfamilies. Cellular osmotic stress has been shown to modulate the activity of signaling pathways involving members of the mitogen-activated Ste20, WNK, and Src kinase families.

1. Mitogen-Activated Protein Kinases

Mitogen-activated protein kinases, or MAPKs, are a large family of eukaryotic serine/threonine kinases that regulate the expression and activity of genes involved in diverse cellular processes, including the cell cycle, cell growth, differentiation, cell death, and multiple stress responses.[169] Activation of MAPKs typically requires a cascade of protein phosphorylation events that includes at least three kinases in series; a MAP kinase kinase kinase (MAPKKK) phosphorylates and activates a MAP kinase kinase (MAPKK), which then phosphorylates and activates one of the MAPKs.[234] In some cases, an upstream MAP kinase kinase kinase kinase (MAPKKKK) or small GTP-binding protein is known to initiate a signaling cascade by activating a MAPKKK. At least 12 MAPKs subfamilies are present in mammals. The best characterized subfamilies are the extra-cellular-signal-regulated kinases 1 and 2 (ERK1/2), stress-activated or *c*-Jun N-terminal kinases (JNK1–3), and p38 MAPKs (α, β, χ, and δ).[169] Members of all three of these subfamilies are activated by hyper- or hypotonic stress.[10,102,107,157,237]

ERK1/2 appear to be ubiquitously activated by hypertonicity in mammalian cells, but their function in cell volume regulation varies with cell type. An early study found no evidence for ERK1/2 regulation of hypertonicity-induced organic osmolyte accumulation in Madin–Darby canine kidney epithelial cells.[108] In contrast, ERK1/2 inhibition reduces organic osmolyte accumulation in inner medullary collecting duct cells[102] and intervertebral disc cells.[213] Because ERK1/2 are activated by growth factors and play a role in cell proliferation, they may also mediate cell survival by suppressing apoptotic signals initiated during hypertonicity.[187]

ERK1/2 are also activated by hypotonicity in some cell types;[157] however, as with hypertonicity, no consensus has been reached on whether ERK1/2 activation contributes to volume regulation. Pharmacological and dominant-negative inhibition of ERK decreases RVD in some cells[150,188] but not in others.[27,52,147,226]

Overexpression of dominant-negative mutant JNKs decreases survival of mouse inner medullary collecting duct cells exposed to hypertonicity but does not alter cell volume recovery or organic osmolyte accumulation.[238] Instead, JNKs regulate hypertonicity-induced expression of

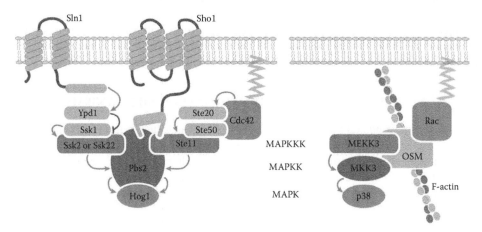

FIGURE 2.5 (See color insert following page 208.) Osmotically activated p38 MAPK signaling cascades in yeast and mammals. Green arrows and red lines indicate activation and inhibition, respectively. Physical interactions are indicated by overlap of components. MAPKKKs, MAPKKs, and MAPKs are colored orange, red, and green, respectively. In yeast, there are two signaling branches upstream from Pbs2, a MAPKK that activates the p38 MAPK Hog1. The Sln1 branch is comprised of the Ypd1 and Ssk1 phosphorelay system that regulates the activity of two redundant MAPKKKs, Ssk2 and Ssk22. Increased osmolality inhibits this phosphorelay system and allows Ssk1 to interact with and activate Ssk2 and Ssk22. Cdc42 in the Sho1 signaling branch is activated by increased osmolality and in turn activates Ste20, a MAPKKKK. Ste20 then activates the MAPKKK Ste11. The efficiency and specificity of the Sho1 cascade is modulated by the scaffolding functions of Sho1, Cdc42, and Ste50. In mammals, Rac, MEKK3, MKK3, and p38 MAPK are thought to comprise a cascade that is homologous to Cdc42, Ste11, Pbs2, and Hog1. OSM acts as a scaffold for Rac, MEKK3, and MKK3.

cyclooxygenase 2,[243] a cytoprotective gene and the Na,K-ATPase,[25] an ion pump essential for intracellular ion homeostasis. Thus, the role of JNKs during hypertonicity may be to promote cell survival instead of controlling cell-volume-regulatory mechanisms. JNKs are also activated by hypotonicity, and pharmological inhibition of kinase activity decreases RVD in mammalian renal and corneal epithelial cells.[27,150] More studies are needed, however, to define the mechanism of JNK-mediated volume regulation and to determine if JNKs are volume sensitive in other cell types.

The best characterized osmotic signaling pathway in eukaryotes is the high-osmolarity glycerol response (HOG) pathway, of the budding yeast *Saccharomyces cerevisiae*.[78,100,178] In yeast, two parallel branches of the HOG pathway are defined by their dependence on one of two transmembrane proteins, Sln1 or Sho1. Within each branch, hypertonic stress activates a MAPK cascade. Both of these cascades terminate on Hog1, the single yeast p38 MAPK homolog (Figure 2.5). Hog1 activates transcription of genes that mediate synthesis of glycerol, the dominant organic osmolyte of *S. cerevisiae*.

Sln1 is a membrane-bound histidine kinase sensor that forms a phosphorelay system together with two other proteins, Ypd1 and Ssk1.[78,100,178] Under stable osmotic conditions, Sln1 is constitutively active and phosphorylates Ypd1, which then transfers its phosphate group to Ssk1. Phosphorylation of Ssk1 is thought to prevent it from interacting with and activating the MAPKKKs Ssk2 and Ssk22. During hypertonic stress, Sln1 is inhibited, Ypd1 and Ssk1 lose their phosphate groups, and Ssk1 interacts with Ssk2 and Ssk22. The interaction with Ssk1 triggers autophosphorylation of the MAPKKKs, which then phosphorylate and activate Pbs2, the MAPKK that activates Hog1.

Sho1 is a fungi-specific protein that contains four transmembrane domains and a C-terminal domain that interacts with downstream signaling components.[78,100,178] Sho1 was originally thought to be an osmosensor like Sln1, but studies of Raitt et al.[168] indicate that it instead functions as a scaffolding protein that provides docking sites for kinases upstream from Hog1. Recent work of Tatebayashi et al.[203] suggests that two transmembrane mucin proteins function upstream of Sho1 as osmosensors.

A recent study of Sho1 signaling demonstrated that Cdc42, a Rho-type GTP-binding protein, becomes activated during hypertonic stress via an unknown mechanism.[204] Activated GTP-bound Cdc42 then binds and activates Ste20, a MAPKKKK.[204] Cdc42 and Ste20 interact with Ste50, a possible cofactor for Ste20. Ste50 also interacts with Ste11, a MAPKKK that is activated by Ste20. Sho1 binds to Ste11 and brings it close to its substrate Pbs2, the MAPKK that activates p38 MAPK.[78,100,178] The specificity and efficiency of the phosphorylation steps of this signaling pathway are enhanced by the scaffolding functions of Cdc42, Ste50, and Sho1[233] (Figure 2.5).

Vertebrate p38 MAPKs are also activated by cellular osmotic stress.[106,187] Studies with pharmacological inhibitors, dominant negative kinases, and siRNA suggest that p38 MAPKs play a role in hypertonicity-induced transcription of genes that mediate organic osmolyte accumulation in cultured mammalian cells.[101,102,135,186,213] Until recently, however, little was known about the upstream signaling mechanisms that activate p38 MAPKs during hypertonicity. Using siRNA, Uhlik and coworkers[217] demonstrated that MEKK3 (mitogen-activated, extracellular-regulated kinase kinase kinase 3), a MAPKKK with homology to yeast Ste11, and MKK3 (mitogen-activated protein kinase kinase 3), a MAPKK, function to activate p38 MAPK. Yeast 2-hybrid screening demonstrated that MEKK3 interacts with a novel gene product termed *osmosensing scaffold for MEKK3* (OSM). OSM also interacts with the actin cytoskeleton, and Rac, a Rho-type GTP-binding protein. Rac is activated by hypertonicity, and a dominant-negative form of the protein inhibits hypertonicity-induced activation of p38 MAPKs. Uhlik and coworkers[217] have proposed a model in which Rac, OSM, MEKK3, and MKK3 activate p38 MAPKs during hypertonic stress similar to Cdc42, Ste20, Sho1, Ste11, and Pbs2 in yeast (Figure 2.5).

The downstream targets of p38 MAPK are not clearly defined. Like Hog1 in yeast, p38 MAPK may regulate the activity of transcription factors that control the expression of genes required for organic osmolyte accumulation. As discussed above, TonEBP regulates the transcription of organic osmolyte synthesis and transporter genes.[130] Dahl et al.[37] demonstrated that TonEBP is phosphorylated in hypertonically stressed MDCK cells. Several studies have suggested that p38 and other MAPKs,[101,135,149,186,213] as well as other types of kinases,[59,82,83,101] play a role in TonEBP activation. Direct phosphorylation of TonEBP has not been demonstrated for any kinase, and the molecular mechanisms by which these kinases regulate TonEBP are unknown.[91] In addition, Lee et al.[113] have demonstrated that truncated versions of TonEBP lacking phosphorylation sites can still induce gene transcription during hypertonicity. Thus, the precise role of kinase signaling in organic osmolyte accumulation in animal cells remains uncertain.

As with ERK1/2 and JNKs, p38 MAPKs are also sometimes activated during hypotonic stress,[150,188,226,227] but their role in cell volume regulation is not well defined. Pharmacological inhibition of p38 MAPKs decreases RVD in trout and rat hepatocytes[52,226] but has no effect on volume regulation in rabbit[150] and human epithelial cells.[188] Interestingly, p38 activation and RVD require the activity of Src kinases and integrins in swollen rat hepatocytes.[227]

Hog1 and p38 MAPKs demonstrate that at least one pathway of osmotic stress signal transduction has been conserved from yeast to vertebrates; however, multiple cell volume signaling pathways appear to have evolved in metazoans that function independently from p38 MAPKs. This diversity of osmotic signal transduction mechanisms may reflect the diversity of cell types or the diversity of cell volume sensor and effector mechanisms that exist in metazoans.

2. Ste20 and WNK Kinases

As discussed above (see Section III.B), swelling-activated K–Cl and shrinkage-activated Na–K–2Cl cotransporters play central roles in RVD and RVI (see Figure 2.1A). Swelling-induced activation and shrinkage-induced inactivation of the K–Cl cotransporter are mediated by serine/threonine dephosphorylation and phosphorylation, respectively. The converse is true for the Na–K–2Cl cotransporter; shrinkage-induced activation is mediated by phosphorylation, and swelling-induced inactivation is brought about by dephosphorylation. Pharmacological, transport, and molecular studies suggest that a type 1 protein phosphatase (PP1) mediates dephosphorylation of both cotransporters[40,50,88,121,191] and that kinase activity and phosphorylation are sensitive to cell-volume changes.[86,87,122] Parker[153] proposed that a common cell volume sensor and signal transduction pathway coregulates K–Cl and Na–K–2Cl cotransporter activity.

Considerable progress has been made recently in identifying components of the volume-sensitive kinase cascade. Delpire and coworkers[165] demonstrated that the Ste20-related kinase PASK and OSR1 interact with the N-termini of both K–Cl and Na–K–2Cl cotransporters. Genetic analysis of mating in the budding yeast *S. cerevisiae* led to the discovery of *ste*, or sterile, genes. Ste20 is the founding member of a large serine/threonine kinase superfamily and, as discussed above (Section V.A.1), functions to regulate hypertonicity-induced glycerol accumulation in yeast.[78,100,178]

The Ste20 kinase superfamily is divided into the PAK (p21-activated kinase) and GCK (germinal center kinase) kinases.[39] These two groups are further subdivided into PAK-I–II and GCK-I–VIII subfamilies. PASK and OSR1 are vertebrate members of the GCK-VI subfamily, which also includes *Drosophila* Fray and *Caenorhabditis elegans* GCK-3. Ste20 kinases regulate numerous fundamental cellular processes, including apoptosis, stress responses, morphogenesis, cytoskeletal architecture, cell cycle, and oocyte meiotic maturation.[39]

Dowd and Forbush[50] provided the first evidence that PASK plays a role in regulating shrinkage-induced activation of Na–K–2Cl cotransport. Subsequently, several groups have demonstrated that both OSR1 and PASK phosphorylate Na–K–2Cl and K–Cl cotransporters and regulate their activity.[3,63,132,222,223]

The initial studies of Piechotta et al.[165] suggested that PASK and OSR1 were not important regulators of cotransporter activity. In addition, Gagnon et al.[63] observed that PASK coexpressed with either K–Cl or Na–K–2Cl cotransporters in *Xenopus* oocytes has no effect on transport activity under basal or osmotic stress conditions; however, when PASK is coexpressed with WNK4, dramatic activation of the Na–K–2Cl cotransporter and inhibition of the K–Cl cotransporter are observed.[65]

WNK4 is a member of the *with no lysine* (K) family of serine/threonine kinases.[221,240] Humans have four WNK kinases, and rare mutations in WNK1 and WNK4 cause pseudohypoaldosteronism type II (PHAII), an autosomal dominant form of hypertension.[235] WNK1 and WNK4 control blood pressure by regulating the activity of ion transport pathways that mediate salt transport in distal renal tubules of mammals.[96,109]

Delpire and coworkers[164] identified WNK4 as a binding partner of PASK. Subsequent biochemical studies have demonstrated that WNK1 and WNK4 bind to, phosphorylate, and activate PASK and OSR1.[3,63,132,222,223] Taken together, these studies indicate that WNK1 and WNK4 function immediately upstream from PASK and OSR1 to reciprocally regulate the activity of both Na–K–2Cl and K–Cl cotransporters (Figure 2.6A).

The mechanism of cell-shrinkage activation of WNK1 was investigated by Zagórska and coworkers.[247] Using HEK 293 cells, they demonstrated that WNK1 kinase activity, measured as OSR1 phosphorylation levels, was specifically stimulated by hypertonicity but not other cell stressors. Phosphorylation of a specific serine in WNK1, S382, activates the kinase, but inhibition of p38 MAPKs, ERK1/2, and JNKs has no effect on WNK1 activation. WNK1 expressed in *Escherichia coli* transautophosphorylates at S382, suggesting that WNK1 may self-regulate its activity in response to cell volume changes. Hypertonicity also stimulates rapid movement of WNK-1 from a diffuse localization to discrete intracellular vesicles. Further work is necessary to determine the exact mechanism of WNK1 activation by cell shrinkage.

Caenorhabditis elegans oocytes express a ClC type of anion channel, CLH-3b, that is activated by swelling and oocyte meiotic maturation.[176] Channel activation is triggered by PP1-mediated serine/threonine dephosphorylation.[177] The PASK/OSR1 homolog GCK-3 interacts with

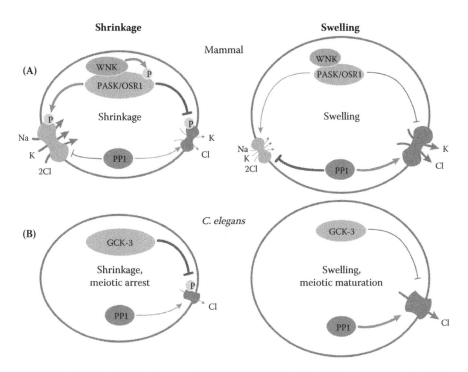

FIGURE 2.6 (See color insert following page 208.) Models of phosphorylation and dephosphorylation events that regulate the activity of volume-sensitive Na–K–2Cl and K–Cl cotransporters in mammalian cells and a volume-sensitive ClC type of anion channel in *Caenorhabditis elegans*. The relative activities of kinases and ion transport pathways are indicated by size. Green arrows and red lines indicate activation and inhibition, respectively. Physical interactions are indicated by overlap of components. (A) During shrinkage in mammalian cells, Na–K–2Cl cotransporters are activated and K–Cl cotransporters are inactivated by phosphorylation via PASK or OSR1. Dephosphorylation via a type 1 protein phosphatase (PP1) inactivates Na–K–2Cl cotransporters and activates K–Cl cotransporters during cell swelling. WNK1 and WNK4 interact with PASK and OSR1. Cell shrinkage activates WNK1 and WNK4 via unknown mechanisms. Activated WNK1 or WNK4 then phosphorylates and activates PASK and OSR1. An unknown phosphatase is thought to dephosphorylate and inactivate PASK and OSR1 during cell swelling. (B) In *C. elegans* oocytes, the ClC channel CLH-3b is inactivated by the PASK/OSR1 homolog GCK-3 by a mechanism that requires kinase activity. CLH-3b is inhibited by GCK-3 during cell shrinkage and meiotic arrest; inhibition by GCK-3 is removed and the channel is activated by the PP1 homologs GLC-7α and GCL-7β during cell swelling and meiotic maturation.

the cytoplasmic C-terminus of the channel. Coexpression of GCK-3 and CLH-3b in HEK293 cells dramatically inhibits channel activity, and RNA interference knockdown of the kinase constitutively activates CLH-3b in worm oocytes (see Figure 2.6B).[46]

Like PASK and OSR1, GCK-3 interacts with the single *Caenorhabditis elegans* WNK homolog WNK1. Both kinases are required for whole animal osmoregulation and for survival during hypertonic stress, and they appear to function in a common signaling pathway.[28] *C. elegans* and mammals are separated by hundreds of millions of years of evolution.[14,15,219] These results as well as those from studies in *C. elegans* oocytes[46,176,177] demonstrate that GCK-VI and WNK kinases play highly conserved and evolutionarily ancient roles in osmoregulation and cellular osmotic stress signal transduction.

WNK and GCK-VI kinases fit the model of a single, common cell volume signal transduction pathway that regulates both shrinkage- and swelling-induced transport pathways as proposed by Parker.[152,153] Future studies are necessary to identify the sensors that activate WNK and to address whether signaling through WNK and GCK-VI kinases is required for cell volume regulation.

3.　Src Kinases

Src kinases are cytoplasmic tyrosine kinases that have well-defined roles in cell growth, differentiation, cell adhesion, carcinogenesis, and immunity.[20,205] These kinases are also emerging as important regulators of cell volume and volume-related cellular processes.[31] The first Src kinase, *v*-Src (viral-sarcoma), was isolated from a cancer-causing retrovirus and shown to be a constitutively active mutant tyrosine kinase oncogene. The wild-type, cellular form of this kinase was named *c*-Src or Src. Activation of Src kinases can be complex but usually involves dephosphorylation of a C-terminal inhibitory tyrosine residue.[31] Interestingly, this residue is missing from the mutant oncogene form of Src. The Src kinases are the largest subfamily of nonreceptor tyrosine kinases and include Src, Yes, Fgr, Blk, Lck, Hck, Lyn, and Yrk kinases.[20,31,205]

Hypotonicity and hypertonicity activate different Src kinases;[31] however, a role for most osmotically sensitive Src kinases in volume regulation has not been established. The best evidence for Src regulation of cell volume is from studies on RVD ion transport pathways; for example, red blood cells from Fgr and Hck knockout mice have threefold greater K–Cl cotransporter activity.[41] De Franceschi et al.[41] have suggested that Fgr and Hck negatively regulate a protein phosphatase that activates the K–Cl cotransporter.

Lck is activated by hypotonicity, and RVD and $I_{Cl,swell}$ activity are inhibited in human T cells that lack functional Lck.[115] In rat hepatocytes, pharmacological inhibition of Src prevented hypotonicity-induced activation of ERK1/2 and p38 MAPKs and reduced RVD, suggesting that Src may regulate RVD via MAPKs.[227] Lck and Lyn may also regulate swelling-induced organic osmolyte efflux from red blood cells of an elasmobranch,[103] suggesting that Src regulation of RVD is conserved among distantly related vertebrates. Finally, studies using pharmacological inhibitors, dominant-negative kinase, and kinase-deficient cell lines suggest that Fyn contributes to hypertonicity-induced activation of TonEBP by a mechanism independent from p38 MAPKs.[101]

As discussed above, TRPV4 is activated by cell swelling.[9,64,117,195,236] Pharmacological studies in cultured rat nociceptors suggest that hypotonicity-induced activation of TRPV4 is mediated by Src kinases.[2] Xu et al.[241] demonstrated that hypotonic stress induces phosphorylation of TRPV4 at tyrosine 253 (Y253). They also showed that Lyn is activated by cell swelling, that the kinase interacts with the channel and mediates phosphorylation, and that mutation of Y253 to phenylalanine blocks hypotonicity-induced channel activation. Other investigators, however, have been unable to reproduce many of these findings. Vriens et al.[228] could find no role for Lyn or Y253 in swelling-induced activation of TRPV4 and instead concluded that volume-dependent channel regulation is mediated by arachidonic acid metabolites (discussed in Section V.B).

B.　Eicosanoids

PLA$_2$ catalyzes the hydrolysis of cellular phospholipids to generate arachidonic acid. Arachidonic acid can be further metabolized by lipoxygenases to generate leukotrienes.[35] As discussed earlier (Section IV.A.2), PLA$_2$ is sensitive to lipid packing density and can be activated by swelling when reconstituted into liposomes.[21,114] Studies in several cell types have shown that both PLA$_2$ and lipoxygenase activity are required for RVD.[85] In addition, arachidonic acid and leukotrienes have been shown to regulate the activity of K$^+$, Cl$^-$, and organic osmolyte efflux pathways.[85,110,112,159,196] Vriens et al[228] have also shown that swelling-induced activation of TRPV4 is inhibited by PLA$_2$ blockers and blockers of cytochrome P450 epoxygenase. They suggest that swelling activates PLA$_2$, leading to arachidonic acid production. Cytochrome P450 epoxygenase converts arachidonic acid into 5′,6′-epoxyeicosatrienoic acid (5′,6′-EET), which in turn activates TRPV4.

C.　Intracellular Ca^{2+}

Intracellular Ca^{2+} ([Ca^{2+}]$_i$) is a ubiquitous second messenger that regulates numerous diverse cellular processes.[12] Swelling-induced increases in [Ca^{2+}]$_i$ and Ca^{2+}-dependent RVD have been observed in several vertebrate cell types.[112,158] In addition, studies in cnidarian[126] and crustacean[11] cells have demonstrated that RVD is Ca^{2+} dependent. The underlying RVD transport pathways that are regulated by $[Ca^{2+}]_i$ are poorly defined. Pasantes-Morales and Mulia[158] have carefully reviewed the literature and noted that swelling-activated Cl^- and organic osmolyte efflux mechanisms are Ca^{2+} independent in most cell types, whereas swelling-activated K^+ channels are commonly regulated by $[Ca^{2+}]_i$ changes. The biophysical properties of these channels suggest that they are BK, or maxi-K^+, channels.

1. Ca^{2+} Entry

Changes in $[Ca^{2+}]_i$ can be brought about by increases in Ca^{2+} influx or release from intracellular stores.[12] In a variety of vertebrate cell types, an extracellular source of Ca^{2+} is required for swelling-induced increases in $[Ca^{2+}]_i$ and for RVD.[119,184,188,207,209] Several recent studies have indicated that TRP channels mediate swelling-induced Ca^{2+} entry. Arniges et al.[4] used siRNA in human airway cells to demonstrate that TRVP4 is required for swelling-induced activation of Ca^{2+}-dependent KCNN4 potassium channels and RVD. Becker et al.[9] have shown that human keratinocytes have a robust RVD that is completely blocked by removal of extracellular Ca^{2+} or by exposure to Gd^{3+}, an inhibitor of TRPV4. Furthermore, they showed that Chinese hamster ovary cells that do not express TRPV4 lack an RVD response; however, RVD can be rescued by heterologous expression of TRPV4. Swelling-induced Ca^{2+} entry via TRPV4 may also regulate RVD in human salivary epithelial cells.[119] In HeLa cells, extracellular Ca^{2+} removal, TRPM7 inhibitors, and TRPM7 siRNA block RVD, suggesting that TRPM7 mediates swelling-induced Ca^{2+} entry.[143] Calcium entry may also be mediated by swelling-induced activation of TRPM3[68] and TRPV2.[134]

2. Intracellular Ca^{2+} Release

Depletion of intracellular Ca^{2+} stores or inhibition of Ca^{2+} store release has been shown to partially inhibit swelling-induced increases in $[Ca^{2+}]_i$ and RVD in several vertebrate cell types.[179,189,209,244] Release of Ca^{2+} from intracellular stores is mediated by inositol 1,4,5-trisphosphate (IP_3) or ryanodine receptor Ca^{2+} channels;[12] however, little direct evidence supports the involvement of these channels in swelling-induced Ca^{2+} signaling and RVD.[158]

D. Rho GTP-Binding Proteins

Rho GTP-binding proteins are a ubiquitous eukaryotic family of Ras-related GTPases that includes 22 members in mammals.[48,84] Rho GTPases are small (~21-kDa) signaling molecules that switch between inactive GDP-bound and active GTP-bound states. In the active GTP-bound state, Rho GTPases interact with multiple proteins and have well-characterized roles in regulating the polymerization of the actin cytoskeleton. The activation state of GTPases is controlled by a large number of regulatory proteins, including guaninine nucleotide exchange factors (GEFs) that replace GTP for GDP, GTPase-activating proteins (GAPs) that stimulate intrinsic GTPase activity, and guanine nucleotide dissociation inhibitors (GDIs) that block spontaneous activation.

As mentioned above, Rac and Cdc42 are Rho GTPases that are activated by hypertonic stress and initiate p38 MAPK cascades in mammalian and yeast cells, respectively.[47,78,100,116,178,217] Rho, another Rho GTPase, is activated in less than 1 minute by hypertonicity in mammalian renal tubule cells.[49] Cell shrinkage and increased intracellular ionic strength can activate Rho GTPases, but the underlying molecular mechanisms by which this occurs are unknown.[48,49] Hypertonic stress-induced activation of Rho GTPases functions to control cytoskeletal remodeling in the cortex, which is thought to help cells withstand physical forces imposed by volume changes.[48] Rho GTPases also function to regulate p38 MAPK cascades in yeast and mammals that control the transcription of genes required for organic osmolyte accumulation.[78,100,178,217]

Rho GTPases may also play a role in controlling RVD. Activation of $I_{Cl,swell}$ has been proposed to be regulated by Rho GTPases in at least three different vertebrate cells types.[56,139,140,206] In NIH3T3 cells, overexpression of constitutively active RhoA increases the rate of RVD, the rate of swelling-activated K^+ and taurine efflux, and the magnitude of $I_{Cl,swell}$.[161]

VI. CONCLUSIONS AND FUTURE PERSPECTIVE

The ability to tightly control solute and water balance during osmotic challenge is an essential prerequisite for cellular life. Cellular osmotic homeostasis is maintained by the regulated accumulation and loss of inorganic ions and organic osmolytes. The effector mechanisms responsible for osmoregulatory solute accumulation and loss in animal cells are generally well understood; however, major gaps exist in our understanding of the signals and signaling pathways by which animal cells detect different types, rates, and magnitudes of volume perturbations[192] and activate volume-regulatory mechanisms.

Over the last few years, some molecular insight into osmotically sensitive signaling mechanisms has been gained. In our opinion, the most significant breakthrough has been the discovery of the role of Ste20 kinase and WNK signaling in regulating volume-sensitive K–Cl and Na–K–2Cl cotransporters in mammals[43,96,194] and a volume-sensitive anion channel[194] and systemic osmotic homeostasis in *Caenorhabditis elegans*.[28] Nematodes and mammals are separated by hundreds of millions of years of evolution,[14,15,219] demonstrating that Ste20/WNK signaling is evolutionarily ancient and likely represents an essential and highly conserved osmosensing pathway in animals.

As detailed in Section V, other osmotically sensitive signaling components have been identified in animals. For the most part, the specific regulatory targets of these components are unclear. In addition, it is unclear whether signaling mechanisms that have been identified or postulated are specific to certain cell types and experimental conditions or whether they represent more universal mechanisms by which cells respond to volume perturbations. We also have little understanding of how volume regulation and various osmotic stress signaling pathways are coordinated with other cellular processes.

Perhaps the most vexing problem is the nature of the signals that indicate to cells that their volume has been perturbed and the sensing mechanisms that detect those signals. It is likely that volume regulation and other osmotic stress responses require the integration of a number of different signals and signal transduction pathways. Considerable understanding of osmosensing has been obtained in bacteria and yeast by forward genetic analysis. Similarly, forward as well as reverse genetic analysis of osmotic stress responses in model organisms such as *Caenorhabditis elegans* should provide important insights into how animals detect cell volume changes. Elucidation of volume-sensing mechanisms and signaling pathways represents the most pressing and significant challenge in the field and is essential for a full, integrative understanding of cell volume control and related cellular processes.

REFERENCES

1. Al Habori, M., Macromolecular crowding and its role as intracellular signalling of cell volume regulation, *Int. J. Biochem. Cell Biol.*, 33, 844–864, 2001.
2. Alessandri-Haber, N., Dina, O. A., Yeh, J. J., Parada, C. A., Reichling, D. B., and Levine, J. D., Transient receptor potential vanilloid 4 is essential in chemotherapy-induced neuropathic pain in the rat, *J. Neurosci.*, 24, 4444–4452, 2004.
3. Anselmo, A. N., Earnest, S., Chen, W., Juang, Y. C., Kim, S. C., Zhao, Y., and Cobb, M. H., WNK1 and OSR1 regulate the Na^+, K^+, $2Cl^-$ cotransporter in HeLa cells, *Proc. Natl. Acad. Sci. USA*, 103, 10883–10888, 2006.
4. Arniges, M., Vazquez, E., Fernandez-Fernandez, J. M., and Valverde, M. A., Swelling-activated Ca^{2+} entry via TRPV4 channel is defective in cystic fibrosis airway epithelia, *J. Biol. Chem.*, 279, 54062–54068, 2004.

5. Arreola, J., Begenisich, T., Nehrke, K., Nguyen, H. V., Park, K., Richardson, L., Yang, B., Schutte, B. C., Lamb, F. S., and Melvin, J. E., Secretion and cell volume regulation by salivary acinar cells from mice lacking expression of the Clcn3 Cl⁻ channel gene, *J. Physiol.*, 545, 207–216, 2002.

6. Auton, M. and Bolen, D. W., Predicting the energetics of osmolyte-induced protein folding/unfolding, *Proc. Natl. Acad. Sci. USA*, 102, 15065–15068, 2005.

7. Bachhuber, T., Konig, J., Voelcker, T., Murle, B., Schreiber, R., and Kunzelmann, K., Cl⁻ interference with the epithelial Na⁺ channel ENaC, *J. Biol. Chem.*, 280, 31587–31594, 2005.

8. Bazopoulou, D. and Tavernarakis, N., Mechanosensitive ion channels in *Caenorhabditis elegans*. In *Current Topics in Membranes, Mechanosensitive Ion Channels, Part B*, Hamill, O. P., Ed., Elsevier, St. Louis, MO, 2007, chap. 3.

9. Becker, D., Blase, C., Bereiter-Hahn, J., and Jendrach, M., TRPV4 exhibits a functional role in cell-volume regulation, *J. Cell Sci.*, 118, 2435–2440, 2005.

10. Berl, T., Siriwardana, G., Ao, L., Butterfield, L. M., and Heasley, L. E., Multiple mitogen-activated protein kinases are regulated by hyperosmolality in mouse IMCD cells, *Am. J. Physiol.*, 272, R305–R311, 1997.

11. Berman, D. M., Pena-Rasgado, C., and Rasgado-Flores, H., Changes in membrane potential associated with cell swelling and regulatory volume decrease in barnacle muscle cells, *J Exp. Zool.*, 268, 97–103, 1994.

12. Berridge, M. J., Bootman, M. D., and Roderick, H. L., Calcium signalling: dynamics, homeostasis and remodelling, *Nat. Rev. Mol. Cell Biol.*, 4, 517–529, 2003.

13. Biemans-Oldehinkel, E., Mahmood, N. A., and Poolman, B., A sensor for intracellular ionic strength, *Proc. Natl. Acad. Sci. USA*, 103, 10624–10629, 2006.

14. Blaxter, M., *Caenorhabditis elegans* is a nematode, *Science*, 282, 2041–2046, 1998.

15. Blaxter, M. L., De Ley, P., Garey, J. R., Liu, L. X., Scheldeman, P., Vierstraete, A., Vanfleteren, J. R., Mackey, L. Y., Dorris, M., Frisse, L. M., Vida, J. T., and Thomas, W. K., A molecular evolutionary framework for the phylum Nematoda, *Nature*, 392, 71–75, 1998.

16. Blumberg, S., Tkachenko, A. V., and Meiners, J. C., Disruption of protein-mediated DNA looping by tension in the substrate DNA, *Biophys. J.*, 88, 1692–1701, 2005.

17. Bond, T., Basavappa, S., Christensen, M., and Strange, K., ATP dependence of the I$_{Cl,swell}$ channel varies with rate of cell swelling: evidence for two modes of channel activation, *J. Gen. Physiol.*, 113, 441–456, 1999.

18. Brigotti, M., Petronini, P. G., Carnicelli, D., Alfieri, R. R., Bonelli, M. A., Borghetti, A. F., and Wheeler, K. P., Effects of osmolarity, ions and compatible osmolytes on cell-free protein synthesis, *Biochem. J.*, 369, 369–374, 2003.

19. Browe, D. M. and Baumgarten, C. M., Stretch of β_1 integrin activates an outwardly rectifying chloride current via FAK and Src in rabbit ventricular myocytes, *J. Gen. Physiol.*, 122, 689–702, 2003.

20. Brown, M. T. and Cooper, J. A., Regulation, substrates and functions of *src*, *Biochim. Biophys. Acta*, 1287, 121–149, 1996.

21. Burack, W. R. and Biltonen, R. L., Lipid bilayer heterogeneities and modulation of phospholipase A₂ activity, *Chem. Phys. Lipids*, 73, 209–222, 1994.

22. Burg, M. B., Renal osmoregulatory transport of compatible organic osmolytes, *Curr. Opin. Nephrol. Hypertens.*, 6, 430–433, 1997.

23. Burg, M. B., Kwon, E. D., and Kültz, D., Regulation of gene expression by hypertonicity, *Ann. Rev. Physiol.*, 59, 437–55, 1997.

24. Cannon, C. L., Basavappa, S., and Strange, K., Intracellular ionic strength regulates the volume sensitivity of a swelling-activated anion channel, *Am. J. Physiol.*, 275, C416–C422, 1998.

25. Capasso, J. M., Rivard, C., and Berl, T., The expression of the gamma subunit of Na–K-ATPase is regulated by osmolality via C-terminal Jun kinase and phosphatidylinositol 3-kinase-dependent mechanisms, *Proc. Natl. Acad. Sci. USA*, 98, 13414–13419, 2001.

26. Chaqour, B., Yang, R., and Sha, Q., Mechanical stretch modulates the promoter activity of the profibrotic factor CCN2 through increased actin polymerization and NF-κB activation, *J. Biol. Chem.*, 281, 20608–20622, 2006.

27. Chiri, S., Bogliolo, S., Ehrenfeld, J., and Ciapa, B., Activation of extracellular signal-regulated kinase ERK after hypo-osmotic stress in renal epithelial A6 cells, *Biochim. Biophy. Acta*, 1664, 224–229, 2004.

28. Choe, K. P. and Strange, K., Evolutionarily conserved WNK and Ste20 kinases are essential for acute volume recovery and survival following hypertonic shrinkage in *Caenorhabditis elegans*, *Am. J. Physiol.*, 293, C915–C927, 2007.

29. Christensen, A. P. and Corey, D. P., TRP channels in mechanosensation: direct or indirect activation?, *Nat. Rev. Neurosci.*, 8, 510–521, 2007.

30. Ciano-Oliveira, C., Thirone, A. C., Szaszi, K., and Kapus, A., Osmotic stress and the cytoskeleton: the R(h)ole of Rho GTPases, *Acta Physiol.*, 187, 257–272, 2006.

31. Cohen, D. M., SRC family kinases in cell volume regulation, *Am. J. Physiol.*, 288, C483–C493, 2005.

32. Colbert, H. A., Smith, T. L., and Bargmann, C. I., OSM-9, a novel protein with structural similarity to channels, is required for olfaction, mechanosensation, and olfactory adaptation in *Caenorhabditis elegans*, *J. Neurosci.*, 17, 8259–8269, 1997.

33. Colclasure, G. C. and Parker, J. C., Cytosolic protein concentration is the primary volume signal in dog red cells, *J. Gen. Physiol.*, 98, 881–892, 1991.

34. Colclasure, G. C. and Parker, J. C., Cytosolic protein concentration is the primary volume signal for swelling-induced [K–Cl] cotransport in dog red cells, *J. Gen. Physiol.*, 100, 1–10, 1992.

35. Cook, J. A., Eicosanoids, *Crit. Care Med.*, 33, S488–S491, 2005.

36. Culham, D. E., Henderson, J., Crane, R. A., and Wood, J. M., Osmosensor ProP of *Escherichia coli* responds to the concentration, chemistry, and molecular size of osmolytes in the proteoliposome lumen, *Biochemistry*, 42, 410–420, 2003.

37. Dahl, S. C., Handler, J. S., and Kwon, H. M., Hypertonicity-induced phosphorylation and nuclear localization of the transcription factor TonEBP, *Am. J. Physiol.*, C248–C253, 2001.

38. Dallos, P. and Fakler, B., Prestin, a new type of motor protein, *Nat. Rev. Mol. Cell Biol.*, 3, 104–111, 2002.

39. Dan, I., Watanabe, N. M., and Kusumi, A., The Ste20 group kinases as regulators of MAP kinase cascades, *Trends Cell Biol.*, 11, 220–230, 2001.

40. Darman, R. B., Flemmer, A., and Forbush, B., Modulation of ion transport by direct targeting of protein phosphatase type 1 to the Na–K–Cl cotransporter, *J. Biol. Chem.*, 276, 34359–34362, 2001.

41. De Franceschi, L., Fumagalli, L., Olivieri, O., Corrocher, R., Lowell, C. A., and Berton, G., Deficiency of Src family kinases Fgr and Hck results in activation of erythrocyte K/Cl cotransport, *J. Clin. Invest.*, 99, 220–227, 1997.

42. Defilippi, P., Di Stefano, P., and Cabodi, S., p130Cas: a versatile scaffold in signaling networks, *Trends Cell Biol.*, 16, 257–263, 2006.

43. Delpire, E. and Gagnon, K. B., SPAK and OSR1, key kinases involved in the regulation of chloride transport, *Acta Physiol.*, 187, 103–113, 2006.

44. DeMali, K. A., Wennerberg, K., and Burridge, K., Integrin signaling to the actin cytoskeleton, *Curr. Opin. Cell Biol.*, 15, 572–582, 2003.

45. Denton, J., Nehrke, K., Rutledge, E., Morrison, R., and Strange, K., Alternative splicing of N- and C-termini of a *C. elegans* ClC channel alters gating and sensitivity to external Cl$^-$ and H$^+$, *J. Physiol.*, 555, 97–114, 2004.

46. Denton, J., Nehrke, K., Yin, X., Morrison, R., and Strange, K., GCK-3, a newly identified Ste20 kinase, binds to and regulates the activity of a cell cycle-dependent ClC anion channel, *J. Gen. Physiol.*, 125, 113–125, 2005.

47. Di Ciano, C., Nie, Z., Szaszi, K., Lewis, A., Uruno, T., Zhan, X., Rotstein, O. D., Mak, A., and Kapus, A., Osmotic stress-induced remodeling of the cortical cytoskeleton, *Am. J. Physiol.*, 283, C850–C865, 2002.

48. Di Ciano-Oliveira, C., Thirone, A. C. P., Szaszi, K., and Kapus, A., Osmotic stress and the cytoskeleton: the R(h)ole of Rho GTPases, *Acta Physiol.*, 187, 257–272, 2006.

49. Di Ciano-Oliveira, C., Sirokmany, G., Szaszi, K., Arthur, W. T., Masszi, A., Peterson, M., Rotstein, O. D., and Kapus, A., Hyperosmotic stress activates Rho: differential involvement in Rho kinase-dependent MLC phosphorylation and NKCC activation, *Am. J. Physiol.*, 285, C555–C566, 2003.

50. Dowd, B. F. and Forbush, B., PASK (proline-alanine-rich STE20-related kinase), a regulatory kinase of the Na–K–Cl cotransporter (NKCC1), *J. Biol. Chem.*, 278, 27347–27353, 2003.

51. Duan, D., Winter, C., Cowley, S., Hume, J. R., and Horowitz, B., Molecular identification of a volume-regulated chloride channel, *Nature*, 390, 417–421, 1997.

52. Ebner, H. L., Fiechtner, B., Pelster, B., and Krumschnabel, G., Extracellular signal regulated MAP-kinase signalling in osmotically stressed trout hepatocytes, *Biochim. Biophy. Acta*, 1760, 941–950, 2006.

53. Ellis, R. J., Macromolecular crowding: obvious but underappreciated, *Trends Biochem. Sci.*, 26, 597–604, 2001.

54. Ellis, R. J. and Minton, A. P., Cell biology: join the crowd, *Nature*, 425, 27–28, 2003.

55. Emma, F., McManus, M., and Strange, K., Intracellular electrolytes regulate the volume set point of the organic osmolyte/anion channel VSOAC, *Am. J. Physiol.*, 272, C1766–C1775, 1997.

56. Estevez, A. Y., Bond, T., and Strange, K., Regulation of $I_{Cl,swell}$ in neuroblastoma cells by G protein signaling pathways, *Am. J. Physiol.*, 281, C89–C98, 2001.

57. Ferraris, J. D. and Burg, M. B., Tonicity-dependent regulation of osmoprotective genes in mammalian cells, *Contrib. Nephrol.*, 152, 125–141, 2006.

58. Ferraris, J. D., Williams, C. K., Jung, K. Y., Bedford, J. J., Burg, M. B., and Garcia-Perez, A., ORE, a eukaryotic minimal essential osmotic response element, *J. Biol. Chem.*, 271, 18318–18321, 1996.

59. Ferraris, J. D., Persaud, P., Williams, C. K., Chen, Y., and Burg, M. B., cAMP-independent role of PKA in tonicity-induced transactivation of tonicity-responsive enhancer/osmotic response element-binding protein, *Proc. Natl. Acad. Sci. USA*, 99, 16800–16805, 2002.

60. ffrench-Constant, C. and Colognato, H., Integrins: versatile integrators of extracellular signals, *Trends Cell Biol.*, 14, 678–686, 2004.

61. Fischmeister, R. and Hartzell, H. C., Volume sensitivity of the bestrophin family of chloride channels, *J. Physiol.*, 562, 477–491, 2005.

62. Gagnon, K. B., England, R., and Delpire, E., Characterization of SPAK and OSR1, regulatory kinases of the Na–K–2Cl cotransporter, *Mol. Cell Biol.*, 26, 689–698, 2006.

63. Gagnon, K. B., England, R., and Delpire, E., Volume sensitivity of cation–Cl⁻ cotransporters is modulated by the interaction of two kinases: Ste20-related proline-alanine-rich kinase and WNK4, *Am. J. Physiol.*, 290, C134–C142, 2006.

64. Gao, X., Wu, L., and O'Neil, R. G., Temperature-modulated diversity of TRPV4 channel gating: activation by physical stresses and phorbol ester derivatives through protein kinase C-dependent and -independent pathways, *J. Biol. Chem.*, 278, 27129–27137, 2003.

65. Garzon-Muvdi, T., Pacheco-Alvarez, D., Gagnon, K. B., Vazquez, N., Ponce-Coria, J., Moreno, E., Delpire, E., and Gamba, G., WNK4 kinase is a negative regulator of K⁺–Cl⁻ cotransporters, *Am. J. Physiol.*, 292, F1197–F1209, 2007.

66. Gemmen, G. J., Millin, R., and Smith, D. E., Tension-dependent DNA cleavage by restriction endo-nucleases: two-site enzymes are "switched off" at low force, *Proc. Natl. Acad. Sci. USA*, 103, 11555–11560, 2006.

67. Gong, Z., Son, W., Chung, Y. D., Kim, J., Shin, D. W., McClung, C. A., Lee, Y., Lee, H. W., Chang, D. J., Kaang, B. K., Cho, H., Oh, U., Hirsh, J., Kernan, M. J., and Kim, C., Two interdependent TRPV channel subunits, inactive and Nanchung, mediate hearing in *Drosophila*, *J. Neurosci.*, 24, 9059–9066, 2004.

68. Grimm, C., Kraft, R., Sauerbruch, S., Schultz, G., and Harteneck, C., Molecular and functional characterization of the melastatin-related cation channel TRPM3, *J. Biol. Chem.*, 278, 21493–21501, 2003.

69. Grunder, S., Thiemann, A., Pusch, M., and Jentsch, T. J., Regions involved in the opening of ClC-2 chloride channel by voltage and cell volume, *Nature*, 360, 759–762, 1992.

70. Grunnet, M., Jespersen, T., MacAulay, N., Jorgensen, N. K., Schmitt, N., Pongs, O., Olesen, S. P., and Klaerke, D. A., KCNQ1 channels sense small changes in cell volume, *J. Physiol.*, 549, 419–427, 2003.

71. Guizouarn, H. and Motais, R., Swelling activation of transport pathways in erythrocytes: effects of Cl⁻, ionic strength, and volume changes, *Am. J. Physiol.*, 276, C210–C220, 1999.

72. Hamill, O. P., Twenty odd years of stretch-sensitive channels, *Pflügers Arch.*, 453, 333–351, 2006.

73. Harrison, B. and Zimmerman, S. B., Polymer-stimulated ligation: enhanced ligation of oligo- and polynucleotides by T4 RNA ligase in polymer solutions, *Nucleic Acids Res.*, 12, 8235–8251, 1984.

74. Harrison, B. and Zimmerman, S. B., Stabilization of T4 polynucleotide kinase by macromolecular crowding, *Nucleic Acids Res.*, 14, 1863–1870, 1986.

75. Harrison, B. and Zimmerman, S. B., T4 polynucleotide kinase: macromolecular crowding increases the efficiency of reaction at DNA termini, *Anal. Biochem.*, 158, 307–315, 1986.

76. Higashijima, T., Ferguson, K. M., and Sternweis, P. C., Regulation of hormone-sensitive GTP-dependent regulatory proteins by chloride, *J. Biol. Chem.*, 262, 3597–3602, 1987.

77. Hogan, E. M., Davis, B. A., and Boron, W. F., Intracellular Cl⁻ dependence of Na–H exchange in barnacle muscle fibers under normotonic and hypertonic conditions, *J. Gen. Physiol.*, 110, 629–639, 1997.

78. Hohmann, S., Osmotic stress signaling and osmoadaptation in yeasts, *Microbiol. Mol. Biol. Rev.*, 66, 300–372, 2002.

79. Ignatova, Z. and Gierasch, L. M., Inhibition of protein aggregation *in vitro* and *in vivo* by a natural osmoprotectant, *Proc. Natl. Acad. Sci. USA*, 103, 13357–13361, 2006.

80. Ignoul, S. and Eggermont, J., CBS domains: structure, function, and pathology in human proteins, *Am. J. Physiol.*, 289, C1369–C1378, 2005.

81. Ingber, D. E., Cellular mechanotransduction: putting all the pieces together again, *FASEB J.*, 20, 811–827, 2006.

82. Irarrazabal, C. E., Burg, M. B., Ward, S. G., and Ferraris, J. D., Phosphatidylinositol 3-kinase mediates activation of ATM by high NaCl and by ionizing radiation: role in osmoprotective transcriptional regulation, *Proc. Natl. Acad. Sci. USA*, 103, 8882–8887, 2006.

83. Irarrazabal, C. E., Liu, J. C., Burg, M. B., and Ferraris, J. D., ATM, a DNA damage-inducible kinase, contributes to activation by high NaCl of the transcription factor TonEBP/OREBP, *Proc. Natl. Acad. Sci. USA*, 101, 8809–8814, 2004.

84. Jaffe, A. B. and Hall, A., Rho GTPases: Biochemistry and biology, *Ann. Rev. Cell Dev. Biol.*, 21, 247–269, 2005.

85. Jakab, M., Furst, J., Gschwentner, M., Botta, G., Garavaglia, M. L., Bazzini, C., Rodighiero, S., Meyer, G., Eichmueller, S., Woll, E., Chwatal, S., Ritter, M., and Paulmichl, M., Mechanisms sensing and modulating signals arising from cell swelling, *Cell Physiol. Biochem.*, 12, 235–258, 2002.

86. Jennings, M. L., Volume-sensitive K⁺/Cl⁻ cotransport in rabbit erythrocytes: analysis of the rate-limiting activation and inactivation events, *J. Gen. Physiol.*, 114, 743–758, 1999.

87. Jennings, M. L. and al-Rohil, N., Kinetics of activation and inactivation of swelling-stimulated K⁺/Cl⁻ transport. The volume-sensitive parameter is the rate constant for inactivation, *J. Gen. Physiol.*, 95, 1021–1040, 1990.

88. Jennings, M. L. and Schulz, R. K., Okadaic acid inhibition of KCl cotransport: evidence that protein dephosphorylation is necessary for activation of transport by either cell swelling or *N*-ethylmaleimide, *J. Gen. Physiol.*, 97, 799–817, 1991.

89. Jensen, H. S., Callo, K., Jespersen, T., Jensen, B. S., and Olesen, S. P., The KCNQ5 potassium channel from mouse: a broadly expressed M-current-like potassium channel modulated by zinc, pH, and volume changes, *Brain Res. Mol. Brain Res.*, 139, 52–62, 2005.

90. Jentsch, T. J., Stein, V., Weinreich, F., and Zdebik, A. A., Molecular structure and physiological function of chloride channels, *Physiol. Rev.*, 82, 503–568, 2002.

91. Jeon, U. S., Kim, J. A., Sheen, M. R., and Kwon, H. M., How tonicity regulates genes: story of TonEBP transcriptional activator, *Acta Physiol.*, 187, 241–247, 2006.

92. Jordt, S. E. and Jentsch, T. J., Molecular dissection of gating in the ClC-2 chloride channel, *EMBO J.*, 16, 1582–1592, 1997.

93. Junankar, P. R. and Kirk, K., Organic osmolyte channels: a comparative view, *Cell Physiol. Biochem.*, 10, 355–360, 2000.

94. Jung, K. and Altendorf, K., Individual substitutions of clustered arginine residues of the sensor kinase KdpD of *Escherichia coli* modulate the ratio of kinase to phosphatase activity, *J. Biol. Chem.*, 273, 26415–26420, 1998.

95. Jung, K., Veen, M., and Altendorf, K., K⁺ and ionic strength directly influence the autophosphorylation activity of the putative turgor sensor KdpD of *Escherichia coli*, *J. Biol. Chem.*, 275, 40142–40147, 2000.

96. Kahle, K. T., Rinehart, J., Ring, A., Gimenez, I., Gamba, G., Hebert, S. C., and Lifton, R. P., WNK protein kinases modulate cellular Cl⁻ flux by altering the phosphorylation state of the Na–K–Cl and K–Cl cotransporters, *Physiology*, 21, 326–335, 2006.

97. Kim, J., Chung, Y. D., Park, D. Y., Choi, S., Shin, D. W., Soh, H., Lee, H. W., Son, W., Yim, J., Park, C. S., Kernan, M. J., and Kim, C., A TRPV family ion channel required for hearing in *Drosophila*, *Nature*, 424, 81–84, 2003.

98. Kirk, K. and Strange, K., Functional properties and physiological roles of organic solute channels, *Ann. Rev. Physiol.*, 60, 719–739, 1998.

99. Kleckner, N., Zickler, D., Jones, G. H., Dekker, J., Padmore, R., Henle, J., and Hutchinson, J., A mechanical basis for chromosome function, *Proc. Natl. Acad. Sci. USA*, 101, 12592–12597, 2004.

100. Klipp, E., Nordlander, B., Kruger, R., Gennemark, P., and Hohmann, S., Integrative model of the response of yeast to osmotic shock, *Nat. Biotechnol.*, 23(8), 975–982, 2005.

101. Ko, B. C. B., Lam, A. K. M., Kapus, A., Fan, L., Chung, S. K., and Chung, S. S. M., Fyn and p38 signaling are both required for maximal hypertonic activation of the osmotic response element-binding protein/tonicity-responsive enhancer-binding protein (OREBP/TonEBP), *J. Biol. Chem.*, 277, 46085–46092, 2002.

102. Kojima, R., Randall, J. D., Ito, E., Manshio, H., Suzuki, Y., and Gullans, S. R., Regulation of expression of the stress response gene Osp94: identification of the tonicity response element and intracellular signalling pathways, *Biochem. J.*, 380, 783–794, 2004.

103. Koomoa, D.-L. T., Musch, M. W., Vaz MacLean, A., and Goldstein, L., Volume-activated trimethylamine oxide efflux in red blood cells of spiny dogfish (*Squalus acanthias*), *Am. J. Physiol.*, 281, R803–R810, 2001.

104. Kruse, E., Uehlein, N., and Kaldenhoff, R., The aquaporins, *Genome Biol.*, 7, 206, 2006.

105. Kuang, K., Yiming, M., Zhu, Z., Iserovich, P., Diecke, F. P., and Fischbarg, J., Lack of threshold for anisotonic cell volume regulation, *J. Membr. Biol.*, 211, 27–33, 2006.

106. Kultz, D. and Burg, M., Evolution of osmotic stress signaling via MAP kinase cascades, *J. Exp. Biol.*, 201, 3015–3021, 1998.

107. Kultz, D. and Avila, K., Mitogen-activated protein kinases are *in vivo* transducers of osmosensory signals in fish gill cells, *Comp. Biochem. Physiol. B*, 129, 821–829, 2001.

108. Kwon, H. M., Itoh, T., Rim, J. S., and Handler, J. S., The MAP kinase cascade is not essential for transcriptional stimulation of osmolyte transporter genes, *Biochem. Biophys. Res. Comm.*, 213, 975–979, 1995.

109. Lalioti, M. D., Zhang, J., Volkman, H. M., Kahle, K. T., Hoffmann, K. E., Toka, H. R., Nelson-Williams, C., Ellison, D. H., Flavell, R., Booth, C. J., Lu, Y., Geller, D. S., and Lifton, R. P., Wnk4 controls blood pressure and potassium homeostasis via regulation of mass and activity of the distal convoluted tubule, *Nat. Genet.*, 38, 1124–1132, 2006.

110. Lambert, I. H., Regulation of the cellular content of the organic osmolyte taurine in mammalian cells, *Neurochem. Res.*, 29, 27–63, 2004.

111. Lamitina, T., Huang, C. G., and Strange, K., Genome-wide RNAi screening identifies protein damage as a regulator of osmoprotective gene expression, *Proc. Natl. Acad. Sci. USA*, 103, 12173–12178, 2006.

112. Lang, F., Busch, G. L., Ritter, M., Volkl, H., Waldegger, S., Gulbins, E., and Haussinger, D., Functional significance of cell volume-regulatory mechanisms, *Physiol. Rev.*, 78, 247–306, 1998.

113. Lee, S. D., Colla, E., Sheen, M. R., Na, K. Y., and Kwon, H. M., Multiple domains of TonEBP cooperate to stimulate transcription in response to hypertonicity, *J. Biol. Chem.*, 278, 47571–47577, 2003.

114. Lehtonen, J. Y. and Kinnunen, P. K., Phospholipase A_2 as a mechanosensor, *Biophys. J.*, 68, 1888–1894, 1995.

115. Lepple-Wienhues, A., Szabo, I., Laun, T., Kaba, N. K., Gulbins, E., and Lang, F., The tyrosine kinase p56lck mediates activation of swelling-induced chloride channels in lymphocytes, *J. Cell Biol.*, 141, 281–286, 1998.

116. Lewis, A., Di Ciano, C., Rotstein, O. D., and Kapus, A., Osmotic stress activates Rac and Cdc42 in neutrophils: role in hypertonicity-induced actin polymerization, *Am. J. Physiol.*, 282, C271–C279, 2002.

117. Liedtke, W., Choe, Y., Marti-Renom, M. A., Bell, A. M., Denis, C. S., Sali, A., Hudspeth, A. J., Friedman, J. M., and Heller, S., Vanilloid receptor-related osmotically activated channel (VR-OAC), a candidate vertebrate osmoreceptor, *Cell*, 103, 525–535, 2000.

118. Liedtke, W., Tobin, D. M., Bargmann, C. I., and Friedman, J. M., Mammalian TRPV4 (VR-OAC) directs behavioral responses to osmotic and mechanical stimuli in *Caenorhabditis elegans*, *Proc. Natl. Acad. Sci. USA*, 100, 14531–14536, 2003.

119. Liu, X., Bandyopadhyay, B., Nakamoto, T., Singh, B., Liedtke, W., Melvin, J. E., and Ambudkar, I., A role for AQP5 in activation of TRPV4 by hypotonicity: concerted involvement of AQP5 and TRPV4 in regulation of cell volume recovery, *J. Biol. Chem.*, 281, 15485–15495, 2006.

120. Lohr, J. W. and Grantham, J. J., Isovolumetric regulation of isolated S_2 proximal tubules in anisotonic media, *J. Clin. Invest.*, 78, 1165–1172, 1986.

121. Lytle, C., Activation of the avian erythrocyte Na–K–Cl cotransport protein by cell shrinkage, cAMP, fluoride, and calyculin-A involves phosphorylation at common sites, *J. Biol. Chem.*, 272, 15069–15077, 1997.

122. Lytle, C., A volume-sensitive protein kinase regulates the Na–K–2Cl cotransporter in duck red blood cells, *Am. J. Physiol.*, 274, C1002–C1010, 1998.

123. MacLeod, R. J. and Hamilton, J. R., Ca^{2+}/calmodulin kinase II and decreases in intracellular pH are required to activate K^+ channels after substantial swelling in villus epithelial cells, *J. Membr. Biol.*, 172, 59–66, 1999.

124. Mahmood, N. A., Biemans-Oldehinkel, E., Patzlaff, J. S., Schuurman-Wolters, G. K., and Poolman, B., Ion specificity and ionic strength dependence of the osmoregulatory ABC transporter OpuA, *J. Biol. Chem.*, 281, 29830–29839, 2006.

125. Manning, G., Plowman, G. D., Hunter, T., and Sudarsanam, S., Evolution of protein kinase signaling from yeast to man, *Trends Biochem. Sci.*, 27, 514–520, 2002.

126. Marino, A. and La Spada, G., Calcium and cytoskeleton signaling during cell volume regulation in isolated nematocytes of *Aiptasia mutabilis* (Cnidaria: Anthozoa), *Comp. Biochem. Physiol. A*, 147, 196–204, 2007.

127. McManus, M. L., Churchwell, K. B., and Strange, K., Regulation of cell volume in health and disease, *N. Engl. J. Med.*, 333, 1260–1266, 1995.

128. Medalia, O., Weber, I., Frangakis, A. S., Nicastro, D., Gerisch, G., and Baumeister, W., Macromolecular architecture in eukaryotic cells visualized by cryoelectron tomography, *Science*, 298, 1209–1213, 2002.

129. Minton, A. P., Colclasure, G. C., and Parker, J. C., Model for the role of macromolecular crowding in regulation of cellular volume, *Proc. Natl. Acad. Sci. USA*, 89, 10504–10506, 1992.

130. Miyakawa, H., Woo, S. K., Dahl, S. C., Handler, J. S., and Kwon, H. M., Tonicity-responsive enhancer binding protein, a Rel-like protein that stimulates transcription in response to hypertonicity, *Proc. Natl. Acad. Sci. USA*, 96, 2538–2542, 1999.

131. Moeckel, G. W., Zhang, L., Chen, X., Rossini, M., Zent, R., and Pozzi, A., Role of integrin $\alpha_1\beta_1$ in the regulation of renal medullary osmolyte concentration, *Am. J. Physiol.*, 290, F223–F231, 2006.

132. Moriguchi, T., Urushiyama, S., Hisamoto, N., Iemura, S., Uchida, S., Natsume, T., Matsumoto, K., and Shibuya, H., WNK1 regulates phosphorylation of cation-chloride-coupled cotransporters via the STE20-related kinases, SPAK and OSR1, *J. Biol. Chem.*, 280, 42685–42693, 2005.

133. Motais, R., Guizouarn, H., and Garcia-Romeu, F., Red cell volume regulation: the pivotal role of ionic strength in controlling swelling-dependent transport systems, *Biochim. Biophys. Acta*, 1075, 169–180, 1991.

134. Muraki, K., Iwata, Y., Katanosaka, Y., Ito, T., Ohya, S., Shigekawa, M., and Imaizumi, Y., TRPV2 is a component of osmotically sensitive cation channels in murine aortic myocytes, *Circ. Res.*, 93, 829–838, 2003.

135. Nadkarni, V., Gabbay, K. H., Bohren, K. M., and Sheikh-Hamad, D., Osmotic response element enhancer activity. Regulation through p38 kinase and mitogen-activated extracellular signal-regulated kinase kinase, *J. Biol. Chem.*, 274, 20185–90, 1999.

136. Neuhofer, W., Woo, S. K., Na, K. Y., Grunbein, R., Park, W. K., Nahm, O., Beck, F. X., and Kwon, H. M., Regulation of TonEBP transcriptional activator in MDCK cells following changes in ambient tonicity, *Am. J. Physiol.*, 283, C1604–C1611, 2002.

137. Niemeyer, M. I., Cid, L. P., Barros, L. F., and Sepulveda, F. V., Modulation of the two-pore domain acid-sensitive K^+ channel TASK-2 (KCNK5) by changes in cell volume, *J. Biol. Chem.*, 276, 43166–43174, 2001.

138. Nilius, B., Prenen, J., Voets, T., Eggermont, J., and Droogmans, G., Activation of volume-regulated chloride currents by reduction of intracellular ionic strength in bovine endothelial cells, *J. Physiol.*, 506, 353–361, 1998.

139. Nilius, B., Prenen, J., Walsh, M. P., Carton, I., Bollen, M., Droogmans, G., and Eggermont, J., Myosin light chain phosphorylation-dependent modulation of volume-regulated anion channels in macrovascular endothelium, *FEBS Lett.*, 466, 346–350, 2000.

140. Nilius, B., Voets, T., Prenen, J., Barth, H., Aktories, K., Kaibuchi, K., Droogmans, G., and Eggermont, J., Role of Rho and Rho kinase in the activation of volume-regulated anion channels in bovine endothelial cells, *J. Physiol.*, 516, 67–74, 1999.

141. Nollen, E. A., Garcia, S. M., van Haaften, G., Kim, S., Chavez, A., Morimoto, R. I., and Plasterk, R. H., Genome-wide RNA interference screen identifies previously undescribed regulators of polyglutamine aggregation, *Proc. Natl. Acad. Sci. USA*, 101, 6403–6408, 2004.

142. Numata, T., Shimizu, T., and Okada, Y., Direct mechano-stress sensitivity of TRPM7 channel, *Cell Physiol. Biochem.*, 19, 1–8, 2007.

143. Numata, T., Shimizu, T., and Okada, Y., TRPM7 is a stretch- and swelling-activated cation channel involved in volume regulation in human epithelial cells, *Am. J. Physiol.*, 292, C460–C467, 2007.

144. O'Hagan, R., Chalfie, M., and Goodman, M. B., The MEC-4 DEG/ENaC channel of *Caenorhabditis elegans* touch receptor neurons transduces mechanical signals, *Nat. Neurosci.*, 8, 43–50, 2005.

145. O'Neill, W. C., Physiological significance of volume-regulatory transporters, *Am. J. Physiol.*, 276, C995–C1011, 1999.

146. Okada, Y., Cell-volume-sensitive chloride channels: phenotypic properties and molecular identity, *Contrib. Nephrol.*, 152, 9–24, 2006.

147. Ollivier, H. L. N., Pichavant, K., Puill-Stephan, E., Roy, S., Calvès, P., Nonnotte, L., and Nonnotte, G., Volume regulation following hyposmotic shock in isolated turbot (*Scophthalmus maximus*) hepatocytes, *J. Comp. Physiol. B*, 176, 393–403, 2006.

148. Pacheco-Alvarez, D., Cristobal, P. S., Meade, P., Moreno, E., Vazquez, N., Munoz, E., Diaz, A., Juarez, M. E., Gimenez, I., and Gamba, G., The Na^+:Cl^- cotransporter is activated and phosphorylated at the amino-terminal domain upon intracellular chloride depletion, *J. Biol. Chem.*, 281, 28755–28763, 2006.

149. Padda, R., Wamsley-Davis, A., Gustin, M. C., Ross, R., Yu, C., and Sheikh-Hamad, D., MEKK3-mediated signaling to p38 kinase and TonE in hypertonically stressed kidney cells, *Am. J. Physiol.*, 291, F874–F881, 2006.

150. Pan, Z., Capo-Aponte, J. E., Zhang, F., Wang, Z., Pokorny, K. S., and Reinach, P. S., Differential dependence of regulatory volume decrease behavior in rabbit corneal epithelial cells on MAPK superfamily activation, *Exp. Eye Res.*, 84, 978–990, 2007.

151. Parker, J. C., Volume-responsive sodium movements in dog red blood cells, *Am. J. Physiol.*, 244, C324–C330, 1983.

152. Parker, J. C., In defense of cell volume?, *Am. J. Physiol.*, 265, C1191–C1200, 1993.

153. Parker, J. C., Coordinated regulation of volume-activated transport pathways. In *Cellular and Molecular Physiology of Cell Volume Regulation*, Strange, K., Ed., CRC Press, Boca Raton, FL, 1994, chap. 18.

154. Parker, J. C. and Colclasure, G. C., Macromolecular crowding and volume perception in dog red cells, *Mol. Cell Biochem.*, 114, 9–11, 1992.

155. Parker, J. C., Dunham, P. B., and Minton, A. P., Effects of ionic strength on the regulation of Na/H exchange and K–Cl cotransport in dog red blood cells, *J. Gen. Physiol.*, 105, 677–699, 1995.

156. Parton, R. G. and Simons, K., The multiple faces of caveolae, *Nat. Rev. Mol. Cell Biol.*, 8, 185–194, 2007.

157. Pasantes-Morales, H., Lezama, R. A., and Ramos-Mandujano, G., Tyrosine kinases and osmolyte fluxes during hyposmotic swelling, *Acta Physiol.*, 187, 93–102, 2006.

158. Pasantes-Morales, H. and Morales Mulia, S., Influence of calcium on regulatory volume decrease: role of potassium channels, *Nephron*, 86, 414–427, 2000.

159. Pasantes-Morales, H., Lezama, R. A., Ramos-Mandujano, G., and Tuz, K. L., Mechanisms of cell volume regulation in hypo-osmolality, *Am. J. Med.*, 119, S4–S11, 2006.

160. Patel, A. J., Honore, E., Maingret, F., Lesage, F., Fink, M., Duprat, F., and Lazdunski, M., A mammalian two pore domain mechano-gated S-like K^+ channel, *EMBO J.*, 17, 4283–4290, 1998.

161. Pedersen, S. F., Beisner, K. H., Hougaard, C., Willumsen, B. M., Lambert, I. H., and Hoffmann, E. K., Rho family GTP binding proteins are involved in the regulatory volume decrease process in NIH3T3 mouse fibroblasts, *J. Physiol.*, 541, 779–796, 2002.

162. Peter, H., Burkovski, A., and Kramer, R., Osmo-sensing by N- and C-terminal extensions of the glycine betaine uptake system BetP of *Corynebacterium glutamicum*, *J. Biol. Chem.*, 273, 2567–2574, 1998.

163. Pheiffer, B. H. and Zimmerman, S. B., Polymer-stimulated ligation: enhanced blunt- or cohesive-end ligation of DNA or deoxyribooligonucleotides by T4 DNA ligase in polymer solutions, *Nucleic Acids Res.*, 11, 7853–7871, 1983.

164. Piechotta, K., Garbarini, N., England, R., and Delpire, E., Characterization of the interaction of the stress kinase SPAK with the Na^+–K^+–$2Cl^-$ cotransporter in the nervous system: evidence for a scaffolding role of the kinase, *J. Biol. Chem.*, 278, 52848–52856, 2003.

165. Piechotta, K., Lu, J., and Delpire, E., Cation chloride cotransporters interact with the stress-related kinases Ste20-related proline-alanine-rich kinase (SPAK) and oxidative stress response 1 (OSR1), *J. Biol. Chem.*, 277, 50812–50819, 2002.

166. Poolman, B., Spitzer, J. J., and Wood, J. M., Bacterial osmosensing: roles of membrane structure and electrostatics in lipid–protein and protein–protein interactions, *Biochim. Biophys. Acta*, 1666, 88–104, 2004.

167. Racher, K. I., Culham, D. E., and Wood, J. M., Requirements for osmosensing and osmotic activation of transporter ProP from *Escherichia coli*, *Biochemistry*, 40, 7324–7333, 2001.

168. Raitt, D. C., Posas, F., and Saito, H., Yeast Cdc42 GTPase and Ste20 PAK-like kinase regulate Sho1-dependent activation of the Hog1 MAPK pathway, *EMBO J.*, 19, 4623–4631, 2000.

169. Raman, M., Chen, W., and Cobb, M. H., Differential regulation and properties of MAPKs, *Oncogene*, 26, 3100–3112, 2007.

170. Rand, R. P., Probing the role of water in protein conformation and function, *Philos. Trans. R. Soc. Lond. B*, 359, 1277–1284, 2004.

171. Rand, R. P., Parsegian, V. A., and Rau, D. C., Intracellular osmotic action, *Cell Mol. Life Sci.*, 57, 1018–1032, 2000.

172. Rubenhagen, R., Morbach, S., and Kramer, R., The osmoreactive betaine carrier BetP from *Corynebacterium glutamicum* is a sensor for cytoplasmic K^+, *EMBO J.*, 20, 5412–5420, 2001.

173. Rubenhagen, R., Ronsch, H., Jung, H., Kramer, R., and Morbach, S., Osmosensor and osmoregulator properties of the betaine carrier BetP from *Corynebacterium glutamicum* in proteoliposomes, *J. Biol. Chem.*, 275, 735–741, 2000.

174. Ruepp, B., Bohren, K. M., and Gabbay, K. H., Characterization of the osmotic response element of the human aldose reductase gene promoter, *Proc. Natl. Acad. Sci. USA*, 93, 8624–8629, 1996.

175. Russell, J. M., Sodium–potassium–chloride cotransport, *Physiol. Rev.*, 80, 211–276, 2000.

176. Rutledge, E., Bianchi, L., Christensen, M., Boehmer, C., Morrison, R., Broslat, A., Beld, A. M., George, A., Greenstein, D., and Strange, K., CLH-3, a ClC-2 anion channel ortholog activated during meiotic maturation in *C. elegans* oocytes, *Curr. Biol.*, 11, 161–170, 2001.

177. Rutledge, E., Denton, J., and Strange, K., Cell cycle- and swelling-induced activation of a *C. elegans* ClC channel is mediated by CeGLC-7α/β phosphatases, *J. Cell Biol.*, 158, 435–444, 2002.

178. Saito, H. and Tatebayashi, K., Regulation of the osmoregulatory HOG MAPK cascade in yeast, *J. Biochem.*, 136, 267–272, 2004.

179. Sanchez, J. C., Danks, T. A., and Wilkins, R. J., Mechanisms involved in the increase in intracellular calcium following hypotonic shock in bovine articular chondrocytes, *Gen. Physiol. Biophys.*, 22, 487–500, 2003.

180. Sawada, Y., Tamada, M., Dubin-Thaler, B. J., Cherniavskaya, O., Sakai, R., Tanaka, S., and Sheetz, M. P., Force sensing by mechanical extension of the Src family kinase substrate p130Cas, *Cell*, 127, 1015–1026, 2006.

181. Schiller, D., Kramer, R., and Morbach, S., Cation specificity of osmosensing by the betaine carrier BetP of *Corynebacterium glutamicum*, *FEBS Lett.*, 563, 108–112, 2004.

182. Schiller, D., Ott, V., Kramer, R., and Morbach, S., Influence of membrane composition on osmosensing by the betaine carrier BetP from *Corynebacterium glutamicum*, *J. Biol. Chem.*, 281, 7737–7746, 2006.

183. Schiller, D., Rubenhagen, R., Kramer, R., and Morbach, S., The C-terminal domain of the betaine carrier BetP of *Corynebacterium glutamicum* is directly involved in sensing K^+ as an osmotic stimulus, *Biochemistry*, 43, 5583–5591, 2004.

184. Sheader, E. A., Brown, P. D., and Best, L., Swelling-induced changes in cytosolic $[Ca^{2+}]$ in insulin-secreting cells: a role in regulatory volume decrease?, *Mol. Cell. Endocrinol.*, 181, 179–187, 2001.

185. Sheikh-Hamad, D., Youker, K., Truong, L. D., Nielsen, S., and Entman, M. L., Osmotically relevant membrane signaling complex: association between HB-EGF, β_1-integrin, and CD9 in mTAL, *Am. J. Physiol.*, 279, C136–C146, 2000.

186. Sheikh-Hamad, D., Di Mari, J., Suki, W. N., Safirstein, R., Watts III, B. A., and Rouse, D., p38 kinase activity is essential for osmotic induction of mRNAs for HSP70 and transporter for organic solute betaine in Madin–Darby canine kidney cells, *J. Biol. Chem.*, 273, 1832–1837, 1998.

187. Sheikh-Hamad, D. and Gustin, M. C., MAP kinases and the adaptive response to hypertonicity: functional preservation from yeast to mammals, *Am. J. Physiol.*, 287, F1102–F1110, 2004.

188. Shen, M.-R., Chou, C.-Y., Browning, J. A., Wilkins, R. J., and Ellory, J. C., Human cervical cancer cells use Ca^{2+} signalling, protein tyrosine phosphorylation and MAP kinase in regulatory volume decrease, *J. Physiol.*, 537, 347–362, 2001.

189. Shinozuka, K., Tanaka, N., Kawasaki, K., Mizuno, H., Kubota, Y., Nakamura, K., Hashimoto, M., and Kunitomo, M., Participation of ATP in cell volume regulation in the endothelium after hypotonic stress, *Clin. Exp. Pharmacol. Physiol.*, 28, 799–803, 2001.

190. Stallkamp, I., Dowhan, W., Altendorf, K., and Jung, K., Negatively charged phospholipids influence the activity of the sensor kinase KdpD of *Escherichia coli*, *Arch. Microbiol.*, 172, 295–302, 1999.

191. Starke, L. C. and Jennings, M. L., K–Cl cotransport in rabbit red cells: further evidence for regulation by protein phosphatase type 1, *Am. J. Physiol.*, 264, C118–C124, 1993.

192. Strange, K., Are all cell volume changes the same?, *News Physiol. Sci.*, 9, 223–228, 1994.

193. Strange, K., Molecular identity of the outwardly rectifying, swelling-activated anion channel: time to re-evaluate pICln, *J. Gen. Physiol.*, 111, 617–622, 1998.

194. Strange, K., Denton, J., and Nehrke, K., Ste20–type kinases: evolutionarily conserved regulators of ion transport and cell volume, *Physiology*, 21, 61–68, 2006.

195. Strotmann, R., Harteneck, C., Nunnenmacher, K., Schultz, G., and Plant, T. D., OTRPC4, a nonselective cation channel that confers sensitivity to extracellular osmolarity, *Nat. Cell Biol.*, 2, 695–702, 2000.

196. Stutzin, A. and Hoffmann, E. K., Swelling-activated ion channels: functional regulation in cell-swelling, proliferation and apoptosis, *Acta Physiol.*, 187, 27–42, 2006.

197. Sukharev, S. and Anishkin, A., Mechanosensitive channels: what can we learn from "simple" model systems?, *Trends Neurosci.*, 27, 345–351, 2004.

198. Sukharev, S. and Corey, D. P., Mechanosensitive channels: multiplicity of families and gating paradigms, *Sci. Signal.*, 219, re4, 2004.

199. Summers, J. C., Trais, L., Lajvardi, R., Hergan, D., Buechler, R., Chang, H., Peña-Rasgado, C., and Rasgado-Flores, H., Role of concentration and size of intracellular macromolecules in cell volume regulation, *Am. J. Physiol.*, 273, C360–C370, 1997.

200. Suzuki, M. and Mizuno, A., A novel human Cl⁻ channel family related to *Drosophila* flightless locus, *J. Biol. Chem.*, 279, 22461–22468, 2004.

201. Syntichaki, P. and Tavernarakis, N., Genetic models of mechanotransduction: the nematode *Caenorhabditis elegans*, *Physiol. Rev.*, 84, 1097–1153, 2004.

202. Takenaka, M., Preston, A. S., Kwon, H. M., and Handler, J. S., The tonicity-sensitive element that mediates increased transcription of the betaine transporter gene in response to hypertonic stress, *J. Biol. Chem.*, 269, 29379–29381, 1994.

203. Tatebayashi, K., Tanaka, K., Yang, H. Y., Yamamoto, K., Matsushita, Y., Tomida, T., Imai, M., and Saito, H., Transmembrane mucins Hkr1 and Msb2 are putative osmosensors in the SHO1 branch of yeast HOG pathway, *EMBO J.*, 26, 3521–3533, 2007.

204. Tatebayashi, K., Yamamoto, K., Tanaka, K., Tomida, T., Maruoka, T., Kasukawa, E., and Saito, H., Adaptor functions of Cdc42, Ste50, and Sho1 in the yeast osmoregulatory HOG MAPK pathway, *EMBO J.*, 25, 3033–3044, 2006.

205. Thomas, S. M. and Brugge, J. S., Cellular functions regulated by Src family kinases, *Ann. Rev. Cell Dev. Biol.*, 13, 513–609, 1997.

206. Tilly, B. C., Edixhoven, M. J., Tertoolen, L. G., Morii, N., Saitoh, Y., Narumiya, S., and de Jonge, H. R., Activation of the osmo-sensitive chloride conductance involves P21rho and is accompanied by a transient reorganization of the F-actin cytoskeleton, *Mol. Biol. Cell*, 7, 1419–1427, 1996.

207. Tinel, H., Kinne-Saffran, E., and Kinne, R., Calcium-induced calcium release participates in cell volume regulation of rabbit TALH cells, *Pflügers Arch.*, 443, 754–761, 2002.

208. Tobin, D., Madsen, D., Kahn-Kirby, A., Peckol, E., Moulder, G., Barstead, R., Maricq, A., and Bargmann, C., Combinatorial expression of TRPV channel proteins defines their sensory functions and subcellular localization in *C. elegans* neurons, *Neuron*, 35, 307–318, 2002.

209. Trischitta, F., Denaro, M. G., and Faggio, C., Cell volume regulation following hypotonic stress in the intestine of the eel, *Anguilla anguilla*, is Ca^{2+} dependent, *Comp. Biochem. Physiol. B*, 140, 359–367, 2005.

210. Trouet, D., Carton, I., Hermans, D., Droogmans, G., Nilius, B., and Eggermont, J., Inhibition of VRAC by *c*-Src tyrosine kinase targeted to caveolae is mediated by the Src homology domains, *Am. J. Physiol.*, 281, C248–C56, 2001.

211. Trouet, D., Hermans, D., Droogmans, G., Nilius, B., and Eggermont, J., Inhibition of volume-regulated anion channels by dominant-negative caveolin-1, *Biochem. Biophys. Res. Commun.*, 284, 461–465, 2001.

212. Trouet, D., Nilius, B., Jacobs, A., Remacle, C., Droogmans, G., and Eggermont, J., Caveolin-1 modulates the activity of the volume-regulated chloride channel, *J. Physiol.*, 520, 113–119, 1999.

213. Tsai, T.-T., Guttapalli, A., Agrawal, A., Albert, T. J., Shapiro, I. M., and Risbud, M. V., MEK/ERK signaling controls osmoregulation of nucleus pulposus cells of the intervertebral disc by transactivation of TonEBP/OREBP, *J. Bone Miner. Res.*, 22, 965–974, 2007.

214. Turner, J.-E., Sedej, S., and Rupník, M., Cytosolic Cl^- ions in the regulation of secretory and endocytotic activity in melanotrophs from mouse pituitary slices, *J. Physiol.*, 566, 443–453, 2005.

215. Uchida, S., Garcia-Perez, A., Murphy, H., and Burg, M., Signal for induction of aldose reductase in renal medullary cells by high external NaCl, *Am. J. Physiol.*, 256, C614–C620, 1989.

216. Uesono, Y. and Toh, E., Transient inhibition of translation initiation by osmotic stress, *J. Biol. Chem.*, 277, 13848–13855, 2002.

217. Uhlik, M. T., Abell, A. N., Johnson, N. L., Sun, W., Cuevas, B. D., Lobel-Rice, K. E., Horne, E. A., Dell'Acqua, M. L., and Johnson, G. L., Rac-MEKK3-MKK3 scaffolding for p38 MAPK activation during hyperosmotic shock, *Nat. Cell Biol.*, 5, 1104–1110, 2003.

218. Ullrich, N., Caplanusi, A., Brone, B., Hermans, D., Lariviere, E., Nilius, B., Van Driessche, W., and Eggermont, J., Stimulation by caveolin-1 of the hypotonicity-induced release of taurine and ATP at basolateral, but not apical, membrane of Caco-2 cells, *Am. J. Physiol.*, 290, C1287–C1296, 2006.

219. Vanfleteren, J. R., Van de, P. Y., Blaxter, M. L., Tweedie, S. A., Trotman, C., Lu, L., Van Hauwaert, M. L., and Moens, L., Molecular genealogy of some nematode taxa as based on cytochrome c and globin amino acid sequences, *Mol. Phylogenet. Evol.*, 3, 92–101, 1994.

220. Vartiainen, M. K., Guettler, S., Larijani, B., and Treisman, R., Nuclear actin regulates dynamic subcellular localization and activity of the SRF cofactor MAL, *Science*, 316, 1749–1752, 2007.

221. Verissimo, F. and Jordan, P., WNK kinases, a novel protein kinase subfamily in multi-cellular organisms, *Oncogene*, 20, 5562–5569, 2001.

222. Vitari, A. C., Deak, M., Morrice, N. A., and Alessi, D. R., The WNK1 and WNK4 protein kinases that are mutated in Gordon's hypertension syndrome phosphorylate and activate SPAK and OSR1 protein kinases, *Biochem. J.*, 391, 17–24, 2005.

223. Vitari, A. C., Thastrup, J., Rafiqi, F. H., Deak, M., Morrice, N. A., Karlsson, H. K., and Alessi, D. R., Functional interactions of the SPAK/OSR1 kinases with their upstream activator WNK1 and downstream substrate NKCC1, *Biochem. J.*, 397, 223–231, 2006.

224. Voets, T., Droogmans, G., Raskin, G., Eggermont, J., and Nilius, B., Reduced intracellular ionic strength as the initial trigger for activation of endothelial volume-regulated anion channels, *Proc. Natl. Acad. Sci. USA*, 96, 5298–5303, 1999.

225. Vollrath, M. A., Kwan, K. Y., and Corey, D. P., The micromachinery of mechanotransduction in hair cells, *Annu. Rev. Neurosci.*, 30, 339–365, 2007.

226. vom Dahl, S., Schliess, F., Graf, D., and Haussinger, D., Role of p38(MAPK) in cell volume regulation of perfused rat liver, *Cell Physiol. Biochem.*, 11, 285–294, 2001.

227. vom Dahl, S., Schliess, F., Reissmann, R., Gorg, B., Weiergraber, O., Kocalkova, M., Dombrowski, F., and Haussinger, D., Involvement of integrins in osmosensing and signaling toward autophagic proteolysis in rat liver, *J. Biol. Chem.*, 278, 27088–27095, 2003.

228. Vriens, J., Watanabe, H., Janssens, A., Droogmans, G., Voets, T., and Nilius, B., Cell swelling, heat, and chemical agonists use distinct pathways for the activation of the cation channel TRPV4, *Proc. Natl. Acad. Sci. USA*, 101, 396–401, 2004.

229. Waldegger, S., Steuer, S., Risler, T., Heidland, A., Capasso, G., Massry, S., and Lang, F., Mechanisms and clinical significance of cell volume regulation, *Nephrol. Dial. Transplant.*, 13, 867–874, 1998.

230. Wang, J., Xu, H., Morishima, S., Tanabe, S., Jishage, K., Uchida, S., Sasaki, S., Okada, Y., and Shimizu, T., Single-channel properties of volume-sensitive Cl⁻ channel in ClC-3-deficient cardiomyocytes, *Jpn. J. Physiol.*, 55, 379–383, 2005.

231. Wehner, F., Cell volume-regulated cation channels, *Contrib. Nephrol.*, 152, 25–53, 2006.

232. Wheeler, J. M. and Thomas, J. H., Identification of a novel gene family involved in osmotic stress response in *Caenorhabditis elegans*, *Genetics*, 174, 1327–1336, 2006.

233. Whitmarsh, A. J. and Davis, R. J., Structural organization of MAP-kinase signaling modules by scaffold proteins in yeast and mammals, *Trends Biochem. Sci.*, 23, 481–485, 1998.

234. Widmann, C., Gibson, S., Jarpe, M. B., and Johnson, G. L., Mitogen-activated protein kinase: conservation of a three-kinase module from yeast to human, *Physiol. Rev.*, 79, 143–180, 1999.

235. Wilson, F. H., Disse-Nicodeme, S., Choate, K. A., Ishikawa, K., Nelson-Williams, C., Desitter, I., Gunel, M., Milford, D. V., Lipkin, G. W., Achard, J. M., Feely, M. P., Dussol, B., Berland, Y., Unwin, R. J., Mayan, H., Simon, D. B., Farfel, Z., Jeunemaitre, X., and Lifton, R. P., Human hypertension caused by mutations in WNK kinases, *Science*, 293, 1107–1112, 2001.

236. Wissenbach, U., Bodding, M., Freichel, M., and Flockerzi, V., Trp12, a novel Trp-related protein from kidney, *FEBS Lett.*, 485, 127–134, 2000.

237. Wojtaszek, P. A., Heasley, L. E., and Berl, T., *In vivo* regulation of MAP kinases in *Ratus norvegicus* renal papilla by water loading and restriction, *J. Clin. Invest.*, 102, 1874–1881, 1998.

238. Wojtaszek, P. A., Heasley, L. E., Siriwardana, G., and Berl, T., Dominant-negative *c*-Jun NH₂-terminal kinase 2 sensitizes renal inner medullary collecting duct cells to hypertonicity-induced lethality independent of organic osmolyte transport, *J. Biol. Chem.*, 273, 800–804, 1998.

239. Wood, J. M., Osmosensing by bacteria, *Sci. Signal.*, 357, pe43, 2006.

240. Xu, B. E., English, J. M., Wilsbacher, J. L., Stippec, S., Goldsmith, E. J., and Cobb, M. H., WNK1, a novel mammalian serine/threonine protein kinase lacking the catalytic lysine in subdomain II, *J. Biol. Chem.*, 275, 16795–16801, 2000.

241. Xu, H., Zhao, H., Tian, W., Yoshida, K., Roullet, J.-B., and Cohen, D. M., Regulation of a transient receptor potential (TRP) channel by tyrosine phosphorylation: SRC family kinase-dependent tyrosine phosphorylation of TRPV4 on tyr-253 mediates its response to hypotonic stress, *J. Biol. Chem.*, 278, 11520–11527, 2003.

242. Yancey, P. H., Organic osmolytes as compatible, metabolic and counteracting cytoprotectants in high osmolarity and other stresses, *J. Exp. Biol.*, 208, 2819–2830, 2005.

243. Yang, T., Huang, Y., Heasley, L. E., Berl, T., Schnermann, J. B., and Briggs, J. P., MAPK mediation of hypertonicity-stimulated cyclooxygenase-2 expression in renal medullary collecting duct cells, *J. Biol. Chem.*, 275, 23281–23286, 2000.

244. Yellowley, C. E., Hancox, J. C., and Donahue, H. J., Effects of cell swelling on intracellular calcium and membrane currents in bovine articular chondrocytes, *J. Cell Biochem.*, 86, 290–301, 2002.

245. Yuan, A., Dourado, M., Butler, A., Walton, N., Wei, A., and Salkoff, L., SLO-2, a K⁺ channel with an unusual Cl⁻ dependence, *Nat. Neurosci.*, 3, 771–779, 2000.

246. Yuan, A., Santi, C. M., Wei, A., Wang, Z. W., Pollak, K., Nonet, M., Kaczmarek, L., Crowder, C. M., and Salkoff, L., The sodium-activated potassium channel is encoded by a member of the Slo gene family, *Neuron*, 37, 765–773, 2003.

247. Zagorska, A., Pozo-Guisado, E., Boudeau, J., Vitari, A. C., Rafiqi, F. H., Thastrup, J., Deak, M., Campbell, D. G., Morrice, N. A., Prescott, A. R., and Alessi, D. R., Regulation of activity and localization of the WNK1 protein kinase by hyperosmotic stress, *J. Cell Biol.*, 176, 89–100, 2007.

248. Zimmerman, S. B. and Harrison, B., Macromolecular crowding increases binding of DNA polymerase to DNA: an adaptive effect, *Proc. Natl. Acad. Sci. USA*, 84, 1871–1875, 1987.

3 The Contractile Vacuole Complex and Cell Volume Control in Protozoa

Richard D. Allen, Takashi Tominaga, and Yutaka Naitoh

CONTENTS

I. INTRODUCTION

Protozoa, or protistans as a whole, are single-celled organisms that live in water or a moist environment. The percentage of solutes dissolved in the water varies from mere trace amounts in freshwater streams to increasing concentrations in sewage treatment plants, in brackish water, and in the ocean. Living in a wide range of environments, protozoa, particularly those that lack cell walls, developed ways of coping with sudden or prolonged changes in their surroundings.

Many wall-less species such as *Paramecium* rely on a contractile vacuole complex (CVC) (Figure 3.1) to maintain their water balance both under normal environmental conditions as well as during dramatic hypoosmotic changes in their environment.[3,6,62,92] This organelle apparently quickly accumulates much of the excess water that passes by osmosis across the plasma membrane of the cell and stores this water briefly in a vacuole before expelling the fluid, along with any accompanying solutes, from the cell. In this way, the CVC provides both a constant water-regulating organelle and a *fast-responding mechanism* for coping with a potentially catastrophic change in environmental osmolarity. For more long-term adaptation, however, and for adjusting the cytosolic

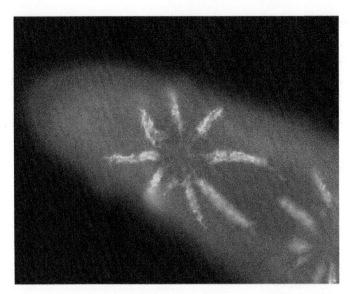

FIGURE 3.1 (See color insert following page 208.) A contractile vacuole complex (CVC) of *Paramecium multimicronucleatum* labeled with monoclonal antibodies (mAbs). Texas red-tagged mAb (red) labels the G4 antigen of the membranes of the smooth spongiome that make up the contractile vacuole (that lies at the hub of this CVC), the radiating collecting canals of the radial arms, and the ampulli. This antibody also cross-reacts with antigens in the membranes of the pellicle of the cell. The flourescein-tagged mAb (green) labels the A4 antigen found only as part of the V-ATPase that is particularly abundant on the decorated tubules of the CVC. The decorated tubules attach peripherally to each radial arm distal to the ampullus.

osmolarity level and the ionic balance to a new plateau, protozoa have other mechanisms that adapt their internal osmolarity to their surroundings. Like other organisms (see Chapter 2), protozoa have pathways located in their plasma membranes that regulate the overall volume of the cell, in both regulated volume decrease (RVD)[54,101] and probably also regulated volume increase (RVI).[54] Transport mechanisms have been identified or postulated for moving osmolytes into the cytosol and out of the cytosol to adjust the osmolarity of the cell relative to the external osmolarity so the interior of the cell will always remain hyperosmotic to the outside. One of the better studied osmoregulatory mechanisms that does not rely in part on a CVC is that of the glycerol regulatory system in yeast cells,[49] which is used for regulated volume decrease (see Chapter 2). Whether such a system is used by protozoa is not known.

Thus, the CVC is a unique, osmolarity-sensitive organelle limited primarily to single-celled algae and protozoa. Although this organelle was eliminated as multicellular organisms evolved, it has survived to the present time in the single-celled zoospore stage of some multicellular fungi and in several kinds of cells (amoebocytes, pinacocytes, and choanocytes) of freshwater sponges.[15] Presumably, a key role for CVCs in wall-less protists, that of ensuring against rupture of their plasma membranes when exposed to the low osmolarities of hypoosmotic environments, was taken over by other cellular or tissue specializations, such as the introduction of the elimination of organic osmolytes across the plasma membrane as in yeast, or it was no longer required by cells of multicellular organisms where individual cells or tissues were protected against large osmolality fluctuations by the surrounding cells or by specialized nephridial organs.

Although the CVC as a separate organelle has not been passed on to higher forms of life, some of its unique membrane structures and functions were, no doubt, retained and now form the basis for how higher organisms deal with their own osmoregulatory challenges. Thus, further study of how primitive cells solved problems they faced in their environments can continue to shed light on how the essential properties of these systems changed and evolved in higher organisms.

Recent studies on the CVC and its relevance to cellular volume regulation in protistans have dealt with (1) the *in vivo* ionic contents of the contractile vacuole (CV), both in standard saline solution and when cells were subjected to different external ionic and osmotic conditions; (2) the regulation of cellular volume and the relative roles of possible transport systems in the plasma membrane vs. the involvement of the CVC in volume regulation; and (3) the involvement of an aquaporin water channel in the swelling of the CV, as well as the role played by acidocalcisomes (accessory vesicles that contain aquaporin) in osmoregulation in some CVCs (see Section II.G). In addition, a number of recent studies on *Dictyostelium* and *Paramecium* have identified, by molecular biological techniques that include forming constructs with green fluorescent protein, several other proteins associated with their CVCs (see Section II.G).

Our own lab has completed studies on the ionic contents of *in vivo* CVs of *Paramecium multimicronucleatum* (see Section II.D). These studies show that the major *in vivo* ions of the CVC, at least in *Paramecium*, are K^+ and Cl^-. The cation presumably enters the CVC as the result of an exchange process occurring across the CVC membrane in which K^+ or other cations are exchanged for protons that have been pumped into the lumen of the CVC as a consequence of the hydrolytic activity of the vast number of proton-translocating $V-H^+$-ATPase enzymes located in the membranes of the CVC. In *Paramecium*, the chloride anion will probably be cotransported with K^+ or will follow through chloride channels attracted by the positive electrical gradient that is formed inside the CVC lumen. Under some conditions, cations other than K^+ can also accumulate in the CV of the *Paramecium*, such as Na^+ and Ca^{2+}, when these ions are present in significant amounts in the external medium. Thus, although the principal osmolytes in the CV are K^+ and Cl^-, the CVC may accumulate other cations if K^+ ions are limited or if other cations in the cytosol such as Ca^{2+} must be eliminated from the cell.

The amount of fluid expelled will usually follow an inverse relationship to the osmolarity of the external medium; at higher osmolarities, less fluid will be expelled from the cell. Even in very high external osmolarities, however, the CV will continue to eliminate fluid at a reduced rate as the osmolarity of the cytosol will always be adjusted upward to maintain it hyperosmotic to the external medium. The CVC is thus an osmolarity-sensitive organelle that accumulates and expels water and (to some extent by default) osmolytes.

The relatively large size of the CVC in *Paramecium* has made it possible to study the functions and contents of this organelle by electrophysiological and biophysical techniques when it has so far not been possible to do so in many smaller protozoa. As may be true of most CVCs, the CVC of *Paramecium* has a two-membrane compartment system. One compartment has proton pumps that set up the electrochemical gradient (positive inside), and a second compartment has membranes that undergo a cycle of spontaneous tension increase followed by relaxation apparently controlled by its own internal molecular components and timing mechanism. *Paramecium* also has the unique ability to keep the K^+ concentration inside the CVC at a relatively constant level of 2.0- to 2.4-fold higher than that in the cytosol over a wide range of external osmolarities and conditions that alter the K^+ activity of the cytosol. How this K^+ ratio is sensed and regulated is not currently understood. By itself, the CVC does not set the level of osmolarity in the cytosol, as this parameter is determined more by the ion transport mechanisms in the plasma membrane, by the free amino acid composition of the cytosol, and by the accumulation or production of other organic osmolytes produced by the cell.

II. CYTOSOLIC OSMOLARITY AND THE CONTRACTILE VACUOLE COMPLEX OF PROTOZOA

A. CYTOSOLIC OSMOLARITY

Freshwater protozoa have a cytosolic osmolarity under normal growth conditions that ranges roughly between 50 and 110 mOsmol/L and is hyperosmotic to the environment.[94,129] In the ciliate *Tetrahymena pyriformis*, the cytosolic osmolarity was reported to rise linearly as cells were adapted to

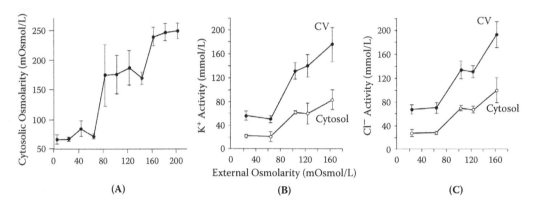

FIGURE 3.2 (A) The cytosolic osmolarity of *Paramecium multimicronucleatum* increases in steps rather than linearly as the external osmolarity increases. Plateaus exist at approximately 75, 160, and 245 mOsmol/L. (From Stock, C. et al., *J. Exp. Biol.*, 204, 291, 2001. With permission.) (B) The K^+ activities of both the cytosol (open circles) and CV (closed circles) also increase in steps as the external osmolarity rises. (C) The cytosolic (open circles) and CV (closed circles) Cl^- activities also increase in steps with increasing external osmolarity. (B and C from Stock, C. et al., *J Cell Sci.*, 115, 2339, 2002. With permission.)

increasing external osmolarities, at least in an external osmolarity range from 40 to 170 mOsmol/kg of cells.[31,112] *Amoeba proteus* was found to have a cytosolic osmolality of ~240 mOsmol/L when cultured in a medium containing 0.7 mg KCl, 0.8 mg $CaCl_2$, and 0.8 mg $MgSO_4$–$7H_2O$ per liter.[86]

Paramecium multimicronucleatum cells living in an external environment having an osmolarity from 4 to 64 mOsmol/L exhibit a relatively constant cytosolic osmolarity of ~75 mOsmol/L.[109] Thus, the cytosol of this cell, like other freshwater protozoa, is also hyperosmotic to its environment; however, as the environmental osmolarity increased beyond 75 mOsmol/L, it was observed that the osmolarity of the cytosol dramatically increased stepwise rather than linearly when three critical cytosolic osmolarities were exceeded.[109] These three key osmolarities were at approximately 75, 160, and 245 mOsmol/L (see Figure 3.2A). When these barriers were approached or crossed, water segregation by the CVC was temporarily disrupted. Once disruption had occurred, time was necessary for water segregation to restart. During this time, the cytosolic osmolarity rapidly increased to the next plateau level of either 160 or 245 mOsmol/L.

B. Structure of CVCs

The contractile vacuole of the CVC appears in wall-less, single-celled organisms to be a single-membrane-lined compartment.[6,92] Based on electron microscopy, the membrane of the CV itself is decorated with neither ribosomes nor an extensive cytosolic coat such as clathrin or the COPI or COPII coats of some membranes of the endocytic and biosynthetic pathways,[5,6,46,47] and they lack, for the most part, the luminal polysaccharide lining such as occurs in lysosomes and some stages of food vacuoles.[5,6] Molecules usually associated with clathrin have been reported to be associated with the CV membrane of *Dictyostelium* in developmental stages[66,88] or under certain experimental treatments.[46] The major specialization noted so far is its tendency to be continuous with a meshwork of tubules or with much smaller vesicles that have a peg-like decoration and, in some cells, an uncharacterized luminal lining. In several cases, the cytosolic decorations are known to be V-type proton-translocating ATPase complexes (V-H+-ATPases).[32,47,76] Genes encoding some of the subunits of the V-ATPase complex have been cloned and sequenced from a few protozoa.[34,124,125]

In the smallest CVs, which appear in the small green algae such as *Chlamydomonas*[69] and in the zoospores of Oomycetes such as *Phytophthora*,[76] the CV is composed of vesicles that are either undecorated or decorated on their cytosolic sides with these pegs. Some of these vesicles will fuse together to form a vacuole as water crosses their membrane. In *Chlamydomonas reinhartii*, this vacuole eventually passes its content to the exterior of the cell through the plasma-membrane-lined flagellar pocket. In this case, the CV may not fuse with the plasma membrane in the conventional way but one or a number of minipores may open between the closely opposed CV membrane and the plasma membrane to allow for exocytosis of the fluid content of the contractile vacuole.[69] Such an exocytic mechanism was first proposed for the CVC of a trypanosomatid protozoan, *Leptomonas collosoma*.[67]

Some algal cells such as *Poterioochromonas* show a particularly clear difference between the CV membrane that is undecorated and the decorated flattened or tubular membranes that bind to and extend from the smooth CV membrane.[126] These tubules do not appear to expand to become part of the CV membrane, as neither their cytosolic pegs nor their luminal fibrous contents are found to be part of the CV membrane proper. In other algae, the decorated tubules are more spherical and, after fusion with the CV membrane, remain as decorated patches protruding from the CV surface into the cytosol.[44,45]

In ciliates, the number of CVs per cell or some of the parameters of the CV cycle (e.g., CV volume and rate of expulsion) vary positively or negatively depending on the size of the cell.[70] Flagellated protozoa usually have one or two small CVs located in a pocket at the base of the flagellum. Amoebae, such as *Amoeba proteus*, have one CV, which is not fixed in place but can lie close to the nucleus and then migrate to the uroid (posterior) region, where it will dock at and fuse with an apparently unspecialized region of the plasma membrane.[1,23] Giant amoebae, such as *Chaos carolinensis*,[98] have many CVs, as does the ciliate *Ichthyophthirius*, the causative agent of white spot disease of freshwater fish.[18] *Dictyostelium discoideum* has an extensively studied CVC system that consists of one or two main bladders, which originally were reported to contain the marker enzyme alkaline phosphatase,[97] but the reliability of this marker has been questioned recently.[21] Extending from these bladders is a network of tubules and smaller expanded compartments that remain close to the cell surface.[47] All of these membranes, when viewed in quick-frozen, deep-etched replicas and in replicas of freeze-dried fragments of disrupted cells, bear V-ATPase pegs of 15 nm; even the largest bladders that fuse with the plasma membrane have pegs.[46,47] Using an antibody specific for the B-subunit of the V_1 subcomplex of the V-ATPase holoenzyme, Heuser et al.[47] showed that these 15-nm pegs clump in the presence of this B-subunit antibody which confirms that they contain components of the V-ATPase complexes. A monoclonal antibody (mAb) specific for the 100-kDa accessory protein of the V-ATPase of *Dictyostelium* and other antibodies such as that to calmodulin were used to fluorescently label the tubules and saccules[33,130] of this CVC in interphase as well as in dividing cells.[131]

The CVC seems to have reached its largest size and complexity in the ciliated protozoa. For electron micrographs of such CVCs in the ciliates, see appropriate chapters in Allen's website (www.pbrc.hawaii.edu/allen): *Tetrahymena*, Chapter 18, Figures 50 and 55; *Paramecium*, Chapter 9; *Nassula*, Chapter 16, Figure 15; *Vorticella*, Chapter 19, Figures 1i and 24 to 27; *Coleps*, Chapter 11, Figure 7a; and *Didinium*, Chapter 13, Figure 9. Most ciliated protozoa have one or two CVs (but occasionally many more) that are composed of a smooth undecorated membrane that is in contact with a three-dimensional spongiome of membranes that encircles or extends from the CV. This peripheral mass of membranes includes both smooth and decorated tubules that are not easily distinguished from each other in transmission electron micrographs of many ciliates. Presumably, the smooth membranes can become part of the CV membrane to provide for CV enlargement while the decorated tubules do not become part of the expanding CV membrane. The decorated or peg-bearing membranes, for most of the time, remain as tubules even when they are continuous with the smooth, undecorated membrane meshwork.[73,74] Under hyperosmotic stress, however, exposure to cold or during cell division all or some of these tubules in *Paramecium* round into vesicles and separate from the smooth membranes of the radial arms.[34]

The CVs of ciliates are not free to move about in the cell but are attached to the surface of the cell, each at an indentation of the plasma membrane called a *CV pore*.[72] This pore (or often several closely spaced pores) is located at a specific location on the surface of the cell. Where the pore is located seems to be unique to different groups of ciliates; for example, in *Tetrahymena* two pores lie near the posterior end of the cell and both are attached to the same CV.[7] In *Vorticella*, several pores are attached to one CV that opens into a pellicle-lined chamber that in turn opens into the peristome region of the cell (see Allen's website, Chapter 19, Figures 1i and 24 to 27). In *Paramecium*, which typically has two CVs per cell, pores are found on the dorsal somatic surface of the cell, one on the anterior-dorsal half and one on the posterior-dorsal half of the cell. Only one pore is associated with each CV. In most ciliates, if not all, the pores are short cylindrical or funnel-shaped indentations of the plasma membrane whose cytosolic surfaces are each supported by one or more helically wrapped microtubules. Five to ten bands of microtubules originate from or against these helically wound pore microtubules and radiate out over the CV membrane where they are bound to and hold the CV against the bottom of the pore. In most ciliates, these bands of microtubules are relatively short and do not extend much beyond the surface of the expanded CV. In *Paramecium*, however, these radiating microtubules extend far from the CV membrane where they pass into the cytosol, remaining near the cell surface, for many micrometers.[8,34] The number of microtubular ribbons determines the number of radial arms (Figure 3.1), as well as the number of collecting canals and thus the overall radial shape of the CVC.

Only in the peniculine ciliates, to which *Paramecium* belongs, are such elaborate CVCs found. Thus the complexity of the CVC seems to have reached its apex in *Paramecium* and its close relatives. In these ciliates, the CVC has a strict spatial separation of smooth membranes from decorated membranes so these two populations of membranes are easily distinguished in transmission electron micrographs as well as in immunologically labeled fluorescence micrographs (Figure 3.1). Only the smooth spongiome is in contact with the microtubular ribbons, and these ribbons permit the formation of the long collecting canals in *Paramecium* by allowing the tubular smooth spongiome to form linear rows of 40-nm circular connections that lie between the subdivisions of the ribbons where the tubules then expand into the canals.[43]

In summary, many CVCs are composed of at least two pools of membrane: (1) a smooth membrane that can expand from a tubular form into fluid reservoirs including the CV, ampulli, and collecting canals, and (2) a decorated form that maintains a tubular or small-vesicle shape with a greatly reduced volume-to-surface area ratio. This latter decorated form can fuse with the smooth form but does not appear to fuse with the plasma membrane and does not contact microtubules. The smooth form can fuse both with the decorated form and with the plasma membrane and, in *Paramecium*, it also binds to microtubules.[120] In *Dictyostelium*, both tubules and bladders have only one type of membrane based on their decoration with pegs,[46,47] although biochemically two types of membranes were reported.[87] *Chlamydomonas* has also been reported to have only one type of membrane.[69] A third pool of membrane, consisting of vesicles called *acidocalcisomes*, has been observed in the smaller miroorganisms, including *Trypanosoma*,[28,75] *Chlamydomonas*,[102] and *Dictyostelium*,[71] that can apparently fuse with some part of the CVC. The spongiome membrane pools, once they are formed in the cell, appear to remain separate from other endomembranes in the cytoplasm, including the endoplasmic reticulum, the Golgi cisternae, the endosomal membrane system, and the digestive vacuole membrane system.[36] CVC membranes must originate from the endoplasmic reticulum and pass through the Golgi stacks as their membranes form and become differentiated, but the process of new CVC membrane formation has not been studied in detail. CVC membranes appear to be slow to break down, as no trace of these membranes has been observed inside autophagosomes, which are most often seen in *Paramecium* to contain the break-down products of mitochondria.

C. Effects of External Osmolarity on CVC Morphology

The rate of fluid segregation and expulsion by the CVC (R_{CVC})[109] varies as the external osmolarity of the bathing solution varies. When the cell is placed in solutions hypoosmotic to the cytosol, the rate of fluid accumulation increases. This may be detected as an increase in the maximum CV diameter before fluid expulsion, as a decrease in time between successive CV expulsions, or by a combination of both parameters. No morphological change in the fine structure of the CVC was noted following relatively small hypoosmotic changes. The usual maximum size of a CV adapted to standard saline in *Paramecium multimicronucleatum* is 13 μm in diameter,[50] while CVs of *Amoeba proteus* can be 27 to 45 μm in diameter.[23,86] The filling and expulsion cycle of *P. multimicronucleatum* growing in 80 mOsmol/L is completed in about 10 sec.[50] In the epimastigotes of *Trypanosoma cruzi*, the CVC cycle lasts 60 to 75 sec;[15] in the zoospores of *Phythophthora*, the cycle is completed in 6 sec;[76] and, in *Dictyostelium*, it is completed in 3 to 4 sec.[37]

Morphological changes in the water segregation system of *Paramecium multimicronucleatum* occur when cell cultures are subjected to pronounced decreases in external osmolarities or when cell cultures have been subjected to a high external Ca^{2+} concentration.[53] These changes involve the production of: (1) longer and bifurcated radial arms, (2) more radial arms per CVC, and (3) the development of additional CVCs per cell. The maximum number of CVCs observed in one *P. multimicronucleatum* cell was seven.[53] Additional CVCs over the usual interphase number had been reported before,[8,57] but this phenomenon is now known to be triggered either by a significant decrease in the external osmolarity or by a significant rise in the external calcium concentration.[53] On the other hand, an equally large increase in external K^+ concentration was not found to lead to extra CVCs.[53]

The response of the CVC to hyperosmotic conditions can also be complex. When the cell is placed in hyperosmotic solutions, the rate of water accumulation will soon fall to zero and the CVC will become inactive. In high hyperosmotic conditions, the decorated tubules around the radial arms, at least in *Paramecium*, will disappear and can no longer be detected immunologically when V-ATPase-specific mAb labeling is used.[39,52] Instead, in these cases, the mAb becomes dispersed throughout the cytosol. Observed by electron microscopy, the decorated tubules will have lost their tubular shape and will have expanded into vesicles of various sizes that are no longer connected to the smooth spongiome.[34,52] If the cell remains in the high hyperosmotic solution long enough, around 8 hours, the decorated tubules will gradually reappear around the radial arms, and CVC activity will be partially restored.[52] If the cell is returned to a hypoosmotic medium, the decorated tubules begin to reappear in 20 min and the CV will return to full activity within 1 hr.[39]

D. Ions and Osmolytes of the Cytosol and CVC

The older literature on inorganic ion concentrations (K^+, Na^+, and Cl^-) in the cytosol of protozoa was summarized by Prusch.[94] K^+ is present in most freshwater protozoa at a concentration of ~25 to 35 mmol/kg wet weight of cells, with amoebae generally containing concentrations near the low end of the range. The marine ciliate *Miamiensis avidus* has a cytosolic K^+ concentration of 74 mmol/kg. The concentration of Na^+ in the cytosol was more variable and generally lower, ranging from 0.5 to 20 mmol/kg, with the exception of the marine ciliate, which had a Na^+ concentration of 88 mmol/kg. Cl^- was low in both freshwater and marine cells, measuring 0.36 to 16 mmol/kg, much lower than the Cl^- concentration in the external medium, which measured up to 550 mmol/L in saltwater. Only in the amoebae *Chaos carolinensis* and *Amoeba proteus* were the Cl^- concentrations in the cytosol considerably higher than in the external medium. In these earlier studies, in no case was enough Cl^- reported in the cytosol to counterbalance all inorganic cation concentrations present.

Studies performed to determine if active transport of ions is occurring across the plasma membrane and if this active transport is coupled reported that Na^+ and K^+ are apparently both transported actively in *Acanthamoeba* but they were not coupled.[63] In *Amoeba proteus*, both Na^+ and K^+ are actively transported across the plasma membrane, K^+ is actively accumulated, and Na^+ is actively eliminated.[96] The ability of the cell to regulate Na^+ was dependent on Ca^{2+} in the external medium. Both Na^+ and K^+ can be actively transported in *Tetrahymena*, whereas Cl^- is apparently distributed passively.[10,30]

Marine and brackish water ciliates, as expected, had much higher cytosolic osmolarities which resulted from a summation of their inorganic ions and free amino acid concentrations, as well as other unspecified intracellular participants that might include organic osmolytes. The marine ciliate *Miamiensis avidus*, a facultative parasite living on the seahorse, has lower concentrations of Na$^+$ and Cl$^-$ in its cytosol than are present in the external environment.[55] These ions vary with changes in salinity, increasing with increased salinity but always remaining lower than their concentrations in the external medium. The cytosolic concentration of K$^+$ was much higher than the exterior K$^+$ concentration, but this was affected less by changes in salinity than were the Na$^+$ and Cl$^-$ concentrations. Osmolarities changed rapidly, within 10 min. A 30-min exposure to a changed external osmolarity caused swelling or shrinkage of the cell under hypoosmotic or hyperosmotic stress, respectively. The cells then returned to their approximate original volume in about 90 min. The total cytosolic osmolarity of *M. avidus* that resulted from both inorganic ions and free amino acids was about 540 mmol/kg cells in seawater. The free amino acid concentration was 317 mmol/kg cells, and alanine, glycine, and proline made up 73% of the free amino acids.[56] The concentration of free amino acids increased and decreased with salinity changes; 25% seawater resulted in a 76% reduction in free amino acids, and 200% seawater resulted in an increase of 22% free amino acids over that in 100% seawater. These changes were completed in 20 min and were apparently the result of the metabolic release of bound amino acids during hyperosmotic stress rather than amino acid uptake from the medium.[56]

The brackish water ciliate *Paramecium calkinsi* can be adapted (over a month) to osmolarities from 10 to 2000 mOsmol/L but it does not divide above 1000 mOsmol/L. It uses both organic and inorganic osmolytes for osmoregulation. Cells exposed to hyperosmotic changes will increase their free amino acid concentrations, particularly alanine and proline, but this requires several hours in the case of alanine and much longer in the case of proline. Upon exposure to hypoosmotic stress, however, much of the free alanine and proline of the cell are released from the cell (within 5 min), and these can be recovered in the external medium. Thus, free amino acids, particularly alanine and proline, play an important role in osmoregulation when the protozoan is subjected to hypoosmotic stress, but the increase of free amino acids in the cytosol would be too slow to quickly reestablish the required hyperosmotic cytosol when the cell is under hyperosmotic stress.[24] *Dictyostelium* also secretes half or more of its load of amino acids, particularly glycine, alanine, and proline, when it encounters hypoosmotic stress.[107]

The *in situ* ionic contents of the CVC have only recently been determined in a living protozoan cell. Earlier attempts to determine CVC osmolarity relied on micropuncture and freezing point depression techniques and, for ionic content, on helium-glow photometry to collect and assay cellular fluids.[98,104] This led to the conclusion that the CV fluid of an amoeba was hypoosmotic to the cytosol. This finding has been difficult to reconcile with generally accepted ideas of water permeability of cellular membranes, based on the obvious accumulation of fluid within the CV, and has required some innovative speculation as to how water could accumulate against a hypoosmotic gradient in the CV, a question still unresolved in cells such as *Dictyostelium*.[46,107] Recently, with the use of ion-selective microelectrodes, it has been possible to actually measure the ionic activities of several major inorganic ions present in living *Paramecium multimicronucleatum*, both in the cytosol of the cell and at the same time in the CV of the cell.[110] In cells adapted to a 24-mOsmol/L standard saline solution that did not contain Na$^+$ (as used by *Paramecium* electrophysiologists[83]), the cytosol had a 22.6 K$^+$ activity (all ionic activities are in mmol/L) compared to 56.0 in the CV, 3.9 of Na$^+$ in the cytosol (presumably carried over in the cell from the previous culture conditions) compared to 4.7 in the CV, and 27.3 of Cl$^-$ in the cytosol compared to 66.5 in the CV. Ca^{2+} activity in the cytosol was too low to measure by this ion-selective microelectrode technique but was measurable at 0.23 in the CV. Thus, the major inorganic ions in the CV of *Paramecium* are K$^+$ and Cl$^-$. These results show that the cytosolic Cl$^-$ activity in freshwater protozoa is actually much higher than the older determinations had reported (see Table 1 in Prusch).[94] Thus, Cl$^-$ can act as the counterbalancing anion for most, if not all, of the free inorganic cations present in the cytosol and

TABLE 3.1
Ratios of K^+ and Cl^- Activities in the CV Fluid Compared to Those in the Cytosol

External Osmolarity (mOsmol/L)	K^+-Containing		Choline-Containing		Ca^{2+}-Containing		Furosemide		DMSO	
	K^+	Cl^-	K^+	Cl^-	K^+	Cl^-	K^+	Cl^-	K^+	Cl^-
24	2.5	2.4	2.4	2.1	2.3		2.4	2.4	2.5	2.5
64	2.4	2.5	—	—	—	—	—	—	—	—
104	2.1	1.9	—	—	—	—	—	—	—	—
124	2.3	2.0	2.5	2.0	2.4	2.0	—	—	—	—
164	2.1	2.3	—	—	—	—	—	—	—	—

Source: Stock, C. et al., *J. Cell Sci.*, 115, 2339, 2002. With permission.

in the CV. It is not yet known, however, if K^+ and Cl^- are the major osmolytes in the CVCs of other cells that have this organelle.

A potentially important finding was that the ratios of both K^+ and Cl^- activities in the CV compared to the K^+ and Cl^- activities in the cytosol were maintained at between a 2.0- and 2.4-fold higher level in the CV over that of the cytosol, and these ratios stayed within this narrow range even as the ion concentrations in the cytosol and CV increased as the external osmolarity was increased from 24 to 164 mOsmol/L (see Table 3.1).[110] This suggests the presence of a mechanism in the cell, probably in the CV membrane, that maintains the K^+ and Cl^- activities of the CV at 2.0 to 2.4 times that of the cytosol. The activities of these two ions within the CV and presumably the overall osmolarity of the CV are therefore regulated by the cytosolic osmolarity and by the individual ionic activities in the cytosol, rather than the CVC determining or regulating the cytosolic ionic composition or the overall cytosolic osmolarity. Thus, the CVC does not control the osmolarity of the cytosol; rather, the cytosol, with a possible contributing role by the CVC membrane, determines the osmolarity of the CV.

As was the case for cytosolic osmolarity (Figure 3.2A), the individual ionic activities of K^+ and Cl^- increased in both the cytosol and CV in steps instead of linearly (Figure 3.2B and C). These steps occurred at the same external osmolarities as the increases in cytosolic osmolarity, at ~75 and ~160 mOsmol/L. It seems likely that either K^+ is actively transported into the cytosol from the external medium together with Cl^- or active K^+ transport is followed by Cl^- moving through Cl^- channels. These two ion species may then be moved together or separately into the CV, and water will enter both the cytosol and the CVC passively by osmosis.

As already mentioned, the presence of different single ion species or a mixture of ion species outside the cell combined with changes in external osmolarity will ultimately affect the cytosolic osmolarity and cytosolic ionic composition as well as the osmolarity and ionic composition of the CVC fluid (Figure 3.3) in a complex way. Single-ion species or a mixture of ion species outside the cell will also affect the rate of fluid expulsion by the CVC (R_{CVC}).[111] When the bathing solution to which *Paramecium* was adapted for 18 hr contained: (1) 2-mmol/L K^+ in MOP-KOH buffer, (2) 2-mmol/L Na^+ in MOP–NaOH buffer, (3) a mixture of 1-mmol/L K^+ and 1-mmol/L Na^+ in MOP–KOH buffer, or (4) 2 mmol/L of the organic cation choline without buffer, it was observed that the fluid segregation (or expulsion) rate (R_{CVC}) was the same in cells adapted to either K^+ or Na^+ alone; however, when K^+ and Na^+ were both present at 1 mmol/L each, thus making a total of 2 mmol/L, R_{CVC} was almost twofold higher (Table 3.2). In the solution enriched in the organic cation choline, R_{CVC} was about half the rate of that in Na^+ or K^+ alone. The sum total of K^+, Na^+, Ca^{2+}, and Cl^- activities in the CV fluid remained about equal in cells adapted to K^+ or Na^+ alone for a particular external osmolarity (e.g., 24 mOsmol/L), but when both K^+ and Na^+ were present

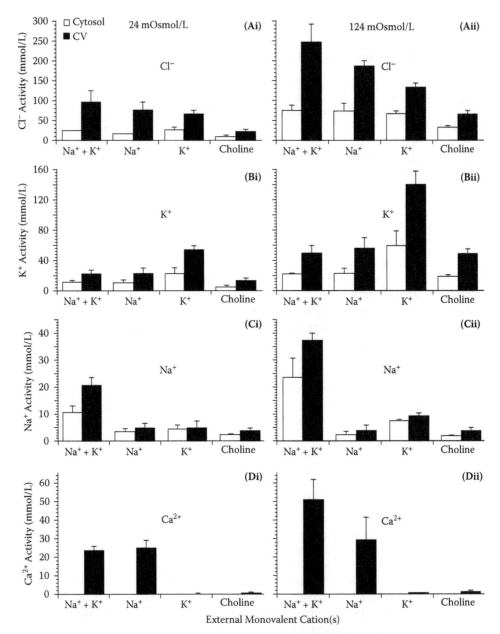

FIGURE 3.3 Ionic activities of (A) Cl⁻, (B) K⁺, (C) Na⁺, and (D) Ca²⁺ in the cytosol (white bars) and CV fluid (black bars) of *Paramecium multimicronucletum* as a function of the external osmolality. The left column is of cells adapted to 24 mOsmol/L (Ai to Di), and the right column to 124 mOsmol/L (Aii to Dii). The two adaptation osmolalities also contained four ionic conditions: (1) 1-mM K⁺ plus 1-mM Na⁺; (2) 2-mM Na⁺ alone; (3) 2-mM K⁺ alone; and (4) 2-mM choline alone as the monovalent cations. (Numerical values for the CV fluid are reported in Table 3.2). Vertical lines represent SD. (From Stock, C. et al., *Eur. J. Cell Biol.*, 81, 505, 2002. With permission.)

together the sum total of their ionic activities within the CV was significantly higher. The presence of both K⁺ and Na⁺ in the adaptation solution did not lead to an increase in the cytosolic osmolarity but did increase the estimated osmolarity of the CV fluid (due mostly to the rise in Na⁺ activity in the CV); consequently, the osmotic gradient between the cytosol and the CV fluid was significantly

TABLE 3.2
Overview of Ionic Activities in Contractile Vacuole, Overall Cytosolic Osmolarity, Estimated Osmotic Gradient across the Contractile Vacuole Membrane, Rate of Fluid Segregation, and Membrane Potential of Contractile Vacuole Complex in *Paramecium multimicronucleatum* Cells Adapted to 24- or 124-mOsmol/L Solutions Containing K⁺ + Na⁺, Na⁺, K⁺, or Choline as the Monovalent Cation

Monovalent Cation Species	Adaptation Solution							
	24 mOsmol/L				124 mOsmol/L			
	K⁺ + Na⁺	Na⁺	K⁺	Choline	K⁺ + Na⁺	Na⁺	K⁺	Choline
$a_{Cl^-_{CV}}$ (mmol/L)	96	77	67	33	244	187	131	66
$a_{Na^+_{CV}}$ (mmol/L)	20	5	5	4	38	4	10	4
$a_{K^+_{CV}}$ (mmol/L)	23	24	56	14	51	57	141	50
$a_{Ca^{2+}_{CV}}$ (mmol/L)	23	25	0.2	0.7	51	29	0.7	1.2
$\Sigma a_{ion_{CV}}$ (mmol/L)	162	131	128.2	51.7	384	277	282.7	122.2
Osm_c (mOsmol/L)	60	56	66	32	168	165	178	132
Osmotic gradient$_{CV}$[a]	102	75	62.2	19.7	216	112	104.7	−9.8
R_{CVC} (fL/sec)	131	67	69	33	34	20	18	3
V_{CVC} (mV)	81	95	84	93	79	87	84	87

[a] Estimated by subtracting Osm_c from $\Sigma a_{ions_{CV}}$.

Note: $a_{ion_{CV}}$, Cl⁻, Na⁺, K⁺, and Ca²⁺ activities in the contractile vacuole; $\Sigma a_{ion_{CV}}$, the sum of all ion activities in the CV; Osm_c, the overall cytosolic osmolarity; R_{CVC}, the rate of fluid segregation; V_{CVC}, the CVC membrane potential. The values shown are rounded numbers.

Source: Stock, C. et al., *Eur. J. Cell Biol.*, 81, 505, 2002. With permission.

higher when both K⁺ and Na⁺ were present together. All of the above values determined for 124-mOsmol/L-exposed cells were higher by equivalent amounts (the external osmolarity had been increased by adding sorbitol). Thus, cells adapted to a mixture of K⁺ and Na⁺ exhibited a higher osmotic gradient across the CVC membrane than if either K⁺ or Na⁺ had been present alone. Water also flowed into the CVC faster which was reflected by the higher R_{CVC} in the mixture of K⁺ and Na⁺.[111]

When adapted to the organic cation choline, most corresponding values were significantly lower—50 to 80% or more lower, except for the cytosolic osmolarity, which was lower by only ~25%. The CVC was still active in cells in choline but its R_{CVC} in an external osmolarity of 124 mOsmol/L was only 3 fL/sec (one femtoliter = 10⁻¹⁵ L) compared to 33 fL/sec when in a 24-mOsmol/L adaptation solution—a 10-fold decrease.[111]

Na⁺ was accumulated significantly in the CVC and the cytosol only when external Na⁺ was present together with K⁺. This probably indicates that a significant part of the Na⁺ in *Paramecium* is cotransported with K⁺ across both membranes. Some Na⁺ cotransport was previously proposed for *Tetrahymena pyriformis*.[10,31] At the higher 124-mOsmol/L osmolarity a higher amount of Na⁺ was present in both the cytosol and the CV when the cells had been adapted to K⁺ solution alone rather than to Na⁺ solution alone. On the other hand, K⁺ was taken up much more rapidly when the cells were adapted to K⁺ alone, in both the cytosol and CV, as opposed to when a mixture of Na⁺ and K⁺ was present (Figure 3.3).

Exposing cells to a medium with high levels of calcium, as would be expected, did not produce a significant rise of calcium in the cytosol but did result in a profound accumulation of calcium activity in the CVC fluid.[110] The calcium activity in the CVC of calcium-exposed cells in an external osmolarity of 24 mOsmol/L was between 20 and 25 mmol/L, whereas in 124-mOsmol/L-exposed cells it was 50 mmol/L. This huge increase over the activity in the cytosol was true only in cells adapted to Na^+ plus K^+ or in Na^+ alone; very little calcium was found in the CVs of K^+-adapted or choline-adapted cells. This indicated that calcium entered the cell in association with Na^+ rather than with K^+, but once in the cytosol Ca^{2+} is quickly transferred to calcium storage sites, such as the alveolar sacs of ciliated protozoa[42,93,108] or it is transferred to the CVC compartments for excretion from the cell.[110,111]

Although each CVC in *Paramecium* is estimated to have millions of $V-H^+-ATPase$ enzymes, the fluid of the CV does not become acid as it does in phagosomes that have far fewer $V-H^+-$ATPases per unit of membrane area.[32,51] By using ion-selective microelectrodes filled with a cocktail sensitive to H^+ activity, the *in vivo* pH of the CV fluid was found in *Paramecium* to be only mildly acid (pH of 6.4), whereas that of the cytosol of the same cell was neutral (pH of 7.0). Altering the external osmolarity from 24 or 124 mOsmol/L had no effect on the pH of either the cytosol or the CV. Thus, most of the protons inside the CVC lumen are either not present in ionic form or they are quickly exchanged for cations (K^+, Na^+, and Ca^{2+}) during the import of these cations into the CVC that help to give rise to some of the +80-mV luminal electrochemical charge across the CV membrane.[118] The primary role of the V-ATPases in the CVC membrane is clearly to energize the CVC membrane rather than to produce a strongly acidic compartment.

E. Electrophysiology of the CVC of *Paramecium*

Electrophysiological techniques were used to determine if the CVC of *Paramecium* had an electrical potential across its membrane, as well as to estimate the membrane fusion and fission events that occur and the resistance/conductance of its membranes to ion flow. A fine-tipped microelectrode was inserted into the living CV and an electrical potential of +80 mV was recorded relative to the cytosol.[118] A continuous recording of several filling and expulsion cycles of the same CV showed that, just before expulsion, the electrical potential dropped precipitously to a level near +10 mV (Figure 3.4). In addition, input capacitance measurements made at the same time indicated that the radial arms had become disconnected from the CV shortly before expulsion of the CV concomitant with the CV undergoing a rounding or contraction process. Thus, at the time of expulsion, the CV was no longer in continuity, with its electrogenic source providing most of the +80-mV electro-chemical potential of the CVC. The electrogenic source was therefore found to apparently reside in the V-type proton pumps arrayed on the decorated tubules that, in turn, are found only along the radial arms. At this point, the CV membrane fused with the plasma membrane at the bottom of the CV pore, and the CV was emptied by cytosolic pressure.[84] The pore then closed by fission, and the CV membrane was resealed against the exterior of the cell. After resealing, the CV registered a brief period of negative potential before the electrical potential quickly returned again to +80 mV as the several radial arms quickly reassociated with the CV. Input capacitance, which was low during the expulsion phase, rapidly returned to its earlier higher plateau value, indicating that the radial arms had once again fused with the membrane of the collapsed CV. These measurements also showed that the conductance (the reciprocal of input resistance) was high when the radial arms were attached to the CV but fell rapidly when the CV was no longer attached to the decorated tubules via the smooth spongiome located along the radial arms. Thus, the CV membrane itself seems to have little electrical conductance compared to the total of all the membranes of the CVC.

By determining the diameter of the CV and the membrane area of the CV, it was evident that the visible CV itself did not suddenly return to its maximum diameter (Figure 3.4). As expected, the microscopically visible CV grew more or less linearly during the filling phase following the initial emptying of the engorged ampulli into the CV;[118] however, the input capacitance measurements showed that the total membrane area of the CVC system had rapidly reconnected with the CV membrane long before the microscopically visible CV had returned to its maximum diameter. This confirmed that the CV during expulsion did not vesiculate into a myriad of individual vesicles but

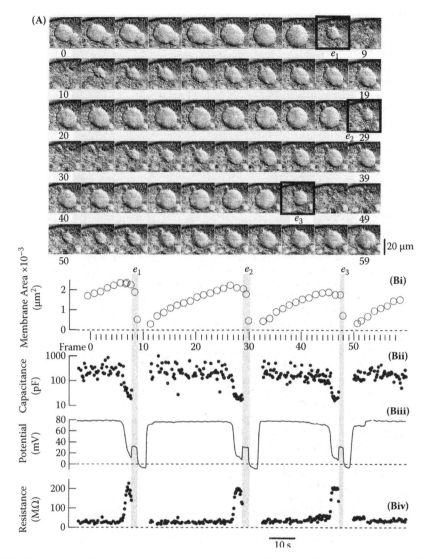

FIGURE 3.4 Electrophysiological parameters of a microelectrode-impaled contractile vacuole of a *Paramecium multimicronucleatum* cell during three successive exocytotic cycles. (A) A series of 60 consecutive images of the contractile vacuole profile taken at 2-sec intervals. Some frames (0–59) are numbered. Black-bordered images correspond to the fluid expulsion phases (e_1–e_3). (Bi) The contractile vacuole membrane area in each frame. (Bii) Input capacitance of the organelle. (Biii) Membrane potential of the organelle with reference to the cytosolic potential. (Biv) Input resistance of the organelle. (From Tominaga, T. et al., *J. Exp. Biol.* 201, 451, 1998. With permission.)

underwent a process of total membrane tubulation where the membrane of the CV reverted into a three-dimensional array of contiguous 40-nm tubules that maintained continuity with a single CV system or with its radial arms (Figure 3.5). Apparently, the only fission occurring along the radial arms was when the membrane of each arm separated as a unit from the CV membrane prior to rounding. During CV expulsion, the collapsed CV could not be detected by light microscopy, so it superficially appeared that the CV had vanished; however, as fluid flowed back into the CV compartment from the reconnected radial arms, the tubules from the collapsed CV reexpanded to form the membrane of the once again microscopically visible CV (see Allen's website, Chapter 9, Video 1).

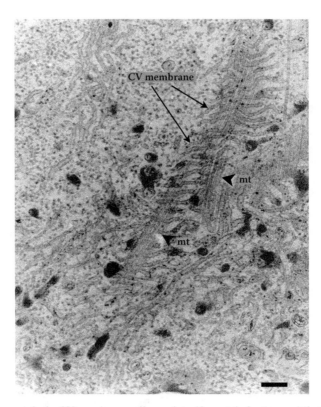

FIGURE 3.5 During systole the CV membrane collapses into 40-nm tubules (arrows) that tend to lie crossways to the microtubular ribbons (arrowheads) that extend from the CV pore. Transmission electron micrograph of chemically fixed, broken cell; electron opaque material entered the disrupted CV prior to or during tubulation. Scale bar, 0.2 μm.

To confirm that the radial arms are the sites of the electrogenic engines of the CVC of *Paramecium*, compressed cells were studied.[39] In these compressed cells, CV potential was found to increase in steps after the expulsion phase (systole), with each step representing the reattachment of one or a few radial arms with the CV. Compressing the cell seemed to interfere with and slow reattachment of the arms. In noncompressed cells, arms seem to reattach simultaneously. A stepwise reduction in CV potential was also observed in these compressed cells at the start of rounding prior to opening of the CV pore.

The number of functional V-ATPases did not directly determine the rate of fluid segregation in the CVC nor did different osmolarities significantly change the CV membrane potential.[39] The CV membrane potential rose to +80 mV or slightly higher and stayed there in cells adapted to external osmolarities from 4 to 124 mOsmol/L. The rate of fluid segregation by the CVC (R_{CVC}) was reduced from 98 to 20 fL/sec as cells were adapted upward through this range of osmolarities. No changes in the final immunologically fluorescent images of the radial arms were observed in cells that had changed from a R_{CVC} of 98 fL/sec to 20 fL/sec, which supported the conclusion that the number of functional V-ATPases did not change significantly. When hypoosmotically adapted cells were changed to a much higher hyperosmotic condition (from 4 mOsmol/L to 124 mOsmol/L) for 30 min and then returned to the original hypoosmotic conditions, it was observed that the fluorescently labeled decorated tubules of the radial arms had disappeared and the potential of the tubulated CV membranes had fallen drastically. This was based on the fact that 20 min after the return of the cell to a hypoosmotic standard saline solution, as radial arm fluorescence began to reappear, the CV potential was now only +44 mV. The potential reached +80 mV by 60 min.

R_{CVC} that had fallen to 0 increased over time from 58 fL/sec at 20 min to ~100 fL/sec at 60 min. Hypoosmotic environments (going directly from 124 mOsmol/L to 4 mOsmol/L) had no effect on the CV membrane potential but did result in an increase of R_{CVC} from 20 fL/sec at 124 mOsmol/L to 103 fL/sec at 4 mOsmol/L. No change was observed in the fluorescent images of the radial arms during this drastic hypoosmotic change.

Exposure of cells to 30 nmol/L of concanamycin B, an inhibitor of V-type ATPases, for 30 min resulted in a 50% decrease from +80 to +40 mV of CV membrane potential in 4-mOsmol/L-adapted cells, and the R_{CVC} decreased 43%.[39] These experiments help to confirm that the membrane potential is generated by voltage-producing processes occurring in the CVC membranes. These processes involve the V-ATPase enzymes that pump protons into the lumen of the decorated tubules. When the cell is placed in a strongly hyperosmotic environment, the V-ATPase holoenzymes fall apart, the CV membrane loses most of its electrochemical potential, and the CVC is no longer functional in eliminating water and electrolyte ions from the cell. It then takes 60 to 120 min for the total population of V-ATPases to reassemble, for the +80-mV CV membrane potential to be restored, and for the fluid segregation rate to return to its normal activity level.[39]

Because the CV potential remains the same at widely different external osmolarities but R_{CVC} changes significantly under this same range (i.e., decreasing with increasing osmolarity), it appears that the CV potential is kept at a maximum to provide for the exchange of K^+ and other cations for protons. Such an exchange may than be followed by the attraction of Cl^- into the CV lumen by the positive electrical gradient of protons, or Cl^- may enter by cotransport of K^+ and Cl^- into the CV lumen. A CV pH of 6.4 argues for proton exchange as the CV only becomes mildly acid.[111] The resultant accumulation of ions in the CVC (K^+ and Cl^- are 2.0 to 2.4 times higher) provides the osmotic gradient that will support the flow of water into the CVC by osmosis. The sum of the activities of the common inorganic ions in the CV is much higher than the activities of these same ions in the cytosol that contribute to the lower osmolarity of the cytosol (see Table 3.2).[111]

F. Membrane Dynamics of the CVC

Because of its apparent periodic contractility, the contractile vacuole has fascinated observers of protozoa from the very early days of microscopy. At the end of a cycle of filling, the CV was observed to fuse with the cell membrane and to disappear from view as its contents were released. It was generally assumed that this contractility was caused by an actin–myosin cytoskeletal system that surrounded the CV; however, no such system has ever been observed either by electron microscopy or by immunological labeling techniques. Only recently has it been possible to begin to understand what precedes fusion of the CV with the plasma membrane and what happens to the CV membrane once fusion occurs. A combination of electron microscopy, electrophysiology, and biophysical techniques has led to a partial understanding of the events that occur at the end of the filling phase (diastole) and during the expulsion phase (systole).

Electron microscopy of *Paramecium* demonstrated that the CV membrane during systole collapses not into a flattened sac but into a meshwork of 40-nm tubules (Figure 3.5), which branch from each other to form a meshwork that remains bound to the noncontractile microtubular ribbons radiating from the cytosolic funnel-shaped surface of the CV pore.[4,85,120] The particular combination of molecular components that make up the bilayer of the CV membrane probably ensures that the membrane returns to a tubular form when the internal hydrostatic pressure of the CV is released. Bending energy is stored in the membrane when the tubules are forced into a more planar shape, and this energy is released when the membrane is allowed again to become tubular.[84] Although this bending energy is not sufficient to account for the rapid expulsion of the fluid from the CV in a living cell, it does seem to be sufficient to expel the fluid at the lower rate observed when a CV was still able to fuse with the plasma membrane of a ruptured cell where the cytosolic pressure had been eliminated.[84]

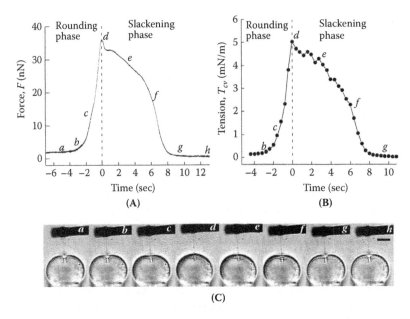

FIGURE 3.6 (A) A trace of the output voltage of a position sensor that follows the tip of a cantilever pressed against an *in vitro* CV. The output voltage varies with the force (*F*) generated by the rounding/relaxing cycle of the *in intro* CV. (B) A single tension (T_{cv})-developing cycle of an *in vitro* CV. (C) Each image labeled by a letter was taken at the time corresponding to each letter beside the trace of (A) *F* and (B) T_{cv}. (Adapted from Tani, T. et al., *J. Cell Sci.*, 114, 785, 2001. With permission.)

Not only does this smooth part of the CV spongiome membrane exist as a highly curved membrane when it is at its lowest energy state, but it also has other unique properties that are revealed during the rounding phase of the CV that occurs just before systole. At this point, an innate timing mechanism is triggered that leads to an increase in tension within the CV membrane of 35 times its resting level (Figure 3.6).[115,116] This timing mechanism is a unique property of the smooth spongiome membrane itself and may involve an enzyme system that is dependent on adenosine triphosphate (ATP) as its energy source. These enzymes may bring about a reversible modification of the membrane structure. Any isolated part of the fragmented smooth spongiome membrane can undergo its own cycle of rounding and relaxing as long as sufficiently undiluted cytosol is still present that contains ATP and other undetermined cofactors.[114,119] No master pacemaker exists to regulate the rounding and relaxing cycles of all vesicles derived from the same CVC to keep their cycles in phase. Even the cycles of the two parts of the same isolated CV, when it is pinched in two by a microneedle, will soon go out of phase with each other (see Allen's website, Chapter 9, Video 3).[114]

The isolated CV of *Amoeba proteus* has also been shown to contract when in the presence of ATP and appropriate ions.[86,95] No fibrous system has been reported to be associated with this CV, not even microtubules. Nishihara et al.[86] showed that isolated CVs from *Amoeba proteus* would suddenly shrink or burst after 2 to 3 min of exposure to 1-mmol/L ATP. Only the ATP nucleotide had this effect. Though these results remain unexplained, the studies suggest that a mechanism such as that observed in *Paramecium* may cause apparent contraction of the CV of *Amoeba proteus* when this organelle is freed from the cell and supplied with ATP.

Thus, what was perceived as contraction in the CV actually is a periodic rapid buildup of membrane tension which has the effect, in *Paramecium*, of causing the CV to round up and to proceed to separate from its attached ampulli and their collecting canal extensions. Just before the CV membrane fuses with the plasma membrane, the CV seems to begin to relax, as its diameter increases slightly, indicating that relaxation of the membrane tension has begun prior to the actual fusion of the CV with the pore membrane.[113] Once the CV and plasma membrane bilayers have completely fused, the natural fluid properties of the bilayer will cause the initial pore opening to expand to the edge of the pore indentation. The contents of the CV will be pushed out of the cell by the cytosolic pressure, which is a product of ongoing osmosis into a cytosol that is hyperosmotic to the environment.[84] This is also presumed to be true in other cells such as *Dictyostelium*.[12] It is at this time that the CV membrane collapses into the meshwork of 40-nm tubules. The meshwork of tubules effectively closes the pore, and this allows the pore membrane to separate from the CV membrane by the fission of the last single, small 40-nm membrane neck that links the CV membrane to the pore membrane.

Fluid does not reenter the CV lumen until the CV has reattached to the ampulli. Fusion probably requires the same complement of fusion proteins as those known to be present in membrane fusion sites universally. The gene coding for the *N*-ethylmaleimide-sensitve factor (NSF)[80] protein in *Paramecium* has been cloned in *P. tetraurelia* and localized by anti-NSF antibody to the junctions between the ampulli and the CV, among other cellular sites.[60] This protein regulates interactions of the soluble NSF attachment protein (SNAP) receptors (SNAREs). Genes for both synaptobrevin-like and syntaxin-like SNAREs that are specific for the CVC of *P. tetraurelia* have also been identified.[61,103] SNAREs from both the vesicle membrane and target membrane must be complexed in *trans* configuration before fusion can occur.[106]

Fluid continues to flow into each ampullus during systole, showing that the entire mechanism required for fluid accumulation is present in each radial arm but not in the CV itself. The one component known to be present in the radial arms that is not in the CV in *Paramecium* is the array of V-ATPase holoenzymes present only in the decorated tubules. The membrane of the decorated tubules differs from the smooth spongiome. Although this membrane forms 50-nm tubules when the V-ATPases are complete and organized into helices, it loses its tubular shape and vesiculates when the cell is placed under hyperosmotic stress or is subjected to cold.[32,52] Under these conditions, the V-ATPases disassemble, at least in part. Membranes of the decorated tubules are thus fundamentally different from the 40-nm tubule-forming, smooth spongiome, because these decorated membranes revert to a spherical shape, not a tubular shape, at their lowest energy state. Tubulation of the membrane of the decorated tubules may depend on the helical associations of the complete V-ATPase holoenzymes, which promote reshaping of the spherical, more planar membrane into bundles of 50-nm tubules, probably one bundle per each large vesicle.[2]

G. COMPARING CVCs

A close comparison of the CVC of *Paramecium* with that of the CVC of *Dictyostelium* is useful and informative, as these two cells, although vastly different in size and structural complexity, have CVCs that share important features and taken together help us understand what is unique about the CVC that sets it apart from other organelles in cells. The CVC of *Dictyostelium* is composed of one or two relatively large 1- to 3-μm-diameter, membrane-limited bladders or CVs that are connected to a meshwork of secondary saccules by membrane tubules that in motile cells lie close to the substrate-facing surface of the cell.[47] All of these structures are studded with V-ATPase pegs, although the observation of a peg in a deep-etch replica does not tell us if the enzyme is, in fact, complete and capable of pumping protons.[46,47] The bladders lie close to the plasma membrane and are associated with this membrane by a layer of palisade-like connections, which, in all likelihood, include the protein drainin,[12] Rab-like GTPase proteins,[17,41] and membrane fusion complexes consisting of, at a minimum, SNAREs, SNAPs, and NSF proteins. Surrounding the cytosolic side of this docking site is an annulus of actin filaments that is in contact with the plasma membrane but does not extend over the CV membrane itself.[46] Fusion of these two membranes is followed by the contraction and emptying of the CV but not, under normal conditions, the mixing of the components of the plasma membrane with the CV membrane.[36,46] The CV first rounds before fusion and then collapses after fusion is completed; its membrane flattens and tubulates during systole but does not vanish into the cytosol. The tubules of the CV can be refilled to restore the bladder.[36,47] The V-ATPase pegs remain with the CV membrane and actin remains only on the luminal surface of the plasma membrane. Only when experimental conditions were used was intermixing of these two membranes seen, and under these conditions clathrin-coated regions arose, and coated vesicles would form that could potentially collect the intramembrane CV components into coated pits to reorganize the CV.[46]

The contraction of the CV during systole, although thought in the past to be actomyosin driven,[27,35] is probably not[46] because cytosolic pressure and the release of bending energy in the CV membrane are sufficient for fluid release, as has been calculated for *Paramecium*.[84,85] No F-actin has ever been seen bound to membranes of the CVs in *Dictyostelium*[46] or *Paramecium*.[5] Heuser and his coworkers[20,22,38,47] concluded that an asymmetrical distribution of phospholipids in the CV membranes may account for the contraction of the CV as its membrane rapidly tubulates during systole. The transient presence of a protein known as LvsA (for large-volume sphere A) in *Dictyostelium* which is related to a protein in mammalian cells that has a sphingomyelinase activating activity can be used in support of this conclusion.[26,37] This protein is associated with the CV only in late stages of the CV cycle and during the return of the more planar CV membrane to a tubular form. LvsA is required for the localization of calmodulin to the CV membrane in *Dictyostelium*.[37]

Thus, contraction of the CV during systole may be, in part, a return of the more planar membrane of the CV to a tubular form; however, as mentioned above, experiments with isolated CVs from *Paramecium* have shown that, during the rounding phase of the CV prior to systole, the tension of the CV membrane increases 35-fold, and this increase in tension is apparently ATP driven.[115] Thus, we know that the CVs of both *Paramecium* and *Dictyostelium* will automatically return to a tubular form during systole (possibly with the aid of phospholipid-altering enzymes) so their underlying CV membrane bilayers seem to be similar in this respect. The V-ATPase pegs on the CVCs of *Dictyostelium* are not organized into helical patterns as they are on the decorated tubules of Paramecium,[2] so this pattern was not required for membrane tubulation of the spongiome in *Dictyostelium*, nor is it required for tubulation of the smooth spongiome of *Paramecium*.

Unfortunately, little is known about the tension in the CV prior to systole in *Dictyostelium* except that the CV has been reported to round up as it does in ciliates.[36,38] The lumen of the bladder then separates from the tubules by membrane fission, or the tubules may simply constrict so their lumens become disconnected from the bladder's lumen.[38] As in *Paramecium*, cytosolic pressure is also thought to be sufficient to produce fluid discharge from the open CV in this cell.[12]

We conclude that contraction of the CV probably has two parts: (1) a rapid buildup of tension immediately prior to systole that results in the rounding of the CV and its detachment from the tubules and the bulk of the CVC, and (2) the rapid return of the CV membrane to a tubular form after CV membrane fusion with the plasma membrane. Both aspects are visible and measurable in *Paramecium*,[114,115] but in *Dictyostelium* a buildup of tension has not yet been documented, only tubulation of the CV. In *Paramecium*, any part of the smooth spongiome when isolated from the rest of the CVC can cycle between a membrane with increased tension (rounding) and one with relaxed tension (tubulation) (see Allen's Web site, Chapter 9, Videos 2 and 5). The CVC membrane of *Dictyostelium* has not yet been studied *in vitro*, but the fact that any part of its fragmented CVC membrane during mitosis or in multinuclear cells can undergo rounding[38] prior to systole which then leads to tubulation during systole strongly suggests that this membrane may undergo the same tension increases followed by relaxation and tubulation that occur in *Paramecium*. Do isolated CV membranes from *Dictyostelium* continue to round and relax *in vitro*, as such membranes do in *Paramecium*, independent of the ability of the vesicle to accumulate additional fluid? A positive answer to this question would establish that the membranes of CV organelles are indeed unique and are strikingly different biophysically from most other membranes of living organisms. Such membranes may fall under the category of the little-studied cubic membranes.[9]

Molecular biological studies of the CVC are more advanced in *Dictyostelium* than in *Paramecium* or other protozoa. Several early studies were focused on the proteins of the individual subunits of the V-ATPase that is enriched in CVC membranes.[68,117,128] Similar studies have now been performed on the V-ATPase subunits of *Paramecium tetraurelia*,[124,125] and one report on *P. multimicronucleatum* has been published.[34] Other studies have dealt with the small GTPases of the Rho family of proteins and its Rac subfamily, as well as on the Rab family of regulatory proteins and the proteins that, in turn, regulate these small GTPases by promoting the release of guanosine diphosphate (GDP) from the GTPase to allow a guanine exchange factor (GEF) to insert a guanosine triphosphate (GTP) in its place. Two Rabs, Rab D and Rab 11, have been identified in the CVC of *Dictyostelium* that presumably act as molecular switches for regulating some aspect of the CV activity.[17,41] Another protein, DRG, which has a GTPase-activating domain (i.e., a GAP), has been identified in *Dictyostelium*; it functions in both the Rac (actin-modifying) and Rab (membrane-trafficking) pathways and appears to be important in CVC regulation.[64] Also, a Rho GDP-dissociation inhibitor (RhoGDI-1) that might also act on both the Rho and Rab pathways has been localized to the CV in *Dictyostelium*.[99]

Copine A (CpnA), another protein associated with the CV of *Dictyostelium*, is a soluble calcium-dependent, membrane-binding protein that may be involved in membrane trafficking pathways or signaling pathways. CpnA has been localized to the CVC as well as other organelles and may only localize to the CVC membrane during a rise in cytosolic calcium concentration.[25] Another protein recently found to associate with the CV of *Dictyostelium* is a protein known as VwkA for its von Willebrand Factor A-like motif that contains a conserved α-kinase catalytic domain that is reported to be present in myosin heavy-chain kinases (MHCKs). This factor may influence myosin II abundance and assembly at the CV membrane of *Dictyostelium*.[14]

These studies have revealed proteins that are for the most part peripherally or transiently associated with the CV of *Dictyostelium*. Currently, their roles seem to be mostly regulatory and are not yet precisely understood. They will likely be shown to be important in CV function or development in the future as more becomes known about the complete pathways involved. Such studies are obviously just beginning, and it is necessary to expand these studies to other cells that have CVCs and to complete the entire pathways in *Dictyostelium*.

Studies of another CVC-possessing organism that is of significant medical interest is that of the parasite *Trypanosoma cruzi*, the causative agent of Chagas disease. Chagas disease is a major problem in Latin America, where it has infected more than 11 million people and 40 million more are at risk.[121] This parasitic flagellate has a CVC adjacent to its flagellar pocket that consists of a vacuole and spongiome. The CVC has a pulsation cycle that lasts 60 to 75 sec.[19] Recent work shows that the membranes of the CVC contain an aquaporin water channel. The gene for this protein was cloned and a polypeptide with a molecular mass of 24.7 kDa (23 residues) was produced.[79] The polypeptide had similarities to other known aquaporins, including the signature Asn–Pro–Ala motif that forms an aqueous channel through the membrane bilayer.[90,127]

Not only was the aquaporin found to be part of the CVC in this cell but it was also present in the membranes of acidocalcisome vesicles.[28,29] These vesicles have a high content of pyrophosphate (PPi), polyphosphate (poly P), calcium, and magnesium, as well as other elements.[105] In addition, membranes of acidocalcisomes also contain two types of proton pumps, a V-H$^+$-ATPase as well as a pyrophosphatase (V-H$^+$-PPase) proton pump, and they also have a vacuolar Ca^{2+}-ATPase. As mentioned above, these vesicles, first described in *Trypanosoma cruzi*, have also been found in *Chlamydomonas reinhardtii*[102] and *Dictyostelium discoideum*[71] and seem to be linked to the functioning of the CVC.[28,100] Placing epimastigotes of *T. cruzi* under hypoosmotic stress caused the acidocalcisomes to migrate to and apparently fuse with the CVC, as fluorescently labeled acidocalcisomes accumulate at and cause the CV to fluoresce more brightly. Acidocalcisomes themselves can swell by 50% when they are exposed to hypoosmotic conditions.[100]

Thus, fusion of acidocalcisomes with a CVC would add free amino acids (mainly arginine and lysine), pyrophosphates, and polyphosphates (that may be reduced to inorganic phosphate), as well as the inorganic ions present in their lumens, along with the integral membrane complexes

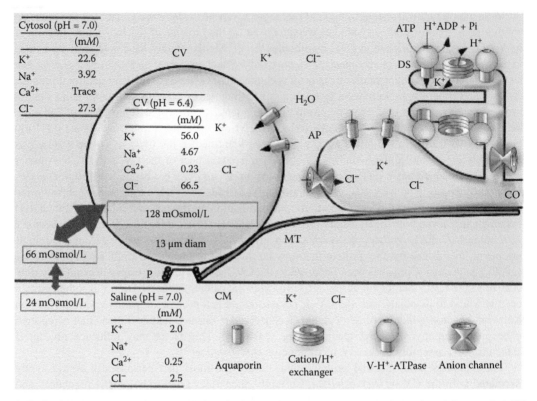

FIGURE 3.7 Summary of the inorganic ions in the standard saline solution, the cytosol, and the rounded CV of *Paramecium multimicronuleatum* just before fluid expulsion (systole). *In situ* cytosolic and CV ionic activities in mmol/L and pH were determined with ion-selective microelectrodes inserted into living cells.[109,110] The osmolarities of the saline solution and cytosol were determined as reported.[108,109] The presence and location of aquaporin, cation/H⁺ exchangers, and anion channels are only speculative. The location of V-H⁺-ATPase was determined by immunofluorescence and electrophysiology.[32,39,117] Abbreviations: CV, contractile vacuole; AP, ampullus; CO, collecting canal; DS, decorated spongiome; P, CV pore; MT, microtubular ribbon; CM, cell membrane.

V-ATPases, V-H⁺-PPases, vacuolar Ca^{2+} pumps, and aquaporins to the CVC. This contribution would favor movement of water osmotically into the CVC; however, how widespread acidocalcisomes are in cells containing CVCs has not been determined. We have never observed electron-opaque vesicles around a CVC nor such bodies fusing with the CVC in electron micrographs of intact *Paramecium* cells.

H. The Role of CVC in Osmoregulation

Cellular osmoregulation involves at least two interacting processes: (1) the acquisition or production of osmolytes that are dissolved in the cytoplasmic fluid phase of the cell, and (2) the balancing of cellular water, which constantly flows into the cell by osmosis to maintain the cytosol at its required hyperosmotic level. Thus, the primary role of the CVC is to facilitate the second process, to sequester excess water from the cell and to expel this water to the exterior of the cell. *Paramecium* does this principally by transferring the inorganic ions K⁺ and Cl⁻ to the CVC at a level 2.0 to 2.4 times higher than that in the cytosol (Figure 3.7). A secondary function is to sequester and excrete cations such as Ca^{2+} and Na⁺ from the cell.

To accomplish this, the CVC membranes must contain mechanisms such as cation/H⁺ exchangers and cotransporters for concentrating K⁺ and Na⁺, Ca^{2+} pumps such as those identified in

Dictyostelium,[77,78] and/or anion channels for the entry of Cl⁻ into their lumens. Although such mechanisms remain to be studied in detail, it is now clear that the membranes of the CVC are highly enriched in V-ATPases. These V-ATPases are not used to form a highly acid compartment but are important for energizing the membrane to establish the +80-mV luminal electrochemical charge and for providing protons that are then available to be exchanged for the cations that increase the osmolarity of the CV and so promote osmosis across the CVC membrane.

Water then enters the CV by osmosis, probably through aquaporin water channels, as we first postulated for *Paramecium*[118] and which is now confirmed by molecular techniques for the CVCs of trypanosomes.[13,29] Periodically, an innate timing mechanism, which in *Paramecium* is not tied to the volume of the CV, triggers the CV to round up and at the same time to separate from its radial arms and proton pumps. This precedes the fusion of the CV membrane with the plasma membrane and the opening of the CV pore so expulsion of the contents of the CV, both osmolytes (such as K⁺ and Cl⁻) and water, occurs. After CV emptying, both the plasma membrane and CV membrane separate and reseal, and the collapsed and tubulated CV membrane will fuse again with each ampullus. The ampulli will empty their accumulated fluid content into the tubules of the collapsed CV which will cause these tubules to swell into a vacuole.

This scheme implies that only those osmolytes that are trapped in the CV during rounding will be expelled from the cell during the expulsion process. Osmolytes remaining in the radial arms will be retained as they are no longer in continuity with the CV during systole. During each cycle the cell will maintain its K⁺ and Cl⁻ levels in the CVC using the energy of the proton gradient produced by the huge number of V-ATPases in the CVC membrane for the import of additional cations. Currently, no experimental evidence has demonstrated a mechanism for retrieving and returning osmolytes from the rounded CV back to the cytosol during the very short period of time that the CV is separated from the CVC and is in the rounded phase. An osmolyte retrieval mechanism in the CV membrane would probably defeat much of the purpose of water segregation and expulsion by the CVC, as water would rapidly flow out of the CV back into the cytosol as the osmolytes are retrieved from the CV, particularly as the CV membrane now seems to contain aquaporins.

In contrast to what was reported for amoebae,[98,104] the osmolarity of the CV of *Paramecium* is not hypoosmotic (hypotonic) to the cytosol.[109,110] Earlier techniques used to measure the osmolarity and ionic contents of the CV in amoebae may have been inadequate to provide reliable results, so these earlier studies should be repeated with improved techniques, preferably on living cells. In *Paramecium* not only do we find K⁺ activity 2.0 to 2.4 times higher in the CV than in the cytosol but we also find Cl⁻ activity equally as high or higher (when the Ca²⁺ concentration is high externally) which can account for most if not all of the counterbalancing anions.

III. CELL VOLUME CONTROL IN *PARAMECIUM* AND PARASITIC PROTOZOA

A. VOLUME ADAPTATION TO THE EXTERNAL OSMOLARITY

1. Adapted Cells Remain Osmotically Swollen

We previously found that the cytosolic osmolarity (C_{cyt}) of a *Paramecium multimicronucleatum* cell changed stepwise at ~75 or 160 mOsmol/L when the adaptation osmolarity (C_{adp}) is continuously changed. That is, C_{cyt} is ~75, ~160, and ~245 mOsmol/L when C_{adp} is (1) less than 75, (2) more than 75 but less than 160, and (3) more than 160, respectively (see Figure 3.2A and Figure 3.8B).[109] This finding implies that an active change in C_{cyt} takes place when the external osmolarity is changed beyond these osmolarities. Hereafter, these two osmolarities (~75 and ~160 mOsmol/L) and also ~245 mOsmol/L (see the legend for Figure 3.8) will each be termed a *critical osmolarity* (C_N), as these cause an active change in C_{cyt} (i.e., an activation of a hypothetical osmolyte-transport mechanism).[54] This finding also implies that C_{cyt} will normally be higher than C_{adp}, and, therefore, an adapted cell will remain osmotically swollen.

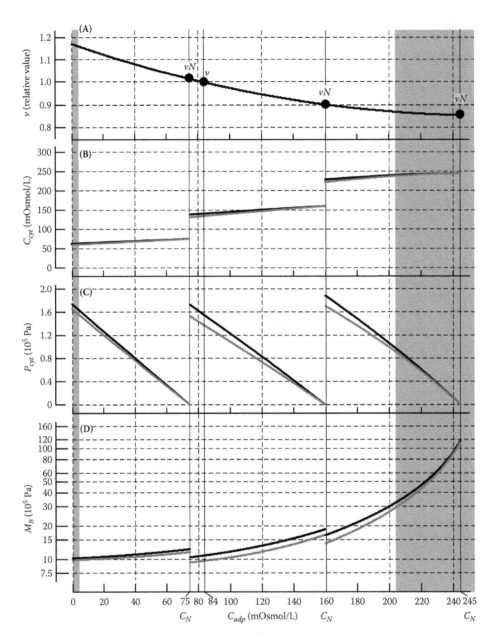

FIGURE 3.8 (A) Cell volume (v); (B) cytosolic osmolarity (C_{cyt}); (C) hydrostatic pressure of the cytosol with reference to the external (adaptation) solution (P_{cyt}); and (D) bulk modulus of the cell (M_B) in *Paramecium multimicronucleatum* cells plotted against the adaptation osmolarity (C_{adp}). A filled circle labeled v corresponds to the reference cell volume (1.0) (i.e., the volume of the cell adapted to 84 mOsmol/L, which is the osmolarity of the axenic culture medium). All cell volumes are presented as the values relative to the reference volume. Three filled circles labeled vN correspond to the natural volumes of the cells for three C_{adp} ranges (C_{adp} < 75, 75 < C_{adp} < 160, and C_{adp} > 160 mOsmol/L). C_N represents the critical osmolarities for three stepwise changes in C_{cyt} (75, 160, and 245 mOsmol/L). In B, C, and D, the values for respective parameters were estimated based on the assumption that the amount of the osmotically nonactive portion of the cell (v_{na}) is 20% (black lines) and 40% (gray lines) of the reference cell volume. The left gray column is for C_{adp} ranging from 0 to 4 mOsmol/L; the right gray column is for C_{adp} ranging from 204 to 245 mOsmol/L. These C_{adp} ranges in gray are nonexperimental, and all the values in these regions were obtained by extrapolation. (Modified from Iwamoto, M. et al., *J. Exp. Biol.*, 208, 523, 2005. With permission.)

The extent of swelling depends on the osmotic pressure of the cytosol with reference to the external solution (π_{cyt}), which is proportional to the difference between C_{cyt} and C_{adp}. π_{cyt} is written as:

$$\pi_{cyt} = \left(C_{cyt} - C_{adp} \right) R \cdot T \qquad (3.1)$$

where R and T are the gas constant and the absolute temperature, respectively.

2. Osmotic Pressure Balances Hydrostatic Pressure in Cytosol in Adapted Cells

In an adapted cell, π_{cyt} equals or balances the hydrostatic pressure in the cytosol with reference to the external solution (P_{cyt}). P_{cyt} is generated when an elastic membrane and its associated cytoskeletal structures, which surround the cell, are expanded as the cell is osmotically swollen. The balance can be written as:

$$\pi_{cyt} = P_{cyt} \qquad (3.2)$$

If either one or both of these pressures become modified, the cell volume will change until a new balance between these pressures is established (Figure 3.9). The cell elasticity, or physical resistance to swelling or shrinking, can be represented as the modulus of volume elasticity—that is, the bulk modulus (M_B). The M_B of a *Paramecium* cell adapted to an osmolarity (C_{adp_n}) is defined as:

$$M_{B_n} = \frac{\pi_{cyt_{n+1}} - \pi_{cyt_n}}{\dfrac{v_{n+1} - v_n}{v_n}} = \frac{\Delta\pi_{cyt_n}}{\dfrac{\Delta v_n}{v_n}} = \frac{v_n \cdot \Delta\pi_{cyt_n}}{\Delta v_n} \qquad (3.3)$$

where n stands for the nth experiment among a series of experiments with varied C_{adp}, and $n+1$ stands for the $(n + 1)$th experiment employing a C_{adp} that is slightly different from that in the preceding nth experiment; v is the volume of the cell in either experiment n or $n + 1$, and π_{cyt_n} can be written as:

$$\pi_{cyt_n} = \left(C_{cyt_n} - C_{adp_n} \right) R \cdot T \qquad (3.1')$$

3. Volume of the Cell Adapted to a New Osmolarity Will Always Change as Adaptation Osmolarity Changes

The volume of a cell (v) that has been adapted for several hours to a given C_{adp} will then continuously change as the C_{adp} is continuously raised in an osmolarity range from 4 to 204 mOsmol/L. The value for v relative to that in the original culture medium employed (84 mOsmol/L) changes by a ratio of ~1.16 to ~0.87 as C_{adp} changes from 4 to 204 mOsmol/L (Figure 3.8A).[54]

4. Estimation of C_{cyt}, π_{cyt}, and M_B of Cells Adapted to Varied C_{adp}

The hypothetical osmolyte-transport mechanism responsible for the stepwise change in C_{cyt} is not activated by a change in C_{adp} within a range where no critical osmolarity (C_N) is crossed, so the number of osmolytes in the cytosol (N) remains unchanged regardless of a change in v due to a change in C_{adp}.[54] Within such a C_{adp} range, C_{cyt_n} can be written as:

$$C_{cyt_n} = \frac{N}{v_n - v_{na}} \qquad (3.4)$$

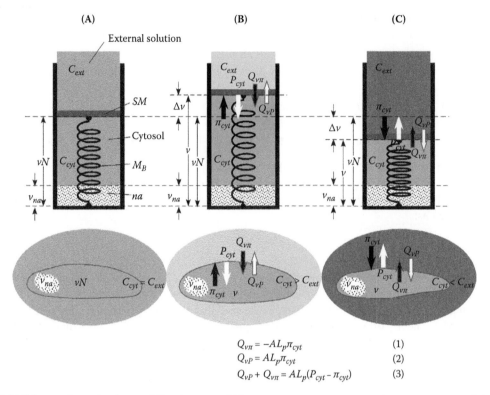

$$Q_{v\pi} = -AL_p\pi_{cyt} \tag{1}$$
$$Q_{vP} = AL_p\pi_{cyt} \tag{2}$$
$$Q_{vP} + Q_{v\pi} = AL_p(P_{cyt} - \pi_{cyt}) \tag{3}$$

FIGURE 3.9 (See color insert following page 208.) Mechanoosmotic model of a *Paramecium* cell in an (A) isosmotic, (B) hypoosmotic, or (C) hyperosmotic solution. The model is drawn as a cylinder with a semipermeable piston (bar labeled *SM*) that is fixed to the bottom of the cylinder by a coil spring (M_B). The piston and the coil spring correspond to the semipermeable cell membrane and its elasticity (the bulk modulus of the cell), respectively. The inside and outside of the cylinder correspond to the cytosol and the external solution, respectively. Medium gray corresponds to the cytosolic osmolarity (C_{cyt}), lighter gray to an osmolarity of the external solution (C_{ext}) lower than C_{cyt}, and darker gray to a C_{ext} higher than C_{cyt}. A dotted area in the cylinder corresponds to an osmotically nonactive portion of the cell (*na*). The corresponding cell shape is shown below each model. When $C_{ext} = C_{cyt}$ (A), no osmotic water flow takes place across the piston, so the coil spring is neither expanded nor compressed; that is, the cell is neither swollen nor shrunken. The length of the coil spring in this situation corresponds to its natural resting length. The corresponding cell volume is thereby termed the *natural cell volume* (*vN*). When C_{ext} is lowered below C_{cyt} (B), water osmotically flows into the cylinder through the piston (water inflow, $Q_{v\pi}$; downward blue-bordered black arrow). $Q_{v\pi}$ is proportional to the osmotic pressure of the cytosol with reference to the external solution so π_{cyt} (upward black arrow) can be obtained from Equation 1, where A is the area of the semipermeable cell membrane and L_p is the hydraulic conductivity of the membrane. A hydrostatic pressure in the cytosol with reference to the external solution P_{cyt} (downward white arrow) is generated as the coil spring is expanded by the water inflow (the cell is osmotically swollen) and causes a water outflow from the cylinder through the piston (upward blue-bordered white arrow labeled Q_{vP}), which is proportional to P_{cyt} (Equation 2). Q_{vP} cancels $Q_{v\pi}$ and the overall water flow across the piston becomes 0 when the cell swells to a level where $P_{cyt} = \pi_{cyt}$ and, therefore, $Q_{vP} = Q_{v\pi}$ (Equation 3). Inversely, when C_{ext} is raised beyond C_{cyt} (C), the water osmotically leaves the cylinder through the piston (the water outflow; upward blue-bordered black arrow labeled $Q_{v\pi}$), so the coil spring is compressed (the cell is osmotically shrunken). A negative hydrostatic pressure with reference to the external solution is thereby generated in the cylinder and causes a water inflow through the piston (downward blue-bordered white arrow labeled Q_{vP}). The overall water flow across the piston becomes 0 when the coil spring is compressed to a level where $P_{cyt} = \pi_{cyt}$ and, therefore, $Q_{vP} = Q_{v\pi}$ (Equation 3). For discussion see Baumgarten and Feher.[11] Abbreviations: v_{na}, volume of the osmotically nonactive portion of the cell (*na*); *v*, cell volume; Δv, volume change after changing the external osmolarity.

where v_{na} is the volume of the osmotically nonactive portion (nonaqueous phase) of the cell (Figure 3.9). For each C_{adp} range ($C_{adp} < 75$, $75 < C_{adp} < 160$, or $160 < C_{adp} < 245$ mmol/L), N can be written as:

$$N = C_N \left(v_N - v_{na} \right) \tag{3.5}$$

because $C_{cyt} = C_N$ and $v = v_N$, when $C_{adp} = C_N$.

The values for C_{cyt_n}, π_{cyt_n}, and M_{B_n} can be estimated from the data in Figure 3.10A and Figure 3.8A according to Equations 3.1 to 3.5 based on the assumption that v_{na} remains constant in the specific C_{adp} range employed, as represented in Figure 3.8 (B, C_{cyt_n}; C, π_{cyt_n}; D, M_{B_n}). The value for v_{na} is not available at present so we employed two different plausible values for v_{na}: 20 and 40% of the cell volume of cells growing in the original culture medium (84 mOsmol/L) for the estimation (black and gray lines in Figure 3.8B, C, and D correspond to 20 and 40%, respectively). Changes in these parameters caused by a change in C_{adp} are essentially the same for the two different v_{na} cases; that is, π_{cyt} is highest (greater than $\sim 1.6 \times 10^5$ Pa) at its lower C_{adp} values in each range, where the difference between C_{cyt} and C_{adp} is largest, and it is almost 0 at the highest C_{adp} values because, at these values, their difference is essentially 0 (Equation 3.1).

That the cell volume decreases continuously as the adaptation osmolarity increases (Figure 3.8A), even though the cytosolic osmolarity increases stepwise (Figure 3.8B), implies that the resistance by the cell to volume change (its M_B) also increases stepwise at each critical osmolarity as C_{adp} increases. M_B is highest at the highest C_{adp} values in each osmolarity range, and the highest value in each C_{adp} range is larger in the higher osmolarity range (~ 12, ~ 17, and $\sim 114 \times 10^5$ Pa at 75, 160, and 245 mOsmol/L for C_{adp}, respectively). For an easier understanding of the mechanoosmotic behaviors of a *Paramecium* cell, a cell model is presented in Figure 3.9 that is consistent with the mechanoosmotic behaviors exhibited by the cell when the external osmolarity is changed.

B. Regulatory Volume Control

1. Time Course of Change in Cell Volume after Changing the External Osmolarity

When a cell adapted to a specific C_{adp} is transferred into another solution with an osmolarity (C_{stm}) different from the C_{adp}, the π_{cyt} changes so the cell volume will change until a new balance for π_{cyt} and P_{cyt} is established (Equations 3.1 and 3.2). The time courses for the changes in volume of 16 groups of cells adapted to different C_{adp} after they were each transferred into a different stimulatory C_{stm} are shown in Figure 3.10. C_{adp} and C_{stm} for each group are visualized by vertical arrows in each corresponding inset. Downward and upward arrows correspond to a decrease and an increase, respectively, in the external osmolarity upon subjection of a cell to the stimulatory C_{stm}. The number at the tail end of each arrow indicates the C_{adp} and the number at the head end indicates C_{stm} in mOsmol/L. The length of each arrow corresponds to the amount of change in the external osmolarity upon subjecting the cell to C_{stm} (black, gray, and light gray arrows correspond to 60, 40, and 20 mOsmol/L of change, respectively). A horizontal bar that crosses an arrow corresponds to a C_N (75 or 160 mOsmol/L) and is placed on the same osmolarity scale with that for the arrows. When the arrow crosses a horizontal bar, the external osmolarity is changed beyond a C_N (Figure 3.10A, C, D, F, H, I); when the arrow does not cross a horizontal bar, the external osmolarity is changed in an osmolarity range where no C_N is crossed (Figure 3.10B, E, G, J). Some representative microscope images of the cells adapted to different C_{adp} before and 15 and 30 min after subjecting the cell to a different C_{stm} are shown in Figure 3.11.

2. Activation of RVD

When the external osmolarity is decreased ($C_{stm} < C_{adp}$; see Figure 3.10A–E, downward arrows), cell volume increases with time to a higher level (osmotic swelling). The cell volume then resumes its initial level only if the osmolarity decrease crosses a C_N (Figure 3.10A, C, D; see also Figure

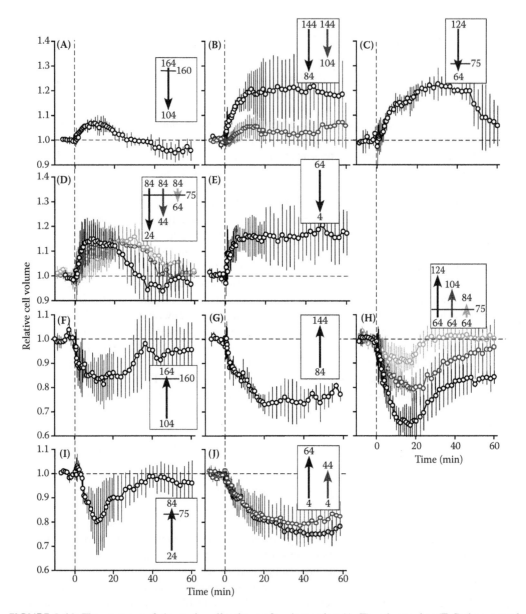

FIGURE 3.10 Time courses of change in cell volume after decreasing (A–E) or increasing (F–J) the external osmolarity in 16 different groups of *Paramecium multimicronucleatum*. Black, gray, and light gray circles correspond to changes in the external osmolarity by 60, 40, and 20 mOsmol/L, respectively. Arrows in the insets show (1) the direction of the osmolarity change (i.e., downward corresponds to its decrease and upward to its increase), and (2) the degree of change in the osmolarity (i.e., the long black, medium gray, and short light gray arrows correspond to 60, 40, and 20 mOsmol/L changes, respectively). The number at the tail end is the adaptation osmolarity (C_{adp}), and that at the head end is the changed osmolarity (C_{stm}) in mOsmol/L. A horizontal bar across the arrow corresponds to a critical osmolarity (C_N), where the change in osmolarity has crossed a C_N. The value for the C_N is shown as a number beside the bar. See the text for details. (Adapted from Iwamoto, M. et al., *J. Exp. Biol.*, 208, 523, 2005. With permission.)

3.11B). By contrast, the cell volume stays at the higher level if no C_N is crossed (Figure 3.10B, E; see also Figure 3.11A). These findings imply that an outward-directed osmolyte-transport mechanism that decreases C_{cyt} (with subsequent cell volume decreases) is activated when C_{stm} is decreased

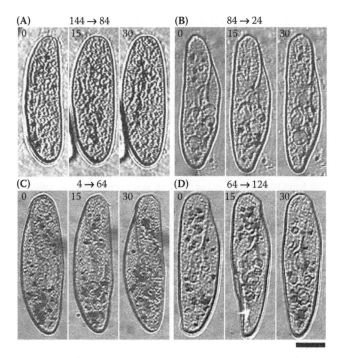

FIGURE 3.11 Four sets of three consecutive pictures of a representative cell each obtained from a different group of *Paramecium multimicronucleatum* cells adapted to one of four different osmolarities taken at 0, 15, and 30 min, respectively, after changing the external osmolarity by 60 mOsmol/L. (A) A 144-mOsmol/L-adapted cell was subjected to 84 mOsmol/L. (B) A 84-mOsmol/L-adapted cell was subjected to 24 mOsmol/L. (C) A 4-mOsmol/L-adapted cell was subjected to 64 mOsmol/L. (D) A 64-mOsmol/L-adapted cell was subjected to 124 mOsmol/L. A white arrowhead points to an indentation of the cell caused by osmotic shrinkage of the cell. The top of each cell image corresponds to the anterior end of the cell. A number on the upper left corner of each picture is the time in minutes, after changing the external osmolarity, when the picture was taken. Scale bar, 50 μm. (From Iwamoto, M. et al., *J. Exp. Biol.*, 208, 523, 2005. With permission.)

beyond a C_N. This osmolyte-transport mechanism corresponds to a regulatory volume decrease (RVD) mechanism (see Chapter 2).

It is clear from the figures that the following are not directly correlated with the activation of RVD: (1) the amount of decrease in the external osmolarity, represented as the length of a downward arrow; (2) the maximum amount of swelling of the cell, represented as a peak or plateau value for cell volume; and (3) the rate of increase in cell volume, which corresponds to the tangent to the time course of cell volume change after decreasing the external osmolarity.

As is shown in Figure 3.8C, P_{cyt} of an adapted cell is highest when C_{adp} is lowest—that is, close to C_N in a given C_{adp} range (as at ~75 in the C_{adp} < 75 mOsmol/L range or ~160 in the 160 < C_{adp} < 245 mOsmol/L range), as the cell is maximally swollen at this C_{adp} value (Figure 3.8A). If a cell adapted to an osmolarity in one of the above C_{adp} ranges is subjected to an external osmolarity lower than the C_N at the lower end of the range, the cell would be expected to swell more than the maximal value observed and P_{cyt} should increase beyond the highest value obtained. In contrast, however, P_{cyt} does not exceed the highest value when the adapted cell is subjected to an osmolarity increase within the specific C_{adp} range. It is therefore concluded that an increase in P_{cyt} beyond the highest value is a primary factor required for activation of RVD.

Currently, we do not know how an increase in P_{cyt} beyond the highest value (the threshold value) will cause RVD to be activated. There might be a pressure sensor or a tension sensor in the cell membrane that is triggered by osmotic swelling of the cell that may, in turn, activate the RVD system to release osmolytes from the cell across its membrane.

The threshold P_{cyt} is approximately1.5 to 1.9×10^5 Pa (Figure 3.8C). The threshold membrane tension estimated from the P_{cyt} value approximates 2 N/m, which is thousands of times larger than the threshold tension required for activation of some mechanosensitive ion channels.[40,81] This high pressure must be countered by the expansion of a cytoskeletal system that lines the plasma membrane or the cell pellicle, which has a high bulk modulus. The tension in the plasma membrane, where the hypothetical RVD mechanism is thought to reside, when expanded to the same degree as the cytoskeleton, would appear to be much lower than that of the estimated tension in the cytoskeletal lining.

3.   Activation of RVI

When the external osmolarity is increased ($C_{stm} > C_{adp}$; Figure 3.10F–J, upward arrows), the cell volume decreases with time to a lower value. It then resumes its initial value when the osmolarity increases beyond a C_N (Figure 3.10F, H, I; see also Figure 3.11D). In contrast, cell volume remains at the lower value when the increase in external osmolarity occurs within a range where no C_N is crossed (Figure 3.10G, J; see also Figure 3.11C). These findings imply that an inward-directed osmolyte-transport mechanism that increases C_{cyt} (cell volume consequently increases) is activated when C_{stm} is increased beyond a C_N. This osmolyte-transport mechanism corresponds to a regulatory volume increase (RVI) mechanism. As is similar to the case for RVD, the amount of increase in the external osmolarity (the length of the upward arrows), the maximum amount of decrease in cell volume (the lower peak or plateau value for cell volume), and the rate of decrease in cell volume (the tangent to the time course of cell volume change) after increasing the external osmolarity are not directly correlated with the activation of RVI.

As shown in Figure 3.8C, P_{cyt} of an adapted cell is almost 0 when C_{adp} is highest in a specific C_{adp} range where no C_N is included (as at ~75 in the $C_{adp} < 75$ mOsmol/L range or ~160 in the 75 $< C_{adp} < 160$ mOsmol/L range), as the cell is neither swollen nor shrunken at a C_{adp} that is close to C_N. If a cell adapted to a given osmolarity is subjected to an external osmolarity higher than C_N at the border of this C_{adp} range that includes the given osmolarity, P_{cyt} should become negative and the cell would shrink. Because P_{cyt} never becomes negative when the external osmolarity is increased within the given osmolarity range, it is therefore concluded that a decrease to 0 or a negative P_{cyt} is the primary factor required for activation of RVI. The cell shrinks and the cell membrane wrinkles when P_{cyt} becomes 0 (Figure 3.11D, white arrowhead). A 0 or negative pressure sensor or a wrinkle-sensitive mechanosensor must be involved in the activation of RVI.

4.   Regulatory Volume Control Involves K⁺ Channels of the Cell Membrane

When a stimulatory solution contains 10 mmol/L tetraethylammonium (TEA) (a potent K⁺ channel inhibitor), adapted cells that would normally be stimulated to undergo RVD or RVI are not so stimulated. In these cells, the volume increases (upon decreasing the external osmolarity) or decreases (upon increasing the external osmolarity) to a new plateau level without returning to their initial volumes. In the presence of 30 mmol/L of K⁺, RVD is also inhibited so the cell volume increases without showing a resumption of the initial volume after decreasing the external osmolarity to a point that RVD should have been activated. On the other hand, RVI is enhanced so cell volume is restored more quickly then normal after increasing the external osmolarity to a point where RVI would normally be activated.[54] These findings strongly suggest that K⁺ channels in the plasma membrane of the cell are involved in regulatory volume control mechanisms in *Paramecium* in both RVD and RVI. Involvement of several kinds of K⁺ channels in regulatory cell volume control has been demonstrated in several cell types.[11,48,81,89] (Also refer to Chapter 2.)

C.   Cell Volume Control and CVC Activity

Four representative time courses of change in the CVC activity presented as R_{CVC} after changing the external osmolarity by 60 mOsmol/L are shown in Figure 3.12. In Figure 3.12A, a 164-mOsmol/L-adapted cell was subjected to 104 mOsmol/L; the external osmolarity was decreased beyond a C_N

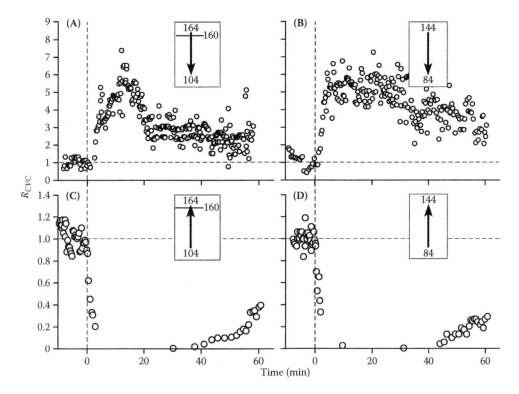

FIGURE 3.12 Four representative time courses of change in the rate of fluid segregation and discharge by the contractile vacuole (R_{CVC}) after changing the external osmolarity by 60 mOsmol/L in *Paramecium multimicronucleatum*. (A) A 164-mOsmol/L-adapted cell was subjected to 104 mOsmol/L. (B) A-144 mOsmol/L-adapted cell was subjected to 84 mOsmol/L. (C) A 104-mOsmol/L-adapted cell was subjected to 164 mOsmol/L. (D) A 84-mOsmol/L-adapted cell was subjected to 144 mOsmol/L. An arrow in each inset shows the direction of change in the external osmolarity; the downward arrow (A, B) corresponds to a decrease, while the upward arrow (C, D) to an increase. A number at the tail end shows the adaptation osmolarity and that at the head end shows the changed external osmolarity in mOsmol/L. A horizontal bar that crosses the arrow corresponds to a C_N, indicating that the change in the external osmolarity crossed a C_N. A number beside the bar is the osmolarity of the C_N. (From Iwamoto, M. et al., *J. Exp. Biol.*, 208, 523, 2005. With permission.)

of 160 mOsmol/L so RVD was activated and the cell volume returned to normal. The corresponding time course of change in cell volume is shown in Figure 3.10A. In Figure 3.12B, a 144-mOsmol/L-adapted cell was subjected to 84 mOsmol/L; the osmolarity change was made without crossing a C_N so RVD was not activated and the cell volume remained higher than normal. The corresponding time course of change in cell volume is shown in Figure 3.10B. In Figure 3.12C, a 104-mOsmol/L-adapted cell was subjected to 164 mOsmol/L; the external osmolarity was changed beyond a C_N of 160 mOsmol/L so RVI was activated and the cell volume approached a normal value. The corresponding time course of change in cell volume is shown in Figure 3.10F. In Figure 3.12D, an 84-mOsmol/L-adapted cell was subjected to 144 mOsmol/L; the osmolarity change was made without crossing a C_N so RVI was not activated and cell volume remained below normal. The corresponding time course of change in cell volume is shown in Figure 3.10G.

In the cases of osmolarity decrease, R_{CVC} appeared to increase in parallel with an increase in cell volume; that is, R_{CVC} increased as cell volume increased after subjection of the cell to a 60-mOsmol/L decrease in the external osmolarity. R_{CVC} then decreased when the cell volume resumed its initial value after activation of RVD (compare Figure 3.12A with Figure 3.10A), while R_{CVC} remained higher when the RVD was not activated; therefore, the cell volume also remained higher (compare Figure 3.12B with Figure 3.10B).

C_{cyt} decreases as the cell osmotically swells. The reduction of C_{cyt}, therefore, appears to be involved in enhancing the R_{CVC} activity. It is highly probable that control by the cell of the ratio of osmolarity in the CV fluid (C_{CVC}) to cytosolic osmolarity (C_{CVC}/C_{cyt}) is a primary factor for determining the R_{CVC} activity. When the ratio increases over the ratio existing before changing the external osmolarity, R_{CVC} will become higher. When the cell resumes its initial volume after activation of RVD (Figure 3.10A), C_{cyt} is assumed to decrease to the same extent as the decrease in external osmolarity. We previously reported that the ionic activities of the K$^+$ and Cl$^-$ ions in the CV fluid are reduced stepwise similar to the overall cytosolic osmolarity that follows RDV activation.[110] Thus, the CV fluid is also affected by RDV activation, and the cell keeps its ratios of K$^+$ and Cl$^-$ activities in the CVC to those in the C_{cyt} more or less constant at ~2.4 (Table 3.1).[110] R_{CVC}, therefore resumes its previous value as the cell volume resumes its initial value (Figure 3.12A). Quantitative analysis of the relationship between the C_{CVC}/C_{cyt} ratio and R_{CVC} is necessary to understand the control mechanism of the rate of water segregation by the CVC (R_{CVC}).

On the other hand, the time course of change in R_{CVC} after increasing the external osmolarity by 60 mOsmol/L was essentially the same, independently of whether RVI was activated or not (compare Figure 3.12C and D with Figure 3.10F and G, respectively), That is, R_{CVC} became 0 immediately after the cell was subjected to an increase in the external osmolarity. It began to recover around 30 to 40 min after increasing the external osmolarity and then gradually increased with time. These findings imply that a decrease in cell volume or a concomitant increase in C_{cyt} or a decrease in C_{CVC}/C_{cyt}, caused by an increase in the external osmolarity, is not directly correlated with the inhibition of R_{CVC}. In fact, as explained earlier, we have determined that the decorated tubules, which bear the electrogenic V-ATPases, immunologically disappear when the cell is subjected to increases in external osmolarity, and it takes 60 to 120 min for the CVC membrane potential to return again to its plateau value of +80 mV.[39] R_{CVC} cannot return to normal until the decorated tubules with their V-ATPases are reattached to the CVC.

It is unlikely that the CVC of *Paramecium* regulates the overall cell volume during either RVD or RVI by extrusion of cytosolic water and osmolytes through the CV. The initial rapid increase in R_{CVC} when the cell is subjected to a decreased external osmolarity will buffer the cell against mechanical disruption that could result from a large initial osmotic swelling. Similarly, the rapid decrease in R_{CVC} to 0 upon subjecting the cell to an increased external osmolarity would eliminate the effects of continued CVC activity upon the cell if excessive osmotic shrinkage were to occur. Dunham and his colleagues[31,112] had earlier suggested a role for the CV in buffering the osmotic changes in cell volume.

D. PARASITIC PROTOZOA

Physiological responses to hypoosmotic stress have been studied in parasitic protozoa, such as *Trypanosoma*,[101] *Giardia*,[91] *Crithidia*,[16] and *Leishmania*.[122] These parasitic protozoa encounter a wide range of fluctuation in external osmolarity as they progress through their life cycles; that is, a single life cycle that begins in the gut of the intermediate host (insects) may end in the cytoplasm of the definitive host (mammals). Docampo and his collaborators[101] demonstrated that *Trypanosoma cruzi* showed RVD in response to hypoosmotic stress in their various life-cycle stages. The major osmolytes responsible for RVD are neutral and anionic amino acids instead of K$^+$ and Cl$^-$, which are the predominant osmolytes responsible for RVD in many cell types, including *Paramecium*.[11,48,65,89,110] The efflux is assumed to be mediated by some putative osmotic swelling-sensitive organic anion channels.[58,59] They also demonstrated that external Ca^{2+} ions were indispensable for triggering RVD in *T. cruzi*. It is generally accepted in many cell types that external Ca^{2+} ions that enter the cytosol through osmotic swelling-sensitive Ca^{2+} channels in the plasma membrane are the mediators of RVD.[65,123] Interestingly, control of the cytosolic Ca^{2+} concentration by using Ca^{2+} ionophores showed little effects on the amino acid efflux responsible for RVD. They concluded that, although Ca^{2+} appears to play a role in modulating the early phase of amino acid efflux, it is not a key determinant of the final outcome of RVD.

IV. FUTURE WORK ON PROTOZOAN OSMOREGULATION AND VOLUME CONTROL

One of the most pressing questions waiting to be answered in CVC research is the osmolarity of the CV in protozoa other than *Paramecium*. Most *Dictyostelium* researchers seem to have accepted the unproven conclusion that the CV of this cell is hypoosmotic to the cytosol.[46,107] This must be confirmed or disproved using *in situ* techniques that can be applied to the living cells of these small protozoa. In addition, the ion channels and osmolyte transport systems of the plasma membrane and CVC membranes must be explored with regard to their possible osmoregulatory and water and osmolyte secretion activities: biochemically, molecularly, and proteomically. Finally, as with the osmolarity of the CV, the number of organisms investigated must be increased to look for differences in the CVC cycle that may be revealed in different species.

Physiological studies on the regulatory volume control in *Paramecium* have only just begun; therefore, various conventional physiological approaches are now required to compare physiological characteristics of this specific cell with other cells for which more detailed physiological studies have already been completed. The comparison will potentially lead to a better understanding of physiological mechanisms underlying cell volume control. Some suggested experiments include the following: (1) Continuous measurement of the cellular membrane potential after changing the external osmolarity would reveal the dynamic change in the cytosolic K^+ concentration associated with RVD or RVI, as the resting membrane potential of *Paramecium* is dependent predominantly on K^+.[83] (2) Continuous monitoring electrically with a Ca^{2+}-sensitive microelectrode[110] or photometrically using a Ca^{2+}-sensitive dye of the cytosolic Ca^{2+} concentration after changing the external osmolarity would reveal the possible involvement of Ca^{2+} in modulating RVD or RVI. (3) Examination of the effects of appropriate channel inhibitors on the time course of cell volume change would reveal the ion channels that directly or indirectly participate in RVD or RVI.

To characterize RVD or RVI, simultaneous monitoring of associated cellular events such as changes in cell volume, cytosolic osmolarity, membrane potential, cytosolic activities of K^+ and Cl^-, etc. are indispensable. We are developing a novel intracellular osmometer.[82] The tip of the probe of the osmometer is ~2 µm in diameter so it can be inserted into the cell together with other microelectrodes for measuring membrane potential and ion activity. The basic principles of the microcapillary osmometer are illustrated in Figure 3.13. At present, the semipermeable $Cu_2Fe(CN)_6$ plug at the tip of the probe is too short lived (~3 min) to be very useful. Development of a long-lasting (at least 1 hr) semipermeable plug is necessary for the practical use of this technique.

The probe without a semipermeable plug can be used for monitoring the change in the hydrostatic pressure of the cytosol after changing the external osmolarity. Measurements of the cytosolic pressure are indispensable for estimation of the volume of the osmotically nonactive portion of the cytoplasm as well as the bulk modulus of the cell.

REFERENCES

1. Akbarieh, M. and Couillard, P., Ultrastructure of the contractile vacuole and its periphery in *Amoeba proteus*: evolution of vesicles during the cycle, *J. Protozool.*, 35, 99–108, 1988.
2. Allen, R.D., Membrane tubulation and proton pumps, *Protoplasma*, 189, 1–8, 1995.
3. Allen, R.D., The contractile vacuole and its membrane dynamics, *BioEssays*, 22, 1035–1042, 2000.
4. Allen, R.D. and Fok, A.K., Membrane dynamics of the contractile vacuole complex of *Paramecium*, *J. Protozool.*, 35, 63–71, 1988.
5. Allen, R.D. and Fok, A.K., Membrane trafficking and processing in *Paramecium*, *Int. Rev. Cytol.*, 198, 277–318, 2000.
6. Allen, R.D. and Naitoh, Y., Osmoregulation and contractile vacuoles of protozoa. In *Molecular Mechanisms of Water Transport Across Biological Membranes*, Zeuthen, T. and Stein, W.D., Eds., Academic Press, New York (reviewed in *Int. Rev. Cytol.*, 215, 351–394, 2002).

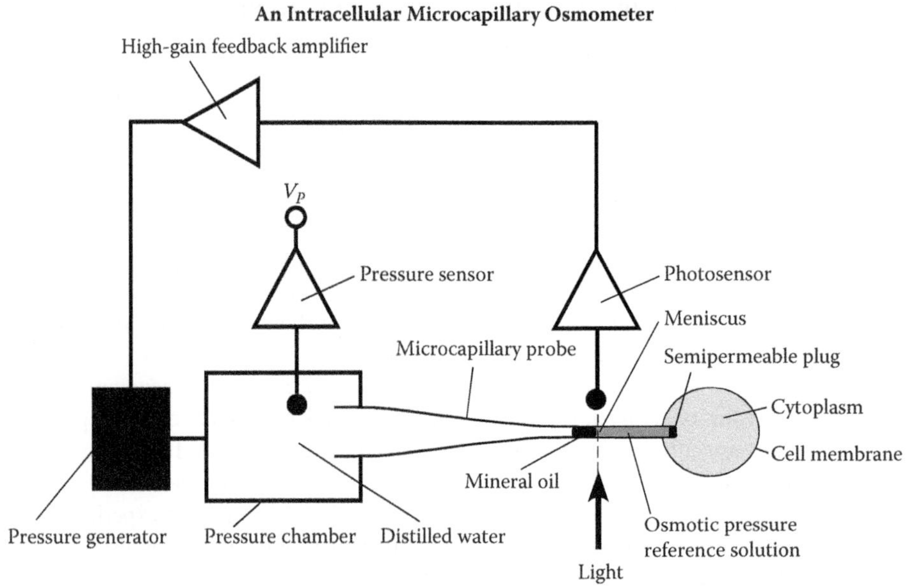

An Intracellular Microcapillary Osmometer

FIGURE 3.13 A novel intracellular microcapillary osmometer that can measure the osmolarity of the cytosol with a time constant of less than 1 sec. The microcapillary probe (the tip of a glass microcapillary similar in shape and dimensions to a conventional glass microcapillary electrode) is first plugged by a semipermeable material such as cupric ferrocyanide. An osmotic pressure reference solution is then introduced into the capillary and mineral oil is introduced to make a meniscus between the reference solution and the mineral oil in the capillary. When the tip of the probe is inserted into a cell, water moves across the plug according to an osmotic pressure difference between the reference solution and the cytosol, causing the meniscus to shift. The shift of the meniscus is detected by a photosensor that produces an electric signal proportional to the shift. The electric signal is amplified and fed into a pressure generator to generate a counter hydrostatic pressure in the pressure chamber to which the probe is connected and prevent a shift of the meniscus. The counter pressure required to prevent a shift in the meniscus is monitored by using a pressure sensor in the pressure chamber. The output of the sensor (V_p) corresponds to the counter pressure. The difference between the counter pressure and the osmotic pressure of the reference solution is the osmotic pressure of the cytosol.

7. Allen, R.D. and Wolf, R.W., Membrane recycling at the cytoproct of *Tetrahymena*, *J. Cell Sci.*, 35, 217–227, 1978.
8. Allen, R.D., Ueno, M.S., Pollard, L.W., and Fok A.K., Monoclonal antibody study of the decorated spongiome of contractile vacuole complexes of *Paramecium*, *J. Cell Sci.*, 96, 469–475, 1990.
9. Almsherqi, Z.A., Kohlwein, S.D., and Deng, Y., Cubic membranes: a legend beyond the *Flatland* of cell membrane organization, *J. Cell Biol.*, 173, 839–844, 2006.
10. Andrus, W. de W. and Giese, A.C., Mechanisms of sodium and potassium regulation in *Tetrahymena pyriformis* W., *J. Cell. Comp. Physiol.*, 61, 17–30, 1963.
11. Baumgarten, C. M. and Feher, J. J., Osmosis and regulation of cell volume. In *Cell Physiology Source Book*, 3rd ed., Sperelakis, N., Ed., Academic Press, San Diego, CA, 2001, pp. 319–355.
12. Becker, M., Matzner, M., and Gerisch, G., Drainin required for membrane fusion of the contractile vacuole in *Dictyostelium* is the prototype of a protein family also represented in man, *EMBO J.*, 18, 3305–3316, 1999.
13. Beitz, E., Aquaporins from pathogenic protozoan parasites: structure, function and potential for chemotherapy, *Biol. Cell*, 97, 373–383, 2005.
14. Betapudi, V., Mason, C., Licate, L., and Egelhoff, T.T., Identification and characterization of a novel α-kinase with a von Willebrand factor A-like motif localized to the contractile vacuole and Golgi complex in *Dictyostelium discoideum*, *Mol. Biol. Cell*, 16, 2248–2262, 2005.

15. Brauer, E.B. and McKanna, J.A., Contractile vacuoles in cells of a fresh water sponge, *Spongilla lacustris*, *Cell Tissue Res.*, 192, 309–317, 1978.
16. Bursell, J. D. H., Kirk, J., Hall, S. T., Gero, A. M., and Kirk, K., Volume-regulatory amino acid release from the protozoan parasite *Crithidia luciliae*, *J. Membr. Biol.*, 154, 131–141, 1996.
17. Bush, J., Temesvari, L., Rodriguez-Paris, J., Buczynski, G., and Cardelli, J., A role for a Rab4-like GTPase in endocytosis and in regulation of contractile vacuole structure and function in *Dictyostelium discoideum*, *Mol. Biol. Cell*, 7, 1623–1638, 1996.
18. Chapman, G.B. and Kern, R.C., Ultrastructural aspects of the somatic cortex and contractile vacuole of the ciliate, *Ichthyophthirius multifiliis* Fouquet, *J. Protozool.*, 30, 481–490, 1983.
19. Clark, T.B., Comparative morphology of four genera of trypanosomatidae, *J. Protozool.*, 6, 227–232, 1959.
20. Clarke, M. and Heuser, J., Water and ion transport. In *Dictyostelium: A Model System for Cell and Developmental Biology*, Maeda, Y., Inouye, K., and Takeuchi, I., Eds., University Academic Press, Tokyo, 1997, pp. 75–91.
21. Clarke, M. and Maddera, L., Distribution of alkaline phosphatase in vegetative *Dictyostelium* cells in relation to the contractile vacuole complex, *Eur. J. Cell Biol.*, 83, 289–296, 2004.
22. Clarke, M., Köhler, J., Arana, Q., Liu, T., Heuser, J., and Gerisch, G., Dynamics of the vacuolar H$^+$-ATPase in the contractile vacuole complex and the endosomal pathway of *Dictyostelium* cells, *J. Cell Sci.*, 115, 2893–2905, 2002.
23. Couillard, P., Forget, J., and Pothier, F., The contractile vacuole of *Amoeba proteus*. III. Vacuolar response to phagocytosis. *J. Protozool.*, 32, 333–338, 1985.
24. Cronkite, D.L. and Pierce, S.K., Free amino acids and cell volume regulation in the euryhaline ciliate *Paramecium calkinsi*, *J. Exp. Zool.*, 251, 275–284, 1989.
25. Damer, C.K., Bayeva, M., Hahn, E.S., Rivera, J., and Socec, C.I., Copine A, a calcium-dependent membrane-binding protein, transiently localizes to the plasma membrane and intracellular vacuoles in *Dictyostelium*, *BMC Cell Biol.*, 6, 46, 2005.
26. De Lozanne, A., The role of BEACH proteins in *Dictyostelium*, *Traffic*, 4, 6–12, 2003.
27. Doberstein, S.K., Baines, I.C., Wiegand, G., Korn, E.D., and Pollard, T.D., Inhibition of contractile vacuole function *in vivo* by antibodies against myosin-I, *Nature*, 365, 841–843, 1993.
28. Docampo, R., de Souza, W., Miranda, K., Rohloff, P., and Moreno, S.N.J., Acidocalcisomes: conserved from bacteria to man, *Nature Rev. Microbiol.*, 3, 251–261, 2005.
29. Docampo, R., Scott, D.A., Vercesi, A.E., and Moreno, S.N.J., Intracellular Ca^{++} storage in acidocalcisomes of *Trypanosoma cruzi*, *Biochem. J.*, 310, 1005–1012, 1995.
30. Dunham, P.B. and Child, F.M., Ion regulation in *Tetrahymena*, *Biol. Bull.*, 121, 129–140, 1961.
31. Dunham, P.B. and Kropp, D.L., Regulation of solutes and water in *Tetrahymena*. In *Biology of Tetrahymena*, Elliott, A.M., Ed., Dowden, Hutchinson and Ross, Stroudsburg, PA, 1973, pp. 165–198.
32. Fok, A.K., Aihara, M.S., Ishida, M., Nolta, K.V., Steck, T.L., and Allen, R.D., The pegs on the decorated tubules of the contractile vacuole complex of *Paramecium* are proton pumps, *J. Cell Sci.*, 108, 3163–3170, 1995.
33. Fok, A.K., Clarke, M., Ma, L., and Allen, R.D., Vacuolar H$^+$-ATPase of *Dictyostelium discoideum*: a monoclonal antibody study, *J. Cell Sci.*, 106, 1103–1113, 1993.
34. Fok, A.K., Yamauchi, K., Ishihara, A., Aihara, M.S., Ishida, M., and Allen, R.D., The vacuolar-ATPase of *Paramecium multimicronucleatum*: gene structure of the B subunit and the dynamics of the V-ATPase-rich osmoregulatory membranes, *J. Eukaryot. Microbiol.*, 49, 185–196, 2002.
35. Furukawa, R. and Fechheimer, M., Differential localization of α-actinin and the 30 kD actin-bundling protein in the cleavage furrow, phagocytic cup and contractile vacuole of *Dictyostelium discoideum*, *Cell Motil. Cytoskel.*, 29, 46–56, 1994.
36. Gabriel, D., Hacker, U., Köhler, J., Müller-Taubenberger, A., Schwartz, J.-M., Westphal, M., and Gerisch, G., The contractile vacuole network of *Dictyostelium* as a distinct organelle: its dynamics visualized by a GFP marker protein, *J. Cell Sci.*, 112, 3995–4005, 1999.
37. Gerald, N.J., Siano, M., and De Lozanne, A., The *Dictyostelium* LvsA protein is localized on the contractile vacuole and is required for osmoregulation, *Traffic*, 3, 50–60, 2002.
38. Gerisch, G., Heuser, J., and Clarke, M., Tubular-vesicular transformation in the contractile vacuole system of *Dictyostelium*, *Cell Biol. Int.*, 26, 845–852, 2002.

39. Grønlien, H.K., Stock, C., Aihara, M.S., Allen, R.D., and Naitoh, Y., Relationship between the membrane potential of the contractile vacuole complex and its osmoregulatory activity in *Paramecium multimicronucleatum*, *J. Exp. Biol.*, 205, 3261–3270, 2002.

40. Gustin, M. C., Mechanosensitive ion channels in yeast: mechanism of activation and adaptation. *Adv. Compar. Environ. Physiol.*, 10, 19–38, 1992.

41. Harris, E., Yoshida, K., Cardelli, J., and Bush, J., A Rab11-like GTPase associates with and regulates the structure and function of the contractile vacuole system in *Dictyostelium*, *J. Cell Sci.*, 114, 3035–3045, 2001.

42. Hauser, K., Pavlovic, N., Kissmehl, R., and Plattner, H., Molecular characterization of a sarco(endo)plasmic reticulum ATPase gene from *Paramecium tetraurelia* and localization of this gene product to subplasmalemmal calcium stores. *Biochem. J.*, 334, 31–38, 1998.

43. Hausmann, K. and Allen, R.D., 1977. Membranes and microtubules of the excretory apparatus of *Paramecium caudatum*, *Cytobiologie (Eur. J. Cell Biol.)*, 15, 303–320, 1977.

44. Hausmann, K. and Patterson, D.J., Involvement of smooth and coated vesicles in the function of the contractile vacuole complex of some cryptophycean flagellates, *Exp. Cell Res.*, 135, 449–453, 1981.

45. Hausmann, K. and Patterson, D.J., Contractile vacuole complexes in algae. In *Compartments in Algal Cells and Their Interaction*, Wiessner, W., Robinson, D., and Starr, R.C., Eds., Springer-Verlag, Berlin, 1984, pp. 139–146.

46. Heuser, J., Evidence for recycling of contractile vacuole membrane during osmoregulation in *Dictyostelium* amoebae: a tribute to Günther Gerisch, *Eur. J. Cell Biol.*, 85, 859–871, 2006.

47. Heuser, J., Zhu, Q., and Clarke, M., Proton pumps populate the contractile vacuoles of *Dictyostelium* amoebae, *J. Cell Biol.*, 121, 1311–1327, 1993.

48. Hoffman, E.K. and Dunham, P.B., Membrane mechanisms and intracellular signaling in cell volume regulation, *Int. Rev. Cytol.*, 161, 173–262, 1995.

49. Hohmann, S., Osmotic adaptation in yeast: control of the yeast osmolyte system, *Int. Rev. Cytol.*, 215, 149–187, 2002.

50. Ishida, M., Aihara, M.S., Allen, R.D., and Fok, A.K., 1993, Osmoregulation in *Paramecium:* the locus of fluid segregation in the contractile vacuole complex, *J. Cell Sci.*, 106, 693–702, 1993.

51. Ishida, M., Aihara, M.S., Allen, R.D., and Fok, A.K., Acidification of the young phagosomes of *Paramecium* is mediated by proton pumps derived from the acidosomes, *Protoplasma*, 196, 12–20, 1997.

52. Ishida, M., Fok, A.K., Aihara, M.S., and Allen, R.D., Hyperosmotic stress leads to reversible dissociation of the proton pump-bearing tubules from the contractile vacuole complex in *Paramecium, J. Cell Sci.*, 109, 229–237, 1996.

53. Iwamoto, M., Allen, R.D., and Naitoh, Y., Hypo-osmotic or Ca^{2+}-rich external conditions trigger extra contractile vacuole complex generation in *Paramecium multimicronucleatum*, *J. Exp. Biol.*, 206, 4467–4473, 2003.

54. Iwamoto, M., Sugino, K., Allen, R.D., and Naitoh, Y., Cell volume control in *Paramecium*: factors that activate the control mechanisms, *J. Exp. Biol.*, 208, 523–537, 2005.

55. Kaneshiro, E.S., Dunham, P.B., and Holtz, G.G., Osmoregulation in a marine ciliate, *Miamiensis avidus*. I. Regulation of inorganic ions and water, *Biol. Bull.*, 136, 63–75, 1969.

56. Kaneshiro, E.S., Holtz, G.G., and Dunham, P.B., Osmoregulation in a marine ciliate, *Miamiensis avidus*. II. Regulation of intracellular free amino acids, *Biol. Bull.*, 137, 161–169, 1969.

57. King, R.L., The contractile vacuole of *Paramecium multimicronucleata*, *J. Morph.*, 58, 555–571, 1935.

58. Kirk, K. Swelling-activated organic osmolyte channels, *J. Membr. Biol.*, 158, 1–16, 1997.

59. Kirk, K. and Strange, K., Functional properties and physiological roles of organic solute channels, *Annu. Rev. Physiol.*, 60, 719–739, 1998.

60. Kissmehl, R., Froissard, M., Plattner, H., Momayezi, M., and Cohen, J., NSF regulates membrane traffic along multiple pathways in *Paramecium, J. Cell Sci.*, 115, 3935–3946, 2002.

61. Kissmehl, R, Schilde, C., Wassmer, T., Danzer, C., Nuehse, K., Lutter, K., and Plattner, H., Molecular identification of 26 syntaxin genes and their assignment to the different trafficking pathways in *Paramecium*, *Traffic*, 8, 523–542, 2007.

62. Kitching, J.A., Contractile vacuoles, ionic regulation and excretion. In *Research in Protozoology*, Vol. 1, Chen, T.-T., Ed., Pergamon Press, Oxford, 1967, pp. 305–336.

63. Klein, R.L., Effects of active transport inhibitors on K movements in *Acanthamoeba* sp., *Exp. Cell Res.*, 34, 231–238, 1964.

64. Knetsch, M.L.W., Schäfers, N., Horstmann, H., and Manstein, D.J., The *Dictyostelium* Bcr/Abr-related protein DRG regulates both Rac- and Rab-dependent pathways, *EMBO J.*, 20, 1620–1629, 2001.

65. Lang. F., Bush, G.L., and Volk, H., The diversity of volume regulatory mechanisms, *Cell. Physiol. Biochem.*, 8, 1–45, 1998.

66. Lefkir, Y., de Chassey, B., Dubois, A., Bogdanovic, A., Brady, R.J., Destaing, O., Bruckert, F., O'Halloran, T.J., Cosson, P., and Letourneur, F., The AP-1 clathrin-adaptor is required for lysosomal enzymes sorting and biogenesis of the contractile vacuole complex in *Dictyostelium* cells, *Mol. Biol. Cell*, 14, 1835–1851, 2003.

67. Linder, J.C. and Staehelin, L.A., A novel model for fluid secretion by the trypanosomatid contractile vacuole apparatus, *J. Cell Biol.*, 83, 371–382, 1979.

68. Liu, T. and Clarke, M., The vacuolar proton pump of *Dictyostelium discoideum*: molecular cloning and analysis of the 100 kDa subunit, *J. Cell Sci.*, 109, 1041–1051, 1996.

69. Luykx, R., Hoppenrath, M., and Robinson, D.G., Structure and behavior of contractile vacuoles in *Chlamydomonas reinhardtii*, *Protoplasma*, 198, 73–84, 1997.

70. Lynn, D.H., Dimensionality and contractile vacuole function in ciliated protozoa, *J. Exp. Zool.*, 223, 219–229, 1982.

71. Marchesini, N., Ruiz, F.A., Vieira, M., and Docampo, R., Acidocalcisomes are functionally linked to the contractile vacuole of *Dictyostelium discoideum*, *J. Biol. Chem.*, 277, 8146–8153, 2002.

72. McKanna, J.A., Fine structure of the contractile vacuole pore in *Paramecium*, *J. Protozool.*, 20, 631–638, 1973.

73. McKanna, J.A., Permeability modulating membrane coats. I. Fine structure of fluid segregation organelles of peritrich contractile vacuoles, *J. Cell Biol.*, 63, 317–322, 1974.

74. McKanna, J.A., Fine structure of the fluid segregation organelles of *Paramecium* contractile vacuoles, *J. Ultrastruct. Res.*, 54, 1–10, 1976.

75. Miranda, K., Benchimol, M., Docampo, R., and de Souza, W., The fine structure of acidocalcisomes in *Trypanosoma cruzi*, *Parasitol. Res.*, 86, 373–384, 2000.

76. Mitchell, H.J. and Hardham, A.R., Characterisation of the water expulsion vacuole in *Phytophthora nicotianae* zoospores, *Protoplasma*, 206, 118–130, 1999.

77. Moniakis, J., Coukell, M.B., and Forer, A., Molecular cloning of an intracellular P-type ATPase from *Dictyostelium* that is up-regulated in calcium-adapted cells, *J. Biol. Chem.*, 270, 28276–28281, 1995.

78. Moniakis, J., Coukell, M.B., and Janiec, A., Involvement of the Ca^{2+}-ATPase PAT1 and the contractile vacuole in calcium regulation in *Dictyostelium discoideum*, *J. Cell Sci.*, 112, 405–414, 1999.

79. Montalvetti, A., Rohloff, P., and Docampo, R., A functional aquaporin co-localizes with the vacuolar proton pyrophosphatase to acidocalcisomes and the contractile vacuole complex of *Trypanosoma cruzi*, *J. Biol. Chem.*, 279, 38673–38682, 2004.

80. Morgan, A. and Burgoyne, R.D., Is NSF a fusion protein? *Trends Cell Biol.*, 5, 335–339, 1995.

81. Morris, E., Mechanosensitive ion channels in eukaryotic cells. In *Cell Physiology Source Book*, 3rd ed., Sperelakis, N., Ed., Academic Press, San Diego, CA, 2001, pp. 745–760.

82. Naitoh, Y., Real time measurement of the osmolarities of the cytosol and the contractile vacuole fluid in *Paramecium* by using a microcapillary osmometer, *Mol. Biol. Cell*, 13(Suppl.), 223a, 2002.

83. Naitoh, Y. and Eckert, R., Electrical properties of *Paramecium caudatum*: modification by bound and free cations, *Z. vergl. Physiol.*, 61, 427–452, 1968.

84. Naitoh, Y., Tominaga, T., Ishida, M., Fok, A.K., Aihara, M.S., and Allen, R.D., How does the contractile vacuole of *Paramecium* expel fluid? Modelling the expulsion mechanism, *J. Exp. Biol.*, 200, 713–721, 1997.

85. Naitoh, Y., Tominaga, T., and Allen, R.D., The contractile vacuole fluid discharge rate is determined by the vacuole size immediately before the start of discharge in *Paramecium multimicronucleatum*, *J. Exp. Biol.*, 200, 1737–1744, 1997.

86. Nishihara, E., Shimmen, T., and Sonobe, S., Functional characterization of contractile vacuole isolated from *Amoeba proteus*, *Cell Struct. Funct.*, 29, 85–90, 2004.

87. Nolta, K.V. and Steck, T.L., Isolation and initial characterization of the bipartite contractile vacuole complex from *Dictyostelium discoideum*, *J. Biol. Chem.*, 269, 2225–2233, 1994.

88. O'Halloran, T.J. and Anderson, R.G.W., Clathrin heavy chain is required for pinocytosis, the presence of large vacuoles, and development in *Dictyostelium*, *J. Cell Biol.*, 118, 1371–1377, 1992.

89. O'Neil, W. C., Physiological significance of volume regulatory transporters, *Am. J. Physiol.*, 276, C995–C1011, 1999.

90. Pao, G.M., Wu, L.-F., Johnson, K.D., Höfte, H., Chrispeels, M.J., Sweet, G., Sandal, N.N., and Saier, Jr., M.H., Evolution of the MIP family of integral membrane transport proteins, *Mol. Microbiol.*, 5, 33–37, 1991.

91. Park, J.H., Schofield, P.J., and Edwards, M.R., The role of alanine in the acute response of *Giardia intestinalis* to hypo-osmotic shock, *Microbiology*, 141, 2455–2462, 1995.

92. Patterson, D.J., Contractile vacuoles and associated structures: their organization and function, *Biol. Rev.*, 55, 1–46, 1980.

93. Plattner, H., Habermann, A., Kissmehl, R., Klauke, N., Majoul, I., and Söling, H.-D., Differential distribution of calcium stores in *Paramecium* cells: occurrence of a subplasmalemmal store with a calsequestrin-like protein, *Eur. J. Cell Biol.*, 72, 297–306, 1997.

94. Prusch, R.D., Protozoan osmotic and ionic regulation. In *Transport of Ions and Water in Animals*, Gupta, B.L., Moreton, R.B., and Oschman, J.L., Eds., Academic Press, London, 1977, chap. 14.

95. Prusch, R.D. and Dunham, P.B., Contraction of isolated contractile vacuoles from *Amoeba proteus*, *J. Cell Biol.*, 46, 431–434, 1970.

96. Prusch, R.D. and Dunham, P.B., Ionic distribution in *Amoeba proteus*, *J. Exp. Biol.*, 56, 551–563, 1972.

97. Quiviger, B., de Chastellier, C., and Ryter, A., Cytochemical demonstration of alkaline phosphatase in the contractile vacuole of *Dictyostelium discoideum*, *J. Ultrastruct. Res.*, 62, 228–236, 1978.

98. Riddick, D.H., Contractile vacuole in the amoeba *Pelomyxa carolinensis*, *Am. J. Physiol.*, 215, 736–740, 1968.

99. Rivero, F., Illenberger, D., Somesh, B.P., Dislich, H., Adam, N., and Meyer, A.-K., Defects in cytokinesis, actin reorganization and the contractile vacuole in cells deficient in RhoGDI, *EMBO J.*, 21, 4539–4549, 2002.

100. Rohloff, P., Montalvetti, A., and Docampo, R., Acidocalcisomes and the contractile vacuole complex are involved in osmoregulation in *Trypanosoma cruzi*, *J. Biol. Chem.*, 279. 52270–52281, 2004.

101. Rohloff, P., Rodrigues, C.O., and Docampo, R., Regulatory volume decrease in *Trypanosoma cruzi* involves amino acid efflux and changes in intracellular calcium, *Mol. Biochem. Parasitol.*, 126, 219–230, 2003.

102. Ruiz, F.A., Marchesini, N., Seufferheld, M., Govindjee, and Docampo, R., The polyphosphate bodies of *Chlamydomonas reinhardtii* possess a proton-pumping pyrophosphatase and are similar to acidocalcisomes, *J. Biol. Chem.*, 276, 46196–46203, 2001.

103. Schilde, C., Wassmer, T., Mansfeld, J., Plattner, H., and Kissmehl, R., A multigene family encoding R-SNAREs in the ciliate *Paramecium tetraurelia*, *Traffic*, 7, 440–455, 2006.

104. Schmidt-Nielsen, B. and Schrauger, C.R., *Amoeba proteus*: studying the contractile vacuole by micropuncture, *Science*, 139, 606–607, 1963.

105. Scott, D.A., Docampo, R., Dvorak, J.A., Shi, S., and Leapman, R.D., *In situ* compositional analysis of acidocalcisomes in *Trypanosoma cruzi*, *J. Biol. Chem.*, 272, 28020–28029, 1997.

106. Söllner, T.H., Regulated exocytosis and SNARE function, *Mol. Memb. Biol.*, 20, 209–220, 2003.

107. Steck, T.L., Chiaraviglio, L., and Meredith, S., Osmotic homeostasis in *Dictyostelium discoideum*: excretion of amino acids and ingested solutes, *J. Eukaryot. Microbiol.*, 44, 503–510, 1997.

108. Stelly, N., Mauger, J.-P., Claret, M., and Adoutte, A., Cortical alveoli of *Paramecium*: a vast submembranous calcium storage compartment, *J. Cell Biol.*, 113, 103–112, 1991.

109. Stock, C., Allen, R.D., and Naitoh, Y., How external osmolarity affects the activity of the contractile vacuole complex, the cytosolic osmolarity and the water permeability of the plasma membrane in *Paramecium multimicronucleatum*, *J. Exp. Biol.*, 204, 291–304, 2001.

110. Stock, C., Grønlien, H.K., Allen, R.D., and Naitoh, Y., Osmoregulation in *Paramecium*: *in situ* ion gradients permit water to cascade through the cytosol to the contractile vacuole, *J. Cell Sci.*, 115, 2339–2348, 2002.

111. Stock, C., Grønlien, H.K., and Allen, R.D., The ionic composition of the contractile vacuole fluid of *Paramecium* mirrors ion transport across the plasma membrane, *Eur. J. Cell Biol.*, 81, 505–515, 2002.

112. Stoner, L.C. and Dunham, P.B., Regulation of cellular osmolarity and volume in *Tetrahymena*, *J. Exp. Biol.*, 53, 391–399, 1970.

113. Sugino, K., Tominaga, T., Allen, R.D., and Naitoh, Y., Electrical properties and fusion dynamics of *in vitro* membrane vesicles derived from separate parts of the contractile vacuole complex of *Paramecium multimicronucleatum*, *J. Exp. Biol.*, 208, 3957–3969, 2005.

114. Tani, T., Allen, R.D., and Naitoh, Y., Periodic tension development in the membrane of the *in vitro* contractile vacuole of *Paramecium*: modification by bisection, fusion and suction, *J. Exp. Biol.*, 203, 239–251, 2000.

115. Tani, T., Allen, R.D., and Naitoh, Y., Cellular membranes that undergo cyclic changes in tension: direct measurement of force generation by an *in vitro* contractile vacuole of *Paramecium multimicronucleatum*, *J. Cell Sci.*, 114, 785–795, 2001.

116. Tani, T., Tominaga, T., Allen, R.D., and Naitoh, Y., Development of periodic tension in the contractile vacuole complex membrane of *Paramecium* governs its membrane dynamics, *Cell Biol. Int.*, 26, 853–860, 2002.

117. Temesvari, L.A., Rodriquez-Paris, J.M., Bush, J.M., Zhang, L., and Cardelli, J.A., Involvement of the vacuolar proton-translocating ATPase in multiple steps of the endo-lysosomal system and the contractile vacuole system of *Dictyostelium discoideum*, *J. Cell Sci.*, 109, 1479–1495, 1996.

118. Tominaga, T., Allen, R.D., and Naitoh, Y., Electrophysiology of the *in situ* contractile vacuole complex of *Paramecium* reveals its membrane dynamics and electrogenic site during osmoregulatory activity, *J. Exp. Biol.*, 201, 451–460, 1998.

119. Tominaga, T., Allen, R.D., and Naitoh, Y., Cyclic changes in the tension of the contractile vacuole complex membrane control its exocytotic cycle, *J. Exp. Biol.*, 201, 2647–2658, 1998.

120. Tominaga, T., Naitoh, Y., and Allen, R.D., A key function of non-planar membranes and their associated microtubular ribbons in contractile vacuole membrane dynamics is revealed by electrophysiologically controlled fixation of *Paramecium*. *J. Cell Sci.*, 112, 3733–3745, 1999.

121. Urbina, J.A. and Docampo, R., Specific chemotherapy of Chagas disease: controversies and advance, *Trends Parasitol.*, 19, 495–501, 2003.

122. Vieira, L. L., Lafuente, E., Gamarro, F., and Cabantchik, Z.I., An amino acid channel activated by hypotonically induced swelling of *Leishmania major* promastigotes, *Biochem. J.*, 319, 691–697, 1996.

123. Vieira, L. L., Lafuente, E., Blum, J., and Cabantchik, Z.I., Modulation of the swelling-activated amino acid channel of *Leishmania major* promastigotes by protein kinases, *Mol. Biochem. Parasitol.*, 90, 449–461, 1997.

124. Wassmer, T., Froissard, M., Plattner, H., Kissmehl, R., and Cohen, J., The vacuolar proton-ATPase plays a major role in several membrane-bounded organelles in *Paramecium*, *J. Cell Sci.*, 118, 2813–2825, 2005.

125. Wassmer, T., Kissmehl, R., Cohen, J., and Plattner, H., Seventeen a-subunit isoforms of *Paramecium* V-ATPase provide high specialization in localization and function, *Mol. Biol. Cell*, 17, 917–930, 2006.

126. Wessel, D. and Robinson, D.G., Studies on the contractile vacuole of *Poterioochromonas malhamensis* Peterfi. I. The structure of the alveolate vesicles, *Eur. J. Cell Biol.*, 19, 60–66, 1979.

127. Wistow, G.J., Pisano, M.M., and Chepelinsky, A.B., Tandem sequence repeats in transmembrane channel proteins, *Trends Biochem. Sci.*, 16, 170–171, 1991.

128. Xie, Y., Coukell, M.B., and Gombos, Z., Antisense RNA inhibition of the putative vacuolar H^+-ATPase proteolipid of *Dictyostelium* reduces intracellular Ca^{2+} transport and cell viability, *J. Cell Sci.*, 109, 489–497, 1996.

129. Zeuthen, T., From contractile vacuole to leaky epithelia: coupling between salt and water fluxes in biological membranes, *Biochim. Biophys. Acta*, 1113, 229–258, 1992.

130. Zhu, Q. and Clarke, M., Association of calmodulin and an unconventional myosin with the contractile vacuole complex of *Dictyostelium discoideum*, *J. Cell Biol.*, 118, 347–358, 1992.

131. Zhu, Q., Liu, T., and Clarke, M., Calmodulin and the contractile vacuole complex in mitotic cells of *Dictyostelium discoideum*, *J. Cell Sci.*, 104, 1119–1127, 1993.

4 Osmotic and Ionic Regulation in Molluscs

Lewis Deaton

CONTENTS

I. MOLLUSCAN HABITATS

A. SALINITY AND THE DISTRIBUTION OF MOLLUSCS IN AQUATIC HABITATS

Living organisms occupy a wide range of aquatic habitats that differ greatly in water and solute chemistry. The primacy of salinity as a determinant of species richness in aquatic habitats is exemplified by Remane's curve (Figure 4.1).[169] The number of species in marine habitats and in freshwaters is large. Fewer species are found in brackish waters, and even fewer in hypersaline environments. Freshwaters can be "hard," with higher concentrations of divalent ions such as calcium, magnesium, and carbonate than monovalent ions such as sodium and chloride, or they can be "soft," with higher concentrations of Na^+ and Cl^- than Ca^{2+} and CO_3^{2-}. Hypersaline waters that support life range from marine salinity (35‰) to 300‰. This diversity of habitats has resulted in the evolution of a variety of mechanisms involved in the maintenance of cellular and organismal salt and water balance in aquatic organisms. Molluscs are found in freshwater, brackish waters, seawater, and hypersaline waters; the numbers of species in the phylum found in each habitat conforms to Remane's curve.[41]

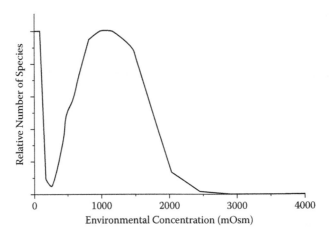

FIGURE 4.1 The relative number of species in aquatic habitats as a function of the osmotic concentration of the environment; 1100 mOsm is equivalent to 35‰, or oceanic salinity. (Adapted from Remane, A. and Schlieper, K., *The Biology of Brackish Water*, Wiley Interscience, New York, 1971.)

B. TERRESTRIAL MOLLUSCS

Gastropods have invaded terrestrial habitats and are widely distributed. Some snails are amphibious, moving between aquatic and terrestrial habitats. Additionally, a few species of bivalves that occur in high intertidal habitats may be out of the water for longer periods of time than they are submerged.[63] These bivalve species are at least semiterrestrial and can breathe air with surprising facility.[36,136]

II. SALT AND WATER BALANCE

In all animals, the cytoplasm and the extracellular fluid are in osmotic equilibrium; however, the ionic composition of the two media are dissimilar in that the cells of all animals have relatively high concentrations of K^+ and low concentrations of Na^+. This ratio is (with a few exceptions) reversed in the extracellular fluid. The low permeability of cell membranes to Na^+ coupled with the high extracellular concentration of Na^+ results in a Donnan effect in the extracellular fluid. This balances the Donnan effect inside the cell due to cytoplasmic proteins that are negatively charged at cytoplasmic pH and cannot cross the plasma membrane. As long as the diffusive movements of Na^+ and K^+ across the plasma membrane are counterbalanced, the cell will maintain its volume and structural integrity. The membrane-bound protein Na^+,K^+-ATPase compensates for the diffusive influx of Na^+ and efflux of K^+; the function of this enzyme represents a very large portion of the energy budget of all cells.[1]

Any change in the osmotic concentration of the extracellular fluid will perturb the diffusion gradients for water and ions across the cell membrane, with deleterious consequences for cellular function. Osmotic regulation, then, has two components: (1) maintenance of cellular water content or volume and (2) maintenance of the ionic composition of the cytoplasm and the extracellular fluid. Kirschner[110] termed the physiological mechanisms involved in osmotic and ionic regulation as either *evasive* or *compensatory*. The former act to minimize the diffusion gradients at exchange surfaces (i.e., cell membranes, body wall); the latter balance diffusive movements of water and ions by transport in the opposite direction. The diversity of habitats occupied by molluscs and the phylogenetic diversity of invasion of dilute and terrestrial habitats in the phylum have led to extensive research on the osmoregulatory physiology of the group.

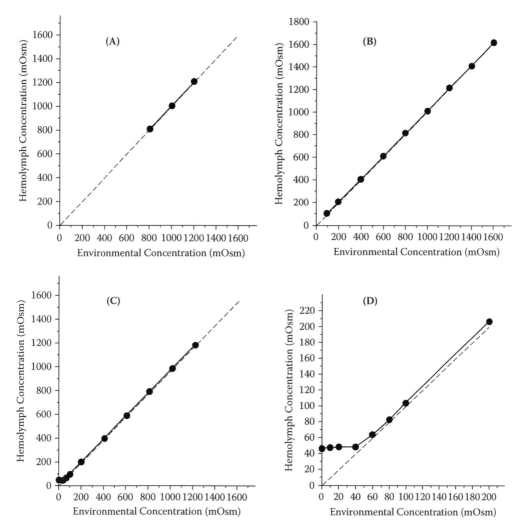

FIGURE 4.2 The relationship between the osmotic concentration of the hemolymph and the osmotic concentration of the environment for typical aquatic molluscs: (A) stenohaline osmotic conformer; (B) euryhaline osmotic conformer; (C) oligohaline animal; (D) freshwater bivalve. The dotted line is the isosmotic line.

A. Marine Osmoconformers

In most marine molluscs, the osmotic concentration of the hemolymph is roughly equal to that of the ambient seawater.[146,194] Osmoconformers that can tolerate only a very narrow range of external salinities are termed *stenohaline*, and species with a wider range of salinity tolerance are *euryhaline* (Figure 4.2A and B). Within the phylum Mollusca, entire classes, such as the cephalopods, solenogasters, monoplacophorans, polyplacophorans, and aplacophorans, contain only marine stenohaline species, whereas other classes (e.g., Bivalvia, Gastropoda) have numerous euryhaline species. The range of salinity tolerance of osmoconforming molluscs can be remarkable.[147,159] In these animals, any change in the external osmotic concentration results in a similar change in the osmotic concentration of the extracellular fluids (Figure 4.2). The animals are also generally isoionic to the external medium. The concentrations of Mg^{2+} and SO_4^{2-} in the hemolymph of osmoconforming molluscs are the same as that of seawater.[112,119,146,174,182] These animals produce a urine that is isosmotic to the hemolymph (see Table 4.6).

TABLE 4.1
Intracellular Concentrations of Ions in Molluscs

Species (Class)	Habitat	Cell	Intracellular Concentration (mM)			Ref.
			Na$^+$	K$^+$	Cl$^-$	
Mytilus edulis (B)	M	Muscle	79	152	94	Potts[162]
Mytilus edulis (B)	M	Nerve	105	206	—	Wilmer[209]
Eledone cirrhosa (C)	M	Muscle	33	167	55	Robertson[171]
Sepia officinalis (C)	M	Muscle	31	189	45	Robertson[171]
Loligo pealei (C)	M	Nerve	50	400	60	Hill et al.[91]
Elysia chlorotica (G)	M	Nerve	300	450	325	Quinn and Pierce[168]
Aplysia californica (G)	M	Nerve	67	232	12	Sato et al.[176]
Acmaea scutum (G)	M	Muscle	46	162	29	Weber and Dehnel[203]
Anodonta cygnea (B)	FW	Muscle	5	21	2	Potts[162]
Viviparus viviparus (G)	FW	Muscle	8	15	4	Little[117]

Note: B, bivalve; C, cephalopod; G, gastropod; M, marine; FW, freshwater.

Evasive strategies found in euryhaline animals include a reduction in the permeability of the body wall to water and ions.[110] Although differences in the permeability of the body wall to water exist among molluscs that inhabit marine, brackish, or freshwaters, they are not large enough to be adaptive.[166] In an osmoconforming animal, a reduction in the osmotic permeability of the epithelium to water is useful only during the initial adjustment to osmotic shock, when it would serve to decrease the rate of water movement across the body wall and give the cells more time to adjust. Indeed, exposure of isolated pieces of mantle tissue from the euryhaline mussel *Geukensia demissa* to 50% seawater for a few hours results in a decrease in diffusional permeability to water.[33] No change is observed in the permeability of the mantle to water and ions in animals acclimated to media of varying salinity for several weeks.[37] This is to be expected, because when the extracellular fluids and ambient medium come into equilibrium, no further osmotic movement of water occurs.

B. Ion Regulation in Osmoconformers: Storage of Ammonia

Because the density of living tissue is higher than that of seawater, maintenance of position in the water column may represent a large cost of energy for a pelagic animal. In at least ten families of squid, large amounts of ammonia are sequestered in the tissues to decrease the density of the animals.[200] The ammonia is stored in specialized coelomic compartments or in vesicles disbursed through the tissues. The pH within the ammonia storage areas is lower than that of the hemolymph, ensuring that the ammonia is in the form of the ammonium ion; concentrations can exceed 500 mM.[200] Molluscs that harbor large numbers of symbiotic algae, such as giant clams of the genus *Tridacna*, take up ammonia from the surrounding medium to provide nitrogen to the symbionts. Young animals with no symbiotic algal cells excrete ammonia, but as the clams grow and acquire a higher density of symbionts they take up ammonia from the medium during the day and excrete ammonia only at night, when the symbionts are not photosynthesizing.[66]

C. Volume Regulation in Osmoconformers: The Cellular Response

Studies have shown that the maximal concentration of K$^+$ in the cytoplasm of animal cells is about 200 to 300 mM.[110] This holds for a variety of cells from marine molluscs (Table 4.1). Other cations and anions increase the total osmotic pressure of the cytoplasm of marine animals to about 500 to

600 mOsm.[110] Because the cytoplasm is in osmotic equilibrium with extracellular fluid that has the same osmotic concentration as seawater (1100 mOsm), there is a seeming deficit in the osmotic concentration of the cytoplasm. In molluscs, this deficit is made up by a mixture of amino acids and quaternary amine compounds. The cells adjust to changes in the osmotic concentration of the extracellular fluids by either increasing or decreasing the size of this pool of organic osmolytes. The mix of organic osmolytes varies greatly among species and even among different populations within a species (Table 4.2),[113] but the amino acids that make up the bulk of the cytoplasmic pool are limited to only a few of the amino acids found in proteins (Table 4.2). Usually, taurine, alanine, glutamic acid, and glycine make up the majority of the pool. These amino acids are thought to be compatible solutes that, even in relatively high concentrations, have little effect on the tertiary structure of proteins.[190,210]

The response of isolated tissues and cells to changes in the ambient osmolality has been studied in a wide variety of molluscs. Cells exposed to increases in the osmotic concentration of the bathing media shrink, and if volume regulation occurs it is accompanied by a rapid increase in the concentration of cytoplasmic osmolytes. An extensive literature on the accumulation of organic osmolytes during regulatory volume increase is available, but data on increases in the concentrations of inorganic ions are few.

Baginski and Pierce[13] have shown that the increase in the cytoplasmic pool of amino acids during prolonged acute hyperosmotic stress is a highly organized process. Both the gills and ventricle of *Geukensia demissa* demonstrate a rapid initial increase in the concentration of alanine; the increase reaches a maximal value after 24 hours and then declines. The amount of proline increases during the first 6 days of hyperosmotic adjustment. A concurrent, but slower, increase in glycine peaks after 2 weeks. The cytoplasmic levels of taurine show a very gradual increase over 2 months. During this latter phase, the concentrations of first proline and then glycine decrease.

A thorough study of the biochemistry involved in the early phases of the increase in the cytoplasmic amino acid pool has been published.[17] The metabolic sources of alanine include synthesis from pyruvate[13] and protein catabolism followed by transamination.[34,79] The slow rate of taurine accumulation is probably a function of limited synthetic capacity, but the biochemical pathways involved in the synthesis of taurine in molluscs have not been studied. Quaternary amines, such as glycine betaine and proline betaine, have also been shown to increase in the tissues of some molluscs during hyperosmotic stress.[39,45,152,158] Whether this response is a general one among molluscs is as yet unknown; relatively few species have been studied. Betaine has been found in cephalopod tissues.[171] Most of the enzymes involved in increasing the size of the amino acid pool in molluscs are localized in the mitochondria, suggesting that these mechanisms may be compartmentalized and therefore regulated in concert.[17] Studies have shown that the enzymes involved in the production of glycine betaine in oysters are also in the mitochondria.[158]

Cells exposed to a decrease in the osmotic concentration of the ambient medium initially swell due to an influx of water; in most cells, a regulatory volume decrease occurs. The mechanism is a reduction in the osmotic concentration of the cytoplasm accomplished by the release of osmolytes to the extracelluar fluid. The osmolytes released from the cells include inorganic ions and the constituents of the cytoplasmic pool of organic molecules.[4,168,189,209]

Although the release of amino acids from isolated tissues is often equated with volume regulation, studies in which the release of osmolytes and actual changes in cell volume have been measured are scarce. Studies of the red blood cells from the clam *Noetia ponderosa* have measured both volume regulation and changes in osmolytes, resulting in a model for the control of the release of cytoplasmic amino acids during hyperosmotic volume regulation.[4,5] Exposure of these cells to hypoosmotic media results in an influx of Ca^{2+}, possibly through a stretch-activated channel.[151] In *N. pondersosa* red cells, the efflux of amino acids and volume regulation are inhibited by phenothiazine inhibitors of calmodulin action.[151] Calmodulin is present in these cells,[154] and hypoosmotic stress induces the phosphorylation of plasma membrane proteins.[148] Taken together, these results suggest the following model: Cellular swelling in response to hypoosmotic stress activates a

TABLE 4.2
Primary Constituents of the Cytoplasmic Amino Acid Pool in Marine Molluscs

Mollusc	π_{ext}	Tissue	Amino Acids (% of Total)	Ref.
Bivalves				
Anadara trapezia	1100	Adductor	Tau (31), Gly (31)	Ivanovici et al.[99]
Arca umbonata	1100	Soft tissue	Tau (79)	Simpson et al.[188]
Noetia ponderosa	840	Adductor	Gly (25), Asp (21)	Amende and Pierce[3]
	840	Foot	Tau (62)	Amende and Pierce[3]
	840	Gill	Tau (71)	Amende and Pierce[3]
Mytilus edulis	1180	Adductor	Gly (52), Pro (18)	Gilles[75]
	1100	Adductor	Tau (34), Gly (29)	Zachariassen et al.[211]
	1060	Adductor	Tau (46), Gly (35)	Shumway et al.[185]
	500	Adductor	Tau (64)	Deaton et al.[43]
Mytilus californianus	980	Gill	Tau (72)	Silva and Wright[186]
Modiolus modiolus	1060	Adductor	Tau (51), Gly (31)	Shumway et al.[185]
Modiolus squamosus	730	Mantle	Tau (79)	Pierce[146]
Geukensia demissa	1190	Mantle	Ala (35), Tau (19)	Pierce[146]
	1160	Ventricle	Tau (57), Gly (23)	Baginski and Pierce[12]
	1090	Gill	Tau (66)	Neufeld and Wright[141]
Saccostrea commercialis	1100	Adductor	Tau (32), β-Ala (23)	Ivanovici et al.[99]
Crassostrea virginica	1000	Mantle	Tau (64)	Heavers and Hammen[86]
Chesapeake Bay	920	Adductor	Ala (60)	Pierce et al.[157]
Atlantic	920	Adductor	Tau (50), Ala (19)	Pierce et al.[157]
Crassostrea gigas	1060	Adductor	Gly (38), Tau (26)	Shumway et al.[185]
Chlamys opercularis	1060	Adductor	Gly (72)	Shumway et al.[185]
Glycymeris glycymeris	1180	Adductor	Tau (56), Asp (11)	Gilles[75]
Mya arenaria	630	Adductor	Ala (37), Gly (35)	DuPaul and Webb[64]
	630	Adductor	Gly (54), Tau (19)	Shumway et al.[185]
	1000	Gill	Gly (42), Tau (17)	DuPaul and Webb[64]
Scrobicularia plana	1060	Adductor	Ala (38), Gly (28)	Shumway et al.[185]
Cardium edule	1060	Adductor	Tau (42), Gly (22)	Shumway et al.[185]
Mercenaria mercenaria	1060	Adductor	Tau (39), Gly (31)	Shumway et al.[185]
Macoma balthica	230	Soft tissue	Ala (36), Glu (13)	Sokolowski et al.[191]
Gastropods				
Pyrazus ebenius	1100	Foot	Tau (25), Ala (22)	Ivanovici et al.[99]
Thais hemastoma	990	Foot	Tau (75)	Kapper et al.[104]
Amphibola crenata	1070	Foot	Gly (14), Glu (12)	Shumway and Freeman[184]
Nassarius obsoletus	1000	Digestive gland	Tau (52), Glu (13)	Kasschau[105]
Purpura lapillus	1100	Foot	Tau (53), Ala (11)	Hoyeaux et al.[97]
Patella vulgata	1100	Foot	Tau (74)	Hoyeaux et al.[97]
Elysia chlorotica	920	Whole animal	Glu (61)	Pierce et al.[159]
Busycon perversum	1100	Soft tissue	Ala (31), Tau (25)	Simpson et al.[188]
Fasciolara distans	1100	Soft tissue	Tau (50), Gly (19)	Simpson et al.[188]
Siphonaria lineolata	1100	Soft tissue	Ala (31), Tau (28)	Simpson et al.[188]
Polynices duplicata	1100	Soft tissue	Tau (65)	Simpson et al.[188]
Oliva sayana	1100	Soft tissue	Tau (42), Ala (21)	Simpson et al.[188]
Hydrobia ulvae	1100	Soft tissue	Gly (37), β-Ala (27)	Negus[140]
Chitons				
Acanthochitona discrepans	1180	Foot	Tau (64)	Gilles[75]
Cephalopods				
Loliguncula brevis	1100	Soft tissue	Tau (59), Gly (21)	Simpson et al.[188]
Lithophaga bisulcata	1100	Soft tissue	Gly (50), Tau (46)	Simpson et al.[188]
Architeuthis sp.	1100	Muscle	Glu (13), Asp (10)	Rosa et al.[172]

calcium-selective. stretch-activated channel in the plasma membrane. The resulting influx of Ca^{2+} binds to calmodulin, and the calcium–calmodulin complex activates a kinase that phosphorylates targets in the plasma membrane. The phosphorylation of these membrane proteins then activates an amino acid efflux pathway, probably a nonselective amino acid–anion channel.

This model, however, is clearly not general to all molluscan cells. Both ventricles from *Geukensia demissa* and *Noetia* red cells have been subjected to identical experiments investigating the control of amino acid release. Differences between the ventricles and the red cells in response to these treatments are summarized in Table 4.3. Calmodulin seems to be involved in the release of amino acids from *Noetia* red cells, but it does not appear to have a role in this process in *Geukensia* ventricles. Also, although phorbol esters increase the amino acid release from ventricles, they have no effect on amino acid release from the red cells but do increase the release of K^+. This result suggests that a mechanism that involves the activation of protein kinase C is involved in the release of amino acids from the ventricles of *Geukensia*. Consistent with this idea is the finding that the amount of diacylglycerol increases in ventricles exposed to hypoosmotic seawater; the turnover of inositol-1,4,5-triphosphate, however, is unaffected.[38] Phorbol esters also potentiate the release of amino acids from gills of *Geukensia* and ventricles of the clam *Mercenaria mercenaria* exposed to hypoosmotic seawater.[35,38]

The physiological function of the ventricle in bivalves would seem to require a mechanism for control of volume regulation that is different from the *Noetia* red cell model. The ventricle is composed of muscle cells that are rhythmically active; some of the Ca^{2+} that activates contraction enters the cell with the action potential.[40] In these cells, the mechanical activity would activate stretch-activated channels, and the action potential would initiate volume regulation with each contraction of the heart, even in isosmotic conditions.

The release of amino acids from ventricles of the clam *Mercenaria mercenaria* is increased by the neurotransmitter 5-hydroxytryptamine and the molluscan neuropeptide FMRFamide.[35] In addition, a neuropeptide isolated from neurons in the abdominal ganglion of the sea hare, *Aplysia californica*, has been shown to affect the water content of the animal.[204] These data suggest that the central nervous system is involved in adjustment to osmotic stress. In summary, the control of volume regulation in molluscs has been investigated in detail in only a few types of cells (and only in bivalves); significant differences have been observed; and nothing is known about the variety of mechanisms that may be present in other types of cells or in other molluscan classes.

TABLE 4.3
Response of *Noetia* Red Cells and *Geukensia* Ventricles to Experimental Treatments

	Result		
Treatment	***Noetia* Red Cells**	***Geukensia* Ventricles**	**Refs.**
Hypoosmotic Ca^{2+} free	Decreased FAA release	Increased FAA release	Amende and Pierce,[5] Pierce and Greenberg[149]
Hypoosmotic high Ca^{2+}	Increased FAA release	Decreased FAA release	Amende and Pierce,[5] Pierce and Greenberg[149]
Hypoosmotic + verapamil	Decreased FAA release	No effect	Amende and Pierce,[5] Deaton (unpublished)
Hypoosmotic + Co^{2+}	Increased FAA release	No effect	Amende and Pierce,[5] Deaton (unpublished)
Hypoosmotic + phorbols	Increased K^+ release	Increased FAA release	Pierce et al.,[155] Deaton[38]
Hypoosmotic + dintrophenol	Decreased FAA release	Increased FAA release	Amende and Pierce,[5] Pierce and Greenberg[149]
Hypoosmotic + trifluoperazine	Decreased FAA release	No effect	Pierce et al.,[155] Deaton (unpublished)

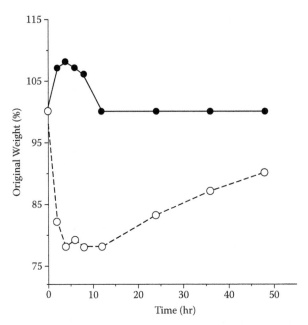

FIGURE 4.3 Volume regulation in *Mytilus edulis*. Animals (with shells held open with pegs) were transferred from 600 mOsm to 300 mOsm (filled circles, solid line) or from 300 mOsm to 600 mOsm (open circles, dotted line) at time 0. (Adapted from Gainey, L.F., *Comp. Biochem. Physiol.*, 87A, 151–156, 1987.)

D. VOLUME REGULATION IN OSMOCONFORMERS: THE ORGANISMAL RESPONSE

An acute change in the ambient salinity presents an osmoconforming animal with an osmotic gradient between the ambient medium and the extracellular fluid. If it has a shell that closes tightly or can be sealed with an operculum, the animal can isolate its soft tissues from the environment by retreating into the shell. This behavior may shield the animal from environmental changes that are short lived, such as those caused by tidal cycles in an estuary. When the change in ambient osmotic concentration is prolonged, despite the capacity for anaerobic metabolism in many molluscs,[49] the animal must eventually emerge or open to feed, to release nitrogenous waste, and to exchange respiratory gases. As the extracellular fluid comes to osmotic equilibrium with the new ambient osmotic concentration, the cells of the animal gain or lose water by osmosis; this results in weight gain or loss by the organism. Molluscs can control the rate of exposure to the altered ambient medium by changing the gape of the shell (bivalve) or degree of withdrawal (gastropod), the diameter of the inhalant and exhalant siphons, and the rate of beating of the cilia on the gills that propel water through the mantle cavity.[29,30,147]

Animals exposed to an increase in the ambient osmotic concentration lose water (Figure 4.3). Depending on the species and the magnitude of the osmotic stress, the animal may or may not regulate its water content.[69,147,194] In animals that are able to regain volume, the mechanism is an increase in the cytoplasmic concentration of osmolytes. The source of organic osmolytes for hyperosmotic volume regulation is an area of some controversy. Studies have shown conclusively that molluscs can take up amino acids from seawater.[126] The critical question is whether the concentration of these molecules in the environment is high enough for the molluscs to accumulate enough of them to contribute to volume regulation. The measurement of amino acids in natural waters is not a straightforward proposition; published estimates can vary over several orders of magnitude, and those on the high end are probably erroneous.[65,68] At present, the role of uptake of amino acids from the medium in volume regulation is unclear, but the evidence suggests that it is negligible.[84] Some studies have suggested that amino acids released into the hemolymph by cells exposed to hypoosmotic stress may be transported back into the cells during a subsequent hyper-osmotic stress, as may occur during an estuarine tidal cycle.[193] Whatever the contributions of these uptake mechanisms are, it is clear that the cells of molluscs exposed to hyperosmotic media produce amino acids by synthesis and by catabolism of proteins.[17]

Animals exposed to a decrease in the ambient osmotic concentration gain water (Figure 4.3). Over time, the volume of the animal returns toward the initial, preosmotic stress value. Both the time course and the degree of completion of this volume regulatory response vary among species and depend on the magnitude of the osmotic shock.[69,147,184] Hypoosomotic volume regulation is accomplished by the release of cytoplasmic osmolytes into the blood and the excretion of osmolytes by the animal. This brings the three compartments (cytoplasm, extracellular fluid, and ambient medium) into osmotic equilibrium.

These osmolyte molecules are not excreted by the animal after they are released from the cells. The rate of excretion of amino acids in molluscs exposed to reduced osmotic concentrations does not show a marked increase.[14,87,124,192] Instead, the rate of excretion of ammonia increases. This suggests that the amino acids released into the extracellular fluid are taken up and deaminated in some tissue in the animals. Bartberger and Pierce[14] have shown that the amino acid content of the mantle of the mussel *Geukensia demissa* does not change when the animals are exposed to decreased external salinity. A taurine transporter in the epidermal tissues of the mussel *Mytilus galloprovincialis* has been cloned; the protein is induced in the mantle by exposure to hypoosmotic media.[94] These observations are consistent with the notion that the amino acids released into the extracellular fluid by other tissues are taken up by the mantle and deaminated; the carbon skeletons of the amino acids are presumably polymerized and conserved while the ammonia is excreted.

The existence of a relationship between the salinity tolerance and capacity for regulation of volume in animals was proposed 50 years ago.[181] In the intervening years, this hypothesis has been explored in a variety of molluscs. Lange[116] found that tissue water content changed less in euryhaline species relative to stenohaline species exposed to a variety of external salinities. The euryhaline clam *Polymesoda caroliniana* has a greater capacity for volume regulation than the stenohaline species *Corbicula manilensis*,[69] and the tissue amino acid pool is larger in the euryhaline mussel *Geukensia demissa* than in the stenohaline mussel *Modiolus americanus*.[146] Gainey and Greenberg[73] linked the capacity for changes in the amino acid pool to salinity tolerance, but more recent studies suggest the available data do not support this hypothesis.[41,72]

E. Volume Regulation in Osmoconformers

Given the complexity of volume regulation at both the cellular and organismal level, it is instructive to ask whether conditions in the natural environment are ever associated with the responses evoked in the typical laboratory experiments described above. A few studies have subjected intact animals or isolated gills to fluctuating changes in ambient osmolality with the periodicity and magnitude of those that are typical for a tidal cycle in an estuarine habitat. The results suggest that when intact animals are exposed to short-term changes in the ambient osmolality, no volume regulatory response occurs.[182,183,185] At the cellular level, results vary among cell types. Ventricles of *Geukensia demissa* do initiate volume regulation in response to a moderate change (100% to 60% seawater) in the ambient medium, but cells in the gills and mantle, as well as circulating hemocytes, do not.[141] The failure of some tissues to respond to a change in the osmolality of the medium may be due to the energetic cost of processing the amino acids that would be released.[84,151]

Many studies of the responses of molluscs to both hypoosmotic and hyperosmotic stress have shown that the volume regulatory responses of bivalves to increases in salinity and to decreases in salinity are not symmetrical.[26,69,71,131,146,181,193] When *Mytilus edulis* was transferred from 600 to 300 mOsm, the osmotic concentration of the hemolymph decreased at a rate of about 10 mOsm/hr.[71] In contrast, the rate of increase in the hemolymph osmotic concentration in mussels transferred from 300 mOsm to 600 mOsm was about 200 mOsm/hr. The volume regulatory response of the animals is also asymmetrical. Transfer to hyposomotic media results in a smaller initial increase in weight and a more rapid return to the original weight than does a hyperosmotic shock of the same magnitude (Figure 4.3). The comparatively slower rate of equilibration between hemolymph and ambient in animals exposed to hypoosmotic rather than hyperosmotic media may be due to a reduction in the permeability of the body wall to water, a decrease in movement of water through the mantle cavity (due to inhibition of the lateral cilia of the gills), and an increase in urine production.[71] Acute changes in the ambient salinity depress the activity of the lateral cilia of isolated gills, but the effects on the cilia are similar for hypoosmotic and hyperosmotic stresses.[143] A comprehensive explanation of the rapid equilibration between medium and extracellular fluids during hyperosmotic stress awaits further research. The slower volume regulatory response is probably due to limits on the rate of accumulation of amino acids in the cytoplasm of the cells of the animals.

Animals that are osmotic and ionic conformers have little need for the kidney to be involved in salt and water balance. Curiously, the rates of urine production in marine molluscs (e.g., *Octopus*, *Haliotis*) are of the same order of magnitude as those found in freshwater bivalves (see Table 4.6). The purpose of such high rates of urine production in marine animals is unclear, but they may be necessary for the excretion of wastes such as heavy metals.

F. Oligohaline Molluscs: Mixed Conformity and Hyperregulation

A select group of molluscs that includes both bivalves and gastropods inhabits brackish waters. These animals cannot live in marine salinities but are osmoconformers in concentrated brackish water and hyperosmotic regulators in dilute brackish waters (Figure 4.2C). These species are capable of regulating volume when exposed to a modest increase or decrease in the ambient osmotic concentration.[59,87,88,129] The environmental salinity that initiates the transition from conformity to hyperregulation is 60 to 70 mOsm and 125 to 150 mOsm for bivalves and gastropods, respectively (Table 4.4). The blood osmotic concentration of oligohaline species acclimated to freshwater is not different from that of freshwater species. The available data suggest that a blood osmotic concentration of about 40 mOsm may be the low limit for molluscs in freshwaters.

In an animal that is maintaining the osmotic concentration of the hemolymph above that of the ambient medium, the influx of water and efflux of ions are continuous. The lower osmotic concentration of the hemolymph of oligohaline bivalves relative to that of oligohaline gastropods when the animals are in very dilute media (Table 4.4) means that diffusive movements of water and ions between the extracellular fluid and the dilute medium are comparatively lower in the bivalves. The filter-feeding habit of bivalves may expose large surface areas of permeable tissues to more water than the respiratory currents in nonpulmonate freshwater snails. Data on the relative size of the respiratory surfaces in gastropods and molluscs provide some support for this idea. Ghiretti[74] cited data for the surface area of the gills in a variety of molluscs. Values for gastropods range between 7 and 9.3 cm^2/g wet weight. The values for bivalve gills are 13.5 cm^2/g. Numerous values are available for ventilation rates among bivalves but there are no comparable data for gastropods. If this hypothesis is correct, gastropods in dilute media would be expected to have lower ventilation rates than bivalves.

All hyperregulating animals must produce a large volume of urine to maintain volume and be capable of taking up ions from the medium by active transport. Because urine production increases the loss of ions, these oligohaline animals presumably produce urine that is hypoosmotic to the extracellular fluids; however, data pertinent to this question are rare (see data for *Assiminea grayana* in Table 4.6). Measurements of sodium and chloride fluxes have not been done in any oligohaline species, but data are available for *Corbicula fluminea*, a freshwater species that can tolerate dilute brackish water.[31] If an animal is in a steady state, comparison of the equilibrium potential for an ion calculated with the Nernst equation to the measured electrical potential across the body wall can be used to assess whether or not the ion is in equilibrium across the body wall. If a substantial difference is found, the ion is being moved across the body wall by active

TABLE 4.4
Osmoregulation in Selected Oligohaline and Freshwater Molluscs

Species	Class	π_{int} in FW	π_{ext} Break	Ref.
Oligohaline animals				
Polymesoda caroliniana	B	48	60	Deaton[31]
Rangia cuneata	B	41	60	Deaton[31]
Mytilopsis leucophaeta	B	40	70	Deaton et al.[42]
Melanopsis trifasciata	G	120	150	Bedford[15]
Potamopyrgus jenkinsi	G	125	125	Duncan[62]
Assiminea grayana	G	180	150	Little and Andrews[123]
Freshwater animals				
Corbicula fluminea	B	53	70	Deaton[31]
Limnoperna fortunei	B	40	70	Deaton et al.[42]
Elliptio lanceolata	B	42	40	Gainey and Greenberg[73]
Ligumia subrostrata	B	47	50	Dietz and Branton[55]
Lampsilis teres	B	50	50	Jordan and Deaton[101]
Lampsilis claibornensis	B	46	50	Deaton[31]
Pomacea bridgesi	G	100	100	Jordan and Deaton[101]
Lymnaea stagnalis	G	95	100	De With[46]
Viviparus viviparus	G	80	95	Little[117]

Note: The external osmotic concentration above which the animal is an osmotic conformer and below which it is a hyperosmotic regulator is the π_{ext} break. The osmotic concentration of the hemolymph of animals acclimated to freshwater is the π_{int}. Osmotic concentrations are in mOsm. B, bivalve; G, gastropod.

transport.[110] The transepithelial potential of the clam *Corbicula fluminea* acclimated to artificial freshwater is –7 mV, whereas the calculated equilibrium potentials for Na^+ and Cl^- are, respectively, –89 and –74 mV.[133] This result indicates that neither ion is in equilibrium across the body wall. The Na^+,K^+-ATPase activity of the mantle and kidney tissue, but not gill, of several oligohaline bivalves increases when the animals are acclimated to dilute media below the breakpoint separating osmotic conformity from osmotic regulation.[32,175] Oligohaline molluscs can live in freshwater but cannot reproduce in this habitat. Unlike freshwater molluscs, these animals are unable to survive long-term exposure to deionized water, presumably because the capacity of their ion uptake mechanisms is too low. Because they are capable of volume regulation over a relatively wide range of salinity (as osmoconformers) and ionic regulation in very dilute media, these species are often considered transitional forms on the evolutionary path from marine habitats to the invasion of freshwaters.[41,73]

G. FRESHWATER MOLLUSCS

Both bivalves and gastropods are well represented in freshwater habitats. Independent invasions of freshwaters by a variety of bivalve and gastropod taxa have occurred numerous times.[2,40,135] In general, the osmotic concentration of the hemolymph of freshwater bivalves is lower than that of freshwater gastropods (Table 4.5). In many of the bivalves, the concentration of bicarbonate ion in the hemolymph is equal to or higher than that of chloride (Table 4.5). Molluscs that are fully adapted to freshwater, like all freshwater animals, produce urine that has a lower osmotic concentration than that of the hemolymph, although the urine/blood ratio is considerably higher than that found in freshwater fishes (Table 4.6).[110] The rates of urine production in freshwater gastropods are generally higher than those reported for freshwater bivalves; this is consistent with the more concentrated hemolymph of gastropods (Table 4.6).

TABLE 4.5
Ionic Composition of the Hemolymph of Freshwater (FW) and Terrestrial (T) Molluscs

Species	Habitat	π	Na⁺	K⁺	Ca²⁺	Cl⁻	HCO₃⁻	Ref.
Bivalves								
Dreissena polymorpha	FW	36	11.5	0.5	5.2	14.5	5.1	Horohov et al.[93]
Corbicula fluminea	FW	53	26.7	1.0	5.2	24.3	2.5	Deaton[31]
Carunculina texasensis	FW	45	15.4	0.5	4.7	11.4	11.6	Dietz[51]
Anodonta grandis	FW	55	19.5	0.5	5.8	16.1	11.2	Dietz[51]
Anodonta cygnea	FW	42	15.6	0.5	8.4	11.7	14.6	Potts[161]
Anodonta woodiana	FW	45	15.8	0.5	—	13.7	—	Matsushima and Kado[130]
Ligumia substrostrata	FW	47	20.6	0.6	3.6	12.5	11.5	Dietz[51]
Lampsilis claibornensis	FW	46	27.1	0.9	3.2	11.7	6.4	Deaton[31]
Margaritifera margaritifera	FW	—	14.4	0.5	7.8	11.4	—	Chaisemartin[21]
Margaritifera hembeli	FW	39	14.6	0.3	5.2	9.3	11.9	Dietz[51]
Sphaerium transversum	FW	45	15.2	0.4	2.8	14.2	9.0	Dietz[53]
Gastropods								
Viviparus viviparus	FW	81	34.0	1.2	5.7	31.0	11.0	Little[117]
Pomacea depressa	FW	140	55.7	3.0	6.6	52.0	19.0	Little[120]
Pomacea lineata	FW	135	49.8	2.4	7.2	41.3	23.4	Little[120]
Theodoxusa fluviatilis	FW	105	45.0	2.2	2.3	32.8	11.3	Little[121]
Lymnaea stagnalis	FW	—	55.3	1.7	4.4	36.2	28.3	De With and Sminia[47]
Lymnaea trunculata	FW	137	49	2.4	8.3	32.1	18.4	Pullin[167]
Pila globosa	FW	—	126	19.3	30.7	191	—	Saxena[177]
Eutrochatella tankervillei	T	74	27	1.2	3.2	24	12.5	Little[121]
Helix pomacea	T	158	59.3	4.5	10.0	49.2	40	Wieser[206]
Arion ater	T	216	62	2.7	2.3	—	—	Roach[170]
Achatina fulica	T	212	65.6	3.3	10.7	72.2	13	Matsumoto et al.[128]
Strophocheilus oblongus	T	166	38	2.4	12.3	53	23.5	DeJorge et al.[44]
Tropidophora cuvierana	T	198	89	3.5	8.0	77	13.5	Rumsey[173]
Tropidophora ligata	T	294	127	5.7	12.2	124	—	Rumsey[173]
Tropidophora fulvescens	T	206	86.0	4.9	9.7	75.5	—	Rumsey[173]
Pomatais elegans	T	280	110	6.0	16.5	106	11.0	Rumsey[173]
Orthalicus undulatus	T	126	35	3.0	7.3	—	—	Burton[19]
Sphincterochila candidissima	T	134	53	3.1	7.7	—	—	Burton[19]
Eobania vermiculata	T	168	71	3.8	5.7	—	—	Burton[19]
Cepaea nemoralis	T	—	88	4.6	2.0	65.6	—	Trams et al.[198]
Maizamia wahlbergi	T	64	26	1.8	4.5	24.5	—	Andrews and Little[9]
Incidostoma impressus	T	80	30	1.2	3.7	24	—	Andrews and Little[9]
Poteria lineata	T	82	31	1.8	5.1	25	13.7	Andrews and Little[9]
Poteria yalluhsensis	T	86	36.8	2.4	8.3	27.3	11.7	Andrews and Little[9]

Freshwater molluscs also take up ions from the dilute medium against an electrochemical gradient. In the unionid mussel *Ligumia subrostrata*, the calculated equilibrium potentials for Na⁺ and Cl⁻ are, respectively, –85 and –65 mV.[133] The measured transepithelial potential is –15 mV, indicating uptake of both ions by an active mechanism.[133] The site of extrarenal uptake of ions is assumed to be the gills, and data comparing the rate of sodium uptake by isolated demibranchs and intact unionid mussels are consistent with this hypothesis; for example, the rate for two demibranchs from *Ligumia subrostrata* accounts for that of the whole animal (Table 4.7). In unionid mussels, the water channel epithelium of the gills contains groups of nonciliated cells that are packed with mitochondria but lack other common organelles (e.g., Golgi, endoplasmic reticulum); these cells also have extensive apical microvilli. This morphology is characteristic of

TABLE 4.6
Urine Production in Molluscs

Species	π_{ext}	Urine/Hemolymph Ratio	Urine Production	Refs.
Bivalves				
Anodonta cygnea	FW	0.67	1.9	Picken,[154] Potts[161]
Margaritana margaritifera	FW	0.09	3.4	Chaisemartin[21]
Hyridella australis	FW	0.60	—	Hiscock[92]
Gastropods				
Strombus gigas	1100	0.96	3	Little[119]
Nerita fulgurans	1100	1.01	—	Little[121]
Haliotis rufescens	1100	1.0	6–21	Harrison[81]
Hydrobia ulvae	500	1.0	—	Todd[197]
Turritella communis	1100	1.0	—	Avens and Sleigh[11]
Buccinium undulatum	1100	1.0	—	Avens and Sleigh[11]
Littorina saxatalis	1100	1.0	—	Avens and Sleigh[11]
Assiminea grayana	1150	0.96	—	Little and Andrews[123]
	FW	0.56	—	Little and Andrews[123]
Viviparus viviparus	FW	0.28	15	Little[118]
Viviparus malleatus	FW	—	42	Monk and Stewart[139]
Lymnaea peregra	FW	0.70	—	Picken[144]
Pomatia lineata	FW	0.23	5.7	Little[120]
Neritina latissima	FW	0.63	—	Little[121]
Potamopyrgus jenkensi	FW	0.83	—	Todd[197]
Achatina fulica	T	0.72	4.4	Martin et al.[127]
Pomatais elegans	T	0.98	—	Rumsey[173]
Tropidophora cuvieriana	T	1.02	—	Rumsey[173]
Tropidophora fulvescens	T	1.03	—	Rumsey[173]
Tropidophora ligata	T	1.10	—	Rumsey[173]
Parachondria angustae	T	1.00	—	Rumsey[173]
Licinia nuttii	T	0.97	—	Rumsey[173]
Annularia sp.	T	1.03	—	Rumsey[173]
Helix pomatia	T	—	10.3	Vorwohl[201]
Archachatina ventricosa	T	—	8.4	Vorwohl[201]
Cephalopods				
Octopus dofleini	1100	1.0	2.6	Harrison and Martin[82]

Note: Urine/hemolymph ratio is calculated from osmolality or Na^+ concentrations; π_{ext} is in mOsm; rates of urine production are μL per g wet weight per hr.

ion-transporting cells in a wide variety of animals and suggests that the mitochondria-rich cells in the water channels of the gill may be responsible for the uptake of ions.[106] The K_m values for the influx of ions are similar in bivalves and gastropods and not different from those for other freshwater animals.[110] Rates of uptake are also roughly similar for bivalves and gastropods; a few clams (*Sphaerium transversum, Corbicula fluminea, Dreissena polymorpha*) have higher rates (Table 4.7). The higher rates of ion flux in the latter two species may reflect their comparatively recent invasion of freshwater.[41]

The uptake of sodium and chloride ions occurs independently in unionid mussels, *Corbicula fluminea*, and freshwater pulmonates.[46,53,133] In *Dreissena polymorpha*, Cl^- uptake is dependent on Na^+.[93] The mechanism of sodium uptake in the unionid mussels involves exchange of H^+ for Na^+ and Na^+,K^+-ATPase. Chloride uptake has been associated with a Cl^-,HCO_3^--ATPase,[54,57] but the data

TABLE 4.7
Ion Influx in Freshwater Molluscs

Species	Ion	Maximum Rate (μmol per g dry weight per hr)	K_m (mM)	Ref.
Bivalves				
Carunculina texasensis	Na$^+$	1.3	0.1	Dietz[52]
Ligumia subrostrata	Na$^+$	2	0.1	Dietz[51]
	Cl$^-$	1	0.1	Dietz and Branton[56]
Ligumia subrostrata (isolated gill)	Cl$^-$	0.5	0.2	Dietz and Hagar[58]
Margaritana margaritifera	Na$^+$	67[b]	0.05	Chaisemartin et al.[22]
Corbicula fluminea	Na$^+$	13	0.05	McCorkle and Dietz[133]
Dreissena polymorpha	Na$^+$	22	—	Horohov et al.[93]
	Cl$^-$	25	—	Horohov et al.[93]
	Ca^{2+}	29	—	Horohov et al.[93]
Sphaerium transversum	Na$^+$	21	—	Dietz[53]
	Cl$^-$	9	—	Dietz[53]
Gastropods				
Lymnaea stagnalis	Na$^+$	3.6[a]	0.1	De With and van der Schors[48]
	Cl$^-$	2.4[a]	0.1	De With and van der Schors[48]
	Ca^{2+}	6.0[a]	0.3	Greenaway[78]
Biomphalria glabrata	Ca^{2+}	2.4[a]	0.3	Thomas and Lough[195]
Ancylastrum fluviatilis	Ca^{2+}	7.2[a]	0.07	Chaisemartin and Videaud[23]

[a] Wet weight converted to dry weight by a factor of 12.[199]
[b] Wet weight converted to dry weight assuming tissue hydration = 85%.[101]

supporting this mechanism are inconclusive. More recent models of the mechanism of Cl$^-$ uptake in freshwater animals implicate Cl$^-$/HCO$_3^-$ ion exchangers and V-type H$^+$-ATPases (see Chapter 3 in this volume). The activity of Na$^+$,K$^+$-ATPase in the gills of freshwater mussels is higher than that of the gills of oligohaline bivalves in freshwater.[57,175] Some species demonstrate evidence of a diurnal rhythm in the uptake of ions, with rates being higher during darkness.[77,134] The uptake of ions by freshwater bivalves is stimulated by serotonin and cAMP; serotonergic synapses are abundant in the gills.[60,61,93,179] Prostaglandins have been shown to inhibit the uptake of sodium ions in unionid mussels.[76] If kidney tissue from freshwater snails acclimated to dilute seawater is incubated in extracts of the visceral ganglion, the spaces between the epithelial cells and the basal infoldings of the cells expand; this morphological change mimics that induced by placing the animals in distilled water.[108] These observations suggest, again, that the components of osmotic regulation are integrated and controlled by messenger molecules released from the nervous system.

Freshwater molluscs have a limited tolerance for brackish water. Populations of freshwater species persist in, for example, dilute brackish water in the Baltic Sea,[169] but such instances are rare. Laboratory studies have shown that long-term exposure of freshwater molluscs to media more concentrated than about 200 mOsm is fatal (Table 4.8). The cells of freshwater molluscs accumulate amino acids in hyperosmotic media; the capacity of this mechanism seems to be lower in freshwater gastropods than in freshwater bivalves.[101] Direct comparison of the rates of increase in the levels of amino acids in isolated tissues from oligohaline and freshwater bivalves (Table 4.9) suggests that unionid mussels have a relatively limited ability to accumulate amino acids. Differences between an oligohaline and a freshwater species of *Corbicula* are small in either the initial rapid rate, when the largest increase is probably alanine, or a longer term rate during which increases in other amino acids occur. For reasons that are unclear, full adaptation to freshwater seems to be accompanied by a loss of tolerance for dilute brackish water.

TABLE 4.8
Upper Limits of Salinity Tolerance for Freshwater Molluscs

Species	Upper Limit (mOsm)	Ref.
Bivalves		
Lampsilis claibornensis	200	Deaton[31]
Lampsilis teres	200	Jordan and Deaton[101]
Corbicula leana	160	Kado and Murata[103]
Corbicula fluminea	400	Deaton[31]
Mytilopsis leucophaeta	200	Deaton et al.[42]
Limnoperna fortunei	200	Deaton et al.[42]
Dreissena polymorpha	100	Komendatov et al.[111]
Anodonta piscinalis	200	Komendatov et al.[111]
Gastropods		
Viviparus viviparus	200	Little[117]
Helisoma duryi	150	Khan and Saleuddin[107]
Pomacea bridgesi	200	Jordan and Deaton[101]

TABLE 4.9
Rates of Accumulation of Ninhydrin-Positive Substances in Isolated Foot Tissues from Selected Bivalves

Species	Habitat	Initial Rate (3 hr)	Cumulative Rate (9 hr)
Corbicula japonica	O	6.3	2.5
Corbicula leana	FW	7.7	2.8
Anodonta woodiana	FW	2.0	0.3

Note: O, oligohaline habitat; FW, freshwater habitat. Rates are μmol per g wet weight per hr.

Source: Adapted from Matsushima, O. et al., *J. Exp. Mar. Biol.*, 109, 93–99, 1987.

H. Hypersaline Molluscs

Hypersaline habitats have salinities of 40‰ or above. Little is known about the physiology of molluscs that live in such habitats. In the Laguna Madre system in Texas and in hypersaline habitats of the northern Yucatan peninsula, several species of molluscs maintain populations in areas where the salinity ranges between 50 and 80‰.[18] These animals are all marine bivalves and doubtless are osmotic conformers. They most probably maintain levels of cytoplasmic organic osmolytes that are higher than those typical for marine animals. In addition, gastropods occur in hypersaline pools near hydrocarbon seeps on the floor of the Gulf of Mexico; nothing is as yet known about the osmoregulatory physiology of these animals.[90]

I. Terrestrial Molluscs

All truly terrestrial molluscs are gastropods, although some species of bivalve can survive out of water for impressive lengths of time (1 to 12 months).[28,50,92] For a terrestrial soft-bodied animal, the primary osmoregulatory problem presented by the habitat is dessication. The relative humidity of air that is in equilibrium with an extracellular fluid osmotic concentration typical of a terrestrial snail (200 to 300 mOsm; see Table 4.5) is 99.5%. The implication is obvious for animals with a

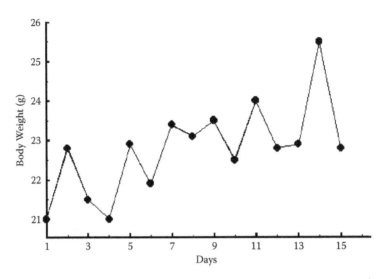

FIGURE 4.4 Fluctuations in the body weight of an individual *Helix pomatia*. The animal was maintained in water-saturated air with food and water available. (Data are from Howes, N.H. and Wells, G.P., *J. Exp. Biol.*, 11, 327–343, 1934.)

body wall that is permeable to water. Routes of water loss in terrestrial molluscs include evaporation across the body wall and the respiratory surface, production of urine, secretion of the mucus (slime) trail, and feces. Sources of water include metabolic water, water drunk or obtained in the food, and water taken up from the environment by osmosis. The concentration of the hemolymph of these animals varies, with some species resembling freshwater gastropods and others having hemolymph at about twice the osmotic concentration of that typical for a freshwater snail (Table 4.5). These numbers, however, should be interpreted with some caution because they can change with the hydration state of the animal.

The body weight of terrestrial gastropods fluctuates by 10 to 50%, even if the animals are maintained in constant conditions (Figure 4.4).[95,96,165] A balance sheet for water in a terrestrial gastropod (Table 4.10) shows that the animals must have sources of water other than that contained in their food. Secreted mucus is 98% water[125] and represents a small, but steady loss of water in an active animal. Whether or not the animals drink water is unclear.[28,114] Numerous studies have shown, however, that terrestrial gastropods can absorb water across the body wall.[126,163] This osmotic uptake occurs through the bottom of the foot via a paracellular pathway,[164] and the rate in dehydrated

TABLE 4.10
Water Balance in a Typical Active Terrestrial Slug

Gains		Losses	
Food	7	Mucus	0.3
Metabolic	0.2	Feces	1.2
Drinking	?	Urine	0.003
Absorbed	Up to 3000	Evaporation	40
Total	42	Total	41.5

Note: Metabolic water assumes a VO_2 of 200 µL/g/hr. Gains and losses are in mL per g body weight per hr.

Source: Data from Machin,[125] Martin et al.,[127] Dainton,[27] and Prior.[163]

animals is high enough to obviate the need for drinking of water (Table 4.10). The osmotic concentration of the hemolymph of the slug *Limax maximus* in a fully hydrated state is 140 mOsm.[163] When loss of water reduces their body weight to 68% of normal (blood π = 200 mOsm), slugs will rehydrate from damp substrates; after termination of the "osmotic drinking" behavior, the hemolymph of rehydrated animals has an osmotic concentration of 117 mOsm, so an "overshoot" of the predehydration concentration has occurred.[163] Snails rehydrating after prolonged dehydration during estivation also overshoot their predehydration weight.[120] This extra water may later be lost as urine to flush out accumulated metabolic wastes.

Even though terrestrial molluscs can survive water losses of up to 90% of their body weight,[125,163] prolonged activity is limited by the availability of water in the habitat. Many species are nocturnal or restrict activity on dry, hot days.[125] Rates of evaporative water loss from active terrestrial pulmonate gastropods range from 2 to 45 mg per g body weight per hr, depending on species and the relative humidity of the air.[125,165] Snails can withdraw into the shell, which reduces water loss by one or two orders of magnitude, and sealing the aperture of the shell with a secreted epiphragm further reduces evaporative loss.[125] Snails in this inactive state can survive for hundreds of days.[125] Slugs, however, lack a shell and cannot live for more than a few days unless they have access to a source of water.[96] Groups of slugs may huddle together to reduce evaporative water loss.[165]

In estivating terrestrial molluscs, the osmotic concentration of the hemolymph increases due to evaporative loss of water. After 200 days of estivation, individuals of the amphibious snail *Pomacea lineata* have lost 50% of their initial body weight, and the osmotic pressure of the hemolymph has increased from 130 mOsm to 240 mOsm.[120] The proportions of Na^+, K^+, and Ca^{2+} in the hemolymph do not change during estivation, but a slight proportional increase in Cl^- and decrease in HCO_3^- occur. In *Helix pomatia*, dehydration is accompanied by changes in the composition of the tissue amino acid pool, but no net increase occurs.[207]

In bivalves that are exposed to air, the physiological responses show some variation in comparison to estivating terrestrial gastropods. Exposure of the unionid clam *Ligumia subrostrata* to air for 7 days resulted in an increase in the osmotic concentration of the hemolymph from 53 mOsm to 92 mOsm.[50] As with estivating snails, the proportions of the cations in the hemolymph did not change, but an increase in chloride was observed (HCO_3^- was not measured).[50] To maintain cellular volume, the intracellular amino acid pool increases in these animals during dehydration.[80] The osmotic concentration of the hemolymph of the clam *Corbicula fluminea* increases from 60 mOsm to 120 mOsm, and 20% of the total body water is lost during exposure to air for 120 hr.[20] In contrast to the response of unionid clams, the only ions that increase in concentration are Ca^{2+} and presumably HCO_3^-, leading to the hypothesis that Na^+, K^+, and Cl^- are transported into the cells to maintain volume.[20] Whether *C. fluminea* tissues accumulate amino acids during immersion is not known. In summary, indirect evidence suggests that evaporative water loss in molluscs results in increases in inorganic ions in some species and increases in amino acids and inorganic ions in others.[59,207]

III. URINE FORMATION IN MOLLUSCS

The anatomy and function of the excretory systems in molluscs have been thoroughly reviewed elsewhere[6,7,122,127,205] and will not be treated in detail here. The phylogenetic diversity of the phylum is reflected in the variability of the morphology of the excretory organs in molluscs. Monopolacophorans have six or seven pairs of excretory organs, chitons have one pair, and many advanced gastropods have only one.[7] In general, molluscs produce urine as an ultrafiltrate of the hemolymph. The walls of the auricles of the heart, the pericardial glands (Keber's organs), or the kidney epithelium act as a filtration membrane, and the hydrostatic pressure that drives filtration of the hemolymph is generated by contraction of the ventricles of the heart. Podocytes have been found to be associated with the heart complex in many molluscs.[6,83,85,115,137,138,142,180,202] In bivalves, the presence of podocytes in both the auricle and the pericardial organs has confused the issue of the site or sites of filtration. Despite some contrary data,[145,196] measurements of hemodynamics in a few molluscs support the idea that the formation of urine occurs by filtration of the hemolymph into the pericardial cavity.[67,89,100] The morphology of the podocytes suggests that these cells may have a secretory function.[7,109] The filtrate may be modified by the kidney. In freshwater animals, ions are removed to produce a urine that is hypoosmotic to the hemolymph (Table 4.6).[122] In some freshwater snails (e.g., *Helisoma duryi*), a ureter contains epithelial cells with apical microvilli, extensive basal infolding, and numerous mitochondria.[106] This specialized structure is not present in freshwater bivalves.[7] Molluscan kidneys contain cells with large vacuoles; these cells are known as *excretory cells* and release the vacuolar contents into the urine.[7]

IV. CALCIUM UPTAKE FOR SHELL FORMATION

Biomineralization in molluscs has also been extensively reviewed elsewhere.[24,208] The shell is secreted by the mantle, a tissue that consists of two epithelia, each a single cell layer thick, that are separated by a space filled with hemolymph. The mantle separates the mantle cavity, which contains the ambient medium, from the extrapallial fluid, which bathes the shell. The concentration of Ca^{2+} in seawater and in the hemolymph of a marine mollusc is 10 mM; no diffusion gradient exists to oppose the uptake of calcium ions from the medium. In marine molluscs the K_m for the uptake and deposition of calcium into the shell is about 7.5 mM.[102] In freshwaters the concentration of Ca^{2+} ranges from 0.1 to 1 mM, considerably lower than the concentrations typical of the hemolymph of freshwater molluscs (Table 4.5). Freshwater molluscs not only can secrete and maintain a large, thick shell (the author has shells of unionid mussels that weigh over 600 g) but can also accumulate additional calcium that is stored as concretions in the tissues and used to provision the shells of developing larvae that are brooded in the gill marsupia.[187] Although some of this calcium is doubtless dietary, an enormous amount must be transported into these animals from the medium against a large concentration gradient. The K_m values for the influx of calcium into freshwater animals are within the range of calcium concentrations found in freshwaters (Table 4.7).

The transepithelial potentials across the body wall of the clams *Corbicula fluminea* and *Ligumia subrostrata* in artificial pond water with a Ca^{2+} concentration of 0.4 mM are, respectively, –7 and –15 mV (hemolymph negative).[55,133] Assuming a water temperature of 22°C, the calculated equilibrium potentials for calcium are –32 and –28 mV in *C. fluminea* and *L. subrostrata*, respectively. These values are not close to the measured transepithelial potentials and indicate that the uptake of Ca^{2+} from the medium occurs by active transport. The partitioning of calcium ion uptake between the gills and mantle is not known, but, once in the hemolymph, calcium is moved across the shell-facing epithelium of the mantle into the extrapallial cavity by a mechanism that is still unclear but does not appear to involve active transport of calcium ions.[25,98]

V. CONCLUSIONS

This review suggests several avenues for future research. Among osmoconformers, the asymmetry in the response of animals to hyper- or hypoosmotic stress seems to be universal. The morphological and physiological diversity among the major classes, such as the bivalves and gastropods, is enormous; the reasons for the differential response are far from clear. The mechanisms involved in changes in the permeability of the body wall to ions and water remain unknown, but modern experimental approaches (such as the use of antibodies to aquaporins) would improve our insight into the question.

Table 4.2 reveals that very little recent work has been done on the role of amino acids in volume regulation in gastropods. What is the fate of amino acids and quaternary amines released into the hemolymph from the tissues? Are the carbon skeletons conserved in all molluscs? If so, which tissue or tissues deaminate these osmolytes? At the cellular level, very few studies have examined the role of inorganic ions in both hyper- and hypoosmotic volume regulation in molluscs. Little is known about the mechanisms involved in the initiation and control of either osmolyte release in response to a decrease in the osmolality of the medium or osmolyte accumulation in response to an increase in the osmolality of the medium. Finally, the reported rates of urine production in marine osmotic conformers are of the same order of magnitude as those of some freshwater animals (Table 4.6). If all of these measurements are accurate, some explanation other than salt and water balance must account for the comparatively high rates of urine production in the marine species.

Many studies have shown that oligohaline animals can tolerate freshwater but, in contrast to freshwater species, cannot survive in deionized water. This may be due to differences in the kinetics of ion uptake mechanisms, but we have no relevant data from oligohaline molluscs. The larval stages of oligohaline molluscs may also be less tolerant of very low salinities. Unionid bivalves brood their larvae, and the form that is released, the parasitic glochidium, attaches to fish gills for a period of maturation. Among more recent colonizers of fresh waters, *Corbicula fluminea* lacks a specialized larval stage but broods the larvae prior to releasing them into the environment; *Dreissena polymorpha* does not even brood its larvae. Studies on the physiology of these larval stages are scarce. For all freshwater molluscs, the upper limit of salinity tolerance is about 200 mOsm (Table 4.8). This inability to tolerate higher external salinities has been ascribed to a decreased capacity for hyperosmotic volume regulation. Whereas this may be true for unionid mussels (Table 4.9), the data are less than conclusive for other freshwater taxa. Experiments on the rates of accumulation of amino acid in the tissues of a variety of freshwater and oligohaline bivalves and gastropods would shed light on the question.

Although the gills of freshwater bivalves are clearly involved in the transport of ions from the medium, the contribution of the mantle, if any, to the uptake of Na^+, Ca^{2+}, and Cl^- in these animals, in freshwater gastropods, and in oligohaline species is not known. Despite numerous studies, the mechanisms involved in the movement of calcium ions from the hemolymph to the extrapallial cavity for shell growth and repair are unclear. No evidence supports the existence of any primary or secondary active transport mechanism involved in the flux of Ca^{2+} across the shell-facing epithelium of the mantle.[98] Some preliminary evidence suggests that hormones are involved in the control of osmoregulatory mechanisms in molluscs, but none has yet been identified. Very little is known, for example, about the link between tissue hydration and the initiation of estivation in terrestrial molluscs. Finally, given their well-known fondness for beer, the question of whether or not pulmonate slugs imbibe water by mouth appears to be a trivial one, but we do not have an unambiguous answer.

REFERENCES

1. Alberts, B., Bray, D., Lewis, J., Raff, M., Roberts, K., and Watson, J. D., *Molecular Biology of the Cell*, Garland Publishing, New York, 1983, 1146 pp.
2. Aldridge, D. W., Physiological ecology of freshwater prosobranchs. In: *Physiology of the Mollusca*. Vol. 6. *Ecology*, Russell-Hunter, W. D., Ed., Academic Press, New York, 1983, pp. 329–358.
3. Amende, L. M. and Pierce, S. K., Hypotaurine: the identity of an unknown ninhydrin-positive compound co-eluting with urea in amino acid extracts of bivalve tissue. *Comp. Biochem. Physiol.*, 59B, 257–261, 1978.
4. Amende, L. M. and Pierce, S. K., Cellular volume regulation in salinity stressed molluscs: the response of *Noetia ponderosa* (Arcidae) red blood cells to osmotic variation. *J. Comp. Physiol.*, 138, 283–289, 1980.
5. Amende, L. M. and Pierce, S. K., Free amino acid mediated volume regulation of isolated *Noetia ponderosa* red blood cells: control by Ca^{2+} and ATP. *J. Comp. Physiol.*, 138, 291–298, 1980.
6. Andrews, E. B., Osmoregulation and excretion in prosobranch gastropods: structure in relation to function. *J. Mollusc. Stud.*, 47, 248–289, 1981.

7. Andrews, E. B., Excretory systems of molluscs. In: *The Mollusca*. Vol. 11. *Form and Function*, Trueman, E. R. and Clarke, M. R., Eds., Academic Press, New York, 1988, pp. 381–448.

8. Andrews, E. B. and Jennings, K. H., The anatomical and ultrastructural basis of primary urine formation in bivalve molluscs. *J. Mollusc. Stud.*, 59, 223–257, 1993.

9. Andrews, E. B. and Little C., Structure and function in the excretory system of some prosobranch snails (Cyclophoridae), *J. Zool.*, 168, 395–422, 1972.

10. Andrews, E. B. and Little, C., Structure and function in the excretory organs of some terrestrial prosobranch snails (Cyclophoridae). *J. Zool. Lond.*, 168, 395–422, 1993.

11. Avens, A. C. and Sleigh, M. A., Osmotic balance in gastropod molluscs. I. Some marine and littoral gastropods. *Comp. Biochem. Physiol.*, 16, 121–141, 1965.

12. Baginski, R. M. and Pierce, S. K., The time course of intracellular free amino acid accumulation in tissues of *Modiolus demissus* during high salinity adaptation. *Comp. Biochem. Physiol.*, 57A, 407–412, 1977.

13. Baginski, R. M. and Pierce, S. K., A comparison of amino acid accumulation during high salinity adaptation with anaerobic metabolism in the ribbed mussel, *Modiolus demissus demissus. J. Exp. Zool.*, 203, 419–428, 1978.

14. Bartberger, C. A. and Pierce, S. K., Relationship between ammonia excretion rates and hemolymph nitrogenous compounds of a euryhaline bivalve during low salinity acclimation. *Biol. Bull.*, 150, 1–14, 1978.

15. Bedford, J. J., Osmoregulation in *Melanopsis trifasciata*. II. The osmotic pressure and the principal ions of the hemocoelic fluid. *Physiol. Zool.*, 45, 143–151, 1972.

16. Bedford, J. J., Osmotic relationships in a freshwater mussel, *Hyridella menziesi* Gray (Lamellibranchia: Unionidae). *Archs. Int. Physiol. Biochim.*, 81, 819–831, 1973.

17. Bishop, S. H., Greenwalt, D. E., Kapper, M. A., Paynter, K. T., and Ellis, L. L., Metabolic regulation of proline, glycine, and alanine accumulation as intracellular osmolytes in ribbed mussel gill tissue. *J. Exp. Zool.*, 268, 151–161, 1994.

18. Britton, J. C. and Morton, B., *Shore Ecology of the Gulf of Mexico*. University of Texas Press, Austin, TX, 1989, 387 pp.

19. Burton, R. F., Concentrations of cations in the blood of some terrestrial snails. *Comp. Biochem. Physiol.*, 39A, 875–878, 1971.

20. Byrne, R. A., McMahon, R. F., and Dietz, T. H., The effects of aerial exposure and subsequent reimmersion on hemolymph osmolality, ion composition, and ion flux in the freshwater bivalve *Corbicula fluminea. Physiol. Zool.*, 62, 1187–1202, 1989.

21. Chaisemartin, C., Place de la fonction renale dans la regulation de l'eau et des sels chez *Margaritana margaranitifera* (Unionidae). *Compte Rendu Seanc. Soc. Biol.*, 162, 1193–1195, 1968.

22. Chaisemartin, C., Martin, P. M., and Bernard, M., Homeoionemie chez *Margaritana margarantifera* L. (Unionides), etudiee a la aide des radioelements ^{22}Na et ^{36}Cl. *Compte Rendu Seanc. Soc. Biol.*, 162, 523–526, 1968.

23. Chaisemartin, C. and Videaud, A., Seuils calciques de l'eau et economie du calcium chez *Ancylastrum fluviatilis* (Gastropodes, Pulmones). *Compte Rendu Seanc. Soc. Biol.*, 165, 2401–2404, 1971.

24. Checa, A., A new model for periostracum and shell formation in the unionidae (Bivalvia, Mollusca). *Tissue Cell*, 32, 405–416, 2000.

25. Coimbra, J., Machado, J., Fernandes, P. L., Ferreira, H. G., and Ferreira, K. G., Electrophysiology of the mantle of *Anodonta cygnea. J. Exp. Biol.*, 140, 65–88, 1988.

26. Costa, C. J. and Pritchard, A. W., The response of *Mytilus edulis* to short duration hypoosmotic stress. *Comp. Biochem. Physiol.*, 64A, 91–95, 1978.

27. Dainton, B. H., The activity of slugs. I. The induction of activity by changing temperatures. *J. Exp. Biol.*, 31, 165–187, 1954.

28. Dance, S. P., Drought resistance in an African freshwater bivalve. *J. Conchol.*, 24, 281–283, 1958.

29. Davenport, J., The isolation response of mussels (*Mytilus edulis* L.) exposed to falling seawater concentrations. *J. Mar. Biol. Assoc. U.K.*, 59, 123–132, 1979.

30. Davenport, J. and Fletcher, J. S., The effects of simulated estuarine mantle cavity conditions upon the activity of the frontal gill cilia of *Mytilus edulis. J. Mar. Biol. Assoc. U.K.*, 58, 671–681, 1978.

31. Deaton, L. E., Ion regulation in freshwater and brackish water bivalve molluscs. *Physiol. Zool.*, 54, 109–121, 1981.

32. Deaton, L. E., Tissue (Na$^+$ K$^+$)-activated adenosinetriphosphatase activities in freshwater and brackish water bivalve molluscs. *Mar. Biol. Lett.*, 3, 107–112, 1982.

33. Deaton, L. E., Epithelial water permeability in the euryhaline mussel *Geukensia demissa*: decrease in response to hypooosmotic media and hormonal regulation. *Biol. Bull.*, 173, 230–238, 1987.

34. Deaton, L. E., Hyperosmotic cellular volume regulation in the ribbed mussel *Geukensia demissa*: inhibition by lysosomal and proteinase inhibitors. *J. Exp. Zool.*, 244, 375–282, 1987.

35. Deaton, L. E., Potentiation of hypoosmotic cellular volume regulation in the quahog, *Mercenaria mercenaria*, by 5-hydroxytryptamine, FMRFamide, and phorbol esters. *Biol. Bull.*, 178, 260–266, 1990.

36. Deaton, L. E., Oxygen uptake and heart rate of the clam *Polymesoda caroliniana* Bosc in air and in seawater. *J. Exp. Mar. Biol. Ecol.*, 147, 1–7, 1991.

37. Deaton, L. E., Osmoregulation and epithelial permeability in two euryhaline bivalve molluscs: *Mya arenaria* and *Geukensia demissa*. *J. Exp. Mar. Biol. Ecol.*, 158, 167–177, 1992.

38. Deaton, L. E., Hypoosmotic volume regulation in bivalves: protein kinase C and amino acid release. *J. Exp. Zool.*, 268, 145–150, 1994.

39. Deaton, L. E., Hyperosmotic volume regulation in the gills of the ribbed mussel, *Geukensia demissa*: rapid accumulation of betaine and alanine. *J. Exp. Mar. Biol. Ecol.*, 260, 185–197, 2001.

40. Deaton, L. E. and Greenberg, M. J., The ionic dependence of the cardiac action potential in bivalve molluscs: systematic distribution. *Comp. Biochem. Physiol.*, 67A, 155–161, 1990.

41. Deaton, L. E. and Greenberg, M. J., The adaptation of bivalve molluscs to oligohaline and fresh waters: phylogenetic and physiological aspects. *Malacol. Rev.*, 24, 1–18, 1991.

42. Deaton, L. E., Derby, J. G. S., Subhedar, N., and Grenberg, M. J., Osmoregulation and salinity tolerance in two species of bivalve mollusc: *Limnoperna fortunei* and *Mytilopsis leucophaeta*. *J. Exp. Mar. Biol. Ecol.*, 133, 67–79, 1989.

43. Deaton, L. E., Hilbish, T. J., and Koehn, R. K., Hyperosmotic volume regulation in the tissues of the mussel *Mytilus edulis*. *Comp. Biochem. Physiol.*, 80A, 571–574, 1985.

44. DeJorge, F. B., Ulhoa Cintra, A. B., Haeser, P. E., and Sawaya, P., Biochemical studies on the snail *Strophocheilus oblongus musculus* (Bequaert). *Comp. Biochem. Physiol.*, 14, 35–42, 1965.

45. De Vooys, C. G. N. and Geenevasen, J. A. J., Biosynthesis and role in osmoregulation of glycine-betaine in the Mediterranean mussel *Mytilus galloprovincialis* LMK. *Comp. Biochem. Physiol.*, 132B, 409–414, 2002.

46. De With, N. D., Evidence for the independent regulation of specific ions in the haemolymph of *Lymnaea stagnalis* (L.). *Proc. Kon. Ned. Akad. Wetenschap. Ser. C*, 80, 144–157, 1977.

47. De With, N. D. and Sminia, T., The effects of the nutritional state and the external calcium concentration on the ionic composition of the haemolymph and on the calcium cells in the pulmonate freshwater snail *Lymnaea stagnalis*. *Proc. K. Ned. Akad. Wet. (C)*, 83, 217–227, 1980.

48. De With, N. D. and van der Schors, R. C., On the interrelations of the Na+ and CT– influxes in the pulmonate freshwater snail *Lymnaea stagnalis*. *J. Comp. Physiol.*, 148, 131–135, 1982.

49. De Zwaan, A., Anaerobic energy metabolism in bivalve molluscs. *Oceanogr. Mar. Biol.*, 15, 103–187, 1977.

50. Dietz, T. H., Body fluid composition and aerial oxygen consumption in the freshwater mussel, *Ligumia subrostrata* (Say): effects of dehydration and anoxic stress. *Biol. Bull.*, 147, 560–572, 1974.

51. Dietz, T. H., Solute and water movement in freshwater bivalve molluscs. In: *Water Relations in Membrane Transport in Plants and Animals*, Jungreis, A., Hodges, T., Kleinzeller, A., and Schultz, S., Eds., Academic Press, New York, 1977, pp. 11–119.

52. Dietz, T. H., Sodium transport in the freshwater mussel, *Caruculina texasensis* (Lea). *Am. J. Physiol.*, 235, R35–R40, 1978.

53. Dietz, T. H., Uptake of sodium and chloride by freshwater mussels. *Can. J. Zool.*, 57, 156–160, 1979

54. Dietz, T. H., Ionic regulation in freshwater mussels: a brief review. *Am. Malacol. Bull.*, 3, 233–242, 1985.

55. Dietz, T. H. and Branton, W. D., Ionic regulation in the freshwater mussel, *Ligumia subrostrata* (Say). *J. Comp. Physiol.*, 104, 19–26, 1975.

56. Dietz, T. H. and Branton, W. D., Chloride transport in freshwater mussels. *Physiol. Zool.*, 52, 520–528, 1979.

57. Dietz, T. H. and Findley, A. M., Ion-stimulated ATPase activity and NaCl uptake in the gills of freshwater mussels. *Can. J. Zool.*, 58, 917–923, 1980.

58. Dietz, T. H. and Hagar, A. F., Chloride uptake in isolated gills of the freshwater mussel *Ligumia subrostrata*. *Can. J. Zool.*, 68, 6–9, 1990.

59. Dietz, T. H., Neufeld, D. H., Silverman, H., and Wright, S. H., Cellular volume regulation in freshwater bivalves. *J. Comp. Physiol.*, 168B, 87–95, 1998.

60. Dietz, T. H., Scheide, J. I., and Saintsing, D. G., Monoamine transmitters and cAMP stimulation of Na transport in freshwater mussels. *Can. J. Zool.*, 60, 1408–1411, 1982.

61. Dietz, T. H., Steffens, W. L., Kays, W. T., and Silverman, H., Serotonin localization in the gills of the freshwater mussel, *Ligumia subrostrata*. *Can. J. Zool.*, 63, 1237–1243, 1985.

62. Duncan, A., Osmotic balance in *Potamopyrgus jekinisi* (Smith) from two Polish populations. *Pol. Arch. Hydrobiol.*, 14, 1–10, 1967.

63. Duobinis-Gray, E. M. and Hackney, C. T., Seasonal and spatial distribution of the Carolina marsh clam *Polymesoda caroliniana* (Bosc) in a Mississippi tidal marsh. *Estuaries*, 5, 102–109, 1982.

64. DuPaul, W. D. and Webb, K. L., Free amino acid accumulation in isolated gill tissue of *Mya arenaria*. *Arch. Int. Physiol. Biochim.*, 79, 327–336, 1971.

65. Ferguson, R. L. and Sunda, W. G., Utilization of amino acids by planktonic marine bacteria: importance of clean technique and low substrate concentrations. *Limnol. Oceanogr.*, 29, 258–274, 1984.

66. Fitt, W. K., Rees, T. A. V., Braley, R. D., Lucas, J. S., and Yellowlees, D., Nitrogen flux in giant clams: size dependency and relationship to zooxanthellae density and clam biomass in the uptake of dissolved organic nitrogen. *Mar. Biol.*, 117, 381–386, 1993.

67. Florey, E. and Cahill, M. A. Hemodynamics in lamellibranch molluscs: confirmation of constant-volume mechanism of auricular and ventricular filling, remarks on the heart as a site for ultrafiltration. *Comp. Biochem. Physiol.*, 57A, 47–52, 1977.

68. Fuhrman, J. A. and Bell, T. M., Biological considerations in the measurement of dissolved free amino acids in seawater and implications for chemical and microbial studies. *Mar. Ecol. Prog. Ser.*, 25, 13–21, 1985.

69. Gainey, L. F., The response of the Corbiculidae (Mollusca: Bivalvia) to osmotic stress: the organismal response. *Physiol. Zool.*, 51, 68–78, 1978.

70. Gainey, L. F., The response of the Corbiculidae (Mollusca: Bivalvia) to osmotic stress: the cellular response. *Physiol. Zool.*, 51, 79–91, 1978.

71. Gainey, L. F., The effect of osmotic hysteresis on volume regulation in *Mytilus edulis*. *Comp. Biochem. Physiol.*, 87A, 151–156, 1987.

72. Gainey, L. F., Volume regulation in three species of marine mussels. *J. Exp. Mar. Biol. Ecol.*, 181, 201–211, 1994.

73. Gainey, L. F. and Greenberg, M. J., Physiological basis of the species abundance–salinity relationship in molluscs: a speculation. *Mar. Biol.*, 40, 41–49, 1977.

74. Ghiretti, F., Respiration. In: *Physiology of Mollusca*, Vol. II, Wilbur, K. M. and Yonge, C. M., Eds., Academic Press, New York, 1966, pp. 175–208.

75. Gilles, R., Osmoregulation in three molluscs: *Acontchitona discrepans* (Brown), *Glycymeris glycymeris* and *Mytilus edulis*. *Biol. Bull.*, 142, 25–35, 1972.

76. Graves, S. Y. and Dietz, T. H., Prostaglandin E_2 inhibition of sodium transport in the freshwater mussel. *J. Exp. Zool.*, 210, 195–202, 1979.

77. Graves, S. Y. and Dietz, T. H., Diurnal rhythms of sodium transport in the freshwater mussel. *Can. J. Zool.*, 58, 1626–1630, 1980.

78. Greenaway, P., Calcium regulation in the freshwater mollusk, *Limnaea stagnalis* (L.) (Gastropoda: Pulmonata). *J. Exp. Biol.*, 54, 100–214, 1971.

79. Greenwalt, D. E. and Bishop, S. H., Effect of aminotransferase inhibitors on the pattern of free amino acid accumulation in isolated mussel hearts subjected to hyperosmotic stress. *Physiol. Zool.*, 53, 262–269, 1980.

80. Hanson, J. A. and Dietz, T. H., The role of free amino acids in cellular osmoregulation in the freshwater bivalve *Ligumia subrostrata* (Say). *Can. J. Zool.*, 54, 1927–1931, 1976.

81. Harrison, F. M., Some excretory processes in the abalone *Haliotis rufescens*. *J. Exp. Biol.*, 39, 179–192, 1962.

82. Harrison, F. M. and Martin, A. W., Excretion in the cephalopod *Octopus dofleini*. *J. Exp. Biol.*, 42, 71–98, 1965.

83. Haszprunar, G. and Fahner, A., Anatomy and ultrastructure of the excretory system of a heart-bearing and heart-less sacoglossan gastropod (Opisthobrancia, Sacoglossa). *Zoomorphology*, 12, 85–93, 2001.

84. Hawkins, A. J. S. and Hilbish, T. J., The costs of cell volume regulation: protein metabolism during hyperosmotic adjustment. *J. Mar. Biol. Assoc. U.K.*, 72, 569–578, 1992.

85. Hawkins, W. E., House, H. D., and Sarphie, T. G., Ultrastructure of the heart of the oyster *Crassostrea virginica*. *J. Submicrosc. Cytol.*, 12, 359–374, 1980.

86. Heavers, B. W. and Hammen, C. S., Fate of endogenous free amino acids in osmotic adjustment of *Crassostrea virginica* (Gmelin). *Comp. Biochem. Physiol.*, 82A, 571–576, 1985.

87. Henry, R. P. and Mangum, C. P., Salt and water balance in the oligohaline clam *Rangia cuneata*. III. Reduction of the free amino acid pool during low salinity adaptation. *J. Exp. Zool.*, 211, 25–32, 1980.

88. Henry, R. P., Mangum, C. P., and Webb, K. L., Salt and water balance in the oligohaline clam, *Rangia cuneata*. II. Accumulation of intracellular free amino acids during high salinity adaptation. *J. Exp. Zool.*, 211, 11–24, 1980.

89. Hevert, F., Urine formation in the lamellibranchs: evidence for ultrafiltration and quantitative description. *J. Exp. Biol.*, 11, 1–12, 1984.

90. Hickman, C. S., Mollusc-microbe mutualisms extend the potential for life in hypersaline systems. *Astrobiology*, 3, 631–644, 2003.

91. Hill, R. W., Wyse, G. A., and Anderson, M., *Animal Physiology*, Sinauer Associates, Saunderland, MA, 2004, 770 pp.

92. Hiscock, I. D., Osmoregulation in Australian freshwater mussels (Lamellibranchiata). *Austral. J. Mar. Freshwater Res.*, 4, 317–329, 1953.

93. Horohov, J., Silverman, H., Lynn, J. W., and Dietz, T. H., Ion transport in the freshwater zebra mussel, *Driessena polymorpha. Biol. Bull.*, 183, 297–303, 1992.

94. Hosoi, M., Takeuchi, K., Sawada, H., and Toyohara, H., Expression and functional analysis of mussel taurine transporter, as a key molecule in cellular osmoconforming. *J. Exp. Biol.*, 208, 4203–4211, 2005.

95. Howes, N. H. and Wells, G. P., The water relations of snails and slugs I. Weight rhythms in *Helix pomatia* L. *J. Exp. Biol.*, 11, 327–343, 1934.

96. Howes, N. H. and Wells, G. P., The water relations of snails and slugs II. Weight rhythms in *Arion ater* L. and *Limax flavus* L. *J. Exp. Biol.*, 11, 344–351, 1934.

97. Hoyeaux, J., Gilles, R., and Jeuniaux, C., Osmoregulation in molluscs of the intertidal zone. *Comp. Biochem. Physiol.*, 53A, 361–365, 1976.

98. Hudson, R. L., Ion transport by the isolated mantle epithelium of the freshwater clam, *Unio complanatus. Am. J. Physiol.*, 263, R76–R83, 1992.

99. Ivanovici, A. M., Rainer, S. F., and Wadly, V. A., Free amino acids in three species of mollusc: responses to factors associated with reduced salinity. *Comp. Biochem. Physiol.*, 70A, 17–22, 1981.

100. Jones, H. D. and Peggs, D., Hydrostatic and osmotic pressures in the heart and pericardium of *Mya arenaria* and *Anodonta cygnea. Comp. Biochem. Physiol.*, 76A, 381–385, 1983.

101. Jordan, P. J. and Deaton, L. E., Osmotic regulation and salinity tolerance in the freshwater snail *Pomacea bridgesi* and the freshwater clam *Lampsilis teres. Comp. Biochem. Physiol.*, 122A, 199–205, 1999.

102. Kado, Y., Studies on shell formation in molluscs. *J. Sci. Hiroshima Univ. Ser. B*, 19, 1–210, 1960.

103. Kado, Y. and Murata, H., Responses of brackish and fresh-water clams, *Corbicula japonica* and *C. leana*, to variations in salinity. *J. Sci. Hiroshima Univ. Ser. B*, 25, 217–224, 1974.

104. Kapper, M. A., Stickle, W. B., and Blakeney, E., Volume regulation and nitrogen metabolism in the muricid gastropod *Thais haemastoma. Biol. Bull.*, 169, 458–475, 1985.

105. Kasschau, M. R., The relationship of free amino acids to salinity changes and temperature–salinity interactions in the mud-flat snail, *Nassarius obsoletus. Comp. Biochem. Physiol.*, 51A, 301–308, 1975.

106. Kays, W. T., Silverman, H., and Dietz, T. H., Water channels and water canals in the gill of the freshwater mussel, *Ligumia subrostrata*: ultrastructure and histochemistry. *J. Exp. Zool.*, 254, 256–269, 1990.

107. Khan, H. R. and Saleuddin, A. S. M., Osmotic regulation and osmotically induced changes in the neurosecretory cells of the pulmonate snail *Helisoma. Can. J. Zool.*, 57, 1371–1383, 1979.

108. Khan, H. R. and Saleuddin, A. S. M., Effects of osmotic changes and neurosecretory extracts on kidney ultrastructure in the freshwater pulmonate *Helisoma. Can. J. Zool.*, 57, 1256–1270, 1979.

109. Khan, H. R., Asthon, M. L., and Saleuddin, A. S. M., A study on the cytoplasmic granules of the pericardial gland cells of some bivalve molluscs. *Tissue Cell*, 20, 587–597, 1988.

110. Kirschner, L. B., Water and ions. In: *Environmental and Metabolic Animal Physiology*, Prosser, C. L., Ed., Wiley-Liss, New York, 1991, pp. 13–107.

111. Komendatov, A. Y., Khlebovich, V. V., and Aladin, H. B., Features of osmotic and ionic regulation in bivalve molluscs as they depend on environmental factors [in Russian]. *Ekologiya*, 5, 39–46, 1985.

112. Krogh, A., *Osmotic Regulation in Aquatic Animals*. Cambridge University Press, Cambridge, U.K., 1939, 242 pp.

113. Kube, S. A., Gerber, A., Jansen, J. M., and Schiede, D., Patterns of organic osmolytes in two marine bivalves, *Macoma balthica* and *Mytilus* spp., along their European distributions. *Mar. Biol.*, 149, 1387–1396, 2006.

114. Kunkel, K., Zur Biologie der Nacktschnecken. *Verh. Zool. Ges.*, 22–31, 1900.

115. Lachtel, D. L., Martin, A. W., Deyrup-Olsen, I., and Boer, H. H., Gastropoda: Pulmonata. In: *Microscopic Anatomy of Invertebrates*. Vol. 6B. *Mollusca II*, Harrison F. W. and Kohn, A. W., Eds., Wiley-Liss, New York, 1997, pp. 459–718.

116. Lange, R., Isosmotic intracellular regulation and euryhalinity in marine bivalves. *J. Exp. Mar. Biol.*, 5, 170–179, 1970.

117. Little, C., Osmotic and ionic regulation in the prosobranch gastropod mollusc, *Viviparus viviparus* Linn. *J. Exp. Biol.*, 43, 23–37, 1965.

118. Little, C., The formation of urine by the prosobranch gasatropod mollusc *Viviparus viviparus* Linn. *J. Exp. Biol.*, 43, 49–54, 1965.

119. Little, C., Ionic regulation in the queen conch, *Strombus gigas* (Gastropoda, Prosobranchia). *J. Exp. Biol.*, 46, 459–474, 1967.

120. Little, C., Aestivation and ionic regulation in two species of *Pomacea* (Gasatropoda, Prosobranchia). *J. Exp. Biol.*, 48, 569–585, 1968.

121. Little, C., The evolution of kidney function in the Neritacea (Gasatropoda, Prosobranchia). *J. Exp. Biol.*, 56, 249–261, 1972.

122. Little, C., Renal adaptations of prosobranchs to the freshwater environment. *Am. Malacol. Bull.*, 3, 223–231, 1985.

123. Little, C. and Andrews, E. B., Some aspects of excretion and osmoregulation in assimineid snails. *J. Moll. Stud.*, 43, 263–285, 1977.

124. Livingstone, D. R., Widdows, J., and Fieth, P., Aspects of nitrogen metabolism of the common mussel *Mytilus edulis*: adaptation to abrupt and fluctuating changes in salinity. *Mar. Biol.*, 53, 41–55, 1979.

125. Machin, J., Water relationships. In: *Pulmonates*, Vol. 1, Fretter, V. and Peake, J., Eds., Academic Press, New York, 1975, pp. 105–163.

126. Manahan, D. T., Wright, S. H., and Stephens, G. C., Simultaneous determination of transport of 16 amino acids into a marine bivalve. *Am. J. Physiol.*, 244, R832–R838, 1986.

127. Martin, A. W., Stewart, D. M., and Harrison, F. M., Urine formation in a pulmonate land snail, *Achatina fulica*. *J. Exp. Biol.*, 42, 99–123, 1965.

128. Matsumoto, M., Morimasa, T., Takeuchi, H., Mori, A., Kohsaka, M., Kobayashi, J., and Morii, F. Amounts of inorganic ions and free amino acids in hemolymph and ganglion of the giant African snail, *Achatina fulica* Ferussac. *Comp. Biochem. Physiol.*, 48A, 465–470, 1974.

129. Matsushima, O., Comparative studies on responses to osmotic stresses in brackish and fresh-water clams. *J. Sci. Hiroshima Univ. Ser. B*, 30, 173–192, 1982.

130. Matsushima, O. and Kado, Y., Hyperosmoticity of the mantle fluid in the freshwater bivalve, *Anodonta woodiana*. *J. Exp. Zool.*, 221, 379–381, 1982.

131. Matsushima, O., Katayama, H., and Yamada, K., The capacity for intracellular osmoregulation mediated by free amino acids in three bivalve molluscs. *J. Exp. Mar. Biol.*, 109, 93–99, 1987.

132. McAlister, R. O., and Fisher, F. M., Responses of the false limpet *Siphonaria pectinata* Linnaeus (Gastropoda, Pulmonata) to osmotic stress. *Biol. Bull.*, 134, 96–117, 1968.

133. McCorkle, S. and Dietz, T. H., Sodium transport in the freshwater asiatic clam *Corbicula fluminea*. *Biol. Bull.*, 159, 325–336, 1980.

134. McCorkle-Shirley, S., Effects of photoperiod on sodium flux in *Corbicula fluminea* (Mollusca: Bivalvia). *Comp. Biochem. Physiol.*, 71A, 325–327, 1982.

135. McMahon, R. F., Physiological ecology of freshwater pulmonates. In: *Physiology of the Mollusca*. Vol. 6. *Ecology*, Russell-Hunter, W. D., Ed., Academic Press, New York, 1983, pp. 359–430.

136. McMahon, R. F., Respiratory response to periodic emergence in intertidal molluscs. *Am. Zool.*, 28, 97–114, 1988.

137. Meyerhofer, E. and Morse, M. P., Podocytes in bivalve molluscs: morphological evidence for ultra-filtration. *Invert. Biol.*, 115, 20–29, 1996.

138. Meyerhofer, E., Morse, M. P., and Robinson, W. E., Podocytes in bivalve molluscs: morphological evidence for ultrafiltration. *J. Comp. Physiol.*, 156B, 151–161, 1985.

139. Monk, C. and Stewart, D. M., Urine formation in a freshwater snail, *Viviparus malleatus*. *Am. J. Physiol.*, 210, 647–651, 1966.

140. Negus, M. R. S., Oxygen consumption and amino acid levels in *Hydrobia ulvae* (Pennant) in relation to salinity and behavior. *Comp. Biochem. Physiol.*, 24, 317–325, 1968.

141. Neufeld, D. S and Wright, S. H., Salinity change and cell volume: the response of tissues from the estuarine mussel *Geukensia demissa*. *J. Exp. Biol.*, 199, 1619–1630, 1996.

142. Okland, S., Ultrastructure of the pericardium in chitons (Mollusca: Polyplacophora) in relation to filtration and contraction mechanisms. *Zoomorphology*, 97, 193–203, 1981.

143. Paparo, A. A. and Dean, R. C., Activity of the lateral cilia of the oyster *Crassostrea virginica* Gmelin: response to changes in salinity and to changes in potassium and magnesium concentration. *Mar. Behav. Physiol.*, 11, 111–130, 1984.

144. Picken, L. E. R., The mechanism of urine formation in invertebrates. II. The excretory mechanism in certain mollusca. *J. Exp. Biol.*, 14, 20–34, 1937.

145. Pierce, S. K., The water balance of *Modiolus* (Mollusca: Bivalvia: Mytilidae): osmotic concentration in changing salinities. *Comp. Biochem. Physiol.*, 36, 521–533, 1970.

146. Pierce, S. K., A source of solute for volume regulation in marine mussels. *Comp. Biochem. Physiol.*, 38A, 619–635, 1971.

147. Pierce, S. K., Volume regulation and valve movements by marine mussels. *Comp. Biochem. Physiol.*, 39A, 103–117, 1971.

148. Pierce, S. K., Osmolyte permeability in molluscan red cells is regulated by Ca^{2+} and membrane protein phosphorylation: the present perspective. *J. Exp. Zool.*, 268, 166–170, 1994.

149. Pierce, S. K. and Greenberg, M. J., The initiation and control of free amino acid regulation of cell volume in salinity-stressed marine bivalves. *J. Exp. Biol.*, 59, 435–446, 1973.

150. Pierce, S. K. and Greenberg, M. J., Hypoosmotic cell volume regulation in marine bivalves: the effects of membrane potential change and metabolic inhibition. *Physiol. Zool.*, 49, 417–424, 1976.

151. Pierce, S. K. and Politis, A. D., Ca^{2+}-activated cell volume recovery mechanisms. *Ann. Rev. Physiol.*, 52, 27–42, 1990.

152. Pierce, S. K. and Rowland, L. M., Proline betaine and amino acid accumulation in sea slugs (*Elysia chlorotica*) exposed to extreme hyperosmotic conditions. *Physiol. Zool.*, 61, 205–212, 1988.

153. Pierce, S. K. and Rowland-Faux, L. M., Ionomycin produces an improved volume recovery by an increased efflux of taurine from hypoosmotically stressed molluscan red blood cells. *Cell Calcium*, 13, 321–327, 1992.

154. Pierce, S. K., Edwards, S. C., Mazzocchi, P. H., Kingler, L. H., and Warren, M. K., Proline betaine: a unique osmolyte in an extremely euryhaline osmoconformer. *Biol. Bull.*, 167, 495–500, 1984.

155. Pierce, S. K., Politis, A. D., Cronkite, D. H., Rowland, L. M., and Smith, L. H., Evidence of calmodulin involvement in cell volume recovery following hypo-osmotic stress. *Cell Calcium*, 10, 159–169, 1989.

156. Pierce, S. K., Politis, A. D., Smith, L. H., and Rowland, L. M., A Ca^{2+} influx in response to hypo-osmotic stress may alter osmolyte permeability by a phenothiazine-sensitive mechanism. *Cell Calcium*, 9, 129–140, 1988.

157. Pierce, S. K., Rowland-Faux, L. M., and O'Brien, S. M., Different salinity tolerance mechanisms in Atlantic and Chesapeake Bay conspecific oysters: glycine betaine and amino acid pool variations. *Mar. Biol.*, 113, 107–115, 1992.

158. Pierce, S. K., Rowland-Faux, L. M., and Crombie, B. N., The mechanism of glycine betaine regulation in response to hyperosmotic stress in oyster mitochondria: a comparative study of Atlantic and Chesapeake Bay oysters. *J. Exp. Zool.*, 271, 161–170, 1995.

159. Pierce, S. K., Warren, M. K., and West, H. H., Non-amino acid mediated volume regulation in an extreme osmoconformer. *Physiol. Zool.*, 56, 445–454, 1983.

160. Pirie, B. J. S. and George, S. C., Ultrastructure of the heart and excretory system of *Mytilus edulis* (L.). *J. Mar. Biol. Assoc. U.K.*, 59, 819–829, 1979.

161. Potts, W. T. W., The inorganic composition of the blood of *Mytilus edulis* and *Anodonta cygnea*. *J. Exp. Biol.*, 31, 376–385, 1954.

162. Potts, W. T. W., The inorganic and amino acid composition of some lamellibranch muscles. *J. Exp. Biol.*, 35, 749–764, 1958.

163. Prior, D. J., Analysis of contact rehydration in terrestrial gastropods: osmotic control of drinking behavior. *J. Exp. Biol.*, 111, 63–73, 1984.

164. Prior, D. J. and Uglem, G. L., Analysis of contact-rehydration in terrestrial gastropods: absorption ^{14}C inulin through the epithelium of the foot, *J. Exp. Biol.*, 111, 75–80, 1984.

165. Prior, D. J., Hume, M., Varga, D., and Hess, S. D., Physiological and behavioral aspects of water balance and respiratory function in the terrestrial slug *Limax maximus*. *J. Exp. Biol.*, 104, 111–127, 1983.

166. Prusch, R. D. and Hall, C., Diffusional water permeability in selected marine bivalves. *Biol. Bull.*, 154, 292–301, 1978.

167. Pullin, R. S. V., Composition of the hemolymph of *Lymnaea trunculata*, the snail host of *Fasciola hepatica*. *Comp. Biochem. Physiol.*, 40A, 617–626, 1971.

168. Quinn, R. H. and Pierce, S. K., The ionic basis of the hypoosmotic depolarization in neurons from the opisthobranch mollusc *Elysia chlorotica*. *J. Exp. Biol.*, 163, 169–186, 1992.

169. Remane, A. and Schlieper, K., *The Biology of Brackish Water*, Wiley Interscience, New York, 1971, 372 pp.

170. Roach, D. K., Analysis of the hemolymph of *Arion ater* L. (Gastropoda: Pulmonata). *J. Exp. Biol.*, 40, 613–623, 1963.

171. Robertson, J. D., Studies on the chemical composition of muscle tissue. III. The mantle muscle of cephalopod molluscs. *J. Exp. Biol.*, 42, 153–175, 1965.

172. Rosa, R., Periera, J., and Nunes, M. L., Biochemical composition of cephalopods with different life strategies, with special reference to a giant squid, *Architeuthis* sp. *Mar. Biol.*, 146, 739–751, 2005.

173. Rumsey, T. J., Osmotic and ionic regulation in a terrestrial snail, *Pomatais elegans* (Gastropoda, Prosobranchia), with a note on some tropical Pomatiasidae. *J. Exp. Biol.*, 57, 20–215, 1972.

174. Rumsey, T. J., Some aspects of osmotic and ionic regulation in *Littorina littorea* (L.) (Gastropoda, Prosobranchia). *Comp. Biochem. Physiol.*, 45A, 327–344, 1973.

175. Saintsing, D. G. and Towle, D. W., Na$^+$ and K$^+$ ATPase in the osmoregulating clam *Rangia cuneata*. *J. Exp. Zool.*, 206, 435–442, 1978.

176. Sato, M., Austen, G., Yai, H., and Maruhashi, J., The ionic permeability changes during acetylcholine-induced responses of *Aplysia* ganglion cells. *J. Gen. Physiol.*, 51, 321–345, 1968.

177. Saxena, B. B., Inorganic ions in the blood of *Pila globosa* (Swainson). *Physiol. Zool.*, 30, 161–164, 1957.

178. Scemes, E. and Cassaola, A. C., Regulatory volume decrease in neurons of *Aplysia brasiliana*. *J. Exp. Zool.*, 272, 329–337, 1995.

179. Scheide, J. I. and Dietz, T. H., Serotonin regulation of gill cAMP production, Na, and water uptake in freshwater mussels. *J. Exp. Zool.*, 240, 309–314, 1986.

180. Schipp, R. and Hevert, F., Ultrafiltration in the branchial heart appendage of dibranchiate cephalopods: a comparative ultrastructual and physiological study. *J. Exp. Biol.*, 92, 23–35, 1981.

181. Schlieper, C. Physiologie des Brackwassers. In: *Die Biologie des Brackwassers. Die Binnengewasser*, Vol. 22, Remane, A. and Schlieper, C., Eds., E. Schweizerbart sche Verl., Stuttgart, Germany, 1958, pp. 219–230.

182. Shumway, S. E., Effect of salinity fluctuation on the osmotic pressure and Na$^+$, Ca^{2+}, and Mg^{2+} ion concentrations in the hemolymph of bivalve molluscs. *Mar. Biol.*, 41, 153–177, 1977.

183. Shumway, S. E., The effect of fluctuating salinity on the tissue water content of eight species of bivalve molluscs. *J. Comp. Physiol.*, 116, 269–285, 1977.

184. Shumway, S. E. and Freeman, R. F. H., Osmotic balance in a marine pulmonate, *Amphibola crenata*. *Mar. Behav. Physiol.*, 11, 157–183, 1984.

185. Shumway, S. E., Gabbott, P. A., and Youngson, A., The effect of fluctuating salinity on the concentrations of free amino acids and ninhydrin-positive substances in the adductor muscles of eight species of bivalve molluscs. *J. Exp. Mar. Biol. Ecol.*, 29, 131–150, 1977.

186. Silva, A. L. and Wright, S. H., Short-term cell volume regulation in *Mytilus californianus* gill. *J. Exp. Biol.*, 194, 47–68, 1994.

187. Silverman, H., Kays, W. T., and Dietz, T. H., Maternal calcium contribution to glochidial shells in freshwater mussels (Eulamellibranchia: Unionidae). *J. Exp. Zool.*, 242, 137–146, 1987.

188. Simpson, J. W., Allen, K., and Awapara, J., Free amino acids in some aquatic invertebrates. *Biol. Bull.*, 117, 371–381, 1959.

189. Smith, L. H. and Pierce, S. K., Cell volume regulation by molluscan erythrocytes during hypoosmotic stress: Ca^{2+} effects on ionic and organic osmolyte fluxes. *Biol. Bull.*, 173, 407–418, 1987.

190. Somero, G. N. and Bowlus, R. D., Osmolytes and metabolic products in molluscs: the design of compatible solute systems. In: *The Mollusca*. Vol. 2. *Environmental Biochemistry and Physiology*, Hochachka, P., Ed., Academic Press, New York, 1983, pp. 77–100.

191. Sokolowski, A., Wolowicz, M., and Hummel, H., Free amino acids in the clam *Macoma balthica* L. (Bivalvia, Mollusca) from brackish waters of the southern Baltic Sea. *Comp. Biochem. Physiol.*, 134A, 579–592, 2003.

192. Stickle, W. B., Kapper, M. A., Blakeney, E., and Bayne, B. L., Effects of salinity on the nitrogen metabolism of the muricid gastropod, *Thais (Nucella) lapillus* (L.) (Mollusca: Prosobranchia). *J. Exp. Mar. Biol. Ecol.*, 91, 1–16, 1985.

193. Strange, K. B. and Crowe, J. H., Acclimation to successive short term salinity changes by the bivalve *Modiolus demissus*. II. Nitrogen metabolism. *J. Exp. Zool.*, 210, 227–236, 1979.

194. Tarr, K. J., An analysis of water-content regulation in osmoconforming limpets (Molluscs: Patellacea). *J. Exp. Zool.*, 201, 259–268, 1977.

195. Thomas, J. D. and Lough, A., The effects of external calcium concentration on the rate of uptake of this ion of *Biomphalria glabrata* (Say). *J. Anim. Ecol.*, 43, 861–872, 1974.

196. Tiffany, W. J., Aspects of excretory ultrafiltration in the bivalved molluscs. *Comp. Biochem. Physiol.*, 43A, 527–536, 1972.

197. Todd, M. E., Osmotic balance in *Hydrobia ulvae* and *Potamopyrgus jenkinsi* (Gastropoda: Hydrobiidae). *J. Exp. Biol.*, 41, 665–677, 1964.

198. Trams, E. G., Lauter, C. J., Bourke, R. S., and Tower, D. B., Composition of *Cepaea nemoralis* hemolymph and tissue extracts. *Comp. Biochem. Physiol.*, 14, 399–404, 1965.

199. Van Aardt, W. J., Quantitative aspects of the water balance in *Lymnaea stagnalis* (L.). *Neth. J. Zool.*, 18, 253–312, 1968.

200. Voight, J. R., Portner, H. O., and O'Dor, R. K., A review of ammonia-mediated buoyancy in squids (Cephalopoda: Teuthoidea). *Mar. Fresh. Behav. Physiol.*, 25, 193–203, 1994.

201. Vorwohl, G., Zur Funktion der exkretionsorgane von *Helix pomatia* L. und *Archachatina ventricosa* Gould. *Z. vergl. Physiol.*, 45, 12–49, 1961.

202. Watts, J. A., Koch, R. A., Greenberg, M. J., and Pierce, S. K., Ultrastructure of the heart of the marine mussel, *Geukensia demissa*. *J. Morphol.*, 170, 301–319, 1981.

203. Weber, H. H. and Dehnel, P. A., Ion balance in the prosobranch gastropod *Acmaea scutum*. *Comp. Biochem. Physiol.*, 25, 49–64, 1968.

204. Weiss, K. R., Bayley, H., Lloyd, P. E., Tenenbaum, R., Gawinowicz Kolks, M. A., Buck, L., Cropper, E. C., Rosen, S. C., and Kupfermann, I., Purification and sequencing of neuropeptides contained in neuron R15 of *Aplysia californica*. *Proc. Nat. Acad. Sci. USA*, 86, 2913–2917, 1989.

205. White, K. M., The pericardial cavity and the pericardial gland of the lamellibranchia. *Proc. Malacol. Soc. Lond.*, 25, 37–88, 1942.

206. Wieser, W., Responses of *Helix pomatia* to anoxia: changes in solute activity and other properties of the hemolymph. *J. Comp. Physiol.*, 141, 503–509, 1981.

207. Wieser, W. and Schuster, M., The relationship between water content, activity and free amino acids in *Helix pomatia* L. *J. Comp. Physiol.*, 98B, 169–181, 1975.

208. Wilbur, K. M. and Saleuddin, A. S. M., Shell Formation. In: *The Mollusca*. Vol. 4. *Physiology*, Part 1, Saleuddin, A. S. M. and Wilbur, K. M., Eds., Academic Press, New York, 1983, pp. 235–287.

209. Wilmer, P. G., Volume regulation and solute balance in the nervous tissue of an osmoconforming bivalve (*Mytilus edulis*). *J. Exp. Biol.*, 77, 157–179, 1978.

210. Yancey, P. H., Organic osmolytes as compatible, metabolic, and counteracting cryoprotectants in high osmolality and other stresses. *J. Exp. Biol.*, 208, 2819–2830, 2005.

211. Zachariassen, K. E., Olsen, A. J., and Aunaas, T., The effect of formaldehyde exposure on the transmembrane distribution of free amino acids in muscles of *Mytilus edulis*. *J. Exp. Biol.*, 199, 1287–1294, 1996.

5 Osmoregulation in Annelids

Robert L. Preston

CONTENTS

I. INTRODUCTION AND OVERVIEW

The organisms that constitute the phylum Annelida, comprised of at least 20,000 species, are remarkably diverse and occupy habitats from open ocean, to estuaries, to freshwater streams and lakes, to soil in terrestrial environments.[3,81] A suite of biological adaptations is necessary for survival in each of these specific niches, but one of the most important adaptations is certainly the capacity to regulate internal osmotic pressure and the composition of cellular and tissue osmolytes. Freshwater species generally maintain comparatively high internal osmotic pressures (150 to 250 mOsm; see below) and must compensate for osmotic water gain. Terrestrial species face transient freshwater challenges and desiccation stress. Saltwater species generally are in near osmotic equilibrium with seawater but of necessity must regulate intracellular composition (as do virtually all living cells). Brackish water species and those saltwater species that migrate into estuarine habitats may regulate or resist osmotic challenge but in most cases are capable of osmotolerance. In addition, parasitic, mutualistic, and commensal species also exist.[6,81] Although some reviews in the area of epithelial transport and osmoregulation have been published, it appears that a review from a more general perspective has not been undertaken for some time.[11,96] This broad approach seeks to honor in some small way the spirit of the classic volume of Potts and Parry that this series commemorates.[66]

Perhaps one of the most interesting aspects of annelid species is that there is a fairly clear correlation among the classification of annelid species and their physiological capabilities to osmoregulate. Recent phylogenetic analyses have split Annelida into Polychaeta and Clitellata.[3,81,82] These analyses take advantage of new molecular approaches as well as reexamination of the morphological bases of classification. It appears that even some phyla previously considered

TABLE 5.1
Typical Osmotic Pressures of Coelomic Fluids in Annelid Species

Class and Species	Medium Osmotic Pressure (mOsm)	Coelomic Fluid (or Blood) Osmotic Pressure (mOsm)	Na (mM)	K (mM)	Cl (mM)	Refs.	Notes
Hirudinea							
Hirudo medicinalis	—	201 (blood)	136	6.0	36	Zerbst-Boroffka[125]	—
Poecilobdella granulosa	—	145 (blood)	67	11.2	41	Ramamurthi[75]	—
Oligochaeta							
Lumbricus terrestris	—	154 ± 2	71 ± 2	4 ± 0	48 ± 1	Dietz and Alvarado[16]	Worm kept in soil
Pheretima posthuma	—	154–167	80.4	5.9	21.7–22.9	Bahl[2]	Fully hydrated
Polychaeta							
Nereis virens	1000	1033	458	14.7	526	Oglesby[62]	—
Glycera dibranchiata	1000	1048	436 ± 4	12.9 ± 2.9	463 ± 2	Stevens and Preston;[114–116] Oglesby;[62] Mead and Preston (unpubl.)	—

Note: The values listed here show the range of osmotic pressures and solute composition for coelomic fluid from selected species. These values are more or less representative of the general case for other species within these classes, with some exceptions. These values are in part taken from earlier reviews that provide a more comprehensive listing of these and other species (see Oglesby[62]). In some cases, the data were recalculated to express all values in mOsm or mM units. These values are intended to reflect those of the worms in their typical natural habitat.

independent (e.g., Pogonophora, Echiura, Sipincula) may be greatly modified annelids.[3,81,82] A comprehensive discussion of this reclassification is beyond the scope of this review, but the significant changes occurring in this field should be kept in mind. That being said, it is convenient for discussion of the osmoregulatory adaptations of annelids to temporarily return to the older classification (classes Hirudinea, Oligochaeta, and Polychaeta), as it correlates reasonably well with the habitats and the physiological mechanisms employed in osmoregulation, although with exceptions. The following section characterizes very briefly the typical niche of these groups in relation to osmoregulatory stress.

Class Hirudinea (leeches) usually frequent freshwater environments, but some species may be marine or terrestrial. Many are free-living ectoparasitic bloodsuckers or scavengers.[6] Most leeches maintain osmotic pressures of 150 to 200 mOsm (Table 5.1) in spite of very large gradients favoring passive osmotic water uptake (freshwater osmotic pressure may be variable but typically may be considered to be about 10 mOsm; typical vertebrate blood osmotic pressure is about 300 mOsm). Most leeches parasitize fishes, amphibians, or other vertebrates and obtain blood meals of high salt, protein, and free amino acid content. The osmotic water gain must be compensated for by active excretion mechanisms (nephridia) and resisted to some lesser extent by hydrostatic pressure developed within the animal that acts as a driving force opposing osmotic water uptake. Almost all annelids, including Hirudinea, have a fairly thick body wall with a cuticle comprised of cellular components, collagen fibers in multiple layers, and muscle that functions as a *hydrostatic skeleton*. The hydrostatic skeleton is necessary for locomotion in annelids—the transmission of forces via the tube-shaped cavities filled with fluid and tissues.[1,38,54,55,73,74,98–100] The extent to which these hydrostatic forces contribute to the driving force for water excretion is open to discussion, but at least a theoretical case may be made for a significant contribution under some circumstances (see further discussion below).

Class Oligochaeta (e.g., earthworms) are commonly terrestrial or freshwater. Less frequently, members of this group can be found in intertidal zones and marine environments. Taken as a model, the earthworm (*Lumbricus* sp.) is a soil processor or detritus feeder. The typical osmotic pressure for lumbricoids is about 150 mOsm.[8,16,60,71,76,77] The terrestrial subsoil environment subjects the animals to potential desiccation, floods of rainwater, and anoxia.[14,46,47] With regard to the osmotic environment, these organisms should minimize drastic osmotic water loss (via desiccation) or gain (flooding by rain water) and take advantage of microenvironments in the soil that moderate osmotic stress. It is well known that earthworms are capable of moving fairly large distances in the soil; therefore, a major component of *behavioral osmoregulation* (seeking optimal external osmotic conditions) may be present.[46] Freshwater oligochaetes face osmotic challenges parallel to those of the Hirudinea: potential excessive passive water gain and the necessity to tolerate it or actively excrete water. It is further obvious that the hydrostatic skeleton plays a role in locomotion and possible water excretion in oligochaetes.

Class Polychaeta is primarily marine, although some species may inhabit brackish water and a few live in freshwater. Among the multitude of marine species, the osmotic pressure of the internal body fluids (blood, coelomic fluid, and extracellular fluid) is generally in near equilibrium with that of the external medium. Assuming that a typical reference value for ocean seawater osmotic pressure is about 1000 mOsm, it is generally the case that coelomic fluid is slightly hypertonic to 1000 mOsm (see below). In bays, estuaries, and areas with significant influx of freshwater from rivers and streams, the external water osmotic pressure may be significantly lower (500 to 900 mOsm), and in those species that survive in these areas the osmotic pressure of the coelomic fluid generally approaches that of the external medium but reaches steady state slightly hypertonic to the external medium. Some estuarine species are osmotolerant, but most do not seem to express the same *osmotic resistance* shown in the Hirudinea and Oligochaeta. This implies that compensatory cellular osmoregulatory mechanisms must play a role in these species. The body wall and cuticle still resemble those of Hirudinea and Oligochaeta, because the hydrostatic skeleton plays a crucial role in locomotion. Water and solute exchange with the environment may also be significant across gills and across the intestine in polychaetes. Gills, in particular, tend to be diverse and sometimes elaborate in polychaetes, and, because they usually live in isotonic media, this presumably confers a selective advantage in oxygen uptake in an isotonic habitat.

II. BIOLOGY AND FORM OF ANNELIDS

A. Body Plans of Annelid Classes

This section provides a brief review of the basic characteristics of the annelid body plan, along with a summary of some of the similarities and differences in the three classes of annelids. Of particular interest are those features that define fluid and solute compartments that are crucial in osmoregulation. The key issues center around the permeability of body surfaces to water and solutes, the organs involved in solute and water uptake, and the organs involved in solute and water excretion.

The most obvious and salient feature of annelids are that they are usually tube-like and bilaterally symmetrical and are composed of repeating segmental units (Figure 5.1A–C). In some species, the segments are very much alike; others may demonstrate specialization for different functions in different areas of the body.[6] The head (prostromium) may be highly specialized with tentacles (antennae, palps, cirri), feeding apparatus, pharynx, gills, and sensory organs, including eyes. In some species, these are very reduced and primitive; in others, they are highly elaborated. In some species, the tail (pygidium) may be elaborated, but usually it is not.

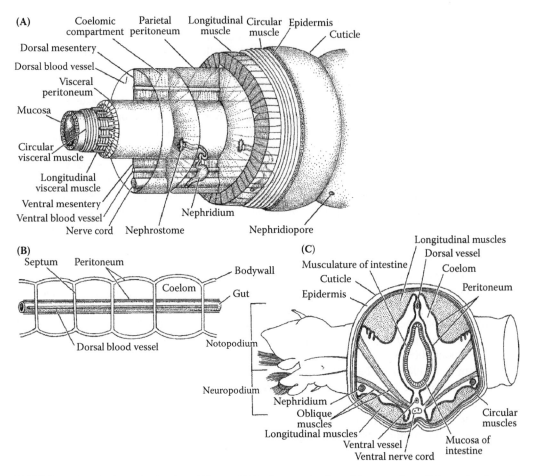

FIGURE 5.1A Polychaete body plan. (A) Annelid body organization; this general condition exists in polychaetes and oligochaetes. (B) Metameric coelom arrangement in a polychaete, seen in dorsal view (the dorsal body wall has been removed). (C) Nereid polychaete (cross-section); note the consolidation of longitudinal muscles into nearly separate bands. (From Brusca, R.C. and Brusca, G.J., *Invertebrates*, 2nd ed., Sinauer, Sunderland, MA, 2003, chap 13. With permission.)

The segments may be physically separated by septa that form distinct compartments. Typically, the coelomic cavity within the segments may be divided dorsal ventrally by mesenteric membranes, producing functionally right and left coelomic cavities. In some species, septa are fenestrate or missing so the coelomic fluid freely circulates among them. The extreme example of this condition occurs in the Glyceridae, which may lack septa altogether. In the marine polychaete *Glycera dibranchiata*, for example, the coelom is completely open, and the coelomic fluid containing hemoglobin-containing coelomocytes (nucleated red blood cells) and varieties of white blood cells moves throughout the entire body by body-wall muscular contraction.[40] During breeding season, the gametes form and develop while floating freely in the coelomic fluid, and they may outnumber other cell types in breeding season.[40]

Typical body plans for the three classes of annelids are shown in Figure 5.1A–C.[6] The general features of potential relevance to osmoregulatory physiology include a body wall surrounded by a cuticle of proteinaceous fibers (typically high in collagen content) and mucopolysaccharide fibers, both deposited by the cells of the epidermal layer (Figure 5.2). The epidermis is usually composed of columnar epithelial cells that may be ciliated in some areas of the body. Elongated microvilli from epidermal cells may penetrate the cuticle and, in polychaetes in particular, contact the external

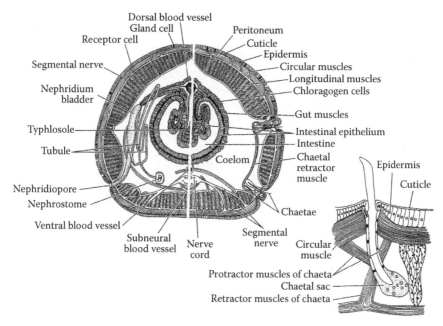

FIGURE 5.1B Oligochaete body plan showing body wall and general internal organization of the earthworm (cross-section). The left side of the illustration depicts a single nephridium so the drawing is a composite of two segments; the right side of the illustration shows a chaeta and its associated musculature. (From Brusca, R.C. and Brusca, G.J., *Invertebrates*, 2nd ed., Sinauer, Sunderland, MA, 2003, chap 13. With permission.)

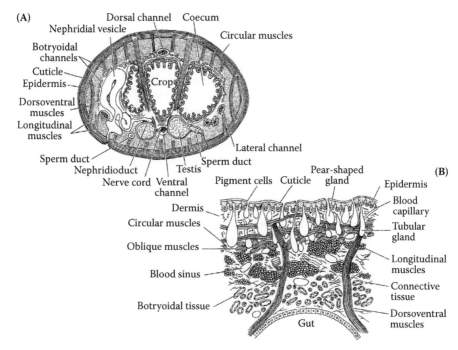

FIGURE 5.1C Hirudinea body plan showing body wall and general internal organization. (A) Cross-section of the leech *Hirudo*, in which the original circulatory system has been lost and replaced by coelomic channels. (B) Body wall of *Hirudo*. Note in both of these illustrations the effectively "acoelomate" body structure resulting from reduction of the coelom. (From Brusca, R.C. and Brusca, G.J., *Invertebrates*, 2nd ed., Sinauer, Sunderland, MA, 2003, chap 13. With permission.)

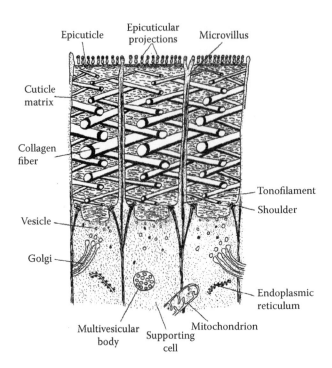

FIGURE 5.2 Annelid cuticle. The cuticle has connective tissue fibers (collagen) reinforcing layers of muscle. Microvilli penetrate the epidermal layers and are exposed to the external medium. (From Richards, K.S., in *Physiology of Annelids*, Mill, P.J., Ed., Academic Press, New York, 1978, pp. 33–62. With permission.)

medium.[78] These microvilli are implicated in polychaete species in the organic solute absorption that is virtually ubiquitous in marine species (Figure 5.2).[10,78] Beneath the epidermis is a layer of connective tissue that binds together layers of muscles that may be oriented circularly (around the worm body circumference), longitudinally or at oblique angles. Various internal bundles of muscles may also be present to control parapodia (in polychaetes), setae, or other structures.

Penetrating the septa and running the length of the body is the digestive track, itself surrounded by thin layers of circular and longitudinal muscle, with an inner absorptive epithelial mucosal cell layer. In addition, a ventral nerve cord communicates with the head region (and with ganglia that function as a brain). In species with closed circulatory systems, a dorsal blood vessel runs the length of the body.

Some structures are distinctly associated with the annelid class, and these structures appear likely to have some impact on body-wall water permeability. Parapodia are present only in polychaetes.[6] Figure 5.1A shows a cross-section of a typical polychaete. Parapodia project from each side of each segment and usually contain bundles of stiff chitinous and scleroprotein bristles (setae), which are connected to internal supporting rods (acicula). Muscle bundles control setae motion directly or are connected to the acicula. The setal surface membranes are usually thin (compared with the body wall); they may have fleshy projections (cirri) and, in some species, gills. The setae may be served by the circulatory system or coelomic fluid may circulate through the internal setae space forced by muscular contraction of the body wall. In some cases, ciliated tracts line the internal setal tissues that assist in circulation of coelomocytes through gills.[40] From the perspective of osmoregulatory physiology, the generally thin setal epidermal tissue is a potential area of water loss or gain. Because most polychaetes are marine and their coelomic fluid isotonic to the external medium, this is not usually a problem. Furthermore, Stephens and coworkers[10,72] have shown that the parapodia are an important route through which dissolved organic molecules may be actively absorbed.

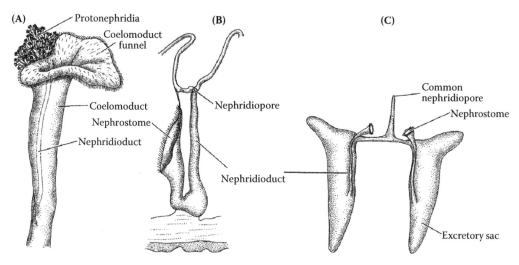

FIGURE 5.3A Structure of polychaete nephridia. (A) Protonephromixium of a phyllodocid. Here, a cluster of solenocytic protonephridia sits atop a nephridioduct that joins with the coelomoduct. (B) Mixonephrium of a spionid. (C) Single pair of nephridia joined to a common duct in a serpulid. (From Brusca, R.C. and Brusca, G.J., *Invertebrates*, 2nd ed., Sinauer, Sunderland, MA, 2003, chap 13. With permission.)

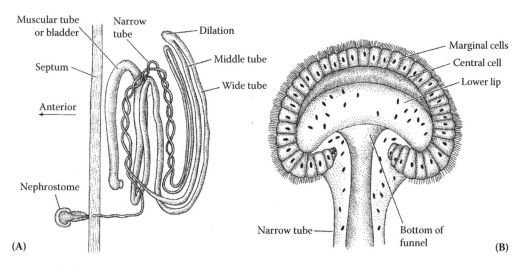

FIGURE 5.3B Structure of oligochaete (*Lumbricus*) nephridia. (A) Single nephridium and its relationship to a septum; (B) details of the nephrostome. Evidence suggests that earthworm nephridia are highly selective excretory and osmoregulatory units. The nephridioduct is regionally specialized along its length. The narrow tube receives body fluid and various solutes, first from the coelom through the nephrostome and then from the blood via capillaries that lie adjacent to the tube. In addition to various forms of nitrogenous wastes (ammonia, urea, uric acid), certain coelomic proteins, water, and ions (Na^+, K^+, Cl^-) are also picked up. Apparently, the wide tube serves as a site of selective reabsorption (probably into the blood) of proteins, ions, and water, leaving the urine rich in nitrogenous wastes. (From Brusca, R.C. and Brusca, G.J., *Invertebrates*, 2nd ed., Sinauer, Sunderland, MA, 2003, chap 13. With permission.)

Oligochaetes and hirudineans lack parapodia (see Figures 5.1B and C). Extensive analyses of locomotion by all three annelid classes suggest that their particular musculature and epidermal/cuticular structures fit their specific habitats reasonably well;[73,74,123] however, these structural differences also seem consistent with the need to minimize osmotic water loss or gain in terrestrial or freshwater environments. Oligochaetes have reduced setae, comparatively large coelomic cavity spaces, and

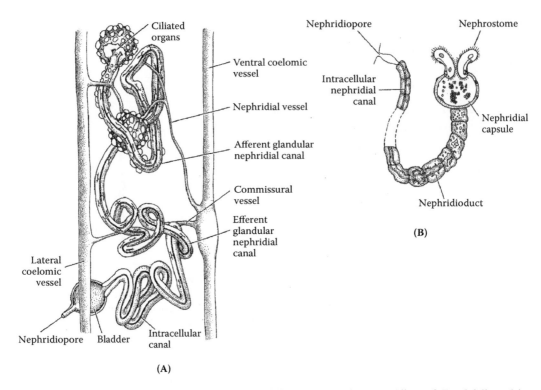

FIGURE 5.3C Structure of hirudinean (leech) nephridia. (A) Complex nephridium of *Erpobdella* and its association with the coelomic channels. (B) Details of a nephridium of an arhynchobdellid leech. (From Brusca, R.C. and Brusca, G.J., *Invertebrates*, 2nd ed., Sinauer, Sunderland, MA, 2003, chap 13. With permission.)

moderately thick cuticles and epidermis. The Hirudinea lack parapodia and setae and have a thick cuticle/epidermal layer. They have a more or less solid body construction (muscle and connective tissue) without large coelomic-fluid-filled cavities, and they lack segmentation by septa (Figure 5.1C). They typically have a large crop and intestinal cavity and a well-elaborated system of nephridia.

All annelids have metanephridia or protonephridia that function as primitive kidney tubules to assist in regulation of solute and water balance (Figure 5.3A–C). Most annelids have metanephridia, a tubular structure composed of absorptive and secretory epithelia, beginning at the nephrostome, which is open to the coelom and collects coelomic fluid. Typically, each segment has two meta-nephridia, although the number may be much reduced in some species. The nephridial tubule may wind through the segment and pass into the next body segment.[6] The length of the tubule seems to correlate to some extent with the habitat, being longer in freshwater than marine species. The tubule opens to the outside through the nephridiopore in the body wall. The nephridiopore opening may be closed or opened by sphincter muscles that respond to the osmotic status of the animal.[6]

Protonephridia are considered somewhat more primitive excretory structures and are present in one form or another in a number of invertebrate species. In general, these structures contain ciliated or flagellated cells that propel coelomic fluid through the fairly short tubules, which open externally through a nephridiopore. They tend to be best developed in those species that reside in freshwater and are important in water export.[6]

B. CLITELLA AND COCOONS

The clitellum is a reproductive structure composed of modified segments that are unique to Oli-gochaetes and Hirudinea, hence their classification together as Clitellata.[6] Polychaetes do not possess this structure. The clitellum forms a sac-like structure in which eggs are deposited, usually following copulation and fertilization. The clitellum usually forms a fairly stable cocoon, which is shed after mating.[81] The cocoon is a protected environment where embryo development occurs and young worms eventually escape when developmentally mature. Most authors suggest that the major function of the clitellum/cocoon is protection of the young from predators.[81] Although this is most certainly true, another obvious function is that the cocoon and clitella protect eggs and embryos from osmotic stress. This correlates nicely with the freshwater and terrestrial habitats of the Clitellata and with the absence of these structures on the predominately marine Polychaeta.

III. OSMOREGULATION: A BRIEF REVIEW OF RELEVANT PRINCIPLES

A. OSMOTIC- AND PRESSURE-DRIVEN WATER FLOW

The fundamentals of osmosis, osmoregulation, and membrane transport and diffusion have been treated by Dawson in an earlier chapter in this volume; however, a few key concepts that have relevance to the physiology of osmoregulation in annelids are worth restating. A succinct summary of these principles is given by Baumgarten and Feher,[4] and the following development is essentially that presented by those authors (with permission).

The osmotic pressure of ideal dilute solutions is given by the van't Hoff equation:

$$\pi = RT\Sigma C_s \tag{5.1}$$

where π is the osmotic pressure, R is the gas constant, T is the temperature in degrees Kelvin, and ΣC_s is the sum of the concentrations of osmotically active particles (*osmotic activities*) that are formed on dissociation of solutes. We use the traditional physiological convention of expressing the units in osmotic concentration terms: osmolar (Osm; osmoles/liter) or osmolal (osmoles/kilogram water). Alternatively, osmolarity may be expressed as pressure units, atmospheres, mmHg, pascals (Pa; N/m^2) or dyn/cm^2. Equation 5.1 gives a reasonable approximation for dilute solutions, but real solutions are not ideal, and at high concentrations the values for osmotic pressure may depart substantially from the ideal state. A correction factor is applied that takes the difference into account, the osmotic coefficient (φ_s). Thus, the van't Hoff equation can be rewritten:

$$\pi = RT\Sigma\varphi_s C_s \tag{5.2}$$

where the terms are as defined previously.

The equivalence of osmotic concentration and pressure is not accidental; osmotic gradients generate real pressures that may drive flow of solvent. Conversely, pressure gradients may be used to do osmotic work. An example of the magnitude of the quantitative relation between osmotic concentration and pressure is worth consideration. A 10-mM ideal solution of glucose or a 5-mM solution of NaCl (complete dissociation into two particles) will have an osmolarity of 10 mOsm. At 37°C, a 10-mOsm solution would have an osmotic pressure of 0.254 atmospheres, 193 mmHg, or 25.7 kilopascals (kPa). In other words, *fairly small osmotic pressure gradients may establish rather significant hydrostatic pressures in physiological systems.*

The formal relationship of osmotic and hydrostatic pressure is shown below.[4] The flow rate across a membrane is linearly related to the osmotic concentration difference across the membrane by Equation 5.3:

$$J_v = -L_p(\pi_i - \pi_o) = -L_p\Delta\pi \tag{5.3}$$

where J_v is the volume flux in cm^3/sec per unit area of membrane, π_i is the internal osmotic concentration, π_o is the external osmotic concentration, and L_p is the hydraulic conductivity (filtration coefficient or hydraulic permeability). The negative sign is required to indicate that the water flow is from low osmotic concentration to high osmotic concentration, and we assume for this analysis that $\pi_i > \pi_o$ (in most cases). The flow across the entire membrane (Q_v; units cm³/sec) is given by:

$$Q_v = -AL_p\Delta\pi \tag{5.4}$$

The flow rates due to hydrostatic pressure in the absence of osmotic concentration gradients are given by similar equations:

$$J_v = L_p (P_i - P_o) = -L_p\Delta P \tag{5.5}$$

$$Q_v = AL_p\Delta P \tag{5.6}$$

where P_i is the internal hydrostatic pressure, and P_o is the external hydrostatic pressure. The L_p was found to have the same value relating pressure driven and osmotically driven flow. Osmotic-driven flow can be nulled by opposing pressure flow, and the equivalent value for L_p allows the following relationship:

$$Q_v = AL_p[(P_i - P_o) - (\pi_i - \pi_o)] \tag{5.7}$$

$$Q_v = AL_p(\Delta P - \Delta\pi) \tag{5.8}$$

This equation describes the net flow in the presence of both hydrostatic and osmotic pressure gradients across a semipermeable membrane.

A final topic must be addressed with regard to this type of osmotic flow—that of the selectivity for solutes (and perhaps solvent) of real membrane vs. ideal membranes. An ideal semipermeable membrane would permit only water to flow, excluding any solute flow. Of course, real biological membranes do not usually behave in an ideal way, and solute also flows to a greater or lesser extent depending on the particular membrane, the particular organism, and perhaps also the adaptation state and expression of particular membrane channel proteins or transporters. If a membrane is partially permeable to solute, the measured osmotic pressure should be less than that predicted by van't Hoff's equation. A second membrane coefficient can be defined that corrects for difference, the reflection coefficient (σ), which is defined as:

$$\sigma = [(\pi_{observed})/(\pi_{theoretical})] = [(\pi_{observed})/(\varphi_s RTC_s)] \tag{5.9}$$

With an ideal semipermeable membrane, $\sigma = 1$, and the calculated osmotic pressure would match the measured osmotic pressure. The parallel permeation of solute would reduce the driving force for osmotic flow and the osmotic pressure, so the volume flow would be given by:

$$Q_v = AL_p\left[(P_i - P_o) - \left(\Sigma_j\sigma_j\pi_{j,i} - \Sigma_j\sigma_j\pi_{j,o}\right)\right] \tag{5.10}$$

where σ_j is the reflection coefficient for solute j, $\pi_{j,i}$ is the osmotic pressure of solute j on the inside, and $\pi_{j,o}$ is the osmotic pressure of solute j on the outside.

It should be recognized that these relationships refer to cellular biological plasma membranes or simple artificial membranes. In the case of whole epithelial membranes, connective tissues layers and muscle, as well as the surface epithelial cell layer, must be considered. In addition, parallel flow pathways through gills and parapodial membranes (in polychaetes) may be present. These same relationships may apply but one needs to recognize that the values for L_p and σ may be aggregates of the properties of these layers and in the best case may represent the rate-limiting step for water and solute movements. Furthermore, one would expect that these epithelia could vary widely from species to species. This is even more problematic for annelid body walls, as they have a cuticular layer of collagen fibers of various thicknesses. In addition, in the intact organism,

TABLE 5.2
Hydrostatic Pressures in Oligochaetes and Hirudinea

Class and Species	Apparent Fluid Compartment Measured	Resting or Baseline Pressure[a] (kPa)	Peak Pressure Active[b] (kPa)	Maximum Pressure[c] (kPa)	Notes	Ref.
Oligochaeta						
Lumbricus terrestris	Coelom	0.49	1.96	7.82	—	Seymour[98–100]
	Coelom	2	10	—	Axial forces	Quillin[74]
	Coelom	20	100	100	Burrowing forces	Quillin[74]
	Coelom	—	46.3 ± 3	—	Axial forces, mean values	Keudel and Schrader[38]
	Coelom	—	72.6 ± 12	—	Radial forces, mean values	Keudel and Schrader[38]
Aporectoda rosa	Coelom	—	72.8	116.5	—	McKenzie and Dexter[54,55]
Glossoscolex gigantea	Dorsal vessel	1.17	1.96	—	—	Johansen and Martin[33]
	Ventral vessel	4.40	5.87	—	—	Johansen and Martin[33]
	—	5.87	9.78	—	Quiet worm	Johansen and Martin[33]
	—	4.89	12.7	—	Active worm	Johansen and Martin[33]
Hirudinea						
Hirudo medicinalis	Coelom	0.196	1.08	—	Crawling	Wilson et al.[123]
	Coelom	0.196	2.93	—	Swimming	Wilson et al.[123]
	Vascular system	0.67	6.40	13.3	—	Krahl and Zerbst-Boroffka[41]

[a] Resting or baseline pressures recorded on an inactive animal.
[b] Peak pressure during normal locomotor activities.
[c] Maximum pressure observed when the animal was maximally stimulated.

nephridial urine output must be considered as a depressurizing flow component, and any apparent L_p and σ estimates for whole body walls would necessarily include this component. Urine formation and output are most likely controlled to some extent by various regulatory processes including neural osmoregulatory peptides (see below).[84–89] Nonetheless, with these reservations in mind, one should be able to characterize apparent L_p and perhaps σ values that may be useful in modeling the osmotic behavior of annelid body walls.

Because maintenance of a positive internal hydrostatic pressure compared to the environment is crucial for normal annelid locomotion, it can be predicted that the normal regulatory setpoint for the osmotic concentration in coelomic fluid (which is presumably in equilibrium with blood and extracellular fluids) for annelids at steady state with the medium (not undergoing transient osmotic challenge or in isomotic media) should be slightly hyperosmotic to the environment. It is obvious that in freshwater oligochaetes and hirudineans, very large gradients exist that favor osmotic water uptake, and maintenance of hydrostatic pressure is not a problem (Table 5.2). The cuticles must be relatively impermeable to limit osmotic water uptake (and efflux), and the nephridial output of urine to compensate for osmotic water gain must be continuous and regulated. Table 5.2 shows some of the observed hydrostatic pressures reported under resting and active conditions for selected oligochaetes and leeches. In terrestrial oligochaetes (earthworms), water conservation may be crucial in desiccating environments.

In polychaetes, on the other hand, where the internal body fluids are rather close to that of the external medium, maintenance of internal hypertonicity is essential. The question that arises is whether the hydrostatic pressures that could be potentially generated by the observed osmotic gradients in polychaetes is sufficient to produce realistic hydrostatic pressures observed in these animals. Unfortunately, direct measurements of polychaete hydrostatic pressures seem to be lacking, but comparisons with the values in Table 5.2 for oligochaetes and leeches should give us a rough estimate of what pressures may be necessary for normal baseline resting activities and locomotory activities.

Oglesby[62] comprehensively reviewed the osmotic relationships and major ion concentrations in coelomic fluid compared with the external medium for 38 species in 18 families of polychaetes (and 2 species of oligochaetes) adapted to high salinities (usually close to 100% seawater, which Oglesby takes as 1033 mOsm) and concluded that most annelids are hyperosmotic regulators. The body fluid/medium ratios ranged from 0.890 to 1.525, but two thirds of the values fell between 1.010 and 1.199. The outlying values may be less trustworthy due to differences in technique or other experimental concerns. As noted above, a 10-mOsm osmotic pressure gradient can generate a hydrostatic pressure of 25.7 kPa. Assuming that the external medium is 1033 mOsm, the range of hydrostatic pressures that could be generated is 26.7 to 528 kPa. This range encompasses that for the resting states listed in Table 5.2 and even potentially reaches the very high values observed in active animals. From the summary data of Oglesby, 28 of 59 values for the internal/external medium ratios fell in the range of 1.002 to 1.095.[62] If we assume that this range may more realistically represent the modal condition (and that perhaps some of the higher and lower values resulted from experimental problems), the calculated average ratio is 1.043 ± 0.005 ($n = 28$). The internal medium would therefore be about 44 mOsm higher in concentration than the external medium, and this could potentially generate a maximum hydrostatic pressure of 114 kPa. This easily encompasses the predicted range of pressures for resting worms (based on Table 5.2).

In a more recent paper, Generlich and Giere[24] summarized the data from a number of studies that included examples from polychaetes, oligochaetes, and leeches (Table 5.3), and this prediction seems to be basically correct. The ratio of the osmotic pressure of the coelomic fluid to the external medium ranges from about 1.075 (for *Hirudinea medicinalis*) to 1.009 (for *Heterochaeta costata*) under nearly isosmotic conditions. Some organisms were reported to be isosmotic under these conditions. In hypotonic media, the ratios increased (Table 5.3), commonly exceeding 1.1 to as high as 15 (*Enchytraeus albidus*, *Hediste diversicolor*).

The arguments presented here are consistent with the notion that the slight hypertonicity of internal body fluid that is routinely measured in polychaetes potentially plays an important physiological role in maintaining a positive hydrostatic pressure that is essential for the locomotory activities of the animals. In addition, this pressure gradient could be a significant driving force favoring nephridial filtration of coelomic fluid and urine formation.[2,18,25–29,83,108,126]

B. Overview of the Pathways for Solute and Water Movement

To put the global picture in perspective, one may describe an ideal annelid for each of the three environments: ocean, land, and freshwater lakes and streams. The ideal polychaete could have a highly water-permeable but *solute-impermeable* body wall. Because it lives in isotonic medium (nature's Ringer's solution), solute exchange and dissolved organic nutrient uptake would be selectively advantageous. The surface of the worm may also have elaborated gills and tentacles for oxygen exchange and food intake. Nutrient uptake occurs via the gut and in most species via surface uptake of dissolved organic nutrients, especially free amino acids, sugars, and the like.[10,13,53,68–70,72,109–116] The surface uptake of dissolved organic nutrients (DOMs) occurs in virtually all soft-bodied marine

TABLE 5.3
Concentration of Body Fluids of Annelids in Different Salinities at Steady State

Species	Concentration of Medium (% Seawater)	Coelomic Fluid/Medium Concentration Ratio	Refs.
A. Polychaeta			
Nereidae			
Hediste diversicolor	1.4	11–15	Hohendorf[31]
	14	2.16	Hohendorf[31]
	29	1.27	Hohendorf[31]
	50	1.199	Schlieper[91]
	50	1.121	Fletcher[23]
	70	1.086	Fletcher[23]
	97	1.022	Fletcher[21]
	100	1.049	DeLeersnyder[15]
	102	1.000	Hohendorf[31]
	106–109	1.026–1.098	Karandeeva[37]
B. Oligochaeta			
Enchytraeidae			
Enchytraeus albidus	2.6	14.96 or 12.44	Generlich and Giere[24]
	31	1.715 or 1.503	Generlich and Giere[24]
	44	1.205	Generlich and Giere[24]
	58	1.146	Generlich and Giere[24]
	94	1.042	Generlich and Giere[24]
	115	1.066	Generlich and Giere[24]
	<75	Hyperosmotic	Schone[97]
	112–120	Isosmotic	Schone[97]
Marionina achaeta	<75	Hyperosmotic	Lasserre[45]
Tubificidae			
Heterochaeta costata	44	1.083	Generlich and Giere[24]
	91	Isosmotic (0.991–1.009)	Generlich and Giere[24]
Clitellio arenarius	47	1.083	Ferraris;[19] Ferraris and Schmidt-Nielsen[20]
	62	1.062	Ferraris;[19] Ferraris and Schmidt-Nielsen[20]
	92	Isosmotic	Ferraris;[19] Ferraris and Schmidt-Nielsen[20]
Naididae			
Nais elinguis	<20	Hyperosmotic	Little[48]
	20	Isosmotic	Little[48]
	20–57	Hyposmotic	Little[48]
Megascolecidae			
Pontodrilus bermudensis	14	2.62	Subba Rao;[117] Subba Rao and Ganapati[118]
	43	0.99	Subba Rao;[117] Subba Rao and Ganapati[118]
	86	1.013	Subba Rao;[117] Subba Rao and Ganapati[118]
Lumbricidae			
Lumbricus terrestris	0.3	62.8	Prusch and Otter[71]
	1	10.6	Ramsay[76]
	1	17.1	Dietz and Alvarado[16]
	13	1.27	Dietz and Alvarado[16]
	15	1.44	Ramsay[76]
	26	1.055	Dietz and Alvarado[16]
	45	1.1	Ramsay[76]
Hirudinea			
Hirudo medicinalis	33	1.075	Nieczaj and Zerbst-Boroffka[57]

Source: Adapted from Generlich, O. and Giere, O., *Hydrobiologia*, 334, 251–261, 1996.

invertebrates, and in some instances can contribute significantly to the nutrition of these animals.[110] Active transport of amino acids has been thoroughly studied, and it has been shown that annelids and other marine invertebrates are capable of accumulating amino acids against very high gradients (approaching under some circumstances gradients of 1 million to 1), primarily by sodium-dependent cotransport mechanisms.[68–70,114–116]

Preston[68–70] reviewed the thermodynamic requirements of such processes, and to accumulate amino acids to such high gradients would require multiple coupling coefficients (two or three sodiums per cotransported amino acid), low cytosolic sodium activities, and electrogenic coupling of influx to the cellular transmembrane potential (which is typically on the order of –60 mV in marine invertebrate tissues). Support for this sort of mechanism has been provided by a number of studies.[68] This capability provides some selective advantage to having an integument that may allow function of these transport systems. In freshwater annelids, very low or no such organic solute transport occurs.[109–113] Obviously, sodium cotransport systems in freshwater environments are impractical, as the sodium gradients favor efflux from the animal, and, further, the integument must be generally quite impermeable to resist osmotic water gain. For marine polychaetes, their environment has no deficit of minerals and the coelomic fluid resembles seawater compositionally, so it might be expected that salt regulation may occur at the tissue level. The nephridia presumably fine-tune the coelomic fluid solute and water content to retain a positive pressure in the hydrostatic skeleton and further conserve organic nutrients that may be present in the coelomic fluid. High surface-water permeability limits the distribution to ocean, intertidal, and mudflat environments, excluding freshwater and terrestrial environments where water gain or loss would be substantial.

The ideal terrestrial oligochaete would balance the water permeability properties of the body wall to resist desiccation but allow water gain from the interstitial moisture in the soil.[124] Minerals and nutrients must be recovered from the food, perhaps with some assistance via surface uptake of salts, especially sodium chloride. Large volumes of soil with low nutrient and ion content must be processed to recover sufficient critical nutrients and ions. Oligochaete intestine should be adapted to avid absorption of both salts and organic molecules.[12] Unlike the polychaetes, organic nutrient uptake by oligochaetes may not occur to any significant extent, presumably because the pathways that permit organic solute uptake may permit adventitious water uptake, as well. This same principle should also apply to the Hirudinea (see below). In fact, Stephens[109–113] has shown in extensive comparative studies that active organic solute uptake occurs across the body surfaces of virtually every soft-bodied marine invertebrate species (excluding the arthropods, which have an impermeant chitinous exoskeleton) and that the uptake of these nutrients is very small or not detectable in freshwater species. Nephridia should ordinarily recover efficiently and highly conserve both salts and organic molecules. The regulation and maintenance of hydrostatic pressure are crucial in the potential variable terrestrial environment, so expression of strong behavioral and hormonal control of the osmotic internal *milieu* would be expected. It might also be expected that in some circumstances salt and water excretion as well as conservation may be necessary.

The Hirudinea (modeled on parasitic freshwater leeches) should have thick highly water impermeant body walls that slow osmotic water uptake. Because the osmotic gradients are very substantial— approximately 10 mOsm for freshwater compared with 220 mOsm for the coelomic fluid—substantial passive water uptake must occur. It might be expected that the body wall should be thick and generally impermeable to solutes. Gills and other surface elaborations that increase the area through which passive water gain may occur should be minimal or absent. The uptake of organic nutrients across the body wall should be very low or nonexistent, but it is possible that sodium uptake can occur across the body surface.[11,39,71,96] The nephridia should function primarily to excrete water gained osmotically and regulate coelomic fluid osmotic pressure (partly to maintain the hydroskeleton) and composition.[120–123,125,126] Parasitic leeches feed periodically on vertebrates, fish, reptiles, amphibians, birds, and mammals, all of which have blood and body fluids comparatively high in Na and Cl (~ 300 mOsm), as well as protein, lipids, and other organic molecules in the cellular fraction. Leeches therefore may experience a potentially hyperosmotic challenge after feeding, and their tissues might be expected to be osmotolerant around the range of typical vertebrate blood osmotic pressures. It is also likely that salt excretion via the nephridia may be necessary on some occasions. The physiological responses to these blood meals might be expected to be tightly regulated by hormonal pathways.

IV. OSMOREGULATION IN ANNELID CLASSES

A. POLYCHAETES AS OSMOCONFORMERS

Most polychaetes are stenohaline and rarely face osmotic challenges; however, some species penetrate estuaries to salinities 50% that of open ocean (~500 mOsm) and apparently thrive there. In at least one case, *Nereis limnicola* can apparently breed in low salinities and survive in freshwater.[58,59] Oglesby reviewed the responses of 20 species of polychaetes in 9 families with regard to their adaptive responses to a wide range of salinities.[62] Oglesby[59] defined the term *critical low salinity*, in which the "internal solute concentrations fall below the plateau level of hyperionic and hyperosmotic regulation." This parameter is an approximate index of the salinity at which the compensatory transport mechanisms for salt transport begin to fail. Of the 20 species listed, the following polychaetes showed significant hyperosmotic regulation to the critical low salinity (as percent seawater, shown in parentheses): *Nereis limnicola* (<1%), *N. diversicolor* (1 to 2%), *N. succinea* (6 to 10%), and *Laeonereis culveri* (<2%). The rest of the species showed osmoconformity and, in general, could not survive in low salinities for prolonged periods. Extensive work on these nereid species by Oglesby and others has provided data on the physiological patterns of polychaete osmoregulation in those species that are not strict osmoconformers.[58–64,105–108] Examples of some of these data are shown in Figure 5.4.

Another example of osmoconformity in polychaetes is shown in Figure 5.5, in which we measured the weight change of whole *Glycera dibranchiata* (bloodworm) at various salinities (Preston et al., unpublished data). Note that water influx is rapid and the weight change approaches steady state after approximately 4 hours. Osmotic pressure measurements of the coelomic fluid of these worms showed values very close to that of the medium (Figure 5.6), although the coelomic fluid may be slightly hypertonic to 50% and 100% seawater. The coelomic fluid in 150% seawater appears to be somewhat hypotonic to the external medium. In nature, *Glycera dibranchiata* may invade estuarine mudflats up to the point where the water osmolarity is about 50% that of open ocean. In our laboratory experiments, prolonged exposure to this salinity usually kills the animals; however, in the field it should be remembered that the interstitial water in mud probably does not exchange rapidly with the water of the estuary and that the mud buffers to some extent the salinity to which the worms are exposed. In some habitats (such as coastal Maine and Canada, where bloodworms are common), these mudflats undergo twice-daily tidal cycles that typically range from 6 to 12 feet. This means that the estuarine substratum may be replenished with salt on a cyclical basis. Furthermore, the animals may exhibit *behavioral regulation* (in this case, *behavioral osmoregulation*); that is, they may detect and seek depths in the mud or seek surface locations that minimize osmotic stress and other factors such as oxygen levels and temperature.

B. OSMOTIC REGULATION BY OLIGOCHAETES

In terrestrial oligochaetes, the prime example being *Lumbricus terrestris* (the common earthworm), the cuticle may be relatively thin compared to muscle layers, presumably because oxygen absorption is considered to be cutaneous.[47] In the subsoil habitat, oxygen levels may be quite variable, but it is commonly observed that earthworms surface periodically. This would cause exposure to high atmospheric oxygen levels. One of the reasons attributed to the commonly observed phenomenon

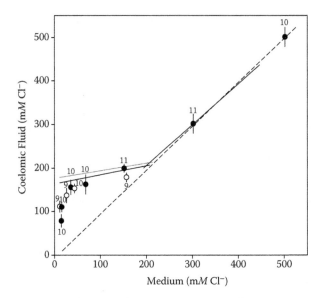

FIGURE 5.4 Osmotic pressure of the coelomic fluid of *Nereis limnicola* adapted to various osmotic pressures as indicated by chloride concentration. This species is capable of osmoregulation at low salinities, although most polychaetes show osmoconformity. Solid circles are values for worms from Schooner Creek, Oregon ($N = 72$; T = 10°C). Open circles are values for worms from the Salinas River estuary in California ($N = 46$; T = 5°C) (data from Table I of Smith[105]). Solid lines are regression lines for worms from Lake Merced, Walker Creek, and the Salinas River estuary in California ($N = 339$; T = 14 to 18°C) (data of R.I. Smith taken from Oglesby[58]). Dotted line is regression line for worms from Lake Merced only. Numbers indicate sample size. Vertical bars represent one standard deviation above and below the mean. Dashed diagonal line is line of equal internal and external chloride concentration. (Adapted from Oglesby, L.C., in *Physiology of Annelids*, Mill, P.J., Ed., Academic Press, New York, 1978, pp. 555–657.)

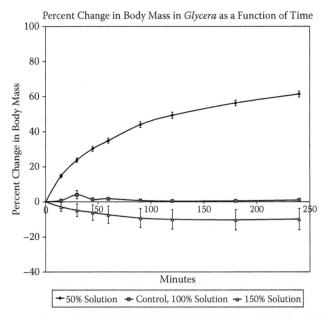

FIGURE 5.5 Osmotic responses of whole *Glycera dibranchiata* to three salinities (50%, 100%, and 150% seawater). Whole worms were immersed in seawater and weighed periodically. Worms were approximately the same size, and the data were normalized for comparison. Values shown are mean ± S.E. ($N = 4$). (Preston, unpublished data.)

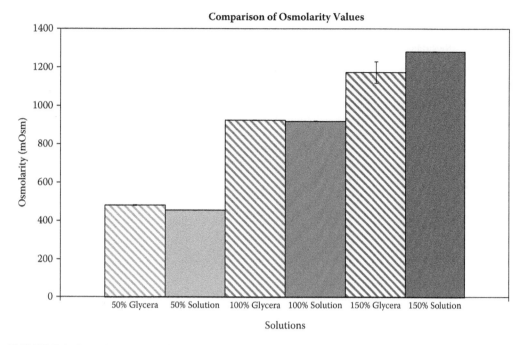

FIGURE 5.6 Osmotic pressure of *Glycera* coelomic fluid after 250 minutes of exposure to 50%, 100%, and 150% seawater measured by vapor pressure osmometry. The values shown are mean ± SE ($N = 4$). Some error bars are not visible in this plot due to their small size. (Preston, unpublished data.)

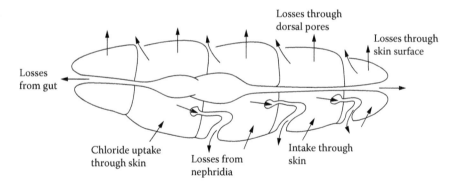

FIGURE 5.7 Routes of water loss or gain in earthworms. (From Laverack, M., *The Physiology of Earthworms*, Macmillan, New York, 1963. With permission.)

of earthworms moving to the surface after heavy rains is that they are driven to the surface by low oxygen levels in their flooded burrows.[47] It is also possible that osmotic stress of a sudden deluge of freshwater may also be a factor. Earthworms may also undergo cycles of dehydration stress. Dry environmental conditions and a fairly water-permeable body wall suggest that potentially significant water loss may occur. In fact, in some older laboratory studies it was stated that earthworms may lose as much as 60% of the body water content for short periods and when rehydrated remain viable.[79,80] This suggests that some unique physiological adaptations may be in place, perhaps enhanced heat shock protein (stress protein) expression or the like. In nature, it is also likely that behavioral responses trigger migration to soil with more moisture to avoid the most severe conditions.

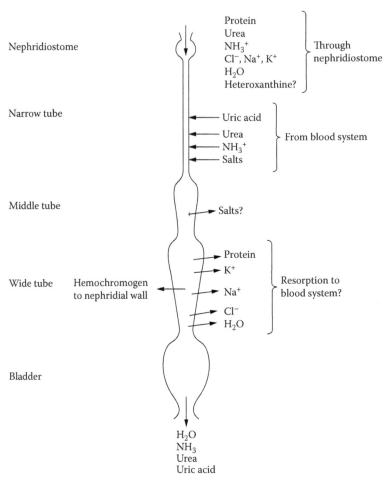

FIGURE 5.8 Functions of *Lumbricus terrestris* nephridia, based on Ramsay.[76,77] (From Laverack, M., *The Physiology of Earthworms*, Macmillan, New York, 1963. With permission.)

Very early work elucidated the basic characteristics of water and salt exchange in *Lumbricus*. Laverack[47] summarized the basic routes of water loss or gain based on these early investigations (Figure 5.7). Maintenance of water balance depends on balancing passive osmotic water gain driven by solute transport (mainly ions) and nephridial loss. Classic studies by Ramsay that are still frequently cited showed that *Lumbricus* nephridia secrete a hypotonic urine and recover ions and organic solutes.[76,77] Figure 5.8 summarizes these data. Note that sodium and chloride are recovered in the wide tube (Ramsay's terminology), and presumably the water permeability of this region must be low enough so hypotonic urine formation is possible. Ramsay also showed the osmotic pressure relationships in *Lumbricus* (Figure 5.9), and these data indicate that hypotonic urine formation begins in the middle tube but is largely formed in the wide tube.[76,77] The typical osmotic pressure of *Lumbricus* coelomic fluid is about 150 mOsm, and the urine may be 10 to 120% of this value.

Dietz and Alvarado[16] measured the ionic composition and osmotic pressure of *Lumbricus terrestris* that had been equilibrated for 1 week in artificial pond water (0.5-m*M* NaCl, 0.05-m*M* KCl, 0.40-m*M* CaCl$_2$, and 0.20-m*M* NaHCO$_3$). This permitted more precise control of the extra-cellular ion composition than in earlier studies using worms equilibrated with moist soil. Figure 5.10 shows that *Lumbricus* regulates coelomic fluid osmotic pressure well in hypotonic media but seems to be an osmoconformer in hypertonic media. They also found a significant change in

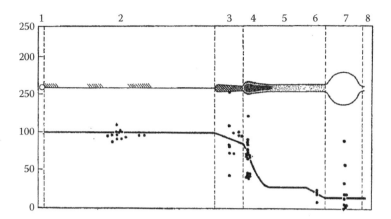

FIGURE 5.9 Osmotic pressure of urine at different levels in the nephridium of *Lumbricus*. The osmotic pressure of the Ringer's solution surrounding the nephridium was equated to 100. (1) Nephridiostome; (2) narrow tube; (3) middle tube; (4) wide tube, proximal; (5) wide tube, middle; (6) wide tube, distal; (7) bladder; (8) exterior. Based on Ramsay.[76,77] (From Laverack, M., *The Physiology of Earthworms*, Macmillan, New York, 1963. With permission.)

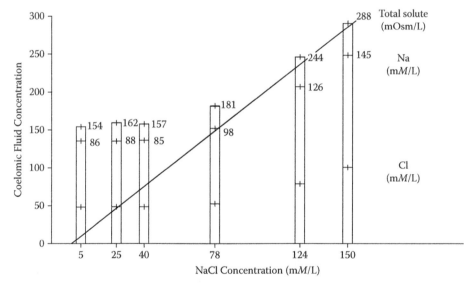

FIGURE 5.10 Coelomic fluid composition of *Lumbricus terrestris* acclimated 10 to 13 days in different NaCl solutions. Sodium is added to chloride in each column. The number adjacent to each bar is the mean concentration of the ion indicated ($N = 8$). Vertical lines indicate ± S.E. The diagonal line represents isosmoticity. (From Dietz, T.H. and Alvarado, R.H., *Biol. Bull.*, 138, 247–261, 1970. With permission.)

coelomic fluid composition for Na and water, but Cl, K, and total solute were not changed (Table 5.4). Water uptake could occur through the digestive tract or the skin. Using high-molecular-weight inulin and dextran markers, they estimated the drinking rate to be about 4.2 µL per 10 g worm per hour ($n = 5$). They measured the clearance rate of inulin from the coelomic fluid (about 75 µL per 10 g worm per hour), which can be used as an indirect estimate of urine flow, although the nephridia may reabsorb some water so this is likely to be an underestimate. Formation of a rectal fluid occurred at an estimated rate of about 22 µL per 10 g worm per hour. They therefore estimated total water excretion of about 100 µL per 10 g worm per hour. If 4 µL per 10 g worm per hour is due to drinking, the remaining influx (at steady state) must arise from water influx across the skin,

TABLE 5.4

Concentrations of Ions in Coelomic Fluid (CF) of *Lumbricus* Acclimated to Soil or Pond Water

Measurements	Units	Soil	Pond Water
Water content	mL/10 g wet weight	8.4 ± 0.1 (18)	8.8 ± 0.0[a] (22)
Total Na+	μEq/10 g wet weight	235 ± 3 (8)	373 ± 15[a] (12)
Total K+	μEq/10 g wet weight	345 ± 15 (8)	365 ± 15 (12)
Total Cl-	μEq/10 g wet weight	172 ± 8 (8)	174 ± 13 (12)
CF Na+	mEq/L	71 ± 2 (14)	75 ± 1 (24)
CF K+	mEq/L	4 ± 0 (14)	3 ± 0 (24)
CF Cl-	mEq/L	48 ± 1 (14)	47 ± 1 (24)
CF total solute	mOsmol/L	154 ± 2 (10)	159 ± 2 (15)

[a] Significantly different from soil animals ($p < 0.05$). Number of observations is in parentheses.

Source: Adapted from Dietz, T.H. and Alvarado, R.H., *Biol. Bull.*, 138, 247–261, 1970.

which would account for 96% of total water influx. Thus, ion transport and obligated passive water influx across the skin must be a very important process in osmoregulation in *Lumbricus*. Dietz and Alvarado[16] concluded that sodium and chloride transport by the skin is the primary route of ion absorption, because the absorption rate changes very little if the mouth and anus are blocked. Measurements of sodium uptake reveal that sodium is absorbed via a saturable transport system with kinetic constants: $V_{max} = 1$ μEq per 10 g worm per hour and $K_m = 1.3$ mM. They also concluded that, "Water balance is achieved when the osmotic force is balanced by hydrostatic force generated by the elasticity of the body wall plus the forces involved in eliminating water in urine and rectal fluid."

Prusch and Otter[71] measured the transepithelial transport of Na and Cl in *Lumbricus terrestris* and in the leech (*Haemopsis grandis*) using Ussing chambers. They measured the transepithelial potential (TEP) across the body wall *in vivo* in whole animals by inserting an agar bridge into the coelomic cavity of restrained animals bathed in artificial pond water (0.5-mM NaCl, 0.05-mM KCl, 0.40-mM CaCl$_2$, and 0.20-mM NaHCO$_3$). They found the TEP was -16 ± 1.8 mV ($n = 8$) inside negative. By comparison, the leech TEP was $+25 \pm 1.9$ mV ($n = 8$) inside positive; the polarity of the TEP reversed from that of the earthworm. The isolated body-wall preparations consisted of body-wall tissue, with septa and organs removed. This was mounted in a Ussing chamber, taking care to avoid the ventrolateral nephridial pores. The outside was bathed in artificial pond water and the inside with annelid Ringer's (116-mM NaCl, 1.9-mM KCl, 1.1-mM CaCl$_2$, and 2.4-mM NaHCO$_3$). The *in vitro* TEPs were -14 ± 1.4 mV ($n = 26$) inside negative. By comparison, the leech TEP was $+22 \pm 1.1$ mV ($n = 33$) inside positive. These values were reasonably close to those of the *in vivo* measurements, indicating that the Ussing chamber measurements were viable. They measured unidirectional fluxes of Na^{24} and Cl^{36} and reported the somewhat curious result that the unidirectional efflux exceeded the unidirectional influx in both earthworms (earthworm influx, $J_i^{Na} = 1.66 \pm 0.13 \times 10^{-9}$ mol/cm^2·min, $n = 7$; efflux, $J_o^{Na} = 6.66 \pm 0.24 \times 10^{-8}$ mol/cm^2·min, $n = 6$) and leeches. For chloride, the fluxes in earthworms were $J_i^{Cl} = 5.90 \pm 0.36 \times 10^{-9}$ mol/cm^2·min, $n = 8$, for influx and $J_o^{Cl} = 7.8 \pm 0.44 \times 10^{-9}$ mol/cm^2·min, $n = 5$, for efflux.

Leeches also showed a net outwardly directed flux and demonstrated saturation kinetics for Na and Cl uptake. Amiloride applied to the outside decreased Na influx by about 50% but did not affect efflux. Externally applied amiloride (10^{-4} M) causes the earthworm TEP to hyperpolarize (13 mV to -25 mV), but this is rapidly reversible upon removal. They also calculated theoretical flux ratios and compared them with the measured flux ratios, which supported the possibility of active uptake. They concluded that, taken together, these data (saturability, effect of inhibitors, and flux ratio measurements) suggest that Na and Cl uptake across the earthworm body wall is electrogenic and is likely to be the result of active processes. It is possible that uptake across the gut may be significant in the total ion balance and that osmoregulatory hormones may change the relative uptake rates above these apparently basal levels.[12,85]

More recent studies on *Lumbricus terrestris* integument conducted by Schnizler et al.[96] using Ussing chambers have provided more detailed information, but these studies were in the context of comparison with extensive studies on leeches by this and other groups. Their measurements of TEP of -10.8 ± 1.2 mV ($n = 14$) compares favorably with the findings of Prusch and Otter[71] described above. The values for transepithelial resistance (R_T) of up to 10 kOhms show that earthworm integument is a tight epithelium, as would be expected for a freshwater organism transporting ions up steep gradients. The net Na transport is sensitive to amiloride, but the large variability in these measurements may be associated with seasonal differences in worms.[96] Apical application of furosemide increased short-circuit current (I_{sc}), although this effect may be indirect because the usual target of furosemide is the Na–K–2Cl transport, which is electroneutral. The basolateral administration of ouabain to earthworm integument shifts the I_{sc} to more negative values, which Schnizler et al.[96] suggested implies a high paracellular resistance (2.5 to 24.5 megOhms); therefore, very little if any paracellular movement of Na occurs.

C. Osmotic Regulation by Hirudinea

Hirudinea are generally freshwater aquatic organisms, and they are under continuous osmotic stress. Many more studies have been done on leeches than other annelids, most probably because of their convenient size and hardiness, as well as the use of *Hirudinea medicinalis* in medical treatments. Reviews by Clauss and Schnizler et al.[11,96] cover the more recent work on leeches with particular emphasis on transepithelial ion transport using Ussing chamber techniques. *Hirudinea medicinalis* controls blood (which is presumably in equilibrium with the coelomic fluid) Na and Cl concentrations and osmotic pressure quite closely in water below 200 mOsm (Figure 5.11). Interestingly, in external medium concentrations above 200 mOsm, the osmotic pressure of blood and urine and the Na and Cl content increase. The setpoint for regulation in hyperosmotic media is clearly about 200 mOsm. Osmotic water gain is compensated for by increased urine flow.[93,120,125,126] After a blood meal, the internal osmotic pressure in the crop may increase to near 300 mOsm, and excess salt load must be excreted by the nephrida.

The Ussing chamber studies of Clauss and coworkers used an integument preparation from which the underlying muscle layers were dissected and then collagenase was used to disintegrate the collagen bundle layers that reinforce the apical surface.[11,119] It was also possible to forego the collagenase treatment and arrive at results very similar to those for the collagenase-treated tissues.[11] They confirmed the basic data of Prusch and Otter that the integument of *Hirudinea medicinalis* is a tight epithelium with a resistance of >1 kOhm·cm^2.[71] The basolateral presence of the Na,K-ATPase was confirmed using ouabain. Weber et al.[119] showed that sodium channels were present (ENaCs) by using amiloride inhibition and noise analysis. They also showed that cAMP stimulated Na absorption and that it was most likely due to an increase in functional apical Na channels.[119] The Na absorption was highest at an external medium concentration of about 20-mM Na.[119] The data further suggest that the Na channel is highly selective, with Na:K ratios of 30:1. These workers also observed stimulatory effects of cGMP on short-circuit current (I_{sc}) and mixed stimulatory and inhibitory effects of adenosine triphosphate (ATP).

Leeches (and other annelids) have a number of signaling peptides secreted by ganglionic neurosecretory cells that affect osmoregulation.[34–36,84,85] A novel peptide, leech osmoregulatory factor (LORF), was characterized by Salzet et al.[88,89] An extensive series of investigations has isolated over 30 neuropeptides in four classes (summarized by Salzet and Stefano;[85] see Table 5.5). Some examples discussed by Salzet and Stefano are given below. The biological activity of the AII-amide isolated from *Erpobdella octoculata* is involved in the control of the leech water balance,

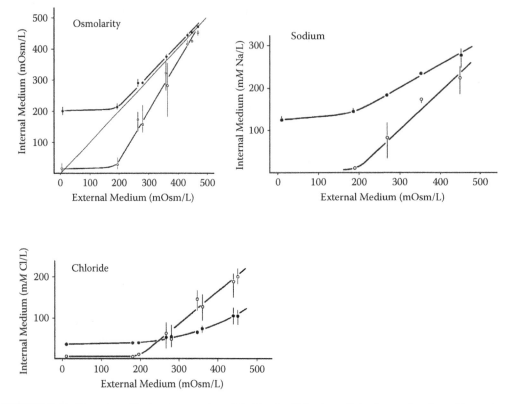

FIGURE 5.11 Osmotic and ionic concentration of the blood of *Hirudo medicinalis* as a function of the external medium concentration. Solid circles are blood; hollow circles, urine. (From Sawyer, R.T., *Leech Biology and Behaviour*, Vol. 1, Clarendon Press, Oxford, 1986, p. 129. With permission.)

exerting a diuretic effect (20% loss of mass on average for 1 nmole of AII). Annetocin is related to oxytocin and vasopressin by sequence homology and acts on osmoregulation via nephridia.[65,86] Lysine-conopressin inhibits the Na amiloride-dependent transitory current and highly stimulates it in *Hirudo medicinalis* stomach or integument preparation. Lysine-conopressin induces egg laying in earthworm like oxytocin does in vertebrates, and this is consistent with the hypothesis that the oxytocin/vasopressin peptide family functions in both osmoregulation and reproduction.[65] RF-amide peptides are probably secreted into the dorsal vessel. The leech *Theromyzon tessulatum* shows weight loss after a GDPFLRF-amide injection and an increase of weight after a FMRF-amide injection.[84,85] GDPFLRF-amide may act as a diuretic hormone and FMRF-amide as an antidiuretic hormone. GDPFLRF-amide also has a stimulating effect on Cl secretion across the caecal epithelium but not Na absorption. Water follows passively, causing water loss from the cells.[11,119] The antidiuretic effect of FMRF-amide might control water balance by direct action on nephridia. Wenning and Calabrese[122] showed that in *H. medicinalis* the nephridial nerve cells, which innervate the nephridia and contact the urine-forming cells, contain RF-amide peptides. They also showed that FMRF-amide leads to hyperpolarization and decreases the rate of firing of the nephridial nerve cells, suggesting autoregulation of peptide release. The leech osmoregulator factor (LORF) is involved in osmoregulation.[88,89] Electrophysiological experiments conducted in *H. medicinalis* revealed an inhibition of the efficacy of Na conductance in leech skin.[88,89]

The transport properties of the gastrointestinal tract in leeches has been investigated to some extent by Milde et al.[56] After a mammalian blood meal (~300 mOsm), which is hypertonic to the blood and coelomic fluid (~200 mOsm), leeches excrete hypertonic urine over 24 hours.[120–122,125,126]

TABLE 5.5
Characterized Annelid Neuropeptides from Leeches

Species	Sequence	Name
Theromyzon tessulatum	SYVMEHFRWDKFGRKIKRRPIKVYPNGAED	ACTH-like
	ESAEAFPLE	Angiotensin I
	DRVYIHPFHLLXWG	Angiotensin II
	DRVYIHPF	Angiotensin III
	RVYIHPF	LORF (leech osmoregulatory factor)
	IPEPYVWD	FMRF-amide
	FMRF-amide	FMRF-amide sulfoxide
	FM(O)RF-amide	—
	FLRF-amide	FLRF-amide
	GDPFLRF-amide	GDPFLRF-amide
	PLG	MIF-1
	YGGFL	Leucine-enkephalin
	YGGFM	Methionine-enkephalin
	YGGFLRKYPK	β-Neoendorphin
	YVMGHFRWDKF-amide	MHS-like peptide
	GSGVSNGGTEMIQLSHIRERQRYWAQDNLR	Leech egg-laying hormone
	RRFLEK-amide	—
Erpobdella octoculata	DRVYIHPF-amide	Angiotensin II-amide
	CFIRNCPKG-amide	Lysine-conopressin
	FMRF-amide; FM(O)RF-amide	FMRF-amide
	GDPFLRF-amide	GDPFLRF-amide
	FLRF-amide	FLRF-amide
	IPEPYVWD; IPEPYVWD-amide	LORF
Hirudo medicinalis	FMRF-amide, FM(O)RF-amide	FMRF-amide
	FLRF-amide	FLRF-amide
	AMGMLRM-amide	Myomoduline-like peptide
Hirudo nipponia	WRLRSDETVRGTRAKCEGEWAIHACLCLG	Leech excitatory peptide
Whitmania pigra	GN-amide	GN-amide

Source: Adapted from Salzet, M. and Stefano, G., *Placebo*, (2)3, 54–72, 2001.

Milde et al.[56] were able to mount the foregut diverticula of *Hirudo medicinalis* in a Ussing chamber and characterized its basic properties. This epithelium was leaky (60 Ohm·cm^2), and the TEP was about −1 mV (lumen negative). The transport rate of Na under short-circuit conditions was about 50 µA·cm^2. This flux was not sensitive to an amiloride or its analogs, and Clauss[11] suggested that the entry pathway is therefore unlikely to be the ENaC channel. The transport showed linear kinetics and was partially blocked (40%) by lanthanum and terbium, suggesting a nonselective cation conductance. Milde et al.[56] also showed that basolateral Na extrusion was ouabain sensitive and probably involved the Na,K-ATPase. Figure 5.12 shows a model developed by Milde et al. that compares Na absorption by leech integument and by leech diverticulum.[56] The apical side of the integument contains an ENaC-like channel that is inhibited by amiloride and stimulated by cAMP. The apical side of the gut diverticulum appears to have another type of non-ENaC sodium channel that may be blocked by certain nonselective cation channel blockers. The effect of cAMP was to stimulate uptake. In both cases, the evidence suggests that Na,K-ATPase is present in the basolateral membrane which maintains the cellular ion gradients. In addition, it is postulated that K channels must be present. This sort of arrangement is typical for many types of epithelia.[11] Presumably little Na flux occurs via the paracellular pathway in the integument, but it is likely to be significant in the diverticulum.

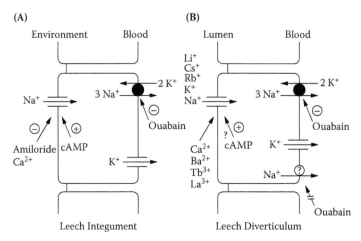

FIGURE 5.12 Proposed models of sodium absorption across (A) leech dorsal skin and (B) leech diverticulum. (Adapted from Clauss, W.G., *Can. J. Zool.*, 79, 192–203, 2001.)

V. TISSUE OSMOTIC REGULATION

The tissues of osmoconforming polychaetes and transiently stressed clitellates express cellular osmoregulatory mechanisms (*cellular volume regulation*).[49,50,101–104] These have been observed to some extent in virtually all tissues and have been extensively studied in mammalian and vertebrate tissues.[9,32] This has been discussed in depth in other sections of this volume, but one should recognize that osmoconforming annelids may be especially dependent on cellular mechanisms of osmoregulation.[49,50] It is generally observed that cells exposed to moderately hypotonic media swell rapidly by osmotic water uptake; however, within a few minutes the cell volume decreases, sometimes approaching the initial cell volume. This is the *regulatory volume decrease* (RVD). The volume decrease results from obligate water efflux coupled with solute efflux. In general, the solutes released from the cell occur via highly regulated and quite specific membrane channels.[9,32,49,50] Some of these osmoregulatory channels have broad solute selectivity, and some are rather specific.[9] The sensing of osmotic stress at the tissue level and the signal transduction pathways that turn these channels off or on have been the subject of intense study for many years (mainly in vertebrate cells).[32] Key osmotic regulatory solutes tend to be potassium and free amino acids (especially taurine), but other solutes, such as sorbitol and betaine, are used in some tissues.[9,32]

In hypertonic medium, the initial cellular response is shrinking followed by an increase in cell volume due to obligated osmotic water flow coupled to solute uptake or synthesis (*regulatory volume increase*, or RVI). This process usually involves distinctly different transporters and metabolic responses than RVD. Potassium gain may occur presumably through action of Na,K-ATPase. Comparatively little work in this area has been done with annelid tissues. With both RVD and RVI, one of the most important initial impacts of increases or decreases in coelomic fluid (or, in vertebrates, blood and extracellular fluid) water content is the rapid change in external potassium ion content.[7] In most cells, the largest component of the cellular membrane resting potential is due to a potassium diffusion potential. A good approximation of the resting potential is given by the Nernst equation:

$$E_{ion} = \frac{RT}{zF} \ln \frac{[Ion_{out}]}{[Ion_{in}]} \tag{5.11}$$

where E_{ion} is the transmembrane equilibrium potential (Nernst potential), R is the gas constant, T is the absolute temperature (Kelvin), z is the charge on the ion, F is the Faraday constant, ln is the natural logarithm, $[Ion_{out}]$ is the ion concentration outside the membrane, and $[Ion_{in}]$ is the ion concentration inside the membrane. A simplified form for standard conditions (assuming that the temperature is 25°C and converting ln to log 10 with appropriate unit conversions):

$$E_{ion} = \frac{59}{z} \log \frac{[Ion_{out}]}{[Ion_{in}]} \tag{5.12}$$

A hypothetical example for a polychaete tissue is as follows: Assuming the coelomic fluid K^+ concentration is 10 mM (taken to be the same as 900-mOsm seawater) and the cytosolic K^+ concentration is 100 mM, the membrane potential would be –59 mV. If the coelomic fluid becomes diluted by 50% (resulting in 5-mM K^+), the resting membrane potential would be –77 mV. If the coelomic fluid is concentrated by 50% (assuming K^+ is 20 mM), the membrane potential would be –41 mV. Obviously, hypotonic dilution of the coelomic fluid may lead to hyperpolarization and hyperexcitability (typically nerve action potentials are triggered after about a +15-mV depolarization). Hypertonic coelomic fluid may lead to the inability to initiate normal action potentials at normal threshold depolarizations. Consequently, voltage-dependent cell function, especially nerve and muscle function, is very sensitive to internal osmotic change. Organismal responses and organ system responses are crucial in vertebrates to maintaining ion homeostasis in the longer term, but the initial RVD and RVI responses are rapid and potentially important in the short term. In annelids, especially osmoconformers, RVD and RVI appear to be the primary responses. In the clitellates, the organismal osmoregulatory responses that confer a degree of homeostatic stability in osmotically stressful environments no doubt provide a considerable selective advantage to these organisms.

VI. MOLECULAR STUDIES

It seems that there should be considerable interest in applying molecular analysis to the presence and expression of membrane transporters and channels involved in osmoregulation in annelids. The diverse habitats and the range of physiological responses should make them prime candidates for in-depth molecular analysis. It is therefore rather surprising that at the time of writing of this review, a search of GenBank revealed only 16 possible nucleotide sequences for organic and ion transporters and channels. In contrast, thousands of ribosomal gene and mitochondrial and metabolic enzyme nucleotide sequences have been used for the many extensive studies of phylogeny of annelids.[82] Of the transporters and channels partially or completely sequenced, eight are voltage-gated Na or K channels in neurons of leeches and one oligochaete. Three transporters are for organic molecules and five for transport ATPases. The complete Na,K-ATPase mRNA sequence for the alpha subunit for *Hirudinea medicinalis* has been completed by Kusche et al.[44] Partial sequences for the Na,K-ATPase mRNA sequence for the alpha subunit are listed for the polychaete *Marenzelleria viridis* and two sequences for P-type ATPase in the polychaete *Platynereis dumerilii*. One partial Na,K-ATPase mRNA sequence for the alpha subunit is listed for *Lumbricus terrestris*. It is certainly obvious that these techniques would be very helpful in analysis of the presence and change of expression during osmotic stress of ion and organic solute transporters in integument, gills, gut, and nephridia. Furthermore, presumptive regulatory factors and pathways may be screened. Much remains to be done that is potentially valuable and exciting.

VII. CONCLUSION

The diversity of habitats occupied by annelids makes them ideal subjects to study osmoregulation and the process of adaptation to diverse environments from marine, freshwater, and terrestrial environments. The core of the studies currently available in the literature harkens back to classic studies that certainly have contributed importantly to our current understanding of osmoregulation in annelids. Of the more recent studies, most have focused on the properties of ion transport by the integumental epithelia of oligochaetes, especially *Lumbricus terrestris*, and of hirudinea that focused on *Hirudo medicinalis*. Some key principles are worth reiterating. Most annelids, even the polychaetes that live in near-isotonic medium, maintain a slightly hypertonic blood and coelomic fluid. This appears to be essential to maintaining the pressure within the hydrostatic skeleton, which is essential for locomotion. The size of the osmotic gradient can be rather small (internal excess of 10 to 40 mOsm), as the equivalent hydrostatic pressure generated by these small osmotic gradients is rather substantial. In freshwater oligochaetes and in the hirudinea, maintenance of a positive internal pressure is usually not a problem, as they are capable of maintaining their osmotic pressure at 150 to 200 mOsm. Water excretion via the nephridia is crucial, and ions and nutritive organic solutes are recaptured during urine formation. It is frequently stated that the fluid flow is driven by the action of cilia at the mouths of the nephridia and this certainly may be true; however, it is also very likely that hydrostatic pressure may be a primary force driving flow into nephridia, and this hydrostatic pressure gradient relies on maintenance of a stable hypertonic gradient, particularly in polychaetes.

The epithelia that absorb and transport ions must be gills (primarily in polychaetes) and intestine and integument, with regulation and recapture by nephridia. The general outline of these processes seems convincing, but a careful balance sheet of the fluxes in these tissues, the losses and changes that occur during osmotic adjustment, has not been done. The work on the leech *Hirudo medicinalis* has probably been the most productive, followed by work on the earthworm *Lumbricus terrestris*. Much research should be done on the signaling pathways and the role of osmoregulatory peptides in these species. The time seems ripe for thorough molecular analysis of the osmoregulatory processes in annelids. Future investigators will find this an interesting and important challenge.

REFERENCES

1. Accoto, D., Castrataro, P., and Dario, P., Biomechanical analysis of Oligochaeta crawling. *J. Theor. Biol.*, 230, 49–55, 2004.
2. Bahl, K. N., Studies on the structure, development, and physiology of the nephridia of Oligochaeta. VI. The physiology of excretion and the significance of the enteronephric type of nephridial system in Indian earthworms. *Q. J. Micr. Sci.*, 85, 343–389, 1945.
3. Bartolomaeus, T. and Purschke, G., Eds., *Morphology, Molecules, Evolution and Phylogeny in Polychaeta and Related Taxa*, Springer, Dordrecht, Netherlands, 2005, p. 387.
4. Baumgarten, C.M. and Feher, J.J., Osmosis and the regulation of cell volume. In *Cell Physiology Source Book*, 2nd ed., Sperelakis, N., Ed., Academic Press, New York, 1998, pp. 253–292.
5. Boroffka, I., Osmo- und Volumenregulation bei *Hirudo medicinalis*. *Z. vergl. Physiol.*, 57, 348–375, 1968.
6. Brusca, R.C. and Brusca, G.J., *Invertebrates*, 2nd ed., Sinauer, Sunderland, MA, 2003, chap 13.
7. Burton, R.F., Cell potassium and the significance of osmolarity in vertebrates. *Comp. Biochem. Physiol.*, 27, 763–773, 1968.
8. Carley, W.W., Water economy of the earthworm *Lumbricus terrestris* L.: coping with the terrestrial environment. *J. Exp. Zool.*, 205, 71–78, 1978.
9. Chamberlin, M. and Strange, K., Anisosmotic cell volume regulation: a comparative view. *Am. J. Physiol.*, 257, 159–173, 1989.
10. Chien, P.K., Stephens, G.C., and Healey, P.L., The role of ultrastructure and physiological differentiation of epithelia in amino acid uptake by the bloodworm. *Glycera, Biol. Bull.*, 142, 219–235, 1972.
11. Clauss, W.G., Epithelial transport and osmoregulation in annelids. *Can. J. Zool.*, 79, 192–203, 2001.
12. Cornell, J.C., Sodium and chloride transport in the isolated intestine of the earthworm, *Lumbricus terrestris* (L.). *J. Exp. Biol.*, 97, 197–216, 1982.
13. Costopulos, J.J., Stephens, G.C., and Wright, S.H., Uptake of amino acids by marine polychaetes under anoxic conditions. *Biol. Bull.*, 157(3), 434, 1979.
14. Dales, R.P., *Annelids*, Hutchinson University Library, London, 1963, p. 200.
15. DeLeersnyder, M., Sur la regulation ionique du milieu interieur de *Nereis diversicolor* O.F. Miller. *Cah. Biol. Mar.*, 12, 49–55, 1971.
16. Dietz, T.H. and Alvarado, R.H., Osmotic and ionic regulation in *Lumbricus terrestris* L. *Biol. Bull.*, 138, 247–261, 1970.

17. Doube, B. and Styan, C., The response of *Aporrectodea rosea* and *Aporrectodea trapezoides* (Oligochaeta: Lubricidae) to moisture gradients in three soil types in the laboratory. *Biol. Fertil. Soils*, 23, 166–172, 1996.

18. El-Duweini, A.K. and Ghabbour, S.I., Nephridial systems and water balance of three oligochaete genera. *Oikos*, 19, 61–70, 1968.

19. Ferraris, J.D., Volume regulation in intertidal *Procephalothrix spiralis* (Nemertina) and *Clitellio arenarius* (Oligochaeta). II. Effects of decerebration under fluctuating salinity conditions. *J. Comp. Physiol.*, 154B, 125–137, 1984.

20. Ferraris, J.D. and Schmidt-Nielsen, B., Volume regulation in an intertidal oligochaete, *Clitellio arenarius* (Miller). I. Short term effects and the influence of the supra- and subesophageal ganglia. *J. Exp. Zool.*, 222, 113–128, 1982.

21. Fletcher, C.R.,. Volume regulation in *Nereis diversicolor*. I. The steady state. *Comp. Biochem. Physiol.*, 47A, 1199–1214, 1974.

22. Fletcher, C. R., Volume regulation in *Nereis diversicolor*. II. The effect of calcium. *Comp. Biochem. Physiol.*, 47A, 1215–1220, 1974.

23. Fletcher, C.R., Volume regulation in *Nereis diversicolor*. III. Adaptation to a reduced salinity. *Comp. Biochem. Physiol.*, 47A, 1221–1234, 1974.

24. Generlich, O. and Giere, O., Osmoregulation in two aquatic oligochaetes from habitats with different salinity and comparison to other annelids. *Hydrobiologia*, 334, 251–261, 1996.

25. Hansen, U., New aspects of the possible sites of ultrafiltration in annelids (Oligochaeta). *Tissue Cell*, 27, 73–78, 1995.

26. Hansen, U., Permeability of the podocytes of *Lumbricus terrestris* (Annelida, Oligochaeta) to different markers. *Zoomorphology*, 117, 63–69, 1997.

27. Haupt, J., Function and ultrastructure of the nephridium of *Hirudo medicinalis* L. II. Fine structure of the central canal and the urinary bladder. *Cell Tissue Res.*, 152, 385–401, 1974.

28. Hildebrandt, J.P. and Zerbst-Boroffka, I., Osmotic and ionic regulation during hypoxia in the medicinal leech, *Hirudo medicinalis* L. *J. Exp. Zool.*, 263, 374–381, 1992.

29. Hoeger, U., Wenning, A., and Greisinger, U., Ion homeostasis in the leech: contribution of organic anions. *J. Exp. Biol.*, 147, 43–51, 1989.

30. Hoeger, U. and Abe, H., Beta-alanine and other free amino acids during salinity adaptation of the polychaete *Nereis japonica*. *Comp. Biochem. Physiol. A*, 137, 161–171, 2004.

31. Hohendorf, K., Der Einfiuß der Temperatur auf die Salzgehaltstoleranz und Osmoregulation von *Nereis diversicolor* O.F. Miller. *Kieler Meeresforsch.*, 19, 196–218, 1963.

32. Hoffmann, E.K. and Dunham, P.B., Membrane mechanisms and intracellular signalling in cell volume regulation. *Int. Rev. Cytol.*, 161, 173–262, 1995.

33. Johansen, K. and Martin, A.W., Circulation in the giant earthworm *Glossoscolex giganteus*. I. Contractile processes and pressure gradients in the large blood vessels. *J. Exp. Biol.*, 43, 333–347, 1965.

34. Kamemoto, F.I., The influence of the brain on osmotic and ionic regulation in earthworms. *Gen. Comp. Endocrinol.*, 17, 420–426, 1964.

35. Kamemoto, F.I., Spalding, A.E., and Keister, S.M., Ionic balance in blood and coelomic fluid of earthworms. *Biol. Bull.*, 122, 228–231, 1962.

36. Kamemoto, F.I., Kato, K.N., and Tucker, L.E., Neurosecretion and salt and water balance in the Annelida and crustacean. *Am. Zool.*, 6, 213–219, 1966.

37. Karandeeva, O.G., Correlation of rates of individual processes participating in the initial osmoregulating reaction in invertebrates. *Byull. Mosk. O.-Va. Ispyt. Prir. Otd. Biol.*, 144–145, 1965.

38. Keudel, M. and Schrader, S., Axial and radial pressures exerted by earthworms of different ecological groups. *Biol. Fertil. Soils*, 29, 262–269, 1999.

39. Kirschner, L.B., Sodium chloride absorption across the body surface: frog skins and other epithelia. *Am. J. Physiol.*, 244, R429–R43, 1983.

40. Klawe, W. and Dickie, L.M., Biology of the bloodworm, *Glycera dibranchiata* Ehlers, and its relation to the bloodworm fishery of the maritime provinces. *Bull. Fish. Res. Board Can.*, 115, 1–136, 1957.

41. Krahl, B. and Zerbst-Boroffka, I., Blood pressure in the leech *Hirudo medicinalis*. *J. Exp. Biol.*, 107, 163–168, 1983.

42. Krumm, S., Goebel-Lauth, S.G., Fronius, M., and Clauss, W., Transport of sodium and chloride across earthworm skin *in vitro*. *J. Comp. Physiol. B*, 175, 601–608, 2005.

43. Kuhl, D.L. and Oglesby, L.C., Reproduction and survival of the pileworm *Nereis succinea* in higher Salton Sea salinities. *Biol. Bull.*, 157, 153–165, 1979.

44. Kusche, K., Bangel, N., Mueller, C., Hildebrandt, J.P., and Weber, W.M., Molecular cloning and sequencing of the Na$^+$/K$^+$-ATPase α-subunit of the medical leech *Hirudo medicinalis* (Annelida) implications for modelling protostomian evolution. *J. Zoolog. Syst. Evol. Res.*, 43, 339–342, 2005.

45. Lasserre, P., Metabolisme et osmoregulation chez une annelide oligochete de la miofaune: Marionina achaeta Lasserre. *Cah. Biol. Mar.*, 16, 765–798, 1975.

46. Lavelle, P., Earthworm activities and the soil system. *Biol. Fertil. Soils*, 6, 237–251, 1988.

47. Laverack, M., *The Physiology of Earthworms*, Macmillan, New York, 1963.

48. Little, C., Ecophysiology of *Nais elinguis* (Oligochaeta) in a brackish-water lagoon. *Estuar. Coast. Shelf. Sci.*, 18, 231–244, 1984.

49. Machin, J. and O'Donnell, M., Volume regulation in the coelomocytes of the blood worm *Glycera dibranchiata. J. Comp. Physiol. B*, 117, 303–311, 1977.

50. Machin, J., Osmotic responses of the bloodworm *Glycera dibranchiata* Ehlers: a graphical approach to the analysis of weight regulation. *Comp. Biochem. Physiol. A*, 52, 49–54, 1975.

51. Macknight, A.D., Principles of cell volume regulation. *Renal Physiol. Biochem.*, 11, 114–141, 1988.

52. Malecha, J., Osmoregulation in Hirudinea Rhynchobdellida *Theromyzon tessulatum* (O.F.M.): experimental localization of the secretory zone of a regulation factor of water balance. [L'osmoregulation chez l'Hirudinee Rhynchobdelle *Theromyzon tessulatum* (O.F.M.): localisation experimentale de la zone secretrice d'un facteur de regulation de la balance hydrique]. *Gen. Comp. Endocrinol.*, 49, 344–351, 1983.

53. Manahan, D.T., Wright, S.H., Stephens, G.C., and Rice, M.A., Transport of dissolved amino acids by the mussel *Mytilus edulis*: demonstration of net uptake from natural seawater. *Science*, 215, 1253–1255, 1982.

54. McKenzie, B. M. and Dexter, A. R., Axial pressures generated by the earthworm *Aporrectodea rosea. Biol. Fertil. Soils*, 5, 323–327, 1988.

55. McKenzie, B. M. and Dexter, A. R., Radial pressures generated by the earthworm *Aporrectodea rosea. Biol. Fertil. Soils*, 5, 328–332, 1988.

56. Milde, H., Weber, W.M., Salzet, M., and Clauss, W., Regulation of Na$^{(+)}$ transport across leech skin by peptide hormones and neurotransmitters. *J. Exp. Biol.*, 204, 1509–1517, 2001.

57. Nieczaj, R. and Zerbst-Boroffka, I., Hyperosmotic acclimation in the leech, *Hirudo medicinalis* L.: energy metabolism, osmotic, ionic and volume regulation. *Comp. Biochem. Physiol.*, 106A, 595–602, 1993.

58. Oglesby, L.C., Steady-state parameters of water and chloride regulation in estuarine nereid polychaetes. *Comp. Biochem. Physiol.*, 14, 621–640, 1965.

59. Oglesby, L.C., Water and chloride fluxes in estuarine nereid polychaetes, *Comp. Biochem. Physiol.*, 16, 437–455, 1965.

60. Oglesby, L.C., Salinity-stress and desiccation in intertidal worms. *Am. Zool.*, 9, 319–331, 1969.

61. Oglesby, L.C., Studies on the salt and water balance of *Nereis diversicolor*. I. Steady-state parameters. *Comp. Biochem. Physiol*, 36, 449–466, 1970.

62. Oglesby, L.C., Salt and water balance. In *Physiology of Annelids*, Mill, P.J., Ed., Academic Press, New York, 1978, pp. 555–657.

63. Oglesby, L.C., Volume regulation in aquatic invertebrates. *J. Exp. Zool.*, 215, 289–301, 1981.

64. Oglesby, L., Mangum, C., Heacox, A., and Ready, N., Salt and water balance in the polychaete *Nereis virens. Comp. Biochem. Physiol. A*, 73, 15–19, 1982.

65. Oumi, T., Ukena, K., Matsushima, O., Ikeda, T., Fujita, T., Minakata, H. et al., Annetocin: an oxytocin-related peptide isolated from the earthworm, *Eisenia foetida. Biochem. Biophys. Res. Comm.*, 198, 393–399, 1994.

66. Potts, W.T.W and Parry, G, *Osmotic and Ionic Regulation in Animals*, Pergamon Press, New York, 1963.

67. Potts, W.T., Osmotic and ionic regulation. *Annu. Rev. Physiol.*, 30, 73–104, 1968.

68. Preston, R.L., Sodium/amino acid cotransport systems in marine invertebrates. In *Comparative Physiology: Comparative Aspects of Sodium Cotransport Systems*, Vol. 7, Kinne, R. Ed., Karger Press, New York, 1990.

69. Preston, R.L., Transport of amino acids by marine invertebrates. *J. Exp. Zool.*, 265, 410–421, 1993.

70. Preston, R.L. and Stevens, B.R., Kinetic and thermodynamic aspects of sodium-coupled amino acid transport by marine invertebrates. *Am. Zool.*, 22, 709–721, 1982.

71. Prusch, R. and Otter, T., Annelid transepithelial ion transport. *Comp. Biochem. Physiol. A*, 57, 87–92, 1977.

72. Qafaiti, M. and Stephens, G.C., Distribution of amino acids to internal tissues after epidermal uptake in the annelid *Glycera dibranchiata*. *J. Exp. Biol.*, 136, 177–191, 1988.

73. Quillin, K.J., Ontogenetic scaling of hydrostatic skeletons: geometric, static stress and dynamic stress scaling of the earthworm *Lumbricus terrestris*. *J. Exp. Biol.*, 201, 1871–1883, 1998.

74. Quillin, K.J., Ontogenetic scaling of burrowing forces in the earthworm *Lumbricus terrestris*. *J. Exp. Biol.*, 203, 2757–2770, 2000.

75. Ramamurthi, R., Studies on the Respiration of Freshwater Poikilotherms in Relation to Osmotic Stress, Ph.D. thesis, Sri Venkateswara University, Tirupati, India, 1962.

76. Ramsay, J.A., The osmotic relations of the earthworm. *J. Exp. Biol.*, 26, 46–56, 1949.

77. Ramsay, J. A., The site of formation of hypotonic urine in the nephridium of *Lumbricus*. *J. Exp. Biol.*, 26, 65–75, 1949.

78. Richards, K.S., Epidermis and cuticle. In *Physiology of Annelids*, Mill, P.J., Ed., Academic Press, New York, 1978, pp. 33–62.

79. Roots, B.I., The water relations of earthworms. I. The activity of the nephridiostome cilia of *Lumbricus terrestris* L. and *Allolobophora chlorotica* Savigny, in relation to the concentration of the bathing medium. *J. Exp. Biol.*, 32, 765–774, 1955.

80. Roots, B.I., The water relations of earthworms. II. Resistance to desiccation and immersion, and behaviour when submerged and when allowed a choice of environment, *J. Exp. Biol.*, 33, 29–44, 1956.

81. Rouse, G. and Pleijel, F., *Reproductive Biology and Phylogeny of Annelida*, Reproductive Biology and Phylogeny Series, Jamieson, B.M., Ed., Science Publishers, Enfield, NH, 2006.

82. Rousset, V., Pleijel, F., Rouse, G.W., Erséus, C., and Siddall, M.E., A molecular phylogeny of annelids. *Cladistics*, 23, 41–63, 2007.

83. Ruppert, E.E. and Smith, P.R., The functional organization of filtration nephridia. *Biol. Rev.*, 63, 231–258, 1988.

84. Salzet, M., The neuroendocrine system of annelids. *Can. J. Zool.*, 79, 175–191, 2001.

85. Salzet, M. and Stefano, G., Biochemical evidence for an annelid neuroendocrine system: evolutionarily conserved molecular mechanisms. *Placebo*, (2)3, 54–72, 2001.

86. Salzet, M. et al., FMRFamide-related peptides in the sex segmental ganglia of the pharyngobdellid leech *Erpobdella octoculata:* identification and involvement in the control of hydric balance. *Eur. J. Biochem. FEBS*, 221, 269–275, 1994.

87. Salzet, M. and Deloffre, L., PLGamide characterization and role in osmoregulation in leech brain. *Brain Res. Molec. Brain Res.*, 76, 161–169, 2000.

88. Salzet, M., Vandenbuckle, F., and Verger-Bocquet, M., Structural characterization of osmoregulator peptides from the brain of the leech *Theromyzon tessulatum*: IPEPYVWD and IPEPYVWD-amide. *Brain Res. Molec. Brain Res.*, 43, 301–310, 1996.

89. Salzet, M. and Verger-Bocquet, M., Biochemical evidence of the sodium influx stimulating related peptide in the brain of the leech *Theromyzon tessulatum*. *Neurosci. Lett.*, 213, 161–164, 1996.

90. Sawyer, R.T., *Leech Biology and Behaviour*, Vol. 1, Clarendon Press, Oxford, 1986, p. 129.

91. Schlieper, C., Uber die Einwirkung niederer Salzkonzentration auf marine Organismen. *Z. vergl. Physiol.*, 9, 478–514, 1929.

92. Schmidt, H. and Zerbst-Boroffka, I., Recovery after anaerobic metabolism in the leech (*Hirudo medicinalis* L.). *J. Comp. Physiol. B Biochem. Syst. Environ. Physiol.*, 163, 574–580, 1993.

93. Schmidt-Nielsen, B. and Laws, D.F., Invertebrate mechanisms for diluting and concentrating the urine. *Annu. Rev. Physiol.*, 25, 631–658, 1963.

94. Schnizler, M., Buss, M., and Clauss, W., Effects of extracellular purines on ion transport across the integument of *Hirudo medicinalis*. *J. Exp. Biol.*, 205, 2705–2713, 2002.

95. Schnizler, M. and Clauss, W., Gd^{3+}-sensitive Na^+ transport across the integument of *Hirudo medicinali*. *Physiol. Biochem. Zool.*, 76, 115–121, 2003.

96. Schnizler, M., Krumm, S., and Clauss, W., Annelid epithelia as models for electrogenic Na^+ transport. *Biochim. Biophys. Acta*, 1566, 84–91, 2002.

97. Schone, C., Uber den Einfluß von Nahrung und Substratsalinitat auf Verhalten, Fortpflanzung und Wasserhaushalt von *Enchytraeus albidus* Henle. *Oecologia (Berlin)*, 6, 254–266, 1971.

98. Seymour, M.K., Locomotion and coelomic pressure in *Lumbricus terrestris* L. *J. Exp. Biol.*, 51, 47–58, 1969.

99. Seymour, M.K., Skeletons of *Lumbricus terrestris* L. and *Arenicola marina* (L.). *Nature*, 228, 383–385, 1970.

100. Seymour, M.K., Coelomic pressure and electromyogram in earthworm locomotion. *Comp. Biochem. Physiol.*, 40A, 859–864, 1971.

101. Skaer, H.L., The water balance of a serpulid polychaete, *Mercierella enigmatica* (Fauvel). I. Osmotic concentration and volume regulation. *J. Exp. Biol.*, 60, 321–330, 1974.

102. Skaer, H.L., The water balance of a serpulid polychaete, *Mercierella enigmatica* (Fauvel). II. Ion concentration. *J. Exp. Biol.*, 60, 331–338, 1974.

103. Skaer, H.L., The water balance of a serpulid polychaete, *Mercierella enigmatica* (Fauvel). III. Accessibility of the extracellular compartment and related studies. *J. Exp. Biol.*, 60, 339–349, 1974.

104. Skaer, H.L., The water balance of a serpulid polychaete *Mercierella enigmatica* (Fauvel). IV. The excitability of the longitudinal muscle cells. *J. Exp. Biol.*, 60, 351–370, 1974.

105. Smith, R.I., A note on the tolerance of low salinities by nereid polychaetes and its relation to temperature and reproductive habit. *L'Annee Biologique*, 33, 93–107, 1957.

106. Smith, R.I., Chloride regulation at low salinities by *Nereis diversicolor* (Annelida, Polychaeta). II. Water fluxes and apparent permeability to water. *J. Exp. Biol.*, 53, 93–100, 1970.

107. Smith, R.I., Exchanges of sodium and chloride at low salinities by *Nereis diversicolor* (Annelida, Polychaeta). *Biol. Bull.*, 151, 587–600, 1976.

108. Smith, R.I., The larval nephridia of the brackish-water polychaete, *Nereis diversicolor*. *J. Morphol.*, 179, 273–289, 1984.

109. Stephens, G.C., Uptake of organic material by aquatic invertebrates. III. Uptake of glycine by brackish-water annelids. *Biol. Bull.*, 126, 150–162, 1964.

110. Stephens, G.C., Dissolved organic matter as a potential source of nutrition for marine organisms. *Am. Zool.*, 8, 95–106, 1968.

111. Stephens, G.C. and Schinske, R.A., Uptake of amino acids by marine invertebrates. *Limnol. Oceanogr.*, 6, 175–181, 1961.

112. Stephens, G.C., Uptake of naturally occurring primary amines by marine annelids, *Biol. Bull.*, 149, 397–407, 1975.

113. Stephens, G.C. and Virkar, R.A., Uptake of organic material by aquatic invertebrates. IV. The influence of salinity on the uptake of amino acids by the brittle star, *Ophiactis arenosa. Biol. Bull.*, 131, 172–185, 1966.

114. Stevens, B.R. and Preston, R.L., The transport of L-alanine by the integument of the marine polychaete *Glycera dibranchiata*. *J. Exp. Zool.*, 212, 119–127, 1980.

115. Stevens, B.R. and Preston, R.L., The effect of sodium on the kinetics of L-alanine influx by the integument of the marine polychaete *Glycera dibranchiate*, Part 1. *J. Exp. Zool.*, 212, 129–138, 1980.

116. Stevens, B.R. and Preston, R.L., The effect of sodium on the kinetics of L-alanine influx by the integument of the marine polychaete *Glycera dibranchiate*, Part 2. *J. Exp. Zool.*, 212, 139–146, 1980.

117. Subba Rao, B.V.S.S.R., Osmotic regulation in a brackish water oligochaete, *Pontodrilus bermudensis* Beddard. *Indian J. Mar. Sci.*, 7, 132–134, 1978.

118. Subba-Rao, B.V.S S.R. and Ganapati, P.N., Regulation of chlorides, sodium and potassium in a brackish-water oligochaete, *Pontodrilusb ermudensis* Beddard. In *Biology of Benthic Marine Organisms: Techniques and Methods as Applied to the Indian Ocean*, Thompson, M.F., Sarojini, R., and Nagabhushanam, R., Eds., Oxford and IBH Publishing, New Delhi, India, 1986, pp. 19–34.

119. Weber, W.M. et al., Ion transport across leech integument. I. Electrogenic Na^+ transport and current fluctuation analysis of the apical Na^+ channel. *J. Comp. Physiol. B*, 163, 153–159, 1993.

120. Wenning, A., Salt and water regulation in *Macrobdella decora* (Hirudinea: Gnathobdelliformes) under osmotic stress. *J. Exp. Biol.*, 131, 337–349, 1987.

121. Wenning, A., Sensory and neurosecretory innervation of leech nephridia is accomplished by a single neurone containing FMRFamide. *J. Exp. Biol.*, 182, 81–96, 1993.

122. Wenning, A. and Calabrese, R.L., An endogenous peptide modulates the activity of a sensory neurone in the leech *Hirudo medicinalis*. *J. Exp. Biol.*, 198, 1405–1415, 1995.

123. Wilson, R.J., Skierczynski, B.A., Blackwood, S., Skalak, R., and Kristan, Jr., W.B., Mapping motor neurone activity to overt behaviour in the leech: internal pressures produced during locomotion. *J. Exp. Biol.*, 199, 1415–1428, 1996.

124. Wolf, A.V., Paths of water exchange in the earthworm, *Physiol. Zool.*, 13, 294–308, 1940.

125. Zerbst-Boroffka, I., Function and ultrastructure of the nephridium in *Hirudo medicinalis* L. III. Mechanisms of the formation of primary and final urine. *J. Comp. Physiol. B*, 100, 307–315, 1975.

126. Zerbst-Boroffka, I., Bazin, B., and Wenning, A., Chloride secretion drives urine formation in leech nephridia. *J. Exp. Biol.*, 200, 2217–2227, 1997.

6 Osmotic and Ionic Regulation in Aquatic Arthropods

*Guy Charmantier, Mireille Charmantier-Daures,
and David Towle*

CONTENTS

I. INTRODUCTION

This chapter on osmoregulation of aquatic arthropods deals mainly with crustaceans. Among the Pancrustacea group, the Hexapoda (formerly insects) will be discussed in Chapter 7. Section VIII of this chapter is devoted to a few other aquatic arthropods. Osmotic and ionic regulations have been studied in crustaceans for over 80 years and have been reviewed in Robertson,[499] Potts and Parry,[465] Shaw,[535] Lockwood,[345] Prosser,[471] Mantel and Farmer,[382] Schoffeniels and Dandrifosse,[519] Péqueux,[447] and Péqueux et al.[448] In this chapter, we briefly summarize the knowledge on crustacean osmoregulation up to the 1980s and 1990s, before concentrating on more recent information in two areas. One is linked to the adaptive function of osmoregulation—that is, the relations between osmoregulation and the ecology of crustaceans, especially during their development. The second deals with the mechanisms of osmoregulation, mainly at the cellular and molecular levels.

II. CRUSTACEAN HABITATS

With a fossil record extending from the Lower Cambrian to Recent, crustaceans have evolved over more than 500 million years.[11,164,520,521] This very long period of evolution and radiation has led to a great variety in size, shape, and occupation of various habitats. Most of the 42,000 contemporary described[60] species live in aquatic habitats, and about 90% of the current species live in the sea or in brackish water. Often noted is the fact that the hemolymph osmolality and ion composition of crustaceans are close to those of seawater, perhaps reminiscent of the original media in which the early crustaceans appeared.

From these ancestral marine habitats, crustaceans have occupied a variety of aquatic habitats where salinity may vary. The bottoms of large oceans are among the most stable environments with regard to temperature and salinity, which usually is in the range of 34 to 35‰. One exception is found in the deep-sea hydrothermal vents that release fluids with particular ionic ratios.[609] The surfaces of oceans, where salinity is relatively stable, are exposed to precipitation, which can cause local changes in salinity. Coastal areas receive continental freshwater from rivers and are thus more exposed to salinity fluctuations. In estuaries, salinity gradients can be modified twice daily in tidal regions. Tidal pools are subjected to high salinity due to evaporation. Lagoons are also the site of important variations of salinity, ranging from low values to occasional saturation of seawater corresponding to a salinity of about 280‰. All of these media host different species of crustaceans that have developed various means of coping with the stress originating from salinity variations. At the other end of the salinity spectrum, several groups of crustaceans have successfully adapted to very low salinities, as low as that found in freshwater. They are able to live in rivers and lakes or in locked bodies of inland water, where the ion concentrations and ratios can be different from those found in rivers. A few other species have become terrestrial.

We have so far considered adult crustaceans with a supposedly limited ability to move between media, but some species are able to move over large distances, such as the spiny[342] and homarid lobsters.[126] Migrations are often related to reproduction and development, and they expose the animals at different ontogenetical stages to variable salinity (see Section VII).[78]

Salinity tolerance varies between species. Crustaceans that cannot tolerate large variations of salinity are designated as *stenohaline*. They live under stable conditions of salinity, usually in marine habitats in a salinity range of about 30 to 38‰. Others are restricted to freshwater. Crustaceans that live in habitats where salinity fluctuates or that migrate between media of different salinities are considered to be *euryhaline* and to have varying amplitudes of salinity tolerance. We agree with Mantel and Farmer (p. 54)[382] that "the dividing line between 'steno' and 'eury' is well nigh impossible to define. These terms are relative and are most useful as comparative, rather than absolute, measures of the animal's capabilities." It is also worth noting that in some species, particularly those undertaking ontogenetical migrations, tolerance to salinity may vary with the developmental stages, often from stenohalinity to euryhalinity.

III. HYDROMINERAL REGULATION: INTRACELLULAR OR EXTRACELLULAR REGULATION?

In animal cells, the cell volume must remain close to constant; thus, the osmolality of the cytosol must be kept almost equal to that of the cell-surrounding medium to prevent water exchanges. In multicellular animals, this medium corresponds to the extracellular fluid, which is in close osmotic equilibrium with the circulating fluid (blood or hemolymph). Most vertebrates are able to tightly regulate their blood osmolality within a range of 280 to 350 mOsm/kg even under highly variable salinity, but in invertebrates, including crustaceans, the ability to osmoregulate, when present, is not as efficient, resulting in variations in hemolymph osmolality when salinity fluctuates. In crustaceans, the cells must cope with such variations.

The corresponding mechanisms are referred to as *intracellular isosmotic regulation*. They have been particularly studied in crustacean cells by the Liège, Belgium, research group.[123,210–214,447,448,518] Experimental evidence strongly points to the involvement of free amino acids as intracellular osmotic effectors. Following a transitional change in cell volume originating from temporary water movement across the cell wall, the intracellular ionic composition is affected, which in turn affects the activity of enzymes involved in the anabolism and catabolism of amino acids, particularly asparagine, glutamine, proline, alanine, glycine, and serine.[1,405,640] At high salinity, the increasing hemolymph osmolality is followed by an increase in the cellular content of free amino acids resulting from higher synthesis and lower catabolism of amino acids. At low salinity, the decrease in intracellular osmolality results from a higher catabolism of free amino acids released to the hemolymph before final deamination and excretion as ammonia to the external medium through the posterior gills.

IV. PATTERNS OF OSMOREGULATION

Since studies conducted during the first part of the 20th century,[382] solutes of the crustacean hemolymph have been known to include organic (proteins, amino acids, carbohydrates, and lipids) and inorganic (ions) compounds. Given their relative concentrations, it clearly appears that ions are the main osmotic effectors, accounting for over 90% of the hemolymph osmolality. Among them, sodium and chloride are dominant; for example, in the lobster *Homarus americanus*, maintained in seawater, the hemolymph Cl^-, Na^+, K^+, Ca^{2+}, and Mg^{2+} concentrations are, respectively, 470, 470, 10, 15.6, and 7 mmol/L.[471] Assuming a complete dissociation, their sum yields a value of 995 mOsm/kg, very close to 1000 mOsm/kg, which is the osmolality of standard 34.3‰ seawater. In addition, sodium and chloride represent 94% of the osmotic effect of these five major ions. This case is representative of most marine crustaceans. Solute regulation in crustacean osmoregulation thus concerns mainly Na^+ and Cl^-. The hemolymph concentration of these two ions is very close to their concentration in seawater and, of course, is much higher than in freshwater. In addition to ion concentration, the water content is also regulated, hence the concept of hydromineral regulation.

Interest in crustaceans serving as models for ecophysiology stems from the wide variety of their habitats and their patterns of osmoregulation. All of the possible types of osmoregulation are represented in this group. In comparison, teleost fishes present a single pattern of osmoregulation. Crustacean osmoregulation is thus a good example of the benefits of a comparative approach, as advocated by Bartholomew.[33]

Data on the patterns of osmoregulation in crustaceans are abundant and are summarized in Figure 6.1. Some crustaceans are osmoconformers, with an isosmotic pattern (Figure 6.1, Pattern 1); a few of them, usually referred to as *hyper-osmoconformers*, maintain a slight positive and constant difference of osmolality with the environment, usually in the range of 10 to 40 mOsm/kg. Other crustaceans are *osmoregulators*. Their type of osmoregulation may be hyperisosmotic, with hyper-regulation at low salinity and isosmotic regulation in salinities close to and higher than seawater (see Figure 6.1, Pattern 2). Freshwater crustaceans display that type, but with lower values of hemolymph osmolality (see Figure 6.1, Pattern 2'). The strongest osmoregulators among crustaceans are hyper–hyporegulators, with an isosmotic point (same osmolality in hemolymph and medium) usually close to or below seawater osmolality (see Figure 6.1, Pattern 3).

In osmoregulators, the ability to hyper- or hypo-osmoregulate at a given salinity varies with the species, the stage of development, and environmental parameters (see Section VI). The level of osmoregulation can be numerically evaluated though the measurement of the osmoregulatory capacity (OC), which is the difference between the hemolymph osmolality and the medium osmolality at a given salinity. Values of OC are positive and negative under conditions of hyper- and hypo-osmoregulation, respectively.

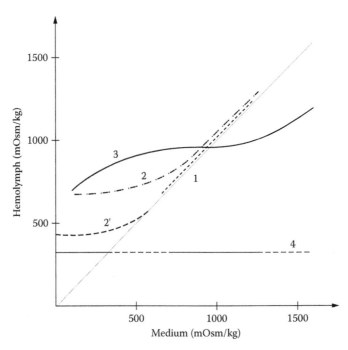

FIGURE 6.1 Patterns of osmoregulation in crustaceans, and teleost fishes: (1) isosmotic; (2) hyper-isosmotic; (2′) hyper-isosmotic in freshwater species; (3) hyper–hypoosmotic; (4) osmoregulation in teleosts.

Comparing these patterns with osmoregulation in teleost fishes leads to several observations. First, although teleost fishes maintain their extracellular osmolality within a narrow range, even the strongest osmoregulators among crustaceans are subjected to variations in their hemolymph osmolality when salinity varies. Second, the value of the isosmotic point is close to 800 to 1000 mOsm/kg in crustaceans, much higher than the 300 mOsm/kg in teleosts. Most crustaceans are marine species, thus living close to isosmoticity with their environment—seawater (30 to 38‰, 880 to 1120 mOsm/kg). Under natural conditions, the probability of exposure to decreasing salinities (lower than seawater) due to precipitation or freshwater flow in coastal environments is higher than the probability of exposure to increased salinities originating from evaporation. Thus, crustaceans are much more likely to hyperosmoregulate than to hypo-osmoregulate, a fact reflected in the relatively more abundant data regarding the former metabolism. In contrast, teleost fishes do not often spend time in an isosmotic medium, because a salinity of 10.2‰ (300 mOsm/kg) is mostly transitory in brackish environments. They consequently either hyperosmoregulate (in freshwater) or hypo-osmo-regulate (in seawater), and both mechanisms are equally well known, as detailed in Chapter 8.

Numerous studies have described the pattern of osmoregulation of crustaceans; they have been reviewed by Mantel and Farmer[382] and Péqueux et al.,[448] and some representative data are given in Table 6.1. In laboratory studies of osmoregulation, the time of exposure to various salinities must be chosen to allow for complete stabilization of the hemolymph osmolality. If insufficient, it results in the measurement of unstable values. For adult crustaceans, a stabilization time of at least 4 days is generally recognized as sufficient for osmotic equilibration. A shorter time (a few hours) is required in small specimens, particularly in larval stages (see Section VII).

In crustaceans, the pattern of osmoregulation is related to the tolerance to salinity and hence to the distribution in natural habitats. Although generalization of the following proposition is hampered by exceptions, it is generally recognized that osmoconformers are stenohaline and that osmoregulators tend to be euryhaline. The former species tend to be restricted to marine zones, while more variable environments tend to be populated by the latter ones; for example, a typical tolerance range for osmoconformers such as majid crabs is about 30 to 36‰, which restricts them

TABLE 6.1
Patterns of Osmoregulation in Crustaceans, Selected Examples

Species	Pattern	Habitat	Media/OC (mOms/kg)	Refs.
Branchiopoda: Calmanostraca: Notostraca				
Triops longicaudatus	2'	FW	FW/135	Horne[273]
Branchiopoda: Sarsostraca: Anostraca				
Branchinella australensis	2	FW	FW/120	Geddes[203]
Artemia salina	3	Salt marshes	150, SW, 5000/70, −670, −4400	Crogham[135]
Branchiopoda: Cladocera				
37 species	2, 2', 3	FW, brackish, hypersaline water		Aladin,[12] Aladin and Potts[13]
Maxillopoda: Copepoda: Calanoida				
Boeckella triarticulata	2	FW	FW, 1/2SW/135, 510	Brand and Bayly[62]
Acartia tonsa	2	Brackish	130, 1/2SW, SW/120, 130, 45	Farmer,[179] Lance[324]
Gladioferens pectinatus	3	Brackish	175, 1/2SW, SW/160, 30, −240	Brand and Bayly[62]
Maxillopoda: Cirripedia: Thoracica				
Balanus crenatus	1	SW, intertidal	1/2SW, SW/20, 40	Foster[193]
Balanus improvisus	2	SW, intertidal	FW, 1/2SW, SW, 1800/90, 10, 10, 0	Fyhn[200]
Malacostraca: Hoplocarida: Stomatopoda				
Squilla empusa	1	SW, intertidal	1/2SW, SW/10, 10	Lee and McFarland[327]
Malacostraca: Eumalacostraca: Peracarida, Mysidacea				
Praunus flexuosus	3	SW	1/2SW, SW, 1200/200, −120, −210	McLusky and Heard[395]
Neomysis integer	3	SW, estuaries	50, 1/2SW, 800, SW/570, 90, 0, −250	McLusky and Heard,[395] Vilas et al.[612]
Amphipoda				
Gammarus pulex	2'	FW	FW, 1/2SW/275, 10	Lockwood,[344] Sutcliffe[572]
Gammarus oceanicus	2	SW	100, 1/2SW, SW/320, 300, 0	Brodie and Halcrow,[65] Werntz[631]
Orchestia gammarellus	3	SW, semiterrestrial	265, 1/2SW, SW/550, 316, −105	Morritt and Spicer[414,416]
Orchestia cavimana	2	SW, FW, semiterrestrial	FW, 1/2SW, 850/390, 100, 50	Morritt and Spicer[416]
Mysticotalitrus cryptus	2	Terrestrial	30, 1/2SW, SW/500, 250, 20	Morritt and Richardson,[411] Morritt and Spicer[416]

(continued)

TABLE 6.1 (cont.)
Patterns of Osmoregulation in Crustaceans, Selected Examples

Species	Pattern	Habitat	Media/OC (mOms/kg)	Refs.
Isopoda				
Saduria entomon	2	FW, low brackish	FW, 1/2SW/520, 120	Crogham and Lockwood,[137] Lockwood and Crogham[346]
Sphaeroma rugicauda	3	SW, intertidal	1/2SW, SW, 1700/100, −140, −400	Harris[242]
Sphaeroma serratum	2–3	SW, coastal, lagoons	120, 1/2SW, SW, 2000/420, 320, 0, −90	Charmantier et al.,[106] Charmantier and Trilles[107]
Malacostraca: Eumalacostraca: Decapoda: Dendrobranchiata				
Metapenaeus bennetae	3	SW, coastal	270, 1/2SW, SW, 1200/295, 110, −300, −460	Dall[140]
Penaeus chinensis	3	SW, coastal	200, 1/2SW, SW, 1100/390, 250, −150, −280	Charmantier-Daures et al.,[111] Chen and Lin[114]
Decapoda: Pleocyemata, Caridea				
Macrobrachium petersi	3	SW, brackish	FW, 1/2SW, SW/460, 0, −180	Read[477]
Palaemonetes varians	3	FW, brackish	FW, 1/2SW, SW/565, 80, −320	Parry,[443] Potts and Parry[466]
Crangon crangon	3	SW, estuaries, lagoons	160, 1/2SW, SW, 1350/325, 160, −130, −300	Cieluch et al.,[122] Hagerman,[234] Spaargaren[560]
Decapoda: Astacidea				
Astacus leptodactylus	2'	FW, seldom brackish	FW, 1/2SW/420, 40	Bielawski,[47] Holdich et al.,[261] Susanto and Charmantier[570]
Nephrops norvegicus	1	SW	SW/−85	Robertson[500]
Homarus americanus	2	SW	1/2SW, SW/125, 10	Charmantier et al.,[98,110] Dall[143]
Decapoda: Palinura				
Panulirus longipes	1	SW	760, SW, 1300/0, 0, 0	Dall[144]
Decapoda: Thalassinidea				
Callianassa jamaicense	2	SW, estuaries	100, 1/2SW, SW, 1400/400, 100, 0, −50	Felder[181]

Decapoda: Anomura

Species	Pattern	Habitat	Values	Reference
Birgus latro	3	Terrestrial, SW at reproduction	250, 1/2SW, SW, 1370/450, 350, 30, –220; field/HI 735–885	Gross,[227] Harms[241]
Clibanarius taeniatus	2	SW, intertidal	120, 1/2SW, SW/360, 220, 70	Dunbar and Coates[169]
Paralithodes camtschatica	1	SW, deep	850, SW/10, 15	Mackay and Prosser[369]

Decapoda: Brachyura

Species	Pattern	Habitat	Values	Reference
Callinectes sapidus	3	SW, brackish, estuaries	FW, 1/2SW, SW, 1880/625, 230, –100, –300	Ballard and Abbott,[28] Cameron,[75] Gifford,[209] Lynch et al.,[366] Mangum and Amende[378]
Carcinus maenas	2	SW, brackish, coastal, lagoons	180, 1/2SW, SW, 1190/360, 250, 0, –10	Lucu et al.,[360] Taylor et al.,[580] Theede[583]
Scylla serrata	2–3	SW, estuaries, mangroves	420, SW, 1300/350, 0, –30	Chen and Chia[113]
Bythograea thermydron	1	SW, deep hydrothermal vents	740, SW, 1200/10, 20, 10	Martinez et al.[389]
Cancer magister	2	SW, coastal	300, 1/2SW, SW/275, 225, 0	Brown and Terwilliger,[67] Hunter and Rudy,[277] Jones[285]
Chionoecetes tanneri	1	SW, deep	860, SW/0, 0	Mackay and Prosser[369]
Gecarcinus lateralis	2	Terrestrial, SW, FW, brackish	FW, 1/2SW, SW/575, 250, 30	Mantel et al.[381]
Potamon edulis	2	FW	FW, 1/2SW/540, 180	Harris and Micaleff[245]
Eriocheir sinensis	3	FW, brackish, SW	FW, 1/2SW, SW, 1440/615, 250, –150, –240	Cieluch et al.,[121] De Leersnyder,[152] Roast et al.[495]
Hemigrapsus nudus	2	SW, intertidal	55, 1/2SW, SW, 1380/495, 300, 150, 165	Dehnel,[154] Dehnel and Stone[157]
Neohelice granulata	3	SW, brackish, semiterrestrial	30, 1/2SW, SW, 1300/610, 250, –120, –310	Castilho et al.,[82] Charmantier et al.,[103] Novo et al.[426]
Sesarma reticulatum	3	SW, estuaries, salt marshes	135, 1/2SW, SW, 1530/630, 320, –130, –265	Foskett[192]
Ucides cordatus	3	SW, coastal, mangrove	260, 760, SW/470, 10, –150	Harris and Santos,[246,247] Santos,[508] Santos and Salomao[509–510]

Chelicerata: Xiphosura: Limulidae

Species	Pattern	Habitat	Values	Reference
Limulus polyphemus	2	Coastal, intertidal	50, 1/2SW, SW, 2000/240, 120, 5, –20	Robertson[501]

Note: Pattern 1, isosmotic; Pattern 2, hyper-isosmotic; Pattern 2′, hyperosmotic in freshwater; Pattern 3, hyper–hypoosmotic (see Figure 6.1). Published or estimated values of the osmoregulatory capacity OC (in mOsm/kg) are given in freshwater (FW, ca. 0.3‰, 5–10 mOsm/kg) or at the lowest used salinity (in mOsm/kg), in half seawater (1/2SW, 17‰, 500 mOsm/kg), in seawater (SW, 34‰, 1000 mOsm/kg), and in some cases in concentrated media (in mOsm/kg). HI, hemolymph osmolality.

Source: Data from Mantel, L.H. and Farmer, L.L., in *Internal Anatomy and Physiological Regulation,* Mantel, L.H., Ed., Academic Press, New York, 1983, pp. 53–161; Péqueux, A. et al., in *Treatise on Zoology—Anatomy, Taxonomy, Biology, The Crustacea,* Forest, J. and von Vaupel Klein, J.C., Eds., Brill Academic Publishers, Leiden, 2006, pp. 205–308.

to marine habitats. Osmoregulators such as grapsid crabs or peneid shrimps can tolerate much wider ranges, about 5 to 45‰, and their habitats include tidal estuaries, lagoons, and mangroves. Among osmoregulating species, the level of osmoregulation can be variable, and it affects their distribution. As an example, among three species of mysids living in the same estuary, one of them (*Neomysis integer*), displaying the highest ability to hyperosmoregulate, is found in the most oligohaline zone.[612] A similar separation in habitat has been reported in four species of *Uca* spp. displaying different abilities to osmoregulate.[340]

Extreme cases of adaptation are well represented by crayfish, which are hyper-isoregulators and fully adapted to freshwater, where they strongly hyperosmoregulate, maintaining a hemolymph osmolality close to 400 mOsm/kg against an external osmolality of 5 to 10 mOsm/kg. At the other end of the adaptive scale, anostracan branchiopods of the *Artemia* group, which are hyper–hypo-osmoregulators, are well known for their very high capacity for hyporegulation in the highly concentrated media found in salt marshes—for example, 580 mOsm/kg in the hemolymph of *Artemia salina* in a 5470-mOsm/kg (186‰) medium.[134–136,548,549]

Except for such striking cases, few crustaceans can be categorized as strong osmoregulators. The strong hyper-hypo-osmoregulating grapsid crabs represent a case in point. The hemolymph osmolality of *Metopograpsus messor* is maintained at a constant 965 mOsm/kg from 25% seawater to full seawater.[292] In *Grapsus grapsus*, it is regulated at 990 ± 30 mOsm/kg from 50 to 125% seawater. The wide array of habitats populated by grapsids results from the exceptional osmoregulatory abilities that make these crabs good models for studying the mechanisms of hyper- and hypo-osmoregulation.

As in grapsids, several families possess a common type of osmoregulation. Most peneid shrimps, for example are hyper-hypo-osmoregulators. The osmoregulatory abilities of crustaceans used in aquaculture, (e.g., *Macrobrachium* spp. and mainly peneid shrimps) have been heavily studied (Table 6.1), as they are related to their salinity tolerance and growth.[24,114,320] Peneid aquaculture is expanding from coastal locations to inland sites with brackish well waters in which ionic ratios differ from those found in seawater, and studies on the osmoregulation of peneid shrimp have revealed the importance of minimum K^+ and sometimes Mg^{2+} concentrations in the water.[468,503,556] The colonization of new habitats by certain species is sometimes linked to a shift in their pattern of osmoregulation that differs from the general pattern of the group to which they belong. Palaemonid shrimps, for example, are mostly euryhaline hyper–hypo-osmoregulators living in lagoons, estuaries, and intertidal coastal areas (Table 6.1), but some species such as *Palaemonetes paludosus*[165] and *P. argentinus*[92] that live in freshwater or in low salinity media are unable to hyporegulate. At salinities above 17 to 20‰, they iso-osmoregulate and cannot tolerate salinities above 30‰. The hydrothermal vent crab *Bythograea thermydron*,[389] a stenohaline osmoconformer, is exposed to stable salinity close to 33 to 35‰.[511] Bythograeidae may have derived from Potamoidae, Portunoideae, or Xanthoideae,[231,232,389] most of which are able to strongly osmoregulate. During their evolution, freshwater palaemonids and *Bythograea thermydron* would have lost part or all their ancestors' osmoregulatory abilities, which had became superfluous in environments where salinity is stable.

V. SITES AND MECHANISMS OF OSMOREGULATION

At salinities below seawater, a hyper-regulating organism is exposed to osmotic influx of water and to ion loss. At higher salinities, a hypo-osmoregulating animal undergoes the reverse passive exchanges. Mechanisms limiting these passive fluxes and compensating for them are localized at four sites: the integument, represented by the cuticle and underlying epithelium in crustaceans; the digestive tract; the excretory urinary organs; and the branchial chambers, including gills. For reasons discussed in Section IV, we will primarily address mechanisms of hyperosmoregulation. They include active uptake of ions through the gills and production of a requisite volume of urine, which can be diluted (hypotonic to the hemolymph) in some freshwater species.

A. Integument

One first adaptation in osmoregulation is the limitation of water and ion fluxes through reduction in the integumental permeability. The complete impermeabilization of the integument is impossible because it would be incompatible with respiratory gas exchanges and excretion of waste products. Early experiments based on changes in weight after blockage of excretory pores showed that osmoregulators are less permeable to water than osmoconformers.[274,522] These observations were later confirmed using radioactive tracers (^3H_2O), which also revealed that an exposure to low salinity could induce a very quick (within 30 sec) decrease in water permeability, such as in the isopod *Sphaeroma serratum*.[587] Water permeability is generally lower in crustaceans living in brackish and freshwater media.[79,238,550] Osmoconformers are more permeable to ions, particularly sodium, than osmoregulators;[227,242,535,573] for example, permeability to ions is 20 to 30 times higher in cancrid and majid crabs than in grapsids and crayfish. Recent studies have shown that the permeability of the cuticle covering the gill epithelium is higher in osmoconformers than in osmoregulators and is generally higher than the permeability of the underlying epithelium.[332–334,448] These variations are not related to cuticle structural or ultrastructural differences and warrant further investigations at the molecular level because the branchial epicuticle of euryhaline decapods might contain specific ion channels.[448]

B. Digestive Tract

1. Anatomy and Cellular Structure

The morphology and structure of the digestive tract of crustaceans have been extensively studied.[83,145] The gut of crustaceans is comprised of three parts. The foregut and the hindgut are ectodermic and lined with cuticle. The midgut has an endodermic origin and is subdivided into one to several diverticula or caeca. The foregut is limited by a monolayered epithelium covered by cuticle.[420] In addition to its function in nutrition,[83] it may also be involved in ion and water movements.[80,380] The midgut is lined by high epithelial cells directly in contact with the lumen content. They present features of transporting cells,[83,420] also present in the midgut caeca of different decapods.[419,420] Typical features include a large cell size, apical microvilli, basolateral infoldings, and numerous mitochondria. These cells might be involved in osmoregulation.[248,271] The hindgut is also lined by transporting cells in several isopods.[262,263] In such cells, Na$^+$,K$^+$-ATPase was detected through immunostaining in the basolateral infolding membranes of the hindgut epithelium of the terrestrial isopod *Armadillo officinalis*[621] and in the midgut of *Homarus gammarus* larvae and post-larvae,[307] which points to their ion-transporting function.

2. Functions

In addition to its role in nutrition, evidence suggests that the digestive tract is involved in osmoregulation. Drinking through the mouth and anus has been reported in several species.[141,142,195,205,246,350,351,373,460,508] It is still unclear whether crustaceans modify their drinking rate according to salinity, as teleosts do,[382] although evidence in some species points to an increase of drinking rate at higher salinities.[220,423] In the mangrove crab *Ucides cordatus*, the drinking rate more than doubles from 26‰ (isosmotic medium) to 34‰ seawater in which the animals hyporegulate.[508] Fluid absorption by the gut of some species has been observed, and, at least in hyperosmoregulators, it appears isosmotic to hemolymph.[2,4,9,205] Ion movements have been reported that potentially could be associated with water uptake.[116,179,262,271,373,380,419,420,422,440,508] In the terrestrial crab *Gecarcinus lateralis*, the foregut is permeable to water and ions and under neuroendocrine control, which is interpreted as an adaptation to conserve water.[380] The midgut and the midgut caeca participate in ion and water regulation in osmoregulating species,[4,9,141,142,145,248,351,577] but their involvement is not significant in osmoconformers.[271,421,422] When present, the osmoregulatory function of the midgut and the caeca may constitute an important adaptation in hypo-osmoregulators that tend to be chronically dehydrated.[2,3,9,10,141,142,204,209,221] In addition, in hyporegulators and in terrestrial species, the gut could be the site of salt transport,[592] probably oriented to salt extrusion.[142,221] More generally, as some terrestrial crabs reingest their urine, these species could rely on the gut for ion reabsorption from the urine (see also Section V.C.2).[6,54] The hindgut seems involved in ion transport in *Corophium volutator*[279] and in water uptake during molting in *Carcinus maenas*,[117] but its involvement in osmoregulation during the intermolt is still uncertain, particularly given the cuticle lining of this part of the digestive tract.[118] Differentiating the digestive and osmoregulatory functions of the digestive tract is difficult and additional research is necessary, particularly in conditions of hypo-osmoregulation.

C. Excretory Organs

1. Anatomy and Cellular Structure

The excretory glands of crustaceans are usually paired organs hypothetically derived from segmented pairs of excretory organs present in ancestral crustaceans.[219] Their structure and functions have been reviewed several times.[382,447,448,463,487,499] The segment location of the excretory organs differs according to the evolutionary position of crustaceans. In most entomostracans—including Branchiopoda, Ostracoda, Cirripedia, Copepoda, and, in the lower Malacostraca, Phyllocarida, Hoplocarida, and some Peracarida (Isopoda in particular)—the excretory ducts open on the ventral face of the second maxilla (sixth) somite, hence we refer to them as *maxillary glands*. In other Peracarida, including Amphipoda, Mysidacea, and Eucarida (Euphausiacea and Decapoda), the excretory organs are *antennal glands*, as their ducts open at the base of the antennae on the antennal (third) somite. In a few species of the first group, antennal glands would develop first in larvae before their replacement by maxillary glands, but generalization is not warranted.[61,519]

The anatomy of the excretory glands is organized according to a common three-part plan comprising an end sac or coelomosac including remnants of the coelom; an excretory canal or tubule, which may include a labyrinth; and an exit duct sometimes differentiated into a bladder (Figure 6.2). This last section is ectodermic and thus lined by cuticle. A valve often separates the end sac from the tubule. The microscopic anatomy of these glands, studied in a limited number of species, shows variations among them.[382,448] In *Artemia salina*, a central sac is surrounded by three coils of tubule ending in the terminal duct without a bladder.[607,608] A similar general organization has been found in *Balanus balanoides* and *B. hameri*,[633] *Corophium volutator*,[279] *Uca mordax*,[516] *Callinectes sapidus*,[284] and *Homarus gammarus*.[306,307] The organization and structure of the antennal glands have been well studied in crayfish.[19,38,199,305,308,321,375,403,454,455,487,490,514,533,538] The excretory canal, proportionally longer than in other species, is divided into a labyrinth and a tubule, sometimes called a *nephridial tubule*, itself including proximal and distal parts, and the exit duct is clearly dilated into a bladder.

The coelomosac is limited by epithelial cells with basal podocytic extensions extending to the basal membrane; slit diaphragms, well observed in crayfish, bridge the gap between adjacent foot processes. Hemolymph, brought to the coelomosac by an antennary artery, is separated from the urinary space inside the sac by the thin process made up of the basal membrane and the slit diaphragms. This type of structure is reminiscent of an ultrafiltration system as in the glomerulus–Malpighian corpuscle of the vertebrate nephron. The end-sac cells also contain numerous vesicles that look like lysosomes and contain residual bodies—the formed bodies of Riegel,[483] also found in *Homarus gammarus*[306] and *Astacus leptodactylus*.[305] Along the excretory canal, cells bordering the proximal part (labyrinth and proximal tubule when they are differentiated) are generally high and present the typical features of ion-transporting cells. Toward the distal part of the canal, endocytic cell vacuoles are more frequent, but the density of the apical microvilli tends to decrease. In the bladder, some or all of the limiting cells present apical microvilli or cytoplasmic extrusions

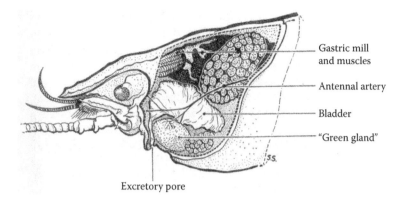

Gastric mill
and muscles

Antennal artery

Bladder

"Green gland"

Excretory pore

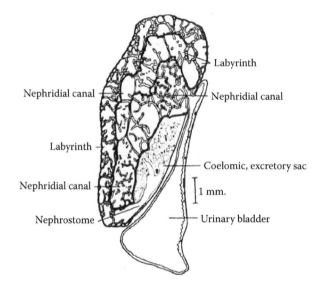

Labyrinth

Nephridial canal

Nephridial canal

Labyrinth

Coelomic, excretory sac

Nephridial canal

1 mm.

Nephrostome

Urinary bladder

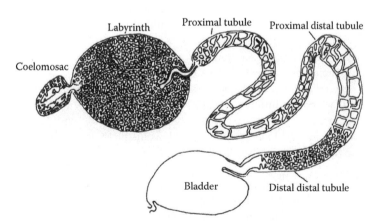

Labyrinth

Proximal tubule

Proximal distal tubule

Coelomosac

Bladder

Distal distal tubule

FIGURE 6.2 Excretory organs. (Top) Location of the "renal organ" in the crayfish *Astacus pallipes*. (From Potts, W.T.W. and Parry, G., *Osmotic and Ionic Regulation in Animals*, Pergamon Press, Oxford, 1963. With permission.) (Middle) Transverse section of the antennal gland of *Astacus fluviatilis*. (From Peters, H., *Z. Morphol. Ökol. Tiere*, 30, 355, 1935. With permission.) (Bottom) Schematic diagram of the antennal gland in *Astacus* spp. (From Riegel, J.A., *Comparative Physiology of Renal Excretion*, Oliver and Boyd, Edinburgh, 1972. With permission.)

into the lumen and numerous mitochondria within basal infoldings, which indicate that the function of the organ is more than storage and release of urine. Immunolocalization of Na^+,K^+-ATPase in *Homarus gammarus*[306,307] and in several cryfish[305,308,455] has shown that the enzyme is absent in the coelomosac cells and present at the basolateral part of the ionocyte-like cells (see Section V.D.1) of the labyrinth, the tubular section, and the bladder.

2. Functions

As clearly indicated by its structure, the coelomosac is involved in ultrafiltration. Solutes with a molecular mass lower than 90,000 Da or smaller than 40 Å can filtrate from the hemolymph to the primary urine formed in the coelomosac. Hemolymph is carried to the coelomosac at low pressure by hemolymph sinuses or, in decapods, by the antennary artery. The driving force for ultrafiltration might be this pressure difference, probably enhanced by the difference in osmotic pressure between the two compartments resulting from the release of the formed bodies content from the coelomosac cells to the primary urine,[484–486,488,489] which is considered to be isosmotic to hemolymph. Its volume and composition are altered during its transit to the excretory pore. Reabsorption of organic molecules such as glucose and amino acids has been observed in the bladder of some crab species.[49–53,228,266] Nitrogenous compounds as ammonia or urea can be excreted to a limited extent through the excretory glands, but their participation in the acid–base balance is limited.[77] Mg^{2+} is secreted into the urine, perhaps in relation to Na^+ and Cl^- transport, which accounts for the low Mg^{2+} hemolymph concentration in crustaceans.[221,229,246,473,491,497,498,508,510] Mg^{2+} transport would be an active process[196] effected through the bladder epithelium in crabs.[265,267,473] The dynamics of modification of the primary urine Na^+ and Cl^- content is for the most part unknown. In most marine and brackish water crustaceans, even exposed to salinity changes, urine is isosmotic to and has similar Na^+ and Cl^- concentrations as hemolymph (see Tables I and II of Mantel and Farmer[382]). However, the presence of ionocytes and of Na^+,K^+-ATPase therein (see above), even in such slight osmoregulators as homarid lobsters and in crab,[270,284] shows that active transport of Na^+ and Cl^- occur in the excretory canal and in the bladder, probably as mediators of secondary transports of other ions and molecules.[306]

In terrestrial anomuran and brachyuran crabs, where water and ion conservation are important, specific adaptations have been reported. Urine, isosmotic to hemolymph, is directed from the nephropores to the branchial chambers.[407] Ion reabsorption from urine to hemolymph occurs across the gills.[224,225,406,408,578,636,637] In some species, urine could also be reingested,[223,225,579,636] giving an ion-regulatory function to the gut.[6,54]

In contrast, some freshwater crustaceans, particularly gammarids and crayfish, produce urine hypotonic to hemolymph with lower Na^+ and Cl^- concentrations (Table 6.2), which contributes to lower ion loss. At higher salinity, urine becomes isosmotic to hemolymph.[512] In freshwater, ionic reabsorption takes place in the different sections of the excretory glands (labyrinth, tubule, and bladder), where ionocytes are present and show a clear expression of Na^+,K^+-ATPase that tends to decrease if salinity increases.[512] In crayfish, the tubule is distinctly longer than in other decapods, and the abundance and activity of Na^+,K^+-ATPase are generally higher than in other sections, particularly the labyrinth.[258,296,455] The distal part of the tubule, where the number of ionocytes and presence of Na^+,K^+-ATPase are highest, might be the primary site of ion absorption in the antennal gland,[488–490,616,632] perhaps linked to other transports such as glucose and amino acid reabsorption or organic acid secretion.[512,616] The density of ionocytes and high expression (although lower than in the distal tubule) of Na^+,K^+-ATPase in the bladder found in *Astacus leptodactylus*[305,308] confirm that this part of the antennal gland is also implicated in ion reabsorption as already suggested.[294,482,483,490] Unlike crayfish, freshwater crabs such as potamids and *Eriocheir sinensis*, which have low cuticle permeability, have not evolved the ability to produce dilute urine, but they release small volumes of isotonic urine.[222,243,410,433]

TABLE 6.2
Hemolymph and Urine Osmolality and Ion Concentrations in Selected Crustaceans Maintained in Freshwater

Species	Osmolality (mOsm/kg)		Na+ Concentration (mEq/L)		Cl- Concentration (mEq/L)		Refs.
	Hemo-lymph	Urine	Hemo-lymph	Urine	Hemo-lymph	Urine	
Gammarus duebeni	480	155	—	—	—	—	Lockwood,[344] Sutcliffe,[574]
Gammarus fasciatus	320	110	—	—	—	—	Werntz[631]
Gammarus pulex	275	50	—	—	—	—	Lockwood,[344] Sutcliffe and Shaw[575]
Macrobrachium australiense	515	25	—	—	—	—	Denne[158]
Austropotamobius pallipes	415	55	205	11	—	—	Riegel[484]
Pacifastacus leniusculus	445	35	200	14.2	195	38	Kerley and Pritchard,[304] Pritchard and Kerley[470]
Astacus leptodactylus	375	182	171	11.4	201	5.6	Khodabandeh et al.[308]
Procambarus clarkii	400	50	—	—	—	—	Sarver et al.[512]
Potamon edulis	540	560	250	295	210	275	Harris and Micaleff[245]
Eriocheir sinensis	615	610	305	325	275	285	De Leersnyder[152]

Source: Adapted from Mantel, L.H. and Farmer, L.L., in *Internal Anatomy and Physiological Regulation*, Mantel, L.H., Ed., Academic Press, New York, 1983, pp. 53–161.

The primary function of the excretory organs related to osmoregulation is the regulation of the volume of hemolymph. When exposed to a dilute medium, most crustaceans get rid of the excess water through an increased volume of urine, a reaction observed in osmoconforming and osmo-regulating crabs.[49,50,156,230,265,502,534,643] The signal triggering the increase in urine production could be the increased volume of hemolymph,[265] but other mechanisms could be involved, including hormonal mediation.[382] As the urine is isosmotic to hemolymph in most marine and brackish water species, the increased urine volume at low salinity translates into a loss of ions. Exceptions are found in a few species, such as *Macrobrachium australiense*,[158] that are able to produce hypotonic urine at low salinity. The same response of hypotonic urine production, found in such freshwater crustaceans as gammarids and crayfish but not in crabs (see above), represents an important adaptation to that medium.

At salinities higher than seawater, most crustaceans, including the semiterrestrial *Uca mordax*,[516] produce isotonic urine; however, in two semiterrestrial crabs, *Ocypode quadrata*[209] and *Uca pugnax*,[221] maintained in seawater and at higher salinities where they hypo-osmoregulate, urine is hypertonic to hemolymph by 100 to 150 mOsm/kg. This fact may be related to the high amount of Na+,K+-ATPase found in the antennal glands of terrestrial crabs[592] that are exposed to dehydration. Further studies are necessary in such models to investigate the possible function of the excretory glands in salt excretion or the absorption of water. The potential expression of aquaporin-like molecules in the excretory glands should also be studied.

In summary, the excretory glands are mainly involved in the maintenance of water balance through the production of urine. As urine is generally isotonic to hemolymph, these glands are less involved in ion balance. A few exceptions that are significant with relation to habitat adaptation can be found in some freshwater and semiterrestrial species in which the production of hypotonic urine contributes to ion regulation. The function of the glands in the hydromineral balance of terrestrial and hypo-osmoregulating species remains to be further investigated.

D. Gills and Branchial Chambers

1. Anatomy and Cellular Structure

Several reviews have considered the anatomy and structure of the gills,[183,382,397,447,448,582] which are the main site for gas and ion exchanges. As diffusional exchanges of gas are proportional to the area of the exchange surface, the shape and organization of gills are primarily dependent on this function. Ionic regulation, if effective, is located in different places according to the groups and species. In small individuals, where the surface/volume ratio is high enough to permit sufficient gas exchanges, the presence of gills is not necessary. In such cases, ionic regulation is effected in specialized areas or organs entirely devoted to this function. The fenestra dorsalis is an example; it is located on the dorsal part of the cephalothorax of some freshwater syncarids.[323] Dorsal or neck organs with similar structure and function have been described in several cladocera.[12,387,401,427,464] In the tanaid *Sinelobus stanfordi*, transporting tissues are located on the gills and along the branchiostegites.[310] In amphipods, extrabranchial sites of osmoregulation have been detected at several locations as sternal and pereopodal disks[311] and blood vessels of the coxal gills.[312] In decapods, ionoregulatory sites, generally located on the gills, are also found at other locations of the branchial chambers (see below for examples). In embryonic or early post-embryonic stages, the existence (temporary, in most cases) of extrabranchial ionoregulatory organs has been reported in several species (see Section VII).

Gills would have first developed as flattened appendages or epipodites of appendages perfused by hemolymph, such as in the branchiopod *Artemia* spp.[134,136] or in the primitive decapod *Anaspides tasmaniae*.[397] Gills are differently located according to the groups. In branchiopods, each thoracic limb bears a respiratory and osmoregulatory epipod.[386] In isopods, the pleopods develop into gills.[618] Most amphipods possess coxal gills on four or more thoracic segments; epipodites and sternal gills are also present in some species.[517] In decapods, each thoracic appendage can bear gills, with variations being observed among species.[582] According to their location on the appendage or the pleura, a theoretical number of three gill types per appendage could occur—podobranch, artho-branch, and pleurobranch located, respectively, on the coxa, the articular membrane between the coxa and precoxa, and the pleura. The number of gill pairs is highest in peneids and homarids (19 and 21) but generally tends to be lower in other decapods with a usual range of 6 to 9 in brachyuran crabs.[29]

A decapod gill is structurally formed of an elongated axis bearing branchial lamellae (comprehensively reviewed in Taylor and Taylor[582]). The axis contains an afferent and an efferent vessel carrying hemolymph into and out of the gill. Three main types of gills have been described according to the shape and structure of lamellae: the phyllobranch, trichobranch, and dendrobranch (Figure 6.1). A phyllobranch is formed of a series of paired flat lamellae perpendicular to the axis, including a marginal canal linking the afferent and efferent vessels for hemolymph circulation. This type is found in carid shrimps, in some anomurans, and in brachyuran crabs, except in Dromiidae. In a trichobranch, found in astacids, lamellae are replaced by several single filaments internally separated into two longitudinal compartments where hemolymph flows. Dendrobranchs, a main feature of Dendrobranchiata, are found in Penaeoidea and Sergestoidea. Their axes bear branched or secondary filaments.

The gills of decapods are enclosed in a pair of branchial chambers formed by the pleurae and lateral extensions of the tergum, the branchiostegites. Each chamber is longitudinally open at its base, providing an entrance for the external water that flows along the gills, usually with the guidance of epipodites separating each gill group; water is then actively ejected from the branchial chamber through an anterior channel by the pumping of the scaphognathites (elongated and flattened exopodite of the second maxilla[393]). On the inside part of the two-compartment system separated by the gill tegument, hemolymph perfuses the gill epithelium through the circulatory system.[393,397] In terrestrial anomuran and brachyuran crabs, gills are specialized in air breathing and are kept wet through several adaptations, including redirection of urine from the nephropores, a process also allowing ion reabsorption (see Section V.C.2).[224,407]

In addition to their respiratory function, gills are involved in ion regulation, and both functions interact with acid–base balance.[259] High Cl⁻ concentrations in cells lining some regions of the cuticle have been revealed early through silver staining. Following the pioneering work of Koch,[316] this technique has been used in various species of crustaceans, including *Artemia salina*[136] and several decapods.[447] Na⁺ transport was demonstrated in isolated gills of *Eriocheir sinensis*.[317,318,450] The cellular structure of the gills of crustaceans is thus related to their two main functions that require very different cell types. Respiration based on gas diffusions is accomplished through thin (2- to 4-μm) poorly differentiated cells. In contrast, ion transports are effected in the so-called *ionocytes* in which specific structural and molecular features dictate a larger size (over 10-μm thickness) not compatible with gas diffusion. These features include apical microvilli, basolateral infoldings, and numerous mitochondria (see below for details). Separate functions result in the separation of the two types of epithelia—respiratory or osmoregulatory—either on the same gill or on different gills. The gill structure, seldom reviewed,[448,582] has been studied in several species of crustaceans.[14,27,30,48,59,68,69,71,120–122,127,131,132,138,153,155,161,170,171,180,182,190,197,206,207,218,224,240,269,284,309,310,335, 336,338,362,371,388,392,402,407,418,433,441,442,445,448,458,517,537,551,552,576,577,586,587,590,617]

Gill cells are separated from hemolymph by a basal membrane and are covered by a thin apical cuticle. Respiratory cells are flat and contain few organelles, among them a small number of mitochondria and vesicles. In contrast, the ionoregulatory cells or ionocytes present typical differentiations. They are usually interlinked through infoldings with or without septate junctions. These junctions may play a dual role in adhesion between neighboring cells and in permeability regulation as suggested in gammarid amphipods.[537] Although they vary between species, ionocytes present several basic features common in many salt-transporting vertebrate and invertebrate tissues such as fish gills, renal tubules, and intestinal cells. Ionocytes can be located in gills but also in other sites of the branchial chamber, including the pleura, the branchiostegite, and the epipodites (see below). The apical and basal sides of ionocytes are different. Apical microvilli, sometimes forming deep channels, increase the surface in contact with the subcuticular space; their number and dimensions vary with salinity, usually expanding when salinity decreases. On the basolateral side, deep infoldings that can extend high in the cell are associated with numerous mitochondria (Figure 6.3). Several transmembrane proteins involved in ion transport and exchanges are basolaterally or apically located (see Section V.D.2). Among them, Na⁺,K⁺-ATPase is found in abundance on the basolateral side.[361] Its presence has been revealed through immunocytochemistry in several decapods (Figure 6.4).[31,32,120–122,335,336,338,388,448,451,458,590,595] Na⁺,K⁺-ATPase was also detected through the same method in the salt gland of *Artemia* spp.[568] and in the calcium-transporting sternal epithelium of the terrestrial isopod *Porcellio scaber*.[647] Ultracytochemical studies (immunogold) have shown that Na⁺,K⁺-ATPase is mainly located along the basolateral membranes of ionocytes.[335,336,448,451,595]

In adult decapods, the partition of respiratory and ionoregulatory cells varies with species and the ability to osmoregulate. During development, the location of the osmoregulatory function may shift between different sites (see Section VII). In osmoconformers, no or few ionocytes are found on the gills or elsewhere in the branchial chamber.[388,448] In osmoregulators, different types of ionocytes vs. respiratory cells partition have been observed. Osmoregulation can be effected on the posterior gills only or on all of the gills by ionocytes separated from respiratory cells. In some decapods, other sites of the branchial chamber can be involved in osmoregulation, such as the epipodites and the inner side of the branchiostegites.

In isopods, gill functions are localized in the pleopods. Among their five pairs, exopodites are respiratory, and endopodites are involved in osmoregulation, as shown by their structure[618] and by the increase in Na⁺,K⁺-ATPase activity at low salinity.[462,585]

In several crabs, a clear separation exists between the anterior gills that are respiratory and the posterior ones, where ionocytes are numerous on the lamellae and thus involved in osmoregulation This organization has been reported in *Callinectes sapidus*,[314,349,352,595] *Carcinus maenas*,[120,254,595]

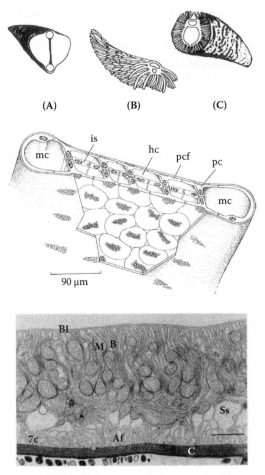

FIGURE 6.3 Gills. (Top) Schematic drawings of the different types of gills in decapods: (A) phyllobranch; (B) trichobranch; (C) dendrobranch. (From McMahon, B.R. and Wilkens, J.L., in *The Biology of Crustacea*, Vol. 5, Mantel L.H., Ed., Academic Press, New York, 1983. With permission.) (Middle) Schematic detail of a lamella from a posterior phyllobranchiate gill in *Macrobrachium olfersii*: is, intralamellar septum; hc, hemolymph channels; mc, marginal canal; pc, pillar cells; pcf, pillar cell flanges. (From Freire C.A. and McNamara, J.C., *J. Crust. Biol.*, 15, 103, 1995. With permission.) (Bottom) Ionocytes in gill epithelium of *Neohelice granulata (Chasmagnathus granulatus)* in seawater: Af, apical folds; B, basolateral membrane; Bl, basal lamina; M, mitochondria; Ss, subcuticular space. Scale bar = 1 μm. (From Genovese, G. et al., *Mar. Biol.*, 144, 111, 2004. With permission.)

Eriocheir sinensis,[30,121,448,590] *Neohelice granulata*,[206,207,362,364] *Dilocarcinus pagei*,[433] and *Pachygrapsus marmoratus*.[447,448,458,540,543] Exposure to low salinity usually results in an increase of the apical infoldings[127,133,188] and of the subcuticular space of ionocytes.[448] Few data are available in crabs able to hyporegulate. When exposed to high salinities, ultrastructural changes seem to indicate a shift from a salt-uptake morphology to salt secretion in *Goniopsis cruentata*[385] and *Uca uruguayensis*.[363] In *Neohelice granulata*, the septate junctions between adjacent ionocytes are shorter at high salinity compared to low salinity, pointing to a possible role of the paracellular pathway in ion secretion.[206,362] In addition, respiratory cells can be also associated with ionocytes on the same posterior gill, as observed in *Carcinus maenas*,[127] *Callinectes sapidus*,[133] and *Gecarcinus lateralis*.[132] In the freshwater crab *Dilocarcinus pagei*, structural and functional asymmetries have been observed in the posterior gills; their lamellae are lined by a proximal thick epithelium and a distal thin one, apparently involved, respectively, in Na+ and Cl− absorption.[433]

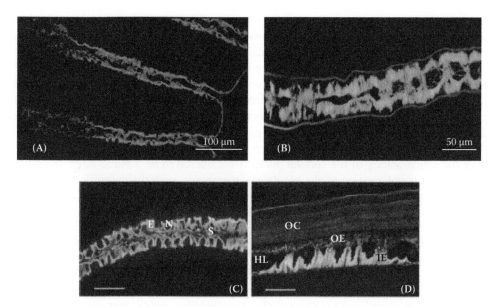

FIGURE 6.4 Immunolocalization of Na+,K+-ATPase in diverse organs of the branchial chamber. (Top) Lamellae of the posterior gill 6 in *Pachygrapsus marmoratus*. (Courtesy of T. Spanings-Pierrot.) (Bottom) Epipodite (C) and branchiostegite (D) in *Palaemon adspersus*: E, epithelium; HL, hemolymph lacuna; IE, inner epithelium; N, nucleus; OC, outer cuticle; OE, outer epithelium; S, septum. Scale bars = 50 μm. (From Martinez, A.-S. et al., *Tissue Cell*, 37, 153, 2005. With permission.)

In carid shrimps, the two types of epithelia also coexist in the gills with a predominance of the respiratory function. The osmoregulatory epithelium, instead of laterally bordering the gill as in brachyurans, is axial.[167,441,442,445] In *Palaemon adspersus*[388] and *Crangon crangon*,[122] H-shaped cells enclose hemolymph lacunae. They present a central shaft, with the features of ionocytes, and thin extensions that would be involved in ion transport and gas exchange. A similar organization has been described in the palaemonid *Macrobrachium olfersii*,[197] but the thick and thin parts correspond to different cells (Figure 6.4). In addition, the epipodites and inner side of the branchiostegites of *Palaemon adspersus* bear numerous ionocytes with a high amount of Na+,K+-ATPase (Figure 6.4).[388]

In the crayfish *Procambarus clarkii*[71,161] and *Astacus leptodactylus*,[31,32,170,338] both respiratory and osmoregulatory epithelia have been found on the same gills; some filaments are responsible for respiration, and others bearing ionocytes are involved in osmoregulation. Other slightly different ionocytes are present on the lamina (equivalent to the epipodites), perhaps involved in Cl– transport.[31,32,170,338] In *Homarus gammarus*, the gills present a poorly differentiated epithelium devoid of Na+,K+-ATPase; in contrast, ionocytes are present on the epithelia lining the epipodites and the inner side of the branchiostegites. Exposure to low salinity results in an increase of the ionocytes thickness and of Na+,K+-ATPase immunostaining. Thus, the slight hyperosmoregulatory capacity of this species is most probably based on these two extra-gill sites.[240,335,336] In peneid shrimps such as *Penaeus aztecus*[194,577] and *P. japonicus*,[59] the gills bear poorly differentiated ionocytes probably more involved in respiration than in osmoregulation, but numerous ionocytes are present on the epipodites, which would be the main site of osmoregulation in *P. japonicus*.[59]

2. Functions

Models of osmoregulatory NaCl transport across crustacean gills are universally based on Na+,K+-ATPase, which, as in other epithelia, is restricted to the basolateral membrane of branchial epithelial cells. Evidence supporting such localization comes from physiological studies as well as via direct determinations. In isolated perfused gills of the blue crab *Callinectes sapidus*, for example, the Na^+,K^+-ATPase inhibitor ouabain has an inhibitory effect on active transepithelial Na^+ flux only when applied in the internal (basolateral) perfusion medium; ouabain in the external (apical) bathing medium has no effect,[73] confirming earlier experiments measuring transport-related potential differences and sodium fluxes across posterior gills of *Carcinus maenas*.[359,543] Direct localization via ultracytochemical or immunohistochemical methods corroborates a basolateral site for Na^+,K^+-ATPase in ionocytes of crustacean gills and branchial chamber tissues, as discussed in detail in the previous section. In this position, due to the directionality of the pumping mechanism, the Na^+,K^+-ATPase is poised to pump Na^+ ions from the cytosol of the ionocyte outwardly across the basolateral infoldings into the hemolymph, in exchange for K^+ or NH_4^+.[594] Because the stoichiometry of the sodium pump is considered to be 3 Na^+:2 K^+,[343] the outcome of this transport is polarization of the basolateral membrane, with the cytosolic side negatively charged with respect to the hemolymph side. This polarization of charge as well as Na^+ concentration provide a source of potential energy to drive secondary active transport processes, to be discussed below.

Many authors have reported that the enzymatic activity of the Na^+,K^+-ATPase measured in gill homogenates or subcellular membrane fractions responds to hypoosmotic stress (low salinity) in a variety of euryhaline marine crustaceans. Highest activities are recorded in posterior gills, corresponding with the predominance of ion-transporting cell structures in epithelia lining the posterior but not anterior gills, as discussed in detail earlier. In *Callinectes sapidus* acclimated to 5‰ salinity, mitochondria-rich areas of posterior gill lamellae exhibit 7-fold higher Na^+,K^+-ATPase specific activity than the lighter colored, mitochondria-poor regions and 14-fold higher activity than anterior gills.[595] Crustacean species in which a salinity effect on branchial Na^+,K^+-ATPase activity has been noted include *Callinectes sapidus*,[383,424,597] *Callinectes similis*,[461] *Carcinus maenas*,[244,357,540] *Neohelice granulata*,[82,206,515] *Cherax destructor*,[645] *Eriocheir sinensis*,[452,590] *Homarus gammarus*,[191,354] *Hemigrapsus* spp.,[623] *Macrobrachium olfersii*,[399] *Macrophthalmus* spp.,[604] *Scylla paramamosain*,[604] *Uca* spp.,[166,268,620] *Ucides cordatus*,[247] and others. In contrast, gills of osmoconforming or weakly osmoregulating species have been shown to possess generally lower Na^+,K^+-ATPase activity that shows little response to salinity stress, as noted in *Calappa hepatica*,[565] *Maja crispata*,[356] *Dromia personata*,[356] and the spiny lobster *Palinurus elephas*.[356]

Short-term changes in gill Na^+,K^+-ATPase activity could result from interaction between its α, β, and γ subunits, from direct effects of cell-signaling processes on one or more of its subunits, or from differential recruitment of the pump protein into the plasma membrane, processes that have been widely noted in other biological systems.[348,584] The amino acid sequence of the Na^+,K^+-ATPase α-subunit from *Artemia franciscana*, *Callinectes sapidus*, *Homarus americanus*, *Scylla paramamosain*, and *Pachygrapsus marmoratus* contains a conserved binding site for protein kinase A,[119,281,367,600] providing a target for regulation by cyclic AMP. Studies with crustacean epithelia, summarized in detail below, show that dopamine and cyclic AMP do indeed affect Na^+,K^+-ATPase activity and the transport properties of gills.

In some but not all crustacean α-subunits, a binding site for 14–3–3 protein resides near the *N*-terminus.[281] Because the 14–3–3 family of proteins regulates the translocation of target proteins between endoplasmic reticulum or cytoplasmic sites and the plasma membrane,[370] the presence of this binding site on the α-subunit affords a mechanism for differential recruitment to the plasma membrane of gill epithelial cells. Longer term changes in Na^+,K^+-ATPase activity are likely to result from the regulation of translational or transcriptional processes within gill epithelial cells.

During embryonic development of the crayfish *Astacus leptodactylus*, mRNA expression of the Na^+,K^+-ATPase α-subunit is closely correlated with the differentiation of tissues involved in hyperosmoregulation.[532] In the brine shrimp *Artemia franciscana*, two transcripts encode different α-subunits of the Na^+,K^+-ATPase.[36,367] During early larval development, mRNA encoding the α_1-subunit is highly expressed in salt gland, antennal gland, and midgut, while expression of the α_2-subunit is restricted to the salt gland;[176] however, the expression of neither transcript appears to respond to altered environmental salinity.[328]

In contrast, in gills of decapod crustaceans, Na^+,K^+-ATPase α-subunit mRNA transcription appears to be quite sensitive to salinity change. In the portunid crab, *Callinectes sapidus*, quantitative PCR analysis showed that α-subunit mRNA abundance in posterior gills increased by 2.5-fold within 4 days following the transfer of crabs from 32‰ to 10‰ seawater, in which the crab strongly hyperosmoregulates.[352] Despite sustained increases in α-subunit protein and Na^+,K^+-ATPase specific activity in low salinity, α-subunit mRNA returned to control levels after 11 days. In the closely related portunid *Scylla paramamosain*, α-subunit mRNA abundance in posterior gill increased 6-fold within 7 days after transfer from 25 5‰ to 5‰ salinity, preceding a 4-fold increase in Na^+,K^+-ATPase activity at 14 days.[119] This transcriptional response is consistent with a central role of the Na^+,K^+-ATPase in supporting hyperosmoregulation in dilute salinities. Although *S. paramamosain*, like *C. sapidus*, is incapable of hypo-osmoregulation in salinities above the isoionic value, α-subunit mRNA expression was enhanced in 45‰ salinity but curiously was not accompanied by changes in enzymatic activity.[119]

In the strongly hyper–hypo-osmoregulating varunid crab *Neohelice granulata*, α-subunit mRNA levels increased dramatically in posterior gills following transfer from 30‰ to either 2‰ or 45‰ salinity.[365] During acclimation to 2‰, α-subunit expression reached 35- to 55-fold higher levels after just 24 hours.[365] Following transfer from 30‰ to 45‰ salinity, a condition in which the crab hypo-osmoregulates effectively, α-subunit mRNA abundance increased by 25-fold but only after 4 days of acclimation.[365] These observations led to the conclusion that the Na^+,K^+-ATPase is likely to be intimately associated with both hyperosmoregulation in low salinities and hypo-osmoregulation in high salinities.

A gill-by-gill study of α-subunit mRNA expression in the hyper–hypo-osmoregulating grapsid crab, *Pachygrapsus marmoratus*, showed a complex pattern of response following transfer of animals from 36‰ to 10‰ salinity. Gills 5 to 9 all showed increased levels of α-subunit mRNA during acclimation but with different time courses; gill 7 responded within 2 hours, whereas gill 9 showed no response until 24 hours.[281] This pattern contrasted with the response following transfer from 36‰ to 45‰ salinity, where this species hypo-osmoregulates; little effect of acclimation on α-subunit expression was observed in any tested gill except for gill 7, in which a significant increase occurred within 4 hours after transfer.[281] Because it is known that gills of other hyper–hypo-osmoregulating crabs exhibit distinct functions with regard to ion uptake or ion excretion,[390] it is tempting to suggest that in *P. marmoratus* all tested gills (5 to 9) may participate in ion uptake and gill 7 is specialized for ion excretion, both processes energized by an increased level of Na^+,K^+-ATPase mRNA and presumably protein. A preliminary examination of the promoter structure upstream of the α-subunit gene in *P. marmoratus* identified more than six putative binding sites for transcription factors, perhaps allowing the complex pattern of transcriptional responses observed following salinity change.[281]

Although Na^+,K^+-ATPase has received the most attention from investigators interested in the osmoregulatory function of crustacean gills, other transporters and associated proteins are also of biological importance. Because Na^+,K^+-ATPase is basolaterally located, in contact with the internal milieu, other membrane proteins must take the role of mediating transport of ions across the apical membrane. To accomplish hyperosmoregulation in dilute salinities, NaCl uptake from the aqueous environment may be achieved via apical Na^+/H^+ exchange, Cl^-/HCO_3^- exchange, $Na^+/K^+/2Cl^-$ cotransport, other ion pumps, or ion channels. To achieve hypo-osmoregulation in high salinities, these transport functions may be redistributed between apical and basolateral membranes, a phenomenon observed with V-type H^+-ATPase during mineralization–remineralization cycles in sternal epithelial cells of the isopod *Porcellio scaber*.[648] Only in a few cases have these transporters been investigated in detail in crustacean gills, and little information is available regarding hypo-osmoregulatory functions in particular.

One of the transporter candidates to receive experimental attention is the Na^+/H^+ exchanger, a family of proteins associated with pH regulation, Na^+ transport, and osmotic responses in a variety of epithelial cells.[374,619] Evidence supporting an osmoregulatory role for an apical amiloride-sensitive Na^+/H^+ exchanger comes from studies of intact animals and isolated gill preparations, as well as partially purified membranes. In the blue crab *Callinectes sapidus*, exposure of crabs to external amiloride (1×10^{-4} mol/L) reduced Na^+ influx rates by 97%, with little effect on Cl^- influx.[76] In the adult water flea *Daphnia magna*, amiloride reduced whole-body Na^+ uptake by about 40%.[46] Amiloride and its derivative ethylisopropyl amiloride inhibited Na^+ influx in intact crayfish (*Procambarus clarkii*), the latter being somewhat more effective in salt-depleted animals.[315] In isolated posterior gills of *C. sapidus* perfused with asymmetric salines, reflecting the transbranchial ion gradient in brackish water, external amiloride (1×10^{-4} mol/L) inhibited net Na^+ uptake by 60%.[73] Earlier studies of isolated gills perfused symmetrically showed similar inhibitory effects of amiloride, notably in the shore crabs *Carcinus aestuarii* and *Carcinus maenas*.[358,541] Subsequently, inhibition of Na^+ fluxes or short-circuit current by amiloride or its derivatives has been demonstrated in perfused gills of *Ucides cordatus*[390] and in isolated split gill lamellae of *Carcinus maenas*[438] and *Eriocheir sinensis*.[430,646] Whether these inhibitory effects were due to actions on a Na^+/H^+ exchanger or on amiloride-sensitive Na^+ channels in the epithelium or possibly on exchange sites in the acellular cuticle[436] remains controversial.

In membrane vesicles isolated from gills of the crayfish *Orconectes limosus*, an amiloride-sensitive Na^+/H^+ exchanger was detected by acridine orange quenching.[566] Exchanger activity could be separated from basolateral Na^+,K^+-ATPase activity by density gradient centrifugation, sedimenting with likely apical membrane markers. The $K_{0.5}$ for Na^+ was estimated at 17×10^{-3} mol/L with a Hill coefficient of approximately 1. In contrast, a similar study using membrane vesicles from posterior gills of *Carcinus maenas* showed sigmoid kinetics for Na^+, with a Hill coefficient of approximately 2, suggesting cooperativity and a 2:1 ratio of Na^+/H^+ exchange, unlike the electroneutral 1:1 ratio universally observed among Na^+/H^+ exchangers of vertebrate species.[536] Experiments with a potential-sensitive dye confirmed the electrogenic nature of the Na^+/H^+ exchanger in vesicles from *Carcinus* gill. The presence of such an exchanger has been confirmed in hepatopancreas and antennal glands of other decapod crustaceans, including the lobster *Homarus americanus*, where it may also function as a Ca^{+2}/H^+ exchanger.[5,7,8] On the basis of its sensitivity to inhibition by Ca^{2+}, the electrogenic $2Na^+/H^+$ exchanger was implicated in whole-body Na^+ uptake by the water flea *Daphnia magna*.[217] Immunocytochemical localization of the $2Na^+/H^+$ exchanger in lobster hepatopancreas showed strong reactivity in the apical membrane; in gills, however, the exchanger was associated with intracellular vacuoles rather than apical or basolateral plasma membranes.[313] No similar localization studies on strongly euryhaline crustaceans have been published.

The electrogenicity of the Na^+/H^+ exchanger in gill membrane vesicles may help to explain the hyperpolarizing effect of amiloride on isolated gills and split gill lamellae noted by several authors.[73,358,431] An amiloride-sensitive conductive pathway, initially suggested to be Na^+ channels, was in fact suggested as an alternative to electrically neutral 1 Na^+/1 H^+ exchange as the first step in Na^+ entry.[542]

Molecular evidence for a Na^+/H^+ exchanger in crustacean gill was obtained through the polymerase chain reaction (PCR) using degenerate primers based on vertebrate cDNA sequences. Starting with RNA prepared from posterior gills of *Carcinus maenas*, a 2595-base-pair cDNA was obtained that contained an open reading frame encoding a 673-amino-acid protein similar to Na^+/H^+ exchangers identified in other species.[598] A BLAST search of GenBank with this sequence showed that it bears strong similarity to the mammalian NHE-3 isoform, known to be localized to the apical membrane of proximal tubule epithelial cells in mammalian kidney and other tissues.[17,374] cRNA transcribed from the cloned *Carcinus* sequence supported Na^+/H^+ exchange across the membrane of injected *Xenopus* oocytes; however, the stoichiometry of exchange could not be measured due to the fragility of the oocytes expressing the transporter. Semiquantitative PCR showed the highest expression of the Na^+/H^+ exchanger in posterior gill, followed closely by anterior gill, with much lower expression levels in nonbranchial tissues.[598] Recently, a putative Na^+/H^+ exchanger nearly identical to the earlier sequence was identified among expressed sequence tags in normalized cDNA libraries derived from *Carcinus maenas* and the copepod *Calanus finmarchicus*[599] (Accession Nos. DV944270, DV943567, EL773341), confirming the presence of this exchanger in crustacean tissues.

Whether expression of this Na^+/H^+ exchanger varies with salinity, as well as its relationship to the electrogenic Na^+/H^+ exchanger described in membrane vesicles, remains to be investigated.

Several investigators have presented evidence supporting the existence in crustacean gills of epithelial Na^+ channels that are sensitive to inhibition by much lower concentrations of amiloride than are needed to inhibit the Na^+/H^+ exchanger. In split gill lamellae of *Eriocheir sinensis*, amiloride inhibited short-circuit current with half-maximal inhibition recorded at 6×10^{-7} mol/L,[646] much lower than the concentration of amiloride usually required to inhibit Na^+/H^+ exchange. By comparison, in perfused whole gills of *Carcinus maenas*, a half-maximal hyperpolarizing effect on transepithelial potential difference was achieved by 4×10^{-5} mol/L amiloride.[542] A subsequent study of split gill lamellae of *Carcinus maenas* suggested that most if not all of the amiloride effect in this species may be explained by its inhibition of ion fluxes through the acellular cuticle.[436]

Sorting through the apparently complex effects of amiloride on osmoregulatory ion transport across crustacean gills awaits further explorations at the molecular level. A voltage-gated sodium channel has been identified in neuronal tissue of *Cancer borealis* (Accession No. EF089568), but no amiloride-sensitive sodium channel has yet been detected using molecular approaches, despite more than 138,000 nucleotide sequences available in GenBank for Crustacea.

An alternative apical sodium transporter that has received attention is the $Na^+/K^+/2Cl^-$ cotransporter, one form of which mediates Na^+ uptake across the apical membrane of vertebrate renal thick ascending limb.[504] Influx rates of Na^+ and Cl^- measured isotopically across isolated gill lamellae of *Carcinus maenas* occurred in a ratio of approximately 1:2.[492] When the basolateral membrane of this preparation was made freely permeable to ions, the apical influx of Cl^- became dependent on the simultaneous presence of Na^+ and K^+, leading the authors to suggest that apical transport of these ions is achieved by a $Na^+/K^+/2Cl^-$ cotransporter. However, known inhibitors of the cotransporter, bumetanide and furosemide, had little effect on Cl^- fluxes, apparently resulting from their limited ability to cross the cuticle boundary between the external medium and the apical membrane, the most likely site of the cotransporter in gills exhibiting net ion uptake.[353,492] In gills of *Uca rapax*, however, apical furosemide did show an inhibitory effect on Na^+ and Cl^- fluxes and the results suggested that the $Na^+/K^+/2Cl^-$ cotransporter is restricted to posterior gills.[644] The role of the $Na^+/K^+/2Cl^-$ cotransporter in Na^+ uptake has been clarified by a pharmacological study in the water flea *Daphnia magna* in which it was shown that both bumetanide and thiazide (an inhibitor of Na^+/Cl^- exchange) reduced whole body Na^+ uptake in neonates but only bumetanide was effective in adults.[46] These results suggest that a Na^+/Cl^- exchanger may be important in Na^+ uptakes in neonates but not in adults.

A cDNA encoding a putative $Na^+/K^+/2Cl^-$ cotransporter has been amplified from gills of *Callinectes sapidus*,[593] *Eriocheir sinensis*,[628] and *Neohelice granulata*[365] and has been identified in expressed sequence tag libraries from *Carcinus maenas* (Accession No. DV467183 and others)[599] and *Calanus finmarchicus* (Accession No. EL697027). Its mRNA expression in *Neohelice granulata* was found to be strongly responsive to salinity challenge, increasing in posterior gills within 6 hours after transfer from 30‰ to 2‰ and reaching a maximum at 48 hours.[365] When crabs were transferred from 30‰ to 45‰ salinity, a medium in which the animal hypo-osmoregulates, cotransporter expression also increased in posterior gills but only after 96 hours, paralleling the observed increase in Na^+,K^+-ATPase mRNA expression and similar to the time course of the Na^+,K^+-ATPase mRNA response noted in *Pachygrapsus marmoratus*.[281] These responses at the transcriptional level suggest that a $Na^+/K^+/2Cl^-$ cotransporter is strongly involved in the osmoregulatory process in gills of *C. granulatus*; however, it is not known whether the cotransporter is associated with transepithelial ion fluxes, particularly in low salinities, or is simply a part of the volume regulatory response in the gill epithelial cells themselves.

An apical Cl^-/HCO_3^- exchanger may be important in mediating Cl^- uptake in some species, perhaps in addition to a $Na^+/K^+/2Cl^-$ cotransporter. The anion exchange inhibitor 4-acetamido-4'-isothiocyanostilbene-2,2'-disulfonate (SITS) reduced Cl^- uptake across perfused gills of *Carcinus maenas* by 28 to 39%, whereas the augmentation of the perfusion media with HCO_3^- enhanced uptake.[353] When gills of *Neohelice granulata* were perfused with Na^+-free medium, SITS reduced the transepithelial potential difference by 45%, indicating an electrogenic pathway involving a Cl^-/HCO_3^- exchanger.[208] In *Eriocheir sinensis* acclimated to freshwater, short-circuit current measurements across split gill lamellae indicated that Cl^- uptake proceeded by a Na^+-independent pathway, perhaps via a Cl^-/HCO_3^- exchanger energized by an outwardly directed gradient of HCO_3^-.[432,435] Na^+-independent uptake of Cl^- has been described in bicarbonate-loaded plasma membrane vesicles prepared from posterior gills of *Callinectes sapidus*.[329] Uptake could be inhibited by SITS but not by furosemide or bumetanide, supporting the presence of a Cl^-/HCO_3^- exchanger. A cDNA highly similar to those encoding anion exchangers in other species has been identified in an expressed sequence tag library of *Carcinus maenas* (Accession No. DV944614),[599] but no information is available regarding its expression in relation to osmotic challenge.

Basolateral transport of Cl^- across crustacean gills appears to be mediated by chloride channels. In perfused gills of *Eriocheir sinensis*, addition of the chloride channel blocker diphenylamine-2-carboxylate to the internal perfusion medium induced depolarization of transepithelial potential.[44] Similar results for this and other chloride channel blockers were observed with perfused posterior gills of *Carcinus maenas*, the most potent blocker being 5-nitro-2-(3-phenylpropylamino)-benzoate.[539] None of these blockers was effective in the external bathing medium but required access to the basolateral membrane of gill epithelial cells. cDNAs encoding a putative calcium-activated epithelial chloride channel have been identified among expressed sequence tags of *Carcinus maenas* (Accession No. DW584526) and *Homarus americanus* (Accession No. CN853980).[599] Whether the transcriptional expression of this channel varies with osmotic stress remains to be investigated.

V-type H^+-ATPase has been implicated as the driving force for Na^+ uptake via epithelial Na^+ channels in fish gills[177] in addition to serving important roles in ion transport and pH regulation in many other systems; it has also been suggested as an important component in crustacean osmoregulation. A specific inhibitor of V-type H^+-ATPase, bafilomycin, has been shown to block at least partially Cl^- influx and the short-circuit current due to Cl^- transport in gills of *Eriocheir sinensis*[434,493] as well as the transepithelial potential difference in gills of *Neohelice granulata* perfused with Na^+-free saline.[208] A membrane vesicle preparation from *E. sinensis* gills was shown to accumulate H^+ via an ATP-dependent pathway that could be blocked by bafilomycin.[434] Concanamycin, also an inhibitor of V-type H^+-ATPase, reduced the short-circuit current developed by a split lamella preparation of posterior gill from the freshwater crab *Dilocarcinus pagei*, being most effective when administered in the internal perfusion fluid.[626] In the water flea *Daphnia magna*, bafilomycin reduced whole-body Na^+ uptake in neonates but not in adults, suggesting that V-type H^+-ATPase may be important to osmoregulation during early stages of development.[46]

A V-type H^+-ATPase B subunit has been amplified and fully sequenced from gills of *Carcinus maenas* and partially sequenced from *Callinectes sapidus*, *Cancer irroratus*, *Neohelice granulata*, *Dilocarcinus pagei*, and *Eriocheir sinensis*.[365,626,629] Its transcriptional expression was found to be generally higher in posterior gills of *E. sinensis* and *D. pagei* than anterior, the converse apparently being the case in *C. maenas*, in which salinity had little effect on mRNA abundance measured semiquantitatively.[626,629] In *C. granulatus*, as measured by quantitative PCR, V-type H^+-ATPase B subunit mRNA showed increases in both anterior and posterior gills by 24 hours after transfer of crabs from 30‰ to 2‰ salinity and by 96 hours in posterior gills after transfer from 30‰ to 45‰ salinity, although sample-to-sample variation was quite high.[365]

Immunocytochemical localization of the V-type H^+-ATPase B subunit showed that it is primarily cytoplasmic in gills of *Carcinus maenas*, quite absent from the apical region.[629] Indeed, its distribution appears to be punctate, reflective of localization in discrete vesicles (Lignot, Weihrauch, and Towle, unpublished data). Because bafilomycin was without effect on transepithelial potentials in perfused gills of *C. maenas*, the authors concluded that V-type H^+-ATPase is not essential to osmoregulatory ion transport in this species.[629] A systematic study of 13 euryhaline crab species varying in their tolerance for freshwater found that gill V-type H^+-ATPase appears to be apical in species that tolerate freshwater but cytoplasmic in at least some species that do not tolerate freshwater.[604] In *Uca formosensis*, specific activity of the apically located V-type H^+-ATPase was higher in 5‰ salinity than 35‰, with no difference observed in Na$^+$,K$^+$-ATPase activity.[604] Although not directly related to osmoregulatory processes, the V-type H$^+$-ATPase of sternal epithelial cells in the isopod *Porcellio scaber* was shown to shift in polarity, from basolateral to apical distribution, during the transition from calcium deposition to calcium resorption.[648] Whether a similar shift in polarity, perhaps from cytoplasmic to apical, can occur in osmoregulatory tissues remains to be investigated.

The possibility that ammonium ion might serve as a counterion in Na$^+$/NH$_4^+$ exchange across crustacean gills has been of interest for many decades.[319] At the subcellular level, it is clear that NH$_4^+$ substitutes effectively for K$^+$, not only in stimulating the ATPase activity of the sodium pump[547] but also in its ability to mediate ATP-dependent Na$^+$ transport across inside-out plasma membrane vesicles.[264,594] Further support for this conclusion has come from studies with perfused *Carcinus* gills demonstrating a clear inhibitory effect of basolateral ouabain on the transbranchial excretion of NH$_4^+$.[355,624,627,630] Although earlier conclusions were based on the assumption that K$^+$ and NH$_4^+$ compete for the same external sites on the Na$^+$,K$^+$-ATPase protein, it now appears that the two sites may not be identical. In the portunid crabs *Callinectes danae* and *Callinectes ornatus*, kinetic analysis indicates that K$^+$ and NH$_4^+$ bind synergistically to different sites on the Na$^+$,K$^+$-ATPase protein.[201,391] Thus, even in the presence of physiological levels of hemolymph K$^+$, in contact with the K$^+$-binding aspect of the pump, NH$_4^+$ may be pumped as well. Active excretion of NH$_4^+$ across *Carcinus* gills requires not only a functioning Na$^+$,K$^+$-ATPase but also the V-type H$^+$-ATPase and intact microtubules, because the process is blocked by bafilomycin, a specific V-type H$^+$-ATPase inhibitor, as well as colchicine and other microtubule inhibitors.[630] A model of active NH$_4^+$ excretion suggests basolateral NH$_4^+$ transport via Na$^+$,K$^+$-ATPase, sequestration of NH$_3$ in vesicles acidified by V-type H$^+$-ATPase, vesicle transport via microtubules, and exocytosis at the apical membrane.[627] The discovery of a Rhesus-related ammonium transporter in crab gills may afford an additional component in such a model.[625] A Rhesus-related protein has been implicated in ammonium transport across both fish gills and mammalian nephron.[175,275] Demonstration of an inhibitory effect of apical amiloride on Na$^+$/NH$_4^+$ exchange across gills of *Callinectes sapidus*[469] and *Petrolisthes cinctipes*[276] suggests that an apical cation exchanger may be involved with this process as well, although other authors suggest that amiloride may have an indirect effect via interference with cation exchange across the gill cuticle rather than the epithelium itself, as noted above.[436]

Although not a membrane transport protein, carbonic anhydrase has been strongly implicated in osmoregulatory processes in crustaceans. In addition to facilitating CO$_2$ excretion by gills, carbonic anhydrase catalyzes the formation of H$^+$ and HCO$_3^-$, counterions for Na$^+$/H$^+$ and Cl$^-$/HCO$_3^-$ exchange. At least two forms of carbonic anhydrase are believed to occur in crustacean gills: a membrane-bound form primarily associated with CO$_2$ excretion[72,250] and a cytoplasmic form that is highly sensitive to environmental salinity, increasing in activity as hyperosmoregulating crabs acclimate to dilute media.[251,252] The salinity-related response of carbonic anhydrase activity is most notable in posterior gills of euryhaline crabs and branchial tissues of other crustaceans and has been reported in *Callinectes sapidus*,[251,253] *Callinectes similis*,[461] *Carcinus maenas*,[57,254] *Neohelice granulata*,[208,347] *Eriocheir sinensis*,[429] *Homarus gammarus*,[446] and others. In the lobster *Homarus gammarus*, salinity-sensitive carbonic anhydrase activity resides mainly in epipodites and branchiostegites rather than in the gills themselves,[446] corresponding with the distribution of Na$^+$,K$^+$-ATPase activity in these tissues.[191,336]

Two forms of carbonic anhydrase have been identified at the molecular level in gills of *Callinectes sapidus*: a cytoplasmic form (Accession No. EF375490) and a form linked to glycosyl-phosphatidylinositol (Accession No. EF375491).[531] Other cDNAs encoding putative carbonic anhydrase sequences have been identified in cDNA libraries from *Carcinus maenas* (Accession Nos. DN202505 and DV467246),[599] *Calanus finmarchicus* (Accession Nos. EL697164 and ES237390), and *Litopenaeus vannamei* (Accession No. BF024146)[226] and by conventional PCR in *Carcinus maenas* using degenerate primers.[255] Transcriptional expression of carbonic anhydrase-encoding mRNA was induced in posterior gills of *C. maenas* within 24 hours after transfer of 32‰-acclimated crabs to 10‰ salinity, preceding by 24 hours a significant increase in carbonic anhydrase activity.[257]

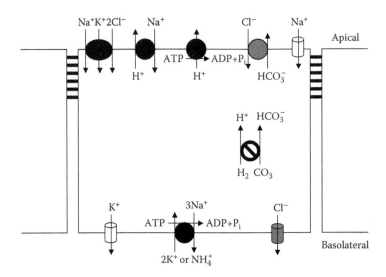

FIGURE 6.5 Hypothetical working model of NaCl uptake across gill epithelial cells of hyperosmoregulating aquatic crustaceans, based on numerous physiological, ultrastructural, and molecular studies. Subcellular localization studies have been accomplished for only a few of the transporters. Basolateral Na^+/K^+-ATPase is thought to generate an electrochemical potential that energizes apical transport processes, including epithelial Na^+ channels, Na^+/H^+ exchangers, and $Na^+/K^+/2Cl^-$ cotransporters. Apical Cl^-/HCO_3^- exchangers may mediate Cl^- uptake along with $Na^+/K^+/2Cl^-$ cotransporters, the HCO_3^- being provided by the action of intracellular carbonic anhydrase. V-type H^+-ATPase may polarize the apical membrane in certain freshwater-adapted species. Basolateral transport of Cl^- may be mediated by epithelial Cl^- channels; K^+ channels may be involved in recycling K^+ across the basolateral membrane. Transporters represented by gray symbols have been identified at the molecular level in crustacean gills; in addition, gene expression data are available for those represented in black. Please see text for supporting references.

By 7 days after the transfer, carbonic anhydrase mRNA abundance had declined, but enzymatic activity continued to increase. When crabs were transferred from 32‰ to 20 or 25‰ salinity, modest increases in carbonic anhydrase mRNA occurred, but upon transfer to 15‰, a 10-fold increase was noted,[257] clearly implicating carbonic anhydrase as a major component of the hyperosmoregulatory response. In *Callinectes sapidus*, mRNA expression of the glycosylphosphatidylinositol-linked form of carbonic anhydrase is induced about 4- to 5-fold following transfer to low salinity, whereas the cytoplasmic form is induced about 100-fold.[531]

Based on physiological and molecular studies of crustacean gill, a model of osmoregulatory ion uptake can be proposed (Figure 6.5). Unfortunately, insufficient information exists regarding mechanisms of ion excretion in species that effectively hypo-osmoregulate. Even in species that hyperosmoregulate, clear differences exist between those that can and cannot tolerate freshwater. It seems to be broadly accepted that Na^+,K^+-ATPase is the major driving force for osmoregulatory ion transport across gills, with the V-type H^+-ATPase augmenting or possibly replacing the driving force function of Na^+,K^+-ATPase in freshwater-tolerant animals.[604] Entry of Na^+ across the apical membrane appears to be mediated by three transporters, perhaps singly or in combination: the Na^+/H^+ exchanger, the $Na^+/K^+/2Cl^-$ cotransporter, or epithelial Na^+ channels. Apical entry of Cl^- may be achieved by the same $Na^+/K^+/2Cl^-$ cotransporter or by a Cl^-/HCO_3^- exchanger, with basolateral transport mediated by Cl^- channels. To permit recycling of K^+, basolateral K^+ channels likely exist. Intracellular carbonic anhydrase is poised to generate the counterions H^+ and HCO_3^- for apical exchange processes. Molecular evidence has been obtained for the existence of several of these transporters, but more research, particularly regarding intracellular localization, is necessary before the functional role of all of the players becomes clear.

VI. EFFECT OF DIFFERENT FACTORS ON OSMOREGULATION

A. MOLT CYCLE

The variability of the hemolymph ionic composition and osmolality throughout a molt cycle has been known for a long time.[35] These two parameters tend to increase in premolt, preceding an uptake of water at ecdysis.[55,112,382] The entry of water causes linear growth by stretching the new tegument.[112,168,428,499,602] The amount of absorbed water compared to pre-ecdysial weight ranges from 6 to 48%.[421,480,481] Total water content shifts from 60 to 70% in intermolt stages to about 75 to 80% in early postmolt. Excess water could be released through dilute urine in postmolt.[235] The variations in hemolymph osmolality throughout a molt cycle also depend on salinity; in *Penaeus monodon* and *P. vannamei*, they are highest at low salinity and in seawater but limited at 15 to 20‰ salinity.[105,186] The mechanism of uptake of water at ecdysis is still being discussed. Although an osmotic process is probably part of it in some species,[152] others do not depend solely on it, as osmolality does not always increase before molting.[233,382,499] In hypo-osmoregulating species, the osmotic gradient is opposed to osmotic water uptake, as shown in *Penaeus vannamei*,[105] which might account for lower growth rates at high salinities.[24] One site of water entry is represented by the gills and the entire body surface, which are covered by the new thin cuticle, as shown on isolated integument of *Maia squinado*.[149] A second major site of water entry is the gut. The drinking rate increases just before ecdysis in *Homarus americanus*[421] and *Panulirus longipes*,[146] and the absorbed water probably moves to hemolymph as its volume increases concomitantly. Water movements related to molting are probably under humoral control. In *Maia squinado*, water uptake through the tegument is changed by the addition of hemolymph from a molting animal.[149] In *Carcinus maenas*[117,118] and *Homarus americanus*,[87,88] an increase in circulating crustacean hyperglycemic hormone (CHH) occurs that seems to mediate the onset of ecdysis through water uptake.

In summary, due in particular to water uptake at molt, the ion composition and osmolality of hemolymph change according to molting stages. As a practical consequence, any study of osmoregulation in crustaceans must take the molt stages in account. As ecdysis implies large water movements across tegumental boundaries, future research might address the possible presence of aquaporins at selected sites.

B. TEMPERATURE AND DISSOLVED OXYGEN

Temperature variations affect ionic and osmotic regulation, thus the capacity to osmoregulate varies with seasons as reported in several branchiopods, amphipods, isopods, and decapods.[74,89,382,426,447] In temperate species, very low (below 5°C) and relatively high (over 20°C) temperatures generally decrease the capacity to osmoregulate, with a generally more pronounced effect of high temperatures.[169] In tropical species such as peneid shrimps, a decrease in osmoregulation occurs only at temperatures over 25°C,[111,635] but the effect of temperature is variable between species. In the coldwater *Homarus americanus*, for example, hyperosmoregulation at low salinity increases from 2 to 11 and 25°C.[104] No clear correlation exists with salinity tolerance, which is maximum at 5 to 12°C.[394] The mechanisms accounting for the temperature effect are not clear. Keeping the American lobster as an example, the main enzymes of metabolism have a thermal optimum close to the mean temperature of the habitat[519] (e.g., 12°C for lactate dehydrogenase in *Homarus gammarus*),[601] but the optimum temperature for the activity of Na$^+$,K$^+$-ATPase is close to 37°C.[354,589] Clearly, further research is necessary in this area. More generally, global change should renew the interest in the effect of temperature. As temperature rises, so will evaporation in selected aquatic areas, leading to increased salinity that should in turn enhance the interest in the mechanisms of hypo-osmoregulation.

Variations in dissolved oxygen affect osmotic and ionic regulation at different levels according to species. In *Carcinus maenas*, exposure to low oxygen concentration is followed by no substantial changes in seawater where the crabs are close to isosmoticity[561] but by a decrease in hemolymph Cl$^-$ levels at low salinity in which they hyperosmoregulate.[283] Hypoxic stress also induces lower hemolymph Cl^- levels in *Palaemon adspersus*[236] and *Crangon crangon*[237] at low salinity. The consequences of hypoxic conditions are of particular importance in aquaculture, as shown in peneid shrimps,[105] for which the oxygen lethal concentration is close to 1 mg O_2 per L (oxygen tension, PO_2, of 2.7 kPa). At the minimum PO_2 of 8 kPa, usually retained to minimize stress that affects molting and growth, hyper- and hypo-osmoregulation are negatively affected in *Penaeus vannamei*, an impairment that increases at lower PO_2 levels.[272] The decrease in osmoregulatory ability originates from several factors, including interference with the respiratory physiology,[377,642] disturbance in the hemolymph acid–base balance,[259,603] or reallocation of available oxygen from active osmoregulatory processes to other vital processes.

C. POLLUTANTS AND OTHER STRESSORS

Pollutants affect salinity tolerance and osmoregulation.[214,337,447] Exposure to waterborne toxics exposes crustaceans to stress.[37] To measure the resulting alteration of osmoregulation, the use of osmoregulatory capacity (OC), either hyper-OC or hypo-OC, has been proposed. The osmoregulatory capacity is the difference between the osmolalities of the hemolymph and the external medium at a given salinity.[93,337]

The effects of several types of pollutants present in water have been tested on osmotic and ionic regulation: oil, pesticides and PCBs, metals, phenols, potassium, ammonia, nitrite.[337] Recent evidence has accumulated that illustrates the generally negative impact of pollutants on osmoregulation, particularly with regard to atrazine,[546] benzene,[508] ammonia,[478] cadmium,[544,545,615,638] copper,[43,66] lead,[16] and zinc,[41,638] although no change in osmoregulation was reported in *Penaeus duorarum* exposed to silver in seawater.[45] In freshwater and diluted media, significant decreases of hyper-OC range from a few percent to usually 20 to 50% and up to 90 to 100%. In seawater and at high salinity, stress exposure usually induces a decrease in hypo-OC. In both cases, Na^+ and Cl^- regulations are also affected. These effects are particularly apparent upon exposure to ammonia, tributyltin oxide (TBTO), oil, benzene, pesticides, and most metals, such as aluminum, cadmium, and copper. Alternately, the uptake and toxicity of metals are influenced by salinity; however, the level of osmoregulation does not alone control metal uptake rates, and the reciprocal relationships among metal toxicity, metal uptake, and osmoregulation remain an open field for research.[495,496,611]

Regarding other stressors, studies of the effect of pH on osmoregulation have been largely triggered by the acidification of large bodies of freshwater due to acid rains.[339] Few marine species have thus far been investigated.[15,396] In freshwater branchiopods and amphipods and in crayfish, low pH between 3.0 and 5.6 consistently results in a reduction of OC as well as Na^+ and Cl^- concentrations, by 10 to 75% depending on the pH level.[184,185,337] As for radioactive emissions, exposure of the isopod *Cyathura polita* to gamma radiation was followed by a 16% decrease in the hypo-OC at 40‰.[301] High levels of turbidity have been shown to reduce OC in *Penaeus japonicus*.[341] Pathogenic agents, such as fungal infections[555] and cyanobacteria mycrocystin toxins,[613,614] can also decrease OC. Ultraviolet radiation (UVR) is an ecologically important parameter in marine ecosystems that is susceptible to increase with global change[330] and to affect planctonic crustaceans including larvae. Few studies on the impact of UVR have been conducted in crustaceans,[148,160,297] but the impact on osmoregulation is worth studying.

In the hydrothermal vent crab *Bythograea thermydron*, exposure to high pressure did not affect hemolymph osmolality at low salinity,[389] but short-term exposure to 50 to 100 bars (1 bar is close to 1 atmosphere) significantly affected the hemolymph ion concentrations in *Carcinus maenas*[449] and Ca^{2+} content in *Eriocheir sinensis*.[524] Regarding intracellular isosmotic regulation, unusual organic osmolytes such as trimethylamine-*N*-oxide (TMAO) replace several amino acids in deep-sea crabs and carid shrimps,[303] perhaps as an adaptation that protects protein stability against pressure.[640]

Ion-transporting organs and cells are generally severely affected by toxicant-induced stress.[337] Exposure to pollutants often results in a blackening of gills that are the site of necrosis and hemocytic congestion. Metals cause such gill cell alterations as fewer and swollen mitochondria, nuclear pycnosis, intracellular vacuolization, fragmentation of the basolateral membrane infoldings, and the occurrence of pseudomyelinic structures. Similar impacts of pollutants have been reported in the epithelial cells of the coelomosac and labyrinth in the antennal excretory glands. Alterations of the basal membrane where Na^+,K^+-ATPase is located, structural changes of mitochondria, and vacuolization indicating a possible failure in the regulation of water content[326] all contribute to the decrease in OC. In addition, Na^+,K^+-ATPase activity and water and ion integument permeability are generally affected by environmental pollutants.[45,66,337,508] Na^+,K^+-ATPase activity rarely increases, indicating a possible temporary compensation effect. In most cases, the enzyme activity tends to decrease, and the passive transtegumental exchanges of water and ions increase due to higher permeability, thus contributing to a lower OC.

In summary, exposure to pollutants and more generally to stress generally results in a disruption of ionic regulation and of osmoregulation. This effect is widespread, as the ability to osmoregulate was affected in 79% of the species reviewed in Lignot et al.[337] In cultured species, osmoregulation was disrupted in 93% of peneid shrimps and 100% of crayfish exposed to various stress. These effects are commonly dose dependent, and they vary among individuals.[115] The changes in iono- and osmoregulation are induced by sublethal doses of stressors, and they are often detectable before any mortality is noticeable in exposed animals. The variations in ion content or OC can thus be considered to be early warnings of sublethal stress in crustaceans. Measuring OC can thus be used as a reliable biomarker to monitor the physiological condition and effect of stressors in osmoregulating crustaceans.[93,337] Given their wide variety of osmoregulatory capacity, crustaceans can also be used as bioindicators of the quality of media.

D. NEUROENDOCRINE CONTROL

The possible existence of a humoral control of salt and water balance in crustaceans was first suggested by weight and size increase in decapods without eyestalks compared to intact animals, primarily because of higher water content.[34,81,444,523] Later, mainly in decapods, neuroendocrine cells were described as clusters in parts of the nervous system including the cerebroid ganglia, the eyestalk complex, and the thoracic and abdominal ganglia. These neuroendocrine centers produce neurohormones that are stored in and released from neurohemal organs, essentially the sinus glands in the eyestalks and the pericardial organs.[85,87,187,289] Other than a few studies showing the involvement of endocrine antennal glands in the control of salt and water balance in isopods,[108,372] most investigations have been conducted in decapods.[98,291,382,448] Whole organism injections of extracts from the central nervous system, thoracic ganglion, or pericardial organs usually stimulate osmoregulation. Other studies have been based on the perfusion of isolated gills[291,353,447,448] or on the use of split gill lamellae mounted in Ussing chambers.[433,437,439,494] Hormones with diuretic and antidiuretic effects have been detected in *Gecarcinus lateralis*.[55] Various factors of regulation, from the thoracic ganglion to pericardial organs, have been shown to control water and ionic movements in the gills.[39,290,291,553,606] Dopamine, one of the catecholamines found in the pericardial organs, might be one of the responsible factors, as shown through its stimulation of Na^+,K^+-ATPase activity[553] and of Na^+ influx.[159,291,404,458] In several species, especially in *Eriocheir sinensis*, cAMP appears involved, probably as a second messenger, in the neuroendocrine control effected by the pericardial organs on Na^+ uptake.[42,404] Additional evidence results from the increase in Na^+,K^+-ATPase activity in posterior gills of *Carcinus maenas* incubated with dBcAMP,[553,554] and the involvement of cAMP in the upregulation of branchial ion pumping seems ubiquitous in aquatic brachyuran species;[159,409] however, serotonergic stimulation of branchial ion uptake, independent of cAMP, has been shown in terrestrial crabs and seems unique to them.[408]

In decapods, another neuroendocrine center, the X organ–sinus gland complex located in the eyestalks, has received much attention. In most tested crayfish and crab species,[98,448] bilateral eyestalk removal lowers the ionic and osmoregulatory capability; reimplantation or injection of eyestalks or sinus glands generally partially restores the initial level of ionic and osmotic regulation. This control has been clearly demonstrated in the strong osmoregulators *Metopograpsus messor*[293,298] and *Uca pugilator*[150,249] and in the slight hyper-regulator *Homarus americanus*.[98,99,110,111] Similar surgical operations have also shown that the eyestalks positively control gill Na$^+$,K$^+$-ATPase activity in *Procambatus clarkii*, *Metopograpsus messor*,[296] and *Callinectes sapidus*.[513] In *Pachygrapsus marmoratus*, sinus gland extracts perfused in isolated gills increase Na$^+$ influx[459] and Na$^+$,K$^+$-ATPase activity in a dose-dependent way.[173] NaCl uptake is also stimulated by eyestalk extracts in split gill lamellae of *Eriocheir sinensis* through the increase in number of open apical Na$^+$ channels and activation of the apical V-type H$^+$-ATPase.[437] Regarding the nature of the eyestalk factors, recent accumulating evidence points to the involvement of CHH, a member of the CHH–MIH–VIH–MOIH family of hormones (crustacean hyperglycemic, molt-inhibiting, vitellogenesis-inhibiting, mandibular-inhibiting hormones), a group of 8- to 9.5-kDa neuropeptides. Synthesized in the X organs and stored in and released from the sinus glands, they are involved in the control of metabolism, reproduction and development.[56,84,85,87,102,151,178,299,300,322,507,558,610]

An additional function of CHH in the control of osmoregulation has been shown experimentally in a few species. In previously destalked lobsters (*Homarus americanus*), injection of one CHH isoform from the sinus glands increased hemolymph osmolality.[110] A similar effect was reported in the crayfish *Astacus leptodactylus* injected with D-Phe³-CHH.[529] CHH polymorphism resulting from posttranslational isomerization of one amino acid residue in position 3 of the amino-terminal fragment from the L to the D configuration has been reported in several crayfish, lobsters, and crabs,[557–559,591] leading to a wide functional diversity of CHH, with the D-enantiomer being involved in the control of osmoregulation.[529,532] The gills appear as one important target of CHH, because in isolated posterior gills of *Pachygrapsus marmoratus* perfusion of CHH isolated from sinus glands induced an increase in Na$^+$ influx.[563] CHH might control the level of Na$^+$,K$^+$-ATPase activity, as incubation of gills in sinus gland extracts increases the enzyme activity.[173] Also noteworthy is the fact that exposure of *Homarus americanus* to low salinity (15‰) results in an increase in circulating CHH titer within 2 hours.[88] As a peptidic hormone, CHH probably acts through a second messenger, which could be cGMP, as shown in *Orconectes limosus* and *Callinectes sapidus*[295,525] and in the gills, hindgut, and midgut gland of *Carcinus maenas*.[118]

In summary, a possible scenario of control of osmoregulation in decapods emerges as follows: CHH produced by the eyestalk X organs would be released through the sinus glands; it would reach the osmoregulating cells of the gills and, through a second messenger, would stimulate Na$^+$,K$^+$-ATPase. Its increased activity would then enhance Na$^+$ uptake and thus hemolymph osmolality. In addition to this possible CHH control route, other possible sites for CHH synthesis have recently been found. Endocrine cells from the foregut and hindgut of *Carcinus maenas* produce CHH that controls water and ion uptake at ecdysis.[117] CHH is also produced in the thoracic ganglion and in sub-esophageal neurons, and it could be released from the pericardial organs where CHH was detected in lobsters.[86,163] In *Carcinus maenas*, the expression of one of the CHH isoforms isolated from the pericardial organs[162] increased following exposure of the crabs to low salinity, suggesting a role in the control of osmoregulation.[600] Similar changes have been reported in *Pachygrapsus marmoratus*.[562,564] These findings show that the current interpretation of the control of osmoregulation must be further studied at various levels: sites of neurohormone production and release; different molecular forms, including the CHH isomers; molecular and cellular mechanisms of actions, including second messengers and activated enzymes and channels other than Na$^+$,K$^+$-ATPase; target organs, including excretory glands (especially in freshwater species) and the digestive tract; and the gills.

One last point of interest lies in the similarities between CHH and the ion transport peptide (ITP) found in the neurohemal corpora cardiaca of several insects.[25,26,118,368] In addition to the molecular proximity between CHH and ITP,[457,558] their physiological activities are also similar, as ITP stimulates ileal Cl$^-$ transport followed by water reabsorption.[456,457] Thus, CHH-like peptides apparently occur widely in the arthropod phylum, with the control of osmoregulation as a possible homologous function. These findings, in addition to recent phylogenetic analyses of molecular sequence data from different genes and newer morphological studies, provide further evidence of the relationship between hexapods, including insects, and crustaceans, a grouping commonly referred to as Pancrustacea.[216]

VII. ONTOGENY

Most studies of crustacean osmoregulation have been conducted in adults; however, natural selection acts on all stages of development,[33,70,109] and salinity is one of the environmental factors yielding a selective pressure on crustaceans during their entire life cycle. Investigations on the ontogeny of osmoregulation are thus necessary for a better understanding of the adaptation of a species to its habitat. In his milestone book on the biology of decapod crustacean larvae, Anger[21] reported the wide variety of developmental strategies of crustaceans. Eggs can be released in the environment or retained by the female; larvae generally spend a variable amount of time in the pelagic environment as a way of dispersal before eventually returning to the benthos as juveniles and recruiting to the adult population. During these phases, individuals are submitted to various regimes of salinity. Various techniques have been used to study osmoregulation, including embryonic and larval culture, nanoosmometry,[198,467,476] histology *sensu lato*, biochemistry, molecular biology, and developmental ecology, all leading to increasing knowledge regarding the ontogeny of osmoregulation.[90,97,448]

A. OSMOREGULATION THROUGHOUT DEVELOPMENT

In those species whose adults osmoregulate, the capacity to do so occurs either in the embryonic or post-embryonic phase, sometimes with a change in the location of the osmoregulatory sites and related variations in the capacity to osmoregulate.

1. Embryonic Phase

Osmoregulation during the embryonic phase has been investigated in a limited number of species with external or internal development (Table 6.3). Among crustaceans, embryonic development is internal in a proportionally low number of species, especially in cladocerans, amphipods, and isopods. The embryos develop inside the body of the female (e.g., in closed brood chambers in cladocerans living in freshwater or at low salinity or in marine or hypersaline continental water). The adult female regulates the osmolality of its hemolymph and of the fluid of the brood chamber to which the embryos are isosmotic. Osmoregulatory organs develop during the embryogenesis, sometimes as temporary neck organs later replaced by epipodites, and the young cladocerans are able to osmoregulate at hatch.[13] In *Sphaeroma serratum*, an isopod living in coastal or lagoon areas, the eggs develop in closed incubating pouches, isosmotic to the female hemolymph, where they acquire the ability to osmoregulate before hatching.[95] Through a comprehensive and elegant set of observations and experiments, similar adaptations have been described in the semiterrestrial amphipod *Orchestia gammarellus*.[413–415,417] The eggs are laid in a semi-closed marsupium where urine, isosmotic to hemolymph, is apparently directed, resulting in control of the local osmolality. The embryos become able to osmoregulate during their development with a switch in the effector site from the embryonic dorsal organ to coxal gills. In these cases, embryos are thus osmotically protected in a specialized part of the female body, where they develop osmoregulatory organs (sometimes later replaced by other organs) leading to acquisition of a physiological competency[416,417] in osmoregulation, which in turns results in a certain level of euryhalinity necessary to cope with salinity variations in their habitat. A similar function of osmoprotection of the brood has been suggested in other similar cases, through marsupial pouches in terrestrial amphipods,[411] isopods,[260,569] and mysids[395] and through ovisacs in freshwater calanoid copepods.[40]

External development is much more common among crustaceans. The developing eggs are directly exposed to the external medium, either kept in open brood pouches or attached to the

TABLE 6.3
Ontogeny of Osmoregulation During the Embryonic Phase

Development	Group	Species	Refs.
Internal development: embryos or eggs in female body	Branchiopoda:		
	Cladocera	Several species	Aladin and Potts[13]
	Malacostraca:		
	Amphipoda	*Orchestia gammarellus*	Morritt and Spicer[413–415,417]
	Isopoda	*Sphaeroma serratum*	Charmantier and Charmantier-Daures[95]
External development: eggs exposed to external medium	Branchiopoda:		
	Cladocera	Several species	Aladin and Potts[13]
	Anostraca	*Artemia* spp.	Conte[128]
	Malacostraca:		
	Amphipoda	*Gammarus duebeni*	Morritt and Spicer[412,413]
	Isopoda	*Cyathura polita*	Kelley and Burbanck[302]
	Decapoda	*Callianassa jamaicense*	Felder et al.[182]
		Homarus americanus	Charmantier and Aiken[91]
		Astacus leptodactylus	Susanto and Charmantier[571]
		Hemigrapsus crenulatus	Seneviratna and Taylor,[528] Taylor and Seneviratna[581]
		Hemigrapsus edwardsii	Taylor and Seneviratna[581]
		Hemigrapsus sexdentatus	Seneviratna and Taylor[528]

Source: Adapted from Charmantier, G. and Charmantier-Daures, M., *Am. Zool.*, 41, 1078–1089, 2001.

pleopods of the female (in most decapods) or freely released in the environment. The egg envelopes, formed by embryonic envelopes surrounded by an outer coat,[215] constitute the only barrier between the embryo and the water. Among the studied species (Table 6.3), most are able to osmoregulate at hatch, meaning that the ability to osmoregulate occurs at some point of the embryonic development. Until then, it is generally recognized that the egg envelopes offer at least partial osmotic protection to the embryo. The development of osmoregulatory organs varies according to species. In cladocerans, an embryonic nuchal or neck organ is later replaced by epipodites.[13] In *Artemia* spp., which have been particularly well studied given their remarkable hypo-osmoregulatory ability, the embryo is first protected by the cyst envelope, then osmoregulation in the embryonic pre-nauplius is effected by a dorsal organ or salt gland which is retained in the nauplius that usually hatches in very high-salinity media, before replacement of the organ by coxal gills.[128,588] In the amphipod *Gammarus duebeni* developing in eggs carried in an open brood chamber, osmoregulation occurs early in the development, apparently based on the function of a temporary dorsal organ later replaced by coxal gills.[412] Similar changes have been observed in the isopod *Cyathura polita*, but osmoregulation in the egg could be effected either by the vitelline and embryonic membranes[302] or by temporary dorsolateral organs.[567] In the thalassinid decapod *Callianassa jamaicense* living in coastal and estuarine waters, Na^+,K^+-ATPase activity increases in late embryos, suggesting their ability to osmoregulate.[182] The activity of Na^+,K^+-ATPase also increases during the embryonic development of the shrimp *Macrobrachium rosenbergii*, resulting in temporary efficient osmoregulation in hatching larvae.[634] In Astacidea, sharp differences in the time of occurrence of osmoregulation seem related to the habitat. In the marine and coastal lobster *Homarus americanus*, the embryos are osmoconformers, as are the early larvae; osmoprotection is provided by the outer egg membrane,[91] although a slight presence of Na^+,K^+-ATPase has been detected through immunocytochemistry in the antennal glands, the intestine, and the epipodotes of late embryos.[307] In contrast, in freshwater crayfish such as *Astacus leptodactylus*, the egg envelopes also appear to osmotically protect embryos during the majority of their development, but the development of gills bearing ionocytes[338,571] and of excretory antennal glands[305,308] shortly before hatching allows efficient hyper-regulation at hatch. In several species of the euryhaline intertidal crabs of the genus *Hemigrapsus*, which are strong hyper-regulators at low salinity, the ontogeny of osmoregulation has been followed in embryos.[528,581] Post-gastrula embryos of *H. crenulatus* actively hyperosmoregulate, with high Na$^+$,K$^+$-ATPase activity, increased at low salinity. The involvement of a dorsal organ is hypothesized (see below).[528]

2. Post-Embryonic Phase

From the available data (Table 6.4) following the early study of Kalber and Costlow in 1966,[287] three patterns of ontogeny of osmoregulation have been recognized. In species belonging to Pattern 1, osmoregulation varies little throughout development; all post-embryonic developmental stages including adults are osmoconformers or weak osmoregulators. This type is found in marine stenohaline species (e.g., majid crabs). In Pattern 2 species, the adults are euryhaline, live in environments where salinity varies or in freshwater, and are hyper-iso- or hyper–hypo-osmoregulators. The adult type of osmoregulation is present in the first post-embryonic stage, and osmoregulatory capacity increases in successive developmental stages. This group includes cladocerans, amphipods, isopods, and decapods such as crayfish and carid shrimps living in low-salinity media or in freshwater. In a third group of species (Pattern 3), the early post-embryonic stages are osmoconformers or they slightly osmoregulate. A shift to increased ability to osmoregulate generally occurs at the metamorphic larva–juvenile transition, along with a change in habitat, sometimes a migration. Peneid shrimps, some carid shrimps, homarid lobsters, and portunid, ocypodid, and grapsid crabs belong to this pattern. In those Pattern 3 species whose adults hyper–hypo-osmoregulate, larvae are first only able to hyper-regulate at low salinity, then, after metamorphosis, this ability increases while the capacity to hypo-osmoregulate at high salinity occurs. Species such as *Crangon crangon*,[122] *Eriocheir sinensis*,[121] *Armases miersii*,[100] *Sesarma curacaoense*,[22] *Neohelice granulata* (Figure 6.6),[103] and *Uca subcylindrica*[474,475] exemplify these changes.

B. Functional Basis

The mechanisms of hydromineral regulation in early stages of development appear close to those described in adult crustaceans with some differences regarding the location of the osmoregulatory sites. Intracellular isosmotic regulation has been reported in several osmoconforming larvae, which rely on it for their limited tolerance to salinity variations. As in adults, it is based on the adjustment of the intracellular free amino acid concentration, as shown in *Menippe mercenaria*,[605] *Penaeus japonicus*,[147,384] and *Homarus gammarus*.[239]

In those early stages that are able to osmoregulate, osmoregulation is based on the regulation of the hemolymph ion content, mainly Na$^+$ and Cl$^-$, through active ion transport. As in adults, Na$^+$,K$^+$-ATPase is a key enzyme, the activity of which tends to increase with the occurrence of Na$^+$ regulation in early developmental stages, as demonstrated in *Artemia* spp.,[128,129] *Callianassa jamaicense*,[182] *Homarus americanus*,[589] *Penaeus japonicus*,[58,59] *Macrobrachium rosenbergii*,[278] and *Hemigrapsus crenulatus*.[528,581] Carbonic anhydrase has also been found in the larvae of some species.[129,589] Ionocytes, with the typical features of ion-transporting cells (see Section V.D.1), are present in early developmental stages, but, in contrast to adults, their distribution includes extra-branchial organs and may vary with time.

The embryonic development often takes place inside an egg, which may offer a level of osmotic protection to the embryo, but the mechanisms of egg and embryonic osmoregulation are still uncertain. The outer egg envelopes are acellular and do not present sites of active ion transport in the species studied so far. The water and ion permeability of the egg envelopes could be low during

TABLE 6.4
Ontogeny of Osmoregulation during the Post-Embryonic Phase

Group	Species	Pattern	Refs.
Branchiopoda			
Cladocera	Several species	2	Aladin and Potts[13]
Anostraca	*Artemia* spp.	2	Conte,[128] Conte et al.,[129,130]
Malacostraca			
Amphipoda	*Gammarus duebeni*	2	Morritt and Spicer[412]
	Orchestia gammarellus	2	Morritt and Spicer[415,417]
Isopoda	*Cyathura polita*	2	Kelley and Burbanck[301,302]
	Sphaeroma serratum	2	Charmantier and Charmantier-Daures[95]
Decapoda	*Penaeus japonicus*	3	Charmantier et al.[101]
	Macrobrachium petersi	2	Read[477]
	Palaemonetes argentinus	2	Charmantier and Anger[92]
	Crangon crangon	3	Cieluch et al.[122]
	Astacus leptodactylus	2	Susanto and Charmantier[570,571]
	Homarus americanus	3	Charmantier et al.[98,101,104]
	Homarus gammarus	3	Charmantier et al.,[104] Thuet et al.[589]
	Callianassa jamaicense	2?	Felder et al.[182]
	Clibanarius vittatus	?	Young[641]
	Hepatus ephiliticus	1	Kalber[286]
	Callinectes sapidus	?	Kalber[286]
	Carcinus maenas	3	Cieluch et al.[120]
	Cancer irroratus	3	Charmantier and Charmantier-Daures[94]
	Cancer borealis	?	Charmantier and Charmantier-Daures[94]
	Cancer magister	?	Brown and Terwilliger[67]
	Rhithropanopeus harrisii	?	Kalber and Costlow[287]
	Chionoecetes opilio	1	Charmantier and Charmantier-Daures[96]
	Libinia emarginata	1	Kalber[286]
	Cardisoma guanhumi	?	Kalber and Costlow[288]
	Eriocheir sinensis	3	Cieluch et al.[121]
	Armases miersii	3	Charmantier et al.[100]
	Sesarma reticulatum	?	Foskett[192]
	Sesarma curacaoense	3	Anger and Charmantier[22]
	Neohelice granulata	3	Charmantier et al.[103]
	Uca subcylindrica	3	Rabalais and Cameron[475]

Note: Pattern 1, all stages weak regulators or osmoconformers; Pattern 2, adult type of efficient osmoregulation present at hatch, adults osmoregulate; Pattern 3, early post-embryonic stages osmoconform or slightly osmoregulate, shift from larval to adult type of osmoregulation during post-embryonic development, often at metamorphosis; ?, insufficient data or no clear pattern.

Source: Adapted from Charmantier, G., *Invert. Reprod. Develop.*, 33, 177–190, 1998; Péqueux, A. et al., in *Treatise on Zoology—Anatomy, Taxonomy, Biology, The Crustacea*, Forest, J. and von Vaupel Klein, J.C., Eds., Brill Academic Publishers, Leiden, 2006, pp. 205–308.

most of the embryonic development, before an increase induces an osmotic uptake of water favoring hatching,[91,215,302,505,506] but complete impermeability of the envelopes would be incompatible with respiratory gas exchanges. Recent evidence in *Hemigrapsus crenulatus* shows that, although water permeability is relatively low, ion exchange across the egg envelopes is high.[528] The presence of charged inorganic molecules could lead to the accumulation of ions according to a Donnan effect.[472] Alternatively, the tensile strength of the envelopes[425] would limit the osmotic uptake of water, resulting in a probably high hydrostatic pressure in the egg (J.-P. Truchot, pers. comm.).

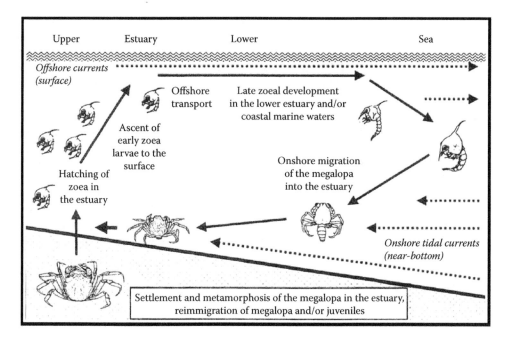

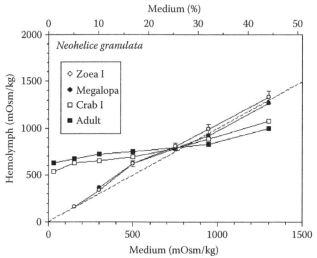

FIGURE 6.6 Relationship between ecology and osmoregulation in *Neohelice granulata*. (Top) Conceptual model of export strategy, with pattern of ontogenetic migration. (From Anger, K., *The Biology of Decapod Crustacean Larvae*, A.A. Balkema, Lisse, 2001. With permission.) (Bottom) Ontogeny of osmoregulation in selected post-embryonic stages (From Charmantier, G. et al., *Mar. Ecol. Progr. Ser.*, 229, 185, 2002. With permission.)

In species that become able to osmoregulate during the embryonic phase, ionocytes differentiate at various locations and times. These osmoregulatory sites, such as dorsal, nuchal, or neck organs, can be temporary and later replaced by definitive organs such as gills. Several cladocerans,[13] *Artemia* spp.,[128] amphipods,[400,412,414–416] and *Cyathura polita*[567] illustrate this pattern. Ionocytes may also appear once at their definitive location (e.g., in gills) with a strong expression of Na⁺,K⁺-ATPase in late embryos of *Astacus leptodactylus*,[307,338] correlating with peaks in mRNA expression of the enzyme.[532] In several species of *Hemigrapsus* spp., mainly *H. crenulatus*, recent evidence has shown that postgastrula embryos become able to hyperosmoregulate.[528] The hypothetical functional model to interpret these data includes osmotic uptake of water balanced by excretion of salts and water via a dorsal organ, as well as salt loss balanced by active uptake through an unknown site. These hypotheses warrant further research, particularly with regard to the dorsal organ of decapods which is represented by a thickening of the extra-embryonic ectoderm in the dorsal midline opposite to the developing embryo[18,189,387] and the possible presence of ionocytes on the embryonic ectoderm, as in fishes.

After hatching, during the post-embryonic phase, dorsal or neck organs can persist in larvae (e.g., *Daphnia magna*[13] and *Artemia* spp.[128]) and in later stages.[12,13,387] Ionocytes may also differentiate sequentially on successive organs—on branchiostegites in larvae then on epipodites and gills in juveniles and adults of peneid shrimps;[59,577] on epipodites in larvae then, in addition, on the branchiostegites of juveniles in homarid lobsters;[335] and on branchiostegites in zoeal larvae then on the posterior gills in the megalopa and first juvenile of *Eriocheir sinensis*.[121] In the carid shrimp *Crangon crangon*, the sequence of localization of ionocytes is particularly complex, as they are located on pleurae and branchiostegites in zoeal larvae, on branchiostegites and epipodites in decapodids and early juveniles, then on gills in addition to these organs in later juveniles.[122] As in peneids and lobsters, metamorphosis marks the time of appearance of the adult type of osmoregulation in several species, in close relation to the occurrence of ionocytes. Ionocytes, for example, first occur at the megalopa stage of *Carcinus maenas* on the posterior gills, where they persist in later stages.[120]

In species such as *Eriocheir sinensis* and *Crangon crangon*, changes in the location of ionocytes are related to the synchronous occurrence of hypo-osmoregulation following metamorphosis. In several of these species (*Homarus gammarus*,[335] *Carcinus maenas*,[120] *Crangon crangon*,[122] and *Eriocheir sinensis*[121]), the functionality of the ionocytes is demonstrated by their structure and the expression of Na^+,K^+-ATPase as revealed through immunofluorescence. Later in development, in juveniles, the relative gill area is greater than in adults, which presents a challenge for water invasion and salt loss at low salinity. As shown in juvenile *Callinectes sapidus*, a partial compensation results from the reduction of the gill permeability and from a large increase in the expression and utilization of the gill Na^+,K^+-ATPase.[331] As in adults, the gut and excretory organs could be involved in osmoregulation during the early development, but information on these sites is still scarce. In *Astacus leptodactylus*, ionocytes have been detected in the labyrinth and in the bladder of the antennary gland in late embryos.[305,308]

A neuroendocrine positive control of osmoregulation has been demonstrated through surgical eyestalk removal and reimplantation in larvae of *Rhithropanopeus harrisii*[287] and in the early juveniles of *Homarus americanus*.[99] In *Astacus leptodactylus* embryos, both L- and D-CHH enantiomers are synthesized in the eyestalk X organ and stored in the sinus glands. The D-Phe^3-CHH, which has been shown to influence osmoregulation in adult crayfish,[529,532] occurs later than the L-CHH, a few days before the onset of the ability to osmoregulate.[571] Thus, D-CHH would begin controlling osmoregulation in late embryos.[530]

C. Ecological Implications

From molecules to the environment, an integrated series of events links osmoregulation to the habitat of a species at each developmental stage. The expressions of specific enzymes, transporters, and ion channels, coordinated with the structural differentiation of ionocytes at several anatomical sites, result in stage-specific levels of osmoregulation; they in turn determine the salinity tolerance of the successive ontogenetical stages which is a parameter of their selection of and adaptation to habitats.

When all developmental stages osmoconform, their salinity tolerance is usually limited despite their possible reliance on intracellular isosmotic regulation, and these stenohaline species are usually restricted to marine habitats during their entire life (Pattern 1). Typical examples are found among majid crabs (Table 6.4). In those species that can tolerate salinity variations during part or all of their development, euryhalinity allows the colonization of media where salinity fluctuates. Before hatching, euryhalinity first originates from an osmoprotection of the embryo by incubating pouches or by the egg envelopes. Then, at some point of the embryonic or post-embryonic development, an autonomous ability to osmoregulate develops. If acquired before hatching, it results in an osmoregulating and euryhaline hatchling being able to cope with ambient variable or extreme salinity. In other cases, the osmoregulatory ability occurs later in post-embryonic stages, often at the metamorphic transition, and it may result in changes in habitats. These possibilities depend on the ontogenetic patterns of osmoregulation and are reflected in the different strategies of adaptations to habitats characterized by their levels of and variations in salinity.

A few Pattern 2 species use a *limited export* strategy; for example, *Macrobrachium petersi* adults live and breed in freshwater, but their larvae temporarily require saline waters. The females migrate downstream, and hatching occurs close to estuaries. Adults and stage 1 larvae are strong hyper–hypo-osmoregulators, able to tolerate fresh and saline waters. Subsequent larval stages lose the ability to hyper-regulate in freshwater and are thus temporarily confined to estuarine waters. Post-metamorphic juveniles regain this ability and migrate upstream back to the adult freshwater habitat.[477]

Most Pattern 2 species have a *retention* strategy; that is, they spend their entire life cycle in a single habitat where salinity can be variable, brackish, very high (hypersaline media), or very low (freshwater). Osmoregulatory adaptations include osmoprotection of the embryos and subsequent development of osmoregulation in embryos, which result in the capacity to osmoregulate in all post-embryonic stages including the hatchlings. Among those exhibiting this pattern are Cladocerans,[13] *Artemia* spp.,[128] *Gammarus duebeni* and *Orchestia gammarellus*,[412–414,417] *Cyathura polita*,[301,302] *Sphaeroma serratum*,[95] *Callianassa jamaicense*,[182] *Palaemonetes argentinus*,[92] and *Astacus leptodactylus*.[570,571]

Pattern 3 species generally use an export strategy, in which ontogenetical stages with different levels of osmoregulatory ability are transported or migrate between habitats with different levels or regimes of salinity. Osmoconforming or poorly regulating stages are restricted to marine environments, whereas osmoregulating stages can cope with variable or extreme salinities. In homarid lobsters, salinity variations in coastal areas where larvae hatch are tolerated first by intracellular isosmotic regulation in the osmoconforming larvae then through the occurrence of a slight hyper-osmoregulation in juveniles.[98,101,104,239] In *Penaeus japonicus*, the osmoconforming larvae hatch in the open sea. They drift to the coast and enter lagoons where they grow into juveniles following their metamorphosis into post-larvae that are able to hyper–hypo-osmoregulate.[58,59,101] In *Carcinus maenas*, the osmoconforming zoeae are exported offshore through several mechanisms, including vertical migrations and tidal transport;[21] following metamorphosis, the increased ability to hyper-osmoregulate allows for a reinvasion of areas with low salt concentrations such as estuaries.[120] Similar relationships between ecology and hydromineral metabolism have been reported in *Crangon crangon*.[122] Striking links between ontogenetic migrations and physiology are exemplified in two grapsid crab species. In *Neohelice granulata* (Figure 6.6), adults live in lagoons or estuaries and are strong hyper–hypo-osmoregulators. At hatching, apparently synchronized by external factors such as tidal cycles,[21,23] zoea I larvae are temporarily able to slightly osmoregulate at low salinity before being exported within a few hours by tidal currents to the sea. The subsequent zoeal stages are osmoconformers and develop in marine waters; the megalopae are reimported into the lagoons for settlement. The change to a hyper–hypo-osmoregulating pattern from the megalopa, which is linked to a rapid increase in euryhalinity in juvenile crabs, is one of the main adaptations allowing a return to conditions of variable salinity.[103] The adults of *Eriocheir sinensis* are known to live in freshwater, where they strongly hyperosmoregulate, and they are also able to slightly hyporegulate. Berried females migrate downstream before hatching occurs in brackish estuarine areas. A temporary, strong hyper-osmotic regulation is used by zoea I larvae that are exported by surface currents to the sea, where later larval stages develop. The increased hyperosmoregulation occurring in megalopae and mainly in juvenile crabs allows for their progressive return to the estuary and the upstream migration.[121] In other species, much wider salinity variations are experienced early in the life cycle, such as in *Uca subcylindrica*[475] and *Sesarma curacaoense*,[22] which breed in landlocked habitats such as temporary water puddles, and in *Armases miersii*,[100] which breeds in supratidal rock pools. Larvae of the latter species are able to hyper-regulate to cope with exposure to frequent low-salinity periods. The additional capacity to hypo-osmoregulate at high salinity is acquired following metamorphosis (i.e., in megalopa or young crabs).[100]

In conclusion, the ontogenetical changes in the ability to osmoregulate are related to the ecology of the species during their development. Several aspects of the ontogeny of osmoregulation, such as salinity tolerance and osmoregulatory capacity, are closely related with the ontogenetic expression of Na^+,K^+-ATPase and the appearance of specialized transporting epithelia in osmoregulating organs, and both are correlated with ontogenetic changes in the habitat of the successive developmental stages.

Future work should include an extension of comparative studies to document other cases of ecological and physiological relationships, particularly in terrestrial and subterrestrial species.[20,21] Among the effector organs, the development of gills is relatively well documented, but the extra-branchial organs, temporary or not, should be further investigated, along with the excretory organs and the digestive tract. The osmoregulatory functions of the incubating pouches and of the egg envelopes should also be studied. As in adults, the mechanisms of hypo-osmoregulation should be investigated, particularly the timing of their occurrence, which often coincides with metamorphosis. At the cellular and molecular level, much work remains to be conducted in conjunction with the developmental biology of crustaceans. The fields open to research include the origin and differentiation of the ionocytes, identification and expression of the enzymes, ion and water channels involved in the osmoregulatory processes, and their regulation during the ontogeny and according to environmental factors, natural or anthropic in origin.

VIII. OTHER ARTHROPODS

Among aquatic arthropods, hydromineral regulation has been heavily studied in crustaceans, and for a lesser part in Hexapoda (former insects; see Chapter 7). The horseshoe crabs constitute a third and original group that has triggered the interest of biologists studying osmoregulation. These arthropods, well known to immunologists, originate from ancestors from the Silurian and Cambrian of the Paleozoic era.[501] Among the Chelicerata, they belong to the Merostomata: Xiphosura: Limulidae. Four living species are recognized: *Limulus polyphemus*, along the eastern North American coast of the Atlantic Ocean from Nova Scotia to the Yucatan, as well as *Carcinoscorpius rotundicauda*, *Trachypleus gigas*, and *T. tridentatus* in coastal habitats of the Indian and Pacific Oceans in Southeast Asia. The adults live in marine waters in coastal and estuarine areas where they can be submitted to varying salinities. They migrate shoreward in spring and summer to spawn intertidally on sandy beaches,[174,526] a time during which they are exposed to salinities as low as 7‰.[124,398] The eggs are deposited below 10 to 20 cm of sand near the waterline in the mid- to upper intertidal areas. While regularly inundated by tides, they are surrounded by wet sediment at low tide during which they may be exposed to high temperatures on sunny days and to low salinities due to precipitation. Following hatching, larvae swim freely for up to 6 days, then settle to the bottom in shallow waters of the intertidal zone, thus they are under conditions of variable salinity. Later, juveniles migrate to deeper waters. All stages are euryhaline, in the range of 10 to 55‰ (5‰ and over 60‰ for short-term periods) in adults and 10 to 70‰ (5 to 90‰ for short-term periods) in larvae of *Limulus polyphemus*.[174,501] Optimal salinity for embryonic development is in the range of 20 to 30‰[282,325,527] or 30 to 40‰.[174]

The osmoregulation of *Limulus polyphemus* has been studied.[125,139,202,398,501,596] Adult and large juvenile horseshoe crabs are hyper-isosmotic (Pattern 2 in Table 6.1). In the experimental salinity range of 5 to 64‰ tested by Robertson,[501] they slightly hyper-regulate from 5 to 21‰ and are osmoconformers at higher salinities.

The osmoregulating sites are represented by the book gills and the coxal glands. The opisthosomatic appendages are swimming legs. Their second to sixth pairs carry gill books formed of gill lamellae[639] morphologically differentiated into thin (peripheral) and thick (central) regions.[376] Although an ultrastructural study of the gills of *Limulus polyphemus* and *Trachypleus tridentatus* showed no structure characteristic of a transport epithelium,[479] other evidence points to the presence of ion-transporting cells in parts of the gills. In *Limulus polyphemus*, electron microscopy suggests that the thin cells of the peripheral region are specialized in respiration, while typical ionocytes are found on the ventral part of the central region.[256] High concentrations of Na^+,K^+-ATPase and of carbonic anhydrase are found in the ionocytes,[256,280] but exposure to low salinity is not followed by an increase in the enzyme activity.[256,596] The osmoregulatory function of horseshoe crab gills is thus spread on all of them, as in astacid crayfish and contrary to brachyuran crabs. The coxal glands are paired organs, each represented in adult horseshoe crabs by four nephridial lobes connected by a stolon terminating in an end sac at the base of the fourth lobe; the end sac is continuous with a convoluted nephric duct up to an excretory pore at the base of the fifth walking leg.[64,639] Each nephridial lobe consists of two cortical layers surrounding a medulla. The cortexes, in which podocytes separate hemolymph lacunae from the urinary space, are most probably the site of hemolymph ultrafiltration.[64] Ionocytes are present in other parts of the glands, particularly in tubules of the medulla, in the stolon, and in the epithelial lining of the end sac.[64,256] Na^+,K^+-ATPase is present in the cells of the coxal glands. Its activity is much higher than in the antennal glands of euryhaline decapod crustaceans, and it increases at low salinity.[256,280,596] Carbonic anhydrase is also found in coxal glands.[256] After exposure to low salinity, the glands produce urine that is hypotonic to hemolymph.[379,596] Thus, at least in *Limulus polyphemus*, both the gills and coxal glands are involved in ion transport and osmoregulation, whereas the coxal glands, where hemolymph ultrafiltration occurs, also probably function in excretion. Horseshoe crabs, however, are characterized by a high water permeability of the carapace and gills, which is about tenfold higher than in decapod crustaceans.[172,238,256] Severe swelling of the articular membranes and of the gills has been observed in *Limulus polyphemus* at low salinity before hemorrhage and death.[172,379,501] At the cellular level, osmotic water uptake and cell swelling occur, but intracellular volume regulation is slow and incomplete.[622] Compared to decapod crustaceans, the intracellular free amino acid content is lower in horseshoe crabs,[63] and their contribution to intracellular isosmotic regulation is less efficient.[622] In summary, the lower limit of chronic salinity tolerance (ca. 10‰) of adult *Limulus polyphemus* may not be set by a limitation in ion transport by the gills and coxal glands but rather by a combination of high water permeability of the tegument, particularly of the gills, and of limited cell volume regulation.

Due to the location of the reproduction sites, early developmental stages are submitted to variations of salinity. When exposed to different salinities, the perivitelline fluid contained within the outer membrane in embryos of *Trachypleus tridentatus*[527] and *Limulus polyphemus*[174] changes rapidly and becomes nearly isosmotic to the medium. Thus, the embryos are not protected from salinity changes by the egg membranes. A pair of embryonic lateral organs has been described in horseshoe crab embryos.[527] Although cauterization of these organs interferes with the weight increase of the embryos, their possible role in osmoregulation is still unclear. Following hatching, larvae and juveniles are also exposed to wide salinity fluctuations, and their euryhalinity has been demonstrated; however, their capacity to osmoregulate is still unknown and should be investigated.

IX. CONCLUSION

In the past 10 to 15 years, several techniques have been utilized to deepen our understanding of the mechanisms and adaptive role of osmoregulation in crustaceans. Molecular approaches have complemented research based on physiology and cellular biology, ecology, and ecophysiology. The cellular and molecular bases of ionic and osmotic regulation are being deciphered, several ion channels and transporters and their functions have been revealed, and the endocrine control of osmoregulation has been further analyzed. The aptitude to osmoregulate has been confirmed as being a key factor for the occupation of a habitat, and the ontogenetic variations of osmoregulation have been linked to different strategies of development.

Old questions remain unresolved, however, and new ones are appearing, leaving many avenues open for future research. Our understanding of the mechanisms of hypo-osmoregulation is still quite poor, compared to the state of knowledge on hyper-regulation in crustaceans and to the wealth of data available in fishes. The study of this metabolism would renew interest in the function of the tegument, of excretory organs and gills, and, particularly, of the digestive tract, which has somehow been neglected so far. At the cellular and molecular level, the search for additional ion channels, their site of expression, and their regulation should also be expanded, as should studies of the open or septate junctions between adjacent cells of osmoregulatory sites. Also worth developing is a search for crustacean aquaporins or AQP-like molecules, the expression of which would be important to allow passive water movements that are vital under conditions of dehydration at high salinity or in terrestrial conditions. Several directions for future research are related to the ontogeny of osmoregulation, as stated at the end of Section VII.C. Osmotic protection and osmoregulation of the embryo, the origin of osmoregulatory sites (including extra-embryonic tissues), the origin of ionocytes (linked to developmental biology), the molecular structure and expression of water and ion exchangers, and the ecophysiological role of osmoregulation in developmental strategies are among the important areas of research to pursue in this field. Rising concerns about global change and anthropic alteration of natural habitats will also warrant an increase in studies related to the effect of pollution and adverse environmental effects (e.g., temperature, ultraviolet radiation) on osmoregulation.

REFERENCES

1. Abe, H., Yoshikawa, N., Sarower, M.G., and Okada, S., Physiological function and metabolism of free D-alanine in aquatic animals, *Biol. Pharm. Bull.*, 28, 1571–1577, 2005.
2. Ahearn, G.A., Allosteric co-transport of sodium, chloride and calcium by the intestine of freshwater prawns, *J. Membr. Biol.*, 42, 281–300, 1978.
3. Ahearn, G.A., Intestinal electrophysiology and transmural ion transport in freshwater prawns, *Am. J. Physiol.*, 239, C1–C10, 1980.
4. Ahearn, G.A., Water and solute transport by crustacean gastrointestinal tract, in *Membrane Physiology of Invertebrates*, Podesta, R.B., Ed., Marcel Dekker, New York, 1982, pp. 261–339.
5. Ahearn, G.A. and Clay, L.P., Kinetic analysis of electrogenic $2Na^+–1H^+$ antiport in crustacean hepatopancreas, *Am. J. Physiol.*, 257, R484–R493, 1989.
6. Ahearn, G.A., Duerr, J.M., Zhuang, Z., Brown, R.J., Aslamkhan, A., and Killebrew, D.A., Ion transport processes of crustacean epithelial cells, *Physiol. Biochem. Zool.*, 72, 1–18, 1999.
7. Ahearn, G.A. and Franco, P., Sodium and calcium share the electrogenic $2Na^+–1H^+$ antiporter in crustacean antennal glands, *Am. J. Physiol.*, 259, F758–F767, 1990.
8. Ahearn, G.A., Franco, P., and Clay, L.P., Electrogenic $2Na^+–1H^+$ exchange in crustaceans, *J. Membr. Biol.*, 116, 215–226, 1990.
9. Ahearn, G.A., Maginniss, L.A., Song, Y.K., and Tornquist, A., Intestinal water and ion transport in freshwater malacostracan prawns (Crustacea), in *Water Relations in Membrane Transports in Plants and Animals*, Jungrens, A.M., Hodges, T.K., Kleihzeller, A., and Shultz, S., Eds., Academic Press, New York, 1977, pp. 129–142.
10. Ahearn, G.A. and Tornquist, A., Allosteric cooperativity during intestinal cotransport of sodium and chloride in freshwater prawns, *Biochim. Biophys. Acta*, 471, 273–279, 1977.
11. Ahyong, S.T. and O'Meally, D., Phylogeny of the Decapoda Reptantia: resolution using three molecular loci and morphology, *Raffles Bull. Zool.*, 52, 673–693, 2004.
12. Aladin, N.V., Salinity tolerance and morphology of the osmoregulation organs in Cladocera with special reference to Cladocera from the Aral sea, *Hydrobiologia*, 225, 291–299, 1991.
13. Aladin, N.V. and Potts, W.T.W., Osmoregulatory capacity of the Cladocera, *J. Comp. Physiol. B*, 164, 671–683, 1995.

14. Aldridge, J.B., Structure and Respiratory Function in the Gills of the Blue Crab, *Callinectes sapidus* (Rathbun), M.A. thesis, University of Texas, 1977.

15. Allan, G.L. and Maguire, G.B., Effects of pH and salinity on survival, growth and osmoregulation in *Penaeus monodon* Fabricius, *Aquaculture*, 107, 33–47, 1992.

16. Amado, E.M., Freire, C.A., and Souza, M.M., Osmoregulation and tissue water regulation in the freshwater red crab *Dilocarcinus pagei* (Crustacea, Decapoda), and the effect of waterborne inorganic lead, *Aquat. Toxicol.*, 79, 1–8, 2006.

17. Amemiya, M., Loffing, J., Lotscher, M., Kaissling, B., Alpern, R.J., and Moe, O.W., Expression of NHE-3 in the apical membrane of rat renal proximal tubule and thick ascending limb, *Kidney Int.*, 48, 1206–1215, 1995.

18. Anderson, D.T., *Embryology and Phylogeny in Annelids and Arthropods* Pergamon Press, Oxford, 1973.

19. Anderson, E. and Beams, H.W., Light and electron microscopic studies on the cells of the labyrinth in the "green gland" of *Cambarus* sp., *Iowa Acad. Sci.*, 63, 681–685, 1956.

20. Anger, K., Developmental biology of *Armases miersii* (Grapsidae), a crab breeding in supratidal rock pools. II. Food limitation in the nursery habitat and larval cannibalism, *Mar. Ecol. Progr. Ser.*, 117, 83–89, 1995.

21. Anger, K., *The Biology of the Decapod Crustacean Larvae*, A.A. Balkema, Lisse, 2001.

22. Anger, K. and Charmantier, G., Ontogeny of osmoregulation and salinity tolerance in a mangrove crab, *Sesarma curacaoense* (Decapoda: Grapsidae), *J. Exp. Mar. Biol. Ecol.*, 251, 265–274, 2000.

23. Anger, K., Spivack, E., Bas, C., Ismael, D., and Luppi, T., Hatching rhythms and dispersion of decapod crustacean larvae in a brackish coastal lagoon in Argentina, *Helgoländer Meeresun.*, 48, 445–466, 1994.

24. Atwood, H.L., Young, S.P., and Tomasso, J.R., Survival and growth of Pacific white shrimp *Litopenaeus vannamei* postlarvae in low-salinity and mixed-salt environments, *J. World Aquacult. Soc.*, 34, 518–523, 2003.

25. Audsley, N., McIntosh, C., and Phillips, J.E., Isolation of a neuropeptide from locust corpus cardiacum which influences ileal transport, *J. Exp. Biol.*, 173, 261–274, 1992.

26. Audsley, N., McIntosh, C., and Phillips, J.E., Actions of ion-transport peptide from locust corpus cardiacum on several hindgut transport processes, *J. Exp. Biol.*, 173, 275–288, 1992.

27. Babula, A., Ultrastructure of gill epithelium in *Mesidothea entomon* (Crustacea, Isopoda), *Acta Med. Polonica*, 18, 281–282, 1977.

28. Ballard, B.S. and Abbott, W., Osmotic accommodation in *Callinectes sapidus* Rathbun, *Comp. Biochem. Physiol.*, 29, 671–687, 1969.

29. Balss, H., Decapoda Lief. 3 and 4, in *Bronn's Kl. Ordn. Tierreichs*, Akademische Verlagsgesellschaft, Leipzig, 1944, pp. 470–561, 562–591.

30. Barra, J.A., Péqueux, A., and Humbert, W., A morphological study on gills of a crab acclimated to fresh water, *Tissue Cell*, 15, 583–596, 1983.

31. Barradas, C., Dunel-Erb, S., Lignon, J., and Péqueux, A., Superimposed morphofunctional study of ion regulation and respiration in single gill filaments of the crayfish *Astacus leptodactylus*, *J. Crust. Biol.*, 19, 14–25, 1999.

32. Barradas, C., Wilson, J.M., and Dunel-Erb, S., Na$^+$,K$^+$-ATPase activity and immunocytochemical labelling in podobranchial filament and lamina of the freshwater crayfish *Astacus leptodactylus* Eschscholtz: evidence for the existence of sodium transport in the filaments, *Tissue Cell*, 31, 523–528, 1999.

33. Bartholomew, G.A., Interspecific comparison as a tool for ecological physiologists, in *New Directions in Ecological Physiology*, Feder, M.E., Bennett, A.F., Burggren, W.W., and Huey, R.B., Eds., Cambridge University Press, Cambridge, U.K., 1987, pp. 11–37.

34. Bauchau, A.G., Phénomènes de croissance et glande sinusaire chez *Eriocheir sinensis*, *Ann. Soc. R. Zool. Belg.*, 79, 73–86, 1948.

35. Baumberger, J.P. and Olmsted, J.M.D., Changes in the osmotic pressure and water content of crabs during the molt cycle, *Physiol. Zool.*, 1, 531–544, 1928.

36. Baxter-Lowe, L.A., Guo, J.Z., Bergstrom, E.E., and Hokin, L.E., Molecular cloning of the Na,K-ATPase α-subunit in developing brine shrimp and sequence comparison with higher organisms, *FEBS Lett.*, 257, 181–187, 1989.

37. Bayne, B.L., Aspects of physiological condition in *Mytilus edulis* L., with respect to the effects of oxygen tension and salinity, in *Proceedings of the Ninth. European Marine Biology Symposium*, Barnes, H., Ed., Allen & Unwin, London, 1975, pp. 213–238.

38. Beams, H.W., Anderson, E., and Press, N., Light and electron microscope studies on the cells of the distal portion of the crayfish nephron tubule, *Cytologia*, 21, 50–57, 1956.

39. Berlind, A. and Kamemoto, F.I., Rapid water permeability changes in eyestalkless euryhaline crabs and in isolated, perfused gills, *Comp. Biochem. Physiol.*, 58A, 383–385, 1977.

40. Bernard, M., Influence de la salinité et de la température sur le développement embryonnaire de *Temora stylifera* (Copépode pélagique). Conséquences pour l'adaptation aux milieux de salinités diverses, *Vie Milieu*, 22, 109–117, 1971.

41. Bianchini, A. and Castilho, P., Effects of zinc exposure on oxygen consumption and gill Na$^+$,K$^+$-ATPase of the estuarine crab *Chasmagnathus granulata* Dana, 1851 (Deacapoda, Grapsidae), *Bull. Environ. Contam. Toxicol.*, 62, 63–69, 1999.

42. Bianchini, A. and Gilles, R., Cyclic AMP as a modulator of NaCl transport in gills of the euryhaline Chinese crab *Eriocheir sinensis*, *Mar. Biol.*, 104, 191–196, 1990.

43. Bianchini, A., Martins, S.E.G., and Barcarolli, I.F., Mechanism of acute copper toxicity in the euryhaline crustaceans: implication for the biotic ligand model, *Int. Congr. Ser.*, 1275, 189–194, 2004.

44. Bianchini, A., Péqueux, A., and Gilles, R., Effects of TAP and DPC on the transepithelial potential difference of isolated perfused gills of the freshwater acclimated crab, *Eriocheir sinensis*, *Comp. Biochem. Physiol.*, 90A, 315–319, 1988.

45. Bianchini, A., Playle, R.C., Wood, C.M., and Walsh, P.J., Mechanism of silver acute toxicity in marine invertebrates, *Aquat. Toxicol.*, 72, 67–82, 2005.

46. Bianchini, A. and Wood, C.M., Sodium uptake in different life stages of crustaceans: the water flea *Daphnia magna* Strauss, *J. Exp. Biol.*, 211, 539–547, 2008.

47. Bielawski, J., Chloride transport and water intake into isolated gills of crayfish, *Comp. Biochem. Physiol.*, 13, 423–432, 1964.

48. Bielawski, J., Ultrastructure and ion transport in gill epithelium of the crayfish *Astacus leptodactylus* Esch., *Protoplasma*, 73, 177–190, 1971.

49. Binns, R., The physiology of the antennal gland of *Carcinus maenas* (L.). I. The mechanism of urine production, *J. Exp. Biol.*, 51, 1–10, 1969.

50. Binns, R., The physiology of the antennal gland of *Carcinus maenas* (L.). II. Urine production rates, *J. Exp. Biol.*, 51, 11–16, 1969.

51. Binns, R., The physiology of the antennal gland of *Carcinus maenas* (L.). III. Glucose reabsoption, *J. Exp. Biol.*, 51, 17–27, 1969.

52. Binns, R., The physiology of the antennal gland of *Carcinus maenas* (L.). IV. The reabsoption of amino acids, *J. Exp. Biol.*, 51, 29–39, 1969.

53. Binns, R., The physiology of the antennal gland of *Carcinus maenas* (L.). V. Some nitrogenous constituents in the blood and urine, *J. Exp. Biol.*, 51, 41–45, 1969.

54. Bliss, D.E., Transition from water to land in decapod crustaceans, *Am. Zool.*, 8, 355–392, 1968.

55. Bliss, D.E., Wang, S.M.E., and Martinez, E.A., Water balance in the land crab, *Gecarcinus lateralis*, during the intermolt cycle, *Am. Zool.*, 6, 197–212, 1966.

56. Böcking, D., Dircksen, H., and Keller, R., The crustacean neuropeptides of the CHH/MIH/GIH family: structures and biological activities, in *The Crustacean Nervous System*, Wiese, K., Ed., Springer Verlag, Berlin, 2002, pp. 84–97.

57. Böttcher, K., Siebers, D., and Becker, W., Carbonic anhydrase in branchial tissues of osmoregulating shore crabs, *Carcinus maenas*, *J. Exp. Zool.*, 255, 251–261, 1990.

58. Bouaricha, N., Charmantier, G., Charmantier-Daures, M., Thuet, P., and Trilles, J.-P., Ontogenèse de l'osmorégulation chez la crevette *Penaeus japonicus*, *Cah. Biol. Mar.*, 32, 149–158, 1991.

59. Bouaricha, N., Charmantier-Daures, M., Thuet, P., Trilles, J.-P., and Charmantier, G., Ontogeny of osmoregulatory structures in the shrimp *Penaeus japonicus* (Crustacea, Decapoda), *Biol. Bull., Woods Hole*, 186, 29–40, 1994.

60. Bowman, T.E. and Abele, L.G., Classification of the recent Crustacea, in *Systematics: The Fossil Record and Biogeography*, Abele, L.G., Ed., Academic Press, New York, 1982, pp. 1–27.

61. Boxshall, G.A., Copepoda, in *Microscopic Anatomy of Invertebrates*, Harrison, F.W. and Humes, A.G., Eds., Wiley-Liss, New York, 1992, pp. 347–384.

62. Brand, G.W. and Bayly, I.A.E., A comparative study of osmotic regulation in four species of calanoid copepods, *Comp. Biochem. Physiol.*, 388, 361–371, 1971.

63. Bricteux-Grégoire, S., Duchâteau-Bosson, G., Jeuniaux, C., and Florkin, M., Les constituants osmotiquement actifs des muscles et leur contribution à le régulation isosmotique intracellulaire chez *Limulus polyphemus*, *Comp. Biochem. Physiol.*, 19, 729–736, 1966.

64. Briggs, R.T. and Moss, B.L., Ultrastructure of the coxal gland of the horseshoe crab *Limulus polyphemus*: evidence for ultrafiltration and osmoregulation, *J. Morphol.*, 234, 233–252, 1997.

65. Brodie, D.A. and Halcrow, K., Hemolymph regulation to hyposaline and hypersaline conditions in *Gammarus oceanicus* (Crustacea: Amphipoda), *Experientia*, 34, 1297–1298, 1978.

66. Brooks, S.J. and Mills, C.L., The effect of copper on osmoregulation in the freshwater amphipod *Gammarus pulex*, *Comp. Biochem. Physiol.*, 135A, 527–537, 2003.

67. Brown, A.C. and Terwilliger, N.B., Developmental changes in ionic and osmotic regulation in the Dungeness crab, *Cancer magister*, *Biol. Bull.*, 182, 270–277, 1992.

68. Bubel, A., Histological and microscopical observations on the effects of different salinities and heavy metal ions, on the gills of *Jaera nordmanni* (Rathke) (Crustacea, Isopoda), *Cell Tissue Res.*, 167, 65–95, 1976.

69. Bubel, A. and Jones, M.B., Fine structure of the gills of *Jaera nordmanni* (Rathke) (Crustacea, Isopoda), *J. Mar. Biol. Assoc. U.K.*, 54, 737–743, 1974.

70. Burggren, W.W., The importance of an ontogenetic perspective in physiological studies: amphibian cardiology as a case study, in *Physiological Adaptations in Vertebrates: Respiration, Circulation, and Metabolism*, Wood, S.C. et al., Eds., Marcel Dekker, New York, 1992, pp. 235–253.

71. Burggren, W.W., McMahon, B.R., and Costerton, J.W., Branchial water- and blood-flow patterns and the structure of the gill of the crayfish *Procambarus clarkii*, *Can. J. Zool.*, 52, 1511–1518, 1974.

72. Burnett, L.E. and McMahon, B.R., Facilitation of CO_2 excretion by carbonic anhydrase located on the surface of the basal membrane of crab gill epithelium, *Respir. Physiol.*, 62, 341–348, 1985.

73. Burnett, L.E. and Towle, D.W., Sodium ion uptake by perfused gills of the blue crab *Callinectes sapidus*: effects of ouabain and amiloride, *J. Exp. Biol.*, 149, 293–305, 1990.

74. Burton, R.F., Ionic regulation in Crustacea: the influence of temperature on apparent set points, *Comp. Biochem. Physiol.*, 84A, 135–139, 1986.

75. Cameron, J.N., NaCl balance in blue crabs, *Callinectes sapidus*, in fresh water, *J. Comp. Physiol.*, 123, 127–135, 1978.

76. Cameron, J.N., Effects of inhibitors on ion fluxes, trans-gill potential and pH regulation in freshwater blue crabs, *Callinectes sapidus* (Rathbun), *J. Comp. Physiol.*, 133, 219–225, 1979.

77. Cameron, J.N. and Batterton, C.V., Antennal gland function in the freshwater blue crab, *Callinecte sapidus*: water, electrolyte, acid–base and ammonia excretion, *J. Comp. Physiol.*, 123, 143–148, 1978.

78. Campbell, A., Growth of tagged American lobsters *Homarus americanus* in the Bay of Fundy, *Can. J. Fish. Aquat. Sci.*, 40, 1667–1675, 1986.

79. Campbell, P.J. and Jones, M.B., Water permeability of *Palaemon longirostris* and other euryhaline caridean prawns, *J. Exp. Biol.*, 150, 145–158, 1990.

80. Cantelmo, A.C., Water Permeability of Isolated Tissues from Three Species of Decapod Crustaceans with Respect to Osmotic Conditions and Effects of Neuroendocrine Factors, Ph.D. thesis, University of New York, 1976.

81. Carlisle, D.B., On the hormonal control of water balance in *Carcinus*, *Publ. Staz. Zool. Napoli*, 27, 227–231, 1955.

82. Castilho, P.C., Martins, I.A., and Bianchini, A., Gill Na^+,K^+-ATPase and osmoregulation in the estuarine crab, *Chasmagnathus granulata* Dana, 1851 (Decapoda, Grapsidae), *J. Exp. Mar. Biol. Ecol.*, 256, 215–227, 2001.

83. Ceccaldi, H.J., The digestive tract: anatomy, physiology, and biochemistry, in *Treatise on Zoology: Anatomy, Taxonomy, Biology, The Crustacea*, Forest, J. and von Vaupel Klein, J.C., Eds., Brill Academic Publishers, Leiden, 2006.

84. Chan, S.M., Gu, P.L., Chu, K.H., and Tobe, S.S., Crustacean neuropeptide genes of the CHH/MIH/GIH family: implications from molecular studies, *Gen. Comp. Endocrinol.*, 134, 214–219, 2003.

85. Chang, E.S., Chemistry of crustacean hormones that regulate growth and reproduction, in *Recent Advances in Marine Biotechnology*, Fingerman, M., Nagabhushanam, R., and Thompson, M.-F., Eds., Science Publishers, Enfield, NH, 1997, pp. 163–178.

86. Chang, E.S., Chang, S.A., Beltz, B.S., and Kravitz, E.A., Crustacean hyperglycemic hormone in the lobster nervous system: localization and release from cells in the suboesophageal ganglion and thoracic second roots, *J. Comp. Neurol.*, 414, 50–56, 1999.

87. Chang, E.S., Chang, S.A., and Mulder, E.P., Hormones in the lives of crustaceans: an overview, *Am. Zool.*, 41, 1090–1097, 2001.

88. Chang, E.S., Keller, R., and Chang, S.A., Quantification of crustacean hyperglycemic hormone by ELISA in hemolymph of the lobster, *Homarus americanus*, following various stresses, *Gen. Comp. Endocrinol.*, 111, 359–366, 1998.

89. Charmantier, G., Variations saisonnières des capacités ionorégulatrices de *Sphaeroma serratum* (Fabricius, 1787) (Crustacea, Isopoda, Flabellifera), *Comp. Biochem. Physiol.*, 50A, 339–345, 1975.

90. Charmantier, G., Ontogeny of osmoregulation in crustaceans: a review, *Invert. Reprod. Develop.*, 33, 177–190, 1998.

91. Charmantier, G. and Aiken, D.E., Osmotic regulation in late embryos and prelarvae of the American lobster *Homarus americanus* H. Milne Edwards, 1837 (Crustacea, Deacapoda), *J. Exp. Mar. Biol. Ecol.*, 109, 101–108, 1987.

92. Charmantier, G. and Anger, K., Ontogeny of osmoregulation in the Palaemonid shrimp *Palaemonetes argentinus*, *Mar. Ecol. Progr. Ser.*, 181, 125–129, 1999.

93. Charmantier, G., Bouaricha, N., Charmantier-Daures, M., Thuet, P., and Trilles, J.-P., Salinity tolerance and osmoregulatory capacity as indicators of physiological state of peneid shrimps, *Eur. Aquac. Soc. Spec. Publ.*, 10, 65–66, 1989.

94. Charmantier, G. and Charmantier-Daures, M., Ontogeny of osmoregulation and salinity tolerance in *Cancer irroratus*: elements of comparison with *C. borealis* (Crustacea, Decapoda), *Biol. Bull.*, 180, 125–134, 1991.

95. Charmantier, G. and Charmantier-Daures, M., Ontogeny of osmoregulation and salinity tolerance in the isopod crustacean *Sphaeroma serratum*, *Mar. Ecol. Progr. Ser.*, 114, 93–102, 1994.

96. Charmantier, G. and Charmantier-Daures, M., Osmoregulation and salinity tolerance in zoeae and juveniles of the snow crab *Chionoecetes opilio*, *Aquat. Liv. Res.*, 8, 171–179, 1995.

97. Charmantier, G. and Charmantier-Daures, M., Ontogeny of osmoregulation in crustaceans: the embryonic phase, *Am. Zool.*, 41, 1078–1089, 2001.

98. Charmantier, G., Charmantier-Daures, M., and Aiken, D.E., Neuroendocrine control of hydromineral regulation in the American lobster *Homarus americanus* H. Milne Edwards, 1837 (Crustacea, Decapoda). I. Juveniles, *Gen. Comp. Endocrinol.*, 54, 8–19, 1984.

99. Charmantier, G., Charmantier-Daures, M., and Aiken, D.E., Neuroendocrine control of hydromineral regulation in the American lobster *Homarus americanus* H. Milne Edwards, 1837 (Crustacea, Decapoda). II. Larval and postlarval stages, *Gen. Comp. Endocrinol.*, 54, 20–34, 1984.

100. Charmantier, G., Charmantier-Daures, M., and Anger, K., Ontogeny of osmoregulation in the grapsid crab *Armases miersii* (Crustacea, Decapoda), *Mar. Ecol. Progr. Ser.*, 164, 285–292, 1998.

101. Charmantier, G., Charmantier-Daures, M., Bouaricha, N., Thuet, P., Aiken, D.E., and Trilles, J.-P., Ontogeny of osmoregulation and salinity tolerance in two decapod crustaceans: *Homarus americanus* and *Penaeus japonicus*, *Biol. Bull.*, 175, 102–110, 1988.

102. Charmantier, G., Charmantier-Daures, M., and Van Herp, F., Hormonal regulation of growth and reproduction in crustaceans, in *Recent Advances in Marine Biotechnology*, Fingerman, M., Nagabhushanam, R., and Thompson, M.-F., Science Publishers, Enfield, NH, 1997, pp. 109–161.

103. Charmantier, G., Gimenez, L., Charmantier-Daures, M., and Anger, K., Ontogeny of osmoregulation, physiological plasticity, and export strategy in the grapsid crab *Chasmagnathus granulata* (Crustacea, Decapoda), *Mar. Ecol. Progr. Ser.*, 229, 185–194, 2002.

104. Charmantier, G., Haond, C., Lignot, J.-H., and Charmantier-Daures, M., Ecophysiological adaptation to salinity throughout a life-cycle: a review in homarid lobsters, *J. Exp. Biol.*, 204, 967–977, 2001.

105. Charmantier, G., Soyez, C., and Aquacop, Effect of molt stage and hypoxia on osmoregulatory capacity in the peneid shrimp *Penaeus vannamei*, *J. Exp. Mar. Biol. Ecol.*, 178, 233–246, 1994.

106. Charmantier, G. and Thuet, P., Recherches écophysiologiques sur deux Sphéromes de l'étang de Thau (Hérault): *Sphaeroma serratum* (Fabricius) et *Sphaeroma hookeri* Leach, *C. R. Acad. Sci. Paris*, 269, 2405–2408, 1969.

107. Charmantier, G. and Trilles, J.-P., Recherches physiologiques chez *Sphaeroma serratum* (Fabricius) (Isopode, Flabellifère). Variations de la teneur en eau de l'organisme et de la teneur en ions Cl⁻, Na⁺, K⁺ et Ca⁺⁺ de l'hémolymphe au cours du cycle d'intermue, *C. R. Acad. Sci. Paris*, 272, 286–288, 1971.

108. Charmantier, G. and Trilles, J.-P., Influence des glandes antennaires sur la régulation ionique, la teneur en eau et éventuellement la mue de *Sphaeroma serratum* (Crustacea, Isopoda, Flabellifera), *Gen. Comp. Endocrinol.*, 31, 295–301, 1977.

109. Charmantier, G. and Wolcott, D.-L., Introduction to the SICB symposium: ontogenetic strategies of invertebrates in aquatic environments, *Am. Zool.*, 41, 1053–1056, 2001.

110. Charmantier-Daures, M., Charmantier, G., Aiken, D.E., Janssen, K.P.C., and Van Herp, F., Involvement of eyestalk factors in the neuroendocrine control of hydromineral metabolism in adult American lobster *Homarus americanus*, *Gen. Comp. Endocrinol.*, 94, 281–293, 1994.

111. Charmantier-Daures, M., Thuet, P., Charmantier, G., and Trilles, J.-P., Tolérance à la salinité et osmorégulation chez les post-larves de *Penaeus japonicus* et *P. chinensis*. Effet de la température, *Aquat. Liv. Res.*, 1, 267–276, 1988.

112. Charmantier-Daures, M. and Vernet, G., Molting autotomy and regeneration, in *Treatise on Zoology: Crustacea*, Forest, J. and von Vaupel Klein, J.C., Eds., Brill Academic, Leiden, 2004, pp. 161–255.

113. Chen, J.-C. and Chia, P.-G., Osmotic and ionic concentrations of *Scylla serrata* (Forskal) subjected to different salinity levels, *Comp. Biochem. Physiol.*, 117A, 239–244, 1997.

114. Chen, J.C. and Lin, J.-L., Responses of hemolymph osmolality and tissue water of *Penaeus chinensis* Osbeck juveniles subjected to sudden change in salinity, *Mar. Biol.*, 120, 115–121, 1994.

115. Chim, L., Bouveret, R., Lemaire, P., and Martin, J.L.M., Tolerance of the shrimp *Litopenaeus stylirostris*, Stimpson 1894, to environmental stress: interindividual variability and selection potential for stress-resistant individuals, *Aquac. Res.*, 34, 629–632, 2003.

116. Chu, K.H., Sodium transport across the perfused midgut and hindgut of the blue crab, *Callinectes sapidus*: the possible role of the gut in crustacean osmoregulation, *Comp. Biochem. Physiol.*, 87A, 21–25, 1987.

117. Chung, J.S., Dircksen, H., and Webster, S.G., A remarkable, precisely timed release of hyperglycemic hormone from endocrine cells in the gut is associated with ecdysis in the crab *Carcinus maenas*, *Proc. Nat. Acad. Sci. USA*, 96, 13103–13107, 1999.

118. Chung, J.S. and Webster, S.G., Binding sites of crustacean hyperglycemic hormone and its second messengers on gills and hindgut of the green shore crab, *Carcinus maenas*: a possible osmoregulatory role, *Gen. Comp. Endocrinol.*, 147, 206–213, 2006.

119. Chung, K.F. and Lin, H.C., Osmoregulation and Na,K-ATPase expression in osmoregulatory organs of *Scylla paramamosain*, *Comp. Biochem. Physiol.*, 144A, 48–57, 2006.

120. Cieluch, U., Anger, K., Aujoulat, F., Buchholz, F., Charmantier-Daures, M., and Charmantier, G., Ontogeny of osmoregulatory structures and functions in the green crab *Carcinus maenas* (Crustacea, Decapoda), *J. Exp. Biol.*, 207, 325–336, 2004.

121. Cieluch, U., Anger, K., Charmantier-Daures, M., and Charmantier, G., Osmoregulation and immunolocalization of Na⁺/K⁺-ATPase during the ontogeny of the mitten crab, *Eriocheir sinensis* (Decapoda, Grapsoidea), *Mar. Ecol. Progr. Ser.*, 329, 169–178, 2007.

122. Cieluch, U., Charmantier, G., Grousset, E., Charmantier-Daures, M., and Anger, K., Osmoregulation, immunolocalization of Na⁺/K⁺-ATPase, and ultrastructure of branchial epithelia in the developing brown shrimp, *Crangon crangon* (Decapoda, Caridea), *Physiol. Biochem. Zool.*, 78, 1017–1025, 2005.

123. Clark, M.E., The osmotic role of amino acids: discovery and function, in *Transport Processes, Iono- and Osmoregulation*, Gilles, R. and Gilles-Baillien, M., Eds., Springer, Berlin, 1985, pp. 412–423.

124. Cohen, J.A. and Brockman, H.J., Breeding activity and mate selection in the horseshoe crab, *Limulus polyphemus*, *Bull. Mar. Sci.*, 33, 274–281, 1983.

125. Cole, W.H., The composition of fluids and sera of some marine animals and of the sea water in which they live, *J. Gen. Physiol.*, 23, 575–584, 1940.

126. Comeau, M. and Savoie, F., Movement of American lobster (*Homarus americanus*) in the southwestern Gulf of St. Lawrence, *Fish. Bull.*, 100, 181–192, 2002.

127. Compère, P., Wanson, S., Péqueux, A., Gilles, R., and Goffinet, G., Ultrastructural changes in the gill epithelium of the green crab *Carcinus maenas* in relation to the external salinity, *Tissue Cell*, 21, 229–318, 1989.

128. Conte, F.P., Structure and function of the crustacean larval salt gland, *Int. Rev. Cytol.*, 91, 45–106, 1984.

129. Conte, F.P., Droukas, P.C., and Ewing, R.D., Development of sodium regulation and *de novo* synthesis of Na⁺,K⁺-activated ATPase in larval brine shrimp, *Artemia salina*, *J. Exp. Zool.*, 202, 339–362, 1977.

130. Conte, F.P., Hootman, S.R., and Harris, P.J., Neck organ of *Artemia salina nauplii*: a larval salt gland, *J. Comp. Physiol.*, 80, 239–246, 1972.

131. Copeland, D.E., A study of salt secreting cells in the brine shrimp (*Artemia salina*), in *Protoplasma*, Porter, K.R., Ed., Springer-Verlag, Vienna, 1967, pp. 363–384.

132. Copeland, D.E., Fine structure of salt and water uptake in the land crab, *Gecarcinus lateralis*, *Am. Zool.*, 8, 417–432, 1968.

133. Copeland, D.E. and Fitzjarrell, A.T., The salt absorbing cells in the gills of the blue crab (*Callinectes sapidus* Rathbun) with notes on modified mitochondria, *Z. Zellforsch.*, 92, 1–22, 1968.

134. Crogham, P.C., The survival of *Artemia salina* (L.) in various media, *J. Exp. Biol.*, 35, 213–218, 1958.

135. Crogham, P.C., The osmotic and ionic regulation of *Artemia salina* (L.), *J. Exp. Biol.*, 35, 219–233, 1958.

136. Crogham, P.C., The mechanism of osmotic regulation in *Artemia salina* (L.): the physiology of the branchiae, *J. Exp. Biol.*, 35, 234–242, 1958.

137. Crogham, P.C. and Lockwood, A.P.M., Ionic regulation of the Baltic and freshwater races of the isopod *Mesidotea* (*Saduria*) *entomon* (L.), *J. Exp. Biol.*, 48, 141–158, 1968.

138. Curra, R.A., Notes on the morphology of the branchiae of the crayfish *Austropotamobius pallipes* (Lereboullet), *Zool. Anzeiger*, 174, 313–323, 1965.

139. Dailey, M.E., Fremont-Smith, F., and Carroll, M.P., The relative composition of sea water and of the blood of *Limulus polyphemus*, *J. Biol. Chem.*, 93, 17–24, 1931.

140. Dall, W., Studies on the physiology of a shrimp, *Metapenaeus mastersii* (Haswell) (Crustacea: Decapoda: Peneidae), *Aust. J. Mar. Freshwater Res.*, 15, 145–161, 1964.

141. Dall, W., Hypo-osmoregulation in Crustacea, *Comp. Biochem. Physiol.*, 21, 653–678, 1967.

142. Dall, W., The functional anatomy of the digestive tract of a shrimp *Metapenaeus bennettae* Racek & Dall (Crustacea: Decapoda: Peneidae), *Aust. J. Zool.*, 15, 699–714, 1967.

143. Dall, W., Osmoregulation in the lobster *Homarus americanus*, *J. Fish. Res. Board Can.*, 27, 1123–1130, 1970.

144. Dall, W., Osmotic and ionic regulation in the western rock lobster *Panulirus longipes* (Milne Edwards), *J. Exp. Mar. Biol. Ecol.*, 15, 97–125, 1974.

145. Dall, W. and Moriarty, D.J.W., Functional aspects of nutrition and digestion, in *Internal Anatomy and Physiological Regulation*, Mantel, L.H., Ed., Academic Press, New York, 1983, pp. 215–261.

146. Dall, W. and Smith, D.M., Water uptake at ecdysis in the western rock lobster, *J. Exp. Mar. Biol. Ecol.*, 35, 165–176, 1978.

147. Dalla Via, G.J., Salinity responses of the juvenile penaeid shrimp *Penaeus japonicus*. II. Free amino acids, *Aquaculture*, 55, 307–316, 1986.

148. Damkaer, D.M. and Dey, D.B., UV damage and photoreactivation potentials of larval shrimp, *Pandalus platyceros*, and adult euphausiids, *Thysanoessa raschii*, *Oecology*, 60, 169–175, 1983.

149. Dandrifosse, G., Absorption d'eau au moment de la mue chez un Crustacé décapode: *Maia squinado* Herbst, *Archs Intern. Physiol. Bioch.*, 74, 329–331, 1966.

150. Davis, C.W., Neuroendocrine control of sodium balance in the fiddler crab, *Uca pugilator*, *Diss. Abstr. Int. B. Sci. Eng.*, 39, 3169, 1979.

151. De Kleijn, D.P.V., De Leeuw, E.P.H., Van Den Berg, M.C., Martens, G.J.M., and Van Herp, F., Cloning and expression of two mRNAs encoding structurally different crustacean hyperglycemic hormone precursors in the lobster *Homarus americanus*, *Mar. Biotechnol.*, 2, 80–91, 1995.

152. De Leersnyder, M., Le milieu intérieur d'*Eriocheir sinensis* H. Milne Edwards et ses variations. II. Etude expérimentale, *Cah. Biol. Mar.*, 8, 295–321, 1967.

153. Debaisieux, P., Appareil branchial de *Crangon vulgaris* (Décapode nageur): anatomie et histologie, *La Cellule*, 19, 64–77, 1970.

154. Dehnel, P.A., Aspects of osmoregulation in two species of intertidal crabs, *Biol. Bull.*, 122, 208–227, 1962.

155. Dehnel, P.A., Gill tissue respiration in the crab *Eriocheir sinensis*, *Can. J. Zool.*, 52, 923–937, 1974.

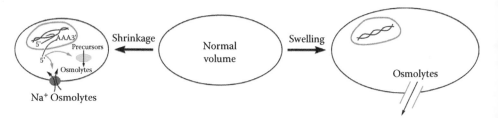

FIGURE 2.2 Mechanisms of organic osmolyte accumulation and loss. Shrinkage-induced organic osmolyte accumulation in animal cells is mediated largely by increased transcription and translation (green arrows) of genes encoding Na^+-coupled membrane transporters or enzymes involved in organic osmolyte synthesis. During swelling, organic osmolytes are lost from animal cells largely by passive efflux through channel-like transport pathways. In addition, cell swelling inhibits the expression of genes involved in organic osmolyte accumulation.

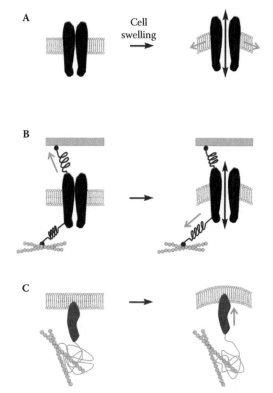

FIGURE 2.3 Possible mechanisms of cellular mechanical force detection and transduction. Panels A and B illustrate mechanisms by which ion channel gating can be modulated by mechanical force; however, the function of any protein embedded plasma membrane may be subject to similar mechanical regulation. (A) Bilayer model of mechanosensitive channel gating; changes in mechanical force generated within the bilayer during cell swelling or shrinkage directly alter channel conformation and gating. (B) Tethered model of mechanosensitive channel gating; the channel is tethered to relatively immobile extracellular matrix and/or intracellular cytoskeletal proteins. Mechanical force is placed on the channel through tether proteins during swelling or shrinkage. (C) Mechanical-force-induced change in conformation of an intracellular macromolecule; in addition to membrane proteins, mechanical force can alter the conformation and hence function of intracellular macromolecules. The illustration shows a cytoplasmic protein (blue) tethered to a membrane-embedded protein and a cytoskeletal network. Cell swelling displaces the membrane protein relative to the cytoskeletal network, thereby stretching the cytoplasmic protein. The resulting conformational change could alter enzyme activity, expose functional domains such as phosphorylation sites, alter protein–protein interactions, etc. Green arrows in all panels indicate direction of applied mechanical force.

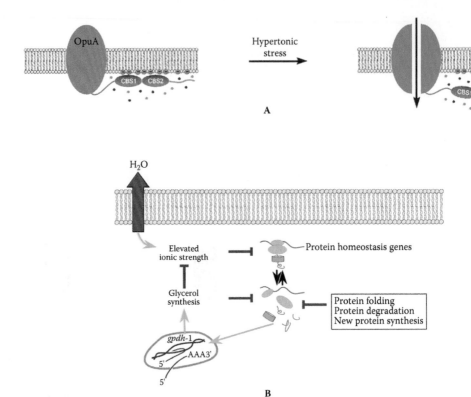

FIGURE 2.4 Examples of osmoregulatory pathways activated by increases in intracellular ionic strength OpuA is a bacterial organic osmolyte transporter activated by hypertonic stress. The cationic surface (sh by red shading) of the CBS domains interacts with anionic membrane lipids (shown in green) and inactiv the transporter. Hypertonic stress and cell shrinkage increase intracellular ionic strength (small red and g circles), which in turn disrupts this electrostatic interaction leading to increased OpuA activity. The ani C-terminus (green) of OpuA is expected to be repelled away from anionic membrane lipids which may mod ionic strength sensitivity by weakening the interaction between the membrane and CBS domains. (B) M for regulation of *C. elegans* osmosensitive gene expression by disruption of protein homeostasis. Hypert stress induced water loss causes elevated cytoplasmic ionic strength which in turn disrupts new protein synth and cotranslational protein folding. Misfolded and incompletely synthesized proteins function as a signa activates *gpdh-1* expression and glycerol synthesis. Glycerol replaces inorganic ions in the cytoplasm functions as a chemical chaperone that aids in the refolding of misfolded proteins. Loss of function of pr homeostasis genes also causes accumulation of damaged proteins and activation of *gpdh-1* expression.

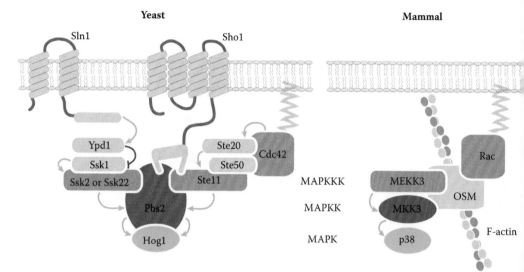

FIGURE 2.5 Osmotically activated p38 MAPK signaling cascades in yeast and mammals. Green arrows and red lines indicate activation and inhibition, respectively. Physical interactions are indicated by overlap of components. MAPKKKs, MAPKKs, and MAPKs are colored orange, red, and green, respectively. In yeast there are two signaling branches upstream from Pbs2, a MAPKK that activates the p38 MAPK Hog1. The Sln1 branch is comprised of the Ypd1 and Ssk1 phosphorelay system that regulates the activity of two redundant MAPKKKs, Ssk2 and Ssk22. Increased osmolality inhibits this phosphorelay system and allows Ssk1 to interact with and activate Ssk2 and Ssk22. Cdc42 in the Sho1 signaling branch is activated by increased osmolality and in turn activates Ste20, a MAPKKKK. Ste20 then activates the MAPKKK Ste11. The efficiency and specificity of the Sho1 cascade is modulated by the scaffolding functions of Sho1, Cdc42, and Ste50. In mammals, Rac, MEKK3, MKK3, and p38 MAPK are thought to comprise a cascade that is homologous to Cdc42, Ste11, Pbs2, and Hog1. OSM acts as a scaffold for Rac, MEKK3, and MKK3.

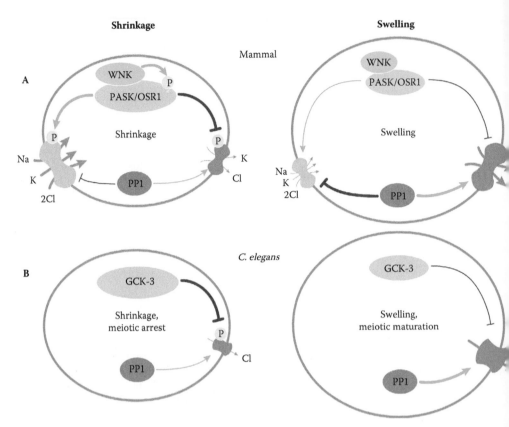

FIGURE 2.6 Models of phosphorylation and dephosphorylation events that regulate the activity of vol[ume] sensitive Na–K–2Cl and K–Cl cotransporters in mammalian cells and a volume-sensitive ClC type of a[nion] channel in *Caenorhabditis elegans*. The relative activities of kinases and ion transport pathways are indi[cated] by size. Green arrows and red lines indicate activation and inhibition, respectively. Physical interaction[s are] indicated by overlap of components. (A) During shrinkage in mammalian cells, Na–K–2Cl cotransporter[s are] activated and K–Cl cotransporters are inactivated by phosphorylation via PASK or OSR1. Dephosphoryl[ation] via a type 1 protein phosphatase (PP1) inactivates Na–K–2Cl cotransporters and activates K–Cl cotranspo[rters] during cell swelling. WNK1 and WNK4 interact with PASK and OSR1. Cell shrinkage activates WNK[1 and] WNK4 via unknown mechanisms. Activated WNK1 or WNK4 then phosphorylates and activates PASK [or] OSR1. An unknown phosphatase is thought to dephosphorylate and inactivate PASK and OSR1 during [cell] swelling. (B) In *C. elegans* oocytes, the ClC channel CLH-3b is inactivated by the PASK/OSR1 hom[olog] GCK-3 by a mechanism that requires kinase activity. CLH-3b is inhibited by GCK-3 during cell shrin[kage] and meiotic arrest; inhibition by GCK-3 is removed and the channel is activated by the PP1 homologs GL[C-7α] and GCL-7β during cell swelling and meiotic maturation.

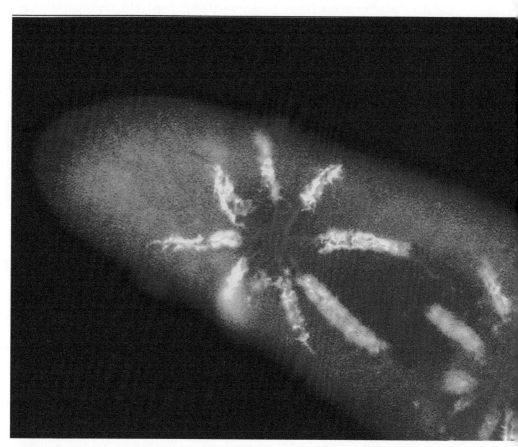

FIGURE 3.1 (A contractile vacuole complex (CVC) of *Paramecium multimicronucleatum* labeled with monoclonal antibodies (mAbs). Texas red-tagged mAb (red) labels the G4 antigen of the membranes of the smooth spongiome that make up the contractile vacuole (that lies at the hub of this CVC), the radiating collecting canals of the radial arms, and the ampulli. This antibody also cross-reacts with antigens in the membranes of the pellicle of the cell. The flourescein-tagged mAb (green) labels the A4 antigen found only as part of the V-ATPase that is particularly abundant on the decorated tubules of the CVC. The decorated tubules attach peripherally to each radial arm distal to the ampullus.

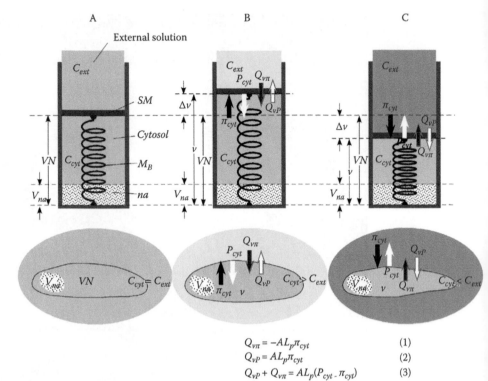

$$Q_{v\pi} = -AL_p\pi_{cyt} \qquad (1)$$
$$Q_{vP} = AL_p\pi_{cyt} \qquad (2)$$
$$Q_{vP} + Q_{v\pi} = AL_p(P_{cyt} - \pi_{cyt}) \qquad (3)$$

FIGURE 3.9 Mechanoosmotic model of a *Paramecium* cell in an (A) isoosmotic, (B) hypoosmotic, or
hyperosmotic solution. The model is drawn as a cylinder with a semipermeable piston (bar labeled
that is fixed to the bottom of the cylinder by a coil spring (M_B). The piston and the coil spring corresp
to the semipermeable cell membrane and its elasticity (the bulk modulus of the cell), respectively.
inside and outside of the cylinder correspond to the cytosol and the external solution, respectively. Med
gray corresponds to the cytosolic osmolarity (C_{cyt}), lighter gray to an osmolarity of the external solu
(C_{ext}) lower than C_{cyt}, and darker gray to a C_{ext} higher than C_{cyt}. A dotted area in the cylinder correspe
to an osmotically nonactive portion of the cell (*na*). The corresponding cell shape is shown below
model. When $C_{ext} = C_{cyt}$ (A), no osmotic water flow takes place across the piston, so the coil sprin
neither expanded nor compressed; that is, the cell is neither swollen nor shrunken. The length of the
spring in this situation corresponds to its natural resting length. The corresponding cell volume is the
termed the *natural cell volume* (*vN*). When C_{ext} is lowered below C_{cyt} (B), water osmotically flows into
cylinder through the piston (water inflow, $Q_{v\pi}$; downward blue-bordered black arrow). $Q_{v\pi}$ is proporti
to the osmotic pressure of the cytosol with reference to the external solution so π_{cyt} (upward black ar
can be obtained from Equation 1, where A is the area of the semipermeable cell membrane and L_p is
hydraulic conductivity of the membrane. A hydrostatic pressure in the cytosol with reference to the exte
solution P_{cyt} (downward white arrow) is generated as the coil spring is expanded by the water inflow
cell is osmotically swollen) and causes a water outflow from the cylinder through the piston (upward t
bordered white arrow labeled Q_{vP}), which is proportional to P_{cyt} (Equation 2). Q_{vP} cancels $Q_{v\pi}$ and
overall water flow across the piston becomes 0 when the cell swells to a level where $P_{cyt} = \pi_{cyt}$ and, there:
$Q_{vP} = Q_{v\pi}$ (Equation 3). Inversely, when C_{ext} is raised beyond C_{cyt} (C), the water osmotically leaves
cylinder through the piston (the water outflow; upward blue-bordered black arrow labeled $Q_{v\pi}$), so the
spring is compressed (the cell is osmotically shrunken). A negative hydrostatic pressure with referenc
the external solution is thereby generated in the cylinder and causes a water inflow through the pi
(downward blue-bordered white arrow labeled Q_{vP}). The overall water flow across the piston becom
when the coil spring is compressed to a level where $P_{cyt} = \pi_{cyt}$ and, therefore, $Q_{vP} = Q_{v\pi}$ (Equation 3).
discussion see Baumgarten and Feher.[11] Abbreviations: v_{na}, volume of the osmotically nonactive portio
the cell (*na*); v, cell volume; Δv, volume change after changing the external osmolarity.

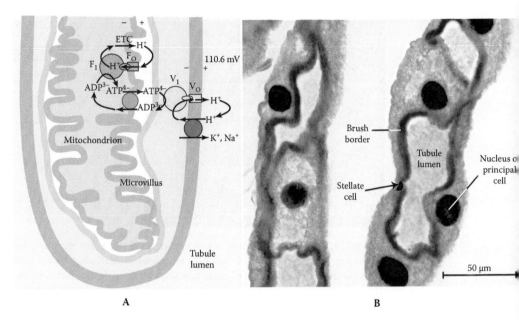

A   B

FIGURE 7.14 The brush border of principal cells in Malpighian tubules of *Aedes aegypti*: (a) Each microvillus contains a mitochondrion. ATP is produced by the F-synthase located in the inner mitochondrial membrane. F_1 and F_0 are, respectively, the catalytic and the proton-translocating complexes of the synthase. ETC is the electron transport chain. V_1 and V_0 are, respectively, the catalytic and proton-translocating complex of the V-type H^+-ATPase (See Figure 7.30 for structural details of the V-type H^+-ATPase). The V_1 complex contains subunit B against which the antibody used in (b) was prepared. (b) Immunolocalization of the B-subunit of the V-type H^+-ATPase in the brush border of principal cells of Malpighian tubules of the yellow fever mosquito (*Aedes aegypti*). A stellate cell (arrow) gives no evidence for the B subunit of the V-type H^+-ATPase. The antibody was kindly provided by Marcus Huss from the University of Osnabrueck, Germany. (Adapted from Beyenbach, K.W., *News Physiol. Sci.*, 16, 145, 2001.)

FIGURE 7.19 Female *Anopheles* mosquito taking a blood meal. Note the urination while feeding. Repeating this experiment on himself, allowing a female, pathogenic-free yellow fever mosquito (*Aedes aegypti*) to take a blood meal, James Williams in our laboratory found that the first urine droplets eliminate the NaCl and water fraction of the blood meal.[152] (Photograph courtesy of Jack Kelly Clark, University of California.)

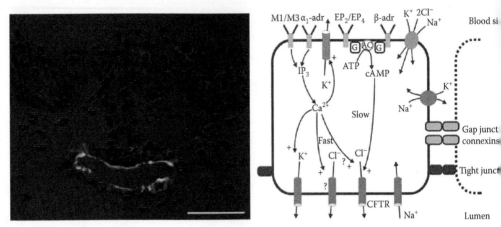

FIGURE 9.7 The fluid-secreting subepidermal gland of anuran skin. (A) Immunofluorescence labelir
AQP-x5 in mucous gland of the toad *Bufo woodhouseii*. AQP-x5, which is homologous to mammalian A
is visible as a narrow light band in the apical plasma membrane (green) of the secretory cells of the mu
gland. AQP-h3BL was similarly immunolocalized in the basolateral membrane (red) of the same cells
granular cells in the epidermis. Nuclei are counterstained with DAPI (blue). Scale bar = 50 μm. (B) M
of the organization of ion transport systems of frog skin acinar cells identified by transepithelial isotope t
and water flow studies, measurements of intracellular ion concentrations, patch clamp electrophysiology
application of pharmacological protocols as explained in the text. (From Sørensen, J.B. and Larsen, I
Pflügers Arch., 439, 101–112, 1999. With permission.)

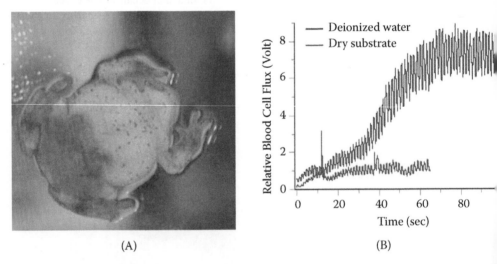

(A)　　　　　　　　　　　　　　　　　　(B)

FIGURE 9.15 (A) The toad *Bufo punctatus* exhibiting the water absorption response. Note that the hindl
are abducted from the body and cutaneous capillaries in the skin are highly perfused. (B) Seat patch cap
blood flow in *Bufo woodhouseii* estimated by laser Doppler flowmetry of relative red blood cell flux. *L*
trace: The seat patch blood flow in a dehydrated toad on a dry substrate is low. *Upper trace*: The blood
is greatly stimulated when the dehydrated toad is exposed (at time zero) to deionized water. (From Vi
A.L. and Hillyard, S.D., *Physiol. Biochem. Zool.*, 78, 394–404, 2005.).

156. Dehnel, P.A. and Malley, D.F., Urine production in the shore crab, *Hemigrapsus nudus*, *Can. J. Zool.*, 58, 1542–1550, 1980.

157. Dehnel, P.A. and Stone, D., Osmoregulatory role of the antennary gland in two species of estuarine crabs, *Biol. Bull.*, 126, 354–372, 1964.

158. Denne, L.B., Some aspects of osmotic and ionic regulation in the prawns *Macrobrachium australiense* (Holthuis) and *M. equidens* (Dana), *Comp. Biochem. Physiol.*, 26, 17–30, 1968.

159. Detaille, D., Trausch, G., and Devos, P., Dopamine as a modulator of ionic transport and glycolytic fluxes in the gills of the Chinese crab, *Eriocheir sinensis*, *Comp. Biochem. Physiol.*, 103C, 521–526, 1992.

160. Dey, D.B., Damkaer, D.M., and Heron, G.A., Dose/dose rate responses of seasonally abundant copepod of Puget Sound, *Oecology*, 76, 321–329, 1988.

161. Dickson, J.S., Dillaman, R.M., Roer, R.D., and Roye, D.B., Distribution and characterization if ion transporting and respiratory filaments of the gills of *Procambarus clarkii*, *Biol. Bull.*, 180, 154–166, 1991.

162. Dircksen, H., Böcking, D., Heyn, U., Chung, J.S., Baggerman, G., Verhaert, P., Daufeldt, S., Plösch, T., Jaros, P.P., Waelkens, E., Keller, R., and Webster, S.G., Crustacean hyperglycaemic hormone (CHH)-like peptides and CHH-precursor-related peptides from pericardial organ neurosecretory cells in the shore crab, *Carcinus maenas*, are putatively spliced and modified products of multiple genes, *Biochem. J.*, 356, 159–170, 2001.

163. Dircksen, H. and Soyez, D., The lobster thoracic ganglia–pericardial organ neurosecretory system: a large source of novel crustacean hyperglycemic hormone-like peptides, in *Proc. of the 19th Conf. of European Comparative Endocrinologists*, Nijmegen, Netherlands, 1998, p. 19.

164. Dixon, C.J., Ahyong, S.T., and Schram, F.R., A new hypothesis of decapod phylogeny, *Crustaceana*, 76, 935–975, 2003.

165. Dobkin, S. and Manning, R.S., Osmoregulation in two species of *Palaemonetes* (Crustacea: Decapoda) from Florida, *Bull. Mar. Sci. Gulf Caribbean*, 14, 149–157, 1964.

166. D'Orazio, S.E. and Holliday, C.W., Gill Na,K-ATPase and osmoregulation in the sand fiddler crab, *Uca pugilator*, *Physiol. Zool.*, 58, 364–373, 1985.

167. Dougthie, D.G. and Rao, K.R., Histopathological and ultrastructural changes in the antennal gland, midgut, hepatopancreas and gill of grass shrimp following exposure to hexavalent chromium, *J. Invertebr. Pathol.*, 43, 89–108, 1984.

168. Drach, P., Mue et cycle d'intermue chez les Crustacés Décapodes, *Ann. Inst. Oceanogr. Monaco*, 19, 103–391, 1939.

169. Dunbar, S.G. and Coates, M., Differential tolerance of body fluid dilution in two species of tropical hermit crabs: not due to osmotic/ionic regulation, *Comp. Biochem. Physiol.*, 137A, 321–337, 2004.

170. Dunel-Erb, S., Barradas, C., and Lignon, J.M., Morphological evidence for the existence of two distinct types of mitochondria rich cells in the gills of the crayfish *Astacus leptodactylus* Eschscholtz, *Acta Zool.*, 78, 195–203, 1997.

171. Dunel-Erb, S., Massabuau, J.C., and Laurent, P., Organisation fonctionnelle de la branchie d'écrevisse, *C. R. Soc. Biol.*, 176, 248–258, 1982.

172. Dunson, W.A., Permeability of the integument of the horseshoe crab, *Limulus polyphemus*, to water, sodium, and bromide, *J. Exp. Zool.*, 230, 495–499, 1984.

173. Eckhardt, E., Pierrot, C., Thuet, P., Van Herp, F., Charmantier-Daures, M., Trilles, J.-P., and Charmantier, G., Stimulation of osmoregulating processes in the perfused gill of the crab *Pachygrapsus marmoratus* (Crustacea, Decapoda) by a sinus gland peptide, *Gen. Comp. Endocrinol.*, 99, 169–177, 1995.

174. Ehlinger, G.S. and Tankersley, R.A., Survival and development of horseshoe crab (*Limulus polyphemus*) embryos and larvae in hypersaline conditions, *Biol. Bull.*, 206, 87–94, 2004.

175. Eladari, D., Cheval, L., Quentin, F., Bertrand, O., Mouro, I., Cherif-Zahar, B., Cartron, J.-P., Paillard, M., Doucet, A., and Chambrey, R., Expression of RhCG, a new putative NH_3/NH_4^+ transporter, along the rat nephron, *J. Am. Soc. Nephrol.*, 13, 1999–2008, 2002.

176. Escalante, R., Garcia-Sáez, A., and Sastre, L., *In situ* hybridization analyses of Na,K-ATPase alpha-subunit expression during early larval development of *Artemia franciscana*, *J. Histochem. Cytochem.*, 43, 391–399, 1995.

177. Evans, D.H., Piermarini, P.M., and Choe, K.P., The multifunctional fish gill: dominant site of gas exchange, osmoregulation, acid–base regulation, and excretion of nitrogenous waste, *Physiol. Rev.*, 85, 97–177, 2005.

178. Fanjul-Moles, M.L., Biochemical and functional aspects of crustacean hyperglycemic hormone in decapod crustaceans: review and update, *Comp. Biochem. Physiol.*, 142C, 390–400, 2006.

179. Farmer, L., Evidence for hyporegulation in the calanoid copepod *Acartia tonsa*, *Comp. Biochem. Physiol.*, 65A, 359–362, 1980.

180. Farrelly, C.A. and Greenaway, P., Morphology and ultrastructure of the gills of terrestrial crabs (Crustacea, Gecarcinidae and Grapsidae): adaptations for air-breathing, *Zoomorphology*, 112, 39–50, 1992.

181. Felder, D.L., Osmotic and ionic regulation in several western Atlantic Callianassidae (Crustacea, Decapoda, Thalassinidea), *Biol. Bull.*, 54, 409–429, 1978.

182. Felder, J.M., Felder, D.L., and Hand, S.C., Ontogeny of osmoregulation in the estuarine ghost shrimp *Callianassa jamaicense louisianensis* Schmitt (Decapoda, Thalassinidea), *J. Exp. Mar. Biol. Ecol.*, 99, 91–105, 1986.

183. Felgenhauer, B.E., Internal anatomy of the Decapoda: an overview, in *Decapod Crustacea: Microscopic Anatomy of Invertebrates*, Harrison, F.W. and Humes, A.G., Eds., Wiley-Liss, New York, 1992, pp. 45–75.

184. Felten, V. and Guerold, F., Hyperventilation and loss of hemolymph Na^+ and Cl^- in the freshwater amphipod *Gamarus fossarum* exposed to acid stress: a preliminary study, *Dis. Aquat. Org.*, 45, 77–80, 2001.

185. Felten, V. and Guerold, F., Haemolymph $[Na^+]$ and $[Cl^-]$ loss in *Gammarus fossarum* exposed *in situ* to a wide range of acidic streams, *Dis. Aquat. Org.*, 61, 113–121, 2004.

186. Ferraris, R.P., Parado-Estepa, F.D., De Jesus, E.G., and Ladja, J.M., Osmotic and chloride regulation in the hemolymph of the tiger prawn *Penaeus monodon* during molting in various salinities, *Mar. Biol.*, 95, 377–385, 1987.

187. Fingerman, M., Glands and secretions, in *Microscopic Anatomy of Invertebrates*, Harrison, F.W., Ed., Willey-Liss, New York, 1992, pp. 345–394.

188. Finol, H.J. and Crogham, P.C., Ultrastructure of the branchial epithelium of an amphibious brackish-water crab, *Tissue Cell*, 15, 63–75, 1983.

189. Fioroni, P., The dorsal organ of arthropods with special reference to Crustacea Malacostraca: a comparative survey, *Zool. Jb. Anat.*, 104, 425–465, 1980.

190. Fisher, J.M., Fine structural observations on the gills of the fresh-water crayfish *Astacus pallipes* Lereboullet, *Tissue Cell*, 4, 287–299, 1972.

191. Flik, G. and Haond, C., Na^+ and Ca^{2+} pumps in the gills, epipodites and branchiostegites of the European lobster *Homarus gammarus*: effects of dilute sea water, *J. Exp. Biol.*, 203, 213–220, 2000.

192. Foskett, J.K., Osmoregulation in the larvae and adults of the grapsid crab *Sesarma reticulatum* Say, *Biol. Bull.*, 153, 505–526, 1977.

193. Foster, B.A., Responses and acclimation to salinity in the adults of some balanomorph barnacles, *Philos. Trans. Roy. Soc. Lond. B*, 256, 377–400, 1970.

194. Foster, C.A. and Howse, H.D., A morphological study on gills of the brown shrimp, *Penaeus aztecus*, *Tissue Cell*, 10, 77–92, 1978.

195. Fox, H.N., Anal and oral intake of water by Crustacea, *J. Exp. Biol.*, 29, 583–599, 1952.

196. Franklin, S.E., Tiensongrusmee, B., and Lockwood, A.P.M., Inhibition of magnesium secretion in the prawn *Palaemon serratus* by ethacrinic acid and by ligature of the eyestalks, in *Comparative Physiology: Water, Ions and Fluid Mechanics*, Schmidt Nielsen, K., Bolis, L., and Maddrell, S.H., Cambridge University Press, Cambridge, U.K., 1978.

197. Freire, C.A. and McNamara, J.C., Fine structure of the gills of the fresh-water shrimp *Macrobrachium olfersii* (Decapoda): effect of acclimation to high salinity medium and evidence for involvement of the lamellar septum in ion uptake, *J. Crust. Biol.*, 15, 103–116, 1995.

198. Frick, J.H. and Sauer, J.R., Examination of a biological cryostat/nanoliter osmometer for use in determining the freezing point of insect hemolymph, *Ann. Entomol. Soc. Am.*, 66, 781–783, 1973.

199. Fuller, E.G., Highison, G.J., Brown, F., and Bayer, C., Ultrastructure of the crayfish antennal gland revealed by scanning and transmission electron microscopy combined with ultrasonic microdissection, *J. Morphol.*, 200, 9–15, 1989.

200. Fyhn, H.J., Holeurhyalinity and its mechanisms in a cirriped crustacean *Balanus improvisus*, *Comp. Biochem. Physiol.*, 53A, 19–30, 1976.

201. Garçon, D.P., Masui, D.C., Mantelatto, F.L., McNamara, J.C., Furriel, R.P., and Leone, F.A., K+ and NH$_4^+$ modulate gill (Na+,K+)-ATPase activity in the blue crab, *Callinectes ornatus*: fine tuning of ammonia excretion, *Comp. Biochem. Physiol.*, 147A, 145–155, 2007.

202. Garrey, W.E., The osmotic pressure of sea water and of the blood of marine animals, *Biol. Bull.*, 8, 257–270, 1905.

203. Geddes, M.C., Salinity tolerance and osmotic and ionic regulation in *Branchinella australiensis* and *B. compacta* (Crustacea: Anostraca), *Comp. Biochem. Physiol.*, 45A, 559–569, 1973.

204. Geddes, M.C., Studies on an Australian brine shrimp *Parartemia zietziana* Sayce (Crustacea: Anostraca). II. Osmotic and ionic regulation, *Comp. Biochem. Physiol.*, 51A, 561–571, 1975.

205. Geddes, M.C., Studies of an Australian brine shrimp *Parartemia zietziana* Sayce (Crustacea: Anostraca). III. The mechanisms of osmotic and ionic regulation, *Comp. Biochem. Physiol.*, 51A, 573–578, 1975.

206. Genovese, G., Luchetti, C.G., and Luquet, C.M., Na+/K+-ATPase activity and gill ultrastructure in the hyper–hyporegulating crab *Chasmagnatgus granulatus* acclimated to dilute, normal, and concentrated seawater, *Mar. Biol.*, 144, 111–118, 2004.

207. Genovese, G., Luquet, C.M., Paz, D.A., Rosa, G.A., and Pellerano, G.N., The morphometric changes in the gills of the estuarine crab *Chasmagnathus granulatus* under hyper- and hyporegulation conditions are not caused by proliferation of specialised cells, *J. Anat.*, 197, 239–246, 2000.

208. Genovese, G., Ortiz, N., Urcola, M.R., and Luquet, C.M., Possible role of carbonic anhydrase, V-H+-ATPase, and Cl−/HCO$_3^-$ exchanger in electrogenic ion transport across the gills of the euryhaline crab *Chasmagnathus granulatus*, *Comp. Biochem. Physiol.*, 142A, 362–369, 2005.

209. Gifford, C.A., Some aspects of osmotic and ionic regulation in the blue crab *Callinectes sapidus*, and the ghost crab *Ocypode albicans*, *Bull. Inst. Mar. Sci.*, 8, 97–125, 1962.

210. Gilles, R., Métabolisme des acides aminés et contrôle du volume cellulaire, *Arch. Intern. Physiol. Bioch.*, 82, 423–589, 1974.

211. Gilles, R., Intracellular organic osmotic effectors, in *Mechanisms of Osmoregulation in Animals*, Gilles, R., Ed., Wiley Interscience, London, 1979, pp. 11–153.

212. Gilles, R., "Compensatory" organic osmolytes in high osmolarity and dehydration stresses: history and perspectives, *Comp. Biochem. Physiol.*, 117A, 279–290, 1997.

213. Gilles, R. and Delpire, E., Variation in salinity, osmolarity and water availability: vertebrates and invertebrates, in *Handbook of Comparative Physiology*, Dantzler, W.H., Ed., Oxford University Press, New York, 1997, pp. 1523–1586.

214. Gilles, R. and Péqueux, A., Interactions of chemical and osmotic regulation with the environment, in *Environmental Adaptations*, Vernberg, F. and Vernberg, W.B., Eds., Academic Press, New York, 1983, pp. 109–177.

215. Glas, P.S., Courtney, L.E., Rayburn, J.R., and Fisher, W.S., Embryonic coat of the grass shrimp *Palaemonetes pugio*, *Biol. Bull.*, 192, 231–242, 1997.

216. Glenner, H., Thomsen, P.F., Hebsgaard, M.B., Sorensen, M.V., and Willerslev, E., The origin of insects, *Science*, 314, 1883–1884, 2006.

217. Glover, C.N. and Wood, C.M., Physiological characterisation of a pH- and calcium-dependent sodium uptake mechanism in the freshwater crustacean, *Daphnia magna*, *J. Exp. Biol.*, 208, 951–959, 2005.

218. Goodman, S.H. and Cavey, M.J., Organization of a phyllobranchiate gill from the green shore crab *Carcinus maenas* (Crustacea, Decapoda), *Cell Tissue Res.*, 260, 495–505, 1990.

219. Goodrich, E.S., The study of the nephridia and the genital ducts since 1895, *Q. J. Microsc. Sci.*, 86, 113–301, 1945.

220. Greco, T.M., Alden, M.B., and Holliday, C.W., Control of hemolymph volume by adjustments in urinary and drinking rates in the crab *Cancer borealis*, *Comp. Biochem. Physiol.*, 84A, 695–701, 1986.

221. Green, J.W., Harsch, H., Barr, L., and Prosser, C.L., The regulation of water and salt by the fiddler crabs *Uca pugnax* and *Uca pugilator*, *Biol. Bull.*, 116, 76–87, 1959.

222. Greenaway, P., Sodium regulation in the freshwater/land crab *Holthuisana transversa*, *J. Comp. Physiol.*, 142B, 451–456, 1981.

223. Greenaway, P., Salt and water balance in field populations of the terrestrial crab *Gecarcoidea natalis*, *J. Crust. Biol.*, 14, 438–453, 1994.

224. Greenaway, P., Terrestrial adaptations in the Anomoura (Crustacea: Decapoda), *Memoirs Museum Vistoria*, 60, 13–26, 2003.

225. Greenaway, P., Taylor, H.H., and Morris, S., Adaptation to a terrestrial existence by the robber crab *Birgus latro*. VI. The role of the excretory system in fluid balance, *J. Exp. Biol.*, 64, 767–786, 1990.

226. Gross, P.S., Bartlett, T.C., Browdy, C.L., Chapman, R.W., and Warr, G.W., Immune gene discovery by expressed sequence tag analysis of hemocytes and hepatopancreas in the Pacific white shrimp, *Litopenaeus vannamei*, and the Atlantic white shrimp, *L. setiferus*, *Dev. Comp. Immunol.*, 25, 565–577, 2001.

227. Gross, W.J., An analysis of response to osmotic stress in selected decapod Crustacea, *Biol. Bull.*, 112, 43–62, 1957.

228. Gross, W.J., Glucose absorption from the urinary bladder in a crab, *Comp. Biochem. Physiol.*, 20, 313–317, 1967.

229. Gross, W.J. and Capen, R.L., Some functions of the urinary bladder in a crab, *Biol. Bull. Woods Hole*, 131, 272–291, 1966.

230. Gross, W.J. and Marshall, L.A., The influence of salinity on the magnesium and water fluxes of a crab, *Biol. Bull., Woods Hole*, 119, 438–446, 1960.

231. Guinot, D., Les crabes des sources hydrothermales de la dorsale du Pacifique oriental (campagne Biocyarise, 1984), *Oceanol. Acta*, 8, 109–118, 1988.

232. Guinot, D., *Austinograea alayseae* sp. nov., Crabe hydrothermal découvert dans le bassin de Lau, Pacifique sud-occidental (Crustacea, Decapoda, Brachyura), *Bull. Mus. Natl. Hist. Nat.*, 11, 879–903, 1990.

233. Haefner, P. and Schuster, D., Length increments during terminal moult of the female blue crab *Callinectes sapidus* in different salinity environments, *Chesapeake Sci.*, 5, 114–118, 1964.

234. Hagerman, L., Osmoregulation and sodium balance in *Crangon vulgaris* (Fabricius) (Crustacea, Natantia) in varying salinities, *Ophelia*, 9, 21–30, 1971.

235. Hagerman, L. and Larsen, M., The urinary flow in *Crangon vulgaris* (Fabr.) (Crustacea, Natantia) during the molt cycle, *Ophelia*, 16, 143–150, 1977.

236. Hagerman, L. and Uglow, R.F., Ventilatory behaviour and chloride regulation in relation to oxygen tension in the shrimp *Palaemon adspersus* Rathke maintained in hypotonic medium, *Ophelia*, 20, 193–200, 1981.

237. Hagerman, L. and Uglow, R.F., Effects of hypoxia on osmotic and ionic regulation in the brown shrimp *Crangon crangon* (L.) from brackish water, *J. Exp. Mar. Biol. Ecol.*, 63, 93–104, 1982.

238. Hannan, J. and Evans, D., Water permeability in some euryhaline decapods and *Limulus polyphemus*, *Comp. Biochem. Physiol.*, 44A, 1199–1213, 1973.

239. Haond, C., Bonnal, L., Sandeaux, R., Charmantier, G., and Trilles, J.-P., Ontogeny of intracellular isosmotic regulation in the European lobster *Homarus gammarus* (L.), *Physiol. Biochem. Zool.*, 72, 534–544, 1999.

240. Haond, C., Flik, G., and Charmantier, G., Confocal laser scanning and electron microscopical studies on osmoregulatory epithelia in the branchial cavity of the lobster *Homarus gammarus*, *J. Exp. Biol.*, 201, 1817–1833, 1998.

241. Harms, J.W., Die Realisation von Genen und die consecutive Adaptation. II. *Birgus latro* L. als Land Krebs und seine Beziehungen den Coenobiten, *Zeitsch. F. Wissensch. Zool.*, 140, 167–290, 1932.

242. Harris, R.R., Aspects of sodium regulation in a brackish-water and a marine species of the isopod genus *Sphaeroma*, *Mar. Biol.*, 12, 18–27, 1972.

243. Harris, R.R., Urine production rate and urinary sodium loss in the freshwater crab *Potamon edulis*, *J. Comp. Physiol.*, 96, 143–153, 1975.

244. Harris, R.R. and Bayliss, D., Gill (Na$^+$+K$^+$)-ATPases in decapod crustaceans: distribution and characteristics in relation to Na$^+$ regulation, *Comp. Biochem. Physiol.*, 90A, 303–308, 1988.

245. Harris, R.R. and Micaleff, H., Osmotic and ionic regulation in *Potamon edulis*, a fresh water crab from Malta, *Comp. Biochem. Physiol.*, 38A, 127–129, 1971.

246. Harris, R.R. and Santos, M.C.F., Ionoregulatory and urinary responses to emersion in the mangrove crab *Ucides cordatus* and the intertidal crab *Carcinus maenas*, *J. Comp. Physiol.*, 163B, 18–27, 1993.

247. Harris, R.R. and Santos, M.C.F., Sodium uptake transport (Na$^+$+K$^+$) ATPase changes following Na+ depletion and low salinity acclimation in the mangrove crab *Ucides cordatus* (L.), *Comp. Biochem. Physiol.*, 105A, 35–42, 1993.

248. Heeg, J. and Cannone, J., Osmoregulation by means of a hitherto unsuspected osmoregulatory organ in two grapsid crabs, *Zoologica Africana*, 2, 127–129, 1966.

249. Heit, M. and Fingerman, M., The role of an eyestalk hormone in the regulation of the sodium concentration of the blood of the fiddler crab, *Uca pugilator*, *Comp. Biochem. Physiol.*, 50A, 277–280, 1975.

250. Henry, R.P., Membrane-associated carbonic anhydrase in gills of the blue crab *Callinectes sapidus*, *Am. J. Physiol.*, 252, R966–R971, 1987.

251. Henry, R.P., Subcellular distribution of carbonic anhydrase activity in the gills of the blue crab, *Callinectes sapidus*, *J. Exp. Zool.*, 245, 1–8, 1988.

252. Henry, R.P., Environmentally mediated carbonic anhydrase induction in the gills of euryhaline crustaceans, *J. Exp. Biol.*, 204, 991–1002, 2001.

253. Henry, R.P. and Cameron, J.N., The distribution and partial characterization of carbonic anhydrase in selected aquatic and terrestrial decapod crustaceans, *J. Exp. Zool.*, 221, 309–321, 1982.

254. Henry, R.P., Garrelts, E.E., McCarthy, M.M., and Towle, D.W., Differential time course of induction for branchial carbonic anhydrase and Na/K-ATPase activity in the euryhaline crab, *Carcinus maenas*, during low salinity acclimation, *J. Exp. Zool.*, 292, 595–603, 2002.

255. Henry, R.P., Gehnrich, S., Weihrauch, D., and Towle, D.W., Salinity-mediated carbonic anhydrase induction in the gills of the euryhaline green crab, *Carcinus maenas*, *Comp. Biochem. Physiol.*, 136A, 243–258, 2003.

256. Henry, R.P., Jackson, S.A., and Mangum, C.P., Ultrastructure and transport-related enzymes of the gill and coxal gland of the horseshoe crab *Limulus polyphemus*, *Biol. Bull.*, 191, 241–250, 1996.

257. Henry, R.P., Thomason, K.L., and Towle, D.W., Quantitative changes in branchial carbonic anhydrase activity and expression in the euryhaline green crab, *Carcinus maenas*, in response to low salinity exposure, *J. Exp. Zool.*, 305, 842–850, 2006.

258. Henry, R.P. and Wheatly, M.G., Dynamics of salinity adaptations in the euryhaline crayfish *Pacifastacus leniusculus*, *Physiol. Zool.*, 61, 260–271, 1988.

259. Henry, R.P. and Wheatly, M.G., Interaction of respiration, ion regulation and acid–base balance in the everyday life of aquatic crustaceans, *Am. Zool.*, 32, 407–416, 1992.

260. Hoese, B. and Janssen, H.H., Morphological and physiological studies on the marsupium in terrestrial isopods, *Monit. Zool. Ital. (N.S.) Monograf.*, 4, 153–173, 1989.

261. Holdich, D.M., Harlioglu, M.M., and Firkins, I., Salinity adaptations of crayfish in British waters with particular reference to *Austrapotamobius pallipes*, *Astacus leptodactylus* and *Pacifastacus leniusculus*, *Estuar. Coast. Shelf Sci.*, 44, 147–154, 1997.

262. Holdich, D.M. and Mayes, K.R., A fine structural re-examination of the so-called midgut of the isopod *Porcellio*, *Crustaceana*, 29, 186–192, 1975.

263. Holdich, D.M. and Ratcliffe, N., A light and electron microscopic study of the hindgut of the herbivorous isopod *Dynamene bidentata* (Crustacea, Pericarida), *Z. Zellforsch.*, 11, 209–227, 1970.

264. Holleland, T. and Towle, D.W., Vanadate but not ouabain inhibits Na+K+-ATPase and sodium transport in tight inside-out native membrane vesicles from crab gill (*Carcinus maenas*), *Comp. Biochem. Physiol.*, 96B, 177–181, 1990.

265. Holliday, C.W., Aspects of Antennal Gland Function in a Crab, *Cancer magister* (Dana), Ph.D. dissertation, University of Oregon, 1978.

266. Holliday, C.W., Glucose absorption by the bladder of the crab *Cancer magister* (Dana), *Comp. Biochem. Physiol.*, 61A, 73–79, 1978.

267. Holliday, C.W., Magnesium transport by the urinary bladder of the crab *Cancer magister*, *J. Exp. Biol.*, 86, 187–201, 1980.

268. Holliday, C.W., Salinity-induced changes in gill Na,K-ATPase activity in the mud fiddler crab, *Uca pugnax*, *J. Exp. Zool.*, 233, 199–208, 1985.

269. Holliday, C.W., Branchial Na+/K+-ATPase and osmoregulation in the Isopod *Idotea wosnesenskii*, *J. Exp. Biol.*, 136, 259–272, 1988.

270. Holliday, C.W. and Miller, D.S., Cellular mechanisms of organic anion transport in crustacean renal tissue, *Am. Zool.*, 24, 275–284, 1984.

271. Holliday, C.W., Mykles, D.L., Terwilliger, N.B., and Dangott, L.J., Fluid secretion by the midgut caeca of the crab, *Cancer magister*, *Comp. Biochem. Physiol.*, 67A, 259–263, 1980.

272. Hopkins, J.S., Stokes, A.D., Browdy, C.L., and Sandifer, P.A., The relationship between feeding rate, paddle-wheel aeration rate and expected dawn dissolved oxygen in intensive shrimp ponds, *Aquacult. Eng.*, 10, 281–290, 1991.

273. Horne, F.R., Some aspects of osmotic regulation in the tadpole shrimp *Triops longicaudatus*, *Comp. Biochem. Physiol.*, 19, 313–316, 1966.

274. Huf, E., Der Einfluss des mechanischen Innendrucks auf die Flûssigkeitsausscheidung bei gepanzerten Süsswasser- und Meereskrebsen, *Pflüg. Arch. Ges. Physiol.*, 237, 240–250, 1936.

275. Hung, C.Y., Tsui, K.N., Wilson, J.M., Nawata, C.M., Wood, C.M., and Wright, P.A., Rhesus glyco-protein gene expression in the mangrove killifish *Kryptolebias marmoratus* exposed to elevated environmental ammonia levels and air, *J. Exp. Biol.*, 210, 2419–2429, 2007.

276. Hunter, K.C. and Kirschner, L.B., Sodium absorption coupled to ammonia excretion in osmoconform-ing marine invertebrates, *Am. J. Physiol.*, 251, R957–R962, 1986.

277. Hunter, K.C. and Rudy, P.P.J., Osmotic and ionic regulation in the Dungeness crab, *Cancer magister* Dana, *Comp. Biochem. Physiol.*, 51A, 439–447, 1975.

278. Huong, D.T.T., Jayasanker, V., Jasmani, S., Saido-Sakanaka, H., Wigginton, A.J., and Wilder, M.N., Na/K-ATPase activity during larval development in the giant freshwater prawn *Macrobrachium rosen-bergii* and the effects of salinity on survival rates, *Fish. Sci.*, 70, 518–520, 2004.

279. Icely, J.D. and Nott, J.A., The general morphology and fine structure of the antennary gland of *Corophium volutator* (Amphipoda: Crustacea), *J. Mar. Biol. Assoc. U.K.*, 59, 745–756, 1979.

280. Jackson, S.A., Branchial ultrastructure and the presence of Na⁺-K⁺-ATPase in the horseshoe crab, *Limulus polyphemus* (Lineaeus), *Va. J. Sci.*, 35, 83, 1984.

281. Jayasundara, N., Towle, D.W., Weihrauch, D., and Spanings-Pierrot, C., Gill-specific transcriptional regulation of Na⁺+K⁺-ATPase α-subunit in the euryhaline shore crab *Pachygrapsus marmoratus*: sequence variants and promoter structure, *J. Exp. Biol.*, 210, 2070–2081, 2007.

282. Jegla, T.C. and Costlow, J.D., Temperature and salinity effects on development and early posthatch stages of *Limulus*, in *Physiology and Biology of Horseshoe Crabs: Studies on Normal and Environ-mentally Stressed Animals*, Bonaventura, J., Bonaventura, C., and Tesh, S., Alan R. Liss, New York, 1982, pp. 103–113.

283. Johnson, I. and Uglow, R.F., The effect of hypoxia on ion regulation and acid–base balance in *Carcinus maenas*, *Comp. Biochem. Physiol.*, 86A, 261–268, 1987.

284. Johnson, P.T., *Histology of the Blue Crab* Callinectes sapidus *Praeger*, New York, 1980.

285. Jones, L.L., Osmotic regulation in several crabs of the Pacific coast of North America, *J. Cell. Comp. Physiol.*, 18, 79–92, 1941.

286. Kalber, F.A., Osmoregulation in decapod larvae as a consideration in culture techniques, *Helgoländer Meeresun.*, 20, 697–706, 1970.

287. Kalber, F.A. and Costlow, J.D., The ontogeny of osmoregulation and its neurosecretory control in the decapod crustacean *Rhithropanopeus harrisii*, *Am. Zool.*, 6, 221–229, 1966.

288. Kalber, F.A. and Costlow, J.D.J., Osmoregulation in larvae of the land crab, *Cardisoma guanhumi* Latreille, *Am. Zool.*, 8, 411–416, 1968.

289. Kamemoto, F.I., Neuroendocrinology of osmoregulation in Decapod Crustacea, *Am. Zool.*, 16, 141–150, 1976.

290. Kamemoto, F.I., Crustacean neuropeptides and osmoregulation, in *Neurosecretion: Molecules, Cells, Systems*, Farmer, D.S. and Lederis, K., Eds., Plenum, New York, 1982, pp. 329–335.

291. Kamemoto, F.I., Neuroendocrinology of osmoregulation in crabs, *Zool. Sci.*, 8, 827–833, 1991.

292. Kamemoto, F.I. and Kato, K.N., The osmotic and chloride regulation capacities of five Hawaiian decapod crustaceans, *Pacific Sci.*, 23, 232–237, 1969.

293. Kamemoto, F.I., Kato, K.N., and Tucker, L.E., Neurosecretion and salt and water balance in the Annelida and Crustacea, *Am. Zool.*, 6, 213–219, 1966.

294. Kamemoto, F.I. and Ono, J.K., Urine flow determinations by continuous collection in the crayfish *Procambarus clarkii*, *Comp. Biochem. Physiol.*, 27, 851–857, 1968.

295. Kamemoto, F.I. and Oyama, S.N., Neuroendocrine influence on effector tissues of hydromineral balance in crustaceans, in *Current Trends in Comparative Endocrinology*, Lofts, B. and Holmes, W.N., Eds., Hong Kong University Press, Hong Kong, 1985, pp. 883–886.

296. Kamemoto, F.I. and Tullis, R.E., Hydromineral regulation in Decapod Crustacea, *Gen. Comp. Endo-crinol.*, 3(Suppl.), 299–307, 1972.

297. Karansas, J.J., van Dyke, H., and Worrest, R.C., Mid-ultraviolet (UV-B) sensitivity of *Acartia clausii* Giesbrecht (Copepoda), *Limnol. Oceanogr.*, 24, 1104–1116, 1979.

298. Kato, K.N. and Kamemoto, F.I., Neuroendocrine involvement in osmoregulation in the grapsid crab *Metopograpsus messor*, *Comp. Biochem. Physiol.*, 28, 665–674, 1969.

299. Keller, R., Crustacean neuropeptides: structures, functions and comparative aspects, *Experientia*, 48, 439–448, 1992.

300. Keller, R., Jaros, P.P., and Kegel, G., Crustacean hyperglycemic neuropeptides, *Am. Zool.*, 25, 207–221, 1985.

301. Kelley, B.J.J. and Burbanck, W.D., Osmoregulation in juvenile and adult *Cyathura polita* (Stimpson) subjected to salinity changes and ionizing gamma irradiation (Isopoda, Anthuridea), *Chesapeake Sci.*, 13, 201–205, 1972.

302. Kelley, B.J.J. and Burbanck, W.D., Responses of embryonic *Cyathura polita* (Stimpson) (Isopoda: Anthuridea) to varying salinities, *Chesapeake Sci.*, 17, 159–167, 1976.

303. Kelly, R.H. and Yancey, P.H., High content of trimethylamine oxide correlating with depth in deep-sea teleost fishes, skates and decapod crustaceans, *Biol. Bull.*, 196, 18–25, 1999.

304. Kerley, D.E. and Pritchard, A.W., Osmotic regulation in the crayfish, *Pacifastacus leniusculus*, stepwise acclimated to dilutions of seawater, *Comp. Biochem. Physiol.*, 20, 101–113, 1967.

305. Khodabandeh, S., Charmantier, G., Blasco, C., Grousset, E., and Charmantier-Daures, M., Ontogeny of the antennal glands in the crayfish *Astacus leptodactylus* (Crustacea, Decapoda): anatomical and cell differentiation, *Cell Tissue Res.*, 319, 153–165, 2005.

306. Khodabandeh, S., Charmantier, G., and Charmantier-Daures, M., Ultrastructural studies and Na⁺,K⁺-ATPase immunolocalization in the antennal urinary glands of the lobster *Homarus gammarus* (Crustacea, Decapoda), *J. Histochem. Cytochem.*, 53, 1203–1214, 2005.

307. Khodabandeh, S., Charmantier, G., and Charmantier-Daures, M., Immunolocalization of Na⁺/K⁺-ATPase in osmoregulatory organs during the embryonic and postembryonic development of the lobster *Homarus gammarus*, *J. Crust. Biol.*, 26, 515–523, 2006.

308. Khodabandeh, S., Kutnik, M., Aujoulat, F., Charmantier, G., and Charmantier-Daures, M., Ontogeny of the antennal glands in the crayfish *Astacus leptodactylus* (Crustacea, Decapoda). Immunolocalization of Na⁺K⁺-ATPase, *Cell Tissue Res.*, 319, 167–174, 2005.

309. Kikuchi, S., The fine structure of the gill epithelium of a fresh-water flea, *Daphnia magna* (Crustacea: Phyllopoda) and changes associated with acclimation to various salinities, *Cell Tissue Res.*, 229, 253–268, 1983.

310. Kikuchi, S. and Matsumasa, M., Two ultrastructurally distinct types of transporting tissues, the branchiostegal and the gill epithelia, in an estuarine tanaid, *Sinelobus stanfordi* (Crustacea, Peracarida), *Zoomorphology*, 113, 253–260, 1993.

311. Kikuchi, S. and Matsumasa, M., Pereopodal disk: a new type of extrabranchial ion-transporting organ in an estuarine amphipod, *Melita setiflagella* (crustacea), *Tissue Cell*, 27, 635–643, 1995.

312. Kikuchi, S., Matsumasa, M., and Yashima, Y., The ultrastructure of the sternal gills form a striking contrast with the coxal gills in a fresh-water amphipod (Crustacea), *Tissue Cell*, 25, 915–928, 1993.

313. Kimura, C., Ahearn, G.A., Busquets-Turner, L., Haley, S.R., Nagao, C., and De Couet, H.G., Immunolocalization of an antigen associated with the invertebrate electrogenic 2Na⁺–1H⁺ antiporter, *J. Exp. Biol.*, 189, 85–104, 1994.

314. Kinsey, S.T., Buda, E., and Nordeen, J., Scaling of gill metabolic potential as a function of salinity in the euryhaline crab, *Callinectes sapidus* Rathbun, *Physiol. Biochem. Zool.*, 76, 105–114, 2003.

315. Kirschner, L.B., Sodium–proton exchange in crayfish, *Biochim. Biophys. Acta*, 1566, 67–71, 2002.

316. Koch, H.J., Essai d'interprétation de la soi-disant réduction vitale des sels d'argent par certains organes d'Arthropodes, *Annales Société Royale des Sciences Médicales et Naturelles de Bruxelles*, 54, Ser. B, 346–361, 1934.

317. Koch, H.J., Cholinesterase and active transport of sodium chloride through the isolated gills of the crab *Eriocheir sinensis* (M. Edw.), in *Recent Developments in Cell Physiology*, Kitching, J.A., Ed., Butterworths, London, 1954, pp. 15–27.

318. Koch, H.J., Evans, J., and Schicks, E., The active absorption of ions by the isolated gills of the crab *Eriocheir sinensis* (M. Edw.), *Meded. Vlaamse Acad. Kl. Wet.*, 16, 3–16, 1954.

319. Krogh, A., The active absorption of ions in some freshwater animals, *Z. Vergleich. Physiol.*, 25, 335–350, 1938.

320. Kumlu, M. and Jones, D.A., Salinity tolerance of hatchery-reared postlarvae of *Penaeus indicus* H. Milne Edwards originating from India, *Aquaculture*, 130, 287–296, 1995.

321. Kummel, G., Das Cölomosäckchen der Antenndrüse von *Cambarus affinis* Say (Decapoda: Crustacea). Eine elektronmikroskopische Untersuchung mit einer Diskussion über die Funktion, *Zool. Beitr.*, 10, 227–252, 1964.

322. Lacombe, C., Grève, P., and Martin, G., Overview of the sub-grouping of the crustacean hyperglycemic hormone family, *Neuropeptides*, 33, 71–80, 1999.

323. Lake, P.S., Swain, R., and Ong, J.E., The ultrastructure of the fenestra dorsalis of the syncarid crustaceans *Allanaspides helonomus* and *Allanaspides hickmani*, *Z. Zellforsch.*, 147, 335–351, 1974.

324. Lance, J., Respiration and osmotic behavior of the copepod *Acartia tonsa* in diluted sea water, *Comp. Biochem. Physiol.*, 14, 155–165, 1965.

325. Laughlin, R., The effects of temperature and salinity on larval growth of the horseshoe crab *Limulus polyphemus*, *Biol. Bull.*, 164, 93–103, 1983.

326. Lawson, S.L., Jones, M.B., and Moate, R.M., Effect of copper on the ultrastructure of the gill epithelium of *Carcinus maenas* (Decapoda: Brachyra), *Mar. Pollut. Bull.*, 31, 63–72, 1995.

327. Lee, B.D. and McFarland, W.N., Osmotic and ionic concentration in the mantis shrimp *Squilla empusa* Say, *Publ. Inst. Mar. Sci. Univ. Texas*, 8, 126–142, 1962.

328. Lee, K.J. and Watts, S.A., Specific activity of Na⁺,K⁺ ATPase is not altered in response to changing salinities during early development of the brine shrimp *Artemia franciscana*, *Physiol. Zool.*, 67, 910–924, 1994.

329. Lee, S.-H. and Pritchard, J.B., Bicarbonate-chloride exchange in gill plasma membranes of blue crab, *Am. J. Physiol.*, 249, R544–R550, 1985.

330. Lesser, M.P. and Barry, T.M., Survivorship, development, and DNA damage in echinoderm embryos and larvae exposed to ultraviolet radiation (290–400 nm), *J. Exp. Mar. Biol. Ecol.*, 292, 75–91, 2003.

331. Li, T., Roer, R.D., Vana, M., Pate, S., and Check, J., Gill area, permeability and Na⁺,K⁺-ATPase activity as a function of size and salinity in the blue crab, *Callinectes sapidus*, *J. Exp. Zool.*, 305A, 233–245, 2006.

332. Lignon, J.M., Structure and permeability of the decapod crustacean cuticle, in *Comparative Physiology of Environmental Adaptation*. Vol. 1. *Adaptation to Salinity and Dehydration*, Kirsch, R. and Lalhou, B., Eds., Karger, Basel, 1987, pp. 178–187.

333. Lignon, J.M. and Péqueux, A., Permeability properties of the cuticle and gill ion exchanges in decapod crustaceans, in *Animal Nutrition and Transport Processes*. Vol. 2. *Transport, Respiration and Excretion: Comparative and Environmental Aspects*, Truchot, J.-P. and Lalhou, B., Eds., Karger, Basel, 1990, pp. 14–27.

334. Lignon, J.M., Péqueux, A., and Gendner, J.-P., Perméabilité cuticulaire branchiale aux ions chez les Crustacés Décapodes et salinité du milieu environnant, *Océanis*, 14, 487–503, 1988.

335. Lignot, J.-H. and Charmantier, G., Immunolocalization of Na⁺,K⁺-ATPase in the branchial cavity during the early development of the European lobster *Homarus gammarus* (Crustacea, Decapoda), *J. Histochem. Cytochem.*, 49, 1013–1023, 2001.

336. Lignot, J.-H., Charmantier-Daures, M., and Charmantier, G., Immuno-localization of Na⁺,K⁺-ATPase in the organs of the branchial cavity of the European lobster *Homarus gammarus* (Crustacea, Decapoda), *Cell Tissue Res.*, 296, 417–426, 1999.

337. Lignot, J.-H., Spanings-Pierrot, C., and Charmantier, G., Osmoregulatory capacity as a tool in monitoring the physiological condition and the effect of stress in crustaceans, *Aquaculture*, 191, 209–245, 2000.

338. Lignot, J.-H., Susanto, G.N., Charmantier-Daures, M., and Charmantier, G., Immunolocalization of Na⁺,K⁺-ATPase in the branchial cavity during the early development of the crayfish *Astacus leptodactylus* (Crustacea, Decapoda), *Cell Tissue Res.*, 319, 331–339, 2005.

339. Likens, G.E., Butker, T.J., and Buso, D.C., Long- and short-term changes in sulphate deposition: effects of the 1990 Clean Air Act Amendments, *Biogeochemistry*, 52, 1–11, 2001.

340. Lin, H.C., Su, Y.C., and Su, S.H., A comparative study of osmoregulation in four fiddler crabs (Ocypodidae: *Uca*), *Zool. Sci.*, 19, 643–650, 2002.

341. Lin, H.P., Charmantier, G., Thuet, P., and Trilles, J.-P., Effects of turbidity on survival, osmoregulation and gill Na⁺-K⁺-ATPase in the shrimp *Penaeus japonicus*, *Mar. Ecol. Progr. Ser.*, 90, 31–37, 1992.

342. Linnane, A., Dimmlich, W., and Ward, T., Movement patterns of the southern rock lobster, *Jasus edwardsii*, off South Australia, *N. Z. J. Mar. Fresh. Res.*, 39, 335–346, 2005.

343. Livengood, D.R., Coupling ratio of the Na–K pump in the lobster cardiac ganglion, *J. Gen. Physiol.*, 82, 853–874, 1983.

344. Lockwood, A.P.M., Osmoregulation in gammarids, *J. Exp. Biol.*, 38, 647–658, 1961.

345. Lockwood, A.P.M., *Aspects of the Physiology of Crustacea*, Oliver & Boyd, London, 1968.

346. Lockwood, A.P.M. and Crogham, P.C., The chloride regulation of the brackish and fresh water races of *Mesidotea entomon* (L.), *J. Exp. Biol.*, 34, 253–258, 1957.

347. López Mañanes, A.A., Magnoni, L.J., and Goldemberg, A.L., Branchial carbonic anhydrase (CA) of gills of *Chasmagnathus granulata* (Crustacea Decapoda), *Comp. Biochem. Physiol.*, 127B, 85–95, 2000.

348. Lopina, O.D., Na$^+$,K$^+$-ATPase: structure, mechanism, and regulation, *Membr. Cell. Biol.*, 13, 721–744, 2000.

349. Lovett, D.L., Colella, T., Cannon, A.C., Lee, D.H., Evangelisto, A., Muller, E.M., and Towle, D.W., Effect of salinity on osmoregulatory patch epithelia in gills of the blue crab *Callinectes sapidus*, *Biol. Bull.*, 210, 132–139, 2006.

350. Lovett, D.L. and Felder, D.L., Ontogeny of kinematics in the gut of the white shrimp *Penaeus setiferus* (Decapoda, Penaeidae), *J. Crust. Biol.*, 10, 53–68, 1990.

351. Lovett, D.L. and Felder, D.L., Ontogenetic changes in enzyme distribution and midgut function in developmental stages of *Penaeus setiferus* (Crustacea, Decapoda, Penaeidae), *Biol. Bull.*, 178, 160–174, 1990.

352. Lovett, D.L., Verzi, M.P., Burgents, J.E., Tanner, C.A., Glomski, K., Lee, J.J., and Towle, D.W., Expression profiles of Na$^+$,K$^+$-ATPase during acute and chronic hypo-osmotic stress in the blue crab *Callinectes sapidus*, *Biol. Bull.*, 211, 58–65, 2006.

353. Lucu, C., Evidence for Cl$^-$ exchanges in perfused *Carcinus* gills, *Comp. Biochem. Physiol.*, 92A, 415–420, 1989.

354. Lucu, C. and Devescovi, M., Osmoregulation and branchial Na$^+$,K$^+$-ATPase in the lobster *Homarus gammarus* acclimated to dilute seawater, *J. Exp. Mar. Biol. Ecol.*, 234, 291–304, 1999.

355. Lucu, C., Devescovi, M., and Siebers, D., Do amiloride and ouabain affect ammonia fluxes in perfused *Carcinus* gill epithelia?, *J. Exp. Zool.*, 249, 1–5, 1989.

356. Lucu, C., Devescovi, M., Skaramuca, B., and Kozul, V.V., Gill Na,K-ATPase in the spiny lobster *Palinurus elephas* and other marine osmoconformers: adaptiveness of enzymes from osmoconformity to hyperregulation, *J. Exp. Mar. Biol. Ecol.*, 246, 163–178, 2000.

357. Lucu, C. and Flik, G., Na$^+$,K$^+$-ATPase and Na$^+$/Ca$_2^+$ exchange activities in gills of hyperregulating *Carcinus maenas*, *Am. J. Physiol.*, 276, R490–R499, 1999.

358. Lucu, C. and Siebers, D., Amiloride-sensitive sodium flux and potentials in perfused *Carcinus* gill preparations, *J. Exp. Biol.*, 122, 25–35, 1986.

359. Lucu, C. and Siebers, D., Linkage of Cl$^-$ fluxes with ouabain sensitive Na/K exchange through *Carcinus* gill epithelia, *Comp. Biochem. Physiol.*, 87A, 807–811, 1987.

360. Lucu, C., Siebers, D., and Sperling, K.R., Comparison of osmoregulation between Adriatic and North Sea *Carcinus*, *Mar. Biol.*, 22, 85–95, 1973.

361. Lucu, C. and Towle, D.W., Na$^+$+K$^+$-ATPase in gills of aquatic crustacea, *Comp. Biochem. Physiol.*, 135A, 195–214, 2003.

362. Luquet, C.M., Genovese, G., Rosa, G.A., and Pellerano, G.N., Ultrastructural changes in the gill epithelium of the crab *Chasmagnathus granulatus* (Decapoda: Grapsidae) in diluted and concentrated seawater, *Mar. Biol.*, 141, 753–760, 2002.

363. Luquet, C.M., Pellerano, G.N., and Rosa, G.A., Salinity-induced changes in the fine structure of the gills of the semiterrestrial estuarine crab *Uca uruguayensis*, *Tissue Cell*, 29, 495–501, 1997.

364. Luquet, C.M., Rosa, G.A., Ferrari, C.C., Genovese, G., and Pellerano, G.N., Gill morphology of the intertidal estuarine crab *Chasmagnathus granulatus* Dana, 1851 (Decapoda, Grapsidae) in relation to habitat and respiratory habits, *Crustaceana*, 73, 53–63, 2000.

365. Luquet, C.M., Weihrauch, D., Senek, M., and Towle, D.W., Induction of branchial ion transporter mRNA expression during acclimation to salinity change in the euryhaline crab *Chasmagnathus granulatus*, *J. Exp. Biol.*, 208, 3627–3636, 2005.

366. Lynch, M.P., Webb, K.L., and Van Engel, W.A., Variations in serum constituents of the blue crab *Callinectes sapidus*: chloride and osmotic concentration, *Comp. Biochem. Physiol.*, 44A, 719–734, 1973.

367. Macías, M.-T., Palmero, I., and Sastre, L., Cloning of a cDNA encoding an *Artemia franciscana* Na/K ATPase α-subunit, *Gene*, 105, 197–204, 1991.

368. Macins, A., Meredith, J., Zhao, Y., Brock, H.W., and Phillips, J.E., Occurrence of ion transport peptide (ITP) and ion transport-like peptide (ITP-L) in orthopteroids, *Arch. Insect Biochem. Physiol.*, 40, 107–118, 1999.

369. Mackay, W.C. and Prosser, C.L., Ionic and osmotic regulation in the king crab and two other North Pacific crustaceans, *Comp. Biochem. Physiol.*, 34, 273–280, 1970.

370. Mackintosh, C., Dynamic interactions between 14–3–3 proteins and phosphoproteins regulate diverse cellular processes, *Biochem. J.*, 381, 329–342, 2004.

371. Maina, J.N., The morphology of the gills of the freshwater African crab *Potamon niloticus* (Crustacea: Brachyura: Potamonidae): a scanning and transmission electron microscope study, *J. Zool. Lond.*, 221, 499–515, 1990.

372. Maissiat, J., Contribution à l'étude du rôle endocrine de la glande antennaire de l'Oniscoïde *Porcellio dilatatus* Brandt. Effet de son ablation sur l'équilibre hydrominéral, *C. R. Séanc. Soc. Biol.*, 166, 916–919, 1972.

373. Malley, D.F., Salt and water balance in the spiny lobster *Panulirus argus*: the role of the gut, *J. Exp. Biol.*, 70, 231–245, 1977.

374. Malo, M.E. and Fliegel, L., Physiological role and regulation of the Na^+/H^+ exchanger, *Can. J. Physiol. Pharmacol.*, 84, 1081–1095, 2006.

375. Maluf, N.S.R., On the anatomy of the kidney of the crayfish and on the absorption of chloride from fresh water by this animal, *Zool. Jahrb. (Allg. Zool. Physiol. Tiere)*, 59, 515–534, 1939.

376. Mangum, C.P., The functions of gills in several groups of invertebrates animals, in *Gills*, Houlihan, D., Rankin, K., and Shuttelworth, T., Eds., Cambridge University Press, Cambridge, U.K., 1982, pp. 77–97.

377. Mangum, C.P., Oxygen transport in the blood, in *The Biology of Crustacea*, Bliss, D.E. and Mantel, L.H., Eds., Academic Press, New York, 1983, pp. 373–429.

378. Mangum, C.P. and Amende, L.M., Blood osmotic concentration of the blue crabs *Callinectes sapidus* found in fresh water, *Chesapeake Sci.*, 13, 318–320, 1972.

379. Mangum, C.P., Booth, C.E., DeFur, P.L., Heckel, N.A., Henry, R.P., Oglesby, L.C., and Polites, G., The ionic environment of hemocyanin in *Limulus polyphemus*, *Biol. Bull.*, 150, 453–467, 1976.

380. Mantel, L.H., The foregut of *Gecarcinus lateralis* as an organ of salt and water balance, *Am. Zool.*, 8, 433–442, 1968.

381. Mantel, L.H., Bliss, D.E., Sheeham, S.W., and Martinez, E.A., Physiology of hemolymph, gut fluid, and hepatopancras of the land crab *Gecarcinus lateralis* (Fréminville) in various neuroendocrine states, *Comp. Biochem. Physiol.*, 51A, 663–671, 1975.

382. Mantel, L.H. and Farmer, L.L., Osmotic and ionic regulation, in *Internal Anatomy and Physiological Regulation*, Mantel, L.H., Ed., Academic Press, New York, 1983, pp. 53–161.

383. Mantel, L.H. and Olson, J.R., Studies on the Na^+,K^--activated ATPase of crab gills, *Am. Zool.*, 16, 223, 1976.

384. Marangos, C., Brogren, C.-H., Alliot, E., and Ceccaldi, H.J., The influence of water salinity on the free amino acid concentration in muscle and hepatopancreas of adult shrimps, *Penaeus japonicus*, *Biochem. Syst. Ecol.*, 17, 589–594, 1989.

385. Martelo, M.J. and Zanders, I.P., Modifications of gill ultrastructure and ionic composition in the crab *Goniopsis cruentata* acclimated to various salinities, *Comp. Biochem. Physiol.*, 84A, 383–389, 1986.

386. Martin, J.W., Branchiopoda, in *Microscopic Anatomy of Invertebrates*, Harrison, F.W. and Humes, A.G., Eds., Wiley-Liss, New York, 1992, pp. 25–224.

387. Martin, J.W. and Laverack, M.S., On the distribution of the crustacean dorsal organ, *Acta Zool.*, 73, 357–368, 1992.

388. Martinez, A.-S., Charmantier, G., Compère, P., and Charmantier-Daures, M., Branchial chambers ionocytes in two caridean shrimps: the epibenthic *Palaemon adspersus* and the deep-sea hydrothermal *Rimicaris exoculata*, *Tissue Cell*, 37, 153–165, 2005.

389. Martinez, A.-S., Toullec, J.-Y., Shillito, B., Charmantier-Daures, M., and Charmantier, G., Hydromineral regulation in the hydrothermal vent crab *Bythograea thermydron*, *Biol. Bull.*, 201, 167–174, 2001.

390. Martinez, C.B., Harris, R.R., and Santos, M.C., Transepithelial potential differences and sodium fluxes in isolated perfused gills of the mangrove crab (*Ucides cordatus*), *Comp. Biochem. Physiol.*, 120A, 227–236, 1998.

391. Masui, D., Furriel, R., McNamara, J., Mantelatto, F., and Leone, F., Modulation by ammonium ions of gill microsomal (Na$^+$,K$^+$)-ATPase in the swimming crab *Callinectes danae*: a possible mechanism for regulation of ammonia excretion, *Comp. Biochem. Physiol.*, 132C, 471–482, 2002.

392. McConnell, F.M., Morphometry of transport tissues in a freshwater crustacean, *Tissue Cell*, 19, 319–349, 1987.

393. McLaughlin, P.A., Internal anatomy, in *Internal Anatomy and Physiological Regulation*, Mantel, L.H., Ed., Academic Press, New York, 1983, pp. 1–52.

394. McLeese, D.W., Effects of temperature, salinity and oxygen on the survival of the American lobster, *J. Fish. Res. Board Can.*, 13, 247–272, 1956.

395. McLusky, D.S. and Heard, V.E.J., Some effects of salinity on the mysid *Praunus flexuosus*, *J. Mar. Biol. Assoc. U.K.*, 51, 709–715, 1971.

396. McMahon, B.R. and Doyle, J.E., Acid exposure in euryhaline environments: ion regulation and acid tolerance in larval and adult *Artemia franciscana*, *Hydrobiologia*, 352, 1–7, 1997.

397. McMahon, B.R. and Wilkens, J.L., Ventilation, perfusion and oxygen uptake, in *Internal Anatomy and Physiological Regulation*, Mantel, L.H., Ed., Academic Press, New York, 1983, pp. 289–392.

398. McManus, J.J., Osmotic relations in the horseshoe crab, *Limulus polyphemus*, *Am. Midl. Nat.*, 81, 569–573, 1969.

399. Mendonça, N.N., Masui, D.C., McNamara, J.C., Leone, F.A., and Furriel, R.P., Long-term exposure of the freshwater shrimp *Macrobrachium olfersii* to elevated salinity: effects on gill (Na$^+$,K$^+$)-ATPase alpha-subunit expression and K$^+$-phosphatase activity, *Comp. Biochem. Physiol.*, 146A, 534–543, 2007.

400. Meschenmoser, M., Ultrastructure of the embryonic dorsan organ of *Orchestia cavimana* (Crustacea, Amphipoda); with a note on localization of chloride and on the change in calcium deposit before the embryonic moult, *Tissue Cell*, 21, 431–442, 1989.

401. Meurice, J.C. and Goffinet, G., Structure et fonction de l'organe nuccal des cladocères marins gymnomères, *C. R. Acad. Sci. Paris*, 295, 693–694, 1992.

402. Milne, D.J. and Ellis, R.A., The effect of salinity acclimation on the ultrastructure of the gills of *Gammarus oceanicus* (Segerstrale, 1947) (Crustacea: Amphipoda), *Z. Zellforsch.*, 139, 311–318, 1973.

403. Miyawaki, M. and Ukeshima, A., On the ultrastructure of the epithelium of the antennal gland of the crayfish, *Procambarus clarkii*, *Kumamoto J. Sci. B*, 8, 57–73, 1967.

404. Mo, J.L., Devos, P., and Trausch, G., Dopamine as a modulator of ionic transport and Na$^+$/K$^+$-ATPase activity in the gills of the Chinese crab *Eriocheir sinensis*, *J. Crust. Biol.*, 18, 442–448, 1998.

405. Moran, W.M. and Pierce, S.K., The mechanism of crustacean salinity tolerance: cell volume regulation by K$^+$ and glycine effluxes, *Mar. Biol.*, 81, 41–46, 1984.

406. Morris, S., Neuroendocrine regulation of osmoregulation and the evolution of air-breathing in decapod crustaceans, *J. Exp. Biol.*, 204, 979–989, 2001.

407. Morris, S., The ecophysiology of air-breathing in crabs with special reference to *Gecarcoidea natalis*, *Comp. Biochem. Physiol.*, 131B, 559–570, 2002.

408. Morris, S. and Ahern, M.D., Regulation of urine reprocessing in the maintenance of sodium and water balance in the terrestrial Christmas Island red crab *Gecarcoidea natalis* investigated under field conditions, *J. Exp. Biol.*, 206, 2869–2881, 2003.

409. Morris, S. and Edwards, T., Control of osmoregulation via regulation of Na$^+$/K$^+$-ATPase activity in the amphibious purple shore crab *Leptograpsus variegatus*, *Comp. Biochem. Physiol.*, 112C, 129–136, 1995.

410. Morris, S. and Van Aardt, W.J., Salt and water relations, and nitrogen excretion in the amphibious freshwater crab *Potamonautes warreni* in water and in air, *J. Exp. Biol.*, 201, 883–893, 1998.

411. Morritt, D. and Richardson, A.M.M., Female control of embryonic environment in a terrestrial amphipod, *Mysticotalitrus cryptus* (Crustacea), *Function. Ecol.*, 12, 351–358, 1998.

412. Morritt, D. and Spicer, J.I., Changes in the pattern of osmoregulation in the brackish water amphipod *Gammarus duebeni* Lilljeborg (Crustacea) during embryonic development, *J. Exp. Zool.*, 273, 271–281, 1995.

413. Morritt, D. and Spicer, J.I., The culture of eggs and embryos of amphipod crustaceans; implications for brood pouch physiology, *J. Mar. Biol. Assoc. U.K.*, 76, 361–376, 1996.

414. Morritt, D. and Spicer, J.I., Developmental ecophysiology of the beachflea *Orchestia gammarellus* (Pallas) (Crustacea: Amphipoda). I. Female control of embryonic environment, *J. Exp. Mar. Biol. Ecol.*, 207, 191–203, 1996.

415. Morritt, D. and Spicer, J.I., Developmental ecophysiology of the beachflea *Orchestia gammarellus* (Pallas) (Crustacea: Amphipoda). II. Embryonic osmoregulation, *J. Exp. Mar. Biol. Ecol.*, 207, 205–216, 1996.

416. Morritt, D. and Spicer, J.I., The physiological ecology of talitrid amphipods: an update, *Can. J. Zool.*, 76, 1965–1982, 1998.

417. Morritt, D. and Spicer, J.I., Developmental ecophysiology of the beachflea *Orchestia gammarellus* (Pallas) (Crustacea: Amphipoda: Talitridae). III. Physiological competency as a possible explanation for timing of hatchling release, *J. Exp. Mar. Biol. Ecol.*, 232, 275–283, 1999.

418. Morse, H.C., Harris, P.J., and Dornfeld, E.J., *Pacifastacus leniusculus*: fine structure of arthobranch with reference to active ion uptake, *Trans. Am. Microsc. Soc.*, 89, 12–27, 1970.

419. Mykles, D.L., The ultrastructure of the posterior midgut caecum of *Pachygrapsus crassipes* (Decapoda, Brachyura) adapted to low salinity, *Tissue Cell*, 9, 681–691, 1977.

420. Mykles, D.L., Ultrastructure of alimentary epithelia of lobsters, *Homarus americanus* and *H. gammarus*, and crab, *Cancer magister*, *Zoomorphology*, 92, 201–215, 1979.

421. Mykles, D.L., The mechanism of fluid absorption at ecdysis in the American lobster, *Homarus americanus*, *J. Exp. Biol.*, 84, 89–101, 1980.

422. Mykles, D.L., Ionic requirements of transepithelial potential difference and net water flux in the perfused midgut of the American lobster *Homarus americanus*, *Comp. Biochem. Physiol.*, 69A, 317–320, 1981.

423. Neufeld, D.S. and Cameron, J.N., Mechanism of the net uptake of water in moulting blue crabs (*Callinectes sapidus*) acclimated to high and low salinities, *J. Exp. Biol.*, 188, 11–23, 1994.

424. Neufeld, G.J., Holliday, C.W., and Pritchard, J.B., Salinity adaptation of gill Na,K-ATPase in the blue crab, *Callinectes sapidus*, *J. Exp. Zool.*, 211, 215–224, 1980.

425. Noblitt, S.B. and Payne, J.F., A comparative study of selected chemical aspects of the eggs of the crayfish *Procambarus clarkii* (Girard, 1852) and *P. zonangulus* Hobbs & Hobbs, 1990 (Decapoda, Cambaridae), *Crustaceana*, 68, 695–704, 1995.

426. Novo, M.S., Miranda, R.B., and Bianchini, A., Sexual and seasonal variations in osmoregulation and ionoregulation in the estuarine crab *Chasmagnathus granulata* (Crustacea, Decapoda), *J. Exp. Mar. Biol. Ecol.*, 323, 118–137, 2005.

427. Olesen, J., External morphology and phylogenetic significance of the dorsal/neck organ in the Conchostraca and the head pores of the cladoceran family Chydoridae (Crustacea, Branchiopoda), *Hydrobiologia*, 330, 213–226, 1996.

428. Olmsted, J.M.D. and Baumberger, J.P., Form and growth of grapsoid crabs: a comparison of the form of three species of grapsoid crabs and their growth at molting, *J. Morphol.*, 38, 279–294, 1923.

429. Olsowski, A., Putzenlechner, M., Böttcher, K., and Graszynski, K., The carbonic anhydrase of the Chinese crab *Eriocheir sinensis*: effects of adaptation from tap to salt water, *Helgoländer Meeresun.*, 49, 727–735, 1995.

430. Onken, H., Active and electrogenic absorption of Na^+ and Cl^- across posterior gills of *Eriocheir sinensis*: influence of short-term osmotic variations, *J. Exp. Biol.*, 199, 901–910, 1996.

431. Onken, H., Active NaCl absorption across split lamellae of posterior gills of Chinese crabs (*Eriocheir sinensis*) adapted to different salinities. *Comp. Biochem. Physiol.*, 123A, 377–384, 1999.

432. Onken, H., Graszynski, K., and Zeiske, W., Na^+-independent, electrogenic Cl^- uptake across the posterior gills of the Chinese crab (*Eriocheir sinensis*): voltage-clamp and microelectrode studies, *J. Comp. Physiol.*, 161, 293–301, 1991.

433. Onken, H. and McNamara, J.C., Hyperosmoregulation in the red crab *Dilocarcinus pagei* (Brachyura, Trichodactylidae): structural and functional asymmetries of the posterior gills, *J. Exp. Biol.*, 205, 165–175, 2002.

434. Onken, H. and Putzenlechner, M., A V-ATPase drives active, electrogenic and Na⁺-independent Cl⁻ absorption across the gills of *Eriocheir sinensis*, *J. Exp. Biol.*, 198, 767–774, 1995.

435. Onken, H. and Riestenpatt, S., NaCl absorption across split gill lamellae of hyperregulating crabs: transport mechanisms and their regulation, *Comp. Biochem. Physiol.*, 119A, 883–893, 1998.

436. Onken, H. and Riestenpatt, S., Ion transport across posterior gills of hyperosmoregulating shore crabs (*Carcinus maenas*): amiloride blocks the cuticular Na⁺ conductance and induces current noise, *J. Exp. Biol.*, 205, 523–531, 2002.

437. Onken, H., Schöbel, A., Kraft, J., and Putzenlechner, M., Active NaCl absorption across split lamellae of posterior gills of the Chinese crab *Eriocheir sinensis*: stimulation by eyestalk extract, *J. Exp. Biol.*, 203, 1373–1381, 2000.

438. Onken, H. and Siebers, D., Voltage-clamp measurements on single split lamellae of posterior gills of the shore crab *Carcinus maenas*, *Mar. Biol.*, 114, 385–390, 1992.

439. Onken, H., Tresguerres, M., and Luquet, C.M., Active NaCl absorption across posterior gills of hyperosmoregulating *Chasmagnathus granulata*, *J. Exp. Biol.*, 206, 1017–1023, 2003.

440. Palackal, T., Faso, L., Zung, J.L., Vernon, G., and Witkus, R., The ultrastructure of the hindgut epithelium of terrestrial isopods and its role in osmoregulation, *Symp. Zool. Soc. Lond.*, 53, 185–198, 1984.

441. Papathanassiou, E., Effects of cadmium ions on the ultrastructure of the gill cells of the brown shrimp *Crangon crangon* (L.) (Decapoda, Caridea), *Crustaceana*, 48, 6–17, 1985.

442. Papathanassiou, E. and King, P.E., Ultrastructural studies on the gills of *Palaemon serratus* (Pennant) in relation to cadmium accumulation, *Aquat. Toxicol.*, 3, 273–284, 1983.

443. Parry, G., Osmoregulation in some fresh water prawns, *J. Exp. Biol.*, 34, 417–423, 1957.

444. Passano, L.M., Neurosecretory control of molting in crabs by the X-organ sinus gland complex, *Physiol. Comp. Oecol.*, 3, 155–189, 1953.

445. Patil, H.S. and Kaliwal, M.B., Histopathological effects of zinc on the gills of prawn *Macrobrachium hendersodyanum*, *Z. Angen. Zool.*, 76, 505–509, 1989.

446. Pavicic-Hamer, D., Devescovi, M., and Lucu, C., Activation of carbonic anhydrase in branchial cavity tissues of lobsters (*Homarus gammarus*) by dilute seawater exposure, *J. Exp. Mar. Biol. Ecol.*, 287, 79–92, 2003.

447. Péqueux, A., Osmotic regulation in crustaceans, *J. Crust. Biol.*, 15, 1–60, 1995.

448. Péqueux, A., Dandrifosse, G., Loret, S., Charmantier, G., Charmantier-Daures, M., Spanings-Pierrot, C., and Scoffeniels, E., Osmoregulation: morphological, physiological, biochemical, hormonal, and developmental aspects, in *Treatise on Zoology: Anatomy, Taxonomy, Biology, The Crustacea*, Forest, J. and von Vaupel Klein, J.C., Eds., Brill Academic Publishers, Leiden, 2006, pp. 205–308.

449. Péqueux, A. and Gilles, R., Control of extracellular fluid osmolality in crustaceans, in *Osmoregulation in Estuarine and Marine Animals*, Péqueux, A., Gilles, R., and Bolis, L., Eds., Springer, Berlin, 1984, pp. 18–34.

450. Péqueux, A. and Gilles, R., The transepithelial potential difference of isolated perfused gills of the Chinese crab *Eriocheir sinensis* acclimated to fresh water, *Comp. Biochem. Physiol.*, 89A, 163–172, 1988.

451. Péqueux, A., Gilles, R., and Marshall, L.A., NaCl transport in gills and related structures. I. Invertebrates, in *Advances in Comparative and Environmental Physiology*, Greger, R., Ed., Springer-Verlag, Berlin, 1988, pp. 1–73.

452. Péqueux, A., Marchal, A., Wanson, S., and Gilles, R., Kinetic characteristics and specific activity of gill (Na⁺+K⁺) ATPase in the euryhaline Chinese crab, *Eriocheir sinensis*, during salinity acclimation, *Mar. Biol. Lett.*, 5, 35–45, 1984.

453. Peters, H., Über den Einfluss des Salzgehaltes im Aussenmedium auf den Bau und die Funktion der Exkretionsorgane decapoder Crustaceen (nach Untersuchungen an *Potamobius fluviatilis* und *Homarus vulgaris*), *Z. Morphol. Ökol. Tiere*, 30, 355–381, 1935.

454. Peterson, D.R. and Loizzi, R.F., Ultrastructure of the crayfish kidney, coelomosac, labyrinth, and nephridial canal, *J. Morphol.*, 142, 241–264, 1974.

455. Peterson, D.R. and Loizzi, R.F., Biochemical and cytochemical investigation of Na⁺,K⁺-ATPase in the crayfish kidney, *Comp. Biochem. Physiol.*, 49A, 763–773, 1975.

456. Phillips, J.E. and Audsley, N., Neuropeptide control of ion and fluid transport across locust hindgut, *Am. Zool.*, 35, 503–514, 1995.

457. Phillips, J.E., Meredith, J., Audsley, N., Richardson, N., Macins, A., and Ring, M., Locust ion transport peptide (ITP): a putative hormone controlling water and ionic balance in terrestrial insects, *Am. Zool.*, 38, 461–470, 1998.

458. Pierrot, C., Physiologie branchiale et osmorégulation chez *Pachygrapsus marmoratus* (Crustacé, Décapode), Ph.D. thesis, Montpellier II, 1994.

459. Pierrot, C., Eckhardt, E., Van Herp, F., Charmantier-Daures, M., Charmantier, G., Trilles, J.-P., and Thuet, P., Effet d'extraits de glandes du sinus sur la physiologie osmorégulatrice des branchies perfusées du crabe *Pachygrapsus marmoratus*, *C. R. Acad. Sci. Paris*, 317, 411–418, 1994.

460. Pillai, R.S., Studies on the shrimp *Caridina laevis* (Heller). I. The digestive system, *J. Mar. Biol. Assoc. India*, 2, 7–74, 1960.

461. Piller, S.C., Henry, R.P., Doeller, J.E., and Kraus, D.W., A comparison of the gill physiology of two euryhaline crab species, *Callinectes sapidus* and *Callinectes similis*: energy production, transport-related enzymes and osmoregulation as a function of acclimation salinity, *J. Exp. Biol.*, 198, 349–358, 1995.

462. Postel, U., Becker, W., Brandt, A., Luck-Kopp, S., Riestenpatt, S., Weihrauch, D., and Siebers, D., Active osmoregulatory ion uptake across the pleopods of the isopod *Idotea baltica* (Pallas): electrophysiological measurements on isolated split endo- and exopodites mounted in a micro-Ussing chamber, *J. Exp. Biol.*, 203, 1141–1152, 2000.

463. Potts, W.T.W., Osmotic and ionic regulation, *Ann. Rev. Physiol.*, 30, 73–104, 1968.

464. Potts, W.T.W. and Durning, C.T., Physiological evolution in the branchiopods, *Comp. Biochem. Physiol.*, 67B, 475–484, 1980.

465. Potts, W.T.W. and Parry, G., *Osmotic and Ionic Regulation in Animals* Pergamon Press, Oxford, 1963.

466. Potts, W.T.W. and Parry, G., Sodium and chloride balance in the prawn *Palaemonetes varians*, *J. Exp. Biol.*, 41, 591–601, 1964.

467. Prager, D.J. and Bowman, R.L., Freezing-point depression: new method of measuring ultra-micro quantities of fluids, *Science*, 142, 237–239, 1963.

468. Prangnell, D.I. and Fotedar, R., Effect of sudden salinity change on *Penaeus latisulcatus* Kishinouye osmoregulation, ionoregulation and condition in inland saline water and potassium-fortified inland saline water, *Comp. Biochem. Physiol.*, 415A, 449–457, 2006.

469. Pressley, T.A., Graves, J.S., and Krall, A.R., Amiloride-sensitive ammonium and sodium ion transport in the blue crab, *Am. J. Physiol.*, 241, R370–R378, 1981.

470. Pritchard, A.W. and Kerley, D.E., Kidney function in the North American crayfish *Pacifastacus leniusculus* (Dana) stepwise acclimated to dilutions of sea water, *Comp. Biochem. Physiol.*, 35, 427–437, 1970.

471. Prosser, C.L., Water: osmotic balance, hormonal regulation: inorganic ions, in *Comparative Animal Physiology*, Prosser, C.L., Ed., Saunders, Philadelphia, PA, 1973, pp. 1–110.

472. Prosser, C.L., *Environmental and Metabolic Animal Physiology*, Wiley-Liss, New York, 1991.

473. Prosser, C.L., Green, J.W., and Chow, T.J., Ionic and osmotic concentration in blood and urine of *Pachygrapsus crassipes* acclimated to different salinities, *Biol. Bull.*, 109, 99–107, 1955.

474. Rabalais, N.N. and Cameron, J.N., Physiological and morphological adaptations of adult *Uca subcylindrica* to semi-arid environments, *Biol. Bull.*, 168, 135–146, 1985.

475. Rabalais, N.N. and Cameron, J.N., The effects of factors important in semi-arid environments on the early development of *Uca subcylindrica*, *Biol. Bull.*, 168, 147–160, 1985.

476. Ramsay, J.A. and Brown, R.H.J., Simplified apparatus and procedure for freezing-point determinations upon small volumes of fluid, *J. Sci. Instrum.*, 32, 372–375, 1955.

477. Read, G.H.L., Instraspecific variation in the osmoregulatory capacity of larval, post-larval, juvenile and adult *Macrobrachium petersi* (Hilgendorf), *Comp. Biochem. Physiol.*, 78A, 501–506, 1984.

478. Rebelo, M.F., Santos, E.A., and Montserrat, J.M., Ammonia exposure of *Chasmagnathus granulata* (Crustacea, Deacapoda) Dana, 1851: accumulation in haemolymph and effects on osmoregulation, *Comp. Biochem. Physiol.*, 122A, 429–435, 1990.

479. Reisinger, P.W.M., Tutter, I., and Welsch, U., Fine structure of the gills of the horseshoe crabs *Limulus polyphemus* and *Trachypleus tridentatus* and of the book lungs of the spider *Eurypelma californicum*, *Zool. Jb. Anat.*, 121, 331–357, 1991.

480. Renaud, L., Le cycle des réserves organiques chez les Crustacés Décapodes, *Ann. Inst. Océanogr. Paris*, 24, 259–357, 1949.

481. Richard, P., Rôle biologique et écologique des acides aminés libres chez quelques Crustacés Décapodes marins, Ph.D. thesis, Aix-Marseille II, 1982.

482. Riegel, J.A., Micropuncture studies of chloride concentrations and osmotic pressure in the crayfish antennal gland, *J. Exp. Biol.*, 40, 487–492, 1963.

483. Riegel, J.A., Analysis of formed bodies in urine removed from the crayfish antennal gland by micropuncture, *J. Exp. Biol.*, 44, 387–395, 1966.

484. Riegel, J.A., Analysis of the distribution of sodium, potassium and osmotic pressure in the urine of crayfishes, *J. Exp. Biol.*, 48, 587–596, 1968.

485. Riegel, J.A., *In vitro* studies of fluid and ion movements due to the swelling of formed bodies, *Comp. Biochem. Physiol.*, 35, 843–856, 1970.

486. Riegel, J.A., A new model of transepithelial fluid movement with detailed application to fluid movement in the crayfish antennal gland, *Comp. Biochem. Physiol.*, 36, 403–410, 1970.

487. Riegel, J.A., *Comparative Physiology of Renal Excretion*, Oliver and Boyd, Edinburgh, Scotland, 1972.

488. Riegel, J.A., Aspects of fluid movement in the crayfish antennal gland, in *Water Relations in Membrane Transports in Plants and Animals*, Jungrens, A.M., Hodges, T.K., Kleihzeller, A., and Shultz, S., Eds., Academic Press, New York, 1977, pp. 121–127.

489. Riegel, J.A., Fluid movement through the crayfish antennal gland, in *Transport of Ions and Water in Animals*, Gupta, B.L., Moreton, R.B., Osshman, J.L., and Wall, B.J., Eds., Academic Press, London, 1977, pp. 613–631.

490. Riegel, J.A. and Cook, M.A., Recent studies of excretion in Crustacea, *Fortschr. Zool.*, 23, 48–75, 1975.

491. Riegel, J.A. and Lockwood, A.P.M., The role of the antennal gland in the osmotic and ionic regulation of *Carcinus maenas*, *J. Exp. Biol.*, 38, 491–499, 1961.

492. Riestenpatt, S., Onken, H., and Siebers, D., Active absorption of Na^+ and Cl^- across the gill epithelium of the shore crab *Carcinus maenas*: voltage-clamp and ion-flux studies., *J. Exp. Biol.*, 199, 1545–1554, 1996.

493. Riestenpatt, S., Petrausch, G., and Siebers, D., Cl^- influx across posterior gills of the Chinese crab (*Eriocheir sinensis*): potential energization by a V-type H^+-ATPase, *Comp. Biochem. Physiol.*, 110A, 235–241, 1995.

494. Riestenpatt, S., Zeiske, W., and Onken, H., Cyclic AMP stimulation of electrogenic uptake of Na^+ and Cl^- across the gill epithelium of the Chinese crab *Eriocheir sinensis*, *J. Exp. Biol.*, 188, 159–174, 1994.

495. Roast, S.D., Rainbow, P.S., Smith, B.D., Nimmo, M., and Jones, M.B., Trace metal uptake by the Chinese crab *Eriocheir sinensis*: the role of osmoregulation, *Mar. Environ. Res.*, 53, 453–464, 2002.

496. Roast, S.D., Widdows, J., and Jones, M.B., Effect of salinity and chemical speciation on cadmium accumulation and toxicity to two mysid species, *Environ. Toxicol. Chem.*, 20, 1078–1084, 2001.

497. Robertson, J.D., Ionic regulation in some marine invertebrates, *J. Exp. Biol.*, 26, 182–200, 1949.

498. Robertson, J.D., Further studies on ionic regulation in marine invertebrates, *J. Exp. Biol.*, 30, 277–296, 1953.

499. Robertson, J.D., Osmotic and ionic regulation, in *The Physiology of Crustacea*. Vol. 1. *Metabolism and Growth*, Waterman, T.H., Ed., Academic Press, New York, 1960, pp. 317–339.

500. Robertson, J.D., Studies on the chemical composition of muscle tissue. II. The abdominal flexor muscle of the lobster *Nephrops norvegicus* (L.), *J. Exp. Biol.*, 38, 707–728, 1961.

501. Robertson, J.D., Osmotic and ionic regulation in the horseshoe crab *Limulus polyphemus* (Linnaeus), *Biol. Bull.*, 138, 157–183, 1970.

502. Robinson, G.D., Water fluxes and urine production in blue crabs (*Callinectes sapidus*) as a function of environmental salinity, *Comp. Biochem. Physiol.*, 71A, 407–412, 1982.

503. Roy, L.A., Davies, D.A., Saoud, I.P., and Henry, R.P., Effects of various levels of aqueous potassium and magnesium on survival, growth, and respiration of the pacific white shrimp, *Litopenaeus vannamei*, reared in low salinity waters, *Aquaculture*, 262, 461–469, 2007.

504. Russell, J.M., Sodium–potassium–chloride cotransport, *Physiol. Rev.*, 80, 211–276, 2000.

505. Saigusa, M., Two kinds of active factor in crab hatch water: ovigerous-hair stripping substance (OHSS) and a proteinase, *Biol. Bull.*, 191, 234–240, 1996.

506. Saigusa, M. and Terajima, M., Hatching of an estuarine crab, *Sesarma haematocheir*: from disappearance of the inner (E3) layer to rupture of the egg case, *J. Exp. Zool.*, 287, 510–523, 2000.

507. Santos, F.H. and Keller, R., Crustacean hyperglycemic hormone (CHH) and the regulation of carbohydrate metabolism: current perspectives, *Comp. Biochem. Physiol.*, 106A, 405–411, 1993.

508. Santos, M.C.F., Drinking and osmoregulation in the mangrove crab *Ucides cordatus* following exposure to benzene, *Comp. Biochem. Physiol.*, 133A, 29–42, 2002.

509. Santos, M.C.F. and Salomao, L.C., Hemolymph osmotic and ionic concentrations in the gecarcinid crab *Ucides cordatus*, *Comp. Biochem. Physiol.*, 81A, 581–583, 1985.

510. Santos, M.C.F. and Salomao, L.C., Osmotic and cationic urine concentrations in hyporegulating *Ucides cordatus*, *Comp. Biochem. Physiol.*, 81A, 895–898, 1985.

511. Sarradin, P.-M., Caprais, J.-C., Briand, P., Gaill, F., Shillito, B., and Desbruyères, D., Chemical and thermal description of the environment of the Genesis hydrothermal vent community (13°N, EPR), *Cah. Biol. Mar.*, 40, 93–104, 1998.

512. Sarver, G.L., Flynn, M.A., and Holliday, C.W., Renal Na,K-ATPase and osmoregulation in the crayfish *Procambarus clarkii*, *Comp. Biochem. Physiol.*, 107A, 349–356, 1994.

513. Savage, J.P. and Robinson, G.D., Inducement of increased gill/Na$^+$-K$^+$ ATPase activity by a hemolymph factor in hyperregulating *Callinectes sapidus*, *Comp. Biochem. Physiol.*, 74A, 65–69, 1983.

514. Schaffner, A. and Rodewald, R., Filtration barriers in the coelomic sac of the crayfish, *Procambarus clarkii*, *J. Ultrastruct. Res.*, 65, 36–47, 1978.

515. Schleich, C.E., Goldemberg, L.A., and López Mananes, A.A., Salinity-dependent Na$^+$-K$^+$ATPase activity in gills of the euryhaline crab *Chasmagnathus granulata*, *Gen. Physiol. Biophys.*, 20, 255–266, 2001.

516. Schmidt-Nielsen, K., Gertz, K.H., and Davis, L.E., Excretion and ultrastructure of the antennal gland of the fiddler crab, *Uca mordax*, *J. Morphol.*, 125, 473–496, 1968.

517. Schmitz, E.H., Amphipoda, in *Microscopic Anatomy of Invertebrates*, Harrison, F.W. and Humes, A.G., Eds., Wiley-Liss, New York, 1992, pp. 443–528.

518. Schoffeniels, E., Biochemical aspects of adaptation: adaptation to salinity, *Biochem. Soc. Symp.*, 41, 179–204, 1976.

519. Schoffeniels, E. and Dandrifosse, G., Osmorégulation: aspects morphologiques et biochimiques, in *Morphologie, Physiologie, Reproduction, Systématique*, Forest, J., Ed., Masson, Paris, 1994, pp. 529–594.

520. Schram, F.R., The fossil record and evolution of Crustacea, in *Embryology, Morphology and Genetics*, Abele, L.G., Ed., Academic Press, New York, 1982.

521. Schram, F.R., Phylogeny of decapods: moving toward a consensus, *Hydrobiologia*, 449, 1–20, 2001.

522. Schwabe, E., Uber die Osmoregulation verschiedener Krebse (Malacostracen), *Z. Vergl. Physiol.*, 19, 183–236, 1933.

523. Scudamore, H.H., The influence of the sinus glands upon molting and associated changes in the crayfish, *Physiol. Zool.*, 20, 187–208, 1947.

524. Sébert, P., Simon, B., and Péqueux, A., Effects of hydrostatic pressure on energy metabolism and osmoregulation in crab and fish, *Comp. Biochem. Physiol.*, 116A, 281–290, 1997.

525. Sedlmeier, D. and Keller, R., The role of action of the crustacean neurosecretory hyperglycemic hormone. 1. Involvement of cyclic nucleotides, *Gen. Comp. Endocrinol.*, 45, 82–90, 1981.

526. Sekiguchi, K., *Biology of the Horseshoe Crabs*, Science House, Tokyo, 1988.

527. Sekiguchi, K., Yamamichi, Y., Seshimo, H., and Sugita, H., Normal development, in *Biology of the Horseshoe Crabs*, Sekiguchi, K., Ed., Science House, Tokyo, 1988, pp. 133–224.

528. Seneviratna, D. and Taylor, H.H., Ontogeny of osmoregulation in embryos of intertidal crabs (*Hemigrapsus sexdentatus* and *H. crenulatus*, Grapsidae, Brachyura): putative involvement of the embryonic dorsal organ., *J. Exp. Biol.*, 209, 1487–1501, 2006.

529. Serrano, L., Blanvillain, G., Soyez, D., Charmantier, G., Grousset, E., Aujoulat, F., and Spanings-Pierrot, C., Putative involvement of crustacean hyperglycemic hormone isoforms in the neuroendocrine mediation of osmoregulation in the crayfish *Astacus leptodactylus*, *J. Exp. Biol.*, 206, 979–988, 2003.

530. Serrano, L., Grousset, E., Charmantier, G., and Spanings-Pierrot, C., Occurrence of L- and D-crustacean hyperglycemic hormone isoforms in the eyestalk X-organ/sinus gland complex during the ontogeny of the crayfish *Astacus leptodactylus*, *J. Histochem. Cytochem.*, 52, 1129–1140, 2004.

531. Serrano, L., Halanych, K.M., and Henry, R.P., Salinity-stimulated changes in expression and activity of two carbonic anhydrase isoforms in the blue crab, *Callinectes sapidus*, *J. Exp. Biol.*, 210, 2320–2332, 2007.

532. Serrano, L., Towle, D.W., Charmantier, G., and Spanings-Pierrot, C., Expression of Na⁺/K⁺-ATPase α-subunit mRNA during embryonic development of the crayfish *Astacus leptodactylus*, *Comp. Biochem. Physiol.*, 2D, 126–134, 2007.

533. Sesma, P., Bayona, C., Villaro, A.C., and Vazquez, J.J., A microscopic study on the antennal gland of *Antrapotamobius ballines* (Crustacea, Decapoda), *Morfologia Normal y Patologica*, 7, 289–301, 1983.

534. Shaw, J., Studies on ionic regulation in *Carcinus maenas*, *J. Exp. Biol.*, 38, 135–152, 1961.

535. Shaw, J., The control of salt balance in Crustacea, *Soc. Exp. Biol. Symp.*, 18, 237–254, 1964.

536. Shetlar, R.E. and Towle, D.W., Electrogenic sodium–proton exchange in membrane vesicles from crab (*Carcinus maenas*) gill, *Am. J. Physiol.*, 257, R924–R931, 1989.

537. Shires, R., Lane, N.J., Inman, C.B.E., and Lockwood, A.P.M., Microtubule systems associated with the septate junctions of the gill cells of four gammarid amphipods, *Tissue Cell*, 27, 3–12, 1995.

538. Shivers, R.R. and Chauvin, W.J., Intercellular junctions of antennal gland epithelial cells in the crayfish *Orconectes virilis*, *Cell Tissue Res.*, 175, 425–438, 1977.

539. Siebers, D., Böttcher, K., Petrausch, G., and Hamann, A., Effects of some chloride channel blockers on potential differences and ion fluxes in isolated perfused gills of shore crabs *Carcinus maenas*, *Comp. Biochem. Physiol.*, 97A, 9–15, 1990.

540. Siebers, D., Leweck, K., Markus, H., and Winkler, A., Sodium regulation in the shore crab *Carcinus maenas* as related to ambient salinity, *Mar. Biol.*, 69, 37–43, 1982.

541. Siebers, D., Lucu, C., Winkler, A., Grammerstorf, U., and Wille, H., Effects of amiloride on sodium chloride transport across isolated perfused gills of shore crabs *Carcinus maenas* acclimated to brackish water, *Comp. Biochem. Physiol.*, 87A, 333–340, 1987.

542. Siebers, D., Wille, H., Lucu, C., and Venezia, L.D., Conductive sodium entry in gill cells of the shore crab, *Carcinus maenas*, *Mar. Biol.*, 101, 61–68, 1989.

543. Siebers, D., Winkler, A., Lucu, C., Thedens, G., and Weichart, D., Na,K-ATPase generates an active transport potential in the gills of the hyperregulating shore crab *Carcinus maenas*, *Mar. Biol.*, 87, 185–192, 1985.

544. Silvestre, F., Trausch, G., and Devos, P., Hyper-osmoregulatory capacity of the Chinese mitten crab (*Eriocheir sinensis*) exposed to cadmium; acclimation during chronic exposure, *Comp. Biochem. Physiol.*, 140C, 29–37, 2005.

545. Silvestre, F., Trausch, G., Péqueux, A., and Devos, P., Uptake of cadmium through isolated perfused gills of the Chinese mitten crab, *Eriocheir sinensis*, *Comp. Biochem. Physiol.*, 137A, 189–196, 2004.

546. Silvestre, F., Trausch, G., Spano, L., and Devos, P., Effects of atrazine on osmoregulation in the Chinese mitten crab, *Eriocheir sinensis*, *Comp. Biochem. Physiol.*, 132C, 385–390, 2002.

547. Skou, J.C., Further investigations on a Mg⁺⁺ + Na⁺-activated adenosinetriphosphatase, possibly related to the active, linked transport of Na⁺ and K⁺ across the nerve membrane, *Biochim. Biophys. Acta*, 42, 6–23, 1960.

548. Smith, P.G., The ionic regulation of *Artemia salina* (L.). I. Measurements of electrical potential difference and resistance, *J. Exp. Biol.*, 51, 727–738, 1969.

549. Smith, P.G., The ionic regulation of *Artemia salina* (L.). II. Fluxes of sodium, chloride and water, *J. Exp. Biol.*, 51, 739–757, 1969.

550. Smith, R.I. and Rudy, P.P., Water-exchange in the crab *Hemigrapsus nudus* measured by use of deuterium and tritium oxides as tracers, *Biol. Bull.*, 143, 234–246, 1972.

551. Soegianto, A., Bambang, Y., Charmantier-Daures, M., Trilles, J.-P., and Charmantier, G., Impact of copper on the structure of gills and epipodites of the shrimp *Penaeus japonicus* (Crustacea, Decapoda), *J. Crust. Biol.*, 19, 209–223, 1999.

552. Soegianto, A., Charmantier-Daures, M., Trilles, J.-P., and Charmantier, G., Impact of cadmium on the structure of gills and epipodites of the shrimp *Penaeus japonicus* (Crustacea: Decapoda), *Aquat. Liv. Res.*, 12, 57–70, 1999.

553. Sommer, M.J. and Mantel, L.H., Effect of dopamine, cyclic AMP and pericardial organs on sodium uptake and Na/K ATPase activity in gills of the green crab *Carcinus maenas*, *J. Exp. Zool.*, 248, 272–277, 1988.

554. Sommer, M.J. and Mantel, L.H., Effects of dopamine and acclimation to reduced salinity on the concentration of cyclic AMP in the gills of the green crab, *Carcinus maenas* (L.), *Gen. Comp. Endocrinol.*, 82, 364–368, 1991.

555. Souheil, H., Vey, A., Thuet, P., and Trilles, J.-P., Pathogenic and toxic effect of *Fusarium oxysporum* (Schelcht) on survival and osmoregulatory capacity of *Penaeus japonicus* (Bate), *Aquaculture*, 178, 209–224, 1999.

556. Sowers, A.D., Young, S.P., Grosell, M., Browdy, C.L., and Tomasso, J.R., Hemolymph osmolality and cation concentrations in *Litopenaeus vannamei* during exposure to artificial sea salt or a mixed-ion solution: relationship to potassium flux, *Comp. Biochem. Physiol.*, 145A, 176–180, 2006.

557. Soyez, C., Larverdure, A.M., Kallen, J., and Van Herp, F., Demonstration of a cell-specific isomerisation of invertebrate neuropeptides, *Neuroscience*, 82, 935–942, 1998.

558. Soyez, D., Occurrence and diversity of neuropeptides from the crustacean hyperglycemic hormone family in arthropods, *Ann. N.Y. Acad. Sci.*, 814, 319–323, 1997.

559. Soyez, D., Toullec, J.-Y., Olliveaux, C., and Géraud, G., L to D amino acid isomerization in a peptide hormone is a late post-translational event occurring in specialized neurosecretory cells, *J. Biol. Chem.*, 275, 37870–37875, 2000.

560. Spaargaren, D.H., Aspects of the osmotic regulation in the shrimps *Crangon crangon* and *Crangon allmani*, *Neth. J. Sea Res.*, 5, 275–335, 1971.

561. Spaargaren, D.H., On the water and salt economy of some decapod crustaceans from the Gulf of Aqaba (Red Sea), *Neth. J. Sea Res.*, 11, 99–106, 1977.

562. Spanings-Pierrot, C., Bisson, L., and Towle, D.W., Expression of a crustacean hyperglycemic hormone isoform in the shore crab *Pachygrapsus marmoratus* during adaptation to low salinity, *Bull. Mt. Desert Isl. Biol. Lab.*, 44, 67–69, 2005.

563. Spanings-Pierrot, C., Soyez, D., Van Herp, F., Gompel, M., Skaret, G., Grousset, E., and Charmantier, G., Involvement of crustacean hyperglycemic hormone in the control of gill ion transport in the crab *Pachygrapsus marmoratus*, *Gen. Comp. Endocrinol.*, 119, 340–350, 2000.

564. Spanings-Pierrot, C. and Towle, D.W., Time course of crustacean hyperglycemic hormone isoforms mRNA levels in the crab *Pachygrapsus marmoratus* following salinity adaptation, *Bull. Mt. Desert Isl. Biol. Lab.*, 46, 10–11, 2007.

565. Spencer, A.M., Fielding, A.H., and Kamemoto, F.I., The relationship between gill Na,K-ATPase activity and osmoregulatory capacity in various crabs, *Physiol. Zool.*, 52, 1–10, 1979.

566. Strauss, O. and Graszynski, K., Isolation of plasma membrane vesicles from the gill epithelium of the crayfish, *Orconectes limosus* Rafinesque, and properties of the Na$^+$/H$^+$ exchanger, *Comp. Biochem. Physiol.*, 102A, 519–526, 1992.

567. Strömberg, J.O., *Cyathura polita* (Crustacea, Isopoda), some embryological notes, *Bull. Mar. Sci. Gulf Caribbean*, 22, 463–483, 1972.

568. Sun, D.Y., Guo, J.Z., Hartmann, H.A., Uno, H., and Hokin, L.E., Na,K-ATPase expression in the developing brine shrimp *Artemia*: immunochemical localization of the alpha- and beta-subunits, *J. Histochem. Cytochem.*, 39, 1455–1460, 1991.

569. Surbida, K.L. and Wright, J.C., Embryo tolerance and maternal control of the marsupial environment in *Armadillidium vulgare* (Isopoda), *Physiol. Biochem. Zool.*, 74, 894–906, 2001.

570. Susanto, G.N. and Charmantier, G., Ontogeny of osmoregulation in the crayfish *Astacus leptodactylus*, *Physiol. Biochem. Zool.*, 73, 169–176, 2000.

571. Susanto, G.N. and Charmantier, G., Crayfish freshwater adaptation starts in eggs: ontogeny of osmo-regulation in embryos of *Astacus leptodactylus*, *J. Exp. Zool.*, 289, 433–440, 2001.

572. Sutcliffe, D.W., Sodium regulation in the fresh-water amphipod *Gammarus pulex* (L), *J. Exp. Biol.*, 46, 499–518, 1967.

573. Sutcliffe, D.W., Sodium regulation and adaptation to fresh water in gammarid crustaceans, *J. Exp. Biol.*, 48, 359–380, 1968.

574. Sutcliffe, D.W., Regulation of water and some ions in gammarids (Amphipoda). I. *Gammarus duebeni* Lilljeborg from brackish water and fresh water, *J. Exp. Biol.*, 55, 325–344, 1971.

575. Sutcliffe, D.W. and Shaw, J., The sodium balance mechanism in the fresh-water amphipod *Gammarus lacustris* Sars, *J. Exp. Biol.*, 46, 339–358, 1967.

576. Takeuchi, I., Matsumasa, M., and Kikuchi, S., Gill ultrastucture and salinity tolerance of *Caprella* spp. (Crustacea: Amphipoda: Caprellidea) inhabiting the *Sargassum* community, *Fish. Sci.*, 69, 966–973, 2003.

577. Talbot, P., Clark, W.H., and Lawrence, A.L., Light and electron microscopic studies on osmoregulatory tissue in the developing brown shrimp, *Penaeus aztecus*, *Tissue Cell*, 4, 271–286, 1972.

578. Taylor, A.C. and Greenaway, P., Osmoregulation in the terrestrial Christmas Island red crab *Gecarcoidea natalis* (Brachyura: Gecarcinidae): modulation of branchial chloride uptake from the urine, *J. Exp. Biol.*, 205, 3251–3260, 2002.

579. Taylor, A.C., Greenaway, P., and Morris, S., Adaptation to a terrestrial existence by the robber crab *Birgus latro*. VIII. Osmotic and ionic regulation on freshwater and saline drinking regimens, *J. Exp. Biol.*, 179, 93–113, 1993.

580. Taylor, E.W., Butler, P.J., and Al-Wassia, A., The effect of a decrease in salinity on respiration, osmoregulation and activity of the shore crab *Carcinus maenas* (L.) at different acclimation temperatures, *J. Comp. Physiol.*, 119, 155–170, 1977.

581. Taylor, H.H. and Seneviratna, D., Ontogeny of salinity tolerance and hyper-osmoregulation by embryos of the intertidal crabs *Hemigrapsus edwardsii* and *Hemigrapsus crenulatus* (Decapoda, Grapsidae): survival of acute hyposaline exposure, *Comp. Biochem. Physiol.*, 140A, 495–505, 2005.

582. Taylor, H.H. and Taylor, E.W., Gills and lungs: the exchange of gases and ions, in *Microscopic Anatomy of Invertebrates*, Harrison, F.W. and Humes, A.G., Eds., Wiley-Liss, New York, 1992, pp. 203–343.

583. Theede, H., Einige neue Aspekte bei der Osmoregulation von *Carcinus maenas*, *Mar. Biol.*, 2, 14–120, 1969.

584. Therien, A.G. and Blostein, R., Mechanisms of sodium pump regulation, *Am. J. Physiol.*, 279, C541–C566, 2000.

585. Thuet, M., Thuet, P., and Phillippot, J., Activité de l'ATPase (Na⁺-K⁺) en fonction de la morphologie et de la structure histologique des pléopodes ainsi que de la concentration du sodium de l'hémolymphe chez *Sphaeroma serratum* (Fabricius), *C. R. Acad. Sci. Paris*, 269, 233–236, 1969.

586. Thuet, P., Etude des flux de diffusion de l'eau en fonction de la concentration du milieu extérieur chez l'isopode *Sphaeroma serratum* (Fabricius) [Influence of environmental salinity on the diffusion fluxes of water in the isopod *Sphaeroma serratum* (Fabricius)], *Arch. Intern. Physiol. Bioch.*, 86, 289–316, 1978.

587. Thuet, P., Transfer of water in the isopod crustacean *Sphaeroma serratum* (Fabricius) as a function of salinity of the environment, *Arch. Intern. Physiol. Bioch.*, 85, 1011–1042, 1978.

588. Thuet, P., Adaptations écophysiologiques d'*Artemia* (Crustacé, Branchiopode, Anostracé) aux variations de salinité, *Bull. Soc. Ecophysiol.*, 7, 203–225, 1982.

589. Thuet, P., Charmantier-Daures, M., and Charmantier, G., Relation entre l'osmorégulation et l'activité d'ATPase NA⁺-K⁺ et d'anhydrase carbonique chez les larves et post-larves de *Homarus gammarus* (L.) (Crustacea, Decapoda), *J. Exp. Mar. Biol. Ecol.*, 115, 249–261, 1988.

590. Torres, G., Charmantier-Daures, M., Chifflet, S., and Anger, K., Effects of long term exposure to different salinities on the location and activity of Na⁺-K⁺-ATPase in the gills of juvenile mitten crab *Eriocheir sinensis*, *Comp. Biochem. Physiol.*, 147A, 460–465, 2007.

591. Toullec, J.-Y., Serrano, L., Lopez, P., Soyez, D., and Spanings-Pierrot, C., The crustacean hyperglycemic hormones from an euryaline crab, *Pchygrapsus marmoratus*, and a freshwater crab, *Potamon ibericum*: eyestalk and pericardial isoforms, *Peptides*, 27, 1269–1280, 2006.

592. Towle, D.W., Role of Na⁺,K⁺-ATPase in ionic regulation by marine and estuarine animals, *Mar. Biol. Lett.*, 2, 107–121, 1981.

593. Towle, D.W., Cloning and sequencing a Na⁺/K⁺/2Cl⁻ cotransporter from gills of the euryhaline blue crab *Callinectes sapidus*, *Am. Zool.*, 38, 114A, 1998.

594. Towle, D.W. and Holleland, T., Ammonium ion substitutes for K⁺ in ATP-dependent Na⁺ transport by basolateral membrane vesicles, *Am. J. Physiol.*, 252, R479–R489, 1987.

595. Towle, D.W. and Kays, W.T., Basolateral localization of Na⁺/K⁺-ATPase in gill epithelium of two osmoregulating crabs, *Callinectes sapidus* and *Carcinus maenas*, *J. Exp. Zool.*, 239, 311–318, 1986.

596. Towle, D.W., Mangum, C.P., Johnson, B.A., and Mauro, N.A., The role of coxal gland in ionic, osmotic and pH regulation in the horseshoe crab, *Limulus polyphemus*, in *Physiology and Biology of Horseshoe Crabs: Studies on Normal and Environmentally Stressed Animals*, Bonaventura, J., Bonaventura, C., and Tesh, S., Alan R. Liss, New York, 1982.

597. Towle, D.W., Palmer, G.E., and Harris III, J.L., Role of gill Na⁺+K⁺-dependent ATPase in acclimation of blue crabs (*Callinectes sapidus*) to low salinity, *J. Exp. Zool.*, 196, 315–322, 1976.

598. Towle, D.W., Rushton, M.E., Heidysch, D., Magnani, J.J., Rose, M.J., Amstutz, A., Jordan, M.K., Shearer, D.W., and Wu, W.S., Sodium–proton antiporter in the euryhaline crab *Carcinus maenas*: molecular cloning, expression and tissue distribution, *J. Exp. Biol.*, 200, 1003–1014, 1997.

599. Towle, D.W. and Smith, C.M., Gene discovery in *Carcinus maenas* and *Homarus americanus* via expressed sequence tags, *Integr. Comp. Biol.*, 46, 912–918, 2006.

600. Townsend, K., Spanings-Pierrot, C., Hartline, D.K., King, S., Henry, R.P., and Towle, D.W., Salinity-related changes in crustacean hyperglycemic hormone (CHH) mRNA in pericardial organs of the shore crab *Carcinus maenas*, *Am. Zool.*, 41, 1608–1609, 2001.

601. Trausch, G., Effect of temperature upon catalytic properties of lactate dehydrogenase in the lobster, *Biochem. Syst. Ecol.*, 4, 65–68, 1976.

602. Travis, D.F., The molting cycle in the spiny lobster *Panulirus argus* Latreille. I. Molting and growth in laboratory maintained individuals, *Biol. Bull. Woods Hole*, 107, 433–450, 1954.

603. Truchot, J.-P., Regulation of acid–base balance, in *Internal Anatomy and Physiological Regulation*, Mantel, L.H., Ed., Academic Press, New York, 1983, pp. 431–471.

604. Tsai, J.-R. and Lin, H.-C., V-type H^+-ATPase and Na^+,K^+-ATPase in the gills of 13 euryhaline crabs during salinity acclimation, *J. Exp. Biol.*, 210, 620–627, 2007.

605. Tucker, R.K., Free amino acids in developing larvae of the stone crab, *Menippe mercenaria*, *Comp. Biochem. Physiol.*, 60A, 169–172, 1978.

606. Tullis, R.E. and Kamemoto, F.I., Separating and biological effects of CNS factors affecting water balance in the decapod crustacean *Thalamita crenata*, *Gen. Comp. Endocrinol.*, 23, 19–28, 1974.

607. Tyson, G.E., The fine structure of the maxillary gland of the brine shrimp *Artemia salina*: the end sac, *Z. Zellforsch. Mikrosk. Anat.*, 86, 129–138, 1968.

608. Tyson, G.E., The fine structure of the maxillary gland of the brine shrimp *Artemia salina*: the efferent duct, *Z. Zellforsch. Mikrosk. Anat.*, 93, 151–163, 1969.

609. Van Dover, C.L., *The Ecology of Deep-Sea Hydrothermal Vents*, Princeton University Press, Princeton, NJ, 2000.

610. Van Herp, F., Molecular, cytological and physiological aspects of the crustacean hyperglycemic hormone family, in *Recent Advances in Arthropod Endocrinology*, Coast, G.M. and Webster, S.G., Eds., Cambridge University Press, Cambridge, U.K., 1998, pp. 53–70.

611. Verslycke, T., Vangheluwe, M., Heijerick, D., Van Sprang, P., and Janssen, C.R., The toxicity of metal mixtures to the estuarine mysid *Neomysis integer* (Crustacea: Mysidacea) under changing salinity, *Aquat. Toxicol.*, 64, 307–315, 2003.

612. Vilas, C., Drake, P., and Pascual, E., Oxygen consumption and osmoregulatory capacity in *Neomysis integer* reduce competition for resources among mysid shrimp in a temperate estuary, *Physiol. Biochem. Zool.*, 79, 866–877, 2006.

613. Vinagre, T.M., Alciati, J.C., Regoli, F., Bocchetti, R., Yunes, J.S., Bianchini, A., and Montserrat, J.M., Effects of microcystin on ion regulation and antioxydant system in gills of the estuarine crab *Charmagnathus granumlata* (Decapoda, Grapsidae), *Comp. Biochem. Physiol.*, 135C, 67–75, 2003.

614. Vinagre, T.M., Alciati, J.C., Yunes, J.S., Richards, J., Bianchini, A., and Montserrat, J.M., Effects of extracts from the cyanobacterium *Microcystis aeruginosa* on ion regulation and gill Na^+,K^+-ATPase and K^+-dependent phosphatase activities of the estuarine crab *Chasmagnathus granulata* (Deacapoda, Grapsidae), *Physiol. Biochem. Zool.*, 75, 600–608, 2002.

615. Vitale, A.M., Montserrat, J.M., Castilho, P., and Rodriguez, E.M., Inhibitory effects of cadmium on carbonic anhydrase activity and ionic regulation of the estuarine crab *Chasmagnathus granulata* (Decapoda, Grapsidae), *Comp. Biochem. Physiol.*, 122C, 121–129, 1999.

616. Vogt, G., Functional anatomy, in *Biology of Freshwater Crayfish*, Holdich, D.M., Ed., Blackwell Science, Oxford, 2002, pp. 53–151.

617. Wägele, J.W., Ultrastructure of the pleopods of the estuarine isopod *Cyathura carinata* (Crustacea: Isopoda: Anthuridea), *Zoomorphology*, 101, 215–226, 1982.

618. Wägele, J.W., Isopoda, in *Microscopic Anatomy of Invertebrates*, Harrison, F.W. and Humes, A.G., Eds., Wiley-Liss, New York, 1992, pp. 529–617.

619. Wakabayashi, S., Shigekawa, M., and Pouyssegur, J., Molecular physiology of vertebrate Na^+-H^+ exchangers, *Physiol. Rev.*, 77, 51–74, 1997.

620. Wanson, S.A., Péqueux, A., and Roer, R.D., Na regulation and Na^+,K^+-ATPase activity in the euryhaline fiddler crab *Uca minax* (La Conte), *Comp. Biochem. Physiol.*, 79A, 673–678, 1984.

621. Warburg, M.R. and Rosenberg, M., Ultracytochemical identification of Na^+,K^+-ATPase activity in the isopodan hindgut epithelium, *J. Crust. Biol.*, 9, 525–528, 1989.

622. Warren, M.K. and Pierce, S.K., Two cell volume regulatory system in the *Limulus* myocardium and interaction of ions and quaternary ammonium compounds, *Biol. Bull.*, 163, 504–516, 1982.

623. Watanabe, K., Osmotic and ionic regulation and the gill sodium potassium ATPase activity in the Japanese shore crab *Hemigrapsus sanguineus*, *Bull. Jpn. Soc. Sci. Fish.*, 48, 917–920, 1982.

624. Weihrauch, D., Becker, W., Postel, U., Riestenpatt, S., and Siebers, D., Active excretion of ammonia across the gills of the shore crab *Carcinus maenas* and its relation to osmoregulatory ion uptake, *J. Comp. Physiol.*, 168B, 364–376, 1998.

625. Weihrauch, D., Marini, A.-M., and Towle, D.W., Cloning and expression of a putative Rh-like ammonium transporter from gills of the shore crab *Carcinus maenas*, *Am. Zool.*, 41, 1621, 2001.

626. Weihrauch, D., McNamara, J.C., Towle, D.W., and Onken, H., Ion-motive ATPases and active, transbranchial NaCl uptake in the red freshwater crab, *Dilocarcinus pagei* (Decapoda, Trichodactylidae), *J. Exp. Biol.*, 207, 4623–4631, 2004.

627. Weihrauch, D., Morris, S., and Towle, D.W., Ammonia excretion in aquatic and terrestrial crabs, *J. Exp. Biol.*, 207, 4491–4504, 2004.

628. Weihrauch, D. and Towle, D.W., Na$^+$/H$^+$ exchanger and Na$^+$/K$^+$/2Cl$^-$ cotransporter are expressed in gills of the euryhaline Chinese crab *Eriocheir sinensis*, *Comp. Biochem. Physiol.*, 126B, S94, 2000.

629. Weihrauch, D., Ziegler, A., Siebers, D., and Towle, D.W., Molecular characterization of V-type H$^+$-ATPase (β-subunit) in gills of euryhaline crabs and its physiological role in osmoregulatory ion uptake, *J. Exp. Biol.*, 204, 25–37, 2001.

630. Weihrauch, D., Ziegler, A., Siebers, D., and Towle, D.W., Active ammonia excretion across the gills of the green shore crab *Carcinus maenas*: participation of Na$^+$/K$^+$-ATPase, V-type H$^+$-ATPase and functional microtubules, *J. Exp. Biol.*, 205, 2765–2775, 2002.

631. Werntz, H.O., Osmotic regulation in marine and freshwater gammarids (Amphipoda), *Biol. Bull. Woods Hole*, 124, 225–239, 1963.

632. Wheatly, M.G. and Gannon, A.T., Ion regulation in crayfish: freshwater adaptations and the problem of molting, *Am. Zool.*, 35, 49–59, 1995.

633. White, K.N. and Walker, G., The barnacle excretory organ, *J. Mar. Biol. Assoc. U.K.*, 61, 529–547, 1981.

634. Wilder, M.N., Huong, D.T.T., Atmomarsono, M., and Yang, W.-J., Ouabain-sensitive Na/K-ATPase activity increases during embryogenesis in the giant freshwater prawn *Macrobrachium rosenbergii*, *Fish. Sci.*, 67, 182–184, 2001.

635. Williams, A.B., The influence of temperature on osmotic regulation in two species of estuarine shrimps (*Penaeus*), *Biol. Bull. Woods Hole*, 119, 560–571, 1960.

636. Wolcott, T., Water and solute balance in the transition to land, *Am. Zool.*, 32, 370–381, 1992.

637. Wolcott, T.G. and Wolcott, D.-L., Ion conservation and reprocessing of urine in the land crab *Gecarcinus lateralis*, *Physiol. Zool.*, 64, 344–361, 1991.

638. Wu, J.P. and Chen, H.C., Effects of cadmium and zinc on oxygen consumption, ammonium excretion, and osmoregulation of the white shrimp (*Litopenaeus vannamei*), *Chemosphere*, 57, 1591–1598, 2004.

639. Yamasaki, T., Makioka, T., and Saito, J., Morphology, in *Biology of the Horseshoe Crabs*, Sekiguchi, K., Ed., Science House, Tokyo, 1988, pp. 69–132.

640. Yancey, P.H., Blake, W.R., and Conley, J., Unusual organic osmolytes in deep-sea animals: adaptations to hydrostatic pressure and other perturbants, *Comp. Biochem. Physiol.*, 133A, 667–676, 2002.

641. Young, A.M., Osmoregulation in three hermit crab species, *Clibanarius vittatus* (Bosc), *Pagurus longicarpus* Say and *P. pollicaris* Say (Crustacea, Decapoda, Anomura), *Comp. Biochem. Physiol.*, 63A, 377–382, 1979.

642. Zainal, K.A.Y., Taylor, A.C., and Atkinson, R.J.A., The effect of temperature and hypoxia on the respiratory physiology of the squat lobsters, *Munida rugosa* and *Munida sarsi* (Anomura, Galatheidae), *Comp. Biochem. Physiol.*, 101A, 557–567, 1992.

643. Zanders, I.P., Ionic regulation in the mangrove crab *Goniopsis cruentata*, *Comp. Biochem. Physiol.*, 60, 293–302, 1978.

644. Zanders, I.P. and Rojas, W.E., Transbranchial potentials and ion fluxes across isolated, perfused gills of *Uca rapax*, *Mar. Biol.*, 125, 307–314, 1996.

645. Zare, S. and Greenaway, P., The effect of molting and sodium depletion on sodium transport and the activities of Na$^+$K$^+$-ATPase and V-ATPase in the freshwater crayfish *Cherax destructor* (Crustacea: Parastacidae), *Comp. Biochem. Physiol.*, 119A, 739–745, 1998.

646. Zeiske, W., Onken, H., Schwarz, H.-J., and Graszynski, K., Invertebrate epithelial Na^+ channels: amiloride-induced current-noise in crab gill, *Biochim. Biophys. Acta*, 1105, 245–252, 1992.

647. Ziegler, A., Immunocytochemical localization of Na^+,K^+-ATPase in the calcium-transporting sternal epithelium of the terrestrial isopod *Porcellio scaber* L. (Crustacea), *J. Histochem. Cytochem.*, 45, 437–446, 1997.

648. Ziegler, A., Weihrauch, D., Hagedorn, M., Towle, D.W., and Bleher, R., Expression and polarity reversal of V-type H^+-ATPase during the mineralization–demineralization cycle in *Porcellio scaber* sternal epithelial cells, *J. Exp. Biol.*, 207, 1749–1756, 2004.

7 Osmotic and Ionic Regulation in Insects

Klaus W. Beyenbach and Peter M. Piermarini

CONTENTS

I. INTRODUCTION

Insects are by far the most numerous animals on Earth, with an estimated 10^{19} individuals alive. The current count of about 900,000 known species accounts for more than 80% of all the species of animals on the planet. Moreover, an additional 2 to 30 million species of insects are thought to exist. Insects are so deeply entrenched in the web of living things that it is difficult to imagine a world of plants and animals without them.[1] No one will doubt the importance of insects in human welfare. Our own lives depend on the relatively few species of pollinating insects that by way of plants nourish us and animals. We all know the taste of honey, the feel of silk, and the fragrance of wax, but we also look with disbelief at the destruction insects can bring on forests, fields, and civilizations; however, that destruction can also facilitate the early conclusion of war.[2]

The building of the Panama Canal illustrates the struggle that can exist between humans and insects and the self-sacrifice humans are capable of making toward the common good. The Panama Canal project was completed only after learning how yellow fever spreads in human populations. Soldiering volunteers provided the definitive conclusion in an experimental protocol that would challenge ethics committees today. The volunteers proved that the bite of the mosquito *Aedes aegypti* distributes the disease in the population. One set of volunteers with the stomach to wear soiled clothes and sleep in the dirty linens of yellow fever patients in rooms isolated from mosquitoes did not get the disease, whereas volunteers wearing clean clothes and sleeping in fresh linen but who were exposed to mosquitoes that had feasted on the blood of a yellow fever patient did get the disease.

The French initiated the Panama Canal project but gave it up in 1893 after losing 22,000 workers to yellow fever and malaria. Five years later, Americans had lost 2450 soldiers in the Spanish–American War: 385 in battle and all the others to diseases such as yellow fever. A casualty list demonstrating 84% of loss of life to tropical disease led the U.S. Army to commission Walter Reed to investigate yellow fever in Cuba, where the physician Carlos Juan Finlay had claimed, as early as 1881, that yellow fever is transmitted by mosquitoes. Reed established the link between yellow fever and mosquitoes in 1900 in the experiment described above. Thereafter, Americans picked up the Panama Canal project, drained the swamps in the vicinity of the construction sites, reduced the population of mosquitoes, and contained the spread of yellow fever. The Panama Canal opened in 1914. We now know that the causative agent of yellow fever is a virus, but we still do not have a cure for the disease. The causative agent of malaria, also transmitted through the saliva of blood-feeding mosquitoes, is a protozoan of the genus *Plasmodium*. Nematodes may also be transmitted through the saliva of mosquitoes, causing heartworm disease in dogs and elephantiasis in humans.

One of the aims of this chapter is to review the salt and water balance in insects—a subject that can fill volumes in view of the enormous diversity of insects and their habitats. No one has yet risen to this challenge; however, the reader will find excellent discussions of this topic in *The Insects*, by Chapman;[3] in *Biology of Disease Vectors*, by Marquardt;[4] and in *The Biology of Mosquitoes*, by Clements.[5] Also see the reviews listed in Table 7.1. A recent book by Chown and Nicolson[6] takes a look at water balance in insects from the perspective of environmental physiology. Dow and Davies[7] discovered putative transport activities in Malpighian tubules beyond osmoregulation by examining the so-called transcriptomes of Malpighian tubules. Gaede[8] examined how peptides couple metabolism to water balance in insects, and Coast[9] has provided the most recent review of the endocrine control of salt and water balance in insect Malpighian tubules and gut.

We approach the review of osmotic and ionic regulation in insects by focusing on challenges at the extremes: conserving water at one end of the spectrum and getting rid of water at the other end. Minimizing water loss is the overall concern of terrestrial insects, but getting rid of water is the concern of larval insects inhabiting freshwater and of insects that gorge on blood or the sap of plants either for nourishment (e.g., *Rhodnius*, *Homalodisca*) or as part of their reproductive cycle (e.g., *Anopheles*, *Aedes*). As will be shown below, the exoskeleton and respiratory, renal, and intestinal/rectal systems all aim to conserve water in terrestrial insects, but the renal system retains the capacity to dump water when water floods the hemolymph.

We begin our review by examining how insects conserve water by way of minimizing cuticular and respiratory water loss. We then focus on Malpighian tubules and their epithelial transport mechanisms and show how these transport mechanisms can be manipulated to accomplish specific homeostatic goals. The review then continues to examine the astonishing feat in some insects of pulling water out of the ambient air. Finally, we conclude this chapter by showing how insects

TABLE 7.1
Guide to Malpighian Tubules of Specific Insects

Insect	Refs.
Diptera:	
Fruit fly (*Drosophila*)	Blumenthal,[262] Dow,[7] O'Donnell, [356] Riegel,[283] Wessing[132]
Mosquitoes (*Aedes*, *Anopheles*)	Beyenbach,[124,207] Bradley,[143,357,358] Coast,[244] Donini,[359] Gill,[360] Patrick,[180] Pullikuth[361]
Orthoptera:	
Cricket (*Acheta*)	Spring,[162] Neufeld,[362] Xu,[363] Hazelton[364]
Weta (*Hemideina*)	Neufeld,[187] Leader[365]
Hemiptera:	
Kissing bug (*Rhodnius*)	Caruso-Neves,[189] Orchard,[192] Gutierrez,[191] Sofia Hernandez,[366] Whittembury,[157] Hazel,[367] Te Brugge,[195] Maddrell,[271] Ianowski[368]
Hymenoptera:	
Ant (*Formica*)	Van Kerkhove,[185] Laenen,[369] Zhang[370]
Coleoptera:	
Beetle (*Tenebrio*)	Holtzhausen,[371] Wiehart[372,373]
Lepidoptera:	
Tobacco hornworm (*Manduca*)	Gaertner,[374] Li,[375] Reagan,[247] Skaer[116]

continue to make important contributions to our understanding of the molecular biology and biochemistry of the vacuolar-type H⁺-ATPase.

Potts and Parry[10] cautioned years ago that, "The insects are a class of specialists and vary so greatly in habit and physiology that no insect can be regarded as typical; *Carausius* [stick insect] examined in detail by Ramsay is as representative as any other." Considerable differences in the composition of the hemolymph both between different orders of insects and even within a single order, exemplify the functional diversity that can be found in the class Insecta.[11] Our focus on Malpighian tubules of mosquitoes (Diptera) in this chapter will also bring out the complexity and functional diversity of Malpighian tubules that not long ago were considered a "simple epithelium."

II. MINIMIZING CUTICULAR WATER LOSS

This complex film undergoes a change of phase with which is associated a greatly increased permeability to water molecules.

Ramsay[12]

Given the small body size of insects, the water household is disadvantaged by: (1) a large surface area-to-volume ratio; (2) a limited storage capacity for water, especially for flying insects; and (3) exposure to high temperature and low relative humidity in terrestrial habitats. Thus, a major challenge to terrestrial insects in general is to manage a water household that is threatened primarily by evaporative water loss through the cuticle and the respiratory system. Transpirative water loss through the cuticle and the respiratory system can account for more than 60% of the body water loss in terrestrial insects,[13,14] and most of that water loss takes place across the cuticle.[14,15] Cuticular water loss would be incompatible with life in most insects if insects were not waterproofed by an external lipid layer; however, the cuticle is not solely specialized for water conservation. It serves other functions as well, such as exoskeletal armor, locomotion, growth, communication, respiration, reproduction, and sensory perception. Thus, the evolution of cuticular waterproofing in different insect species must have been met with tradeoffs to other functions of the cuticle.

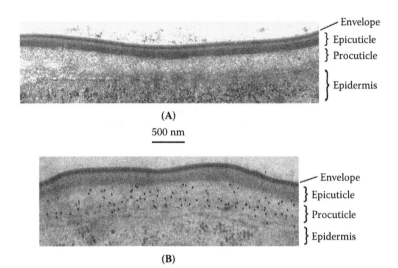

FIGURE 7.1 Transmission electron micrograph of the cuticle in *Drosophila melanogaster* embryo: (A) relative thickness of cuticular layers; (B) black dots represent gold particles conjugated to wheat-germ agglutinin, which selectively binds to and identifies chitin. (From Moussian, B. et al., *Arthropod Struct. Dev.*, 35, 137, 2006. With permission.)

A. STRUCTURE OF THE CUTICLE

The structure of the insect integument has been described in several textbooks and reviews.[3,13,14,16–19] Together with the underlying epidermis and a basal lamina, the cuticle forms the integument of insects. Classically, the cuticle is described as consisting of two main layers: the procuticle and epicuticle (Figure 7.1). Locke,[19] however, has proposed that the thin cuticulin component of the epicuticle is an envelope analogous to that enclosing bacterial cells and that it should be considered an additional main layer. The envelope is extremely thin, approximating the thickness of a plasma membrane (Figure 7.1). It forms the external boundary of the cuticle, and its chemical composition is unknown.[19]

The epicuticle (0.1 to 3 μm thick) lies immediately below the envelope (Figure 7.1 and Figure 7.2). The epicuticle is enriched with protein and lipid; the latter contributes to waterproofing. Underlying the epicuticle is the procuticle, which may be several hundred micrometers thick. The presence of chitin distinguishes the procuticle from the other cuticular layers (Figure 7.1B). In adult insects, the procuticle can be delineated into two sublayers: the endocuticle and the exocuticle (Figure 7.2A). Exo- and endocuticle consist of a protein matrix reinforced with chitin fibers. The proteins of the exocuticle are heavily sclerotized (i.e., cross-linked), which makes it stiffer than the endocuticle. The procuticle is not considered a major barrier to evaporative water loss, but it is thought to shape and strengthen the exoskeleton.

Essential to waterproofing are a series of canals and ducts that arise from the epidermis and traverse the cuticular layers to reach the outer surface of the envelope. Wax or pore canals leaving the epidermis proliferate into a fine meshwork of filaments in the epicuticle (Figure 7.2). The filaments deposit a layer of wax on the outer surface of the envelope (Figure 7.2B). This wax consists primarily of waterproofing lipids. In some insects, dermal gland ducts extend from secretory dermal glands of the epidermis to the envelope (Figure 7.2A). The ducts secrete a layer of cement containing proteins and waterproofing lipids on top of the wax layer (Figure 7.2B). In the cockroach *Periplaneta americana*, cuticular water loss increases when the dermal gland ducts are open. They close during dehydration stress to limit transpirative losses.[20,21]

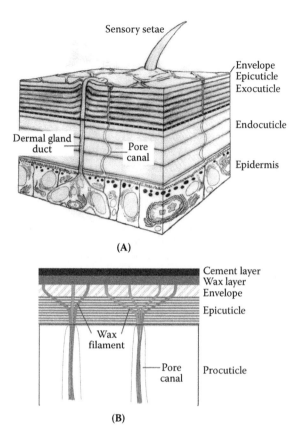

FIGURE 7.2 Diagram of the arthropod cuticle: (A) Subdivision of the procuticle into the exocuticle and endocuticle. (From Hadley, N.F., *J. Exp. Zool.*, 222, 239, 1982. With permission.) (B) Pore canals traverse the cuticle and form a meshwork of filaments in the epicuticle that deposit the wax layer on top of the envelope. (From Klowden, M.J., *Physiological Systems in Insects*, 2nd ed., Academic Press, New York, 2007. With permission.)

B. CLASSICAL EVIDENCE FOR CUTICULAR WATERPROOFING

The study of cuticular waterproofing has been motivated and shaped by two pioneers of insect physiology, J.A. Ramsay and V.B. Wigglesworth, as well as by one of their contemporaries, J.W.L Beament.

1. Ramsay

The first evidence that cuticular lipids were responsible for waterproofing was provided over 70 years ago by Ramsay.[12] In his classic paper on evaporative water loss in the cockroach *Periplaneta americana* (see Reference 22 for a succinct and enjoyable history), Ramsay made three critical observations. First, he found that the rate of water loss from the body surface of dead roaches (with sealed spiracles) suddenly and dramatically increased when the ambient temperature rose above 30°C (Figure 7.3A). The finding confirmed and revised an earlier study on living cockroaches by Gunn,[23] who concluded that the water loss at the higher temperature was due to increased respiration, which was obviously not occurring in Ramsay's dead subjects. Second, Ramsay observed that small drops of water deposited on the cuticle of roaches did not evaporate as readily as those on glass and other surfaces. Third, Ramsay observed with a microscope that the drops of water on the surface of roaches appeared to be covered by a film that resisted puncture of the drop with a fine,

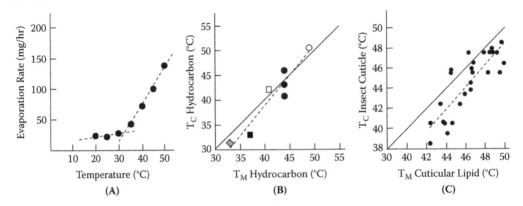

FIGURE 7.3 Cuticular waterproofing in insects: (A) Ramsay's discovery of the critical temperature (T_C), above which evaporative water loss suddenly increases in the cockroach.[12] (B) Lipids applied to a synthetic model membrane, demonstrating nearly perfect correlation between the melting temperature (T_M) and T_C of known lipids.[28] Each symbol represents a unique saturated hydrocarbon (*n*-alkane). Solid line indicates the line of identity; dashed line depicts the measured correlation. (C) Correlation between T_M and T_C measured in the grasshopper (*Melanoplus sanguinipes*); the solid line indicates the line of identity, and the dashed line depicts the measured correlation. (Adapted from Rourke, B.C. and Gibbs, A.G., *J. Exp. Biol.*, 202, 3255, 1999.)

glass needle. Much of that resistance disappeared at temperatures above 30°C, which led Ramsay to conclude that: (1) the exoskeleton of the roach is covered with a "film of fatty substance" that decreases evaporative water loss, and (2) the fatty film melts at temperatures above 30°C, exacerbating the loss of body water. Wigglesworth[24] later referred to critical temperature (T_C) as the temperature at which an insect's evaporative water losses abruptly increase.

2. Wigglesworth

To extend Ramsay's original findings, Wigglesworth[24] conducted similar experiments but studied a number of insect species and different life stages and further examined the lipid nature of the surface film. He found that values of T_C vary among species; the values are from 20 to 30°C higher in insects from drier habitats than those for insects from moist habitats. Similarly, values of T_C were lower in larval insects inhabiting moist soils than in the pupae or adult insects exposed to relatively dry air. Negative correlations between values of T_C and the availability of environmental water led Wigglesworth to suggest that the chemical and physical properties of the cuticle match the waterproofing needs of the insect.[24] Importantly, Wigglesworth found that evaporative water loss was greatly increased by: (1) mild abrasions of the cuticle, and (2) the application of nonpolar solvents to the exoskeleton. Wigglesworth's findings demonstrated the delicate, hydrophobic nature of the cuticular waterproofing mechanism, which he attributed to a thin layer of highly ordered lipids that becomes disordered at temperatures above the T_C.

3. Beament

Published in the same journal issue and perhaps overshadowed by Wigglesworth's paper, Beament[25] found that artificial cuticles prepared from lipids extracted from molted exoskeletal casts showed responses to temperature and abrasion similar to those observed in dead insects. Beament[25] hypothesized that the cuticle of insects was sealed by a highly ordered monolayer of lipids. He envisioned hydrophobic heads of the monolayer interacting with the underlying cuticle and hydrophobic tails forming the water seal at the surface. Although Beament further developed a complex biophysical model to explain this attractive hypothesis,[26] it was largely discounted after biochemical studies found that lipids expected to form such monolayers (e.g., phospholipids, saturated alcohols) are not abundant in insect epicuticle.

4. Back to the Future

The most important and prevailing hypothesis to emerge from the above studies is that cuticular waterproofing is mediated by an external surface of structured lipids that becomes disrupted or melts at the T_C. The hypothesis went unchallenged for more than 60 years because the technology to accurately and precisely measure the melting temperature (T_M) of cuticular lipids was not available. T_M is the temperature at which 50% of the lipid melts.[27] It was not until the sensitive and precise technique of Fourier transform infrared (FTIR) spectroscopy was established that the lipid melting hypothesis was tested on the grasshopper *Melanoplus sanguinipes*.[28] The authors first showed that the T_M value of a pure lipid (i.e., hydrocarbons) coating a model membrane matched the T_C value (Figure 7.3B). The authors then demonstrated that T_C values measured in intact grasshoppers closely matched T_M values of cuticular lipid extracts (Figure 7.3C), providing convincing evidence in support of the lipid melting hypothesis. Ramsay[12] "got the critical temperature story right the first time," notes Gibbs.[22]

C. BIOCHEMICAL COMPOSITION AND BIOPHYSICAL PROPERTIES OF CUTICULAR LIPIDS

A thorough summary of the studies describing the biochemistry and biophysics of epicuticular lipids in insects could easily encompass a chapter in itself. For extensive reviews on this subject, we direct the reader to References 29 to 35.

1. Biochemical Composition

The heterogeneity of lipids detected in the cuticle of insects is remarkable. The epicuticle of the house fly *Musca domestica* may contain more than 100 different hydrophobic compounds.[36,37] The list of lipid classes detected in cuticles includes, but is not limited to, saturated hydrocarbons (straight-chained *n*-alkanes and branched methylalkanes), unsaturated hydrocarbons (*n*-alkenes and alkadienes), fatty acids, wax esters, ketones, and sterols. Notably, phospholipids and acylglycerols (e.g., triglycerides) are absent from the epicuticle or detectable only in low amounts. In view of the diversity of insects and the habitats they select, generalizations about the lipid composition of cuticles are difficult; however, saturated hydrocarbons (*n*-alkanes and methylalkanes) containing 20 to 40 carbons are usually the most abundant. Next are unsaturated hydrocarbons (*n*-alkenes) containing 20 to 30 carbons and saturated wax esters containing at least 30 carbon atoms (Figure 7.4).

2. Lipid Biophysics and Waterproofing

For a cuticular lipid to be effective at waterproofing, it must be in a solid or semisolid phase at the temperatures experienced by an insect; thus, the T_M value of cuticular lipids should be greater than the insect's ambient temperature. The T_M of any lipid, and consequently its waterproofing ability, is ultimately dependent on its molecular packing and size. The molecular packing of a lipid, which is determined by its biochemical structure, is the most influential parameter (Figure 7.4 and Figure 7.5).

In view of their long, slender hydrocarbon chains, the *n*-alkanes are able to pack the tightest and thus are considered the most effective at waterproofing (Figure 7.4A). Moreover, the *n*-alkanes exhibit the highest melting points, with T_M values of at least 50°C (Figure 7.5).[38] In terms of structure, the methylalkanes are similar to *n*-alkanes but contain a single methyl branch on one of their carbons. The degree of molecular packing and waterproofing for methylalkanes is highly dependent on the location of that methyl branch; for example, if the branch occurs near the beginning of the carbon chain, the melting temperature is slightly lower than its *n*-alkane equivalent, because the branch does not greatly affect the linear structure (e.g., compare *n*-alkane with 2-methylalkane in Figure 7.4A). As the branch moves closer to the center of the molecule, however, the linear

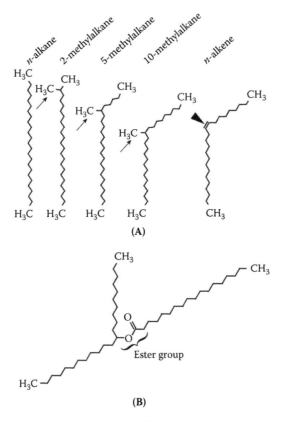

(A)

(B)

FIGURE 7.4 Molecular structures of cuticular lipids: (A) Saturated and unsaturated hydrocarbons. Note how the position of the methyl branch (arrow) influences the structure of the methylalkane; also note the kink in the structure of the *n*-alkene caused by the double-bonded carbon (arrow head). (B) Saturated wax ester. Note the ester linkage resulting in a nonlinear branched structure.

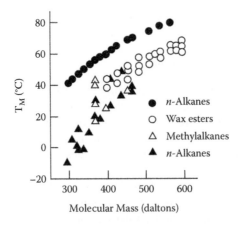

FIGURE 7.5 Effect of molecular mass on the melting temperature (T_M) of hydrocarbons. The linear *n*-alkanes are characterized by the highest T_M values, followed by the wax esters. Methylalkanes of similar molecular mass (e.g., ~375 daltons) can have different T_M values depending on the location of their methyl branch (see Figure 7.4A). The *n*-alkenes typically exhibit the lowest T_M values, because of their kinked structure (see Figure 7.4A). Every 100 daltons approximates 7 carbon atoms of a hydrocarbon chain. (Adapted from Gibbs, A.G., *J. Insect Physiol.*, 48, 391, 2002.)

structure becomes more disrupted along with a decrease in T_M (Figure 7.4A). Thus, methylalkanes of a similar molecular mass can exhibit different T_M values depending on the location of the methyl branch (Figure 7.5).[38]

Saturated wax esters contain an ester group that slightly disrupts molecular packing (Figure 7.4B), but these compounds usually exhibit relatively high T_M values of at least 40°C (Figure 7.5) and thus are effective at waterproofing.[39] In contrast, double-bonded carbons in the n-alkenes kink the hydrocarbon chain and greatly disrupt molecular packing (Figure 7.4A). As shown in Figure 7.5, these compounds can exhibit T_M values below 0°C and thus are the least effective at water-proofing.[40]

Waterproofing hydrocarbons likely exist on the cuticular envelope as mixtures that may or may not be miscible with one another. Gibbs[41] showed that when an n-alkane is mixed with another n-alkane or with a methylalkane the mixture exhibited a single T_M value close to that predicted and melted over a broader range of temperatures than the pure components. Mixtures of n-alkanes and saturated wax esters melted at T_M values only a few degrees lower than predicted.[39] These findings indicated that saturated hydrocarbons and wax esters of insect cuticles are miscible and display unique biophysical properties not found in the pure components alone.[41] The broader temperature range in which the mixtures melt suggests that the lipids may exist as a semisolid at the physiological temperatures of the insect. Although solid-phase lipids would be most effective at waterproofing, a partially melted layer of lipids may have the advantage of a lower viscosity, thus enabling the lipids to spread over the cuticular surface area.

A more complex picture evolves when an unsaturated n-alkene is mixed with a saturated n-alkane, because the two compounds are immiscible.[30,40,41] At physiological temperatures, the n-alkene exists as a fluid distinct from the mostly solid n-alkane. The two hydrocarbons would only become miscible if the temperature reached the T_M of the n-alkane.[30,40] Thus, in insects containing high proportions of n-alkenes on their cuticle, such as fruit flies and house flies, the lipids may be part of a dynamic surface that approaches a fluid mosaic in which solid patches of alkanes float in n-alkenes. The implications of such a dynamic surface are discussed below.

D. Cuticular Lipids and Rates of Cuticular Water Loss

Insects living in xeric habitats typically exhibit lower rates of cuticular water loss or higher T_C values than those living in hygric or mesic habitats which indicates a greater degree of cuticular waterproofing in dry habitats.[13,14] Cuticular waterproofing changes also during metamorphosis as the insect passes from larval, to pupal, and adult stages.[14] Numerous studies have attempted to correlate the biochemical properties of cuticular lipids to: (1) the waterproofing needs of an insect in a particular habitat or at a particular life stage, or (2) the rates of cuticular water loss. In most insects examined—including several orthopterans and coleopterans, the stonefly *Pteronarcys californica*, the fleshflies (*Sarcophaga* sp.), and a mosquito (*Culex pipiens*)—the properties of or alterations to cuticular lipids are consistent with the waterproofing needs of an insect. In brief, enhanced cuticular waterproofing (or lower rates of cuticular water loss) appears to be associated with at least one of the following changes in the cuticle: (1) an increase in the amount of hydrocarbons,[28,42–48] (2) an increase in the proportion of saturated or unbranched hydrocarbons,[46,49–52] and (3) an increase in the T_M values of lipids.[27,28,47,51,53]

An increase in the quantity of hydrocarbons is expected to increase the thickness of the wax or cement layer on top of the cuticular envelope, thereby increasing the diffusion distance and reducing evaporative water loss. One of the more spectacular examples of this mechanism of reducing cuticular water loss is the so-called wax bloom in tenebrionid beetles that inhabit the Sonora Desert of the United States and the Namib Desert of Africa. When exposed to a low relative humidity, the beetles secrete a thick (~20 μm) meshwork of lipid filaments and protein onto the cuticular surface via dermal gland ducts.[45,54,55] The meshwork gives the beetle the appearance of wearing a blue exoskeleton. More importantly, it reduces transpirative losses via the cuticle by

~22%.[45] Thus, upon exposure to low relative humidity, some tenebrionid beetles are able to rapidly secrete an additional waterproofing coat onto their cuticular envelope, which is an obvious osmo-regulatory advantage for these desert dwellers.

One of the best-studied examples of how cuticular lipids acclimate to the habitat of the insect is found in the grasshopper *Melanoplus sanguinipes*. When this species is reared at a temperature of 34°C, the T_M of epicuticular lipids increases on average by 3°C compared to grasshoppers raised at 27°C.[51] The increase in T_M correlates with an increase in the proportion of *n*-alkanes and with a decrease in the proportion of methylalkanes in the cuticular lipids. Field studies confirm these laboratory findings: Grasshoppers collected from relatively warm habitats with low water availability have (1) increased amounts of cuticular lipids,[47] (2) higher T_M values of cuticular lipids,[47,53] and (3) lower rates of body water loss[47] compared to grasshoppers collected from relatively cool habitats with high water availability.

Although the trends described above are tantalizing in that they make good physiological sense consistent with lipid melting, Gibbs[40] warns that comprehensive studies are rare that measure in the same species: (1) rates of cuticular water loss or T_C values, (2) biochemical composition of cuticular lipids, and (3) the T_M values of cuticular lipids. Moreover, the above trends should not be considered dogmatic for all insects, because exceptions exist.

Two glaring exceptions are fruit flies and house flies. Desiccation-resistant lines of *Drosophila melanogaster* exhibit rates of evaporative water loss that are ~40% lower than control lines.[56] The amounts of cuticular hydrocarbons per individual, the composition of epicuticular lipids, and the T_M values of cuticular lipids are very similar between the two lines, however, and what small differences exist do not account for the large differences in rates of water loss.[56] In other fruit flies[57–60] and in the housefly *Musca domestica*,[36,61] correlations between the biochemical properties of cuticular lipids and the temperature or water availability of their habitat or their rates of evaporative water loss are tenuous at best.

One reason why fruit flies and house flies are so exceptional may be the high *n*-alkene composition of the cuticle. The *n*-alkenes exhibit the lowest T_M values and are not miscible with the saturated alkanes at physiological temperatures; thus, the hydrocarbon surface of fruit flies and house flies may exist in two distinct phases which may obscure correlations between the biochemical composition of cuticular lipids and their T_M in relation to water loss.[40] Because *n*-alkenes are important for chemical communication in insects, Gibbs[40] proposed that insects with high proportions of *n*-alkenes in their cuticle risk enhanced water loss in exchange for more receptive chemical communication. Although *n*-alkenes would not provide effective waterproofing, they may provide a pathway for unsaturated pheromones to permeate the cuticle and reach receptors in the underlying epidermis. The *n*-alkenes are necessary, because saturated alkanes provide a barrier not only to water loss but also to pheromones. The two-phase model is consistent with studies on house flies demonstrating that sexual maturation results in increased rates of cuticular water loss and proportions of *n*-alkenes among cuticular lipids.[36,61]

III. MINIMIZING RESPIRATORY WATER LOSS

Respiratory surfaces in animals have evolved to maximize the diffusion of O_2 and CO_2. The strategy is to increase the area for diffusion and to decrease the diffusion distance for O_2 and CO_2; however, the same strategy also invites diffusive exchanges of water and solutes between an animal's environment and its blood or hemolymph. For terrestrial insects, the primary threat to water balance is osmotic desiccation from their moist tracheal system to the relatively dry air that they breathe. Faced with the vital need to breathe, terrestrial insects have devised their own mechanisms to minimize respiratory water loss. The following summary is offered with the caveat that no insect can be representative of other insects. Indeed, Figure 7.6 illustrates what all is possible in insectan designs of respiratory systems.

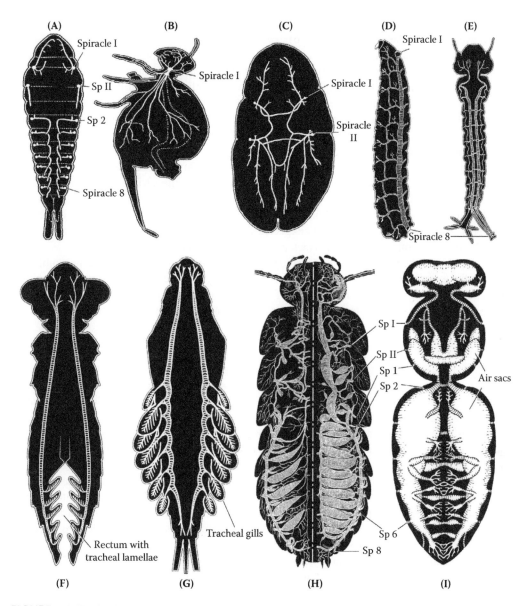

FIGURE 7.6 Tracheal systems in aquatic and terrestrial insects: (A) Separate, segmental tracheal system in machilids (silverfish); (B) secondarily reduced tracheal system in the collembolan *Sminthurus* (a springtail) with a single spiracle (sp); (C) secondarily reduced tracheal system in the flightless scale insect; (D) amphipneustic system with one anterior and one posterior opening in the larval housefly; (E) metapneustic system with only one opening in aquatic larval mosquitoes; (F) rectal tracheal lamellae in the aquatic larval dragonfly; (G) external tracheal gills in the larval mayfly; (H) tracheal system in the adult cockroach; (I) air sacs of the tracheal system in the honeybee. Roman and Arabic numbers identify thoracic and abdominal spiracles respectively. (Adapted from Wasserthal, L.T., in *Lehrbuch der Entomologie*, 2nd ed., Dettner, K.P. and Peters, W. Eds., Spektrum Verlag, Heidelberg, 2003, p. 165.)

A. PULMONARY AND TRACHEAL SYSTEMS

The human lung is a compact structure in the thorax that facilitates diffusive gas exchange between two flowing media across a large alveolar surface approximately the size of a tennis court. Air flows on one side of the surface and blood on the other. The two media exchange gases across a thin barrier that maximizes the diffusion of O_2 and CO_2. In contrast, the "lung" of insects can be found in every part of the insect (Figure 7.6).

Whereas a single nasopharyngeal opening admits access to the vertebrate lung, the insect "lung" has up to 20 openings at the body surface. These openings lead to a network of ducts (tracheae and tracheoles) so extensive and deep as to reach each metabolizing cell of the insect. The major function of the tracheal system is the delivery of O_2 to each cell and the removal of CO_2 from it. O_2 must first dissolve in water to be useful for metabolism, and CO_2 is produced in the aqueous environment of cells. Accordingly, the respiratory handling of O_2 and CO_2 is intricately coupled to the respiratory handling of water.

As illustrated in Figure 7.6, the tracheal system begins at several openings in the body wall, at so-called spiracles, and terminates at blind-ended tracheoles at the surface or inside cells.[3] The basic body plan provides for two pairs of thoracic and eight pairs of abdominal spiracles; however, this body plan is not subscribed by all insects (Figure 7.6). The number of terminal tracheoles has not yet been counted in any insect but is expected to be somewhat less than the total number of cells in the insect, as one tracheole provides gas exchange for more than one cell.

In some insects, the primary tracheae are expanded to form air sacs that increase the ventilation volume at the expense of hemolymph volume (Figure 7.6H,I).[62] The air sacs are more compliant than tracheae, forming balloon-like reservoirs of air. In terrestrial insects, these reservoirs are useful when an insect closes its spiracles to conserve water during evaporative water stress in dry air.

B. Spiracles

Most insects have spiracles that can open and close. The gating structures are located at the surface of the body wall or slightly below comb-like filters and an atrium.[63] Spiracles are under peripheral (local) and central (neural) neuromuscular control. Relaxing of the closer muscle in response to increasing CO_2 is an example of local control. Neural commands open and close spiracles in coordination with the pumping actions of the exoskeleton that ventilate the tracheal system and cause hemolymph to circulate.[64] Between open (O) periods and between ventilation (V) periods, spiracles can close (C) or flutter (F) in cycles of discontinuous gas exchange cycle (DGC).[65–68] Flutter can be described as repetitive openings and closures with a frequency as high as that of a tremoring muscle. It is intuitive that open spiracles pose the greatest threat to respiratory water loss and that closed spiracles prevent water loss. Partially open spiracles or fluttering spiracles tend to conserve water quite effectively. Serving as gates to the tracheal system, spiracles are therefore considered the primary controllers of respiratory water loss.[69]

Spiracles control the gas exchange between the ambient air and tracheal manifolds. Tracheal manifolds turn into segmental tracheal tubes that branch to give rise to tracheoles. Figure 7.7 illustrates the branching of tracheae on their way to muscle and epithelial cells of the gut of a tobacco hornworm, the larva of *Manduca sexta*. Terminal tracheoles often lie on top of cells for the direct transfer of O_2 and CO_2 (Figure 7.8A). Tracheoles are intracellular tubes inside thin, flat tracheolar cells (Figure 7.8B,C). Tracheoles may indent cells without penetrating their plasma membranes which brings tracheoles in close proximity to mitochondria, especially in very active cells such as those of flight muscle.[69,70]

C. Fluid-Filled Terminal Tracheoles

At the level of metabolizing cells, the terminal end of tracheoles may be filled with fluid.[71] Wigglesworth was first to observe that the column of this fluid falls in tracheoles during muscular contractions. The fluid column also shortens with increasing temperature and decreasing O_2 content of the ambient air. Mitochondria are never far from terminal tracheoles (Figure 7.8). The proximity minimizes the diffusion distance for O_2 and CO_2. Fluid within the terminal tracheoles impedes the axial diffusion of O_2 and CO_2. Thus, diffusion of O_2 and CO_2 can be enhanced by filling the terminal

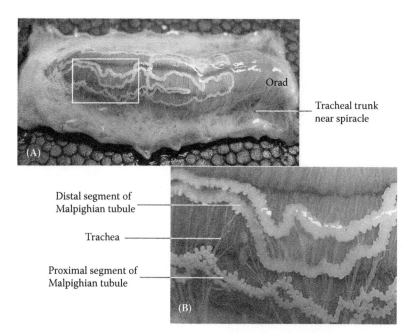

Orad

Tracheal trunk
near spiracle

(A)

Distal segment of
Malpighian tubule

Trachea

Proximal segment of
Malpighian tubule

(B)

FIGURE 7.7 Tracheolar branching in the 5th instar larva of *Manduca sexta*: (A) Most trachea lead to muscular and epithelial cells of the gut. Note the trachea radiating from a tracheal trunk near a spiracle. (B) Cut-out to illustrate tracheal branching. (Photographs courtesy of O. Vitavska, University of Osnabrueck, Germany.)

tracheoles with air. In some insects, such as *Drosophila*, the tracheal cells contain hemoglobin,[72] which assists in the delivery of O_2 to metabolizing cells, because the fluid column in the tracheoles is bypassed. The bypass may be sufficient to support metabolism in the resting insect when spiracles are most likely closed. In contrast, an active insect must reduce the fluid column in terminal tracheoles to maximize the diffusion of O_2 from tracheole to mitochondria. It is unknown how water enters and leaves terminal tracheoles, but aquaporin water channels have been identified in tracheoles.[73]

D. Taenidia

Taenidia are coil-like structures in tracheal and tracheolar walls that prevent the physical collapse of air ducts while allowing axial volume changes (Figure 7.8B). The change in volume can derive from a change in tracheal pressure[74] or hemolymph pressure;[62,67] for example, pressure in the tracheal tree decreases as O_2 is removed from it when spiracles are closed. Changes in hemolymph pressure can derive from the cyclical muscular contractions of the body wall associated with ventilation and the circulation of hemolymph.[62] Body movements associated with mere physical activity of the insect also cause changes in hemolymph pressure that autoventilate the tracheal tree.[75]

E. Diffusion and Convection

Respiratory systems subscribe to diffusion and convection that may be supplemented by molecular carriers such as hemoglobin. Diffusion is the most utilized transport mechanism in plants and animals. It is automotive transport that stems from the molecular mobility of solutes, water, and gases, all in fields of thermal energy from the sun. Students of biology tend to think that active transport is the most utilized transport mechanism; however, no biological system could muster all the energy that would be required if the transport of O_2, CO_2, water, ions, and nutrients were all active and dependent on the hydrolysis of adenosine triphosphate (ATP).

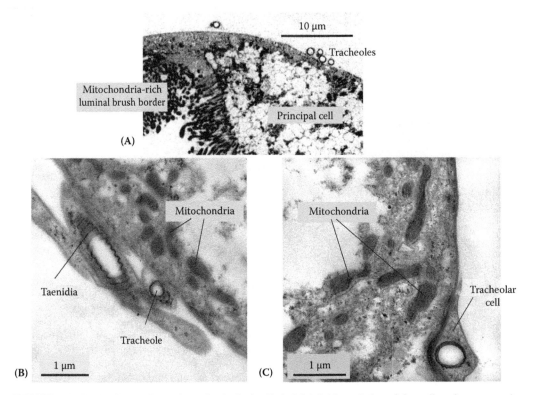

FIGURE 7.8 Tracheoles on the surface of principal cells in Malpighian tubules of the yellow fever mosquito (*Aedes aegypti*). (A) Bundle of tracheoles close to a trachea; other tracheoles were lost during the isolation of the tubule; (B) taenidium in a tracheole; (C) tracheolar cell with tracheole. Note the close proximity of the tracheole and mitochondria.

For gases, the general diffusion equation illustrates that the rate of diffusion (F) is inversely proportional to the length (L) of the diffusion path and directly proportional to the area (A) of diffusion, the effective diffusion coefficient (D'_x) of gas x, the capacitance coefficient (β_x) of gas x, and the driving force for diffusion—the partial pressure difference (ΔP_x) of gas x. In the case of $x = CO_2$, D'_x is the effective biological diffusion coefficient of CO_2 that includes CO_2 coefficients in air, water, and lipid:

$$F_x = \frac{AD'_x\beta_x}{L}\Delta P_x \tag{7.1}$$

The capacitance coefficient (β_x) includes a physical capacitance (solubility of CO_2 in water), a chemical capacitance (CO_2 in the form of HCO_3^-), and a biological capacitance (CO_2 bound to macromolecules). If $x = O_2$, the biological capacitance is particularly large in the presence of respiratory pigments; for example, hemoglobin raises β_{O_2} to overcome the low physical O_2 solubility in water.

Equation 7.1 shows that the rate of diffusion is high for short distances and low for long distances. An increase in the driving force ΔP_x can extend the range of diffusion when, for example, the partial pressure of CO_2 in tracheolar air behind the spiracle may reach 6.5 kPa compared to 0.035 kPa in ambient air (at sea level, 1 atm = 760 mmHg = 29.9 inHg = 101.325 kPa). Indeed, insects use this specific partial pressure strategy for water retention during discontinuous gas exchange when they build up large partial pressure differences for O_2 and CO_2 between the tracheal lumen and the ambient air, which increases rates of diffusion when spiracles do open.[76]

A second mechanism for overcoming the limits of diffusion distance is to increase the area of diffusion, which is accomplished in mammals via progressive branching; for example, one bronchus in the human thorax eventually branches into about 500 million alveoli.[77] Progressive branching of the tracheal tree in insects amplifies the area for diffusive gas exchange with the tissues.

A theoretical analysis of insect tracheal systems by August Krogh (who received the Nobel Prize in Physiology or Medicine in 1920) suggested that pure diffusion of O_2 through the tracheal system is adequate for supporting metabolism in resting insects up to 3.4 g in weight.[78] Diffusion is also sufficient during flight of small insects such as *Drosophila*, where diffusion distances in the tracheal system are very short.[63,79] Krogh, however, did allow that ventilation (i.e., convective gas exchange across open spiracles) supports the diffusion of O_2 in the tissues.[80] These observations led to the classical theory that diffusion can meet oxygen demands in resting insects less than 3.4 g in size whereas larger insects require support by ventilation, even at rest.[69]

When the advantages afforded by minimizing diffusion distance and maximizing diffusion area are exhausted, a third mechanism for increasing transport is convection. Convection moves the whole medium and the diffusible elements it contains in a heteromotive way. Convection is bulk transport or mass flow, the kind of transport that sweeps away driftwood and algae in rivers and erythrocytes in blood. Convective transport in the context of respiratory physiology is known as *ventilation*. It brings air into our lungs, and it blows off CO_2 in humans and whales alike.

F. VENTILATION

What the movement of the diaphragm accomplishes for mammalian lungs, the movement of a compliant exoskeleton in insects accomplishes for the tracheal system. Ventilation includes inhalation and exhalation. Whereas pulmonary exhalation is normally passive in vertebrates (relaxing the diaphragm), it is an active neuromuscular process in insects. In some insects, abdominal segments telescope; in other insects, the abdomen flattens.[81] The pressure that these geometric changes exert on the abdominal hemolymph increases the tracheal pressure at closed spiracles[67] and drives convective flow (exhalation) when spiracles open after tracheal pressure has increased somewhat (delayed opening). Relaxation of the compliant exoskeleton mixes tracheal air at closed spiracles but drives inhalation as spiracles open after the tracheal pressure has dropped a little (delayed opening). In many adult insects, the flow of air can be (1) unidirectional, as in birds, but through separate inflow and outflow orifices (Figure 7.6), or (2) tidal, as in human lungs. One-way air flow became possible in insects after developing longitudinal intersegmental tracheal trunks that provide a pathway from one spiracle to another.[82–85] Accordingly, inhalation may occur at thoracic spiracles and exhalation at abdominal spiracles, or *vice versa* (Figure 7.6H,I).

G. DISCONTINUOUS GAS EXCHANGE

Respiration in many insects, but not all insects, is thought to be unique in what has been termed the discontinuous gas exchange cycle (DGC)[86] or classically cyclic CO_2 release.[67] To the uninitiated, the term *discontinuous gas exchange* can be confounding, because DGC describes a cyclical event that includes a period of no gas exchange with the ambient environment when spiracles are closed (C). As illustrated in Figure 7.9, other phases of the cycle include flutter (F), open (O), and ventilation (V). When spiracles open, CO_2 escapes to the ambient environment in a burst because of the large partial pressure that has been built up in trachea during C and F periods (Figure 7.9). Water also escapes in a burst because the tracheal mixture of gases is saturated with the water of hemolymph. Note that in the absence of ventilation the bursts of CO_2 and H_2O escape are driven by diffusion alone.

The closed phase (C) identifies the discontinuation of gas exchange between the tracheae and ambient air. It occurs only at low metabolic rates (and low temperatures). In general, pupae and some adult insects such as ants exhibit CFO and FO cycles,[65] where diffusion drives respiratory

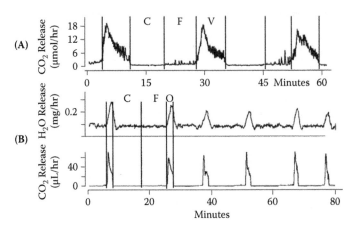

FIGURE 7.9 Discontinuous gas exchange (DGC) in two insects: (A) DGC with a period of ventilation (V) in the resting cockroach (*Periplaneta americana*). (Data from Kestler, P., in *Environmental Physiology and Biochemistry of Insects*, Hoffmann, K.H., Ed., Springer-Verlag, Berlin, 1985, p. 137.) (B) DGC with an open period (O) of spiracles in the ant *Pogonomyrmex rugosus*. C, closed spiracle; F, fluttering spiracle. (Data from Lighton, J.R., *Physiol. Zool.*, 67, 142, 1994.) (Figure adapted from Wasserthal, L.T., in *Lehrbuch der Entomologie*, 2nd ed., Dettner, K.P. and Peters, W. Eds., Spektrum Akademischer Verlag, Heidelberg, 2003, p. 165.)

gas exchange across spiracles. Most resting adult insects exhibit CFV cycles at temperatures below 27°C and FV cycles above 27°C.[66] During the C phase, spiracles are tightly shut for up to an hour and longer, blocking external respiration between the insect and ambient air,[87] but internal respiration between tracheoles and metabolizing cells continues. As a result, the partial pressure of O_2 decreases and the partial pressure of CO_2 increases in the tracheal system. O_2 removed from the tracheoles is not immediately replaced with CO_2 because hemolymph is a large sink for CO_2. The removal of O_2 from the tracheal system—faster than the addition of CO_2—lowers the tracheal pressure below atmospheric pressure. Thus, when spiracles open slightly and briefly during the flutter (F) period, they bring about a passive suction ventilation.[67,74,75] The principal benefits of the F period are (1) the import of O_2 via diffusion and some convection, and, importantly, (2) the negligible loss of H_2O from the tracheal system as illustrated in Figure 7.9B.

H. Setting Physiological Priorities: Gas Exchange vs. Water Conservation

The metabolic requirement for gas exchange, on the one hand, and the need for water conservation, on the other hand, pose conflicting challenges for plants and animals alike. Rather than settle on compromise, insects alternate between physiological priorities. They save water whenever possible but sacrifice water when other activities are more important. Respiratory patterns reflect the change in setting physiological priority.

1. Saving Water at Rest

Good agreement can be found in the literature that water conservation is a benefit of closing the spiracles when metabolic rates are low in resting insects or diapausing pupae.[6,67,88,89] In general, spiracles are closed as long as possible in the resting insect. They open partially during periods of flutter to admit O_2 by diffusion supplemented with suction ventilation and then open fully to allow the release of CO_2 and consequently H_2O (Figure 7.9). CFO and CFV cycles minimize respiratory loss by the closing strategy in C and partial repetitive openings in F when metabolic rates are low (e.g., during rest, low ambient temperatures, in diapausing pupae).

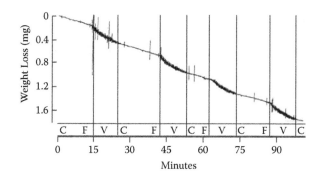

FIGURE 7.10 Weight loss from a resting cockroach (*Periplaneta americana*) during discontinuous gas exchange cycles consisting of closed (C), fluttering (F), and ventilating (V) spiracles at 24.8°C and 76% relative humidity. The large spikes reflect antennal movements. (Adapted from Kestler, P., in *Environmental Physiology and Biochemistry of Insects*, Hoffmann, K.H., Ed., Springer-Verlag, Berlin, 1985, p. 137.)

Studies of the American cockroach at rest during CFV cycles illustrate barely any difference in the water loss during the C and F periods, as a cockroach (weighing approximately 1,030,000 µg) loses water at a rate of 12 µg/min (i.e., less than 2% of its body weight per day), primarily through the cuticle (Figure 7.10). The rate of H_2O loss increases to 25 µg/min, or 4% of the body weight per day, as the cockroach ventilates the tracheal system through open spiracles (Figure 7.10). Not all spiracles may open at rest. Only two spiracles may be active in the insect at rest, opening and closing rhythmically, whereas all eight pairs of spiracles will stay open during activity.[90]

It is intuitive that water loss increases during ventilation, especially during expiration. Less intuitive is the substantial water loss by diffusion when spiracles are open but not ventilating. In general, the diffusion coefficient D (Equation 7.1) is inversely proportional to molecular size. Because H_2O is smaller than O_2 than CO_2, it follows that H_2O diffuses faster than O_2 and CO_2 through the open, nonventilating spiracle; therefore, an open spiracle threatens primarily with diffusive water loss made worse by the water saturation of the tracheal gas mixture. High rates of diffusive water loss can be diminished by opening spiracles only partially during the F period in both resting and active insect.[91] The diffusion equation (Equation 7.1) indicates that this strategy reduces the area of diffusion via the partial opening; however, the diffusive flux through the partially open spiracle can be quite high because the length of the diffusion path through the spiracle orifice is negligibly small compared to the tracheal path to cells. For this reason, it is thought that limiting the open time of spiracles is superior to controlling diffusion area as a water-saving measure of fluttering in FO and FV cycles.[76] Indeed, Figure 7.9 shows that the open period is shortest of all periods during DGC in the resting insect. Open time can be even further reduced by coordinating the opening of the spiracle with abdominal pumping.[66,67]

2. Dealing with Respiratory Water Loss During Activity

Under conditions of activity, metabolic demands override water conservation, and the insect loses water, especially during flight when the saturation partial pressure of water in trachea rises with body temperature.[92–94] Ventilations are now deep and frequent, with spiracles more open than closed or always open, which further increases diffusive water loss. As a result, respiratory water loss increases dramatically, tenfold and more.[68,94,95] Insect flight is energetically the most demanding activity of all animals;[96] however, one consequence of the high fuel requirements in flying insects is the metabolic generation of water when the usual flight fuel, trehalose, is burned in the presence of oxygen. Metabolic water replaces up to 75% of the water lost in the flight of *Drosophila*.[97] Bumble bees produce even more metabolic water than they lose by respiratory diffusion and convection, which prompts them to eliminate the excess water in the urine.[6,98] Other adaptations to reduce water loss may exist. One reason why locusts fly at altitude may be the lower temperatures and consequently the reduced partial pressure of water in trachea.[99]

I. THE PHYSICS OF SAVING WATER IN A TRACHEAL SYSTEM

The most rigorous treatment of water balance as it relates to external respiration in insects is an analysis by Kestler.[67] Starting with first principles, Kestler produces a physical model that couples diffusion to convection. Predictions from that model show that, to minimize respiratory water loss, the spiracles: (1) should be mostly closed during rest and activity until dwindling oxygen levels or a rising CO_2 level force their opening, (2) should have a geometry that maximizes the inflow of air consistent with a subatmospheric pressure in the tracheal system to minimize water loss (suction ventilation), and (3) should open only briefly during bursts of respiratory CO_2 release. Further analysis of this mathematical model shows that convective gas exchange is advantageous for respiratory water conservation in all insects, but especially for small insects,[100] because convective gas exchange reduces the open time of spiracles. Experimental observations in intact insects confirm that the biology follows the physics. In the case of conflicting physiological interests (i.e., gas exchange vs. water conservation), the physics of the tracheal system advises the elimination of CO_2 during short, strong bursts of exhalation. To enhance uptake of O_2 and reduce diffusive losses of CO_2 and H_2O, the physics suggests long periods of inhalation at subatmospheric pressures (suction ventilation). Not surprisingly, Kestler has observed these phenomena in the cockroach.[66,67]

IV. RENAL MECHANISMS OF SALT AND WATER BALANCE

> The secretion of potassium (together with some anion) into the tubule will set up an osmotic pressure, which in its turn will promote a passive inward diffusion of water.
>
> **Ramsay**[101]

A. HOMEOSTASIS OF THE EXTRACELLULAR FLUID COMPARTMENT

Multicellular animals have two major fluid compartments: (1) an intracellular fluid compartment that houses the mechanisms of metabolism of cells, and (2) an extracellular fluid compartment that bathes and supports these cells. The extracellular fluid compartment in insects is the hemolymph. The constancy of the extracellular fluid compartment in both volume and composition is known as *homeostasis*. The homeostasis of the extracellular fluid is the primary function of the kidneys. Turning over extracellular fluid at high rates, the kidneys can rapidly correct changes in extracellular volume and composition that result from the diverse activities of the individual and the unpredictable changes in the external environment, the habitat. In vertebrates, the renal turnover of extracellular fluid begins with the filtration of plasma water and its dissolved constituents. Filtration is possible because of a closed circulatory system capable of producing high filtration pressures in glomerular capillaries. In the absence of blood vessels, insects circulate the hemolymph at pressures too low for filtration. Accordingly, insects must initiate the renal turnover of extracellular fluid by a different mechanism—namely, epithelial secretion that delivers salt and water into the lumen of renal tubules.

Renal tubules in insects are known as *Malpighian tubules*, named after the Italian physician Marcello Malpighi (1628–1694). Malpighian tubules secrete fluid via mechanisms conceptually similar to the secretions of human salivary, sweat, and tear glands. After secreting fluid into the lumen of distal (blind-ended) Malpighian tubules, solutes and water essential to life may be reabsorbed as secreted fluid flows downstream through the proximal Malpighian tubule, the hindgut, and the rectum. The reabsorption of life-essential solutes and water leaves other solutes (in excess or toxic) behind for excretion from the animal. Thus, insects subscribe to the general two-step extracellular fluid homeostasis observed widely in multicellular animals: tubular secretion (or glomerular filtration) followed by tubular reabsorption. If much water is reabsorbed, the remaining solutes may reach precipitating concentrations which allows even greater water reabsorption and conservation.

Our current understanding of Malpighian tubules rests on a sizeable group of biologists who have found their study fascinating. The most popular insects inhabiting laboratories around the world and donating Malpighian tubules for their study are listed in Table 7.1. Malpighian tubules in *Drosophila* have received most of the attention because of the genetic information that has been available for this species for some time; however, integrative physiological and genetic studies on Malpighian tubules from a wider variety of insect species will soon be possible due to the current wave of genomic biology. Since publication of the *Drosophila melanogaster* genome,[102] the genomes of 11 other *Drosophila* species,[103,104] of the malaria mosquito (*Anopheles gambiae*),[105] of the honey bee (*Apis mellifera*),[106] of the silkworm (*Bombyx mori*),[107,108] and of the yellow fever mosquito (*Aedes aegypti*)[109] have been published. Among the genomes currently under analysis are those of the flour beetle (*Tribolium castaneum*), the house mosquito (*Culex pipiens*), the pea aphid (*Acyrtho-siphon pisum*), the human body louse (*Pediculus humanus*), the kissing bug (*Rhodnius prolixus*), the squinting brush brown butterfly (*Bicyclus anynana*), the Glanville fritillary butterfly (*Melitaea cinxia*), and three species of the parasitoid wasps *Nasonia*. The reader can follow the progress of these genome projects and discover new insect genomes under study by accessing http://www.ncbi.nlm.nih.gov/sites/entrez?db=genome and searching for "insecta."

B. Malpighian Tubules in *Drosophila* and *Aedes*

The number of Malpighian tubules in insects varies from zero to several hundred. Aphids (plant lice, Hemiptera) do not have any Malpighian tubules at all,[110,111] which could lift the ink from these pages were it not for the minority status of this contrary group. But, it says a great deal about the functional plasticity of other organs in insects. All other insects studied to date apparently do have Malpighian tubules. The fruit fly has four Malpighian tubules,[112] the yellow fever mosquito has five,[113] the cockroach 150,[114] and the locust 233.[115] The number of Malpighian tubules does not correlate well with the size of the insect, because the larva of the tobacco hornworm (*Manduca sexta*, Latin for "glutton sixfold"; Lepidoptera), one of the biggest insects, has only six Malpighian tubules (Figure 7.7). What this larval moth lacks in number of tubules, however, it makes up for in tubule length: 25 cm for an insect less than 5 cm long.[116] Thus, epithelial mass appears more important to extracellular fluid homeostasis rather than the number of tubules.

In *Drosophila*, Malpighian tubules drain their secretions into a ureter, which then empties into the gut (Figure 7.11). The five Malpighian tubules of *Aedes* drain their secretions directly into the gut at the junction of the midgut and hindgut. We have observed no functional difference between the five Malpighian tubules of the same female mosquito, suggesting that the Malpighian tubule is the functional equivalent of the insect kidney.[113]

Malpighian tubules are much larger in the female mosquito than in the male (Figure 7.12A). The sexual dimorphism[117] extends to functional differences that are more quantitative than qualitative. Malpighian tubules of female mosquitoes secrete salt and water *in vitro* at six times the rate of male Malpighian tubules.[118] Extra mass and added capacity for salt and water transport serve not only the bigger size and metabolism of the female but also the large salt and water loads she occasionally acquires when feeding on blood as part of the reproductive cycle.[119]

Malpighian tubules are formed by a single layer of epithelial cells (Figure 7.12B). When viewed under the light microscope, two types of epithelial cells—principal cells and stellate cells—can be readily observed in Malpighian tubules of *Drosophila* and *Aedes*.[120,121] Principal cells mediate the active transport of Na^+ and K^+ from the hemolymph into the lumen of the Malpighian tubule, whereas stellate cells and the paracellular pathway provide for transepithelial secretion of Cl^-.[122–125] Principal cells are five times more numerous and are much larger than stellate cells. Principal cells account for more than 90% of the tubule mass in female mosquitoes of *Aedes aegypti* (Figure 7.12).

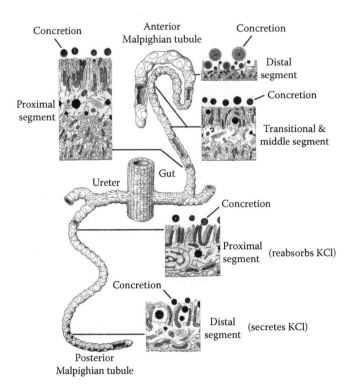

FIGURE 7.11 Malpighian tubules in *Drosophila melanogaster*. Two pairs of Malpighian tubules, an anterior pair and a posterior pair, empty their secretion into the gut via a "ureter." Distal segments form the blind end of the tubule and secrete primarily KCl and water into the tubule lumen. Proximal segments reabsorb KCl and water.[149] Proximal segments merge to form the ureter in *Drosophila*; proximal segments enter the gut directly in *Aedes*. Every tubule segment produces concretions, which are mineralized structures that may be expelled into the tubule lumen. (Adapted from Wessing, A. and Zierold, K., *Cell Tissue Res.*, 272, 491, 1993.)

The focus on macroscopic cell types can be misleading, as molecular/genetic studies in *Drosophila* indicate several functional domains along the length of the Malpighian tubule and epithelial transport systems far more numerous than the number of epithelial cell types suggests.[126–128] In one insect, the New Zealand glow-worm, one segment of the Malpighian tubule glows in the dark.[129] The luciferin glow in the larvae attracts prey that is subsequently ensnarled in mucous trap lines for consumption, whereas the glow in the adult attracts the opposite sex for reproductive consummation. Other insects use Malpighian tubules in the larva to accumulate calcium that later is used to thicken and strengthen the shell of the pupa. Still other insects use Malpighian tubules to produce silk. No vertebrate renal epithelium can match the functional diversity of insect Malpighian tubules.

1. Intracellular Concretions

One striking feature of Malpighian tubules in flies, mosquitoes, and possibly other insects is the presence of dense bodies in the cytoplasm of principal cells. Dense bodies are far more numerous in female Malpighian tubules than in male Malpighian tubules. As a result, female tubules appear opaque and male tubules transparent (Figure 7.12A). Because chemical analyses have revealed the mineralized nature of dense bodies, they are more commonly called *concretions* (Figure 7.12 and Figure 7.13). If the concretions are not properly fixed for microscopic examination, empty spaces will appear in electron micrograph sections of the cytoplasm (Figure 7.12B).

The analysis of concretions has elucidated compositions that depend on the diet, on environmental factors, and on physiological activity. When maintained on a calcium-rich diet, *Rhodnius*

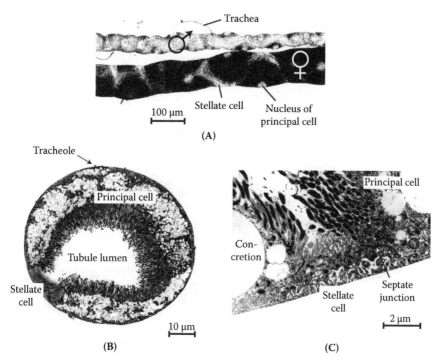

FIGURE 7.12 Malpighian tubules of the yellow fever mosquito (*Aedes aegypti*): (A) The sexual dimorphism displayed by Malpighian tubules may reflect osmotic and ionic challenges that are greater in the female than in the male.[118] Intracellular concretions in principal cells of female tubules make these cells opaque and dark against a bright background. Principal cells of male Malpighian tubules contain far less intracellular concretions than female tubules. The thin stellate cells are devoid of concretions and are therefore transparent. (B) Cross-section through a female Malpighian tubule. A single principal cell curling to fold upon itself forms the tubule lumen. Where the lateral edges of the cell touch, a septate junction normally seals the tubule. In the cross-section shown, a stellate cell is wedged between the lateral boundaries of the principal cell. Note the tall microvilli of principal cells. Virtually every microvillus of the principal cell houses a long and slender mitochondrion. Empty, vesicle-like spaces in the cytoplasm of the principal cell are artifacts stemming from the loss of concretions during the fixation step. (C) Stellate cell embedded between two principal cells. Note the short microvilli that lack mitochondria. The basolateral membrane shows extensive infoldings. Septate junctions at the lateral edges of the cell trace the paracellular pathway from tubule lumen to hemolymph.

accumulates calcium in concretions.[130] Some concretions store calcium and magnesium in a matrix of proteoglycans, and other concretions accumulate potassium.[131–133] Concretions may also contain uric acid which are absent in Malpighian tubules of transgenic *Drosophila* knock-outs lacking a critical subunit of the V-type H^+-ATPase.[134,135] Apparently, the proton pump is involved in the formation of concretions. Concretions in Malpighian tubules of the housefly contain phosphorus, sulfur, chlorine, calcium, iron, zinc, and copper.[136,137]

Studies in the alkali fly (*Ephydra hians*) illustrate the effect of habitat and development on concretions.[138] Larvae of this fly are able to inhabit alkaline lakes such as Mono Lake in California with a pH above 10 and a combined CO_3^{2-} and HCO_3^- concentration approaching 500 mM. Intracellular concretions in one pair of Malpighian tubules are so numerous as to give the tubule the appearance of a gland: a lime gland in particular, because the concretions consist of nearly pure $CaCO_3$.[138] Perfectly round and smooth concretions that range from less than 1 to 10 μm in diameter suggest their layered growth as the larva passes through three instars (Figure 7.13). Before larvae commence puparition, the epithelial cells of the lime gland discharge their concretions into the tubule lumen for excretion, suggesting exocytosis as the mechanism for expelling

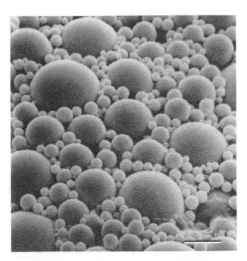

FIGURE 7.13 Scanning electron micrograph of concretions from the lumen of the lime gland (Malpighian tubule) of larval alkali fly (*Ephydra hians*). Bar, 5 μm. The larvae were maintained in filtered water of Mono Lake, CA. The molal concentrations of the major solutes in Mono Lake water at pH 9.8 are Na^+, 1.432; Cl^-, 0.537; SO_4^{2-}, 0.114; CO_3^{2-}, 0.295; and HCO_3^-, 0.054.[352] These solutes add up to an osmotic pressure of 2432 mOsm/L. (From Herbst, D.B. and Bradley, T.J., *J. Exp. Biol.*, 145, 63, 1989. With permission.)

concretions from the cytoplasm.[131,132] In the face fly (*Musca autumnalis*), calcium and phosphorus are accumulated and stored as intracellular concretions in Malpighian tubules during the prepupal period. The concretions are mobilized and moved for deposition in the cuticle during the pupal period.[139–141] The Ca^{2+} mineralization of the cuticle turns the pupa white and as brittle as an egg shell.

Concretions may serve useful functions such as: (1) the storage of calcium, metals, and trace metals, similar to the function of vertebrate bone; (2) the removal of metals and potentially heavy metals from the hemolymph, thereby supporting homeostasis of the extracellular fluid compartment; and (3) the renal excretion of excess ions in precipitates which greatly reduces the water a flying animal must carry.

2. Mitochondria and the V-Type H⁺-ATPase in Microvilli of the Brush Border

Next to intracellular concretions, another striking feature of Malpighian tubules is the presence of mitochondria in microvilli of the apical brush border of principal cells (Figure 7.14A and Figure 7.12B,C). Mitochondria can be observed to move into and out of microvilli, respectively with increased and decreased secretory activity of the tubule.[142,143] Because mitochondria generate ATP, the presence of mitochondria in microvilli suggests an ATP-dependent activity. Indeed, physiological and molecular evidence indicates that the brush border is home to an ATP-driven proton pump, the V-type H⁺-ATPase, as illustrated in Figure 7.14.[144] It will be shown later that the V-type H⁺-ATPase not only energizes ion transport across the apical (plasma) membrane of microvilli but also ion transport through the paracellular pathway and across the basolateral membrane of principal cells. Antibodies against the B-subunit of the V-type H⁺-ATPase localize the pump to the brush border of principal cells in *Aedes* Malpighian tubules (Figure 7.14B). No immunoreactivity is observed in stellate cells (Figure 7.14B), which do not house mitochondria in their microvilli (Figure 7.12C). In addition to the B-subunit of the V-type H⁺-ATPase, principal cells in *Drosophila* Malpighian tubules express the transcripts that encode all the other subunits of this proton pump.[127,135]

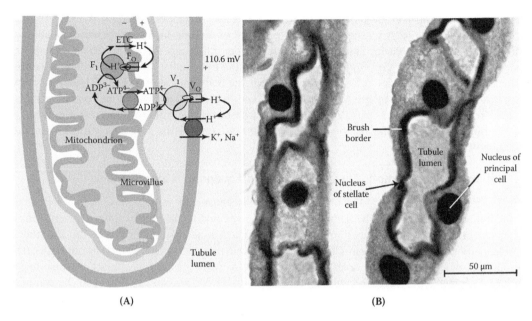

(A)   (B)

FIGURE 7.14 (See color insert following page 208.) The brush border of principal cells in Malpighian tubules of *Aedes aegypti*: (A) Each microvillus contains a mitochondrion. ATP is produced by the F-synthase located in the inner mitochondrial membrane. F_1 and F_0 are, respectively, the catalytic and the proton-translocating complexes of the synthase. ETC is the electron transport chain. V_1 and V_0 are, respectively, the catalytic and proton-translocating complex of the V-type H^+-ATPase (See Figure 7.30 for structural details of the V-type H^+-ATPase). The V_1 complex contains subunit B against which the antibody used in Part B was prepared. (B) Immunolocalization of the B subunit of the V-type H^+-ATPase in the brush border of principal cells of Malpighian tubules of the yellow fever mosquito (*Aedes aegypti*). A stellate cell (arrow) gives no evidence for the B subunit of the V-type H^+-ATPase. The color insert shows immunoperoxidase staining (red) for subunit B of the V-type H^+-ATPase in the brush border of principal cells. A hematoxylin counterstain (blue) labels the nuclei of principal and stellate cells. The antibody was kindly provided by Marcus Huss from the University of Osnabrueck, Germany. (Adapted from Beyenbach, K.W., *News Physiol. Sci.*, 16, 145, 2001.)

3. Primary Urine and Renal Detoxification

Recent transcriptome analyses of Malpighian tubules from *Drosophila* have provided a long list of putative transporters for inorganic and organic solutes[127,128] at odds with the transport physiology displayed by distal (blind-ended) Malpighian tubules *in vitro*—namely, the secretion of NaCl, KCl, and water. To be sure, distal tubules do not represent the whole tubule and its transcriptome; however, another explanation illuminates the paradox: Ringer solutions typically lack the organic solutes the tubule might secrete *in vivo*. If Ringer solutions do not offer organic substrates for secretion, transport systems handling these solutes are silent and consequently not observed *in vitro*. Hence, models of transepithelial electrolyte transport presented in this chapter should be considered minimal transport models that can account for the transepithelial secretion of NaCl, KCl, and water but fall short in acknowledging the organic solute transport systems the tubule might possess.

Transport systems that secrete organic acids and bases into the tubule lumen may play an important role in the detoxification of the hemolymph and in pesticide resistance;[145–147] however, before toxins can be moved from the hemolymph into the tubule lumen, it would seem advantageous to have water already present in the tubule lumen. Without such a primary urine in the lumen, luminal toxin concentrations could reach levels that might kill the epithelial cells that have transported them. Thus, the spontaneous transepithelial secretion of NaCl and KCl and water in minimal Ringer solution *in vitro* may serve *in vivo* to dilute secreted toxins, thereby increasing the rate of renal toxin excretion and the clearance of toxins from the circulation. Nicolson[148] came to this conclusion when investigating the paradoxical presence of diuretic factors in an insect inhabiting the desert, where the need for diuresis might rarely occur. She suggests that diuretic hormones in insects may also serve to clear toxins from the circulation by stimulating the secretion of primary urine. Consistent with such a role of primary urine is the reabsorption of NaCl, KCl, and water in proximal segments of the tubule near the gut[149,150] and in the hindgut and rectum before urine is voided from the body.[9] As early as 1981, Maddrell discussed the effect of urine flow on the efficiency of clearing solutes from the hemolymph of insects, introducing the idea of clearance without mention of the word.[151]

4.   Primary Urine, Isosmotic Fluid Secretion, and Aquaporin Water Channels

The most abundant ions in the hemolymph of most insects are Na^+ and Cl^-, resembling the extra-cellular fluid of vertebrates.[3] The hemolymph K^+ concentration is maintained at low concentrations as in vertebrates, but it can reach 50 mM in some insects. Amino acids and trehalose can contribute substantially to the hemolymph osmotic pressure. We have measured a hemolymph osmotic pressure of 354 mOsm/kg in the yellow fever mosquito.[152] The osmotic pressure of primary urine secreted by distal segments of *Aedes* Malpighian tubules *in vitro* is 340 mOsm/kg, similar to the osmotic pressure of 330 mOsm/kg of the peritubular Ringer.[153] Because these two osmotic pressures fall within the experimental error of the measurement, it can be assumed that primary urine is essentially isosmotic to the peritubular bath in the experiment and to the hemolymph in the animal.

Isosmotic secretion reflects a high water permeability of the epithelium. One advantage of a high water permeability is that little osmotic pressure difference is necessary to drive osmosis across the epithelium.[154] Both transcellular[150,155] and paracellular pathways[156,157] for water flow into the tubule lumen have been suggested. Aquaporin water channels are known to render biological membranes permeable to water.[158] Malpighian tubules of adult *Drosophila* express transcripts for five aquaporin-like genes. Three transcripts (DRIP, Aqp17664, and Aqp4019) are significantly enriched in the tubule relative to their expression in the whole body.[127,128] Stellate cells exclusively express the mRNA for DRIP, and principal cells express transcripts for Aqp17664 and Aqp4019;[159] however, only DRIP has been shown to have water channel activity.[159] Thus, DRIP stands out as the likeliest candidate for transcellular water transport through stellate cells in *Drosophila* Malpighian tubules. Molecular and immunochemical evidence from Malpighian tubules of other insects also supports a transcellular route for secreting water into the tubule lumen.[159–162] Still, a paracellular route for the transepithelial osmosis of water cannot be excluded.[156,157,163] The problem is our experimental inability to distinguish transcellular water flow from paracellular water flow.

Although the osmotic pressures of the peritubular Ringer solution and the fluid secreted into the tubule lumen are nearly identical, ionic compositions are markedly different. Invariably, the primary urine secreted into the tubule lumen of distal segments of the *Aedes* Malpighian tubule reveals Na^+ concentrations that are lower and K^+ concentrations that are higher than respective concentrations in the peritubular Ringer solution (Figure 7.15). In contrast, peritubular and luminal Cl^- concentrations are very close. Because the transepithelial voltage is lumen positive by about 53 mV, it follows that the cations K^+ and Na^+ are secreted into the tubule lumen against their electrochemical potentials[164] (i.e., by active transport). Cl^- moves across the epithelium down its electrochemical potential via passive transport.

5.   *In Vitro* Study of Transepithelial Transport in Malpighian Tubules

The data presented in Figure 7.15 were obtained from isolated Malpighian tubules studied by two methods. The first method, the method of Ramsay, measures the rate of fluid secretion by the tubule *in vitro* (Figure 7.16A). The method also allows the collection of secreted fluid for the analysis of its ionic composition.[165,166] The second method is an adaptation of the method of Burg for the *in vitro* microperfusion of renal tubules.[167] In brief, the tubule lumen is cannulated with a perfusion

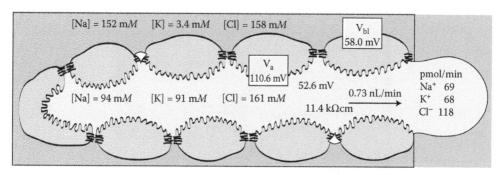

FIGURE 7.15 Spontaneous generation of primary urine in isolated Malpighian tubules of the yellow fever mosquito (*Aedes aegypti*). Next to the indicated concentrations of Na^+, K^+, and Cl^-, the peritubular Ringer solution contained (in mM) 25 HEPES, 5 glucose, 1.8 $NaHCO_3$, 1.7 $CaCl_2$, 1 $MgSO_4$, and 1.7 $CaCl_2$ at pH 7. In the presence of this peritubular Ringer solution, the tubule secretes primary urine at a rate of 0.73 nL/min.[153] Transepithelial secretion rates of Na^+, K^+, and Cl^- are calculated as the product of the volume secretion rate and the concentrations of Na^+, K^+, and Cl^- in primary urine (nL and pmol are, respectively, 10^{-9} L and 10^{-12} mol). The transepithelial voltage is 52.6 mV (lumen positive) across a tubule wall with an electrical resistance of 11.4 KΩcm (normalized to a tubule 1 cm long). The apical membrane voltage ($V_a = 110.6$ mV, cell negative) stems largely from the activity of the electrogenic proton pump, the V-type H^+-ATPase located in the microvillar apical membrane (Figure 7.14). The basolateral membrane voltage (V_{bl}) is 58.0 mV (cell negative).

pipette so the solutions on both sides of the epithelium can be controlled (Figure 7.16B). Ions can be added to or removed from either side of the epithelium to evaluate effects on transepithelial electrolyte transport, fluid secretion, and tubule electrophysiology.[165,167,168] In addition, the effects of potential hormones, stimulators, and inhibitors can be evaluated. To investigate transport steps across the basolateral and apical membranes of epithelial cells, a principal cell can be impaled with conventional microelectrodes to measure membrane voltages or with ion-selective microelectrodes to measure intracellular ion concentrations of interest.[165,169]

6. Definitions of Active and Passive Transport

The Ramsay method and the Burg method have allowed us to elucidate the mechanism and the regulation of transepithelial ion transport in Malpighian tubules of the yellow fever mosquito. Fundamental in this elucidation was measurement of the transepithelial electrochemical potential differences for Na^+, K^+, and Cl^- as the thermodynamic evidence of active and passive transport mechanisms. Active transport can be primary or secondary. Primary active transport is mediated by an ATP-driven pump, such as H^+ transport by the V-type H^+-ATPase in the apical (microvillus) membrane of principal cells of Malpighian tubules (Figure 7.14A and Figure 7.17C). The classical Na,K-ATPase of the eukaryotic cell membrane, which produces an inward Na^+ electrochemical potential and an outward K^+ electrochemical potential, is another example of primary active transport (Figure 7.17C). Primary active transport is immediately dependent on the hydrolysis of ATP that yields the energy for translocating an ion or organic solute.

Secondary active transport is not immediately but ultimately dependent on the hydrolysis of ATP. It is mediated by a carrier that uses the energy of one electrochemical potential to generate another electrochemical potential (Figure 7.17B). As an example, the Na/H exchanger of the NHE (Na/H exchanger) family of transporters utilizes the energy of the inward Na^+ electrochemical potential to drive H^+ out of the cell. The inward Na^+ electrochemical potential is utilized by a cotransporter of the SLGT (sodium-linked glucose transporter) family to bring glucose into the cell (Figure 7.17B). In both cases, the inward Na^+ electrochemical potential was first generated by the Na,K-ATPase.

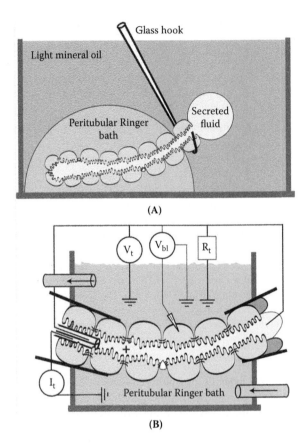

FIGURE 7.16 Basic methods for investigating transepithelial transport in isolated Malpighian tubules: (A) Ramsay method of fluid secretion. The isolated Malpighian tubule is bathed in a 50-μL droplet of Ringer solution under oil. The open end of the tubule is pulled into the oil with a glass hook so fluid secreted by the tubule accumulates as a droplet separate from the peritubular Ringer bath. (B) *In vitro* microperfusion of a Malpighian tubule. A short segment of the isolated Malpighian tubule is cannulated with a double-barrel perfusion pipette for (1) perfusion of the tubule lumen, (2) measurement of the transepithelial voltage (V_t), and (3) measurement of the transepithelial resistance (R_t) upon the injection of current (I) into the tubule lumen. A glass microelectrode impaling one principal cell provides measurements of the basolateral membrane voltage (V_{bl}).

Passive transport is down the electrochemical potential as in diffusion. Ion channels mediate the passive transport of ions (Figure 7.17A). Carriers may mediate diffusion as well, which in this case is called *facilitated diffusion*. Osmosis exemplifies the passive transport of water; it is the diffusion of water from a high water concentration to a low water concentration; the concentration of water is greater in a dilute solution (of solute and water) than in a concentrated solution.

7. Minimal Model of Transepithelial Electrolyte Secretion by Malpighian Tubules of *Aedes aegypti*

The transcellular transport of solute across an epithelium often encompasses both passive and active transport steps. As illustrated in Figure 7.18, the entry of K^+ across the basolateral membrane of principal cells is passive and mediated by K^+ channels that dominate the electrical conductance of the basolateral membrane.[170,171] But, the extrusion of K^+ from the cell on the apical side is apparently by secondary active transport, mediated by a hypothetical NHE (Figure 7.18). Here, the inward H^+ electrochemical potential that was generated by the V-type H^+-ATPase drives the extrusion of K^+

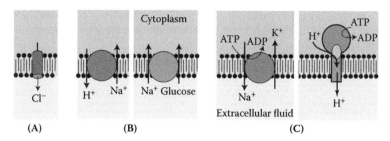

FIGURE 7.17 Some basic transport steps across biological membranes: (A) Passive transport through a membrane channel. Channels mediate the diffusion of noncharged solutes or the electrodiffusion of solutes carrying electrical charge. Diffusion through the lipid bilayer is not shown. (B) secondary active transport. Under the usual physiological conditions of low intracellular and high extracellular Na$^+$ concentrations, the inward Na$^+$ electrochemical potential delivers the energy for extruding H$^+$ from the cell via exchange transport that is electroneutral and voltage independent in the example shown. Another carrier couples the energy of the inward Na$^+$ electrochemical potential to the uptake of glucose via cotransport that is electrogenic and voltage sensitive in this case. (C) Primary active transport. The Na,K-ATPase generates transmembrane concentration differences for both Na$^+$ and K$^+$. The exchange of 3 Na$^+$ ions for 2 K$^+$ ions per pump cycle makes the pump electrogenic and contributes to the cell-negative membrane voltage. The V-type H$^+$-ATPase is purely electrogenic in the transport of H$^+$ ions alone, bringing about high voltages across membranes inhabited by this proton pump.

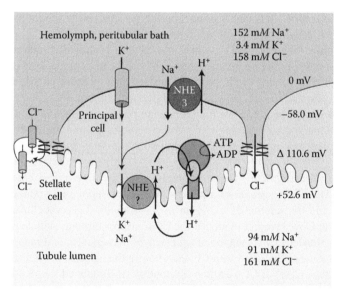

FIGURE 7.18 Minimal model of transepithelial NaCl and KCl secretion by Malpighian tubules of the yellow fever mosquito (*Aedes aegypti*) under control conditions. Principal cells provide the transcellular active transport pathways for secreting the cations Na$^+$ and K$^+$ into the tubule lumen. Cl$^-$ passes passively into the tubule lumen through septate junctions or stellate cells. Transepithelial secretion of NaCl and KCl is energized by the V-type H$^+$-ATPase located in the brush-border apical membrane. Aquaporin water channels are not shown. By translocating H$^+$ ions from the cytoplasm to the tubule lumen, the V-type H$^+$-ATPase generates electrical current that must return to the cytoplasmic face of this proton pump. Current returns to the peritubular bath (or hemolymph) taking a pathway outside principal cells, through the paracellular pathway or stellate cells. Current returning from the tubule lumen to the peritubular bath is carried by Cl$^-$ passing from bath to lumen as the mechanism of transepithelial secretion (passive transport). Current passing from the peritubular bath to the cytoplasm is carried by K$^+$ entering the cell through K$^+$ channels in the basolateral membrane.

(as well as Na$^+$) from cell to tubule lumen. Theoretically, secondary active transport systems are reversible. Whether they operate in forward or reverse mode depends on the net electrochemical potential of participating solutes or ions.

In *Drosophila* Malpighian tubules, as many as 30 transcripts encoding K$^+$ channels have been identified, with those encoding inward rectifiers scoring the highest enrichment.[127] Inward rectification allows the entry of K$^+$ from hemolymph into the cytoplasm with greater ease than the exit of K$^+$ from cytoplasm into the hemolymph (Figure 7.18). Na$^+$ enters the cell via an electroneutral mechanism that most likely is mediated by an NHE isoform that resembles the NHE3 of mammals, as illustrated in Figure 7.18.[172–175] The NHE3 cDNA has been cloned from *Aedes* Malpighian tubules,[174,175] and NHE3 immunoreactivity occurs in basal and cytoplasmic regions of principal cells from proximal and distal segments, as well as in the apical regions of principal cells from median segments.[174] Another potential pathway for Na$^+$ entry is NDAE1 (Na-driven anion exchanger), which localizes to the basolateral region of *Drosophila* Malpighian tubules.[176]

Both Na$^+$ and K$^+$ are thought to exit the cell across the apical membrane via exchange transport with H$^+$ (Figure 7.18). Kang'ethe et al.[177] suggest that the recently cloned NHE8 isoform from *Aedes* Malpighian tubules accepts both Na$^+$ and K$^+$ for electroneutral exchange transport with H$^+$. The proton electrochemical potential that would drive the uphill transport of Na$^+$ and K$^+$ from the cytoplasm to the tubule lumen is generated by the V-type H$^+$-ATPase located in the apical membrane (Figure 7.18).

Measurements of intracellular and luminal K$^+$ and H$^+$ concentrations[178] and known apical membrane voltages (110.6 mV; Figure 7.14) in Malpighian tubules of *Aedes aegypti* present the argument—on the basis of thermodynamics—that H$^+$/K$^+$ exchange must be electrogenic, transporting more than one H$^+$ ion (n) for each K$^+$ ion.[178] Such a nH$^+$/K$^+$ exchanger would derive its primary driving force from the apical membrane voltage generated by the V-type H$^+$-ATPase. An electrogenic transporter nH$^+$/cation of the NHA family of transporters has been proposed for the apical membrane of intestinal epithelial cells of the *Anopheles* mosquito.[179]

Although subunits of the Na,K-ATPase are detectable in Malpighian tubules of *Aedes aegypti* by reverse transcription–polymerase chain reaction (RT-PCR) and immunohistochemistry,[180] we do not detect ATPase activity of the Na/K pump.[144] Instead, we measure an ATPase activity that is sensitive to bafilomycin and nitrate, consistent with the operation of a V-type H$^+$-ATPase. Because the V-type H$^+$-ATPase translocates H$^+$ without replacing positive charge, the proton pump is highly electrogenic, producing some of the highest membrane voltages in animal cells, on average 111 mV across the apical membrane of Malpighian tubules of *Aedes aegypti* (Figure 7.14A, Figure 7.15, and Figure 7.18).

Because the apical membrane voltage is nearly twice as large as the basolateral membrane voltage, the V-type H$^+$-ATPase generates a lumen-positive transepithelial voltage that may serve as a driving force for the transepithelial secretion of Cl$^-$ through the paracellular pathway (Figure 7.18). O'Donnell and Dow[125] report evidence for Cl$^-$ transport through stellate cells in *Drosophila* Malpighian tubules. Indeed, Cl$^-$ channels are present in the apical membrane of stellate cells in *Aedes* Malpighian tubules consistent with Cl$^-$ secretion in this species.[181] Stellate cells may be the primary route for transepithelial Cl$^-$ secretion in control, unstimulated *Aedes* Malpighian tubules; however, the passage through the paracellular pathway dominates in tubules stimulated by the diuretic hormone leucokinin.[182] Despite the uncertainty regarding the magnitude of Cl$^-$ fluxes through trans- and paracellular pathways, it is generally agreed that principal cells do not mediate the transepithelial secretion of Cl$^-$.[122]

The transport model illustrated in Figure 7.18 is a minimal model that accounts for the transepithelial secretion of NaCl and KCl in isolated Malpighian tubules of *Aedes aegypti*.[183] The model is minimal because additional mechanisms for Na$^+$ and K$^+$ entry across the basolateral membrane of principal cells have been reported. Furthermore, the model reflects the minimal composition of the Ringer solution bathing the isolated Malpighian tubule. It consists of the usual salts of Na$^+$, K$^+$, Mg^{2+}, and Ca^{2+}; a buffer; and some glucose but no organic solutes the tubules might secrete *in vivo*.

The transport model elucidated in Malpighian tubules of *Aedes aegypti* (Figure 7.18) is similar to that proposed for Malpighian tubules of *Drosophila*, *Formica*, and *Hemideina*.[184–187] On first inspection, the model does not appear to apply to Malpighian tubules of the kissing bug (*Rhodnius prolixus*; Hemiptera), as these tubules generate lumen-negative transepithelial voltages,[188] use the Na,K-ATPase and secondary active transport of Cl- to produce primary urine,[185,189] employ serotonin as diuretic hormone,[190–193] and apparently do not use kinin-like proteins as diuretic agents.[194–196] Accordingly, the generic transport model in Hemiptera (bugs) may differ from that in Diptera (flies and mosquitoes), Hymenoptera (ants), and Orthoptera (crickets, wetas). On second inspection, evidence for the V-type H^+-ATPase at the apical membrane of *Rhodnius* Malpighian tubules has been reported,[188] and kinin-like peptides have also been detected in the central nervous system and in neurohemal sites of *Rhodnius*.[194]

Clearly, additional studies are required to establish real differences between Malpighian tubules from different species. At present, it appears that Malpighian tubules share a large functional repertoire that is variably expressed in species. The transcriptome of *Drosophila* Malpighian tubules lists over 1000 genes significantly enriched in Malpighian tubules, and of the top 200 genes less than half can be associated with known functions.[127] Thus, many transport functions have yet to be observed in *Drosophila* Malpighian tubules, the most widely studied tubule of all insects. Even more transport functions await their detection in Malpighian tubules of other species.

8. Renal Responses to the Osmoregulatory Challenges of the Blood Meal

About 14,000 known species of insects feed on blood. Some insects, such as the kissing bug (*Rhodnius prolixus*), are exclusively hematophagous. The kissing bug can go for weeks without a meal but, upon finding a donor, it can take on a volume 12 times its own body weight.[197] Feeding on blood is not as obligatory in other insects. In fact, nectar meals are important for most mosquitoes because the low glucose content of mammalian blood is inadequate for fueling flight.[198] Only the female gender of the yellow fever mosquito (*Aedes aegypti*) feeds on blood and then only during the reproductive period as a convenient source of proteins and nutrients for egg production.[199] Because mammalian blood consists largely of NaCl, KCl, and water, the blood meal presents excess quantities of NaCl from ingested plasma and KCl from ingested blood cells as well as the weight of unwanted water.[152,153] The average female mosquito of *Aedes aegypti* in our laboratory weighs 1.3 mg. When she consumes about 3 μL of blood in a single meal,[119,200] her body weight increases to 4.3 mg. Thus, her take-off weight is more than three times her empty weight. By comparison, the take-off weight of a passenger airplane may not exceed twice its empty weight, and most of that payload is fuel, not passengers or cargo.

As shown in Figure 7.19, the mosquito deals with excess cargo by triggering a diuresis before she has finished her meal. The initial diuresis excretes urine that consists primarily of NaCl and water. KCl is excreted later.[152] It has been suggested that the first fluid droplets excreted from the rectum (Figure 7.19) do not stem from Malpighian tubules but from the gut, where ingested plasma is passed on to the hindgut for excretion while the blood cells are retained in the midgut for digestion.[5,9] This intestinal mechanism for excreting unwanted salt and water bypasses the kidneys and has therefore been termed *prediuresis* in *Anopheles gambiae* (Figure 7.19). The prediuresis apparently does not take place in the mosquito *Aedes aegypti*, because Trypan blue and [144]Ce added to the blood prior to ingestion do not appear in the first excreted droplets.[200,201]

C. Diuresis and Antidiuresis in Insects

Diuresis is a Latin term meaning increased excretion of urine from the body. It may come about in insects by: (1) increasing the rate of transepithelial secretion of salt and water in distal segments of Malpighian tubules, or (2) decreasing the reabsorption of salt and water in proximal segments of Malpighian tubules, in the hindgut, and in the rectum. A natriuresis is marked by increased Na^+

FIGURE 7.19 (See color insert following page 208.) Female *Anopheles* mosquito taking a blood meal. Note the urination while feeding. Repeating this experiment on himself, allowing a female, pathogenic-free yellow fever mosquito (*Aedes aegypti*) to take a blood meal, James Williams in our laboratory found that the first urine droplets eliminate the NaCl and water fraction of the blood meal.[152] (Photograph courtesy of Jack Kelly Clark, University of California.)

excretion rates; a kaliuresis signifies increased K^+ excretion; and a chloruresis indicates increased Cl^- excretion. In contrast, an antidiuresis minimizes urinary excretion to conserve water for the animal. It may come about by: (1) reducing the rate of fluid secretion in distal segments of the Malpighian tubule, or (2) increasing the rate of water reabsorption in proximal segments of the tubule, hindgut, and rectum.[202]

Both diuresis and antidiuresis can be triggered by natural and synthetic agents that target specific transport systems. Table 7.2 lists samples of the major classes of diuretic and antidiuretic peptides that so far have been discovered in insects. The mechanisms of action are best understood for peptide hormones such as the kinins, calcitonin-like and CRF-related diuretic peptides (CRF, corticotrophin releasing factor), and the biogenic amine serotonin.

The mechanism and regulation of diuresis in insects have been recently reviewed by Maddrell et al.,[203] Torfs et al.,[204] and Gaede et al.[205] The diuresis in the kissing bug *Rhodnius* has been reviewed by Coast,[9] Orchard,[192] Maddrell et al.,[193] Te Brugge and Orchard,[194] and Te Brugge et al.[195] Diuresis in the desert beetle, the cricket, and the mosquito has been reviewed, respectively, by Nicolson,[148] Spring and Clark,[206] and Beyenbach.[124,207] Below we discuss some of the different classes of diuretic and antidiuretic agents in insects, their effects on Malpighian tubule function, and their mechanisms of action.

1. The CRF-Related Diuretic Peptides

The CRF-related diuretic peptides structurally resemble vertebrate corticotropin-releasing factor (CRF). The first CRF-related diuretic peptide in insects was isolated from heads of the adult moth *Manduca sexta*.[208] The hormonal status of this diuretic peptide has been established in the locust on the basis of radioimmunoassay determination of the peptide in tissues and its circulation in the hemolymph[8,209] and by the block of diuretic activity by an antiserum against the peptide.[210] CRF-related diuretic peptides are common in insects. They have been found in locusts, crickets, cockroaches, termites, beetles, bugs, moths, mosquitoes, and flies.[8]

In the locust, CRF-related diuretic hormones are synthesized by neurosecretory cells in the brain. Axonal transport delivers the hormones to the corpora cardiaca, from where they are released

TABLE 7.2

Representative Diuretic and Antidiuretic Peptides in Insects

Family/Peptide	Code Name	Primary Sequence/Structure	Primary Second Messenger	Refs.
Diuretic				
CRF-related peptides	Manse-DH	PMPSLSIDLPMSVLRQKLS-LEKERKVHALRAAANRN-FLNDIamide	cAMP	Kataoka et al.[208] Cabrero et al.[284] Furuya et al.[376]
Insect kinins	Leucokinin VIII Aedeskinin 2	GDAFYSWGamide NPFHAYFSAWGamide	Ca^{2+}	Holman et al.[222] Veenstra,[377] O'Donnell et al.[122]
Calcitonin-like peptide	Dippu-DH$_{31}$	GLDLGLSRGFSGSQAAKH-LMGLAAANYAGGGPamide	cAMP	Furuya et al.[376]
Cardioaccelerator peptide	Manse-CAP$_{2b}$	pQLYAFPRV-NH$_2$	NO/cGMP	Davies et al.,[289] Huesmann et al.,[378] Kean et al.[379]
Serotonin	—	5-Hydroxytryptamine	cAMP	Orchard[192]
Tyramine	—	Tyramine	Ca^{2+}	Blumenthal[262]
Antidiuretic				
Unknown	Tenmo-ADFa	VVNTPGHAVSYHVY-OH	cGMP	Eigenheer et al.[273]

into the hemolymph. Small groups of posterolateral neurosecretory cells in the abdominal ganglia also synthesize CRF-related diuretic hormone and release it from neurohemal sites of abdominal nerves (G. Coast, pers. comm.). Of interest is that CRF-related peptide colocalizes with kinins (another family of diuretic peptides) in mesothoracic ganglia in the kissing bug (*Rhodnius prolixus*).[211] Figure 7.20 illustrates the location of neurosecretory cells and hormone storage sites in the mosquito *Aedes aegypti*. The medial and lateral neurosecretory cells together with the corpus cardiacum are considered analogous to the hypothalamus in vertebrates (M. Brown, pers. comm.).

In the tobacco hornworm (*Manduca sexta*), cricket (*Acheta domesticus*), and fruit fly (*Drosophila melanogaster*), CRF-related diuretic peptides bind to G-protein-coupled receptors[212–214] and increase intracellular cAMP concentrations.[212] In isolated Malpighian tubules, both CRF-related diuretic hormone and cAMP stimulate fluid secretion, suggesting that cAMP serves as a second messenger of CRF-related diuretic peptides.[208,215–217] One target of cAMP in Malpighian tubules of the cricket and the mosquito is a Na/K/2Cl cotransporter in the basolateral membrane which is upregulated.[218,219] A second target of cAMP is the V-type H$^+$-ATPase at the apical membrane where the nucleotide leads to the assembly of the catalytic V$_1$ complex and the proton-translocating V$_0$ complex to form the holoenzyme.[220] The assembled holoenzyme is now capable of coupling the hydrolysis of ATP to the translocation of protons from cytoplasm to tubule lumen (see Figure 7.24). Thus, cAMP affects transport systems at both basolateral and apical membranes. Matching cation entry into the cell across the basolateral membrane with cation exit across the apical membrane maintains the cell in a steady state.

2. Insect Kinins

Mark Holman is credited for discovering the kinins, which are now known to be widely distributed in invertebrates. He isolated eight kinins from the cockroach *Leucophaea* on the basis of their ability to stimulate the contractions of the cockroach hindgut—hence, the designation leucokinin, the kinin of *Leucophaea*.[221,222] Insect kinins are small peptides of no more than 15 amino acids.[223] They share the core C-terminal pentapeptide sequence $FX_1X_2WGamide$ required for biological activity, where X_1 can be Y, F, S, or H and X_2 can be S or P. Kinins are synthesized in neurosecretory cells and released from the corpora cardiacum or abdominal neurosecretory cells.[224]

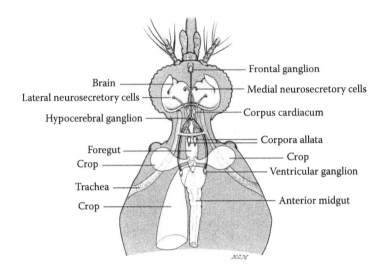

FIGURE 7.20 Neurosecretory cells and neuropeptides storage sites (corpus cardiacum and corpora allata) in the female yellow fever mosquito (*Aedes aegypti*). According to Arden Lea and Mark Brown (who have kindly provided the diagram), the composite medial and lateral neurosecretory cells and the corpus cardiacum are analogous to the hypothalamus in vertebrate brains. In the yellow fever mosquito, the corpora cardiaca form one structural unit with the aorta and should therefore be referred to in the singular: corpus cardiacum. The corpus cardiacum receives the axon terminals from medial and lateral neurosecretory cells in the brain and from 8 to 10 neurosecretory cells (not shown) immediately posterior to the corpus cardiacum. Medial cells are the source of ecdysteroidogenic hormone[353] and insulin-like peptides.[354] The products of lateral neurosecretory cells are unknown. Neurosecretory cells near the corpus cardiacum are the source of adipokinetic hormones.[355]

It is not uncommon for hormones or neuropeptides that influence the motility of epithelial structures to also affect epithelial transport. Because leucokinin increases the contractions of the hindgut—thereby facilitating excretion from the hindgut—we examined the effects of leucokinin on an epithelium further upstream, the Malpighian tubule. We found that synthetic leucokinin increased the secretion of fluid in isolated Malpighian tubules of the yellow fever mosquito.[225] The compositional analysis of secreted fluid revealed the nonselective stimulation of both NaCl and KCl, suggesting an effect on transepithelial Cl⁻ transport (Figure 7.21). Electrophysiological studies confirmed this hypothesis: the effects of leucokinin on the transepithelial voltage, resistance, and Cl⁻ diffusion potentials all pointed to the increase in the Cl⁻ conductance of a paracellular transport pathway.[182] As illustrated in Figure 7.22, it was the first demonstration in any epithelium that a hormone exerted its effect on a paracellular transport pathway.[207]

The stimulation of Cl⁻ transport through stellate cells was ruled out in *Aedes* Malpighian tubules by studying the effects of leucokinin on very short tubule segments that did not include these cells.[120] In stellate-cell-free tubule segments, leucokinin induced the transepithelial Cl⁻ conductance as quickly and reversibly as in the presence of stellate cells, confirming the effect on the septate junctional Cl⁻ pathway while documenting the presence of the kinin receptor on the basolateral membrane of principal cells (Figure 7.22).

Paracellular transport, the transepithelial transport between epithelial cells, is defined by the permselective properties of the paracellular pathway—i.e., septate junctions in invertebrates and tight junctions in vertebrates. Probing the permselectivity of the septate junction in the *Aedes* Malpighian tubule, we found the permeability sequence of halides in free solution (I⁻ > Br⁻ > Cl⁻ > F⁻) under control conditions. In free solution, ions are surrounded by shells of water which slows down their diffusion. In the presence of leucokinin the permeability sequence shifted to Br⁻ > Cl⁻ > I⁻ > F⁻, reflecting the selection of small halides solely on the basis of size and charge (i.e., halide

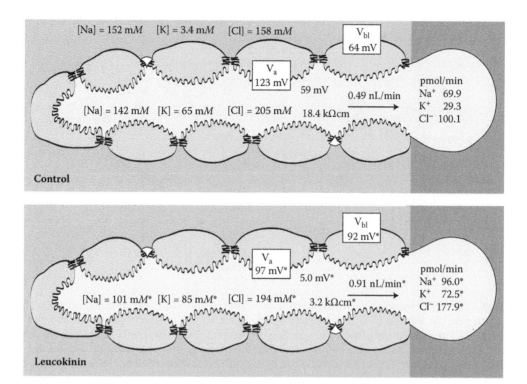

FIGURE 7.21 Leucokinin VIII stimulates the transepithelial section of NaCl, KCl, and water. Asterisk indicates significant difference from the control. Leucokinin increases the transepithelial secretion of Na^+, K^+, Cl^-, and water together with large reductions in transepithelial voltage and resistance. V_{bl} and V_a are cell negative with respect to ground in the hemolymph and tubule lumen, respectively. V_t is lumen positive with respect to ground in the hemolymph. V, voltage; a, apical membrane; bl, basolateral membrane. (Data from Pannabecker, T.L. et al., *J. Membr. Biol.*, 132, 63, 1993; Beyenbach, K.W., *Curr. Opin. Nephrol. Hypertens.*, 12, 543, 2003.)

ions without their water shells). The shift from hydrated to dehydrated halides suggests that in the presence of leucokinin a channel-like structure in the septate junction becomes accessible to hydrated halide ions, allowing coulombic interactions with the channel. In the process of these interactions, halide ions lose their water shells, making them small enough for channel permeation (Figure 7.22).

Remarkably, the on/off effects of leucokinin on the septate junctional Cl^- conductance proceed with switch-like speed, indicating posttranslational modifications of septate junctional proteins.[182,207] Such rapid changes in the junctional Cl^- conductance, which changes nearly tenfold in response to adding and removing leucokinin, is consistent with channel-like structures residing in the septate junction. Proteins of the claudin family are known to define the barrier and permselectivity properties of the paracellular pathway in vertebrate epithelia.[226–231] Two claudin-like proteins, sinuous and megatrachea, have been found in *Drosophila*. Both localize to septate junctions in tracheal tubes, where they provide a barrier function, like claudin;[230,232,233] however, claudin-like proteins have not yet been observed in Malpighian tubules of any insect.

Figure 7.22 illustrates the signaling pathway of leucokinin in *Aedes* Malpighian tubules.[234,235] Leucokinin binds to a G-protein-coupled receptor at the basolateral membrane of principal cells, thereby activating phospholipase C to yield inositol trisphosphate.[236,237] Subsequently, inositol trisphosphate triggers the release of Ca^{2+} from intracellular stores. The depletion of these stores activates a Ca^{2+} channel in the basolateral membrane, allowing the entry of Ca^{2+} into principal cells.[207,234,235] How Ca^{2+} proceeds to increase the Cl^- conductance of the paracellular, septate junctional pathway is an intriguing question (Figure 7.22). It probably involves posttranslational modifications such as phosphorylations, glycosylations, or palmitylations that have yet to be

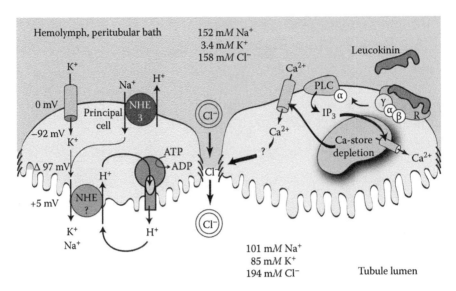

FIGURE 7.22 Signal transduction of leucokinin in principal cells of Malpighian tubules of *Aedes aegypti*. α, β, and γ are the subunits of G protein; PLC, phospholipase C; IP₃, inositol trisphosphate. The proteins defining the Cl⁻ permselectivity of the septate junction are unknown. The permeation of Cl⁻ through the septate junction is hypothetical and thought to involve a conformational change in septate junctional proteins that can interact with hydrated Cl⁻. The interaction strips Cl⁻ of its hydration shell which admits the ion for permeation through the septate junction. Emerging on the other side of the junction, Cl⁻ hydrates again. (Data from Yu, M.J. and Beyenbach, K.W., *J. Exp. Biol.*, 207, 519, 2004; Yu, M. and Beyenbach, K.W., *J. Insect Physiol.*, 47, 263, 2001; Yu, M.J. and Beyenbach, K.W., *Am. J. Physiol.*, 283, F499, 2002.)

elucidated. Equally important is identification of the proteins that define the Cl⁻ permselectivity of the septate junction in the presence of leucokinin.

Mechanisms are different in Malpighian tubules of *Drosophila*, where stellate cells signal and mediate the transepithelial secretion of Cl⁻. O'Donnell et al.[125] have found Cl⁻ channels in stellate cells and transepithelial Cl⁻-dependent current sinks in the vicinity of stellate cells, consistent with a transcellular route for transepithelial Cl⁻ secretion. Moreover, leucokinin increases intracellular Ca²⁺ concentration primarily in stellate cells,[223,236] and the receptor for drosokinin (the kinin of *Drosophila*) is expressed by stellate cells.[236,238] Altogether, these observations provide strong evidence that kinins increase Cl⁻ transport through stellate cells in *Drosophila* Malpighian tubules. Malpighian tubules of the cricket *Acheta domesticus* do not have stellate cells,[162] yet the tubules respond to achetakinin, the kinin of *Acheta*. Surprisingly, all five achetakinins decreased fluid secretion in the distal (blind-ended) tubules, but two achetakinins increased fluid secretion in mid-tubule segments.[239] Thus, kinins can trigger diuretic and antidiuretic activity, even in the absence of stellate cells.

The above sojourn into the comparative physiology of kinins in *Aedes*, *Drosophila*, and *Acheta* brings back to mind Potts and Parry's warning that "no insect can be regarded as typical." Malpighian tubules offer both transcellular and paracellular routes for secreting Cl⁻ into the tubule lumen. The two routes are used to varying degrees in different species. Clearly, a single epithelial transport model cannot be assigned to the mechanism of action of kinins in insect Malpighian tubules.

3. Calcitonin-Like Peptides

In contrast to the nonselective stimulation of transepithelial NaCl and KCl secretion by leucokinin, the mosquito natriuretic peptide selectively stimulates the secretion of NaCl.[153,164] The secretion of K⁺ is not affected, as the fourfold increase in water secretion dilutes the K⁺ concentration in the tubule lumen fourfold (Figure 7.23). Mosquito natriuretic peptide hyperpolarizes the transepithelial

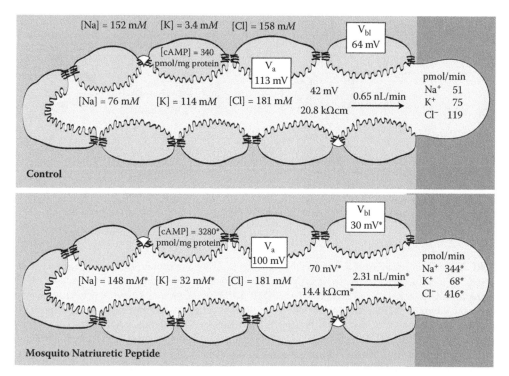

FIGURE 7.23 Selective stimulation of transepithelial NaCl and water secretion by the mosquito natriuretic factor. Mosquito natriuretic factor is a peptide[241] that Coast et al. have identified as one of the calcitonin-like diuretic hormones in insects.[244] Note the nearly tenfold increase in intracellular cAMP concentration in the presence of mosquito natriuretic peptide. Asterisk indicates significant difference from the control. For voltage polarities, see legend for Figure 7.21. (Data from Williams, J.C. and Beyenbach, K.W., *J. Comp. Physiol. B*, 154, 301, 1984; Petzel, D.H. et al., *Am. J. Physiol. Regul. Integr. Comp. Physiol.*, 249, R379, 1985; Beyenbach, K.W. and Petzel, D.H., *News Physiol. Sci.*, 2, 171, 1987; Petzel, D.H. et al., *Am. J. Physiol. Regul. Integr. Comp. Physiol.*, 253, R701, 1987.)

voltage from 42 mV to 70 mV due to the depolarization of the basolateral membrane voltage of principal cells. Because the Na^+ concentration in primary urine increases from 76 mM to 148 mM and the transepithelial voltage becomes more lumen positive, it is clear that mosquito natriuretic peptide targets the stimulation of active transepithelial transport of Na^+ (Figure 7.23).

Years ago, we extracted mosquito natriuretic peptide from several thousand heads of the yellow fever mosquito.[240,241] At that time we thought that mosquito natriuretic peptide was a CRF-related diuretic peptide because it increases intracellular concentrations of cAMP (Figure 7.23), and cAMP perfectly duplicates the effects of mosquito natriuretic peptide on the transepithelial electrolyte secretion and electrophysiology of the tubule.[123,124,240–243] It now appears, however, that mosquito natriuretic peptide is a member of the calcitonin-like diuretic peptides. Calcitonin-like peptides mediate their effects also via cAMP.[244] Anoga-DH31, the calcitonin-like diuretic hormone of *Anopheles gambiae*, has natriuretic activity in isolated Malpighian tubules of both *Anopheles* and *Aedes* via the elevation of intracellular cAMP.[244] In parallel, cAMP depolarizes the basolateral membrane voltage of principal cells to the same extent it hyperpolarizes the transepithelial voltage in Malpighian tubules of both *Anopheles*[244] and *Aedes*.[245] Moreover, the amino-acid sequence of the predicted calcitonin-like diuretic peptide in *Aedes* (Accession No. XP_001658868) is identical to that of Anoga-DH31 (pers. observ.). Thus, the mosquito natriuretic peptide isolated from *Aedes* is likely a calcitonin-like diuretic peptide, perhaps to be named Aedae-DH31. The amino-acid sequencing of the mosquito natriuretic peptide purified from *Aedes* will tell for sure.

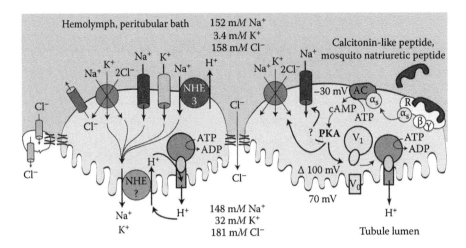

FIGURE 7.24 The mechanism of action and the signaling pathway of mosquito natriuretic peptide. Mosquito natriuretic peptide increases intracellular cAMP concentrations by stimulating adenylate cyclase (AC) presumably via a G-protein-coupled receptor. The cAMP activation of protein kinase A (PKA) is thought to assemble the V-type H^+-ATPase at the apical membrane, thereby activating ATP hydrolysis and the translocation of H^+ from the cytoplasm to the tubule lumen. At the basolateral membrane, cAMP activates: (1) a Na^+ conductance, presumably a Na^+ channel, and (2) a bumetanide-sensitive transport system, presumably Na/K/2Cl cotransport. Whether the basolateral membrane channel and cotransporter are activated via phosphorylation by PKA is unknown. A hypothetical Cl^- channel is added to the basolateral membrane to allow the exit of Cl^- that has entered the cell via Na/K/2Cl cotransport. Cl^- may pass through stellate cells and/or septate junctions for transepithelial secretion into the tubule lumen. Whether Cl^- passes through the septate junction as hydrated or dehydrated ion is unknown.

Both mosquito natriuretic peptide and cAMP depolarize the basolateral membrane voltage of principal cells by activating a Na^+ conductance in that membrane.[242,245] The molecular correlate to this Na^+ conductance could be a Na^+ channel as illustrated in Figure 7.24, an electrogenic Na^+-dependent cotransporter, or an electrogenic exchange transporter. In addition, cAMP activates an electroneutral, bumetanide-sensitive transport system in the basolateral membrane, presumably a Na/K/2Cl cotransporter related to the SLC12 (solute-linked carrier) family of cation-coupled Cl^- cotransporters.[219] A SLC12-like transcript has been cloned from a cDNA library prepared from the gut and Malpighian tubules of *Aedes aegypti*, but neither expression of the transcript nor immunoreactivity of the protein could be demonstrated in Malpighian tubules.[246] A SLC12-like transporter has been identified in Malpighian tubules of *Manduca sexta*,[247,248] and physiological evidence suggests Na/K/2Cl cotransport in *Formica*,[185] *Rhodnius*,[249] and *Drosophila*.[250]

The entry of Na^+ into the principal cell via conductive and carrier-mediated transport is expected to raise the intracellular Na^+ concentration, improving its competition with K^+ for extrusion across the apical membrane via the H^+/cation exchanger (Figure 7.24). The H^+/cation exchanger may be related to the NHE family of transporters, such as NHE8 identified in the *Aedes* Malpighian tubule.[177] Kinetic studies of the *Aedes* NHE8 transporter in proteoliposomes revealed that it can accept Na^+ as well as K^+ for exchange transport with H^+. It is unknown whether cAMP affects the expression or activity of NHE8 in the principal cells.

Whereas Figure 7.24 summarizes measured values of voltages and electrolyte concentrations in the presence of mosquito natriuretic peptide (or cAMP), the signaling pathway is hypothetical. Calcitonin-like peptide is thought to bind to a G-protein-coupled receptor located in principal cells of Malpighian tubules.[224,251,252] Activation of G protein stimulates adenylate cyclase with the effect of increasing intracellular cAMP concentrations. The binding of cAMP to the regulatory subunits of protein kinase A releases the catalytic kinase subunits setting off the phosphorylation of transport systems in basolateral and apical membranes. Phosphorylation of the epithelial Na/K/2Cl transporter is known to increase its activity.[253] Likewise, phosphorylation of the epithelial Na^+ channel (ENaC) is known to increase Na^+ transport through this channel.[254] Best understood in the transport model illustrated in Figure 7.24 are the effects of cAMP on the V-type H^+-ATPase,[220,255] which are discussed in Section VI.

4. Serotonin

Maddrell discovered the diuretic effect of serotonin in Malpighian tubules of *Rhodnius prolixus* as early as 1969.[256] Serotonin (5-hydroxytryptamine) is a common neurotransmitter in vertebrates and invertebrates that can serve (1) synaptic transmission and the stimulation of tissues innervated by serotonergic nerves, and (2) as a hormone in the hemolymph.[192,193] In *Rhodnius*, serotonergic nerves activated by the ingestion of a blood meal stimulate salivary glands, the dorsal aorta (heart), and the gut. Serotonin released into the hemolymph from neurohemal areas of abdominal nerves increases the secretory activity of Malpighian tubules more than 1000-fold.[192]

In *Rhodnius*, serotonin binds to a G_s-protein-coupled receptor, activates adenylate cyclase, and increases intracellular cAMP concentrations, not unlike the effect of calcitonin-like peptide (Figure 7.24). Because serotonin inhibits the Na,K-ATPase in *Rhodnius* Malpighian tubules, cAMP is thought to activate a protein kinase that inhibits the Na/K pump.[190] The inhibition of the Na,K-ATPase by ouabain is known to stimulate transepithelial Na^+ secretion in *Rhodnius* Malpighian tubules.[257] The inhibition is expected to increase the intracellular concentration of Na^+, thereby out-competing K^+ for transport into the tubule lumen via H^+/cation exchange (Figure 7.24). Rates of transepithelial Na^+ and fluid secretion increase as a result; thus, serotonin appears to mimic the effect of ouabain, bringing about diuresis in distal segments of *Rhodnius* Malpighian tubules.

At the apical membrane of *Rhodnius* Malpighian tubules, serotonin increases the volume of microvilli nearly threefold and causes mitochondria to move from the cell cortex into the microvilli.[142] Thus, the suppliers of ATP become situated adjacent to the ion-transporting apical plasma membrane, which is commensurate with the stimulation of transepithelial ion transport.[142] In addition, serotonin and cAMP—as well as the blood meal—all activate aquaporin water channels to increase the water permeability of the tubule.[161]

Whereas serotonin is clearly the primary diuretic agent in *Rhodnius*, the amine appears to have only secondary roles in Malpighian tubules of other insects. In Malpighian tubules of larval *Aedes aegypti*, serotonin stimulates fluid secretion via cAMP.[258,259] Serotonin also increases fluid secretion in Malpighian tubules of adult *Aedes aegypti*,[260] but the stimulation is small compared to that induced by mosquito natriuretic peptide and leucokinin (unpubl. observ.). Because serotonin and calcitonin-like peptide have additive if not synergistic effects in some insects, it has been suggested that serotonin may modulate the effects of diuretic peptides.[261]

5. Tyramine

The biogenic amine tyramine has recently joined the list of diuretic agents in insect Malpighian tubules. Blumenthal made this discovery in *Drosophila* Malpighian tubules after first noting that the common amino acid tyrosine stimulated transepithelial fluid secretion by increasing a trans-epithelial Cl^- conductance.[262] In principal cells of the tubule, tyrosine decarboxylase converts tyrosine to tyramine. Tyramine is then exported from principal cells and binds in paracrine fashion to a G-protein-coupled tyramine receptor on nearby stellate cells. Blumenthal has cloned the cDNA encoding this tyramine receptor in *Drosophila* Malpighian tubules. Receptor binding elevates intracellular Ca^{2+} concentrations in stellate cells. Calcium goes on to increase a transepithelial Cl^- conductance as the principal mechanism for stimulating transepithelial fluid secretion. The mechanism of action of tyramine is strikingly similar to that of leucokinin in *Aedes* Malpighian tubules (Figure 7.22), except that tyramine and kinin signaling in *Drosophila* takes place in stellate cells and not in principal cells as in *Aedes* Malpighian tubules.

In a follow-up study, Blumenthal made the interesting observation that tyramine and leucokinin signaling in *Drosophila* Malpighian tubules is significantly affected by the osmotic pressure in the peritubular bath of the tubule.[263] An increase in osmotic pressure reduced the diuretic effects of tyramine and leucokinin, which Blumenthal interprets as protecting the fruit fly during dehydration stress. In the insect, dehydration stress is expected to increase the osmotic pressure of the hemolymph.

Increased osmotic pressures also decreased the frequency of spontaneous transepithelial voltage oscillations in *Drosophila* Malpighian tubules. Spontaneous transepithelial voltage oscillations stem from cyclical changes in transepithelial Cl^- conductance[264] that Blumenthal attributes to paracrine tyramine signaling. By decreasing the frequency of transepithelial Cl^- conductance changes, elevated peritubular osmotic pressures limit Cl^- secretion, thereby reducing fluid secretion. Blumenthal suggests further that the conversion of extracellular osmotic pressure (amplitude modulated, AM) into a frequency domain of oscillating Cl^- conductances (frequency modulated, FM) may stabilize transepithelial fluid secretion rates, especially in the case of small changes in extracellular osmolality.

6.   Signaling in the Intact Animal

To establish the relevance of our observations in isolated Malpighian tubules to mechanisms taking place in the intact mosquito, we demonstrated that: (1) in the intact yellow fever mosquito, the initial diuresis that begins during feeding on blood (Figure 7.19) excretes a NaCl-rich urine;[152] (2) the initiation and maintenance of the diuresis in the mosquito requires the presence of the head on the insect;[242,265] (3) mosquito natriuretic peptide injected into the hemolymph of decapitated mosquitoes triggers diuresis *in vivo*;[242,265] (4) hemolymph collected from blood-fed mosquitoes stimulates fluid secretion in isolated Malpighian tubules, but hemolymph from unfed or blood-fed and then decapitated mosquitoes does not;[242,265] and (5) Malpighian tubules isolated from blood-fed mosquitoes during the peak NaCl diuresis reveal elevated cAMP concentrations.[243] These observations establish the physiological connections between the blood meal, the release of mosquito natriuretic peptide, and the role of cAMP as second messenger. Stretch receptors in the abdomen are thought to sense the distension of the gut and to cause the release of diuretic hormones.[200,266] Klowden provides an engaging review of the links between the two frequently tandem activities of eating and mating, this time in the mosquito.[119]

7.   Synergism of Diuretic Peptides

The signaling cascade that regulates the release of kinin diuretic peptides is largely unknown, but it would appear that it does not differ substantially from the release of other diuretic peptides. One reason is that the colocalization of kinins and CRF-related peptides in abdominal neurosecretory cells of the tobacco hornworm is consistent with the release of both peptides when the need for diuresis arises.[211,267] Because kinins increase transepithelial Cl^- secretion via intracellular Ca^{2+} (Figure 7.22) and because CRF-related peptides increase transepithelial cation secretion via cAMP, these separate controls of transepithelial anion and cation secretion allow at least additive if not synergistic effects consequent to the release of both diuretic peptides. Indeed, the kinin and CRF-related peptide of the locust act cooperatively to increase the rate of fluid secretion in Malpighian tubules by more than the sum of their separate responses.[268] Likewise, the kinin and CRF-related peptide of the house fly act synergistically to increase fluid secretion in Malpighian tubules,[269] even though the two peptides do not colocalize to single neurons or neurohemal release sites.[270] Also, evidence in the Malpighian tubules of the kissing bug (*Rhodnius prolixus*) suggests synergistic effects on fluid secretion of the amine serotonin and an unidentified diuretic hormone.[195,271,272]

8. Antidiuretic Peptides

The laboratory of Schooley was the first to isolate an antidiuretic peptide that reduced fluid secretion in Malpighian tubules.[273] The peptide was isolated from heads of the mealworm *Tenebrio molitor* (a tenebrionid beetle) and named Tenmo-ADFa. With an EC_{50} of about 10^{-14} M it is a potent inhibitor of fluid secretion in mealworm Malpighian tubules. Tenmo-ADFa uses cGMP as second messenger. The laboratory of Schooley isolated a second antidiuretic peptide, Tenmo-ADFb, a year later.[274] The antidiuretic effect of ADFb is also mediated via cGMP.

We tested the effects of Tenmo-ADFa in Malpighian tubules of the yellow fever mosquito. The peptide inhibited the spontaneous secretion of fluid without significant effects on the electrophysiology of the tubule and principal cells.[275] Cyclic GMP duplicated the inhibition of fluid secretion without electrophysiological correlates. These observations suggest that the antidiuresis is mediated by inhibiting an electroneutral transport system.

Because cAMP stimulates and cGMP inhibits fluid secretion in mealworm Malpighian tubules, Wiehart and coworkers[276] investigated the interaction of these two nucleotides. When both cAMP and cGMP were added to the peritubular bath of isolated mealworm Malpighian tubules, the stimulatory effect of cAMP was not observed. Similarly, the stimulatory effect of the CRF-related diuretic hormone Tenmo-DH(37) was reversed by the antidiuretic peptide Tenmo-ADFa.[276] Antagonism between diuretic and antidiuretic peptides and their second messengers was thus demonstrated.

Cyclic GMP is known to stimulate cAMP phosphodiesterase activity, thereby reducing the intracellular concentration of cAMP.[277,278] Diminishing intracellular cAMP concentrations are therefore expected to decrease fluid secretion which may explain in part the antagonism between cGMP and cAMP. The antagonism may be useful *in vivo* for terminating the diuresis at the site of primary urine generation.[279]

At the level of the intestine, antidiuretic peptides stimulate reabsorptive transport which diminishes excretory output. One hormone that stimulates fluid absorption is the ion-transport peptide (ITP) that first was isolated in the desert locust.[280] ITP increases fluid absorption in the ileum of the locust by stimulating the absorption of Cl⁻ using cAMP as second messenger.[281] A second hormone to produce antidiuretic effects is the chloride transport-stimulating hormone (CTSH), also identified in the locust.[280,282] CTSH stimulates fluid reabsorption in the rectum by increasing the reabsorption of Cl⁻, again using cAMP as second messenger.[282]

9. Caveat

Biological mechanisms are hardly as straightforward as the research suggests or as presented above. First, although it is true that cAMP stimulates and cGMP inhibits fluid secretion in most Malpighian tubules, exceptions exist. Cyclic GMP, for example, stimulates fluid secretion in *Drosophila* Malpighian tubules. Like cAMP, cGMP activates the V-type H^+-ATPase.[122,186,272,283] In *Drosophila* Malpighian tubules, fluid secretion can be stimulated via cAMP by the binding of Drome-DH44 (a CRF-related peptide) to a G-protein-coupled receptor.[284] Fluid secretion in tubules can also be stimulated via a nitric oxide–cGMP pathway in response to cardiac accelerator peptide CAP_{2b}.[285–287] In both cases, a rise in intracellular Ca^{2+} is thought to increase the activity of apical mitochondria, which would elevate ATP levels in the vicinity of the V-type H^+-ATPase, thereby stimulating proton transport and transepithelial fluid secretion.[288] Thus, two different diuretic peptides, Drome-DH44 and CAP_{2b}, working through different intracellular signaling pathways, may still converge on the same target—namely, the V-type H^+-ATPase to increase transepithelial fluid secretion.

Second, whether a peptide is diuretic or antidiuretic depends on the insect and tissue; for example, CAP_{2b} has diuretic activity in *Drosophila* Malpighian tubules[289] but antidiuretic activity in *Rhodnius* Malpighian tubules.[279] Likewise, kinins have diuretic activity in most insect Malpighian tubules but antidiuretic activity in Malpighian tubules of the cricket.[239] In the same insect, serotonin stimulates K^+ secretion in distal Malpighian tubules and K^+ absorption in proximal Malpighian tubules.[290,291]

Third, the specific signaling pathway activated by a ligand may be dose dependent; for example, low concentrations of synthetic CRF-related diuretic peptide of the mosquito *Culex* caused a mild diuresis in isolated Malpighian tubules of *Aedes aegypti*, apparently via Ca^{2+}-mediated effects on paracellular Cl^- conductance (see Figure 7.22). High concentrations of *Culex* CRF-related diuretic peptide caused a strong diuresis via cAMP-mediated effects on transcellular Na^+ secretion, similar to the effect of mosquito natriuretic peptide (Figure 7.24).[292,293] Thus, CRF-related diuretic peptide may engage both Ca^{2+} and cAMP signaling pathways in *Aedes* Malpighian tubules and in *Anopheles* Malpighian tubules (G. Coast, pers. comm.).

Fourth, cAMP may exert different effects in the same tissue. The CRF-related diuretic hormone Anoga-DH44 and the calcitonin-like hormone Anoga-DH31 use cAMP as second messenger.[9] Whereas cAMP activated by Anoga-DH31 brings about a natriuresis, cAMP activated by Anoga-DH44 does not. Accordingly, a second messenger signaling pathway (cAMP, cGMP, Ca^{2+}) may lead to a particular target, but that signaling pathway is part of an intracellular signaling network that may reroute and alter the message.

V. WATER ABSORPTION BY THE RECTAL COMPLEX

The precise significance of the arrangement is not known; perhaps this serves to add the absorptive powers of the Malpighian tubules to those of the rectal epithelium.

Wigglesworth[294]

In some larval insects and rarely in adult insects, the distal (blind) ends of Malpighian tubules are not suspended in the abdominal hemolymph but instead are encased by a sheath that keeps them closely associated with the rectum. The resulting structure is referred to as the *cryptonephridial* or *rectal complex*. The complex occurs in several coleopterans, including the tenebrionid beetles (e.g., the mealworm, *Tenebrio molitor*), lepidopterans (e.g., the tobacco hornworm, *Manduca sexta*), and at least one dipteran (i.e., the glow-worm, *Arachnocampa luminosa*). The rectal complex is an osmoregulatory adaptation to terrestrial life that helps minimize excretory water losses by reabsorbing water from the feces and the fluid Malpighian tubules have emptied into the gut.[295,296] The rectal complex is absent or poorly developed in coleopterans and lepidopterans with aquatic or semiaquatic larval stages.[297] In larvae of tenebrionid beetles, the rectal complex can also absorb water vapor from air of high relative humidity.[298,299]

A. STRUCTURE OF THE RECTAL COMPLEX

The rectal complex consists of three major structures: the rectum, the distal Malpighian tubules (now referred to as *perinephric tubules*), and the perinephric membrane (Figure 7.25). The latter is the most distinguishing feature of the rectal complex, because the perinephric membrane isolates both the perinephric tubules and the rectum from the hemolymph. Importantly, the membrane is not a lipid bilayer as the term might imply, but instead it consists of several layers of compressed epithelial cells. In lepidopteran larvae,[300] the rectal complex occurs as three longitudinal bands along the rectum (Figure 7.26), whereas in larvae of tenebrionid beetles[295,297,301] and at least one dipteran[302] the perinephric membrane encloses the entire circumference of the rectum (Figure 7.27). Thus, in lepidopterans, portions of the rectal epithelium are bathed by hemolymph and are referred to as the *normal rectal epithelium*, in contrast to the *perinephric rectal epithelium*, which is bathed by fluid occupying the *perinephric space* (Figure 7.26).

In tenebrionid beetles and the glow-worm, the entire rectal epithelium is enclosed within the complex and is bathed by fluid within the perinephric space (Figure 7.27). The perinephric tubules meander in an anterior direction until they reach the anterior end of the perinephric membrane, where they become free of the rectum and together form a common trunk still enclosed by perinephric membrane.[295,301] Once free of the trunk, the Malpighian tubules—now bathed by hemolymph—separate and continue toward the midgut/hindgut junction, where they empty their secretions.

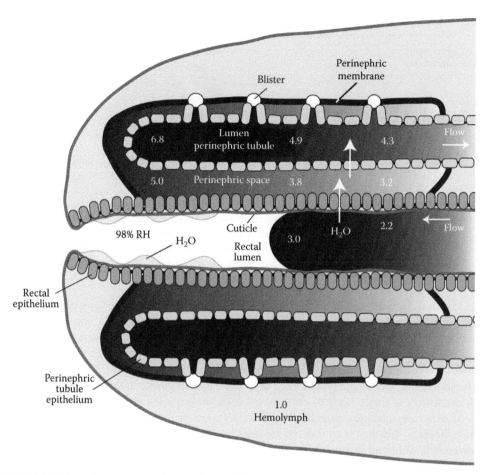

FIGURE 7.25 Osmotic pressure profiles in the rectal complex of the mealworm *Tenebrio molitor*. Numbers indicate osmotic pressures in Osm/kg. Arrows indicate direction of water flow. (Adapted from Machin, J., *Am. J. Physiol.*, 244, R187, 1983.)

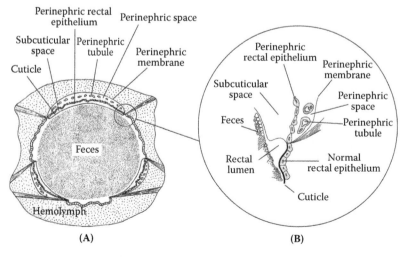

FIGURE 7.26 Rectal complex of a lepidopteran larva: (A) Transverse section through the posterior rectum; (B) enlargement of the region encircled in Part A. (From Ramsay, J.A., *Philos. Trans. R. Soc. Ser. B*, 274, 203, 1976. With permission.)

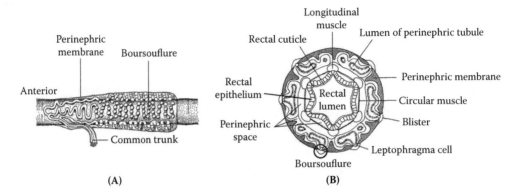

FIGURE 7.27 Rectal complex of the mealworm *Tenebrio molitor*: (A) The perinephric membrane is removed from one tubule to reveal the boursouflures of a perinephric tubule. (From Grimstone, A.V. et al., *Philos. Trans. R. Soc. Ser. B*, 253, 343, 1968. With permission.) (B) Transverse section; circled region of the boursouflure is detailed in Figure 7.28B. (From O'Donnell, M.J. and Machin, J., *J. Exp. Biol.*, 155, 375, 1991. With permission.)

As shown in Figure 7.27B, the rectal epithelium is composed of a single layer of cells that form a tightly packed, uniform cell layer surrounded by circular and longitudinal muscle layers.[301,303] The luminal surface of the rectal epithelium is covered by cuticle and is exposed to either feces or air. Bathing the basolateral surface of the rectal epithelium is the fluid of the perinephric space, which also bathes the basolateral surface of the perinephric tubules.

The perinephric membrane is composed of an outer and an inner cellular sheath.[301,304] The outer sheath consists of a single layer of compressed cells, whereas the inner sheath contains several layers of extremely compressed cells (Figure 7.28B). Given the densely packed cell layers that compose the inner sheath, it is considered the main barrier to osmotic and ionic exchanges between the perinephric fluid and hemolymph;[301] however, exchanges of solutes between the perinephric fluid and the hemolymph might occur in the anterior region of the complex where the inner sheath is thinner compared to that of the posterior region.[301]

In the rectal complex of tenebrionid beetle larvae, the perinephric tubules form *boursouflures* at regular intervals along their length.[295,301] Boursouflures are nodule-like structures that form when the wall of a perinephric tubule bulges out toward the perinephric sheath and the tubule lumen forms two or three diverticulae (Figure 7.27 and Figure 7.28). A morphometric analysis indicates that the boursouflures increase the luminal surface area of the perinephric tubules, thereby providing a greater area for active transport which is hypothesized to enhance water absorption from the rectum.[305] Boursouflures, however, are not essential for rectal water absorption, because the perinephric tubules of lepidopterans and glow-worms lack these structures.[297,300,302]

At the apex of a boursouflure (usually at the most central diverticulum; see Figure 7.28A) is a relatively small, hyaline cell that wedges between the principal cells, not unlike stellate cells in Malpighian tubules of *Aedes aegypti* (Figure 7.12). The cell forms an extremely thin window, or *leptophragma*, into the tubule lumen. As shown in Figure 7.28, the cell body of the leptophragma cell hangs down into the lumen of the diverticulum,[295,297,301,304] and then flattens to bridge a gap in the inner sheath of the perinephric membrane. In the vicinity of leptophragma cells, the outer sheath thins out to a basement membrane, which forms the blister that, according to Grimstone et al.,[301] is filled with "amorphous material" (Figure 7.28B). On the other side, the inner sheath has also thinned out to its basement membrane in contact with the thin portion of the leptophragma cell (Figure 7.28B). In some histological preparations of rectal complexes, the blisters are not noticeable due to shrinkage artifacts.[295,301] Thus, the conclusion that blisters are absent from the rectal complex of the beetle[305] and the glow-worm[302] should be reexamined, because leptophragma cells are present in these two insects.

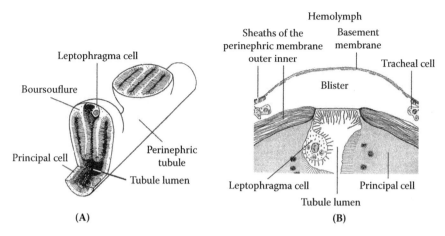

FIGURE 7.28 Boursouflures and blisters of the mealworm rectal complex: (A) Schematic of two boursouflures that are cut open to show the orientation of the diverticulae; note the location of the leptophragma cell. (B) Detail of a blister and a leptophragma cell (circled region in Figure 7.27B); note the relative thickness of the outer and inner sheaths of the perinephric membrane. (From Grimstone, A.V. et al., *Philos. Trans. R. Soc. Ser. B*, 253, 343, 1968. With permission.)

The function of the leptophragma cells (and associated blisters) is enigmatic, but the dramatic reduction in the thickness of perinephric membrane at these sites suggests a transport pathway between the hemolymph and the lumen of the perinephric tubules. Given the high osmotic pressures that perinephric tubules generate in the tubule lumen, it is unlikely that the blisters or leptophragma cells allow water to enter from the hemolymph. Leptophragma cells may provide a pathway for hemolymph Cl^- to enter the tubule lumen, complementing the transcellular secretion of K^+, similar to stellate cells in Malpighian tubules of *Drosophila*. In addition, leptophragma cells may mediate the exchange of metabolic fuels or waste products between the perinephric tubules and hemolymph.[305,306] Whatever their function, leptophragma cells are not essential for the reabsorption of rectal water from excretory material, because these cells are not found in the perinephric tubules of lepidopterans and some coleopterans.[297,300]

Both inner portions of the rectal complex and the perinephric membrane are well supplied by the tracheal system,[295] suggesting high transport activities. In the posterior region of the rectal complex, tracheoles form numerous branches in the space between the inner and outer sheaths of the perinephric membrane (Figure 7.28B). Within this space, the tracheolar cells usually associate with the external surface of the inner sheath.[301] Furthermore, the tracheolar cells often become flattened and contain long peripheral projections,[301] which likely increases their functional surface area for delivering oxygen to the nearby perinephric tubules.

B. FUNCTIONS OF THE RECTAL COMPLEX

The digestion and passage of food through the gut is a wet process. Not only is water required for the activity of hydrolytic enzymes, but it is also needed to make the products of digestion transportable by dissolving them first in water. Only then can channels, carriers, and pumps translocate the products of digestion. Furthermore, water in the intestinal lumen maintains the fluidity of the gut contents and lubricates the easy passage to more distal sites. In insects, the midgut epithelium secretes water into the gut.[307–309] Water also arrives in the intestinal lumen via Malpighian tubules that empty their secretions into the gut. For insects living in dry environments or eating dry foods, preventing intestinal losses of water is critical for maintaining water balance. In coleopteran and lepidopteran larvae, the rectal complex contributes to water conservation by reabsorbing intestinal water. In tenebrionid beetles, the rectal complex is also able to absorb water vapor from air but only at high relative humidity.

1. Tenebrionid Beetles

The stunning feat of absorbing water from air (and from fecal matter) has made tenebrionid beetles the favorite models for studying the rectal complex. Digested food material leaving the midgut of beetle larvae contains a high water content; it is nearly isosmotic with the hemolymph (~0.5 Osm/kg).[13,295] As the rectal complex reabsorbs water it dries the fecal pellets to almost a powder.[295] Assuming that the water content of freshly expelled fecal pellets is in equilibrium with the water vapor pressure of the rectum, Ramsay[295] determined that the relative humidity (RH) in the rectal chamber of *Tenebrio molitor* is 90%. A RH of 90% is expected above a solution with an osmotic pressure of 6.2 Osm/kg, whereas a relative humidity of 99% is expected above a solution with an osmotic pressure of 0.5 Osm/kg, as calculated from the following equation:

$$a_w = \frac{55.5}{55.5 + P_{osm}}$$ (7.2)

Where a_w is the water activity (i.e., RH divided by 100), 55.5 is the number of moles of water in 1 kg of water, and P_{osm} is the osmotic pressure in Osm/kg (see Edney[13] and Kiss and Hansson[310]).

To increase the osmotic pressure of the feces from 0.5 Osm/kg to 6.2 Osm/kg by the abstraction of water alone, the rectal complex must reabsorb 92% of the water. In extreme cases, the RH of the rectal chamber can drop to 75%,[295] reflecting a fecal osmotic pressure of 18.5 Osm/kg. The rectal complex now reabsorbs as much as 97% of the water from the rectal chamber.

The relationship between relative humidity and the osmotic pressure of solutions (see Equation 7.2) suggests that the rectal complex can absorb atmospheric water vapor. Indeed, Nobel-Nesbitt[311] and Machin[299] showed that starved larvae of *Tenebrio molitor* increase both their body mass and body water content if held at RHs above 88%. Blocking the anus with wax prevents the gains of mass and water, indicating that the rectum is the likely site of water vapor absorption.[299,311] The threshold value of 88% RH for vapor absorption is consistent with the mean RH of 90% in the rectal chamber of *T. molitor*.[295,312] In tenebrionid beetle larvae (*Onymacris* sp.) that inhabit the Namib Desert, water vapor absorption can occur at ambient RHs as low as 83%.[298,313]

The rate of water vapor absorption increases as ambient RH increases above the threshold RH for water absorption.[298,299,314] Rates of rectal water uptake can be as high as 5 to 7% of their body mass per day.[315] It is unlikely that the threshold RH for rectal water uptake is reached in *T. molitor* living in stored grain or flour. Here, the rectal complex returns most of the intestinal water to the hemolymph;[316] however, in larvae of *Onymacris* sp. of the Namib Desert, rectal absorption of water vapor may be relevant during the oceanic fogs that moisten the desert air. During the peak intervals of these fogs, the larvae move toward the surface of the sand, where the RH may exceed 80%.[298]

2. Lepidoptera

The rectal complex of lepidopterans was initially considered to play only a minor role in rectal water reabsorption, because larvae of the large white butterfly (*Pieris brassicae*) could increase the osmotic pressure of the rectal milieu only to 1.4 Osm/kg, consistent with a RH of 97.5%.[300] Furthermore, lepidopteran larvae typically eat foods with high water content, such as succulent plant material. Some studies suggest, however, that the rectal complex—along with the normal rectal epithelium—is important for recycling water to help maintain the high water content of the midgut. If *Manduca sexta* larvae are fed a relatively dry diet, the rate of rectal water absorption increases three- to sixfold.[296] The reabsorbed water is presumably recycled back to the midgut.[296,309] The diuretic hormone Mas-DH enhances fluid secretion by the perinephric tubules of *M. sexta* as it does in Malpighian tubules in insects without a rectal complex. The stimulation of fluid secretion in perinephric tubules increases water reabsorption by the rectal complex two- to threefold.[317] The same function of the rectal complex *in vivo* would recycle renal water, consistent with a clearance role of diuretic hormones.[148]

C. Mechanisms of Water Absorption from the Rectal Lumen

The mechanisms for absorbing water by the rectal complex have been most extensively studied in the mealworm *Tenebrio molitor*, especially for absorbing water from air. It can be assumed that the same mechanisms apply to the reabsorption of water from fecal material. Figure 7.25 illustrates osmotic pressure profiles in the rectal complex of a mealworm larva. In this case, air with a RH of 98% (above a solution of 1.1 Osm/kg) fills the posterior half of the rectal lumen, whereas the anterior portion is filled with fecal matter from which some water has already been reabsorbed. At 98% RH, it can be expected that water condenses on the surface of the rectal cuticle,[318] making it available for absorption (Figure 7.25).

The uptake of water from the rectum is driven by radial osmotic pressure gradients established by active transport in the perinephric tubules.[318,319] As illustrated in Figure 7.25, measured osmotic pressures in the lumen of the blind-ended segment of the perinephric (Malpighian) tubule can be as high as 6.8 Osm/kg,[318] and can reach 9.0 Osm/kg in perinephric tubules of *Onymacris* sp.[315] As the fluid in the perinephric tubule lumen flows downstream, it gains water by osmosis from the perinephric space and ultimately from the rectal chamber. The addition of water dilutes solutes and, consequently, reduces the osmotic pressure of the fluid flowing down the perinephric tubule. Because flow down the perinephric lumen is counter-current to flow in the rectum, it would appear that water is picked from the rectum along the entire length of the rectal complex (Figure 7.25). When the perinephric tubule emerges from the rectal complex, osmotic inflow of water across the epithelium of the Malpighian tubule renders the luminal contents isosmotic with the hemolymph;[312,316,318] consequently, the luminal volume and flow toward the tubule/gut junction increase.

In the larvae of tenebrionid beetle species (*Tenebrio molitor*, *Onymacris* sp.), the perinephric tubules generate high luminal osmolalities by actively secreting K^+ into the lumen, with Cl^- following passively.[305,306] Although Na^+ and H^+ can also be transported actively into the lumen,[305,306] K^+ is by far the more important cation, because: (1) the concentration of K^+ in secreted fluid is between four and eight times greater than that of Na^+,[305,306] and (2) the fluid reabsorption by the isolated rectal complex is dependent on the presence of K^+ but not Na^+ in the saline hemolymph bathing the perinephric membrane.[312] The sites of K^+ and Cl^- transport along perinephric tubules have been a matter of debate. Grimstone and colleagues[301] proposed that K^+ is actively transported from the hemolymph into the lumen of perinephric tubules at the blisters via the leptophragma cells, with Cl^- from the hemolymph following passively. Active transport of K^+ by leptophragma cells, however, is inconsistent with their ultrastructure (Figure 7.28B), which more closely resembles stellate cells than principal cells (Figure 7.12). A study of the electrochemical potentials for K^+ and Cl^- in the rectal complex indicates that K^+ is secreted by active transport from the perinephric space into the lumen of perinephric tubules (Figure 7.25), most likely by principal cells of the tubule.[305,306] The active transport of K^+ then provides a favorable electrochemical gradient for Cl^- to enter the tubule lumen passively from the perinephric fluid via a paracellular route or from the hemolymph, possibly via a transcellular pathway through leptophragma cells (Figure 7.25). The hypothesis of transcellular Cl^- transport through leptophragma cells is attractive, as O'Donnell and colleagues[125] have suggested a similar function for stellate cells in Malpighian tubules of *Drosophila*.

If K^+ is secreted into the lumen of the perinephric tubule, then the question arises as to what replaces K^+ in the perinephric fluid of the rectal complex. The first hypothesis to come to mind is reabsorption from the rectum, because Malpighian tubules empty their K^+ rich secretions into the gut; however, if the rectal lumen of the *Tenebrio molitor* rectal complex is filled with air or concentrated sucrose solution, then water absorption still occurs[312] and concentrations of K^+ in the perinephric space and the lumen of perinephric tubules are not affected.[306] O'Donnell and Machin[306] suggest that, as the perinephric tubules deplete K^+ from the perinephric fluid, it is replaced by electrodiffusion from the hemolymph across the anterior portions of perinephric membrane, which are more permeable to solutes than the posterior portions. Tracer studies are needed to identify the routes and mechanisms K^+ and Cl^- take for getting into the perinephric space.

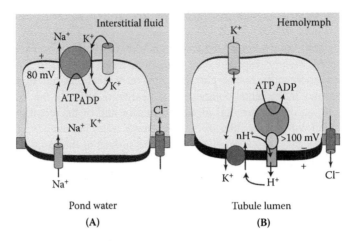

FIGURE 7.29 Paradigms of transepithelial transport: (A) The Ussing model of active Na$^+$ transport across frog skin. Central to the model is the Na,K-ATPase (pump) located at the basolateral membrane facing the interstitial fluid/blood. Bringing K$^+$ from the interstitial fluid into the cell (active transport), the pump elevates intracellular K$^+$ concentrations far above extracellular concentrations. K$^+$ diffusing out of the cell (passive transport) generates a voltage that is used to pull in Na$^+$ from pond water through Na$^+$ channels located in the apical membrane (passive transport). Na$^+$ entering the cell across the apical membrane is extruded from the cell across the basolateral membrane by the Na,K-ATPase (active transport). (B) Wieczorek–Harvey model of active K$^+$ secretion in Malpighian tubules. Here, the ATPase is a proton pump of the vacuolar type, the V-type H$^+$-ATPase. The ATP-driven extrusion of H$^+$ ions (protons) across the apical membrane carries current. Current returns to the cytoplasmic side of the pump passing through the paracellular pathway and the basolateral membrane. Cl$^-$ carries current through the paracellular pathway as the mechanism for secreting Cl$^-$ into the tubule lumen. K$^+$ carries current across the basolateral membrane as the mechanism for K$^+$ entry through K$^+$ channels. K$^+$ leaves the cell via hypothetical exchange transport with H$^+$. A hypothetical stoichiometry of 2H$^+$ for each K$^+$ transported could take advantage of the large membrane voltage (>100 mV) generated across the apical membrane by the proton pump.

In the rectal complex of the lepidopteran *Manduca sexta*, pharmacological evidence indicates that the activity of a V-type H$^+$-ATPase and a Na/H exchanger (presumably in the perinephric tubules) contribute to the basal rate of fluid reabsorption[320] and that a Na/K/2Cl cotransporter is involved with the stimulation of fluid reabsorption mediated by cAMP.[317] Comprehensive molecular and immunochemical studies on the expression of ion transporters and channels have yet to be undertaken on the rectal complex of any insect. Accordingly, completion of the genome of the tenebrionid beetle *Tribolium castaneum* promises to shed new light on the old question of drinking water from air.

VI. EPITHELIAL TRANSPORT POWERED BY THE V-TYPE H$^+$-ATPase

We conclude this chapter with a focus on the V-type H$^+$-ATPase, because insects have figured importantly in establishing the physiological significance of this proton pump that for some time was thought unique to membranes of intracellular vacuoles.[321] Studies of an insect midgut have demonstrated that the V-type H$^+$-ATPase can also reside in the plasma membrane of cells where it can energize the transport of diverse solutes other than H$^+$.[322,323] Prior to this discovery, Na,K-ATPase was considered the sole energizer of cell membranes and epithelial transport systems according to the famous Ussing model that for nearly 50 years has dominated our understanding of transepithelial transport (Figure 7.29). The Wieczorek–Harvey model of transepithelial transport built around a proton pump added a new paradigm to the physiology of epithelia, but it did not replace the Ussing paradigm of epithelial transport (Figure 7.29).

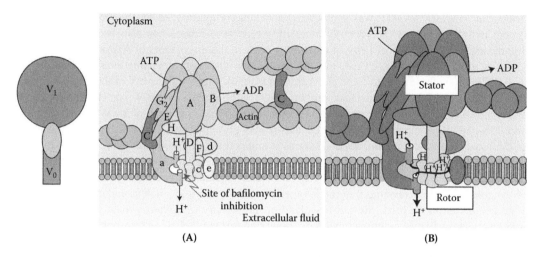

FIGURE 7.30 Molecular and mechanical models of the V-type H⁺-ATPase. The figure illustrates (A) the constituent protein subunits of the V_0 and V_1 complexes, and (B) their assembly as a mechanical stator and a rotor. (Adapted from Beyenbach, K.W. and Wieczorek, H., *J. Exp. Biol.*, 209, 577, 2006.)

In vertebrates, the V-type H⁺-ATPase is now known to be essential for the transport of Na⁺ and Cl⁻ transport across the gills of freshwater animals.[324–326] In mammals, the V-type H⁺-ATPase is needed for reabsorbing bone,[327] for maturing sperm,[328,329] for reabsorbing HCO_3^- to secrete H⁺ in the kidney,[330,331] and for regulating the pH in the inner ear.[332] Cancer cells are thought to use the V-type H⁺-ATPase for acidifying and digesting tissue in advance of the growing tumor.[333]

Parallel to the interest in the physiological roles of the V-type H⁺-ATPase, biophysicists have ingeniously elucidated the function of F-ATPases, which share many features with the V-type H⁺-ATPase[334] (see Figure 7.14 and Figure 7.30). Whereas F-ATPases use a proton concentration difference across the mitochondrial membrane to generate ATP, V-type H⁺-ATPases use ATP to produce a H⁺ electrochemical potential difference across a membrane: endosomal, vacuolar, or plasma.[335,336]

The V-type H⁺-ATPase, the holoenzyme, consists of more than 10 polypeptides assembled in two ring structures connected by two stalks (Figure 7.30). The peripheral ring structure, the so-called V_1 complex, is invariably found in the cytoplasm. It is the site of ATP hydrolysis mediated by subunits A and B. The other ring structure, the so-called V_0 complex, is located in the membrane and consists of six or more c-subunits. It is the site of proton translocation. One stalk is peripheral. It holds the V_1 complex in place by anchoring it to actin and to subunit a of the V_0 complex via subunit C (Figure 7.30). The other stalk is central and consists of subunits D and F (Figure 7.30).

The functional model views the V-type H⁺-ATPase holoenzyme as a molecular motor composed of a stator and a rotor (Figure 7.30). The rotor consists of subunits D and F and the ring of subunit c. The remaining subunits of the holoenzyme form the stator. The hydrolysis of ATP energizes the rotation of the rotor which translocates protons from one side of the membrane to the other, interacting with two hypothetical half channels in the plasma membrane.[337] The inner half of the channel delivers H⁺ to rotating c subunits, each capable of binding one proton. Protein–protein interaction of the c subunit with the outer half channel releases H⁺ to the outside.

The above mechanical model of the V-type H⁺-ATPase (Figure 7.30) rests on rotary models proposed for F-ATPases[338,339] and V-type H⁺-ATPases.[337,340] In one ingenious experiment, the catalytic ring of the F-ATPase was immobilized upside down on a glass surface, and a fluorescent actin filament was attached to the central stalk.[341] The addition of ATP triggered rotation of the actin propeller. What is more, the reversibility of this motor could be demonstrated! A magnetic bead, instead of an actin filament, was attached to the central stalk, and luciferin and luciferase were added to the medium. Cranking the bead with an external magnet gave evidence of a "molecular sparkler" as ATP was synthesized and quickly hydrolyzed by the luciferin/luciferase system emitting photons. The rotary nature of V-type H^+-ATPases has also been visualized.[342] Significantly, the 1997 Nobel Prize in Chemistry went to the discoverers of ATP-synthesizing and -utilizing proteins: to Boyer (UCLA) and Walker (Cambridge) for elucidating the synthesis of ATP by F-ATPase (synthase) and to Skou (U. Aarhus, Denmark) for his discovery of the Na,K-ATPase.

The most widely studied mechanism that regulates the activity of the V-type H^+-ATPase is the reversible assembly and disassembly of this proton pump. The dissociation of the V-type H^+-ATPase into its V_1 and V_0 complexes, which inactivates proton pumping, was first observed in the midgut of the tobacco hornworm (*Manduca sexta*) as it stopped feeding with the onset of molt.[343] The withdrawal of glucose leads to the disassembly of the V-type H^+-ATPase in yeasts, and restoring glucose reassembled the proton pump.[344] Whereas these two examples exemplify nutritional controls of the assembly and disassembly of the V-type H^+-ATPase, the insect salivary gland demonstrates a hormonal control. In brief, the salivary gland of the blowfly *Calliphora vicina* secretes salt and water, very much like Malpighian tubules do, powering transepithelial ion transport with the V-type H^+-ATPase (Figure 7.14 and Figure 7.18). The hormone serotonin stimulates salivary secretion via an increase in intracellular cAMP, and both serotonin and cAMP cause the catalytic V_1 complex of the proton pump to dock at the V_0 complex in the apical membrane of glandular epithelial cells, thereby initiating V-type H^+-ATPase activity and increasing transepithelial ion and water secretion.[220,345] One subunit of the proton pump, subunit C, figures importantly in the cAMP-induced docking of the V_1 complex to the V_0 complex (Figure 7.24 and Figure 7.30). The subunit becomes phosphorylated by protein kinase A in the presence of cAMP.[255] The phosphorylation of subunit C is then thought to enhance its binding to the V_1 and V_0 complex, thereby assembling the holoenzyme.[346] Additional binding of subunit C to actin filaments may stabilize the assembled holoenzyme in the plasma membrane.[347,348] The physiological and molecular similarities of the cAMP stimulation of electrolyte secretion in salivary glands and Malpighian tubules suggest that the phosphorylation of subunit C is an essential, if not canonical, event in activating the V-type H^+-ATPase via the assembly of V_0 and V_1 complexes (Figure 7.24).

ACKNOWLEDGMENTS

One of us (Beyenbach) is old enough to acknowledge the role that the Potts and Parry book had in shaping his research interest in biology. A course I took from Prof. Francis Horne (Southwest Texas State University, San Marcos) in the late 1960s led me to the fascinating subject of "osmotic and ionic regulation in animals," made even more captivating by the clear and pleasing English of Potts and Parry. It embarked me on a life-long interest in the subject. Paging through the 1963 book today, I see how Potts and Parry influenced my decision to study the mechanisms of salt and water homeostasis. "The importance of homeostatic mechanisms in a living animal cannot be overestimated" is the first sentence in their book. The book goes on to examine homeostatic mechanisms from several perspectives: the chemistry of solutions, environmental challenges faced by animals in different habitats, and mechanisms to deal with these challenges in vertebrates and invertebrates. What emerges is a synthesis that illuminates the astonishing number of ways animals can make a living in diverse environments. The book exemplifies the creative contributions of comparative physiology written in the heyday of comparative physiology. The final paragraph of Potts and Parry is oracular in its brief mention of diuretic hormones in insects. The student reader I was in the 1960s would have never imagined that 40 years later I would have the privilege of updating the story on diuretic hormones in insects. It is my hope that readers who know the old Potts and Parry have detected traces of their influence in my pages.

Thank you, David Schooley, Lutz Wasserthal, and David Evans, for proofreading these pages. The authors are particularly grateful to Paul Kestler for deep discussions of respiratory water loss and to Geoffrey Coast for sharing his good knowledge of Malpighian tubules with us.

REFERENCES

1. Hickman, Jr., C. P., Roberts, L. S., and Hickman, F. M., *Integrated Principles of Zoology*, 7th ed., Times Mirror/Mosby College Publishing, St. Louis, MO, 1984.
2. Miller, G. M. and Petersen, R., *Insects, Disease, and History*, http://entomology.montana.edu/historybug/, 2008.
3. Chapman, R. F., *The Insects: Structure and Function*, 4th ed., Cambridge University Press, Cambridge, U.K., 1998.
4. Marquardt, W. C., *Biology of Disease Vectors*, Academic Press, Burlington, MA, 2005.
5. Clements, A. N., *The Biology of Mosquitoes*, 1st ed., Chapman & Hall, London, 1992.
6. Chown, S. L. and Nicolson, S. W., *Insect Physiological Ecology*, Oxford University Press, Oxford, U.K., 2004.
7. Dow, J. A. and Davies, S. A., The Malpighian tubule: rapid insights from post-genomic biology, *J. Insect Physiol.*, 52, 365, 2006.
8. Gaede, G., Regulation of intermediary metabolism and water balance of insects by neuropeptides, *Annu. Rev. Entomol.*, 49, 93, 2004.
9. Coast, G., The endocrine control of salt balance in insects, *Gen. Comp. Endocrinol.*, 152, 332, 2007.
10. Potts, W. T. W. and Parry, G., *Osmotic and Ionic Regulation*, Pergamon Press, London, 1963.
11. Sutcliffe, D. W., The chemical composition of haemolymph in insects and some other arthropods, in relation to their phylogeny, *Comp. Biochem. Physiol.*, 9, 121, 1963.
12. Ramsay, J. A., The evaporation of water from the cockroach, *J. Exp. Biol.*, 12, 373, 1935.
13. Edney, E. B., *Water Balance in Land Arthropods*, Springer, Berlin, New York, 1977.
14. Hadley, N. F., *Water Relations in Terrestrial Arthropods*, Academic Press, San Diego, CA, 1994.
15. Nation, J. L., *Insect Physiology and Biochemistry*, CRC Press, Boca Raton, FL, 2002.
16. Hadley, N. F., Cuticle ultrastructure with respect to the lipid waterproofing layer, *J. Exp. Zool.*, 222, 239, 1982.
17. Hadley, N. F., Integumental lipids of plants and animals: comparative function and biochemistry, *Adv. Lipid Res.*, 24, 303, 1991.
18. Klowden, M. J., *Physiological Systems in Insects*, 2nd ed., Academic Press, New York, 2007.
19. Locke, M., The Wigglesworth Lecture: insects for studying fundamental problems in biology, *J. Insect Physiol.*, 47, 495, 2001.
20. Machin, J., Smith, J., and Lampert, G., Evidence for hydration-dependent closing of pore structures in the cuticle of *Periplaneta americana*, *J. Exp. Biol.*, 192, 83, 1994.
21. Smith, J. J., Machin, J., and Lampert, G. J., An electrical model for *Periplaneta americana* pronotal integument: an epidermal location for hydration-dependent resistance, *J. Exp. Biol.*, 198, 249, 1995.
22. Gibbs, A. G., Waterproof cockroaches: the early work of J. A. Ramsay, *J. Exp. Biol.*, 210, 921, 2007.
23. Gunn, D., The temperature and humidity relations of the cockroach *Blatta orientalis*, *J. Exp. Biol.*, 10, 274, 1933.
24. Wigglesworth, V. B., Transpiration through the cuticle of insects, *J. Exp. Biol.*, 21, 97, 1945.
25. Beament, J. W. L., The cuticular lipoids of insects, *J. Exp. Biol.*, 21, 115, 1945.
26. Beament, J. W. L., The active transport and passive movement of water in insects. In *Advances in Insect Physiology*, Beament, J. W. L., Treherne, J. E., and Wigglesworth, V. B., Eds., Academic Press, London, 1964, p. 67.
27. Gibbs, A. and Crowe, J. H., Intra-individual variation in cuticular lipids studied using Fourier transform infrared spectroscopy, *J. Insect Physiol.*, 37, 743, 1991.
28. Rourke, B. C. and Gibbs, A. G., Effects of lipid phase transitions on cuticular permeability: model membrane and *in situ* studies, *J. Exp. Biol.*, 202, 3255, 1999.
29. Blomquist, G. J., Nelson, D. R., and De Renobales, M., Chemistry biochemistry and physiology of insect cuticular lipids, *Arch. Insect Biochem. Physiol.*, 6, 227, 1987.
30. Gibbs, A. G., Water-proofing properties of cuticular lipids, *Am. Zool.*, 38, 471, 1998.
31. Hadley, N. F., Lipid water barriers in biological systems, *Prog. Lipid Res.*, 28, 1, 1989.
32. Lockey, K. H., Insect hydrocarbon classes implications for chemotaxonomy, *Insect Biochem.*, 21, 91, 1991.
33. Lockey, K. H., Lipids in the insect cuticle: origin, composition and function, *Comp. Biochem. Physiol. B*, 89, 595, 1988.

34. Noble-Nesbitt, J., Cuticular permeability and its control. In *Physiology of the Insect Epidermis*, Binnington, K. and Retnakaran, A. Eds., CSIRO Publications, East Melbourne, Australia, 1991, p. 252.

35. Renobales, M. D., Nelson, D. R., and Blomquist, G. J., Cuticular lipids. In *Physiology of the Insect Epidermis*, Binnington, K. and Retnakaran, A. Eds., CSIRO Publications, East Melbourne, Australia, 1991, p. 240.

36. Gibbs, A., Kuenzli, M., and Blomquist, G. J., Sex- and age-related changes in the biophysical properties of cuticular lipids of the housefly, *Musca domestica*, *Arch. Insect Biochem. Physiol.*, 29, 87, 1995.

37. Nelson, D. R., Dillwith, J. W., and Blomquist, G. J., Cuticular hydrocarbons of the house fly, *Musca domestica*, *Insect Biochem.*, 11, 187, 1981.

38. Gibbs, A. and Pomonis, J. G., Physical properties of insect cuticular hydrocarbons: the effects of chain length, methyl-branching and unsaturation, *Comp. Biochem. Physiol. B*, 112, 243, 1995.

39. Patel, S., Nelson, D. R., and Gibbs, A. G., Chemical and physical analyses of wax ester properties, *J. Insect Sci.*, 1, 1, 2001.

40. Gibbs, A. G., Lipid melting and cuticular permeability: new insights into an old problem, *J. Insect Physiol.*, 48, 391, 2002.

41. Gibbs, A., Physical properties of insect cuticular hydrocarbons: model mixtures and lipid interactions, *Comp. Biochem. Physiol. B*, 112, 667, 1995.

42. Benoit, J. B. and Denlinger, D. L., Suppression of water loss during adult diapause in the northern house mosquito, *Culex pipiens*, *J. Exp. Biol.*, 210, 217, 2007.

43. Hadley, N. F., Epicuticular lipids of the desert tenebrionid beetle, *Eleodes armata:* seasonal and acclimatory effects on composition, *Insect Biochem.*, 7, 277, 1977.

44. Hadley, N. F., Cuticular lipids of adults and nymphal exuviae of the desert cicada, *Diceroprocta apache* (Homoptera, Cicadidae), *Comp. Biochem. Physiol. B*, 65, 549, 1980.

45. Hadley, N. F., Wax secretion and color phases of the desert tenebrionid beetle *Cryptoglossa verrucosa* (LeConte), *Science*, 203, 367, 1979.

46. Hadley, N. F. and Schultz, T. D., Water loss in three species of tiger beetles (*Cicindela*): correlations with epicuticular hydrocarbons, *J. Insect Physiol.*, 33, 677, 1987.

47. Rourke, B. C., Geographic and altitudinal variation in water balance and metabolic rate in a California grasshopper, *Melanoplus sanguinipes*, *J. Exp. Biol.*, 203, 2699, 2000.

48. Yoder, J. A. and Denlinger, D. L., Water balance in flesh fly pupae and water vapor absorption associated with diapause, *J. Exp. Biol.*, 157, 273, 1991.

49. Baker, J. E., Cuticular lipids of larvae of *Attagenus megatoma*, *Insect Biochem.*, 8, 287, 1978.

50. Baker, J. E., Nelson, D. R., and Fatland, C. L., Developmental changes in cuticular lipids of the black carpet beetle, *Attagenus megatoma*, *Insect Biochem.*, 9, 335, 1979.

51. Gibbs, A. and Mousseau, T. A., Thermal acclimation and genetic variation in cuticular lipids of the lesser migratory grasshopper (*Melanoplus sanguinipes*): effects of lipid composition on biophysical properties, *Physiol. Zool.*, 67, 1523, 1994.

52. Hadley, N. F., Cuticular permeability of desert tenebrionid beetles correlations with epicuticular hydrocarbon composition, *Insect Biochem.*, 8, 17, 1978.

53. Gibbs, A., Mousseau, T. A., and Crowe, J. H., Genetic and acclimatory variation in biophysical properties of insect cuticle lipids, *Proc. Natl. Acad. Sci. USA*, 88, 7257, 1991.

54. Hanrahan, S. A., McClain, E., and Gernecke, D., Dermal glands concerned with production of wax blooms in desert tenebrionid beetles, *S. Afr. J. Sci.*, 80, 176, 1984.

55. Hanrahan, S. A., McClain, E., and Warner, S. J. C., Protein component of the surface wax bloom of a desert tenebrionid, *Zophosis testudinaria*, *S. Afr. J. Sci.*, 83, 495, 1987.

56. Gibbs, A. G., Chippindale, A. K., and Rose, M. R., Physiological mechanisms of evolved desiccation resistance in *Drosophila melanogaster*, *J. Exp. Biol.*, 200, 1821, 1997.

57. Gibbs, A. G., Louie, A. K., and Ayala, J. A., Effects of temperature on cuticular lipids and water balance in a desert *Drosophila*: is thermal acclimation beneficial?, *J. Exp. Biol.*, 201, 71, 1998.

58. Gibbs, A. G. and Matzkin, L. M., Evolution of water balance in the genus *Drosophila*, *J. Exp. Biol.*, 204, 2331, 2001.

59. Markow, T. A. and Toolson, E. C., Temperature effects on epicuticular hydrocarbons and sexual isolation in *Drosophila mojavensis*. In *Ecological and Evolutionary Genetics of Drosophila*, Barker, J. S. F., Starmer, W. T., and MacIntyre, R. J. Eds., Plenum Press, New York, 1990.

60. Toolson, E. C., Effects of rearing temperature on cuticle permeability and epicuticular lipid composition in drosophila-pseudoobscura, *J. Exp. Zool.*, 222, 249, 1982.

61. Montooth, K. L. and Gibbs, A. G., Cuticular pheromones and water balance in the house fly, *Musca domestica*, *Comp. Biochem. Physiol. A Mol. Integr. Physiol.*, 135(3), 457, 2003.

62. Wasserthal, L. T., Haemolymphe und Haemolymphtransport. In *Lehrbuch der Entomologie*, 2nd ed., Dettner, K. P. and Peters, W. Eds., Spektrum-Verlag, Heidelberg, 2003, p. 185.

63. Wasserthal, L. T., Atemsystem. In *Lehrbuch der Entomologie*, 2nd ed., Dettner, K. P. and Peters, W. Eds., Spektrum-Verlag, Heidelberg, 2003, p. 165.

64. Wasserthal, L. T., Interaction of circulation and tracheal ventilation in holometabolous insects, *Adv. Insect Physiol.*, 26, 297, 1996.

65. Schneiderman, H. A., Discontinuous gas exchange in insects: role of the spiracles, *Biol. Bull.*, 119, 494, 1960.

66. Kestler, P., *Die diskontinuierliche Ventilation bei Periplanata americana L. und anderen Insekten*, Julius-Maximillian-Universitaet, 1971.

67. Kestler, P., Respiration and respiratory water loss. In *Environmental Physiology and Biochemistry of Insects*, Hoffmann, K. H., Ed., Springer-Verlag, Berlin, 1985, p. 137.

68. Lighton, J. R. B., Simultaneous measurement of oxygen uptake and carbon dioxide emission during discontinuous ventilation in the tok-tok beetle, *Psammodes striatus*, *J. Insect Physiol.*, 34, 361, 1988.

69. Miller, P. L., Respiration-aerial gas transport. In *The Physiology of Insecta*, Rockstein, M., Ed., Academic Press, New York, 1964, p. 557.

70. Wigglesworth, V. B. and Lee, W. M., The supply of oxygen to the flight muscles of insects: a theory of tracheole physiology, *Tissue Cell*, 14, 501, 1982.

71. Wigglesworth, V. B., A theory of tracheal respiration in insects, *Proc. R. Soc. Lond. B*, 106, 229, 1930.

72. Hankeln, T., Jaenicke, V., Kiger, L. et al., Characterization of *Drosophila* hemoglobin: evidence for hemoglobin-mediated respiration in insects, *J. Biol. Chem.*, 277, 29012, 2002.

73. Pietrantonio, P. V., Jagge, C., Keeley, L. L., and Ross, L. S., Cloning of an aquaporin-like cDNA and *in situ* hybridization in adults of the mosquito *Aedes aegypti* (Diptera: Culicidae), *Insect Mol. Biol.*, 9, 407, 2000.

74. Herford, G. M., Tracheal pulsation in the flea, *J. Exp. Biol.*, 15, 327, 1938.

75. Miller, P. L., Respiration-aerial gas transport. In *The Physiology of Insecta*, Rockstein, M., Ed., Academic Press, New York, 1974, p. 345.

76. Kestler, P., Physiological gas exchange strategies for spiracular control, *J. Comp. Biochem. Physiol.*, 134A, S73, 2003.

77. Ochs, M., Nyengaard, J. R., Jung, A., Knudsen, L., Voigt, M., Wahlers, T., Richter, J., and Gundersen, H. J., The number of alveoli in the human lung, *Am. J. Respir. Crit. Care Med.*, 169, 120, 2004.

78. Krogh, A., Studien ueber Tracheenrespiration. II. Ueber Gasdiffusion in den Tracheen, *Pflügers Arch. Gesamte Physiol. Menschen Tiere*, 179, 95, 1920.

79. Weis-Fogh, T., Diffusion in insect wing muscle, the most active tissue known, *J. Exp. Biol.*, 41, 229, 1964.

80. Krogh, A., Studien ueber Tracheenrespiration. III. Die Kombination von mechanischer Ventilation mit Gasdiffusion nach Versuchen an *Dytiscus*larven, *Pflügers Arch. Gesamte Physiol. Menschen Tiere*, 179, 113, 1920.

81. Harrison, J. F., Ventilatory mechanism and control in grasshoppers, *Am. Zool.*, 37, 73, 1997.

82. Bailey, L., The respiratory currents in the tracheal system of the adult honey-bee, *J. Exp. Biol.*, 31, 589, 1954.

83. Miller, P. L., Ventilation in active and inactive insects. In *Locomotion and Energetics in Arthropods*, Herreid, C. F. and Fourtner, C. R., Eds., Plenum Press, New York, 1981, p. 367.

84. Weis-Fogh, T., Respiration and tracheal ventilation in locusts and other flying insects: a theory of tracheole physiology, *Tissue Cell*, 14, 501, 1967.

85. Wasserthal, L. T., Flight-motor-driven respiratory air flow in the hawkmoth *Manduca sexta*, *J. Exp. Biol.*, 204, 2209, 2001.

86. Lighton, J. R. B., Discontinuous gas exchange in insects, *Annu. Rev. Entomol.*, 41, 309, 1996.

87. Bridges, C. R., Kestler, P., and Scheid, P., Tracheal volume in the pupa of the Saturniid moth *Hyalophora cecropia* determined with inert gases, *Respir. Physiol.*, 40, 281, 1980.

88. Lighton, J. R. B., Garrigan, D. A., Duncan, F. D., and Johnson, R. A., Spiracular control of respiratory water loss in female alates of the harvester ant *Pogonomyrmex rugosus*, *J. Exp. Biol.*, 179, 233, 1993.

89. Chown, S. L. and Davis, A. L. V., Discontinuous gas exchange and the significance of respiratory water loss in scarabaeine beetles, *J. Exp. Biol.*, 206, 3547, 2003.

90. Wigglesworth, V. B., *The Principles of Insect Physiology*, 2nd ed., Methuen, London, 1942.

91. Lehmann, F. O., Matching spiracle opening to metabolic need during flight in *Drosophila*, *Science*, 294, 1926, 2001.

92. Nicolson, S. W. and Louw, G. N., Simultaneous measurement of evaporative water loss oxygen consumption and thoracic temperature during flight in a carpenter bee *Xylocopa capitata*, *J. Exp. Zool.*, 222, 287, 1982.

93. Edney, E. B., Water balance in arthropods, *Science*, 156, 1059, 1967.

94. Harrison, J. F. and Roberts, S. P., Flight respiration and energetics, *Annu. Rev. Physiol.*, 62, 179, 2000.

95. Hadley, N. F. and Quinlan, M. C., Discontinuous carbon dioxide release in the Eastern lubber grasshopper *Romalea guttata* and its effect on respiratory transpiration, *J. Exp. Biol.*, 177, 169, 1993.

96. Ziegler, R., Metabolic energy expenditure and its hormonal regulation. In *Environmental Physiology and Biochemistry of Insects*, Hoffmann, K. H., Ed., Springer-Verlag, Berlin, 1985, p. 94.

97. Lehmann, F. O., Dickinson, M. H., and Staunton, J., The scaling of carbon dioxide release and respiratory water loss in flying fruit flies (*Drosophila* spp.), *J. Exp. Biol.*, 203, 1613, 2000.

98. Bertsch, A., Foraging in male bumble bees *Bombus lucorum* maximizing energy or minimizing water load, *Oecologia*, 62, 325, 1984.

99. Weis-Fogh, T., Metabolism and weight economy in migrating animals, particularly birds and insects. In *Insects and Physiology*, Beament, J. W. L. and Treherne, J. E., Eds., Oliver and Boyd, London, 1967, p. 143.

100. Kestler, P., Weshalb gerade kleine Insekten im Flug einen Gasaustausch durch Diffusion verhindern muessen. In *BIONA-Report*, Nachtigal, W. Ed., Fischer, Stuttgart, 1983, p. 135.

101. Ramsay, J. A., Active transport of water by the Malpighian tubules of the stick insect, *Dixippus morosus* (Orthoptera: Phasmidae), *J. Exp. Biol.*, 31, 104, 1954.

102. Adams, M. D., Celniker, S. E., Holt, R. A. et al., The genome sequence of *Drosophila melanogaster*, *Science*, 287, 2185, 2000.

103. Clark, A. G., Eisen, M. B., Smith, D. R. et al., Evolution of genes and genomes on the *Drosophila* phylogeny, *Nature*, 450, 203, 2007.

104. Richards, S., Liu, Y., Bettencourt, B. R. et al., Comparative genome sequencing of *Drosophila pseudoobscura*: chromosomal, gene, and *cis*-element evolution, *Genome Res.*, 15, 1, 2005.

105. Holt, R. A., Subramanian, G. M., Halpern, A. et al., The genome sequence of the malaria mosquito *Anopheles gambiae*, *Science*, 298, 129, 2002.

106. Honey Bee Genome Sequencing Consortium, Insights into social insects from the genome of the honeybee *Apis mellifera*, *Nature*, 443, 931, 2006.

107. Mita, K., Kasahara, M., Sasaki, S. et al., The genome sequence of silkworm, *Bombyx mori*, *DNA Res.*, 11, 27, 2004.

108. Xia, Q., Zhou, Z., Lu, C. et al., A draft sequence for the genome of the domesticated silkworm (*Bombyx mori*), *Science*, 306, 1937, 2004.

109. Nene, V., Wortman, J. R., Lawson, D. et al., Genome sequence of *Aedes aegypti*, a major arbovirus vector, *Science*, 316, 1718, 2007.

110. Gersch, M., Verteilung und Ausscheidung von Fluorescein bei Aphiden, *Z. vergl. Physiol.*, 29, 506, 1942.

111. Downing, N., The regulation of sodium, potassium and chloride in an aphid subjected to ionic stress, *J. Exp. Biol.*, 87, 343, 1980.

112. Wessing, A. and Eichelberg, D., Malpighian tubules, rectal papillae and excretion. In *The Genetics and Biology of Drosophila*, Ashburner, A. and Wright, T. F., Eds., Academic Press, London, 1978, p. 1.

113. Beyenbach, K. W., Oviedo, A., and Aneshansley, D. J., Malpighian tubules of *Aedes-aegypti*: five tubules, one function, *J. Insect Physiol.*, 39, 639, 1993.

114. Wall, B. J., Oschman, J. L., and Schmidt, B. A., Morphology and function of Malpighian tubules and associated structures in the cockroach, *Periplaneta americana*, *J. Morphol.*, 146, 265, 1975.

115. Garrett, M. A., Bradley, T. J., Meredith, J. E., and Phillips, J. E., Ultrastructure of the Malpighian tubules of *Schistocerca gregaria*, *J. Morphol.*, 195, 313, 1988.

116. Skaer, N. J., Nassel, D. R., Maddrell, S. H., and Tublitz, N. J., Neurochemical fine tuning of a peripheral tissue: peptidergic and aminergic regulation of fluid secretion by Malpighian tubules in the tobacco hawkmoth *M. sexta*, *J. Exp. Biol.*, 205, 1869, 2002.

117. de Sousa, R. C. and Bicudo, H. E., Morphometric changes associated with sex and development in the Malpighian tubules of *Aedes aegypti*, *Cytobios*, 102, 173, 2000.

118. Plawner, L., Pannabecker, T. L., Laufer, S., Baustian, M. D., and Beyenbach, K. W., Control of diuresis in the yellow fever mosquito *Aedes aegypti*: evidence for similar mechanisms in the male and female, *J. Insect Physiol.*, 37, 119, 1991.

119. Klowden, M. J., Blood, sex and the mosquito, *Bioscience*, 45, 326, 1995.

120. Yu, M. J. and Beyenbach, K. W., Effects of leucokinin VIII on *Aedes* Malpighian tubule segments lacking stellate cells, *J. Exp. Biol.*, 207, 519, 2004.

121. Satmary, W. M. and Bradley, T. J., The distribution of cell types in the Malpighian tubules of *Aedes taeniorhynchus* Diptera Culicidae, *Int. J. Insect Morphol. Embryol.*, 13, 209, 1984.

122. O'Donnell, M. J., Dow, J. A., Huesmann, G. R., Tublitz, N. J., and Maddrell, S. H., Separate control of anion and cation transport in Malpighian tubules of *Drosophila melanogaster*, *J. Exp. Biol.*, 199, 1163, 1996.

123. Beyenbach, K. W., Mechanism and regulation of electrolyte transport in Malpighian tubules, *J. Insect Physiol.*, 41, 197, 1995.

124. Beyenbach, K. W., Transport mechanisms of diuresis in Malpighian tubules of insects, *J. Exp. Biol.*, 206, 3845, 2003.

125. O'Donnell, M. J., Rheault, M. R., Davies, S. A., Rosay, P., Harvey, B. J., Maddrell, S. H., Kaiser, K., and Dow, J. A., Hormonally controlled chloride movement across *Drosophila* tubules is via ion channels in stellate cells, *Am. J. Physiol.*, 274, R1039, 1998.

126. Sozen, M. A., Armstrong, J. D., Yang, M., Kaiser, K., and Dow, J. A., Functional domains are specified to single-cell resolution in a *Drosophila* epithelium, *Proc. Natl. Acad. Sci. USA*, 94, 5207, 1997.

127. Wang, J., Kean, L., Yang, J., Allan, A. K., Davies, S. A., Herzyk, P., and Dow, J. A., Function-informed transcriptome analysis of *Drosophila* renal tubule, *Genome Biol.*, 5, R69, 2004.

128. Chintapalli, V. R., Wang, J., and Dow, J. A., Using FlyAtlas to identify better *Drosophila melanogaster* models of human disease, *Nat. Genet.*, 39, 715, 2007.

129. Green, L. F. B., The fine structure of the light organ of the New Zealand glow-worm *Arachnocampa luminosa* (Diptera, Mycetophilidae), *Tissue Cell*, 11, 457, 1979.

130. Maddrell, S. H., Whittembury, G., Mooney, R. L., Harrison, J. B., Overton, J. A., and Rodriguez, B., The fate of calcium in the diet of *Rhodnius prolixus*: storage in concretion bodies in the Malpighian tubules, *J. Exp. Biol.*, 157, 483, 1991.

131. Wessing, A. and Zierold, K., Metal-salt feeding causes alterations in concretions in *Drosophila* larval Malpighian tubules as revealed by x-ray microanalysis, *J. Insect Physiol.*, 38, 623, 1992.

132. Wessing, A. and Zierold, K., The formation of type-I concretions in *Drosophila* Malpighian tubules studied by electron microscopy and x-ray microanalysis, *J. Insect Physiol.*, 45, 39, 1999.

133. Wessing, A., Zierold, K., and Hevert, F., Two types of concretions in *Drosophila* Malpighian tubules as revealed by x-ray microanalysis: a study on urine formation, *J. Insect Physiol.*, 38, 543, 1992.

134. Davies, S. A., Goodwin, S. F., Kelly, D. C., Wang, Z., Sozen, M. A., Kaiser, K., and Dow, J. A., Analysis and inactivation of *vha55*, the gene encoding the vacuolar ATPase B-subunit in *Drosophila melanogaster*, reveals a larval lethal phenotype, *J. Biol. Chem.*, 271, 30677, 1996.

135. Allan, A. K., Du, J., Davies, S. A., and Dow, J. A., Genome-wide survey of V-ATPase genes in *Drosophila* reveals a conserved renal phenotype for lethal alleles, *Physiol. Genom.*, 22, 128, 2005.

136. Sohal, R. S., Peters, P. D., and Hall, T. A., Fine structure and x-ray microanalysis of mineralized concretions in the Malpighian tubules of the housefly, *Musca domestica*, *Tissue Cell*, 8, 447, 1976.

137. Sohal, R. S., Peters, P. D., and Hall, T. A., Origin, structure, composition and age-dependence of mineralized dense bodies (concretions) in the midgut epithelium of the adult housefly, *Musca domestica*, *Tissue Cell*, 9, 87, 1977.

138. Herbst, D. B. and Bradley, T. J., A Malpighian tubule lime gland in an insect inhabiting alkaline salt lakes, *J. Exp. Biol.*, 145, 63, 1989.

139. Grodowitz, M. J. and Broce, A. B., Calcium storage in face fly (Diptera: Muscidae) larvae for puparium formation, *Ann. Entomol. Soc. Am.*, 76, 418, 1983.

140. Krueger, R. A., Broce, A. B., Hopkins, T. L., and Kramer, K. J., Calcium transport from Malpighian tubules to puparial cuticle of *Musca autumnalis*, *J. Comp. Physiol. B*, 158, 413, 1988.

141. Roseland, C. R., Grodowitz, M. J., Kramer, K. J., Hopkins, T. L., and Broce, A. B., Stabilization of mineralized and sclerotized puparial cuticle of muscid flies, *Insect Biochem.*, 15, 521, 1985.

142. Bradley, T. J. and Satir, P., 5-Hydroxytryptamine-stimulated mitochondrial movement and microvillar growth in the lower Malpighian tubule of the insect, *Rhodnius prolixus*, *J. Cell Sci.*, 49, 139, 1981.

143. Bradley, T. J. and Snyder, C., Fluid secretion and microvillar ultrastructure in mosquito Malpighian tubules, *Am. J. Physiol.*, 257, R1096, 1989.

144. Weng, X. H., Huss, M., Wieczorek, H., and Beyenbach, K. W., The V-type H^+ ATPase in Malpighian tubules of *Aedes aegypti*: localization and activity, *J. Exp. Biol.*, 206, 2211, 2003.

145. Maddrell, S. H. and Gardiner, B. O., Excretion of alkaloids by Malpighian tubules of insects, *J. Exp. Biol.*, 64, 267, 1976.

146. Maddrell, S. H. and Gardiner, B. O., Induction of transport of organic anions in Malpighian tubules of *Rhodnius*, *J. Exp. Biol.*, 63, 755, 1975.

147. Maddrell, S. H. and Casida, J. E., Mechanism of insecticide-induced diuresis in *Rhodnius*, *Nature*, 231, 55, 1971.

148. Nicolson, S. W., Diuresis or clearance: is there a physiological role for the diuretic hormone of the desert beetle *Onymacris*?, *J. Insect Physiol.*, 37, 447, 1991.

149. O'Donnell, M. J. and Maddrell, S. H. P., Fluid reabsorption and ion transport by the lower Malpighian tubules of adult female *Drosophila*, *J. Exp. Biol.*, 198, 1647, 1995.

150. Bradley, T. J., Functional design of microvilli in the Malpighian tubules of the insect *Rhodnius prolixus*, *J. Cell Sci.*, 60, 117, 1983.

151. Maddrell, S. H. P., The functional design of the insect excretory system, *J. Exp. Biol.*, 90, 1, 1981.

152. Williams, J. C., Hagedorn, H. H., and Beyenbach, K. W., Dynamic changes in flow rate and composition of urine during the post-bloodmeal diuresis in *Aedes aegypti* (L.), *J. Comp. Physiol. B*, 153, 257, 1983.

153. Williams, J. C. and Beyenbach, K. W., Differential effects of secretagogues on Na and K secretion in the Malpighian tubules of *Aedes aegypti* (L.), *J. Comp. Physiol. B*, 149, 511, 1983.

154. O'Donnell, M. J., Aldis, G. K., and Maddrell, S. H. P., Measurements of osmotic permeability in the Malpighian tubules of an insect *Rhodnius prolixus*, *Proc. R. Soc. Lond. B*, 216, 267, 1982.

155. O'Donnell, M. J., Maddrell, S. H., and Gardiner, B. O., Passage of solutes through walls of Malpighian tubules of *Rhodnius* by paracellular and transcellular routes, *Am. J. Physiol.*, 246, R759, 1984.

156. Hernandez, C. S., Gonzalez, E., and Whittembury, G., The paracellular channel for water secretion in the upper segment of the Malpighian tubule of *Rhodnius prolixus*, *J. Membr. Biol.*, 148, 233, 1995.

157. Whittembury, G., Paz-Aliaga, A., Biondi, A., Carpi-Medina, P., Gonzalez, E., and Linares, H., Pathways for volume flow and volume regulation in leaky epithelia, *Pflügers Arch.*, 405(Suppl. 1), S17, 1985.

158. Agre, P., King, L. S., Yasui, M., Guggino, W. B., Ottersen, O. P., Fujiyoshi, Y., Engel, A., and Nielsen, S., Aquaporin water channels: from atomic structure to clinical medicine, *J. Physiol.*, 542, 3, 2002.

159. Kaufmann, N., Mathai, J. C., Hill, W. G., Dow, J. A., Zeidel, M. L., and Brodsky, J. L., Developmental expression and biophysical characterization of a *Drosophila melanogaster* aquaporin, *Am. J. Physiol. Cell Physiol.*, 289, C397, 2005.

160. Echevarria, M., Ramirez-Lorca, R., Hernandez, C. S., Gutierrez, A., Mendez-Ferrer, S., Gonzalez, E., Toledo-Aral, J. J., Ilundain, A. A., and Whittembury, G., Identification of a new water channel (Rp-MIP) in the Malpighian tubules of the insect *Rhodnius prolixus*, *Pflügers Arch.*, 442, 27, 2001.

161. Martini, S. V., Goldenberg, R. C., Fortes, F. S., Campos-de-Carvalho, A. C., Falkenstein, D., and Morales, M. M., *Rhodnius prolixus* Malpighian tubule's aquaporin expression is modulated by 5-hydroxytryptamine, *Arch. Insect Biochem. Physiol.*, 57, 133, 2004.

162. Spring, J. H., Robichaux, S. R., Kaufmann, N., and Brodsky, J. L., Localization of a *Drosophila* DRIP-like aquaporin in the Malpighian tubules of the house cricket, *Acheta domesticus*, *Comp. Biochem. Physiol. A*, 148, 92, 2006.

163. Hernandez, C. S., Gutierrez, A. M., Vargas-Janzen, A., Noria, F., Gonzalez, E., Ruiz, V., and Whittembury, G., Fluid secretion in *Rhodnius* upper Malpighian tubules (UMT): water osmotic permeabilities and morphometric studies, *J. Membr. Biol.*, 184, 283, 2001.

164. Williams, J. C. and Beyenbach, K. W., Differential effects of secretagogues on the electrophysiology of the Malpighian tubules of the yellow fever mosquito, *J. Comp. Physiol. B*, 154, 301, 1984.

165. Beyenbach, K. W. and Dantzler, W. H., Comparative kidney tubule sources, isolation, perfusion, and function, *Meth. Enzymol.*, 191, 167, 1990.

166. Ramsay, J. A., Active transport of potassium by the Malpighian tubules of insects, *J. Exp. Biol.*, 93, 358, 1953.

167. Burg, M., Grantham, J., Abramow, M., and Orloff, J., Preparation and study of fragments of single rabbit nephrones, *Am. J. Physiol.*, 210, 1293, 1966.

168. Helman, S. I., Determination of electrical resistance of the isolated cortical collecting tubule and its possible anatomical location, *Yale J. Biol. Med.*, 45, 339, 1972.

169. Beyenbach, K. W. and Fromter, E., Electrophysiological evidence for Cl secretion in shark renal proximal tubules, *Am. J. Physiol.*, 248, F282, 1985.

170. Beyenbach, K. W. and Masia, R., Membrane conductances of principal cells in Malpighian tubules of *Aedes aegypti*, *J. Insect Physiol.*, 48, 375, 2002.

171. Masia, R., Aneshansley, D., Nagel, W., Nachman, R. J., and Beyenbach, K. W., Voltage clamping single cells in intact Malpighian tubules of mosquitoes, *Am. J. Physiol.*, 279, F747, 2000.

172. Hegarty, J. L., Zhang, B., Carroll, M. C., Cragoe, E. J. J., and Beyenbach, K. W., Effects of amiloride on isolated Malpighian tubules of the yellow fever mosquito (*Aedes aegypti*). *J. Insect Physiol.*, 38, 329, 1992.

173. Petzel, D. H., Na$^+$/H$^+$ exchange in mosquito Malpighian tubules, *Am. J. Physiol. Regul. Integr. Comp. Physiol.*, 279, R1996, 2000.

174. Pullikuth, A. K., Aimanova, K., Kang'ethe, W., Sanders, H. R., and Gill, S. S., Molecular characterization of sodium/proton exchanger 3 (NHE3) from the yellow fever vector, *Aedes aegypti*, *J. Exp. Biol.*, 209, 3529, 2006.

175. Hart, S. J., Knezetic, J. A., and Petzel, D. H., Cloning and tissue distribution of two Na$^+$/H$^+$ exchangers from the Malpighian tubules of *Aedes aegypti*, *Arch. Insect Biochem. Physiol.*, 51, 121, 2002.

176. Sciortino, C. M., Shrode, L. D., Fletcher, B. R., Harte, P. J., and Romero, M. F., Localization of endogenous and recombinant Na$^+$-driven anion exchanger protein NDAE1 from *Drosophila melanogaster*, *Am. J. Physiol. Cell Physiol.*, 281, C449, 2001.

177. Kang'ethe, W., Aimanova, K. G., Pullikuth, A. K., and Gill, S. S., NHE8 mediates amiloride-sensitive Na$^+$/H$^+$ exchange across mosquito Malpighian tubules and catalyzes Na$^+$ and K$^+$ transport in reconstituted proteoliposomes, *Am. J. Physiol. Renal Physiol*, 292, F1501, 2007.

178. Petzel, D. H., Pirotte, P. T., and Van Kerkhove, E., Intracellular and luminal pH measurements of Malpighian tubules of the mosquito *Aedes aegypti*: the effects of cAMP, *J. Insect Physiol.*, 45, 973, 1999.

179. Rheault, M. R., Okech, B. A., Keen, S. B., Miller, M. M., Meleshkevitch, E. A., Linser, P. J., Boudko, D. Y., and Harvey, W. R., Molecular cloning, phylogeny and localization of AgNHA1: the first Na$^+$/H$^+$ antiporter (NHA) from a metazoan, *Anopheles gambiae*, *J. Exp. Biol.*, 210, 3848, 2007.

180. Patrick, M. L., Aimanova, K., Sanders, H. R., and Gill, S. S., P-type Na$^+$/K$^+$-ATPase and V-type H$^+$-ATPase expression patterns in the osmoregulatory organs of larval and adult mosquito *Aedes aegypti*, *J. Exp. Biol.*, 209, 4638, 2006.

181. O'Connor, K. R. and Beyenbach, K. W., Chloride channels in apical membrane patches of stellate cells of Malpighian tubules of *Aedes aegypti*, *J. Exp. Biol.*, 204, 367, 2001.

182. Pannabecker, T. L., Hayes, T. K., and Beyenbach, K. W., Regulation of epithelial shunt conductance by the peptide leucokinin, *J. Membr. Biol.*, 132, 63, 1993.

183. Beyenbach, K. W., Energizing epithelial transport with the vacuolar H$^+$-ATPase, *News Physiol. Sci.*, 16, 145, 2001.

184. O'Donnell, M. J. and Spring, J. H., Modes of control of insect Malpighian tubules: synergism, antagonism, cooperation and autonomous regulation, *J. Insect Physiol.*, 46, 107, 2000.

185. Van Kerkhove, E., Cellular mechanisms of salt secretion by the Malpighian tubules of insects, *Belg. J. Zool.*, 124, 73, 1994.

186. Dow, J. A. T., Davies, S. A., and Sozen, M. A., Fluid secretion by the *Drosophila* Malpighian tubule, *Am. Zool.*, 38, 450, 1998.

187. Neufeld, D. S. and Leader, J. P., Electrochemical characteristics of ion secretion in Malpighian tubules of the New Zealand alpine weta (*Hemideina maori*), *J. Insect Physiol.*, 44, 39, 1998.

188. Ianowski, J. P. and O'Donnell, M. J., Transepithelial potential in Malpighian tubules of *Rhodnius prolixus*: lumen-negative voltages and the triphasic response to serotonin, *J. Insect Physiol.*, 47, 411, 2001.

189. Caruso-Neves, C. and Lopes, A. G., Sodium pumps in the Malpighian tubule of *Rhodnius* sp., *An. Acad. Bras. Cienc.*, 72, 407, 2000.

190. Grieco, M. A. B. and Lopes, A. G., 5-Hydroxytryptamine regulates the (Na$^+$+K$^+$)ATPase activity in Malpighian tubules of *Rhodnius prolixus*: evidence for involvement of G-protein and cAMP-dependent protein kinase, *Arch. Insect Biochem. Physiol.*, 36, 203, 1997.

191. Gutierrez, A. M., Hernandez, C. S., and Whittembury, G., A model for fluid secretion in *Rhodnius* upper Malpighian tubules (UMT), *J. Membr. Biol.*, 202, 105, 2004.

192. Orchard, I., Serotonin: a coordinator of feeding-related physiological events in the blood-gorging bug, *Rhodnius prolixus*, *Comp. Biochem. Physiol. A Mol. Integr. Physiol.*, 144, 316, 2006.

193. Maddrell, S. H., Herman, W. S., Mooney, R. L., and Overton, J. A., 5-Hydroxytryptamine: a second diuretic hormone in *Rhodnius prolixus*, *J. Exp. Biol.*, 156, 557, 1991.

194. Te Brugge, V. A. and Orchard, I., Evidence for CRF-like and kinin-like peptides as neurohormones in the blood-feeding bug, *Rhodnius prolixus*, *Peptides*, 23, 1967, 2002.

195. Te Brugge, V. A., Schooley, D. A., and Orchard, I., The biological activity of diuretic factors in *Rhodnius prolixus*, *Peptides*, 23, 671, 2002.

196. Te Brugge, V. A., Nassel, D. R., Coast, G. M., Schooley, D. A., and Orchard, I., The distribution of a kinin-like peptide and its co-localization with a CRF-like peptide in the blood-feeding bug, *Rhodnius prolixus*, *Peptides*, 22, 161, 2001.

197. Adams, T. S., Hematophagy and hormone release, *Ann. Entomol. Soc. Am.*, 92, 1, 1999.

198. Gary, Jr., R. E. and Foster, W. A., *Anopheles gambiae* feeding and survival on honeydew and extra-floral nectar of peridomestic plants, *Med. Vet. Entomol.*, 18, 102, 2004.

199. O'Meara, G. F. and Evans, D. G., Blood feeding requirements of the mosquito geographical variation in *Aedes taeniorhynchus*, *Science*, 180, 1291, 1972.

200. Stobbart, R. H., The control of the diuresis following a blood meal in females of the yellow fever mosquito *Aedes aegypti* (L.), *J. Exp. Biol.*, 69, 53, 1977.

201. Boorman, J. P. T., Observations on the feeding habits of the mosquito *Aedes (Stegomyia) aegypti* (Linnaeus): the loss of water after a blood-meal and the amount of blood taken during feeding, *Ann. Trop. Med. Parasit.*, 54, 8, 1960.

202. Spring, J. H., Endocrine regulation of diuresis in insects, *J. Insect Physiol.*, 36, 13, 1990.

203. Maddrell, S. H., O'Donnell, M. J., and Caffrey, R., The regulation of haemolymph potassium activity during initiation and maintenance of diuresis in fed Rhodnius prolixus, *J. Exp. Biol.*, 177, 273, 1993.

204. Torfs, P., Nieto, J., Veelaert, D. et al., The kinin peptide family in invertebrates, *Ann. N.Y. Acad. Sci.*, 897, 361, 1999.

205. Gaede, G., Hoffmann, K.-H., and Spring, J. H., Hormonal regulation in insects: facts, gaps, and future directions, *Physiol. Rev.*, 77, 963, 1997.

206. Spring, J. H. and Clark, T. M., Diuretic and antidiuretic factors which act on the Malpighian tubules of the house cricket, *Acheta domesticus*, *Prog. Clin. Biol. Res.*, 342, 559, 1990.

207. Beyenbach, K. W., Regulation of tight junction permeability with switch-like speed, *Curr. Opin. Nephrol. Hypertens.*, 12, 543, 2003.

208. Kataoka, H., Troetschler, R. G., Li, J. P., Kramer, S. J., Carney, R. L., and Schooley, D. A., Isolation and identification of a diuretic hormone from the tobacco hornworm, *Manduca sexta*, *Proc. Natl. Acad. Sci. USA*, 86, 2976, 1989.

209. Audsley, N., Goldsworthy, G. J., and Coast, G. M., Quantification of *Locusta* diuretic hormone in the central nervous system and corpora cardiaca: influence of age and feeding status, and mechanism of release, *Reg. Pept.*, 69, 25, 1997.

210. Patel, M., Hayes, T., and Coast, G., Evidence for the hormonal function of a CRF-related diuretic peptide (Locusta-DP) in *Locusta migratoria*, *J. Exp. Biol.*, 198, 793, 1995.

211. Thompson, K. S., Rayne, R. C., Gibbon, C. R., May, S. T., Patel, M., Coast, G. M., and Bacon, J. P., Cellular colocalization of diuretic peptides in locusts: a potent control mechanism, *Peptides*, 16, 95, 1995.

212. Reagan, J. D., Expression cloning of an insect diuretic hormone receptor: a member of the calcitonin/secretin receptor family, *J. Biol. Chem.*, 269, 9, 1994.

213. Reagan, J. D., Molecular cloning and function expression of a diuretic hormone receptor from the house cricket, Acheta domesticus, *Insect Biochem. Mol. Biol.*, 26, 1, 1996.

214. Johnson, E. C., Bohn, L. M., and Taghert, P. H., *Drosophila* CG8422 encodes a functional diuretic hormone receptor, *J. Exp. Biol.*, 207, 743, 2004.

215. Kay, I., Coast, G. M., Cusinato, O., Wheeler, C. H., Totty, N. F., and Goldsworthy, G. J., Isolation and characterization of a diuretic peptide from *Acheta domesticus*: evidence for a family of insect diuretic peptides, *Biol. Chem. Hoppe Seyler*, 372, 505, 1991.

216. Lehmberg, E., Ota, R. B., Furuya, K., King, D. S., Applebaum, S. W., Ferenz, H. J., and Schooley, D. A., Identification of a diuretic hormone of *Locusta migratoria*, *Biochem. Biophys. Res. Commun.*, 179, 1036, 1991.

217. Coast, G. M., The neuroendocrine regulation of salt and water balance in insects, *Zoology*, 103, 179, 2001.

218. Coast, G. M., Orchard, I., Phillips, J. E., and Schooley, D. A., Insect diuretic and antidiuretic hormones. In *Advances in Insect Physiology*, Evans, P. D., Ed., Academic Press, London, 2002, p. 279.

219. Hegarty, J. L., Zhang, B., Pannabecker, T. L., Petzel, D. H., Baustian, M. D., and Beyenbach, K. W., Dibutyryl cAMP activates bumetanide-sensitive electrolyte transport in Malpighian tubules., *Am. J. Physiol. Cell Physiol.*, 261, C521, 1991.

220. Dames, P., Zimmermann, B., Schmidt, R., Rein, J., Voss, M., Schewe, B., Walz, B., and Baumann, O., cAMP regulates plasma membrane vacuolar-type H$^+$-ATPase assembly and activity in blowfly salivary glands, *Proc. Natl. Acad. Sci. USA*, 103, 3926, 2006.

221. Holman, G. M., Cook, B. J., and Nachman, R. J., Isolation, primary structure, and synthesis of two neuropeptides from *Leucophaea maderae*: members of a new family of cephalomyotropins, *Comp. Biochem. Physiol. C*, 84, 205, 1986.

222. Holman, G. M., Cook, B. J., and Nachman, R. J., Isolation, primary structure and synthesis of leucokinin VII and VIII: the final members of the new family of cephalomyotropic peptides isolated from head extracts of *Leucophaea maderae*, *Comp. Biochem. Physiol.*, 88, 31, 1987.

223. Terhzaz, S., O'Connell, F. C., Pollock, V. P., Kean, L., Davies, S. A., Veenstra, J. A., and Dow, J. A. T., Isolation and characterization of a leucokinin-like peptide of *Drosophila melanogaster*, *J. Exp. Biol.*, 202, 3667, 1999.

224. Muren, J. E., Lundquist, C. T., and Naessel, D. R., Quantitative determination of myotropic neuropeptide in the nervous system of the cockroach *Leucophaea maderae*: distribution and release of leucokinins, *J. Exp. Biol.*, 179, 289, 1993.

225. Hayes, T. K., Pannabecker, T. L., Hinckley, D. J., Holman, G. M., Nachman, R. J., Petzel, D. H., and Beyenbach, K. W., Leucokinins, a new family of ion transport stimulators and inhibitors in insect Malpighian tubules, *Life Sci.*, 44, 1259, 1989.

226. Yu, A. S. L., Claudins and epithelial paracellular transport: the end of the beginning, *Curr. Opin. Nephrol. Hypertens.*, 12, 503, 2003.

227. Van Itallie, C. M. and Anderson, J. M., Claudins and epithelial paracellular transport, *Annu. Rev. Physiol.*, 68, 403, 2006.

228. Schneeberger, E. E. and Lynch, R. D., The tight junction: a multifunctional complex, *Am. J. Physiol. Cell Physiol.*, 286, C1213, 2004.

229. Tsukita, S. and Furuse, M., Occludin and claudins in tight-junction strands: leading or supporting players?, *Trends Cell Biol.*, 9, 268, 1999.

230. Furuse, M. and Tsukita, S., Claudins in occluding junctions of humans and flies, *Trends Cell Biol.*, 16, 181, 2006.

231. Sawada, N., Murata, M., Kikuchi, K., Osanai, M., Tobioka, H., Kojima, T., and Chiba, H., Tight junctions and human diseases, *Med. Electron Microsc.*, 36, 147, 2003.

232. Behr, M., Riedel, D., and Schuh, R., The claudin-like megatrachea is essential in septate junctions for the epithelial barrier function in *Drosophila*, *Dev. Cell*, 5, 611, 2003.

233. Wu, V. M., Schulte, J., Hirschi, A., Tepass, U., and Beitel, G. J., Sinuous is a *Drosophila* claudin required for septate junction organization and epithelial tube size control, *J. Cell Biol.*, 164, 313, 2004.

234. Yu, M. and Beyenbach, K. W., Leucokinin and the modulation of the shunt pathway in Malpighian tubules, *J. Insect Physiol.*, 47, 263, 2001.

235. Yu, M. J. and Beyenbach, K. W., Leucokinin activates Ca^{2+}-dependent signal pathway in principal cells of *Aedes aegypti* Malpighian tubules, *Am. J. Physiol.*, 283, F499, 2002.

236. Radford, J. C., Davies, S. A., and Dow, J. A., Systematic G-protein-coupled receptor analysis in *Drosophila melanogaster* identifies a leucokinin receptor with novel roles, *J. Biol. Chem.*, 277, 38810, 2002.

237. Pietrantonio, P. V., Jagger, C., Taneja-Bageshwar, S., Nachman, R. J., and Barhoumi, R., The mosquito *Aedes aegypti* (L.) leucokinin receptor is a multiligand receptor for the three *Aedes* kinins, *Insect Mol. Biol.*, 14, 55, 2005.

238. Radford, J. C., Terhzaz, S., Cabrero, P., Davies, S. A., and Dow, J. A., Functional characterisation of the *Anopheles* leucokinins and their cognate G-protein-coupled receptor, *J. Exp. Biol.*, 207, 4573, 2004.

239. Spring, J. H. and Kim, I., Differential effects of neuropeptides on the distal and mid-tubules of the house cricket, *Arch. Insect Biochem. Physiol.*, 29, 11, 1995.

240. Petzel, D. H., Hagedorn, H. H., and Beyenbach, K. W., Preliminary isolation of mosquito natriuretic factor, *Am. J. Physiol. Regul. Integr. Comp. Physiol.*, 249, R379, 1985.

241. Petzel, D. H., Hagedorn, H. H., and Beyenbach, K. W., Peptide nature of two mosquito natriuretic factors, *Am. J. Physiol. Regul. Integr. Comp. Physiol.*, 250, R328, 1986.

242. Beyenbach, K. W. and Petzel, D. H., Diuresis in mosquitoes: role of a natriuretic factor, *News Physiol. Sci.*, 2, 171, 1987.

243. Petzel, D. H., Berg, M. M., and Beyenbach, K. W., Hormone-controlled cAMP-mediated fluid secretion in yellow-fever mosquito, *Am. J. Physiol. Regul. Integr. Comp. Physiol.*, 253, R701, 1987.

244. Coast, G. M., Garside, C. S., Webster, S. G., Schegg, K. M., and Schooley, D. A., Mosquito natriuretic peptide identified as a calcitonin-like diuretic hormone in *Anopheles gambiae* (Giles), *J. Exp. Biol.*, 208, 3281, 2005.

245. Sawyer, D. B. and Beyenbach, K. W., Dibutyryl-cAMP increases basolateral sodium conductance of mosquito Malpighian tubules, *Am. J. Physiol. Regul. Integr. Comp. Physiol.*, 248, R339, 1985.

246. Filippov, V., Aimanova, K., and Gill, S. S., Expression of an *Aedes aegypti* cation-chloride cotransporter and its *Drosophila* homologues, *Insect Mol. Biol.*, 12, 319, 2003.

247. Reagan, J. D., Molecular cloning of a putative Na^+-K^+-$2Cl^-$ cotransporter from the Malpighian tubules of the tobacco hornworm, *Manduca sexta*, *Insect Biochem. Mol. Biol.*, 25, 875, 1995.

248. Gillen, C. M., Blair, C. R., Heilman, N. R., Somple, M., Stulberg, M., Thombre, R., Watson, N., Gillen, K. M., and Itagaki, H., The cation-chloride cotransporter, masBSC, is widely expressed in *Manduca sexta* tissues, *J. Insect Physiol.*, 52, 661, 2006.

249. Ianowski, J. P., Christensen, R. J., and O'Donnell, M. J., Na^+ competes with K^+ in bumetanide-sensitive transport by Malpighian tubules of *Rhodnius prolixus*, *J. Exp. Biol.*, 207, 3707, 2004.

250. Ianowski, J. P. and O'Donnell, M. J., Basolateral ion transport mechanisms during fluid secretion by *Drosophila* Malpighian tubules: Na^+ recycling, Na^+:K^+:$2Cl^-$ cotransport and Cl^- conductance, *J. Exp. Biol.*, 207, 2599, 2004.

251. Johnson, E. C., Shafer, O. T., Trigg, J. S., Park, J., Schooley, D. A., Dow, J. A., and Taghert, P. H., A novel diuretic hormone receptor in *Drosophila*: evidence for conservation of CGRP signaling, *J. Exp. Biol.*, 208, 1239, 2005.

252. Hill, C. A., Fox, A. N., Pitts, R. J. et al., G-protein-coupled receptors in *Anopheles gambiae*, *Science*, 298, 176, 2002.

253. Kurihara, K., Nakanishi, N., Moore-Hoon, M. L., and Turner, R. J., Phosphorylation of the salivary Na^+-K^+-$2Cl^-$ cotransporter, *Am. J. Physiol. Cell Physiol.*, 282, C817, 2002.

254. Shimkets, R. A., Lifton, R., and Canessa, C. M., *In vivo* phosphorylation of the epithelial sodium channel, *Proc. Natl. Acad. Sci. USA*, 95, 3301, 1998.

255. Voss, M., Vitavska, O., Walz, B., Wieczorek, H., and Baumann, O., Stimulus-induced phosphorylation of vacuolar H^+-ATPase by protein kinase A, *J. Biol. Chem.*, 282, 33735, 2007.

256. Maddrell, S. H. P., Pilcher, D. E. M., and Gardiner, B. O. C., Stimulatory effect of 5-hydroxytryptamine serotonin on secretion by Malpighian tubules of insects, *Nature*, 222, 784, 1969.

257. Maddrell, S. H. and Overton, J. A., Stimulation of sodium transport and fluid secretion by ouabain in an insect Malpighian tubule, *J. Exp. Biol.*, 137, 265, 1988.

258. Clark, T. M., Integrative aspects of epithelial transport in larval *Aedes aegypti*, *Am. Zool.*, 40, 975, 2000.

259. Clark, T. M. and Bradley, T. J., Additive effects of 5-HT and diuretic peptide on *Aedes* Malpighian tubule fluid secretion, *Comp. Biochem. Physiol. A*, 119, 599, 1998.

260. Veenstra, J. A., Effects of 5-hydroxytryptamine on the Malpighian tubules of *Aedes aegypti*, *J. Insect Physiol.*, 34, 299, 1988.

261. Te Brugge, V. A., Lombardi, V. C., Schooley, D. A., and Orchard, I., Presence and activity of a Dippu-DH31-like peptide in the blood-feeding bug, *Rhodnius prolixus*, *Peptides*, 26, 29, 2005.

262. Blumenthal, E. M., Regulation of chloride permeability by endogenously produced tyramine in the *Drosophila* Malpighian tubule, *Am. J. Physiol. Cell Physiol.*, 284, C718, 2003.

263. Blumenthal, E. M., Modulation of tyramine signaling by osmolality in an insect secretory epithelium, *Am. J. Physiol. Cell Physiol.*, 289, C1261, 2005.

264. Beyenbach, K. W., Aneshansley, J. D., Pannabecker, T. L., Masia, R., Gray, D., and Yu, M. J., Oscillations of voltage and resistance in Malpighian tubules of *Aedes aegypti*, *J. Insect Physiol.*, 46, 321, 2000.

265. Wheelock, G. D., Petzel, D. H., Gillett, J. D., Beyenbach, K. W., and Hagedorn, H. H., Evidence for hormonal control of diuresis after a blood meal in the mosquito *Aedes aegypti*, *Arch. Insect Biochem. Physiol.*, 7, 75, 1988.

266. Maddrell, S. H., Excretion in the blood-sucking bug, *Rhodnius prolixus* Stal. III. The control of the release of the diuretic hormone, *J. Exp. Biol.*, 41, 459, 1964.

267. Chen, Y., Veenstra, J. A., Hagedorn, H., and Davis, N. T., Leucokinin and diuretic hormone immunoreactivity of neurons in the tobacco hornworm, *Manduca sexta*, and co-localization of this immunoreactivity in lateral neurosecretory cells of abdominal ganglia, *Cell Tissue Res.*, 278, 493, 1994.

268. Coast, G. M., Synergism between diuretic peptides controlling ion and fluid transport in insect Malpighian tubules, *Reg. Pept.*, 57, 283, 1995.

269. Holman, G. M., Nachman, R. J., and Coast, G. M., Isolation, characterization and biological activity of a diuretic myokinin neuropeptide from the housefly, *Musca domestica*, *Peptides*, 20, 1, 1999.

270. Iaboni, A., Holman, G. M., Nachman, R. J., Orchard, I., and Coast, G. M., Immunocytochemical localisation and biological activity of diuretic peptides in the housefly, *Musca domestica*, *Cell Tissue Res.*, 294, 549, 1998.

271. Maddrell, S. H. P., Herman, W. S., Farndale, R. W., and Riegel, J. A., Synergism of hormones controlling epithelial fluid transport in an insect, *J. Exp. Biol.*, 174, 65, 1993.

272. Coast, G. M., Webster, S. G., Schegg, K. M., Tobe, S. S., and Schooley, D. A., The *Drosophila melanogaster* homologue of an insect calcitonin-like diuretic peptide stimulates V-ATPase activity in fruit fly Malpighian tubules, *J. Exp. Biol.*, 204, 1795, 2001.

273. Eigenheer, R. A., Nicolson, S. W., Schegg, K. M., Hull, J. J., and Schooley, D. A., Identification of a potent antidiuretic factor acting on beetle Malpighian tubules, *Proc. Natl. Acad. Sci. USA*, 99, 84, 2002.

274. Eigenheer, R. A., Wiehart, U. M., Nicolson, S. W., Schoofs, L., Schegg, K. M., Hull, J. J., and Schooley, D. A., Isolation, identification, and localization of a second beetle antidiuretic peptide, *Peptides*, 24, 27, 2003.

275. Massaro, R. C., Lee, L. W., Patel, A. B., Wu, D. S., Yu, M. J., Scott, B. N., Schooley, D. A., Schegg, K. M., and Beyenbach, K. W., The mechanism of action of the antidiuretic peptide Tenmo ADFa in Malpighian tubules of *Aedes aegypti*, *J. Exp. Biol.*, 207, 2877, 2004.

276. Wiehart, U. I., Nicolson, S. W., Eigenheer, R. A., and Schooley, D. A., Antagonistic control of fluid secretion by the Malpighian tubules of *Tenebrio molitor*: effects of diuretic and antidiuretic peptides and their second messengers, *J. Exp. Biol.*, 205, 493, 2002.

277. Geoffroy, V., Fouque, F., Nivet, V., Clot, J. P., Lugnier, C., Desbuquois, B., and Benelli, C., Activation of a cGMP-stimulated cAMP phosphodiesterase by protein kinase C in a liver Golgi-endosomal fraction, *Eur. J. Biochem.*, 259, 892, 1999.

278. Day, J. P., Dow, J. A., Houslay, M. D., and Davies, S. A., Cyclic nucleotide phosphodiesterases in *Drosophila melanogaster*, *Biochem. J.*, 388, 333, 2005.

279. Quinlan, M. C., Tublitz, N. J., and O'Donnell, M. J., Anti-diuresis in the blood-feeding insect *Rhodnius prolixus* Stal: the peptide CAP2b and cyclic GMP inhibit Malpighian tubule fluid secretion, *J. Exp. Biol.*, 200, 2363, 1997.

280. Phillips, J. and Audsley, N., Neuropeptide control of ion and fluid transport across locust hindgut, *Am. Zool.*, 35, 503, 1995.

281. King, D. S., Meredith, J., Wang, Y. J., and Phillips, J. E., Biological actions of synthetic locust ion transport peptide (ITP), *Insect Biochem. Mol. Biol.*, 29, 11, 1999.

282. Phillips, J. E., Wiens, C., Audsley, N., Jeffs, L., Bilgen, T., and Meredith, J., Nature and control of chloride transport in insect absorptive epithelia, *J. Exp. Zool.*, 275, 292, 1996.

283. Riegel, J. A., Farndale, R. W., and Maddrell, S. H. P., Fluid secretion by isolated Malpighian tubules of *Drosophila melanogaster* Meig: effects of organic anions, quinacrine and a diuretic factor found in the secreted fluid, *J. Exp. Biol.*, 202, 2339, 1999.

284. Cabrero, P., Radford, J. C., Broderick, K. E., Costes, L., Veenstra, J. A., Spana, E. P., Davies, S. A., and Dow, J. A., The Dh gene of *Drosophila melanogaster* encodes a diuretic peptide that acts through cyclic AMP, *J. Exp. Biol.*, 205, 3799, 2002.

285. Davies, S. A., Stewart, E. J., Huesmann, G. R., Skaer, N. J. V., Maddrell, S. H. P., Tublits, N. J., and Dow, J. A. T., Neuropeptide stimulation of the nitric oxide signaling pathway in *Drosophila melanogaster* Malpighian tubules, *Am. J. Physiol.*, 273, R823, 1997.

286. Rosay, P., Davies, S. A., Yu, Y., Sozen, A., Kaiser, K., and Dow, J. A., Cell-type specific calcium signalling in a *Drosophila* epithelium, *J. Cell Sci.*, 110, 1683, 1997.

287. Dow, J. A., Maddrell, S. H., Davies, S. A., Skaer, N. J., and Kaiser, K., A novel role for the nitric oxide-cGMP signaling pathway: the control of epithelial function in *Drosophila*, *Am. J. Physiol.*, 266, R1716, 1994.

288. Terhzaz, S., Southall, T. D., Lilley, K. S., Kean, L., Allan, A. K., Davies, S. A., and Dow, J. A., Differential gel electrophoresis and transgenic mitochondrial calcium reporters demonstrate spatiotemporal filtering in calcium control of mitochondria, *J. Biol. Chem.*, 2006.

289. Davies, S. A., Huesmann, G. R., Maddrell, S. H., O'Donnell, M. J., Skaer, N. J., Dow, J. A., and Tublitz, N. J., CAP_{2b}, a cardioacceleratory peptide, is present in *Drosophila* and stimulates tubule fluid secretion via cGMP, *Am. J. Physiol.*, 269, R1321, 1995.

290. Maddrell, S. H. P. and Phillips, J. E., Active transport of sulfate ions by the Malpighian tubules of larvae of the mosquito *Aedes campestris*, *J. Exp. Biol.*, 62, 367, 1975.

291. Phillips, J. E. and Maddrell, S. H. P., Secretion of hypoosmotic fluid by the lower Malpighian tubules of *Rhodnius prolixus*, *J. Exp. Biol.*, 62, 671, 1975.

292. Clark, T. M., Hayes, T. K., and Beyenbach, K. W., Dose-dependent effects of CRF-like diuretic peptide on transcellular and paracellular transport pathways, *Am. J. Physiol.*, 274, F834, 1998.

293. Clark, T. M., Hayes, T. K., Holman, G. M., and Beyenbach, K. W., The concentration-dependence of CRF-like diuretic peptide: mechanisms of action, *J. Exp. Biol.*, 201, 1753, 1998.

294. Wigglesworth, V. B., *Insect Physiology*, Methuen, London, 1934.

295. Ramsay, J. A., The rectal complex of the mealworm *Tenebrio molitor* L. (Coleoptera, Tenebrionidae), *Philos. Trans. R. Soc. Lond. B Biol. Sci.*, 248, 279, 1964.

296. Reynolds, S. E. and Bellward, K., Water balance in *Manduca sexta* caterpillars: water recycling from the rectum, *J. Exp. Biol.*, 141, 33, 1989.

297. Saini, R. S., Histology and physiology of the cryptonephridial system of insects, *Trans. R. Entomol. Soc. Lond.*, 116, 347, 1964.

298. Coutchie, P. A. and Crowe, J. A., Transport of water vapor by tenebrionid beetles, *Physiol. Zool.*, 52, 67, 1979.

299. Machin, J., Water balance in *Tenebrio molitor* L. larvae; the effect of atmospheric water absorption, *J. Comp. Physiol. B*, 101, 121, 1975.

300. Ramsay, J. A., The rectal complex in the larvae of Lepidoptera, *Philos. Trans. R. Soc. Ser. B*, 274, 203, 1976.

301. Grimstone, A. V., Mullinger, A. M., and Ramsay, J. A., Further studies on the rectal complex of the mealworm *Tenebrio molitor* L. (Coleoptera, Tenebrionidae), *Philos. Trans. R. Soc. Ser. B*, 253, 343, 1968.

302. Green, L. F. B., Cryptonephric Malpighian tubule system in a dipteran larva, the New Zealand glow-worm, *Archnocampa luminosa* (Diptera; Mycetophilidae): a structural study, *Tissue Cell*, 12, 141, 1980.

303. Noble-Nesbitt, J., Cellular differentiation in relation to water vapor absorption in the rectal complex of the mealworm, *Tenebrio molitor*, *Tissue Cell*, 22, 925, 1990.

304. Koefoed, B. M., Ultrastructure of the cryptonephridial system in the mealworm, *Tenebrio molitor*, *Z. Zellforsch. Mikrosk. Anat.*, 116, 487, 1971.

305. Machin, J. and O'Donnell, M. J., Rectal complex ion activities and electrochemical gradient in larvae of the desert beetle *Onymacris*: comparisons with *Tenebrio*, *J. Insect Physiol.*, 37, 829, 1991.

306. O'Donnell, M. J. and Machin, J., Ion activities and electrochemical gradients in the mealworm rectal complex, *J. Exp. Biol.*, 155, 375, 1991.

307. Moffett, D. F., Recycling of K^+, acid–base equivalents, and fluid between gut and hemolymph in lepidopteran larvae, *Physiol. Zool.*, 67, 68, 1994.

308. Zerahn, K., Water transport across the short-circuited midgut of the American silkworm, *J. Exp. Biol.*, 116, 481, 1985.

309. Reynolds, S. E., Nottingham, S. F., and Stephens, A. E., Food and water economy and its relation to growth in 5th-instar larvae of the tobacco hornworm *Manduca sexta*, *J. Insect Physiol.*, 31, 119, 1985.

310. Kiss, G. and Hansson, H. C., Application of osmolality for the determination of water activity and the modelling of cloud formation, *Atmos. Chem. Phys. Discuss.*, 4, 7667, 2004.

311. Noble-Nesbitt, J., Water uptake from subsaturated atmospheres: its site in insects, *Nature*, 225, 753, 1970.

312. Tupy, J. H. and Machin, J., Transport characteristics of the isolated rectal complex of the mealworm, *Tenebrio molitor*, *Can. J. Zool.*, 63, 1897, 1985.

313. Coutchie, P. A. and Machin, J., Allometry of water vapor absorption in two species of tenebrionid beetle larvae, *Am. J. Physiol.*, 247, R230, 1984.

314. Machin, J., Passive exchanges during water vapor absorption in mealworms, *Tenebrio molitor*: a new approach to studying the phenomenon, *J. Exp. Biol.*, 65, 603, 1976.

315. Machin, J., Water compartmentalization in insects, *J. Exp. Zool.*, 215, 327, 1981.

316. Nicolson, S., Excretory function in *Tenebrio molitor* fast tubular secretion in a vapour-absorbing insect, *J. Insect Physiol.*, 38, 139, 1992.

317. Audsley, N., Coast, G. M., and Schooley, D. A., The effects of *Manduca sexta* diuretic hormone on fluid transport by the Malpighian tubules and cryptonephric complex of *Manduca sexta*, *J. Exp. Biol.*, 178, 231, 1993.

318. Machin, J., Compartmental osmotic pressures in the rectal complex of *Tenebrio* larvae: evidence for a single tubular pumping site, *J. Exp. Biol.*, 82, 123, 1979.

319. Machin, J., Water vapor absorption in insects, *Am. J. Physiol.*, 244, R187, 1983.

320. Liao, S., Audsley, N., and Schooley, D. A., Antidiuretic effects of a factor in brain/corpora cardiaca/ corpora allata extract on fluid reabsorption across the cryptonephric complex of *Manduca sexta*, *J. Exp. Biol.*, 203, 605, 2000.

321. Beyenbach, K. W. and Wieczorek, H., The V-type H^+ ATPase: molecular structure and function, physiological roles and regulation, *J. Exp. Biol.*, 209, 577, 2006.

322. Wieczorek, H., Putzenlechner, M., Zeiske, W., and Klein, U., A vacuolar-type proton pump energizes K^+/H^+ antiport in an animal plasma membrane, *J. Biol. Chem.*, 266, 15340, 1991.

323. Wieczorek, H., Brown, D., Grinstein, S., Ehrenfeld, J., and Harvey, W. R., Animal plasma membrane energization by proton-motive V-ATPases, *Bioessays*, 21, 637, 1999.

324. Kirschner, L. B., The mechanism of sodium chloride uptake in hyperregulating aquatic animals, *J. Exp. Biol.*, 207, 1439, 2004.

325. Weihrauch, D., McNamara, J. C., Towle, D. W., and Onken, H., Ion-motive ATPases and active, transbranchial NaCl uptake in the red freshwater crab, *Dilocarcinus pagei* (Decapoda, Trichodactyl- idae), *J. Exp. Biol.*, 207, 4623, 2004.

326. Evans, D. H., Piermarini, P. M., and Choe, K. P., The multifunctional fish gill: dominant site of gas exchange, osmoregulation, acid–base regulation, and excretion of nitrogenous waste, *Physiol. Rev.*, 85, 97, 2005.

327. Schlesinger, P. H., Blair, H. C., Teitelbaum, S. L., and Edwards, J. C., Characterization of the osteoclast ruffled border chloride channel and its role in bone resorption, *J. Biol. Chem.*, 272, 18636, 1997.

328. Breton, S., Smith, P., Lui, B., and Brown, D., Acidification of male reproductive tract by a bafilomycin- sensitive H^+ ATPase, *Nat. Med.*, 2, 470, 1996.

329. Garcia-MacEdo, R., Rosales, A. M., Hernandez-Perez, O., Chavarria, M. E., Reyes, A., and Rosado, A., Effect of bafilomycin A1, a specific inhibitor of vacuolar (V-type) proton ATPases, on the capac- itation of rabbit spermatozoa, *Andrologia*, 33, 113, 2001.

330. Al Awqati, Q., Plasticity in epithelial polarity of renal intercalated cells: targeting of the H^+ ATPase and band 3, *Am. J. Physiol. Cell Physiol.*, 270, C1571, 1996.

331. Wagner, C. A., Finberg, K. E., Breton, S., Marshansky, V., Brown, D., and Geibel, J. P., Renal vacuolar H^+-ATPase, *Physiol. Rev.*, 84, 1263, 2004.

332. Stankovic, K. M., Brown, D., Alper, S. L., and Adams, J. C., Localization of pH regulating proteins H^+ ATPase and Cl^-/HCO_3^- exchanger in the guinea pig inner ear, *Hear. Res.*, 114, 21, 1997.

333. De Milito, A. and Fais, S., Tumor acidity, chemoresistance and proton pump inhibitors, *Future Oncol.*, 1, 779, 2005.

334. Nishi, T. and Forgac, M., The vacuolar H^+-ATPases: nature's most versatile proton pumps, *Nat. Rev. Mol. Cell. Biol.*, 3, 94, 2002.

335. Hirata, T., Iwamoto-Kihara, A., Sun-Wada, G. H., Okajima, T., Wada, Y., and Futai, M., Subunit rotation of vacuolar-type proton pumping ATPase: relative rotation of the G and C subunits, *J. Biol. Chem.*, 278, 23714, 2003.

336. Hirata, T., Nakamura, N., Omote, H., Wada, Y., and Futai, M., Regulation and reversibility of vacuolar H^+-ATPase, *J. Biol. Chem.*, 275, 386, 2000.

337. Murata, T., Yamato, I., Kakinuma, Y., Leslie, A. G. W., and Walker, J. E., Structure of the rotor of the V-type Na$^+$-ATPase from *Enterococcus hirae*, *Science*, 308, 654, 2005.

338. Feniouk, B. A., Kozlova, M. A., Knorre, D. A., Cherepanov, D. A., Mulkidjanian, A. Y., and Junge, W., The proton-driven rotor of ATP synthase: ohmic conductance (10 fS), and absence of voltage gating, *Biophys. J.*, 86, 4094, 2004.

339. Junge, W., Lill, H., and Engelbrecht, S., ATP synthase: an electrochemical transducer with rotatory mechanics, *Trends Biochem. Sci.*, 22, 420, 1997.

340. Grabe, M., Wang, H., and Oster, G., The mechanochemistry of V-ATPase proton pumps, *Biophys. J.*, 78, 2798, 2000.

341. Noji, H., Yasuda, R., Yoshida, M., and Kinosita, K. J., Direct observation of the rotation of F$_1$-ATPase, *Nature*, 386, 299, 1997.

342. Imamura, H., Nakano, M., Noji, H., Muneyuki, E., Ohkuma, S., Yoshida, M., and Yokoyama, K., Evidence for rotation of V$_1$-ATPase, *Proc. Natl. Acad. Sci. USA*, 100, 2312, 2003.

343. Sumner, J. P., Dow, J. A., Earley, F. G., Klein, U., Jäger, D., and Wieczorek, H., Regulation of plasma membrane V-ATPase activity by dissociation of peripheral subunits, *J. Biol. Chem.*, 270, 5649, 1995.

344. Kane, P. M., Disassembly and reassembly of the yeast vacuolar H$^+$-ATPase *in vivo*, *J. Biol. Chem.*, 270, 17025, 1995.

345. Zimmermann, B., Dames, P., Walz, B., and Baumann, O., Distribution and serotonin-induced activation of vacuolar-type H$^+$-ATPase in the salivary glands of the blowfly *Calliphora vicina*, *J. Exp. Biol.*, 206, 1867, 2003.

346. Inoue, T. and Forgac, M., Cysteine-mediated cross-linking indicates that subunit C of the V-ATPase is in close proximity to subunits E and G of the V$_1$ domain and subunit a of the V$_0$ domain, *J. Biol. Chem.*, 280, 27896, 2005.

347. Vitavska, O., Merzendorfer, H., and Wieczorek, H., The V-ATPase subunit C binds to polymeric F-actin as well as monomeric G-actin and induces cross-linking of actin filaments, *J. Biol. Chem.*, 280, 1070, 2005.

348. Vitavska, O., Wieczorek, H., and Merzendorfer, H., A novel role for subunit C in mediating binding of the H$^+$-V-ATPase to the actin cytoskeleton, *J. Biol. Chem.*, 278, 18499, 2003.

349. Moussian, B., Seifarth, C., Muller, U., Berger, J., and Schwarz, H., Cuticle differentiation during *Drosophila* embryogenesis, *Arthropod Struct. Dev.*, 35, 137, 2006.

350. Lighton, J. R., Discontinuous ventilation in terrestrial insects, *Physiol. Zool.*, 67, 142, 1994.

351. Wessing, A. and Zierold, K., Heterogeneous distribution of elemental contents in the larval Malpighian tubules of *Drosophila hydei*: x-ray microanalysis of freeze-dried cryosections, *Cell Tissue Res.*, 272, 491, 1993.

352. Jellison, R., Macintyre, S., and Millero, F. J., Density and conductivity properties of Na–CO$_3$–Cl–SO$_4$ brine from Mono Lake, California, USA, *Int. J. Salt Lake Res.*, 8, 41, 1999.

353. Brown, M. R., Graf, R., Swiderek, K. M., Fendley, D., Stracker, T. H., Champagne, D. E., and Lea, A. O., Identification of a steroidogenic neurohormone in female mosquitoes, *J. Biol. Chem.*, 273, 3967, 1998.

354. Riehle, M. A., Fan, Y., Cao, C., and Brown, M. R., Molecular characterization of insulin-like peptides in the yellow fever mosquito, *Aedes aegypti*: expression, cellular localization, and phylogeny, *Peptides*, 27, 2547, 2006.

355. Kaufmann, C. and Brown, M. R., Adipokinetic hormones in the African malaria mosquito, *Anopheles gambiae*: identification and expression of genes for two peptides and a putative receptor, *Insect Biochem. Mol. Biol.*, 36, 466, 2006.

356. O'Donnell, M. J., Ianowski, J. P., Linton, S. M., and Rheault, M. R., Inorganic and organic anion transport by insect renal epithelia, *Biochim. Biophys. Acta*, 1618, 194, 2003.

357. Bradley, T. J., Physiology of osmoregulation in mosquitoes, *Ann. Rev. Entomol.*, 32, 439, 1987.

358. Bradley, T. J., Membrane dynamics in insect Malpighian tubules, *Am. J. Physiol.*, 257, R967, 1989.

359. Donini, A., Patrick, M. L., Bijelic, G., Christensen, R. J., Ianowski, J. P., Rheault, M. R., and O'Donnell, M. J., Secretion of water and ions by Malpighian tubules of larval mosquitoes: effects of diuretic factors, second messengers, and salinity, *Physiol. Biochem. Zool.*, 79, 645, 2006.

360. Gill, S. S., Chu, P. B., Smethurst, P., Pietrantonio, P. V., and Ross, L. S., Isolation of the V-ATPase A and c subunit cDNAs from mosquito midgut and Malpighian tubules, *Arch. Insect Biochem. Physiol.*, 37, 80, 1998.

361. Pullikuth, A. K., Filippov, V., and Gill, S. S., Phylogeny and cloning of ion transporters in mosquitoes, *J. Exp. Biol.*, 206, 3857, 2003.
362. Neufeld, D. S., Kauffman, R., and Kurtz, Z., Specificity of the fluorescein transport process in Malpighian tubules of the cricket *Acheta domesticus*, *J. Exp. Biol.*, 208, 2227, 2005.
363. Xu, W. and Marshall, A. T., Control of ion and fluid transport by putative second messengers in different segments of the Malpighian tubules of the black field cricket *Teleogryllus oceanicus*, *J. Insect Physiol.*, 46, 21, 2000.
364. Hazelton, S. R., Townsend, V. R., Felgenhauer, B. E., and Spring, J. H., Membrane dynamics in the Malpighian tubules of the house cricket, *Acheta domesticus*, *J. Membr. Biol.*, 185, 43, 2002.
365. Leader, J. P. and Neufeld, D. S., Electrochemical characteristics of ion secretion in Malpighian tubules of the New Zeeland alpine weta (*Hemideina maori*), *J. Insect Physiol.*, 44, 39, 1997.
366. Sofia Hernandez, C., Gonzalez, E., and Whittembury, G., The paracellular channel for water secretion in the upper segment of the Malpighian tubule of *Rhodnius prolixus*, *J. Membr. Biol.*, 148, 233, 1995.
367. Hazel, M. H., Ianowski, J. P., Christensen, R. J., Maddrell, S. H., and O'Donnell, M. J., Amino acids modulate ion transport and fluid secretion by insect Malpighian tubules, *J. Exp. Biol.*, 206, 79, 2003.
368. Ianowski, J. P. and O'Donnell, M. J., Electrochemical gradients for Na^+, K^+, Cl^- and H^+ across the apical membrane in Malpighian (renal) tubule cells of *Rhodnius prolixus*, *J. Exp. Biol.*, 209, 1964, 2006.
369. Laenen, B., De Decker, N., Steels, P., Van Kerkhove, E., and Nicolson, S., An antidiuretic factor in the forest ant: purification and physiological effects on the Malpighian tubules, *J. Insect Physiol.*, 47, 185, 2001.
370. Zhang, S. L., Leyssens, A., Van Kerkhove, E., Weltens, R., Van Driessche, W., and Steels, P., Electrophysiological evidence for the presence of an apical H^+-ATPase in Malpighian tubules of *Formica polyctena*: intracellular and luminal pH measurements, *Pflügers Arch.*, 426, 288, 1994.
371. Holtzhausen, W. D. and Nicolson, S. W., Beetle diuretic peptides: the response of mealworm (*Tenebrio molitor*) Malpighian tubules to synthetic peptides, and cross-reactivity studies with a dung beetle (*Onthophagus gazella*), *J. Insect Physiol.*, 53, 361, 2007.
372. Wiehart, U. I., Nicolson, S. W., and Van Kerkhove, E., The effects of endogenous diuretic and antidiuretic peptides and their second messengers in the Malpighian tubules of *Tenebrio molitor*: an electrophysiological study, *J. Insect Physiol.*, 49, 955, 2003.
373. Wiehart, U. I., Nicolson, S. W., and Van Kerkhove, E., K^+ transport in Malpighian tubules of *Tenebrio molitor* L: a study of electrochemical gradients and basal K^+ uptake mechanisms, *J. Exp. Biol.*, 206, 949, 2003.
374. Gaertner, L. S., Murray, C. L., and Morris, C. E., Transepithelial transport of nicotine and vinblastine in isolated Malpighian tubules of the tobacco hornworm (*Manduca sexta*) suggests a P-glycoprotein-like mechanism, *J. Exp. Biol.*, 201, 2637, 1998.
375. Li, H., Wang, H., Schegg, K. M., and Schooley, D. A., Metabolism of an insect diuretic hormone by Malpighian tubules studied by liquid chromatography coupled with electrospray ionization mass spectrometry, *Proc. Natl. Acad. Sci. USA*, 94, 13463, 1997.
376. Furuya, K., Milchak, R. J., Schegg, K. M., Zhang, J., Tobe, S. S., Coast, G. M., and Schooley, D. A., Cockroach diuretic hormones: characterization of a calcitonin-like peptide in insects, *Proc. Natl. Acad. Sci. USA*, 97, 6469, 2000.
377. Veenstra, J. A., Isolation and identification of three leucokinins from the mosquito *Aedes aegypti*, *Biochem. Biophys. Res. Commun.*, 202, 715, 1994.
378. Huesmann, G. R., Cheung, C. C., Loi, P. K., Lee, T. D., Swiderek, K. M., and Tublitz, N. J., Amino acid sequence of CAP2b, an insect cardioacceleratory peptide from the tobacco hawkmoth *Manduca sexta*, *FEBS Lett.*, 371, 311, 1995.
379. Kean, L., Cazenave, W., Costes, L., Broderick, K. E., Graham, S., Pollock, V. P., Davies, S. A., Veenstra, J. A., and Dow, J. A., Two nitridergic peptides are encoded by the gene capability in *Drosophila melanogaster*, *Am. J. Physiol. Regul. Integr. Comp. Physiol.*, 282, R1297, 2002.

8 Osmotic and Ionic Regulation in Fishes

David H. Evans and James B. Claiborne

CONTENTS

I. PALEOECOLOGY OF CHORDATE EVOLUTION

Fishes are the first vertebrates, descendents of the early chordates* that evolved from marine, deuterostome invertebrates (related to modern echinoderms). The extant lancelets (cephalochordates) and tunicates (urochordates), both marine and stenohaline, are modern invertebrate chordates, with the lancelets generally accepted as closest to the early fish lineage, based on molecular data.†[221] The recent discovery of fossil chordates (e.g., genus *Haikouella*) from early Cambrian marine deposits (ca. 520 million years before present) that resemble lancelets in many details[310] supports this conclusion. Modern jawless fishes (hagfishes and lampreys, termed *agnathans* or *cyclostomes*) are sister groups to a large assemblage of marine, jawless fishes (sometimes termed *ostracoderms*)

---

* For the most recent appraisal of fossil and modern fish systematics, see Nelson.[365] For an analysis of fish genome evolution, see Volff.[517]

† A genomic study, however, suggests that the urochordates may be the basal stock,[90] and urochordate embryos apparently have neural crest cells, which are generally considered to be a vertebrate trait.[237]

found in Silurian and Devonian deposits that are 440 to 360 million years old. The earliest fossils of both hagfishes and lampreys are found in what appear to be marine deposits, but despite some superficial similarities the two modern, agnathan lineages are thought to have been separate for at least 500 million years.[147,148,236] The hagfishes may be more properly placed in the subphylum Craniata[94,365] to distinguish them from the true Vertebrata (lampreys and jawed fishes), although this proposition is still debated.[313,376,377,492] Modern hagfishes (e.g., *Myxine* or *Eptatretus*) are marine and stenohaline,[339] while modern lampreys are anadromous (marine adults but breeding in freshwater, such as *Petromyzon marinus*, *Lampetra fluviatilis*, and *Geotria australis*) with some landlocked populations in freshwater (e.g., *Petromyzon*, *Lampetra*).[189] The chondrichthyan fishes (holocephalans and sharks, skates, and rays) are thought to have evolved some 450 million years ago in estuarine environments[511] and now are predominantly marine, although some 10% of modern species can enter freshwater (e.g., bull shark, stingrays), and at least one group is stenohaline in freshwater (Potamotrygonid rays).[56,188] The earliest bony fishes—Actinopterygii (ray-finned fishes) and Sarcopterygii (lungfishes and coelacanths)—are found in marine and freshwater deposits that are nearly 400 million years old.[187] The least derived, extant groups of actinopterygians are either freshwater (birchir, paddlefish, gar, bowfin) or marine and anadromous (sturgeon), while the lungfishes are freshwater and the coelacanth marine. The more derived actinopterygians (teleosts, some 25,000 species) are found in all salinities, with many groups being euryhaline.[119,197,365]

Thus, despite an incomplete and often contradictory fossil record, it is apparent that (with the exception of the hagfishes) early fishes evolved in low salinity or freshwater environments.[187] Consistent with such a zoogeographic history, the blood ionic concentrations of all modern fishes (indeed, vertebrates), except hagfishes, are significantly below that of seawater (hypoionic) and above that of freshwater (hyperionic) (Table 8.1). Given the physiological constraints of gas exchange across the fish gill, such as a large surface area and thin epithelium (see Evans et al.[127] for a more complete discussion), these ionic and osmotic gradients across a permeable surface must have produced significant osmoregulatory problems early in fish evolution. One might argue that the evolution of physiological solutions to these osmoregulatory problems was a key element in the ecological success of fishes, as well as their descendent vertebrate groups.

II. HAGFISHES

Despite probable specialization during at least 400 million years of evolution, 70 species of modern hagfishes[365] represent the best model available for any osmoregulatory mechanisms that may have been present in the basal vertebrate clade. The general biology of this interesting group of marine fishes was reviewed most recently by Jorgensen et al.,[241] and general aspects of hagfish osmoregulation have been reviewed in the past 30 years.[114,120,246]

Although it is isotonic to seawater, hagfish plasma appears to have significantly more Na^+ but lower K^+, Ca^{2+}, Mg^{2+}, Cl^-, and SO_4^{2-} than seawater (Table 8.1).[114] No published measurements of the transepithelial electrical potential (TEP) across hagfishes are available, so (for the present) we assume that each of these ions is out of electrochemical equilibrium. Some of the ionic gradients seen, however, could be accounted for by a TEP of only a few millivolts (inside negative for Na^+ and Cl^- and positive for K^+). Potassium, Ca^{2+}, Mg^{2+}, and SO_4^{2-} are thought to be excreted in the urine and mucous.[359] More data on these putative ionic gradients and excretion pathways are needed, but this finding suggests that the early vertebrate kidney was primarily a divalent excretory system, as it is in extant invertebrates (see Chapters 5 and 6).

A. KIDNEY

The hagfish, opisthonephric kidney has 30 to 40 segmental glomeruli, connected to short, nonciliated neck segments that drain into paired archinephric ducts that are structurally similar to proximal tubules in other vertebrates.[93,132,133] The hagfish glomerulus has the same basic structure as other vertebrates, with a capillary endothelium, mesengial cells, basement membrane, and podocyte[201] forming the putative filtration surface. That surface has been calculated to be 1.8 mm^2 per glomerulus for *Myxine glutinosa*, substantially greater than that calculated for the marine flounder (0.023 mm^2 per glomerulus), freshwater carp (0.064 mm^2 per glomerulus), or marine skate (0.340 mm^2 per glomerulus).[133] On the other hand, measured glomerular pressures are so low that filtration is debatable,[134,435,436] suggesting that the primary urine is produced by secretion.[437] Urine formation by proximal tubular solute secretion has been described for both elasmobranchs and teleosts (see Sections IV.A.1 and V.A.2),[26] so it is likely that this process preceded glomerular filtration to form primary urine in the early vertebrate kidney. Because Mg^{2+} and SO_4^{2-} transport plays a role in secretion of urine in other fish groups[26] and the urine/plasma (U/P) ratios in *Eptatretus stoutii* urine are 1.23 and 8.59, respectively,[359] it seems likely that these divalents are the driving force in hagfish urine production, especially as the inulin U/P ratio is 1.0, signifying a lack of water reabsorption. Secretion of organic molecules may also play a role, as it does in other fish kidneys.[26] Urine/plasma ratios for Na^+ and Cl^- also are approximately 1.0 in both *M. glutinosa*[1] and *E. stoutii*,[359] so neither monovalent ion appears to be secreted in the renal tubules, contrary to the situation in elasmobranchs and teleosts.[26]

The hagfish archinephric duct is composed of columnar epithelial cells that possess an apical brush border, prominent Golgi, and smooth endoplasmic reticulum but relatively minor rough endoplasmic reticulum (ER), few mitochondria, and no basal labyrinth, in contrast to other vertebrate proximal tubules.[133] Despite some similarity to the typical vertebrate proximal tubule, it is apparent that the hagfish archinephric duct is unable to absorb Na^+ from the urine,[133,347,359] which is generally considered to be a functional requirement for osmoregulation in dilute salinities. In fact, hagfishes are stenohaline, able to control body weight only at salinities above 80% seawater by reduction in plasma ionic concentrations (to remain isotonic to the new salinity),[68,346] presumably by renal and diffusional salt loss. Interestingly, hagfishes apparently have an extremely high water permeability[445] and extremely low apparent ionic permeability,[123] a condition also found in elasmobranchs but not teleosts.[114] The high water permeability may be associated with an AQP4-like aquaporin in the gill pavement cells.[366] Whether this differential water and ionic permeability is a primitive or derived condition remains to be determined, but it could mean that entry of early agnathan fishes into reduced salinities was limited by both an inability to reabsorb ions from the urine and high branchial water permeability. In addition, it is not clear whether hagfishes can either initiate or increase glomerular filtration secondary to increased vascular pressures,[1,434] although water excretion must have increased to account for the return to control weight in the experiments where *Myxine glutinosa* was transferred to 80% seawater.[346]

B. Gill

Maintenance of plasma Na^+ and Cl^- levels in reduced salinities requires gill ionic uptake mechanisms as first proposed for modern freshwater fishes (and other organisms) by August Krogh,[268] so one might suppose that these ionoregulatory pathways are missing in hagfish gill epithelia, limiting euryhalinity. It would follow, therefore, that evolution of the first freshwater fishes was predicated upon evolution of some ionic uptake mechanisms. Specifics of these pathways will be discussed later in this chapter and they have been reviewed for fishes recently,[127,326] but it is clear that NaCl uptake by fishes in hypoionic salinities is mediated by chemical or electrical coupling with the excretion of H^+ (possibly NH_4^+) and HCO_3^-, both exchanges vital to acid–base regulation, even in seawater.[72,111] Surprisingly, early studies demonstrated that acid and base excretion by *Myxine glutinosa* was dependent on external Na^+ and Cl^-,[117] and it was suggested that the mitochondrion-rich cells (MRCs) in the hagfish branchial epithelium are the sites of these ionic exchange mechanisms,[311,312] as they are found in both elasmobranches and teleosts.[127,530]

The hagfish gill morphology is very different from that found in lampreys and jawed piscine vertebrates.[15,127,530] Rather than gill arches built upon a branchial skeleton (e.g., lampreys), the hagfish gill consists of multiple (5 to 14) pairs of gill pouches that are medial to the much-reduced

TABLE 8.1

Representative Blood Chemistry Data for Various Fish Groups

Salinity or Species	Total Osmolarity (mOsm/L)	Na	Cl	K	Ca	Mg	SO₄	Urea	TMAO	Species (Ref.)
		(mM/L or mM/kg)								
Seawater[a]	1050	439	513	9.3	9.6	50	26	—	—	(Robertson[441])
Hagfish	1035	486	508	8.2	5.1	12	3	—	—	Myxine glutinosa (Robertson[441])
Anadromous lamprey	333	156	159	5.6	3.5	7.0	—	—	—	Petromyzon marinus (Beamish et al.,[22] Morris[357])
Euryhaline bull shark	1067	289	296	5.8	4.4	1.8	—	370	47	Carcharhinus leucas (Pillans and Franklin,[403] Pillans et al.[404])
Dogfish sharks	1118	255	241	6.0	5.0	3.0	0.5	441	72	Scyliorhinus canicula, Squalus acanthias (Forster et al.,[150] Payan and Maetz[386])
Euryhaline Atlantic stingray	953	319	295	—	—	—	—	330	—	Dasyatis sabina (Piermarini and Evans[399])
Elephant fish	1068	292	288	—	—	—	—	465	—	Callorhinchus milli (Hyodo et al.[228])
Euryhaline sturgeon	294	152	149	2.9	2.1	1.6	—	—	—	Acipenser oxyrinchus (Altinok et al.,[3] Holmes and Donaldson[222])
Anglerfish	452	180	196	5.1	2.8	2.5	2.7	—	—	Lophius piscatorius (Holmes and Donaldson,[222] Smith[472])
Euryhaline steelhead trout[b]	325	153	135	4.0	1.4	0.65	—	—	—	Oncorhynchus mykiss (Liebert and Schreck[286])
Euryhaline puffer	363	179	144	2.8	3.1	1.8	—	2.6	—	Takifugu obscurus (Kato et al.[248])
Coelacanth	931	197	187	5.8	4.9	5.3	—	377	122	Latimeria chalumnae (Griffith[172])
Freshwater (soft)	1.0	0.25	0.23	0.01	0.07	0.04	0.05	—	—	(Potts and Parry[418])

Common name	Species								
Anadromous lamprey	*Petromyzon marinus* (Pickering and Morris[398])	—	112	99.6	2.3	1.8	1.5	—	—
Landlocked lamprey	*Lampetra fluviatilis* (Pickering and Morris[398])	272	120	104	3.9	2.5	2.0	—	—
Euryhaline bull shark	*Carcharhinus leucas* (Pillans and Franklin,[403] Pillans et al.[404])	595	221	220	4.2	3.0	1.3	151	19
Euryhaline Atlantic stingray	*Dasyatis sabina* (Piermarini and Evans[399])	621	212	209	—	—	—	196	—
Freshwater stingray	*Potamotrygon* sp. (Wood et al.[541])	319	178	146	—	—	—	1.22	—
Euryhaline sturgeon	*Acipenser oxyrinchus* (Althoff et al.,[2] Holmes and Donaldson[222])	260	136	107	2.9	2.3	1.5	—	—
Gar	*Lepisosteus osseus* (Holmes and Donaldson,[222] Sulya et al.[487])	—	159	133	4.2	6.1	0.3	—	—
Bowfin	*Amia calva* (Butler and Youson[49])	279	133	110	1.5	—	—	—	—
Carp	*Cyprinus carpio* (Holmes and Donaldson[222])	274	130	125	2.9	2.1	1.2	—	—
Lake whitefish	*Coregonus clupoides* (Robertson[440])	—	141	117	3.8	2.7	1.7	2.3	—
Euryhaline steelhead trout	*Oncorhynchus mykiss* (Liebert and Schreck[286])	260	153	133	3.8	1.4	0.5	—	—
Euryhaline puffer	*Takifugu obscurus* (Kato et al.[248])	346	166	128	3.0	3.5	1.3	—	—
Lungfish	*Protopterus aethiopicus* (Smith[473])	238	99	44	8.2	2.1	—	0.6	—
	Protopterus dolloi (Wilkie et al.[527])	—	99	88	—	—	—	—	—

[a] The solute levels in this seawater are somewhat below open ocean seawater but similar to what is often used when working with experimental animals.

[b] Data estimated from graphs.

Note: The most complete databases remain the early reviews by Holmes[222] and Evans.[114]

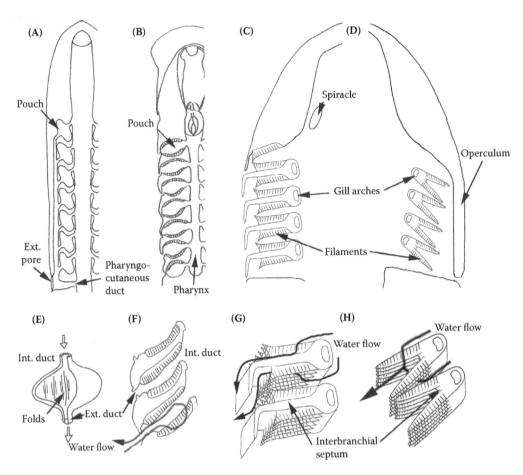

FIGURE 8.1 Simplified anatomy of gills and associated pouches and arches in hagfishes (A and E), lampreys (B and F), elasmobranchs (C and G), and teleosts (D and H) oriented anterior to posterior with oral opening at top. (From Evans, D.H. et al., *Physiol. Rev.*, 85, 97–177, 2005. With permission.)

branchial skeleton (cartilaginous plates associated with the gill pores) (Figure 8.1A). The lens-shaped gill pouches are internal and connected to the pharynx by an incurrent duct (Figure 8.1E and Figure 8.2A). The excurrent ducts from each pouch converge and lead to a single gill pore in the myxinid hagfishes; multiple gill pores are present in eptatretid hagfishes.[15] The branchial epithelium is also unique in hagfishes. The walls of the pouch are expanded into folds that run parallel to the axis of the pouch. These folds vary in height, with only a few bordering the central water channel, and each fold, in turn, is subdivided into smaller (second- to sixth-order) folds (Figure 8.2B). The primary folds are considered to be equivalent to the filaments forming the hemibranchs on the gill arches of other fishes; the smaller folds are considered to be equivalent to the lamellae (see below). In hagfishes, the lamellae run parallel to the filaments; in other fishes, the lamellae are at right angles to the filaments. Individual pouches are perfused via an afferent branchial artery that feeds the afferent circular artery that surrounds the excurrent duct (Figure 8.2A). Afferent radial arteries perfuse each primary fold (filament) via "cavernous tissue" and subdivide into afferent lamellar arterioles within the lamellae. Like other fishes, pillar cells in the lamellae appear to contain contractile elements and may function to maintain lamellar structure. Efferent lamellar arterioles drain blood into efferent "cavernous tissue," which feeds blood to an efferent circular artery that surrounds the incurrent duct (Figure 8.3).[104] Thus, as in other vertebrates, blood flow in the hagfish gill is counter-current to water flow across the branchial epithelium.

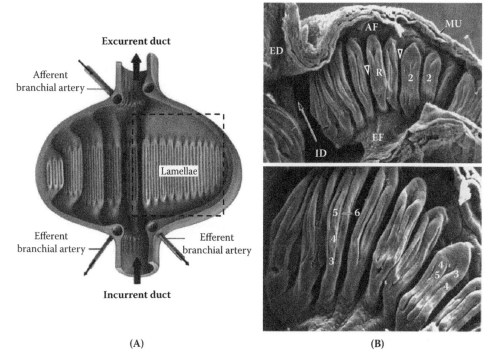

(A)																																									(B)

FIGURE 8.2 Anatomy of hagfish gills. (A) Schematic of a longitudinal cut through a gill pouch from the Atlantic hagfish, with a lateral perspective of a primary gill fold (filament) and its lamellae (boxed area). Note radial arrangement of additional filaments around the pouch. Large arrows indicate direction of water flow; small arrows indicate direction of blood flow. (Adapted from Elger, M., *Anat. Embryol. (Berl.)*, 175(4), 489–504, 1987.) (B) Scanning electron micrographs of a gill filament from the Pacific hagfish, comparable to boxed area in Part A. The upper micrograph (30×) shows an overview of a filament with afferent (AF) and efferent (EF) regions and respiratory lamellae (R) with second-order folds (2). MU indicates muscular layer around the pouch. Arrow indicates flow of water through pouch from incurrent duct (ID) to excurrent duct (ED); arrowheads indicate flow of blood across filament. Bottom panel (70×) reveals higher order folds of the lamellae: third-order (3), fourth-order (4), fifth-order (5), and sixth-order (6). (From Evans, D.H. et al., *Physiol. Rev.*, 85, 97–177, 2005. With permission.)

In hagfishes, the MRCs are on the lateral half of the gill folds and the lateral wall of the gill pouches. They are single and separated by pavement cells, and they display extensive tight junctions.[127] The cells have a subapical vesiculotubular system, numerous large mitochondria, and basolateral infoldings, which are continuous with an intracellular tubular system. The apical membrane contains microvilli, but deep apical crypts (which characterize teleost MRCs) usually are not found (Figure 8.4).[530] Both Na^+,K^+-ATPase (NKA) and carbonic anhydrase (CA) have been localized to the MRCs in hagfish,[77] although both NKA protein expression and enzymatic activity are below levels commonly found in teleosts.[61,311,505] More recently, the Na^+/H^+ exchanger (NHE) has been localized in gill tissue from *Myxine glutinosa*[65,100] and in MRCs in the gill of *Eptatretus stoutii*, along with V-type H^+-ATPase (V-HAT),[505] and expression of NHE is upregulated by induced acidosis in both species.[100,378] In *E. stoutii*, alkalosis is associated with an upregulation of MRC V-HAT protein (western blot) and a downregulation of NKA protein, and the authors suggest that this is due to differential insertion of the respective proteins in the basolateral membrane.[506] Tonic activity of NHE (for acid excretion) may account for the slight hypernatric condition commonly found in hagfish plasma (Table 8.1).[114] These acid transporters in the hagfish MRCs may be associated with the apical, rod-shaped particles that have been described in these

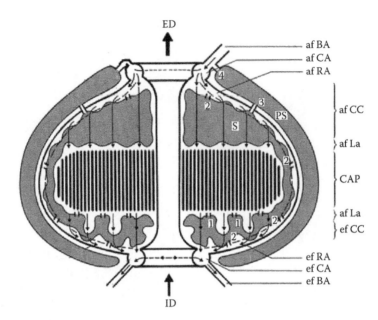

FIGURE 8.3 Schematic of arterioarterial and arteriovenous vasculature in the gill pouch and filament of hagfishes. The arterioarterial vasculature includes afferent (af) and efferent (ef) branchial arteries (BA), circular arteries (CA), radial arteries (RA), cavernous tissue (CC), and lamellar arterioles (La). Also noted are the lamellar sinusoids or capillaries (CAP). The arteriovenous vasculature includes the peribranchial sinus (PS) and a sinusoid system (S). The numbers 1, 2, and 4 indicate sites of arteriovenous anastomoses. Number 3 marks an anastomosis between the venous sinusoid system and the venous peribranchial sinus. Thin arrows indicate direction of blood flow, and thick arrows indicate direction of water flow. (From Evans, D.H. et al., *Physiol. Rev.*, 85, 97–177, 2005. With permission.)

cells[13,14] and similar cells in the lamprey gill (see below), as well as acid–base regulating, intercalated cells in the mammalian collecting duct.[166] Interestingly, heterologous antibodies have localized the cystic fibrosis transmembrane conductance regulator (CFTR), which is important in salt extrusion in teleost gill MRCs (see Section V.A.3), in the gill epithelium of *M. glutinosa* but in cells that are distinct from those that express NKA (K.A. Hyndman and D.H. Evans, unpublished data). In teleosts, CFTR and NKA are expressed in the same MRCs that mediate salt extrusion (see Section V.A.3),[127] so one might hypothesize that the lack of colocalization of these two transport proteins in the same cell is associated with the apparent inability of the hagfishes to hypoosmoregulate.[346]

In summary, it is apparent that two mechanisms that are vital to entry into freshwater (a glomerular kidney and gill ionic uptake mechanisms) were probably actually present in the basal, marine, agnathan vertebrates; however, it seems likely euryhalinity is limited by the inability of either of these pathways to upregulate sufficiently to balance water gain or salt loss in salinities below about 80% seawater.[119] Moreover, it is not clear if mechanisms for renal reabsorption of NaCl existed which are critical for salt balance in low salinities. More work on these interesting agnathan fishes is needed.

III. LAMPREYS

The earliest lamprey fossils are found in marine or estuarine deposits,[163] but the fact that modern lampreys (38 species) are either freshwater or marine and anadromous suggests that members of this agnathan group were the first vertebrates to enter freshwater. Thus, study of extant lampreys can give us insight into the physiological strategies that allowed osmoregulation during this

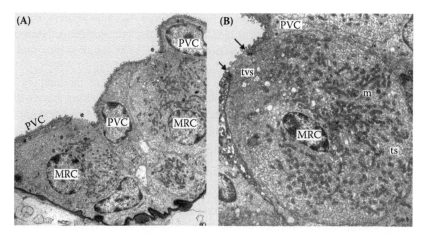

FIGURE 8.4 Transmission electron micrographs of the gill epithelium from a hagfish (Atlantic hagfish). In Part A, note the large MRCs intercalated between pavement cells (PVCs) (original magnification, 3740×). Asterisks indicate exposed apical membranes of MRCs. (Unpublished micrograph generously provided by Dr. Helmut Bartels, University of Munchen.) Part B shows a higher magnification micrograph of a MRC (original magnification, 6630×), with a basolateral tubular system (ts) that closely associates with mitochondria (m). Also note the subapical tubulovesicular system (tvs) and the deep intercellular junctions (black arrows) between the MRC and neighboring PVCs. (From Evans, D.H. et al., *Physiol. Rev.*, 85, 97–177, 2005. With permission.)

evolutionary transition. The general biology of lampreys has been reviewed by Hardisty and Potter,[189] and their osmoregulation was reviewed by Morris[357] and Evans;[114] more recent reviews include Evans,[120] Karnaky,[246] and Rankin.[423]

A. Freshwater Species

Presumably, the reduced NaCl concentration that is characteristic of lamprey plasma (Table 8.1) is the result of ionic loss during the evolutionary entry into freshwater. The fact that this reduced ionic content is maintained demonstrates that modern (and presumably ancestral) lampreys can osmoregulate in hypoionic media, as can members of all descendent vertebrate clades. No transepithelial electrical potential measurements have been published for lampreys, but they would have to be substantial ($> \pm 100$ mV) to account for the chemical gradients displayed in Table 8.1 for the lamprey in freshwater.

What evolved in lampreys that allowed osmoregulation in reduced salinities? Like marine hagfishes, freshwater lampreys appear to maintain a relatively low branchial permeability to ions (although direct measurements have not been published), as the radioisotopic efflux of both Na^+ and Cl^- is low, and renal efflux can only account for <10% of the total.[114,482] Lamprey urine flow rates are substantial (>10 mL/kg/hr)[42,357] and even higher than freshwater teleosts (see Section V.B.1), suggesting that they have retained the high osmotic permeability that may have been present in their marine, isotonic ancestors (see Section II.A), and that they can increase glomerular filtration as needed, contrary to hagfishes (see above). The most critical evolutionary addition, however, appears to be a distal renal tubule[93,105,201,263,299] that allows the production of a dilute urine.[357]

1. Kidney

The kidney of larval (ammocoete) and adult lampreys contains distinct glomeruli perfused from the dorsal aorta, podocytes, fenestrated endothelial cells, mesangial cells, and the microscopic structural elements characteristic of the glomeruli of all other vertebrates.[93,263] Contrary to the architecture in other vertebrate nephrons, lamprey afferent arterioles may perfuse more than one glomerulus, and a single glomerulus may be perfused by more than one afferent arteriole.[201] This may be the structural basis for the apparent lack of glomerular recruitment or intermittency that generally has been found in lamprey renal function studies.[42,424] The lamprey proximal tubule cell possesses microvilli, as well as numerous mitochondria and basolateral infoldings, both of which are missing in hagfishes.[201] The proximal tubule cell comprises 50% of the lamprey nephron.[357] The distal segments are characterized by cuboidal epithelial cells lacking microvilli but with numerous mitochondria and a smooth endoplasmic reticulum. More than one distal segment drains into a series of common collecting ducts. The distal parts of the proximal tubules, the distal tubules, and the collecting ducts are arranged in a series of ascending and descending loops, adjacent to blood vessels running in the same direction.[357] Such a looped architecture is also found in the elasmobranch nephron.[200,202]

The relatively high urine flow rate found in lampreys (15 to 20 mL/kg/hr) is correlated with a substantial single-nephron glomerular filtration rate (SNGFR) and total GFR.[300,424] In fact, all of the individual glomeruli appear to filter in freshwater.[42] Some 40 to 60% of the filtered urine is reabsorbed, presumably secondary to the reabsorption of approximately 90% of the Na and Cl in the distal tubule and collecting duct.[300,357,424] We are not aware of any investigations of the cellular sites or mechanisms of this ionic uptake, but it seems likely that they are similar to those described for other vertebrate renal salt reabsorption pathways, with the proviso that they are in distal rather than proximal renal segments (see below and other chapters). No evidence published suggests that the lamprey proximal tubule can secrete NaCl, as has been described for both teleosts and elasmobranchs,[26] but such secretory pathways seem likely. Modern molecular techniques will now allow a more thorough investigation of putative salt transport pathways in the lamprey nephron.

2. Gill Salt Uptake

Despite substantial renal NaCl reabsorption and apparent relatively low gill ionic permeability, freshwater lampreys presumably lose a net amount of salt via the urine, as well as by diffusion across the branchial epithelium. Thus, lampreys must either replace this salt by ingestion or possess branchial ionic uptake systems to maintain plasma NaCl concentrations above freshwater levels. As far as we know, no data on the role of ingestion of salt via food have been published. It was shown nearly 50 years ago, however, that the river lamprey (*Lampetra fluviatilis*) is able to reduce the Na^+, Cl^-, and K^+ concentration of the freshwater medium, presumably via some branchial uptake mechanism.[356] Moreover, the uptake of radiolabeled Na^+ was saturable, suggesting that some transport process was involved.[357]

The lamprey gill shares many more morphological characters with the gills of jawed fishes (especially elasmobranchs) than with the gills of the hagfishes.[127,530] Gill filaments extend from each gill arch as holobranchs, with somewhat concave interbranchial septa. These produce cranial and caudal hemibranchs of adjacent holobranchs formed into pouch-like structures through which water travels from the pharynx to the outside (Figure 8.1B). The filaments on each hemibranch express numerous lamellar folds at right angles to the axis of the filament. Each filament is perfused by afferent filamental arteries (AFAs), which arise from the afferent branchial artery in each gill arch. The AFA on each filament subdivides into afferent lamellar arterioles, which perfuse individual lamellae. Blood flow through a lamella is broken up into "sheet flow" through lamellar sinusoids produced by the presence of pillar cells, which separate the lateral and medial epithelial sheets comprising the lamellae. Thus, blood flow is approximately counter-current to water flow across the outside of the lamellae. Blood exits individual lamellae via efferent lamellar arterioles and is collected into efferent filamental arteries, which enter the efferent branchial arteries in the gill arches. Like hagfishes, lampreys possess MRCs in their branchial epithelium in the interlamellar and afferent filamental spaces, but more than one type has been described, and salinity affects their distribution (Figure 8.5).[18,19,127,530]

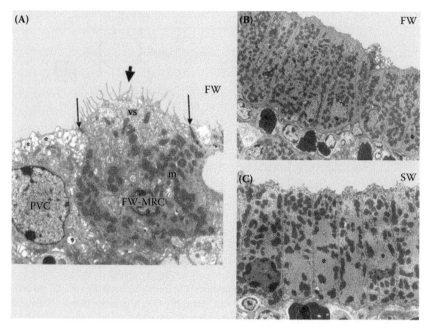

FIGURE 8.5 Transmission electron micrographs of the gill epithelium from freshwater (FW) and seawater (SW) lampreys. Part A shows a PVC and a FW-MRC from a freshwater brook lamprey (*Lampetra appendix*) (original magnification, 3740×). In the PVC, note the subapical secretory vesicles (asterisks) and relatively flat apical membrane. In the FW-MRC, note the numerous mitochondria (m), subapicalvesicular system (vs), and extensive apical membrane microprojections (short arrow). Intercellular junctions between PVCs and FW-MRCs are extensive (long arrows). (Unpublished micrograph generously provided by Drs. Helmut Bartels and John Youson, University of Toronto.) Part B and Part C show cross-sections through the gill filament of freshwater (original magnification, 2750×) and seawater (original magnification, 5000×) pouched lampreys to show the SW-MRCs lined up next to one another. Note the more extensive and organized tubular system (asterisks) between mitochondria in the SW-MRCs of seawater lampreys relative to freshwater lampreys. (Unpublished micrograph generously provided by Dr. Helmut Bartels, University of Toronto.) (From Evans, D.H. et al., *Physiol. Rev.*, 85, 97–177, 2005. With permission.)

The ammocoete larvae express a unique MRC that is found in groups on and between gill lamellae and disappears during metamorphosis. In addition to numerous mitochondria, the cell is characterized by globular particles in the apical membrane (shown by freeze fracture), as well as short, apical microvilli or microplicae. Its function is unknown, but it is probably not a precursor of the adult MRC, as it becomes apoptotic during metamorphosis (Bartels and Potter, pers. comm.). A second MRC type (fwMRC, termed *intercalated mitochondria-rich cell* by Bartels and Potter[18]) is found between the lamellae and at the base of the filament in both adults and ammocoetes. The fwMRC is typically single, cuboidal, and 10 to 15 μm in diameter and lies on the surface epithelium between ammocoete MRCs in the larvae and between pavement cells or between pavement cells and other MRCs in the adult.[19,530] Its apical membrane has elaborate microplicae and numerous small vesicles and tubules in the cytoplasm, although the extensive, basolateral infoldings, which produce the cytoplasmic tubular system that characterizes teleost and marine lamprey MRCs (see below), are missing. Freeze-fracture studies of these fwMRCs have found rod-shaped particles in either the apical or basolateral membranes. The fwMRC is the only MRC type in adult, freshwater lampreys, and it degenerates as the fish enter seawater but regenerates when they reenter freshwater on their spawning runs.[18]

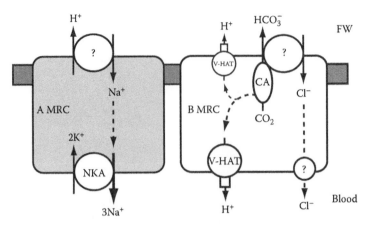

FIGURE 8.6 Working model of ion and acid–base exchange mechanisms in freshwater adult lamprey gills. In the A-type MRC, basolateral Na$^+$,K$^+$-ATPase (NKA) creates a low intracellular [Na$^+$], which drives the entry of Na$^+$ from the environmental water through an apical Na$^+$ transporter (possibly an NHE). In the second MRC type (B-type), carbonic anhydrase (CAII) is expressed in the apical membrane, along with V-type H$^+$-ATPase (V-HAT) in the basolateral or apical membranes. CAII may be providing HCO$_3^-$ to an apical anion exchanger as part of a metabolon secreting HCO$_3^-$ and absorbing Cl$^-$. A question mark (?) indicates proteins that have not yet been described in lampreys but have been in other taxa (see Sections IV and V). (Adapted from Choe, K.P. et al., *J. Exp. Zool.*, 301, 654–665, 2004.)

Initial freeze-fracture studies of fwMRCs suggested that there are actually two subtypes, depending on whether the rod-shaped particles seen in the electron micrographs are apical (A-type MRCs or IMRC-A) or basolateral (B-type MRCs or IMRC-B). It has been suggested that the A-type MRC is associated with Na$^+$ uptake and the B-type MRC is associated with Cl$^-$ uptake,[18,19] just as A- and B-type intercalated cells mediate these transports in the mammalian collecting duct.[166] A recent immunohistochemical study (using heterologous antibodies) of the *Geotria australis* gill epithelium has corroborated this idea by identifying two populations of fwMRCs. One type of fwMRC expressed NKA protein on the basolateral surface; the second type stained for carbonic anhydrase on the apical membrane and V-HAT diffuse in the cytoplasm.[66] A similar cellular distribution for NKA, CA, and V-HAT has been described for the transforming juvenile *Petromyzon marinus*, as they acclimated to increased salinities.[428] This diffuse distribution of V-HAT (cytoplasmic vesicles?) mimics that described for base-secreting cells in the elasmobranch gill epithelium[401,535] which have also been shown to express an apical Cl$^-$/HCO$_3^-$ exchanger (pendrin[402]) (see Section IV.B.3). Choe et al.[66] proposed that the NKA-expressing fwMRCs in the lamprey gill probably also express apical NHE and thereby function in net acid secretion (A-type MRCs), in parallel to the base-secreting B-type MRCs that express apical CA and probably an apical Cl$^-$/HCO$_3^-$ exchanger (pendrin or AE) and V-ATPase, primarily on the basolateral membrane (Figure 8.6). It is noteworthy that the A-type MRCs in the lamprey (and elasmobranch) gill epithelium are more like proximal tubule acid-secreting cells in the mammalian kidney than collecting-duct A-type intercalated cells, which extrude protons via an apical V-ATPase.[166] Thus, despite some uncertainty about the specifics of cellular localization of the putative transport proteins, it appears that the freshwater lamprey gill expresses the ionic pathways necessary for acid–base regulation and, coincidentally, ionic uptake in these dilute salinities. It also appears that the cells mediating these vital transports evolved from hagfish branchial cells that presumably mediated acid–base regulation, but not ionic balance,[19] in the marine ancestors.

Because freshwater lamprey plasma contains substantially less K$^+$, Ca^{2+}, and Mg^{2+} than the environment (Table 8.1), the assumed branchial and renal loss of these ions must be balanced by oral or branchial uptake. To our knowledge, no data have been published that suggest the relative roles of these putative pathways nor mechanisms for K$^+$, Ca^{2+}, or Mg^{2+} transport across the relevant epithelia.

B. Marine Species

Like all other vertebrates, except hagfishes, marine lampreys (e.g., *Petromyzon marinus*) have plasma ionic concentrations distinctly below those in the surrounding seawater (Table 8.1), although few data are published because of the difficulty in capturing and maintaining lampreys as they migrate from the ocean into brackish and freshwaters (termed *fresh run*), and fresh-run lampreys captured in freshwater or landlocked lampreys do not osmoregulate very well in high salinities (Beamish;[21] pers. observ.). Hypoosmoregulation requires a reduction in urine flow, increased urinary salt excretion, and two, presumably new, strategies: oral ingestion of the medium (with attending intestinal salt and water uptake) and extrarenal salt extrusion if the kidney is unable to produce a urine that is hypertonic to the plasma (which is the case, except in birds and mammals). We might presume that the marine lamprey gill epithelium has retained the relatively low ionic permeability that has been calculated for the freshwater lamprey (see above). No measurements have been published, however, and the type of MRC that characterizes the marine lamprey gill epithelium has shallow tight junctions between cells,[16,17] much like those in the relatively ionic-leaky epithelium of the marine teleost gill (see below).[245] The following sections demonstrate the critical need for more data on the mechanisms of osmoregulation in lampreys in seawater.

1. Kidney

Few data on lamprey renal function in seawater have been published, and these are often over 25 years old. Nevertheless, it appears that when *Lampetra fluviatilis* is acclimated for 2 weeks to seawater, its kidney can reduce urine flows by 95% by doubling distal tubule and collecting duct water reabsorption[301] and reducing the single nephron glomerular filtration rate (SNGFR), rather than reducing the number of filtering glomeruli.[42,424] The authors suggested that the fall in SNGFR is secondary to reduced renal blood flow,[301] and subsequent measurements demonstrated a correlation between reduced vascular pressure and SNGFR in *L. fluviatilis* transferred from freshwater to isosmotic brackish water.[348] Seawater-acclimated *L. fluviatilis* had urine Na+ concentrations of approximately 50 mM/L, significantly below plasma levels, suggesting that renal Na+ reabsorption continues to exceed water reabsorption in hypertonic salinities; urine Cl- concentrations were equivalent to plasma levels, indicating that NaCl reabsorption drives the necessary water reabsorption.[301,423] Urinary Mg^{2+} concentration was five times what would be expected by 95% tubular water reabsorption,[301] suggesting that the marine lamprey tubules secrete this divalent ion, as do marine invertebrate kidneys and all other vertebrate kidneys. Especially noteworthy is the finding that the urine of *L. fluviatilis* in seawater appears to be significantly hypertonic to the plasma (50 to 100 mOsm/L),[301,423,424] a measurement that certainly needs to be confirmed. The hypertonicity of the final urine is largely due to very high Mg^{2+} and SO_4^{2-} concentrations, but the mechanisms for final tubular water reabsorption are unknown, and it is tempting to suggest that the looped renal tubules may play a role.[301,423] Despite the limited database, it appears that lampreys in seawater have the appropriate renal mechanisms for water conservation and divalent ion excretion, but renal NaCl loss is not sufficient to balance the assumed diffusional gain of these ions.

2. Oral Ingestion

Hyporegulating marine vertebrates must drink seawater to offset osmotic and renal loss of fluid, so it is not surprising that *Lampetra fluviatilis* ingests the medium in 50% seawater,[355] as well as 100% seawater.[423] The rate of drinking in seawater (approximately 1% body weight per hour)[423] is above that described for many teleosts in seawater,[114,392] so it appears that lampreys in hypertonic salinities may have retained the relatively high branchial water permeability that has been described for freshwater lampreys and marine hagfishes (see above). Approximately 70 to 80% of the ingested fluid is absorbed in the intestine,[355,423] similar to what has been found in the teleost intestine (see Table 8.2).[203] Likewise, ingested divalent ions are either left in the intestinal fluids[398,423] and probably excreted rectally or absorbed and excreted in the urine (see above).[355,398,423]

3. Gill Salt Extrusion

Marine lampreys and freshwater lampreys moving downstream to seawater express a third type of MRC, the swMRC or chloride cell,[18] which is very similar to the MRCs found in teleost gills and presumably functions in salt secretion (see below). They are columnar or flask shaped, 20 to 23 μm in length, with extensive basolateral infoldings; they occur in groups and extend below the surface epithelium.[387] Apical microvilli, present in fwMRCs, are missing in swMRCs.[18] The swMRC disappears when marine lampreys migrate upstream for reproduction.[18] As indicated earlier, these swMRCs (like their marine teleost equivalents) have much reduced tight junctions between the cells,[16,17] suggesting relatively greater ionic permeability in the marine gill vs. the freshwater gill of lampreys, as is found in teleosts.[244] Based on common morphology with the marine teleost gill MRC (see below), Bartels and Potter[18] suggested that the lamprey swMRC is responsible for secreting NaCl to balance the salt gained by ingestion of the medium and diffusion, via the mechanisms outlined in Figure 8.16 (see Section V.A.3). Although logical, no physiological or histochemical data have been published to support this hypothesis; however, at least partial sequences for cDNA of the relevant genes (NKA, NKCC1, and CFTR) can be found by "blasting" sequences from other fish species on the Web site for the preassembly of the genome for *Petromyzon marinus* (http://pre. ensembl.org/Petromyzon_marinus/index.html) (K.A. Hyndman, unpublished data).

The foregoing suggests that modern lampreys display the osmoregulatory strategies that allowed the earliest vertebrates to enter freshwater and then reenter the marine environment of their agnathan ancestors. Critical data are lacking, but it appears that the strategies of hyporegulation in seawater are basically the same in lampreys and marine teleosts, suggesting either independent, convergent evolution over a span of 400 million years or that a common ancestor possessed these pathways over 400 million years ago.

IV. CARTILAGINOUS FISHES

The cartilaginous fishes (Chondrichthyes) include both the elasmobranchs (937 species of sharks, skates, and rays) and the much-less studied elephant fishes and chimeras (holocephalans, comprised of 33 species). The osmoregulatory hallmark of chondrichthyan fishes is the retention of urea in the plasma and cytoplasm to levels that would be lethal to most other vertebrates (Table 8.1). The potential denaturing effects of such high urea concentrations in chondrichthyan plasma and tissue are countered by either changes in enzyme kinetics or the addition of counteracting solutes, such as trimethylamine oxide (TMAO) (Table 8.1).[217,546,547] Recent comparative work has demonstrated that other methylamines (betaine and sarcosine) and β-amino acids (β-alanine and taurine) also may play a modulatory role in elasmobranch muscles.[502] The vast majority of elasmobranchs and all holocephalan fishes are entirely marine, but at least 171 species of elasmobranchs are able to enter brackish waters and freshwaters. Furthermore, one group of elasmobranchs (Potamotrygonid rays) are restricted to freshwater of the Amazon and Orinoco basins.[333] The biology and physiology of elasmobranchs have been reviewed by Shuttleworth,[462] Hamlett,[188] and Carrier et al.,[56] each including chapters dealing with osmoregulation.[126,198,274,374,463] Additional, relevant reviews include Anderson et al.,[6] Evans,[114,116,120] Evans et al.,[127] Hazon et al.,[191,193] Karnaky,[246] and Marshall and Grosell.[326]

A. MARINE SPECIES

In both elasmobranchs and holocephalans, plasma Na[+], Cl[-], K[+], and Ca[2+] levels are distinctly below those in seawater (Table 8.1). Few measurements of transepithelial electrical potentials have been published for marine elasmobranchs, but they appear to be −2 to −5 mV (plasma negative to seawater)[114] below what would account for the ionic gradients measured between the plasma and the external medium in either marine or freshwater species. Because of the structural complexity of the gills, calculation of true branchial permeabilities to ions, water, and urea are probably impossible, but early radioisotopic flux data suggested that elasmobranchs have retained the low ionic but high osmotic permeabilities found in both hagfishes and lampreys.[114,463] The apparent low ionic permeabilities may be associated with the presence of what appear to be multistrand tight junctions in the gill epithelium (see Section IV.A.3).[127] Another study (using an isolated perfused head preparation) suggests, however, that the shark gill water permeability may not be substantially greater than other epithelia, including teleost gills,[381] and this hypothesis is supported by measurement of water fluxes across apical and basolateral membrane vesicles from the spiny dogfish (*Squalus acanthias*) gill.[206] Unfortunately, no molecular studies have been reported on putative gill aquaporin (AQP) channels, as have been published for hagfish and teleost gill tissue,[80,366] although partial AQP cDNA sequences have been obtained from dogfish rectal gland and bull shark (*Carcharhinus leucas*) kidney.[78] The urea permeability of the elasmobranch gill appears to be very low,[35,206,344,381,385] consistent with the maintenance of substantial plasma urea concentrations, despite the extremely large urea gradient favoring urea loss across the gills (Table 8.1). The apical membrane of the gill epithelial cell appears to be the effective barrier to urea loss,[206,381] and some evidence suggests that the basolateral membrane contains a urea transporter that transports urea back into the blood from the branchial cells (in exchange for blood Na^+).[138] In addition, an apparent urea transporter (UT)[448] has been cloned from elasmobranch kidney,[470] and expression was localized to the gill tissue (also liver, blood, intestine, and rectal gland) by northern blot. Unfortunately, this preliminary study of a putative elasmobranch gill urea transporter has not been pursued.[344]

Given the osmotic and ionic gradients across the gills of chondrichthyan fishes (Table 8.1), osmoregulation in seawater entails excretion of the osmotically gained water, renal and branchial retention of urea, and excretion of excess ions. Entry into brackish or freshwater by euryhaline species (and residency in freshwater by the potamotrygonid and dasyatid stingrays) requires increased renal loss of water (and possibly a reduction in the urinary loss of ions) and either ingestion of needed salts by feeding or branchial extraction of ions from the surrounding hypotonic medium. As one might expect, a recent proteomic analysis (two-dimensional gels) of spiny dogfish tissues has shown that osmoregulatory tissues (kidney, intestine, gill, rectal gland) are more similar in their overall proteomes than non-osmoregulatory tissues (heart and brain).[282] Despite the fact that oral ingestion of the medium is not a necessary osmoregulatory strategy in marine elasmobranches, various studies have shown that ingestion of the medium can be stimulated by reduction in blood volume (or injection of angiotensin II) or by transfer to a higher salinity,[5,6] so this might be an important osmoregulatory response in euryhaline species as they reenter seawater.[6]

1. Kidney

The elasmobranch kidney is arguably one of the most complex among the vertebrates, rivaling that found in the mammals.[93,126,198–201,273,274,326] The elaborate arrangement of each nephron and its vasculature gives rise to bundle zones (also termed *lateral bundles*[2]) in the dorsolateral part of the kidney and sinus zones (also termed *mesial tissue*) in the ventromedial region (Figure 8.7 and Figure 8.8). Renal corpuscles (containing Bowman's capsules and glomeruli) are found at the bundle–sinus boundary. Blood from the dorsal aorta perfuses the kidneys via intercostal arteries, which subdivide into renal arteries. These renal arteries, in turn, subdivide into afferent glomerular arterioles and bundle arteries, which perfuse the bundle zones via interstitial capillaries. Efferent glomerular arterioles perfuse the sinus zones, which are blood sinuses that surround the distal tubule and portions of the collecting tubule. The sinus zone also receives blood from the bundle interstitial capillaries and renal portal veins, the latter coming from the vasculature in the caudal portions of the body. Efferent intrarenal veins drain the sinus zones and join the systemic venous circulation via an afferent renal vein (Figure 8.7). The nephron itself is composed of a glomerulus followed by five tubular segments, although specific nomenclature and location are still debated in the

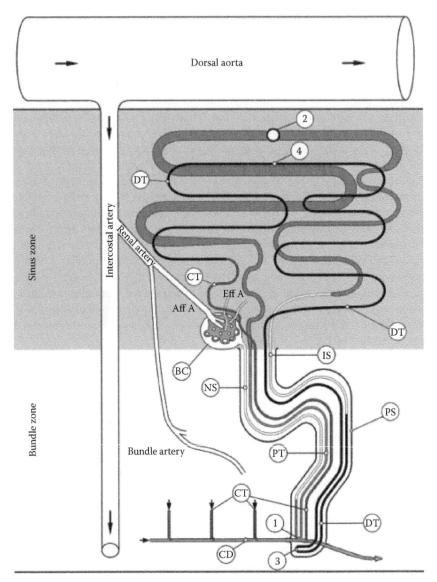

FIGURE 8.7 Schematic of an elasmobranch nephron and renal arterial blood flow. The neck segment (NS) arises from the distal end of Bowman's capsule and extends into the bundle zone, where it becomes the proximal tubule (PT). The PT continues into the bundle zone but then takes a sharp turn (loop 1) toward the sinus zone. Deep in the sinus zone, the PT turns back toward the bundle zone (loop 2), and before reaching the bundle zone the nephron transforms into the intermediate segment (IS). The IS extends into the bundle zone, where it transitions into the distal tubule (DT). The DT continues through the bundle zone but takes a sharp turn (loop 3) toward the sinus zone. The DT progresses deep into the sinus zone, where it turns back (loop 4) toward the bundle zone. Before reaching the bundle zone, the DT transforms into the collecting tubule (CT), which continues through the bundle zone and empties into a collecting duct (CD). The CDs eventually empty into a ventral ureter, which carries urine to the cloaca for excretion. Aff A, afferent arteriole; PS, peritubular sheath. Arrows indicate direction of blood and urine flow. (From Evans, D.H. et al., in *Biology of Sharks and Their Relatives*, Carrier, J. et al., Eds., CRC Press, Boca Raton, FL, 2004. With permission.)

literature and may be species specific. The tubular subdivisions are generally termed the neck, proximal, intermediate, and distal segments, followed by collecting tubules, which drain into a collecting duct at the base of the bundle zone (Figure 8.7). The most important structural components

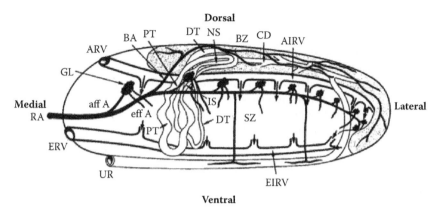

Dorsal

Ventral

FIGURE 8.8 Schematic of a cross-section through a skate (*Raja erinacea*) kidney showing general arrangement of blood vessels and a single nephron. Arterial circulation (including capillaries and glomeruli) is solid black. Venous circulation is thick-lined structures. Kidney tubules (including nephron, collecting duct, and ureter) are thin-lined structures. RA, renal artery; aff A, afferent arteriole; GL, glomerulus; eff A, efferent arteriole; BA, bundle artery; ARV, afferent renal vein; AIRV, afferent intrarenal vein; EIRV, efferent intrarenal vein; ERV, efferent renal vein; NS, neck segment; PT, proximal tubule; IS, intermediate segment; DT, distal tubule; CD, collecting duct; UR, ureter; BZ, bundle zone; SZ, sinus zone. (From Evans, D.H. et al., in *Biology of Sharks and Their Relatives*, Carrier, J. et al., Eds., CRC Press, Boca Raton, FL, 2004. With permission.)

(with potential functional consequences) are four loops in the nephron tubules: early proximal segment (loop 1) and early distal segment (loop 3) tubular loops in the bundle zone and late proximal segment (loop 2) and late distal segment (loop 4) loops in the sinus zone (see Figure 8.7 for details). The tubular structure in loop 3 suggests that it is homologous to the thick ascending limb (TAL) of the mammalian loop of Henle.[155,200] At the bundle and sinus zone boundaries, the intermediate segment and collecting tubule run parallel to the proximal and distal segments. These loops and parallel segments may provide potential counter-current exchange of various tubular contents, especially those in the bundle zone, which are surrounded by a cellular, peritubular sheath which isolates them from the blood in the sinus zone (Figure 8.7).

Unfortunately, few studies of elasmobranch renal function have been published (presumably because of animal size and mobility), but an *in situ*, perfused kidney preparation from the lesser spotted dogfish (*Scyliorhinus canicula*) has been described,[522,523] so advances are expected. In this preparation, as well as in the spiny dogfish in an experimental chamber, the glomerular filtration rate (GFR) is approximately 1 mL/kg/hr,[23,523] of the same order as that described for freshwater teleosts and substantially above that described for marine teleosts (see below). Presumably, this high GFR is correlated with the osmotic influx of water across the gills, secondary to the osmotic gradient produced by plasma urea levels (Table 8.1). The associated urinary flow rate (UFR) is 30 to 50% of the GFR in the spiny dogfish[23,523] as well as the Atlantic stingray (*Dasyatis sabina*),[234] indicating that tubular absorption of filtered urine takes place. Elasmobranch urine osmolarity is below that of the plasma, largely due to the reabsorption of the filtered organic osmolytes. Urine urea levels are very low, with a fractional excretion equal to <1%, demonstrating nearly complete tubular reabsorption of filtered urea.[198] The concentration of TMAO in the urine is 10% of that in the plasma, suggesting that this important solute also is reabsorbed in the elasmobranch renal tubules.[198] Urine monovalent ion concentrations approximate plasma levels, but urine divalent ion (e.g., Mg^{2+}, SO_4^{2-}) concentrations are above plasma levels (and above what could be produced by urinary water absorption), suggesting tubular secretion.[198]

Like the proximal tubules of bony fishes (see below), the isolated proximal tubule of the spiny dogfish can secrete fluid, secondary to the active extrusion of monovalent salts across the tubular epithelium.[26,451] Interestingly, the fluid secretion rate of the isolated, proximal tubules from this marine shark is quite similar to that described in proximal tubules isolated from teleosts, such as the euryhaline, glomerular killifish (*Fundulus heteroclitus*); the marine, glomerular flounder (*Pleuronectes americanus*); and the marine, aglomerular toadfish (*Opsanus tau*). This suggests that this NaCl-driven, proximal tubule secretory urine formation is a general phenomenon in the piscine vertebrates, and one might wonder what role similar pathways may play in tetrapod urine formation.[26] The cellular pathways driving this fluid secretion are thought to be basolateral Na^+–K^+–$2Cl^-$ cotransport (NKCC) coupled to an apical Cl^- channel (CFTR) and paracellular leakage of Na^+. The basolateral Na^+ gradient is maintained by the ubiquitous NKA, and K^+ exits the basolateral membrane via a K channel.[28,451] This secretory transport suite is identical to that described for the elasmobranch rectal gland and marine teleost gill, both of which extrude NaCl (see Sections IV.A.2 and V.A.3).[126,127] In the spiny dogfish preparation, tubular Mg^{2+} secretion is apparently not present,[451] contrary to what has been suggested for other elasmobranch species,[198] so the mechanisms of the putative divalent ion secretion in the elasmobranch proximal tubule are unknown, contrary to teleosts.[26]

The tubular reabsorption of NaCl (and urea) by the elasmobranch nephron is especially difficult to study because of the complex anatomy, but early studies in the spiny dogfish demonstrated a good correlation between salt and urea reabsorption[454] and that infusion of the diuretic furosemide (NKCC inhibitor) produced both natriuresis and inhibition of urea reabsorption.[360] In an elegant series of experiments, Friedman and Hebert[155] used isolated, perfused segments from the loop 3 region in spiny dogfish kidney (Figure 8.7) and demonstrated that Cl^- was reabsorbed against its electrochemical gradient. Moreover, the Cl^- transport was dependent on perfusate Na^+ and inhibited by furosemide, suggesting cotransport via NKCC, as has been described for the TAL in mammals. Corroborating these data, the presence of shark-specific NKCC mRNA has been demonstrated in the elasmobranch nephron by northern blots[544] and localized to the apical membrane of distal segment cells by immunohistochemistry.[29] Based on their studies, Friedman and Hebert[155] proposed that reabsorption of NaCl from the lumen of the loop 3 tubule produces extratubular hypertonicity that osmotically withdraws urine from the subsequent loop 4 in the sinus zone (Figure 8.7). The resulting increase in urea concentration in the urine as it enters the collecting tubule in the bundle zone (designated as "distal tubule" in Friedman and Hebert[155]) favors reabsorption of urea into the surrounding extratubular fluids within the bundle zone. Thus, urea reabsorption in the collecting tubule is coupled to Na^+ reabsorption in the distal tubule, which runs counter-current in the bundle zone, analogous to the TAL and collecting duct morphology in the mammalian nephron (see Chapter 12). Such a functional linkage supports the seminal proposition by Boylan[36] that the counter-current arrangement of the elasmobranch nephron facilitates the nearly total reabsorption of critical urea. Transcripts of the mRNA for an elasmobranch urea transporter have been measured in kidney extracts from the spiny dogfish,[470] Atlantic stingray,[233] little skate (*Raja erinacea*),[353] and Japanese dogfish (*Triakis scyllium*).[229] In the Japanese dogfish, the protein for the urea transporter was localized only to the collecting tubules in the bundle zone by immunohistochemistry, consistent with the model proposed by Friedman and Hebert.[155] In addition, phloretin-sensitive, nonsaturable urea transport has been described in brush-border vesicles isolated from the bundle zone in the little skate kidney.[353] Importantly, the same study described phloretin-sensitive, Na^+-linked urea transport that did show Michaelis–Menton saturation kinetics (with a low K_m) in vesicles isolated from the sinus zone.[354] These data suggest that the sinus zone may contain tubules (loops 2 and 4) that reabsorb urea via a Na^+-urea carrier, a transporter that could account for the functional link between Na^+ and urea reabsorption that was initially described.[454] Thus, the relative importance of these alternative transporters in the reabsorption of urea (and Na^+) by the elasmobranch kidney remains to be quantified. Pathways and mechanisms for the reabsorption of other osmolytes such as TMAO have not been described, despite the fact that early studies suggested that a relatively specific TMAO pathway was present in the shark kidney.[76] A bacterial TMAO transporter has been described,[426] but the uptake in the elasmobranch kidney tubules may be passive, via channels that are volume activated, as they are in erythrocytes of the little skate.[265]

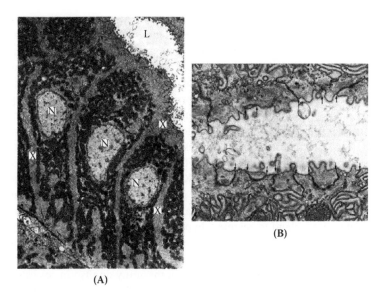

FIGURE 8.9 (A) Electron micrograph of secretory tubule cells from spiny dogfish rectal gland (original magnification, 4800×). Note the nuclei (N) surrounded by clusters of mitochondria and extensive basolateral membrane infoldings (X). L, tubule lumen. (B) Electron micrograph of apical regions of secretory tubule cells (original magnification, 20,000×). Arrows indicate shallow tight junctions between tubule cells. (From Evans, D.H. et al., in *Biology of Sharks and Their Relatives*, Carrier, J. et al., Eds., CRC Press, Boca Raton, FL, 2004. With permission.)

2. Rectal Gland Salt Extrusion

The best studied osmoregulatory organ in the elasmobranchs is the digital rectal gland, which secretes a fluid that contains more Na$^+$ and Cl$^-$ (>524 mM)[126] than either the plasma or seawater into a duct that drains into the shark intestine and exits via the cloaca.[47,246,326,374,463,466,467] It is noteworthy that nearly all of the functional data have been generated from two species: *Squalus acanthias* and *Scyliorhinus canicula*. The rectal gland is encapsulated by connective tissue (and circumferential smooth muscle that can contract[125]), perfused by a single (rarely multiple) rectal artery, and drained by a single vein. This simple vasculature, and the fact that a single duct carries the secreted fluid, provides an organ that is relatively easy to perfuse with specific solutions while collecting the secreted fluid for analysis.[466] The gland also can be studied *in situ*,[479,480] and both isolated tubules[171] and cultured epithelia[92,512] have been utilized. The secretory tubules (which empty into a central lumen) are arranged radially in sharks and in discrete lobules in batoids and are surrounded by capillary beds perfused by arterioles distal to circumferential arteries, which arise from the rectal gland artery (from the posterior mesenteric artery). The secretory tubules are composed of a columnar epithelium characterized by mitochondrion-rich cells (MRCs) with extensive basolateral infoldings (Figure 8.9) and relatively shallow tight junctions between adjacent cells (Figure 8.9B).[149] Subapical, membrane vesicles are often present. The most complete review of glandular ultrastructure is by Olson.[374]

In vivo, the shark rectal gland produces about 500 μL/kg/hr of fluid; so, given fluid NaCl concentrations that are above plasma and seawater (see above), the Na$^+$ and Cl$^-$ excretory rates are of the order of 200 μM/kg/hr.[463] Interestingly, the salt excretory rate in isolated, perfused glands is approximately 10% of these values and only attains *in vivo* rates after stimulation with secretagogs that stimulate intracellular cyclic AMP production.[374,483] The molecular pathways mediating tubular NaCl secretion are well known[126,374,438,463] and thought to be identical to those described for NaCl secretion by the marine teleost gill (see Figure 8.15 and Section V.A.3) and intestinal epithelium (and other secretory epithelia[156]), which contains an MRC that is structurally similar to the rectal gland MRC.[127] Early evidence supporting this model includes: (1) inhibition of secretion by removal of Na$^+$ or addition of either ouabain or furosemide to the perfusate,[467] (2) intracellular Cl$^-$ activities far above what could be predicted by the electrical potential across the basolateral membrane,[524] and (3) a lumen-negative electrical gradient between the perfusate and duct fluids during secretion.[464] Specific studies on the mechanisms of fluid transport into the gland tubule are lacking, but one must suppose that fluid enters the lumen either paracellularly through tight junctions[149] or transcellularly via cellular AQP water channels. Isolated basolateral and apical rectal gland membrane vesicles display relatively low water permeabilities, suggesting that aquaporins may not be present,[555] but cDNAs for aquaporins have now been isolated from rectal gland tissue.[78] Complete or partial sequences for cDNAs for rectal gland NKA,[307] NKCC,[544] CFTR,[314] and a K$^+$ channel[519] have been published, and immunochemical studies of rectal gland cells have localized protein expression of NKCC and NKA to the basolateral membrane[306] and CFTR to the apical membrane.[284,314]

The ability of the rectal gland to secrete a hypersaline solution and the inability of the kidney to secrete a hypersaline urine suggest that the rectal gland would be vital for osmoregulation. Surprisingly, this is not the case, because extirpation of the gland is followed by survival and maintenance of slightly elevated blood NaCl levels in the spiny dogfish in seawater.[46,124,533] Under these conditions, urinary Cl$^-$ excretion increases threefold (diuresis without an increase in urine Cl$^-$ concentration),[46] which may compensate for removal of the rectal gland. The elasmobranch kidney might be able to excrete the excess NaCl gained by diffusion and oral ingestion, as long as the osmotic gain of water across the gill epithelium is greater or equal to the diffusional or oral gain of salt. Thus, the kidney does not have to produce a urine that is hypernatric or hyperchloric to the plasma,[127] especially when an extrarenal salt gland is present (see below). Interestingly, the *in vivo* gland responds to a volume load but not a salt load,[478] and rectal gland NKA activity is stimulated by feeding,[307] so control of rectal gland secretion appears to be much more complex than merely for osmoregulation. In addition, because osmoregulation appears to be possible without a rectal gland, one might suggest that the gill provides another site for salt secretion (see below).

3. Gill Salt Extrusion?

The morphology of the elasmobranch gill is basically intermediate between those of the lamprey and bony fishes.[126,127,530] The individual filaments on the four or more holobranchs are more elaborate than those on the lamprey holobranch, and individual hemibranchs are separated by a distinct interbranchial septum. The water flowing across adjacent posterior and anterior hemibranchs exits the branchial chamber by individual gill slits (Figure 8.1C). Blood flow through the filaments and lamellae is as described for lampreys and teleosts (see Sections III.A.2 and V.A.3), including the presence of lamellar sheet flow of blood around pillar cells. The elasmobranch gill epithelium is composed mostly (90%) of squamous pavement cells (PVCs) characterized by apical microvilli or microplicae, few mitochondria, and no basolateral infoldings. Adjacent pavement cells share deep tight junctions.[127,530] Relatively large, ovoid MRCs are also present on the gill filaments (usually single and interlamellar), as well as on the lamellae.[62,503] The elasmobranch MRCs are characterized by relatively complicated apical surfaces (microvilli on concave or convex membranes), basolateral membrane infoldings into a basal labyrinth, and a tubulovesicular system near the apical surface (Figure 8.10). Adjacent PVCs and MRCs share deep tight junctions,[530] contrary to the relatively shallow tight junctions between MRCs and pavement cells in the teleost gill epithelium.[127] Shark MRCs express NKA,[533] but evidence for the other putative transport proteins that are found in the shark rectal gland and teleost gill (NKCC, CFTR, K channel) has not been published. Moreover, removal of the rectal gland did not produce any morphological (size, number, fine structure) or functional (NKA activity) change in the branchial epithelium of *Squalus acanthias*.[533] In fact, acclimation of freshwater Atlantic stingrays (*Dasyatis sabina*) to seawater actually reduced the number of MRCs and NKA activity.[400] Thus, no evidence suggests that the elasmobranch gill

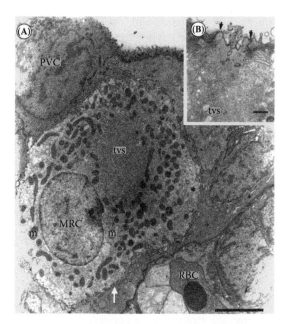

FIGURE 8.10 Transmission electron micrographs of the gill epithelium from a seawater elasmobranch (spiny dogfish). In Part A, note the large size of the MRC with its numerous mitochondria (m), complex basolateral membrane infoldings (white arrow), and extensive tubulovesicular system (tvs), relative to the neighboring PVC. The apical membrane of this MRC is not visible, but Part B shows the apical surface of another MRC which forms deep intercellular junctions (black arrows) with adjacent PVCs and has an extensive subapical-tubulovesicular system. In Part A, bar = 5 mm and RBC indicates a red blood cell. In Part B, bar = 0.5 mm. (From Evans, D.H. et al., *Physiol. Rev.*, 85, 97–177, 2005. With permission.)

epithelium plays a vital role in osmoregulation in seawater, in contrast to its importance in marine bony fish osmoregulation (see Section V.A.3). The marine elasmobranch gill MRCs, however, are probably involved in systemic acid–base regulation, as various physiological experiments have demonstrated branchial acid or base extrusion,[72] and recent studies have localized acid–base relevant proteins (i.e., V-HAT and NHE) to the MRCs in spiny dogfish in seawater.[62,503,507,508]

B. EURYHALINE AND FRESHWATER SPECIES

Euryhaline and freshwater elasmobranches maintain their blood ionic concentrations above those in the freshwater environment (Table 8.1). Despite the fact that a surprising number of elasmobranch species (about 171[333]) are able to enter brackish or freshwater, the resulting osmoregulatory changes have been examined in only a few species. Of particular interest is the euryhaline bull shark (*Carcharhinus leucas*), which is found in rivers throughout the world (and was described 3800 km up the Mississippi River in Alton, Illinois[496]); euryhaline members of the stingray family (Dasyatidae), which also occur in rivers worldwide (including a reproductive population in lakes in Central Florida[240]); and the stenohaline Potamotrygonid stingrays that occur only in the Amazon and Orinoco basins.[498] All three genera reduce their plasma urea levels and osmolarity significantly in freshwater (Table 8.1). Importantly, the Potamotrygonid stingrays have reduced their plasma urea levels to near zero and have actually lost the ability to increase urea levels when transferred to higher salinities.[162,497] In fact, they cannot tolerate salinities above approximately 65% seawater[174] and are restricted to salinities of <10% seawater.[38] The extremely low plasma urea concentration appears to be due to a relatively low rate of urea synthesis combined with a relatively high rate of urea loss, secondary to both high branchial permeability and relatively low rates of urea reabsorption by the renal tubules.[167] These truly freshwater rays, therefore, have reduced the osmotic gradient across their gill epithelium by 50% compared with the euryhaline species, presumably reducing the osmotic influx of water that must be excreted by the kidneys. Early studies indicated that *Potamotrygon* has whole-body water and ion permeabilities equivalent to those of other elasmobranchs,[54] but a more recent study suggests that the ionic permeability of *Potamotrygon* might be only 10% that of marine elasmobranchs.[541]

Despite a reduction in plasma osmolarity, euryhaline and freshwater elasmobranchs must osmoregulate in a hypotonic external medium, as do all other freshwater vertebrates (see above and other chapters in this volume). This requires an increase in urinary excretion of water, and uptake (or ingestion) of needed salts to balance diffusional loss across the gills, renal loss, and possibly loss via the rectal gland, which increases flow in response to a volume load.[478] Moreover, if they reenter the marine environment (bull sharks, for example), they must turn off these osmoregulatory strategies and restart the marine osmoregulatory strategies, including possibly oral ingestion of the medium as their blood volume falls before their plasma regains the high urea (and TMAO) levels characteristic of marine species.[6]

1. Kidney

The renal structure of the euryhaline Atlantic stingray is characteristic of elasmobranchs in general (see above),[274] but the kidney of *Potamotrygon* has obvious structural modifications. In this genus, the bundle zone and sinus zone are replaced by a peripheral "complex zone" and central "sinus zone," respectively, with the renal corpuscles within the complex zone. The most distinguishing characteristics are the absence of loops 3 and 4 of the intermediate and distal segments and the lack of a peritubular sheath surrounding the tubular segments in the complex zone.[274] It is generally assumed that the lack of the distal loops is functionally associated with the inability of Potamotrygonid stingrays to reabsorb urea from the urine.[167,200] Unfortunately, no studies have been published on renal function in this very interesting group, other than the finding that the GFR is 8 mL/kg/hr,[167] nearly 10 times that described for marine elasmobranchs.

We know much more about renal function in euryhaline elasmobranchs. When *Dasyatis sabina* is acclimated to freshwater, the urine flow rate is 10 mL/kg/hr,[235] which is equivalent to the rate described for *Pristis microdon* (sawfish) in freshwater,[476] slightly below that described for a freshwater lamprey (see above) but more than 10 times that published for marine elasmobranchs[23] and approximately 5 to 10 times that described for bony fishes in freshwater.[235] In addition, the stingray urine osmolarity is only 10% that of the plasma, generating a free water clearance that is nearly equivalent to the urine flow rate.[235] These data are consistent with the proposition that, even in freshwater, the elasmobranch gill retains the relatively high osmotic permeability that may have been a primitive vertebrate characteristic (see above). Despite tubular reabsorption of approximately 90% of the filtered urea, it is the dominant urinary solute (20 mM/L), followed by Na^+ (8 mM/L) and Cl^- (2 mM/L).[235] Similar results have been published for a study of renal function in *D. sabina* in dilute seawater (850 mOsm) vs. brackish water (440 mOsm);[234] in this case, GFR was also shown to increase threefold in the lower salinity. In elasmobranchs, increases in GFR may be secondary to either increases in single-nephron GFR or glomerular recruitment.[198] In the study of stingrays in 50% seawater,[234] the fractional urea excretion increased fivefold in the lower salinity, which may be due to downregulation of the renal urea transporter that another study reported.[353] Excretion of such a dilute urine must involve production of a distal tubular osmotic gradient that may exceed that produced by other vertebrate kidneys.[235] Unfortunately, no data have been published on the actual tubular gradients or osmotic permeabilities in the distal renal segments of the elasmobranch kidney. Presumably, this reabsorption of ions takes place in the distal loops, as described for the marine dogfish shark.[155] Of note, transport across the isolated loop 4 was inhibited by ouabain, suggesting that NKA was involved. Consistent with this proposition, acclimation of the euryhaline bull shark to freshwater is associated with increased renal NKA activity.[404] Clearly, much remains to be discovered about renal function in euryhaline and freshwater elasmobranchs.

2. Rectal Gland

As one might expect, early studies determined that the rectal gland of elasmobranchs in freshwater is much reduced compared to conspecifics in seawater,[370] and the rectal gland of *Potamotrygon* is quite atrophied,[499] even considering that batoids have much reduced rectal glands compared to sharks. More recently, Piermarini (unpublished) has found that the rectal gland of Atlantic stingrays living in freshwater in central Florida is 30% (corrected for body weight) as large as the gland in marine populations. On the other hand, rectal gland size and putative secretion rate (assumed to be correlated with NKA activity) may not always vary together. For example, acclimation of members of the freshwater population of Atlantic stingrays to seawater (1 week, after a gradual 7-day salinity change) was not associated with any significant change in rectal gland mass,[400] and the rectal gland mass of bull shark captured along a salinity gradient from 0 to 33‰ did not change with salinity.[403] In both studies, however, the enzymatic activity of NKA in the rectal gland tissue doubled in the higher salinity, and in the stingray the immunoreactivity of the NKA protein increased in seawater to a level equivalent to that measured in members of the marine population.[400] These data support the hypothesis that rectal gland activity is reduced in lower salinities in euryhaline elasmobranchs, but no physiological data have been published to support this idea. Moreover, entry of euryhaline elasmobranchs into reduced salinities probably stimulates rectal gland function initially, because a volume load in the spiny dogfish stimulates a fourfold increase in both duct flow and Cl^- secretion, presumably to aid in the reduction in plasma osmolarity.[478] One might suppose that long-term acclimation to lowered salinities would be associated with a reduction in rectal gland salt secretion (either by reduced flow of plasma-hyperionic solutions or by reabsorption of salt in the gland tubules), but no data have been published. Like the kidney, the rectal gland of euryhaline and freshwater elasmobranchs requires more study, but, even if rectal gland and renal salt excretion is reduced to very low levels, the elasmobranch in freshwater (or any hypoosmotic salinity) must extract needed salts either from the external medium or from ingested food.

3. Gill Salt Uptake

The influx of radioisotopes of both Na^+ and Cl^- into *Potamotrygon* shows saturation kinetics,[541] suggesting a carrier-mediated process. Attempts to discriminate putative ionic exchange vs. ionic channel pathways via addition of inhibitors to the external medium were largely unsuccessful in this study, and one has to merely assume that uptake is across the gills. Based on immunoreactivity for NKA and V-HAT, two obvious MRC populations are present in the gill epithelium of the Atlantic stingray in freshwater,[64,400,401] but no cytological data have been published to determine if ultrastructural changes have taken place during the evolution of this freshwater population. As might be expected, NKA-expressing cells can be immunolocalized in the gill epithelium of *Potamotrygon* (P.M. Piermarini, unpublished data).

The gill NKA activity increases in the gill epithelium of both the bull shark and Atlantic stingray in freshwater,[400,404] contrary to what is often seen in the gill epithelium in teleosts (see below) and the reduction in the rectal gland NKA activity in these species in freshwater (see above). The expression of mRNA for the $\alpha 1$-subunit of NKA also is higher in freshwater-acclimated stingrays,[64] and the number of NKA-rich MRCs is higher in the freshwater population of the stingray, largely due to increased MRCs on the gill lamellae.[64,400] These data suggest that the NKA-rich MRCs must be important for hyperosmoregulation in freshwater elasmobranchs. Supporting this hypothesis is a recent study that cloned a putative Na^+/H^+ exchanger (NHE3) from the gill of the stingray and localized its expression (both mRNA and protein) to NKA-expressing cells.[64] Like NKA, expression of this exchanger is greater (and more lamellar) in the freshwater population vs. the seawater population or individuals acclimated to seawater.[64] As stated above, a basolateral V-HAT has also been localized (by immunohistochemistry) in putative MRCs of the Atlantic stingray gill epithelium

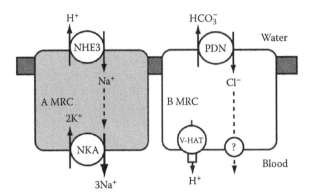

FIGURE 8.11 Working model of NaCl uptake mechanisms proposed for the freshwater Atlantic stingray. One type of MRC (A MRC) expresses Na^+,K^+-ATPase (NKA) on its basolateral membrane and draws in Na^+ across the apical surface in exchange for cytoplasmic H^+ (NHE3). The other MRC (B MRC) expresses V-type H^+-ATPase (V-HAT) on its basolateral membrane and draws Cl^- into the cell via an apical Cl/HCO_3 exchanger (pendrin, PDN). The pathway for basolateral Cl^- movement is unknown. (From Evans, D.H. et al., *Physiol. Rev.*, 85, 97–177, 2005. With permission.)

that do not express NKA.[401] Gill V-HAT immunoreactivity and the number of these V-HAT-expressing MRCs (in both the interlamellar and lamellar regions) were higher in stingrays from the freshwater population than from individuals acclimated to seawater or from the marine population.[401] These same cells also express immunologically identifiable pendrin (a putative Cl^-/HCO_3^- exchanger)[444] on the apical surface, and expression of this protein varies with salinity in the same way as V-HAT.[402]

Thus, it appears that at least the freshwater stingray gill epithelium contains two separate cells that provide putative pathways for the uptake of needed NaCl (Figure 8.11). The NHE–NKA cell (A-type MRC) mediates Na^+ uptake (and proton secretion) via coupled pathways similar to those in the proximal tubules of the mammalian kidney; the pendrin V-HAT cell (B-type MRC) provides for Cl^- uptake (and base secretion) via coupled pathways very similar to those expressed by HCO_3^- - secreting, intercalated cells in the mammalian collecting duct (see Chapter 12). Recent evidence suggests that an H^+,K^+-ATPase (HKA) is also expressed in the A-type MRC in stingray gills, which is upregulated in freshwater,[67] so this transporter may play a role in K^+ balance. It should be obvious that these transport systems (NHE, V-HAT, HKA, pendrin) also provide putative pathways for acid–base regulation in elasmobranchs that are important in all salinities. Indeed, V-HAT-expressing cells have been described for the stenohaline spiny dogfish in seawater,[503,529] as have NHE-expressing cells.[62,70]

4. Limits to Elasmobranch Euryhalinity?

If the marine elasmobranch kidney is able to respond to a volume load by increasing GFR and urine flow (while decreasing urine salt concentration), and the NaCl uptake pathways are present in both freshwater and marine elasmobranchs (used for ion regulation and acid–base regulation in the former and acid–base regulation in the latter), what limits euryhalinity in most elasmobranchs? Presumably the failure is quantitative rather than qualitative. Entry into freshwater requires a significant reduction in salt efflux (turn off the rectal gland, reduce urine salt concentration, and possibly reduce gill permeability), increase in excretion of water (increased GFR, reduced renal tubular water reabsorption), and stimulation of NaCl uptake via carriers with sufficient affinity to extract salt from very low salinities. Presumably, failure occurs in control at the tissue (hormones, nerves) and cellular (phosphorylation and protein–protein interactions) level.[62,193] This is certainly a fertile area for future research.

V. BONY FISHES

The bony fishes (Euteleostomi; formerly Osteichthyes) include two classes: Actinopterygii and Sarcopterygii. The actinopterygians incorporate the basal bichirs, sturgeons, paddlefishes, gar, and bowfins—all freshwater or anadromous (sturgeon)—as well as the more derived teleosts, the largest fish group (approximately 27,000 species[365]). The sarcopterygians include two species of extant, marine coelacanths[220] and six species of lungfishes, all freshwater.[365] The osmolarity and specific ionic concentration of the plasma in all bony fishes studied is intermediate between seawater and freshwater (Table 8.1). Osmoregulation by non-teleostean fishes has only been studied rarely (see Section V.D), but multiple reviews of teleost osmoregulation have been published.[114,116,119,120,127,129,215,246,319,326,540]

A. MARINE SPECIES

Because their plasma is hypotonic (and hypoionic) to seawater (Table 8.1), marine teleosts must compensate for the osmotic loss of water and diffusional gain of ions (most particularly NaCl), across the gill epithelium and possibly the skin. As noted above, accurate measurements of true gill epithelial water and ionic permeabilities are difficult in all fish groups because of the structural complexity of the tissue, but early, whole-animal, isotopic flux studies suggested that marine teleosts display relatively higher NaCl permeabilities and lower water permeabilities than agnathan and chondrichthyan fishes.[109,114,119,232,384] In fact, the estimated water permeability of the teleost gill (1 to 2×10^{-5} cm/sec)[481] is significantly below that of other epithelia (10^{-4} cm/sec),[206] but more recent study of water permeabilities of elasmobranch and teleost gill membrane vesicles found somewhat higher permeabilities in both species ($\sim 10 \times 10^{-4}$ cm/sec).[206] This may be correlated with the substantial expression of an aquaporin (AQP3) that has been described for the European eel gill.[80,288] The expression of this water channel appears to be regulated, because expression of mRNA for AQP3 is reduced by up to 97% in seawater animals,[80] and the protein expression is reduced by 65%, compared to freshwater-acclimated eels.[288] This reduction in water channel abundance in seawater corroborates early water flux data that indicated that both diffusional and osmotic water permeability was higher in freshwater teleosts than marine teleosts.[109,358] Protein expression of AQP3 is found throughout the gill epithelium, including on both the apical and basolateral membranes of mitochondrion-rich cells that also express NKA.[288] On the other hand, it has been suggested that other factors, such as ventilation/perfusion mismatches, mucin diffusion barriers, and regulation of ionic gradients across the gill epithelium, may play a major role in determining functional gill water permeability.[206] Some evidence suggests that external medium Ca^{2+} concentrations may affect at least the ionic permeability of the gill epithelium,[55] but, somewhat surprisingly, it appears that gill ionic permeability is significantly lower in freshwater than marine teleosts.[114,384]

It is generally assumed that Cl^- (but not Na^+) is out of electrochemical equilibrium across the teleost gill in seawater[414] and hence is actively secreted across the gill. Indeed, the current model for passive Na^+ transport contains this assumption (see below), but a significant number of trans-gill electrical potential (TGP) measurements (reviewed in Evans[115] and Marshall and Grosell[326]) suggest that Na^+ actually may be out of electrochemical equilibrium in some species (e.g., toadfish, seahorse, cod). One must assume that these atypical measurements of TGP are in error or that the standard model for NaCl excretion across the teleost gill (see Section V.A.3) may not be applicable to all species.

To compensate for the osmotic loss of water, marine teleosts ingest seawater and desalinate the ingested fluid to move needed water across the intestine. The intestinal ionic uptake (primarily NaCl) adds to the net salt gain produced by the ionic influx across the gill epithelium. Because the teleost kidney cannot produce urine that is hypertonic to the plasma (see below), the excess salts are excreted across the gill epithelium. To conserve water, urine flows are minimal, sometimes approaching zero. Homer Smith first proposed this suite of compensatory pathways,[474] and a substantial database exists that substantiates this hypothesis (see reviews above). Thus, osmoregulatory organs in marine teleosts include the gut, kidney, and gill.

TABLE 8.2
Ionic and Water Balance Sheet for the Southern Flounder

Ion	Swallowed[a]	Intestinal Absorption[b]	Rectal Excretion[b]	Renal Excretion[c]	Extrarenal Excretion[c]
Na^+	1956	98.8	1.2	0.13	99.9
Cl^-	2281	93.9	6.1	1.05	99.0
K^+	41.5	98.0	2.0	0.71	99.3
Ca^{2+}	42.6	68.5	31.5	11.4	88.6
Mg^{2+}	226	15.5	84.5	100	—
SO_4^{2-}	128	11.3	88.7	100	—
Water	4.6[d]	76	24	5.2	94.8

[a] Expressed in μmol/kg/hr.
[b] Percent of swallowed.
[c] Percent of absorbed.
[d] Expressed in mL/kg/hr.

Source: Hickman, Jr., C. P., *Can. J. Zool.*, 46, 457–466, 1968. With permission.

1. Gut

Initial measurements of drinking by marine teleosts found rates of 0.3 to 20 mL/kg/hr,[114] and more recent data are in the same range of 1 to 5 mL/kg/hr.[326] Given the average concentration of NaCl in seawater (Table 8.1), this ingestion presents a NaCl load of approximately 5 mM/kg/hr to an animal that is already facing a diffusional influx of salt. The subsequent processing of the ingested seawater was first described by two studies,[203,459] and a recent review[326] has summarized the data from more modern, but often less complete, investigations. Table 8.2 is from Hickman's work on the southern flounder (*Paralichthys lethostigma*), which remains the most complete balance sheet of intestinal processing of ingested seawater.* Of note is that, although NaCl is nearly completely absorbed (and excreted primarily via extrarenal pathways), Mg^{2+} and SO_4^{2-} are mostly left in the gut contents and excreted rectally. The Mg^{2+} and SO_4^{2-} that *are* absorbed by the gut epithelium are excreted by the kidneys. Calcium handling appears to be intermediate: 30% is not absorbed and excreted rectally, and 70% is absorbed and excreted mostly extrarenally (see below, however). Somewhat surprisingly, only 76% of the ingested water is absorbed (see Figure 8.13). In terms of sequential processing, the esophageal epithelium desalinates the ingested seawater so the osmolarity of the fluid is reduced from approximately 1000 mOsm/L to approximately 350 mOsm/L, isotonic to the plasma.[203] Because the esophageal epithelium apparently has an extremely low permeability to water,[207,380] the volume of the ingested fluid does not change appreciably. Somewhat unexpectedly, aquaporin 1 and 3 (AQP1, APQ3) expression has been demonstrated in the eel esophagus, but their roles are unclear.[82,336] The uptake of NaCl was initially assumed to be passive, but studies on the winter flounder (*Pseudopleuronectes americanus*) demonstrated that an active component is also present.[380] Because the Na^+ influx is inhibited by removal of Cl^- or addition of ouabain and amiloride, but not furosemide, it appears that basolateral NKA is involved and possibly apical Na channels or NHE but not apical NKCC.[380] Similar data have been described for the Japanese eel.[7] Clearly, molecular and immunohistochemical studies on this esophageal NaCl transport system are warranted.

---

* Drinking rates and urinary and rectal outputs were measured. Absorption was calculated as the difference between intake and rectal output; extrarenal excretion was calculated as the difference between the absorption rate and the renal output.

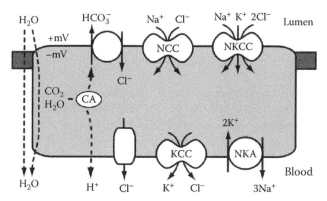

FIGURE 8.12 Working model of NaCl uptake mechanisms proposed for the marine teleost intestine. Uptake of water (either transcellular or paracellular) is driven by uptake of Na^+ and Cl^- via apical Na^+–K^+–Cl^- or Na^+–Cl^- cotransporters (NKCC/NCC), and Cl^-/HCO_3^- exchangers (AE) which are driven by basolateral Na^+,K^+-ATPase (NKA). Basolateral K^+ + Cl^- uptake is via a cotransporter (KCC) and Cl^- channel. The mechanisms of basolateral H^+ uptake are unknown but presumably via either NHE or V-HAT. (Adapted from Marshall, W.S. and Grosell, M., in *The Physiology of Fishes*, 3rd ed., Evans, D.H. and Claiborne, J.B., Eds., CRC Press, Boca Raton, FL, 2006, pp. 177–230.)

The stomach epithelium is not thought to play a significant role in uptake of either NaCl or water,[207] but the intestine appears to be the major site of water absorption, secondary to uptake of NaCl.[7,326,452,469] A variety of studies (reviewed by Marshall and Grosell[326]) have demonstrated that apical NaCl uptake is via parallel cotransporters: NKCC and NCC (Na^+ + Cl^-), with basolateral NKA providing the electrochemical gradient for apical Na^+ uptake (Figure 8.12). Expression levels of NKCC decline along the intestine,[81] and some evidence exists that NCC is only expressed in the posterior intestine.[81,160] Interestingly, Na^+ uptake stops when Cl^- is removed from the apical bathing medium in *in vitro* experiments using anterior, mid-, or posterior intestine (but not rectum), but Cl^- uptake persists when Na^+ is removed,[180] suggesting a component of Cl^- uptake that is not via NKCC. It is now clear that this additional uptake of Cl^- from the intestinal contents is via a Cl/HCO_3 exchanger (AE),[182] which may account for 50% of the total Cl^- uptake.[176] Recently, the apical Cl/HCO_3 exchanger has been identified as Slc26a6 in the euryhaline pufferfish (*Takifugu obscurus*) intestine.[272] Because the secreted HCO_3^- can precipitate intestinal Ca^{2+} (lowering intestinal osmolarity and thereby facilitating water uptake),[537] as well as alkalinize the intestinal contents, it is becoming increasingly clear that apical intestinal AE may play a significant role in teleost osmoregulation, ion regulation (including Ca^{2+}), acid–base balance, and even in the global carbon cycle.[175,177,178,181,326,494,520a,536] Moreover, the precipitated $CaCO_3$ may account for the relatively low intestinal Ca^{2+} concentrations that have been measured, which led to the conclusion that intestinal Ca^{2+} uptake was balanced primarily by extrarenal Ca^{2+} extrusion (Table 8.2).[203] Sodium that enters the enterocytes via the apical transporters exits the cell via basolateral NKA, and Cl^- extrusion appears to be via both Cl^- channels, possibly CFTR[305,332] and a K–Cl cotransporter.[186] Basolateral bicarbonate uptake appears to be via a Na–HCO_3 cotransporter (NBC1).[272] Cellular K^+ leaves the cell via basolateral and apical K^+ channels.[185] As one might expect, NKA mRNA expression and enzyme activity increase in the intestine of euryhaline fishes in seawater, as does expression of NKCC mRNA.[326]

Unfortunately, the specifics of intestinal uptake of Ca^{2+} and Mg^{2+} by the marine fish intestine appear to remain unstudied, but Mg^{2+} uptake by the intestine of the freshwater tilapia (*Oreochromis mossambicus*) is inhibited by ouabain and bumetanide.[514] The apparent coupling to Na^+ transport may be indirect, however, because further experiments determined that Mg^{2+} transport across enterocyte vesicles from the basolateral membrane of tilapia was coupled with an electrically neutral anion symport mechanism.[30,31] Some evidence suggests that enterocytes may cycle SO_4^{2-}. A recent study has demonstrated the presence of an intestinal $2Cl^-/SO_4^{2-}$ exchanger that might facilitate Cl^- (and water because of the 2:1 solute stoichiometry) uptake and SO_4^{2-} secretion.[388] It appears that the intestine also may secrete Cl^- under certain conditions (e.g., increase in intracellular Ca^{2+} and cAMP), via apical CFTR.[328]

Intestinal water uptake is usually thought to be via cellular and paracellular pathways[303,326] and in the Japanese eel (*Anguilla japonica*) is greater in the posterior segment of the intestine.[8] Aquaporin 1 has been sequenced and localized to the apical membrane of enterocytes in the intestine (predominantly posterior) of two eels, *A. japonica* and *A. anguilla*; in both studies, the mRNA expression increased after acclimation to seawater.[8,335] Expression of AQP1 protein also increased in seawater *A. anguilla*, with the posterior intestine displaying 40-fold more protein expression than the anterior intestine.[335] In the gilthead seabream (*Sparus aurata*), two homologs of AQP1 (SaAqp1a and SaAqp1b) have now been sequenced and localized to the duodenum, hindgut, and rectum. The expression was differential, with SaAqp1a predominating in the duodenum and hindgut, and SaAqp1b being the most abundant in the rectum.[422] Immunoreactivity of AQP3 in the intestine of *A. anguilla* is so low that it appears that this AQP does not play a major role in the intestine (contrary to the gill; see Section V.A.3).[288]

2. Kidney

The typical marine teleost kidney has the usual vertebrate complement of glomerulus, neck segment, proximal tubule (usually divided into "early" and "late" or PTI and PTII, characterized by a brush border), and collecting duct, but it usually lacks a distal tubule and, in approximately 30 species, even a glomerulus and PTI.[26,93,201,204,326] These are structural modifications that are consistent with the need for reduced water loss and increased ionic excretion in a hyporegulating fish, but they are not required, as evidenced by the presence of a glomerulus in most marine and euryhaline species and a distal tubule in some euryhaline species (e.g., eel, flounder, salmon).[204] Surprisingly, some aglomerular fishes are euryhaline (e.g., toadfish, *Opsanus tau*).[20,277] Even glomerular species may alter the number of functioning glomeruli in seawater (5%) vs. freshwater (45%).[40] The renal vascular system is similar to that in elasmobranchs, having renal and intercostal arteries from the dorsal aorta.[93] Afferent arterioles lead into glomerular capillaries, which drain into efferent arterioles that lead into a network of sinusoids and peritubular capillaries. In marine species, these capillaries also receive blood from branches of the caudal and segmental veins, giving rise to a renal portal system. Most freshwater teleosts appear to lack a renal portal system.[204] For general discussions of teleost renal morphology and physiology, see Ditrich,[93] Drummond,[96] Hentschel and Elger,[200,201] Kamunde and Kisia,[242] and Marshall and Grosell,[326] although Hickman's 1969 review[204] remains the most complete description of fish renal structure and function.

The glomerular filtration rate of marine teleosts is approximately 0.5 mL/kg/hr, about 50% the GFR measured in marine elasmobranchs (Section IV.A.1) and 10% that measured in freshwater teleosts.[204] Urine flows for glomerular teleosts in seawater are equivalent to the GFR, consistent with the absence of a distal tubule and the lack of net water reabsorption in the proximal tubule or collecting duct/bladder. Somewhat surprisingly, the urine flow in aglomerular marine teleosts is equivalent to that of glomerular teleosts, and the urine ionic composition is similar (approximately isotonic to plasma), thus indicating the primacy of tubular secretion.[26,326]

It is now clear that both aglomerular and glomerular marine teleosts (and perhaps all other vertebrates) secrete various ions across the epithelium of the proximal tubule (PTII) into the urine.[25,26] As described for the elasmobranch proximal tubule (see Section IV.A.1), fluid secretion in the glomerular and aglomerular teleost proximal tubule is primarily driven by NaCl secretion via the combination of basolateral NKCC and apical Cl^- channel (CFTR?), with Na^+ driven into the lumen through a paracellular route by the electrochemical gradient.[26,74,75] The expression of NKCC has now been demonstrated in renal extracts from the European eel.[81] Specific immunolocalizations of most of the putative transport proteins have not been published, but a recent study has found NKCC expression in the basolateral membrane of PII in the killifish kidney (F. Katoh and G.G. Goss, pers. comm.). It is important to note that this proposed salt secretory pathway is identical to that described for the shark rectal gland (see Section IV.A.2) and the teleost gill (see Section V.A.3). It is noteworthy that the fluid secretion rate of proximal tubules isolated from the aglomerular toadfish, two glomerular teleosts (flounder and killifish), and an elasmobranch (spiny dogfish) are all approximately 37 pL/min/mm tubule length.[26] The fluid movement is presumably via aquaporin channels, as AQP1 been localized in the brush border of renal tubules in the European eel.[334] Measurement of the Mg^{2+} and SO_4^{2-} electrochemical gradients across isolated proximal tubules from teleosts suggests that these divalent ions are also actively secreted but play a smaller role in fluid secretion.[26] Beyenbach has proposed that Mg^{2+} enters the tubular cell via a basolateral channel and exits the apical membrane via either a Mg/Na or Mg/H exchanger, the latter driven by an apical V-HAT.[26] Some evidence also suggests that apical Mg^{2+} secretion may be via vesicular exocytosis, as has been proposed for the dogfish PII segments.[59] Sulfate is thought to cross the basolateral membrane via a $2OH/SO_4$ exchanger and enter the tubular lumen via an apical anion/SO_4 exchanger, with intracellular carbonic anhydrase playing a pivotal role.[389,430,431] Consistent with this model, the cDNA for an anion/SO_4^{2-} exchanger (SLC26A1) has been cloned and its mRNA localized to the apical border in PTI of the rainbow trout, along with V-HAT.[252] The teleost proximal tubule also transports a suite of organic anions via relatively well-defined carriers and exchange systems.[26,430]

Surprisingly, these proximal tubular secretory pathways also are apparently present in fresh-water-acclimated euryhaline fishes,[74,75] despite the fact that ion conservation is necessary in hyper-regulating fishes. These studies provide an explanation for the earlier observation that freshwater eels sometimes display urine flow rates greater than the GFR.[453] The attending secretion of fluid into the proximal tubule presumably adds to the necessary excretion of water in these fishes. This fluid and Na^+ secretion can be calculated to be 3.5 times the measured GFR or Na^+ filtration rate![26] This consideration is especially important in calculations of fractional water and Na^+ reabsorption in glomerular, marine teleost kidneys. Adding these secreted fluid and Na^+ inputs into the kidney tubules to the GFR and Na^+ filtration rates increases the calculated, distal tubular reabsorption of water from 40% to 87% and the distal Na^+ reabsorption rate from 89% to 97%, more in line with other vertebrate nephrons.[26] Beyenbach proposes that this proximal solute secretion is the only way to produce urine in aglomerular teleosts, it increases tubular volume in glomerular teleosts, and it provides a mechanism for increasing the renal excretion of unwanted divalent ions, organic acids, and xenobiotics.[26]

Much less is known about the sites and mechanisms of ionic reabsorption in the more distal tubules of the marine teleost kidney, but the extremely high concentrations of Mg^{2+} (140 mM) and SO_4^{2-} (80 mM) in the urine are consistent with substantial late proximal or distal/collecting duct/bladder absorption of water.[326] In the late proximal tubule, it is generally assumed that Na^+ uptake is coupled to luminal membrane glucose or amino acid uptake or else Na/H exchange (NHE), and that Cl^- uptake is via Cl/HCO_3 exchange. Basolateral transport into the blood probably involves NKA in parallel with a Cl/HCO_3 exchanger, but more data are certainly needed.[326] Interestingly, NKCC expression has been demonstrated in renal tissue from the European eel[81] and localized to the apical membranes of the distal and collecting tubules in the killifish and rainbow trout (F. Katoh and G.G. Goss, unpublished data), suggesting a role for the cotransporter in tubular reabsorption as well as secretion (in the proximal tubule). Some evidence also suggests that Mg^{2+} may be reabsorbed in the collecting duct.[59]

When present, the urinary bladder in marine teleosts appears to play a major role in water conservation,[318] reabsorbing 60% of the urine volume in the toadfish (*Opsanus tau*);[225] however, it is important to note that some euryhaline fishes (e.g., the killifish) lack a urinary bladder. The water permeability of the urinary bladder of the starry flounder (*Platichthys stellatus*) in seawater is six times that in the freshwater-acclimated fish,[91] presumably maximizing necessary water reabsorption.

In the winter flounder, bladder NaCl uptake is coupled[429] and inhibited by thiazide but not by bumetanide.[484] The cDNA of a unique class of NaCl cotransporters (NCC, to be distinguished from NKCC) was, in fact, first cloned and localized primarily to the bladder epithelium in the winter flounder.[160] An additional, amiloride-sensitive, electrogenic Na^+ uptake was described in the urinary bladder of the winter flounder[429] and mudsucker goby (*Gillichthys mirabilis*),[304] but the relative importance of these two putative Na^+ uptake mechanisms remains unstudied. The apical Na^+ electrochemical gradient is probably maintained by NKA, which has been localized to the baso-lateral membrane,[432] and the Cl^- that is taken up at the apical surface is thought to exit the cell via a basolateral diphenylamine-2-carboxylic acid (DPC)-sensitive Cl^- channel.[60,85]

3. Gill Salt Extrusion

As indicated in an earlier review,[127] one of the earliest descriptions of active Cl^- transport across an epithelium came from Ancel Keys' studies of the European eel in seawater,[253] and his subsequent paper described the "chloride cell" (now MRC) as the site of this salt secretion.[254] The marine teleost gill has four holobranchs on each side, with no interbranchial septum and, therefore, two hemibranchs (Figure 8.1D), so the direction of irrigating water is counter-current to the blood flow through the lamellae.[127,530] Similar to what has been described for lamprey and elasmobranches (see above), each branchial arch contains numerous gill filaments that are perfused through afferent filamental arteries (on the trailing edge of the filament). Individual lamellae on each filament are perfused by afferent lamellar arterioles, and sheet blood flow through each lamella is controlled by the distribution of pillar cells, which have been demonstrated to be contractile.[164,350,471] Substantial blood flow also occurs through the outer marginal channels on the periphery of individual lamellae. Blood exiting the lamellae through efferent lamellar arterioles enters either the efferent filamental arteries, leading into the efferent branchial arteries and the systemic circulation, or into series of interlamellar and collateral vessels or sinuses that are thought to provide nutrients to the filaments. In some species (e.g., spiny dogfish and European eel), this latter pathway may also be fed by prelamellar arteriovenous anastomoses, providing a nonrespiratory shunt (termed *arteriovenous*) directly back to the branchial venous drainage to the heart.[127,375]

The filamental epithelium is composed of cuboidal and squamous pavement cells (PVCs), mucous cells, accessory cells (ACs), and numerous large MRCs. The lamellar epithelium also contains PVCs and, rarely, MRCs and mucous cells.[127,530] As in the elasmobranchs, PVCs cover approximately 90% of the filamental surface. MRCs are most dense on the trailing edge of the filament (over the afferent filamental artery), as well as in the interlamellar region.[249,513] In addition to numerous mitochondria, the marine teleost MRC is characterized by an apical crypt and an extensive intracellular network of tubules that are intertwined with the mitochondria and derived from basolateral invaginations, and not from the endoplasmic reticulum (Figure 8.13).[405] In fact, the MRC cytoplasm is so packed with tubules and mitochondria that microelectrode determinations of apical vs. basolateral electrical potentials and conductances are technically extremely difficult.[552] Individual MRCs are always associated with adjacent accessory cells, forming multicellular com-plexes that are usually in crypts below the epithelial surface containing PVCs, opening via distinct pores (Figure 8.14).[249,530] At least one study has demonstrated that crypt density (and epithelial conductance) is proportional to salinity.[84] Accessory cells contain numerous mitochondria but have a less developed tubular system.[223] It is unclear if they are a discrete cell type[410] or merely immature MRCs.[525] The extensive interdigitations between MRCs and ACs have shallow tight junctions, so they are generally considered to be the site of the relatively high ionic permeability that characterizes the marine teleost gill epithelium.[245,450] It is noteworthy that the opercular epithelium of the killifish[243] and tilapia[152] and the jaw skin of the mudsucker[330] contain relatively high densities of MRCs that are structurally and functionally identical to those in the branchial epithelium. These tissues have provided experimental approaches (see below) that are impossible with the complex, branchial epithelium.[88,153,316]

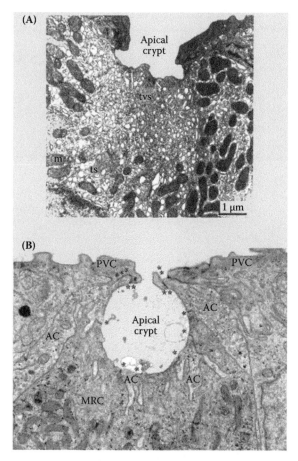

FIGURE 8.13 Transmission electron micrographs of MRCs from the gills of seawater teleosts. (A) MRC from sole (*Solea solea*) containing numerous mitochondria (m), a tubular system (ts), a subapical tubulovesicular system (tvs), and an apical crypt. (B) Apical region of a MRC from the killifish (*Fundulus heteroclitus*). This MRC forms deep tight junctions (**) with surrounding PVCs and shallow tight junctions (*) with surrounding accessory cells (ACs), which share an apical crypt with the MRC. (From Evans, D.H. et al., *Physiol. Rev.*, 85, 97–177, 2005. With permission.)

The molecular biology of ionic transport across seawater teleost MRCs has been extensively reviewed recently[79,127,215,226,319,322] and will only be summarized here. High activity of NKA in marine gill tissue was measured 40 years ago[106] and led to the suggestion that Na^+ extrusion was via apical Na/K exchange.[122,308] This hypothesis, however, was not supported by the subsequent demonstration of NKA (via ouabain-binding autoradiography) on the basolateral membrane of the MRC,[247] which has been confirmed repeatedly using immunohistochemistry for the α-subunit of NKA (α5) in marine and freshwater teleosts, as well as elasmobranchs.[208,292,390,400,531–534] Indeed, α5 staining is now used to delineate MRCs.[211,227,250] Basolateral expression of NKA is consistent with the fact that injection of ouabain into the American eel (*Anguilla rostrata*) completely inhibits NKA activity and both Na^+ and Cl^- effluxes (measured isotopically), with a much smaller effect on tritiated water efflux (indirect measurement of cardiovascular or permeability effects).[465] Consistent with concurrent studies of a variety of other tissues displaying Na^+-coupled Cl^- transport,[156] it was proposed that the marine teleost MRC expresses the suite of transport proteins (basolateral NKA, K^+ channel, and NKCC, as well as an apical CFTR) that were concurrently described for the shark rectal gland (see Section IV.A.2) and, subsequently, the teleost proximal tubule (see Section V.A.2). The proposed model (Figure 8.15) was supported by electrophysiological studies of isolated operculae

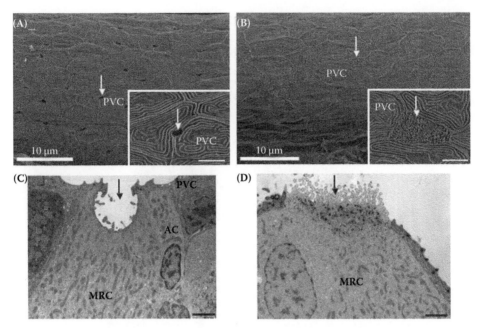

FIGURE 8.14 Scanning electron and transmission electron micrographs of gill filaments from killifish in seawater (A and C) and killifish transferred from seawater to freshwater after 30 days (B and D). Note transformation of the apical region of MRCs (arrows) from a smooth, concave crypt that is recessed below the PVCs (A and C), to a convex surface studded with microvilli that extend above the surrounding PVCs (B and D). Also, note that an accessory cell (AC) is not associated with MRCs from freshwater-acclimated killifish (C and D) and that the distinct, whorl-like microridges on the surfaces of PVCs do not change with salinity (A and C). Bar = 1 mm, except where noted. (From Evans, D.H. et al., *Physiol. Rev.*, 85, 97–177, 2005. With permission.)

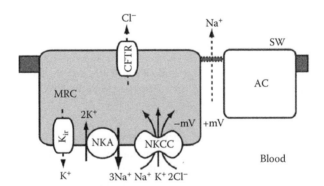

FIGURE 8.15 Working model for the extrusion of NaCl by the marine teleost gill epithelium. Plasma Na$^+$, K$^+$, and Cl$^-$ enter the cell via basolateral NKCC; Na$^+$ is recycled back to the plasma via Na$^+$,K$^+$-ATPase (NKA) and K$^+$ via a K$^+$ channel (K$_{ir}$). Cl$^-$ is extruded across the apical membrane via a Cl$^-$ channel (CFTR). The transepithelial electrical potential across the gill epithelium (plasma positive to seawater) drives Na$^+$ across the leaky tight junctions between the MRC and the AC. (From Evans, D.H. et al., *Physiol. Rev.*, 85, 97–177, 2005. With permission.)

and jaw epithelia (see above) that contain numerous MRCs (60% in killifish operculum) and generate a short-circuit current that is equivalent to the net Cl$^-$ secretion and inhibited by basolateral ouabain, furosemide, or Na$^+$ substitution.[88,151,316] Moreover, a distinct Cl$^-$ current could be measured directly over the MRC,[153] and a cDNA of CFTR was cloned from the killifish gill.[468] More recently, molecular studies have localized NKCC to the basolateral membrane and CFTR to the apical membrane of the marine MRC.[214a,329,343,390,534] Complementary DNAs for these transport proteins have been cloned and sequenced in a variety of marine teleosts, and their mRNA expression (or that of the relevant protein) is upregulated in seawater and downregulated in freshwater.[63,127,135,212,214a,215,226,289,290,292,457,509] Recent work suggests that basolateral membrane lipid content can modulate NKA function in seawater Arctic char (*Salvelinus alpinus*).[50] Interestingly, activity and expression of NKA in some euryhaline species are not always increased in seawater or may be increased in freshwater,[50a,63,127,212,296,297,322] but this may be due to the fact that different α-subunit isoforms (α1a vs. α1b) are upregulated and downregulated during salinity transfer[50a,51,433] or that the kinetics of Na^+ vs. K^+ activation are altered by salinity.[291] Moreover, because NKA is important in ionic uptake in freshwater (and acid–base regulation; see Sections IV.B.3, V.B.2, and V.C), no specific reason exists to think that NKA expression or activity would be higher in seawater vs. freshwater, and at least one study has shown that it increases in both high and low salinities relative to intermediate salinities.[278] The basolateral K^+ channel (Figure 8.15) is suggested by the fact that basolateral Ba^{2+} inhibits the short-circuit current across the killifish opercular epithelium,[87] and the cDNA of an inward-rectifying K^+ channel has now been sequenced and shown to have high expression in the basolateral tubules of the MRCs of the Japanese eel, and the expression is increased in seawater.[488]

B. Freshwater Species

In contrast to marine teleosts (but similar to freshwater lampreys), hyperregulating bony fishes must compensate for osmotic water gain and diffusional ion loss. Water absorbed across the gut with food and salts lost renally also add to the osmoregulatory requirements for these animals. Thus, water gained must be excreted via the kidneys with typical high flow rates of dilute urine (when compared to the seawater state). Ions lost to the environment must be replaced by active uptake across the gills. The regulatory sites (gut, kidneys/bladder, and gills) face the opposite challenge (conserve ions, void water) from those described for marine fishes (excrete ions, conserve water). See also several recent reviews on freshwater osmoregulation.[215,226,326,397]

Branchial water permeability in freshwater teleosts may be reduced from the seawater state,[450] and upregulation of cell-junction claudin proteins may play a role in the decreasing tight junction permeability in freshwater.[500] Osmotic permeability is also affected by the temperature of adaptation and membrane cholesterol concentration.[439] A lower gill ion permeability can also reduce salt loss down a blood-to-water concentration (and electrochemical) gradient of several orders of magnitude (Table 8.1); for example, intraspecies differences in freshwater survival have been noted between northern and southern populations (in North America) of the euryhaline killifish.[455] Southern animals exhibit a higher mortality when transferred from brackish water to freshwater, and this correlates with a fall in plasma chloride, an increase in chloride efflux, and a threefold higher gill paracellular leak (measured as extrarenal clearance of polyethylene glycol, presumably across the gills) when compared to northern populations. Northern animals are better able to convert from seawater-type MRCs to freshwater type (with tighter cell–cell junctions), and this likely contributes to the measured chloride efflux differences. The loss of base equivalents (HCO_3^-) across the gills (perhaps due to the inability to decrease gill ion permeability) has also been observed in two euryhaline species—longhorn sculpin (*Myoxocephalus octodecemspinosus*) and the toadfish (*Opsanus tau*)—when exposed to dilutions of the ambient water beyond their normal tolerance limits.[71,73] Thus, regulatory adjustments to decrease branchial ion permeability are a likely requirement for freshwater osmoregulation.[119]

1. Gut

Most freshwater teleosts ingest very little of the ambient water, except for water taken in while feeding; for example, the basal drinking rate for freshwater juvenile Atlantic salmon (*Salmo salar* L.) is only ~0.13 mL/kg/hr compared to 0.3 to 20 mL/kg/hr for marine fish,[119] but water ingestion increases by 5 times following feeding.[99] Natural foods may consist of ~75% water. When dry food is ingested, the volume of water contained in additional gastric fluid secreted (to provide moisture for digestion) by the stomach is on the same order as the volume ingested with the food.[267] In rainbow trout (*Oncorhynchus mykiss*), water secreted by the stomach during the early digestive process is reabsorbed along the intestine, and ultimately a net water uptake from the gastrointestinal tract may occur when the transfers across all regions are summed.[45] Plasma Na^+ levels spike for several hours following the meal. Bucking and Wood[44] showed that, in the first few hours following a meal, the rate of Na^+ absorption from the stomach was even higher than the reported branchial influx.[477] Little net Na^+ uptake was noted, however, when the entire gastrointestinal tract was taken into account, as secretion of Na^+ took place in the anterior intestine (likely due to the Na^+ content of hepatic bile secretion). In contrast, 85 to 90% of ingested K^+ and Cl^- was reabsorbed by the gastrointestinal tract. Also, at least in freshwater-adapted flounder, the Cl^- uptake may be coupled to HCO_3^- secretion, just as it is in seawater-adapted fish.[495] When trout are fed a dietary salt load (5% of body Na^+ content), branchial Na^+ influx rates are reduced to compensate for the large Na^+ uptake across the stomach.[477] Higher levels of salt included in the meal induced an elevation in plasma [NaCl] and increased gill efflux of Na^+ (perhaps due to adjustments to gill permeability) along with decreased branchial Na^+ uptake. Overall, the salt ions absorbed across the gastrointestinal tract following a meal is a relatively small fraction of the total ion taken up by most freshwater fishes (on the same order as the branchial Na^+ influx over a single hour), but this amount still normally compensates for ions lost in the urine (see review by Marshall and Grosell[326]). In species living in very dilute or acidic freshwater (<50 μM Na^+),[168] where gill ion uptake is even more limited, the gastrointestinal absorption of salt is clearly important.[83]

2. Kidney

As would be expected, urine flow rates (and GFR*) in freshwater species are ~5 to 10 times higher than those in marine fishes; for example, European eel (*Anguilla anguilla*) urine flow ranged from 1.1 to 3.5 mL/kg/hr in freshwater-adapted animals vs. 0.25 to 0.6 mL/kg/hr in seawater-adapted conspecifics (reviewed by Hickman and Trump[204]). Surprisingly, freshwater trout still maintain about 40% of nephrons in reserve,[26] as the glomeruli are perfused but not filtering.[40] In parallel with higher urinary flow rates, freshwater fishes also produce a dilute urine with NaCl concentrations in the range of 10 to 20 mM (for some species, <5 mM).[205] Thus, the kidney tubule must actively reabsorb 90% or more of luminal salt after filtration across the glomerulus. Reabsorption takes place across the proximal (PT), distal (DT), and collecting (CT) tubules, followed by the urinary bladder (UB). As in mammals, reabsorption of filtered salt, glucose,[43] and HCO_3^- [161] occurs in the PT. Although the renal tubules have a relatively low water permeability, especially the DT and CT,[367] some water is reabsorbed with the solutes in the PT so the urine flow rate is less than the GFR.[204] Interestingly, secretion (rather than reabsorption) of electrolytes and water occurs in approximately 10% of the proximal tubules in freshwater-adapted (8 to 180 days in freshwater) killifish,[75] perhaps to provide tubular flow even when some glomeruli are not filtering.[40] Few fish-specific data are available on the cellular or transporter mechanisms of tubular ion uptake, but presumably the majority of ions are absorbed across the DT and CT segments via transfers homologous to other vertebrates. A chloride channel (CLC-K), for example, is coexpressed with NKA on the basolateral membranes of distal tubule cells in freshwater-adapted, but not seawater-adapted, tilapia.[351] In trout, luminal furosemide (NKCC inhibition) decreases DT ion transport, and this segment exhibits very low water permeability.[368] Taken together, these data suggest that the distal tubule may have transporter mechanisms homologous to those thought to drive salt reabsorption from the mammalian thick ascending limb of the loop of Henle.[136]

The urinary bladder also reabsorbs ions before final excretion of the dilute urine to the water. In sea bass adapted to low salinities, bladder ion reabsorption begins early in development to produce a hypoosmotic urine with bladder (and kidney) transporting cells rich in NKA.[363,364] Na^+

---

* Although nearly all freshwater fishes possess glomeruli, several species of freshwater, presumably aglomerular, toadfish have been reported.[17]

and Cl⁻ uptake in salmonids is electroneutral and may or may not be coupled to the luminal concentration of the other ion, depending on the species measured.[48,154,317] The trout bladder may also take up urea via a facilitated transporter.[345] An amiloride-sensitive apical Na⁺/H⁺ exchange is present to drive Na⁺ uptake and may also serve to secrete NH_4^+ into the urine.[321]

3. Gill Salt Uptake

To extract Na⁺ and Cl⁻ from the water into the plasma, both ions must be moved against the existing electrochemical gradient. As proposed by Krogh,[269] the influx of NaCl may also be tied to the excretion of acid–base relevant ions (H⁺ and HCO_3^-). Two mechanisms have been postulated to provide apical Na⁺ uptake: the electroneutral Na⁺/H⁺ antiporter (NHE; see Hirata et al.[209]) and passive uptake of Na⁺ through sodium channels (ENaC) down the electrochemical gradient established by V-HAT-driven proton efflux (see, for example, Perry et al.[393]). It has been suggested that Cl⁻ uptake in freshwater takes place via apical Cl⁻/HCO_3^- exchangers (AE1[531] or SLC26a4/pendrin[402]), NKCC,[211] and thiazide-sensitive NaCl cotransporter (NCC).[226] Gill transport cells are generally thought to appear early in embryological development and are necessary for osmoregulation, even before secondary lamellae for respiration have formed.[443] Over the past decade, new molecular and immunological approaches have provided a broad dataset defining the potential transport mechanisms involved. The following is a summary of some of the most recent work; see also recent reviews.[127,226,326,394,504]

Na⁺ uptake from the water in exchange for intracellular H⁺ (or NH_4^+) via the Na⁺/H⁺ antiporter[37] was supported by early work,[309] but this model fell out of favor on the grounds that the passive NHE could not function against the Na⁺ (and H⁺) concentration gradients thought to exist between the gill cell and the ambient freshwater.[415] Renewed interest in this process, however, has resulted from recent data indicating that NHE may still be important in some freshwater species. Genome sequencing and EST databases indicate that fishes (pufferfish, zebrafish, stickleback) have nine or more isoforms of NHE that are homologous to mammalian paralogs. The cDNA for eight NHE isoforms in zebrafish have now been cloned.[545] Genome duplication events that have occurred during the evolution of the teleost clade[69] have also resulted in multiple copies of some NHEs (and likely other important membrane transporters) that may also play a role. NHE2 and NHE3 have been the isoforms described in the gills of freshwater species to date.[101,103,209,531] Hirata et al.[209] were the first to clone NHE3 from a freshwater species, the Osorezan dace (*Tribolodon hakonensis*), and demonstrate both immunologically and functionally a role for NHE3 in Na⁺ uptake and H⁺ excretion. They hypothesized that the apical NHE3 could function even against large apparent gradients for both Na⁺ and H⁺ excretion when dace were exposed to a water pH of 3.5. By using a subtraction cDNA library approach to isolate transcripts that were altered following acidosis, gill NHE3 mRNA was found to dramatically increase within a day after dace were transferred to low-pH water. The NHE3 was colocalized in MRCs that also contained basolateral NKA, NBC1, aquaporin (AQP3), and cytoplasmic carbonic anhydrase (CA-II). Interestingly, little change in mRNA or protein expression for H⁺-ATPase was noted. The authors proposed that Na⁺ uptake from the water via NHE3 can be driven by the gradients established by NKA and NBC on the basolateral membrane. The H⁺ and HCO_3^- are provided by the hydration of intracellular CO_2 enhanced by cellular CA.[209]

NHE3 has also been immunolocalized in freshwater-adapted trout,[101] and apical NHE2 was detected in tilapia.[531] The mRNA for gill NHE2 is expressed in both MRC and pavement cell fractions of trout gill.[361] Na⁺ uptake in adult zebrafish adapted to hardwater (~1.5-mM ambient [Na⁺]) is inhibited by approximately 50% with external amiloride,[33] and Na⁺ accumulation in skin-surface MRCs in zebrafish larvae is nearly abolished by amiloride (at 1×10^{-4} M, a concentration that effectively blocks NHE) or ethylisopropylamiloride (EIPA), a specific NHE inhibitor.[108] Amiloride analogs also inhibit Na⁺ influx in goldfish.[419] Edwards et al.[103] showed that gill NHE2 protein expression is present in killifish adapted to freshwater but is not consistently detected in seawater animals. Transcription of mRNA for NHE2 is also upregulated following transfer of the animal to freshwater.[456] These data supported earlier ion flux experiments that also suggested Na^+/H^+ exchange was occurring in freshwater-adapted killifish.[383]

Initial evidence for Na^+ uptake via the V-HAT/Na^+ channel model was provided by a variety of work in salmonids[9,264,393,486] and has been described in recent reviews.[127,397] V-type (vacuolar) H^+-ATPase[190] is thought to be the driving force for the proton excretion[293] that in turn creates an electrochemical gradient for Na^+ uptake even in freshwater with micromolar Na^+ concentrations. Suppression of Na^+ uptake with bafilomycin (a specific inhibitor of V-HAT) has been demonstrated in zebrafish,[33] tilapia, carp,[137] and trout[427] MRCs. Phenamil, an ENaC blocker in mammals,[262] added to the water decreases *in vivo* Na^+ influx in goldfish[419] and Na^+ uptake in isolated trout MRCs.[379,427] A new model for the study of ionocytes has recently been described in the skin of zebrafish embryos.[294] Basolateral NKA was detected in MRCs, and a second cell type with fewer mitochondria expressed apical V-HAT. Using a proton-sensitive ion probe passed over the skin (strongest readings were over the pericardial cavity and yolk sac), individual H^+ pumping cells could be functionally detected, and these often correlated with the V-HAT-expressing cells. The drop in pH near the apical surface of the cell could be inhibited with bafilinomycin. Interestingly, a second acid-secreting cell type was not affected by V-HAT inhibition, nor was V-HAT detected immunologically in these cells. Injection of antisense morpholinos against V-HAT mRNA induced a suppression of V-HAT expression and decreased proton efflux across the V-HAT cell type.[224] Na^+ and Ca^{2+} content of the embryos exposed to low-ionic-strength water was decreased following V-HAT inhibition. Agreeing with earlier work on zebrafish,[33] Yan et al.[545] suggested a role for *both* NHE and V-HAT in zebrafish gill cells as NHE3 mRNA is detected in V-HAT-rich cells but not NKA cells. Immunolocalization using antibodies against dace NHE3[209] and killifish V-HAT[250] showed that both transporters are expressed apically in these cells. mRNA for NHE3 is upregulated following transfer of the animal to low Na^+ water, at the same time that mRNA for V-HAT decreases. In contrast, exposure to low pH causes a decrease in NHE3 and an elevation in V-HAT. Thus, as suggested by the variation in NHE expression in freshwater- and seawater-adapted killifish,[103] zebrafish may utilize a combination of both NHE and V-HAT systems for osmoregulation and acid–base regulation, depending on the external salinity.

Less is known about the mechanisms of Cl^- uptake from the water. Gill Cl^-/HCO_3^- exchange has long been postulated to be involved.[309] Antibodies against trout erythrocyte AE1[52] have been used to immunolabel the apical membranes of gill cells (often also containing NKA signal) in freshwater-adapted tilapia.[531] The mRNA for a pendrin-type (SLC26) anion exchanger has been cloned from the trout kidney,[252] and homologs are found in several fish genome databases.[521] Pendrin has been immunolocalized in the gills of freshwater-acclimated Atlantic stingray,[402] but this experiment has not been repeated in teleosts. For both of these apical Cl^-/HCO_3^- exchangers, uptake of external Cl^- at freshwater salinities would require significant cytoplasmic HCO_3^- concentrations. Tresguerres et al.[504] have proposed that if V-HAT is located on basolateral membrane invaginations in close proximity to apical Cl^-/HCO_3^- exchangers, the generation of H^+ and HCO_3^- by cytoplasmic carbonic anhydrase in this microenvironment could power HCO_3^- transfer to the water in exchange for Cl^- uptake. Cl^- would then move across the basolateral membrane through Cl^- channels in parallel to H^+ transfers via the H^+-ATPase.

Na^+–K^+–$2Cl^-$ cotransporters (NKCC) are found in both a secretory and absorptive form in the mammalian nephron.[184] Apical NKCC could drive Cl^- uptake across the gill in a similar fashion. Gill cells from adult freshwater-adapted tilapia[543] and MRCs from yolk sac membranes[211] express apical NKCC with basolateral NKA.[214] These data suggest a role for NKCC in Cl^- uptake in this species. In contrast, basolateral NKCC has been observed immunologically, coexpressed in NKA cells[212] in three species of freshwater-adapted salmonids: lake trout (*Salvelinus namaycush*), brook trout (*Salvelinus fontinalis*), and Atlantic salmon (*Salmo salar* L.). Seawater adaptation further increased NKCC protein abundance in both of the trout species, while salmon expressed high levels in freshwater and seawater. As described above (Section V.A.3), basolateral NKCCs are thought to participate in gill salt excretion in seawater-adapted fishes, so expression in freshwater salmonids could indicate that either a basal level is maintained if the euryhaline animal encounters higher salinities or the NKCC has another function yet to be described.[212]

The thiazide-sensitive NaCl cotransporter (NCC) is another potential pathway for gill salt uptake. It is apically expressed in mammalian distal tubule cells and is responsible for Na$^+$ absorption (5 to 10% of filtered load).[413] NCC isoforms were found in fish genome databases and were strongly expressed in eel kidney and intestine when the animals were adapted to either freshwater or seawater, but little NCC mRNA signal was detected in the eel gill.[81] Other work has suggested that NCC in MRCs in freshwater tilapia gill and zebrafish embryos may be involved with Cl$^-$ uptake (described by Hwang and Lee[226]). Most recently, NCC mRNA and protein have been localized to one class of MRCs (that also express basolateral NKA) in the yolk sac and gill epithelium of tilapia, and the NCC mRNA content increases upon acclimation to freshwater.[214a] Once absorbed apically, NaCl is ultimately moved to the extracellular space as Cl$^-$ moves out through basolateral Cl$^-$ channels[319,493] in parallel with Na$^+$, shuttled by NKA or an electrogenic transfer hypothesized to be decoupled from K$^+$ uptake.[298]

4. Correlation of Gill Cell Types with Transport

The branchial cell type involved with the ion transfers has been studied since Keys and Wilmer[254] observed mitochondria-rich "chloride cells" within the epithelium that had a morphology distinct from the respiratory pavement cells (see reviews by Laurent[279] and Wilson and Laurent[530]). Recent work has provided evidence for several MRC subtypes that are likely the sites for ion transfers. Pisam et al.[410] studied the progressive development of MRCs in the gills of embryos and fry of brown trout in freshwater. They documented three distinct MRC types. β cells appeared first and were typically found in the interlamellar region in the gills beginning in prehatch embryos. Accessory cells (A cells) then developed, nearly always adjacent to the β cells in post-hatch embryos. Finally, after the yolk sac had been reabsorbed, a third type of MRCs (α cells) appeared on the lamellae of the fry. These lamellar (and sometimes interlamellar in other species[460]) cells are thought to be specifically involved in ion uptake in low salinities, because they exhibit hypertrophy when the fish is exposed to deionized water and regress in euryhaline fishes when the animals are transferred to higher salinities.[281] Laurent (pers. comm.) has suggested that the α cells are the freshwater-type chloride cells (providing Cl$^-$ uptake in freshwater), while the β cells[410] may be a precursor to the seawater-type chloride cells (driving Cl$^-$ efflux in seawater). At least two MRC types have been reported in several other species.[95,408,411,461,538]

Apical binding with peanut lectin agglutinin (PNA) has been used as a marker to discriminate between different cell types in the mammalian renal tubule. Goss and coworkers[169] utilized Percoll density-gradient separation of gill MRCs followed by PNA binding assays to demonstrate two distinct cell populations (PNA+ and PNA–) in the freshwater trout gill. Approximately 40% of the MRCs were PNA+. The PNA+ and PNA– MRC cell populations also exhibited ultrastructural differences, with the PNA– cells lacking the normal intracellular vesiculotubular network found in the PNA+ cells. Further separation of PNA subtypes and analysis of NKA and V-HAT expression with western blots showed that NKA activity per cell in control fish was approximately threefold higher in the PNA+ cells. In contrast, V-HAT expression was approximately twofold higher in PNA– cells. Hypercapnic acidosis increased the cellular detection of both transporters in the PNA– cells relative to the PNA+ cells. Interestingly, alkalosis by base infusion also increased the relative NKA activity in PNA– cells.[159]

Further work showed that Na+ uptake following acidosis in isolated PNA– cells was sensitive to phenamil.[4,427] The authors suggested that the PNA+ and PNA– subtypes (1% and 6% of the total gill homogenates, respectively) are analogous to the β-type and α-type intercalated cells of the mammalian collecting duct.[520] These are also presumably the same cells that had been previously termed *freshwater chloride cells* and *MR PVCs*, respectively.[397] PNA– cells would express the

ENaC channels, allowing Na^+ entry as intracellular protons are pumped out. Recent work by Parks et al.[379] observed phenamil sensitivity in a subfraction of MRCs (~77%) that also exhibited Na^+-induced intracellular acidification. These cells likely correlate with the $Na–HCO_3$ cells above. Their data and those of Hirata et al.[209,210] suggested that basolateral $Na^+–HCO_3^-$ cotransporter NBC1 or homologs in combination with (or in lieu of) basolateral NKA could ultimately drive the transfer of intracellular Na^+ to the blood.

Thus, the freshwater chloride cells (PNA+) and MR PVCs (PNA–) cells may correspond with the renal mammalian model for intercalated cells in the distal tubule.[518] Functionally, the PNA+ MRCs of trout may be the base-excreting/Cl^- uptake cells and the PNA– MRCs may be the acid-secreting/Na^+ uptake cells.[427] As pointed out by Kirschner,[259] some questions remain to be clarified however; for example, the phenamil Na^+ channel had been presumed to be an ENaC but it has not been cloned in fishes to date, and homologous genes for ENaC are not present in the pufferfish and zebrafish genomes. The data from Reid and coworkers[427] demonstrate that several other Na^+ entry pathways are also present, as significant Na^+ uptake in control (non-acid) PNA+ and PNA– cells was not affected by phenamil. Inhibition of NHE, NKCC, and NKA was required to expose the phenamil sensitivity of Na^+ uptake even in the PNA– cells. Inhibition of V-HAT with bafilomycin also decreased Na^+ uptake in not only PNA– but also PNA+ cell subtypes. This was attributed to the general depolarization of cell membrane potential (and decreased electrochemical driving force for Na^+ entry) observed when V-HAT is inhibited[24] but also suggests a caveat that must be considered with experiments that inhibit the rather ubiquitous V-HAT. Moreover, recent data demonstrate that PNA+ cells in the trout gill epithelium express NHE2/NHE3 protein and that mRNA for NHE2 is upregulated under hypercapnic acidosis.[232a] It also remains to be seen how this model fits the observation that Na^+ uptake in trout does not change following hypercapnia when measured *in vivo*.[170]

Although freshwater trout have served as a customary model to study the cellular location of ion uptake across the gills, findings from other species such as the dace,[209] killifish,[250] tilapia,[283,543] zebrafish,[108,224] and other euryhaline species[296] have suggested additional variations on the theme.[226] Katoh et al.,[250] for example, found that gill V-HAT in freshwater-adapted killifish appears in well-defined, large MRCs and is colocalized with NKA. In contrast to the trout, the V-HAT is expressed on the basolateral membrane of the MRCs. It was proposed that these were Na^+ uptake MRCs and that basolateral excretion of H^+ via V-HAT, along with Na^+ by NKA, would hyperpolarize the cell and provide the electrochemical gradient for Na^+ apical entry via Na^+ channels (although no data exist for these in killifish).

The freshwater-adapted killifish, like the European eel,[179] is unusual in that it appears to take up only Na^+ when measured *in vivo*.[382,539] Cl^- uptake is near zero, and Cl^- necessary to compensate for diffusive loss is provide by the diet and opercular uptake.[280,319] In light of the lack of gill Cl^- uptake, Laurent et al.[280] suggested that the examination of gill cell types in killifish provides a method to better define the cells responsible for Na^+ uptake, as freshwater-type chloride cells (PNA+ MRCs) driving Cl^- uptake presumably are either not present or not active in this species. On transfer from 10% seawater to freshwater, killifish decreased the number and apical exposure of the seawater-type chloride cell (see Section V.A.3) within 3 hours by necrosis/apoptosis and pavement cell covering of apical pits. Freshwater-type chloride cells did not appear even after one week of freshwater adaptation. In contrast, cuboidal-shaped cells, often with mitochondria near the apical aspect, appeared in the filament epithelium among adjacent pavement cells. The cells made up ~10% of the total PVC population within 3 hours of freshwater transfer. This increase was correlated with an enhancement of epithelia cell mitosis noted in the first day following freshwater transfer.[280] Thus, the cuboidal cells noted in killifish may be the same cell type as trout PNA–, MR PVCs, or α-MRCs noted by other authors. Interestingly, whole-animal inhibitor,[383] molecular,[456] and immunological[103] studies all indicate that apical Na^+/H^+ exchangers may be the mechanism for Na^+ entry across the cuboidal cells in killifish. Reevaluation of earlier data may indicate that cuboidal cells are also active in the bullhead catfish (*Ictalurus nebulosus*) and freshwater-adapted European eel.[280]

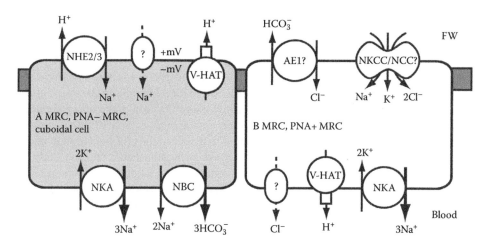

FIGURE 8.16 Working model for the uptake of NaCl by the freshwater teleost gill epithelium. One type of MRC (A MRC; PNA– MRC, cuboidal cell) expresses Na^+,K^+-ATPase (NKA) and a Na–HCO_3 cotransporter (NBC) on its basolateral membrane and draws in Na^+ across the apical surface either in exchange for cytoplasmic H^+ (NHE2/3) or through a Na^+ channel electrically coupled to an apical V-type H^+-ATPase (V-HAT). Whether these alternative, apical pathways are in the same cell or two cells remains to be determined and may be species specific. The other MRC (B MRC; PNA+ MRC) expresses NKA, a Cl^- channel, and V-HAT on its basolateral membrane and draws Cl^- into the cell via an apical Cl/HCO_3 exchanger (AE1?) or an apical Na^+–K^+–Cl^- or Na^+–Cl^- cotransporter (NKCC/NCC?), neither of which has been definitively identified. See text for details and supporting evidence. (Adapted from Evans, D.H. et al., *Physiol. Rev.*, 85, 97–177, 2005.)

It is clear that the cellular site and specific mechanisms for Na^+ and Cl^- uptake by freshwater teleosts is still unsettled. Figure 8.16 presents a working model for this system. It is yet to be determined if two or more cells are involved and if differences may be species specific.

C. What Limits Teleost Euryhalinity?

Early studies have demonstrated that Na^+ uptake by the euryhaline sailfin molly (*Poecilia latipinna*) was saturable in both freshwater and seawater, suggesting that the uptake was not merely by diffusion.[110] In addition, the Na^+ uptake by four species of marine teleosts was inhibited by addition of NH_4^+ to the external medium but stimulated by injection of ammonium. Moreover, NH_4^+ efflux was partially dependent on external Na^+.[112] These studies prompted the suggestion that even marine teleosts might extract NaCl from the external medium, associated with acid–base equivalent or nitrogen excretion.[111] Subsequent studies have demonstrated similar pathways in the gills of stenohaline hagfishes (see Section II.B) and euryhaline and stenohaline elasmobranchs (see Sections IV.A.3 and IV.B.3), as well as the importance of the gill epithelium in acid–base regulation and nitrogen excretion in both marine and freshwater teleosts.[72,194–196,526] More recent, molecular studies have demonstrated the presence of the relevant proteins (i.e., V-HAT, NHE) in the marine teleost gill epithelium,[58,72,103] so it seems relatively clear that Na^+ (and probably Cl^-) uptake mechanisms are resident in the gill epithelium of even stenohaline, marine teleosts. What, then, limits euryhalinity for most marine teleosts? The question is still largely unanswered but probably relates to the control of branchial ionic and osmotic permeability relative to the efficacy of ionic uptake by the branchial epithelium and ionic reabsorption in the renal tubules.[119] How important the presence of a distal renal tubule is for euryhalinity is still not known, because, although it is common in freshwater species (e.g., goldfish, bluegill, swordtail, bullhead) and some euryhaline marine species (e.g., eel, salmon, southern flounder), other euryhaline species (e.g., killifish, three-spine stickleback) do not have a distal tubule.[204]

D. Nonteleost Bony Fishes

The literature on osmoregulation in nonteleost bony fishes (e.g., coelacanth, lungfish, sturgeon, and gar or bowfin) is extremely limited and has not been reviewed for nearly 30 years.[114] The marine coelacanth is the only bony fish that, like the chondrichthyan fishes, stores urea and TMAO to raise plasma osmotic pressure, but it appears that the plasma is still hypotonic to the external medium (Table 8.1). In addition, the coelacanth has a postanal gland[349] that is structurally similar to the elasmobranch rectal gland, including the presence of what appear to be MRCs that express high activity of NKA.[173,285] This suggests a function similar to that described for the elasmobranch rectal gland (see Section IV.A.2), but no functional studies have been published. Only one study of renal structure in the coelacanth has been published,[275] and the complex structure that characterizes the kidney of elasmobranches is not seen[485] (see Section IV.A.1). No studies of the cellular structure of the coelacanth gill have been published. Much is left to be learned about this evolutionarily important fish, but its rarity limits what we might expect in the future.

The other sarcopterygian fishes, the freshwater lungfishes, have been studied in much more detail, but more research is needed. Their blood ionic concentrations appear to be below those of freshwater teleosts (Table 8.1). Recent measurement of radioisotopic and tritiated water fluxes across an African lungfish (*Protopterus dolloi*) indicate a relatively reduced whole-body ionic and diffusional water permeability (gills plus skin),[527] which may be correlated with the reduced gill surface area of the obligate air-breathing African lungfishes.[395] Such a reduced ion and water permeability would be adaptive both in freshwater and during estivation (see below). Surprisingly, some evidence[276] exists for a cloacal excretory gland in *Protopterus* that shares some microscopic similarities with the rectal gland of sharks and the coelacanth, and the authors propose that it may be used for salt excretion during estivation. This interesting gland needs to be reexamined, using more modern immunohistochemical techniques. The kidney of *Protopterus* contains the usual freshwater fish glomerulus, neck segment, proximal tubule, intermediate segment, distal tubule, collecting duct, and collecting tubule.[372] The proximal tubule can be subdivided into PTI and PTII and the distal tubule into DTI and DTII, as described for these tubules in other fish species.[200,201] The PT, DT, collecting duct, and tubule all display cells and nuclei that are considerably larger (two to three times) than those described for these tubules in other fish species. The collecting duct expresses principal cells[372] similar to those found in the CT in other fishes and the vertebrates in general (see Section V.B.1), which are probably the site of salt (and possibly water) reabsorption. Unfortunately, no histochemical or molecular studies have been published on the lungfish kidney. Both the gills (and opercular epithelium) and skin of *Protopterus* contain distinctive MRCs,[485] as defined by staining with an antibody against an epitope on the inner mitochondrial membrane, staining with DASPEI (a vital stain for mitochondria),[330] and other structural features. Two MRC types, equivalent to the α- and β-type of some freshwater teleosts,[407,412] are found in the gill, but only the α-type is found in the skin.[485] On the gill, both types of MRCs are found in the interlamellar space on the filaments, as well as on the lamellae themselves. Both the gill and skin MRCs stain using an antibody against Ca-ATPase, a characteristic of MRCs in teleosts (see Section VII). The occurrence of MRCs in the skin and gill of *Protopterus* suggests the presence of the same NaCl uptake mechanisms that have been described for the freshwater teleost MRCs (see Section V.B.2), and cDNAs for both V-HAT and NBC1 recently have been cloned from *Protopterus* gill tissue,[165] but immunohistochemical localization has not been published. The presence of α-type MRCs in the skin and gill is especially intriguing, as this type has been proposed to be a precursor of the MRC that characterizes the seawater teleost gill epithelium,[406] and lungfishes are never found in seawater. Localization of putative seawater transport proteins (e.g., NKCC and CFTR) in these cells would be especially interesting.

The African lungfishes estivate in mud cocoons during the dry season, and this is usually associated with increased plasma Na⁺, Cl⁻, and urea levels,[89,239,475] presumably secondary to evaporative water loss and the need to detoxify ammonia. A more recent study, however, suggests that moist substrate may reduce osmoregulatory problems during the dry season. During up to 20 weeks of laboratory estivation of *Protopterus dolloi* (mucous cocoon covering the dorsolateral body surface, ventral surface in a film of water = 0.4 mm depth, gills out of water), body mass, blood osmolality, and NaCl concentrations changed little, but plasma and muscle urea increased approximately sevenfold.[527] The increase in plasma urea was less than that described in other species of estivating *Protopterus*,[89,231] which may have been secondary to the carrier-mediated urea excretion across the skin of this species (under moist conditions) that was described in an earlier study.[542] The authors were unable to detect TMAO in the plasma under aquatic or estivation conditions, and they suggest that the lack of this counterbalancing solute might be associated with inhibition of metabolic enzymes by urea, which might be important in the metabolic downregulation associated with estivation.[527] In addition, relatively high concentrations of urea in the muscles would raise intracellular osmotic concentration, which could potentially move water into the muscle cells to act as a water reservoir during prolonged estivation.[527] During estivation, the tritiated water efflux fell, but ionic fluxes (both efflux and influx) remained the same. In both cases, one must propose that the fluxes were across the ventral skin in contact with the small volume of water under the fish, because the gills were not in contact with the water film. The authors propose that the ionic uptake was via the skin MRCs that had been described earlier in *Protopterus annectens*.[485] Thus, it appears that lungfish may have ventral skin similar to the pelvic patch in amphibians that has been shown to be important in ion and water uptake.[516] Of special interest is the finding that ammonia excretion continues in *P. dolloi* under these moist estivation conditions,[302,542] suggesting that urea retention has been selected for water conservation, rather than ammonia detoxification, in estivating lungfish.[527]

Sturgeons are primitive actinopterygians that are either freshwater or anadromous, but their protected status in some areas and natural history (migratory habits, etc.) have limited research. Their plasma concentrations are in the same range as teleosts in seawater or freshwater (Table 8.1), and some species (e.g., *Acipenser naccarii*) are able to maintain relatively consistent plasma osmolarity after acclimation to seawater.[337,338] Early isotopic flux studies suggested that the permeability of the sturgeon to ions and water was basically the same as that found in marine and freshwater teleosts.[114] After transfer to seawater, the shortnose and Atlantic sturgeons (*A. brevirostrum* and *A. oxyrhynchus*) both maintain their plasma ions (and total osmolarity) below seawater levels by drinking the external medium and removing Na^+ and water and concentrating Ca^{2+} and Mg^{2+}.[266] A recent study has demonstrated that bicarbonate secretion (probably via AE) takes place in the intestine of *A. baerii* acclimated to 50% seawater for 2 weeks.[494] As described for teleosts (see Section V.A.1), the secreted bicarbonate presumably precipitates Ca^{2+}, thereby aiding in intestinal water uptake and Ca^{2+} balance.[494] Also similar to teleosts, the sturgeon urine Mg^{2+} concentration increases to above plasma levels (~50 mM) and, surprisingly, the urine Na^+ concentrations appear to reach 450 mM.[266] This interesting finding must be confirmed, because it might have been the result of experimental contamination of collected urine with seawater. During acclimation to seawater, the glomeruli of both *A. naccarii* and *A. brevirostrum* are reduced in size by 20 to 35%, consistent with the expected reduction in GFR,[57,266] although another anatomical study suggests that the GFR may be relatively low in freshwater *A. naccarii*.[371] The gill epithelium of *A. brevirostrum* and *A. naccarii* contain distinct MRCs, with the usual subcellular morphology (e.g., many mitochondria, basolateral invaginations).[57,266,338] Acclimation of *A. naccarii* to seawater is associated with reduction in number of lamellar MRCs[57] and the formation of an apical crypt and subapical vesicles,[53] which are characteristic of the α-type MRCs.[412] To date, nothing has been published on the expression of putative transport proteins in sturgeon MRCs, except that NKA activity increases when *A. naccarii* is acclimated to seawater[338] or juvenile *A. baerii* are acclimated to approximately 40% seawater.[442] Much is to be learned about the basic patterns and molecular biology of osmoregulation by the commercially important and globally distributed sturgeons.

The other major group of nonteleostean actinopterygians, the gar and bowfin, are also largely unstudied. They are freshwater fishes (but some gar may be moderately euryhaline)[183] that maintain the usual pattern of plasma electrolytes (Table 8.1). The kidney of the bowfin contains a glomerulus, neck segment, first and second proximal and first and second distal tubules, and collecting segments.[548] The GFR and UFR of the bowfin are 8.2 and 5.3 mL/kg/hr, respectively,[49] both in the range described for teleosts in freshwater (see Section V.B.1). Because the free water clearance (4.9 mL/kg/hr) and GFR show a linear relationship in this species, variations in urine flow are apparently due to variations in GFR rather than variations in tubular reabsorption.[49] Urine Na^+ and Cl^- concentrations and osmolarity are approximately 7%, 4%, and 11% of plasma levels, respectively, so significant tubular ion reabsorption takes place,[49] as would be expected in this hyperregulating species. The gill filamental and lamellar structure and underlying vascular network of the bowfin are basically like those of teleosts, with the exception that an interfilamental support bar is fused to the outer margins of lamellae of adjacent filaments and the presence of subepithelial sinusoids in the filaments. In addition, the postbranchial circulation of arches III and IV perfuses the air bladder.[373] Early measurement of the rate of radioisotopic fluxes of Na^+, Cl^-, and water suggest that the Florida spotted gar (*Lepisosteus platyrhincus*) maintains the same low ionic and diffusional water permeability (skin + gills)[554] that has been described in freshwater teleosts (see Section V.B).[114] The urine flow was found to be 6.6 mL/kg/hr, with urine Na^+ and Cl^- concentrations of 30 mM and 24 mM, respectively,[554] significantly above levels described for the bowfin (9.6 mM and 4.9 mM)[49] and freshwater teleosts (see Section V.B.1).[204] The causes for this discrepancy are unknown and should be confirmed. Clearly, both bowfin and gar need to be reexamined with more modern molecular and physiological techniques to determine if these nonteleostean bony fishes can extract needed salts from the environment via gill epithelial pathways similar to those described for freshwater teleosts (see Section V.B.2).

VI. ENERGETICS OF OSMOTIC AND IONIC REGULATION

The energetic cost of osmoregulation (and ionic regulation) is of some interest, especially when considering euryhalinity or the evolution of hyporegulation (marine teleosts) vs. ureotelism (elasmobranchs). For example, does osmotic or ionic regulation require sufficient adenosine triphosphate (ATP) production to limit migration into other salinities (historically or currently) or can the energetic costs be offset by increases in metabolic rates or reduction in other functions, such as reproduction? Various techniques have attempted to measure the true metabolic cost of osmoregulation, with varying degrees of success. Respirometry of fishes in various salinities (measuring changes in oxygen uptake relative to those measured in isotonic salinities) generates estimates of costs from near zero in *Ambassis interrupta* (Asiatic glassfish)[369] and killifish[256] to 20 to 50% of the routine metabolic rate for osmoregulation in freshwater species, such as trout,[425] *Tilapia nilotica*,[130] and catfish (*Ictalurus* sp.).[158] Other studies, however, have actually found a decline in routine metabolic rate in salinities other than those that are isotonic to the blood for the milkfish (*Chanos chanos*)[489] and *Tilapia mossambica*.[238] In a tilapia hybrid, Febry and Lutz[131] found that osmoregulation cost approximately 16% of the total metabolic rate in freshwater (corrected for swimming activity) and 12% in seawater.

Using rates of ionic fluxes, ionic permeabilities, and transepithelial electrical potentials, one can calculate the putative energetic cost of osmoregulation.[416] Such calculations suggest that the energetic cost of either freshwater or seawater osmoregulation is more of the order of 1.0% and 0.5%, respectively.[97,98] Comparing the routine metabolic rate with the calculated cost of ATP use by gill, intestinal, and renal ionic transport (i.e., the metabolic method) suggests that 2.5% of the energy can be ascribed to osmoregulation in the trout in freshwater[261] and 7.5% in seawater.[260] Similar calculations indicate that the metabolic cost of osmoregulation by the killifish is <1% in freshwater and 9.8% in seawater.[257] Using another approach, Morgan and Iwama[352] measured the oxygen consumption of isolated, perfused trout gills before and after addition of transport inhibitors—ouabain for NKA and bafilomycin for V-HAT. They calculated that transport-specific, gill metabolism was 3.9% in freshwater and 2.4% in seawater.

The single study (using the metabolic method) that compared the calculated energetic costs of marine teleostean osmoregulation (hyporegulation) vs. elasmobranch osmoregulation (ureotelic regulation) determined that the costs are approximately the same: 7 to 10%.[260] This suggests that neither alternative offers a significant metabolic advantage over the other mode, but this interesting question should be studied further.

Despite a relatively large database, it is clear that a definitive statement about the metabolic costs of osmoregulation in freshwater vs. seawater fishes is not possible. But, the most recent data suggest that the cost is less than 10% of the total metabolic rate of the fish; whether this cost is physiologically or evolutionarily relevant remains to be determined.

VII. OTHER IONS

Regulation of Na^+ and Cl^- content of plasma is the most significant and best-studied component of osmoregulation in fishes, but plasma levels of other ions (e.g., K^+, Ca^{2+}, Mg^{2+}, and SO_4^{2-}) also must be controlled in either seawater or freshwater. In marine fishes, or euryhaline fishes in seawater, the plasma concentration of all of these ions is far below their concentrations in seawater (Table 8.1),[114] although more modern measurements are needed. Because the Nernst equilibrium potential for K^+ is approximately +17 mV (plasma positive to medium), which is near that measured across various marine teleost species,[115,326] it appears that K^+ may be in equilibrium across the gill. Presumably, ingested K^+ is absorbed in the intestine via NKCC (see Section V.A.1) and is excreted renally, although the roles of these pathways have not been quantified to date. Basolateral uptake of K^+ via NKCC in the gill epithelial MRCs presumably does not provide an excretory pathway for K^+ because of basolateral recycling of K^+ via an inward-rectifying K^+ channel (see Section V.A.3).[488] Branchial permeability to divalent ions in seawater is presumably nearly zero (but not studied), and what little is known about the pathways for processing of ingested divalents (secondary to oral ingestion of seawater for osmoregulation) has been described earlier (see Sections V.A.1 and V.A.2).

In freshwater fishes, K^+, Ca^{2+}, Mg^{2+}, and SO_4^{2-} are definitely maintained at higher concentrations than in the external medium (Table 8.1). Presumably, ingested food provides ionic intake, while low branchial permeability and high renal reabsorption lower ionic loss. Choe[67] has recently cloned a putative H^+,K^+-ATPase from the stomach of the euryhaline stingray (*Dasyatis sabina*) and demonstrated expression of the protein in NKA-rich cells in the branchial epithelium. This suggests that some K^+ uptake may be coupled to proton excretion. Interestingly, HKA expression was not upregulated when the stingray was made hypercapnic but it was after acclimation to freshwater, suggesting that the ionic exchange may play a role in K^+ balance in low salinities, rather than acid–base regulation.

If divalent ion intake from food is less than branchial/renal loss, then branchial active uptake must balance the net loss; for example, branchial uptake of Mg^{2+} may account for 30% of uptake in freshwater tilapia fed a low Mg^{2+} diet.[32] Unfortunately, studies on the mechanisms of Mg^{2+} uptake have not been published. On the other hand, putative branchial Ca^{2+} uptake mechanisms have been relatively well studied.[127,144–146,319,322] Teleostean bone is acellular,[287] so external Ca^{2+} must be the source for bone growth and remodeling. Because gill cell Ca^{2+} concentrations are presumably below those in even soft freshwater (<1 μM), entry of Ca^{2+} from freshwater is presumably via an apical Ca^{2+} channel (ECaC). A fish ortholog of ECaC has now been cloned from cDNA from the gill of the pufferfish (*Fugu rubripes*)[421] and the rainbow trout,[458] and the expressed pufferfish protein can mediate Ca^{2+} entry into Madin–Darby canine kidney cells. In the trout, localization of ECaC mRNA via *in situ* hybridization and protein via immunohistochemistry indicates that the channel is expressed apically in both MRCs and PVCs in the gill epithelium,[458] suggesting that both cells may be the site of Ca^{2+} uptake. This finding is in contrast to earlier studies that found that Ca^{2+} uptake was proportional to the density of MRCs in tilapia,[342] rainbow trout,[325] and killifish.[323]

Presumed electrochemical gradients suggest that the basolateral membrane of the MRCs must transport Ca^{2+} up its electrochemical gradient, and Ca^{2+} transport across membrane vesicles and the killifish opercular membrane is Na^+ dependent.[143,515] A putative Na/Ca exchanger (NCX) has now been cloned from tilapia heart (AY283779), and a partial sequence of a high-affinity Ca^{2+}-activated ATPase (PMCA) also has been published for the same species (AF236669), but gill cellular localization of either transporter has not been published.

VIII. CONTROL MECHANISMS

The neuroendocrine control of fish osmoregulation (generally elasmobranchs and teleosts) has been well reviewed in the past decade,[11,12,121,127,192,230,320,340,341,420,446,490,491,501,510,549,550] and space does not permit another review of the substantial literature in this chapter. In recent years, however, it has become increasingly clear that external salinity itself can control a suite of gill intracellular molecules that can mediate rapid responses important both for protection of the gill cells and osmoregulation. The Ca^{2+} receptor protein (CaR) that was first described in mammalian parathyroid and kidney[39] has now been cloned from the dogfish shark kidney and localized (via RT-PCR) in the kidney, rectal gland, stomach, intestine, and gill (as well as the olfactory epithelia and brain).[362] The authors also found similar tissue distributions in two teleosts (winter flounder and salmon), including localization in the MRCs of the gill and in the urinary bladder epithelium. The Ca^{2+} sensitivity of shark CaR (expressed in human embryonic kidney cells) was responsive to external Na^+ concentrations, so the authors propose that this receptor may mediate information about salinity changes that may be sensed by the olfactory, intestinal, gill, or renal epithelial cells.[362] In fact, feeding freshwater rainbow trout a salt-enriched diet prompts gill remodeling to the seawater type and upregulation of gill NKA, NKCC1, and CFTR, even if the external salinity is not changed.[396] These data provide a mechanism for the earlier finding that feeding a salt-enriched diet improves the survival of three species of salmonids when transferred from freshwater to seawater.[391,447,553]

In addition, it has become evident that the gill epithelium itself can respond rapidly to changes in external salinity, thereby decreasing potentially damaging volume changes. The short-circuit current (Isc) across the opercular epithelium of the killifish is doubled within 10 minutes after a 100-mOsm/L increase in the osmolarity in the basolateral (but not apical) Ringer's solution[551] and decreased by 50% by a fall in osmolarity of 50 mOsm/L, which is equivalent to the measured fall in plasma Na^+ 6 hours after transfer of this species from seawater to freshwater.[324] In both experiments, the tissue was mounted *in vitro* (in Ussing chambers), so the responses were cellular, not via neuroendocrine signaling. This rapid response may partially account for the time course of the change in radioactive Na^+ efflux that was measured 30 years before in intact killifish after similar transfers.[417] The stimulation of Isc by increased basolateral osmolarity could be inhibited by various Cl^- and K^+ channel blockers, suggesting that the key step is activation of the basolateral NKCC transporter.[219] The stimulation was also strongly inhibited by a protein kinase (PKC) inhibitor, as well as a myosin light-chain kinase (MLCK) inhibitor, indicating that both of these kinases may be involved, but inhibition of protein kinase A (PKA) had no effect.[219] Importantly, an inhibitor of serine/threonine phosphatases of the PP-1 and PP-2A type stimulated the steady-state Isc by the opercular epithelium.[219] These authors also described a serine-phosphorylated, 190-kDa protein that was upregulated in gill cells by seawater acclimation and suggested that this protein was involved in the phosphorylation/dephosphorylation processes involved in seawater acclimation.[219] Notably, the protein tyrosine kinase (PTK) inhibitor genistein inhibited control Isc, but the effect was not additive with the hypotonic response, suggesting that tyrosine phosphorylation of some intracellular protein is also involved in stimulation of Isc across the opercular membrane, and its inhibition is a component of the hypotonic response.[324] The Hoffman and Marshall group has recently proposed a model (Figure 8.17) for rapid control of NKCC in MRCs, based on the effect of a series of relatively specific inhibitors as well as western blots of expressed proteins after salinity stress. Their model includes stimulation by phosphorylation of NKCC by a suite of protein kinases

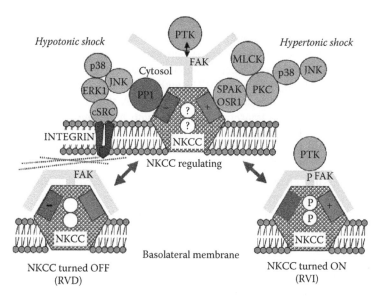

FIGURE 8.17 A hypothetical model of the regulation of NKCC in the basolateral membrane of chloride-secreting cells in isolated killifish opercular epithelia. Extracellular side is below; cytosolic components above. (Right) The kinases that have been demonstrated to be near the basolateral membrane and that are activated by hypertonic stimuli and ultimate NKCC activation associated with regulatory volume increase (RVI): JNK, MLCK, and P38 MAPK. Also included are kinases that are colocalized with NKCC (FAK, OSR1, and SPAK). Because SPAK and OSR1 coimmunoproecipitate with NKCC in other systems, they are placed nearest to NKCC. The order of the others in the cascade is unknown. (Left) The inhibitory cascade terminating with PP1, a phosphatase that in other systems has been shown to coimmunoprecipate with NKCC. On this side are some of the known players (JNK and p38 MAPK), but also included is the involvement of integrin, cSRC, and ERK, which in other systems are associated with detection and mediation of hypotonic stress. The model proposes the following conditions: an unstable regulating phase that can result in NKCC becoming phosphorylated (i.e., a pseudo-stable phase of stimulation during RVI) or becoming dephosphorylated and entering a pseudo-stable inactive state during RVD. The reason FAK is shown occluding phosphorylation sites is because of the unusual effect of the PTK genistein on the system, causing a decrease in Isc when it is high as well as increasing Isc when it is low. The PTK depicted separately from FAK could be FAK itself. (From Hoffman, E.K. et al., *Comp. Biochem. Physiol. A*, 148, 29–43, 2007. With permission.)

such as cAMP/PKA, PKC, MLCK, a p38 mitogen-activated protein kinase (p38 MAPK), a stress protein kinase (SPAK), and an oxidation stress response kinase (OSR1). NKCC is deactivated by hypotonic swelling, by Ca^{2+}, and by an unidentified protein phosphatase, and NKCC is also controlled by PTK acting on a focal adhesion kinase (pFAK), which colocalizes with NKCC on the basolateral membrane of the MRC. Activation of apical CFTR by intracellular cAMP and PKA also may be important,[218,331] and recent evidence suggests that transfer of the killifish to seawater stimulates rapid (15 to 30 minutes) expression of SGK1 and insertion of CFTR into the apical membrane.[458a] The hypothesis that MAPK plays a role in NKCC activation in high salinities corroborated an earlier study that demonstrated increased expression of a MAPK protein (SAPK2) in gills from killifish acclimated to seawater.[270] In addition, the cDNA for a 14–3–3 protein has been localized to the gill tissue of the killifish and shown to be upregulated 2- to 4-fold 24 hours after transfer from seawater to freshwater.[271] These proteins are nearly ubiquitous regulators of cellular function via binding and regulation of protein kinases, phosphatases, and other phophoproteins.[157] Rapid responses also can be demonstrated at the level of transcription factors. Two hours after tilapia are transferred from freshwater to seawater, the mRNA of two transcription factors (OSTF1 and TFIIB) is increased 4- to 6-fold. Four hours after transfer, the protein levels have increased 7.5 to 9-fold.[141] It is of interest that the expression of the homolog of OSTF1 (GILZ) is induced in mammalian cortical collecting duct cells by corticosteroids,[141] because clear evidence is now available that cortisol is important in seawater acclimation in teleosts,[121] and glucocorticoid receptors are upregulated in the gill of tilapia after seawater transfer.[86] A subsequent study, however, demonstrated that the glucocorticoid receptor agonist dexamethazone did not stimulate OSTF1 levels in primary cultures of tilapia gill cells.[140] The authors then demonstrated that the upregulation of OSTF1 by hypertonicity was mediated by transient mRNA stabilization, which is a regulatory mechanism common in inducible transcription factors with high rates of mRNA turnover[10] and involves a reduction in the degradation of the mRNA.[140] See Chapter 2 for a more complete study of intracellular proteins involved in cellular osmoregulation, and see a recent review of osmosensing by Fiol and Kültz.[142]

Most recently, the Kültz group has used suppression subtractive hybridization (SSH) to identify genes that are activated early during acclimation of tilapia to seawater and have found that more than 50% of the identified immediate hyperosmotic stress genes interact within a signaling network that consists of six molecular processes: stress response signal transduction, compatible organic osmolyte accumulation, energy metabolism, lipid transport and cell membrane protection, actin-based cytoskeleton dynamics, and protein and mRNA stability.[139] A similar, transcriptomic approach (using SSH) was used to examine the effect of salinity acclimation on gill and intestinal transcripts in the euryhaline sea bass (*Dicentrarchus labrax*).[34] In this study, specific genes for proteins involved in osmoregulation, cell-cycle regulation, cytoskeleton, energy metabolism, protein regulation, cell communication, detoxification, and nucleic acid regulation were upregulated in seawater in the gill, and genes for cytoskeleton, energy metabolism, protein regulation, lipid metabolism, cell communication, and detoxification were upregulated in the intestine after seawater acclimation. Other genes in the same categories were upregulated after freshwater acclimation. Importantly, however, in each case >70% of sequences for the transcripts that were upregulated were not identifiable from current sequence databases. Despite the current limitations, it is clear that such transcriptomic, proteomic, and systems biology approaches will provide new insights into the signaling and intracellular processing attending salinity changes in fishes.

IX. SUMMARY AND QUESTIONS REMAINING

It is clear that much has been learned about fish osmotic and ionic regulation in the 44 years since Potts and Parry's *Osmotic and Ionic Regulation in Animals* was published, especially with the recent advent of molecular techniques for the identification, localization, and quantification of relevant mRNAs and proteins. The emerging fish genome projects and mRNA knock-down techniques (morpholinos, siRNAs) offer pathways to answer old questions and open up new lines of research. What follows is a short list of what we feel are the most important knowns and unknowns in fish osmoregulation:

1. It appears that descendents of marine hagfishes entered freshwater and gave rise to subsequent vertebrate evolution, but the fossil record, especially for the largely marine elasmobranchs, is unclear and incomplete. Can the fossil record ever give us a clear picture of early vertebrate evolution?

2. Hagfishes are isosmotic but not isoionic to seawater. Are these slight ionic gradients due to transepithelial electrical potentials or secondary to other processes, such as acid–base regulation? Can renal excretion account for the reduced divalent ion concentrations in the hagfish plasma, or is the slime really a means of excretion?

3. It is not clear if there is any net filtration pressure across the hagfish glomerulus. What are the relative roles of filtration vs. secretion in the hagfish kidney? Is it true that the hagfish cannot increase GFR secondary to increased vascular pressures? Is the hagfish archinephric duct really unable to absorb ions?

4. It appears that the hagfish gill epithelium possesses mechanisms for Na$^+$ uptake, secondary to the necessary secretion of acid or ammonia. Is the expected anionic exchanger (Cl/HCO$_3$) also present? What is the relationship between the kinetics of these putative ion uptake transporters and the diffuisonal loss of ions in lowered salinities? Does this limit hagfish euryhalinity, or is that secondary to limited renal responses?

5. Cells in the hagfish gill appear to express CFTR and NKA, although in different cells. Does the hagfish gill express NKCC and in what cell? If so, why can't the hagfish acclimate to hypersaline solutions? Is it a renal limitation?

6. The lamprey distal tubule can dilute the urine in freshwater. What are the cellular mechanisms of salt reabsorption? Can the lamprey proximal tubule secrete salts to form urine, as has been described for both teleosts and elasmobranchs?

7. The freshwater lamprey gill epithelium apparently possesses molecular mechanisms for both Na$^+$ and Cl$^-$ uptake. What is the specific cellular localization of these transporters, and what are their molecular pathways?

8. Our knowledge about osmoregulation by marine lampreys is especially limited, largely because of the difficulty in capturing specimens in seawater, before their entry into brackish water. Does the marine lamprey kidney secrete divalent ions? Is the urine actually hyperosmotic to the plasma? How is water conserved by the kidney tubules?

9. Lampreys appear to have higher drinking rates than marine teleosts. Is this due to a relatively higher, branchial water permeability? What are the mechanisms of ionic uptake by the lamprey intestine. Is the pattern of monovalent vs. divalent uptake vs. excretion similar to that described for teleosts?

10. The marine lamprey gill possesses cells very similar to the MRCs of the teleost gill. Does the epithelium use the same molecular mechanisms for salt extrusion as have been described for the teleost gill (e.g., NKA, NKCC, CFTR)?

11. The elasmobranch gill appears to retain the relatively high water permeability and low ionic permeability found in hagfishes and lampreys. For all three groups, is this actually a function of the structure of their gill epithelia or secondary to gill perfusion patterns? What is the structural basis for the extremely low branchial urea permeability?

12. Why is the osmoregulatory strategy of marine elasmobranchs so different from that of marine teleosts? Is one strategy more energetically efficient?

13. The elasmobranch kidney is extremely complex. What are the sites and mechanisms for urea, NaCl, and water reabsorption?

14. The spiny dogfish rectal gland appears to respond more to a volume load than a salt load. Is volume rather than plasma salt concentration the general stimulant for rectal gland function? How does water enter the rectal gland tubule subsequent to salt secretion?

15. Numerous studies have demonstrated that removal of the rectal gland does not significantly impair osmoregulation in seawater. Under these circumstances, is the excess salt secreted by the urine or is diffusional salt uptake decreased by changes in gill permeability?

16. Evidence suggests that the MRCs in the marine elasmobranch gill epithelium express salt uptake mechanisms that are involved in acid–base regulation rather than salt secretory pathways. Is this actually the case? Are these pathways upregulated in euryhaline species in freshwater?

17. Despite relatively high plasma osmolarities, some elasmobranchs are euryhaline. What limits the euryhalinity of other elasmobranch species?

18. Evidence suggests that both elasmobranchs and teleosts express a Cl/HCO$_3$ exchanger in their gill epithelium. What specific anion exchanger is most common?

19. The most complete balance sheet for intestinal salt absorption and renal, rectal, and gill excretion is 40 years old. These data should be extended to other species to get a more general picture of the pathways involved in monovalent vs. divalent ion balance.

20. How general and how important is proximal tubular salt secretion in the formation of urine in various fish groups? What are the relative roles of divalent vs. monovalent secretion?

21. What are the mechanisms of salt and water reabsorption in the distal segments of the teleost kidney? Is a distal tubule vital for osmoregulation in low salinities?

22. Some euryhaline teleosts lack a urinary bladder (e.g., killifish). What is the relative role in salt reabsorption of the distal tubules vs. the urinary bladder? What are the most important mechanisms for salt uptake by the bladder epithelium?

23. Is the model for marine teleost NaCl secretion that is based largely on the killifish opercular epithelium generally applicable to most teleosts, even those that display a negative transepithelial electrical potential?

24. Some evidence exists for a basolateral K^+ channel in the teleost MRC. What is the molecular mechanism for this K^+ channel?

25. Evidence suggests two mechanisms for Na^+ uptake and acid excretion across the freshwater teleost gill epithelium, but little evidence exists for Cl^- uptake and base excretion. What are the cellular sites and molecular pathways for these uptakes, and are differences species specific? Are one or two cells involved? What are the mechanisms for basolateral Na^+ and Cl^- transport? How is apical Na/NH_4 exchange thermodynamically possible in low salinities?

26. Some evidence exists for a role of apical NKCC in NaCl uptake in the freshwater gill epithelium. Is this an important uptake pathway?

27. Evidence suggests that neither the killifish nor freshwater eel extracts Cl^- from the external medium, only Na^+. Is this a more general phenomenon?

28. What limits euryhalinity in most teleost species? Is there a difference in this limitation between primarily freshwater vs. marine species?

29. What is the structure and function of the coelacanth postanal gland? Does the coelacanth gill have MRCs that secrete salt? Do these cells also express the NaCl uptake proteins that have been described in other groups of fishes?

30. Do the MRCs in the lungfish gill express these NaCl uptake proteins? What is the relative role of gill vs. skin in NaCl uptake? What is the structure and function of the cloacal excretory gland in the African lungfish? Where are the sites of ion reabsorption in the lungfish nephron?

31. What is the relative role of the intestine, kidney, and gill epithelium in osmoregulation in sturgeons, gars, and bowfin? What are the molecular mechanisms involved in the relevant epithelia?

32. What is the real cost of osmoregulation in seawater vs. freshwater?

33. What is the balance sheet for K^+, Mg^{2+}, and SO_4^{2-} in seawater vs. freshwater teleosts? Are there branchial vs. renal and rectal mechanisms for transport of these ions?

34. How do euryhaline fishes sense external or internal salinity changes to stimulate appropriate transport pathways?

35. What are the most important hormones or paracrines that control osmoregulation via effects on the intestine, kidney, and gill epithelia? Are specific hormones or paracrines limiting euryhalinity?

36. What are the intracellular mechanisms that control intestinal, renal, or branchial cell volume regulation in the face of substantial, transcellular ionic and osmotic gradients?

ACKNOWLEDGMENTS

We would like to thank Drs. Helmut Bartels, Klaus Beyenbach, Frank Chapman, Keith Choe, Susan Edwards, George Kidder, Bruce MacFadden, Frank Nordlie, Peter Piermarini, and Ian Potter for their reviews of subsections of this chapter. Their comments were extremely helpful, but any errors that remain are ours. We are especially indebted to Dr. Keith Choe, who drew the "working models" in Figures 8.6, 8.11, 8.12, 8.15, and 8.16. We are the academic offspring (son and grandson) of Bill Potts, so the original "Potts and Parry" was instrumental in our careers. We have been funded by the National Science Foundation for much of our respective careers, and the writing of this chapter was supported by IOB-0519579 to DHE and IOB-06-061687 to JBC.

REFERENCES

1. Alt, J., M., Stolte, H., Eisenbach, G. M., and Walvig, F., Renal electrolyte and fluid excretion in the Atlantic hagfish *Myxine glutinosa*, *J. Exp. Biol.*, 91, 323–330, 1981.
2. Althoff, T., Hentschel, H., Luig, J., Schutz, H., Kasch, M., and Kinne, R. K., Na$^+$-D-glucose cotransporter in the kidney of *Squalus acanthias*: molecular identification and intrarenal distribution, *Am. J. Physiol.*, 290(4), R1094–R1104, 2006.
3. Altinok, I., Galli, S. M., and Chapman, F. A., Ionic and osmotic regulation capabilities of juvenile Gulf of Mexico sturgeon, *Acipenser oxyrinchus* de sotoi, *Comp. Biochem. Physiol. A*, 120(4), 609–616, 1998.
4. Alvarez de la Rosa, D., Canessa, C., Fyfe, G., and Zhang, P., Structure and regulation of amiloride-sensitive sodium channels, *Annu. Rev. Physiol.*, 62, 573–594, 2000.
5. Anderson, W. G., Takei, Y., and Hazon, N., Osmotic and volaemic effects on drinking rate in elasmobranch fish, *J. Exp. Biol.*, 205, 1115–1122, 2002.
6. Anderson, W. G., Taylor, J. R., Good, J. P., Hazon, N., and Grosell, M., Body fluid volume regulation in elasmobranch fish, *Comp. Biochem. Physiol. A*, 148, 3–13, 2006.
7. Ando, M., Mukuda, T., and Kozaka, T., Water metabolism in the eel acclimated to sea water: from mouth to intestine, *Comp. Biochem. Physiol. B*, 136(4), 621–33, 2003.
8. Aoki, M., Kaneko, T., Katoh, F., Hasegawa, S., Tsutsui, N., and Aida, K., Intestinal water absorption through aquaporin 1 expressed in the apical membrane of mucosal epithelial cells in seawater-adapted Japanese eel, *J. Exp. Biol.*, 206, 3495–505, 2003.
9. Avella, M. and Bornancin, M., A new analysis of ammonia and sodium transport through the gills of the freshwater rainbow trout (*Salmo gairdneri*), *J. Exp. Biol.*, 142, 155–176, 1989.
10. Bakheet, T., Frevel, M., Williams, B. R., Greer, W., and Khabar, K. S., ARED: human AU-rich element-containing mRNA database reveals an unexpectedly diverse functional repertoire of encoded proteins, *Nucleic Acids Res.*, 29, 246–54, 2001.
11. Balment, R. J., Lu, W., Weybourne, E., and Warne, J. M., Arginine vasotocin a key hormone in fish physiology and behaviour: a review with insights from mammalian models, *Gen. Comp. Endocrinol.*, 147, 9–16, 2006.
12. Balment, R. J., Song, W., and Ashton, N., Urotensin II: ancient hormone with new functions in vertebrate body fluid regulation, *Ann. N.Y. Acad. Sci.*, 1040, 66–73, 2005.
13. Bartels, H., Assemblies of linear arrays of particles in the apical plasma membrane of mitochondria-rich cells in the gill epithelium of the Atlantic hagfish (*Myxine glutinosa*), *Anat. Rec.*, 211, 229–238, 1985.
14. Bartels, H., Intercellular junctions in the gill epithelium of the Atlantic hagfish, *Myxine glutinosa*, *Cell Tissue Res.*, 254, 573–83, 1988.
15. Bartels, H., The gills of hagfishes. In *The Biology of Hagfishes*, Jorgensen, J. M., Lomholt, J. P., Weber, R. E., and Malte, H., Eds., Chapman & Hall, London, 1998, pp. 205–222.
16. Bartels, H., Moldenhauer, A., and Potter, I. C., Changes in the apical surface of chloride cells following acclimation of lampreys to seawater, *Am. J. Physiol.*, 270, R125–R33, 1996.
17. Bartels, H. and Potter, I. C., Structural changes in the zonulae occludentes of the chloride cells of young adult lampreys following acclimation to seawater, *Cell Tissue Res.*, 265, 447–458, 1991.
18. Bartels, H. and Potter, I. C., Cellular composition and ultrastructure of the gill epithelium of larval and adult lampreys: implications for osmoregulation in fresh and seawater, *J. Exp. Biol.*, 207, 3447–3462, 2004.
19. Bartels, H., Potter, I. C., Pirlich, K., and Mallatt, J., Categorization of the mitochondria-rich cells in the gill epithelium of the freshwater phases in the life cycle of lampreys, *Cell Tissue Res.*, 291, 337–349, 1998.
20. Baustian, M. D., Wang, S. Q., and Beyenbach, K. W., Adaptive responses of aglomerular toadfish to dilute sea water, *J. Comp. Physiol. B*, 167, 61–70, 1997.

21. Beamish, F. W., Osmoregulation in juvenile and adult lampreys, *Can. J. Fish. Aquat. Sci.*, 37, 1739–1750, 1980.

22. Beamish, F. W. H., Strachan, P. D., and Thomas, E., Osmotic and ionic performance of the anadromous sea lamprey, *Petromyzon marinus*, *Comp. Biochem. Physiol. A*, 60, 435–443, 1978.

23. Benyajati, S. and Yokota, S. D., Renal effects of atrial natriuretic peptide in a marine elasmobranch, *Am. J. Physiol.*, 258, R1201–R1206, 1990.

24. Beyenbach, K., Pannabecker, T., and Nagel, W., Central role of the apical membrane H^+-ATPase in electrogenesis and epithelial transport in Malpighian tubules, *J. Exp. Biol.*, 203, 1459–68, 2000.

25. Beyenbach, K. W., Secretory NaCl and volume flow in renal tubules, *Am. J. Physiol.*, 250, R753–R763, 1986.

26. Beyenbach, K. W., Kidneys sans glomeruli, *Am. J. Physiol.*, 286, F811–F827, 2004.

27. Deleted at proofs.

28. Beyenbach, K. W. and Fromter, E., Electrophysiological evidence for Cl secretion in shark renal proximal tubules, *Am. J. Physiol.*, 248, F282–295, 1985.

29. Biemesderfer, D., Payne, J. A., Lytle, C. Y., and Forbush, 3rd, B., Immunocytochemical studies of the Na–K–Cl cotransporter of shark kidney, *Am. J. Physiol.*, 270, F927–F936, 1996.

30. Bijvelds, M. J., Kolar, Z. I., Wendelaar Bonga, S. E., and Flik, G., Magnesium transport across the basolateral plasma membrane of the fish enterocyte, *J. Membr. Biol.*, 154, 217–225, 1996.

31. Bijvelds, M. J., Velden, J. A., Kolar, Z. I., and Flik, G., Magnesium transport in freshwater teleosts, *J. Exp. Biol.*, 201, 1981–1990, 1998.

32. Bijvelds, M. J. C., Flik, G., Kolar, Z. I., and Wendelaar Bonga, S. E., Uptake, distribution and excretion of magnesium in *Oreochromis mossambicus*: uptake, distribution and excretion in diet and water, *Fish Physiol. Biochem.*, 15, 287–298, 1996.

33. Boisen, A. M., Amstrup, J., Novak, I., and Grosell, M., Sodium and chloride transport in soft water and hard water acclimated zebrafish (*Danio rerio*), *Biochim. Biophys. Acta*, 1618, 207–218, 2003.

34. Boutet, I., Long Ky, C. L., and Bonhomme, F., A transcriptomic approach of salinity response in the euryhaline teleost, *Dicentrarchus labrax*, *Gene* 379, 40–50, 2006.

35. Boylan, J., Gill permeability in *Squalus acanthias*. In *Sharks, Skates, and Rays*, Gilbert, P. W., Mathewson, R. F., and Rall, D. P., Eds., The Johns Hopkins University Press, Baltimore, MD, 1967, pp. 197–206.

36. Boylan, J. W., A model for passive urea reabsorption in the elasmobranch kidney, *Comp. Biochem. Physiol. A*, 42, 27–30, 1972.

37. Brett, C. L., Donowitz, M., and Rao, R., Evolutionary origins of eukaryotic sodium/proton exchangers, *Am. J. Physiol.*, 288, C223–C239, 2005.

38. Brooks, D. R., Thorson, T. B., and Mayes, M. A., Fresh-water stingrays (Potamotrygonidae) and their helminth parasites: testing hypotheses of evolution and co-evolution. In *Advances in Cladistics, Proceedings of the First Meeting of the Willi Hennig Society*, Funk, V. A. and Brooks, D. R., Eds., The New York Botanical Garden, Bronx, NY, 1981, pp. 147–175.

39. Brown, E. M., Pollak, M., Seidman, C. E., Seidman, J. G., Chou, Y. H., Riccardi, D., and Hebert, S. C., Calcium-ion-sensing cell-surface receptors, *N. Engl. J. Med.*, 333(4), 234–240, 1995.

40. Brown, J. A., Oliver, J. A., Henderson, I. W., and Jackson, B. A., Angiotensin and single nephron glomerular function in the trout *Salmo gairdneri*, *Am. J. Physiol.*, 239(5), R509–R514, 1980.

41. Deleted at proofs.

42. Brown, J. A. and Rankin, J. C., Lack of glomerular intermittency in the river lamprey, *Lampetra fluviatilis*, acclimated to sea water and following acute transfer to iso-osmotic brackish water, *J. Exp. Biol.*, 202, 939–946, 1999.

43. Bucking, C. and Wood, C. M., Renal regulation of plasma glucose in the freshwater rainbow trout, *J. Exp. Biol.*, 208, 2731–2739, 2005.

44. Bucking, C. and Wood, C. M., Gastrointestinal processing of Na^+, Cl^-, and K^+ during digestion: implications for homeostatic balance in freshwater rainbow trout, *Am. J. Physiol.*, 291, R1764–R1772, 2006.

45. Bucking, C. and Wood, C. M., Water dynamics in the digestive tract of the freshwater rainbow trout during the processing of a single meal, *J. Exp. Biol.*, 209, 1883–1893, 2006.

46. Burger, J. W., Roles of the rectal gland and kidneys in salt and water excretion in the spiny dogfish, *Physiol. Zool.*, 38, 191–196, 1965.

47. Burger, J. W. and Hess, W. N., Function of the rectal gland of the spiny dogfish, *Science*, 131, 670–671, 1960.

48. Burgess, D. W., Miarczynski, M. D., O'Donnell, M. J., and Wood, C. M., Na$^+$ and Cl$^-$ transport by the urinary bladder of the freshwater rainbow trout (*Oncorhynchus mykiss*), *J. Exp. Zool.*, 287, 1–14, 2000.

49. Butler, D. G. and Youson, J. H., Kidney function in the bowfin (*Amia calva* L.), *Comp. Biochem. Physiol. A*, 89, 343–345, 1988.

50. Bystriansky, J. S. and Ballantyne, J. S., Gill Na$^+$-K$^+$-ATPase activity correlates with basolateral membrane lipid composition in seawater- but not freshwater-acclimated Arctic char (*Salvelinus alpinus*), *Am. J. Physiol.*, 292, R1043–R1051, 2007.

50a. Bystriansky, J. S., Frick, N. T., Richards, J. G., Schulte, P. M., and Ballantyne, J. S., Wild Arctic char (*Salvelinus alpinus*) upregulate gill Na$^+$,K$^+$-ATPase during freshwater migration, *Physiol. Biochem. Zool.*, 80(3), 270-282, 2007.

51. Bystriansky, J. S., Richards, J. G., Schulte, P. M., and Ballantyne, J. S., Reciprocal expression of gill Na$^+$/K$^+$-ATPase alpha-subunit isoforms alpha1a and alpha1b during seawater acclimation of three salmonid fishes that vary in their salinity tolerance, *J. Exp. Biol.*, 209, 1848–1858, 2006.

52. Cameron, B. A., Perry, S. F., Wu, C., Ko, K., and Tufts, B. L., Bicarbonate permeability and immunological evidence for an anion exchanger-like protein in the red blood cells of the sea lamprey, *Petromyzon marinus*, *J. Comp. Physiol. B*, 166, 197–204, 1996.

53. Carmona, R., Garcia-Gallego, M., Sanz, A., Domezain, A., and Ostos-Garrido, M. V., Chloride cells and pavement cells in gill epithelia of *Acipenser naccarii*: ultrastructural modifications in seawater-acclimated specimens, *J. Fish Biol.*, 64, 553–566, 2004.

54. Carrier, J. C. and Evans, D. H., Ion and water turnover in the fresh-water elasmobranch *Potamotrygon* sp., *Comp. Biochem. Physiol. A*, 45, 667–670, 1973.

55. Carrier, J. C. and Evans, D. H., The role of environmental calcium in freshwater survival of the marine teleost *Lagodon rhomboides*, *J. Exp. Biol.*, 65, 529–538, 1976.

56. Carrier, J. C., Musick, J. A., and Heithaus, M. R., *Biology of Sharks and Their Relatives*, CRC Press, Boca Raton, FL, 2004.

57. Cataldi, E., Ciccotti, E., Dimarco, P., Disanto, O., Bronzi, P., and Cataudella, S., Acclimation trials of juvenile Italian sturgeon to different salinities: morphophysiological descriptors, *J. Fish Biol.*, 47, 609–618, 1995.

58. Catches, J. S., Burns, J. M., Edwards, S. L., and Claiborne, J. B., Na$^+$/H$^+$ antiporter, V-H$^+$-ATPase and Na$^+$/K$^+$-ATPase immunolocalization in a marine teleost (*Myoxocephalus octodecemspinosus*), *J. Exp. Biol.*, 209, 3440–3447, 2006.

59. Chandra, S., Morrison, G. H., and Beyenbach, K. W., Identification of Mg-transporting renal tubules and cells by ion microscopy imaging of stable isotopes, *Am. J. Physiol.*, 273, F939–F948, 1997.

60. Chang, W. and Loretz, C. A., DPC blockade of transepithelial chloride absorption and single anion channels in teleost urinary bladder, *Am. J. Physiol.*, 265, R66–R75, 1993.

61. Choe, K. P., Edwards, S., Morrison-Shetlar, A. I., Toop, T., and Claiborne, J. B., Immunolocalization of Na$^+$/K$^+$-ATPase in mitochondrion-rich cells of the Atlantic hagfish (*Myxine glutinosa*) gill, *Comp. Biochem. Physiol. A*, 124, 161–168, 1999.

62. Choe, K. P., Edwards, S. L., Claiborne, J. B., and Evans, D. H., The putative mechanism of Na$^+$ absorption in euryhaline elasmobranchs exists in the gills of a stenohaline marine elasmobranch, *Squalus acanthias*, *Comp. Biochem. Physiol. A*, 146, 155–162, 2007.

63. Choe, K. P., Havird, J., Rose, R., Hyndman, K., Piermarini, P., and Evans, D. H., COX2 in a euryhaline teleost, *Fundulus heteroclitus*: primary sequence, distribution, localization, and potential function in gills during salinity acclimation, *J. Exp. Biol.*, 209(Pt. 9), 1696–1708, 2006.

64. Choe, K. P., Kato, A., Hirose, S., Plata, C., Sindic, A., Romero, M. F., Claiborne, J. B., and Evans, D. H., NHE3 in an ancestral vertebrate: primary sequence, distribution, localization, and function in gills, *Am. J. Physiol.*, 289, R1520–R1534, 2005.

65. Choe, K. P., Morrison-Shetlar, A. I., Wall, B. P., and Claiborne, J. B., Immunological detection of Na$^+$/H$^+$ exchangers in the gills of a hagfish, *Myxine glutinosa*, an elasmobranch, *Raja erinacea*, and a teleost, *Fundulus heteroclitus*, *Comp. Biochem. Physiol. A*, 131, 375–385, 2002.

66. Choe, K. P., O'Brien, S., Evans, D. H., Toop, T., and Edwards, S. L., Immunolocalization of Na$^+$/K$^+$-ATPase, carbonic anhydrase II, and vacuolar H$^+$-ATPase in the gills of freshwater adult lampreys, *Geotria australis*, *J. Exp. Zool.*, 301, 654–665, 2004.

67. Choe, K. P., Verlander, J. W., Wingo, C. S., and Evans, D. H., A putative H$^+$-K$^+$-ATPase in the Atlantic stingray *Dasyatis sabina*: primary sequence and expression in gills, *Am. J. Physiol.*, 287, R981–R991, 2004.

68. Cholette, C., Gagnon, A., and Germain, P., Isosmotic adaptation in *Myxine glutinosa* L. I. Variations of some parameters and role of the amino acid pool of the muscle cells, *Comp. Biochem. Physiol.*, 33, 333–346, 1970.

69. Christoffels, A., Koh, E. G., Chia, J. M., Brenner, S., Aparicio, S., and Venkatesh, B., *Fugu* genome analysis provides evidence for a whole-genome duplication early during the evolution of ray-finned fishes, *Mol. Biol. Evol.*, 21, 1146–1151, 2004.

70. Claiborne, J. B., Choe, K. P., Morrison-Shetlar, A. I., Weakley, J. C., Havird, J., Freiji, A., Evans, D. H., and Edwards, S. L., Molecular detection and immunological localization of gill Na+/H+ exchanger (NHE2) in the dogfish (*Squalus acanthias*), *Am. J. Physiol.*, 294, R1092–R1102, 2007.

71. Claiborne, J. B., Compton-McCullough, D., and Walton, J. S., Branchial acid–base transfers in the euryhaline oyster toadfish (*Opsanus tau*) during exposure to dilute seawater, *J. Fish Biol.*, 56, 1539–1544, 2000.

72. Claiborne, J. B., Edwards, S. L., and Morrison-Shetlar, A. I., Acid–base regulation in fishes: cellular and molecular mechanisms, *J. Exp. Zool.*, 293, 302–319, 2002.

73. Claiborne, J. B., Perry, E., Bellows, S., and Campbell, J., Mechanisms of acid excretion across the gills of a marine fish, *J. Exp. Zool.*, 279, 509–520, 1997.

74. Cliff, W. H. and Beyenbach, K. W., Fluid secretion in glomerular renal proximal tubules of freshwater-adapted fish, *Am. J. Physiol.*, 254, R154–R158, 1988.

75. Cliff, W. H. and Beyenbach, K. W., Secretory renal proximal tubules in seawater- and freshwater-adapted killifish, *Am. J. Physiol.*, 262, F108–F116, 1992.

76. Cohen, J. J., Krupp, M. A., Chidsey, I., C.A., and Blitz, C. L., Effect of TMA and its homologues on renal conservation of TMA-oxide in the spiny dogfish *Squalus acanthias*, *Am. J. Physiol.*, 194, 229–235, 1959.

77. Conley, D. M. and Mallatt, J., Histochemical localization of Na+K+ATPase and carbonic anhydrase activity in the gills of 17 fish species, *Can. J. Zool.*, 66, 2398–2405, 1988.

78. Cutler, C. P., Cloning and identification of four aquaporin genes in the dogfish shark (*Squalus acanthias*), *Bull. Mt. Desert Isl. Biol. Lab.*, 46, 19, 2007.

79. Cutler, C. P. and Cramb, G., Molecular physiology of osmoregulation in eels and other teleosts: the role of transporter isoforms and gene duplication, *Comp. Biochem. Physiol. A*, 130, 551–564, 2001.

80. Cutler, C. P. and Cramb, G., Branchial expression of an aquaporin 3 (AQP-3) homologue is down-regulated in the European eel *Anguilla anguilla* following seawater acclimation, *J. Exp. Biol.*, 205, 2643–2651, 2002.

81. Cutler, C. P. and Cramb, G., Differential expression of absorptive cation–chloride cotransporters in the intestinal and renal tissues of the European eel (*Anguilla anguilla*), *Comp. Biochem. Physiol. B*, 149, 63–73, 2008.

82. Cutler, C. P., Martinez, A. S., and Cramb, G., The role of aquaporin 3 in teleost fish, *Comp. Biochem. Physiol. A*, 148, 82–91, 2007.

83. D'Cruz, L. M. and Wood, C. M., The influence of dietary salt and energy on the response to low pH in juvenile rainbow trout, *Physiol. Zool.*, 71, 642–657, 1998.

84. Daborn, K., Cozzi, R. R., and Marshall, W. S., Dynamics of pavement cell–chloride cell interactions during abrupt salinity change in *Fundulus heteroclitus*, *J. Exp. Biol.*, 204, 1889–1899, 2001.

85. Dawson, D. C. and Frizzell, R. A., Mechanism of active K+ secretion by flounder urinary bladder, *Pflügers Arch.*, 414, 393–400, 1989.

86. Dean, D. B., Whitlow, Z. W., and Borski, R. J., Glucocorticoid receptor upregulation during seawater adaptation in a euryhaline teleost, the tilapia (*Oreochromis mossambicus*), *Gen. Comp. Endocrinol.*, 132, 112–118, 2003.

87. Degnan, K. J., The role of K+ and Cl− conductances in chloride secretion by the opercular epithelium, *J. Exp. Zool.*, 236, 19–25, 1985.

88. Degnan, K. J., Karnaky, Jr., K. J., and Zadunaisky, J., Active chloride transport in the *in vitro* opercular skin of a teleost (*Fundulus heteroclitus*), a gill-like epithelium rich in chloride cells, *J. Physiol. Lond.*, 271, 155–191, 1977.

89. DeLaney, R. G., Lahiri, S., Hamilton, R., and Fishman, P., Acid–base balance and plasma composition in the aestivating lungfish (*Protopterus*), *Am. J. Physiol.*, 232, R10–R17, 1977.

90. Delsuc, F., Brinkmann, H., Chourrout, D., and Philippe, H., Tunicates and not cephalochordates are the closest living relatives of vertebrates, *Nature*, 439, 965–968, 2006.

91. Demarest, J. R., Ion and water transport by the flounder urinary bladder: salinity dependence, *Am. J. Physiol.*, 246, F395–F401, 1984.

92. Devor, D. C., Forrest, Jr., J. N., Suggs, W. K., and Frizzell, R. A., cAMP-activated Cl⁻ channels in primary cultures of spiny dogfish (*Squalus acanthias*) rectal gland, *Am. J. Physiol.*, 268, C70–C79, 1995.

93. Ditrich, H., *Renal Structure and Function in Vertebrates*, Science Publishers, Enfield, NH, 2005.

94. Donoghue, P. C., Forey, P. L., and Aldridge, R. J., Conodont affinity and chordate phylogeny, *Biol. Rev. Camb. Philos. Soc.*, 75, 191–251, 2000.

95. Doyle, N. and Gorecki, D., The so-called chloride cell of the fish gill, *Physiol. Zool.*, 34, 81–85, 1961.

96. Drummond, I. A., Zebrafish kidney development, *Meth. Cell Biol.*, 76, 501–530, 2004.

97. Eddy, F. B., The effect of calcium on gill potentials and on sodium and chloride fluxes in the goldfish *Carassius auratus*, *J. Comp. Physiol.*, 96, 131–142, 1975.

98. Eddy, F. B., Osmotic and ionic regulation in captive fish with particular reference to salmonids, *Comp. Biochem. Physiol. B*, 73, 125–142, 1982.

99. Eddy, F. B., Drinking in juvenile Atlantic salmon (*Salmo salar* L.) in response to feeding and activation of the endogenous renin–angiotensin system, *Comp. Biochem. Physiol. A*, 148, 23–28, 2007.

100. Edwards, S. L., Claiborne, J. B., Morrison-Shetlar, A. I., and Toop, T., Expression of Na⁺/H⁺ exchanger mRNA in the gills of the Atlantic hagfish (*Myxine glutinosa*) in response to metabolic acidosis, *Comp. Biochem. Physiol. A*, 130, 81–91, 2001.

101. Edwards, S. L., Tse, C. M., and Toop, T., Immunolocalisation of NHE3-like immunoreactivity in the gills of the rainbow trout (*Oncorhynchus mykiss*) and the blue-throated wrasse (*Pseudolabrus tetrious*), *J. Anat.*, 195, 465–469, 1999.

102. Edwards, S. L., Wall, B. P., Morrison-Shetlar, A., Sligh, S., Weakley, J. C., and Claiborne, J. B., The effect of environmental hypercapnia and salinity on the expression of NHE-like isoforms in the gills of a euryhaline fish (*Fundulus heteroclitus*), *J. Exp. Zool.*, 303, 464–475, 2005.

103. Deleted at proofs.

104. Elger, M., The branchial circulation and the gill epithelia in the Atlantic hagfish, *Myxine glutinosa* L., *Anat. Embryol. (Berl.)*, 175(4), 489–504, 1987.

105. Ellis, L. C. and Youson, J. H., Ultrastructure of the pronephric kidney in upstream migrant sea lamprey, *Petromyzon marinus* L., *Am. J. Anat.*, 185, 429–443, 1989.

106. Epstein, F., Katz, A. I., and Pickford, G. E., Sodium- and potassium-activated adenosine triphosphatase of gills: role in adaptation of teleots to salt water, *Science*, 156, 1245–1247, 1967.

107. Ernst, S. A., Hootman, S. R., Schreiber, J. H., and Riddle, C. V., Freeze-fracture and morphometric analysis of occluding junctions in rectal glands of elasmobranch fish, *J. Membr. Biol.*, 58(2), 101–114, 1981.

108. Esaki, M., Hoshijima, K., Kobayashi, S., Fukuda, H., Kawakami, K., and Hirose, S., Visualization in zebrafish larvae of Na(+) uptake in mitochondria-rich cells whose differentiation is dependent on foxi3a, *Am. J. Physiol. Regul. Integr. Comp. Physiol.*, 292(1), R470–R480, 2007.

109. Evans, D. H., Studies on the permeability to water of selected marine, freshwater and euryhaline teleosts, *J. Exp. Biol.*, 50, 689–703, 1969.

110. Evans, D. H., Sodium uptake by the sailfin molly, *Poecilia latipinna*: kinetic analysis of a carrier system present in both fresh-water-acclimated and sea-water-acclimated individuals, *Comp. Biochem. Physiol. A*, 45(3), 843–850, 1973.

111. Evans, D. H., Ionic exchange mechanisms in fish gills, *Comp. Biochem. Physiol. A*, 51, 491–495, 1975.

112. Evans, D. H., Further evidence for Na/NH₄ exchange in marine teleost fish, *J. Exp. Biol.*, 70, 213–220, 1977.

113. Evans, D. H., Fish. In *Osmotic and Ionic Regulation in Animals*, Vol. 1, Maloiy, G. M. O., Ed., Academic Press, London, 1979, pp. 305–390.

114. Evans, D. H., Fish. In *Comparative Physiology of Osmoregulation in Animals*, Vol. 1, Maloiy, G. M. O., Ed., Academic Press, Orlando, 1979, pp. 305–390.

115. Evans, D. H., Kinetic studies of ion transport by fish gill epithelium, *Am. J. Physiol.*, 238(3), R224–R230, 1980.

116. Evans, D. H., Osmotic and ionic regulation by freshwater and marine fish. In *Environmental Physiology of Fishes*, Ali, M. A., Ed., Plenum Press, New York, 1981, pp. 93–122.

117. Evans, D. H., Gill Na/H and Cl/HCO₃ exchange systems evolved before the vertebrates entered fresh water, *J. Exp. Biol.*, 113, 464–470, 1984.

118. Deleted at proofs.

119. Evans, D. H., The roles of gill permeability and transport mechanisms in euryhalinity. In *Fish Physiology*, Hoar, W. S. and Randall, D. J., Eds., Academic Press, New York, 1984, pp. 239–283.

120. Evans, D. H., Osmotic and ionic regulation. In *The Physiology of Fishes*, Evans, D. H., Ed., CRC Press, Boca Raton, FL, 1993, pp. 315–341.

121. Evans, D. H., Cell signaling and ion transport across the fish gill epithelium, *J. Exp. Zool.*, 293(3), 336–347, 2002.

122. Evans, D. H. and Cooper, K., The presence of Na–Na and Na–K exchange in sodium extrusion by three species of fish, *Nature*, 259(5540), 241–242, 1976.

123. Evans, D. H. and Hooks, C., Sodium fluxes across the hagfish, *Myxine glutinosa*, *Bull. Mt. Desert Isl. Biol. Lab.*, 23, 61–62, 1983.

124. Evans, D. H., Oikari, A., Kormanik, G. A., and Mansberger, L., Osmoregulation by the prenatal spiny dogfish, *Squalus acanthias*, *J. Exp. Biol.*, 101, 295–305, 1982.

125. Evans, D. H. and Piermarini, P. M., Contractile properties of the elasmobranch rectal gland, *J. Exp. Biol.*, 204(Pt. 1), 59–67, 2001.

126. Evans, D. H., Piermarini, P. M., and Choe, K. P., Homeostasis: osmoregulation, pH regulation, and nitrogen excretion. In *Biology of Sharks and Their Relatives*, Carrier, J., Musick, J., and Heithaus, J., Eds., CRC Press, Boca Raton, FL, 2004.

127. Evans, D. H., Piermarini, P. M., and Choe, K. P., The multifunctional fish gill: dominant site of gas exchange, osmoregulation, acid–base regulation, and excretion of nitrogenous waste, *Physiol. Rev.*, 85(1), 97–177, 2005.

128. Deleted at proofs.

129. Evans, D. H., Piermarini, P. M., and Potts, W. T. W., Ionic transport in the fish gill epithelium, *J. Exp. Zool.*, 283, 641–652, 1999.

130. Farmer, G. J. and Beamish, F. W. H., Oxygen consumption of *Tilapia nilotica* in relation to swimming speed and salinity, *J. Fish. Res. Bd. Can.*, 26, 2807–2821, 1969.

131. Febry, R. and Lutz, P., Energy partitioning in fish: the activity-related cost of osmoregulation in a euryhaline cichlid, *J. Exp. Biol.*, 128, 63–85, 1987.

132. Fels, L. M., Kastner, S., and Stolte, E. H., The hagfish kidney as a model to study renal physiology and toxicology. In *The Biology of Hagfishes*, Jorgensen, J. M., Lomholt, J. P., Weber, R. E., and Malte, H., Eds., Chapman & Hall, London, 1998.

133. Fels, L. M., Raguse-Degener, G., and Stolte, H., The archinephron of *Myxine glutinosa* L. (Cyclostomata). In *Structure and Function of the Kidney*, Kinne, R. K. H., Ed., Karger, Basel, 1989, pp. 73–102.

134. Fels, L. M., Sanz-Altamira, P. M., Decker, B., Elger, B., and Stolte, H., Filtration characteristics of the single isolated perfused glomerulus of *Myxine glutinosa*, *Ren. Physiol. Biochem.*, 16(5), 276–284, 1993.

135. Feng, S. H., Leu, J. H., Yang, C. H., Fang, M. J., Huang, C. J., and Hwang, P. P., Gene expression of Na^+-K^+-ATPase alpha 1 and alpha 3 subunits in gills of the teleost *Oreochromis mossambicus*, adapted to different environmental salinities, *Mar. Biotechnol. (N.Y.)*, 4(4), 379–391, 2002.

136. Fenton, R. A. and Knepper, M. A., Mouse models and the urinary concentrating mechanism in the new millennium, *Physiol. Rev.*, 87(4), 1083–1112, 2007.

137. Fenwick, J. C., Wendelaar Bonga, S. E., and Flik, G., *In vivo* bafilomycin-sensitive Na^+ uptake in young freshwater fish, *J. Exp. Biol.*, 202, 3659–3666, 1999.

138. Fines, G. A., Ballantyne, J. S., and Wright, P. A., Active urea transport and an unusual basolateral membrane composition in the gills of a marine elasmobranch, *Am. J. Physiol. Regul. Integr. Comp. Physiol.*, 280(1), R16–R24, 2001.

139. Fiol, D. F., Chan, S., and Kultz, D., Identification and pathway analysis of immediate hyperosmotic stress responsive molecular mechanisms in tilapia (*Oreochromis mossambicus*) gill, *Comp. Biochem. Physiol. D*, 1, 344–356, 2006.

140. Fiol, D. F., Chan, S. Y., and Kultz, D., Regulation of osmotic stress transcription factor 1 (Ostf1) in tilapia (*Oreochromis mossambicus*) gill epithelium during salinity stress, *J. Exp. Biol.*, 209(Pt. 16), 3257–3265, 2006.

141. Fiol, D. F. and Kultz, D., Rapid hyperosmotic coinduction of two tilapia (*Oreochromis mossambicus*) transcription factors in gill cells, *Proc. Natl. Acad. Sci. USA*, 102(3), 927–932, 2005.

142. Fiol, D. F. and Kultz, D., Osmotic stress sensing and signaling in fishes, *FEBS Lett.*, 274(22), 5790–4798, 2007.

143. Flik, G., Kaneko, T., Greco, A. M., Li, J., and Fenwick, J. C., Sodium-dependent ion transporters in trout gills, *Fish Physiol. Biochem.*, 17, 385–396, 1997.

144. Flik, G., Klaren, P. H. M., Schoenmakers, T. J. M., Bijvelds, M. J. C., Verbost, P. M., and Bonga, S. E. W., Cellular calcium transport in fish: unique and universal mechanisms, *Physiol. Zool.*, 69(2), 403–417, 1996.

145. Flik, G. and Verbost, P. M., Calcium transport in fish gills and intestine, *J. Exp. Biol.*, 184, 17–29, 1993.

146. Flik, G., Verbost, P. M., and Wendelaar Bonga, S. E., Calcium transport processes in fishes. In *Cellular and Molecular Approaches to Fish Ionic Regulation*, Wood, C. M. and Shuttleworth, T. J., Eds., Academic Press, San Diego, CA, 1995, pp. 317–342.

147. Forey, P. L. and Janvier, P., Agnathans and the origin of jawed vertebrates, *Nature*, 361(6408), 129–134, 1993.

148. Forey, P. L., Agnathans recent and fossil, and the origin of jawed vertebrates, *Rev. Fish Biol. Fish.*, 5, 267–303, 1995.

149. Forrest, Jr., J. N., Boyer, J. L., Ardito, T. A., Murdaugh, Jr., H. V., and Wade, J. B., Structure of tight junctions during Cl secretion in the perfused rectal gland of the dogfish shark, *Am. J. Physiol.*, 242(5), C388–C392, 1982.

150. Forster, R. P., Goldstein, L., and Rosen, J. K., Intrarenal control of urea reabsorption by renal tubules of the marine elasmobranch, *Squalus acanthias*, *Comp. Biochem. Physiol. A*, 42(1), 3–12, 1972.

151. Foskett, J. K., Bern, H. A., Machen, T. E., and Conner, M., Chloride cells and the hormonal control of teleost fish osmoregulation, *J. Exp. Biol.*, 106, 255–281, 1983.

152. Foskett, J. K., Logsdon, C. D., Turner, T., Machen, T. E., and Bern, H. A., Differentiation of the chloride extrusion mechanisms during seawater adaptation of a teleost fish, the cichlid *Sarotherodon mossambicus*, *J. Exp. Biol.*, 93, 209–224, 1981.

153. Foskett, J. K. and Machen, T. E., Vibrating probe analysis of teleost opercular epithelium: correlation between active transport and leak pathways of individual chloride cells, *J. Membr. Biol.*, 85(1), 25–35, 1985.

154. Fossat, B. and Lahlou, B., The mechanism of coupled transport of sodium and chloride in isolated urinary bladder of the trout, *J. Physiol.*, 294, 211–222, 1979.

155. Friedman, P. A. and Hebert, S. C., Diluting segment in kidney of dogfish shark. I. Localization and characterization of chloride absorption, *Am. J. Physiol.*, 258(2, Pt. 2), R398–R408, 1990.

156. Frizzell, R. A., Field, M., and Schultz, S. G., Sodium-coupled chloride transport by epithelial tissues, *Am. J. Physiol.*, 236(1), F1–F8, 1979.

157. Fu, H., Subramanian, R. R., and Masters, S. C., 14–3–3 proteins: structure, function, and regulation, *Annu. Rev. Pharmacol. Toxicol.*, 40, 617–647, 2000.

158. Furspan, P., Prange, H. D., and Greenwalt, L., Energetics and osmoregulation in the catfish, *Ictalurus nebolusus* and *I. punctatus*, *Comp. Biochem. Physiol. A*, 77, 773–778, 1984.

159. Galvez, F., Reid, S. D., Hawkings, G., and Goss, G. G., Isolation and characterization of mitochondria-rich cell types from the gill of freshwater rainbow trout, *Am. J. Physiol. Regul. Integr. Comp. Physiol.*, 282(3), R658–R668, 2002.

160. Gamba, G., Saltzberg, S. N., Lombardi, M., Miyanoshita, A., Lytton, J., Hediger, M. A., Brenner, B. M., and Hebert, S. C., Primary structure and functional expression of a cDNA encoding the thiazide-sensitive, electroneutral sodium-chloride cotransporter, *Proc. Natl. Acad. Sci. USA*, 90(7), 2749–2753, 1993.

161. Georgalis, T., Gilmour, K. M., Yorston, J., and Perry, S. F., Roles of cytosolic and membrane-bound carbonic anhydrase in renal control of acid–base balance in rainbow trout, *Oncorhynchus mykiss*, *Am. J. Physiol. Ren. Physiol.*, 291(2), F407–F421, 2006.

162. Gerst, J. W. and Thorson, T. B., Effects of saline acclimation on plasma electrolytes, urea excretion, and hepatic urea biosynthesis in a freshwater stingray, *Potamotrygon* sp. Garman, 1877, *Comp. Biochem. Physiol. A*, 56, 87–93, 1977.

163. Gess, R. W., Coates, M. I., and Rubidge, B. S., A lamprey from the Devonian period of South Africa, *Nature*, 443(7114), 981–984, 2006.

164. Gettex-Galland, M. and Hughes, G. M., Contractile filamentous material in the pillar cells of fish gills, *J. Cell Sci.*, 13, 359–370, 1973.

165. Gilmour, K. M., Euverman, R. M., Esbaugh, A. J., Kenney, L., Chew, S. F., Ip, Y. K., and Perry, S. F., Mechanisms of acid–base regulation in the African lungfish *Protopterus annectens*, *J. Exp. Biol.*, 210(Pt. 11), 1944–59, 2007.

166. Gluck, S., Underhill, D., Iyori, M., Holliday, S., Kostrominova, T., and Lee, B., Physiology and biochemistry of kidney vacuolar H+-ATPase, *Annu. Rev. Physiol.*, 58, 427–445, 1996.

167. Goldstein, L. and Foster, R. P., Urea biosynthesis and excretion in freshwater and marine elasmobranchs, *Comp. Biochem. Physiol.*, 398, 415–421, 1971.

168. Gonzalez, R. J., Wilson, R. W., Wood, C. M., Patrick, M. L., and Val, A. L., Diverse strategies for ion regulation in fish collected from the ion-poor, acidic Rio Negro, *Physiol. Biochem. Zool.*, 75(1), 37–47, 2002.

169. Goss, G. G., Adamia, S., and Galvez, F., Peanut lectin binds to a subpopulation of mitochondria-rich cells in the rainbow trout gill epithelium, *Am. J. Physiol. Regul. Integr. Comp. Physiol.*, 281(5), R1718–R1725, 2001.

170. Goss, G. G., Perry, S. F., Wood, C. M., and Laurent, P., Mechanisms of ion and acid–base regulation at the gills of freshwater fish, *J. Exp. Zool.*, 263(2), 143–159, 1992.

171. Greger, R. and Schlatter, E., Mechanism of NaCl secretion in the rectal gland of spiny dogfish (*Squalus acanthias*). I. Experiments in isolated *in vitro* perfused rectal gland tubules, *Pflügers Arch.*, 402(1), 63–75, 1984.

172. Griffith, R. W., Chemistry of the body fluids of the coelacanth, *Latimeria chalumnae*, *Proc. R. Soc. Lond. B Biol. Sci.*, 208(1172), 329–347, 1980.

173. Griffith, R. W. and Burdick, C. J., Sodium–potassium activated adenosine triphosphatase in coelacanth tissues: high activity in rectal gland, *Comp. Biochem. Physiol. B*, 54(4), 557–559, 1976.

174. Griffith, R. W., Pang, P. K. T., Srivastava, A. K., and Pickford, G. E., Serum composition of freshwater stingrays (Potamotrygonidae) adapted to fresh and dilute sea water, *Biol. Bull.*, 144, 304–320, 1973.

175. Grosell, M., Intestinal anion exchange in marine fish osmoregulation, *J. Exp. Biol.*, 209(Pt. 15), 2813–2827, 2006.

176. Grosell, M., De Boeck, G., Johannsson, O., and Wood, C. M., The effects of silver on intestinal ion and acid–base regulation in the marine teleost fish *Parophrys vetulus*, *Comp. Biochem. Physiol. C Pharmacol. Toxicol. Endocrinol.*, 124(3), 259–270, 1999.

177. Grosell, M. and Genz, J., Ouabain-sensitive bicarbonate secretion and acid absorption by the marine teleost fish intestine play a role in osmoregulation, *Am. J. Physiol., Regul. Integr. Comp. Physiol.*, 291(4), R1145–R1156, 2006.

178. Grosell, M., Gilmour, K. M., and Perry, S. F., Intestinal carbonic anhydrase, bicarbonate, and proton carriers play a role in the acclimation of rainbow trout to seawater, *Am. J. Physiol. Regul. Integr. Comp. Physiol.*, 293(5), R2099–R2111, 2007.

179. Grosell, M., Hogstrand, C., Wood, C. M., and Hansen, H. J., A nose-to-nose comparison of the physiological effects of exposure to ionic silver versus silver chloride in the European eel (*Anguilla anguilla*) and the rainbow trout (*Oncorhynchus mykiss*), *Aquat. Toxicol.*, 48(2–3), 327–342, 2000.

180. Grosell, M., Laliberte, C. N., Wood, S. J., Jensen, F. B., and Wood, C. M., Intestinal HCO$_3$ secretion in marine teleost fish: evidence for an apical rather than a basolateral Cl/HCO$_3$ exchanger, *Fish Physiol. Biochem.*, 24, 81–95, 2001.

181. Grosell, M. and Taylor, J. R., Intestinal anion exchange in teleost water balance, *Comp. Biochem. Physiol. A*, 148(1), 14–22, 2007.

182. Grosell, M., Wood, C. M., Wilson, R. W., Bury, N. R., Hogstrand, C., Rankin, C., and Jensen, F. B., Bicarbonate secretion plays a role in chloride and water absorption of the European flounder intestine, *Am. J. Physiol. Regul. Integr. Comp. Physiol.*, 288(4), R936–R946, 2005.

183. Gunter, G., A revised list of euryhaline fishes of North and Middle America, *Am. Midl. Nat.*, 56, 345–354, 1956.

184. Haas, M. and Forbush, B., The Na–K–Cl cotransporter of secretory epithelia, *Annu. Rev. Physiol.*, 62, 515–534, 2000.

185. Halm, D. R., Krasny, Jr., E. J., and Frizzell, R. A., Potassium transport across the intestine of the winter flounder: active secretion and absorption, *Prog. Clin. Biol. Res.*, 126, 245–255, 1983.

186. Halm, D. R., Krasny, Jr., E. J., and Frizzell, R. A., Electrophysiology of flounder intestinal mucosa. I. Conductance properties of the cellular and paracellular pathways, *J. Gen. Physiol.*, 85, 843–864, 1985.

187. Halstead, L. B., The vertebrate invasion of freshwater, *Philos. Trans. R. Soc. Lond. B*, 309, 243–258, 1985.

188. Hamlett, W. C., *Sharks, Skates, and Rays*, The Johns Hopkins University Press, Baltimore, MD, 1999.

189. Hardisty, M. W. and Potter, I. C., *The Biology of Lampreys*, Academic Press, London, 1972.

190. Harvey, B. J., Physiology of V-ATPases, *J. Exp. Biol.*, 172, 1–17, 1992.

191. Hazon, N., Tierney, M. L., Anderson, G., Mackenzie, S., Cutler, C., and Cramb, G., Ion and water balance in elasmobranch fish. In *Ionic Regulation in Animals: A Tribute to Professor W.T.W. Potts*, Hazon, N., Eddy, F. B., and Flik, G., Eds., Springer, Berlin, 1997, pp. 70–86.

192. Hazon, N., Tierney, M. L., and Takei, Y., Renin–angiotensin system in elasmobranch fish: a review, *J. Exp. Zool.*, 284(5), 526–534, 1999.

193. Hazon, N., Wells, A., Pillans, R. D., Good, J. P., Gary Anderson, W., and Franklin, C. E., Urea-based osmoregulation and endocrine control in elasmobranch fish with special reference to euryhalinity, *Comp. Biochem. Physiol. B*, 136, 685–700, 2003.

194. Heisler, N., Regulation of the acid–base status in fishes. In *Environmental Physiology of Fishes*, Ali, M. A., Ed., Plenum Press, New York, 1980, pp. 123–162.

195. Heisler, N., Transepithelial transfer processes as mechanisms for fish acid–base regulation in hypercapnia and lactacidosis, *Can. J. Zool.*, 60, 1108–1123, 1982.

196. Heisler, N., Acid–base regulation in fishes. In *Fish Physiology*, Hoar, W. S. and Randall, D. J., Eds., Academic Press, New York, 1984, pp. 315–401.

197. Helfman, G. S., Collette, B. B., and Facey, D. E., *The Diversity of Fishes*, Blackwell Science, London, 1997.

198. Henderson, I. W., O'Toole, L. B., and Hazon, N., Kidney function. In *Physiology of Elasmobranch Fishes*, Shuttleworth, T. J., Ed., Springer-Verlag, Berlin, 1988, pp. 201–214.

199. Hentschel, H., Renal blood vascular system in the elasmobranch *Raja erinacea* Mitchill in relation to kidney zones, *Am. J. Anat.*, 183, 130–147, 1988.

200. Hentschel, H. and Elger, M., The distal nephron in the kidney of fishes, *Adv. Anat. Embryol. Cell Biol.*, 108, 1–151, 1987.

201. Hentschel, H. and Elger, M., Morphology of glomerular and aglomerular kidneys. In *Structure and Function of the Kidney*, Kinne, R. K. H., Ed., Karger, Basel, 1989, pp. 1–72.

202. Hentschel, H., Mahler, S., Herter, P., and Elger, M., Renal tubule of dogfish, *Scyliorhinus caniculus*: a comprehensive study of structure with emphasis on intramembrane particles and immunoreactivity for H^+-K^+-adenosine triphosphatase, *Anat. Rec.*, 235, 511–532, 1993.

203. Hickman, Jr., C. P., Ingestion, intestinal absorption and elimination of sea water and salts in the southern flounder, *Paralichthys lethostigma*, *Can. J. Zool.*, 46, 457–466, 1968.

204. Hickman, Jr., C. P., The kidney. In *Fish Physiology*, Hoar, W. S. and Randall, D. J., Eds., Academic Press, New York, 1969, pp. 91–239.

206. Hill, W. G., Mathai, J. C., Gensure, R. H., Zeidel, J. D., Apodaca, G., Saenz, J. P., Kinne-Saffran, E., Kinne, R., and Zeidel, M. L., Permeabilities of teleost and elasmobranch gill apical membranes: evidence that lipid bilayers alone do not account for barrier function, *Am. J. Physiol.*, 287, C235–C242, 2004.

207. Hirano, T. and Mayer-Gostan, N., Eel esophagus as an osmoregulatory organ, *Proc. Natl. Acad. Sci. USA*, 73(4), 1348–1350, 1976.

208. Hirata, T., Kaneko, T., Ono, T., Nakazato, T., Furukawa, N., Hasegawa, S., Wakabayashi, S., Shigekawa, M., Chang, M. H., Romero, M. F., and Hirose, S., Mechanism of acid adaptation of a fish living in a pH 3.5 lake, *Am. J. Physiol. Regul. Integr. Comp. Physiol.*, 284(5), R1199–R1212, 2003.

210. Hirata, T., Ono, T., Nakazato, T., Y. Saruta, T. Kaneko, Hirano, T., and Hirose, S., Maintenance of acid–base balance by Na^+/H^+ exchanger (NHE3) and Na^+/HCO_3^- cotransporter (NBC) in the gill of dace adapted to extremely acidic conditions (pH 3.5). In *Control and Diseases of Sodium-Dependent Transport Proteins and Ion Channels*, Suketa, Y., Carafoli, E., Lazdunski, M., Mikoshiba, K., Okada, Y., and Wright, E. M., Eds., Elsevier Science, Amsterdam, 2000, pp. 127–128.

211. Hiroi, J., McCormick, S., Ohtani-Kaneko, R., and Kaneko, T., Functional classification of mitochondrion-rich cells in euryhaline Mozambique tilapia (*Oreochromis mossambicus*) embryos, by means of triple immunofluorescence staining for Na^+/K^+-ATPase, $Na^+/K^+/2Cl^-$ cotransporter and CFTR anion channel, *J. Exp. Biol.*, 208, 2023–2036, 2005.

212. Hiroi, J. and McCormick, S. D., Variation in salinity tolerance, gill Na$^+$/K$^+$-ATPase, Na$^+$/K$^+$/2Cl$^-$ cotransporter and mitochondria-rich cell distribution in three salmonids: *Salvelinus namaycush*, *Salvelinus fontinalis*, and *Salmo salar*, *J. Exp. Biol.*, 210, 1015–1024, 2007.

214. Hiroi, J., Miyazaki, H., Katoh, F., Ohtani-Kaneko, R., and Kaneko, T., Chloride turnover and ion-transporting activities of yolk-sac preparations (yolk balls) separated from Mozambique tilapia embryos and incubated in freshwater and seawater, *J. Exp. Biol.*, 208, 3851–3858, 2005.

214a. Hiroi, J., Yasumasu, S., McCormick, S., Hwang, P., and Kaneko, T., Evidence for an apical NaCl cotransporter involved in ion uptake in a teleost fish, *J. Exp. Biol.*, 211, 2584–2599, 2008.

215. Hirose, S., Kaneko, T., Naito, N., and Takei, Y., Molecular biology of major components of chloride cells, *Comp. Biochem. Physiol. B*, 136, 593–620, 2003.

217. Hochachka, P. W. and Somero, G. N., *Biochemical Adaptation: Mechanism and Process in Physiological Evolution*, Oxford University Press, Oxford, U.K., 2002.

218. Hoffman, E. K., Schettino, T., and Marshall, W. S., The role of volume-sensitive ion transport systems in regulation of epithelial transport, *Comp. Biochem. Physiol. A*, 148, 29–43, 2007.

219. Hoffmann, E. K., Hoffmann, E., Lang, F., and Zadunaisky, J. A., Control of Cl$^-$ transport in the operculum epithelium of *Fundulus heteroclitus*: long- and short-term salinity adaptation, *Biochim. Biophys. Acta Biomembr.*, 1566, 129–139, 2002.

220. Holder, M. T., Erdmann, M. V., Wilcox, T. P., Caldwell, R. L., and Hillis, D. M., Two living species of coelacanths?, *Proc. Natl. Acad. Sci. USA*, 96, 12616–12620, 1999.

221. Holland, N. D. and Chen, J., Origin and early evolution of the vertebrates: new insights from advances in molecular biology, anatomy, and palaeontology, *Bioessays*, 23(2), 142–151, 2001.

222. Holmes, W. N. and Donaldson, E. M., The body compartments and the distribution of electrolytes. In *Fish Physiology*, Hoar, W. S. and Randall, D. J., Eds., Academic Press, New York, 1969, pp. 1–89.

223. Hootman, S. R. and Philpott, C. W., Accessory cells in teleost branchial epithelium, *Am. J. Physiol.*, 238, R199–R206, 1980.

224. Horng, J., Lin, L., Huang, C., Katoh, F., Kaneko, T., and Hwang, P., Knockdown of V-ATPase subunit A (atp6v1a) impairs acid secretion and ion balance in zebrafish (*Danio rerio*), *Am. J. Physiol.*, 292, R2068–R2076, 2007.

225. Howe, D. and Gutknecht, J., Role of urinary bladder in osmoregulation in marine teleost, *Opsanus tau*, *Am. J. Physiol.*, 235, R48–R54, 1978.

226. Hwang, P. P. and Lee, T. H., New insights into fish ion regulation and mitochondrion-rich cells, *Comp. Biochem. Physiol. A*, 148, 479–497, 2007.

227. Hyndman, K. A., Choe, K. P., Havird, J. C., Rose, R. E., Piermarini, P. M., and Evans, D. H., Neuronal nitric oxide synthase in the gill of the killifish, *Fundulus heteroclitus*, *Comp. Biochem. Physiol. B*, 144, 510–519, 2006.

228. Hyodo, S., Bell, J. D., Healy, J. M., Kaneko, T., Hasegawa, S., Takei, Y., Donald, J. A., and Toop, T., Osmoregulation in elephant fish *Callorhinchus milli* (Holocephalii), with special reference to the rectal gland, *J. Exp. Biol.*, 210, 1303–1310, 2007.

229. Hyodo, S., Katoh, F., Kaneko, T., and Takei, Y., A facilitative urea transporter is localized in the renal collecting tubule of the dogfish *Triakis scyllia*, *J. Exp. Biol.*, 207, 347–356, 2004.

230. Inoue, K. and Takei, Y., Molecular evolution of the natriuretic peptide system as revealed by comparative genomics, *Comp. Biochem. Physiol. D*, 1, 69–76, 2006.

231. Ip, Y. K., Yeo, P. J., Loong, A. M., Hiong, K. C., Wong, W. P., and Chew, S. F., The interplay of increased urea synthesis and reduced ammonia production in the African lungfish *Protopterus aethiopicus* during 46 days of aestivation in a mucus cocoon, *J. Exp. Zool.*, 303A, 1054–1065, 2005.

232. Isaia, J., Water and nonelectrolyte permeation. In *Fish Physiology*, Hoar, W. S. and Randall, D. J., Eds., Academic Press, New York, 1984, pp. 1–38.

232a. Ivanis, G., Esbaugh, A., and Perry, S., Branchial expression and localization of SLC9A2 and SLC9A3 sodium/hydrogen exchangers and their possible role in acid–base regulation in freshwater rainbow trout (*Oncorhynchus mykiss*), *J. Exp. Biol.*, 211, 2467–2477, 2008.

233. Janech, M. G., Fitzgibbon, W. R., Chen, R., Nowak, M. W., Miller, D. H., Paul, R. V., and Ploth, D. W., Molecular and functional characterization of a urea transporter from the kidney of the Atlantic stingray, *Am. J. Physiol.*, 284, F996–F1005, 2003.

234. Janech, M. G., Fitzgibbon, W. R., Ploth, D. W., Lacy, E. R., and Miller, D. H., Effect of low environmental salinity on plasma composition and renal function of the Atlantic stingray, a euryhaline elasmobranch, *Am. J. Physiol.*, 291, F770–F780, 2006.

235. Janech, M. G. and Piermarini, P. M., Renal water and solute excretion in the Atlantic stingray in fresh water, *J. Fish Biol.*, 60, 1–5, 2002.

236. Janvier, P., Catching the first fish, *Nature*, 402, 21–22, 1999.

237. Jeffery, W. R., Strickler, A. G., and Yamamoto, Y., Migratory neural crest-like cells form body pigmentation in a urochordate embryo, *Nature*, 431, 696–699, 2004.

238. Job, S. W., The respiratory metabolism of *Tilapia mossambica* (Teleostei). I. The effect of size, temperature and salinity, *Mar. Biol.*, 2, 121–126, 1969.

239. Johansen, K., Lykkeboe, G., Weber, R. E., and Maloiy, G. M. O., Respiratory properties of blood in awake and estivating lungfish, *Protopterus amphibius*, *Resp. Physiol.*, 27(3), 335–345, 1976.

240. Johnson, M. R. and Snelson, F. F., Reproductive life history of the Atlantic stingray, *Dasyatis sabina* (Pisces, Dasyatidae), in the freshwater St. Johns River, Florida, *Bull. Mar. Sci.*, 59, 74–88, 1996.

241. Jorgensen, J. M., Lomholt, J. P., Weber, R. E., and Malte, H., *The Biology of Hagfishes*, Chapman & Hall, London, 1998.

242. Kamunde, C. N. and Kisia, S. M., Fine structure of the nephron in the euryhaline teleost *Oreochromis niloticus*, *Acta Biol. Hung.*, 45(1), 111–121, 1994.

243. Karnaky, Jr., K. G. and Kinter, W. B., Killifish opercular skin: a flat epithelium with a high density of chloride cells, *J. Exp. Zool.*, 199, 355–364, 1977.

244. Karnaky, Jr., K. G., Structure and function of the chloride cell of *Fundulus heteroclitus* and other teleosts, *Am. Zool.*, 26, 209–224, 1986.

245. Karnaky, Jr., K. G., Teleost osmoregulation: changes in the tight junction in response to the salinity of the environment. In *Tight Junctions*, Cereijido, M., Ed., CRC Press, Boca Raton, FL, 1992, pp. 175–185.

246. Karnaky, Jr., K. G., Osmotic and ionic regulation. In *The Physiology of Fishes*, 2nd ed., Evans, D. H., Ed., CRC Press, Boca Raton, FL, 1998, pp. 157–176.

247. Karnaky, Jr., K. G., Kinter, L. B., Kinter, W. B., and Stirling, C. E., Teleost chloride cell. II. Autoradiographic localization of gill Na,K-ATPase in killifish *Fundulus heteroclitus* adapted to low and high salinity environments, *J. Cell Biol.*, 70, 157–177, 1976.

248. Kato, A., Doi, H., Nakada, T., Sakai, H., and Hirose, S., *Takifugu obscurus* is a euryhaline *Fugu* species very close to *Takifugu rubripes* and suitable for studying osmoregulation, *BMC Physiol.*, 5, 18, 2005.

249. Katoh, F., Hasegawa, S., Kita, J., Takagi, Y., and Kaneko, T., Distinct seawater and freshwater types of chloride cells in killifish, *Fundulus heteroclitus*, *Can. J. Zool.*, 79, 822–829, 2001.

250. Katoh, F., Hyodo, S., and Kaneko, T., Vacuolar-type proton pump in the basolateral plasma membrane energizes ion uptake in branchial mitochondria-rich cells of killifish, *Fundulus heteroclitus*, adapted to a low ion environment, *J. Exp. Biol.*, 206, 793–803, 2003.

252. Katoh, F., Tresguerres, M., Lee, K. M., Kaneko, T., Aida, K., and Goss, G. G., Cloning of rainbow trout SLC26A1: involvement in renal sulfate secretion, *Am. J. Physiol.*, 290, R1468–R1478, 2006.

253. Keys, A., Chloride and water secretion and absorption by the gills of the eel, *Z. vergl. Physiol.*, 15, 364–389, 1931.

254. Keys, A. B. and Willmer, E. N., "Chloride-secreting cells" in the gills of fishes with special reference to the common eel, *J. Physiol. Lond.*, 76, 368–378, 1932.

256. Kidder, 3rd, G. W., Petersen, C. W., and Preston, R. L., Energetics of osmoregulation: I. Oxygen consumption by *Fundulus heteroclitus*, *J. Exp. Zool.*, 305A, 309–317, 2006.

257. Kidder, 3rd, G. W., Petersen, C. W., and Preston, R. L., Energetics of osmoregulation. II. Water flux and osmoregulatory work in the euryhaline fish, *Fundulus heteroclitus*, *J. Exp. Zool. A Comp. Exp. Biol.*, 305(4), 318–327, 2006.

258. Deleted at proofs.

259. Kirschner, L. B., The mechanism of sodium chloride uptake in hyperregulating aquatic animals, *J. Exp. Biol.*, 207, 1439–1452, 2004.

260. Kirschner, L. B., The energetics of osmotic regulation in ureotelic and hypoosmotic fishes, *J. Exp. Zool.*, 267, 19–26, 1993.

261. Kirschner, L. B., Energetics of osmoregulation in fresh water vertebrates, *J. Exp. Zool.*, 271, 243–252, 1995.

262. Kleyman, T. and Cragoe, E., Amiloride and its analogs as tools in the study of ion transport, *J. Membr. Biol.*, 105, 1–21, 1988.

263. Kluge, B. and Fischer, A., The pronephros of the early ammocoete larva of lampreys (Cyclostomata, Petromyzontes): fine structure of the external glomus, *Cell Tiss. Res.*, 260, 249–259, 1990.

264. Klungsoyr, L., Magnesium ion activated ATPase and proton transport in vesicles from the gill of rainbow trout (*Salmo gairdneri*), *Comp. Biochem. Physiol. B*, 88, 1125–1134, 1987.

265. Koomoa, D. L., Musch, M. W., MacLean, A. V., and Goldstein, L., Volume-activated trimethylamine oxide efflux in red blood cells of spiny dogfish (*Squalus acanthias*), *Am. J. Physiol.*, 281, R803–R810, 2001.

266. Krayushkina, L. S., Gerasimov, A. A., and Smirnov, A. V., Hypoosmotic regulation in anadromous marine sturgeon, with special reference to the structure and function of their kidneys and gill chloride cells, *Dokl. Biol. Sci.*, 378, 210–212, 2001.

267. Kristiansen, H. R. and Rankin, J. C., Discrimination between endogenous and exogenous water sources in juvenile rainbow trout fed extruded dry feed, *Aquat. Living Resour.*, 14(6), 359–366, 2001.

268. Krogh, A., Osmotic regulation in freshwater fishes by active absorption of chloride ions, *Z. vergl. Physiol.*, 24, 656–666, 1937.

269. Krogh, A., The active absorption of ions in some freshwater animals, *Z. vergl. Physiol.*, 25, 335–350, 1938.

270. Kultz, D. and Avila, K., Mitogen-activated protein kinases are *in vivo* transducers of osmosensory signals in fish gill cells, *Comp. Biochem. Physiol. B*, 129, 821–829, 2001.

271. Kultz, D., Chakravarty, D., and Adilakshmi, T., A novel 14–3–3 gene is osmoregulated in gill epithelium of the euryhaline teleost *Fundulus heteroclitus*, *J. Exp. Biol.*, 204, 2975–2985, 2001.

272. Kurita, Y., Nakada, T., Kato, A., Doi, H., Mistry, A. C., Chang, M. H., Romero, M. F., and Hirose, S., Identification of intestinal bicarbonate transporters involved in formation of carbonate precipitates to stimulate water absorption in marine teleost fish, *Am. J. Physiol.*, 294(4), R1402–R1412, 2008.

273. Lacy, E. R. and Reale, E., Functional morphology of the elasmobranch nephron and retention of urea. In *Cellular and Molecular Approaches to Fish Ionic Regulation*, Wood, C. M. and Shuttleworth, T. J., Eds., Academic Press, San Diego, CA, 1995, pp. 107–146.

274. Lacy, E. R. and Reale, E., Urinary system. In *Sharks, Skates, and Rays*, Hamlett, W. C., Ed., The Johns Hopkins University Press, Baltimore, MD, 1999, pp. 353–397.

275. Lagios, M. D., Granular epithelioid (juxtaglomerular) cell and renovascular morphology of the coelacanth *Latimeria chalumnae* Smith (Crossopterygii) compared with that of other fishes, *Gen. Comp. Endocrinol.*, 22, 296–307, 1974.

276. Lagios, M. D. and McCosker, J. E., A cloacal excretory gland in the lungfish *Protopterus*, *Copeia*, 1977, 176–178, 1977.

277. Lahlou, B., Henderson, I. W., and Sawyer, W. H., Renal adaptations by *Opsanus tau*, a euryhaline aglomerular teleost, to dilute media, *Am. J. Physiol.*, 216, 1266–1272, 1969.

278. Laiz-Carrion, R., Guerreiro, P. M., Fuentes, J., Canario, A. V., Martin Del Rio, M. P., and Mancera, J. M., Branchial osmoregulatory response to salinity in the gilthead sea bream, *Sparus auratus*, *J. Exp. Zool.*, 303A, 563–576, 2005.

279. Laurent, P., Gill internal morphology. In *Fish Physiology*, Hoar, W. S. and Randall, D. J., Eds., Academic Press, New York, 1984, pp. 39–63.

280. Laurent, P., Chevalier, C., and Wood, C., Appearance of cuboidal cells in relation to salinity in gills of *Fundulus heteroclitus*, a species exhibiting branchial Na$^+$ but not Cl$^-$ uptake in freshwater, *Cell Tissue Res.*, 325, 481–492, 2006.

281. Laurent, P. and Dunel, S., Morphology of gill epithelia in fish, *Am. J. Physiol.*, 238(3), R147–R159, 1980.

282. Lee, J., Valkova, N., White, M. P., and Kultz, D., Proteomic identification of processes and pathways characteristic of osmoregulatory tissues in spiny dogfish (*Squalus acanthias*), *Comp. Biochem. Physiol. D*, 1, 328–343, 2006.

283. Lee, T. H., Feng, S. H., Lin, C. H., Hwang, Y. H., Huang, C. L., and Hwang, P. P., Ambient salinity modulates the expression of sodium pumps in branchial mitochondria-rich cells of Mozambique tilapia, *Oreochromis mossambicus*, *Zool. Sci.*, 20, 29–36, 2003.

284. Lehrich, R. W., Aller, S. G., Webster, P., Marino, C. R., and Forrest, Jr., J. N., Vasoactive intestinal peptide, forskolin, and genistein increase apical CFTR trafficking in the rectal gland of the spiny dogfish, *Squalus acanthias*. Acute regulation of CFTR trafficking in an intact epithelium, *J. Clin. Invest.*, 101, 737–745, 1998.

285. Lemire, M. and Lagios, M. D., Ultrastructure of the secretory parenchyma of the postanal gland of the coelacanth, *Latimeria chalumnae* Smith, *Acata Anat. (Basel)*, 104, 1–15, 1979.

286. Liebert, A. M. and Schreck, C. B., Effects of acute stress on osmoregulation, feed intake, IGF-1, and cortisol in yearling steelhead trout (*Oncorhynchus mykiss*) during seawater adaptation, *Gen. Comp. Endocrinol.*, 148, 195–202, 2006.

287. Liem, K. F., Bemis, W., Walker, W. F., and Grande, L., *Functional Anatomy of the Vertebrates: An Evolutionary Perspective*, 3rd ed., Brooks/Cole, Florence, KY, 2001.

288. Lignot, J. H., Cutler, C. P., Hazon, N., and Cramb, G., Immunolocalisation of aquaporin 3 in the gill and the gastrointestinal tract of the European eel *Anguilla anguilla* (L.), *J. Exp. Biol.*, 205, 2653–2663, 2002.

289. Lima, R. N. and Kültz, D., Laser scanning cytometry and tissue microarray analysis of salinity effects on killifish chloride cells, *J. Exp. Biol.*, 207, 1729–1739, 2004.

290. Lin, C.-H., Huang, P.-P., Yang, C.-H., Lee, T.-H., and Hwang, P.-P., Time-course changes in the expression of Na,K-ATPase and the morphometry of mitochondria-rich cells in gills of euryhaline tilapia (*Oreochromis mossambicus*) during freshwater acclimation, *J. Exp. Zool.*, 301A, 85–96, 2004.

291. Lin, C. H. and Lee, T. H., Sodium or potassium ions activate different kinetics of gill Na,K-ATPase in three seawater- and freshwater-acclimated euryhaline teleosts, *J. Exp. Zool.*, 303A, 57–65, 2005.

292. Lin, C. H., Tsai, R. S., and Lee, T. H., Expression and distribution of Na,K-ATPase in gill and kidney of the spotted green pufferfish, *Tetraodon nigroviridis*, in response to salinity challenge, *Comp. Biochem. Physiol. A*, 138, 287–295, 2004.

293. Lin, H. and Randall, D. J., H+-ATPase activity in crude homogenates of fish gill tissue: inhibitor sensitivity and environmental and hormonal regulation, *J. Exp. Biol.*, 180, 163–174, 1993.

294. Lin, L. Y., Horng, J. L., Kunkel, J. G., and Hwang, P. P., Proton pump-rich cell secretes acid in skin of zebrafish larvae, *Am. J. Physiol.*, 290, C371–C378, 2006.

295. Deleted at proofs.

296. Lin, Y. M., Chen, C. N., and Lee, T. H., The expression of gill Na,K-ATPase in milkfish, *Chanos chanos*, acclimated to seawater, brackish water and fresh water, *Comp. Biochem. Physiol. A*, 135, 489–497, 2003.

297. Lin, Y. M., Chen, C. N., Yoshinaga, T., Tsai, S. C., Shen, I. D., and Lee, T. H., Short-term effects of hyposmotic shock on Na+/K+-ATPase expression in gills of the euryhaline milkfish, *Chanos chanos*, *Comp. Biochem. Physiol. A*, 143, 406–415, 2006.

298. Lingwood, D., Harauz, G., and Ballantyne, J., Decoupling the Na+,K+-ATPase *in vivo*: a possible new role in the gills of freshwater fishes, *Comp. Biochem. Physiol. A*, 144, 451–457, 2006.

299. Logan, A. G., Moriarty, R. J., Morris, R., and Rankin, J. C., The anatomy and blood system of the kidney in the river lamprey, *Lampetra fluviatilis*, *Anat. Embryol. (Berl.)*, 158, 245–252, 1980.

300. Logan, A. G., Moriarty, R. J., and Rankin, J. C., A micropuncture study of kidney function in the river lamprey, *Lampetra fluviatilis*, adapted to fresh water, *J. Exp. Biol.*, 85, 137–147, 1980.

301. Logan, A. G., Morris, R., and Rankin, J. C., A micropuncture study of kidney function in the river lamprey, *Lampetra fluviatilis*, adapted to sea water, *J. Exp. Biol.*, 88, 239–247, 1980.

302. Loong, A. M., Hiong, K. C., Lee, S. M., Wong, W. P., Chew, S. F., and Ip, Y. K., Ornithine–urea cycle and urea synthesis in African lungfishes, *Protopterus aethiopicus* and *Protopterus annectens*, exposed to terrestrial conditions for six days, *J. Exp. Zool.*, 303A, 354–65, 2005.

303. Loretz, C. A., Electrophysiology of ion transport in teleost intestinal cells. In *Cellular and Molecular Approaches to Fish Ionic Regulation*, Wood, C. M. and Shuttleworth, T. J., Eds., Academic Press, San Diego, CA, 1995, pp. 25–56.

304. Loretz, C. A. and Bern, H. A., Ion transport by the urinary bladder of the gobiid teleost, *Gillichthys mirabilis*, *Am. J. Physiol.*, 239(5), R415–R423, 1980.

305. Loretz, C. A. and Fourtner, C. R., Functional characterization of a voltage-gated anion channel from teleost fish intestinal epithelium, *J. Exp. Biol.*, 136, 383–403, 1988.

306. Lytle, C., Xu, J. C., Biemesderfer, D., Haas, M., and Forbush, 3rd, B., The Na–K–Cl cotransport protein of shark rectal gland. I. Development of monoclonal antibodies, immunoaffinity purification, and partial biochemical characterization, *J. Biol. Chem.*, 267, 25428–25437, 1992.

307. MacKenzie, S., Cutler, C. P., Hazon, N., and Cramb, G., The effects of dietary sodium loading on the activity and expression of Na,K-ATPase in the rectal gland of the European dogfish (*Scyliorhinus canicula*), *Comp. Biochem. Physiol. B*, 131, 185–200, 2002.

308. Maetz, J., Sea water teleosts: evidence for a sodium–potassium exchange in the branchial sodium-excreting pump, *Science*, 166, 613–615, 1969.

309. Maetz, J., Fish gills: mechanisms of salt transfer in fresh water and sea water, *Philos. Trans. R. Soc. Lond. B*, 262, 209–251, 1971.

310. Mallatt, J. and Chen, J. Y., Fossil sister group of craniates: predicted and found, *J. Morphol.*, 258(1), 1–31, 2003.

311. Mallatt, J., Conley, D. M., and Ridgway, R. L., Why do hagfish have gill "chloride cells" when they need not regulate plasma NaCl concentration?, *Can. J. Zool.*, 65, 1956–1965, 1987.

312. Mallatt, J. and Paulsen, C., Gill ultrastructure of the Pacific hagfish *Eptatretus stouti*, *Am. J. Anat.*, 177(2), 243–269, 1986.

313. Mallatt, J. and Sullivan, J., 28S and 18S rDNA sequences support the monophyly of lampreys and hagfishes, *Mol. Biol. Evol.*, 15, 1706–1718, 1998.

314. Marshall, J., Martin, K. A., Picciotto, M., Hockfield, S., Nairn, A. C., and Kaczmarek, L. K., Identification and localization of a dogfish homolog of human cystic fibrosis transmembrane conductance regulator, *J. Biol. Chem.*, 266, 22749–22754, 1991.

316. Marshall, W. S., Sodium dependency of active chloride transport across isolated fish skin (*Gillichthys mirabilis*), *J. Physiol.*, 319, 165–178, 1981.

317. Marshall, W. S., Independent Na$^+$ and Cl$^-$ active transport by urinary bladder epithelium of brook trout, *Am. J. Physiol.*, 250, R227–R234, 1986.

318. Marshall, W. S., Transport processes in isolated teleost epithelia: opercular epithelium and urinary bladder. In *Cellular and Molecular Approaches to Fish Ionic Regulation*, Wood, C. M. and Shuttleworth, T. J., Eds., Academic Press, San Diego, CA, 1995, pp. 1–23.

319. Marshall, W. S., Na$^+$, Cl$^-$, Ca^{2+} and Zn^{2+} transport by fish gills: retrospective review and prospective synthesis, *J. Exp. Zool.*, 293, 264–283, 2002.

320. Marshall, W. S., Rapid regulation of NaCl secretion by estuarine teleost fish: coping strategies for short-duration freshwater exposures, *Biochim. Biophys. Acta*, 1618, 95–105, 2003.

321. Marshall, W. S. and Bryson, S. E., Intracellular pH regulation in trout urinary bladder epithelium: Na$^+$-H$^+$(NH$_4^+$) exchange, *Am. J. Physiol.*, 261, R652–R658, 1991.

322. Marshall, W. S. and Bryson, S. E., Transport mechanisms of seawater teleost chloride cells: an inclusive model of a multifunctional cell, *Comp. Biochem. Physiol. A*, 119(1), 97–106, 1998.

323. Marshall, W. S., Bryson, S. E., Burghardt, J. S., and Verbost, P. M., Ca^{2+} transport by opercular epithelium of the fresh water adapted euryhaline teleost, *Fundulus heteroclitus*, *J. Comp. Physiol. B*, 165, 268–277, 1995.

324. Marshall, W. S., Bryson, S. E., and Luby, T., Control of epithelial Cl$^-$ secretion by basolateral osmolality in the euryhaline teleost *Fundulus heteroclitus*, *J. Exp. Biol.*, 203, 1897–1905, 2000.

325. Marshall, W. S., Bryson, S. E., and Wood, C. M., Calcium transport by isolated skin of rainbow trout, *J. Exp. Biol.*, 166, 297–316, 1992.

326. Marshall, W. S. and Grosell, M., Ion transport, osmoregulation and acid–base balance. In *The Physiology of Fishes*, 3rd ed., Evans, D. H. and Claiborne, J. B., Eds., CRC Press, Boca Raton, FL, 2006, pp. 177–230.

328. Marshall, W. S., Howard, J. A., Cozzi, R. R., and Lynch, E. M., NaCl and fluid secretion by the intestine of the teleost *Fundulus heteroclitus*: involvement of CFTR, *J. Exp. Biol.*, 205(Pt. 6), 745–758, 2002.

329. Marshall, W. S., Lynch, E. M., and Cozzi, R. R., Redistribution of immunofluorescence of CFTR anion channel and NKCC cotransporter in chloride cells during adaptation of the killifish *Fundulus heteroclitus* to sea water, *J. Exp. Biol.*, 205, 1265–1273, 2002.

330. Marshall, W. S. and Nishioka, R. S., Relation of mitochondria-rich chloride cells to active chloride transport in the skin of a marine teleost, *J. Exp. Zool.*, 214, 147–156, 1980.

331. Marshall, W. S., Ossum, C. G., and Hoffmann, E. K., Hypotonic shock mediation by p38 MAPK, JNK, PKC, FAK, OSR1, and SPAK in osmosensing chloride secreting cells of killifish opercular epithelium, *J. Exp. Biol.*, 208, 1063–1077, 2005.

332. Marshall, W. S. and Singer, T. D., Cystic fibrosis transmembrane conductance regulator in teleost fish, *Biochim. Biophys. Acta*, 1566, 16–27, 2002.

333. Martin, R. A., Conservation of freshwater and euryhaline elasmobranchs, *J. Mar. Biol. Assoc. U.K.*, 85, 1049–1073, 2005.

334. Martinez, A. S., Cutler, C. P., Wilson, G. D., Phillips, C., Hazon, N., and Cramb, G., Cloning and expression of three aquaporin homologues from the European eel (*Anguilla anguilla*): effects of seawater acclimation and cortisol treatment on renal expression, *Biol. Cell*, 97, 615–627, 2005.

335. Martinez, A. S., Cutler, C. P., Wilson, G. D., Phillips, C., Hazon, N., and Cramb, G., Regulation of expression of two aquaporin homologs in the intestine of the European eel: effects of seawater acclimation and cortisol treatment, *Am. J. Physiol.*, 288, R1733–R1743, 2005.

336. Martinez, A. S., Wilson, G., Phillips, C., Cutler, C., Hazon, N., and Cramb, G., Effect of cortisol on aquaporin expression in the esophagus of the European eel, *Anguilla anguilla*, *Ann. N.Y. Acad. Sci.*, 1040, 395–398, 2005.

337. Martinez-Alvarez, R. M., Hidalgo, M. C., Domezain, A., Morales, A. E., Garcia-Gallego, M., and Sanz, A., Physiological changes of sturgeon *Acipenser naccarii* caused by increasing environmental salinity, *J. Exp. Biol.*, 205, 3699–3706, 2002.

338. Martinez-Alvarez, R. M., Sanz, A., Garcia-Gallego, M., Domezain, A., Domezain, J., Carmona, R., del Valle Ostos-Garrido, M., and Morales, A. E., Adaptive branchial mechanisms in the sturgeon *Acipenser naccarii* during acclimation to saltwater, *Comp. Biochem. Physiol. A*, 141, 183–190, 2005.

339. Martini, F. H., The ecology of hagfishes. In *The Biology of Hagfishes*, Jorgensen, J. M., Lomholt, J. P., Weber, R. E., and Malte, H., Eds., Chapman & Hall, London, 1998, pp. 57–77.

340. McCormick, S. D., Endocrine control of osmoregulation in teleost fish, *Am. Zool.*, 41, 781–794, 2001.

341. McCormick, S. D. and Bradshaw, D., Hormonal control of salt and water balance in vertebrates, *Gen. Comp. Endocrinol.*, 147, 3–8, 2006.

342. McCormick, S. D., Hasegawa, S., and Hirano, T., Calcium uptake in the skin of a freshwater teleost, *Proc. Natl. Acad. Sci. USA*, 89, 3635–3638, 1992.

343. McCormick, S. D., Sundell, K., Bjornsson, B. T., Brown, C. L., and Hiroi, J., Influence of salinity on the localization of Na⁺/K⁺-ATPase, Na⁺/K⁺/2Cl⁻ cotransporter (NKCC) and CFTR anion channel in chloride cells of the Hawaiian goby (*Stenogobius hawaiiensis*), *J. Exp. Biol.*, 206, 4575–4583, 2003.

344. McDonald, M. D., Smith, C. P., and Walsh, P. J., The physiology and evolution of urea transport in fishes, *J. Membr. Biol.*, 212, 93–107, 2006.

345. McDonald, M. D., Walsh, P. J., and Wood, C. M., Transport physiology of the urinary bladder in teleosts: a suitable model for renal urea handling?, *J. Exp. Zool.*, 292(7), 604–617, 2002.

346. McFarland, W. N. and Munz, F. W., Regulation of body weight and serum composition by hagfish in various media, *Comp. Biochem. Physiol.*, 14, 393–398, 1965.

347. McInerney, J. E., Renal sodium reabsorption in the hagfish, *Eptatretus stouti*, *Comp. Biochem. Physiol.*, 49, 273–280, 1974.

348. McVicar, A. J. and Rankin, J. C., Dynamics of glomerular filtration in the river lamprey, *Lampetra fluviatilis* L., *Am. J. Physiol.*, 249, F132–F138, 1985.

349. Millot, J. and Anthony, J., Le glande post-anale de *Latimeria*, *Ann. Sci. Natur. Zool. Paris 12e Ser.*, 14, 305–318, 1960.

350. Mistry, A. C., Kato, A., Tran, Y. H., Honda, S., Tsukada, T., Takei, Y., and Hirose, S., FHL5, a novel actin fiber-binding protein, is highly expressed in gill pillar cells and responds to wall tension in eels, *Am. J. Physiol.*, 287, 1141–1154, 2004.

351. Miyazaki, H., Kaneko, T., Uchida, S., Sasaki, S., and Takei, Y., Kidney-specific chloride channel, OmClC-K, predominantly expressed in the diluting segment of freshwater-adapted tilapia kidney, *Proc. Natl. Acad. Sci. USA*, 99, 15782–15787, 2002.

352. Morgan, J. D. and Iwama, G. K., Energy cost of NaCl transport in isolated gills of cutthroat trout, *Am. J. Physiol.*, 277, R631–R639, 1999.

353. Morgan, R. L., Ballantyne, J. S., and Wright, P. A., Regulation of a renal urea transporter with reduced salinity in a marine elasmobranch, *Raja erinacea*, *J. Exp. Biol.*, 206, 3285–3292, 2003.

354. Morgan, R. L., Wright, P. A., and Ballantyne, J. S., Urea transport in kidney brush-border membrane vesicles from an elasmobranch, *Raja erinacea*, *J. Exp. Biol.*, 206, 3293–3302, 2003.

355. Morris, R., The mechanism of marine osmoregulation in the lampern, *Lampetra fluviatilis* L., and the cause of its breakdown during the spawning migration, *J. Exp. Biol.*, 35, 649–665, 1958.

356. Morris, R., General problems of osmoregulation with special reference to cyclostomes, *Symp. Zool. Soc. Lond.*, 1, 1–16, 1960.

357. Morris, R., Osmoregulation. In *The Biology of Lampreys*, Hardisty, M. W. and Potter, I. C., Eds., Academic Press, London, 1972, pp. 193–239.

358. Motais, R., Isaia, J., Rankin, J. C., and Maetz, J., Adaptive changes of water permeability of the teleostean gill epithelium in relation to external salinity, *J. Exp. Biol.*, 51, 529–546, 1969.

359. Munz, F. W. and McFarland, W. N., Regulatory function of a primitive vertebrate kidney, *Comp. Biochem. Physiol.*, 13, 381–400, 1964.

360. Myers, J. D., Murdaugh, H. V., Davis, B., Blumentals, A., Eichenholz, A., Ragn, M. V., and Murdaugh, E. W., Effects of diuretic drugs on renal function in *Squalus acanthias*, *Bull. Mt. Desert Isl. Biol. Lab.*, 11, 71–71, 1971.

361. Nawata, C. M., Hung, C. C., Tsui, T. K., Wilson, J. M., Wright, P. A., and Wood, C. M., Ammonia excretion in rainbow trout (*Oncorhynchus mykiss*): evidence for Rh glycoprotein and H^+-ATPase involvement, *Physiol. Genom.*, 31, 463–474, 2007.

362. Nearing, J., Betka, M., Quinn, S., Hentschel, H., Elger, M., Baum, M., Bai, M., Chattopadyhay, N., Brown, E. M., Hebert, S. C., and Harris, H. W., Polyvalent cation receptor proteins (CaRs) are salinity sensors in fish, *Proc. Natl. Acad. Sci. USA*, 99, 9231–9236, 2002.

363. Nebel, C., Negre-Sadargues, G., Blasco, C., and Charmantier, G., Morphofunctional ontogeny of the urinary system of the European sea bass *Dicentrarchus labrax*, *Anat. Embryol. (Berl.)*, 209, 193–206, 2005.

364. Nebel, C., Romestand, B., Negre-Sadargues, G., Grousset, E., Aujoulat, F., Bacal, J., Bonhomme, F., and Charmantier, G., Differential freshwater adaptation in juvenile sea-bass *Dicentrarchus labrax*: involvement of gills and urinary system, *J. Exp. Biol.*, 208, 3859–3871, 2005.

365. Nelson, J. S., *Fishes of the World*, 4th ed., John Wiley & Sons, Hoboken, NJ, 2006.

366. Nishimoto, G., Sasaki, G., Yaoita, E., Nameta, M., Li, H., Furuse, K., Fujinaka, H., Yoshida, Y., Mitsudome, A., and Yamamoto, T., Molecular characterization of water-selective AQP (EbAQP4) in hagfish: insight into ancestral origin of AQP4, *Am. J. Physiol.*, 292, R644–R651, 2007.

367. Nishimura, H. and Imai, M., Control of renal function in freshwater and marine teleosts, *Fed. Proc.*, 41(8), 2355–2360, 1982.

368. Nishimura, H., Imai, M., and Ogawa, M., Sodium chloride and water transport in the renal distal tubule of the rainbow trout, *Am. J. Physiol.*, 244(3), F247–F254, 1983.

369. Nordlie, F. G., The influence of environmental salinity on respiratory oxygen demands in the euryhaline teleost, *Ambassis interrupta* Bleeker, *Comp. Biochem. Physiol. A*, 59, 271–274, 1978.

370. Oguri, M., Rectal glands of marine and fresh-water sharks: comparative histology, *Science*, 144, 1151–1152, 1964.

371. Ojeda, J. L., Icardo, J. M., and Domezain, A., Renal corpuscle of the sturgeon kidney: an ultrastructural, chemical dissection, and lectin-binding study, *Anat. Rec.*, 272, 563–573, 2003.

372. Ojeda, J. L., Icardo, J. M., Wong, W. P., and Ip, Y. K., Microanatomy and ultrastructure of the kidney of the African lungfish *Protopterus dolloi*, *Anat. Rec.*, 288, 609–625, 2006.

373. Olson, K. R., Morphology and vascular anatomy of the gills of a primitive air-breathing fish, the bowfin (*Amia calva*), *Cell Tissue Res.*, 218, 499–517, 1981.

374. Olson, K. R., Rectal gland and volume homeostasis. In *Sharks, Skates, and Rays*, Hamlett, W. C., Ed., The Johns Hopkins University Press, Baltimore, MD, 1999, pp. 329–352.

375. Olson, K. R., Vascular anatomy of the fish gill, *J. Exp. Zool.*, 293(3), 214–231, 2002.

376. Ota, K. G., Kuraku, S., and Kuratani, S., Hagfish embryology with reference to the evolution of the neural crest, *Nature*, 446, 672–675, 2007.

377. Ota, K. G. and Kuratani, S., The history of scientific endeavors towards understanding hagfish embryology, *Zool. Sci.*, 23, 403–418, 2006.

378. Parks, S. K., Tresguerres, M., and Goss, G. G., Blood and gill responses to HCl infusions in the Pacific hagfish (*Eptatretus stoutii*), *Can. J. Zool.*, 85(8), 855–862, 2007.

379. Parks, S. K., Tresguerres, M., and Goss, G. G., Interactions between Na^+ channels and Na^+-HCO_3^- cotransporters in the freshwater fish gill MR cell: a model for transepithelial Na^+ uptake, *Am. J. Physiol.*, 292, C935–C944, 2007.

380. Parmelee, J. T. and Renfro, J. L., Esophageal desalination of seawater in flounder: role of active sodium transport, *Am. J. Physiol.*, 245, R888–R893, 1983.

381. Pärt, P., Wright, P. A., and Wood, C. M., Urea and water permeability in dogfish (*Squalus acanthias*) gills, *Comp. Biochem. Physiol. A*, 119, 117–123, 1998.

382. Patrick, M., Part, P., Marshall, W., and Wood, C., Characterization of ion and acid–base transport in the fresh-water-adapted mummichog (*Fundulus heteroclitus*), *J. Exp. Zool.*, 279, 208–219, 1997.

383. Patrick, M. L. and Wood, C. M., Ion and acid–base regulation in the freshwater mummichog (*Fundulus heteroclitus*): a departure from the standard model for freshwater teleosts, *Comp. Biochem. Physiol. A*, 122, 445–456, 1999.

384. Payan, P., Girard, J. P., and Mayer-Gostan, N., Branchial ion movements in teleosts: the roles of respiratory and chloride cells. In *Fish Physiology*, Hoar, W. S. and Randall, D. J., Eds., Academic Press, Orlando, FL, 1984, pp. 39–63.

385. Payan, P., Goldstein, L., and Forster, R. P., Gills and kidneys in ureosmotic regulation in euryhaline skates, *Am. J. Physiol.*, 224, 367–372, 1973.

386. Payan, P. and Maetz, J., Balance hydrique chez les Elasmobranches: arguments en faveur d'un controle endocrinien, *Gen. Comp. Endocrinol.*, 16, 535–554, 1971.

387. Peek, W. D. and Youson, J. H., Ultrastructure of chloride cells in young adults of the anadromous sea lamprey, *Petromyzon marinus* L., in fresh water and during adaptation to sea water, *J. Morphol.*, 160, 143–164, 1979.

388. Pelis, R. M., Edwards, S. L., Kunigelis, S. C., Claiborne, J. B., and Renfro, J. L., Stimulation of renal sulfate secretion by metabolic acidosis requires Na^+/H^+ exchange induction and carbonic anhydrase, *Am. J. Physiol. Ren. Physiol.*, 289, F208–F216, 2005.

389. Pelis, R. M. and Renfro, J. L., Role of tubular secretion and carbonic anhydrase in vertebrate renal sulfate excretion, *Am. J. Physiol.*, 287, R491–R501, 2004.

390. Pelis, R. M., Zydlewski, J., and McCormick, S. D., Gill Na^+-K^+-$2Cl^-$ cotransporter abundance and location in Atlantic salmon: effects of seawater and smolting, *Am. J. Physiol.*, 280, R1844–R1852, 2001.

391. Pellertier, D. and Besner, M., The effect of salty diets and gradual transfer to sea water on osmotic adaptation, gill Na^+,K^+-ATPase activation, survival of brook char, *Salvelinus fontinalis* Mitchill, *J. Fish Biol.*, 141, 791–803, 1992.

392. Perrot, M. N., Grierson, C. E., Hazon, N., and Balment, R., Drinking behaviour in sea water and fresh water teleosts, the role of the renin-angiotensin system, *Fish Physiol. Biochem.*, 10, 161–168, 1992.

393. Perry, S. F., Beyers, M. L., and Johnson, D. A., Cloning and molecular characterisation of the trout (*Oncorhynchus mykiss*) vacuolar H^+-ATPase B subunit, *J. Exp. Biol.*, 203, 459–470, 2000.

394. Perry, S. F. and Gilmour, K. M., Acid–base balance and CO_2 excretion in fish: unanswered questions and emerging models, *Respir. Physiol.*, 154, 199–215, 2006.

395. Perry, S. F., Gilmour, K. M., Swenson, E. R., Vulesevic, B., Chew, S. F., and Ip, Y. K., An investigation of the role of carbonic anhydrase in aquatic and aerial gas transfer in the African lungfish *Protopterus dolloi*, *J. Exp. Biol.*, 208, 3805–3815, 2005.

396. Perry, S. F., Rivero-Lopez, L., McNeill, B., and Wilson, J., Fooling a freshwater fish: how dietary salt transforms the rainbow trout gill into a saltwater phenotype, *J. Exp. Biol.*, 209, 4591–4596, 2006.

397. Perry, S. F., Shahsavarani, A., Georgalis, T., Bayaa, M., Furimsky, M., and Thomas, S. L. Y., Channels, pumps, and exchangers in the gill and kidney of freshwater fishes: their role in ionic and acid–base regulation, *J. Exp. Zool.*, 300A, 53–62, 2003.

398. Pickering, A. D. and Morris, R., Osmoregulation of *Lampetra fluviatilis* L. and *Petromyzon marinus* (Cyclostomata) in hyperosmotic solutions, *J. Exp. Biol.*, 53, 231–243, 1970.

399. Piermarini, P. M. and Evans, D. H., Osmoregulation of the Atlantic stingray (*Dasyatis sabina*) from the freshwater Lake Jesup of the St. Johns River, Florida, *Physiol. Zool.*, 71, 553–560, 1998.

400. Piermarini, P. M. and Evans, D. H., Effects of environmental salinity on Na^+/K^+-ATPase in the gills and rectal gland of a euryhaline elasmobranch (*Dasyatis sabina*), *J. Exp. Biol.*, 203, 2957–66, 2000.

401. Piermarini, P. M. and Evans, D. H., Immunochemical analysis of the vacuolar proton-ATPase B-subunit in the gills of a euryhaline stingray (*Dasyatis sabina*): effects of salinity and relation to Na^+/K^+-ATPase, *J. Exp. Biol.*, 204, 3251–3259, 2001.

402. Piermarini, P. M., Verlander, J. W., Royaux, I. E., and Evans, D. H., Pendrin immunoreactivity in the gill epithelium of a euryhaline elasmobranch, *Am. J. Physiol.*, 283, R983–R992, 2002.

403. Pillans, R. D. and Franklin, C. E., Plasma osmolyte concentrations and rectal gland mass of bull sharks *Carcharhinus leucas*, captured along a salinity gradient, *Comp. Biochem. Physiol. A*, 138, 363–371, 2004.

404. Pillans, R. D., Good, J. P., Anderson, W. G., Hazon, N., and Franklin, C. E., Freshwater to seawater acclimation of juvenile bull sharks (*Carcharhinus leucas*): plasma osmolytes and Na⁺/K⁺-ATPase activity in gill, rectal gland, kidney and intestine, *J. Comp. Physiol. B*, 175, 37–44, 2005.

405. Pisam, M., Membranous systems in the "chloride cell" of teleostean fish gill; their modifications in response to the salinity of the environment, *Anat. Rec.*, 200, 401–414, 1981.

406. Pisam, M., Auperin, B., Prunet, P., Rentier-Delrue, F., Martial, J., and Rambourg, A., Effects of prolactin on alpha and beta chloride cells in the gill epithelium of the saltwater adapted tilapia *Oreochromis niloticus*, *Anat. Rec.*, 235, 275–284, 1993.

407. Pisam, M., Boeuf, G., Prunet, P., and Rambourg, A., Ultrastructural features of mitochondria-rich cells in stenohaline freshwater and seawater fishes, *Am. J. Anat.*, 187, 21–31, 1990.

408. Pisam, M., Lemoal, C., Auperin, B., Prunet, P., and Rambourg, A., Apical structures of "mitochondria-rich" a and b cells in euryhaline fish gill: their behaviour in various living conditions, *Anat. Rec.*, 241, 13–24, 1995.

409. Deleted at proofs.

410. Pisam, M., Massa, F., Jammet, C., and Prunet, P., Chronology of the appearance of beta, A, and alpha mitochondria-rich cells in the gill epithelium during ontogenesis of the brown trout (*Salmo trutta*), *Anat. Rec.*, 259, 301–311, 2000.

411. Pisam, M., Prunet, P., and Rambourg, A., Accessory cells in the gill epithelium of freshwater rainbow trout *Salmo gairdneri*, *Am. J. Anat.*, 184, 311–320, 1989.

412. Pisam, M. and Rambourg, A., Mitochondria-rich cells in the gill epithelium of teleost fishes: an ultrastructural approach, *Int. Rev. Cytol.*, 130, 191–232, 1991.

413. Plotkin, M. D., Kaplan, M. R., Verlander, J. W., Lee, W. S., Brown, D., Poch, E., Gullans, S. R., and Hebert, S. C., Localization of the thiazide sensitive NaCl cotransporter, rTSC1, in the rat kidney, *Kidney Int.*, 50, 174–183, 1996.

414. Potts, W. T. W., Transepithelial potentials in fish gills. In *Fish Physiology*, Hoar, W. S. and Randall, D. J., Eds., Academic Press, New York, 1984, pp. 105–128.

415. Potts, W. T. W., Kinetics of sodium uptake in freshwater animals: a comparison of ion-exchange and proton pump hypotheses, *Am. J. Physiol.*, 266, R315–R320, 1994.

416. Potts, W. T. W. and Eddy, F. B., An analysis of the sodium and chloride fluxes in the winter flounder *Platichthys flesus*, *J. Comp. Physiol.*, 87, 21–28, 1973.

417. Potts, W. T. W. and Evans, D. H., Sodium and chloride balance in the killifish *Fundulus heteroclitus*, *Biol. Bull.*, 133, 411–425, 1967.

418. Potts, W. T. W. and Parry, G., *Osmotic and Ionic Regulation in Animals*, Macmillan, New York, 1964.

419. Preest, M. R., Gonzalez, R. J., and Wilson, R. W., A pharmacological examination of Na⁺ and Cl⁻ transport in two species of freshwater fish, *Physiol. Biochem. Zool.*, 78, 259–272, 2005.

420. Prunet, P., Sturm, A., and Milla, S., Multiple corticosteroid receptors in fish: from old ideas to new concepts, *Gen. Comp. Endocrinol.*, 147, 17–23, 2006.

421. Qiu, A. and Hogstrand, C., Functional characterisation and genomic analysis of an epithelial calcium channel (ECaC) from pufferfish, *Fugu rubripes*, *Gene*, 342, 113–123, 2004.

422. Raldua, D., Otero, D., Fabra, M., and Cerda, J., Differential localization and regulation of two aquaporin-1 homologs in the intestinal epithelia of the marine teleost *Sparus aurata*, *Am. J. Physiol.*, 294(3), R993–R1003, 2008.

423. Rankin, C., Osmotic and ionic regulation in cyclostomes. In *Ionic Regulation in Animals: A Tribute to Professor W.T.W. Potts*, Hazon, N., Eddy, F. B., and Flik, G., Eds., Springer, Berlin, 1997, pp. 50–69.

424. Rankin, J. C., Logan, A. G., and Moriarty, R. J., Changes in kidney function in the river lamprey, *Lampetra fluviatilis* L., in response to changes in external salinity. In *Epithelial Transport in the Lower Vertebrates*, Lahlou, B., Ed., Cambridge University Press, Cambridge, U.K., 1980, pp. 171–184.

425. Rao, G. M. M., Oxygen consumption of rainbow trout *Salmo gairdneri* in relation to activity and salinity, *Can. J. Zool.*, 46, 781–786, 1968.

426. Raymond, J. and Plopper, G., A bacterial TMAO transporter, *Comp. Biochem. Physiol. B*, 133, 29, 2002.

427. Reid, S. D., Hawkings, G. S., Galvez, F., and Goss, G. G., Localization and characterization of phenamil-sensitive Na⁺ influx in isolated rainbow trout gill epithelial cells, *J. Exp. Biol.*, 206, 551–559, 2003.

428. Reis-Santos, P., McCormick, S. D., and Wilson, J. M., Ionoregulatory changes during metamorphosis and salinity exposure of juvenile sea lamprey (*Petromyzon marinus* L.), *J. Exp. Biol.*, 211, 978–988, 2008.

429. Renfro, J. L., Interdependence of Active Na^+ and Cl^- transport by the isolated urinary bladder of the teleost, *Pseudopleuronectes americanus*, *J. Exp. Zool.*, 199, 383–390, 1977.

430. Renfro, J. L., Recent developments in teleost renal transport, *J. Exp. Zool.*, 283, 653–661, 1999.

431. Renfro, J. L., Maren, T. H., Zeien, C., and Swenson, E. R., Renal sulfate secretion is carbonic anhydrase dependent in a marine teleost, *Pleuronectes americanus*, *Am. J. Physiol.*, 276, F288–F294, 1999.

432. Renfro, J. L., Miller, D. S., Karnaky, Jr., K. J., and Kinter, W. B., Na-K-ATPase localization in teleost urinary bladder by [^{3}H]ouabain autoradiography, *Am. J. Physiol.*, 231, 1735–1743, 1976.

433. Richards, J. G., Semple, J. W., Bystriansky, J. S., and Schulte, P. M., Na^+/K^+-ATPase alpha-isoform switching in gills of rainbow trout (*Oncorhynchus mykiss*) during salinity transfer, *J. Exp. Biol.*, 206, 4475–4486, 2003.

434. Riegel, J. A., The absence of an arterial pressure effect on filtration by perfused glomeruli of the hagfish, *Eptatretus stouti* (Lockington), *J. Exp. Biol.*, 126, 361–374, 1986.

435. Riegel, J. A., An analysis of the function of the glomeruli of the hagfish mesonephric kidney. In *The Biology of Hagfishes*, Jorgensen, J. M., Lomholt, J. P., Weber, R. E., and Malte, H., Eds., Chapman & Hall, London, 1998.

436. Riegel, J. A., Analysis of fluid dynamics in perfused glomeruli of the hagfish *Eptatretus stouti* (Lockington), *J. Exp. Biol.*, 201, 3097–3104, 1998.

437. Riegel, J. A., Secretion of primary urine by glomeruli of the hagfish kidney, *J. Exp. Biol.*, 202, 947–955, 1999.

438. Riordan, J. R., Forbush, 3rd, B., and Hanrahan, J. W., The molecular basis of chloride transport in shark rectal gland, *J. Exp. Biol.*, 196, 405–418, 1994.

439. Robertson, J. C. and Hazel, J. R., Influence of temperature and membrane lipid composition on the osmotic water permeability of teleost gills, *Physiol. Biochem. Zool.*, 72, 623–632, 1999.

440. Robertson, J. D., The chemical composition of the blood of some aquatic chordates, including members of the Tunicata, Cyclostomata, and Osteichthyes, *J. Exp. Biol.*, 31, 424–442, 1954.

441. Robertson, J. D., Osmotic constituents of the blood plasma and parietal muscle of *Myxine glutinosa*. In *Contemporary Studies in Marine Science*, Barnes, H., Ed., George Allen & Unwin, London, 1966, pp. 631–644.

442. Rodriguez, A., Gallardo, M. A., Gisbert, E., Santilari, S., Ibarz, A., Sanchez, J., and Castello-Orvay, F., Osmoregulation in juvenile Siberian sturgeon (*Acipenser baerii*), *Fish Physiol. Biochem.*, 26, 345–354, 2002.

443. Rombough, P., The functional ontogeny of the teleost gill: which comes first, gas or ion exchange?, *Comp. Biochem. Physiol. A*, 148, 732–742, 2007.

444. Royaux, I. E., Wall, S. M., Karniski, L. P., Everett, L. A., Suzuki, K., Knepper, M. A., and Green, E. D., Pendrin, encoded by the Pendred syndrome gene, resides in the apical region of renal intercalated cells and mediates bicarbonate secretion, *Proc. Natl. Acad. Sci. USA*, 98(7), 4221–4226, 2001.

445. Rudy, P. P. and Wagner, R. C., Water permeability in the Pacific hagfish, *Polistotrema stoutii*, and the staghorn sculpin, *Leptocottus armatus*, *Comp. Biochem. Physiol.*, 34, 399–403, 1970.

446. Sakamoto, T. and McCormick, S. D., Prolactin and growth hormone in fish osmoregulation, *Gen. Comp. Endocrinol.*, 147, 24–30, 2006.

447. Salman, N. A. and Eddy, F. B., Increased sea-water adaptability of non-smolting rainbow trout by salt feeding, *Aquaculture*, 86, 259–270, 1990.

448. Sands, J. M., Mammalian urea transporters, *Annu. Rev. Physiol.*, 65, 543–566, 2003.

449. Deleted at proofs.

450. Sardet, C., Pisan, M., and Maetz, J., The surface epithelial of teleostean fish gills: cellular and tight junctional adaptations of the chloride cell in relation to salt adaptations, *J. Cell Biol.*, 80, 96–117, 1979.

451. Sawyer, D. B. and Beyenbach, K. W., Mechanism of fluid secretion in isolated shark renal proximal tubules, *Am. J. Physiol.*, 249, F884–F890, 1985.

452. Schettino, T. and Lionetto, M. G., Cl^- absorption in European eel intestine and its regulation, *J. Exp. Zool.*, 300A, 63–68, 2003.

453. Schmidt-Nielsen, B. and Renfro, J. L., Kidney function of the American eel *Anguilla rostrata*, *Am. J. Physiol.*, 228, 420–431, 1975.

454. Schmidt-Nielsen, B., Truniger, B., and Rabinowitz, L., Sodium-linked urea transport by the renal tubule of the spiny dogfish *Squalus acanthias*, *Comp. Biochem. Physiol. A*, 42, 13–25, 1972.

455. Scott, G., Rogers, J., Richards, J., Wood, C., and Schulte, P., Intraspecific divergence of ionoregulatory physiology in the euryhaline teleost *Fundulus heteroclitus*: possible mechanisms of freshwater adaptation, *J. Exp. Biol.*, 207, 3399–3410, 2004.

456. Scott, G. R., Claiborne, J. B., Edwards, S. L., Schulte, P. M., and Wood, C. M., Gene expression after freshwater transfer in gills and opercular epithelia of killifish: insight into divergent mechanisms of ion transport, *J. Exp. Biol.*, 208, 2719–2729, 2005.

457. Scott, G. R., Richards, J. G., Forbush, B., Isenring, P., and Schulte, P. M., Changes in gene expression in the gills of the euryhaline killifish *Fundulus heteroclitus* after abrupt salinity transfer, *Am. J. Physiol.*, 287, 300–309, 2004.

458. Shahsavarani, A., McNeill, B., Galvez, F., Wood, C. M., Goss, G. G., Hwang, P. P., and Perry, S. F., Characterization of a branchial epithelial calcium channel (ECaC) in freshwater rainbow trout (*Oncorhynchus mykiss*), *J. Exp. Biol.*, 209(Pt. 10), 1928–1943, 2006.

458a. Shaw, J. R., Sato, J. D., VanderHeide, J., LaCasse, T., Stanton, C. R., Lankowski, A., Stanton, S. E., Chapline, C., Coutermarsh, B., Barnaby, R., Karlson, K., and Stanton, B. A., The role of SGK and CFTR in acute adaptation to seawater in *Fundulus heteroclitus*, *Cell. Physiol. Biochem.*, 22, 69–78, 2008.

459. Shehadeh, Z. H. and Gordon, M. S., The role of the intestine in salinity adaptation of the rainbow trout, *Salmo gairdneri*, *Comp. Biochem. Physiol.*, 30, 397–418, 1969.

460. Shikano, T. and Fujio, Y., Immunolocalization of Na+/K+-ATPase in branchial epithelium of chum salmon fry during seawater and freshwater acclimation, *J. Exp. Biol.*, 201, 3031–3040, 1998.

461. Shirai, N. and Utida, S., Development and degeneration of the chloride cell during seawater and freshwater adaptation of the Japanese eel, *Anguilla japonica*, *Z. Zellforsch. Mikrosk. Anat.*, 103, 247–264, 1970.

462. Shuttleworth, T. J., *Physiology of Elasmobranch Fishes*, Springer-Verlag, Berlin, 1988, p. 324.

463. Shuttleworth, T. J., Salt and water balance: extrarenal mechanisms. In *Physiology of Elasmobranch Fishes*, Shuttleworth, T. J., Ed., Springer-Verlag, Berlin, 1988, pp. 171–199.

464. Siegel, N. J., Schon, D. A., and Hayslett, J. P., Evidence for active chloride transport in dogfish rectal gland, *Am. J. Physiol.*, 230(5), 1250–1254, 1976.

465. Silva, P., Solomon, R., Spokes, K., and Epstein, F., Ouabain inhibition of gill Na,K-ATPase: relationship to active chloride transport, *J. Exp. Zool.*, 199, 419–426, 1977.

466. Silva, P., Solomon, R. J., and Epstein, F. H., Shark rectal gland, *Meth. Enzymol.*, 192, 754–766, 1990.

467. Silva, P., Stoff, J., Field, M., Fine, L., Forrest, J. N., and Epstein, F. H., Mechanism of active chloride secretion by shark rectal gland: role of Na-K-ATPase in chloride transport, *Am. J. Physiol.*, 233(4), F298–F306, 1977.

468. Singer, T. D., Tucker, S. J., Marshall, W. S., and Higgins, C. F., A divergent CFTR homologue: highly regulated salt transport in the euryhaline teleost *F. heteroclitus*, *Am. J. Physiol.*, 274, C715–C723, 1998.

469. Skadhauge, E., The mechanism of salt and water absorption in the intestine of the eel (*Anguilla anguilla*) adapted to waters of various salinities, *J. Physiol.*, 204, 135–158, 1969.

470. Smith, C. P. and Wright, P. A., Molecular characterization of an elasmobranch urea transporter, *Am. J. Physiol.*, 276, R622–R626, 1999.

471. Smith, D. G. and Chamley-Campbell, J., Localisation of smooth-muscle myosin in branchial pillar cells of snapper (*Chrysophys auratus*) by immunofluorescence histochemistry, *J. Exp. Zool.*, 215, 121–124, 1981.

472. Smith, H. W., The excretion of ammonia and urea by the gills of fish, *J. Biol. Chem.*, 81, 727–742, 1929.

473. Smith, H. W., Metabolism of the lungfish *Protopterus aethiopicus*, *J. Biol. Chem.*, 88, 97–130, 1930.

474. Smith, H. W., The absorption and excretion of water and salts by marine teleosts, *Am. J. Physiol.*, 93, 480–505, 1930.

475. Smith, H. W., Observations of the African lung-fish, *Protopterus aethiopicus*, *Ecology* 12, 164–181, 1931.

476. Smith, H. W., The absorption and excretion of water and salts by the elasmobranch fishes. I. Fresh water elasmobranchs, *Am. J. Physiol.*, 98, 279–295, 1931.

477. Smith, N., Eddy, F., and Talbot, C., Effect of dietary salt load on transepithelial Na+ exchange in freshwater rainbow trout (*Oncorhynchus mykiss*), *J. Exp. Biol.*, 198, 2359–2364, 1995.

478. Solomon, R., Taylor, M., Sheth, S., Silva, P., and Epstein, F. H., Primary role of volume expansion in stimulation of rectal gland function, *Am. J. Physiol.*, 248, R638–R640, 1985.

479. Solomon, R. J., Taylor, M., Stoff, J. S., Silva, P., and Epstein, F. H., *In vivo* effect of volume expansion on rectal gland function. I. Humoral factors, *Am. J. Physiol.*, 246, R63–R66, 1984.

480. Solomon, R. J., Taylor, M., Rosa, R., Silva, P., and Epstein, F. H., *In vivo* effect of volume expansion on rectal gland function. II. Hemodynamic changes, *Am. J. Physiol.*, 246, R67–R71, 1984.

481. Steen, J. B. and Stray-Pederson, S., The permeability of fish gills with comments on the osmotic behavior of cellular membranes, *Acta Physiol. Scand.*, 95, 6–20, 1975.

482. Stinson, C. M. and Mallatt, J., Branchial ion fluxes and toxicant extraction efficiency in lamprey (*Petromyzon marinus*) exposed to methylmercury, *Aquat. Toxicol.*, 15, 237–252, 1989.

483. Stoff, J. S., Silva, P., Field, M., Forrest, J., Stevens, A., and Epstein, F. H., Cyclic AMP regulation of active chloride transport in the rectal gland of marine elasmobranchs, *J. Exp. Zool.*, 199, 443–448, 1977.

484. Stokes, J. B., Sodium chloride absorption by the urinary bladder of the winter flounder: a thiazide-sensitive, electrically neutral transport system, *J. Clin. Invest.*, 74, 7–16, 1984.

485. Sturla, M., Masini, M. A., Prato, P., Grattarola, C., and Uva, B., Mitochondria-rich cells in gills and skin of an African lungfish, *Protopterus annectens*, *Cell Tissue Res.*, 303(3), 351–358, 2001.

486. Sullivan, G. V., Fryer, J. N., and Perry, S. F., Localization of mRNA for proton pump (H$^+$-ATPase) and Cl$^-$/HCO$_3^-$ exchanger in rainbow trout gill, *Can. J. Zool.*, 74, 2095–2103, 1996.

487. Sulya, L. L., Box, B. E., and Gunther, G., Distribution of some blood constituents in fish from the Gulf of Mexico, *Am. J. Physiol.*, 199, 1177–1180, 1969.

488. Suzuki, Y., Itakura, M., Kashiwagi, M., Nakamura, N., Matsuki, T., Sakuta, H., Naito, N., Takano, K., Fujita, T., and Hirose, S., Identification by differential display of a hypertonicity-inducible inward rectifier potassium channel highly expressed in chloride cells, *J. Biol. Chem.*, 274, 11376–11382, 1999.

489. Swanson, C., Interactive effects of salinity on metabolic rate, activity, growth and osmoregulation in the euryhaline milkfish (*Chanos chanos*), *J. Exp. Biol.*, 201, 3355–3366, 1998.

490. Takei, Y. and Hirose, S., The natriuretic peptide system in eels: a key endocrine system for euryhalinity?, *Am. J. Physiol. Regul. Integr. Comp. Physiol.*, 282, R940–R951, 2002.

491. Takei, Y. and Loretz, C. A., Endocrinology. In *The Physiology of Fishes*, 3rd ed., Evans, D. H. and Claiborne, J. B., Eds., CRC Press, Boca Raton, FL, 2006, pp. 271–318.

492. Takezaki, N., Figueroa, F., Zaleska-Rutczynska, Z., and Klein, J., Molecular phylogeny of early vertebrates: monophyly of the agnathans as revealed by sequences of 35 genes, *Mol. Biol. Evol.*, 20, 287–292, 2003.

493. Tang, C. H. and Lee, T. H., The effect of environmental salinity on the protein expression of Na$^+$/K$^+$-ATPase, Na$^+$/K$^+$/2Cl$^-$ cotransporter, cystic fibrosis transmembrane conductance regulator, anion exchanger 1, and chloride channel 3 in gills of a euryhaline teleost, *Tetraodon nigroviridis*, *Comp. Biochem. Physiol. A*, 147, 521–528, 2007.

494. Taylor, J. R. and Grosell, M., Evolutionary aspects of intestinal bicarbonate secretion in fish, *Comp. Biochem. Physiol. A*, 143, 523–529, 2006.

495. Taylor, J. R., Whittamore, J. M., Wilson, R. W., and Grosell, M., Postprandial acid–base balance and ion regulation in freshwater and seawater-acclimated European flounder, *Platichthys flesus*, *J. Comp. Physiol. B*, 177, 597–608, 2007.

496. Thomerson, J. E. and Thorson, T. B., The bull shark, *Carcharhinus leucas*, from the upper Mississippi river near Atlon, Illinois, *Copeia*, 1977(1), 166–168, 1977.

497. Thorson, T. B., Fresh water stingrays, *Potamotrygon* spp.: failure to concentrate urea when exposed to saline medium, *Life Sci.*, 9, 893–900, 1970.

498. Thorson, T. B., Cowan, C. M., and Watson, D. E., *Potamotrygon* spp.: elasmobranchs with low urea content, *Science*, 158, 375–377, 1967.

499. Thorson, T. B., Wotton, R. M., and Georgi, T. A., Rectal gland of freshwater stingrays, *Potamotrygon* spp. (Chondrichthyes: Potamotrygonidae), *Biol. Bull.*, 154, 508–516, 1978.

500. Tipsmark, C. K., Baltzegar, D. A., Ozden, O., Grubb, B. J., and Borski, R. J., Salinity regulates claudin mRNA and protein expression in the teleost gill, *Am. J. Physiol. Regul. Integr. Comp. Physiol.*, 294, R1004–R1014, 2008.

501. Toop, T. and Donald, J. A., Comparative aspects of natriuretic peptide physiology in non-mammalian vertebrates: a review, *J. Comp. Physiol. B*, 174, 189–204, 2004.

502. Treberg, J. R., Speers-Roesch, B., Piermarini, P. M., Ip, Y. K., Ballantyne, J. S., and Driedzic, W. R., The accumulation of methylamine counteracting solutes in elasmobranchs with differing levels of urea: a comparison of marine and freshwater species, *J. Exp. Biol.*, 209, 860–870, 2006.

503. Tresguerres, M., Katoh, F., Fenton, H., Jasinska, E., and Goss, G. G., Regulation of branchial V-H$^+$-ATPase, Na$^+$/K$^+$-ATPase and NHE2 in response to acid and base infusions in the Pacific spiny dogfish (*Squalus acanthias*), *J. Exp. Biol.*, 208, 345–354, 2005.

504. Tresguerres, M., Katoh, F., Orr, E., Parks, S., and Goss, G., Chloride uptake and base secretion in freshwater fish: a transepithelial ion-transport metabolon?, *Physiol. Biochem. Zool.*, 79, 981–996, 2006.

505. Tresguerres, M., Parks, S. K., and Goss, G. G., V-H(+)-ATPase, Na(+)/K(+)-ATPase and NHE2 immunoreactivity in the gill epithelium of the Pacific hagfish (*Epatretus stoutii*), *Comp. Biochem. Physiol. A*, 145(3), 312–321, 2006.

506. Tresguerres, M., Parks, S. K., and Goss, G. G., Recovery from blood alkalosis in the Pacific hagfish (*Eptatretus stoutii*): involvement of gill V-H$^+$-ATPase and Na$^+$/K$^+$-ATPase, *Comp. Biochem. Physiol. A*, 148, 133–141, 2007.

507. Tresguerres, M., Parks, S. K., Katoh, F., and Goss, G. G., Microtubule-dependent relocation of branchial V-H$^+$-ATPase to the basolateral membrane in the Pacific spiny dogfish (*Squalus acanthias*): a role in base secretion, *J. Exp. Biol.*, 209, 599–609, 2006.

508. Tresguerres, M., Parks, S. K., Wood, C. M., and Goss, G. G., V-H$^+$-ATPase translocation during blood alkalosis in dogfish gills: interaction with carbonic anhydrase and involvement in the postfeeding alkaline tide, *Am. J. Physiol.*, 292, R2012–R2019, 2007.

509. Tse, W. K., Au, D. W., and Wong, C. K., Characterization of ion channel and transporter mRNA expressions in isolated gill chloride and pavement cells of seawater acclimating eels, *Biochem. Biophys. Res. Commun.*, 346, 1181–1190, 2006.

510. Tsukada, T. and Takei, Y., Integrative approach to osmoregulatory action of atrial natriuretic peptide in seawater eels, *Gen. Comp. Endocrinol.*, 147, 31–38, 2006.

511. Turner, S. and Miller, R. F., New ideas about old sharks, *Am. Sci.*, 93, 244–252, 2004.

512. Valentich, J. D. and Forrest, Jr., J. N., Cl$^-$ secretion by cultured shark rectal gland cells. I. Transepithelial transport, *Am. J. Physiol.*, 260, C813–C823, 1991.

513. Van Der Heijden, A. J. H., Verbost, P. M., Eygensteyn, J., Li, J., Bonga, S. E. W., and Flik, G., Mitochondria-rich cells in gills of tilapia (*Oreochromis mossambicus*) adapted to fresh water or sea water: quantification by confocal laser scanning microscopy, *J. Exp. Biol.*, 200, 55–64, 1997.

514. Van der Velden, J. A., Groot, J. A., Flik, G., Polak, P., and Kolar, Z. I., Magnesium transport in fish intestine, *J. Exp. Biol.*, 152, 587–592, 1990.

515. Verbost, P. M., Bryson, S. E., Bonga, S. E., and Marshall, W. S., Na$^+$-dependent Ca^{2+} uptake in isolated opercular epithelium of *Fundulus heteroclitus*, *J. Comp. Physiol. B*, 167, 205–212, 1997.

516. Viborg, A. L. and Hillyard, S. D., Cutaneous blood flow and water absorption by dehydrated toads, *Physiol. Biochem. Zool.*, 78, 394–404, 2005.

517. Volff, J. N., Genome evolution and biodiversity in teleost fish, *Heredity*, 94, 280–294, 2005.

518. Wagner, C. A. and Geibel, J. P., Acid–base transport in the collecting duct, *J. Nephrol.*, 15(Suppl. 5), S112–S127, 2002.

519. Waldegger, S., Fakler, B., Bleich, M., Barth, P., Hopf, A., Schulte, U., Busch, A. E., Aller, S. G., Forrest, Jr., J. N., Greger, R., and Lang, F., Molecular and functional characterization of s-KCNQ1 potassium channel from rectal gland of Squalus acanthias, *Pflügers Arch.*, 437, 298–304, 1999.

520. Wall, S., Recent advances in our understanding of intercalated cells, *Curr. Opin. Nephrol. Hypertens.*, 14, 480–484, 2005.

520a. Walsh, P., Blackwelder, P., Gill, K. A., Danulat, E., and Mommsen, T. P., Carbonate in the marine fish intestine: a new source of sediment production, *Limnol. Oceanogr.*, 36, 1227–1232, 1991.

521. Weber, T., Gopfert, M., Winter, H., Zimmermann, U., Kohler, H., Meier, A., Hendrich, O., Rohbock, K., Robert, D., and Knipper, M., Expression of prestin-homologous solute carrier (SLC26) in auditory organs of nonmammalian vertebrates and insects, *Proc. Natl. Acad. Sci. USA*, 100, 7690–7695, 2003.

522. Wells, A., Anderson, W. G., Cains, J. E., Cooper, M. W., and Hazon, N., Effects of angiotensin II and C-type natriuretic peptide on the *in situ* perfused trunk preparation of the dogfish, *Scyliorhinus canicula*, *Gen. Comp. Endocrinol.*, 145, 109–115, 2006.

523. Wells, A., Anderson, W. G., and Hazon, N., Development of an *in situ* perfused kidney preparation for elasmobranch fish: action of arginine vasotocin, *Am. J. Physiol.*, 282, R1636–R1642, 2002.

524. Welsh, M. J., Smith, P. L., and Frizzell, R. A., Intracellular chloride activities in the isolated perfused shark rectal gland, *Am. J. Physiol.*, 245, F640–F644, 1983.

525. Wendelaar Bonga, S. E., Flik, G., Balm, P. H. M., and van der Meij, J. C. A., The ultrastructure of chloride cells in the gills of the teleost *Orechromis mossambicus* during exposure to acidified water, *Cell Tissue Res.*, 259, 575–585, 1990.

526. Wilkie, M. P., Ammonia excretion and urea handling by fish gills: present understanding and future research challenges, *J. Exp. Zool.*, 293, 284–301, 2002.

527. Wilkie, M. P., Morgan, T. P., Galvez, F., Smith, R. W., Kajimura, M., Ip, Y. K., and Wood, C. M., The African lungfish (*Protopterus dolloi*): ionoregulation and osmoregulation in a fish out of water, *Physiol. Biochem. Zool.*, 80(1), 99–112, 2007.

528. Wilson, J. and Laurent, P., Fish gill morphology: inside out, *J. Exp. Zool.*, 293, 192–213, 2002.

529. Wilson, J., Randall, D. J., Vogl, A. W., and Iwama, G. K., Immunolocalization of proton-ATPase in the gills of the elasmobranch, *Squalus acanthias*, *J. Exp. Zool.*, 278, 78–86, 1997.

531. Wilson, J. M., Laurent, P., Tufts, B. L., Benos, D. J., Donowitz, M., Vogl, A. W., and Randall, D. J., NaCl uptake by the branchial epithelium in freshwater teleost fish: an immunological approach to ion-transport protein localization, *J. Exp. Biol.*, 203, 2279–2296, 2000.

533. Wilson, J. M., Morgan, J. D., Vogl, A. W., and Randall, D. J., Branchial mitochondria-rich cells in the dogfish *Squalus acanthias*, *Comp. Biochem. Physiol. A*, 132, 365–374, 2002.

534. Wilson, J. M., Randall, D. J., Donowitz, M., Vogl, A. W., and Ip, A. K. Y., Immunolocalization of ion-transport proteins to branchial epithelium mitochondria-rich cells in the mudskipper (*Periophthalmodon schlosseri*), *J. Exp. Biol.*, 203, 2297–2310, 2000.

535. Wilson, J. M., Randall, D. J., Vogl, A. W., and Iwama, G. K., Immunolocalization of proton-ATPase in the gills of the elasmobranch, *Squalus acanthias*, *J. Exp. Zool.*, 278, 78–86, 1997.

536. Wilson, R. W. and Grosell, M., Intestinal bicarbonate secretion in marine teleost fish-source of bicarbonate, pH sensitivity, and consequences for whole animal acid–base and calcium homeostasis, *Biochim. Biophys. Acta*, 1618, 163–174, 2003.

537. Wilson, R. W., Wilson, J. M., and Grosell, M., Intestinal bicarbonate secretion by marine teleost fish: why and how?, *Biochim. Biophys. Acta*, 1566, 182–193, 2002.

538. Wong, C. and Chan, D., Chloride cell subtypes in the gill epithelium of Japanese eel *Anguilla japonica*, *Am. J. Physiol.*, 277, R517–R522, 1999.

539. Wood, C. and Laurent, P., Na^+ versus Cl^- transport in the intact killifish after rapid salinity transfer, *Biochim. Biophys. Acta*, 1618, 106–119, 2003.

540. Wood, C. M. and Marshall, W. S., Ion balance, acid–base regulation and chloride cell function in the common killifish, *Fundulus heteroclitus*: a freely euryhaline estuarine teleost, *Estuaries*, 17, 34–52, 1994.

541. Wood, C. M., Matsuo, A. Y., Gonzalez, R. J., Wilson, R. W., Patrick, M. L., and Val, A. L., Mechanisms of ion transport in *Potamotrygon*, a stenohaline freshwater elasmobranch native to the ion-poor blackwaters of the Rio Negro, *J. Exp. Biol.*, 205, 3039–3054, 2002.

542. Wood, C. M., Walsh, P. J., Chew, S. F., and Ip, Y. K., Greatly elevated urea excretion after air exposure appears to be carrier mediated in the slender lungfish (*Protopterus dolloi*), *Physiol. Biochem. Zool.*, 78, 893–907, 2005.

543. Wu, Y. and Lee, T., $Na^+,K^+,2Cl^-$ cotransporter: a novel marker for identifying freshwater- and seawater-type mitochondria-rich cells in gills of the euryhaline tilapia, *Oreochromis mossambicus*, *Zool. Stud.*, 42, 186–192, 2003.

544. Xu, J. C., Lytle, C., Zhu, T. T., Payne, J. A., Benz, Jr., E., and Forbush, 3rd, B., Molecular cloning and functional expression of the bumetanide-sensitive Na-K-Cl cotransporter, *Proc. Natl. Acad. Sci. USA*, 91, 2201–2205, 1994.

545. Yan, J. J., Chou, M. Y., Kaneko, T., and Hwang, P. P., Gene expression of Na^+/H^+ exchanger in zebrafish H^+-ATPase-rich cells during acclimation to low-Na^+ and acidic environments, *Am. J. Physiol.*, 293, C1814–C1823, 2007.

546. Yancey, P. H., Nitrogen compounds as osmolytes. In *Nitrogen Excretion*, Wright, P. A. and Anderson, P. M., Eds., Academic Press, San Diego, CA, 2001, pp. 309–341.

547. Yancey, P. H. and Somero, G. N., Methylamine osmoregulatory solutes of elasmobranch fishes counteract urea inhibition of enzymes, *J. Exp. Zool.*, 212, 205–213, 1980.

548. Youson, J. H. and Butler, D. G., Morphology of the kidney of adult bowfin, *Amia calva*, with emphasis on "renal chloride cells" in the tubule, *J. Morphol.*, 196, 137–156, 1988.

549. Yuge, S., Inoue, K., Hyodo, S., and Takei, Y., A novel guanylin family (guanylin, uroguanylin, and renoguanylin) in eels: possible osmoregulatory hormones in intestine and kidney, *J. Biol. Chem.*, 278(25), 22726–22733, 2003.

550. Yuge, S., Yamagami, S., Inoue, K., Suzuki, N., and Takei, Y., Identification of two functional guanylin receptors in eel: multiple hormone-receptor system for osmoregulation in fish intestine and kidney, *Gen. Comp. Endocrinol.*, 149, 10–20, 2006.

551. Zadunaisky, J. A., Cardona, S., Au, L., Roberts, D. M., Fisher, E., Lowenstein, B., Cragoe, Jr., E. J., and Spring, K. R., Chloride transport activation by plasma osmolarity during rapid adaptation to high salinity of *Fundulus heteroclitus*, *J. Membr. Biol.*, 143, 207–217, 1995.

552. Zadunaisky, J. A., Curci, S., Schettino, T., and Scheide, J. I., Intracellular voltage recordings in the opercular epithelium of *Fundus heteroclitus*, *J. Exp. Zool.*, 247, 126–130, 1988.

553. Zaugg, W. S., Roley, D. D., Prentice, E. F., Gores, K. X., and Waknitz, F., Increased seawater survival and contribution to the fishery of Chinook salmon (*Oncorhynchus tshawytscha*) by supplemental dietary salt, *Aquaculture*, 32, 183–188, 1983.

554. Zawodny, J. F., Osmoregulation in the Florida Spotted Gar, *Lepisosteus platyrhincus*, M.S. thesis, University of Miami, 1975.

555. Zeidel, J. D., Mathai, J. C., Campbell, J. D., Ruiz, W. G., Apodaca, G. L., Riordan, J., and Zeidel, M. L., Selective permeability barrier to urea in shark rectal gland, *Am. J. Physiol.*, 289, F83–F89, 2005.

Stanley D. Hillyard, Nadja Møbjerg,
Shigeyasu Tanaka, and Erik Hviid Larsen

CONTENTS

I. INTRODUCTION

Amphibians were the first vertebrates to emerge from aquatic habitats, and modern species have evolved a variety of mechanisms to regulate water and electrolyte homeostasis. Fossil evidence for terrestrial vertebrates first appears in the upper Devonian (360–380 mya) in the form of tetrapods such as *Ichthyostega* and *Ancanthostega*. The fossil record has a large gap before the emergence of the major tetrapod lineages in the Upper Carboniferous with numerous amphibian taxa that are primarily associated with freshwater deposits. The earliest fossils that can be attributed to modern amphibians first appear in the early Triassic, and their relationship with the primitive forms is speculative.[67,78] Modern amphibians are collectively termed Lissamphibia (smooth skinned) and include three orders: Anura (frogs and toads), Caudata/Urodela (newts and salamanders), and Gymnophiona (legless wormlike animals also known as apodans or caecilians). All three orders have species that occupy a wide range of habitats from purely aquatic to highly terrestrial. The phylogeny of Amphibia has recently been reevaluated by Frost et al.[143] to include both morphological and molecular parameters. Many of the traditional genera have been divided to reflect a more detailed description of evolutionary relationships; for example, the North American frog *Rana pipiens* is now classified as *Lithobates pipiens*, whereas the European frog, *R. temporaria* remains in the genus *Rana*. We have elected to retain the traditional generic names because they are historically embedded in the literature cited in this chapter and still widely used by contemporary researchers in the field of ionic and osmotic regulation.

The literature on amphibian physiology is heavily biased toward anurans. Frost et al.[143] recognized 5227 species in 32 families of anurans but only 548 species in 10 families of the caudata and 173 species in 6 families of the apoda. In addition, a relatively small number of anuran species have historically been used as model organisms for the study of ionic and osmotic processes collectively termed as *amphibian*. The three orders of modern amphibians are considered to be monophyletic, and transport mechanisms characterized in anurans generally apply to urodeles and apodans, as well. Jørgensen[240] summarized the historical development of our understanding of water balance mechanisms in amphibians, dating from the English naturalist Robert Townson, who, in 1795, observed the roles of the skin, kidneys, and urinary bladder in the water economy of frogs. The literature on ionic and osmotic regulation by aquatic and terrestrial amphibians has been reviewed by Boutilier et al.[44] and Shoemaker et al.[442] In this review, we include more recent studies on mechanisms of ionic and osmotic regulation and also provide a historical perspective of studies leading to the current description of the biological processes required for the amphibious life style and how they have contributed to our understanding of basic physiological mechanisms in other animal phyla.

II. VOLUME AND COMPOSITION OF THE BODY FLUIDS IN RELATION TO ENVIRONMENTAL CONDITIONS

A. PHYSIOLOGICAL FLUID COMPARTMENTS

Representative data on the water content and ionic composition of amphibian body fluids are listed in Table 9.1 and Table 9.2. Body water ranges from 70 to 80% of the body mass and is distributed between the intracellular and interstitial compartment, the lymph space, and the blood plasma. Anurans are unique among vertebrates in having large subcutaneous lymph sacs that may constitute a large fraction of the extracellular fluid volume, whereas the skin of urodeles and apodans is firmly attached to the underlying tissues.[248,272,342] In addition, the large bladder capacity of many terrestrial anuran and urodele species is an aspect that must be considered when evaluating body fluid composition. Generally, the body fluid composition is evaluated relative to a hydrated animal whose bladder has been emptied, a term that Ruibal[414] referred to as the *standard body mass*; however, residual urine may remain in the bladder, and hydration states are variable depending on a variety

TABLE 9.1
Examples of Amphibian Body Fluid Volumes

Species	Total Body Water (TBW) (% of Total Body Mass)	Extracellular Volume and Marker	Plasma Volume and Marker	Hematocrit (% Blood Cells)	Remarks
Rana pipiens	78.9[a]	23.6–29.8[b] Thiocyanate	5.6–9.3[b] T-1824	—	
Rana catesbeiana	—	29.8–47.9[b] Thiocyanate	7.2–8.8[b] T-1824	30 ± 5.3[c]	
Hyla cinerea	80.1[a]	—	—	—	
Scaphiopus hammondii	80.0[a]	—	—	—	
Scaphiopus couchi	69.8–81.6[d]	—	—	—	
Aneides flavipunctatus	71–78[e]	—	—	—	
Bufo bufo	81.3 ± 1.0[f]	32.0 ± 1.6[g] ^{14}C-inulin	—	—	TBW of lean body mass; empty urinary bladder
Bufo marinus	—	—	—	37 ± 1.7[c] 26.4 ± 2.2[i]	
Bufo viridis	72.7 ± 0.9[h]	35[h] ^{14}C-inulin	6.3[h] Evans blue	—	Free access to tapwater; empty urinary bladder

[a] Thorson and Svihla.[485]
[b] Prosser and Weinstein.[393]
[c] Hillman and Withers.[189]
[d] McClanahan.[318]
[e] Ray.[397]
[f] Nielsen and Jørgensen.[355]
[g] Jørgensen et al.[355]
[h] Hoffman and Katz.[204]
[i] Konno et al.[270]

of factors, including natural conditions (e.g., season, breeding status) or husbandry in the laboratory (e.g., temperature, water availability). These variables, discussed below, may contribute to the variability of values for total extracellular fluid volume that can range from 24 to 48% of the body mass and variation of the relative plasma volume as well (Table 9.1). Water and solutes from the plasma are filtered through the capillary endothelium and enter the interstitial fluid from which respiratory gases, ions, nutrients, and metabolites are exchanged with the intracellular compartment. The relative flow of fluid between the extracellular compartments is very fast relative to other vertebrates and may result in a daily lymph production of one to two times the animal's body mass in frogs and toads, as observed by Isayama.[217,218] The interstitial fluid drains into the lymphatic spaces and returns to the blood via the great lymph trunks and sinuses energized by the lymph hearts. These issues are discussed in further detail in Section VII.H at the end of the chapter.

The ion concentrations follow the general vertebrate pattern, with Na$^+$ and Cl$^-$ being dominant in the extracellular fluid (Table 9.2) and K$^+$ above and Na$^+$ significantly below thermodynamic equilibrium in the cell water (Table 9.3). Intracellular Cl$^-$ is near its equilibrium concentration in striated muscle fibers and significantly above equilibrium in the exocrine gland and in the epithelial cells of the skin. With a pH of ~7.8 and with [HCO$_3^-$]/pCO$_2$ as the major extracellular buffer system, the concentration of HCO$_3^-$ is about 22 mM at an arterial pCO$_2$ of ~12 mmHg (~1.6 kPa) in amphibians with pulmonary respiration.[45] Adult amphibians with cutaneous gas exchange and tadpoles with gill respiration have a significantly lower arterial pCO$_2$ of ~3 to 5 mmHg (0.4 to 0.7 kPa). The extracellular pH of about 7.8 is here maintained by an extracellular [HCO$_3^-$] of ~8 mM.[63] Although the data presented in Table 9.2 are from studies on anuran amphibians supposedly in a normally hydrated state, the concentrations vary somewhat both between species and laboratories. The general trend, nevertheless, is that extracellular concentrations of Na$^+$ and Cl$^-$ are lower than those of other vertebrate groups.

TABLE 9.2
Total Osmotic Concentration and Concentration of Small Diffusible Ions and Urea in the Extracellular Fluid of Some Amphibians

Species	Δ (mOsm/ kg H_2O)	Blood Plasma or Lymph					Remarks
		Na⁺ (mM)	K⁺ (mM)	Cl⁻ (mM)	HCO_3^- (mM)	Urea (mM)	
Rana cancrivora[a]	290 ± 10	125 ± 17	9 ± 1	98 ± 10	—	40 ± 1	In freshwater
Rana ridibunda[b]	247 ± 12	115 ± 5	6 ± 1	83 ± 6	—	11 ± 2	In tapwater
Scaphiopus couchi[c]	225–390	141–210	3.6–14.4	84–152	—	37–173	Foraging in water collected from temporary pools
Bufo viridis[d]	253–346	120–174	3.6–5.7	92–125	—	18–51	In tapwater
Bufo bufo[e]	275 ± 5	120 ± 6	4.0 ± 0.4	87 ± 2	—	17 ± 2	In tapwater; lymph samples; mean ± SD
Bufo bufo[f]	—	119 ± 2	3.0 ± 0.25	79 ± 2	—	—	Terrestrial habitat with free access to tapwater; lymph samples
	—	105.5 ± 1.0	—	85.4 ± 2.6	24.1	—	
Bufo marinus[g]	234.3 ± 10.2	107.9 ± 6.8	2.4 ± 0.4	76.1 ± 4.9	—	—	Mean ± SD
Bufo marinus[h]	241.2 ± 3.4	99.8 ± 2.6	4.6 ± 0.1	72.6 ± 3.8	—	19.6 ± 3.2	
Ascaphus truei[i]	172 ± 5	106 ± 3	—	81 ± 3	—	—	Aquatic, dilute high-altitude streams; collected in the field
Rana pipiens[i]	193 ± 8	112 ± 3	—	68 ± 3	—	—	Semiaquatic; kept in about 0.2-mM NaCl
Hyla regilla[i]	218 ± 10	110 ± 3	—	78 ± 5	—	—	Terrestrial; collected in the field
Bufo boreas[i]	235 ± 15	109 ± 3	—	77 ± 5	—	—	Terrestrial; collected in the field

[a] Gordon et al.[161]
[b] Katz.[251]
[c] McClahanan.[318]
[d] Shpun and Katz.[443]
[e] Nielsen and Jørgensen.[355]
[f] Jensen et al.[225]
[g] Stinner and Hartzler.[465]
[h] Konno et al.[270]
[i] Mullen and Alvarado.[341]

Note: Mean ± SEM or range, unless otherwise indicated.

TABLE 9.3
Intracellular Concentrations of Small Diffusible Ions and Membrane Potential of Amphibian Tissue Cells

Tissue	Na$^+$ (mmol/ kg H$_2$O)	K$^+$ (mmol/ kg H$_2$O)	Cl$^-$ (mmol/ kg H$_2$O)	HCO$_3^-$ (mmol/ kg H$_2$O)	V_m (mV)
Striated muscle	10.4[a]	124[a]	1.5[a]	12.4[a]	−95[b]
Acinus of subepidermal gland	11.5[c]	155[c]	55[c]	—	−69.5 ± 0.7[d]
Principal cell compartment of epidermis	12[e]	153[e]	47[e]	—	−108 ± 2[f]

[a] Conway.[85]

[b] Hodgkin and Horowicz.[199]

[c] Electron microprobe analysis corrected for a mean dry mass of 18.6 g/100 g (Mills et al.[328]).

[d] Basolateral membrane potential of resting gland cells (Sørensen and Larsen[448]).

[e] Electron microprobe analysis corrected for a mean dry mass of 25 g/100 g (Rick et al.[406]).

[f] Basolateral membrane potential (Nagel[345]).

B. PHYSIOLOGICAL VARIATIONS OF BODY FLUID VOLUMES AND THEIR COMPOSITION

The extreme tolerance of amphibians to hydration and dehydration is associated with the large and variable extracellular volume and rapid fluid exchange between plasma and lymph space which secure the blood flow despite large variations of total body water content.[188,191,239] Thus, blood pressure and hematocrit values were not significantly affected in the dehydration-tolerant *Bufo marinus* and *Scaphiopus couchi* when dehydrated by 20% of their body mass,[190,318] nor did a 5% loss of body mass by bleeding affect blood flow.[19,189,325] During 14 weeks on soil with water potential of approximately −5 atm, *Bufo viridis* lost about 5% of its body water, which was accounted for by a decrease of extracellular volume with no significant change in plasma volume.[204] Finally, Conklin[81–83] increased the body mass of *Rana pipiens* by 20% by intravenous injection of Ringer's solution, which was passed into the lymphatic space, without affecting blood pressure.

The European toad *Bufo bufo* acclimated in a simulated terrestrial habitat experiences spontaneous variations in hydration of up to ±5% of the standard body mass.[235] With free access to a pool of water, the toads spent most of the time in the dry environment, where they lost about 3% of the body mass per day by evaporation at 16°C. Interestingly, the toad visited the water bath for cutaneous drinking in a hydrated state and generally before the urinary bladder was empty. This drinking behavior characterized by water intake by toads in a hydrated state was termed *anticipatory drinking* as opposed to *emergency drinking* by dehydrated toads[235,236,239] (see Section VII.G). Regulated displacements of body fluid volumes and osmotic concentrations have been observed in response to feeding or changes of the environmental temperature, in hibernating animals, during the breeding season, and during acclimation to changes of environmental salinity.

1. Feeding

Food intake initiates drinking behavior in the terrestrial *Bufo bufo*. Following a meal, the toad visits a water source and takes up water at an amount exceeding the mass of the food eaten. The water intake is proportional to the size of the meal (mealworms) and may amount to as much as 15% of the body mass measured with empty urinary bladder.[237] Eventually, the excess water increases the water store in the bladder. This significant increase in water turnover secures secretion of digestive fluids but may lead to a temporary disturbance of extracellular Cl$^-$ balance and alkalosis, as has been described for another carnivorous vertebrate, the alligator (reviewed in Taplin[478]). The spadefoot toad (*Scaphiopus couchi*) is only active above ground for a few days in a year and may consume

55% of its body mass as food in a single feeding.[105] At the same time, the animals must attain a hydration status that permits survival when burrowed between annual rainfall periods.[415] The balance between food consumption and water gain remains an interesting question in this and other seasonally active species.

2. Temperature

When frogs (*Rana esculenta* and *R. catesbeiana*) and toads (*Bufo bufo*) kept in tapwater at room temperature are transferred to low temperatures (2 to 4°C), they accumulate fluid within 1 to 2 days corresponding to a 5 to 10% gain in body mass. This is caused by both high cutaneous water inflow and reduced urine production.[234,245,327,426] This fast water gain occurs without salt accumulation, and it is rapidly eliminated upon retransfer to room temperature. In *B. bufo* transferred from 20°C to 4°C, the inulin space increased from 32 to 38% of body mass (with an empty urinary bladder), with an associated drop in lymph concentration of Na^+ from 112.4 ± 0.8 to 94.6 ± 2.0 mM. By comparing these changes with the simultaneously measured body mass increase, it was calculated that about 90% of the accumulated water was partitioned to the extracellular space.[245]

3. Hibernation

Frogs and toads emerging from hibernation are edematous, with fairly large amounts of fluid accumulated in the lymphatic system.[76] Fluid accumulation during hibernation is associated with a much more complex change in the water and salt balance of the animal than the above fast (and passing) response to low temperature. Following a few days in tapwater at 4°C in a simulated hibernation, toads begin to accumulate Na^+ at a rate corresponding to 1% of the original Na^+ pool per day, which is accompanied by further water retention. Following about 3 months of NaCl and water accumulations, the uptake through the skin and the urinary elimination of NaCl become about equal, and a steady state is approached where the body Na^+ pool is increased by 60 to 75%.[245,355] The initial fast water volume gain is associated with decreases in the concentrations of Na^+ and Cl^-, but the concentrations of these two ions increase during the subsequent period of further fluid accumulation, with the concentrations rising above their respective values at 20°C. Both the rate and the net accumulation of Cl^- were found to exceed those of Na^+. The lymph concentration of K^+ showed the opposite changes by decreasing from ~3 mM at 20°C to ~1.9 mM after 3 days at 4°C. This lower value was maintained during the following 5 months at 4°C.[355] It appears, therefore, that the body pools of the three diffusible ions Na^+, Cl^-, and K^+ are subject to different regulations in a simulated hibernation.

The above studies also indicate that an overall electroneutral ion exchange is achieved by additional as yet unidentified charge transfers. In sum, at steady state during hibernation in the laboratory, the extracellular and intracellular spaces of *Bufo bufo* are expanded by about 70% and 10%, respectively. Of the accumulated fluid, about 80% is partitioned to the extracellular space, amounting to 40% of the body mass with a new steady state in the turnover of salt and water. At retransfer to 20°C, elimination of salt and water begins immediately at rates that are the reverse of the initial water and ion accumulations at 4°C. Thus, initially water excretion dominates with a temporary increase in lymph concentrations of Na^+ and Cl^-, which is followed by a period of dominating salt elimination.[355] Tadpoles and frogs of *Rana muscosa*, which overwinter in ice-covered lakes and streams, also accumulate water and ions during hibernation in the laboratory.[46] During the first month of simulated hibernation, water content increased by 5.7% in frogs and 14% in tadpoles, with the latter group losing ions to the surroundings as well. Thus, initially, the extracellular fluid became diluted. Interestingly, after this initial period of exposure to 4°C, both frogs and tadpoles began to accumulate salt without losing more water. During a prolonged period at 4°C (7 to 12 months), water content increased again and ions were lost, resulting in dilution of the body fluids.

Jørgensen and coworkers showed that handling of the animals, catheterization of the urinary bladder, and hypophysectomy prior to lowering the environmental temperature partly or fully abolished salt accumulation, confirming that the cold-induced alterations discussed above reflect regulated physiological processes.[245,355] Further studies are necessary to provide detailed information on this and to investigate the physiological significance of the fluid and ion accumulations during hibernation.

4. Breeding

Expanded extracellular volume with distended lymphatic space is characteristic of posthibernating anurans during the breeding season irrespective of the environmental temperature. The explanation for this is unknown, but it illustrates a striking difference between salt and water accumulation in hibernating anurans and anurans acclimated to low temperature in the laboratory, because in the laboratory the animals eliminate the accumulated salt and water immediately upon transfer from 4°C to room temperature.[236]

5. Salinity

Although most amphibians live in or near freshwater, the ability to tolerate brackish water is often overlooked. Brackish water is defined by the European Environment Agency as having a salinity range of 5 to 18 parts per thousand (ppt) NaCl, which corresponds to concentrations of 85 to 308 mM and osmotic concentrations of 159 to 572 mOsm/kg (assuming an osmotic coefficient of 0.93 for NaCl). This ranges from mildly hypoosmotic to considerably hyperosmotic relative to the body fluids of hydrated animals (Table 9.2). Nonetheless, Neill[354] identified 52 species and subspecies of amphibian that have been observed in habitats in which elevated salinity might be encountered.

Bufo calamita, for example, inhabits estuarine regions of northern Europe and is able to tolerate salinities of 16 to 17 ppt (~290 mM NaCl) for up to 4 days.[314] Even the aquatic species *Xenopus laevis* (Pipidae) has been reported to tolerate environmental NaCl concentrations equivalent to 300 to 400 mOsm/kg.[257,408] Among the urodeles, the slender salamander *Batrachoseps attenuatus* (Plethodontidae) has been acclimated to NaCl solutions as high as 600 mOsm and *Ambystoma tigrinum* (Ambystomatidae) to solutions having osmotic concentrations as high as 450 mOsm.[231,409] In both cases, gradual acclimation was required to obtain survival at the higher salt concentrations, and it was also necessary to feed *A. tigirinum*.

Balinsky[16] reviewed the literature on amphibians capable of regulating the internal environment in response to raised external osmotic salinities by maintaining body fluids hyperosmotic to the external bathing solution. As a general rule, saline-tolerant species are able to survive elevated NaCl concentrations in the extracellular fluid and retain urea that arises from protein catabolism, thus the need for gradual acclimation and feeding to obtain maximal survival. A similar mechanism to overcome desiccating environments has been exploited by species that avoid desiccation by burrowing (see Section III.B.1); for example, spadefoot toads (*Scaphiopus*) remain hyperosmotic to the environment and maintain hematocrit and hemoglobin concentrations constant during estivation in the desert throughout the dry season by storing urea in the body fluids.[432,440] In desiccating environments, body volume is reduced by 10 to 20% and is maintained at a lower value with the increased amounts of osmotically active solutes, urea and NaCl, the concentrations of which are regulated by cutaneous transports, urea synthesis, and the kidneys. The intracellular water being little affected and the body water being hyperosmotic relative to the environment indicate that the lower body water content reflects an osmoregulatory response associated with tolerated downward-shifted extracellular volume.

Two anuran species, *Rana cancrivora*[161] and *Bufa viridis*,[159,250] have exploited this general mechanism to maintain their body fluids hypertonic relative to salinities near that of seawater and are able to inhabit estuarine or even marine environments. This extraordinary ability has generated a considerable number of studies, which are summarized below.

a. Rana cancrivora

Even before the revision of Frost et al.,[143] this semiterrestrial crab-eating frog was classified as *Limnonectes cancrivorus*. These frogs live in coastal lowland habitats of Southeast Asia, where they breed in water-filled ditches of varying salinity. In the laboratory, slowly acclimated adult frogs and tadpoles tolerate salinities as high as 80% and 108% seawater, respectively. Over the entire range of tolerated external salinities, the adult frog becomes hyperosmotic relative to the environment, with NaCl and urea contributing about 40 and 60%, respectively, to the raised osmotic concentration of blood plasma. Whereas plasma Na^+ and Cl^- concentrations were found to be above those of the external medium at low salinities, the opposite was found to be true in animals acclimated to salinities above 30%.[161] Following transfer to a relatively more concentrated external medium, the frog lost water with a time course of about 24 hours, at which time a near steady state was reached, with up to 20% decreased body mass. This state of lower body water volume was maintained for a week (the period of observation), during which body water approached a steady state with the production of a hypotonic urine of a concentration, which increased almost linearly to the plasma concentration.[161] Urine production did not stop at higher salinities but was reduced significantly at an external salinity of 600 mOsm, when it amounted to no more than ~1% of the urine flow measured in freshwater acclimated frogs. The antidiuresis was caused by both reduced glomerular filtration rate and increased tubular reabsorption of water.[427] The urea concentration of the urine equilibrated with that of plasma, but, due to the reduced urine flow, the total urinary urea loss decreased to small amounts at the highest external salinities.[427] Muscle inulin space did not change significantly during salinity acclimation of the frogs between freshwater and 80% seawater, and muscle cell volume decreased by only 7% in frogs exhibiting a threefold increase in plasma osmotic concentration, indicating uptake or cellular production of osmotically active solutes. The urea contribution was estimated to be about 60%, the small diffusible ions contributed another 10%, and synthesis of α-amino nitrogen compounds seemed to have contributed the remaining amount of increased osmotically active solutes of the cell water. This latter pool was dominated by taurine, glycine, and β-alanine at the highest external salinities.[160] The increasing urea accumulation in the body fluids with external salinity is due to upregulation of the hepatic urea synthesis, with variation among different tissues of the relative contribution of newly synthesized urea and amino acids (muscle, 119 vs. 38 mM; liver, 132 vs. 3 mM).[548]

b. Bufo viridis

The terrestrial green toad *Bufo viridis* is widely distributed in the old world with its southern boundary in North Africa (Sahara) and eastern boundary in central Asia (Tibet). Among known breeding habitats are coastal rockpools, ditches, small freshwater lakes, estuaries, and salt meadows with salinities of up to 8 ppt (standard seawater has a salinity of 35 ppt). In nature, the adult toad tolerates salinities of ~20 ppt. Katz[250] showed that in the laboratory regularly fed *B. viridis* can be acclimated to live in salinities as high as 800 mOsm (by NaCl). Acclimation is a time-consuming process that depends on the season in which the animal is caught. Thus, a toad experiencing a large change of environmental salinity in one step could die within 3 to 4 days, and winter toads were not able to survive in salinities above 500 mOsm. Salinity acclimation of *B. viridis* is associated with loss of water; for example, in freshwater the water content amounted to 79% of total body mass as compared to 72% in 800 mOsm. Blood plasma is always significantly hypertonic with relative contributions of about 2/3 NaCl and 1/3 urea in animals acclimated at salinities greater than 200 mOsm. In the same study, urine concentrations of both Na^+ and urea increased significantly with external salinity in such a way that the urine was hypoosmotic below 400 mOsm and near isosmotic at higher external salinities. In its natural habitat, *B. viridis* responds to water restriction by burrowing[97] like the desert spadefoot toads.[318,433] Under burrowing conditions in the laboratory in soil containing 8 to 14% water, plasma osmolality increased slowly due to urea accumulation, with smaller extracellular changes occurring in concentrations of Na^+ and Cl^- and paralleled by similar concentration changes of these solutes in bladder urine. Urine was not voided, and whole-body water volume and tissue water were maintained fairly constant even after 70 to 80 days in burrows, provided the animals were allowed contact with the soil.[254] In another laboratory study under similar conditions, a loss of body water was recorded initially that stabilized at 85% of the hydrated body mass during the subsequent period of 2 to 3 months.[203] These and further studies[205] suggested that the increased plasma urea concentration maintains an inward driving force for osmotic cutaneous water uptake from the soil, that urea loss is prevented by recycling across the skin, and that tissue hydration is maintained by equilibration of urea between the extra- and intracellular body compartments.[254,532] In salinity-acclimated toads, urea and K^+ retentions were concluded to be caused by specific tubular processes.[443]

c. Larval Stages

i. Anurans

Embryos and larvae of anuran species that live in estuarine environments show a range of tolerances to salinity. Survival of *Rana temporaria* larvae declines from 80% in NaCl solutions having an osmolarity of 56 mOsm (2.3 ppt) to about 20% for 130-mOsm (4.5-ppt) NaCl solutions.[516] The greatest salinity where oviposition was observed in natural habitats was 0.9 ppt, indicating that the frogs are able to detect dilute NaCl solutions and select sites for oviposition well within the larval salinity tolerance. In contrast, *Bufo calamita* are able to successfully metamorphose in salinities of 10 ppt (~300 mOsm), although larval size is smaller and time to metamorphosis is longer.[157,158] Larval *B. calamita* taken from saline habitats were able to tolerate higher salinities than those from freshwater breeding sites and accumulated a small amount of urea but not enough to be osmotically effective.[157] *R. cancrivora* larvae tolerate salinities up to or even higher than seawater by increasing plasma Na^+ and Cl^- concentrations and maintaining plasma osmolality that is dilute relative to seawater.[160] During climax stages of metamorphosis, *R. cancrivora* express urea cycle enzymes and accumulate urea so the body fluid osmolality slightly exceeds seawater, as is the case in adults. Some larval anurans deposit eggs on land and the larvae accumulate urea. This will be discussed in Section VI.

ii. Urodeles

Larval *Ambystoma tigrinum* reared in the laboratory develop successfully at salinities of 1.37× normal Ringer's (137 m*M* NaCl) but not at 220 m*M* NaCl. During exposure to 137-m*M* NaCl the plasma concentrations of Na^+ and Cl^- increase to keep the plasma hypertonic to the external medium by 22 mOsm.[264] Unlike the adults, urea is not an osmolyte in the larvae. Taylor[480] investigated a reproducing population of *A. subsalsum* in Lake Alchichica in Mexico which has a reported salinity in excess of 5.2 ppt NaCl plus appreciable concentrations of bicarbonate, carbonate, and sulfate that raise the osmotic concentration to an estimated value of about 200 mOsm.

III. STRUCTURE AND PHYSIOLOGY OF THE SKIN

A. Structure and Morphological Cell Types

Except for the presence of dermal scales in some apodan species, all amphibians have a similar arrangement of dermal and epidermal structures (Figure 9.1A). The dermis consists of an inner stratum compactum and an outer stratum spongiosum containing glands that secrete a near isosmotic fluid that may contain mucus, lipids, or toxins. The epidermis consists of an inner layer of cells, the stratum germanitivum, which is attached to the dermis by a basement membrane. Cells of the s. germanitivum undergo mitotic cell division and propagate to form three distinct layers. The first layer above the s. germanitivum is the s. spinosum, and above that is the s. granulosum, which is the outermost living cell layer. Cells of the s. granulosum are connected by zonulae occludens, or tight junctions, that separate the apical plasma membrane, which faces the cornified cells, from the plasma membrane lining the lateral intercellular space. The apical plasma membrane is the limiting barrier for transcellular solute and water transport across the epidermis, although evidence suggests

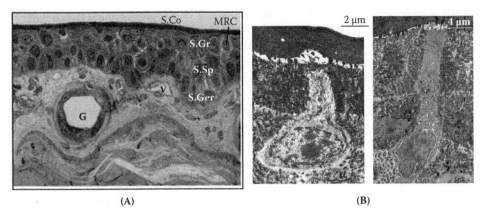

FIGURE 9.1 (A) Cross-section of skin from *Rana pipiens* (see Farquhar and Palade[128]). The epidermis consists of the outermost stratum corneum (S.Co), a dead cell layer that is not a permeability barrier for small ions and water except in some cocoon forming anuran species. The apical membrane of the outermost living cell layer, the stratum granulosum (S.Gr), is the limiting barrier for salt and water transport. The stratum granulosum becomes cornified and replaced by cells of the stratum spinosum (S.Sp) that are renewed by division of cells in the stratum germanitivum (S.Ger). Mitochondria-rich cells (MRs) are interspersed among the principal cells. The dermis consists of an outer layer—the stratum spongiosum, which contains glands (G) and blood vessels (V)—and an inner layer, the stratum compactum. (B) Two mitochondria-rich cells from the same toad skin preparation. On top is seen the dead cornified layer. The mitochondria-rich cells expose a relatively small plasma membrane area to the subcorneal space. (From Willumsen, N.J. et al., *Biochim. Biophys. Acta*, 1566, 28–43, 2002. With permission.)

that the tight junctions are regulated and paracellular transport occurs as well.[144,196,406,501,522] The basolateral plasma membranes of the s. granulosum form extensive connections, both desmosomes and gap junctions, with the plasma membranes of the s. spinosum and the s. spinosum with the s. germanitivum.[435] The cells of these layers form a functional syncytium denoted the *principal cell compartment*, with functions resembling those of principal cells of the urinary bladder and the single-layered collecting tubules and collecting ducts of the kidney. Trans-cellular transport thus requires passage across the apical plasma membrane of the s. granulosum, diffusion through the cytosol of principal cells and through gap junctions between them, and transport across the basolateral plasma membrane, which includes the plasma membranes of the s. spinosum and s. germanitivum.[406,445]

Cells of the stratum granulosum accumulate keratin, die, and are replaced by cells migrating from the s. spinosum. The dead cells form a flattened, one- to two-cell-thick layer, termed the s. corneum. With the exception of some cocoon-forming fossorial species, the s. corneum is generally not a limiting barrier for either evaporative water loss or absorption of water and solutes. The s. corneum is shed at regular intervals—the molting cycle, which is about 1 week at 20°C.[59,60,292] During molting, the skin becomes leaky to electrolytes and water.[233] *In vitro* studies have indicated that the cells replacing the old s. granulosum are not fully functionally differentiated until the slough is formed and shed.[282,362]

In addition to principal cells are Merckel cells of poorly known function and bottle-shaped mitochondria-rich (MR) cells (also called *flask cells*), which extend into the s. granulosum and s. spinosum and connect to the outer surface of the skin, with a narrow neck projecting between the s. granulosum cells to which they are linked by tight junctions (Figure 9.1B). Their apical plasma membrane faces the subcorneal space, and they connect with neighboring cells by desmosomes.[537] MR cell histochemistry[553] and immunoreactivity[61,178,259,451] are different from those for the principal cells. MR cells are selectively stained by silver and methylene blue, revealing a density on the order of 10^5 MR cells per cm^2 of epithelial surface area. The density is species dependent, often with regional differences, and it varies with the prehistory of the animal.[61,116,216,260,542] The area of the apical plasma membrane of MR cells is enlarged by 0.5- to 1-μm microvillar ridges and is critical for acid–base secretion and Cl⁻ uptake (discussed in Section III.F).

B. Water Exchange across the Skin Epithelium

1. Evaporative Water Loss

Evaporation of water from the animal occurs through the respiratory epithelium of the lungs (the nassobuccopharyngal cavity) and through the skin. Amphibian olfaction is closely associated with buccopharyngeal ventilation, by which they smell and react to chemical cues in air and water (reviewed in Jørgensen[242]). The shallow oscillatory movement of the buccal floor that serves olfactory purposes results in evaporative water loss, but its quantitative contribution to the total water loss on land is not known. Adolph[5] observed that skinned frogs (*Rana pipiens*) lost water at rates comparable to intact animals and concluded that the skin forms no appreciable barrier for evaporation. This was for many years the assumption for amphibians in general and may still apply to urodeles and apodans. More recent studies have shown a variety of mechanisms that reduce evaporation across the anuran skin, including those of "waterproof" frogs with evaporative water losses comparable to some reptiles. These mechanisms occur independently among many anuran families that also have species with the more typical high rates of evaporation. Spotila and Berman[453] and Lillywhite[296] quantified the resistance of the integument to evaporative water loss among amphibians and other vertebrates in terms of an evaporative resistance coefficient of the skin (R_S; sec/cm), which is the reciprocal of the rate of evaporation per cm² per unit of vapor pressure difference between the animal and the surrounding air at the prevailing temperature. Resistance values for semiaquatic *Rana* species are predictably low, near zero, which is the resistance for evaporation from a free water surface. Those of the more terrestrial North American desert amphibians *Bufo cognatus* and *Scaphiopus couchi* are about 5× greater but are still reflective of high evaporative water loss. In the 1970s, it was found that the African Rhacophorid *Chiromantis xerampelina* had an evaporative skin resistance of 400 to 900 sec/cm, which is comparable to that of many reptilian species.[301]

This intrinsic property of the skin is as yet not fully understood but has been suggested to be related to the lipid composition of the apical plasma membranes. Reed frogs (Hyperolidae) are also able to tolerate long periods in desiccating environments.[546] The skin of *Hyperolius nasutus* has an evaporative resistance coefficient in excess of 100 to 200 sec/cm depending on the movement of air (flowing vs. still air). Thus, some compensation for a low intrinsic evaporative skin resistance is achieved if the animal stays motionless. An alternative mechanism for reducing evaporative water loss includes wax secretions from skin glands by some members of the South and Central American genus *Phyllomedusa* (Hylidae). The frogs wipe the secretions over the entire skin area to provide very high resistance coefficients of over 300 sec/cm.[441] Other Hylid frogs and arboreal frogs in the genus *Pelopedates* (Rhacophoridae) show similar wiping behaviors and skin secretions and may exhibit resistance values comparable to *Phyllomedusa*.[298] In all cases, wiping is followed by the animals remaining motionless to avoid disruption of the waxy layer. It should be noted that resistance values given in the literature depend on temperature (according to the definition of R_S; see above) and often also on conditions of air flow because an unstirred air layer on the skin surface reducing the rate of diffusion has not always been corrected for. This should be kept in mind in comparative studies and when laboratory data are applied to the natural habitat of the species.

Many species burrow to avoid evaporation. Some fossorial species, such as *Scaphiopus couchi* (Pelobatidae), and many Bufonids accumulate electrolytes and urea to maintain the water potential of the body fluids in equilibrium with the soil.[241] Even species commonly thought of as aquatic, such as *Xenopus laevis* and the urodele *Ambystoma tigrinum*, inhabit ponds that may dry up, requiring the animals to burrow in the mud to accumulate electrolytes and urea.[99,240] Burrowed *S. couchi* accumulate urea more rapidly when burrowed in soils with a lower water potential[319] and routinely estivate for up to a year between summer activity periods that coincide with seasonal summer rainfall. Different soil types have different water-retaining properties, so selection of burrowing sites is important for survival during dry periods.[192] Water absorption from soil is covered below. In addition to accumulating solutes, anurans from several families accumulate many (60 to over 200) layers of shed epidermis to form a watertight cocoon. *Cyclorana australis* (Hylidae) has a skin resistance to evaporative water loss of $R_S = 2.4$ sec/cm, which increases to 60 to 214 sec/cm when a cocoon forms. Cocooned *Neobatrachus aquilonius* (Myobatrachidae) uncovered 1.5 years after a recorded rainfall had urine that was nearly isosmotic with the plasma, and it was estimated they had a 2-year tolerance to dehydration under the ambient soil moisture conditions.[69] Of interest, *N. aquilonius* burrowed in a moist soil did not form a cocoon, indicating that the animals are able to assess the water potential of their surroundings and respond accordingly.

2. Water Absorption

In 1795, Townson[487] observed (his capital letters): "THESE TAKE IN THEIRS THROUGH THE SKIN ALONE: ALL OF THE AQUEOUS FLUID WHICH THEY TAKE IN BEING ABSORBED BY THE SKIN AND ALL THEY REJECT BEING TRANSPIRED THROUGH IT." This was reestablished in the paper "Do Frogs Drink?" by Bentley and Yorio.[31] Frogs placed in hypersaline baths have been observed to drink, but *Rana cancrivora* in near-seawater does not. One frog for which drinking is significant is *Phyllomedusa sauvagii*, which becomes waterproof when it wipes itself with waxy skin secretions. During a rain, these frogs will allow water falling on the head to flow into their mouths[321] and may regain 20% of their body mass per hour, but when not waxed they are able to absorb water across their skin.

a. Comparative Studies of Cutaneous Water Uptake

Comparative measurements of water absorption have been made with intact animals and with isolated skin. The latter has been a popular tissue for studying osmotic water movement because it is long lived when mounted in a chamber consisting of a tube with the skin tied to form a membrane at one end. The tube is filled with one solution and the skin end is immersed in a second solution. This arrangement, termed an *osmometer*,[112] was applied to frog skin by Matteucci.[315] Osmotic water movement can be quantified either as the change in fluid volume of the osmometer or as the mass gain by the intact animal. With body mass (M) expressed in grams, Mullen and Alvarado[341] used the equations $A = 8.9 \cdot M^{0.56}$ cm^2 to estimate area-specific water absorption by *Bufo boreas* and $A = M^{0.56}$ cm^2 for *Ascaphus trueii*, *Hyla regilla*, and *Rana pipiens*. Katz and Ben-Sasson[252] applied the formula $A = 6.3 \cdot M^{0.63}$ cm^2 for *B. viridis*. Dehydration and subsequent rehydration, as they relate to the osmotic concentration of the body fluids, can then be evaluated relative to the standard mass, which, as discussed above, depends on both environmental and physiological conditions. Rehydration is commonly expressed in units of mL/g/hr or extrapolated for the whole surface area based on body mass. The skin of aquatic species of anurans,[29,341] urodeles,[30,187] and apodans[222] has a lower water permeability than the skin of more terrestrial forms. This reduces the tendency of aquatic species to absorb excessive water and permits terrestrial species to rehydrate more rapidly when they return to water.[21] Absorption by the more terrestrial *Bufo* was observed to be greater than for the more aquatic anuran species. Among the urodeles, the terrestrial species *Notopthalmus viridescens* (Salamandridae) and *Aneides lugubris* (Plethodontidae) are able to rehydrate as quickly as terrestrial anurans.[54,187]

Smaller animals have a greater surface area-to-volume ratio. With a given area-specific rate of water absorption, smaller animals will recover a greater percentage of their body mass within a given period of time. In addition, many species of terrestrial anuran have a region in the ventral

TABLE 9.4
Osmotic Permeabilities of Amphibian Skin

Species	Mean Body Mass (g)	J_v (μL/cm²/hr)	P_f (cm/sec)	Remarks
Ascaphus truei	2.5 ± 0.2	3.98 ± 0.19	3.59	*In vivo*
Rana pipiens	39.7 ± 5.0	4.80 ± 0.42	3.83	*In vivo*
Hyla regilla	2.6 ± 0.3	5.60 ± 0.27	3.96	*In vivo*
Bufo boreas	61.1 ± 5.2	15.80 ± 2.1	10.38	*In vivo*

Source: Data from Mullen, T.E. et al., *Am. J. Physiol.*, 231, 501–508, 1976.

skin, termed the *seat patch* or *pelvic patch*, that is specialized for water absorption.[28,75,320] Its water permeability can be greatly stimulated in dehydrated animals and may account for 85% of the water absorbed.[311] Thus, area-specific rates based on body mass will underestimate the actual water flux across the seat patch. Finally, many species have a system of grooves and channels in the skin, termed *epidermal sculpturing*, that can amplify the effective surface area for water absorption[94,297,374] and has been suggested to account for the difference in water permeability between Bufonidae and Ranidae.[75] Examples of osmotic water uptakes in different anuran species under laboratory conditions together with calculated osmotic permeabilities (P_f) are given in Table 9.4 (for definition of P_f, see discussion below). This study found a fairly large difference in P_f between *Bufo boreas* and the other three anurans, probably due to the greater osmotic permeability of the seat patch region of the ventral skin.

b. Water Uptake from Soil Water

Many anuran and urodele and most apodan species are partially or completely fossorial and depend on soil moisture as a hydration source. Soil water potential depends on the adsorption of water to soil particles (*matric potential*, $\pi_M \leq 0$), the solutes dissolved in the soil water (*osmotic potential*, $\pi_{SW} \leq 0$), the elevation in the gravitational field (gravitational potential, which may be negative, zero, or positive), and the external applied pressure (pressure potential, likewise negative, zero, or positive). When discussing water movement between soil water and a burrowing amphibian, only the matric potential and the osmotic potential of the soil need be considered. The matric potential can be measured with the ceramic plate extraction method,[444] in which water-saturated soil is placed on a ceramic plate in a pressure vessel. An imposed nitrogen pressure forces water from the particle matrix to a collection tube until the gas pressure equilibrates with water retained by the soil. For a given pressure, a certain percentage of water is retained by the soil. By measuring soil water content over a range of pressures, one can characterize the water-retention characteristics of soil in a given burrowing site as a function of its water content. The force of water retention thus obtained, expressed as the negative value of the pressure forcing water out, defines the *matric potential* (π_M), which is determined by soil particle size, porosity of particles, and organic content. The osmotic potential (osmotic pressure) of pure water is defined as zero and becomes more negative as solute is added. The accumulation of solutes in the body fluids reduces the osmotic potential relative to that of pure water ($\pi_{BW} < 0$) and can be determined by a freezing point or vapor pressure osmometer. Thermodynamically, osmotic water flow proceeds with a negative free-energy change. Thus, burrowing animals can maintain fluid balance when the osmotic potential of the animal's body fluids is equal to or more negative than the soil water potential, given by the sum of its matric and osmotic potential—that is, when $\pi_{BW} \leq \pi_M + \pi_{SW}$. Historically, π has been expressed in units of bars (1 bar = 1.013 atmospheres), but more recently the SI unit kPa has been used (1 bar = 100 kPa).

c.   Osmotic and Diffusional Permeability

i.   Osmotic Permeability

Experimental measurements of water absorption across the amphibian skin contributed significantly to the quantitative formulation of the two major physiological concepts of epithelial water transport: the osmotic permeability and the diffusional permeability. The osmotic pressure (π) of a diluted solution of an osmolyte (S) with the molar volume concentration of C_S is given by van't Hoff's equation:*

$$\pi = R \cdot T \cdot C_S \tag{9.1}$$

π is in the unit of Pa when $R = 8.31$ Joule/mol·K, T is in Kelvin, and C_S is in mol/m³ ($\equiv$ mM). Considering a membrane with hydraulic conductance L_P and solute reflection coefficient σ_S, the water volume flow (J_V) across the membrane is:

$$J_V = L_P \cdot R \cdot T \cdot \sigma_S \cdot \Delta C_S \tag{9.2}$$

where ΔC_S is the transmembrane concentration difference, and the reflection coefficient is defined such that $\sigma_S = 1$ for a membrane that is permeable to water but impermeable to S (a semipermeable membrane), and $\sigma_S = 0$ for a membrane that cannot discriminate between solvent and solute. The osmotic permeability (P_f) of a membrane is given by:[131]

$$P_f = \frac{RT}{\overline{V}_W} L_P \tag{9.3}$$

which, combined with Equation 9.2 for water flow, gives:

$$J_V = P_f \cdot \overline{V}_W \cdot \sigma_S \cdot \Delta C_S \tag{9.4}$$

In physiological literature, J_V is in cm³/cm²/sec ($\equiv$ cm/sec) when P_f is in cm/sec, $\overline{V}_W$ is in cm³ per mol H$_2$O (~18 cm³/mol at 20°C), and C_S is in mol/cm³. The osmotic coefficients of the external bath and extracellular body fluid can be quite different, which can be taken into account by measuring the osmotic concentrations with an osmometer. In many studies, it is tacitly assumed that $\sigma_S = 1$, which may be a reasonable approximation for tight epithelia such as amphibian skin and urinary bladder with high-resistance tight junctions.

ii.   Diffusional Permeability

Another method that has been used in studies of water exchange across membranes is to add an isotopic tracer for water (^{3}H$_2$O, D$_2$O, H$_2$O^{18}) and measure unidirectional diffusion fluxes across the membrane. If the unidirectional tracer flux per unit area of membrane is proportional to the water concentration in the solution to which the tracer is added ([H$_2$O]$_o$ and [H$_2$O]$_i$), the net diffusion flux would be the difference between influx and efflux as indicated by *Fick's law*:

$$J_{H_2O}^{in} = P_{dw} \cdot [H_2O]_o, \quad J_{H_2O}^{out} = P_{dw} \cdot [H_2O]_i$$

$$J_{H_2O} = P_{dw} \cdot ([H_2O]_o - [H_2O]_i) \tag{9.5}$$

where P_{dw} is the water diffusion permeability based on the assumption that there is no isotopic effect (the tracer behaves exactly as H$_2$O). If the water concentration in the membrane is so small that the water molecules do not interact with each other, it can be shown[131] that the osmotic

---

* The van't Hoff equation[504] applies to *ideal* solutions. The body fluid cannot be considered to be an *ideal* solution. Thus, in the physiological range of osmotic concentrations more correctly, $\pi = \phi \cdot R \cdot T \cdot C_S$, where ϕ is a nondimensional, empirically determined osmotic coefficient with $\phi = 0.9$ for the amphibian Ringer's solution.

permeability given by Equation 9.4 will be equal to the diffusion permeability given by Equation 9.5. In a study of frogs *in vivo*, the cutaneous water permeability obtained from the rate of osmotic water uptake was compared with that obtained by the rate of diffusion of D_2O.[186] It was found that the osmotic permeability was several times higher than the permeability calculated from the diffusion of heavy water assuming that the ratio between the influx and efflux is equal to the ratio of the water concentrations on the two sides of the skin (see Equation 9.5). Koefoed-Johnsen and Ussing[266] pointed out that this latter assumption is not valid if the diffusion of water is superimposed on a convective flow of water through pores in the membrane. They developed appropriate equations for testing this hypothesis in experiments on isolated frog skin stimulated by neurohypophyseal hormone. They found that the hormone increased the osmotic net water flux 2.3× when estimated as water volume uptake but by no more than 10% when the influx was estimated with the D_2O isotope tracer method.[266] From these observations water movement was proposed to result from mass flow through pores rather than from thermal motions of single water molecules dissolved in the membrane phase.

3. Amphibian Aquaporins

Biophysical and physiological studies on the water permeability of biological membranes including amphibian skin[266] and urinary bladder[181,400] predicted water movement to take place through pores. Water channel proteins were initially identified by freeze fracture electron microscopy as intramembrane particle aggregates.[50,72,246] During the early 1990s, such water channel proteins were indeed discovered[390] and given the name *aquaporins* (AQPs) which have subsequently been identified in many organisms, ranging from bacteria to plants and animals.[219,378] AQPs form membrane pores that are selectively permeable to water, and a subfamily of AQPs, termed *aquaglyceroporins*, are comprised of membrane pores that are permeated by water and certain small solutes, such as glycerol and urea. A common feature of AQPs is a sequence of two Asn–Pro–Ala (NPA) motifs that form the water-selective pathway and the ability of Hg^{2+} and other mercurial compounds to inhibit water transport. Specific aquaporins in the skin and urinary bladder are stimulated by the amphibian antidiuretic hormone arginine vasotocin (AVT), thereby facilitating water absorption and retention in most anuran species and in many urodeles.

At the present time, the full-length sequences of 17 AQP cDNAs have been elucidated in anurans. A phylogenetic analysis of AQP proteins from anuran amphibians and mammals suggested that anuran AQPs can be divided into six clusters: types 1, 2, 3, and 5 and two anuran-specific types designated as a1 and a2 (the letter "a" represents anuran) (Figure 9.2). Types 1, 2, 3, and 5 correspond to mammalian AQP1, AQP2, AQP3, and AQP5, respectively.[470] The cluster of type a1 AQPs is composed of AQPxlo from *Xenopus laevis* oocytes[518] and another *X. laevis* AQP (BC090201). The cluster of type a2 AQPs contains AQP-h2[178] and AQP-h3[477] from the tree frog *Hyla japonica* and AQP-t2 (AF02622) and AQP-t3 (AF020622) from the toad *Bufo marinus*. It is of interest that AQP2 and type a2 clusters belong to different groups, although all of these AQPs can be stimulated by AVT (see below).

AQP-h1 has higher homology to toad AQP-t1, Rana FA-CHIP (AQP1),[1] and rat AQP1, whereas AQP-h3 and AQP-h2 have higher homology to mammalian AQP2 than to mammalian AQP3. AQP-h1 showed a ubiquitous tissue distribution, whereas AQP-h3 displayed a specific distribution that was restricted to the ventral pelvic skins. AQP-h2 was expressed in the ventral pelvic skin and urinary bladder but not in the kidney.[178,477] Recently, two aquaporins were added to the amphibian AQP family: AQP-h2K was isolated from the kidney of *Hyla japonica*, which is a homolog to AQP2 (HC-2) from Cope's tree frog (*Hyla chrysoscelis*),[559] and *Hyla* AQP-h3BL was identified by molecular cloning.[6] This AQP-h3BL showed a high amino acid sequence similarity to mammalian AQP3 and, like the mammalian isoform, is predominately expressed in the basolateral plasma membrane of several osmoregulatory tissues, including skin, mucous glands, urinary bladder, and kidney.

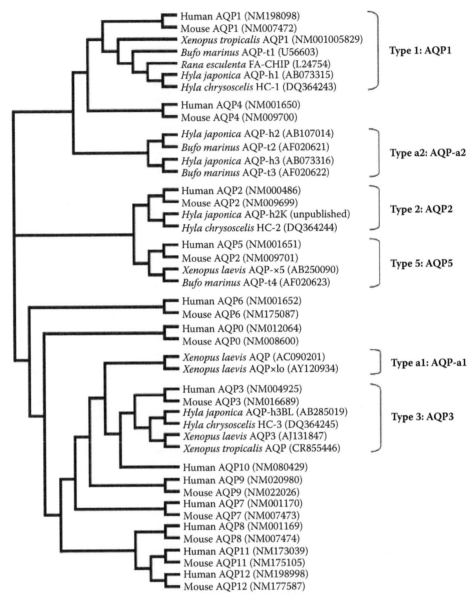

FIGURE 9.2 Phylogenetic and sequence analysis of amphibian AQPs. Note that the AQP-a2 cluster includes the –h2, –t2, –h3, and –t3 isoforms specific for anurans that, like the AQP2 cluster (including mammals), is stimulated by AVT. Note also that the Type 3 AQP3 cluster includes the mammalian AQP3 which, like the amphibian AQP-h3BL and HC-3 isoforms, is located in the basolateral membranes. (From Ogushi, Y. et al., *Endocrinology*, 148, 5891–5901, 2007. With permission.)

a.   Aquaporins of Amphibian Skin

It is the granulosum cell layer (previously denoted the *first-reacting cell layer*) that plays a key role in controlling water transport.[304,521] By immunofluorescence staining, both AQP-h2 and AQP-h3 were localized in two or three layers of principal cells.[178,477] These immunolabels were strongest along the plasma membranes, and AQP-h2 appeared to be colocalized with AQP-h3.[178] No signal was found in the mitochondria-rich cells. AQP-h3BL was constitutively expressed in the basolateral membrane of principal cells in the stratum granulosum of the whole frog body. AQP-h2 and AQP-h3

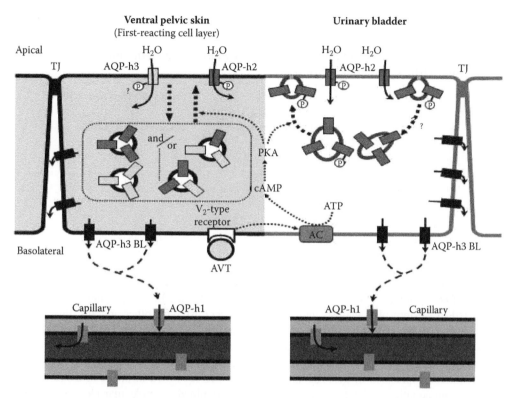

FIGURE 9.3 AVT-controlled signaling pathways for apical membrane insertion of *Hyla* AQP-h2 and AQP-h3 of the outermost granular cells of ventral pelvic skin (left) and urinary bladder (right). AVT increases water permeability by binding to V_2-type AVT receptors on the basolateral plasma membrane of the granular cells. The ligand-bound AVT receptors activate adenylyl cyclase (AC) and the intracellular cAMP level increases, in turn leading to activation of protein kinase A (PKA). The PKA phosphorylates AQP-h2 and AQP-h3 in the ventral pelvic skin and AQP-h2 in the urinary bladder, promoting translocation of the AQP-bearing vesicles to the apical plasma membrane. Transmembrane osmotic water flows are indicated by black and dotted arrows. AQP-h3BL is located at the basolateral membrane of the granular cells, thereby promoting water flows from the cytoplasm into the connective tissues and the spaces between granular cells. Apically located high-resistance tight junctions (TJ) join together the granular cells. AQP-h1 channels are located at the plasma membrane of the endothelial cells of capillaries where they mediate osmotic transport of absorbed water into the bloodstream.

are specifically expressed in the ventral pelvic skins, where water absorption occurs,[6,178,477] and are translocated to the apical plasma membrane in response to treatment with the amphibian antidiuretic hormone arginine vasotocin (AVT), which results in enhancement of the cutaneous water permeability. Water absorbed into the cells exits across the basolateral membrane through the AQP-h3BL proteins and moves to subepidermal capillaries (Figure 9.3). It is an interesting issue to determine whether AQP-h3BL, AQP-h2, and AQP-h3 are colocalized on the plasma membrane and intracellular vesicles. Further studies will be required to define the roles of each AQP in the water absorption of the ventral pelvic skins.

b. Comparative Studies of Amphibian Aquaporins

The response to antidiuretic hormone tends to be greater in the anuran species normally occupying drier habitats than in those from wet habitats.[23,25] To examine the putative involvement of AQPs for this functional differentiation, immunoblots and immunohistochemical analyses have been performed for ventral pelvic skins of five anurans. According to their habitats, they are divided into four groups (Table 9.1): *X. laevis* is aquatic; *R. japonica*, *R. catesbeiana*, and *R. nigromaculata* are

TABLE 9.5
Phylogenetic Distribution of AQP Expression in the Abdominal Skin and Urinary Bladder of Anurans

Species	Habitat	Ventral-Type AQP Pelvic Skin	Urinary Bladder–Type AQP Pelvic Skin	Urinary Bladder–Type AQP Urinary Bladder
Hyla japonica	Arboreal	+	+	+
Bufo japonica	Terrestrial	+	+	+
Rana nigromaculata	Semiterrestrial	+	−	+
Rana japonica	Semiterrestrial	+	−	?
Rana catesbeiana	Semiterrestrial	+	−	+
Xenopus laevis	Aquatic	+	−	+

Source: Suzuki, M. et al., *Comp. Biochem. Physiol. A*, 148 72–81, 2007. With permission.

semiterrestrial; *B. japonicus* explores drier terrestrial habitats and is classified as a terrestrial; and the arboreal *H. japonica* constitutes the fourth group. Because the antibody against *Hyla* AQP-h2 was utilized in western blot analysis and immunohistochemistry for other anurans, this antibody was applied. AQP-h2 or AQP-h2-like protein was detected in the urinary bladder of all of the species examined (Table 9.5). The AQP-h2 homolog was detected in the pelvic skin of the terrestrial toad *Bufo japonica* as in the tree-adapting frog *H. japonica*, but not in the other species (Table 9.1). In contrast, expression of the AQP-h3-like cDNA was identified in the ventral skin of all of the frogs, from aquatic species to terrestrial dwellers, examined by molecular cloning; therefore, AQP-h2 seems to be a urinary-bladder-type AQP, and AQP-h3 seems to be a ventral-skin-type AQP. It may be speculated that, as anurans adapted to drier terrestrial environments, the urinary-bladder-type AQP became expressed in the pelvic skin, as well, resulting in water absorption from the environment via both AQP-h2 and AQP-h3.

C. EPIDERMAL ION TRANSPORT: BACKGROUND

The literature on ion transport in amphibian skin can roughly be divided between papers on general and on comparative physiological issues. As examples of the first group of studies, frog skin was used extensively by physiologists in the 19th and 20th centuries as a model for investigating the nature of bioelectricity and membrane transport (reviewed in Jørgensen[240] and Ussing et al.[503]). DuBois-Reymond[107,108] observed that frog skin generates an electric current (*Froschhautstromme*), which Galeotti[146] showed to be dependent on external Na^+ (or Li^+), and forwarded the interesting hypothesis that the skin is more permeable for the two alkali metal ions in the inward direction than in the outward direction. A number of more speculative theories were also forwarded before Huf[211] demonstrated cutaneous ion uptake in closed sacs from leg skin filled with a Ringer's. The sacs were immersed in the same solution and gained mass by cutaneous water uptake that was abolished by cyanide. Analysis of the contents of the bag revealed an accumulation of Cl^-, which he concluded to be the result of an active transport in the inward direction. A similar conclusion was reached already in 1890 by Reid,[398] who studied fluid transport across the isolated frog skin and concluded "that in the living skin of the frog there is at work some form of absorptive force dependent on protoplasmic activity and exerted in a direction from the external towards the internal surface" (p. 346) (see also Section III.H).

The concept of *active transport* was defined by Krogh as a transport that occurs against a concentration gradient. In his first study, Krogh[277,278] raised two fundamental questions which became major themes in general and comparative physiology. First, can transport of an ion take place against its concentration gradient—that is, in the direction opposite of that predicted by Fick's law? Second, what is the biological significance of active ion transport? These questions were analyzed in experiments with a variety of freshwater animals that maintain large concentration gradients between body fluids and the external bath. It was shown that *Rana esculenta* can absorb Cl⁻ from freshwater and from NaCl solutions as dilute as 10^{-5} *M*, and that Na⁺ and Cl⁻ could be taken up independently of each other.[279] With respect to the physiological significance of cutaneous ion transport, Krogh emphasized that he could demonstrate Cl⁻ accumulation in frogs only if they had been forced into a negative NaCl balance by being kept starved and sprayed for several weeks by distilled water. This treatment resulted in a fairly large loss of body water and extracellular Cl⁻, induced active cutaneous ion uptake, and reduced significantly the cutaneous and renal loss of Cl⁻. Krogh suggested that ion balance is normally maintained by uptake of ions from food, and the uptake via the skin is of physiological significance only during hibernation at the bottom of ponds.[277] As discussed in Section II.B.3, significant amounts of NaCl and water are accumulated by anurans via cutaneous uptake during hibernation in an artificial terrestrial habitat, which is also the case during hibernation in nature. Another function of cutaneous ion uptake in well-fed animals might be to generate solute-coupled water uptake (see Section III.H). It should also be noted that the active Na⁺ transport together with a parallel anion conductance serve a chemosensory function for detecting osmotically available water (see Section VII.G).

Krogh's studies[277] also showed that the skin becomes quite tight to passive ion loss during periods of forced negative whole-body ion balance, which was the first indication of physiological regulation of the passive permeability of the skin. More recent measurements of cutaneous fluxes with radioactive tracers have shown aquatic species to have a higher affinity for salt uptake and lower loss to dilute solutions than do terrestrial species,[164,341] indicating that selection for high-affinity salt uptake and reduced passive loss has occurred in natural habitats.

D. Functional Organization of the Epithelium

To obtain a rigorous theoretical tool in the study of mechanisms of ion transport, Ussing[496] derived the flux ratio equation for passive transport by electrodiffusion which applies to any membrane independent of its complexity:

$$\frac{J_j^{in}}{J_j^{out}} = \frac{(j)_o}{(j)_i} \exp\left[\frac{-z_j \cdot F \cdot V_T}{R \cdot T}\right] \qquad (9.6a)$$

Thus, for passive transport of the ion *j* with charge z_j the ratio of influx to efflux of *j* (left-hand side of the equation) is equal to the ratio of the ion activity in the solutions bathing the membrane, $(j)_o$ and $(j)_i$, multiplied by an exponential function that takes into account the electrical potential difference across the membrane, V_T ($= \psi_i - \psi_o$). In Equation 9.6a, z_j is the valency of the ion, *F* is the Faraday constant, *R* is the universal gas constant, and *T* is the temperature in K.[461] Ussing and Zerahn[499] were the first to conclusively demonstrate active Na⁺ transport by mounting isolated frog skin between two chambers with identical Ringer's bathing either side of the tissue. An external short-circuit current (I_{SC}) was passed to regulate the transepithelial potential difference resulting from active ion transport to 0 mV (considered to be the *Ussing chamber technique*). Under the standard conditions where $(j)_o$ and $(j)_i$ are equal and V_T is clamped to 0 mV (short-circuited), a net flux in the inward direction (J_j^{in}/J_j^{out} greater than unity) defines active transport. Dividing I_{SC} by Faraday's constant provided a measure of net transepithelial ion flux that was shown to be about the same as the net flux of Na⁺ calculated from unidirectional fluxes of radioactive Na⁺ isotopes.

The *two-membrane model* of active sodium transport proposed by Koefoed-Johnsen and Ussing[267] depicted the frog skin epithelium as a single functionally polarized unit with an apical plasma membrane that is permeable to Na⁺ and a basolateral plasma membrane permeable to K⁺, with an active transport step (the Na⁺–K⁺ pump) also in the basolateral plasma membrane (Figure

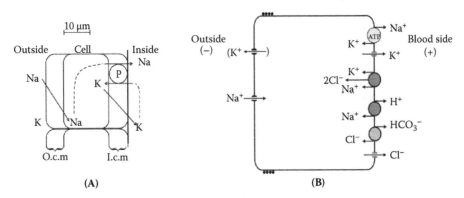

FIGURE 9.4 Functional organization of principal cell compartment. (A) Model for transepithelial Na⁺ transport as proposed by Koefoed-Johnsen and Ussing.[267] Na⁺ enters across the apical (outside facing) membrane down an electrochemical gradient that is maintained by active Na⁺ transport out of the cell by the ATP-consuming basolateral Na⁺–K⁺ pump (P). The relatively high K⁺ permeability of the basolateral membrane secures passive back leak of K⁺ into the interstitial fluid and a negative membrane potential. (B) A more recent model applied to the syncytium of principal cells which also contains NKCC cotransporters and Cl⁻ channels in the basolateral membrane, which displays transport systems for intracellular pH regulation: Na⁺/H⁺ and Cl⁻/HCO₃⁻ exchangers, respectively. The K⁺ channels of the apical membrane (shown in parentheses) are normally downregulated but constitute important pathways for eliminating a body load of K⁺, in which case they become activated.

9.4A). Numerous studies of epithelial transport have shown that this model, with a passive entry step and a primary active exit mechanism, is a general principle for active transepithelial Na⁺ transports that applies to several other types of electrolyte absorbing epithelia, as well.[400]

Jørgensen et al.[243] investigated the existence of active Cl⁻ uptake by living anurans in freshwater taking into consideration the transepithelial potential difference as putative driving force (Equation 9.6a). They measured the Cl⁻ influx with radioactive isotopes and the net Cl⁻ flux by titration of the external bath. The electrical potential across the skin *in vivo* was measured with a Ringer–agar bridge deeply inserted into the dorsal lymph sac and with the reference electrode of appropriate composition dipped into the bath. With some variation between the animals but independent of water uptake (solvent drag), they found that Cl⁻ is actively transported inward across the skin of *B. bufo*, *R. esculenta*, and *R. temporaria*. In studies with radioisotopes of both Na⁺ and Cl⁻ and in studies on selectively ion-depleted animals, they confirmed that the Na⁺ and Cl⁻ are transported in the inward direction by independent membrane mechanisms. Because of the very large electrostatic forces required to separate ions of opposite charge, Krogh proposed that the independent uptake of Na⁺ and Cl⁻ involved exchange of an ion of the same sign, *in casu* (e.g., Cl⁻ for HCO₃⁻), which was in agreement with measured HCO₃⁻ excretion in some of his experiments and because "the CO₂ produced by metabolism and excreted through the skin and gills is probably sufficient to serve in exchanged for Cl⁻ absorbed."[279] The cation exchanging for Na⁺ was not identified in frog skin, but his suggestion that "NH₃ liberated at the outer surface [from NH₄⁺] of an absorbing cell may partly diffuse back and become utilized over again"[279] would imply that Na⁺ was exchanged for H⁺. Similar observations were made with the axolotl *Ambystoma mexicanum* and a large number of freshwater teleosts, indicating that the uptake of Na⁺ and Cl⁻ is common among freshwater organisms.[278,280]

Garcia-Romeu et al.[149] extended previous *in vivo* studies on NaCl uptake by freshwater animals by measuring quantitative relationships of cutaneous Na⁺, Cl⁻, H⁺, and HCO₃⁻ (base) exchanges in the South American frog *Calyptocephalella gayi*. With salt-depleted frogs, they confirmed Na⁺ and Cl⁻ uptake in near 1:1 proportion by animals kept in artificial freshwater. Selectively Cl⁻-depleted frogs displayed a stationary Cl⁻ uptake from NaCl solutions accompanied by excretion of base

(HCO_3^-) without a concomitant uptake of Na^+. In a similar way, Na^+-depleted animals exposed to low concentrations of NaCl absorbed Na^+ with no uptake of Cl^-. As a novel observation, it was found that in selectively Na^+-depleted animals Na^+ exchanged with H^+ in a 1:1 proportion. Both types of exchanges were depressed by a carbonic anhydrase inhibitor.

These and other similar observations were discussed in a review by Motais and Garcia-Romeu,[339] in which they forwarded the general hypothesis that in the skin of amphibians and the gills of freshwater teleosts the exchanges of Cl^- for HCO_3^- and of Na^+ for H^+ were obligatory, with a fixed stoichiometry of 1:1 for both mechanisms. This hypothesis conformed to early observations that anuran skin acidifies the external solution[133,213,496] and to subsequent studies confirming obligatory apical anion exchange.[113] A contemporary study of the isolated skin confirmed that anuran skin displays a proton-secreting mechanism that is dependent on cellular energy metabolism and can be suppressed by an inhibitor of carbonic anhydrase.[120] When considering the energetic requirement, however, a proton pump rather than a 1:1 Na^+/H^+ seems to be demanded for ion uptake from diluted NaCl solutions, and H^+ secretion has turned out to be associated with Cl^- uptake rather than Na^+ uptake (see Section III.F.3.b).

Numerous studies on anuran skin over a period of 40 to 50 years have extended the model with additional transport systems of the principal cells, added a paracellular pathway, and shown that transepithelial Cl^-, HCO_3^-, and H^+ fluxes are governed by flask-shaped intercalated mitochondria-rich cells. These studies are discussed below.

E. Principal Cell Compartment

Two regulated transepithelial ion fluxes are known to flow through the principal cell compartment (see Figure 9.4B). The active uptake of Na^+ and an active secretion of K^+ are both fueled by hydrolysis of adenosine triphosphate (ATP) by the basolateral Na^+,K^+-ATPase. At the expense of metabolic energy these pumps maintain the intracellular $[K^+]$ above thermodynamic equilibrium and the intracellular $[Na^+]$ below (Table 9.2).[148,352,359,406] Experiments with diluted external solutions on the outside indicated that the apical plasma membrane potential is governed by the K^+ distribution across the basolateral membrane with a K^+ equilibrium potential more negative than −100 mV.[346] The combination of low cellular Na^+ activity and negative apical membrane potential draws Na^+ passively across the apical membrane even if the external Na^+ activity is as low as 100 μM.[175] Radioautographic analysis of the sodium pump inhibitor ^{3}H-ouabain showed that the Na^+–K^+ pumps are localized in the plasma membranes of the stratum granulosum and in the plasma membranes of deeper cell layers that interface with the intercellular spaces of the epithelium. In contrast, the serosal plasma membranes facing the basement membrane was not labeled by ^{3}H-ouabain.[329] Due to the small Na^+ dissociation constant of the pump[283,359] and because the principal cells are connected by gap junctions allowing Na^+ to diffuse into cells of the s. spinosum and s. germinativum, the basolateral sodium pumps effectively drive Na^+ into the relatively large lateral intercellular space while maintaining low cellular Na^+ concentrations.

Lindemann and Van Driessche[299] analyzed microscopic amiloride-induced fluctuations in I_{SC} and demonstrated the pathway for apical Na^+ entry to be channels that fluctuate between open, closed, and amiloride-blocked states (reviewed in Van Driessche[507]). The single-channel current is a function of the conductance of individual channels (γ_{Na}) and the driving force of apical Na^+ entry, $i_{Na} = \gamma_{Na} \cdot (V_M - E_{Na})$, where V_M is the apical membrane potential and E_{Na} is the equilibrium (Nernst) potential of the Na^+ distribution across the apical plasma membrane. I_{Na}, measured as the amiloride-sensitive component of I_{SC}, is then a function of the single channel current (i_{Na}), the density of channels in the apical membrane (N), and the probability that a given channel is in an open state (P_o): $I_{Na} = i_{Na} \cdot N \cdot P_o$. N and P_o can be regulated physiologically by channel insertion/retrieval and by controlling the gating of channels resident in the membrane, respectively. A number of hormones, including AVT, stimulate Na^+ transport in the skin, kidney, and urinary bladder to maintain the Na^+ concentration and extracellular fluid volume.[15,184] Baker and Hillyard[15] observed a fivefold increase in amiloride-sensitive I_{SC} following AVT stimulation of isolated toad skin which correlated significantly with the density of electrically active Na^+ channels. Apical membrane area evaluated from the change in capacitance increased significantly by a factor of about 18%. These results might be explained by insertion of vesicles with a high density of channels[152] or by a combination of insertion and activation of channels already in the membrane.

1. The Apical Sodium Channel and Its Regulation: ENaC

Canessa et al.[65] showed that amiloride-blockable Na^+ channels are related to mechanosensitive cation channels in the body wall of the worm *Caenorhabditus elegans*. Subsequent studies have characterized a large family of epithelial sodium channels (ENaCs) that perform a number of transport functions, including Na^+ absorption by amphibian skin, urinary bladder, and renal tubules.[152,261,434] ENaC is a multimeric membrane protein made of three homolog units, α, β, and γ, with a conductance of ~5 pS for Na^+ and ~9 pS for Li^+.[66,394] The functional channel is probably a tetramer of 2α, 1β, and 1γ.[13] The voltage dependence of the Na^+ current is described by the Goldman–Hodgkin–Katz electrodiffusion equation; however, in frog skin the apical Na^+ permeability (P_{Na}) decreases with an increase in the external sodium concentration.[145] The downregulation of P_{Na} predicts Michaelis–Menten-like saturation kinetics of the Na^+ influx with increasing external Na^+ concentration in agreement with direct observations.[34,262,411,503] Supposedly, binding of Na^+ to an external Na^+ selective site closes the adjacent channel, leaving the remaining channels unaffected. This hypothesis was confirmed by fluctuation analysis of Na^+ currents in the presence of small concentrations of amiloride on the external side of the membrane. It was found that the number of amiloride-accessible Na^+ channels decreased with increasing external Na^+ concentration, while the single-channel currents increased linearly with increasing Na^+ concentration.[508] The assumed function of this phenomenon is to reduce Na^+ entry from elevated external Na^+ concentrations and preserve cell volume. Also of physiological significance is downregulation of the apical Na^+ permeability at increasing intracellular concentrations of H^+ and Ca^{2+}, respectively.[49,174]

2. Apical Potassium Channels

Van Driessche and Zeiske[509] used stationary current fluctuation analysis to demonstrate the presence of apical K^+ channels in the skin of *Rana temporaria* that provide the mechanism of cutaneous secretion of K^+ measured in living K^+-loaded frogs.[142,506] The potassium conductance of the apical membrane is usually very small and not easy to detect unless the K^+ concentration in the outside bath is raised[349] or Cl^- of the bathing solutions is replaced by the impermeable gluconate, which induces an ouabain-inhibitable active secretion of K^+ via Ba^{2+}-blockable K^+ channels in the apical plasma membrane.[360] Fluctuation analysis has also revealed an apical cation channel that operates in parallel with ENaC but is not blocked by amiloride. It is poorly selective for Na^+ vs. K^+, and its function is not understood.[510]

3. Basolateral Membrane and Cell Volume Regulation

The Na^+–K^+ pump is located in the membranes lining the labyrinth of lateral intercellular spaces.[329] As in other cells, the pump in frog skin is rheogenic with a cation stoichiometry of 3Na:2K.[348,357,358] One ATP is hydrolyzed per pump cycle with a free energy of hydrolysis (ΔG_{ATP}) = –60 kJ/mol.[77,123] The major conductance of the basolateral membrane is K^+ selective, which can be inhibited by Ba^{2+}.[347] The K^+ channels of the basolateral membrane serve several functions, as they recycle K^+ into the lateral intercellular space after it has been pumped into the cells via the adjacent Na^+–K^+ pumps,[495] control the electrical driving force for Na^+ uptake across the apical Na^+ selective membrane,[174] and recycle K^+ into the interstitial space that has been taken up by the Na^+–K^+–$2Cl^-$ exchange protein (NKCC) of the basolateral membrane.[106] Possibly these functions are governed by different populations of K^+ channels, but little is known about this. The set of basolateral

Cl^-/HCO_3^- and Na^+/H^+ exchange mechanisms (Figure 9.4B) serves to control intracellular pH,[109,110,466] which Harvey[174,176] has shown plays a significant role in regulating the coordinated activity of apical Na^+ and basolateral K^+ channels, denoted as *cross talk*.[430]

Studies on the principal cell compartment of the isolated epithelium with double-barreled Cl^--sensitive microelectrodes have shown that the intracellular activity of 20 to 50 mM for this ion is far above thermodynamic equilibrium in both frog and toad skin (Table 9.2).[155,175,542] The apical membrane was found to be tight to Cl^-,[36,542] so the intracellular Cl^- space and thereby the cell volume is controlled by the NKCC transporters and Cl^- channels in the basolateral cell membrane.[106,130,155,497,498] With the apical Na^+ channels of the skin (*Bufo bufo*) blocked by amiloride, the ratio of the basolateral K^+ conductance and basolateral Cl^- conductance was $G_K/G_{Cl} = 0.87$, confirming macroscopic K^+ selectivity of the basolateral membrane in the resting state.[544] Increasing the volume of this large epidermal cell compartment by diluting the serosal bath activated the basolateral Cl^- conductance, whereby the volume decreased toward its control value.[497] Thus, the cell volume of the principal cells seems to be controlled by mechanisms similar to those of nonpolarized body cells.[206]

F. Mitochondria-Rich Cells

Mitochondria-rich cell volume is about 500 μm^3, and with a density on the order of 10^5 cells per cm^2 they occupy about 50 nL/cm^2 as compared to a total epithelial cell volume of about 7000 nL/cm^2. Due to their small size and inaccessibility with glass pipette microelectrodes, MR cells have not been easy to study, and for several years their role in Cl^- absorption was simply overlooked.[522] As with principal cells, MR cells exhibit high K^+ and low Na^+ concentrations in the cytosol that are regulated by apical ENaCs and a K^+-selective basolateral membrane that also expresses Na^+–K^+ pumps.[173,288,406]

Because of the different intracellular Cl^- concentrations,[288,304,403] different responses of the intracellular Na^+ concentration to antidiuretic hormone,[404] and different volume responses to external electrolyte and osmotic perturbations[288,454,497] it is assumed that MR cells and principal cells are not coupled via conducting gap junctions. With respect to the whole-body functions of MR cells, they have been divided into three subpopulations (Figure 9.5). The α-type is specialized for H^+ secretion in animals experiencing a body acid load, and the β-type is specialized for eliminating a surplus of base as HCO_3^- associated with an active electroneutral uptake of Cl^-. These properties of α- and β-cells (also denoted A- and B-cells) were first characterized in urinary bladder epithelium of turtle and in the collecting duct of mammalian kidney.[156,305,459,460] Studies by Larsen and collaborators[224,288,289,446,542] have indicated the existence of a third type, the γ-type MR cell, that unlike the α- and β-type MR cells displays a passive Cl^- conductance together with a H^+ pump and a Cl^-/HCO_3^- exchange mechanism in the apical plasma membrane. The coupled fluxes of Cl^- and HCO_3^- are fueled by a V-type H^+-ATPase in all three MR cell types (see Figure 9.5).

1. α-Type Mitochondria-Rich Cells

The frog *Rana pipiens*, imposed with metabolic acidosis, increased the rate of cutaneous acid secretion[511] associated with an increased number of MR cells.[375] Likewise, the isolated skin of *R. esculenta* responded to metabolic acidosis by an increased H^+ secretion that was associated with stimulated Na^+ uptake. The density of MR cells was increased together with an increased pit area of their apical membrane, which was taken to indicate that this membrane is the site of the proton pump.[114] Acidification of the solution bathing the epidermal side of the skin was correlated with serosal alkalanization and was suppressed if the serosal Cl^- was replaced by a nonpermeating anion.[110] Taken together, these experiments provide the evidence for an MR cell type with the proton pump in the apical membrane and the Cl^-/HCO_3^- exchanger in the basolateral membrane (Figure 9.5A). Immunostaining with a monoclonal antibody directed against the 31-kDa subunit E of the bovine renal V-ATPase has provided the evidence for expression of a V-type H^+ ATPase in

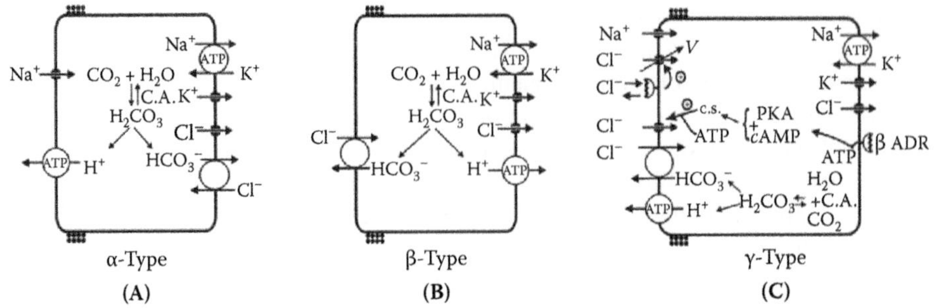

FIGURE 9.5 The current model of anuran skin epithelium also includes three types of intercalated mitochondria-rich cells, here indicated with their apical membranes turned toward the left. (A) The α-type is a proton-secreting cell similar to the A-type of distal renal epithelia. It has been suggested that this type is dominant in animals experiencing an acid load. With Na^+ channels in the apical membrane, the cutaneous elimination of the acid load may be associated with uptake of Na^+ by the same cell (see Ehrenfeld and Harvey[114] and Harvey[173]). (B) The β-type is configured for eliminating a base load like the B-type of distal renal epithelia. The anion exchange across the apical membrane is electroneutral, as is the exit of Cl^- together with H^+ across the basolateral membrane. Thus, this cell type is also configured for electroneutral active uptake of Cl^- fueled by the basolateral proton ATPase as suggested by Steinmetz.[460] (C) Transport mechanisms associated with γ-type MR cells (see Larsen[284] and Willumsen et al.[545]). The apical membrane contains the large depolarization-activated Cl^- channel, the small CFTR-like Cl^- channel, an amiloride-blockable Na^+ channel, and a set of Cl^-/HCO_3^- exchangers and H^+ pumps, which couple cellular energy metabolism to active uptake of Cl^- from freshwater and diluted salt solutions. Active uptake of Na^+ is energized via the basolateral Na^+–K^+ pump. At very low external [NaCl] (<100 μM) the apical rheogenic H^+ pump hyperpolarizes the apical membrane potential as suggested by Ehrenfeld et al.[117] In this mode, the transepithelial electrical potential difference is reversed with the serosal side being relatively negative (discussed in Jensen et al.[225]).

the α-type mitochondria-rich cells in the skin of *R. esculenta*.[115,265] Harvey[173] proposed that a secondary role of the apical H^+ pump of this cell type would be to energize the Na^+ uptake via principal cells in the absence of a permeant anion on the corneal side under open circuit conditions. Such a mechanism would also account for the 1:1 exchange of Na^+ and H^+ observed in selectively Na^+-depleted frogs with no transcutaneous anion fluxes.[149]

2. β-Type Mitochondria-Rich Cells

Bicarbonate-induced alkalosis of *Rana pipiens* evokes cutaneous base secretion,[511] probably through β-type MR cells, which in turtle urinary bladder and mammalian collecting ducts display the H^+ pump in the basolateral membrane and the Cl^-/HCO_3^- antiporter in the apical membrane.[306,462] With this configuration of the pump and the anion exchanger, HCO_3^- secretion would be coupled to active nonrheogenic cutaneous Cl^- uptake energized by the basolateral proton pump (Figure 9.5B). Thus, this cell type has the capacity to account for the 1:1 exchange of Cl^- and HCO_3^- in the absence of transcutaneous cation fluxes observed in selectively Cl^--depleted frogs.[149]

3. γ-Type Mitochondria-Rich Cells

Under normal physiological conditions, the transport properties attributed to the γ-type cell shown in Figure 9.5C suggest that it is the most abundant of the three types of MR cells, and it is probably this cell that displays the active Cl^- mechanism discovered by Krogh[277] and subsequently studied in unperturbed or NaCl-depleted animals by Jørgensen et al.,[243] Garcia-Romeu et al.,[149] and Mullen and Alvarado.[341] It is also the γ-type MR cell that is differentiated for highly regulated passive transepithelial Cl^- uptake via apical chloride channels under conditions where the Cl^- concentration exceeds 3 to 5 mM. To distinguish this cell type from the two other MR cell types discussed above, it was denoted as the γ-type MR cell.[284,289]

a. Apical Chloride/Bicarbonate Exchange

The skin of members of the Ranidae and Bufonidae displays a saturating net uptake of Cl⁻ with a half-maximum saturation concentration of external Cl⁻ of 0.1 to 0.5 mM.[9,58,113,116] Cl⁻ uptake is inhibited by diamox, an inhibitor of carbonic anhydrase that, along with a band-3-related protein, has been immunolocalized in the apical membrane of MR cells of *Bufo viridis*,[256] which would be the expected site of the anion exchanger of the γ-type and the β-type MR cells (Figure 9.5B and C). The coupling of Cl⁻ influx and HCO_3^- efflux is fixed,[113] and the Cl⁻ influx from low concentrations of its salt is unaffected by shifting V_T between positive and negative values,[113,223,273] in agreement with an electroneutral anion uptake mechanism at low [Cl⁻]$_o$.

With cutaneous respiration in freshwater, the cytosolic concentrations of HCO_3^- and Cl⁻ are supposed to be low and of similar magnitude.[225] Thus, the coupled efflux of HCO_3^- and influx of Cl⁻ at low external Cl⁻ concentration[113,149] indicates that the cytosolic binding site of the carrier has higher affinity for HCO_3^- than for Cl⁻. For a symmetrical carrier, this would be the case for the external binding site as well. By keeping the HCO_3^- concentration low at the outer surface of the skin, the apical proton pump would facilitate binding of Cl⁻ to the external binding site and inward transport of this ion even when the external Cl⁻ concentration is low.

b. Active Chloride Uptake Energized by a Proton Pump

In the absence of external halide ions, the proton efflux generates a pH gradient in the external unstirred layer which is eliminated by suppression of the cellular energy metabolism, in agreement with an active mechanism located in the apical membrane.[224,289] The rheogenic nature of the proton efflux mechanism of the skin was indicated by the following types of experiments: First, in the skin of *Rana esculenta*, proton secretion depends on both V_T and externally imposed pH clamps with a significantly decreased proton efflux at an external pH of 5.8 or for $V_T = -80$ mV (inside of the skin negative).[176] Second, the ratio of the CO_2-stimulated hydrogen ion efflux multiplied by the Faraday ($F \cdot \Delta J_H$) and the associated change of amiloride-insensitive short-circuit current (ΔI_{SC}) is not significantly different from unity, with $-\Delta I_{SC}/F \cdot J_H = 0.90 \pm 0.09$ in *R. esculenta*[176] and $-\Delta I_{SC}/F \cdot J_H = 0.96 \pm 0.07$ in *Bufo bufo*.[223] In other words, the proton flux is generated by a rheogenic mechanism. Third, the proton efflux is inhibited by the specific V-ATPase H⁺ pump inhibitor concanamycin A.[223,265] The active fluxes of H⁺ and Cl⁻ are opposite in direction but numerically of similar magnitude in the skin of both *B. bufo* and *R. esculenta*, and quantitatively similar components of the proton efflux and the ³⁶Cl⁻ influx are inhibited by concanamycin A.[225] These observations provide the evidence that the active Cl⁻ uptake is energized by the apical H⁺ pump (see Figure 9.5C). Subsequently, with an immunohistochemical method, it was confirmed that in the γ-type MR cell of toad skin the pump is also a V-type H⁺-ATPase.[225]

At steady state, for each anion transport cycle driven by the apical exit of H⁺ a negative charge is moved in the inward direction across the cell, which is carried by Cl⁻ across both the apical and the basolateral plasma membrane (the simultaneous exit of H⁺ via the pump and HCO_3^- via the exchanger carries no net charge; see Figure 9.5C). This indicates that the γ-cell also accounts for the negative short-circuit current associated with active uptake of Cl⁻ as observed in *Leptodactyllus ocellatus*,[554] *Bufo bufo*,[58] and *Bufo arenarium*.[33]

c. The γ-Type MR Cell Displays a Dynamic Apical Cl⁻ Conductance Activated by [Cl⁻]$_o$ and Transcellular Hyperpolarization

A large number of studies have indicated that even a small increase in the external chloride concentration, such as [Cl⁻]$_o$ = 3 to 6 mM, stimulates a passive Cl⁻ conductance (G_{Cl}) of frog and toad skin.[172,223,263,341] A dependence of G_{Cl} on external [Cl⁻] over and above that expected from the increase in [Cl⁻]$_o$ was first suggested by Linderholm[300] and later exploited by Koefoed-Johnsen and Ussing[267] in their study of chloride-tight skins resulting in the two-membrane model. More detailed studies of the skin of *Rana temporaria*[275] and *Bufo bufo*[172] with radioactive tracers indicated that G_{Cl} activation by [Cl⁻]$_o$ results from binding to an external site of high affinity for Cl⁻ and Br⁻, but not for other anions, and that the activated G_{Cl} exhibits poor anion selectivity with the following relative permeability sequence: $SCN^-\rightarrow Br^-\rightarrow Cl^-\rightarrow I^- = 1.7\rightarrow1.3\rightarrow1\rightarrow0.8$. The above G_{Cl} is also voltage dependent by being slowly activated (seconds) by transepithelial hyperpolarization (inside of the skin relatively more positive) and deactivated when V_T is reversed (inside of the skin negative). This has been found for a number of anurans investigated (*B. bufo*,[58,285] *R. esculenta* and *R. temporaria*,[276] *B. viridis*,[253] *B. marinus*,[288] and *Hyla arborea*[260]). The voltage sensitivity of G_{Cl} results in a steady-state transepithelial I_{Cl}-V_T relationship that is strongly rectified with large currents carried by an inward flux of Cl^- and vanishingly small Cl^- currents in the opposite direction.[58,253,542] The dynamic G_{Cl} is reversibly antagonized by α_1-adrenoceptors.[351] A notable exception to the above anurans is the frog *Xenopus laevis*, the skin of which does not display a dynamic G_{Cl}; that is, the passive Cl^- permeability remains low independent of $[Cl^-]_o$ and V_T.[260]

i.   Localization of the Dynamic G_{Cl}

The passive Cl^- flux (measured as I_{Cl} or with $^{36}Cl^-$) is positively correlated with the density of MR cells in the skin of *Rana esculenta*,[522] *Bufo bufo*,[542] and *B. marinus*.[102] Other studies confirmed that the dynamic G_{Cl} increases or decreases with protocols that stimulate or reduce, respectively, the density of MR cells (*B. viridis*,[255,253] *B. bufo*[61]). In a comparative study[260] of anuran Amphibia it was found that, although the abdominal skin of *Hyla arborea* has a high density of MR cells, this cell type is virtually absent in the dorsal skin. Only the abdominal skin displayed the dynamic Cl^- conductance.* With the estimated density of MR cells of *B. bufo*, the fully activated dynamic G_{Cl} was calculated to correspond to 20 to 30 nS/cell.[61,542] This very high conductance of the relatively small bottle-shaped cell was directly verified by recording transcellular currents in isolated voltage clamped MR cells.[291,545] With the self-referencing vibrating probe, currents of similar magnitudes were sampled above a number of MR cells *in situ* (*R. pipiens*[135] and *B. viridis*[353]).

ii.   Patch Clamp Studies of the Voltage-Activated Cl^- Channel in MR Cells

Fluctuation analysis of depolarization-activated stationary Cl^- currents of single isolated MR-cells in whole-cell patch clamp mode indicated a single-channel conductance of 150 to 300 pS.[286] A subsequent cell-attached patch clamp study of the apical membrane of isolated cells similarly reported channels of this magnitude that were active only upon membrane depolarization, confirming that the transcellular Cl^- conductance is controlled by depolarization-activated apical Cl^- channels.[446] Altogether, three to four types of Cl^- channels were observed, of which a small (~8-pS) channel was the most abundant (discussed below). The basolateral membrane also contains different types of anion-selective channels. The two most frequently observed anion channels displayed single-channel conductances of 10 and 30 pS.[543]

d.   Relationship between Active and Passive Cutaneous Cl^- Uptake

In the low range of environmental concentrations, the uptake of Cl^- by the γ-type MR cell is accomplished by H^+-pump-driven active transport, which is sensitive to the carbonic anhydrase inhibitor diamox (Figure 9.6A). With the high Cl^- affinity of the apical anion exchanger, this would be the prevailing mechanism in freshwater with simultaneous uptake of Na^+ and Cl^-.[149,243,277,341] In this range of concentrations, the passive back flux is suppressed because the dynamic G_{Cl} of the γ-type MR cells is downregulated; however, at larger external Cl^- concentrations, G_{Cl} becomes activated (Figure 9.6B), thus allowing much larger Cl^- fluxes to pass the MR cells provided the electrical driving force is inwardly directed.[172,285] The combination of these two regulations (i.e., $[Cl^-]_o$ and the apical membrane potential) secures high permeability for passive Cl^- uptake and prevents cutaneous loss of Cl^- if the transepithelial electrochemical potential difference of Cl^- reverses. This is illustrated by Figure 9.6C, where the results of unidirectional Cl^- flux studies have been collected and presented as a flux ratio analysis with following rewriting of Equation 9.6a:

---

* Other studies reported no correlation between MR cell density and transepithelial Cl^- conductance (*R. esculenta* and *R. pipiens*[350] and *B. viridis*[413]), which would indicate the presence of MR cell types that do not display the dynamic G_{Cl} (i.e., the α-type or β-type MR cells) or that Cl^- flowed through other pathways (e.g., extracellular leaks).

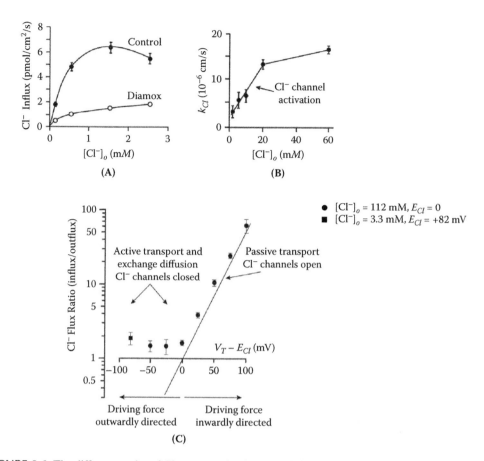

FIGURE 9.6 The different modes of Cl⁻ transport by the γ-type MR cells demonstrated by experiments with preparations of the European toad, *B. bufo*. (A) Active transport at low external Cl⁻ concentrations indicating the very high affinity to Cl⁻ of the external binding site of the apical Cl⁻/HCO$_3^-$ exchanger. In this low range of [Cl⁻]$_o$ the apical Cl⁻ channels are closed, and Cl⁻ uptake takes place by active transport that is inhibited by the carbonic anhydrase inhibitor diamox (data from Bruus et al.[58]). (B) Dependence of the rate coefficient for passive Cl⁻ uptake on [Cl⁻]$_o$. The rate coefficient was calculated as $k_{Cl} = J_{Cl}^{in} / [Cl^-]_o$ with the Cl⁻ influxes measured by the radioactive ³⁶Cl⁻ isotope and V_T clamped at +80 mV with the inside of the skin positive (experimental data from Harck and Larsen[172]). (C) Flux ratio analysis performed according to Equation 9.6b. The experimental data indicated by symbols ● and ■ are from Willumsen and Larsen[542] and Jensen et al.,[223] respectively. The flux ratio follows the line predicted for passive electrodiffusion if the driving force of Cl⁻ (i.e., $V_T - E_{Cl}$) is in the inward direction at elevated external Cl⁻ concentrations where the apical Cl⁻ channels of the γ-MR cells are activated (right-hand side). In contrast, if $V_T - E_{Cl}$ is in the outward direction (left-hand side), the apical chloride channels are closed, and active Cl⁻ transport fueled by the apical H⁺-ATPase and Cl⁻:Cl⁻ exchange diffusion become the dominating modes of Cl⁻ transport.

$$\frac{J_{Cl}^{in}}{J_{Cl}^{out}} = \exp\left[\frac{F \cdot (V_T - E_{Cl})}{R \cdot T}\right]$$

$$E_{Cl} = \frac{RT}{-F} \ln \frac{f_o[Cl^-]_o}{f_i[Cl^-]_i} \tag{9.6b}$$

where the transepithelial equilibrium potential for Cl⁻ (E_{Cl}) is given by Nernst's equation, and f_o and f_i are the activity coefficients of the solutions bathing the two sides of the epithelium. On the right-hand site of the diagram of Figure 9.6C, the driving force on chloride ions is inwardly directed (V_T

$- E_{Cl} > 0$), apical Cl⁻ channels are open (i.e., G_{Cl} is activated), and the flux ratio obeys Equation 9.6b as indicated by the straight line. The flux ratio, however, does not follow the theoretical relationship for passive electrodiffusion at low $[Cl^-]_o$ with outwardly directed driving forces (left-hand side of Figure 9.6C). Here, with the Cl⁻ channels closed (deactivated G_{Cl}), ATP-dependent active uptake and exchange diffusion are the dominating mechanisms of Cl⁻ transport. This type of regulation of G_{Cl} via apical Cl⁻ channels implies that it is the active Na⁺ flux through the principal cells that controls the passive flow of Cl⁻ through mitochondria-rich cells. When the active Na⁺ uptake is stimulated, the transepithelial potential difference hyperpolarizes, and via current loops through the parallel MR cells their apical membrane becomes depolarized. In turn, if $[Cl^-]_o$ is sufficiently high, the apical Cl⁻ channels open. Thus, the active flux of Na⁺ through principal cells both generates the driving force for passive Cl⁻ uptake through MR cells and controls the Cl⁻ conductance of this cell type.[284]

e. The Receptor-Coupled and cAMP-Activated Cl⁻ Channel of MR Cells

The chloride conductance of the apical membrane is also controlled by β-adrenergic receptors via a cAMP-dependent signaling pathway.[96,544] The independence of channel activity on membrane potential and the single-channel conductance of γ_{Cl} = ~8 pS,[446] together with its selectivity (Cl⁻→ Br⁻→NO₃⁻→I⁻ = 1→0.70→0.53→0.18) and pharmacology, indicate that at the molecular level the channel is the toad cystic fibrosis transmembrane regulator (CFTR[407]).[11,545] This channel is active during hormone-stimulated, solute-coupled cutaneous water uptake (see Section III.H).

G. PARACELLULAR TRANSPORT

It is likely that in freshwater a major function of tight junctions is to prevent passive leakage of small diffusive ions from the extracellular fluid to the environment. In a comparative study, these leaks were estimated by Mullen and Alvarado,[341] who calculated the paracellular electrodiffusion permeability by recording the accumulation of ions in the bath. To prevent ion effluxes via exchange pathways, the animals were submerged in distilled water. The permeabilities were calculated by the Goldman–Hodgkin–Katz flux equation. Their data are collected in Table 9.6, together with electrodiffusion permeabilities obtained from isotope tracer flux studies in isolated preparations mounted in Ussing

TABLE 9.6
Estimated Paracellular Ion Permeabilities of Anuran Skin

Species	P_{Na} (10^{-8} cm/sec)	P_{Cl} (10^{-8} cm/sec)	Remarks
Ascaphus truei[a]	2.1	0.9	In vivo, aquatic
Rana pipiens[a]	11.1	6.0	In vivo, semiaquatic
Hyla regilla[a]	3.9	1.3	In vivo, terrestrial
Bufo boreas[a]	10.4	9.3	In vivo, terrestrial
Rana pipiens	3.5[b]	5[c]	Isolated whole skin
Rana temporaria	0.8 and 0.15[d]	22[e]	In vitro
Bufo bufo[f]	1.7	0.97	Isolated whole skin

[a] Mullen and Alvarado.[341]

[b] Helman and Miller.[183]

[c] Biber et al.[35]

[d] Isolated epithelium: bilateral NaCl Ringer's solution and 1/10 Ringer's solution outside, respectively (Eskesen and Ussing[123]).

[e] Whole skin (Kristensen[274]).

[f] Na–gluconate outside (Bruus et al.[58]).

Note: The estimated P_{Cl} in both the in vivo and the in vitro studies are based on effluxes measured into a Cl⁻-free external bath.

chambers. All permeability values are low, and no obvious difference exists between aquatic and terrestrial species (e.g., compare the aquatic *A. truei* and the terrestrial *H. regilla*). The data listed in Table 9.6 also do not indicate much difference in the estimates obtained from *in vivo* and *in vitro* studies. A general tendency for all species studied *in vivo*, however, is that the tight junction is slightly cation selective. This is seen also in the only *in vitro* estimates where the two permeabilities were from the same study. During molting, the selectivity is lost and ions leak out of the animal.[233] The electrical conductance of the junctions can be reversibly increased by exposing the external side of the skin to hypertonic Ringer's solution,[500] which increases the paracellular permeability for Na^+ and SO_4^{2-} [501] and lanthanum.[122] Salt depletion of frogs results in reduced cutaneous ion loss,[277] which suggests physiological regulation of tight-junction ion permeability, but little is known about this.

H. Solute-Coupled Water Transport

Early studies discovered that anuran skin is capable of transporting water in the absence of a transepithelial osmotic concentration difference both *in vitro*[399] and *in vivo*,[111] which was denoted *nonosmotic* water uptake. The osmolarity of the transported fluid depends on anuran species and experimental conditions, and hyperosmotic,[212,361] near isosmotic,[194,363] and hypoosmotic[363] transport has been reported. The nonosmotic water uptake was stimulated by insipidin (*Bufo bufo*[247]), by arginine vasotocin (*Rana esculenta*[361]), and by the β-adrenergic agonist isoproterenol (*B. bufo*[363]). The skin is also capable of water uptake from a hyperosmotic outside solution—that is, of uphill water transport (*R. esculenta*,[361] *B. bufo*[363]). Because the Na^+–K^+ pumps on all plasma membranes line the lateral intercellular space (lis), the prevailing theories on solute-coupled water transport assume coupling of ion and water flows in lis which would be hyperosmotic and hyperbaric at transepithelial osmotic equilibrium conditions. As a general principle governing solute-coupled water transport across such a system, Curran[90] pointed out that the direction of water flow would be given by the relative magnitude of solute reflection coefficients at the two barriers delimiting the coupling compartment. With the reflection coefficient of the apical barriers being larger than that of the interspace basement membrane, water flows in the inward direction, resulting in absorption of fluid. Diamond and Bossert[103] analyzed water transport by this *local osmosis model* and suggested that a longitudinal concentration gradient is built up in lis in such a way that the solute concentration difference is large across the apical boundary and zero across the interface between lis and the serosal bath, denoted as the *standing-gradient theory*. This eliminates diffusion of solutes across the latter boundary (Fick's law) so convection only drives solutes out of lis, which was shown to be a necessary condition for truly isosmotic absorption.

The alternative *Na+ recirculation theory* suggested by Ussing takes into account diffusion fluxes across the interface basement membrane that are different from zero and assumes isosmotic transport to be achieved by a regulated back flux of Na^+ and other ions into lis via cotransporters in the basal plasma membrane and Na^+–K^+ pumps and ion channels in the lateral plasma membranes. Whereas the standing-gradient theory excludes transjunctional water transport into lis, and thus paracellular solvent drag, the Na^+ recirculation theory covers both transjunctional and translateral water uptake and thus paracellular solvent drag. The Na^+ recirculation theory also accounts for the observations that the ratio of the transepithelial-active Na^+ flux and the associated oxygen uptake spans the range from below to above that of the Na^+ pump itself, which is 18 mol Na^+ per mol O_2. In a straightforward way, both theories explain uphill water uptake, but only the Na^+ recirculation theory has the capacity to generate a hyposmotic transport at transepithelial osmotic equilibrium conditions.[287,290]

I. Skin Glands

Besides the ion and water uptake systems of the epithelium discussed above, anuran skin contains secretory subepidermal glands that secrete fluid driven by a secondary active transport of Cl^- when activated by catecholamines.[268] The function of glandular secretion has been suggested to be associated with evaporative cooling because fluid secretion in the dorsal surface of *R. catesbeiana* was increased

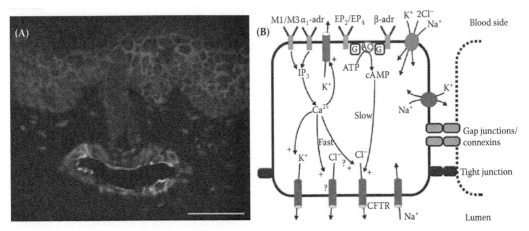

FIGURE 9.7 (See color insert following page 208.) The fluid-secreting subepidermal gland of anuran skin. (A) Immunofluorescence labeling of AQP-x5 in mucous gland of the toad *Bufo woodhouseii*. AQP-x5, which is homologous to mammalian AQP5, is visible as a narrow light band in the apical plasma membrane (green) of the secretory cells of the mucous gland. AQP-h3BL was similarly immunolocalized in the basolateral membrane (red) of the same cells and granular cells in the epidermis. Nuclei are counterstained with DAPI (blue). Scale bar = 50 μm. (B) Model of the organization of ion transport systems of frog skin acinar cells identified by transepithelial isotope tracer and water flow studies, measurements of intracellular ion concentrations, patch-clamp electrophysiology, and application of pharmacological protocols as explained in the text. (From Sørensen, J.B. and Larsen, E.H., *Pflügers Arch.*, 439, 101–112, 1999. With permission.)

when the animals were exposed to elevated temperatures.[295] Of special notice, skin glands are particularly abundant on the ventral surface, especially in the seat patch region of terrestrial bufonids that is specialized for cutaneous water absorption. In this regard, skin glands in the toad *Bufo woodhouseii* contain AQP-X5-like aquaporin in the apical membrane which is homologous with mammalian AQP5 located in the apical membrane of salivary glands of animals that drink orally and rely on salivary secretion to sensitize taste cells in the tongue (Figure 9.7A). Nagai and coworkers[343] have suggested a chemosensory function for toad skin with respect to the salt content of hydration sources, which may be influenced by isosmotic glandular secretion.

The acinar cells of the frog skin gland are electrically coupled.[448] Binding of isoproterenol to β-adrenergic receptors or of prostaglandin E_2 to EP_2/EP_4 receptors stimulates the formation of an isosmotic secretion.[39,328,483,484] The response is mediated by cAMP with no effect on cellular free [Ca^{2+}]. In contrast, carbachol stimulation of muscarinic M1/M2 receptors results in an increase in free [Ca^{2+}] with no effect on cellular cAMP.[356] Frog skin glands also display α_1-adrenergic receptors, the activation of which also elicits Cl^- secretion but via a transient increase in the concentrations of IP_3 and Ca^{2+}.[168] Using preparations of collagenase-stripped glands, patch clamp studies of the apical plasma membrane revealed the presence of CFTR Cl^- channels, depolarization- and Ca^{2+}-stimulated maxi K^+ channels, and a small ENaC-like Na^+ channel.[447,448,450,449] A previous immunocytochemical study of *Xenopus laevis* also indicated expression of CFTR in the luminal membrane.[121] Maxi K^+ channels are expressed in the basolateral membrane, as well, and they recirculate K^+ back into the serosal solution, which has been taken up by the acinar cells via the Na^+–K^+ pumps and the NKCC transporters.[14] Figure 9.7B shows the updated exocrine gland cell model, which integrates the above experimental findings. With respect to the apical membrane, it is likely that the depolarization induced by activation of the CFTR Cl^- channels activates the coexpressed maxi K^+ channels and that this mechanism is responsible for the active K^+ efflux during secretion that results in a [K^+] of the secreted fluid much above that of plasma.[450] Such interplay between apical Cl^- and apical K^+ channels was predicted to enhance the secretory rate in receptor-activated exocrine glands.[86]

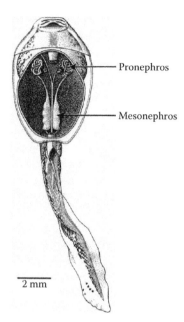

FIGURE 9.8 Anuran tadpole (*Bufo viridis*) with ventral body wall and intestinal system removed to reveal the two kidney systems: the pronephroi and the mesonephroi. The kidneys are located on each side of the dorsal aorta, which anteriorly is divided into two branches. (Adapted from Møbjerg, N. et al., *J. Morphol.*, 245, 177–195, 2000.)

IV. STRUCTURE AND FUNCTION OF THE KIDNEY

Within the amphibian life cycle, two kidney systems are present and functional (Figure 9.8). The pronephroi are the functional kidneys of amphibian larvae. They begin to form soon after fertilization and are fully functional around the time of hatching. The mesonephroi, which will become the functional kidneys of the adult, gradually replace pronephric function, and around the time of metamorphosis the pronephroi undergo regression and apoptosis.[139,153,323,331,402] The amphibian nephron consists of a filtration unit and a renal tubule. Primary urine is produced by filtration of blood across the glomerular filtration barrier of the filtration unit, and the filtrate is subsequently modified by selective reabsorption and secretion of ions and organic molecules across the renal tubule.

A. PRONEPHROS

The paired pronephroi are located retroperitoneally immediately behind the gill region (Figure 9.8). The most obvious function of this early kidney system in amphibian freshwater larvae is to expel water entering the body by osmosis and at the same time minimize ion loss. In anurans and urodeles, each pronephros is a very simple organ, essentially a single nephron composed of a filtration unit and a pronephric tubule which communicates with the exterior through the pronephric duct.[74,137,220,331,517,519] In these taxa the pronephric filtration unit lacks a capsule of Bowman and is formed by an external glomerulus or glomus. The capillary networks forming amphibian pronephric glomeruli and glomera are derived from the dorsal aorta. The efferent arteriole leaves the pronephric filtration unit and enters a sinus of the posterior cardinal vein, surrounding the pronephric tubules. Ultrafiltration of blood occurs across a filtration barrier that, on the ultrastructural level, resembles that of the adult mesonephros (described below). The filtrate enters the coelom before it is taken up by the tubular part of the nephron with the aid of cilia movement. The pronephric tubule (a single convoluted tubule) opens into the coelom via ciliated nephrostomes. The number of nephrostomes varies between species, with three being the typical number for anurans and two to five being reported from urodeles. In caecilians, the pronephros is an elongated organ extending over approximately 10 body segments in *Ichthyophis kohtaoensis*;[549] up to 12 renal tubules (8 to 9 functional) have been reported for *Hypogeophis rostratus*.[47] In *I. kohtaoensis*, glomeruli form a single, large glomus that is partly internalized; the nephrocoel still communicates with the coelom via ciliated funnels.[549] Interfering with cilia formation impairs fluid movement into the nephrostomes and leads to edema formation.[489]

Based on light and electron microscopic investigations, the pronephric tubule can be divided into ciliated and proximal tubule branches corresponding to the number of nephrostomes, a common proximal tubule, and a distal tubule, which continues as the pronephric duct.[74,138,140,154,331] A ciliated intermediate segment is present between the proximal and distal tubule in urodele amphibians.[74]

Physiological investigations on the amphibian pronephros are sparse. Howland[209] demonstrated that bilateral excisions of the pronephric rudiment in embryos of *Ambystoma punctatum* led to the formation of edema and expansion of the pericardial and abdominal cavities in the larvae, followed by death. More recently, Zhou and Vize[556] have shown that fluorescence-labeled macromolecules introduced into the circulation of *Xenopus laevis* tadpoles are filtered by the glomus and subsequently appear in the pronephric tubules; considerable amounts of the filtered molecules are subsequently taken up by the epithelium of the proximal tubule, presumably by endocytosis.

Gene expression assays suggest that proximal as well as distal tubules can be subdivided into early and late segments based on differential expression of transporter proteins in these subdomains.[489,556] Several studies report a high level of expression of the Na^+,K^+-ATPase (α-, β-, and γ-subunits) in both tubules and duct of the maturing pronephros.[118,494,556] The presence of a Na^+,K^+-ATPase in the basolateral cell membrane of pronephric epithelial cells would provide the driving force for luminal uptake of Na^+ through channels or transporters coupling Na^+ uptake to the uptake of, for example, amino acids, glucose, or inorganic ions. Expression of several solute carriers has been reported from primarily the pronephric proximal tubule, including two sodium-dependent solute carriers that may be involved in glucose reabsorption as well as amino acid transporters.[119,344,556,557]

The search for pronephric markers in *X. laevis* has also revealed expression of several channels and cotransporters within the distal nephron and pronepric duct that may participate in reabsorption of NaCl. The Na^+–K^+–$2Cl^-$ cotransporter (NKCC2) is highly expressed in the pronephric early distal tubule of *X. laevis*.[489,556] In the latter kidney generations, this transporter provides the molecular basis for urine dilution through luminal uptake of Na^+, K^+, and Cl^- across the amphibian mesonephric early distal tubule and the mammalian metanephric thick ascending limb of the loop of Henle (see below). In addition, the thiazide-sensitive NaCl cotransporter (NCC) seems to be expressed in late distal tubule and pronephric duct, and high levels of expression of a Cl^- channel (ClC-K) have been reported in both early and late distal tubules and the pronephric duct.[489,520,556] Tran and coworkers[489] furthermore reported that the K^+ channel (ROMK) that mediates the secretion of potassium across the apical plasma membrane of thick ascending limbs and cortical collecting duct principal cells of the mammalian kidney is highly expressed in the early distal tubule and pronephric duct. Moreover, expression of carbonic anhydrase (CAII) and sodium–bicarbonate cotransporters (NBC) has been reported in the early proximal as well as late distal tubule, suggesting that H^+ may be secreted and HCO_3^- reabsorbed in these segments.[489,558] Taken together, these investigations suggest that the amphibian pronephric nephron contains several functional distinct cell types, including types necessary for urine dilution, and that the pronephros plays a key role in regulating the ion and water as well as acid–base balance of amphibian larvae.

B. AMPHIBIAN MESONEPHROS

As holds for amphibian larvae, juvenile and adult amphibians in a freshwater environment maintain the osmolality of their body fluids well above those of the surroundings. In terrestrial environments amphibians tend to lose water and are therefore faced with the problem of dehydration. Amphibians cannot concentrate their urine, and under terrestrial conditions body water is conserved by a drastic reduction in glomerular filtration. In terrestrial anurans, the urinary bladder works in conjunction with the mesonephros and functions as a water reservoir, enabling the animal to remain hydrated on land in the face of high evaporative water loss across the highly water permeable integument[240] (see Section III.B).

The paired mesonephroi are located retroperitoneally in the dorsal wall of the body cavity on each side of the aorta and vena cava. Renal arteries originating from the aorta supply the mesonephric filtration units and, in addition, the mesonephros receives blood from a renal portal vein.[281,338,373,555] Blood containing solutes and water reabsorbed by the renal tubules drains into renal efferent veins that open into vena cava. A number of studies have dealt with the structure of the mesonephros and mesonephric nephron in *anurans*,[18,127,126,129,330,479,493] *urodeles*,[37,79,198,416,457] and *caecilians*.[70,333,417–419,524,533] Uchiyama and Yoshizawa[491] found a positive correlation between mesonephric kidney mass and body mass in anuran amphibians. To estimate nephron numbers, they furthermore counted glomeruli and reported that the number of nephrons in each kidney varied from about 60 in the African dwarf frog (*Hymenochirus boettgeri*; body mass, 1.3 g) to about 8000 in the Colorado River toad (Sonoran Desert toad) (*Bufo alvarius*; body mass, 215 g).

1. The Mesonephric Filtration Unit, Ciliated Segments, and Peritoneal Funnels

The filtration unit in the mesonephric kidney, the Malpighian corpuscle, is formed by two structures: the vascular loops of the glomerulus and the capsule of Bowman surrounding the glomerulus. The capsule of Bowman consists of a visceral layer composed of podocytes, which encircle the glomerular capillaries, and a parietal layer, which is continuous with the epithelium of the renal tubule. The space between the visceral and parietal layers, the nephrocoel or urinary space, is continuous with the lumen of the tubule. Structural investigations on the mesonephric filtration unit in species from all three amphibian orders have revealed that the filtration barrier in the Malpighian corpuscle consists of a fenestrated endothelium of the glomerular capillaries, a glomerular basement membrane, and slit diaphragms bridging podocyte foot processes. The pores of the endothelium are relatively large (100 to 350 nm) and probably serve to limit filtration of cellular elements. The three-layered basement membrane is thick, up to ~1 μm, due to a wide subendothelial space containing abundant microfibrils, collagen fibrils, and occasionally cellular processes of glomerular mesangial cells. Podocyte foot processes are separated by filtration slits bridged by slit diaphragms. These diaphragms are unique cell junctions between the interdigitating foot processes of adjacent podocytes. In amphibians, the slit diaphragms seem to constitute the principal filtration barrier to plasma proteins.[424]

The first studies to clearly demonstrate that formation of primary urine in vertebrate glomerular nephrons relies on ultrafiltration of plasma and to examine the roles of hydrostatic and colloid osmotic pressures within the glomerular capillaries and capsule of Bowman were performed using the micropuncture technique on the mesonephric kidneys of *Necturus maculosus* and *Rana pipiens*.[180,531,536] The kidney micropuncture technique was developed on amphibians in the laboratory of A.N. Richards in the early 1920s, 20 years before the first publications of similar studies in mammals were presented.

The formation of primary urine is controlled by the permeability of the filtration barrier, the available filtration surface, and the balance of hydrostatic and osmotic forces across this barrier (Starling's forces). The net glomerular filtration pressure (P_{GF}) is determined by the hydrostatic pressure difference between the glomerular capillaries (P_{GC}) and Bowman's space (P_{BS}) and by the oncotic pressure exerted by plasma proteins in the glomerular capillaries (π_{GC}). Assuming that the osmotic force due to proteins in Bowman's space is insignificant, the net glomerular filtration pressure is:

$$P_{GF} = P_{GC} - P_{BS} - \pi_{GC}$$

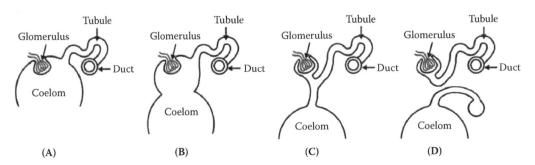

FIGURE 9.9 Schematic illustration of the evolutionary events that may have led to the development of the vertebrate nephron. These events are represented by kidney systems present in extant animal taxa. (A) Invertebrate metanephridium with an external glomerulus situated in the body cavity (coelom); a renal tubule opens into the cavity via a nephrostome and empties into a renal duct at its distal end. (B) The pronephros of anuran larvae; a small coelomic chamber (nephrocoel) surrounds the external glomerulus, and the renal tubule opens into this chamber via a nephrostome. (C) Mesonephric nephron found in caecilian and urodele amphibians; the nephrocoel has become the capsule of Bowman and the glomerulus has consequently been internalized so the two structures now form a Malpighian corpuscle. The renal tubule opens into the coelom via a ciliated peritoneal funnel. The nephrostomes in Parts A and B are homologs with the neck segments of mesonephric renal tubules (compare to Figure 9.10). (D) The nephron has lost its connection to the coelom. In anuran amphibians the peritoneal funnels now open into peritubular blood vessels. (From Møbjerg, N. et al., *J. Morphol.*, 262, 583–607, 2004. With permission.)

The ultrafiltrate has the same composition as plasma with regard to small molecules such as inorganic ions and glucose. The filtration barrier, however, reflects large proteins and the cellular components of the blood.

The available filtration surface is a function of the total length of the filtration area determined by the number of filtering glomeruli and the size and differentiation of the glomerular tufts. In addition, the area and permeability of the filtration barrier may be regulated by the presence of cellular processes from mesangial cells in the glomerulus. The glomerular filtration rate (GFR) in amphibians varies from approximately 10 to 100 mL/kg/hr.[44,92] This reflects variations between species, but noticeably variation also occurs within single individuals. In freshwater, amphibians generally produce copious and very dilute urine; however, under terrestrial conditions or during exposure to hyperosmotic saline, GFR and urine flow drops drastically, while reabsorption of water by the renal tubule may increase.[426] The GFR is regulated by AVT, which binds to V1 receptors in the afferent arterioles to constrict glomerular blood flow and cause a reduction in the filtration rate of single nephrons.[492] In addition, AVT seems to increase the clearance by peritoneal funnels in anurans.[337] In caecilians and salamanders, neck segments of a subpopulation of nephrons communicate with the coelom via ciliated peritoneal funnels. In anurans, these funnels are detached from the tubule and open into the venous system (Figure 9.9). Thus, as holds for the pronephros, not only ultrafiltrate from the filtration unit but also coelomic fluid may be drawn into the lumen of the renal tubule (caecilians and urodeles) or into the renal veins (anurans). Several studies have shown that macromolecules injected into the coleom are taken up by the funnels and either enter peritubular vessels or nephrons.[167,337]

2. Structure and Function of the Mesonephric Tubule

The mesonephric renal tubule is composed of a single-layered epithelium and extends from the Malpighian corpuscle to the junction with the renal duct (the Wolffian duct or accessory ureters). Based on developmental, morphological, and physiological properties it can be divided into six distinct sections: ciliated neck segment, proximal tubule, ciliated intermediate segment, early distal tubule, late distal tubule, and finally the collecting tubule, which opens into collecting ducts that

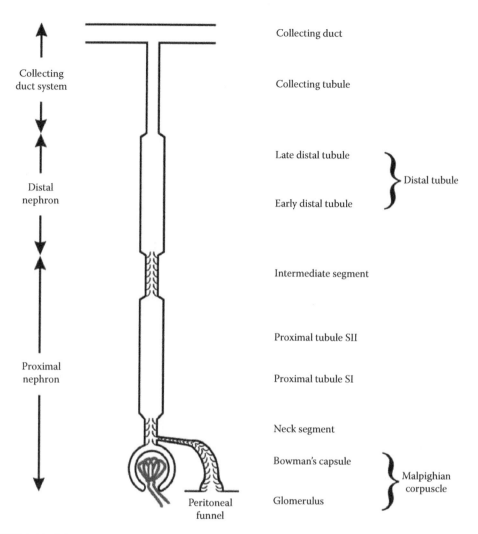

FIGURE 9.10 Schematic illustration of the vertebrate nephron. The nephron can be divided into two major parts based on ontogenetic studies: the proximal nephron and the distal nephron. The proximal nephron is comprised of the Malpighian corpuscle, a ciliated neck segment, the proximal tubule, and a ciliated intermediate segment. The ciliated neck segment may open into the coelom via a ciliated tubule. The distal nephron is comprised of early and late distal tubules. The late distal tubule opens into the collecting tubule, which is the first unbranched portion of the collecting duct system. This model of the nephron is present in salamander and caecilian mesonephric kidneys. (Adapted from Møbjerg, N. et al., *J. Morphol.*, 262, 583–607, 2004.)

lead the urine to the ureter (Figure 9.10). In a series of experiments with punctures of Bowman's capsule and various sites along the nephron in live *Necturus maculosus* and *Rana pipiens*, Richards and collaborators provided essential information on the role of the filtration unit and the renal tubule in urine formation.[401,527,525,526,528] They showed that primary urine contains glucose and Cl⁻ in concentrations similar to that of plasma, whereas the ureter/bladder concentrations were very low, providing evidence for tubular reabsorption. It was also found that the concentration of glucose diminished rapidly as the filtrate moved along the proximal tubule, reaching levels on the order of those measured in urine from the ureter, showing that the site of glucose reabsorption is the proximal tubule. Figure 9.11 is a representation of original charts from Walker and Hudson[525] and Walker et al.[528] showing the glucose, the total osmotic, and the Cl⁻ concentration of the urine relative to that of plasma at different puncture sites along the nephron. The latter study demonstrated that the

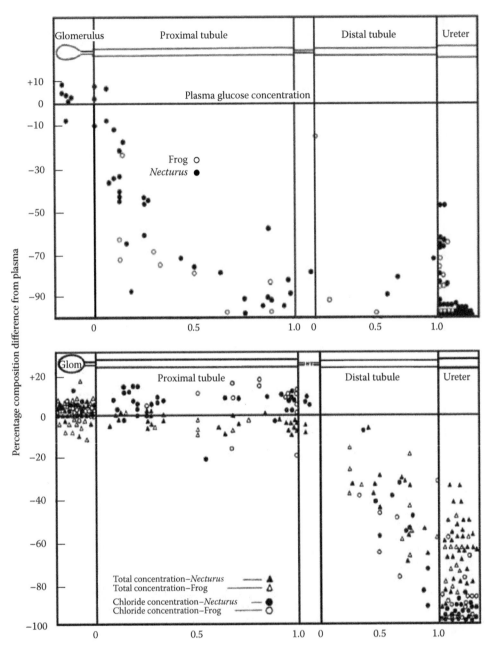

FIGURE 9.11 The percentage deviation in glucose (top) and chloride and total molar concentration of osmolytes (bottom) of the filtrate relative to that of plasma at various puncture sites along the nephron of *Necturus maculosus* and *Rana pipiens*. The ordinate values indicate percentage difference from plasma. The values on the abscissae indicate the relative distance along the proximal and distal tubule, respectively. The glomerular filtrate entering the capsule of Bowman is essentially identical to plasma with regard to glucose, chloride, and molar concentration of osmolytes. Glucose is reabsorbed in the proximal tubule. In the distal tubule, a significant reabsorption of osmolytes occurs, resulting in dilution of the filtrate. (From Walker, A.M. and Hudson, C.L., *Am. J. Physiol.*, 118, 130–143, 1936; Walker, A.M. et al., *Am. J. Physiol.*, 118, 121–129, 1936. With permission.)

most obvious function of the distal nephron is to reabsorb osmolytes and thereby dilute the urine. Another functional aspect of the amphibian nephron was also shown at this time—namely, that the urine was acidified in the distal nephron.[336] In 1962, Bott[43] presented data that were obtained from micropuncture and inulin clearance experiments in *Necturus*. His results showed that up to 50% (on average 25 to 30%) of the primary urine was isosmotically reabsorbed in the proximal tubule. One puncture site in the collecting duct indicated that as much as 75% of the water could be reabsorbed by the end of the distal nephron. In agreement with the results of Walker et al.,[528] Bott found that the Na^+ and Cl^- concentration decreased markedly at the onset of the distal nephron. Whereas K^+ was isotonically absorbed in the proximal tubule, great variation was seen in K^+ concentrations along the distal nephron, indicating that this ion could be reabsorbed and secreted in different distal tubule segments.[43]

In the 1970s and 1980s, much effort was devoted to the study of the amphibian mesonephric nephron, which was used as a model for vertebrate tubular transport. This resulted in a growing knowledge of the function of proximal and distal tubules.[104,170,214,540] As was already noticed in Richards' laboratory almost 60 years earlier, the amphibian kidney represented an organ ideal for vertebrate kidney studies. The advantage of amphibian tissue is the possibility to work at room temperature and the relatively good viability of amphibian renal tubules. In addition, the presence of a renal portal circulation allows for double perfusion of the isolated kidney through the aorta and renal portal veins, respectively. The large tubule diameter and large cell size of urodele renal tubules resulted in a focus on transport characteristics of the mesonephric nephron from this amphibian group, and experiments were especially conducted on either obligate aquatic species or animals held under aquatic conditions. More recent studies have focused on hormonal regulation and on nephric transport mechanisms associated with adaptations to terrestrial and hypersaline environments.[269–271,332,334,335]

a. Proximal Tubule

The cells of the amphibian mesonephric proximal tubule are specialized for the uptake of an isotonic absorbate, as well as macromolecules by receptor-mediated endocytosis. Hence, these cells possess a luminal brush border, apical endocytotic apparatus, and lysosomal system, as well as conspicuous lateral intercellular spaces and a basal labyrinth.[79,316,317,330,418] The mesonephric proximal tubule may be subdivided in two segments, based primarily on cellular and brush-border height, as well as on the extent of the basal labyrinth.[129,333,385,479,493] An intercalated cell type (bald-headed cell) of unknown function is present in the proximal tubules of some caecilians and salamanders.[317,333,418]

b. Distal Nephron

The amphibian mesonephric distal nephron is a complex structure with several subdivisions, each of which has its own structural and functional characteristics. The distal nephron consists of three different segments: the early distal tubule, the late distal tubule, and the collecting tubule. The early distal tubule is composed of a single cell type characterized by a large number of apical junctional complexes and a well-developed basolateral labyrinth, which together with the palisade arrangement of the large number of mitochondria give the cell a striated appearance (Figure 9.12). This segment is also called the *diluting segment*.[467] The late distal tubule can be further subdivided into three morphologically different sections. The late distal tubule section I is distinctly defined; the cells comprising this section have a large nucleus and well-developed lateral and basal labyrinths. The late distal tubule sections II and III represent the gradual transition between section I and the distinctly defined heterocellular collecting tubule. The collecting tubule and duct are composed of principal and intercalated cells. Distal tubule length seems to be greater and late distal tubule segmentation more pronounced in terrestrial anurans.[330,491]

i. Early Distal Tubule: The Diluting Segment

In isolated and perfused renal tubules from urodele and anuran amphibians, Stoner[467] measured transepithelial potential difference (V_T) and collected and subsequently analyzed the ion composition and water content of the tubular fluid. He showed that the early distal tubule has a lumen-positive V_T and possesses transport properties strikingly similar to those of the thick ascending limb of Henle's loop in the mammalian nephron. Stoner therefore named this segment the *diluting segment*.

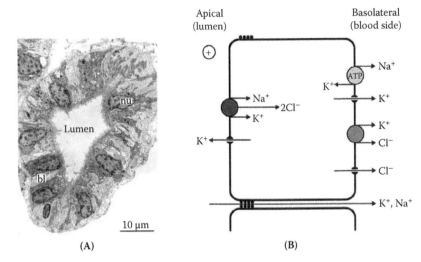

FIGURE 9.12 The amphibian early distal tubule—the diluting segment—reabsorbs NaCl and has low water permeability. (A) Transmission electron microscopy (TEM) of early distal tubule from the mesonephros of *Bufo bufo*. The tubule is composed of a single cell type characterized by a well-developed basolateral labyrinth (bl), which together with the palisade arrangement of a large number of mitochondria give the segment a striated appearance (nu, nucleus). (B) Model illustrating the cellular mechanisms involved in NaCl reabsorption. Na^+,K^+-ATPases in the basolateral cell membrane pump K^+ into and Na^+ out of the cell, thereby providing the driving force for Na^+ uptake at the apical cell membrane. NaCl enters the cell via apical $Na^+–K^+–2Cl^-$ cotransporters, and Cl^- leaves the cell across the basolateral cell membrane via Cl^- channels and $K^+–Cl^-$ cotransporters. Both apical and basolateral cell membranes possess K^+ conductances, which allow K^+ to recycle for $Na^+–K^+–2Cl^-$ cotransporters and Na^+,K^+-ATPases. K^+ movement into the lumen of the tubule and basolateral Cl^- movement out of the cell generate a lumen-positive transepithelial potential difference, which in addition drives Na^+ reabsorption across the cation selective paracellular shunt.

Like its mammalian counterpart, the amphibian diluting segment reabsorbs NaCl and has low water permeability, thus diluting the urine.[165,170] In mammals, the interstitial hypertonicity generated by the NaCl reabsorption in this segment is used to withdraw water from the collecting ducts during antidiuresis, resulting in the production of highly concentrated urine. A similar mechanism is not found in amphibians, which lack a loop of Henle and are unable to form concentrated urine. In the basolateral cell membranes of the diluting segment a Na^+,K^+-ATPase, energized by ATP hydrolysis, maintains a low intracellular Na^+ concentration and thereby provides the driving force for Na^+ uptake at the apical cell membrane (Figure 9.12). NaCl enters the cell via a furosemide-sensitive $Na^+–K^+–2Cl^-$ cotransporter, and Cl^- leaves the cell across the basolateral cell membrane via Cl^- channels and $K^+–Cl^-$ cotransporters. Both the apical and basolateral cell membranes possess K^+ conductances, through which K^+ is recycled for the $Na^+–K^+–2Cl^-$ cotransporter and the Na^+,K^+-ATPase, respectively. K^+ movement into the lumen of the tubule and Cl^- movement out of the cell across the basolateral cell membrane probably generate the current responsible for the lumen-positive transepithelial potential difference. The paracellular pathway is cation selective, and movement of cations across the shunt thus closes the current loop. As a result, Na^+ is reabsorbed across the paracellular pathway in addition to cellular reabsorption.

Two functional cell types with different basolateral K^+ and Cl^- conductances have been described for the *Amphiuma* early distal tubule.[169] In low-conductance cells, most KCl leaves the cell via the basolateral $K^+–Cl^-$ cotransporter, whereas the ions in high-conductance cells preferentially leave via the K^+ and Cl^- channels. Under normal physiological conditions, a small reabsorption of K^+ occurs in the amphibian early distal tubule, whereas K^+ is secreted in animals experiencing a K^+ load.[170,368,482,541] Using the double-perfused kidney from K^+-loaded *Amphiuma* sp., *Rana esculenta*, and *R. pipiens* and with microelectrode recordings, Oberleithner and collaborators[370, 369] showed that Na^+ reabsorption may also occur via luminal amiloride-sensitive Na^+/H^+ exchange; as a consequence, the early distal tubule may participate in urinary acidification. It was suggested that, in these K^+-loaded animals, HCO_3^- leaves the cell at the basolateral cell membrane by electrogenic cotransport with Na^+.[170,529] Cooper and Hunter[87] measured intracellular pH in isolated and perfused early distal tubules from K^+-loaded *R. temporaria*. They could not, however, confirm the presence of an apical Na^+/H^+ exchanger. Instead, the authors presented evidence for the presence of such a transporter on the basolateral cell membrane and suggested that the contradictory results may be a result of species difference or differences in experimental approach.

In a recent study, Konno and coworkers[269] cloned a urea transporter belonging to the UT-A2 family of facilitative urea transporters from the kidney of *Bufo marinus*. The expression of this transporter in the kidney as well as urinary bladder was significantly increased in toads exposed to dry or hypersaline conditions. Within the nephron, the transporter seems to be localized on the luminal membrane of early distal tubule cells, suggesting that this segment may be involved in urea reabsorption.

ii.   Late Distal Tubule

Obviously, a significant amount of salt reabsorption occurs in the diluting segment. In addition, in the amphibians studied so far, reabsorption of NaCl occurs in the late distal tubule and collecting duct system. Whereas the early distal tubule and collecting tubules are relatively easy to identify at the light microscopic level in fresh tissue, the heterogeneity of the late distal tubule makes the identification and thereby the study of the functional characteristics of this nephron segment difficult. Stanton et al.[457] described the ultrastructure of the distal nephron in *Amphiuma means* and subdivided it into early and late distal tubules and a collecting tubule. In this description, the late distal tubule is clearly defined and corresponds to the late distal tubule section I as described above. The transition between this section and the collecting tubule was termed the *transition region*. The late distal tubule of *Amphiuma means*, like its mammalian counterpart, the distal convoluted tubule, reabsorbs Na^+ and Cl^- through electrically neutral mechanisms.[104,455,456] Furthermore, it secretes H^+, which is in agreement with the observations of Montgomery and Pierce.[336] The transepithelial voltage is ~0 mV, and the apical cell membrane has no significant rheogenic pathways, whereas the basolateral cell membrane is conductive to K^+ and Cl^-. NaCl enters the cell across the apical cell membrane via a thiazide-sensitive Na^+–Cl^- cotransporter. In addition, NaCl may be reabsorbed through parallel operation of Na^+/H^+ and Cl^-/HCO_3^- exchangers. Na^+ leaves the cell via the basolateral Na^+,K^+-ATPase, while Cl^- exists down its electrochemical gradient through a basolateral channel of unknown molecular identity. The basolateral cell membrane moreover contains a Cl^-/HCO_3^- exchanger. According to Stanton and collaborators, a second cell type, having the above-mentioned transporters but lacking the luminal Cl^-/HCO_3^- exchanger, accounts for H^+ secretion. In apparent contradiction to the transepithelial measurements in *Amphiuma* by Stanton and collaborators, a lumen-negative transepithelial potential has been observed in the late distal tubule of *Triturus*, *Necturus*, and *Ambystoma*.[12,208,384,467,481] The difference in transepithelial potential values may reflect measurements made in different portions of the late distal tubule, with increasing negative luminal potentials toward the collecting tubule.[481]

iii.   Collecting Duct System

The amphibian collecting tubules, constituting the most terminal part of the nephron, open into the collecting ducts. The epithelium constituting the collecting tubules and ducts is heterocellular, consisting of principal and intercalated (mitochondria-rich) cells, the latter constituting approximately 1/3 of the total cell number in the collecting tubule of *Bufo bufo*.[330] The collecting tubule has been shown to express a lumen-negative voltage, to have low water permeability, to actively reabsorb NaCl, and to secrete K^+ and H^+.[98,104,207,215,271,467,468,541,551,552] In freshwater, the primary function of the collecting duct system is to contribute to urine dilution. A basolateral Na^+,K^+-ATPase provides the driving force for apical uptake of Na^+ through the epithelial sodium channel

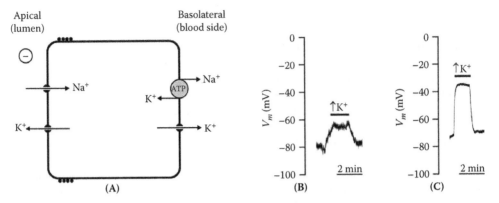

FIGURE 9.13 The principal cell of the mesonephric collecting duct system. (A) Model illustrating cellular transport mechanisms of the principal cell. In aquatic amphibians, the collecting duct system is important for urine dilution, and a large apical Na^+ conductance (ENaC channels) in the principal cells provides the first step for transcellular Na^+ reabsorption. Na^+ leaves the cells through basolateral Na^+,K^+-ATPases. In addition, these cells secrete K^+ as revealed by microelectrode impalements on isolated and perfused tubules of the collecting duct system in the terrestrial anuran *Bufo bufo* (see Møbjerg et al.[331]). K^+ is actively taken up across the basolateral cell membrane by the Na^+,K^+-ATPases. K^+ channels in the apical cell membrane provide a route by which this ion can diffuse into the lumen down its electrochemical gradient, and K^+ channels in the basolateral cell membrane recycle this ion for the Na^+,K^+-ATPases. Voltage recordings from single principal cells in isolated and perfused collecting tubules in Parts B and C demonstrate the effect of luminal and bath $[K^+]$ steps from 3 to 20 m*M* on the membrane potential (V_m). (B) Membrane potential of cell in isolated perfused collecting tubule (V_m) depolarizes in response to an increase in luminal K^+ concentration from 3 m*M* to 20 m*M*, indicating the presence of a K^+ conductance in the apical cell membrane. (C) Raising the K^+ concentration in the bath solution by the same amount also depolarizes V_m, revealing the presence of a large basolateral K^+ conductance.

(ENaC) in the principal cells (Figure 9.13). Electrophysiological studies failed to reveal an apical K^+ conductance in the aquatic urodele *Amphiuma*, and K^+ secretion in the amphibian collecting duct system was therefore presumed to occur across the paracellular pathway.[104,207,215] An apical K^+ conductance, however, is clearly present in principal cells of the collecting duct system from the terrestrial anuran *Bufo bufo*.[332] K^+ secretion through apical K^+ channels is probably a major task of the collecting duct system of this terrestrial amphibian (Figure 9.13). In the terrestrial environment amphibians produce less dilute urine, and NaCl transport by the collecting duct system is downregulated.[334] In addition to fine-tuning urine Na^+, K^+, and Cl^- levels and urine pH, the collecting duct system probably participates in regulating amphibian inorganic phosphate homeostasis.[335]

Heterocellularity in the collecting duct system is a highly advanced feature of the vertebrate meso- and metanephric nephrons. It has no counterpart in the pronephros.[331] Heterocellularity and the presence of principal and mitochondria-rich cells, however, are features closely related to the amphibian lifestyle and characteristic of other osmoregulatory epithelia in amphibians (e.g., skin and urinary bladder).[61,73,284,523,537] Mitochondria-rich cells are characterized by a significant carbonic anhydrase activity. In the skin, they occur in three versions (α, β, γ) that are specialized for active transport of H^+, HCO_3^-, and Cl^-, respectively (see Section III.F). It remains to be shown how many functional types are present in the mesonephros.

c. Aquaporins in the Kidney

A complete picture of aquaporins in the different segments of the amphibian kidney remains to be established and, as with the skin and bladder, comparisons can be made with the mammalian kidney, where six AQPs have been identified at specific segments of the nephrons and other components:

AQP1 and AQP7 at the proximal tubule; AQP1 at the descending thin limb of Henle's loop; AQP2, AQP3, AQP4, and AQP6 at the collecting duct; AQP3 at the renal pelvis; and AQP1 at the vasa recta.[364,476] AQP1 is localized at both apical and basolateral membrane of the epithelial cells in the proximal tubule and the descending thin limb of Henle's loop and is involved in the transcellular movement of water. AQP2 is the vasopressin-dependent water channel that facilitates water reabsorption in response to the antidiuretic hormone vasopressin by translocating from intracellular vesicles to the apical plasma membrane of collecting duct principal cells. Both AQP3 and AQP4 are expressed at the basolateral membrane in the principal cells of the mammalian collecting duct. They are involved in the transcellular pathway for reabsorption of water in concert with apical AQP2. AQP6 is considered to be an anion channel and is expressed in the acid-secreting intercalated cells of the collecting duct. AQP2 mutations and disruption of the AQP2 gene cause nephrogenic diabetes insipidus, a disease characterized by a massive loss of water through the kidney.

In the tree frog, immunolabels for *Hyla* AQP-h3BL were present among the principal cells of the collecting ducts and a portion of late distal tubules of the kidney.[6] The presence of an AVT-dependent AQP in the apical plasma membrane has not yet been demonstrated; however, some reports suggest significant water reabsorption by the renal tubule in response to AVT[421] or noradrenaline.[147,377] Uchiyama[490] provided evidence for selective expression of V_2-type AVT receptors in microdissected collecting tubules of bullfrog by showing increased cAMP production in this segment in response to AVT. Recently, Zimmerman et al.[559] cloned a cDNA encoding an AQP-like 2 protein (HC-2) from Cope's tree frog (*H. chrysoscelis*), and Ogushi et al.[371] cloned a similar AQP (*Hyla* AQP-h2K) from the kidney of *H. japonica*. These AQPs are homologous with mammalian AQP2 isoforms. The amphibian mesonephros, however, does not have a zonation equivalent to the mammalian metanephros. Notably, the amphibian nephron lacks the loop of Henle with its parallel arrangement of vasa recta. The amphibian kidney cannot, therefore, build up the corticomedullary osmotic gradient, which is essential for the AQP2-mediated water withdrawal from collecting ducts during antidiuresis in mammals. As mentioned above, amphibians in terrestrial environments conserve water by drastically reducing GFR and urine flow. Aquaporins present in amphibian renal tubules may under such conditions serve to equilibrate the urine with the interstitial fluid, leading to the production of isosmotic urine. During rehydration by terrestrial species large volumes of dilute urine pass through the ureters into the bladder, where it accumulates and can be reabsorbed by the bladder epithelium.[458]

V. STRUCTURE AND FUNCTION OF THE URINARY BLADDER

In contrast with mammals that conserve water by forming a small volume of concentrated urine, frogs (*Rana pipiens*) cease to form urine when water is deprived.[4] When water is available the amphibian kidney forms a large volume of dilute urine that is stored in a large urinary bladder that is generally a bilobed diverticulum of the cloaca. Adolph[4] also showed that the amount of water stored in the urinary bladder corresponds with the body mass gain when dehydrated frogs are immersed in water (i.e., urine formation corresponds closely with cutaneous absorption). Steen[458] repeated these experiments and further showed that water loss during dehydration corresponds with reabsorption of stored urine. The use of the urinary bladder as a water storage organ is seen in all three amphibian orders and is most important in fossorial and arboreal species (Table 9.7). Bladder capacities reported in the literature[21,54,187,414,436,440] are often averages obtained from animals in the laboratory with water available *ad lib*. In other cases, the largest value recorded is taken as a representative maximal value. For example, Ruibal[414] found that, in the laboratory, *Bufo cognatus* stored 19 to 31% of their hydrated body mass as bladder water, although Shoemaker et al.[440] reported excavating a burrowed *B. cognatus* with bladder water storage in excess of the hydrated body mass. Van Beurden[505] similarly found water storage by burrowed *Cyclorana platycephalus* to be 130% of the hydrated body mass.

TABLE 9.7
Representative Bladder Capacities of Amphibians in Fossorial and Arboreal Species

Species	% Body Mass	Remarks
Anura		
Cyclorana platycephalus	57 (130)	Arid, fossorial, Australia[a,b]
Neobatrachus wilsmorei	50	Arid, fossorial, Australia[a]
Notaden nicholsi	50	Semiarid, fossorial, Australia[a]
Bufo cognatus	31 (103)	Arid/semiarid, fossorial, North America[c,d]
Bufo marinus	25	Tropical, terrestrial, Central America[e]
Scaphiopus couchi	16–33 (60)	Arid, fossorial, North America[d]
Hyla moorei	20–30	Wet, arboreal, Australia[a]
Xenopus laevis	1	Aquatic, Africa[f]
Urodela		
Aniedes lugubris	50	Temperate, arboreal, North America[g]
Salamandra maculosa	34	Temperate, terrestrial, South Europe[h]
Ambystoma tigrinum	20–30	Temperate, fossorial, North America[i]
Notopthalmus viridescens	15–20	Temperate, terrestrial (efts), North America[j]
Triturus cristatus	2	Temperate, mostly aquatic, Europe[k]
Necturus maculosus	5	Temperate, aquatic, North America[k]
Apoda		
Icthyophis kohtaoensis	—	Terrestrial[l]
Typhlonectes compressicauda	—	Aquatic[m]

[a] Main and Bentley.[309]   [h] Bentley and Heller.[27]
[b] van Beurden.[505]   [i] Alvarado.[7]
[c] Ruibal.[414]   [j] Brown and Brown.[54]
[d] Shoemaker et al.[440]   [k] Bentley and Heller.[26]
[e] Shoemaker.[436]   [l] Jared et al.[222]
[f] Bentley.[21]   [m] Wake.[524]
[g] Hillman.[187]

Note: Range or maximal values were obtained in laboratory observations. Values in parentheses indicate extremely high bladder water reserves recorded from animals in the field. References to bladder content of apodans did not calculate bladder volume.

A. Anatomy of Urinary Bladder

Because the transport of water and Na$^+$ across the toad bladder (experiments primarily with *Bufo marinus*) was stimulated by AVT and aldosterone, the tissue became a model for cellular processes in the collecting duct of the mammalian kidney.[89,293,294,422] This resulted in an extensive literature on the anatomy and physiology of the bladder. *B. marinus* bladders are lined by an epithelium that contains three cell types.[381] The most abundant are squamous epithelial cells characterized by apical plasma membranes having small microvilli and ridges. These structures form nearly all of the surface area for salt and water transport from urine stored in the bladder. These cells are termed *granular cells*. The apical membranes of adjacent cells are connected by occluding junctions so transport is mostly transcellular in nature. It was observed that cells swelled when the bladder lumen contained a dilute solution and AVT was added to the serosal solution, indicating that the apical membrane is the limiting barrier for water movement. In contrast, cells became swollen when the serosal solution was dilute, in the absence of AVT. The mechanisms for regulation of water permeability are described below. Interspersed among the granular cells are mitochondria-

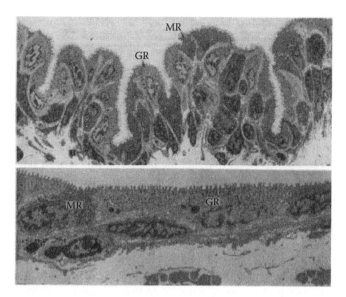

FIGURE 9.14 Bullfrog (*Rana catesbeiana*) bladder fixed for microscopy in the nondistended (upper) vs. distended state (lower). Note that the granular (GR) and MR cells have apical microvilli and are supported by a layer of basal cells (B) that attach to the basement membrane. (From Strum, J.M. and Danon, D., *Anat. Rec.*, 178, 15–40, 1973. With permission.)

rich cells that, like MR cells in the skin, have a narrow neck region that extends to the apical surface and appear to have little absorptive area. Scott and Saperstein[431] found that MR cells made up approximately 15% of the total number of cells and contain enzymes associated with acid–base balance and AVT stimulation of salt and water transport. These observations suggest that the number of MR cells might be regulated, as was observed with the skin (see below). The third type of cell, termed *goblet cells*, are similarly interspersed among the granular cells and contain mucous granules to be secreted over the apical membrane surface of the bladder. The epithelium is supported by a basal lamina that contains collagen fibers, blood vessels, and smooth muscle.

The urinary bladder of bullfrogs (*Rana catesbeiana*) has also been used as a model for studying endocrine control of salt and water transport across the mammalian nephron (Figure 9.14).[469] The granular cells are the most abundant cells forming the outer barrier for salt and water transport. MR cells make up approximately 25% of the epithelium, and both cell types and are sealed by occluding junctions at their apical boundaries. Unlike the toad bladder, goblet cells are lacking and the granular cells are underlain by a layer of basal cells that are attached to the basal lamina. Figure 9.14 also shows the change in appearance when the bladder is distended.

B. Aquaporins in Urinary Bladder

The effects of dehydration on urine reabsorption *in vivo* could be duplicated by posterior pituitary extracts,[125] and Bentley[20] showed a similar effect with isolated bladders. As noted above, stored bladder water may allow fossorial species to endure 1 to 3 years of drought. The discovery of aquaporins provides a mechanism for this process. Granular cells of the urinary bladder are characterized by having specific structures with tubular or vesicular profiles and granular inclusions. In the nonstimulated bladder, immunolabeling of total AQP-h2 protein was exclusively observed in tubular or spherical vesicles in the granular cells but rarely detected in the plasma membrane. On the other hand, in the AVT-stimulated urinary bladder, labeled total AQP-h2 protein was seen in the apical plasma membrane in addition to the cytoplasmic vesicles, and the labeling density in the apical plasma membrane was higher than that in the cytoplasm. As noted for the skin, the insertion of AVT-stimulated AQPs into the apical membrane requires phosphorylation of specific serine residues. Using a specific antibody against phosphorylated AQP-h2, immunoreaction was found in only a small number of vesicles in the granular cells of the nonstimulated urinary bladders, whereas in the AVT-stimulated specimens labeling was found not only in tubular or spherical vesicles but also in the apical membrane of the granular cells.[177]

The AQP-h2 protein, which was translocated after stimulation, was examined for its phosphorylated form prior to and after AVT stimulation by western blot analysis using a specific antibody against phosphorylated AQP-h2 (ST-160). In the nonstimulated condition, the specific purified antibody detected no bands. On the other hand, in the AVT-stimulated condition, a clear band was observed after 2-min incubation with AVT, and the band was still seen in samples prepared after 15-min incubation with it. These observations provide additional evidence that the phosphorylated AQP-h2 protein is translocated from cytoplasmic pools to the apical plasma membranes of the granular cells in the bladder, thereby reabsorbing the water into the body. Taken together, these results suggest that frog AQP-h2 in the bladder and AQP-h2 plus AQP-h3 in the skin are AVT-regulated AQPs and involved in the regulation of body water balance of the frog. QP-h3BL protein was constitutively expressed in the basolateral plasma membrane of the principal cells in the urinary bladder; these cells could be distinguished from mitochondria-rich cells based on the expression of the vacuolar type, proton-pumping ATPase (V-ATPase) E-subunit in the latter. Thus, water reabsorbed from the lumen of the urinary bladder via AQP-h2 in the apical membrane of granular cells will move through the AQP-h3BL to the connective tissues and then to the capillaries (Figure 9.3). Antibodies against these isoforms in *Hyla japonica* were distributed similarly to those in the urinary bladder of *Bufo woodhouseii*.[6]

C. Ion Transport and Acid Secretion by Urinary Bladder

In addition to storing dilute urine, the bladder actively transports Na+ from the lumen to the serosal surface of the epithelium[294] via an amiloride-sensitive pathway.[26] The model for Na+ transport across granular cells is basically that described for frog skin (Figure 9.4A). During hormone stimulation of the ion transport, changes of the Na+ concentration in both granular and basal cells were observed, indicating that the latter cell type also participates in Na+ absorption.[405] Na+,K+-ATPase and K+ channels in the basolateral plasma membrane generate an electrochemical gradient for Na+ entry across the apical plasma membrane via ENaCs. In living toads (*Bufo marinus*) that have been salt restricted, bladder urine is more dilute than ureteral urine, indicating that active Na+ transport enables the urinary bladder to retain salt that was not reabsorbed by the kidney.[324] In general, aquatic anurans and urodeles (e.g., *Xenopus* and *Necturus*) have small urinary bladders. An exception is *Amphiuma means*, which has a large bladder and amiloride-sensitive Na+ transport that is proposed to retain urinary salts that are not reabsorbed by the kidneys.[340] In contrast to toad and frog bladders, Na+ transport across *A. means* bladders is not stimulated by hormones (e.g., antidiuretic hormones and aldosterone) that stimulate Na+ transport in bladders of terrestrial amphibians. The aquatic apodan *Typhlonectes* also has an exceptionally large urinary bladder[524] that might serve to conserve urinary Na+.

The mechanism for Cl- transport across the urinary bladder has not been firmly established. The apical membrane of granular cells is impermeable to Cl-, and it has been suggested that Cl- absorption is paracellular and driven by the transepithelial electrical potential difference generated by the active Na+ flux.[293] MR cells are present in the toad bladder, and their primary function has been associated with urinary acidification. Living toads given an acid load respond by increasing the acidity of the urine in conjunction with an increase in MR cell density.[141] In isolated bladders, inhibition of carbonic anhydrase by acetazolamide inhibits the capacity for acidification of the urine and the enzyme is localized in the MR cells. Populations of *Bufo marinus* collected from Colombia or the Dominican Republic appear to vary. Colombian toads have a high level of carbonic anhydrase and MR cells and can acidify the urine, while those from the Dominican Republic have few MR cells and cannot acidify. It remains to be determined whether this was the result of environmental conditions in their natural habitat or was caused by the laboratory conditions at which the animals were kept. In addition to having carbonic anhydrase, V-ATPase clusters have been localized at the outer surface of the apical membrane of MR cells.[51] To date, only the acidifying α-type MR cell[460] (Figure 9.5A) has been identified in amphibian urinary bladder.

VI. NITROGENOUS WASTES

Fully aquatic amphibians and larvae excrete nitrogen largely in the form of ammonia while ureotelism predominates in the terrestrial environment.[302] Ammonia excretion predominantly occurs via gills, whereas the kidney and bladder play a key role in urea excretion and retention as seen in amphibians acclimatized to hyperosmotic saline or desiccating conditions (see Forster et al.[134] and Section II.B.5). Aquatic larvae are ammoniotelic and use the gills as a primary surface for elimination. Urea was identified in bladder urine of Ranid and Bufonid species in 1821 vs. uric acid in the urine of reptiles and birds.[93] Urea cycle enzymes of *Rana catesbeiana* appear in the liver during the climax stages of metamorphosis, and urea becomes the primary nitrogenous waste at about the same time that the front legs appear.[52] Although this may be the general case for the more commonly studied genera (e.g., *Rana* and *Bufo*), variation occurs that can be related to habitat and reproductive strategy; for example, early development of *Leptodactylus bufonius* larvae occurs in foam nests deposited in burrows made by the adults. Larvae await rainfall to flood the nest. The primary route for nitrogen excretion is urea, and larvae can tolerate plasma urea concentrations as high as 400 mM.[438] Martin and Cooper[313] observed that the anuran *Crinia victoriana* deposits eggs on land vs. aquatic or foam nests, and the larvae excrete 86% of their nitrogenous waste as urea. *Eleutherodactylus coqui* also lays eggs on land, and froglets emerge directly from a protective gelatinous coat about 3 weeks after fertilization.[486] The urea cycle enzyme arginase is detectable prior to hatching and increases after hatching, with maximal levels observed when the yolk sac was completely reabsorbed.[64] Representatives from all three amphibian orders are known to deposit eggs on land,[221,303,312,365,389] but it is not known if all become ureotelic at an early stage

Urea is relatively nontoxic and allows accumulation of nitrogenous wastes when urine formation decreases in terrestrial species as they venture away from a hydration source. As described earlier, anuran and urodele species are able to tolerate reduced water potentials due to salinity or dry soil conditions by accumulating urea in the body fluids (summarized in Jørgensen[241]). This is the result of increased expression of urea cycle enzymes and retention of urea by transport mechanisms in the kidney, urinary bladder, and, in some species, the skin. In the aquatic frog *Xenopus laevis*, adults remain ammoniotelic unless presented with a hyperosmotic environment, in which case they express urea cycle enzymes and increase the urea concentration of the extracellular fluid to remain hyperosmotic to their surroundings.[17] In moderately concentrated (300-mOsm NaCl) solutions, carbamoylphosphate synthase (CPS) levels rise, while in more concentrated solutions (600-mOsm NaCl) both CPS and arginosuccinate lyase (ASL) are elevated. *Bufo viridis* acclimated to 600-mOsm NaCl solutions showed increased activity of all urea cycle enzymes with particular elevations of CPS and ASL.[16] Similarly, *Rana cancrivora* acclimated to 800-mOsm NaCl increased the activity of all urea cycle enzymes with a particular increase in ASL. Recall, however, that larval *R. cancrivora* can tolerate full seawater without increasing urea accumulation (see Section II.B.5).

Urea is secreted into the urine with urine/plasma (U/P) ratios in excess of unity when water is available. Schmid[425] sampled eight species of North American anurans shortly after capture from their natural habitat and found the U/P ratio to be as high as 17 for *Rana septentrionalis* and as low as 1.53 for *Bufo hemiophrys*. Associated with this, species with a lower U/P ratio had higher plasma urea concentrations. In laboratory experiments, the U/P ratio of *Rana esculenta* in water was 7.7 but declined to 1.1 when the frogs were acclimated to 240 mOsm saline.[3] The U/P ratio for *B. marinus* declined from 2.5 in the hydrated state to about 1 when the toads were dehydrated.[308] Jørgensen[241] suggested that animals facing hyperosmotic or dehydrating conditions reduce urine production and reabsorb urea from the kidney and urinary bladder so plasma and blood equilibrate. In this regard, a facilitative urea transport protein (UT) has been identified in the apical membrane of cells in the early distal tubule and urinary bladder of *B. marinus* that is homologous with the mammalian UT-A2 urea transporter found in the ascending thin loop of Henle.[269] The expression of the UT-A2 is increased in the kidneys of rats given the mammalian antidiuretic hormone and in cultured kidney cells treated with hyperosmotic NaCl–urea solutions.[420] Messenger RNA for this transport protein is enhanced in both dehydrated toads and toads immersed in a 300-mOsm NaCl solution.[269] Flux ratio analysis of radiolabeled urea indicated active urea transport in the inward direction in the skin of *B. bufo*.[58,502] Subsequent studies with frogs (*R. esculenta*) indicated passive cutaneous transport in hydrated animals and active transport in the skin of animals exposed to saline solutions or following dehydration, whereas toads (*B. bufo, B. marinus*, and *B. viridis*) displayed active transport across the skin even in a hydrated state.[150,258] The molecular mechanism has not been identified, and the possibility exists that the asymmetrical unidirectional urea fluxes in the studies above are caused by solvent drag on urea in the inward direction.

At least two genera of xeric-adapted anurans (*Chiromantis* from Africa and *Phylomedusa* from South America) have been shown to excrete nitrogenous wastes in the form of uric acid.[301,441] As noted earlier, these animals also have very high resistance to evaporative water loss due to either wax secretions of the skin (*Phylomedusa*) or intrinsic properties of the skin itself (*Chiromantis*). Uric acid minimizes the water loss that is needed for nitrogen excretion. Unlike ureotelic species, such as *Bufo boreas*, that reduce GFR to near zero during periods of dehydration, *P. sauvagii* continue to form urine but reabsorb 98 to 99% of the filtered Na^+, Cl^-, and urea.[439] The renal clearance of uric acid is equal to that of paraaminohippuric acid, indicating secretion of all uric acid delivered to the kidney. As with other terrestrial species, the bladder of *P. sauvagii* serves as a water storage reservoir.

VII. INTEGRATING THE ORGAN FUNCTIONS: ENDOCRINE AND AUTONOMIC CONTROL

Isolated amphibian skin and urinary bladder have been used extensively as model tissues to study factors that regulate epithelial salt and water transport.[284] The stimulation of Na^+ transport is termed the *natriferic response*, and the stimulation of water permeability is termed the *hydroosmotic response*. Historically, the effects of hormones have often been evaluated on isolated tissues treated with pharmacologic doses and the responses extrapolated to the integrated control of ion and water balance by the whole animal, in an environmental context. Interpretation of these studies must be done with caution and should be evaluated, where possible, in conjunction with experiments on whole animals. Plasma levels of some hormones have been measured and related to their physiological effects in the context of the natural history of particular species, as have the effects of agonists and antagonists of the sympathetic nervous system. Finally, vascular perfusion of the skin *in situ* is necessary for the transfer of salt and water into the circulation, and behavioral mechanisms enable the animal to select a favorable hydration source.

A. AVT

The role of arginine vasotocin (AVT) in stimulating water absorption across the skin and reabsorption from the bladder has been discussed in some detail as it relates to the insertion of aquaporin 2 isoforms into the apical membrane and specialized regions for water absorption (see Section III.B.3). Historically, the ability of mammalian posterior pituitary extracts to stimulate cutaneous water absorption by intact animals was observed by Brunn,[57] and stimulation of bladder absorption *in vivo* was observed by Ewer.[125] The amphibian antidiuretic hormone arginine vasotocin was identified in 1959 in a series of related articles[249,383,423] and found to be more potent in stimulating Na^+ transport and water permeability than mammalian isoforms.[22] This capacity, coupled with the reduction of urine formation by lowering GFR, has been termed the *antidiuretic* or *water balance response*[182,437] and is highly developed in terrestrial anuran and urodele species that rely on bladder reserves when foraging away from water and rapid rehydration when returning to water.[21] Experimentally, the antidiuretic response is initiated by forced dehydration, which will increase plasma osmolality and, if sufficient, decrease plasma volume. The antidiuretic response can also be elicited by injection of hyperosmotic NaCl solutions without depleting plasma volume or by depletion of plasma volume by hemorrhage with no change in osmolality.[437] Elevated AVT levels have been measured in anuran species subject to either hyperosmotic or hypovolemic stimuli.[270,366,367] Konno and coworkers[270] also showed that toads maintained with water *ad lib* had plasma osmolarity and AVT levels that were similar to those of toads immersed in tapwater, which supports the concept of *anticipatory drinking* proposed by Jørgensen,[239] in which toads with water available *ad lib* will retain a hydrated state in the absence of elevated AVT levels (to be discussed in more detail below). The simultaneous stimulation of Na^+ transport by these tissues appears to ensure that water and solute absorption maintain the composition and volume of the extracellular fluid.

The effect of AVT on water permeability and Na^+ transport across the skin of aquatic anurans and urodeles shows a wide range of responses that can be related to phylogeny and habitat. The hydroosmotic response of *Xenopus laevis* skin to AVT is lacking but the natriferic response is present. The aquatic, neotenic urodele *Necturus maculosis* (Proteiidae) has low cutaneous water permeability and no measureable active Na^+ transport.[30] As with larval frogs,[8] the gills appear to be the primary site of salt and water exchange. The aquatic urodeles *Siren lacertina* (Sirenidae) and *Amphiuma means* (Amphiumidae) have amiloride-sensitive Na^+ transport across the isolated skin but no AVT-stimulated hydroosmotic response.[24] Larval *Ambystoma tigrinum* (Ambystomatidae) are entirely aquatic, but the adults may be found in highly terrestrial environments.[99] The larvae have gills, and the skin transports Na^+ at a rate one tenth that of the adult. The natriferic response to AVT is absent in the larval skin but present in the adult. In living animals, Jørgensen et al.[232] found neurohypophyseal extracts to stimulate Na^+ uptake by *Ambystoma mexicanum*. Water permeability of *A. tigrinum* skin is comparably low in both adult and larval skin and not stimulated by AVT.[32] It should be noted that Spight[452] observed that *Ambystoma opacum* dehydrated by 20% rehydrated at a greater rate than predicted from the osmotic gradient and suggested that mechanisms for increased water permeability other than AVT might be involved.

Newts (Salamandridae) initially metamorphose from a larval into a terrestrial eft stage and return to the water in an aquatic phase. An AVT-stimulated hydroosmotic response is observed in both eft and aquatic phases of *Notophthalmus viridescens*,[54] *Taricha torosa*,[56] and *Triturus vittatus*[530] and may be comparable to that of terrestrial anurans. In contrast, water permeability of *Salamandra maculosa* skin is not stimulated by AVT; rather, water retention is the result of AVT stimulation of water reabsorption across the urinary bladder.[27] A similar lack of AVT-stimulated water permeability has been observed across the skin of *Triturus alpestris* and *T. cristatus*.[26] Unlike *S. salamandra*, water retention is due to a reduction in urine formation rather than reabsorption from the urinary bladder. Both species do show AVT-stimulation of Na^+ transport across the skin, as does the skin of *Taricha granulosa*,[53] indicating that this is a common feature of the Salamandridae. Lungless salamanders (Plethodontidae) have a slender body and high surface area for water absorption. Spight[452] found that dehydrated *Plethodon jordani* rehydrated at a rate similar to a value reported by Cohen[80] for *Aniedes lugubris*, at that time the greatest rehydration rate reported for urodeles. Hillman[187] later showed the rate of water absorption by *A. lugubris* was stimulated by AVT, as was Na^+ transport across the skin.

The natriferic and hydroosmotic effects of AVT can be duplicated by the addition of cyclic adenosine monophosphate (cAMP) to isolated tissue preparations,[372,396] and the tissue level of cAMP is elevated following AVT treatment of the isolated skin.[226,227,311] The current model is that AVT binds to a G_s-protein-coupled receptor (V_2) to stimulate adenylate cyclase. The resulting increase in cAMP activates protein kinase A, which phosphorylates putative proteins associated with the activation of ENaCs in the apical membrane. It remains to be seen if the phosphorylated protein is an ENaC subunit resident in the membrane or if phosphorylation of cytoskeletal proteins promotes insertion or inhibits withdrawal of vesicles containing ENaCs, as is seen with aquaporins. In contrast, the reduction of GFR during dehydration is mediated by AVT binding to a G_q-protein-coupled receptor (V_1) that activates phospholipase C to form diacylglycerol (DAG) and inositol triphosphate (IP_3).[10]

The concept that has been developed from the laboratory studies discussed above is that AVT plays a key role in regulating the water balance of amphibians. This view is in line with the experimental findings that strong dehydration leads to an increased plasma concentration of AVT and that the hormone enhances the water uptake across the skin and the urinary bladder wall, accompanied by reduced urine excretion. In a review of the literature, Jørgensen[238] pointed out that the above renal and extrarenal responses are accomplished within the range of normal hydration of amphibians (i.e., without raised plasma AVT concentrations). Thus, *in vivo* reabsorption of urine from the bladder takes place in the absence of increased osmolality of the body fluids, toads exhibit anticipatory drinking behavior,[239] and surgical elimination of pars nervosa function has no clear effect on the water balance of the organism nor does it affect the increased cutaneous water permeability caused by dehydration.[244] The role of AVT in amphibian water balance vs. other neural and hormonal factors, discussed below, remains to be established under the environmental conditions to which different species are adapted.

B. HYDRINS

A second group of neurohypophyseal peptides, hydrins 1 and 2, have been isolated from anurans[2] but not urodeles.[71] Hydrins appear to result from less complete processing of provasotocin:

> AVT: Cys–Tyr–Ile–Gln–Asn–Cys–Pro–Arg–Gly
> Hydrin 1: Cys–Tyr–Ile–Gln–Asn–Cys–Pro–Arg–Gly–Gly
> Hydrin 2: Cys–Tyr–Ile–Gln–Asn–Cys–Pro–Arg–Gly–Gly–Lys–Arg

Unlike AVT, hydrins stimulate water permeability of the skin and bladder with little effect on GFR. Acher et al.[2] proposed that the primary effect of AVT at physiological concentrations is antidiuresis via binding to V_1 receptors, whereas hydrins have a higher affinity for V_2 receptors in the skin and bladder to stimulate rehydration and utilization of bladder water. Rehydrating toads are able to rapidly rehydrate via cutaneous absorption and restore bladder water storage via increased urine formation;[458] however, hydrin 1 is present in *Xenopus laevis*, which lacks a hydroosmotic response,[412] so the relative importance of hydrins vs. AVT in amphibian water balance remains to be determined.

C. PROLACTIN

As noted earlier, newts (Salamandridae) have three life stages: Aquatic larvae metamorphose and emerge into a terrestrial eft stage with a dry skin. During the breeding season sexually mature efts return to water and develop an aquatic form with moist skin. The larval stage is prolonged by treatment with prolactin and thyroxine stimulates metamorphosis.[163,162] Prolactin also stimulates the return to water, termed the *water drive*, and the change from the eft to the aquatic form. Although the hydroosmotic response to AVT is comparable between the eft and aquatic phases of *Notopthalmus viridescens*,[54] prolactin did reduce the hydroosmotic effect of AVT in the terrestrial eft stage of *Taricha torosa*,[55] suggesting a reduction in water uptake when the efts return to water. Prolactin has been shown to reduce or prevent the decrease in plasma Na^+ concentration following hypophysectomy in larval *Ambystoma mexicanum*[538,539] and in the aquatic salamander *Necturus maculosus*.[376] This effect seems to be one of reducing efflux of Na^+ into dilute media by more aquatic species, as is the case for the effect of prolactin on gills of euryhaline fish transferred to freshwater.[307]

Exogenous prolactin has also been shown to prolong the larval stage of anurans.[53,124] Early experiments showing delay of metamorphosis were done with ovine prolactin. More recently, Huang and Brown[210] created transgenic *Xenopus laevis* with overexpression of genes for both ovine and *Xenopus* prolactin and observed no delay in developmental time. These authors question a role for prolactin as a juvenile hormone, at least in *Xenopus*. Prolactin receptors are present in the gills during the larval stage and increase rapidly in the kidneys at metamorphosis, suggesting that the osmoregulatory role of prolactin shifts from the gills to the kidneys during metamorphosis. Receptor density in the kidneys is stimulated by thyroxin, which is also increasing during metamorphosis.[535]

The mechanism for prolactin stimulation of Na^+ transport has been studied in isolated skin from premetamorphic larvae and newly metamorphosed bullfrogs. With larval skin, treatment with either aldosterone or glucocorticoids induces the appearance of functional ENaCs and amiloride-sensitive Na^+ transport.[473,474] Prolactin inhibits these effects and promotes the continued expression of a larval-type cation channel that is poorly selective for Na^+ and stimulated by amiloride.[475] In the skin of newly metamorphosed bullfrogs, prolactin, like aldosterone, stimulates Na^+ transport by increasing the density of functional, amiloride-inhibited, ENaCs in the apical membrane and Na^+–K^+ pump activity in the basolateral membrane.[472] The ability of prolactin to stimulate Na^+ transport via increased apical ENaC density has also been shown for adult tree frogs (*Hyla japonica*).[471] This may be caused by changes in receptor expression in the skin as has been proposed for the kidney.[535]

D. INSULIN

Insulin has been shown to stimulate Na^+ transport across isolated skin and urinary bladder in Ussing-type experiments[88,185,429] by increasing the density of ENaCs in the apical membrane.[42] Insulin levels are presumed to be elevated following a meal to facilitate nutrient assimilation and storage. As noted earlier, toads increase water absorption following a meal,[237] so an increase in Na^+ transport may allow the animals to maintain a hydromineral balance at this time.

E. SYMPATHETIC NERVOUS SYSTEM

The β-adrenergic agonist isoproterenol has been shown to stimulate water absorption by the hydrated toads *Bufo cognatus*, *B. bufo*, and *Scaphiopus couchi* (spadefoot toad).[193,513,550] The level of stimulation is equal to or greater than that of exogenous AVT or hyperosmotic NaCl injection. Furthermore, the β-adrenergic antagonist propranolol depressed the rate of water absorption in hydrated *B. cognatus* and *B. bufo*.[513,550] Clearly, the sympathetic nervous system plays an important role in the water balance response. The effects of isoproterenol on hydrated *B. bufo* and propranolol on dehydrated toads were observed, respectively, in conjunction with increased and decreased capillary perfusion in the seat patch,[513] indicating a coordination of vascular perfusion and epithelial transport (discussed below). Similarly, *Bufo arenarum* immersed in an isosmotic NaCl solution rapidly reduced the rate of urine formation, and the reduction is inhibited by guanethidine, which inhibits norepinephrine release from postganglionic neurons.[382]

Studies with the isolated skin and urinary bladder have produced a variety of responses to epinephrine, norepinephrine, sympathomimetics, and inhibitors of adrenergic receptors.[136] Consistent with the whole-animal experiments, low concentrations of norepinephrine stimulated water permeability across the isolated epidermis, and the effect was blocked by propranolol.[395] Like AVT, this response is mediated by increased synthesis of cAMP but appears to be independent of AVT.[101]

The response to higher concentrations of norepinephrine is more variable. When the water permeability of the isolated skin was stimulated with exogenous cAMP and the β-adrenergic receptors were blocked with propranolol, norepinephrine reduced water permeability. The α-adrenergic blocker phentolamine prevented this inhibition, indicating that both α- and β- adrenergic receptors are present in the skin with α-receptors (presumably α_2) inhibiting cAMP synthesis.[395] In the isolated epidermis, β-adrenergic stimulation enhances solute-coupled water absorption associated with prior stimulation of the active uptake of Na^+ by the principal cells and the passive uptake of Cl^- by mitochondria-rich cells.[363]

The bladder receives motor innervation that is cholinergic[62] and presumably controls vascular and bladder smooth muscle tone. Catecholamines have also been localized in the bladder,[322] indicating both sympathetic and parasympathetic control of bladder function. Three lines of evidence suggest sensory innervation that allows toads to detect the presence of bladder water: (1) Hydrated *Bufo woodhouseii* with empty bladders display a water absorption response (WR) behavior and absorb water to a greater extent than toads allowed to retain bladder contents.[488] (2) When presented with water, hydrated *B. alvarius* with empty bladders increase seat patch blood flow to a level similar to that of dehydrated toads.[515] (3) Water uptake across the seat patch region of *B. marinus* is greater in toads with an empty bladder.[379]

F. Renin, Angiotensin, and Aldosterone

Isolated preparations of amphibian skin, urinary bladder, and colon have long been used as models for the study of aldosterone-stimulation of epithelial Na^+ transport.[89,386,492] In the current model, aldosterone binds to a mineralocorticoid receptor that serves as a transcription factor to stimulate the synthesis of proteins that result in a greater number of conducting ENaCs in the apical membranes of absorbing epithelia and also greater activity of Na^+,K^+-ATPase in the basolateral membranes.[84,380] One of the proteins identified is a kinase that increases ENaC density by inhibiting their removal from the apical membrane.

Whole-animal studies have documented elevated plasma aldosterone levels under conditions of plasma Na^+ depletion and reduced plasma volume. *Bufo marinus* maintained for 3 weeks in deionized water (DI; 0 mOsm) had significantly lower plasma osmolality, lower Na^+ concentrations, and elevated aldosterone levels compared with toads maintained in 9% NaCl (285 mOsm), as predicted from the known effect on isolated tissues.[151] In a more recent study, *B. marinus* immersed for 7 days in dilute tapwater had reduced plasma osmolality and Na^+ concentration relative to toads in 300-mOsm NaCl; however, in toads exposed to either tapwater or 300-mOsm NaCl, aldosterone levels were depressed to values below those for toads maintained with water *ad lib*.[270] Aldosterone levels were elevated significantly only in toads that were dehydrated, and the concentration of aldosterone correlated significantly with the increase in hematocrit, suggesting that plasma volume was the stimulus for hormone release. Differences between the experiments include the nutritional status and time of exposure to the dilute solution. It would appear that aldosterone levels are elevated more rapidly in response to plasma volume depletion than to reduced Na^+ concentration. This is consistent with observations that aldosterone levels of *B. japonicus* in the field were highest during the summer period when the toads ventured away from water to their terrestrial habitats and presumably would encounter dehydrating conditions.[230]

The regulation of plasma aldosterone has been assumed to resemble that of mammals, where a decrease in plasma volume is detected by the kidneys. Grill et al.[166] demonstrated that renin from amphibian kidney extracts catalyzes the formation of angiotensin I from angiotensinogen. Hasegawa et al.[179] showed that angiotensin I is converted to angiotensin II (AII) by a converting enzyme, and AII stimulates aldosterone release from the amphibian adrenocortical (interrenal) tissue. Konno et al.[270] found parallel increases in AII and aldosterone in dehydrated *Bufo marinus*. In contrast with the mammalian model, DeRuyter and Stiffler[95] found that adrenocorticotrophic hormone (ACTH) was able to stimulate aldosterone release and restore plasma Na^+ concentration in larval salamanders (*Ambystoma tigrinum*).

In addition, questions remain to be answered regarding the role of the aldosterone and the mineralocorticoid receptor vs. the glucocorticoids and the glucocorticoid receptor in regulating epithelial Na^+ transport in amphibians. The reduced level of Na^+ transport across the skin of frogs (*Rana pipiens*) whose interrenal tissues had been removed could be partially restored by either hydrocortisone or aldosterone, suggesting similar effects of both glucocorticoids and mineralocorticoid.[38] Similarly, hyponatremia produced by adenohypophysectomy in *Bufo marinus* was correctable by glucocorticoids.[326] Furthermore, Schmidt et al.[428] found that stimulation of Na^+ transport in cultured cells from the amphibian kidney by aldosterone was mediated by a glucocorticoid receptor rather than a mineralocorticoid receptor. In this regard, Johnston and Jungreis[229] proposed that aldosterone has glucocorticoid activity and stimulated gluconeogenesis and urea synthesis of *R. pipiens* subject to saline dehydration in the summer but not winter months. This is consistent with the observation by Konno et al.[270] that urea levels were elevated in dehydrated toads that displayed elevated plasma aldosterone concentrations and the field observations of Jolivet-Jaudet et al.[230] that plasma aldosterone concentrations were greatest during the summer when the toads are primarily terrestrial. The issue of mineralocorticoid vs. glucocorticoid function for aldosterone and seasonal variation in hormone responsiveness remain confounding issues that should be considered when examining literature citations or designing experiments.

G. Water Absorption Behavior

For the physiological mechanisms of water absorption to be effective, amphibians must be able to assess their hydration status, perceive the presence of water, and initiate behaviors that place the skin in contact with rehydration sources. Amphibians tend to be secretive and nocturnal, so relating laboratory observations to hydration behaviors in the field is difficult. Shoemaker et al.[442] described a variety of studies that report greater activity of both anuran and urodele species following a rainfall. Stille[463] observed toads (*Bufo woodhouseii*) that routinely moved from daily burrows to sandy beach areas of Lake Michigan, where they were nocturnally active on dry evenings and engaged in rehydration behavior. Toads were weighed at various intervals in their nightly activity cycle and were found to dehydrate by approximately 13% of their hydrated body mass before seeking water. Rehydration behavior included walking to moist sand at the water line and placing the seat patch in contact with the wet surface. From these observations it was concluded that the toads were able to detect osmotically available water with receptors on their feet. Laboratory experiments with *B. woodhouseii* and five other anuran species showed a similar pattern of behavior that Stille[464] termed the *water absorption response* (WR), in which the hindlimbs are abducted and the ventral skin is pressed to a moist surface (Figure 9.15A; see review by Hillyard et al.[195]). Brekke et al.[48] observed that the desert toad *B. punctatus* initiated WR behavior when dehydrated by less than 1% of their standard body mass. WR behavior could only be suppressed by immersing the toads for 1 to 2 hours in water prior to exposure to a rehydration surface. This was attributed to opportunistic drinking, which is an intrinsic behavior due to the scarcity of water in their habitat and may be comparable to the anticipatory drinking observed in *B. bufo*.[239]

In addition to stimulating aldosterone secretion, AII formed during periods of reduced plasma volume stimulates thirst and oral drinking in mammals, birds, and reptiles.[132] Exogenously administered AII similarly increases oral drinking in these vertebrate classes. Hoff and Hillyard[202] ensured a fully hydrated state in *B. punctatus* by immersion for 1 to 2 hours in deionized water and showed that intraperitoneal injection of angiotensin II (AII) stimulated WR behavior. Similar results have been obtained with *B. bufo*, *B. cognatus*, and *Scaphiopus couchi*,[391,512] indicating that AII stimulates cutaneous drinking in addition to oral drinking by other vertebrate classes.[132] Plasma AII levels were elevated in *B. marinus* dehydrated by approximately 20% of their standard body mass and correlated with a decrease in plasma volume and increase in osmolality. Toads in hypersaline (300-mOsm NaCl) solutions decreased plasma AII in association with an increase in both plasma volume and osmolality.[270] Both treatments produced an increase in plasma AVT levels, indicating that plasma AII is primarily sensitive to plasma volume and AVT release is sensitive to plasma osmolality, as well. Propper et al.[392] also showed that AII injection into the cerebral ventricles stimulated WR in *S. couchi*, indicating an AII-sensitive pathway within the blood–brain barrier, in addition

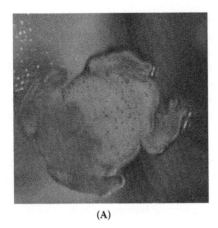
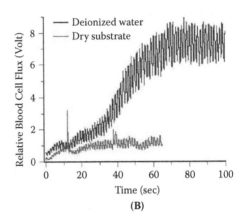

(A)   (B)

FIGURE 9.15 (A) The toad *Bufo punctatus* exhibiting the water absorption response. Note that the hindlimbs are abducted from the body and cutaneous capillaries in the skin are highly perfused. (B) Seat patch capillary blood flow in *Bufo woodhouseii* estimated by laser Doppler flowmetry of relative red blood cell flux. *Lower trace*: The seat patch blood flow in a dehydrated toad on a dry substrate is low. *Upper trace*: The blood flow is greatly stimulated when the dehydrated toad is exposed (at time zero) to deionized water. (From Viborg, A.L. and Hillyard, S.D., *Physiol. Biochem. Zool.*, 78, 394–404, 2005.).

to the circumventricular region of the hypothalamus that is perfused by the circulation. In mammals, the dipsogenic effects of AII are mediated by AII type I receptors[132] that are inhibited by the AII antagonist saralasin. In toads (*B. bufo*), saralasin was shown to be an AII agonist, stimulating hydration behavior, water absorption, and bladder water storage.[512] It should be noted that *S. couchi* dehydrated by 15% of their standard body mass showed WR behavior without measurable increases in brain or plasma AII[228] and that *B. bufo* housed with water available *ad lib* will maintain a constant hydration status (e.g., anticipatory drinking[239]). Both observations suggest that intrinsic factors other than AII also regulate hydration behavior.

Hydration behavior is also sensitive to external stimuli, including barometric pressure changes that might affect rainfall. Numerous anecdotal reports have described toads in closed environments calling during a rainstorm outside the building. Hoff and Hillyard[201] found that dehydrated *B. punctatus* initiate WR behavior and regain their hydrated body mass when the barometric pressure is steady or rising. A fall in barometric pressure by less than 0.5 kPa increased the toads' level of activity and reduced WR behavior and the recovery from dehydration. Saralasin inhibited WR and water gain when the barometric pressure was steady or rising but stimulated WR and water gain when barometric pressure was falling. These results suggest that perception of barometric pressure can modify behavior modulated by AII.

Terrestrial species must also be able to detect osmotically available water, especially species living in brackish water habitats or those that rely on moist soil. Dehydrated toads reject hyperosmotic solutions of urea, NaCl, and KCl as hydration sources.[200,310,343] Amiloride can partially restore the initiation of WR behavior on NaCl but not KCl, suggesting that the active transport of Na[+] serves a chemosensory function like that of the lingual epithelium of mammals that imbibe water orally.[100,197] Hyperosmotic solutions also open tight junctions in the stratum granulosum of the epidermis.[501] Toads are more tolerant of hyperosmotic salt solutions with impermeant anions (e.g., sodium–gluconate vs. NaCl), suggesting that paracellular transport contributes to the chemosensory function of the skin.[196] As noted earlier in this chapter, *Rana temporaria* can detect more dilute salt concentrations for oviposition and fossorial species respond to soil water potentials. Sensory mechanisms for these processes remain to be described. Thirst and water seeking remain interesting subjects for research and are particularly important in regions where climate change may result in reduced rainfall and water availability.

Urodeles lack a clearly identifiable seat patch. As noted previously, water absorption by *Notophthalmus viridescens* was equally high in both the eft and newt stage and was stimulated by injection of either AVT or oxytocin.[54] Water movement across isolated ventral skin was less sensitive to AVT but was greater than that calculated per unit area for the whole animal, which suggests a specialized region for water absorption.

H. Lymphatics and Circulation

Rehydrating *Bufo punctatus* absorb water across their skin at rates as high as 30% of their body mass per hour.[201] The skin is highly vascularized and overlies a network of lymphatic spaces that become engorged when animals rehydrate.[68,248] Unlike other tetrapods, capillary ultrafiltration into the lymphatic spaces is very rapid (reviewed in Hillman et al.[191]). Humans, for example, filter their approximate plasma volume in a day,[171] whereas amphibians filter their plasma volume in an hour. Lymphatic fluid moves to dorsal lymphatic spaces where lymph hearts, derived from skeletal muscle, return fluid to the venous circulation. The question arises as to whether water absorbed across the skin is taken directly into cutaneous capillaries or into the lymphatic spaces. In support of direct absorption, AVT stimulation of osmotic water movement across isolated seat patch skin was only equivalent to that of living toads when the cutaneous vasculature was perfused,[75] removal of lymphatic fluid from rehydrating toads did not account for the body mass gain, and the lymphatic fluid was not diluted as would be predicted from the lymphatic route.[547] Toads rehydrating from a source with ^3H_2O do accumulate the isotope in the lymphatic fluid,[534] so some combination of the two routes is possible.

Rehydrating toads show a large increase in blood flow in cutaneous capillaries of the seat patch[514,513,515] (Figure 9.15B), and the magnitude of blood flow is greater in the more terrestrial species. This is consistent with anatomical observations by Czopek[91] and Roth[410] that the seat patch is more richly vascularized in terrestrial species. The increase in cutaneous blood flow in dehydrated toads requires contact with a moist surface and may be as large as 6 to 8 times the precontact value (i.e., dehydration alone does not stimulate blood flow). Blood flow in hydrated toads can be stimulated by isoproterenol to values similar to those of dehydrated animals, indicating that it is a sympathetic reflex initiated by water potential receptors in the skin.[513]

VIII. EPILOGUE

As indicated in the introductory section of this chapter, we have elected to retain the more traditional genus and species names to be consistent with a considerable literature accumulated prior to and since the original edition of the Potts and Parry book was published in 1964.[387] When possible, we compared physiological mechanisms at the family level with the acknowledgment that the literature includes a limited survey of the diversity among the three orders of the Amphibia. The more recent classification scheme of Frost et al.[143] provides a more rigorous analysis of evolutionary relationships that underlie the physiological adaptations we have described, and we hope that it will stimulate research to better understand how amphibians cope with their environments.

This is particularly significant today because many amphibian populations are undergoing rapid declines and extinctions dating to the 1980s at a time when global climate change was not seen as a significant issue by the general public.[40] In the January 12, 2006, issue of *Nature*,[41] it was acknowledged that the Global Amphibian Assessment (www.globalamphibians.org) found that approximately one third of amphibian species are classified as threatened, and the group has linked many of the declines to human activities, including climate change. In this capacity, amphibians are seen as sentinel species for environmental disruptions that may eventually impact human populations.

Amphibian sensitivity to climate change can, in many ways, be attributed to their utilization of the skin as a surface for osmotic, ionic, and respiratory gas exchange with their environment. The rapid decline of harlequin frogs in Central America has been attributed to global warming, which favors a chytrid fungus that infects their delicate skin.[388] In addition, changes in rainfall patterns may extend the duration of droughts beyond the capacity of seasonal species to survive, and rising temperatures will increase the rate of evaporation across the water-permeable skin and limit terrestrial activity. Anticipated increases in sea levels will project brackish water farther into freshwater drainages, thus exceeding the salinity tolerances of eggs, larvae, or adults. In addition, increased ultraviolet-B radiation as a result of stratospheric ozone depletion may cause DNA damage to amphibian eggs and embryos, leading to malformations and increased mortality rates.

The adaptations discussed in this chapter are the results of gradual climate changes and evolution of remarkable specializations for aquatic and terrestrial species. The study of physiological processes and behavior associated with these specializations will provide insights into the capacity of amphibians to adapt to rapid changes and assist conservation efforts to preserve and restore populations.

REFERENCES

1. Abrami, L., Gobin, R., Berthonaud, V., Thanh, H. L. C. J., Ripoche, P., and Beerbavatz, J. M., Localization of the FA-CHIP water channel in frog urinary bladder, *Eur. J. Cell Biol.*, 73, 215–221, 1994.
2. Acher, R., Chauvet, J., and Rouille, A., Adaptive evolution of water homeostasis regulation in amphibians: vasotocin and hydrins, *Biol. Cell*, 89, 283–291, 1997.
3. Ackrill, P., Hornby, R., and Thomas, S., Responses of *Rana temporaria* and *Rana esculenta* to prolonged exposure to a saline environment, *Comp. Biochem. Physiol.*, 38, 1317–1329, 1969.
4. Adolph, E. F., The excretion of water by the kidneys of frogs, *Am. J. Physiol.*, 81, 315–324, 1927.
5. Adolph, E. F., The vapor tension relations of frogs, *Biol. Bull.*, 62, 112–125, 1932.
6. Akabane, G., Ogushi, Y., Hasegawa, T., Suzuki, M., and Tanaka, S., Gene cloning and expression of an aquaporin (AQP-h3BL) in the basolateral membrane of water-permeable epithelial cells in osmoregulatory organs of the tree frog, *Am. J. Physiol.*, 292, R2340–R2351, 2007.
7. Alvarado, R. H., The effects of dehydration on water and electrolytes in *Ambystoma tigrinum*, *Physiol. Zool.*, 45, 43–53, 1972.
8. Alvarado, R. H. and Moody, A., Sodium and chloride transport in tadpoles of the bullfrog *Rana catesbeiana*, *Am. J. Physiol.*, 218, 1510–1516, 1970.
9. Alvarado, R. H., Dietz, T. H., and Mullen, T. L., Chloride transport across isolated skin of *Rana pipiens*, *Am. J. Physiol.*, 229, 869–876, 1975.
10. Ammar, A., Rajerison, R. M., Roseau, S., Bloch-Faure, M., and Butlen, D., Frog glomerular vasotocin receptors resemble mammalian V1b receptors, *Am. J. Physiol.*, 267, R1198–R1208, 1994.
11. Amstrup, J., Frøslev, J., Willumsen, N. J., Møbjerg, N., Jespersen, A., and Larsen, E. H., Expression of cystic fibrosis transmembrane conductance regulator in the skin of the toad, *Bufo bufo*: possible role for Cl⁻ transport across the heterocellular epithelium, *Comp. Biochem. Physiol. A*, 130 539–550, 2001.
12. Anagnostopoulos, T. and Planelles, G., Cell and luminal activities of chloride, potassium, sodium and protons in the late distal tubule of *Necturus* kidney, *J. Physiol.*, 393, 73–89, 1987.
13. Anatharam, A. and Palmer, L. G., Determination of epithelial Na⁺ channel subunit stoichiometry from single channel conductance, *J. Gen. Physiol.*, 130, 55–70, 2007.
14. Andersen, H. K., Urbach, V., Van Kerkhove, E., Prosser, E., and Harvey, B. J., Maxi K⁺ channels in the basolateral membrane of the exocrine frog skin gland regulated by intracellular calcium and pH, *Pflügers Arch.*, 431, 52–65, 1995.
15. Baker, C. A. and Hillyard, S. D., Capacitance, short-circuit current and osmotic water flow across different regions of the isolated toad skin, *J. Comp. Physiol. B*, 162, 707–713, 1992.
16. Balinsky, J. B., Adaptation of nitrogen metabolism to hyperosmotic environment in Amphibia, *J. Exp. Zool.*, 215, 335–350, 1981.
17. Balinsky, J. B. and Baldwin, E., The mode of excretion of ammonia and urea in *Xenopus laevis*, *J. Exp. Biol.*, 38, 695–705, 1961.
18. Bargmann, W. and Welch, U., Über Kanalchenzellen und dunkle Zellen im Nephron von Anuren, *Z. Zellforschung und mikroskopische Anat.*, 114, 193–204, 1972.

19. Baustian, M., The contribution of lymphatic pathways during recovery from hemorrhage in the toad, *Bufo marinus*, *Physiol. Zool.*, 61, 555–563, 1988.

20. Bentley, P. J., The effects of neurohypophyseal extracts on water transfer across the wall of the isolated urinary bladder of the toad *Bufo marinus*, *J. Endocrinol.*, 17, 201–209, 1958.

21. Bentley, P. J., Adaptations of Amphibia to arid environments, *Science*, 152, 619–623, 1966.

22. Bentley, P. J., Neurohypophyseal hormones in Amphibia: a comparison of their actions and storage, *Gen. Comp. Endocrinol.*, 13, 39–44, 1969.

23. Bentley, P. J., Actions of neurohypophyseal peptides in amphibians, reptiles and birds. In *Handbook of Physiology*. Vol. IV. *Endocrinology*, Geiger, S.R., Ed., American Physiological Society, Washington D.C., 1974, pp. 545–563.

24. Bentley, P. J., The electrical P.D. across the integument of some neotenous urodele amphibians, *Comp. Biochem. Physiol. A*, 50, 639–643, 1975.

25. Bentley, P. J., The Amphibia. In *Endocrines and Osmoregulation: A Comparative Account in Vertebrates*, Springer-Verlag, Berlin, 2002.

26. Bentley, P. J. and Heller, H., The action of neurohypophysial hormones on the water and sodium metabolism of urodele amphibians, *J. Physiol.*, 171, 434–453, 1964.

27. Bentley, P. J. and Heller, H., The water retaining action of vasotocin on the fire salamander (*Salamandra maculosa*): the role of the urinary bladder, *J. Physiol.*, 181, 124–129, 1965.

28. Bentley, P. J. and Main, A. R., Zonal differences in permeability of the skin of some anuran amphibia, *Am. J. Physiol.*, 223, 361–363, 1972.

29. Bentley, P. J. and Yorio, T., The passive permeability of the skin of anuran amphibia, *J. Physiol.*, 261, 603–615, 1976.

30. Bentley, P. J. and Yorio, T., The permeability of the skin of a neotenous urodele amphibian, the mudpuppy *Necturus maculosus*, *J. Physiol.*, 265, 537–547, 1977.

31. Bentley, P. J. and Yorio, T., Do frogs drink?, *J. Exp. Biol.*, 79, 41–46, 1979.

32. Bentley, P. J. and Baldwin, G. F., Comparison of transcutaneous permeability in skins from larval and adult salamanders (*Ambystoma tigrinum*), *Am. J. Physiol.*, 239, R505–R508, 1980.

33. Berman, D. M., Soria, M. O., and Coviello, A., Reversed short-circuit current across the isolated skin of the toad *Bufo arenarum*, *Pflügers Arch.*, 409, 616–619, 1987.

34. Biber, T. U. L. and Curran, P. F., Direct measurement of the uptake of sodium at the outer surface of the frog skin, *J. Gen. Physiol.*, 56, 83–99, 1970.

35. Biber, T. U. L., Walker, T. C., and Mullen, T. L., Influence of extracellular Cl concentration on Cl transport across isolated skin of *Rana pipiens*, *J. Membr. Biol.*, 56, 81–92, 1980.

36. Biber, T. U. L., Drewnowska, K., Baumgarten, C. M., and Fisher, R. S., Intracellular Cl⁻ activity changes of frog skin, *Am. J. Physiol.*, 249, F432–F438, 1985.

37. Biemesderfer, D., Stanton, B., Wade, J. B., Kashgarian, M., and Giebisch, G., Ultrastructure of *Amphiuma* distal nephron: evidence for cellular heterogeneity, *Am. J. Physiol.*, 256, 849–857, 1989.

38. Bishop, W. R., Mumbach, M. W., and Scheer, B. T., Interrenal control of sodium transport across frog skin, *Am. J. Physiol.*, 200, 451–453, 1961.

39. Bjerregaard, H. F. and Nielsen, R., Prostaglandin E$_2$-stimulated glandular ion and water secretion in isolated frog skin (*Rana esculenta*), *J. Membr. Biol.*, 97, 9–19, 1987.

40. Blaustein, A. R. and Wake, D. B., The puzzle of declining amphibian populations, *Sci. Am.*, 272, 52–57, 1995.

41. Blaustein, A. R., and Dobson, A., A message from the frogs, *Nature*, 439, 143–144, 2006.

42. Blazer-Yost, B. L., Liu, X., and Helman, S. I., Hormonal regulation of ENaCs: insulin and aldosterone, *Am. J. Physiol.*, 274, C1373–C1379, 1998.

43. Bott, P. A., Micropuncture study of renal excretion of water, K, Na, and Cl in *Necturus*, *Am. J. Physiol.*, 203, 662–666, 1962.

44. Boutilier, R. G., Stiffler, D. F., and Toews, D. P., Exchange of respiratory gasses, ions, and water in amphibious and aquatic amphibians. In *Environmental Physiology of the Amphibia*, Feder, M. E. and Burggren, W. W., Eds., University of Chicago Press, 1992, pp. 81–124.

45. Boutilier, R. G., Randal, D. J., Shelton, D. P., and Toews, D. P., Acid–base relationships in blood of the toad *Bufo marinus*, *J. Exp. Biol.*, 82, 331–334, 1979.

46. Bradford, D. F., Water and osmotic balance in overwintering tadpoles and frogs, *Rana muscosa*, *Physiol. Zool.*, 57, 474–480, 1984.

47. Brauer, A., Beiträge zur Kenntniss der Entwicklung und Anatomie der Gymnophionen. III. Die Entwicklung der Excretionsorgane, *Zool. Jahrb. Abt. Anat. Ontog. Tiere*, 16, 1–176, 1902.

48. Brekke, D. R., Hillyard, S. D., and Winokur, R. M., Behavior associated with the water absorption response by the toad, *Bufo punctatus*, *Copeia*, 1991, 393–401, 1991.

49. Brodin, B., Rytved, K. A., and Nielsen, R., An increase in [Ca²⁺]ᵢ activates basolateral chloride channels and inhibits apical sodium channels in frog skin epithelium, *Pflügers Arch. Eur. J. Physiol.*, 433, 16–25, 1996.

50. Brown, D., Membrane recycling and epithelial cell function, *Am. J. Physiol.*, 256, F1–F12, 1989.

51. Brown, D., Gluck, S., and Hartwig, J., Structure of the novel membrane coating material in proton-secreting epithelial cells and identification as an H⁺ ATPase, *J. Cell Biol.*, 105, 1637–1648, 1987.

52. Brown, G. W., Brown, W. R., and Cohen, P. P., Comparative biochemistry of urea synthesis. II. Levels of urea cycle enzymes in metamorphosing *Rana catesbeiana* tadpoles, *J. Biol. Chem.*, 234, 1775–1780, 1959.

53. Brown, P. S. and Frye, B. E., Effects of prolactin and growth hormone on growth and metamorphosis of tadpoles of the frog, *Rana pipiens*, *Gen. Comp. Endocrinol.*, 13, 126–138, 1969.

54. Brown, P. S. and Brown, S. C., Water balance responses to dehydration and neurohypophyseal peptides in the newt, *Notophthalmus viridescens*, *Gen. Comp. Endocrinol.*, 31, 189–201, 1977.

55. Brown, P. S. and Brown, S. C., Effects of hypophysectomy and prolactin on the water balance response of the newt, *Taricha torosa*, *Gen. Comp. Endocrinol.*, 46, 7–12, 1982.

56. Brown, S. C. and Brown, P. S., Water balance in the California newt, *Taricha torosa*, *Am. J. Physiol.*, 238, R113–R118, 1980.

57. Brunn, F., Beitrag zur Kenntnis der Wirkung von Hypophysenextracten auf den Wasserhaushalt des Frosches, *Z. für die gesammte experimentelle Medizin*, 25, 120–125, 1921.

58. Bruus, K., Kristensen, P., and Larsen, E. H., Pathways for chloride and sodium transport across toad skin, *Acta Physiol. Scand.*, 97, 31–47, 1976.

59. Budtz, P. E. and Larsen, L. O., Structure of the toad epidermis during the moulting cycle. I. Light microscopic observations in *Bufo bufo* (L.), *Z. für Zellforschung*, 144, 353–368, 1973.

60. Budtz, P. E. and Larsen, L. O., Structure of the toad epidermis during the moulting cycle, *Cell Tissue Res.*, 159, 459–483, 1975.

61. Budtz, P. E., Christoffersen, B. C., Johansen, J. S., Spies, I., and Willumsen, N. J., Tissue kinetics, ion transport and recruitment of mitochondria-rich cells in the skin of the toad (*Bufo bufo*) in response to exposure to distilled water, *Cell Tissue Res.*, 280, 65–75, 1995.

62. Burnstock, G., O'Shay, J., and Wood, M., Comparative physiology of the vertebrate autonomic nervous system. I. Innervation of the urinary bladder of the toad (*Bufo marinus*), *J. Exp. Biol.*, 40, 403–419, 1963.

63. Busk, M., Larsen, E. H., and Jensen, F. B., Acid–base regulation in tadpoles of *Rana catesbeiana* exposed to environmental hypercapnia, *J. Exp. Biol.*, 200, 2507–2512, 1997.

64. Callery, E. M. and Elinson, R. P., Developmental regulation of the urea-cycle enzyme arginase in the direct developing frog *Eleutherodactylus coqui*, *J. Exp. Zool.*, 275, 61–66, 1976.

65. Canessa, C. M., Horisberger, J. D., and Rossier, B. C., Epithelial sodium channel related to proteins involved in neurodegeneration, *Nature*, 361, 467–470, 1993.

66. Canessa, C. M., Schild, L., Buell, G., Thorens, B., Gautschi, I., Horisberger, J. D., and Rossier, B. C., Amiloride-sensitive Na⁺ channel is made of three homologous subunits, *Nature*, 367, 463–467, 1994.

67. Carroll, R. L., The origin and early radiation of terrestrial vertebrates, *J. Paleontol.*, 75, 1202–1215, 2001.

68. Carter, D. B., Structure and function of the subcutaneous lymph sacs in the Anura, *J. Herpetol.*, 13, 321–327, 1979.

69. Cartledge, V. A., Withers, P. C., McMaster, K. A., Thompson, G. G., and Bradshaw, S. D., Water balance of field-excavated aestivating Australian desert frogs, the cocoon-forming *Neobatrachus aquilonius* and the non-cocooning *Notaden nichollsi* (Amphibia: Myobatrachidae), *J. Exp. Biol.*, 209, 3309–3321, 2006.

70. Carvalho, E. T. C. and Junqueira, L. C. U., Histology of the kidney and urinary bladder of *Siphonops annulatus* (Amphibia-Gymnophiona), *Arch. Histol. Cytol.*, 62, 39–45, 1999.

71. Chauvet, J., Rouille, A., Ouedraogo, Y., and Acher, R., Adaptive differential processing of provasotocin in amphibians: occurrence of hydrin 2 (vasotocinyl-glycine) in Anura but not in Urodela, *C. R. Acad. Sci. Paris*, 313(Series III), 351–358, 1991.

72. Chevalier, J., Bourguet, J., and Hugon, J. S., Membrane associated particles: distribution in frog urinary bladder epithelium at rest and after oxytocin treatment, *Cell Tissue Res.*, 152, 129–140, 1974.

73. Choi, J. K., The fine structure of the urinary bladder of the toad, *Bufo marinus*, *J. Cell Biol.*, 16, 53–72, 1963.

74. Christensen, A. K., The structure of the functional pronephros in larvae of *Ambystoma opacum* as studied by light and electron microscopy, *Am. J. Anat.*, 115, 257–278, 1964.

75. Christensen, C. U., Adaptation in the water economy of some anuran amphibia, *Comp. Biochem. Physiol.*, 47A, 1035–1049, 1974.

76. Churchill, E. D., Nakazawa, F., and Drinker, C. D., The circulation of body fluids in the frog, *J. Physiol.*, 63, 304–308, 1927.

77. Civan, M. M., Peterson-Yatorno, K., DiBona, D. R., Wilson, D. F., and Erecinska, M., Bioenergetics of Na transport across frog skin, *Am. J. Physiol.*, 245, F691–F700, 1983.

78. Clack, J. A., *Gaining Ground: The Origin and Evolution of Tetrapods*, Indiana University Press, Bloomington, 2002.

79. Clothier, R. H., Worley, R. T. S., and Balls, M., The structure and ultrastructure of the renal tubule of the urodele amphibian, *Amphiuma means*, *J. Anat.*, 127, 491–504, 1978.

80. Cohen, N. W., Comparative rates of dehydration and hydration in some California salamanders, *Ecology*, 33, 462–479, 1952.

81. Conklin, R. E., The formation and circulation of lymph in the frog. I. The rate of lymph production, *Am. J. Physiol.*, 95, 79–90, 1930.

82. Conklin, R. E., The formation and circulation of lymph in the frog. II. Blood volume and pressure, *Am. J. Physiol.*, 95, 91–97, 1930.

83. Conklin, R. E., The formation and circulation of lymph in the frog. III. The permeability of the capillaries to protein, *Am. J. Physiol.*, 95, 98–110, 1930.

84. Connell, J. M. C. and Davies, E., The new biology of aldosterone, *J. Endocrinol.*, 186, 1–20, 2005.

85. Conway, E. J., Nature and significance of concentration relations of potassium and sodium ions in skeletal muscle, *Physiol. Rev.*, 37, 84–132, 1957.

86. Cook, D. I. and Young, J. A., Effect of K^+ channels in the apical plasma membrane on epithelial secretion based on secondary active Cl^- transport, *J. Membr. Biol.*, 110, 139–146, 1989.

87. Cooper, G. J. and Hunter, M., Na^+-H^+ exchange in frog early distal tubule: effect of aldosterone on the set point, *J. Physiol.*, 479, 423–432, 1994.

88. Cox, M. and Singer, I., Insulin-mediated Na^+ transport in the toad urinary bladder, *Am. J. Physiol.*, 232, F270–F277, 1977.

89. Crabbé, J. and DeWeer, P., Action of aldosterone on the bladder and skin of the toad, *Nature*, 202, 298–299, 1964.

90. Curran, P. F., Na, Cl, and water transport by rat ileum *in vitro*, *J. Gen. Physiol.*, 43, 1137–1148, 1960.

91. Czopek, J., Quantitative studies on the morphology of respiratory surfaces in amphibians, *Acta Anat.*, 62, 296–323, 1965.

92. Dantzler, W. H., Comparative aspects of renal function. In *The Kidney: Physiology and Pathophysiology*, 2nd ed., Seldin, D. W. and Giebisch, G., Eds., Raven Press, New York, 1992, pp. 885–942.

93. Davy, J., An account of the urinary organs and the urine of two species of the genus *Rana*, *Philos. Trans. R. Soc. Lond.*, 111, 95–100, 1821.

94. de Brito-Gitirana, L. and Azevedo, R. A., Morphology of *Bufo ictericus* integument (Amphibia, Bufonidae), *Micron*, 36, 532–538, 2005.

95. De Ruyter, M. L. and Stiffler, D. F., Interrenal function in larval *Ambystoma tigrinum*. II. Control of aldosterone secretion and electrolyte balance by ACTH, *Gen. Comp. Endocrinol.*, 62, 298–305, 1985.

96. De Wolf, I., Van Driessche, W., and Nagel, W., Forskolin activates gated Cl^- channels in frog skin, *Am. J. Physiol.*, 256, C1239–C1249, 1989.

97. Degani, G., Silanikov, N., and Shkolnik, A., Adaptation of green toad (*Bufo viridis*) to terrestrial life by urea accumulation, *Comp. Biochem. Physiol.*, 77A, 585–587, 1984.

98. Delaney, R. and Stoner, L. C., Miniature Ag-AgCl electrode for voltage clamping of the *Ambystoma* collecting duct, *J. Membr. Biol.*, 64, 45–53, 1982.

99. Delson, J. and Whitford, W. G., Adaptation of the tiger salamander, *Ambystoma tigrinum*, to arid habitats, *Comp. Biochem. Physiol.*, 46A, 631–638, 1973.

100. DeSimone, J. A., Heck, G. L., Miersen, S., and DeSimone, S. K., The active ion transport properties of canine lingual epithelium *in vitro*, *J. Gen. Physiol.*, 83, 633–656, 1984.

101. DeSousa, R. C. and Grosso, A., Osmotic water flow across the abdominal skin of the toad, *Bufo marinus*: effect of vasopressin and isoprenaline, *J. Physiol.*, 329, 281–296, 1982.

102. Devuyst, O., Beaujean, V., and Crabbé, J., Effect of environmental conditions on mitochondria-rich cell density and chloride transport in toad skin, *Pflügers Arch.*, 417, 577–581, 1991.

103. Diamond, J. D. and Bossert, W. H., Standing gradient osmotic flow: a mechanism for coupling of water and solute transport in epithelia, *J. Gen. Physiol.*, 50, 2062–2083, 1967.

104. Dietl, P. and Stanton, B. A., The amphibian distal nephron. In *New Insights in Vertebrate Kidney Function*, Brown, J. A., Balment, R. J., and Rankin, J. C., Eds., Cambridge University Press, Cambridge, U.K., 1993, pp. 115–134.

105. Dimmitt, M. and Ruibal, R., Exploitation of food resources by spadefoot toads (*Scaphiopus*), *Copeia*, 1980, 854–862, 1980.

106. Dörge, A., Rick, R., Beck, F., and Thurau, K., Cl transport across the basolateral membrane in frog skin epithelium, *Pflügers Arch.*, 405(Suppl. 1), S8–S11, 1985.

107. Du Bois-Reymond, E., *Undersuchungen über Thierische Electricität (Book 1)*, Georg Reimer, Berlin, 1848.

108. Du Bois-Reymond, E., *Untersuchungen über Thierische Electricität (Book 2)*, Georg Reimer, Berlin, 1884.

109. Duffy, M. E., Kelepouris, E., Peterson-Yan-Torno, K., and Civan, M. M., Microelectrode study of intracellular pH in frog skin: dependence on serosal pH, *Am. J. Physiol.*, 251, F468–F474, 1986.

110. Duranti, E., Ehrenfeld, J., and Harvey, B. J., Acid secretion through the *Rana esculenta* skin: involvement of an anion-exchange mechanism at the basolateral membrane, *J. Physiol.*, 378, 195–211, 1986.

111. Durig, A., Wassergehalt und Organfunktion, *Pflügers Arch.*, 85, 401–504, 1901.

112. Dutrochet, H., *Endosmosis*. In *The Cyclopaedia of Anatomy and Physiology*, Sherwood, Gilbert, Piper, London, 1839.

113. Ehrenfeld, J. and Garcia-Romeu, F., Coupling between chloride absorption and base excretion in isolated skin of *Rana esculenta*, *Am. J. Physiol.*, 235, F33–F39, 1978.

114. Ehrenfeld, J. and Harvey, B. J., The key role of the mitochondria-rich cells in Na^+ and H^+ transport across the frog skin epithelium, *Pflügers Arch.*, 414, 59–67, 1989.

115. Ehrenfeld, J. and Klein, U., The key role of the H^+ V-ATPase in acid–base balance and Na^+ transport processes in frog skin, *J. Exp. Biol.*, 200, 247–256, 1997.

116. Ehrenfeld, J., Masoni, A., and Garcia-Romeu, F., Mitochondria-rich cells of frog skin in transport mechanisms: morphological and kinetic studies of transepithelial excretion of methylene blue, *Am. J. Physiol.*, 231, 120–126, 1976.

117. Ehrenfeld, J., Garcia-Romeu, F., and Harvey, B. J., Electrogenic active proton pump in *Rana esculenta* skin and its role in sodium transport, *J. Physiol.*, 359, 331–355, 1985.

118. Eid, S. R. and Brandli, A. W., *Xenopus* Na,K-ATPase: primary sequence of the beta2 subunit and *in situ* localization if alpha1, beta1 and gamma expression during pronephric kidney development, *Differentiation*, 68, 115–125, 2001.

119. Eid, S. R., Terrettaz, A., Nagata, K., and Brandli, A. W., Embryonic expression of *Xenopus* SGLT-1L, a novel member of the solute carrier family 5 (SLC5), is confined to tubules of the pronephric kidney, *Int. J. Dev. Biol.*, 46, 177–184, 2002.

120. Emilio, M. G., Machado, M. M., and Menano, H. P., The production of a hydrogen ion gradient across the isolated frog skin: quantitative aspects and the effect of acetazolamide, *Biochim. Biophys. Acta*, 203, 394–409, 1970.

121. Engelhardt, J. F., Smith, S. S., Allen, E., Yankaskas, J. R., Dawson, D. C., and Wilson, J. M., Coupled secretion of chloride and mucus in skin of *Xenopus laevis*: possible role for CFTR, *Am. J. Physiol.*, 267, C491–C500, 1994.

122. Erlij, D. and Martinez-Palomo, A., Opening of tight junctions in frog skin by hypertonic urea solutions, *J. Membr. Biol.*, 9, 229–240, 1972.

123. Eskesen, K. and Ussing, H. H., Determination of the electromotive force of active sodium transport in frog skin epithelium (*Rana temporaria*) from presteady-state flux ratio experiments, *J. Membr. Biol.*, 86, 105–111, 1984.

124. Etkin, W. and Gona, A. G., Antagonism between prolactin and thyroid hormone in amphibian development, *J. Exp. Zool.*, 165, 249–268, 1967.

125. Ewer, R. F., The effect of pituitrin on fluid distribution in *Bufo regularis* Reuss, *J. Exp. Biol.*, 29, 173–177, 1952.

126. Farias, A., Hermida, G. N., and Fiorito, L. E., Structure of the kidney of *Bufo arenarum*: intermediate segment, distal tubule and collecting tubule, *Biocell*, 27, 19–28, 2003.

127. Farias, A., FiorFarias, A., Fiorito, L. E., and Hermida, G. N., Structure of the *Bufo arenarum* kidney: renal corpuscle, neck segment and proximal tubule, *Biocell*, 22, 187–196, 1998.

128. Farquhar, M. G. and Palade, G. E., Cell junctions in amphibian skin, *J. Cell Biol.*, 26, 263–291, 1964.

129. Fenoglio, C., Vaccarone, R., Chiari, P., and Gervaso, M. V., An ultrastructural and cytochemical study of the mesonephros of *Rana esculenta* during activity and hibernation, *Eur. J. Morphol.*, 34, 107–121, 1996.

130. Ferreira, K. T. G. and Ferreira, H. G., The regulation of volume and ion composition in frog skin, *Biochim. Biophys. Acta*, 646, 193–202, 1981.

131. Finkelstein, A., *Water Movement Through Lipid Bilayers, Pores and Plasma Membranes: Theory and Reality*, John Wiley & Sons, New York, 1987.

132. Fitzsimons, J. T., Angiotensin, thirst and sodium appetite, *Physiol. Rev.*, 78, 583–686, 1998.

133. Fleming, W. R., On the role of hydrogen ion and potassium ion in the active transport of sodium across the isolated frog skin, *J. Cell. Comp. Physiol.*, 49, 129–152, 1957.

134. Forster, R. P., Schmidt-Nielsen, B., and Goldstein, L., Relation of renal tubular transport of urea to its biosynthesis in metamorphosing tadpoles, *J. Cell. Comp. Physiol.*, 61, 293–244, 1963.

135. Foskett, J. K. and Ussing, H. H., Localization of chloride conductance to mitochondria-rich cells in frog skin epithelium, *J. Membr. Biol.*, 91, 252–258, 1986.

136. Foster, R. W., Adrenoceptors and salt and water movement in epithelia, *Naunyn-Schmiedeberg's Arch. Pharmacol.*, 281, 315–326, 1974.

137. Fox, H., The amphibian pronephros, *Q. Rev. Biol.*, 38, 1–25, 1963.

138. Fox, H., Tissue degeneration: an electron microscopic study of the pronephros of *Rana temporaria*, *J. Embryol. Exp. Morphol.*, 24, 139–157, 1970.

139. Fox, H., *Amphibian Morphogenesis*, Humana Press, Totowa, NJ, 1984.

140. Fox, H. and Hamilton, L., Ultrastructure of diploid and haploid cells of *Xenopus laevis* larvae, *J. Embryol. Exp. Morphol.*, 26, 81–98, 1971.

141. Frazier, L. W., Cellular changes in the toad urinary bladder in response to metabolic acidosis, *J. Membr. Biol.*, 40, 165–177, 1978.

142. Frazier, L. W. and Vanatta, J. C., Excretion of K^+ by frog with rate varying with K^+ load, *Comp. Biochem. Physiol. A*, 69, 157–160, 1981.

143. Frost, D. R., Grant, T., Faivovich, J., Bain, R. H., Haas, A., Hadad, C. F. B., De Sa, R. O., Channing, A., Wilkinson, M., Donnellan, S. C., Raxworthy, C. J., Campbell, J. A., Blotto, B. L., Moler, P., Drewes, R. C., Nussbaum, R. A., Lynch, J. D., Green, D. M., and Wheeler, W. C., The amphibian tree of life, *Bull. Am. Mus. Nat. Hist.*, 297, 1–370, 2006.

144. Fuchs, W., Gebhardt, U., and Lindemann, B., Delayed voltage responses to fast changes of $(Na)_o$ at the outer surface of frog skin epithelium, *Biomembranes*, 3, 483–498, 1972.

145. Fuchs, W., Larsen, E. H., and Lindemann, B., Current–voltage curve of sodium channels and concentration dependence of sodium permeability of frog skin, *J. Physiol.*, 267, 137–166, 1977.

146. Galeotti, G., Concerning the EMF which is generated at the surface of animal membranes in contact with different electrolytes, *Z. Phys. Chem.*, 49, 542–562, 1904.

147. Gallardo, R., Pang, P. K. T., and Sawyer, W. H., Neural influences on bullfrog renal functions, *Proc. Soc. Exp. Biol. Med.*, 165, 233–240, 1980.

148. Garcia-Diaz, J. F., Baxendale, L. M., Klemperer, G., and Essig, A., Cell K in frog skin in the presence and absence of cell current, *J. Membr. Biol.*, 85, 143–158, 1985.

149. Garcia Romeu, F., Salibian, A., and Pezzani-Hernandez, S., The nature of *in vivo* sodium and chloride uptake mechanisms through the epithelium of the Chilean frog, *Calyptocephallela gayi* (Dumeril and Bibron, 1841), *J. Gen. Physiol.*, 53, 616–835, 1969.

150. Garcia Romeu, F., Masoni, A., and Isaia, J., Active urea transport through isolated skins of frog and toad, *Am. J. Physiol.*, 241, R114–R123, 1981.

151. Garland, H. O. and Henderson, I. W., Influence of salinity on renal and adrenocortical function in the toad, *Bufo marinus*, *Gen. Comp. Endocrinol.*, 37, 136–143, 1975.

152. Garty, H. and Palmer, L. G., Epithelial sodium channels: structure, function, regulation, *Physiol. Rev.*, 77, 359–396, 1997.

153. Gealekman, O. M. and Warburg, M. R., Changes in numbers and dimensions of glomeruli during metamorphosis of *Pelobates syriacus* (Anura; Pelobatidae), *Eur. J. Morphol.*, 38, 80–87, 2000.

154. Gibley, C. W. and Chang, J. P., Fine structure of tubule cells in the functional pronephros of *Rana pipiens*, *Am. Zool.*, 6, 610, 1966.

155. Giraldez, F. and Ferreira, K. T. G., Intracellular chloride activity and membrane potential in stripped frog skin, *Biochim. Biophys. Acta*, 769, 625–628, 1984.

156. Gluck, S., Cannon, C., and Al-Awqati, Q., Exocytosis regulates urinary acidification in turtle bladder by rapid insertion of H+ pumps into the luminal membrane, *Proc. Natl. Acad. Sci. USA*, 79, 4327–4331, 1982.

157. Gomez-Maestre, I. and Tejedo, M., Local adaptation of a larval amphibian to osmotically stressful environments, *Evolution*, 57, 1889–1899, 2003.

158. Gomez-Maestre, I., Tejedo, M., Ramayo, E., and Estepa, J., Developmental alterations and osmoregulatory physiology of a larval anuran under osmotic stress, *Physiol. Biochem. Zool.*, 77, 267–274, 2004.

159. Gordon, M. S., Osmotic regulation in the green toad (*Bufo viridis*), *J. Exp. Biol.*, 39, 261–270, 1962.

160. Gordon, M. S. and Tucker, V. A., Further observations on the physiology of salinity adaptation in the crab-eating frog (*Rana cancrivora*), *J. Exp. Biol.*, 49, 185–193, 1968.

161. Gordon, M. S., Schmidt-Nielsen, K., and Kelly, H. M., Osmotic regulation in the crab-eating frog (*Rana cancrivora*), *J. Exp. Biol.*, 38, 659–678, 1961.

162. Grant, W. C. J. and Grant, J. A., Water drive studies on hypophysectomized efts of *Diemyctylus viridescens*. Part I. The role of the lactogenic hormone, *Biol. Bull.*, 114, 1–9, 1958.

163. Grant, W. C. J. and Cooper, G. I., Behavioral and integumentary changes associated with induced metamorphosis in *Diemyctilus*, *Biol. Bull.*, 129, 510–522, 1965.

164. Greenwald, L., Sodium balance in amphibians from different habitats, *Physiol. Zool.*, 45, 229–237, 1972.

165. Greger, R., Ion transport mechanisms in thick ascending limb of Henle's loop of mammalian nephron, *Physiol. Rev.*, 65, 760–797, 1985.

166. Grill, G., Granger, P., and Thurau, K., The renin-angiotensin system of amphibians. I. Determination of the renin content of amphibian kidneys, *Pflügers Arch.*, 331, 1–12, 1972.

167. Gross, M. L., Hanke, W., Koch, A., and Zeibart, H., Intraperitoneal protein injection in the axolotl: the amphibian nephron as a novel model to study tubulointerstitial activation, *Kidney Int.*, 62, 51–59, 2002.

168. Gudme, C. N., Nielsen, M. S., and Nielsen, R., Effect of α_1-adrenergic stimulation of Cl- secretion and signal transduction in exocrine glands (*Rana esculenta*), *Acta Physiol.*, 169, 173–182, 2000.

169. Guggino, W. B., Functional heterogeneity in the early distal tubule of the *Amphiuma* kidney: evidence for two modes of Cl- and K+ transport across the basolateral cell membrane, *Am. J. Physiol.*, 250, F430–F440, 1986.

170. Guggino, W. B., Oberleithner, H., and Giebisch, G., The amphibian diluting segment, *Am. J. Physiol.*, 254, F615–F627, 1988.

171. Guyton, A. C. and Hall, J. E., *Textbook of Medical Physiology*, Elsevier, Philadelphia, PA, 2006.

172. Harck, A. F. and Larsen, E. H., Concentration dependence of halide fluxes and selectivity of the anion pathway in toad skin, *Acta Physiol. Scand.*, 128, 289–304, 1986.

173. Harvey, B. J., Energization of sodium absorption by the H+-ATPase pump in mitochondria-rich cells in frog skin, *J. Exp. Biol.*, 172, 289–309, 1992.

174. Harvey, B. J., Cross talk between sodium and potassium channels in tight epithelia, *Kidney Int.*, 48, 1191–1199, 1995.

175. Harvey, B. J. and Kernan, R. B., Intracellular ion activities in frog skin in relation to external sodium and effects of amiloride and/or ouabain, *J. Physiol.*, 349, 501–517, 1984.

176. Harvey, B. J. and Ehrenfeld, J., Role of Na+/H+ exchange in control of intracellular pH and cell membrane conductances in frog skin epithelium, *J. Gen. Physiol.*, 92, 795–810, 1988.

177. Hasegawa, T., Suzuki, M., and Tanaka, S., Immunocytochemical studies on translocation of phosphorylated aquaporin-h2 protein in granular cells of the frog urinary bladder before and after stimulation with vasotocin, *Cell Tissue Res.*, 322, 407–415, 2005.

178. Hasegawa, T., Tanii, H., Suzuki, M., and Tanaka, S., Regulation of water absorption in the frog skins by two vasotocin-dependent water-channel aquaporins, AQP-h2 and AQP-h3, *Endocrinology*, 144, 4087–4096, 2003.

179. Hasegawa, Y., Watanabe, T. X., Sokabe, H., and Nakajima, T., Chemical structure of angiotensin in the bullfrog *Rana catesbeiana*, *Gen. Comp. Endocrinol.*, 50, 75–90, 1983.

180. Hayman, J. M. J., Estimations of afferent arteriole and glomerular capillary pressures in the frog kidney, *Am. J. Physiol.*, 79, 389–409, 1927.

181. Hays, R. M. and Leaf, A., Studies on the movement of water through the isolated toad bladder and its modification by vasopressin, *J. Gen. Physiol.*, 45, 905–919, 1962.

182. Heller, H., History of neurohypophyseal research. In *Handbook of Physiology*. Section 7. *Endocrinology*. Vol. IV. *The Pituitary Gland and Its Neuroendocrine Control, Part 1*, Knobil, E. and Sawyer, W. H., Eds., American Physiological Society, Washington D.C., 1974, pp. 103–117.

183. Helman, S. I. and Miller, D. A., Edge damage effect on measurements of urea and sodium in frog skin, *Am. J. Physiol.*, 228, 1198–1203, 1974.

184. Helman, S. I., Cox, T. C., and Van Driessche, W., Hormonal control of apical membrane Na transport in epithelia: studies with fluctuation analysis, *J. Gen. Physiol.*, 82, 201–220, 1983.

185. Herrera, F. C., Effect of insulin on short-circuit current across toad urinary bladder, *Am. J. Physiol.*, 209, 819–824, 1965.

186. Hevesy, G. V., Hofer, E., and Krogh, A., The permeability of the skin of frogs to water as determined by D_2O and H_2O, *Skand. Arch. Physiol.*, 72, 199–214, 1935.

187. Hillman, S. S., The effect of arginine vasopressin on water and sodium balance in the urodele amphibian *Aniedes lugubris*, *Gen. Comp. Endocrinol.*, 24, 74–82, 1974.

188. Hillman, S. S., Physiological correlates of differential dehydration tolerance in anurans, *Copeia*, 125–129, 1980.

189. Hillman, S. S. and Withers, P. C., The hemodynamic consequences of hemorrhage and hypernatremia in two amphibians, *J. Comp. Physiol. B*, 157, 807–812, 1988.

190. Hillman, S. S., Zygmunt, A. A., and Baustian, M., Transcapillary fluid forces during dehydration in two amphibians, *Physiol. Zool.*, 60, 339–345, 1987.

191. Hillman, S. S., Hedrick, M. S., Withers, P. C., and Drewes, R. C., Lymph pools in the basement, sump pumps in the attic: dilemma for lymph movement, *Physiol. Biochem. Zool.*, 77, 2004.

192. Hillyard, S. D., The movement of soil water across the isolated amphibian skin, *Copeia*, 1976, 314–320, 1976.

193. Hillyard, S. D., The effect of isoproterenol on the amphibian water balance response, *Comp. Biochem. Physiol.*, 62C, 93–95, 1979.

194. Hillyard, S. D. and Larsen, E. H., Lymph osmolality and rehydration from NaCl solutions by toads, *Bufo marinus*, *J. Comp. Physiol. B*, 171, 283–292, 2001.

195. Hillyard, S. D., Hoff, K., and Propper, C., The water absorption response: a behavioral assay for physiological processes in terrestrial amphibians, *Physiol. Zool.*, 71, 127–138, 1998.

196. Hillyard, S. D., Goldstein, J., Tuttle, W., and Hoff, K., Transcellular and paracellular elements of salt chemosensation in toad skin, *Chem. Senses*, 29, 755–762, 2004.

197. Hillyard, S. D., Viborg, A., Nagai, T., and Hoff, K. V., Chemosensory function of salt and water transport by the amphibian skin, *Comp. Biochem. Physiol.*, 148A, 44–54, 2007.

198. Hinton, D. E., Stoner, L. C., Burg, M., and Trump, B. F., Heterogeneity in the distal nephron of the salamander (*Ambystoma tigrinum*): a correlated structure function study of isolated tubule segments, *Anat. Rec.*, 204, 21–32, 1982.

199. Hodgkin, A. L. and Horowicz, P., The influence of potassium and chloride ions on the membrane potential of single muscle fibres, *J. Physiol.*, 148, 127–160, 1959.

200. Hoff, K. and Hillyard, S. D., Toads taste sodium with their skin: sensory function in a transporting epithelium, *J. Exp. Biol.*, 183, 347–351, 1993.

201. Hoff, K. and Hillyard, S. D., Inhibition of cutaneous water absorption in dehydrated toads by saralasin is associated with changes in barometric pressure, *Physiol. Zool.*, 66, 89–98, 1993.

202. Hoff, K. V. and Hillyard, S. D., Angiotensin II stimulates cutaneous drinking in the toad, *Bufo punctatus*, *Physiol. Zool.*, 64, 1165–1172, 1991.

203. Hoffman, J. and Katz, U., The ecological significance of burrowing behavior in the toad (*Bufo viridis*), *Oecologia*, 81, 510–513, 1989.

204. Hoffman, J. and Katz, U., Tissue osmolytes and body fluid compartments in the toad, *Bufo viridis*, under simulated terrestrial conditions, *J. Comp. Physiol. B*, 161, 433–439, 1991.

205. Hoffman, J., Katz, U., and Eylath, U., Urea accumulation in response to water restriction in burrowing toads (*Bufo viridis*), *Comp. Biochem. Physiol. A*, 97, 423–426, 1990.

206. Hoffmann, E. K., Schettino, T., and Marshall, W. S., The role of volume-sensitive ion transport systems in regulation of epithelial transport, *Comp. Biochem. Physiol. A*, 148, 29–43, 2007.

207. Horisberger, J., Hunter, M., Stanton, B., and Giebisch, G., The collecting tubule of *Amphiuma*. II. Effects of potassium adaptation, *Am. J. Physiol.*, 253, F1273–F1282, 1987.

208. Hoshi, T., Suzuki, Y., and Itoi, K., Differences in functional properties between the early and late segments of the distal tubule of amphibian (*Triturus*) kidney, *Jpn. J. Nephrol.*, 23, 889–896, 1981.

209. Howland, R. B., Experiments on the effect of removal of the pronephros of *Amblystoma punctatum*, *J. Exp. Zool.*, 32, 355–396, 1921.

210. Huang, H. and Brown, D. D., Prolactin is not a juvenile hormone in *Xenopus laevis* metamorphosis, *Proc. Natl. Acad. USA*, 97, 195–199, 2000.

211. Huf, E., Über den anteil vitaler kräfte bei der resorption von flüssigkeit durch die froschhaut, *Pflügers Arch.*, 236, 1–19, 1935.

212. Huf, E., Über aktiven wasser- und saltztransport durch die froshhaut, *Pflügers Arch.*, 237, 143–166, 1936.

213. Huf, E. G., Parrish, J., and Weatherford, C., Active salt and water uptake by isolated frog skin, *Am. J. Physiol.*, 167, 137–142, 1951.

214. Hunter, M., Patch clamp studies of the amphibian nephron, *Ren. Physiol. Biochem.*, 13, 94–111, 1990.

215. Hunter, M., Horisberger, J.-D., Stanton, B., and Giebisch, G., The collecting tubule of *Amphiuma* I. Electrophysiological characterization, *Am. J. Physiol.*, 253, F1263–F1272, 1987.

216. Ilic, V. and Brown, D., Modification in mitochondria-rich cells in different ionic conditions: changes in cell morphology and cell number in the skin of *Xenopus laevis*, *Anat. Rec.*, 196, 153–161, 1980.

217. Isayama, S., Über die stromung der lymphe bei den amphibian, *Z. Biol.*, 82, 90–100, 1924.

218. Isayama, S., Über die geschwindigheit des Flussigkeitsaustausches zwischen blut und gewebe, *Z. Biol.*, 82, 101–106, 1924.

219. Ishibashi, K., Kuwahara, M., and Sasaki, S., Molecular biology of aquaporins, *Physiol. Biochem. Pharmacol.*, 141, 1–32, 2000.

220. Jaffee, O. C., Morphogenesis of the pronephros of the leopard frog (*Rana pipiens*), *J. Morphol.*, 95, 109–123, 1954.

221. Jameson, D. L., Life history and phylogeny of the salientians, *Syst. Zool.*, 6, 75–78, 1957.

222. Jared, C., Navas, C. A., and Toledo, R. C., An appreciation of the physiology and morphology of the Caecelians (Amphibia: Gymnophiona), *Comp. Biochem. Physiol. A*, 123, 313–328, 1999.

223. Jensen, L. J., Willumsen, N. J., and Larsen, E. H., Proton pump activity is required for active uptake of chloride in isolated amphibian skin exposed to freshwater, *J. Comp. Physiol. B*, 172, 503–511, 2002.

224. Jensen, L. J., Sørensen, J. N., Larsen, E. H., and Willumsen, N. J., Proton pump activity of mitochondria-rich cells: the interpretation of external proton concentration gradients, *J. Gen. Physiol.*, 109, 73–91, 1997.

225. Jensen, L. J., Willumsen, N. J., Amstrup, J., and Larsen, E. H., Proton pump-driven cutaneous chloride uptake in anuran amphibia, *Biochim. Biophys. Acta*, 1618, 120–132, 2003.

226. Johnsen, A. and Nielsen, R., Effects of antidiuretic hormone, arginine vasotocin, theophylline, filipin, A23187, on cyclic AMP in isolated frog skin epithelium (*Rana temporaria*), *Acta Physiol. Scand.*, 102, 281–289, 1978.

227. Johnsen, A. and Nielsen, R., Correlation between c-AMP in isolated frog skin epithelium and stimulation of sodium transport and osmotic water flow by antidiuretic hormone and phosphodiesterase inhibitors, *Gen. Comp. Endocrinol.*, 54, 144–153, 1984.

228. Johnson, W. E. and Propper, C. R., Effects of dehydration on plasma osmolality, thirst-related behavior, and plasma and brain angiotensin concentrations in Couch's spadefoot toad, *Scaphiopus couchi*, *J. Exp. Zool.*, 286, 572–584, 2000.

229. Johnston, J. W. and Jungreis, A. W., Urea production in *Rana pipiens*: effects of aldosterone on urea cycle enzymes, *J. Comp. Physiol.*, 134, 359–366, 1979.

230. Jolivet-Jaudet, G., Inoue, M., Takada, K., and Ishii, S., Circannual changes in plasma aldosterone levels in *Bufo japonicus formosus*, *Gen. Comp. Endocrinol.*, 53, 163–167, 1984.

231. Jones, R. M. and Hillman, S. S., Salinity adaptation in the salamander *Batrachoseps*, *J. Exp. Biol.*, 76, 1–10, 1978.

232. Jørgensen, C. B., On the influence of the neurohypophyseal principles on the sodium metabolism in the axolotl (*Amblystoma mexicanum*), *Acta Physiol. Scand.*, 12, 350–371, 1946.

233. Jørgensen, C. B., Permeability of the amphibian skin. II. Effect of moulting of the skin of anurans on the permeability to water and electrolytes, *Acta Physiol. Scand.*, 18, 171–180, 1949.

234. Jørgensen, C. B., Osmotic regulation in the frog, *Rana esculenta* (L.) at low temperature, *Acta Physiol. Scand.*, 20, 46–55, 1950.

235. Jørgensen, C. B., Water economy in the life of a terrestrial anuran, the toad, *Bufo bufo*, *Biologiske Skrifter Det Kongelige Danske Videnskaberenes Selskab*, 39, 1–30, 1991.

236. Jørgensen, C. B., Water and salt balance at low temperature in a cold temperate zone anuran, the toad, *Bufo bufo*, *Comp. Biochem. Physiol. A*, 100, 377–384, 1991.

237. Jørgensen, C. B., Relationships between feeding, digestion and water balance in a carnivorous vertebrate, the toad *Bufo bufo*, *Comp. Biochem. Physiol. A*, 101, 157–160, 1992.

238. Jørgensen, C. B., Role of pars nervosa of the hypophysis in amphibian water economy: a reassessment, *Comp. Biochem. Physiol. A*, 104, 1–21, 1993.

239. Jørgensen, C. B., Water economy of a terrestrial toad (*Bufo bufo*) with special reference to cutaneous drinking and urinary bladder function, *Comp. Biochem. Physiol. A*, 109, 325–334, 1994.

240. Jørgensen, C. B., 200 Years of amphibian water economy: from Robert Townson to the present, *Biol. Rev.*, 72, 153–237, 1997.

241. Jørgensen, C. B., Urea and amphibian water economy, *Comp. Biochem. Physiol. A*, 117, 161–170, 1997.

242. Jørgensen, C. B., Amphibian respiration and olfaction and their relationships: from Robert Townson (1794) to the present, *Biol. Rev.*, 75, 297–345, 2000.

243. Jørgensen, C. B., Levi, H., and Zerahn, K., On active uptake of sodium and chloride ions in anurans, *Acta Physiol. Scand.*, 30, 178–190, 1954.

244. Jørgensen, C. B., Rosenkilde, P., and Wingstrand, K. G., Role of the preoptic-neurohypophyseal system in the water economy of the toad, *Bufo bufo* L., *Gen. Comp. Endocrinol.*, 12, 91–98, 1969.

245. Jørgensen, C. B., Brems, K., and Geckler, P., Volume and osmotic regulation in the toad, *Bufo bufo bufo* (L.) at low temperature with special reference to amphibian hibernation. In *Osmotic and Volume Regulation: Proceedings of the Alfred Benzon Symposium XI*, Jørgensen, C. B. and Skadhauge, E., Eds., Munksgaard, Copenhagen, Denmark, 1978, pp. 62–74.

246. Kachadorian, W. A., Wade, J. B., and Di Scala, V. A., Vasopressin induced structural change in toad bladder luminal membrane, *Science*, 190, 67–69, 1975.

247. Kalman, S. M. and Ussing, H. H., Active sodium uptake by the toad and its response to antidiuretic hormone, *J. Gen. Physiol.*, 38, 361–370, 1955.

248. Kampmeier, O. F., *Evolution and Comparative Morphology of the Lymphatic System*, Charles C Thomas, Springfield, IL, 1969.

249. Katsoyannis, P. G. and du Vigneaud, V., Active principles of the neurohypophysis of cold-blooded vertebrates: arginine vasotocin, *Nature*, 184, 1465, 1959.

250. Katz, U., Studies on the adaptation of the toad *Bufo viridis* to high salinities: oxygen consumption, plasma concentration and water content of the tissues, *J. Exp. Biol.*, 58, 785–796, 1973.

251. Katz, U., NaCl adaptation in *Rana ridibunda* and a comparison with the European euryhaline toad *Bufo viridis*, *J. Exp. Biol.*, 63, 763–773, 1975.

252. Katz, U. and Ben-Sasson, Y., A possible role of the kidney and urinary bladder in urea conservation of *Bufo viridis* under high salt acclimation, *J. Exp. Biol.*, 109, 373–377, 1984.

253. Katz, U. and Larsen, E. H., Chloride transport in toad skin (*Bufo viridis*): the effect of salt adaptation, *J. Exp. Biol.*, 109, 353–371, 1984.

254. Katz, U. and Gabbay, S., Water retention and plasma and urine composition in toads (*Bufo viridis*, Laur.) under burrowing conditions, *J. Comp. Physiol. B*, 156, 735–740, 1986.

255. Katz, U. and Gabbay, S., Mitochondria-rich cells and carbonic anhydrase content of toad skin epithelium, *Cell Tissue Res.*, 251, 425–431, 1988.

256. Katz, U. and Gabbay, S., Band 3 protein and ion transfer across epithelia: mitochondria-rich cells in amphibian skin epithelium, *Funktionsanalyse Biologischer Systeme*, 23, 75–82, 1993.

257. Katz, U. and Hanke, Mechanisms of hyperosmotic acclimation in *Xenopus laevis* (salt, urea or mannitol), *J. Comp. Physiol.*, 163, 189–195, 1993.

258. Katz, U., Garcia Romeu, F., Masoni, A., and Isaia, J., Active transport of urea across the skin of the euryhaline toad, *Bufo viridis*, *Pflügers Arch.*, 390, 299–300, 1981.

259. Katz, U., Zaccone, G., Fasulo, S., Mauceri, A., and Gabbay, S., Lectin binding pattern and band 3 localization in the toad skin epithelium, and the effect of salt acclimation, *Biol. Cell*, 89, 141–152, 1997.

260. Katz, U., Rozman, A., Zaccone, G., Fasulo, S., and Gabbay, S., Mitochondria-rich cells in anuran amphibia: chloride conductance and regional distribution over the body surface, *Comp. Biochem. Physiol. A*, 125, 131–139, 2000.

261. Kellenberger, S. and Schild, L., Epithelial sodium channels/degenerin family of ion channels: a variety of functions for a shared structure, *Physiol. Rev.*, 82, 735–767, 2002.

262. Kirschner, L. B., On the mechanism of active sodium transport across the frog skin, *J. Cell. Comp. Physiol.*, 45, 61–87, 1955.

263. Kirschner, L. B., The study of NaCl transport in aquatic animals, *Am. Zool.*, 10, 365–376, 1970.

264. Kirschner, L. B., Kerstetter, T., and Porter, D., Adaptation of larval *Ambystoma tigrinum* to concentrated environments, *Am. J. Physiol.*, 220, 1814–1819, 1971.

265. Klein, U., Timme, M., Zeiske, W., and Ehrenfeld, J., The H^+ pump in frog skin (*Rana esculenta*): identification and localization of a V-ATPase, *J. Membr. Biol.*, 157, 117–126, 1997.

266. Koefoed-Johnsen, V. and Ussing, H. H., The contribution of diffusion and flow to the passage of D_2O through living membranes: effect of neurohypophyseal hormone on isolated anuran skin, *Acta Physiol. Scand.*, 28, 60–76, 1953.

267. Koefoed-Johnsen, V. and Ussing, H. H., The nature of the frog skin potential, *Acta Physiol. Scand.*, 42, 298–308, 1958.

268. Koefoed-Johnsen, V., Ussing, H. H., and Zerahn, K., The origin of the short-circuit current in the adrenalin-stimulated frog skin, *Acta Physiol. Scand.*, 27, 38–48, 1952.

269. Konno, N., Hyodo, S., Matsuda, K., and Uchiyama, M., Effect of osmotic stress on expression of putative facilitative urea transporter in the kidney and urinary bladder of the marine toad, *Bufo marinus*, *J. Exp. Biol.*, 209, 1207–1216, 2006.

270. Konno, N., Hyodo, S., Takei, Y., Matsuda, K., and Uchiyama, M., Plasma aldosterone, angiotensin II and arginine vasotocin concentrations in the toad, *Bufo marinus*, following osmotic treatments, *Gen. Comp. Endocrinol.*, 140, 86–93, 2005.

271. Konno, N., Hyodo, S., Yamada, T., Matsuda, K., and Uchiyama, M., Immunolocalization and mRNA expression of the epithelial Na^+ channel alpha subunit in the kidney and urinary bladder of the marine toad, *Bufo marinus*, under hyperosmotic conditions, *Cell Tissue Res.*, 328, 583–594, 2007.

272. Kotai, M., New concepts related to the evolution of lymphatics. In *Progress in Lymphology, XII*, Nishii, M., Uchino, S., and Yabuki, S., Eds., Elsevier Science, New York, 1990.

273. Kristensen, P., Chloride transport across isolated frog skin, *Acta Physiol. Scand.*, 84, 338–346, 1972.

274. Kristensen, P., The effect of amiloride on chloride transport across amphibian epithelia, *J. Membr. Biol.*, 40S, 167–185, 1978.

275. Kristensen, P., Chloride transport in frog skin. In *Chloride Transport in Biological Membranes*, Zadunaisky, J., Ed., Academic Press, New York, 1982, pp. 319–332.

276. Kristensen, P., Exchange diffusion, electrodiffusion and rectification in the chloride transport pathway of frog skin, *J. Membr. Biol.*, 72, 141–151, 1983.

277. Krogh, A., Osmotic regulation in the frog (*R. esculenta*) by active absorption of chloride ions, *Skand. Arch. Physiol.*, 76, 60–64, 1937.

278. Krogh, A., Osmotic regulation in fresh water fishes by active absorption of chloride ions, *Z. vergleichende Physiol.*, 24, 656–666, 1937.

279. Krogh, A., The active absorption of ions in some freshwater animals, *Z. vergleichende Physiol.*, 25, 335–350, 1938.

280. Krogh, A., *Osmotic Regulation in Aquatic Animals*, Cambridge University Press, Cambridge, U.K., 1939.

281. Lametschwandtner, A., Albrecht, U., and Adam, H., The vascularization of the kidneys in *Bufo bufo* (L.), *Bombina variegata* (L.), *Rana ridibunda* (L.), and *Xenopus laevis* (L.) as revealed by scanning electron microscopy of vascular corrosion casts, *Acta Zool.*, 59, 11–23, 1978.

282. Larsen, E. H., Effect of aldosterone and oxytocin on the active sodium transport across isolated toad skin in relation to loosening of the *stratum corneum*, *Gen. Comp. Endocrinol.*, 17, 543–553, 1971.

283. Larsen, E. H., NaCl in amphibian skin. In *Advances in Comparative and Environmental Physiology*, Greger, R., Ed., Springer Verlag, Berlin, 1988, pp. 189–248.

284. Larsen, E. H., Chloride transport by high-resistance heterocellular epithelia, *Physiol. Rev.*, 71, 235–283, 1991.

285. Larsen, E. H. and Kristensen, P., Properties of a conductive cellular chloride pathway in the skin of the toad (*Bufo bufo*), *Acta Physiol. Scand.*, 102, 1–21, 1978.

286. Larsen, E. H. and Harvey, B. J., Chloride currents of single mitochondria-rich cells of toad skin epithelium, *J. Physiol.*, 478, 7–15, 1994.

287. Larsen, E. H. and Møbjerg, N., Na$^+$ recirculation and isosmotic transport, *J. Membr. Biol.*, 212, 1–15, 2006.

288. Larsen, E. H., Ussing, H. H., and Spring, K. R., Ion transport by mitochondria-rich cells in toad skin, *J. Membr. Biol.*, 99, 25–40, 1987.

289. Larsen, E. H., Willumsen, N. J., and Christoffersen, B. C., Role of proton pump of mitochondria-rich cells for active transport of chloride ions in toad skin epithelium, *J. Physiol.*, 450, 203–216, 1992.

290. Larsen, E. H., Møbjerg, N., and Nielsen, R., Application of the Na$^+$ recirculation theory to ion coupled water transport in low- and high resistance osmoregulatory epithelia, *Comp. Biochem. Physiol. A*, 148, 101–116, 2007.

291. Larsen, E. H., Kristensen, P., Nedergaard, S., and Willumsen, N. J., Role of mitochondria-rich cells for passive chloride transport, with a discussion of Ussing's contributions to our understanding of shunt pathways in epithelia, *J. Membr. Biol.*, 184, 247–254, 2001.

292. Larsen, L. O., Physiology of moulting. In *Physiology of the Amphibia*, Vol. 3, Lofts, B., Ed., Academic Press, London, 1976, pp. 53–100.

293. Leaf, A., From toad bladder to kidney, *Am. J. Physiol.*, 242, F103–F111, 1982.

294. Leaf, A., Anderson, J., and Page, L. B., Active sodium transport by the isolated bladder, *J. Gen. Physiol.*, 41, 657–688, 1958.

295. Lillywhite, H. B., Thermal modulation of cutaneous mucus discharge as a determinant of evaporative water loss in the frog, *Rana catesbeiana*, *Z. vergleichende Physiol.*, 73, 84–104, 1971.

296. Lillywhite, H. B., Water relations of tetrapod integument, *J. Exp. Biol.*, 209, 202–226, 2006.

297. Lillywhite, H. B. and Licht, P., Movement of water over toad skin: functional role of epidermal sculpturing, *Copeia*, 1974, 165–170, 1974.

298. Lillywhite, H. B., Mittal, A. K., Garg, T. K., and Agrawal, N., Wiping behavior and its ecological significance in the Indian tree frog *Polypedates maculatus*, *Copeia*, 1997, 88–100, 1997.

299. Lindemann, B. and Van Driessche, W., Sodium specific membrane channels of frog skin are pores: current fluctuations reveal high turnover, *Science*, 195, 292–294, 1977.

300. Linderholm, K., On the behaviour of the "sodium pump" in frog skin at various concentrations of Na ions in the solution on the epithelial side, *Acta Physiol. Scand.*, 31, 36–61, 1954.

301. Loveridge, J. P., Observations on nitrogenous excretion and water relations of *Chiromantis xerampelina* (Amphibia, Anura), *Arnoldia (Rhodesia)*, 5, 1–6, 1970.

302. Loveridge, J. P., Nitrogenous excretion in the amphibia. In *New Insights in Vertebrate Kidney Function*, Brown, J. A., Balment, R. J., and Rankin, J. C., Eds., Cambridge University Press, Cambridge, U.K., 1993, pp. 135–143.

303. Lutz, B., Ontogenetic evolution in frogs, *Evolution*, 2, 29–39, 1948.

304. MacRobbie, E. A. C. and Ussing, H. H., Osmotic behaviour of the epithelial cells of frog skin, *Acta Physiol. Scand.*, 53, 348–365, 1961.

305. Madsen, K. M. and Tisher, C. C., Structural-functional relationships along the distal nephron, *Am. J. Physiol.*, 250, F1–F15, 1986.

306. Madsen, K. M., Clapp, W. L., and Verlander, J. W., Structure and function of the inner medullary collecting duct, *Kidney Int.*, 34, 441–454, 1988.

307. Maetz, J., Sawyer, W. H., Pickford, G. E., and Mayer, N., Evolution de la balance du sodium chez Fundulus heteroclitus au cours du transfert d'eau de mer en d'eau douce: effets de l'hypophysectomie et de la prolactine, *Gen. Comp. Endocrinol.*, 8, 169–176, 1967.

308. Maffly, R. H., Hays, R. M., Lamdin, E., and Leaf, A., The effect of neurohypophyseal hormones on the permeability of the toad bladder to urea, *J. Clin. Invest.*, 39, 630–641, 1960.

309. Main, A. R. and Bentley, P. J., Water relations of Australian burrowing frogs and tree frogs, *Ecology*, 45, 379–382, 1975.

310. Maleek, R., Sullivan, P., Hoff, K., Baula, V., and Hillyard, S. D., Salt sensitivity and hydration behavior of the toad, *Bufo marinus*, *Physiol. Behav.*, 67, 739–745, 1999.

311. Marrero, M. B. and Hillyard, S. D., Differences in cyclic AMP levels in epithelial cells from pelvic and pectoral regions of toad skin, *Comp. Biochem. Physiol.*, 82C, 69–73, 1985.

312. Martin, A. A., Australian anuran life histories: some evolutionary and ecological aspects, in *Australian Inland Waters and their Fauna*, Weatherly, A. H., Ed., Australian National University Press, Canberra, 1967, pp. 175–191.

313. Martin, A. A. and Cooper, A. K., The ecology of terrestrial anuran eggs, genus *Crinia* (Leptodactylidae), *Copeia*, 1972, 163–168, 1972.

314. Mathias, J. H., The Comparative Ecologies of Two Species of Amphibia (*Bufo bufo* and *Bufo calamita*) on the Ainsdale Sand Dunes Natural Nature Reserve, Ph.D. thesis, University of Manchester, Manchester, U.K., 1971.

315. Matteucci, C. and Cima, A., Memoire sur l'endosmos, *Ann. Chim. Phys.*, 13, 63–86, 1845.

316. Maunsbach, A. B., Ultrastructure of the proximal tubule. In *Renal Physiology, Handbook of Physiology*, Vol. 8, Geiger, S. R., Ed., American Physiological Society, Washington, D.C., 1973, pp. 31–79.

317. Maunsbach, A. B. and Boulpaep, E. L., Quantitative ultrastructure and functional correlates in proximal tubule of *Amphiuma* and *Necturus*, *Am. J. Physiol.*, 246, F710–F724, 1984.

318. McClanahan, Jr., L. L., Adaptations of the spadefoot toad, *Scaphiopus couchi* to desert environments, *Comp. Biochem. Physiol.*, 20, 73–79, 1967.

319. McClanahan, L. L., Changes in body fluids of burrowed spadefoot toads as a function of soil water potential, *Copeia*, 1972, 209–216, 1972.

320. McClanahan, L. L. and Baldwin, R., Rate of water uptake through the integument of the desert toad, *Bufo punctatus*, *Comp. Biochem. Physiol.*, 28, 381–389, 1969.

321. McClanahan, L. L. and Shoemaker, V. H., Behavior and thermal relations of the arboreal frog, *Phyllomedusa sauvagii*, *Natl. Geogr. Res.*, 3, 11–21, 1987.

322. McClean, G. J. and Burnstock, G., Histochemical localization of catecholamines in the urinary bladder of the toad (*Bufo marinus*), *J. Histochem. Cytochem.*, 14, 538–548, 1966.

323. Meseguer, J., Garcia-Ayala, A., López-Ruiz, A., and Esteban, M. A., Structure of the amphibian mesonephric tubule during ontogenesis in *Rana ridibunda* L. tadpoles: early ontogenetic stages, renal corpuscle formation, neck segment and peritoneal funnels, *Anat. Embryol.*, 193, 397–406, 1996.

324. Middler, S. A., Kleeman, C. R., and Edwards, E., The role of the urinary bladder in salt and water metabolism of the toad, *Bufo marinus*, *Comp. Biochem. Physiol.*, 26, 57–68, 1968.

325. Middler, S. A., Kleeman, C. R., and Edwards, E., Lymph mobilization following acute loss in the toad, *Bufo marinus*, *Comp. Biochem. Physiol.*, 24, 343–353, 1968.

326. Middler, S. A., Kleeman, C. R., Edwards, E., and Brody, D., Effect of adenohypophysectomy on salt and water metabolism of the toad *Bufo marinus* with studies on hormonal replacement, *Gen. Comp. Endocrinol.*, 12, 290–304, 1969.

327. Miller, D. A., Standish, M. L., and Thurman, A. E., Effects of temperature on water and electrolyte balance in the frog, *Physiol. Zool.*, 41, 500–506, 1968.

328. Mills, J. W., Ion transport across the exocrine glands of frog skin, *Pflügers Arch.*, 405(Suppl. 1), S44–S49, 1985.

329. Mills, J. W., Ernst, S. A., and DiBona, D. R., Localization of Na^+-pump sites in frog skin, *J. Cell Biol.*, 73, 88–110, 1977.

330. Møbjerg, N., Larsen, E. H., and Jespersen, Å., Morphology of the nephron in the mesonephros of *Bufo bufo* (Amphibia, Anura, Bufonidae), *Acta Zool. (Stockholm)*, 79, 31–51, 1998.

331. Møbjerg, N., Larsen, E. H., and Jespersen, Å., Morphology of the kidney in larvae of *Bufo viridis* (Amphibia, Anura, Bufonidae), *J. Morphol.*, 245, 177–195, 2000.

332. Møbjerg, N., Larsen, E. H., and Novak, I., K^+ transport in the mesonephric collecting duct system of the toad, *Bufo bufo*: microelectrode recordings from isolated and perfused tubules, *J. Exp. Biol.*, 205, 897–904, 2002.

333. Møbjerg, N., Jespersen, Å., and Wilkinson, M., Morphology of the nephron in the mesonephros of *Geotrypetes seraphini* (Amphibia, Gymnophiona, Caeciliaidae), *J. Morphol.*, 262, 583–607, 2004.

334. Møbjerg, N., Larsen, E. H., and Novak, I., Ion transport mechanisms in the mesonephric collecting duct system of the toad *Bufo bufo*, as revealed by microelectrode recordings from isolated and perfused tubules, *Comp. Biochem. Physiol. A*, 137, 585–595, 2004.

335. Møbjerg, N., Werner, A., Hansen, S. M., and Novak, I., Physiological and molecular mechanisms of inorganic phosphate handling in the toad *Bufo bufo*, *Pflügers Arch.*, 454, 101–113, 2007.

336. Montgomery, H. and Pierce, J. A., The site of acidification of the urine within the renal tubule in Amphibia, *Am. J. Physiol.*, 118, 144–152, 1937.

337. Morris, J. L., Structure and function of ciliated peritoneal funnels in the toad kidney (*Bufo marinus*), *Cell Tissue Res.*, 217, 599–610, 1981.

338. Morris, J. L. and Campbell, G., Renal vascular anatomy of the toad (*Bufo marinus*), *Cell Tissue Res.*, 189, 501–514, 1978.

339. Motais, R. and Garcia Romeu, F., Transport mechanisms in teleostean gill and amphibian skin, *Annu. Rev. Physiol.*, 34, 141–176, 1972.

340. Mullen, T. E., Kashgarian, M., Biemesderfer, D., Giebisch, G., and Biber, T. U. L., Ion transport and function of urinary bladder epithelium of *Amphiuma*, *Am. J. Physiol.*, 231, 501–508, 1976.

341. Mullen, T. L. and Alvarado, R. H., Osmotic and ionic regulation in amphibians, *Physiol. Zool.*, 49, 11–23, 1976.

342. Muller, J., On the existence of hearts, having regular pulsations, connected with the lymphatic system in certain amphibious animals, *Abstracts of Papers Printed in the Royal Society of London (1830–1837)*, 3, 165–166, 1833.

343. Nagai, T., Koyama, H., Hoff, K., and Hillyard, S. D., Desert toads discriminate salt taste with chemosensory function of their ventral skin, *J. Comp. Neurol.*, 408, 125–136, 1999.

344. Nagata, K., Hori, N., Sato, K., Ohta, K., and Tanaka, H., Cloning and functional expression of an SGLT-1-like protein from the *Xenopus laevis* intestine, *Am. J. Physiol.*, 276, G1251–G1259, 1999.

345. Nagel, W., The intracellular electrical potential profile of the frog skin epithelium, *Pflügers Arch.*, 365, 135–143, 1976.

346. Nagel, W., The dependence of the electrical potentials across the membranes of the frog skin upon the concentration of sodium in the mucosal solution, *J. Physiol.*, 249, 777–796, 1977.

347. Nagel, W., Inhibition of potassium conductance by barium in frog skin epithelium, *Biochim. Biophys. Acta*, 552, 346–357, 1979.

348. Nagel, W., Rheogenic sodium transport in a tight epithelium, the amphibian skin, *J. Physiol.*, 302, 281–295, 1980.

349. Nagel, W. and Hirschmann, W., K^+ permeability of the outer border of frog skin (*R. temporaria*), *J. Membr. Biol.*, 62, 107–113, 1980.

350. Nagel, W. and Dörge, A., Analysis of anion conductance in frog skin, *Pflügers Arch.*, 416, 53–61, 1990.

351. Nagel, W. and Katz, U., α_1-Adrenoceptors antagonize activated chloride conductance of amphibian skin epithelium, *Pflügers Arch. Eur. J. Physiol.*, 436, 863–870, 1998.

352. Nagel, W., Garcia-Diaz, J. F., and Armstrong, W. M., Intracellular ionic activities in frog skin, *J. Membr. Biol.*, 61, 127–134, 1981.

353. Nagel, W., Somieski, P., and Katz, U., The route of passive chloride movement across amphibian skin: localization and regulatory mechanisms, *Biochim. Biophys. Acta*, 1566, 44–54, 2002.

354. Neill, W. T., The occurrence of amphibians and reptiles in saltwater areas, and a bibliography, *Bull. Marine Sci. Gulf Caribbean*, 8, 1–97, 1958.

355. Nielsen, K. H. and Jørgensen, C. B., Salt and water balance during hibernating anurans, *Forschritte der Zoologie*, 38, 333–349, 1990.

356. Nielsen, M. S. and Nielsen, R., Effect of carbachol and prostaglandin E_2 on chloride secretion and signal transduction in the exocrine glands of frog skin (*Rana esculenta*), *Pflügers Arch.*, 438, 732–740, 1999.

357. Nielsen, R., A 3 to 2 coupling of the Na–K pump responsible for transepithelial Na transport in frog skin disclosed by the effect of Ba, *Acta Physiol. Scand.*, 107, 189–191, 1979.

358. Nielsen, R., Coupled transepithelial sodium and potassium transport across isolated frog skin: effect of ouabain, amiloride and the polyene antibiotic filipin, *J. Membr. Biol.*, 51, 161–184, 1979.

359. Nielsen, R., Effect of ouabain, amiloride and antidiuretic hormone on the sodium transport pool in isolated epithelia from frog skin (*Rana temporaria*), *J. Membr. Biol.*, 65, 221–226, 1982.

360. Nielsen, R., Active transepithelial transport in frog skin via specific potassium channels in the apical membrane, *Acta Physiol. Scand.*, 120, 287–296, 1984.

361. Nielsen, R., Correlation between transepithelial Na^+ transport and transepithelial water movement across isolated frog skin (*Rana esculenta*), *J. Membr. Biol.*, 159, 61–69, 1997.

362. Nielsen, R. and Tomlison, R. W. S., The effect of amiloride on sodium transport in normal and moulting frog skin, *Acta Physiol. Scand.*, 79, 238–243, 1970.

363. Nielsen, R. and Larsen, E. H., Beta-adrenergic activation of solute coupled water uptake by toad skin epithelium results in near-isosmotic transport, *Comp. Biochem. Physiol. A*, 148, 64–71, 2007.

364. Nielsen, S., Frøkiær, J., Marples, D., Kwon, T.-H., Agre, P., and Knepper, M. A., Aquaporins in the kidney: from molecules to medicine, *Physiol. Rev.*, 82, 205–244, 2002.

365. Noble, G. K., The value of life history data in the study of the evolution of the amphibia, *Ann. N.Y. Acad. Sci.*, 30, 31–128, 1927.

366. Nouwen, E. J., and Kuhn, E. R., Volumetric control of arginine vasotocin and mesotocin release in the frog (*Rana ridibunda*), *J. Endocrinol.*, 105, 371–377, 1985.

367. Nouwen, E. K. and Kuhn, E., R., Radioimmunoassay of arginine vasotocin and mesotocin in serum of the frog *Rana ridibunda*, *J. Comp. Endocrinol.*, 50, 242–251, 1983.

368. Oberleithner, H., Guggino, W. B., and Giebisch, G., Potassium transport in the early distal tubule of *Amphiuma* kidney: effects of potassium adaptation, *Pflügers Arch.*, 396, 185–191, 1983.

369. Oberleithner, H., Lang, F., Messner, G., and Wang, W., Mechanism of hydrogen ion transport in the diluting segment of frog kidney, *Pflügers Arch.*, 402, 272–280, 1984.

370. Oberleithner, H., Lang, F., Wang, W., Messner, G., and Deetjen, P., Evidence for an amiloride sensitive Na^+ pathway in the amphibian diluting segment induced by K^+ adaptation, *Pflügers Arch.*, 399, 166–172, 1983.

371. Ogushi, Y., Mochida, H., Nakakura, T., Suzuki, M., and Tanaka, S., Immunocytochemical and phylogenetic analyses of an arginine vasotocin-dependent aquaporin, AQP-h2K, specifically expressed in the kidney of the tree frog, *Hyla japonica*, *Endocrinology*, 148, 5891–5901, 2007.

372. Orloff, J. and Handler, J. S., The similarity of the effects of vasopressin adenosine 3′,5′ monophosphate (cyclic AMP), and theophylline on the toad bladder, *J. Clin. Invest.*, 41, 702–709, 1962.

373. Othani, O. and Naito, I., Renal microcirculation of the bullfrog (*Rana catesbeiana*): a scanning electron microscope study of vascular casts, *Arch. Histol. Japon.*, 4, 319–330, 1980.

374. Overton, E., Neunundereissing Thesen uber Wasserokonomie der Amphibien und die osmotischen Eigenshaften der Amphibienhaut, *Verhandlungen der physikalish medicinischen Gesellschaft in Würtzburg*, 36, 277–296, 1904.

375. Page, R. D. and Frazier, L. W., Morphological changes in the skin of *Rana pipiens* in response to metabolic acidosis, *Proc. Soc. Exp. Biol. Med.*, 184, 416–422, 1987.

376. Pang, P. K. T. and Sawyer, W. H., Effects of prolactin on hypophysectomized mud puppies, *Necturus maculosus*, *Am. J. Physiol.*, 226, 458–462, 1974.

377. Pang, P. K. T., Furspan, P. B., and Sawyer, W. H., Evolution of neurohypophyseal hormone function in vertebrates, *Am. Zool.*, 23, 655–662, 1983.

378. Park, J. H. and Saier, M. H. J., Phylogenetic characterization of the MIP family of transmembrane channel proteins, *J. Membr. Biol.*, 153, 171–180, 1996.

379. Parsons, R. H., McDevitt, V., Aggerwal, V., LeBlang, T., Manley, K., Lopez, J., and Kenedy, A., Regulation of pelvic patch water flow in *Bufo marinus*: role of bladder volume and angiotensin II, *Am. J. Physiol.*, 264, R686–R689, 1993.

380. Pascual, L. E., Tallec, L., and Lombes, M., The mineralocorticoid receptor: a journey exploring its diversity and specificity of action, *Mol. Endocrinol.*, 19, 2211–2221, 2005.

381. Peachey, L. D. and Rasmussen, H., The structure of the toad's urinary bladder as related to its physiology, *J. Biophys. Biochem. Cytol.*, 10, 529–553, 1961.

382. Petriella, S., Reboreda, J. C., Otero, M., and Segura, E., Antidiuretic responses to osmotic cutaneous stimulation in the toad, *Bufo arenarum*: a possible adaptive control mechanism for urine production, *J. Comp. Physiol. B*, 159, 91–95, 1989.

383. Pickering, B. T. and Heller, H., Active principles of the neurohypophysis of cold-blooded vertebrates: chromatographic and biological characteristics of fish and frog neurohypophyseal extracts, *Nature*, 184, 1463–1464, 1959.

384. Planelles, G. and Anagnostopoulos, T., Electrophysiological properties of amphibian late distal tubule *in vivo*, *Am. J. Physiol.*, 255, F186–F166, 1988.

385. Pons, G., Guardabassi, A., and Pattono, P., The kidney of *Hyla arborea* (L.) (Amphibia Hylidae) in autumn, winter and spring: histological and ultrastructural observations, *Monitore Zoologico Italiano*, 16, 261–281, 1982.

386. Porter, G. A. and Edelman, I. S., The action of aldosterone and related corticosteroids on sodium transport across the toad bladder, *J. Clin. Invest.*, 43, 611–620, 1964.

387. Potts, W. T. W. and Parry, G., *Osmotic and Ionic Regulation in Animals*, Pergamon Press, Oxford, U.K., 1964.

388. Pounds, J. A., Bustamante, M. R., Coloma, L. A., Consuegra, J. A., Fogden, M. P., Foster, P. N., LaMarca, E., Masters, K. L., Merino-Viteri, A., Puschendorf, R., Ron, S. R., Sanchez-Azofeifa, G. A., Still, C. J., and Young, B. E., Widespread amphibian extinctions from epidemic disease driven by global warming, *Nature*, 439, 161–167, 2006.

389. Poynton, J. C., The amphibia of southern Africa, *Ann. Natl. Mus.*, 17, 1–334, 1964.

390. Preston, G. M. and Agre, P., Isolation of the cDNA for erythrocyte integral membrane protein of 28 kilodaltons: member of an ancient channel family, *Proc. Natl. Acad. Sci. USA*, 88, 11110–11114, 1991.

391. Propper, C. R. and Johnson, W. E., Angiotensin II induces water absorption behavior in two species of desert anurans, *Hormones Behav.*, 28, 41–52, 1994.

392. Propper, C. R., Hillyard, S. D., and Johnson, W. E., Central angiotensin II induces thirst-related responses in an amphibian, *Hormones Behav.*, 29, 74–84, 1995.

393. Prosser, C. L. and Weinstein, S. J. F., Comparison of blood volume in animals with open or closed circulatory systems, *Physiol. Zool.*, 23, 113–124, 1950.

394. Puoti, A., May, A., Canessa, M., Horisberger, J.-D., Schild, L., and Rossier, B. C., The highly selective low-conductance epithelial Na channel in *Xenopus laevis* A6 kidney cells, *Am. J. Physiol.*, 269, C188–C197, 1995.

395. Rajerison, R. M., Montegut, M., Jard, S., and Morel, F., The isolated frog skin epithelium: presence of α- and β-adrenergic receptors regulating active sodium transport and water permeability, *Pflügers Arch.*, 332, 313–331, 1972.

396. Rajerison, R. M., Montegut, M., Jard, S., and Morel, F., The isolated frog skin epithelium: permeability characteristics and responsiveness to oxytocin, cyclic AMP and theophylline, *Pflügers Arch.*, 332, 302–312, 1972.

397. Ray, C., Vital limits and rates of desiccation in salamanders, *Ecology*, 39, 75–83, 1958.

398. Reid, E. W., Osmosis experiments with living and dead membranes, *J. Physiol.*, 11, 312–351, 1890.

399. Reid, E. W., Experiments upon "absorption without osmosis," *Br. Med. J.*, 1, 323–326, 1892.

400. Reuss, L., Ussing's two-membrane hypothesis: the model and half a century of progress, *J. Membr. Biol.*, 184, 211–217, 2001.

401. Richards, A. N. and Walker, A. M., Methods of collecting fluid from known regions of the renal tubules of Amphibia and perfusing the lumen of a single tubule, *Am. J. Physiol.*, 118, 111–120, 1936.

402. Richter, S., The opisthonephros of *Rana esculenta* (Anura). I. Nephron development, *J. Morphol.*, 226, 173–187, 1995.

403. Rick, R., Intracellular ion concentrations in the isolated frog skin epithelium: evidence for different types of mitochondria-rich cells, *J. Membr. Biol.*, 127, 227–236, 1992.

404. Rick, R., Short-term bromide uptake in skins of *Rana pipiens*, *J. Membr. Biol.*, 138, 171–179, 1994.

405. Rick, R., Spancken, G., and Dörge, A., Differential effects of aldosterone and ADH on intracellular electrolytes in the toad urinary bladder epithelium, *J. Membr. Biol.*, 101, 275–282, 1988.

406. Rick, R., Dörge, A., Arnim, E. V., and Thurau, K., Electron microprobe analysis of frog skin: evidence for a syncytial sodium transport compartment, *J. Membr. Biol.*, 39, 313–331, 1978.

407. Riordan, J. R., Rommens, J. M., Kerem, B.-S., Alon, N., Rozmehel, R., Grzelczak, Z., Zielenski, J., Lok, S., Plavsic, N., Chou, J.-L., Drumm, M. L., Iannuzzi, M. C., and Tsui, L.-C., Identification of the cystic fibrosis gene: cloning and characterization of complimentary DNA, *Science*, 245, 1066–1073, 1989.

408. Romspert, A. C., Osmoregulation of the African clawed frog, *Xenopus laevis*, in hypersaline media, *Comp. Biochem. Physiol.*, 54A, 207–210, 1976.

409. Romspert, A. L. and McClanahan, L. L., Osmoregulation in the terrestrial salamander, *Ambystoma tigrinum*, in hypersaline media, *Copeia*, 1981, 400–405, 1981.

410. Roth, J. J., Vascular supply to the ventral pelvic skin of anurans related to water balance, *J. Morphol.*, 140, 443–460, 1973.

411. Rotunno, C. A., Vilallonga, F. A., Fernandez, M., and Cereijido, M., The penetration of sodium into the epithelium of the frog skin, *J. Gen. Physiol.*, 55, 716–735, 1970.

412. Rouille, Y., Michel, G., Chauvet, M. T., Chauvet, J., and Acher, R., Hydrins, hydroosmotic neurohypophysial peptides, osmoregulatory processing in amphibians through vasotocin precursor processing, *Proc. Natl. Acad. Sci. USA*, 86, 5272–5275, 1997.

413. Rozman, A., Gabbay, S., and Katz, U., Chloride conductance across toad skin: effects of ionic acclimations and cyclic AMP and relationship to mitochondria-rich cell density, *J. Exp. Biol.*, 203, 2039–2045, 2000.

414. Ruibal, R., The adaptive value of bladder water in the toad, *Bufo cognatus*, *Physiol. Zool.*, 35, 218–223, 1962.

415. Ruibal, R., Tevis, L., and Roig, V., The terrestrial ecology of the spadefoot toad *Scaphiopus hammondi*, *Copeia*, 1969, 571–584, 1969.

416. Sakai, T. and Kawahara, K., The structure of the kidney in the Japanese newt, *Triturus (Cynops) pyrrhogaster*, *Anat. Embryol.*, 166, 31–52, 1983.

417. Sakai, T., Billo, R., and Kritz, W., The structural organization of the kidney of *Typhlonectes compressicaudus* (Amphibia, Gymnophiona), *Anat. Embryol.*, 174, 243–252, 1986.

418. Sakai, T., Billo, R., and Kritz, W., Ultrastructure of the kidney of the South American caecilian, *Typhlonectes compressicaudus* (Amphibia, Gymnophiona). I. Renal corpuscle, neck segment, proximal tubule and intermediate segment, *Cell Tissue Res.*, 252, 589–600, 1988.

419. Sakai, T., Billo, R., Nobiling, R., Gorgas, K., and Kritz, W., Ultrastructure of the kidney of the South American caecilian, *Typhlonectes compressicaudus* (Amphibia, Gymnophiona). II. Distal tubule, connecting tubule, collecting duct and Wolffian duct, *Cell Tissue Res.*, 252, 601–610, 1988.

420. Sands, J. M., Mammalian urea transporters, *Annu. Rev. Physiol.*, 65, 543–566, 2003.

421. Sawyer, W. H., The antidiuretic action of neurohypophyseal hormones in Amphibia. In *The Neurohypophysis*, Heller, H., Ed., Butterworths, London, 1957.

422. Sawyer, W. H. and Schisgall, R. M., Increased permeability of the frog bladder to water in response to dehydration and neurohypophyseal extracts, *Am. J. Physiol.*, 187, 312–314, 1956.

423. Sawyer, W. H., Munsick, R. A., and van Dyke, H. B., Active principles of the neurohypophysis of cold-blooded vertebrates: pharmacologic evidence for the presence of arginine vasotocin and oxytocin in neurohypophyseal extracts of cold blooded vertebrates, *Nature*, 184, 1464–1465, 1959.

424. Schaffner, A. and Rodewald, R., Glomerular permeability in the bullfrog *Rana catesbeiana*, *J. Cell Biol.*, 79, 314–328, 1978.

425. Schmid, W. D., Natural variations in nitrogen excretion of amphibians from different habitats, *Ecology*, 49, 180–185, 1968.

426. Schmidt-Nielsen, B. and Forster, R. P., The effect of dehydration and low temperature on renal function in the bullfrog, *J. Cell. Comp. Physiol.*, 44, 233–246, 1954.

427. Schmidt-Nielsen, K. and Lee, P., Kidney function of the crab eating frog (*Rana cancrivora*), *J. Exp. Biol.*, 39, 167–177, 1962.

428. Schmidt, T. J., Husted, R. F., and Stokes, R. B., Steroid stimulation of Na^+ transport in A6 cells is mediated via glucocorticoid receptors, *Am. J. Physiol.*, 264, C875–C884, 1993.

429. Schoen, H. F. and Erlij, D., Insulin action on electrophysiological properties of apical and basolateral membranes of frog skin, *Am. J. Physiol.*, 252, CC411–C417, 1987.

430. Schultz, S. G., Homocellular regulatory mechanisms in sodium-transporting epithelia: avoidance of extinction by "flush through," *Am. J. Physiol.*, 241, F579–F590, 1981.

431. Scott, W. N., Saperstein, V. S., and Yoder, J., Partition of tissue functions an epithelial cell localization of enzymes in "mitochondria-rich" cells of toad urinary bladder, *Science*, 184, 797–800, 1974.

432. Seymour, R. S., Gas exchanges in spadefoot toads beneath the ground, *Copeia*, 1973, 452–460, 1973.

433. Seymour, R. S., Energy metabolism of dormant spadefoot toads (*Scaphiopus*), *Copeia*, 1973, 435–445, 1973.

434. Shane, M. A., Nofziger, C., and Blazer-Yost, B., Hormonal regulation of the epithelial Na^+ channel: from amphibians to mammals, *Gen. Comp. Endocrinol.*, 147, 85–92, 2006.

435. Sharin, S. H. and Blankemeyer, J. T., Demonstration of gap junctions in frog skin epithelium, *Am. J. Physiol.*, 257, C658–C664, 1989.

436. Shoemaker, V. H., The effects of dehydration on electrolyte concentrations in a toad, *Bufo marinus*, *Comp. Biochem. Physiol.*, 13, 261–271, 1964.

437. Shoemaker, V. H., The stimulus for the water-balance response to dehydration in toads, *Comp. Biochem. Physiol.*, 15, 81–88, 1965.

438. Shoemaker, V. H. and McClanahan, L. L., Nitrogen excretion in the larvae of a land-nesting frog (*Leptodactylus bufonius*), *Comp. Biochem. Physiol.*, 44A, 1149–1156, 1973.

439. Shoemaker, V. H. and Bickler, P. E., Kidney and bladder function in a uricotelic frog (*Phyllomedusa sauvagii*), *J. Comp. Physiol. B*, 133, 211–218, 1979.

440. Shoemaker, V. H., McClanahan, L. L., and Ruibal, R., Seasonal changes in body fluids in a field population of spadefoot toads, *Copeia*, 1969, 585–591, 1969.

441. Shoemaker, V. H., Balding, D., Ruibal, R., and McClanahan, L. J., Uricotelism and low evaporative water loss in a South American frog, *Science*, 175, 1018–1020, 1972.

442. Shoemaker, V. H., Hillman, S. S., Hillyard, S. D., Jackson, D. G., McClanahan, L. L., Withers, P. C., and Wygoda, M. L., Exchange of water, ions, and respiratory gasses in terrestrial amphibians. In *Environmental Physiology of the Amphibia*, Feder, M. E. and Burggren, W. W., University of Chicago Press, 1992.

443. Shpun, S. and Katz, U., Renal function at steady state in a toad (*Bufo viridis*) acclimated in hyperosmotic NaCl and urea solutions, *J. Comp. Physiol. B*, 164, 646–652, 1995.

444. Slatyer, R. O., *Plant–Water Relationships*, Academic Press, New York, 1967.

445. Smith, P. G., The low frequency impedance of the isolated frog skin, *Acta Physiol. Scand.*, 81, 355–366, 1971.

446. Sørensen, J. B. and Larsen, E. H., Heterogeneity of chloride channels in the apical membrane of isolated mitochondria-rich cells, *J. Gen. Physiol.*, 108, 421–433, 1996.

447. Sørensen, J. B. and Larsen, E. H., Patch clamp on the luminal membrane of exocrine gland acini from frog skin (*Rana esculenta*) reveals the presence of cystic fibrosis transmembrane conductance regulator-like Cl⁻ channels activated by cyclic AMP, *J. Gen. Physiol.*, 112, 19–31, 1998.

448. Sørensen, J. B. and Larsen, E. H., Membrane potential and conductance of frog skin gland acinar cells in resting conditions and during stimulation with agonists of macroscopic secretion, *Pflügers Arch.*, 439, 101–112, 1999.

449. Sørensen, J. B., Nielsen, M. S., Nielsen, R., and Larsen, E. H., Luminal ion channels involved in isotonic secretion by Na⁺-recirculation in exocrine gland-acini, *Biologiske Skrifter Det Kongelige Danske Videnskabernes Selskab*, 49, 179–191, 1998.

450. Sørensen, J. B., Nielsen, M. S., Gudme, C. N., Larsen, E. H., and Nielsen, R., Maxi K⁺ channels co-localized with CFTR in the apical membrane of an exocrine gland acinus: possible involvement in secretion, *Pflügers Arch.*, 442, 1–11, 2001.

451. Spies, I., Immunolocalization of mitochondria-rich cells in the epidermis of the common toad, *Bufo bufo* L., *Comp. Biochem. Physiol.*, 118B, 285–291, 1997.

452. Spight, T. M., The water economy of salamanders: water uptake after dehydration, *Comp. Biochem. Physiol.*, 20, 767–771, 1967.

453. Spotila, J. R. and Berman, E. N., Determination of skin resistance and the role of the skin in controlling water loss in amphibians and reptiles, *Comp. Biochem. Physiol.*, 55A, 407–411, 1976.

454. Spring, K. R. and Ussing, H. H., The volume of mitochondria-rich cells of frog skin epithelium, *J. Membr. Biol.*, 92, 21–26, 1986.

455. Stanton, B., Electroneutral NaCl transport by distal tubule: evidence for Na⁺/H⁺-Cl⁻/HCO₃⁻ exchange, *Am. J. Physiol.*, 254, F80–F86, 1988.

456. Stanton, B., Omerovic, A., Koeppen, B., and Giebisch, G., Electroneutral H⁺ secretion in distal tubule of *Amphiuma*, *Am. J. Physiol.*, 252, F691–F699, 1987.

457. Stanton, B., Biemesderfer, D., Stetson, D., Kashgarian, M., and Giebisch, G., Cellular ultrastructure of *Amphiuma* distal nephron: effects of exposure to potassium, *Am. J. Physiol.*, 247, C204–C216, 1984.

458. Steen, W. B., On the permeability of the frog's bladder to water, *Anat. Rec.*, 43, 215–220, 1929.

459. Steinmetz, P. R., Characteristics of hydrogen ion transport in urinary bladder of water turtle, *J. Clin. Invest.*, 46, 1531–1540, 1967.

460. Steinmetz, P. R., Cellular organization of urinary acidification, *Am. J. Physiol.*, 251, F173–F187, 1986.

461. Sten-Knudsen, O., *Biological Membranes. Theory of Transport, Potentials and Electric Impulses*, Cambridge University Press, Cambridge, U.K., 2002.

462. Stetson, D. L., Beauwens, R., Palmosano, J., Mitchell, P. P., and Steinmetz, P. R., A double membrane model for urinary bicarbonate secretion, *Am. J. Physiol.*, 249, F546–F552, 1985.

463. Stille, W. T., The nocturnal amphibian fauna of the southern Lake Michigan beach, *Ecology*, 33, 149–162, 1952.

464. Stille, W. T., The water absorption response of an anuran, *Copeia*, 1958, 217–218, 1958.

465. Stinner, J. N. and Hartzler, L. K., Effect of temperature and pH and electrolyte concentration in air-breathing ectotherms, *J. Exp. Biol.*, 203, 2065–2074, 2000.

466. Stoddard, J. S., Jacobson, E., and Helman, S. I., Basolateral membrane chloride transport in isolated epithelia of frog skin, *Am. J. Physiol.*, 249, C318–C329, 1985.

467. Stoner, L., Isolated perfused amphibian renal tubules: the diluting segment, *Am. J. Physiol.*, 233, F438–F444, 1977.

468. Stoner, L. C. and Viggiano, S. C., Environmental KCl causes an upregulation of apical membrane maxi K and ENaC channels in everted *Ambystoma* collecting tubule, *J. Membr. Biol.*, 162, 107–116, 1998.

469. Strum, J. M. and Danon, D., Fine structure of the urinary bladder of the Bullfrog (*Rana catesbeiana*), *Anat. Rec.*, 178, 15–40, 1973.

470. Suzuki, M., Hasegawa, Y., Ogushi, Y., and Tanaka, S., Amphibian aquaporins and adaptation to terrestrial environments: a review, *Comp. Biochem. Physiol. A*, 148 72–81, 2007.

471. Takada, M. and Kasai, M., Prolactin increases open-channel density of epithelial Na^+ channels in adult frog skin, *J. Exp. Biol.*, 206, 1319–1323, 2003.

472. Takada, M. and Hokari, S., Prolactin increases Na^+ transport across adult bullfrog skin via stimulation of both ENaC and Na^+/K^+-pump, *Gen. Comp. Endocrinol.*, 151 325–331, 2007.

473. Takada, M., Yai, H., and Takayama-Arita, K., Corticoid-induced differentiation of amiloride-blockable active Na^+ transport across larval bullfrog skin *in vitro*, *Am. J. Physiol.*, 268, C218–C226, 1995.

474. Takada, M., Yai, H., and Takayama-Arita, K., Prolactin inhibits corticoid-induced differentiation of active Na^+ transport across cultured frog tadpole skin, *Am. J. Physiol.*, 269, C1326–C1331, 1995.

475. Takada, M., Yai, H., Takayama-Arita, K., and Komazaki, S., Prolactin enables normal development of Ach-stimulated current in cultured larval bullfrog skin, *Am. J. Physiol.*, 271, C1059–C 1063, 1996.

476. Takata, K., Matsuzaki, T., and Tajika, Y., Aquaporins: water channel proteins of the cell membrane, *Prog. Histochem. Cytochem.*, 39, 1–83, 2004.

477. Tanii, H., Hasegawa, T., Hirakawa, N., Suzuki, M., and Tanaka, S., Molecular and cellular characterization of a water-channel protein, AQP-h3, specifically expressed in the frog ventral skin, *J. Membr. Biol.*, 188, 43–53, 2002.

478. Taplin, L. E., Osmoregulation in crocodilians, *Biol. Rev.*, 63, 333–377, 1988.

479. Taugner, R., Schiller, A., and Ntokalou-Knittel, S., Cells and intercellular contacts in glomeruli and tubules of the frog kidney: a freeze-fracture and thin-section study, *Cell Tissue Res.*, 226, 589–608, 1982.

480. Taylor, E. H., A new ambystomid salamander adapted to brackish water, *Copeia*, 1943, 151–156, 1943.

481. Teulon, J. and Anagnostopoulos, T., The electrical profile of the distal tubule in *Triturus* kidney, *Pflügers Arch.*, 395, 138–144, 1982.

482. Teulon, J., Froissart, P., and Anagnostopoulos, T., Electrochemical profile of K^+ and Na^+ in the amphibian early distal tubule, *Am. J. Physiol.*, 248, F266–F271, 1985.

483. Thompson, I. G. and Mills, J. W., Isoproterenol induced current in glands of frog skin, *Am. J. Physiol.*, 241, C250–C257, 1981.

484. Thompson, I. G. and Mills, J. W., Chloride transport in glands of frog skin, *Am. J. Physiol.*, 244, C221–226, 1983.

485. Thorson, T. B. and Svihla, A., Correlation of the habitats of amphibians with their ability to survive the loss of body water, *Ecology*, 24, 374–381, 1943.

486. Townsend, D. S., Stewart, M. M., and Pough, F. H., Male parental care and its adaptive significance in a neotropical frog, *Anim. Behav.*, 32, 421–431, 1984.

487. Townson, R., *Physiological observations on the amphibia. Dissertation the Second. Respiration continued; with a fragment upon the subject of absorption*, Göttingen, 1795.

488. Tran, D.-Y., Hoff, K., and Hillyard, S. D., Effects of angiotensin II and bladder condition on hydration behavior and water uptake in the toad, *Bufo woodhouseii*, *Comp. Biochem. Physiol. A*, 103, 127–130, 1992.

489. Tran, U., Pickney, N. M., Ozpolat, B. D., and Wessley, O., *Xenopus* bicaudal-C is required for the differentiation of the amphibian pronephros, *Dev. Biol.*, 307, 152–164, 2007.

490. Uchiyama, M., Sites of action of arginine vasotocin in the nephron of the bullfrog kidney, *Gen. Comp. Endocrinol.*, 94, 366–373, 1994.

491. Uchiyama, M. and Yoshizawa, H., Nephron structure and immunohistochemical localization of ion pumps and aquaporins in the kidneys of frogs inhabiting different environments. In *Osmoregulation and Drinking in Vertebrates, Symposium of the Society of Experimental Biology*, Hazon, N. and Flik, G., Eds., BIOS, Oxford, U.K., 2002, pp. 109–128.

492. Uchiyama, M. and Konno, N., Hormonal regulation of ion and water transport in anuran amphibians, *Gen. Comp. Endocrinol.*, 147, 54–61, 2006.

493. Uchiyama, M., Murakami, T., Yoshizawa, H., and Wakasugi, C., Structure of the kidney in the crab-eating frog, *Rana crancrivora, J. Morphol.*, 204, 147–156, 1990.

494. Uochi, T., Takahashi, S., Ninomiya, H., Fukui, A., and Asashima, M., The Na^+,K^+-ATPase alpha subunit requires gastrulation in the *Xenopus* embryo, *Dev. Growth Differ.*, 39, 571–580, 1997.

495. Urbach, V., Van Kerkhove, E., and Harvey, B. J., Inward rectifier potassium channels in basolateral membranes of frog skin epithelium, *J. Gen. Physiol.*, 103, 583–604, 1994.

496. Ussing, H. H., The distinction by means of tracers between active transport and diffusion, *Acta Physiol. Scand.*, 19, 43–56, 1949.

497. Ussing, H. H., Volume regulation in frog skin epithelium, *Acta Physiol. Scand.*, 114, 363–369, 1982.

498. Ussing, H. H., Volume regulation and basolateral membrane cotransport of sodium, potassium and chloride ions in frog skin epithelium, *Pflügers Arch.*, 405(Suppl. 1), S1–S7, 1985.

499. Ussing, H. H. and Zerhan, K., Active transport of sodium as the source of electric current in the short-circuited isolated frog skin, *Acta Physiol. Scand.*, 23, 110–127, 1951.

500. Ussing, H. H. and Andersen, B., The relation between solvent drag and active transport of ions. In *Proceedings of the 3rd Congress of Biochemistry*, pp. 434–440, Brussels, 1956.

501. Ussing, H. H. and Windhager, E. E., The nature of the shunt path and active transport through frog skin epithelium, *Acta Physiol. Scand.*, 23, 110–127, 1964.

502. Ussing, H. H. and Johansen, B., Anomalous transport of sucrose and urea in toad skin, *Nephron*, 6, 317–328, 1969.

503. Ussing, H. H., Kruhøffer, P., Thaysen, J. H., and Thorn, N. A., The alkali metal ions in biology. In *Handbuch der experimentellen Pharmakologie 13*, Eichler, O. and Farah, A., Eds., Springer-Verlag, Berlin, 1960.

504. van't Hoff, J. H., Lois de l'equilibre chimique dans l'etat dilud on dissous, *Kungliga Svenska Vetenskaps-Akademiens Handlingar XXI*, 1885.

505. Van Beurden, E. K., Survival strategies of the water-holding frog, *Cyclorana platycephalus*. In *Arid Australia*, Cogger, H. G. and Cameron, E. E., Eds., Australian Museum, Sydney, 1984, pp. 223–234.

506. Van Driessche, W., Physiological role of apical potassium ion channels in frog skin, *J. Physiol.*, 356, 79–95, 1984.

507. Van Driessche, W., Noise and impedance analysis. In *Methods in Membrane and Transporter Research*, Shafer, J. A., Giebisch, G., Kristensen, P., and Ussing, H. H., Eds., Landes, Georgetown, TX, 1994, pp. 22–80.

508. Van Driessche, W. and Lindemann, B., Concentration-dependence of currents through single sodium-selective pores in frog skin, *Nature*, 283, 519, 1979.

509. Van Driessche, W. and Zeiske, W., Spontaneous fluctuations of potassium channels in apical membrane of frog skin, *J. Physiol.*, 299, 101–116, 1980.

510. Van Driessche, W. and Zeiske, W., Ca^{2+}-sensitive, spontaneously fluctuating cation channels in the apical membrane of the adult frog skin, *Pflügers Arch.*, 405, 250–259, 1985.

511. Vanatta, J. C. and Frazier, L. W., The epithelium of *Rana pipiens* secretes H^+ and NH_4^+ in acidosis and HCO_3^- in alkalosis, *Comp. Biochem. Physiol. A*, 68, 511–513, 1981.

512. Viborg, A. L. and Rosenkilde, P., Angiotensin II elicits water seeking behavior and the water absorption response in the toad *Bufo bufo*, *Hormones Behav.*, 39, 225–231, 2001.

513. Viborg, A. L. and Rosenkilde, P., Water potential receptors in the skin regulate blood perfusion in the ventral pelvic patch of toads, *Physiol. Biochem. Zool.*, 77, 39–49, 2004.

514. Viborg, A. L. and Hillyard, S. D., Cutaneous blood flow and water absorption by dehydrated toads, *Physiol. Biochem. Zool.*, 78, 394–404, 2005.

515. Viborg, A. L., Wang, T., and Hillyard, S. D., Cardiovascular and behavioral changes during water absorption in toads, *Bufo alvarius* and *Bufo marinus*, *J. Exp. Biol.*, 209, 834–844, 2006.

516. Viertel, B., Salt tolerance of *Rana temporaria*: spawning site selection and survival during embryonic development (Amphibia, Anura), *Amphibia-Reptilia*, 20, 161–171, 1999.

517. Viertel, B. and Richter, S., Anatomy. In *Tadpoles: The Biology of Anuran Larvae*, McDiarmid, R. W. and Altig, R., Eds., The University of Chicago Press, 1999, pp. 132–140.

518. Virkki, L. V., Franke, C., Somieski, P., and Boron, W. F., Cloning and functional characterization of a novel aquaporin from *Xenopus laevis* oocytes, *J. Biol. Chem.*, 277, 40610–40616, 2002.

519. Vize, P. D., Embryonic kidneys and other nephrogenic models. In *The Kidney: From Normal Development to Congenital Disease*, Vize, P. D., Woolf, A. S., and Bard, J. B. L., Eds., Academic Press, London, 2003, pp. 1–6.

520. Vize, P. D., The chloride conductance channel ClC–K is a specific marker for the *Xenopus* pronephric distal tubule and duct, *Gene Expr. Patterns*, 3, 347–350, 2003.

521. Voûte, C. L. and Ussing, H. H., Some morphological aspects of active sodium transport: the epithelium of the frog skin, *J. Cell Biol.*, 36, 625–639, 1968.

522. Voûte, C. L. and Meier, W., The mitochondria-rich cell of frog skin as hormone sensitive "shunt path," *J. Membr. Biol.*, 40, 141–165, 1978.

523. Wade, J. B., Membrane structural specialization of the toad urinary bladder revealed by the freeze fracture technique. II. The mitochondria-rich cell, *J. Membr. Biol.*, 29, 111–126, 1975.

524. Wake, M. H., Evolutionary morphology of the caecilian urogenital system. II. The kidneys and urogenital ducts, *Acta Anat.*, 75, 321–358, 1970.

525. Walker, A. M. and Hudson, C. L., The reabsorption of glucose from the renal tubule in Amphibia and the action of phlorhizin upon it, *Am. J. Physiol.*, 118, 130–143, 1936.

526. Walker, A. M. and Hudson, C. L., The role of the tubule in the excretion of urea by the amphibian kidney, *Am. J. Physiol.*, 118, 153–166, 1936.

527. Walker, A. M. and Hudson, C. J., The role of the tubule in the excretion of inorganic phosphate by the amphibian kidney, *Am. J. Physiol.*, 118, 167–173, 1936.

528. Walker, A. M., Hudson, C. L., Findley, T. J., and Richards, A. N., The total molecular concentration and the chloride concentration of fluid from different segments of the renal tubule of Amphibia, *Am. J. Physiol.*, 118, 121–129, 1936.

529. Wang, W., Dietl, P., and Oberleithner, H., Evidence for Na^+ dependent rheogenic HCO_3^- transport in fused cells of frog distal tubules, *Pflügers Arch.*, 408, 291–299, 1987.

530. Warburg, M. R. and Goldenberg, S., The changes in osmoregulatory effects of prolactin during the life cycle of two urodeles, *Comp. Biochem. Physiol.*, 61A, 321–324, 1978.

531. Wearn, J. T. and Richards, A. N., Observations on the composition of glomerular urine, with particular reference to the problem of reabsorption in the renal tubules, *Am. J. Physiol.*, 71, 209–227, 1924.

532. Weissberg, J. and Katz, U., Effect of osmolality and salinity adaptation on cellular composition and potassium uptake of erythrocytes from the euryhaline toad, *Bufo viridis*, *Comp. Biochem. Physiol.*, 52A, 165–169, 1975.

533. Welsch, U. and Storch, V., Elektronmikroskopische Beobachtungen am Nephron adulter Gymnophionen (*Ichthyophis kohtaoensis* Taylor), *Zoologisches Jahrbuch der Anatomie*, 90, 311–322, 1973.

534. Wentzel, L. A., McNeil, S. A., and Toews, D. P., The role of the lymphatic system in water balance processes in the toad, *Bufo marinus* (L), *Physiol. Zool.*, 66, 307–321, 1993.

535. White, B. A. and Nicol, C. S., Prolactin receptors in *Rana catesbeiana* during development and metamorphosis, *Science*, 204, 851–853, 1979.

536. White, H. L., Observations on the nature of glomerular activity, *Am. J. Physiol.*, 90, 689–704, 1929.

537. Whitear, M., Flask cells and epidermic dynamics in frog skin, *J. Zool.*, 175, 107–149, 1975.

538. Whittouck, P. J., Modification du retention du sodium chez *Ambystoma mexicanum* (axolotl) intact et hypophysectomise sous l'effect du prolactine, *Arch. Int. Physiol. Biochem.*, 80, 373–381, 1972.

539. Whittouck, P. J., Intensification par la prolactine de l'absorption d'ions sodium au niveau des branchies isolees de larves d' *Ambystoma mexicanum*, *Arch. Int. Physiol. Biochem.*, 80, 825–827, 1972.

540. Wiederholt, M. and Hansen, L. L., *Amphiuma* kidney as a model for distal tubular transport studies, *Contrib. Nephrol.*, 19, 28–32, 1980.

541. Wiederholt, M., Sullivan, W. J., and Giebisch, G., Potassium and sodium transport across single distal tubules of *Amphiuma*, *J. Gen. Physiol.*, 57, 495–525, 1971.

542. Willumsen, N. J. and Larsen, E. H., Membrane potentials and intracellular Cl^- activity of toad skin epithelium in relation to activation and deactivation of the transepithelial conductance, *J. Membr. Biol.*, 94, 173–190, 1986.

543. Willumsen, N. J. and Larsen, E. H., Identification of anion selective channels in the basolateral membrane of mitochondria-rich epithelial cells, *J. Membr. Biol.*, 157, 255–269, 1997.

544. Willumsen, N. J., Vestergaard, L., and Larsen, E. H., Cyclic-AMP and β-agonist activated chloride conductance of a toad skin epithelium, *J. Physiol.*, 449, 641–643, 1992.

545. Willumsen, N. J., Amstrup, J., Møbjerg, N., Jespersen, Å., Kristensen, P., and Larsen, E. H., Mitochondria-rich cells as experimental model in studies of epithelial chloride channels, *Biochim. Biophys. Acta*, 1566, 28–43, 2002.

546. Withers, P. C., Hillman, S. S., Drewes, R. C., and Sokol, O. M., Water loss and nitrogen excretion in sharp-nosed tree frogs (*Hyperolius nausatus*, Anura, Hyperoliidae), *J. Exp. Biol.*, 97, 335–343, 1982.

547. Word, J. M. and Hillman, S. S., Osmotically absorbed water preferentially enters the cutaneous capillaries of the seat patch in the toad, *Bufo marinus*, *Physiol. Biochem. Physiol.*, 78, 40–47, 2005.

548. Wright, P. A., Nitrogen excretion: three end products, many physiological roles, *J. Exp. Biol.*, 198, 273–281, 2004.

549. Wrobel, K.-H. and Süß, F., The significance of rudimentary nephrostomial tubules for the origin of the vertebrate gonad, *Anat. Embryol.*, 201, 273–290, 2000.

550. Yokota, S. D. and Hillman, S. S., Adrenergic control of the anuran cutaneous hydroosmotic response, *Gen. Comp. Endocrinol.*, 53, 309–314, 1984.

551. Yucha, C. B. and Stoner, L. C., Bicarbonate transport by amphibian nephron, *Am. J. Physiol.*, 251, F865–F872, 1986.

552. Yucha, C. B. and Stoner, L. C., Bicarbonate transport by initial collecting tubule of aquatic- and land-phase amphibia, *Am. J. Physiol.*, 253, F310–F317, 1987.

553. Zaccone, G., Fasulo, S., Lo Cascio, P., and Licata, A., Enzyme cytochemical and immunocytochemical studies of flask cells in the amphibian epidermis, *Histochemistry*, 84, 5–9, 1986.

554. Zadunaisky, J. A., Candia, O. A., and Chiarandini, D. J., The origin of the short-circuit current in the isolated skin of the South American frog *Leptodactyllus ocellatus*, *J. Gen. Physiol.*, 47, 393–402, 1963.

555. Zemanova, Z. and Gambaryan, S., Blood supply to the amphibian mesonephros, *Funct. Dev. Morphol.*, 3, 231–236, 1993.

556. Zhou, X. and Vize, P. D., Proximo-distal specialization of epithelial transport processes within *Xenopus* pronephric kidney tubules, *Dev. Biol.*, 27, 322–338, 2004.

557. Zhou, X. and Vize, P. D., Amino acid cotransporter SLC3A2 is selectively expressed in the early proximal segment of *Xenopus* pronephric kidney nephrons, *Gene Expr. Patterns*, 5, 774–777, 2005.

558. Zhou, X. and Vize, P. D., Pronephric regulation of acid–base balance, co-expression of carbonic anhydrase type 2 and sodium bicarbonate cotransporter-1 in the late distal segment, *Dev. Dyn.*, 233, 142–144, 2005.

559. Zimmerman, S. L., Frisbie, J., Goldstein, D. L., West, J., Rivera, K., and Krane, C. M., Excretion and conservation of glycerol and expression of aquaporins and glyceroporins during cold acclimation in Cope's gray tree frog, *Hyla chrysoscelis*, *Am. J. Physiol.*, 292, R544–R555, 2007.

10 Osmotic and Ionic Regulation in Reptiles

William H. Dantzler and S. Donald Bradshaw

CONTENTS

I. THE OSMOTIC ANATOMY OF THE REPTILES

A. CONCEPT OF HOMEOSTASIS

Claude Bernard was the first to advance the concept of a protected *milieu intérieur* composed of the physical elements within the tissues which are maintained at levels and concentrations different from those found in the external environment by the operation of homeostatic mechanisms.[23] Bernard did not use the term *homeostasis*; Walter Cannon coined it in 1929.[64] The notion of homeostatic mechanisms regulating the functional behavior of machines is now commonplace and has been a powerful organizing paradigm in biology. It is important to note, however, one important difference between machines and living organisms. Homeostasis in engineered artifacts results from inbuilt design constraints, whereas that in animals is more often the result of purposive behavior, which may be modulated by specific chemicals secreted internally in the form of hormones. Evidence of homeostasis in animals may thus be sought at two levels: in the temporal stability of the chemical structure of the tissues themselves and by the identification of hierarchical control systems impacting on the composition of the tissues through the maintenance of set points.

TABLE 10.1
Reported Variation in Fluid Distribution and Plasma Electrolyte Concentrations in Reptiles

Group	TBW	ICFV	ECFV	PV	[Na⁺]	[K⁺]
Crocodiles	75.5 ± 0.9 (9)	58.0 ± 0.3 (3)	14.9 ± 0.2 (3)	3.5 ± 0.1 (3)	145.8 ± 2.9 (27)	4.2 + 0.2 (23)
Turtles	67.7 ± 1.3 (19)	46.8 ± 1.9 (9)	22.2 ± 1.3 (13)	6.2 ± 0.5 (18)	144.2 ± 4.3 (26)	5.8 ± 0.7 (20)
Lizards	71.5 ± 1.8 (34)	41.8 ± 1.9 (20)	27.9 ± 1.4 (21)	5.6 ± 0.3 (20)	167.5 ± 3.2 (52)	4.8 ± 0.2 (47)
Snakes	70.9 ± 0.5 (3)	53.8 ± 0.5 (2)	31.1 ± 8.4 (4)	4.1 ± 0.1 (2)	169.1 ± 6.5 (22)	6.5 ± 0.6 (20)

Note: TBW, total body water content; ICFV, intracellular fluid volume; ECFV, extracellular fluid volume; PV, plasma volume (all in mL/100 g); [Na⁺], plasma sodium concentration; [K⁺], plasma potassium concentration (both in mmol/L). Data are expressed as mean ± SE, and numbers in parentheses are the number of animals.

Source: Bradshaw, S.D., *Homeostasis in Desert Reptiles*, Springer, Heidelberg, 1997, pp. 1–213. With permission.

B. A COMPARATIVE ACCOUNT

Bentley[15] briefly reviewed the chemical composition of reptilian plasma and intracellular fluid in relation to other vertebrates in 1971 and again, in more detail, in 1976.[16] Bradshaw[44] specifically examined the question of homeostasis in reptiles and drew together published data on vertebrate electrolyte concentrations and tissue fluid distributions in the various reptilian taxa. This comparison suggested that plasma sodium concentrations are less constant in lower vertebrates (fish, amphibians, and reptiles) than in birds and mammals. When data are available from more than a single season or a single milieu, the variations for fish, amphibians, and reptiles fall in the range of 30 to 60%, whereas seasonal variations for birds and mammals are much lower, of the order of 7%. This trend is not universal, however, as with the case of one desert lizard from Australia (*Ctenophorus nuchalis*) for which the variation in plasma sodium concentration recorded between spring and summer was less than 1%.[224]

C. WATER AND ELECTROLYTES IN REPTILES AS A GROUP

Comparing published values on body water distribution as well as plasma electrolyte concentrations from a wide range of reptiles is a very crude way of investigating homeostasis, but it does provide some idea of the extent to which different reptilian species maintain a constant *milieu intérieur* and whether this is linked in any way with the particular environment the species inhabits. Bradshaw[44] made such a comparison, and we summarize the principal results in the following paragraphs and Table 10.1. Crocodiles appear to differ from chelonians in having significantly higher total body water content (TBW). The extracellular fluid volume (ECFV) of lizards is significantly larger than that of crocodiles and chelonians, and, as would be expected, their intracellular fluid volume (ICFV) is correspondingly smaller. Plasma volume (PV), however, does not differ significantly between the groups (see Bradshaw[44] for details of the statistical treatments).

Plasma potassium concentrations are very much higher at 6.5 ± 0.6 mmol/L in snakes than in either lizards or crocodiles. Plasma potassium levels of Chelonia are also significantly higher than those of crocodiles. Corresponding values for plasma sodium levels show that the mean for lizards (167.5 ± 3.2 mmol/L) is not significantly different from that of snakes but much higher than the levels found in both crocodiles and chelonians (145.8 ± 2.8 and 144.1 ± 4.3 mmol/L, respectively). Plasma sodium levels in snakes are also significantly higher than in both crocodiles and Chelonia, and the two terrestrial groups thus differ from the two aquatic taxa (with $F_{3,123} = 10.38$, $P < 0.0001$ for the ANOVAR), although it is unclear whether this results from dietary or habitat differences. From these data we can draw a number of simple conclusions:

- Crocodiles have high TBW and an ICFV expanded at the expense of the ECFV and PV, which is evident in both freshwater and terrestrial species.
- Chelonians have more normal fluid volumes but high and quite variable plasma K^+ concentrations, with a coefficient of variation (CV) of 54.2%.[44]
- Lizards and snakes appear to have a high ECFV and PV that are expanded at the expense of the ICFV when compared with the crocodiles and chelonians. Large interspecific variation and small sample sizes, however, limit statistical significance to the comparison between lizards and crocodiles ($F_{2,33} = 4.92$, $P = 0.0135$).
- Plasma Na^+ levels are much higher in snakes and lizards than in both crocodilians and chelonians and more variable in snakes than in lizards, with a CV of 19.4% vs. 13.7%.
- Plasma K^+ levels are significantly higher in both snakes and chelonians when compared with crocodilians and lizards. Snakes, lizards, and chelonians all have very variable plasma K^+ concentrations, with CVs ranging from 32.8 to 54.2%.

If one searches, however, for habitat correlations in an effort to explain the above patterns, it becomes clear that the literature on lizards is quite biased toward desert species, with only five species in the dataset analyzed coming from mesic habitats and three from tropical environments.[44] Another obvious lacuna is just how few snakes have been studied from this point of view, regardless of their origins and habitat. The differences observed may thus be real or, instead, simply the result of: (1) vagaries in the literature on reptilian osmoregulation, with its inherent sampling errors and biases due to different techniques employed, and (2) variations due to factors such as season and sex ratio.

The fact that crocodiles have a high TBW and ICFV relative to the other groups may be related to their aquatic habitat, although a similar trend is not apparent in those few species of aquatic turtles in which fluid distributions have been measured. Minnich[213] suggested that TBW (based on total body mass) is lower in turtles than in other reptiles because the calculation in the case of the former takes into account the chelonians' massive bony shell. The shell mass varies between 7 and 30% of the total body mass in chelonians[124,232] and correcting for this would certainly increase the calculated TBW.

Despite the lack of any obvious explanations for the documented differences, it is clear that reptiles as a group do not share a common *milieu intérieur*, and this may vary quite significantly in some groups from what is considered the typical vertebrate model. Another consideration, however, is the fundamental question of the systematic viability of the taxon Reptilia itself. There is a growing realization that chelonians, crocodiles, snakes, lizards, the tuatara, and amphisbaenians are probably not members of a monophyletic group. From a rigorously cladistic viewpoint, reptiles do not exist, being definable only as those amniotes that are neither birds nor mammals.[216] Recent cladistic analyses have also made it very likely that, within the Reptilia, a group such as the Lacertilia, which is currently made up of lizards, is in fact a paraphyletic group that has given rise to snakes and amphisbaenians.[245] Following strict cladistic principles, this name should no longer be retained as a taxon, or, if it is, it should also include both the snakes and Amphisbaenia and would then become a synonym for the squamates.

Cladists differ in how intransigent one should be in eliminating a paraphyletic taxon with such obvious utility as lizards, and an analogous situation arises with the Sauria, which some systematists argue should also include the birds. Such problems will, we hope, be resolved in the coming years with the greater application of cladistic methodologies to the question of the systematic relatedness of the various reptilian groups. For our purposes, however, we will continue to believe in the existence of an ancient group of vertebrates known as the Reptilia, which gave rise to the birds and the mammals, and will study the means by which they regulate their *milieu intérieur* in the belief that this will genuinely give us insight into the evolution of those more complex systems that characterize the latter groups.

TABLE 10.2
Examples of Cutaneous Evaporative Water Loss Rates in Living Reptiles in Dry Air

Species	Normal Habitat	Cutaneous Water Loss (mg/cm²/day)	Species (% TEWL)
Crocodilia			
Caiman sclerops	Semiaquatic, freshwater	32.9 ± 2.45 (8)	87 ± 2.1 (8)
Testudinea			
Pseudemys scripta	Semiaquatic, freshwater	12.2 ± 1.44 (6)	78 ± 2.7 (8)
Terrapene carolina	Terrestrial, mesic	5.3 ± 0.41 (6)	76 ± 3.4 (6)
Squamata: Sauria			
Iguana iguana	Terrestrial, mesic	4.8 ± 0.50 (8)	72 ± 4.3 (8)
Saurmalus obesus	Terrestrial, xeric	1.3 ± 0.10 (6)	66 ± 2.0 (6)

Note: Values are means ± SE, and numbers in parentheses are the number of animals; TEWL, total evaporative water loss.

Source: Data from Bentley, P.J. and Schmidt-Nielsen, K., *Science*, 151, 1547–1549, 1966.

II. FORM AND FUNCTION OF PRIMARY EXCHANGE SITES FOR WATER AND ELECTROLYTES

A. SKIN

Although reptilian skin was once thought to be largely impermeable to water, it is now well known that the integument forms the principal route for evaporative water loss (about 70 to 95% of the total) in terrestrial or semiaquatic reptiles (Table 10.2).[20,67] The integument is also highly permeable to water in aquatic reptiles.[120,126,255] The rate of evaporative water loss under identical conditions, and thus integument permeability, decreases as the aridity of the habitat increases (often also given as increasing skin resistance to water loss with increasing aridity).[20,110,146,192,203,204,252,253] This pattern is found across orders and across species within a single order (Table 10.2 and Table 10.3)[20,252,253] (see also Table 2 in Lillywhite[192] for a more complete list). Indeed, this relationship has even been shown within isolated populations of a single species.[111] Moreover, the skin of some species has been shown to increase its resistance to evaporative water loss with exposure to an arid environment.[169,175] In terrestrial ophidian reptiles, the epidermal resistance to water loss also increases markedly following the first postnatal ecdysis.[307]

Among largely aquatic reptiles, the permeability of the integument to water generally decreases as the salinity of the normal aqueous habitat increases; again, this relationship is found across orders and across species within a single order (Table 10.4).[120,255] Two species of sea snakes (*Hydrophis ornatus* and *H. inornatus*), however, have an integument nearly as permeable to water as that of freshwater snakes (Table 10.4). Because these species maintain osmotic and ionic composition similarly to other marine species, the physiological significance of this relatively high water permeability is not clear.[120] Substantial sodium flux can also occur across the skin of aquatic reptiles, but among ophidian species, at least, the sodium permeability is much greater in freshwater species than in estuarine and marine species.[120] The high rate of sodium influx in freshwater species appears to be correlated with their intolerance of being placed in seawater.[120]

The epidermis of reptiles forms the limiting barrier for exchange of water and, in aquatic species, ions with the environment. Within the epidermis, lipids form the principal permeability barrier for diffusion of water in all those vertebrates studied.[157,158] This was first clearly demonstrated for reptiles by Roberts and Lillywhite,[253] who found that extraction of lipids from shed skins of

TABLE 10.3
Examples of Water Loss Rates through Shed Skin of Reptiles

Species	Normal Habitat	EWL (mg/cm²/hr) Untreated	EWL (mg/cm²/hr) Extracted	Lipid Content (% Dry Weight)
Ophidia				
Acrochordus javanicus	Aquatic, freshwater	0.50 ± 0.08 (10)	1.30 ±0.13 (8)	2.43
Nerodiar hombifera	Semiaquatic, freshwater	0.41 ± 0.13 (7)	2.62 ± 0.50 (9)	4.30
Elaphe obsoleta	Terrestrial, mesic	0.22 ± 0.01 (39)	2.53 ± 0.34 (44)	5.91
Crotalus adamanteus	Terrestrial, mesic	0.22 ± 0.07 (17)	2.59 ± 0.33 (21)	5.89
Crotalus viridis	Terrestrial, semiarid	0.14 ± 0.04 (4)	2.92 ± 0.38 (6)	7.40
Crotalus cerastes	Terrestrial, xeric	0.16 ± 0.04 (8)	2.40 ± 0.40 (10)	8.61
Sauria				
Iguana iguana	Terrestrial, mesic	1.16 ± 0.09 (5)	1.98 ± 0.12 (4)	

Note: Values are means ± SE, and numbers in parentheses are the number of measurements; EWL, evaporative water loss.

Source: Data from Roberts, J.B. and Lillywhite, H.B., *J. Exp. Zool.*, 228, 1–9, 1983; Roberts, J.B. and Lillywhite, H.B., *Science*, 207, 1077–1079, 2007.

TABLE 10.4
Examples of Efflux and Influx of Water in Reptiles in Seawater

Species	Normal Habitat	Water Efflux (mL/100 g/hr)	Water Influx (mL/100 g/hr)
Testudinea[a]			
Chrysemys picta	Semiaquatic, freshwater	—	0.72 ± 0.11 (3)
Malaclemys terrapin	Semiaquatic, estuarine	0.16 ± 0.05 (11)	0.17 ± 0.03 (11)
Squamata: Ophidia[b]			
Nerodia sipedon	Semiaquatic, freshwater	1.54 ± 0.54 (5)	1.33 ± 1.19 (5)
Nerodia fasciata pictiventris	Semiaquatic, freshwater	2.84 ± 2.00 (4)	2.54 ± 2.49 (4)
Acrochordus granulatus	Semiaquatic, estuarine	0.49 ± 0.04 (4)	0.47 ± 0.19 (4)
Laticauda laticauda	Aquatic, seawater	0.20 ± 0.06 (4)	0.17 ± 0.05 (4)
Hydrophis belcheri	Aquatic, seawater	0.61 ± 0.06 (2)	0.54 ± 0.21 (2)
Hydrophis ornatus	Aquatic, seawater	1.26 ± 0.27 (8)	1.08 ± 0.32 (8)
Hydrophis inornatus	Aquatic, seawater	1.18 ± 0.52 (4)	1.19 ± 0.28 (4)

[a] Data from Robinson, G.D. and Dunson, W.A., *J. Comp. Physiol.*, 105, 129–152, 1976.

[b] Data from Dunson, W.A., *Am. J. Physiol.*, 235, R151–R159, 1978.

Note: Values are means ± SD, and numbers in parentheses are the number of measurements. Data were determined from isotopic fluxes.

ophidian and saurian species eliminated the barrier to evaporative water loss (Table 10.3), whereas denaturation of proteins had little effect. These findings were confirmed in later studies on the skin of ophidian reptiles by these authors and others.[252,293]

Roberts and Lillywhite[253] also demonstrated that the lipids involved in water permeation are located in the mesos layer of the squamate epidermis. This epidermal layer consists of cells derived from the α keratins that lie below it, as well as of the extracellular, laminated lipids that form the water barrier.[192,193] Layers of β keratin, which apparently serve a structural function, overlie the mesos layer.[192,193] Lamellar granules within epidermal cells (usually only in the mesos layer but also in the α layer in *Sphenodon punctatus* and some ophidians)[3,307] extrude the lipids into the extracellular compartment of the mesos layer, where they form multiple bilayer sheets.[178,181,192] These sheets are composed primarily of highly saturated, unbranched, long-chain ceramides.[316] In addition, the epidermal lipids of reptiles, like those of birds, include unusual glycolipids, glucosylsterol, and acylglucosylsterol.[2] In the reptilian skin, the glucosylsterol consists of glucose attached to cholesterol, whereas the acylglucosylsterol also includes palmitic, stearic, or oleic fatty acid ester-linked to the carbon 6 of glucose.[2] Because detailed studies of the lipid composition of reptilian epidermis have been made on only a few species, it is not known what differences in lipids, if any, occur between species from arid and mesic environments, with adaptation of a single mesic species to a more arid environment or with growth and development. Roberts and Lillywhite,[252] however, found that the lipid content of the ophidian epidermis increases as the water permeability decreases (Table 10.3). It may be that the quantity of the lipids, not the composition, is the primary factor determining differences or changes in water permeability.

Recently, aquaporin 3 (AQP3) water channels have been found in mammalian epidermis,[290] although these are apparently involved in maintaining hydration of the epidermis, not in transepidermal water permeability. Whether a functioning ortholog of AQP3 is found in the reptilian epidermis is unknown.

How inorganic ions permeate the epidermis of aquatic reptiles is not clearly understood. Inorganic ions do not easily cross lipid membranes. In general, transmembrane ion movements require appropriate protein channels or transporters; however, as noted above, significant sodium movement can occur across the skin of aquatic reptiles or even across the skin of terrestrial ophidian reptiles when the shed skin is maintained in an aqueous medium.[293] Moreover, sodium permeation can vary, perhaps with environment, because whereas the skin of all freshwater snakes studied shows relatively high sodium exchange with the aqueous medium, that of at least one sea snake (*Pelamis platurus*) does not.[126,293] Moreover, Stokes and Dunson[293] found that sodium flux across ophidian skin increases dramatically when lipids are extracted but only modestly or not at all when proteins are extracted. They also reported that there was a directional asymmetry in the tracer fluxes of sodium and water that was also largely abolished with lipid extraction.[293] Although this asymmetry has yet to be confirmed and remains controversial, the data indicate that lipids play a significant role in the permeability of the skin to ions as well as water.

It is possible that vascular perfusion could play some role in the magnitude of diffusion of water and ions across the epidermis of living reptiles. This would, of course, depend on the relationship between the vascular delivery rate of water and ions and the permeability of the epidermis to these substrates. The role of perfusion in epidermal water and ion fluxes has yet to be studied in any detail. Finally, it must be emphasized that most of the data on the permeability of the reptilian integument discussed above are from studies on squamates, particularly ophidians; very few are from studies on chelonians or crocodilians.

B. KIDNEY

1. Introduction

In reptiles, as in other vertebrates, the kidneys play a critical role in regulating the composition of the *milieu intérieur* by controlling the urinary excretion of water, ions, and nitrogenous wastes; however, in some reptile species salt glands help to excrete ions, and in all reptile species structures distal to the kidneys (colon, cloaca, or bladder) can modify the ionic and aqueous composition of the urine before it is finally excreted. Nevertheless, it is reasonable to say that the kidneys have a greater degree of control and are quantitatively more important than any other structure in regulating the output of ions and water and, thus, the composition of the body fluids. Determination of the urinary output of ions and water involves regulation of both filtration at the renal glomerulus and reabsorption and secretion along the renal tubules.

2. Morphology

a. General Form of Kidneys and Nephrons

The external morphology of reptilian kidneys varies a great deal because of the marked variation in body form within the class Reptilia; for example, the kidneys of lizards tend to be compact, somewhat triangular-shaped structures joined at the posterior end, whereas those of snakes are long and thin and those of turtles are constrained by the shape of the carapace[45] (W.H. Dantzler, pers. observ.). These varying kidney shapes also constrain the gross arrangement of the nephrons, the actual functional units of the kidneys; for example, in snake kidneys, the nephrons lie side by side in neat parallel rows and attach at roughly right angles to major collecting ducts, whereas in lizard kidneys the nephrons branch off the collecting ducts more obliquely and are arranged in compact bunches.[229] The number of nephrons in the kidneys of different species also varies. Despite these differences in arrangement, the basic components of the nephrons in all reptiles are glomerulus, short ciliated neck segment, proximal tubule, short ciliated thin intermediate segment, and distal tubule.[86] A few nephrons without glomeruli have been described in the kidney of one species of lizard[229] and in some snakes.[247] No reptilian nephrons have the long loops of Henle between the proximal and distal tubules arranged parallel to collecting ducts that are found in avian and mammalian nephrons. Despite the general differences in nephron arrangement among the orders of living reptiles, all nephrons examined are arranged in such a fashion that the beginning of the distal tubule is closely apposed to the vascular pole, apparently primarily the afferent arteriole, of its own glomerulus[164,229] (S.D. Yokota, R.A. Wideman, and W.H. Dantzler, unpubl. observ.).

b. General Arrangement of Blood Vessels

The number and arrangement of arterial vessels supplying reptilian kidneys vary greatly among the orders and among species within orders, and the arterial blood flow to the kidneys probably varies accordingly. The arterial supply, for example, appears to be extensive in garter snakes (*Thamnophis* spp.) and freshwater snakes (*Nerodia* spp.) (W.H. Dantzler, pers. observ.) but minimal in at least one agamid lizard species (*Ctenophorus ornatus*).[229] In all species examined, however, each glomerulus is supplied by an afferent arteriole that breaks up into the glomerular capillary network. As noted by Bowman[34] over 150 years ago, in reptiles the glomerular capillaries are larger and fewer, the anastomoses are fewer, and, thus, the capillary network is less complex than in mammals.[4,229,236] This relatively simple network is probably required to permit passage of the relatively large and rigid nucleated red blood cells found in reptiles. The glomerular capillaries unite at their efferent end to form a single efferent arteriole that leaves the glomerulus and breaks up into another capillary network surrounding and supplying the renal tubules. Of particular significance, all reptiles have a renal venous portal system that contributes vessels to this peritubular capillary network and supplies it with blood from the posterior regions of the body. The capillaries in this network eventually unite to form the renal veins that drain the kidneys.

c. Detailed Structure

i. Glomerulus

The reptilian glomerulus, as in other vertebrates, is composed of the capillary network described above, a central region of mesangial cells, the parietal layer of Bowman's capsule covering the outside of the capillary filtration surface and the podocytes, and the visceral layer of Bowman's capsule, which is continuous with the epithelium of the proximal tubule.[4,236] The overall ultrastructure of the filtration barrier is also similar to that of other vertebrates and consists of capillary endothelium, basement membrane, and visceral epithelial cells (podocytes) with filtration slits covered by filtration slit diaphragms.[4,236] The podocytes cover more and the filtration slits cover less filtration surface area in reptiles than in mammals,[236] and the fenestrae in the glomerular capillary endothelium are much more numerous in reptiles than in hagfish, lampreys, elasmobranchs, and teleosts but much less numerous than in mammals.[4,85,236] Also, as in other vertebrates, the endothelial cells on the nonfiltering side of the glomerular capillaries rest not on a basement membrane and covering podocytes but on mesangial cells,[85] which contain myofibrils[236] that may play some role in regulating the area available for filtration.[85]

ii.   Proximal Tubule

Reptilian proximal tubule cells, as in mammals, have a distinct brush border of microvilli and numerous mitochondria; however, in contrast to mammals, reptiles show no consistent changes in cell type along the length of the proximal tubule, even when functional changes occur[81] (W.H. Dantzler and R.B. Nagle, unpubl. observ.). Also, reptilian proximal tubule cells do not show the deep basal interdigitations with mitochondria arranged linearly between them observed in the cells of early segments of mammalian proximal tubules[92,237,315] (W.H. Dantzler and R.B. Nagle, unpubl. observ.). Significant amplification of the basolateral cell membrane does occur, however, in the proximal tubule cells of some but not all reptiles, and the tight junctions are always short[92,237,315] (W.H. Dantzler and R.B. Nagle, unpubl. observ.).

iii.   Intermediate Segment

This segment, where studied, is made up of low columnar cells that bear a moderate number of cilia and a few microvilli on the apical surface. The cells also have short tight junctions and uniform intercellular spaces with a few lateral interdigitations.[237]

iv.   Early Distal Tubule

A few reptiles (e.g., gecko, *Hemidactylus* sp.) have early distal tubule cells displaying the features—deep basolateral infoldings with elongated mitochondria occupying the spaces within them—common to this segment in other vertebrates, but most (e.g., horned lizard, *Phrynosoma cornutum*; Galapagos lizard, *Tropidurus* sp.; blue spiny lizard, *Sceloporus cyanogenys*; crocodile, *Crocodylus acutus*; garter snakes, *Thamnophis sirtalis*) do not.[97,98,237,254] Instead, early distal tubule cells in these species have large, often irregular nuclei, ovoid or spherical mitochondria, and extensive lateral interdigitations. The particular type of cell in any given species does not appear to be related to the ability to dilute the urine (see below).

v.   Late Distal and Collecting Tubules

Reptilian nephrons, like those of other vertebrates, have no clear distinctions in cellular structure between the distal portions of the distal tubule (often referred to as *late distal tubule*) and the collecting tubules or collecting ducts into which they empty. The late distal tubule and the connecting tubule of at least one saurian species (*Sceloporus cyanogenys*) have light and dark cells somewhat similar in appearance to the principal and intercalated cells described in other vertebrate species, although the light cells in this lizard appear to secrete mucus.[98]

vi.   *Juxtaglomerular Apparatus*

The arrangement of the early distal tubule of each nephron close to the vascular pole of its own glomerulus (see above) suggests the presence of a juxtaglomerular apparatus (JGA) like that found in mammals.[86,229] Although renin-producing cells in the glomerular arterioles and extraglomerular mesangial cells are present in reptilian nephrons, there is no evidence of a macula densa region in the early distal tubule.[86,227,289] Thus, a complete JGA (at least as defined in mammals) is not present.

3.   Glomerular Function

a.   Introduction

The initial process in urine formation is the delivery of water and solutes to the lumen of the proximal tubule of each nephron. In glomerular nephrons (almost all nephrons in reptiles), this involves ultrafiltration of plasma across the glomerular capillaries, a process first documented in reptiles by Bordley and Richards,[33] who found a protein-free filtrate in Bowman's space outside the glomerular capillaries. The rate at which such an ultrafiltrate is formed at the individual glomerulus equals the rate at which fluid is delivered to the corresponding proximal tubule and, therefore, is the initial determinant of the volume and composition of the urine. Because filtration relies on arterial hydrostatic pressure maintained for other purposes, it is well suited to excreting large volumes of fluid without expending additional energy specifically for that purpose. Moreover, because filtration can be readily altered in reptiles (see below), it can play an important role in regulating water excretion. If the few aglomerular nephrons observed in some species actually function, they must do so by the secretion of ions and water into the proximal tubule (see below), but this process is clearly not of primary importance in the initial formation of urine in reptiles.

b. Filtration Process

As in capillary beds in general, the rate of ultrafiltration at the glomerulus of a single nephron (SNGFR) is the product of the ultrafiltration coefficient (K_f) and the net ultrafiltration pressure, as expressed in the equation:

$$\text{SNGFR} = K_f[(P_{GC} - P_{BS}) - \pi_{GC}]$$

In this equation, P_{GC} is the outwardly directed hydrostatic pressure in the glomerular capillaries, P_{BS} is the inwardly directed hydrostatic pressure in Bowman's space, and π_{GC} is the colloid osmotic pressure in the capillaries that opposes filtration. The sum of these pressures is the net ultrafiltration pressure (PUF). Because proteins are retained within the capillaries, primarily as a result of filtration barriers at the basement membrane and the filtration slits, essentially no protein is found in the ultrafiltrate in Bowman's space and, therefore, no colloid osmotic pressure outside the capillaries to favor filtration. Recently, Yokota measured (in mmHg) a mean arterial pressure of 38, a mean P_{GC} of 22, a mean P_{BS} of 2, and a mean π_{GC} of 17 (afferent arteriole) in anesthetized garter snakes (*Thamnophis* spp.), the first and only such measurements made in reptiles (S.D. Yokota, pers. comm.). These data indicate that mean PUF is 3 mmHg at the afferent end of the glomerular capillary network. Because the measurements of P_{GC} were made randomly in the glomerular capillaries, it can probably be assumed that, as in mammals, P_{GC} remains almost constant along the length of the glomerular capillaries. This relatively constant P_{GC} is maintained by the resistances in the arterioles at both ends of the capillary network. As filtration progresses along the length of the capillaries and the protein in the capillaries is concentrated, π_{GC} rises. Filtration will, of course, continue only so long as the hydrostatic pressure gradient exceeds the opposing colloid osmotic pressure. It is unknown whether or not π_{GC} increases sufficiently to equal the hydrostatic pressure (i.e., PUF becomes zero) so filtration ceases along the length of the glomerular capillaries (i.e., filtration equilibrium is reached) in garter snakes (or in other reptiles). Given the low initial value of PUF, however, this seems likely. If filtration equilibrium is reached along the length of the glomerular capillaries (i.e., not all of the capillary network is used for filtration), then the SNGFR will be particularly sensitive to changes in renal plasma flow.

As noted above, the rate of glomerular ultrafiltration is a function, not only of PUF, but also of K_f. K_f has two components: the surface area of the capillaries available for filtration (*A*) and the hydraulic conductivity (L_p) or the aqueous permeability of the capillary wall. An accurate measurement of either *A* or L_p would permit determination of the other from that value, the PUF, and the SNGFR. Independent determinations of either L_p or *A* would be very useful in gauging the importance of glomerular filtration in regulating solute and water excretion in various reptile species; however, we have no direct way to determine L_p in the glomerular capillaries, and accurately determining *A* also poses many problems.

Approximations of the total surface area available for filtration per glomerulus, based on glomerular diameters, are shown for a wide range of reptile species in Table 10.5. It is apparent that the filtration surface, calculated in this manner, can vary as much as sevenfold within a reptile order. Even within an order, no clear relationship exists between glomerular size or filtration surface area and habitat, despite the suggestion that glomerular size might be related to the availability of water. Moreover, insufficient data are available to determine the extent to which these variations in glomerular size or filtration surface area might simply relate to body size.[147] Aside from the

TABLE 10.5
Glomerular Diameter and Total Surface Available for Filtration

Species	Glomerular Diameter (μm)	Filtration Surface Area per Glomerulus ($mm^2 \times 10^{-2}$)
Testudinea:		
Trionyx sinensis (FW)	154	7.45
Emys orbicularis (FW)	71	1.58
Testudo horsfieldi (D)	96	2.89
Crocodilia:		
Caiman crocodylus (FW)	138	5.98
Squamata:		
Sauria		
Teratoscincus scincus (D)	62	1.21
Agama sanguinolenta (D)	87	2.38
Phrynocephalus mystaceus (D)	114	4.08
Eremias grammica (D)	68	1.45
Eremias aguta (SD)	61	1.17
Lacerta vivipara (M)	88	2.43
Lacerta trilineata (M)	62	1.21
Eumeces schneideri (SD)	63	1.25
Mabuya aurata (SD)	58	1.06
Varanus griseus (D)	155	7.55
Ophisaurus apodus (M)	109	3.73
Ophidia		
Typhlops vermicularis (M)	78	1.91
Eryx miliaris (D)	80	2.01
Eryx jaculus (SD)	85	2.27
Natrix natrix (SA)	152	7.26
Natrix tesselata (SA)	156	7.64
Psammophis lineolatus (D)	63	1.25
Eirenis collaris (SD)	122	4.67
Eirenis punctatolineatu (SD)	102	3.27
Coluber najadum (SD)	90	2.54
Coluber ravergieri (SD)	73	1.67
Naja oxiana (D)	94	2.78
Vipera lebetina (D)	134	5.64
Vipera berus (M)	103	3.33
Thamnophis sirtalis (M)	60[a]	1.13

[a] Data from Peek, W.D. and McMillan, D.B., *Am. J. Anat.*, 155, 83–102, 1979.

Note: Value for glomerular diameter of each species is mean of 30 measurements (2 animals per species with 15 glomeruli analyzed from each). The total filtration surface area per glomerulus was estimated from the relationship $S = \pi d^2$, suggested by Renkin and Gilmore.[249] Habitat code: FW, freshwater; SA, semiaquatic; M, mesic; SD, semidesert; D, desert.

Source: Adapted from Gambaryan, S.P., *J. Morphol.*, 219, 319–339, 1994.

variations in glomerular size, these estimates of filtration surface area are not true determinations of the actual capillary surface per glomerulus; however, even a direct measurement of capillary surface in a given glomerulus would only give the area available for filtration, not the area actually

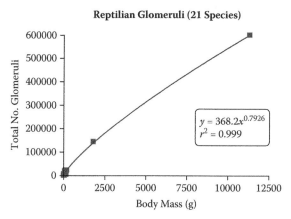

FIGURE 10.1 Relationship between body mass and number of glomeruli. The regression equation for this relationship is shown (see text).

used. The area actually used is determined by whether filtration ceases along the capillary network and, if it does, the capillary area used for filtration up to that point.

Because no accurate measurements of L_p or A can be made for any species, it is only possible to calculate their product (K_f) from PUF and SNGFR. In the case of garter snakes, the only reptile species for which data are available, SNGFRs of 0.6 to 5 nL/min[33] (S.D. Yokota and W.H. Dantzler, unpubl. observ.) and an average PUF of 3 mmHg yield K_f values of 0.2 to 1.7 nL/min/mmHg. The larger value is about the same as that in river lampreys (*Lampetra fluviatilis*) in freshwater[205] but less than one fourth that in the amphibian Congo eel (*Amphiuma means*)[240] and only about one third that in Munich–Wistar rat.[55,102]

c. Changes in Whole-Kidney GFR

The whole-kidney glomerular filtration rate (GFR), which is the sum of all the SNGFRs, cannot be measured over time in the living animal by summing all the individual SNGFRs for a kidney. This is impractical for a number of reasons. First, the number of glomeruli varies among species and even from kidney to kidney in a single animal, and no easy way exists to even approximate this number without direct counting. An attempt to develop an allometric equation that would provide a reasonable estimate of the number of glomeruli for a given body mass, with data from 13 species, was derived by Yokota et al.[320] We have added data from a further 8 species and derived the following equation: $y = 368.2x^{0.7926}$, $r^2 = 0.999$, where y = total number of glomeruli in both kidneys and x = body mass in grams (see Figure 10.1). It is of interest that some species deviate quite significantly from this regression as seen by an examination of residuals. The Australian skink *Tiliqua rugosa*, with a body mass of 156 g but only 4740 glomeruli,[13] falls furthest below the curve for all the species. The lizards *Ctenophorus ornatus* and *Sceloporus cyanogenys*, the chelonian *Clemmys japonica*, and the snake *Natrix sipedon* also all appear to have a low number of glomeruli relative to their body mass. Second, glomerular size varies, not only among species but also within a given kidney;[145,229] SNGFR varies with glomerular size; and there is no practical way to determine differences in glomerular size in the living kidney. Third, even for glomeruli of the same size, SNGFRs can apparently differ at any given time (see below). Finally, the only method currently available for measuring SNGFRs throughout a single kidney[100] is too difficult to permit measurement of all SNGFRs and requires sacrificing the animal, thereby providing only one time point.

Fortunately, whole-kidney GFRs can be measured readily by clearance methods in living animals. Such measurements have been made on representatives of the four extant orders of reptiles during acute changes in hydration or during intravenous administration of a salt load (hyperosmotic sodium chloride, usually 1 mol/L) or a water load (usually a hypoosmotic glucose solution). These measurements (Table 10.6) indicate that whole-kidney GFRs tend to decrease with dehydration or

TABLE 10.6
Changes in Whole-Kidney Glomerular Filtration Rate

Species	Habitat	Condition	GFR (mL/kg/hr)	Refs.
Crocodilia:				
Crocodylus johnsoni	Freshwater, semiaquatic	Control	6.0 ± 1.5	Schmidt-Nielsen and Davis[266]
		Dehydration	1.9 ± 0.2	
		Water load	3.3 ± 1.1	
Crocodylus acutus	Freshwater and seawater, semiaquatic	Control	9.6 ± 1.0	Schmidt-Nielsen and Skadhauge[268]
		Dehydration	6.1 ± 0.6	
		Salt load	7.3 ± 0.6	
		Water load	15.2 ± 2.0	
Crocodylus porosus	Saltwater, semiaquatic	Control	1.5 ± 0.2	Schmidt-Nielsen and Davis[266]
		Salt load	2.8 ± 0.9	
		Water load	18.8 ± 2.3	
Testudinea:				
Gopherus agassizii (desert tortoise)	Arid, terrestrial	Control	4.7 ± 0.60	Dantzler and Schmidt-Nielsen[94]
		Salt load (no urine flow when plasma osmolality increased 100 mOsm)	2.9 ± 0.91	
		Water load	15.1 ± .64	
Pseudemys scripta (freshwater turtle)	Freshwater, semiaquatic	Control	4.7 ± 0.69	Dantzler and Schmidt-Nielsen[94]
		Salt load (no urine flow when plasma osmolality increased 20 mOsm)	2.8 ± 0.90	
		Water load	10.3 ± 2.00	
Squamata:				
Sauria				
Tiliqua scincoides (blue-tongued lizard)	Terrestrial	Control	15.9 ± 1.0	Schmidt-Nielsen and Davis[266]
		Dehydration	0.7	
		Salt load	14.5 ± 0.5	
		Water load	24.5 ± 2.0	

Species	Habitat	Condition	Value	Reference
Phrynosoma cornutum (horned lizard)	Arid, terrestrial	Control	3.5 ± 0.323	Roberts and Schmidt-Nielsen[254]
		Dehydration	2.1 ± 0.20	
		Salt load	1.7 ± 0.40	
		Water load	5.5 ± 0.54	
Varanus gouldii (sand goanna)	Arid, terrestrial	Dehydration	10.99 ± 0.88	Bradshaw and Rice[48]
		Salt load	5.51 ± 1.10	
		Water load	15.89 ± 1.35	
Hemidactylus sp. (Puerto Rican gecko)	Moist, terrestrial	Control	10.4 ± 0.77	Roberts and Schmidt-Nielsen[254]
		Dehydration	3.3 ± 0.37	
		Salt load	11.0 ± 2.18	
		Water load	24.3 ± 1.67	
Ophidia:				
Pituophis melanoleucus (bull snake)	Arid, terrestrial	Salt load	16.1 ± 1.06	Komadina and Solomon[176]
		Water load	10.9 ± 1.07	
Nerodia sipedon (freshwater snake)	Freshwater, semiaquatic	Salt load (no urine flow when plasma osmolality 50 increased mOsm)	13.1 ± 1.26	Dantzler[74,75]
		Water load	22.8 ± 1.75	
Aipysurus laevis (olive sea snake)	Saltwater, aquatic	Control	0.78 (0.49–2.78)	Yokota et al.[321]
		Salt load	2.24 (1.41–6.42)	
		Chronic intraperitoneal seawater load	7.05 ± (6.26–7.83)	
		Water load	0.17 ± (0.03–0.35)	
		Chronic intraperitoneal water load	5.67 ± (4.40–6.20)	
Rhynchocephalia:				
Sphenodon punctatum	Moist terrestrial	Control	3.9	Schmidt-Nielsen and Schmidt[267]
		Dehydration	3.4	
		Water load	4.8	

Note: Values are means or means ± SE, except for sea snakes, for which the data did not show a normal distribution. The values are given as medians and interquartile ranges. All means with SE and medians are for 4 or more values.

Source: Adapted from Dantzler, W.H., in *Structure and Function of the Kidney: Comparative Physiology*, Vol. 1, Kinne, R.K.H., Ed., Karger, Basel, 1989, pp. 143–193. With permission.

a salt load and increase with a water load; however, considerable variation exists among species, especially in response to a salt load, which frequently leads to an increase rather than a decrease in whole-kidney GFR. Although the experiments were not identical in the hands of different investigators, the salt load was always given in an attempt to increase plasma osmolality, especially when it was difficult to study dehydration. Unfortunately, in well-hydrated animals this infusion would initially tend to lead to plasma expansion, increased renal blood flow, and increased whole-kidney GFR. With continued infusion, there would be water and thus volume depletion, increased plasma osmolality, and decreased whole-kidney GFR, but most experiments were not run with a long-term infusion.

In addition, among the species shown in Table 10.6, the crocodilian species and the wholly aquatic sea snakes have an extrarenal route (lingual salt gland) for the excretion of sodium chloride. If these structures remove sodium chloride rapidly enough, then the hyperosmotic salt load might actually be equivalent to at least an isosmotic plasma expansion or perhaps an actual hypoosmotic fluid load; therefore, some increase, rather than a decrease, in whole-kidney GFR might have been expected in these animals. In several studies, renal function was observed to cease when plasma osmolality increased sufficiently (Table 10.6). This appeared to occur at higher plasma osmolalities in the desert tortoise than in semiaquatic or mesic species, suggesting that this desert species can tolerate much greater increases in plasma osmolality than other species and supporting the concept of Nagy and Medica[225] that these animals do not attempt to control tightly their plasma osmolality (see below). Despite the differences among species, in reptiles, in contrast to mammals, marked physiological increases or decreases in whole-kidney GFR can certainly occur, especially with acute changes in hydration; therefore, changes in GFR may play a significant physiological role in regulating the volume and composition of the final urine (see below).

d. Changes in Number of Filtering Nephrons

The observed changes in whole-kidney GFR apparently result primarily from changes in the number of nephrons filtering, although changes in the filtration rates of individual nephrons probably play some role as well.[74,94,266,322] Several lines of evidence, both indirect and direct, support the concept of changes in the number of filtering nephrons as the primary process underlying the changes in whole-kidney GFR. First, in those species studied (freshwater turtles, *Pseudemys scripta*; desert tortoises, *Gopherus agassizii*; water snakes, *Nerodia sipedon*), the maximum rate of tubule transport (T_m) of *para*-aminohippurate (PAH) varies directly with whole-kidney GFR.[74,94] This supports the concept of nephron intermittency because if changes in whole-kidney GFR resulted from changes in the SNGFRs but all nephrons continued to filter, the T_m for PAH transport would not have changed because the mass of tubular tissue secreting PAH and contributing to the final urine would not have changed. Second, histological studies of the kidneys of a number of species (green turtle, *Chelonia mydas*; sea snake, *Laticauda colubrine*; blue-tongued lizard, *Tiliqua scincoides*; freshwater crocodile, *Crocodylus johnsoni*; and saltwater crocodile, *Crocodylus porusus*) showed that the ratio of the number of open to closed proximal tubule lumina correlates roughly with the whole-kidney GFR.[266] Because a proximal tubule lumen tends to collapse when its glomerulus stops filtering, these studies also support the concept that changes in whole-kidney GFR reflect changes in the number of nephrons filtering. Finally, and of greatest importance, direct quantitative measurements of blood flow rates in single glomeruli in kidneys of garter snakes (*Thamnophis sirtalis*) confirm the presence of intermittent blood flow and, presumably, intermittent filtration and indicate that the fraction of glomeruli with intermittent blood flow correlates directly with plasma osmolality (Figure 10.2)[322] (see below for more detail on glomerular blood flow). Changing whole-kidney GFR by changing the number of filtering nephrons appears to be a practical adaptation in reptiles in which nephrons are not arranged to function in concert to produce a urine hyperosmotic to the plasma. Moreover, as stressed above, all reptiles have a renal portal system that can continue to nourish the cells of nonfiltering nephrons in the absence of a postglomerular arterial blood supply.

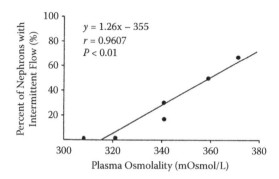

FIGURE 10.2 Relation of percent of nephrons with intermittent flow to plasma osmolality in individual garter snakes (*Thamnophis sirtalis*). Each point represents a single animal. Plasma osmolality was measured in six animals. Line was fitted to data by linear regression. (From Yokota, S.D. and Dantzler, W.H., *Am. J. Physiol.*, 258, R1313–R1319, 1990. With permission.)

4. Tubule Function

a. Introduction

The renal tubules modify the initial glomerular filtrate by the processes of reabsorption and secretion, thereby playing a major role in determining the volume, osmolality, and ionic composition of the ureteral urine. This composition may, of course, be further modified by the colon, cloaca, or bladder. The current discussion is limited to the tubule transport of sodium, potassium, water, and those end products of nitrogen metabolism that have a significant influence on the volume and osmolality of the urine.

b. Inorganic Ion Transport

i. Sodium

Sodium is the major cation of the extracellular fluid and, with its accompanying anions, is responsible for maintaining the extracellular fluid volume. As noted above, neither the extracellular concentration of sodium nor the extracellular fluid volume is maintained as constant in reptiles as in mammals.[42] Nevertheless, maintaining the total amount of sodium in the animal within some broad general range, largely by regulating excretion, is critical to survival.

Reptilian renal tubules reabsorb anywhere from about 50 to 98% of the filtered sodium, depending on the species as well as the specific requirements for conservation of sodium at the time of measurement.[86,294] Under most circumstances in most reptilian species, at least 10% or more of the filtered sodium appears in the ureteral urine, far more than in mammals, where less than 1% of the filtered sodium escapes reabsorption by the renal tubules. When evaluating the fractional reabsorption by the renal tubules of reptiles, two factors must be kept in mind. First, the available measurements come primarily from clearance studies, which only supply net transport for the entire kidney. Although these studies show net reabsorption, they do not permit any distinction between reabsorption only by the renal tubules and the combination of secretion in one portion of the renal tubules and greater reabsorption in another portion of the renal tubules. Indeed, net secretion of sodium may occur in the renal tubules of some marine snakes and in aglomerular nephrons of a few other reptiles (see below). Second, these clearance studies do not include filtered sodium that may be contained in urate precipitates leaving the ureters in uricotelic reptiles (see below). In so far as this occurs, the actual fraction of filtered sodium reabsorbed may be even less than that measured.[82]

In any case, it is apparent that the fraction of filtered sodium reabsorbed by reptilian renal tubules is generally low, when compared to mammalian renal tubules. Additional reabsorption, depending on the requirements of the animal, occurs in the colon, cloaca, or bladder (see below). The degree of such reabsorption must be integrated with the renal reabsorption in the maintenance of overall ionic balance.

TABLE 10.7
Sodium and Chloride Reabsorption by Tubule Segment

Tubule Segment and Species	Sodium J_{Net} (pmol/min/mm)	Sodium (% Filtered Load)	Ref.
Proximal:			
Garter snake (*Thamnophis* spp.)	130.5	45	Dantzler and Bentley[87]
Blue spiny lizard (*Sceloporus cyanogenys*)	27.8	37	Stolte et al.[294]
Late distal and collecting tubule:			
Blue spiny lizard (*Sceloporus cyanogenys*)	61.7	21	Stolte et al.[294]
Collecting duct:			
Blue spiny lizard (*Sceloporus cyanogenys*)	—	37	Stolte et al.[294]

Note: Values are means taken directly or calculated from the references. J_{Net}, net transepithelial reabsorption.

Source: Adapted from Dantzler, W.H., in *Structure and Function of the Kidney: Comparative Physiology*, Vol. 1, Kinne, R.K.H., Ed., Karger, Basel, 1989, pp. 143–193. With permission.

In reptiles, again in contrast to mammals, the proximal tubule is not the primary site of sodium reabsorption. *In vivo* micropuncture studies of lizard tubules and *in vitro* microperfusion studies of snake renal tubules indicate that only some 35 to 45% of the filtered sodium is reabsorbed along the proximal tubule (Table 10.7).[87,294] Some 50 to 70% is reabsorbed along the distal tubules and collecting ducts (Table 10.7).[86,294] The small amount of information available about the rate of sodium reabsorption per unit renal tubule length in reptiles suggests that it is less than or equal to that for mammals in the proximal tubule and about equal to that for mammals in the distal tubule segments (Table 10.7).[86,87,294] The fraction of filtered sodium actually reabsorbed along any given tubule segment must reflect the balance between the filtered amount reaching that segment, the reabsorption rate per unit segment length, and the length of the segment.

Little is known about the mechanism by which sodium is reabsorbed by the proximal tubules of reptiles; however, a very small, lumen-negative transepithelial potential has been observed in the proximal tubules of one species (garter snake, *Thamnophis* spp.) during sodium reabsorption (Table 10.8), indicating that transepithelial sodium transport occurs against an electrochemical gradient.[90] Also, because the fractional reabsorption of chloride in the proximal tubule is essentially the same as that of sodium,[87] transepithelial reabsorption appears to involve active sodium and passive chloride transport, at least in this species and perhaps all reptiles.

Virtually nothing is known about the luminal and basolateral membrane steps in the proximal transepithelial transport of sodium in reptiles. Because, as in other vertebrates, the inside of the proximal cells is negative compared to either the lumen or the basolateral side (Table 10.8),[86,173] sodium certainly can enter the cells from the lumen down an electrochemical gradient. Also, as in other vertebrates, this step probably involves primarily sodium–hydrogen exchange.[86] A functional sodium–hydrogen exchanger has been identified in brush-border membrane vesicles from garter snake (*Thamnophis* spp.) kidneys.[95] The concept of sodium reabsorption by this exchanger is indirectly supported by the observation that the administration of carbonic anhydrase inhibitors to water snakes (*Nerodia sipedon*) results in an alkaline ureteral urine and an increased ureteral excretion of sodium and potassium.[75] Although most luminal sodium entry probably involves some form of sodium–hydrogen exchange, coupled entry with glucose, amino acids, phosphate, and other molecules is certainly involved as well.[8,21,86] Sodium must then be transported out of the cells across the basolateral membrane against an electrochemical gradient in the transepithelial reabsorptive process. As in other vertebrates,[86] this process almost certainly involves Na$^+$,K$^+$-ATPase, which is located in the basolateral membrane of proximal tubules in snakes[21,77] and presumably other reptiles.

TABLE 10.8
Electrical Properties by Tubule Segment

Tubule Segment and Species	V_{BL} (mV)	V_T (mV)	R_T (kΩ·cm)	R_T (Ω·cm²)	Refs.
Proximal:					
Garter snake (*Thamnophis* spp.)	−60.1 ± 1.9 (13)	−0.49 ± 0.155 (14)	—	—	Dantzler and Bentley;[90] Kim and Dantzler[173]
Late distal:					
Garter snake (*Thamnophis* spp.)	—	−34.9 ± 2.1 (27)	23.4 ± 1.6 (27)	83.1	Beyenbach et al.[30]

Note: Values are means or means ± SE; they are taken directly or calculated from the references. Numbers in parentheses indicate number of determinations. V_{BL}, basolateral membrane potential difference (sign indicates inside of cell negative); V_T, transepithelial potential difference (sign indicates lumen negative relative to peritubular side); R_T, transepithelial resistance.

Source: Adapted from Dantzler, W.H., in *Structure and Function of the Kidney: Comparative Physiology*, Vol. 1, Kinne, R.K.H., Ed., Karger, Basel, 1989, pp. 143–193. With permission.

With regard to sodium reabsorption in the distal portions of the nephron, information with particular significance for the intrinsic tubule regulation is available from *in vitro* microperfusion studies on the late distal tubule of snake (*Thamnophis* spp.) nephrons.[27,29,30] These studies revealed a large lumen-negative transepithelial potential (Table 10.8) that is inhibited by amiloride in the luminal perfusate or ouabain in the basolateral bathing medium. This lumen-negative transepithelial voltage and the calculated short-circuit current (presumably representing sodium reabsorption), although dependent on the presence of sodium in the lumen, both decay rapidly when its concentration exceeds 30 mmol/L. The decays in voltage and short-circuit current apparently represent an increase in resistance to sodium transport through the active transport pathway. Because sodium that enters the cells passively across the luminal membrane at this time cannot be extruded rapidly enough to keep up with the entry, the cells swell. This intrinsic response to an excessive sodium load may prevent the distal tubules from reabsorbing too much sodium when there is a need for additional sodium excretion. Moreover, a transport system poised to operate effectively only at low sodium concentrations will enhance the dilution of luminal tubule fluid in which the sodium concentration is already low. Whether such an intrinsic regulatory process operates in the late distal tubules of other reptiles is unknown.

ii.   Potassium

Potassium, the major cation of the intracellular fluids in all vertebrates, is particularly critical for the function of excitable cells. Although the extracellular levels of potassium are not maintained as constant as in mammals (see above), the total quantity is still critical and is regulated largely by renal and extrarenal routes of excretion. Micropuncture studies on one species (blue spiny lizard, *Sceloporus cyanogenys*)[294] and clearance studies on numerous species[81] indicate that either net reabsorption or net secretion by the renal tubules may occur. In uricotelic species, some potassium in the tubule fluid may be combined with urates (see below);[82] therefore, measurements on the aqueous phase of the urine in the clearance and micropuncture studies cited above may not have provided accurate values for the magnitude or even the direction of net transport.[82]

The magnitude and sites of potassium transport along the renal tubules of reptiles are not yet well described; however, the micropuncture studies on *Sceloporus cyanogenys*[294] suggest that about 25 to 35% of the filtered potassium is normally reabsorbed along the proximal tubules. Another 10% of the filtered potassium may be reabsorbed along the distal tubules, but this may change to net secretion of as much as 180% of the filtered load. This general pattern appears to be the same as that observed in mammals and those other nonmammalian vertebrates in which it has been studied;[86] however, the exact distal tubule site (or sites) and the cell types (perhaps the light cells described in this species; see above) involved in determining net reabsorption or net secretion are unknown. Presumably, the distal shifts between net reabsorption and net secretion are related to the need to retain or eliminate dietary potassium. Some additional potassium reabsorption can occur in the collecting ducts of this species, but this does not appear to be very significant under normal conditions.[294] As in the case of urinary sodium excretion, additional modification of potassium excretion may occur by transport in structures distal to the kidney.

Essentially nothing is known about the cellular mechanisms involved in renal tubule transport of potassium. Perhaps, as in mammals, much of the net reabsorption occurs via a paracellular route. Net secretion almost certainly involves enhanced transport into the cells from the blood via basolateral Na^+,K^+-ATPase, but other cellular steps are unknown. Alkalosis enhances potassium excretion, at least in water snakes (*Nerodia sipedon*),[75] but whether this involves stimulation of Na^+,K^+-ATPase or some direct exchange for sodium, or a combination of these, is unknown.

c. Water Transport

i. Introduction

As noted above, changes in GFR play a particularly important role in regulating the excretion of water as well as ions in reptiles, but tubule transport of water, primarily reabsorption, is also of critical importance in maintaining fluid volume, even if this volume is not as rigidly controlled in reptiles as in mammals (see above). Water reabsorption in reptiles sometimes occurs at the same rate as solute reabsorption (as an isosmotic fluid) and sometimes lags behind solute reabsorption, depending on the need to excrete or conserve water.

ii. Water Reabsorption

Water reabsorption is generally measured as total filtered fluid (both water and solutes) reabsorbed. In reptiles, as in other tetrapods, a large fraction of the filtrate is reabsorbed by the renal tubules; however, such reabsorption never approaches, much less exceeds, 99%, even during dehydration, as it does in mammals.[86] Moreover, micropuncture or microperfusion of renal tubules indicates that only about 35 to 45% of the filtered fluid is reabsorbed along the proximal tubules of snakes and lizards (Table 10.9). More indirect studies indicate that this is about the same (30 to 50%) for the proximal tubules of turtles and crocodilians.[94,268] These values are well below the two thirds of the filtered fluid reabsorbed by the proximal tubules of mammals.[86] Instead, a substantial fraction of the filtered fluid, perhaps as much as 45% in some species, can be reabsorbed along the distal portions of the nephrons (Table 10.9).[81] In some, but not all species, the amount of water reabsorbed in these distal segments depends on the need to conserve water and the action of antidiuretic hormone (see below). In addition to reabsorption by the renal tubules, substantial filtered water, like solute, can also be reabsorbed by the colon, cloaca, or bladder (see below), depending on the requirements of the animals.

The details of the process by which filtered fluid is reabsorbed in the proximal tubule are not understood, although *in vivo* micropuncture studies of lizard proximal tubules and *in vitro* microperfusion studies of snake proximal tubules indicate that during reabsorption the filtered fluid in the lumen remains isosmotic with the plasma (Table 10.9).[87,294] Although these studies clearly indicate that sodium and water can be reabsorbed at osmotically equivalent rates, they do not prove that fluid reabsorption by the proximal tubules of reptiles must always be dependent on sodium reabsorption, as it appears to be in other vertebrates. In fact, in the *in vitro* perfusion studies of snake proximal tubules, substitutions for sodium or chloride or both in the solutions used for perfusing and bathing the tubules suggest that neither one of these ions is essential for normal fluid reabsorption.[87] When sodium in the perfusate is replaced with choline (an organic cation), net fluid reabsorption almost ceases, as would be expected; however, when sodium in the bathing medium is also replaced with choline so both solutions are identical, net fluid reabsorption

TABLE 10.9
Reabsorption of Filtered Fluid by Tubule Segment

Tubule Segment and Species	J_v (nL/min/mm)	J_v (% Filtered Load)	Osmolal TF/P	Ref.
Proximal:				
Garter snake (*Thamnophis* spp.)	0.87 ± 0.04 (127)	45	1.00 ± 0.01 (12)	Dantzler and Bentley[87]
Blue spiny lizard (*Sceloporus cyanogenys*)	0.18	35	0.99 (8)	Stolte et al.[294]
Late distal:				
Garter snake (*Thamnophis* spp.)	0.07 ± 0.04 (14)	—	—	Beyenbach[27]
Blue spiny lizard (*Sceloporus cyanogenys*)	0.05	2.5	0.85	Stolte et al.[294]
Collecting duct:				
Blue spiny lizard (*Sceloporus cyanogenys*)	—	36	—	Stolte et al.[294]

Note: Values are means or means ± SE; they are taken directly or calculated from the references. Numbers in parentheses indicate the number of determinations. J_v, net transepithelial fluid reabsorption; osmolal TF/P, ratio of the osmolality in fluid collected from the tubule lumen (TF) to the osmolality in plasma (P).

Source: Adapted from Dantzler, W.H., in *Structure and Function of the Kidney: Comparative Physiology*, Vol. 1, Kinne, R.K.H., Ed., Karger, Basel, 1989, pp. 143–193. With permission.

returns to the control rate. When sodium in the perfusate alone is replaced with lithium (another inorganic alkali metal cation), the rate of net fluid reabsorption is unchanged. Thus, lithium can substitute for sodium in the transepithelial transport process. From these studies on garter snake proximal tubules, it appears that isosmotic fluid reabsorption can proceed at control rates when lithium replaces sodium in the perfusate alone or when some other substance (e.g., choline, tetramethylammonium, methylsulfate, sucrose) replaces sodium or chloride or both in the perfusate and bathing medium simultaneously. Even when sodium is present, net fluid reabsorption is not inhibited by the removal of potassium from the bathing medium or by treatment with ouabain or other cardiac glycosides, as would be expected if fluid reabsorption was driven primarily by sodium reabsorption via Na^+,K^+-ATPase.[87,88] Moreover, net fluid reabsorption with sodium present is not dependent on the nature of the buffer (bicarbonate, phosphate, or Tris) used.[87] Finally, with sodium present, net fluid reabsorption is reduced 18 to 25% by the removal of colloid from the peritubular bathing medium.[87]

Although these observations on snake proximal renal tubules suggest that isosmotic fluid reabsorption can occur in the absence of both sodium and chloride, they do not provide any information on the mechanism underlying this process. Quantitative structural studies on these isolated, perfused tubules[92] have revealed that within minutes after substitution of choline for sodium in both the perfusate and bathing medium significant morphological changes take place. Tubule cells double in size and intercellular spaces nearly quintuple. At the same time, the areas of the lateral and apical cell membranes approximately double, but their surface densities are essentially unchanged; therefore, although the cells are larger in the absence of sodium, they have proportionately larger surface areas so their volume-to-surface ratio remains constant. This rapid increase in membrane area most likely involves incorporation of preformed membrane segments, possibly from intracellular vesicles, although the exact source is unknown. Regardless of the mechanism involved in these morphological changes, they are correlated with the maintenance of a control level of net fluid reabsorption. Perhaps they permit a small, previously unimportant driving force, such as the colloid osmotic pressure in the peritubular fluid (or plasma in the intact animal), to produce a control level of net fluid reabsorption.

TABLE 10.10
Examples of Ranges of Osmolal Urine-to-Plasma Ratios

Species	Habitat	Osmolal Urine-to-Plasma Ratio (Approximate Maximum Range)	Ref.
Crocodilia:			
Crocodile (*Crocodylus acutus*)	Freshwater and marine, semiaquatic	0.55–0.95	Schmidt-Nielsen and Skadhauge[268]
Testudinea:			
Desert tortoise (*Gopherus agassizii*)	Arid, terrestrial	0.3–0.7	Dantzler et al.[94]
Freshwater turtle (*Pseudemys scripta*)	Freshwater, semiaquatic	0.3–1.0	Dantzler et al.[94]
Squamata:			
Sauria			
Blue spiny lizard (*Sceloporus cyanogenys*)	Arid, terrestrial	0.3–0.7	Stolte et al.[294]
Sand goanna (*Varanus gouldii*)	Arid, terrestrial	0.4–1.0	Bradshaw and Rice[48]
Horned lizard (*Phrynosoma cornutum*)	Arid, terrestrial	0.8–1.0	Roberts and Schmidt-Nielsen[254]
Ophidia			
Bull snake (*Pituophis melanoleucus*)	Arid, terrestrial	0.5–1.0	Komadina and Solomon[176]
Freshwater snake (*Nerodia sipedon*)	Freshwater, semiaquatic	0.1–1.0	Dantzler[74]
Olive sea snake (*Aipysurus laevis*)	Marine, aquatic	0.8–1.2	Dantzler[84]

Note: The values are from measurements on ureteral urine.

Source: Adapted from Dantzler, W.H., in *Structure and Function of the Kidney: Comparative Physiology*, Vol. 1, Kinne, R.K.H., Ed., Karger, Basel, 1989, pp. 143–193. With permission.

iii. Water Secretion

In sea snakes (*Aipysurus laevis*), indirect clearance studies suggest that net fluid secretion can sometimes occur.[321] Secretion of fluid has been directly demonstrated in isolated proximal tubules of both glomerular and aglomerular fish[26,28] and in isolated inner medullary collecting ducts of mammals.[313] In some species of glomerular fish, the rate of secretion of fluid by a proximal tubule is about equal to its corresponding SNGFR,[26,28] but in these sea snakes, as in mammalian collecting ducts,[313] it is apparently much lower.[321] Moreover, fluid secretion in sea snakes only appears to be significant when GFR is very low.[321] The tubule site of such secretion and the mechanism involved in the process are unknown, but clearance studies[321] suggest that, as in the proximal tubules of fish[28] and the collecting ducts of mammals,[316] it is dependent on the secretion of sodium and chloride. Although the physiological significance of net fluid secretion in these sea snakes is unclear, the basic process of fluid secretion would be essential to the function of any aglomerular nephrons found in reptiles.

iv. Dilution and Concentration

Reptiles are incapable of producing urine significantly more concentrated than the plasma, although many species, but not all, are capable of varying ureteral urine osmolality from significantly hypoosmotic to the plasma to isosmotic with the plasma (Table 10.10). This variation in urine osmolality between hypoosmotic and isosmotic is the major renal tubule mechanism by which the kidneys of reptiles adjust the amount of solute-free filtered water delivered to the ureters, and the process is generally well regulated, primarily by antidiuretic hormone (see below). The most dilute urine (about one tenth the osmolality of the plasma) is generally produced by freshwater species with a major need to excrete excess water (Table 10.10). A few species (e.g., desert tortoise,

Gopherus agassizii; blue spiny lizard, *Sceloporus cyanogenys*) always produce ureteral urine hypoosmotic to the plasma, and others (e.g., horned lizard, *Phrynosoma cornutum*; olive sea snake, *Aipysurus laevis*) always produce ureteral urine close to isosmotic with the plasma. The observations appear appropriate for the horned lizard and olive sea snake, which rarely have excess water to excrete; however, they appear inappropriate for the desert tortoise and blue spiny lizard, which also rarely have excess water. It is evident that further modification of the urine osmolality occurs in the bladder or cloaca (see below).

Although reptiles are not capable of producing a urine sufficiently hyperosmotic to the plasma to be physiologically significant in the conservation of water, a few species (sea turtle, *Chelonia mydas*; lizard, *Amphibolurus maculosus*; marine snake, *Aipysurus laevis*) have been observed to produce urine slightly hyperosmotic to the plasma (osmolal urine-to-plasma ratio of 1.2 to 1.3).[54,244,321] In the case of the sea turtle and the lizard, production of hyperosmotic urine may involve modification of ureteral urine in the bladder or cloaca (see below) because ureteral urine was not collected directly. This is not the case, however, for *Aipysurus laevis*, in which ureteral urine was found to be slightly hyperosmotic to the plasma at low urine flows (Table 10.10).[321] This hyperosmolality may reflect tubule secretion of solutes (i.e., sodium, potassium, magnesium, or ammonium) into a small volume of tubule fluid.[321] Secretion of ions may be important in regulating their excretion, especially for these marine snakes, although production of urine only slightly hyperosmotic to the plasma cannot be of physiological significance for conserving water in marine species because the plasma osmolality is far below that of seawater.

Formation of urine hypoosmotic to the plasma requires the reabsorption of filtered solutes (probably primarily sodium and chloride) without water accompanying them at some site along the renal tubules. In the xerophilic lizard species *Sceloporus cyanogenys*, micropuncture studies indicate that dilution can occur at least by the early distal tubule and can continue throughout the length of the collecting duct during all states of hydration and during administration of antidiuretic hormone (see below).[294] In this species, further regulation of urine osmolality almost certainly occurs distal to the kidney. Preliminary perfusions of isolated renal tubules from garter snakes (*Thamnophis* spp.) suggest that the thin intermediate segment may have low water permeability and be a site of significant solute reabsorption[45] (S.D. Yokota and W.H. Dantzler, unpubl. observ.). In addition, as discussed above, the late distal tubules of these animals may be specialized to permit additional dilution of luminal fluid that already has a very low sodium concentration.[27,29,30] The maintenance and regulation of the urine osmolality are discussed below.

d. End Products of Nitrogen Metabolism

i. Introduction

The three major end products of nitrogen metabolism excreted in the urine are ammonia, urea, and uric acid. Ammonia is both highly soluble and highly toxic and is generally the primary compound for excretion of nitrogen only in animals in a completely aquatic environment, where it can be rapidly removed from the animal. Urea, although less toxic than ammonia, can still denature proteins and is also highly soluble; therefore, it is generally the primary excretory end product of nitrogen metabolism in animals with ample access to water or the ability to produce urine markedly hyperosmotic to the plasma. Uric acid, even in its usual form of a urate salt, is very poorly soluble and can be removed with very little water. Thus, it is most commonly the primary compound for nitrogen excretion in animals with little access to water or no ability to produce urine significantly hyperosmotic to the systemic plasma (see reviews in Walsh and Wright[314] for more information on comparative aspects of nitrogen metabolism and excretion). Moreover, it can also play an important role in the excretion of inorganic ions (see below).

ii. Ammonia

Apparently very little ammonia is normally present in the systemic blood of reptiles; therefore, it is not filtered to any significant extent. Instead, as in other vertebrates, almost all of the ammonia appearing in reptilian urine is formed within the tubule cells and secreted into the tubule lumen.

TABLE 10.11
Approximate Percents of Total Urinary Nitrogen as Urates, Urea, and Ammonia

Reptile	Percent of Total Urinary Nitrogen as:			Ref.
	Urates	Urea	Ammonia	
Crocodilia	70	0–5	25	Khalil and Haggag[170]
Testudinea:				
Wholly aquatic	5	20–25	20–25	Moyle[218]
Semiaquatic	5	40–60	6–15	Baze and Horne[9]
Wholly terrestrial				
Mesic environment	7	30	6	
Xeric environment	50–60	10–20	5	
Desert tortoise (*Gopherus agassizii*)	20–25	15–50	3–8	Dantzler and Schmidt-Nielsen[94]
Freshwater turtle (*Pseudemys scripta*)	1–24	45–95	4–44	Dantzler and Schmidt-Nielsen[94]
Squamata:				
Sauria	90	0–8	Insignificant to highly significant	Dantzler[86]
Ophidia	98	0–2	Insignificant to highly significant	Dantzler[86]
Rhynchocephalia:				
Sphenodon punctatum	65–80	10–28	3–4	Hill and Dawbin[161]

Source: Adapted from Dantzler, W.H., in *Structure and Function of the Kidney: Comparative Physiology*, Vol. 1, Kinne, R.K.H., Ed., Karger, Basel, 1989, pp. 143–193. With permission.

The tubule sites of production and secretion in reptiles are unknown but probably at least involve the proximal tubule.

In most vertebrates, the primary function of renal ammonia production and excretion is to aid in acid–base balance by permitting the excretion of acid or, more accurately, the production of an equivalent amount of bicarbonate. For this reason, production and excretion tend to increase with an acid load. Except for a slight suggestion that an acid load may increase ammonia excretion in freshwater snakes (*Nerodia* spp.),[81] no information is available on the role of ammonia in acid–base balance in reptiles.

Whatever the role of ammonia in acid–base balance, it is a major excretory end product of nitrogen metabolism in crocodilians (e.g., alligators, *Alligator mississippiensis*) and in semiaquatic and aquatic chelonians (e.g., freshwater turtles, *Pseudemys scripta*) (Table 10.11).[68,82] In semiaquatic and aquatic chelonians, excreted nitrogen is often distributed about equally between ammonia and urea, although much variation exists and urea usually tends to predominate (Table 10.11). In crocodilians, ammonia accounts for about 25% and urates for about 75% of the total nitrogen excreted (Table 10.11).[82,170] Moreover, the absolute amount of ammonia excreted in the urine of alligators on a standard meat diet is greater than that recorded for any other vertebrate species.[69,174]

Although the exact tubule sites of ammonia production are unknown and the total production process has not been studied in reptiles, Lemieux and colleagues[188] have studied some of the steps in ammonia production in alligator kidneys. They assumed that, as in mammals, glutamine is taken up by the renal tubule cells, enters the mitochondria, and is deaminated by glutaminase I to form glutamate and NH_4^+. Glutamate is then presumably deaminated via glutamate dehydrogenase to form α-ketoglutarate and NH_4^+. These authors found that both phosphate-dependent glutaminase I and glutamate dehydrogenase are present in the mitochondria of the alligator kidney at levels of activity suitable for ammonia production. They also found that alanine aminotransferase activity is high and that isolated fragments of alligator renal tubules are capable of producing ammonia from both glutamine and alanine *in vitro*. Finally, they found that glutamine synthetase, but none of the enzymes involved in ammoniagenesis, is found only in the liver, suggesting to them that this organ may be a source of glutamine for ammonia production by the kidney.

When alligators are dehydrated, renal ammonia production decreases and more nitrogen appears in the urine as uric acid.[174] Decreasing renal blood flow during dehydration would decrease delivery of amino acids to the kidneys and thereby reduce the amount of substrate available for ammonia production. In addition, King and Goldstein[174] suggest that the accumulation of ammonia in renal tissue during low urine flow might drive the reversible deamination reactions in the direction of amino acid formation; however, this proposal has not been examined directly.

No experimental data are available on the mechanism of tubule ammonia secretion in reptiles. It could involve nonionic diffusion of free ammonia (NH_3) across the luminal cell membrane and trapping of ammonium ion (NH_4^+) in the lumen in those reptile species in which the pH of the luminal fluid is relatively low. Carrier-mediated secretion of NH_4^+ appears to be necessary in alligators, which normally produce alkaline urine containing large amounts of bicarbonate.[68,70] This process could involve substitution of ammonium ions for hydrogen on the luminal sodium–hydrogen exchanger in proximal tubules. Because alligators maintain a rather low systemic blood pH (~7.1), it is also possible that the initial filtrate is actually below the pH of the early proximal tubule cells and that, therefore, some ammonia can be secreted by nonionic diffusion and diffusion trapping in that portion of the proximal tubule.[187] Neither means of excretion explains why ammonia secreted in the proximal portions of the nephron in these animals is not passively reabsorbed in distal regions where the luminal pH becomes highly alkaline.[68,70] It is possible that, if ammonia secretion into the lumen definitely involves a carrier-mediated process, the luminal membrane of renal tubule cells has a much lower passive permeability for ammonia in alligators than in other species.[86]

iii. Urea

Urea is the dominant excretory end product of nitrogen metabolism only in chelonian reptiles from aquatic, semiaquatic, and mesic terrestrial habitats (Table 10.11); however, it also accounts for a significant fraction of the nitrogen excreted in the urine of chelonian reptiles from arid terrestrial habitats and of the one living rhynchocephalian species (*Sphenodon punctatum*) (Table 10.11). Urea is freely filtered and, in most species studied (including those chelonian species in which it is the major form of nitrogen excretion), undergoes a variable amount of apparently passive tubular reabsorption.[86] In general, the amount of reabsorption varies with urine flow rate, being more obvious with dehydration and low urine flow rate.[86,267,317] Net tubular secretion has also been observed during clearance studies on a few lizard species (*Lacerta viridis* and *Sceloporus cyanogenys*) and Sphenodon.[86,239,267] Net secretion, however, is only observed during extreme diuresis; therefore, in all reptilian species studied, urea excretion is determined primarily by filtration and tubular reabsorption. The sites and mechanisms involved in tubular transport are unknown. Although reabsorption is apparently passive, it almost certainly is carrier mediated. Moreover, any net tubular secretion must involve carrier mediation, but no urea transporters have yet been identified in reptilian kidneys.

iv. Urate

Urates are the primary chemical form in which nitrogen is excreted in all reptiles, regardless of habitat, except for chelonians from aquatic, semiaquatic, and mesic terrestrial habitats (Table 10.11). In all reptiles studied, urate is freely filtered at the glomerulus and is net secreted by the renal tubules.[86] Most of the information available on the sites and mechanism of tubular transport in reptiles derives from studies with isolated, perfused snake (*Thamnophis* spp.) renal tubules. These studies revealed that net secretion from peritubular fluid to lumen against a concentration gradient occurs throughout the proximal tubule but not in the distal tubule.[78,80] No evidence has been found for net urate reabsorption in these tubules, but a passive unidirectional reabsorptive flux can occur throughout the proximal tubule.

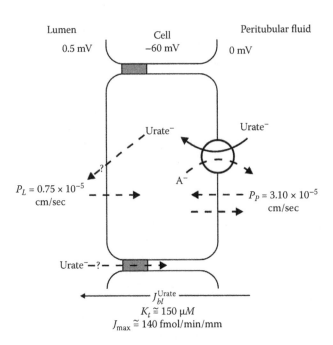

FIGURE 10.3 Model for net tubular secretion of urate based on studies with snake (*Thamnophis* spp.) proximal renal tubules Solid circle with solid arrow indicates either primary or secondary active transport. For countertransport, solid arrow indicates movement against electrochemical gradient; broken arrow, movement down electrochemical gradient. Broken arrows with question marks indicate possible passive movements. A$^-$ indicates anion of unspecified nature. Apparent permeabilities of luminal (P_L) and peritubular (P_P) membranes are shown. Apparent K_t and J_{max} for net secretion are shown at the bottom of figure.

All the snake tubule studies indicate that net secretion of urate occurs by a pathway independent of that involved in net secretion of other organic anions.[86] Net urate secretion involves transport into the cells against an electrochemical gradient at the basolateral membrane, followed by movement from the cells into the lumen down an electrochemical gradient (Figure 10.3). The effects of disulfonic stilbenes suggest that the transport step into the cells at the basolateral membrane involves countertransport for another anion (Figure 10.3).[219] This transport step has no dependence on sodium,[246] which indicates that countertransport cannot involve a dicarboxylate such as α-ketoglutarate (αKG) that is countertransported for other organic anions (e.g., p-aminohippurate, PAH) after entering the cells via the sodium dicarboxylate cotransporter.[66] Moreover, preloading snake tubules with numerous mono-, di-, and tricarboxylates does not stimulate urate uptake (Y.K. Kim and W.H. Dantzler, unpubl. observ.) the way αKG stimulates PAH uptake.[66] Thus, the anion that might drive urate uptake by countertransport remains unknown.

A number of other distinctive features of urate transport not only differentiate it from transport for other organic anions but also may be of adaptive significance with regard to fluid balance. First, the apparent passive permeability of the basolateral membrane is much greater than that of the luminal membrane (Figure 10.3).[80] These findings, which are the opposite of those for the transport of other organic anions, suggest that this is a very inefficient system for net secretion, because urate that is transported into the cells across the basolateral membrane will have a tendency to leak back into the peritubular fluid more readily than to move into the lumen. The basolateral transport step is always working against a large backleak. Second, the basolateral transport step appears to be dependent on the presence of an artificial perfusate (or, *in vivo*, the glomerular filtrate) flowing along the tubule lumen.[78] Third, net transepithelial secretion in these tubules varies directly with the rate of luminal perfusion, suggesting the presence of significant transepithelial backdiffusion from lumen to bath at low perfusion rates.[79] Indeed, significant unidirectional flux from lumen to bath that varies with perfusion rate has been demonstrated in these tubules.[78] Moreover, the transepithelial permeability determined directly from this measured flux (about 2.4×10^{-5} cm/sec) is about four times that (0.60×10^{-5} cm/sec) calculated from the independently measured permeability values for the luminal and peritubular membranes shown in Figure 10.3.[80] This observation suggests that much of the apparent backdiffusion from lumen to bath at low perfusion rates occurs between the cells (Figure 10.3). Fourth, the apparent K_t for net transepithelial urate secretion (about 150 µmol/L) (Figure 10.3) is well below the normal plasma concentration (400 to 500 µmol/L) in these garter snakes.[86] This observation suggests that the net transepithelial secretory system is normally saturated (or nearly saturated) and that changes in plasma urate concentrations have little effect on net urate secretion. Instead, it appears likely that the rate of flow through the lumen and the backdiffusion just described may be particularly important in determining net urate secretion and, thus, excretion.[86] As discussed above, nephrons filter intermittently in reptiles, the number filtering decreasing with dehydration and increasing with hydration. The relatively high passive permeability of the basolateral membrane, the apparent dependence of basolateral transport into the cells on filtrate in the lumen, and the apparent large paracellular backleak at low luminal flow rates may function to reduce urate accumulation in the lumens or cells of nephrons that are not filtering.

Of additional interest with regard to transepithelial urate secretion is the lack of convincing evidence that urate movement from cells to lumen is carrier mediated. Neither the flux of radiolabeled urate from cells to lumen nor the apparent urate permeability of the luminal membrane in perfused snake tubules is affected in any way by unlabeled urate, probenecid, or disulfonic stilbenes.[89,219] Moreover, no evidence of carrier-mediated transport could be found in isolated brush border membrane vesicles from snake kidneys.[21] These negative findings are compatible with simple passive diffusion across the luminal membrane (Figure 10.3), but they do not prove it. Moreover, given the relatively large urate flux across this membrane during net secretion, simple passive diffusion appears quite unlikely.

As noted above, the low solubility of both uric acid itself (0.384 mmol/L) and the modest solubilities of its urate salts (e.g., 6.76 mmol/L for sodium urate and 12.06 mmol/L for potassium urate)[154] mean that almost no pure uric acid and relatively small amounts of the urate salts can exist in true solution in the aqueous phase of the urine. Nevertheless, some urate salts are present. These are generally the sodium and potassium salts, the predominant cation being determined by whether the animal is a carnivore or herbivore. In alligators, in which large amounts of ammonia are excreted in alkaline urine,[68,70] ammonium urate may be present in the aqueous phase. In some reptile species (e.g., turtles), the concentrations of sodium, potassium, ammonium, and urate in the aqueous phase of the urine may be low enough to permit all the urate to be present as salts in true solution;[94] however, in some lizards (e.g., *Dipsosaurus dorsalis*) and snakes (e.g., *Thamnophis* spp.), and probably many other reptiles as well, the concentrations of sodium, potassium, and urate in the aqueous phase of the ureteral urine are above those at which simple urate salts can remain in true solution. In these cases, the urates in this aqueous phase probably exist as, in addition to small amounts of dissolved urate salts, relatively small lyophobic colloid particles and much larger lyophilic colloids produced by the adsorption of the lyophobic colloids to lyophilic proteins.[10,209,243]

Most of the urate excreted by reptiles is in the form of precipitates that do not obligate water excretion. These precipitates develop not only in the cloaca or bladder, where urate may be precipitated from the aqueous phase, thereby permitting additional water reabsorption, but also in the ureters and collecting ducts[214] (W.H. Dantzler, unpubl. observ.). In the ureters and collecting ducts, these precipitates exist as small spheres (2 to 10 µm in diameter), whereas in the cloaca of the same animals they exist as both spheres and sharp-edged uric acid crystals.[214] The small spheres can move along the collecting ducts and ureters without causing damage. How they are formed, however, is unknown. Minnich[209] suggested that the lyophilic proteins could be involved and that precipitation could be initiated by "salting out" of lyophilic colloidal urates. In any case, some forms of proteins appear to be involved in the process.

Although the exact chemical form of the urate precipitates has not yet been determined, they generally contain, in some arrangement, significant amounts of inorganic cations, especially sodium and potassium, but also sometimes calcium, magnesium, or ammonium.[82,208] The predominant cation probably depends not only on dietary intake, as noted above, but also on the ionic or acid–base requirements of the individual species.[82,208] Regardless of the chemical structure of these urate precipitates, the inorganic cations combined with them are excreted without water and without contributing to the osmotic pressure of the urine; therefore, the inability of reptilian kidneys to produce concentrated urine does not necessarily relate to their capacity to excrete inorganic cations. It is also possible that the combination of sodium with urate precipitates in the distal nephrons of reptiles may keep the concentration of free sodium low enough to permit continued reabsorption by the sodium transport system (see above) and maximum dilution of the urine. Finally, in many species, a portion of the inorganic cations combined with urate precipitates in ureteral urine may be reclaimed in the colon, cloaca, or bladder where water can be reabsorbed (see below). In these regions, less complex urate precipitates may be formed and urate salts may be converted to uric acid, thereby freeing the cations for reabsorption. This process may also be enhanced by acidification, which has been demonstrated in the bladders of some turtles and the cloacae of some lizards.[150,207,292]

C. Bladder, Cloaca, and Colon

A urinary bladder occurs only sporadically among reptiles. It is absent in all crocodilians and snakes but is found in the Tuatara and a number of lizards. Turtles all possess a urinary bladder. Embryologically, the tetrapod bladder develops as an evagination from the posterior part of the gut (allantois) and first appears phylogenetically in the Amphibia, where it plays an important role in water balance.[7] Perrault[238] first dissected and described the bladder of the now-extinct Indian giant tortoise *Testudo indica*. Darwin[96] later noted that the giant tortoises of the Galapagos Islands stored large volumes of water in their bladders and suggested that this might be reabsorbed, as in frogs.[167]

The tortoise bladder is osmotically permeable when studied *in vitro*,[14,58,59] and water reabsorption has been demonstrated *in vivo* in the desert tortoise *Gopherus agassizii*[94] and the Tuatara (*Sphenodon punctatus*).[267] Water introduced into the bladder of the gopher tortoise was absorbed at the rate of 20 mL/hr,[94] but, interestingly, bladder permeability in *Testudo graeca* was not affected by arginine vasotocin (AVT).[14] The bladder of desert tortoises functions as an important water-storage organ during periods of drought[167,223] and also as a sink for dietary-derived electrolytes such as potassium and also nitrogen in the form of urea and urates.[211]

The urinary bladders of turtles and tortoises are the site of aldosterone-sensitive active sodium reabsorption, which also enhances water reclamation.[14,58,186] During wet periods, water is stored in the bladder of desert tortoises in the form of very dilute urine that becomes increasing concentrated when the animal is faced with water deprivation,[210,211] eventually becoming isosmotic with the plasma.[213,215] As summer progresses, the tortoises reduce their activity, spending much of their time in estivation. They do, however, emerge to drink when there are infrequent thunderstorms, and they construct small depressions in which the rain collects.[206] The advent of rain results in a dramatic fall in the osmolality of the urine, and a smaller decrease in plasma osmolality.[225] Nagy and Medica[225] speculate that desert tortoises relinquish maintenance of internal homeostasis on a daily basis during most of the year and tolerate large imbalances in their water, energy, and salt budgets. This strategy apparently allows them to exploit resources that are only available periodically while balancing their water and salt budgets on an annual basis and showing an overall energetic profit. Peterson[241] has coined the term *anhomeostasis* to refer to this tendency of the desert tortoises to osmoregulate opportunistically and tolerate significant deviations of the *milieu intérieur* during late spring and summer. In this way, they are able to lay eggs each year, even during droughts.[308]

Little is known of the functions of the bladder in the Tuatara, or in those species of lizards where one is found. The bladder of the large skink *Tiliqua rugosa* was reported to be water permeable[16] but little net water movement occurs in the Tuatara.[267] Stegbauer[291] found that the bladder of a number of species of desert lizards were poorly vascularized and with a transitional epithelium of relatively undifferentiated cells. Rather than being reabsorbed from the lizard bladder, urine may be directed to the cloacal–colonic complex as the bladder duct in species such as *Hemidactylus flaviviridis* has a well-developed sphincter and valvular fold.[273]

The morphology of the cloacal–colonic complex has been described in only a very small number of lizards. The cloaca in *Uromastix hardwickii* is differentiated into three compartments—a coprodeum, urodeum and a proctodeum[274]—but this degree of specialization does not appear to be common. Skadhauge and Duvdevani[288] give some details for *Agama stellio*, and Bentley and Bradshaw[18] describe the colon and cloaca of the agamid lizards *Amphibolurus ornatus* and *A. inermis* (now *Ctenophorus ornatus* and *C. nuchalis*, respectively). In the two latter species, which lack a bladder, urine flows retrograde into the colon from the small cloaca, and up to 2 mL of fluid (i.e., 10% of the lizard's body mass) can be sequestered in this portion of the hindgut at any one time. A well-delineated muscular sphincter prevents the urine from flowing further up into the lower intestine.

A number of early indirect observations suggested that the cloacal–colonic complex is an important reabsorptive site for both water and electrolytes in reptiles. Junqueira et al.[168] exteriorized the ureters in snakes (*Xenodon* sp.) and found that they lost 3.7% of their body mass per day and died after 5 days, with no effect in sham-operated animals. Schmidt-Nielsen and Skadhauge[268] compared the composition of ureteral and cloacal urine in *Crocodylus acutus* and inferred cloacal function from the difference. Similarly, Bradshaw[36,37] compared the composition of ureteral urine with that of voided urine in the lizard *Ctenophorus* (=*Amphibolurus*) *ornatus* and found that the relative osmolar clearance (C_{OSM}/C_{IN}) fell from 29.2% to 1.0%, respectively, indicating massive postrenal reabsorption of osmolytes.

This difference technique was used by Bradshaw[37] to estimate transmural fluxes of water and electrolytes from the colon of *Ctenophorus ornatus* under conditions of both water and saline diuresis. These fluxes were some 20-fold larger than those estimated from *in vitro* measurements with isolated tissues[18,94] and probably reflect the lack of vascularization and adequate mixing. Braysher and Green[53] also used an *in vivo* coprodeum-isolation technique in an attempt to measure rates of fluid and electrolyte reabsorption in the large varanid lizard *Varanus gouldii*. The main conclusion from the studies by Bradshaw[37,42] was that the cloacal–colonic complex routinely reabsorbs approximately 90% of the fluid presented to it, under conditions of either water or saline diuresis, the latter with a much lower glomerular filtration rate. The rate of sodium reabsorption fell significantly, however, from 99% of that presented to the colon to 63% with saline diuresis, and potassium reabsorption also fell from 93% to 35%, indicating the presence of some form of regulation.

A major technical advance was the application to reptiles of the *in situ* perfusion technique of the coprodeum developed by Skadhauge for birds.[284,285,287] Skadhauge and Duvdevani[288] reported that both dehydration and saline loading were associated with increased water reabsorption and decreased rates of sodium transport from the cloacal–colonic complex of the lizard *Agama stellio* (but note that the units in their Table IV should be µEq/kg/hr, not mEq/kg/hr as listed).

Bradshaw and Rice[48] used this same technique to measure transmural fluxes of water and electrolytes from the colon of conscious unanesthetized *Varanus gouldii* held at their preferred body temperature. Estimates for these parameters are given in Table 10.12, where the perfusion rate was adjusted to approximate ureteral flow rates under these same conditions. Sodium loading was accompanied by a decrease in the absolute rates at which sodium, chloride, and water are retained by the colon, and potassium secretion was reduced. When the data were expressed fractionally (i.e., relative to the presentation rate for each component, which is a function of the perfusion rate for each treatment), water reabsorption increased significantly from 23.7% to 40.6% with salt loading, and sodium reabsorption fell from 33.7% to 22.4%. Fractional reabsorption of chloride was unchanged and potassium secretion was reduced.

Surprisingly, colon function in dehydrated *Varanus gouldii* was found to be essentially identical with that of saline-loaded individuals, with an increased rate of reabsorption of water and decreased fractional reabsorption of sodium ions (see Table 10.12). The electrolyte concentration of the

TABLE 10.12
Cloacal Parameters and Rates of Water and Electrolyte Reabsorption and Secretion in
Varanus gouldii

Parameter	Hydration ($n = 6$)		Dehydration ($n = 6$)		Salt-Loaded ($n = 6$)
Perfusion rate (mL/kg/hr)	6.52	—	3.61	—	2.16
Reabsorbate (mL/kg/hr)	1.46 ± 0.15*	NS	1.67 ± 0.25	$P < 0.05$	0.89 ± 0.11*
Na^+ (μmol/kg/hr)	123.9 ± 12.6*	$P < 0.01$	39.49 ± 7.35	NS	27.58 ± 4.08*
K^+ (μmol/kg/hr)	−11.29 ± 2.95*	NS	−6.03 ± 0.94	$P < 0.05$	−2.19 ± 0.54*
Cl^- (μmol/kg/hr)	75.27 ± 9.12*	NS	50.71 ± 10.56	$P < 0.05$	24.77 ± 5.50*
FR_{Na} (%)	33.72 ± 3.94*	$P < 0.01$	20.68 ± 1.39	NS	22.39 ± 3.34*
FR_{Cl} (%)	21.61 ± 1.62	NS	27.51 ± 5.45	NS	22.67 ± 5.34
FR_{H2O} (%)	23.70 ± 2.66*	$P < 0.01$	41.29 ± 4.31	NS	40.61 ± 1.66*
Electrolyte concentration of reabsorbate (in mmol/L of NaCl)	140.3 ± 13.6*	$P < 0.01$	61.5 ± 5.3	NS	63.4 ± 13.9*

Note: Mean ± SE is reported; statistics, ANOVA-SNK test with fractional values arcsine transformed. *$P < 0.01$ for hydrated vs. salt-loaded comparisons within rows.

Source: Bradshaw, S.D. and Rice, G.E., *Gen. Comp. Endocrinol.*, 44, 82–93, 1981. With permission.

reabsorbate has been calculated (in mmol/L of NaCl) with these three treatments; it is hypoosmotic with both dehydration and salt loading but hyperosmotic in water-loaded lizards. This suggests the presence of some solute-linked water flow as found with the cloaca of the crocodile,[268] the intestine of the eel,[286] and the human rectum.[129]

The transport mechanisms underpinning the reabsorption of water and electrolytes from the cloacal–colonic complex are the subject of some debate. Electrical potential differences (PD), with the mucosa negative, have been recorded across colon–cloacal preparations in a number of Brazilian snakes,[168] the tortoise *Testudo graeca*,[14] the crocodilian *Caiman crocodylus*,[19] and the lizards *Ctenophorus ornatus*, *C. nuchalis*, and *Tiliqua rugosa*.[18] Such a PD, which is blocked by mecholyl,[265] is consistent with the occurrence of active sodium transport from the lumen of the colon to the blood (mucosa to serosa) and this may carry water with it as has been demonstrated in the coprodeum and large intestine of the chicken.[31] Differences in colloid osmotic pressure across the cloaca of the desert iguana *Dipsosaurus dorsalis* can also facilitate water reabsorption, even in the absence of an overall osmotic gradient or active sodium transport.[221] Such reabsorption can be prevented by raising the intracloacal colloid osmotic pressure with an infused protein solution, and Lange and Staaland[179] speculate that a similar effect may occur in the colon of the rat.

The ability of the mammalian gall bladder to transport water isosmotically has been well established and depends on the creation of standing sodium gradients in the lateral intercellular spaces.[103,198] Similar intercellular spaces closely associated with mitochondrion-rich cells were described in the anterior rectum of the agamid lizard *Ctenophorus* (=*Amphibolurus*) *maculosus* by Braysher[52] and figured by Minnich,[213] but the ambitious claim that this species is capable of elaborating a hyperosmotic urine (see above) remains to date unverified.[54]

D. SALT GLANDS

Extrarenal specialized glands capable of elaborating hyperosmotic saline solutions occur throughout the vertebrates, are especially well-developed in marine birds,[233] and were first identified in marine reptiles by Schmidt-Nielsen and Fänge.[270] Whereas those of birds are only able to secrete a solution of sodium chloride, the cephalic salt-secreting glands of reptiles have been found to have a surprisingly varied secretory repertoire.[42,119] Some significant differences can be identified in their morphology and embryological origin when compared with avian glands. In birds, the gland is located in a bony recess above the eye (supraorbital), whereas in reptiles it may be an external nasal gland,[143] a lachrymal gland, or a premaxillary, sublingual, or lingual gland, depending on the taxonomic group. An early claim by Dunson and Taub[127] that the Harderian gland of sea snakes of the genus *Laticauda* could be added to this list was later found to be an error, with the true salt gland being located beneath the tongue.[125] Dunson et al.[123] also have described an unusual lateral nasal gland in the Montpellier snake (*Malpolon monspessulanus*), but its function has yet to be established. Comparative data on reptilian salt glands, including embryological origin and secretory characteristics, are summarized in Table 10.13, and histological and cytological information on the external nasal glands of Australian lizards are reviewed in Saint-Girons and Bradshaw.[258] Salt glands have not been found in terrestrial snakes and have only been described in a single terrestrial chelonian, *Testudo carbonaria*.[234] They are also absent in alligators, geckoes, and the one species of pygopodid lizard that has been studied.[258]

An analysis of the data in Table 10.13 reveals a number of clear habitat and diet correlations. Regardless of the embryological origin of the glands, marine species secrete a solution with sodium as the primary electrolyte and Na^+/K^+ ratios ranging from 6 in *Amblyrhynchus cristatus* to as high as 46 in *Crocodylus porosus*. Rates of excretion of sodium are also very high in the few marine species where they have been measured, varying from 730 µmol/kg/hr in the sea snake *Laticauda semifasciata* to 2550 µmol/kg/hr in the Galapagos iguana. Terrestrial tortoises and lizards produce a salt solution that is rich in potassium, rather than sodium, with the Na^+/K^+ ratio typically less than 1.0, reflecting their herbivorous diet. Some exceptions are the two species of varanid lizards, *Varanus semiremex* and *V. salvator*, that feed on crustaceans and other animals in mangrove swamps and littoral habitats and have high intakes of salt water. The Galapagos land iguana (*Conolophus subcristatus*) feeds preferentially on *Opuntia* cactus, and this may explain the high reported Na^+/K^+ ratio of its nasal gland secretion of 3.2.

Comparatively little study of the ultrastructure and vasculature of reptilian salt glands has been done, compared with what is known for bird salt glands.[163] The gross histology and ultrastructural morphology of the lachrymal salt gland in chelonians has been studied in most detail,[1,12,71–73,131,132,148] and Abel and Ellis[1] describe a rich vasculature that is arranged counter-current to the flow of secretion down the tubules, as in marine birds.[233] Cholinesterase-containing nerve fibers form a dense plexus around the secretory tubules, and Abel and Ellis[1] suggest the presence of an additional adrenergic sympathetic nerve supply. Dual cholinergic and adrenergic innervation of the lingual salt glands of the estuarine crocodile *Crocodylus porosus* has also been described recently.[141] The degree of specialization of the cells in the secretory tubules is also similar to that found in the salt glands of marine birds, with small unspecialized cells at the blind end of the tubule that progressively become differentiated into mitochondrion-rich secretory principal cells as one moves down the tubule. The major difference from birds is that the basal membrane in the reptilian lachrymal gland shows little infolding, but, in contrast, the very extensive folding of the lateral intercellular membranes forms complex intercellular spaces. These complex foldings of the lateral membrane are shown very clearly in Figure 10.4 from the nasal salt gland of the varanid lizard *Varanus gouldii*.[260]

Cowan[73] described large amounts of mucopolysaccharide, which has been shown by Farber[135] to function as an ion exchange resin, in the lateral intercellular spaces of the lachrymal gland of *Malaclemys terrapin*. This observation led to the suggestion by Bennett[11] that negative charges on these mucopolysaccharides could attract cations to the absorptive surface of cells. van Lennep and Komnick[310] and van Lennep and Young[311] also called attention to the presence of mucopolysaccharide in the lateral intercellular spaces of the agamid lizard *Uromastix acanthinurus* which was also noted by Lemire[189] and Lemire et al.[195]

Gabe and Saint-Girons[142] compared the histology of the salt glands from 36 species of lizards, representing 16 families, and identified functional secretory units by the presence of principal cells, which they described as *cellules striées* because of their striated appearance under the light microscope.

TABLE 10.13
Comparative Data on Reptilian Salt Glands

Taxon and Species	Embryologic Origin of Gland	Electrolyte Concentration (mmol/L)			Secretion Rate (μmol/kg/hr)		Habitat	Refs.
		Na⁺	K⁺	Na⁺/K⁺	Na⁺	K⁺		
Chelonia:								
Chelonidae								
Caretta caretta	Lachrymal	730–878	18–31	28–40	—	—	Marine	Schmidt-Nielsen and Fänge[270]
Chelonia mydas		685	21	33	—	—	Marine	Holmes and McBean[162]
Lepidochelys olivacea		713	29	25	—	—	Marine	Dunson[116]
Emydidae								
Malaclemys centrata		682	32	21	—	—	Estuarine	Dunson[117]
Testudinae								
Testudo carbonaria		0.1–6	233–260	0.03	70	—	Terrestrial	Peaker[234]
Lacertilia:	External nasal							
Iguanidae								
Amblyrhynchus cristatus		1434	235	6	2550	510	Marine	Dunson[115,116]
Dipsosaurus dorsalis		494–1032	640–1387	0.7–0.8	22 0	31 183 200	Desert	Dunson[117] Hazard,[159] Schmidt-Nielsen et al.,[269] Shoemaker et al.[276]
Sauromalus obesus		82–150	490–1102	0.1–0.2	3	27	Desert	Nagy,[222] Templeton[298]
Ctenosaurus similes		78–475	220–527	0.9	—	—	Terrestrial	Templeton[298]
Iguana iguana		507–728	290–497	1.5–1.7	—	—	Terrestrial	Dunson,[116] Schmidt-Nielsen et al.[269]
Conolophus subcristatus		692	214	3.2	—	—	Terrestrial	Dunson[116]
Uta stansburiana		—	—	—	46	71	Terrestrial	Hazard et al.[160]
Uta tumidarostra		—	—	—	387	113	Intertidal	Hazard et al.[160]

Family / Species	Gland location						Habitat	Reference
Agamidae								
Uromastix aegyptus		639	1398	0.5	—	—	Desert	Schmidt-Nielsen et al.[269]
Uromastix acanthinurus		—	—	0.03–0.6	6	45	Desert	Bradshaw et al.,[47] Lemire et al.[190]
Scincidae								
Tiliqua rugosa		167	433	0.4	—	—	Terrestrial	Bradshaw et al.[50]
Varanidae								
Varanus semiremus		654	54	12.1	—	—	Intertidal	Dunson[118]
Varanus salvator		307	23.2	13.2	—	—	Semiaquatic	Minnich[212]
Ophidia:								
Hydrophiidae	Posterior lingual							
Laticaudia semifasciata		686	57	12	730	33	Marine	Dunson and Taub[127]
Pelamis platurus		620	28	22	2180	92	Marine	Dunson[114]
Aipysurus laevis		798	28	28.5	—	—	Marine	Dunson and Dunson[121]
Lapemis hardwickii		676	23	29.4	—	—	Marine	Dunson and Dunson[121]
Hydrophis elegans		509	20	25.5	—	—	Marine	Dunson and Dunson[121]
Achrochordidae	Posterior lingual							
Acrochordus granulatus		483	15	32.2	—	—	Marine	Dunson and Dunson[121]
Homolapsidae	Premaxillary							
Cerberus rhynchops		414	55	7.5	—	—	Marine	Dunson and Dunson[122]
Crocodilia:	Lingual							
Crocodylus porusus		386–740	10–16	38–46	—	—	Estuarine	Taplin and Grigg[296]
Osteolaemis tetraspsis		545	15	36	—	—	Freshwater	Taplin[295]

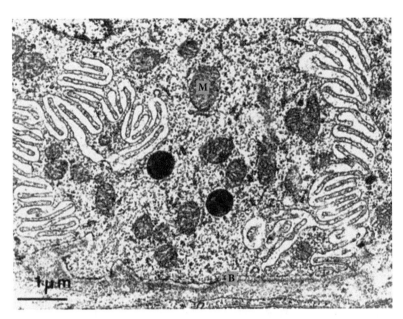

FIGURE 10.4 Basal region of a principal cell (*cellule striée*) of the nasal salt gland of the varanid lizard *Varanus gouldii* showing extensive folding of the intercellular plasma membranes. B, basement membrane.

In species such as *Iguana iguana*, principal cells and mucoserous cells are intermixed in the tubules. A similar situation in the scincid lizard *Tiliqua rugosa* led to some confusion over whether this lizard is capable of secreting a hyperosmotic solution from its nasal gland. Initial claims by Braysher[52] were questioned by Saint-Girons et al.[259] on the basis of the intermixed histology of the gland but finally confirmed by Bradshaw et al.[50] in a physiological investigation.

Early studies suggested that the salt glands of both birds and reptiles were controlled primarily by the activity of cholinergic nerves, with secretion being evoked either by nerve stimulation or injection of parasympathomimetic drugs such as mecholyl.[235] More recent evidence, however, has shown that an adenylate cyclase–cyclic AMP pathway is involved in both avian salt glands and the lachrymal gland of *Malaclemys*[195,278] and that secretion is also stimulated by vasoactive intestinal peptide (VIP),[140,196] probably acting through the adenylate cyclase pathway.[277,276] It is generally accepted that, in birds, the key process is a secondary active secretion of chloride involving a basolateral Na^+–$2Cl^-$–K^+-cotransporter mechanism and apical chloride and basolateral potassium channels—with the whole process being energized by the ion gradients generated by a basolaterally localized electrogenic Na^+,K^+-ATPase pump as shown in Figure 10.5.[277] That this same process is involved in those reptilian salt glands that secrete primarily sodium is supported by the observations that bumetanide (a blocker of Na^+–$2Cl^-$–K^+-cotransporter) and ouabain (a blocker of Na^+/K^+-ATPase) both block stimulation of secretion by isolated lachrymal gland tissue from *Malaclemys terrapin*.[278]

A detailed study of potassium secretion by the nasal gland of the desert lizard *Sauromalus obesus* by Shuttleworth et al.[279] in which blood flow through the gland was measured using microspheres labeled with [86]Sc has revealed important differences between reptilian and avian salt glands. After two potassium loads, the Na^+,K^+-ATPase activity in the salt gland tissue had doubled, but the residual ATPase activity and the total weight of the glands were unchanged. Specific increases in Na^+,K^+-ATPase activity following salt loading are well documented in avian salt glands, but in these cases they are invariably associated with an increase in the size and weight of the gland.[138] Shuttleworth and Hildebrandt[277] note that the fact that *Sauromalus obesus* is capable of producing a secretion with potassium concentrations in excess of 1000 mmol and the virtual absence of any sodium suggest a unique secretory mechanism quite different from that of marine birds.

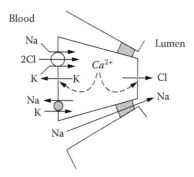

FIGURE 10.5 Diagram illustrating the proposed mechanism of secretion in the avian salt gland. Increases in cytosolic Ca^{2+} activate secretion by increasing the opening of basolateral Ca^{2+}-activated K^+ channels and apical Ca^{2+}-activated Cl^{-1} channels. (From Shuttleworth, T.J. and Hildebrandt, J.-P., *J. Exp. Zool.*, 283, 689–701, 1999. With permission.)

Early work on the control of electrolyte excretion by the nasal salt gland of the north American desert iguana *Dipsosaurus dorsalis* served to highlight the extraordinary secretory repertoire of its salt gland, which is capable of secreting three cations (sodium, potassium, and rubidium) associated with three different anions (acetate, succinate, and bicarbonate).[276] Even more surprising, osmotic loading with sucrose or mannitol did not elicit secretion in this lizard and Shoemaker et al.[276] speculated that alkali metal ions (sodium or potassium) were needed to initiate secretion. More recent work by Hazard[159] with *Dipsosaurus dorsalis* has shown that sodium acetate does not stimulate secretion, whereas histidine chloride and potassium acetate do. These results suggest that *Dipsosaurus* uses ion-specific receptors for potassium and chloride, rather than osmotic or volume receptors that have been described in birds,[235] to detect and respond to an ion load.

Little is known of the actual mechanism by which a hyperosmotic salt solution is generated by the reptilian nasal gland. Following cytochemical and autoradiographic studies showing that Na^+,K^+-ATPase was localized in the lateral intercellular plasma membranes, and not the luminal membranes of the secretory cells,[133] the *standing gradient hypothesis* was proposed to account for the hyperosmotic secretion.[104] This hypothesis required that the plasma membranes of the principal cells be relatively impermeable to water and that NaCl extruded laterally into the intercellular spaces be recycled and not come from the lumen. It also required that Na^+ be pumped *into* cells from the basal and lateral membranes, which would be unprecedented. Ellis et al.[134] resolved this paradox with the conclusion that the cell junctions (zonulae occludentes) between the principal cells and the secretory tubules were not tight but very leaky, with sodium being secreted across the cell membrane at the lumen in a one-stage process as proposed by Peaker and Linzell.[235] Their model envisages that isotonic fluid secreted by the principal cells at the terminal end of the tubules flows along the lumen of the tubule toward the central canal. Water is then reabsorbed from this fluid through the leaky junctions between the principal cells and passes into the intercellular channels, where a standing gradient of Na^+ is maintained by sodium pumps in the lateral membranes.

Direct measurements of intracellular and luminal ion concentrations in the lachrymal salt gland of the sea turtle *Chelonia mydas*[200] seriously disagree with all the proposed mechanisms for salt secretion. X-ray microanalysis of the secretory tubule lumina and central canal in frozen-hydrated lachrymal glands did not reveal the high concentrations of Na^+ and Cl^- that are found in the final secretion. The fluid in the central canals was essentially isosmotic (269 to 380 mOsm/kg), suggesting that the final concentration step occurs not in the gland itself but in the highly vascularized duct system.[202] Another paper using the same method has reported just this situation in the duck nasal gland,[201] with the concentration of the secretion only beginning to increase as it enters the main duct (see Figure 10.6).

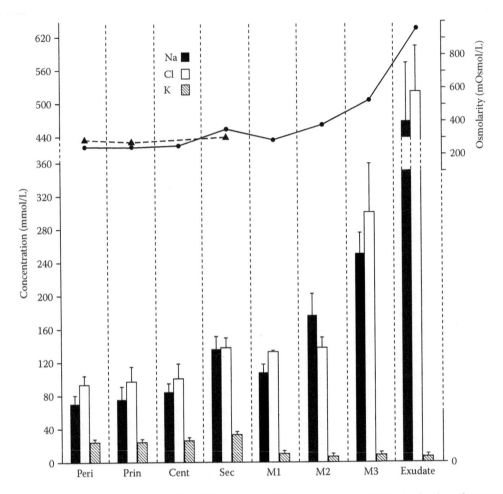

FIGURE 10.6 Concentrations of sodium (Na), chloride (Cl), and potassium (K) in the lumina of secretory tubules of the Pekin duck (*Anas platyrhynchos*), measured by x-ray microanalysis. Concentrations are shown at the position of peripheral (Peri) and principal cells (Prin) in the central canals (Cent), secondary ducts (Sec), and successive positions along the main collecting duct (M1–M3), as well as for the exudate. Note that the concentration of the fluid does not increase until it reaches the main duct system. (From Marshall, A.T. et al., *J. Comp. Physiol.*, 156B, 213–227, 1985. With permission.)

A fascinating insight into the evolution of reptilian salt glands is provided by the description by Grismer[152] of three new species of side-blotched lizards of the genus *Uta* living in intertidal areas of the Baja California. They feed on salt-rich crustaceans (isopods) and all have greatly enlarged nasal salt glands located in protruding bony recesses that are nearly five times larger than those of other mainland species of *Uta* that feed on insects. The author speculates that this evolutionary change has been brought about by a change in diet, forced on these lizards by their island habitat being progressively covered by guano deposited by roosting seabirds. The physiology of the gland of *Uta tumidarostra* was studied by Hazard et al.,[160] and the daily rate of sodium excretion was estimated at 9.3 µmol/g/day compared with a maximal rate of only 1.1 µmol/g/day for *Uta stansburiana*. This rate of sodium excretion from the salt gland of *Uta tumidarostra* compares favorably with the figure of 14.5 µmol/g/day estimated by Shoemaker and Nagy[275] for the Galapagos marine iguana *Amblyrhynchus cristatus*, which possesses one of the most active sodium-excreting glands of any reptile.

III. CONTROL MECHANISMS

A. BLOOD FLOW AND GFR IN THE KIDNEY

Blood flow in the kidney is critical to the formation of glomerular filtrate and to the maintenance of tubular function (see above). In reptiles, the only direct quantitative measurements of renal blood flow and its control have been made on afferent glomerular arterioles of garter snakes (*Thamnophis sirtalis*).[319,322] These studies indicated that blood flow to superficial glomeruli of ophidian kidneys varies considerably, not only between individual animals but also between glomeruli in the same kidneys. The differences between individual animals do not appear to be related to the mean systemic arterial pressure. As noted above, intermittent glomerular blood flow and, presumably, intermittent filtration occur, but they occur most frequently in glomeruli with low blood flow rates. The relationship between intermittency and mean glomerular blood flow rate is not linear, but all glomeruli with blood flow rates below ~5 nL/min apparently show intermittency, whereas only 10% of those glomeruli with average blood flow rates of ~24 nL/min show intermittency. As discussed above, the fraction of glomeruli with intermittent blood flow in each kidney correlates directly with the plasma osmolalities of the individual animals (Figure 10.2). Because plasma osmolality increases with dehydration, this correlation supports the concept that glomerular intermittency plays a major role in the decrease in whole-kidney GFR observed with dehydration (see above). Also, because the release of arginine vasotocin (AVT), the reptilian antidiuretic hormone, from the neurohypophysis is stimulated by increases in plasma osmolality (see below), this correlation supports a role for AVT in mediating glomerular intermittency during dehydration, at least in ophidian reptiles. Moreover, in further support of this concept, an infusion of AVT causes glomerular blood flow to cease in garter snakes[322] (Figure 10.7) just as it produces a decrease in whole-kidney GFR in water snakes (*Natrix sipedon*).[74]

Apparently, decreases in glomerular blood flow with or without exogenous AVT result from constrictions of the afferent arteriole (Figure 10.7).[322] When the diameter of the afferent arteriole becomes sufficiently small, the rigid, nucleated red blood cells cannot pass and occlude the vessel completely, causing cessation of flow.[322] It should be noted, however, that filtration at a given glomerulus may actually cease before blood flow ceases or may not even occur during periods of low continuous blood flow. This would occur if the hydrostatic pressure in the glomerular capillaries fell to a level that equaled the sum of the opposing plasma colloid osmotic pressure and the pressure in Bowman's space (i.e., the net ultrafiltration pressure went to zero; see above). This seems very likely, at least in garter snakes, because of the very low average net ultrafiltration pressure; thus, glomerular intermittency may occur in these reptiles and perhaps in other species without actual cessation of glomerular blood flow.

Prolactin may be involved in regulation of glomerular blood flow. The administration of prolactin to some freshwater turtle species produces a significant increase in whole-kidney GFR in intact animals (definitely in *Chrysemys picta* and possibly in *Pseudemys scripta*) and appears to reverse a decrease in whole-kidney GFR produced by hypophysectomy.[56] These studies suggest that prolactin could help regulate glomerular blood flow by relaxing the afferent arteriole and that it might play a role in determining the increase in GFR observed with a water load, but more detailed studies on the pharmacological vs. physiological effects of this hormone and more direct observations of its effect on regulation of glomerular circulation are needed.

Neural regulation of glomerular blood flow and, thus, filtration, may be more significant in reptiles than generally assumed. First, preliminary observations indicate that nerve endings exist near the glomerular arterioles of garter snakes (*Thamnophis* spp.) (S.D. Yokota, R.A. Wideman, and W.H. Dantzler, unpubl. observ.). Second, the α-adrenergic inhibitors phentolamine and phenoxybenzamine block the decrease in whole-kidney GFR observed with high plasma concentrations of potassium in sea snakes (*Aipysurus laevis*) and garter snakes[22,321] (S. Benyajati, S.D. Yokota, and W.H. Dantzler, unpubl. observ.). These observations suggest that α-adrenergic agonists, whose release is stimulated

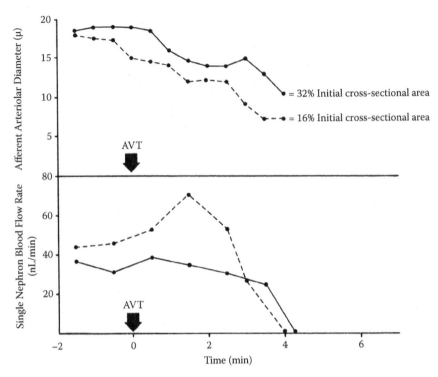

FIGURE 10.7 Simultaneous relationship of afferent arteriolar diameter and blood flow for two representative nephrons in snake (*Thamnophis sirtalis*) kidney during continuous infusion of arginine vasotocin (AVT). Arrows mark start of AVT infusion at a rate of 17 pg/100 g/min. (From Yokota, S.D. and Dantzler, W.H., *Am. J. Physiol.*, 258, R1313–R1319, 1990. With permission.)

by high concentrations of potassium, may play a role in regulating resistance at the afferent glomerular arterioles[22,321] (S. Benyajati, S.D. Yokota, and W.H. Dantzler, unpubl. observ.), but, as in the case of prolactin, more direct studies are needed to determine actual effects on glomerular circulation.

Autoregulation, the process whereby the rates of renal blood flow and glomerular filtration are maintained relatively independent of mean systemic arterial pressure by mechanisms intrinsic to the kidney, has yet to be examined in reptiles. In mammals, this process, which has been studied extensively,[226,271] involves two mechanisms: (1) myogenic control and (2) tubuloglomerular feedback control. Myogenic control is a process intrinsic to the smooth muscle cells of the afferent arterioles whereby increased arterial hydrostatic pressure leads to constriction and decreased pressure leads to relaxation. Tubuloglomerular feedback control involves the macula densa cells sensing an increased load of solute (probably sodium chloride) delivered to the early distal tubule, leading, in turn, via a paracrine factor (apparently adenosine[271]) to constriction of the adjacent afferent glomerular arteriole. This mechanism is poised to act primarily in response to an increased sodium chloride load to produce a decrease in GFR.

Autoregulation may appear unlikely in reptiles with their variable mean arterial pressure, variable SNGFRs, glomerular intermittency, and apparent lack of macula densa cells (despite the apposition of the early distal tubule to the vascular pole of its own glomerulus); however, it may be that certain glomeruli in an individual kidney always have either high or low blood flow and, thus, either high or low GFR. This blood flow may be regulated in some specific fashion relative to the mean systemic arterial pressure. Indeed, as noted above, Yokota and Dantzler[322] found that differences in mean single nephron blood flow rates between individual garter snakes were not related to differences in mean systemic arterial pressures. Direct studies are needed to determine if autoregulation, or a similar process, operates in reptilian kidneys.

B. Blood Flow and Autonomic Control in Salt Glands

As discussed above, cholinergic nerves apparently play a role in controlling salt gland secretion in reptiles as well as birds. In reptiles, this control is evident in species in which the gland secretes primarily sodium (e.g., *Crocodylus porosus* and *Malaclemys terrapin*)[278,296,297] and in species in which the gland secretes primarily potassium (e.g., *Sauromalus obesus*).[279] Presumably this action, as in birds, occurs via the phosphatidylinositol pathway with changes in the cytoplasmic calcium concentration. This pathway is supported by the increased transport activity in salt gland tissue from *Malaclemys* produced by A23187, a calcium ionophore that increases the cytoplasmic calcium concentration.[278]

Although cholinergic nerves apparently regulate salt secretion by the glandular tissue itself, they can also influence blood flow through the glands. This effect was demonstrated in the study by Shuttleworth et al.,[279] noted above, in which blood flow through the nasal salt gland of *Sauromalus* was measured with radiolabeled microspheres. In this study, administration of a cholinomimetic (methacholine) not only stimulated potassium secretion by the salt gland itself but also reduced the blood flow through the gland. Thus, the increased rate of potassium secretion required an effectively greater extraction of potassium from the blood by the gland than would have been the case if the blood flow had not decreased. Indeed, the role of blood flow regulation in determining salt gland secretion is far from clear. For example, despite the remarkable ability of the nasal salt gland in *Sauromalus* to extract 70% or more of the potassium from the blood passing through it, the maximum rate of *in vivo* secretion would require about a fourfold increase in the control blood flow rate. Regulation of blood flow and its integration with the regulation of cellular transport remains to be determined.

C. Hormonal Control: Kidney

1. Introduction

Information has grown in recent years with regard to the extent to which pituitary and adrenal hormones modulate the activity of those organs primarily responsible for controlling rates of water and electrolyte excretion in reptiles: the kidneys, the cloacal–colon complex, and the cephalic salt-secreting glands. The topic has been reviewed periodically over the years,[36,37,42,44,51,93,212,228] but the data are still very fragmentary and garnered from only a very small number of species. Initial studies suggested that the reptilian kidney was unresponsive to corticosteroid hormones exposure, and it was even seriously entertained in the 1960s that reptiles might lack an adrenocorticotrophic hormone (ACTH).[197] This problem was resolved when the importance of studying reptiles at or near their preferred body temperature (PBT) was realized, which in the case of many lizards is closer to 37°C than to room temperature.[5] The erroneous classification of the pituitary corticotroph (the gamma cells) as a luteotroph by cytologists at the time also did not help matters.[257]

2. Arginine Vasotocin

In contrast to the real or suspected role of the adrenal glands, reptiles have long been assumed to have an antidiuretic hormone of pituitary origin, analogous to that of mammals.[15,264] This was early on identified as arginine vasotocin (AVT) based on its presence in the *pars nervosa* of a variety of reptilian species and its pronounced antidiuretic effects when injected.[220,263,264] The antidiuresis is brought about primarily through a reduction in glomerular filtration rate in the kidney.[42,76,81,83,91] This response results from a constriction of glomerular afferent arterioles following the binding of AVT to a V_1-type receptor.[51,128,165] Although reptiles lack a renal-concentrating mechanism in the kidney, evidence suggests that AVT also acts on the renal tubule through a V_2-like receptor, modifying both salt and water reabsorption. The fractional reabsorption of filtrate is enhanced and relative free-water clearance (C_{H_2O}/C_{IN}) declines following administration of physiological doses

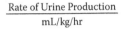

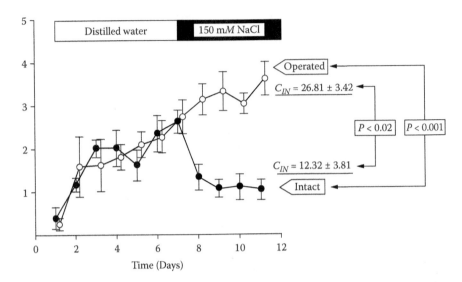

FIGURE 10.8 Rates of urine production and glomerular filtration rate (C_{IN}) with prolonged water and saline diuresis in intact and hypothalamic-tract-lesioned (operated) agamid lizard *Ctenophorus* (=*Amphibolurus*) *ornatus*. All fluid loads = 10 mL 100 g/day, and data are presented as mean ± SE. (From Bradshaw, S.D., *Gen. Comp. Endocrinol.*, 25, 230–248, 1975. With permission.)

of AVT in the freshwater snake *Natrix sipedon*,[74] the freshwater turtles *Chrysemys picta belli* and *Pseudemys scripta*,[60,317] the arid-living lizards *Ctenophorus ornatus*[37] and *Varanus gouldii*,[48] but not in the desert lizard *Sceloporus cyanogenys*.[294] The effect of AVT on relative free-water clearance is usually interpreted as resulting from an increase in tubular permeability to water.[130] Fractional reabsorption of total solutes (i.e., relative osmolar clearance C_{OSM}/C_{IN}) has been shown to increase following AVT injections in a number of species but not in all.[41,43,60,81,85,93] Usually, however, the reabsorption of both water and salts cooperate to form the antidiuretic response, and the localization of these responses is dependent on the permeability to water of the tubular segment and the presence of salt-transporting organelles. There has thus been for some time strong circumstantial evidence for AVTs being the physiological ADH in reptiles but, in an early review, Dantzler and Holmes[93] commented that AVT is "generally considered to be the natural antidiuretic hormone ... [but that] this has never been completely documented by the production of hormone deficiencies by ablation of the hypothalamic neurosecretory cells and correction of the defect by hormone replacement."

Preliminary results of such a study in the agamid lizard *Ctenophorus* (=*Amphibolurus*) *ornatus* were reported by Bradshaw,[37] and full details of the effect of hypothalamic lesions on renal and postrenal function in this species were published later.[38,40] Figure 10.8 shows the effect of the lesions on the ability of the lizards to produce an antidiuresis when challenged with saline loading and the attendant difference in whole-kidney GFR. As would be expected, exogenous AVT provokes a sustained antidiuresis when given to these tract-lesioned individuals.[42]

A direct approach to the question of whether AVT is the physiological ADH in reptiles was taken by Rice,[250] who optimized a sensitive radioimmunoassay originally developed by Rosenbloom and Fisher,[256] to measure changes in circulating levels of this hormone in the large varanid lizard *Varanus gouldii*. Renal function with three treatments (chronic water loading, salt loading, or dehydration) was compared by Bradshaw and Rice,[48] who also monitored concomitant changes in levels of the adrenal corticosteroids aldosterone and corticosterone. Changes in the rate of urine production (V) and the GFR and mean plasma levels of AVT with the three treatments are shown

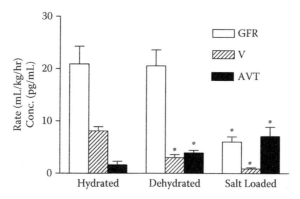

FIGURE 10.9 Renal function and circulating levels of arginine vasotocin (AVT) in the varanid lizard *Varanus gouldii* with water-loading, dehydration, and salt-loading regimes. The glomerular filtration rate (GFR) was measured as the clearance of inulin and the rate of urine production (V), both expressed in mL/kg/hr; AVT concentration is in pg/mL. (Data from Bradshaw, S.D. and Rice, G.E., *Gen. Comp. Endocrinol.*, 44, 82–93, 1981; Bradshaw, S.D., *Homeostasis in Desert Reptiles*, Springer, Heidelberg, 1997, pp. 1–213.)

in Figure 10.9 with AVT increasing significantly with both dehydration and salt loading. Both these treatments provoked an antidiuresis, but this was solely tubular in origin in the case of dehydration; GFR only fell significantly after chronic salt loading. The effects of exogenous AVT at a dosage of 25 ng/kg were blocked by injections of probenecid (an inhibitor of organic acid transport) at a dosage of 100 mg/kg, and a highly significant positive correlation between AVT levels and plasma osmolality was found, as shown in Figure 10.10.

Although the case for AVT functioning as a physiological ADH in *Varanus gouldii* would appear to be strong, plasma levels have only been reported in eight species of reptile to date: one turtle, two snakes, four lizards, and the tuatara.[48,136,137,139,155,177,281,282] The statistical relationship between circulating levels and plasma osmolality was investigated in five of these species but was found to be positive and significant in only three of them (*Varanus gouldii*, *Pogona minor*, and the snake *Notechis scutatus*). The case with the snake *Bothrops jararaca* is problematic, as Silveira et al.[281] argue strongly that no correlation exists between plasma AVT and osmolality in this species, in contrast to that reported by Ladyman et al.[177] in the elapid snake *Notechis scutatus*. A careful examination of their data, however, reveals that plasma AVT levels were undetectable in 47% of

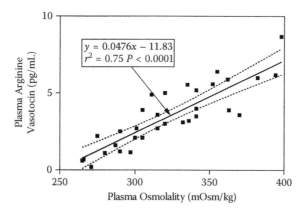

FIGURE 10.10 Correlation between plasma osmolality (in mOsm/kg) and circulating levels of arginine vasotocin (AVT) in the varanid lizard *Varanus gouldii*. (Adapted from Bradshaw, S.D., *Homeostasis in Desert Reptiles*, Springer, Heidelberg, 1997, pp. 1–213.)

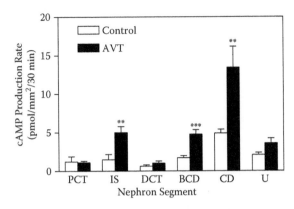

FIGURE 10.11 Localization of arginine vasotocin (AVT)-sensitive adenylate cyclase activity along the nephron of the agamid lizard *Ctenophorus ornatus*. Data are expressed per mm² of the outer membrane of the tubule. PCT, proximal convoluted tubule; IS, thin intermediate segment; DCT, distal convoluted tubule; BCD, branched collecting duct; CD, collecting duct; U, ureter. The statistical significance of differences between the mean of control and AVT addition treatments is shown with $**P = 0.01$ and $***P = 0.001$. (Adapted from Bradshaw, S.D., *Homeostasis in Desert Reptiles*, Springer, Heidelberg, 1997, pp. 1–213.)

the samples assayed, even though the assay sensitivity is given as 0.5 pg/mL. The values reported in their study are also highly variable, and no significant differences were found when all treatment groups were analyzed together; however, the correlation coefficient (r) for plasma AVT levels and osmolality are reported as 1.0 and 0.7 for salt-loaded and acute salt-loaded groups, respectively (see Table 2 in Silveira et al.[281]). No P values are given for these r values, but they appear to belie the conclusion of the authors that AVT levels are not correlated with changes in plasma osmolality.

Bradshaw and Bradshaw[35] used a radioassay of second messenger, developed by Morel,[217] for the tubular localization of hormonal receptors in the mammalian kidney. Working with the agamid lizard *Ctenophorus ornatus*, they detected AVT-stimulated production of cyclic AMP in various segments of the nephron (Figure 10.11). The V_2-type AVT receptors identified by this approach are localized in two segments: the thin intermediate segment (IS) and in all sections of the collecting duct system (LDT, BCD, and CD) but are not found in the distal convoluted tubule (DCT). The site and mode of action of AVT in the reptilian kidney has recently been reviewed by Bradshaw and Bradshaw,[45] who point out that, although the case for AVTs being the physiological ADH in species such as *Varanus gouldii* is strong, hormone ablation studies using neurohypophysial tract lesions have yet to be undertaken. Similarly, although such lesion studies with *Ctenophorus ornatus* are consistent with the hypothesis, plasma AVT levels have yet to be measured in this small 20-g lizard.

3. Adrenal Corticosteroids

Aldosterone and corticosterone (compound B) are the major secretory products of the reptilian adrenal,[261,312] although traces of other steroids such as 18-OH-corticosterone (18-OH-B) and deoxycorticosterone (DOC) have been reported in the literature, along with cortisol (compound F). As is the case with most lower vertebrates, however, the nature of their action on the kidney is not well understood.[17] Partly, this is because of early difficulties experienced in demonstrating any effect of exogenous aldosterone injections on renal function in lizards[36,49,183,299,301] and the failure of all experiments using spironolactone, which is a potent aldosterone antagonist in mammals.[49,61,130] Studies involving adrenalectomy, which is the classical way in which the impact of adrenal hormones has been codified in mammals, have also been singularly uninformative (see Callard and Callard[63] for an early review), primarily because of the technical difficulties associated with this operation in reptiles where the adrenals, particularly the right, are closely adhered to the vena cavae.

LeBrie and Elizondo[184] approached this problem in an original fashion and induced adrenal insufficiency in the water snake *Natrix cyclopion* by vascular occlusion, rather than attempting the far more difficult operation of surgical adrenalectomy, and they found that these animals displayed an increased rate of renal loss of sodium and chloride ions when compared with intact controls. They were able to reduce this loss with injections of aldosterone, which increased reabsorption of both sodium and water by approximately 24%, primarily, they thought, through an action on the proximal rather than the distal segment of the nephric tubule.

The effects of hypophysectomy and dexamethasone blockade on renal function were investigated in the lizards *Ctenophorus ornatus* and *Dipsosaurus dorsalis* by Bradshaw,[36] Bradshaw et al.,[49] and Chan et al.[65] These procedures were associated with an increase, rather than a decrease, in sodium reabsorption, which is inconsistent with the common action of aldosterone as a mineralocorticoid. Depressing adrenal function either by hypophysectomy or dexamethasone blockade also reduces the rate of secretion of other steroids, such as corticosterone, to negligible levels, and replacement studies in *Ctenophorus ornatus* with both corticosterone and ACTH suggested that this steroid is natriuretic, or salt excreting, in this species. Circulating levels of corticosterone also increase with chronic saline diuresis in both *Ctenophorus ornatus* and *Dipsosaurus dorsalis*, which is consistent with this interpretation.[4,36] Chan et al.[65] found that the accumulation of salt and water that they observed in hypophysectomized *Dipsosaurus dorsalis* could be rectified with injections of corticosterone plus prolactin, but not by prolactin alone, lending further support to the notion that corticosterone functions as a natriuretic hormone in these lizards.[37] Hypophysectomy and dexamethasone both affect renal function in the agamid lizard *Ctenophorus ornatus*, with both treatments leading to a significant increase in the fractional reabsorption of sodium ions (FR_{Na}).

Circulating levels of aldosterone were first measured in a reptile by Bradshaw and Grenot,[46] who found a significant negative correlation with increasing plasma sodium concentrations in the lizards *Uromastix acanthinurus* and *Tiliqua rugosa*, although the slope of the regression was much lower and barely significant in the latter, non-desert, species. Bradshaw and Rice[48] found a similar negative correlation for aldosterone in *Varanus gouldii* which contrasts with a strong positive correlation between plasma levels of corticosterone and sodium. The response of *V. gouldii* to sodium loading is thus the same as that of *Ctenophorus ornatus* and *Dipsosaurus dorsalis* (i.e., an increase in the rate of secretion of corticosterone). The opposite changes in plasma aldosterone and corticosterone levels with variation in plasma sodium concentration in the varanid lizard *Varanus gouldii* are shown in Figure 10.12.

It is evident that the homeostatic response of the kidney of *Varanus gouldii*, with an increased rate of excretion of sodium ions following salt loading, could be a result of the increasing levels of corticosterone—if that steroid hormone is natriuretic—or, instead, it could be the result of the falling levels of aldosterone—if that hormone is natriferic (i.e., sodium reabsorbing) in its actions. Rice et al.[185] reasoned that hormonal function is associated with elevated, rather than depressed, levels in the blood and that kidney function in adrenalectomized *Varanus gouldii* would be unchanged with salt loading if aldosterone were responsible for the homeostatic response but compromised during water loading (where the absence of the natriferic aldosterone hormone would prevent excessive loss of sodium). Similarly, if corticosterone were responsible due to its natriuretic action, its absence would be evident under a salt-loading régime.

Figure 10.13 shows clearly that the fractional reabsorption of sodium (FR_{Na}) is significantly reduced in adrenalectomized animals when confronted with a water load (falling from over 95% to 76%) but is not affected in salt-loaded individuals. This confirms that aldosterone is indeed the hormone responsible for the homeostatic response to saline loading in *Varanus gouldii* and, when it was injected into adrenalectomized individuals, FR_{Na} increased significantly from 81.6 ± 3.8% to 92.1 ± 0.8% and potassium excretion increased.[42,185] Surprisingly, corticosterone, when injected into adrenalectomized lizards was also natriferic, although only at much higher dosages than used with aldosterone.[185]

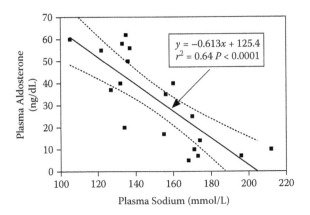

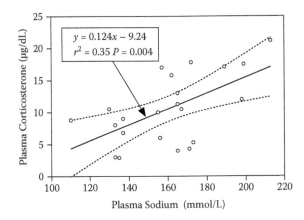

FIGURE 10.12 Variation in plasma aldosterone and corticosterone concentrations with change in plasma sodium concentration in the varanid lizard *Varanus gouldii*. Dotted lines show 95% confidence limits of the regressions. Aldosterone levels are in ng/dL, corticosterone in μg/dL. (Adapted from Bradshaw, S.D., *Homeostasis in Desert Reptiles*, Springer, Heidelberg, 1997, pp. 1–213.)

What is clearly required is further work with species such as *Ctenophorus ornatus* and *Dipsosaurus dorsalis* to clarify the renal effects of corticosterone, which would appear to be the opposite of those observed in *Varanus gouldii*. It would be unwise, too, to assume that all reptiles, or even all lizards for that matter, share a common system of hormonal control of their renal function, and agamids and iguanids may well differ radically from other lizards such as varanids. Levels of aldosterone reported by Duggan and Lofts[112,113] in the sea snake *Hydrophis cyanocinctus* are many times higher than those reported by Bradshaw and Grenot[46] and Bradshaw and Rice[48] for lizards. Those reported by Hadj-Bekkouche et al.[156] and Uva et al.[309] in the tortoises *Testudo hermanni* and *Testudo mauritanica* are also much lower, and the actions of this hormone at the level of the kidney may be totally different in such species.

4. Other Hormones

Early reports that the reptilian *pars nervosa* also contains oxytocin[220] have not been substantiated,[182] but all species that have been examined do contain 8-isoleucine oxytocin, or mesotocin (MT), which is known to be a diuretic in amphibians.[144,230] Both oxytocin and MT provoke a glomerular antidiuresis in the water snake *Natrix sipedon*,[74] but only at dosages that are probably pharmacological (20 to 40 mU/kg). A recent study by Butler and Snitman[62] compared the renal effects of

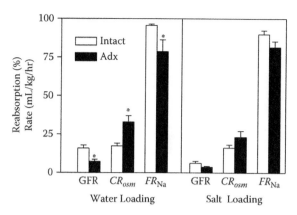

FIGURE 10.13 Renal response to water and salt loading of intact and previously adrenalectomized (Adx) *Varanus gouldii* lizards. The glomerular filtration rate (GFR) is in mL/kg/hr; osmolar clearance (C_{OSM}) and fractional reabsorption of sodium (FR_{Na}) are both in percent, with data expressed as mean ± SE. (Adapted from Rice, G.E. et al., *Gen. Comp. Endocrinol.*, 47, 182–189, 1982.)

MT and AVT in the painted turtle (*Chrysemys picta*). They found that only very high doses of MT (300 to 500 ng/kg) evoked a very small reduction in GFR, whereas AVT was effective at a physiological dose of only 5 ng/kg. We are not aware of any published studies of the effect of thyroid hormones or thyroidectomy on renal function in reptiles but would anticipate some interaction given the general depression of physiological activity noticed in lizards following thyroidectomy.[147]

Atrial natriuretic peptide (ANP)[101] is another hormone that may affect renal function in reptiles. ANP has not been isolated in birds[306] but has been isolated from the heart of the turtle *Amyda japonica*, and a related peptide (BNP) has been found in the atria of the freshwater turtle *Pseudemys scripta*.[248] A BNP cDNA has also been isolated from the atria of the saltwater crocodile (*Crocodylus porosus*) and the longneck tortoise (*Chelodina longicollis*), but no ANP cDNAs were isolated from either species.[306] Several studies have demonstrated the presence of ANP binding sites in the kidney, brain, gastrointestinal tract, adrenal glands, and epididymis of the freshwater turtle *Amyda japonica*,[171,172] but physiological studies with this peptide have yet to be reported in reptiles.

D. Hormonal Control: Cloaca, Colon, and Bladder

The information by Bradshaw and Rice[48] presented in Table 10.12 (see above) on rates of water and electrolyte exchange in the perfused cloacal–colonic complex of the varanid lizard *Varanus gouldii* show that dehydration and salt loading are associated with homeostatic changes in these parameters. Saline loading is associated with a significant reduction in the absolute rates of reabsorption of sodium, chloride, and water from the colon, and potassium secretion is reduced. When the data are expressed fractionally—that is, relative to the presentation rate for each component, which is a function of the rate of perfusion—it may be seen that water reabsorption increases significantly from 23.7% to 40.6% with salt loading, and sodium reabsorption falls from 33.7% to 22.4%. These data raise the interesting question of whether colon function is under some form of hormonal control, as Bradshaw[40] had reported that fractional rates of reabsorption of both sodium and potassium were significantly reduced in *Ctenophorus ornatus* bearing electrolytic lesions in the base of the hypothalamus and seemed intended to inhibit AVT secretion.

Changes in plasma levels of AVT, aldosterone, and corticosterone (Figure 10.14) do not support this contention. Sodium reabsorption from the colon decreases significantly with dehydration (from 33.7% to 20.7%), but both plasma aldosterone and corticosterone levels are unchanged. Salt loading is associated with a significant increase in both AVT and corticosterone levels and a significant fall

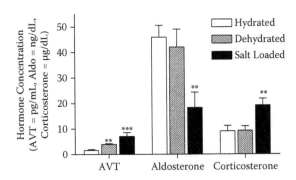

FIGURE 10.14 Changes in plasma levels of arginine vasotocin (AVT), aldosterone, and corticosterone in hydrated, dehydrated, and salt-loaded lizards (*Varanus gouldii*). AVT levels are in pg/mL, aldosterone in ng/dL, and corticosterone in µg/dL. Statistical comparisons for each hormone are with the hydrated state; $**P < 0.01$ and $***P < 0.001$. (Data from Bradshaw, S.D. and Rice, G.E., *Gen. Comp. Endocrinol.*, 44, 82–93, 1981; Bradshaw, S.D., *Homeostasis in Desert Reptiles*, Springer, Heidelberg, 1997, pp. 1–213.)

in aldosterone levels when compared with dehydrated animals, but no significant alteration in any of the transport parameters occurs when compared with dehydrated individuals. Levels of AVT increase from a mean of 3.9 ± 0.3 pg/mL in dehydrated animals to 7.1 ± 1.5 pg/mL with salt loading (see Figure 10.14), but no concomitant increase in the rate of water reabsorption from the colon occurs. Equally, any possible natriferic effect of AVT is negated by the data, as rates of reabsorption of sodium decrease rather than increase with the rise in blood levels of AVT. The rate of sodium reabsorption from the colon also seems to be independent of circulating levels of aldosterone, as it is low and equal in both salt-loaded and dehydrated individuals, despite the fact that aldosterone levels are maximally elevated in dehydrated animals and fall to a minimum with salt loading (see Figure 10.14).

Braysher and Green[53] presented data to suggest that AVT at a dosage of 100 ng/kg enhanced both sodium and water reabsorption from the cloaca of *Varanus gouldii*, but this involved a static procedure with a cannulated cloaca and, in the light of the dynamic perfusion data from the same species, these early conclusions must be treated with some caution. It is true that the increased rate of water reabsorption from the cloaca of salt-loaded and dehydrated *Varanus gouldii* occurs in the face of a decreasing osmotic gradient (the U/P_{OSM} increases from 0.40 with hydration to 0.98 with salt loading), and this does suggest the involvement of hormonal mechanisms. No overwhelming evidence from either study, however, implicates neurohypophysial hormones in the control of postrenal water and electrolyte transfer in this species, and Murrish and Schmidt-Nielsen[221] have argued that at least water reabsorption from the cloaca may be passive in lizards (but see Skadhauge[283] for a critique of this suggestion).

Moreover, despite the well-known response of the amphibian bladder to AVT,[7,17] the reptilian bladder seems generally unresponsive to neurohypophysial hormones, at least for the chelonians *Testudo graeca* and *Pseudemys scripta*.[14,59] In addition, Gilles-Baillien[149] failed to observe any effect of AVT on water permeability of the bladder of *Testudo hermanni*. It would be most interesting, however, to investigate the hormonal responsiveness of the bladder of the desert tortoise *Gopherus agassizii*, as the ecophysiological study of Peterson[242] shows clearly that this species depends on its bladder to both store and reclaim water. An interesting paper by Beuchat et al.[24] also shows how the bladder in neonatal, but not adult, *Sceloporus jarrovi* lizards may function as an extrarenal osmoregulatory organ that can buffer water compartments against osmotic perturbation, and hormonal control may be involved.

An extensive research program on the avian lower intestine by Erik Skadhauge and his colleagues has shown that the colon of normal and high salt-acclimated hens expresses sodium-linked glucose and amino acid cotransporters, whereas the coprodeum is relatively inactive. Following

TABLE 10.14

Sodium Transport by the Colon of the Lizard *Gallotia galloti*

Treatment	PD (mV)	SCC (μmol/cm^2/hr)	Na Flux$_{NET}$ (μmol/cm^2/hr)
Control	1.58 ± 0.37	0.56 ± 0.14	0.79 ± 0.33
Aldosterone (acute; 100 μg/kg)	4.75 ± 0.63**	1.50 ± 0.12**	3.22 ± 0.21***
Aldosterone (chronic)	4.81 ± 1.05**	1.60 ± 0.25**	3.71 ± 0.61***

Note: Data expressed as mean ± S.E. **$P < 0.01$; ***$P < 0.005$.

Source: Adapted from Diaz, M. et al., *Comp. Biochem. Physiol.*, 91A, 71–77, 1988.

acclimation to low-salt diets, however, both colon and coprodeum shift to a pattern of high expression of electrogenic sodium channels, and the colonic cotransporter activity is simultaneously downregulated, with these changes being primarily controlled by aldosterone.[180,303–305] One might thus expect to see some effects of aldosterone in reptilian systems such as the cloacal/colonic complex, including the bladder.

Aldosterone has been shown to affect the rate of sodium transport in the bladder of the chelonians *Testudo graeca*[14] and *Chrysemys picta*.[186] *In vitro* electrophysiological studies[6] on the isolated colon of the lizard *Gallotia galloti* from the Canary Islands have shown that this tissue exhibits the classical properties of a leaky epithelium: low potential difference and short-circuit current, high tissue conductance, and relatively high unidirectional sodium and chloride fluxes compared with net movements. The lizard colon thus appears unique, as most colonic epithelia exhibit discrete passive permeabilities to water and ions, as well as low tissue conductance.[272] The predominant sodium transport mechanism in the colon of this lizard appears to be an electroneutral mechanism, mediated by the presence of an amiloride-sensitive Na^+–H^+ exchange process coupled to a Cl^-/HCO_3^- antiport in the apical membrane of colonocytes.[6,32] In a series of papers, Diaz and Lorenzo[105,107] and Diaz et al.[109] reported that acute or chronic administration of aldosterone brings about a substantial increase in transmural potential difference, short-circuit current, and net sodium transport by the colon of this species, showing that this hormone can also act on leaky as well as tight epithelia, as found in birds.[153,318] These effects of aldosterone on transport parameters are summarized in Table 10.14.

All of the effects of aldosterone on the colon were blocked by administration of amiloride, which also abolished the net flux in control tissues but did not affect the short-circuit current, confirming the electrically silent nature of sodium absorption under basal conditions. Diaz and Lorenzo[107,108] point out that chronic treatment with aldosterone was necessary to achieve maximal effect and suggest that the hormone promotes a morphological transformation or differentiation of new cells in the colonic epithelium containing amiloride-sensitive sodium channels, as suggested by Grubb and Bentley[153] for the avian ileum.

As mentioned above, efforts to show any clear effect of AVT on rates of water reabsorption from the bladder and colon of reptiles have not been particularly successful, and Bentley[14] found this hormone did not change water permeability of the colon of the tortoise *Testudo graeca*, nor did it have any effect on the colon of two species of *Ctenophorus* (=*Amphibolurus*) lizards.[18] Lorenzo et al.,[194] however, reported that both short-circuit current and transmural potential difference (PD) were decreased by arginine vasopressin (AVP) and increased by cAMP in the isolated colon of *Gallotia galloti*. Vasopressin also enhanced the net absorption of water across the colon and the mucosal-to-serosal flux of sodium and chloride across the short-circuited colon but had no effect on absorption from the ileum. Rates of production of cAMP were not altered by AVP but were increased by theophylline, and Diaz and Lorenzo[106] and Lorenzo et al.[194] concluded that the effects of AVP on this lizard colon are mediated by a non-cAMP mechanism.[32] It would be of considerable interest to repeat these experiments with AVT instead of AVP.

TABLE 10.15

Estimated Absolute and Fractional Rates of Water and Electrolyte Reabsorption from the Colon of Intact and Tract-Operated *Ctenophorus ornatus* during Chronic Saline Diuresis

Group	Fluid (mL/kg/hr)	Free Water (mL/kg/hr)	Sodium (μmol/kg/hr)	Potassium (μmol/kg/hr)
Intact	9.0 ± 1.2 (91.6)	4.2 ± 0.6 (93.4)	871.9 ± 33.4 (88.8)	62.9 ± 13.3 (68.3)
Tract-operated	13.1 ± 1.3 (78.2)	9.4 ± 2.7 (87.4)	969.3 ± 59.3 (73.2)*	20.3 ± 4.3*** (33.7)**

Note: Data presented as mean ± S.E. Fractional rates of reabsorption are expressed as percentages and given in parentheses. $*P < 0.05$; $**P < 0.02$; $***P = 0.001$ when compared with corresponding value for intact animals.

Source: Adapted from Bradshaw, S.D., *Gen. Comp. Endocrinol.*, 29, 285, 1976.

The situation at the moment is thus unclear as to whether pituitary or adrenal hormones are involved in the control of homeostatic changes seen in transmural rates of water and electrolyte exchange in the reptilian colon. Studies on the *in situ* perfused colon of the varanid lizard *Varanus gouldii* are not supportive, but the placement of electrolytic lesions in the hypothalamus of *Ctenophorus ornatus*,[40] resulting in the loss of the ability of this lizard to respond antidiuretically to salt loads (see Figure 10.8), also had an effect on rates of sodium and potassium reabsorption from the colon, as seen in Table 10.15. Changes in the rate of fluid reabsorption from the colon of the tract-operated *Ctenophorus ornatus* were not statistically significant, but those for both sodium and potassium were, and the effect of the lesions is thus to reduce the rate of electrolyte reclamation from the urine. This effect is exactly the reverse of the same lesions in the same individuals at the level of the nephron[44] and demonstrates why the hypothalamic lesions have no overall impact on salt excretion in this lizard: They increase the fractional reabsorption of sodium ions by the kidney but decrease their rate of reabsorption in the colon. The two effects thus cancel one another, and the lesions have no net effect on rates of sodium excretion in this lizard.

Changes such as these after lesioning the hypothalamic region are very suggestive of the operation of hormones, but it is difficult to identify what they may be at this stage, as such lesions may affect both anterior and posterior pituitary function. As mentioned earlier, prolactin can influence GFR in the freshwater turtles *Chrysemys picta* and *Testudo graeca*[56,57] and is also known to increase rates of fluid and NaCl absorption in rat, hamster, and guinea-pig ileum.[199] In the absence of further research it would thus be premature to conclude that the reabsorptive activity of such postrenal sites in reptiles is not influenced by pituitary and adrenal hormones.[39,283]

E. HORMONAL CONTROL: SALT GLANDS

Early reviews highlighted the paucity of knowledge regarding any possible hormonal control of reptilian salt glands,[262] although Holmes and McBean[162] first noted the sensitivity of the lachrymal salt gland of the green sea turtle (*Chelonia mydas*) to adrenocortical hormones such as corticosterone (B) in 1964. Aldosterone was also reported to inhibit sodium loss from the nasal gland of adrenalectomized desert iguanas (*Dispsosaurus dorsalis*) in an early short abstract by Templeton et al.[302] Bilateral adrenalectomy in this species was followed by a marked increase in the rate of excretion of sodium ions from the nasal salt gland and injections of aldosterone were able to correct this loss.[300,301]

The response of the nasal salt gland of this same species to injections of a range of adrenal steroids was investigated by Shoemaker et al.,[276] who maintained the lizards at their preferred body temperature in small metabolism cages in a dry atmosphere and collected the nasal salt formed daily. Animals responded within 24 hours to salt loading by excreting a concentrated salt solution composed

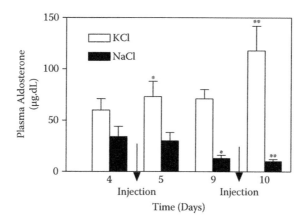

FIGURE 10.15 Changes in mean plasma levels of aldosterone in the agamid lizard *Uromastix acanthinurus* following two intraperitoneal injections of either KCl or NaCl spaced a few days apart (arrows) showing an increase with KCl injections and a fall following NaCl injections. Statistical comparisons in all cases are with the day 4 level of aldosterone in either the NaCl- or KCl-injected group with *$P < 0.05$ and **$P < 0.01$. (Adapted from Bradshaw, S.D., *Homeostasis in Desert Reptiles*, Springer, Heidelberg, 1997, pp. 1–213.)

of a mixture of NaCl and KCl, regardless of the nature of the salt injected. Repetitive injections of a given salt improved the fidelity of the secretion elaborated and K^+/Na^+ ratios were found to vary from 3.0 to 0.3. Daily injections of 15 μg of aldosterone virtually abolished sodium excretion by the gland but had no effect on potassium, suggesting that the action of this hormone on the nasal salt gland is the same as on the kidney and colon (i.e., natriferic and kaliuretic). The effects of other hormones such as corticosterone, cortisol, ACTH, and dexamethasone on the nasal secretion of *D. dorsalis* were all similar to that of aldosterone, although only dexamethasone was as potent.

Another lizard with a well-developed nasal salt gland that was first described by Grenot[151] and that has been studied extensively by Lemire et al.[191] is the Saharan herbivorous agamid *Uromastix acanthinurus*. In studies on the hormonal control of this nasal gland, Bradshaw et al.[47] measured circulating levels of aldosterone and corticosterone in salt-injected animals and assessed the effects of exogenous injections of aldosterone and dexamethasone. The response of the gland to two injections of 1 mmol KCl, spaced several days apart, was very rapid and differed from that of *Dipsosaurus dorsalis* in elaborating an almost pure KCl secretion. Aldosterone levels increase significantly following injections of KCl and decrease following injections of NaCl, as seen in Figure 10.15. The potential role of aldosterone as a classic mineralocorticoid (i.e., natriferic and kaliuretic) is highlighted by calculations showing that total salt excretion correlated positively with aldosterone levels under a KCl injection regime but negatively under a NaCl injection schedule.

These data on changes in aldosterone levels in the plasma provide a clue as to why none of the hormone treatments used by Shoemaker et al.[276] actually stimulated salt secretion by the nasal gland of *Dipsosaurus dorsalis*. Injections of KCl lead, within a few days, to maximal plasma concentrations of aldosterone in *Uromastix acanthinurus* and, if this hormone is indeed controlling the secretory activity of the gland, injections of exogenous aldosterone are unlikely to have any further effect on a gland already secreting to capacity. The only effect likely to be observed is an inhibitory one on sodium excretion, and this is what is seen with both *Dipsosaurus dorsalis* and *Uromastix acanthinurus*. Dexamethasone, again, was as potent as aldosterone in inhibiting sodium excretion in *Uromastix acanthinurus* and spironolactone, which acts as an aldosterone antagonist in mammals,[231] was without effect.[47]

Plasma aldosterone levels showed much less variation with change in plasma sodium concentration in the Australian skink *Tiliqua rugosa*, however, and the differing response of this species and *Uromastix acanthinurus* is shown in Figure 10.16. Both regressions are statistically significant

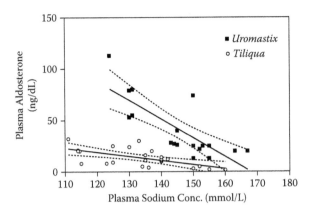

FIGURE 10.16 Variation in plasma aldosterone levels (in ng/dL) with change in plasma sodium concentration (in mmol/L) in the Saharan agamid lizard *Uromastix acanthinurus* and the Australian skink *Tiliqua rugosa*. Regressions for both species shown with 95% confidence limits, as indicated by dotted lines. (Adapted from Bradshaw, S.D., *Homeostasis in Desert Reptiles*, Springer, Heidelberg, 1997, pp. 1–213.)

but the coefficient of determination (r^2) in the case of *Tiliqua rugosa* is only 0.40, compared with 0.61 for *Uromastix acanthinurus*. Bradshaw et al.[50] measured high-affinity binding of both aldosterone and corticosterone in cytosolic preparations from the external nasal gland of *Tiliqua rugosa*, and a Scatchard plot showed high binding for aldosterone, with $K_d = 12.9 \times 10^{-9}\ M$, compared with only $5.2 \times 10^{-9}\ M$ for corticosterone. These data suggest that, if a steroid hormone controls the secretory activity of the nasal salt gland of *Tiliqua rugosa*, it is more likely to be corticosterone than aldosterone. It is well to recall, however, that, as noted earlier, the external nasal salt gland of this species is unusual in being composed of both principal cells and mucoserous secretory cells and is by no means a classical salt-secreting gland.

Vasoactive intestinal peptide (VIP) may be involved in regulating salt gland secretory activity. Franklin et al.[140] have reported that VIP stimulates secretion of NaCl from the lingual glands of the saltwater crocodile (*Crocodylus porosus*) as has also been found in the avian salt gland.[196] Shuttleworth and Thorndyke[280] also reported that a VIP-like peptide stimulated secretion of the rectal gland of the elasmobranch *Scyliorhinus canicula* and the recent isolation of the receptor for VIP confirms that this peptide is a secretagog for the rectal gland.[25] We are not aware, however, of any studies to date that have measured plasma VIP levels in reptiles in relation to salt-gland function. A recent paper by De Falco et al.[99] has reported that VIP alters catecholamine secretion by the adrenal glands of the lizard *Podarcis sicula*, and its possible effect on steroid hormone secretion warrants investigation. The pituitary peptide α-MSH has also been reported to stimulate sodium excretion by the salt gland of the duck,[166] but there are no reports to date of any studies in a reptile.

IV. GENERAL CONCLUSIONS

Reptiles as a group do not share a common osmotic *milieu intérieur*. Moreover, even within a given taxon, they do not regulate that *milieu* as tightly as generally considered typical for vertebrates. Nevertheless, substantial regulation of output of ions and water does occur via the kidneys, with modifications by bladder, cloaca, and colon, and via salt glands. Although we have been able to make a few generalizations about the regulatory processes, these have been limited not simply by incomplete studies on a given process but also by the small number of species studied and by bias in the studied sample toward a given taxon or toward a single environment. In the survey of the overall osmotic anatomy of reptiles, for example, we have already noted that the studies on lizards are biased toward desert species and that few studies have been made on snakes. In terms of water and ion loss across the skin, most studies have been made on snakes, with a modest number on lizards and only a few on turtles and crocodiles. Studies of renal regulation of water and ion excretion are biased toward snakes and lizards, with the most detailed studies on basic renal function being limited to a small number of species of snakes and the most thorough studies on hormonal regulation limited to a few species of desert lizards. Most studies on bladder function have been limited to turtles and those on cloaca and colon to lizards. Studies on salt glands have, of necessity, been limited to those species that have them, but within that group most studies have been made on a few lizard species. From this brief survey of species bias and number, it is clear that truly meaningful generalizations about osmotic regulation in reptiles as a group, for a given taxon, or for a taxon in relation to environment require studies on many more species.

ACKNOWLEDGMENTS

We dedicate this chapter to those scientists who served as our mentors: Wilbur H. Sawyer and Bodil Schmidt-Nielsen (WHD) and Peter J. Bentley (SDB).

REFERENCES

1. Abel, J.H. and Ellis, R.A., Histochemical and electron microscopic observations on the salt secreting lachrymal glands of marine turtles, *Am. J. Anat.*, 118, 337–358, 1966.
2. Abraham, W., Wertz, P.W., Burken, R.R., and Downing, D.T., Glucosylsterol and acylglucosylsterol of snake epidermis: structure determination, *J. Lipid Res.*, 28, 446–449, 1987.
3. Alibardi, L. and Maderson, P.F.A., Observations on the histochemistry and ultrastructure of the epidermis of the Tuatara, *Sphenodon punctatus* (Spenodotida, Lepidosauria, Reptilia): a contribution to an understanding of the Lepidosaurian epidermal generation and evolutionary origin of the squamate shedding complex, *J. Morphol.*, 256, 111–133, 2003.
4. Anderson, E., The ultramicroscopic structure of a reptilian kidney, *J. Morphol.*, 106, 205–240, 1960.
5. Avery, R.A., Field studies of body temperatures and thermoregulation. In *Biology of the Reptilia*, Gans, C. and Pough, F.H., Eds., Academic Press, New York, 1982, pp. 93–166.
6. Badia, P., Gomez, T., Diaz, M., and Lorenzo, A., Mechanism of transport of Na and Cl in the lizard colon, *Comp. Biochem. Physiol. A*, 87, 883–887, 1987.
7. Bakker, H.R. and Bradshaw, S.D., The effect of hypothalamic lesions on water metabolism of the toad *Bufo marinus*, *J. Endocrinol.*, 75, 161–172, 1977.
8. Barfuss, D.W. and Dantzler, W.H., Glucose transport in isolated perfused proximal tubules of snake kidney, *Am. J. Physiol.*, 231, 1716–1728, 1976.
9. Baze, W.B. and Horne, F.R., Ureogenesis in Chelonia, *Comp. Biochem. Physiol.*, 34, 91–100, 1970.
10. Beck, T.R., Hassid, A., and Dunn, M.J., Desamino-D-arginine vasopressin induces fatty acid cyclo-oxygenase activity in the renal medulla of diabetes insipidus rats, *J. Pharmacol. Exp. Ther.*, 221, 269–274, 1982.
11. Bennet, H.S., Morphological aspects of extracellular polysaccharides, *Histochem. Cytochem.*, 11, 14–23, 1967.
12. Benson, G.K., Phillips, J.G., and Holmes, W.N., Observations on the histological structure of the nasal gland of the turtle, *J. Anat.*, 98, 290, 1964.
13. Bentley, P.J., Studies on the water and electrolyte metabolism of the lizard *Trachysaurus rugosus* (Gray), *J. Physiol.*, 145, 37–47, 1959.
14. Bentley, P.J., Studies on the permeability of the large intestine and urinary bladder of the tortoise (*Testudo graeca*), *Gen. Comp. Endocrinol.*, 2, 323–328, 1962.
15. Bentley, P.J., *Endocrines and Osmoregulation*, Springer-Verlag, Berlin, 1971, pp. 1–300.
16. Bentley, P.J., Osmoregulation. In *Biology of the Reptilia*. Vol. 5. *Physiology A*, Gans, C. and Dawson, W.R., Eds., Academic Press, New York, 1976, pp. 365–412.
17. Bentley, P.J., *Endocrines and Osmoregulation*, Springer, Heidelberg, 2002, pp. 1–292.

18. Bentley, P.J. and Bradshaw, S.D., Electrical potential difference across the cloaca and colon of the Australian lizards *Amphibolurus ornatus* and *A. inermis*, *Comp. Biochem. Physiol. A*, 42, 465–471, 1972.

19. Bentley, P.J. and Schmidt-Nielsen, K., Permeability to water and sodium of the crocodilian, *Caiman sclerops*, *J. Cell. Comp. Physiol.*, 66, 303–310, 1965.

20. Bentley, P.J. and Schmidt-Nielsen, K., Cutaneous water loss in reptiles, *Science*, 151, 1547–1549, 1966.

21. Benyajati, S. and Dantzler, W.H., Enzymatic and transport characteristics of isolated snake renal brush-border membranes, *Am. J. Physiol.*, 255, R52–R60, 1988.

22. Benyajati, S., Yokota, S.D., and Dantzler, W.H., Regulation of glomerular filtration by plasma K in reptiles. In *Proc. of the IUPS XXIXth Int. Congr. Physiol. Sci.*, Sydney, Australia, 1983.

23. Bernard, C., *Leçons sur les Phénomènes de la Vie communs aux Animaux et aux Végétaux*, J.-B. Baillière et Fils, Paris, 1878.

24. Beuchat, C.A., Vleck, D., and Braun, E.J., Role of the urinary bladder in osmotic regulation of neonatal lizards, *Physiol. Zool.*, 59, 539–551, 1986.

25. Bewley, M.S., Pena, J.T., Plesch, F.N., Decker, S.E., Weber, G.J., and Forrest, J.N., Shark rectal gland vasoactive intestinal peptide receptor: cloning, functional expression, and regulation of CFTR chloride channels, *Am. J. Physiol.*, 291, R1157–R1164, 2006.

26. Beyenbach, K.W., Direct demonstration of fluid secretion by glomerular renal tubules in a marine teleost, *Nature*, 299, 54–56, 1982.

27. Beyenbach, K.W., Water-permeable and -impermeable barriers of snake distal tubules, *Am. J. Physiol.*, 246, F290–F299, 1984.

28. Beyenbach, K.W., Secretory NaCl and volume flow in renal tubules, *Am. J. Physiol.*, 250, R753–R763, 1986.

29. Beyenbach, K.W. and Dantzler, W.H., Generation of transepithelial potentials by isolated perfused reptilian distal tubules, *Am. J. Physiol.*, 234, F238–F246, 1978.

30. Beyenbach, K.W., Koeppen, B.M., Dantzler, W.H., and Helman, S.I., Luminal Na concentration and the electrical properties of the snake distal tubule, *Am. J. Physiol.*, 239, F412–F419, 1980.

31. Bindslev, N. and Skadhauge, E., Sodium chloride absorption and solute-linked water flow across the epithelium of the coprodeum and large intestine in the normal and dehydrated fowl (*Gallus domesticus*): *in vivo* perfusion studies, *J. Physiol.*, 216, 753–768, 1971.

32. Bolanos, A., Diaz, M., Gomez, T., and Lorenzo, A., Electrolyte transport in lizard intestine: cellular mechanisms and regulatory aspects, *Trends Comp. Biochem. Physiol.*, 1, 229–243, 1993.

33. Bordley, J. and Richards, A.N., Quantitative studies of the composition of glomerular urine. VIII. The concentration of uric acid in glomerular urine of snakes and frogs, determined by an ultramicroadaptation of Folin's method, *J. Biol. Chem.*, 101, 193–221, 1933.

34. Bowman, W., On the structure and use of the Malpighian bodies of the kidney, with observations on the circulation through that gland, *Philos. Trans. R. Soc. Lond.*, 132, 57–80, 1842.

35. Bradshaw, F.J. and Bradshaw, S.D., Arginine vasotocin: locus of action along the nephron of the ornate dragon lizard, *Ctenophorus ornatus*, *Gen. Comp. Endocrinol.*, 103, 281–289, 1996.

36. Bradshaw, S.D., The endocrine control of water and electrolyte metabolism in desert reptiles, *Gen. Comp. Endocrinol.*, Suppl. 3, 360–373, 1972.

37. Bradshaw, S.D., Osmoregulation and pituitary–adrenal function in desert reptiles, *Gen. Comp. Endocrinol.*, 25, 230–248, 1975.

38. Bradshaw, S.D., Effect of hypothalamic lesions on kidney and adrenal function of the lizard *Amphibolurus ornatus*, *Gen. Comp. Endocrinol.*, 29, 285, 1976.

39. Bradshaw, S.D., Aspects of hormonal control of osmoregulation in desert reptiles. In *Comparative Endocrinology*, Gaillard, P.J. and Boer, H.H., Eds., Elsevier, Amsterdam, 1978, pp. 213–216.

40. Bradshaw, S.D., Volume regulation in desert reptiles and its control by pituitary and adrenal hormones. In *Osmotic and Volume Regulation*, Jorgensen, C.B. and Skadhauge, E., Eds., Munksgaard, Copenhagen, 1978, pp. 38–39.

41. Bradshaw, S.D., Hydro-mineral balance in reptiles and its hormonal control. In *Current Trends in Comparative Physiology*, Vol. 2, Lofts, B. and Holmes, W.N., Eds., Hong Kong University Press, Hong Kong, 1985, pp. 939–943.

42. Bradshaw, S.D., *Ecophysiology of Desert Reptiles*, Academic Press, Orlando, FL, 1986, pp. 1–324.

43. Bradshaw, S.D., Hormonal mechanisms and survival in desert reptiles. In *Endocrine Regulations as Adaptive Mechanisms to the Environment*, Assenmacher, I. and Boissin, J., Eds., Centre National de la Recherche Scientifique (CNRS), Bordeaux, 1986, pp. 415–440.

44. Bradshaw, S.D., *Homeostasis in Desert Reptiles*, Springer, Heidelberg, 1997, pp. 1–213.

45. Bradshaw, S.D. and Bradshaw, F.J., Arginine vasotocin: site and mode of action in the reptilian kidney, *Gen. Comp. Endocrinol.*, 126, 7–13, 2002.

46. Bradshaw, S.D. and Grenot, C.J., Plasma aldosterone levels in two reptilian species, *Uromastix acanthinurus* and *Tiliqua rugosa*, and the effect of several experimental treatments, *J. Comp. Physiol. B*, 111, 71–76, 1976.

47. Bradshaw, S.D., Lemire, M., Vernet, R., and Grenot, C.J., Aldosterone and the control of secretion by the nasal salt gland of the North African desert lizard, *Uromastix acanthinurus*, *Gen. Comp. Endocrinol.*, 54, 314–323, 1984.

48. Bradshaw, S.D. and Rice, G.E., The effects of pituitary and adrenal hormones on renal and postrenal reabsorption of water and electrolytes in the lizard *Varanus gouldii* (Gray), *Gen. Comp. Endocrinol.*, 44, 82–93, 1981.

49. Bradshaw, S.D., Shoemaker, V.H., and Nagy, K.A., The role of adrenal corticosteroids in the regulation of kidney function in the desert lizard *Dipsosaurus dorsalis*, *Comp. Biochem. Physiol. A*, 43, 621–635, 1972.

50. Bradshaw, S.D., Tom, J.A., and Bunn, S.E., Corticosteroids and control of nasal salt gland function in the lizard *Tiliqua rugosa*, *Gen. Comp. Endocrinol.*, 54, 308–313, 1984.

51. Braun, E.J. and Dantzler, W.H., Mechanisms of hormone actions on renal function. In *Vertebrate Endocrinology: Fundamentals and Biomedical Implications*, Vol., II, Pang, P.K.T. and Schreibman, M.P., Eds., Academic Press, New York, 1987, pp. 189–210.

52. Braysher, M., The structure and function of the nasal salt gland from the Australian sleepy lizard *Trachysaurus* (formerly *Tiliqua*) *rugosus*: family Scincidae, *Physiol. Zool.*, 44, 129–136, 1971.

53. Braysher, M. and Green, B., Absorption of water and electrolytes from the cloaca of an Australian lizard, *Varanus gouldii* (Gray), *Comp. Biochem. Physiol.*, 35, 607–614, 1970.

54. Braysher, M.I., The excretion of hyperosmotic urine and other aspects of the electrolyte balance in the lizard *Amphibolurus maculosus*, *Comp. Biochem. Physiol.*, 54, 341–345, 1976.

55. Brenner, B.M., Troy, J.L., and Daugharty, T.M., The dynamics of glomerular ultrafiltration in the rat, *J. Clin. Invest.*, 50, 1776–1780, 1971.

56. Brewer, K.J. and Ensor, D.M., Hormonal control of osmoregulation in the Chelonia. I. The effects of prolactin and interrenal steroids in freshwater chelonians, *Gen. Comp. Endocrinol.*, 42, 304–309, 1980.

57. Brewer, K.J. and Ensor, D.M., Hormonal control of osmoregulation in the Chelonia. II. The effects of prolactin and corticosterone on *Testudo graeca*, *Gen. Comp. Endocrinol.*, 42, 310–314, 1980.

58. Brodsky, W.A. and Schilb, T.P., Electrical and osmotic characteristics of the isolated turtle bladder, *J. Clin. Invest.*, 39, 974–975, 1960.

59. Brodsky, W.A. and Schilb, T.P., Osmotic properties of isolated turtle bladder, *Am. J. Physiol.*, 208, 46–57, 1965.

60. Butler, D.G., Antidiuretic effect of arginine vasotocin in the Western painted turtle (*Chrysemys picta belli*), *Gen. Comp. Endocrinol.*, 18, 121–125, 1972.

61. Butler, D.G. and Knox, W.H., Adrenalectomy of the painted turtle (*Chrysemis picta belli*): effect on ionoregulation and tissue glycogen, *Gen. Comp. Endocrinol.*, 14, 551–566, 1970.

62. Butler, D.G. and Snitman, F.S., Renal responses to mesotocin in western painted turtles compared with the antidiuretic response to arginine vasotocin, *Gen. Comp. Endocrinol.*, 144, 101–109, 2005.

63. Callard, I.P. and Callard, G.V., The adrenal gland in Reptilia. Part 2. Physiology. In *General, Comparative and Clinical Endocrinology of the Adrenal Cortex*, Chester Jones, I. and Henderson, I.W., Eds., Academic Press, London, 1978, pp. 370–418.

64. Cannon, W.B., Homeostasis, *Physiol. Rev.*, 9, 399–431, 1929.

65. Chan, D.K.O., Callard, I.P., and Chester Jones, I., Observations on the water and electrolyte composition of the iguanid lizard *Dipsosaurus dorsalis* (Baird and Girard), with special reference to the control by the pituitary gland and the adrenal cortex, *Gen. Comp. Endocrinol.*, 15, 374–387, 1970.

66. Chatsudthipong, V. and Dantzler, W.H., PAH-α-KG countertransport stimulates PAH uptake and net secretion in isolated snake renal tubules, *Am. J. Physiol.*, 261, F858–F867, 1991.

67. Chessman, B.C., Evaporative water loss from three South-Eastern Australian species of freshwater turtle, *Aust. J. Zool.*, 32, 649–655, 1984.

68. Coulson, R.A. and Hernandez, T., *Biochemistry of the Alligator*, Louisiana State University Press, Baton Rouge, 1964, pp. 3–138.

69. Coulson, R.A. and Hernandez, T., Nitrogen metabolism and excretion in the living reptile. In *Comparative Biochemistry of Nitrogen Metabolism*. Vol. 2. *The Vertebrates*, Campbell, J.W., Ed., Academic Press, New York, 1970, pp. 640–710.

70. Coulson, R.A. and Hernandez, T., Alligator metabolism: studies on chemical reactions *in vivo*, *Comp. Biochem. Physiol.*, 74, i–182, 1983.

71. Cowan, F.B.M., Comparative studies on the cranial glands of turtles, *Am. Zool.*, 7, 810, 1967.

72. Cowan, F.B.M., Gross and microscopic anatomy of the orbital glands of *Malaclemys* and other emydine turtles, *Can. J. Zool.*, 47, 723–729, 1969.

73. Cowan, F.B.M., The ultrastructure of the lachrymal "salt" gland and the Harderian gland in the euryhaline *Malaclemys* and some closely related stenohaline emydines, *Can. J. Zool.*, 49, 691–697, 1971.

74. Dantzler, W.H., Glomerular and tubular effects of arginine vasotocin in water snakes *(Natrix sipedon)*, *Am. J. Physiol.*, 212, 83–91, 1967.

75. Dantzler, W.H., Effect of metabolic alkalosis and acidosis on tubular urate secretion in water snakes, *Am. J. Physiol.*, 215, 747–751, 1968.

76. Dantzler, W.H., Kidney function in desert vertebrates. In *Hormones and the Environment*, Benson, G.K. and Phillips, J.G., Eds., Cambridge University Press, London, 1970, pp. 157–190.

77. Dantzler, W.H., Effects of incubations in low potassium and low sodium media on Na-K-ATPase activity in snake and chicken kidney slices, *Comp. Biochem. Physiol. B*, 41, 79–88, 1972.

78. Dantzler, W.H., Characteristics of urate transport by isolated perfused snake proximal renal tubules, *Am. J. Physiol.*, 224, 445–453, 1973.

79. Dantzler, W.H., PAH transport by snake proximal renal tubules: differences from urate transport, *Am. J. Physiol.*, 226, 634–641, 1974.

80. Dantzler, W.H., Comparison of uric acid and PAH transport by isolated, perfused snake renal tubules. In *Amino Acid Transport and Uric Acid Transport Symposium*, Silbernagl, S., Lang, F., and Greger, R., Eds., Georg Thieme, Stuttgart, 1976, pp. 169–180.

81. Dantzler, W.H., Renal function (with special emphasis on nitrogen excretion). In *Biology of Reptilia*, Vol. 5, Gans, C. and Dawson, W.R., Eds., Academic Press, New York, 1976, pp. 447–503.

82. Dantzler, W.H., Urate excretion in nonmammalian vertebrates. In *Handbook of Experimental Pharmacology: Uric Acid*, Vol. 51, Kelley, W.N. and Weiner, I.M., Eds., Springer-Verlag, Berlin, 1978, pp. 185–210.

83. Dantzler, W.H., Renal mechanisms for osmoregulation in reptiles and birds. In *Animals and Environmental Fitness*, Gilles, R., Ed., Pergamon Press, Oxford, 1980, pp. 91–110.

84. Dantzler, W.H., Significance of comparative studies for renal physiology, *Am. J. Physiol.*, 238, F437–F444, 1980.

85. Dantzler, W.H., *Comparative Physiology of the Vertebrate Kidney*, Springer-Verlag, New York, 1989, pp. 1–198.

86. Dantzler, W.H., The nephron in reptiles. In *Structure and Function of the Kidney: Comparative Physiology*, Vol. 1, Kinne, R.K.H., Ed., Karger, Basel, 1989, pp. 143–193.

87. Dantzler, W.H. and Bentley, S.K., Fluid absorption with and without sodium in isolated perfused snake proximal tubules, *Am. J. Physiol.*, 234, F68–F79, 1978.

88. Dantzler, W.H. and Bentley, S.K., Lack of effect of potassium on fluid absorption in isolated, perfused snake proximal renal tubules, *Renal Physiol.*, 1, 268–274, 1978.

89. Dantzler, W.H. and Bentley, S.K., Effects of inhibitors in lumen on PAH and urate transport by isolated renal tubules, *Am. J. Physiol.*, 236, F379–F386, 1979.

90. Dantzler, W.H. and Bentley, S.K., Effects of chloride substitutes on PAH transport by isolated perfused renal tubules, *Am. J. Physiol.*, 241, F632–F644, 1981.

91. Dantzler, W.H. and Braun, E.J., Comparative nephron function in reptiles, birds, and mammals, *Am. J. Physiol.*, 239, R197–R213, 1980.

92. Dantzler, W.H., Brokl, O.H., Nagle, R.B., Welling, D.J., and Welling, L.W., Morphological changes with Na^+-free fluid absorption in isolated perfused snake tubules, *Am. J. Physiol.*, 251, F150–F155, 1986.

93. Dantzler, W.H. and Holmes, W.N., Water and mineral metabolism in reptiles. In *Chemical Zoology*, Florkin, M. and Scheer, B.T., Eds., Academic Press, New York, 1974, pp. 277–336.

94. Dantzler, W.H. and Schmidt-Nielsen, B., Excretion in freshwater turtle *(Pseudemys scripta)* and desert tortoise *(Gopherus agassizii)*, *Am. J. Physiol.*, 210, 198–210, 1966.

95. Dantzler, W.H., Wright, S.H., and Brokl, O.H., Tetraethylammonium transport by snake renal brush-border membrane vesicles, *Pflügers Arch.*, 418, 325–332, 1991.

96. Darwin, C., *Journal of Researches into the Natural History and Geography of the Countries Visited During the Voyage of H.M.S., Beagle Around the World*, Routledge, London, 1839 (reprinted 1905), pp. 1–384.

97. Davis, L.E. and Schmidt-Nielsen, B., Ultrastructure of the crocodile kidney *(Crocodylus porosus)* with special reference to electrolyte and fluid transport, *J. Morphol.*, 121, 255–276, 1967.

98. Davis, L.E., Schmidt-Nielsen, B., and Stolte, H., Anatomy and ultrastructure of the excretory system of the lizard, *Sceloporus cyanogenys*, *J. Morphol.*, 149, 279–326, 1976.

99. De Falco, M., Sciarrillo, R., Capaldo, A., Laforgia, V., Varano, L., Cottone, G., and De Luca, A., Shift from noradrenaline to adrenaline production in the adrenal gland of the lizard, *Podarcis sicula*, after stimulation with vasoactive intestinal peptide (VIP), *Gen. Comp. Endocrinol.*, 131, 325–337, 2003.

100. De Rouffignac, C., Deiss, S., and Bonvalet, J.P., Détermination du taux individuel de filtration glomérulaire des néphrons accessibles et inaccessibles á la microponction, *Pflügers Arch.*, 315, 273–290, 1970.

101. Debold, A.J., Borenstein, H.B., Veress, A.T., and Sonnenberg, H., A rapid and potent natriuretic response to intravenous injection of atrial myocardial extract in rats, *Life Sci.*, 28, 89–94, 1981.

102. Deen, W.M., Troy, J.L., Robertson, C.R., and Brenner, B.M., Dynamics of glomerular ultrafiltration in the rat. IV. Determination of the ultrafiltration coefficient, *J. Clin. Invest.*, 52, 1500–1508, 1973.

103. Diamond, J.M., Water-solute coupling and ion selectivity in epithelia, *Philos. Trans. R. Soc. Lond. B*, 262, 141–151, 1971.

104. Diamond, J.M. and Tormey, J.M., Role of long extracellular channels in fluid transport across epithelia, *Nature*, 210, 817–820, 1966.

105. Diaz, M. and Lorenzo, A., Modulation of active potassium transport by aldosterone, *Comp. Biochem. Physiol. A*, 95, 79–85, 1989.

106. Diaz, M. and Lorenzo, A., Coexistence of absorptive and secretory NaCl processes in isolated lizard colon: effects of cyclic AMP, *Zool. Sci.*, 8, 477–484, 1991.

107. Diaz, M. and Lorenzo, A., Regulation of amiloride-sensitive Na^+ absorption in the lizard *(Gallotia galloti)* colon by aldosterone, *Comp. Biochem. Physiol. A*, 100, 63–68, 1991.

108. Diaz, M. and Lorenzo, A., Aldosterone regulation of active sodium chloride transport in the lizard colon *(Gallotia galloti)*., *J. Comp. Physiol. B*, 162, 189–196, 1992.

109. Diaz, M., Lorenzo, A., Badia, P., and Gomez, T., The role of aldosterone in water and electrolyte transport across the colonic epithelium of the lizard, *Gallotia galloti*, *Comp. Biochem. Physiol. A*, 91, 71–77, 1988.

110. Dmi'el, R., Skin resistance to evaporative water loss in viperid snakes: habitat aridity versus taxonomic status, *Comp. Biochem. Physiol. A*, 121, 1–6, 1998.

111. Dmi'el, R., Perry, G., and Lazell, J., Evaporative water loss in nine insular populations of the lizard *Anolis cristatellus* group in the British Virgin Islands, *Biotropica*, 29, 111–116, 1997.

112. Duggan, R.T. and Lofts, B., Adaptation to fresh water in the sea snake hydrophis cyanocinctus: tissue electrolytes and peripheral corticosteroids, *Gen. Comp. Endocrinol.*, 36, 510–520, 1978.

113. Duggan, R.T. and Lofts, B., The pituitary-adrenal axis in the sea snake, *Hydrophis cyanocinctus* Daudin, *Gen. Comp. Endocrinol.*, 38, 374–383, 1979.

114. Dunson, W.A., Salt gland secretion in the pelagic sea snake *Pelamis*, *Am. J. Physiol.*, 215, 1512–1517, 1968.

115. Dunson, W.A., Electrolyte excretion by the salt gland of the Galapagos marine iguana, *Am. J. Physiol.*, 216, 995–1002, 1969.

116. Dunson, W.A., Reptilian salt glands. In *Exocrine Glands*, Botelho, S.V., Brooks, F.P., and Shelley, W.B., Eds., University of Pennsylvania Press, Philadelphia, 1969, pp. 83–103.

117. Dunson, W.A., Some aspects of electrolyte and water balance in three estuarine reptiles, the diamond-back terrapin, American and "salt water" crocodiles, *Comp. Biochem. Physiol.*, 32, 161–174, 1970.

118. Dunson, W.A., Salt gland secretion in a mangrove monitor lizard, *Comp. Biochem. Physiol. A*, 47, 1245–1255, 1974.

119. Dunson, W.A., Salt glands in reptiles. In *Biology of the Reptilia*. Vol. 5. *Physiology A*, Gans, C. and Dawson, W.R., Eds., Academic Press, New York, 1976, pp. 413–445.

120. Dunson, W.A., Role of the skin in sodium and water exchange of aquatic snakes placed in seawater, *Am. J. Physiol.*, 235, R151–R159, 1978.

121. Dunson, W.A. and Dunson, M.K., Interspecific differences in fluid concentration and secretion rate of sea snake salt glands, *Am. J. Physiol.*, 277, 430–438, 1974.

122. Dunson, W.A. and Dunson, M.K., A possible new salt gland in a marine homalopsid snake (*Cerberus rhynchops*), *Copeia*, 1979, 661–672, 1979.

123. Dunson, W.A., Dunson, M.K., and Keith, A.D., The nasal gland of the Montpellier snake *Malpolon monspessulanus*: fine structure, secretion composition, and a possible role in reduction of dermal water loss, *J. Exp. Zool.*, 203, 461–474, 1978.

124. Dunson, W.A. and Heatwole, H., Effect of relative shell size in turtles on water and electrolyte composition, *Am. J. Physiol.*, 250, R1133–R1137, 1986.

125. Dunson, W.A., Packer, R.K., and Dunson, M.K., Sea snakes: an unusual salt gland under the tongue, *Science*, 173, 437–441, 1971.

126. Dunson, W.A. and Robinson, G.D., Sea snake skin: permeable to water but not to sodium, *J. Comp. Physiol.*, 108, 303–311, 1976.

127. Dunson, W.A. and Taub, A.M., Extrarenal salt excretion in sea snakes (*Laticauda*), *Am. J. Physiol.*, 213, 975–982, 1967.

128. Dworkin, L.D., Ichikawa, I., and Brenner, B.M., Hormonal modulation of glomerular function, *Am. J. Physiol.*, 244, F95–F104, 1983.

129. Edmonds, C.J. and Pilcher, D., Sodium transport mechanisms of the large intestine. In *Transport Across the Intestine*, Burland, W.L. and Samuels, P., Eds., Churchill-Livingstone, Edinburgh, 1969, pp. 43–57.

130. Elizondo, R.S. and Lebrie, S.J., Adrenal–renal function in water snakes *Natrix cyclopion*, *Am. J. Physiol.*, 217, 419–425, 1969.

131. Ellis, R.A. and Abel, J.H., Electron microscopy and cytochemistry of the salt gland of sea turtles, *Anat. Rec.*, 148, 278, 1964.

132. Ellis, R.A. and Abel, J.H., Intercellular channels in the salt-secreting glands of marine turtles, *Science*, 144, 1340–1343, 1964.

133. Ellis, R.A. and Goertemiller, Jr., C.C., Cytological effects of salt-stress and localization of transport adenosine triphosphate in the lateral nasal glands of the desert iguana, *Dipsosaurus dorsalis*, *Anat. Rec.*, 180, 285–298, 2007.

134. Ellis, R.A., Goertemiller, Jr., C.C., and Stetson, D.L., Significance of extensive "leaky" cell junctions in the avian salt gland, *Nature*, 268, 555–556, 1977.

135. Farber, S.J., Mucopolysaccharides and sodium metabolism, *Circulation*, 21, 941–953, 1960.

136. Fergusson, B. and Bradshaw, S.D., Plasma arginine vasotocin, progesterone and luteal development during pregnancy in the viviparous lizard, *Tiliqua rugosa*, *Gen. Comp. Endocrinol.*, 82, 140–151, 1991.

137. Figler, R.A., MacKenzie, D.S., Owens, D.W., Licht, P., and Amoss, M.S., Increased levels of arginine vasotocin and neurophysin during nesting in sea turtles, *Gen. Comp. Endocrinol.*, 73, 223–232, 1989.

138. Fletcher, G.L., Stainer, I.M., and Holmes, W.N., Sequential changes in the adenosine triphosphatase activity and the electrolyte excretory capacity of the nasal glands of the duck (*Anas platyrhynchos*) during the period of adaptation to hypertonic saline, *J. Exp. Biol.*, 47, 375–391, 1967.

139. Ford, S.S. and Bradshaw, S.D., Kidney function and the role of arginine vasotocin (AVT) in three agamid lizards from differing habitats in Western Australia, *Gen. Comp. Endocrinol.*, 147, 62–69, 2006.

140. Franklin, C.E., Holmgren, S., and Taylor, G.C., A preliminary investigation of the effects of vasoactive intestinal peptide on secretion from the lingual salt glands of *Crocodylus porosus*, *Gen. Comp. Endocrinol.*, 102, 74–78, 1996.

141. Franklin, C.E., Taylor, G., and Cramp, R.L., Cholinergic and adrenergic innervation of lingual salt glands of the estuarine crocodile, *Crocodylus porosus*, *Aust. J. Zool.*, 53, 345–351, 2005.

142. Gabe, M. and Saint-Girons, H., Polymorphisme des glandes nasales externes des sauriens, *C. R. Acad. Sci. Paris*, 272, 1275–1278, 1971.

143. Gabe, M. and Saint-Girons, H., Contributions à la morphologie comparée des fosses nasales et de leurs annexes chez les lépidosauriens, *Mem. Mus. Nat. Hist. (Paris)*, A98, 1–87, 1976.

144. Galli-Gallardo, S.M., Oguro, C., and Pang, P.K.T., Renal responses of the Chilean toad, *Calyptocephalella caudiverbera*, and the mudpuppy, *Necturus maculosus*, to mesotocin, *Gen. Comp. Endocrinol.*, 37, 134–136, 1979.

145. Gambaryan, S.P., Microdissectional investigation of the nephrons in some fishes, amphibians, and reptiles inhabiting different environments, *J. Morphol.*, 219, 319–339, 1994.

146. Gans, C., Krakauer, T., and Paganelli, C.V., Water loss in snakes: interspecific and intraspecific variability, *Comp. Biochem. Physiol.*, 27, 747–761, 1968.

147. Gerwein, R.W. and John-Alder, H.B., Growth and behavior in thyroid-deficient lizards, *Gen. Comp. Endocrinol.*, 87, 312–324, 1992.

148. Gerzeli, G., Observazioni e considerazioni morfo-funzionali comparate sulle ghiandole lacrimali dei cheloni, *Arch. Zool. Ital.*, 11, 37–47, 1967.

149. Gilles-Baillien, M., Seasonal changes in the permeability of the isolated vesical epithelium of *Testudo hermanni hermanni* Gmelin, *Biochim. Biophys. Acta*, 193, 129–136, 1969.

150. Green, B., Aspects of renal function in the lizard *Varanus gouldii*, *Comp. Biochem. Physiol. A*, 43, 747–756, 1972.

151. Grenot, C.J., Sur l'excrétion nasale de sels chez le lézard saharien *Uromastix acanthinurus*, *C. R. Acad. Sci. Paris*, 226, 1871–1874, 1968.

152. Grismer, L.L., Three new species of intertidal side-blotched lizards (genus *Uta*) from the Gulf of California, Mexico, *Herpetologica*, 50, 451–474, 1994.

153. Grubb, B. and Bentley, P.J., Aldosterone-induced amiloride-inhibitable short-circuit current in the avian ileum, *Am. J. Physiol.*, 253, G211–G2161987.

154. Gudzent, F., Physikalisch-chemusche untersuchungen uber das verhakten der harnsauren salze losungen, *Hoppe Seylers Zeitschrift fur Physiol. Chemie*, 56, 150–179, 1908.

155. Guillette, L.J., Propper, C.R., Cree, A., and Dores, R.M., Endocrinology of oviposition in the tuatara (*Sphenodon punctatus*). II. Plaasma arginine vasotocin concentrations during natural nesting, *Comp. Biochem. Physiol. A*, 100, 819–822, 1991.

156. Hadj-Bekkouche, F., Cherifa, A., Cherifa, D., and Saidi, A., Variations nychthémérales de l'aldostérone et de la corticostérone chez la tortue terrestres, *Testudo mauritanica*, Dumer (Chelonia, Testudinidae), unpublished manuscript, 2007 (available on request).

157. Hadley, N.F., Lipid barriers in biological systems, *Prog. Lipid Res.*, 28, 1–33, 1989.

158. Hadley, N.F., Integumental lipids of plants and animals: comparative function and biochemistry, *Adv. Lipid Res.*, 24, 303–320, 1991.

159. Hazard, L.C., Ion secretion by salt glands of desert iguanas (*Dipsosaurus dorsalis*), *Physiol. Biochem. Zool.*, 74, 22–31, 2001.

160. Hazard, L.C., Shoemaker, V.H., and Grismer, L.L., Salt gland secretion by an intertidal lizard, *Uta tumidarostra*, *Copeia*, 1998, 231–234, 1998.

161. Hill, L. and Dawbin, W.H., Nitrogen excretion in the tuatara *Sphenodon punctatus*, *Comp. Biochem. Physiol.*, 31, 453–468, 1969.

162. Holmes, W.N. and McBean, R.L., Some aspects of electrolyte excretion in the green turtle, *Chelonia mydas mydas*, *J. Exp. Biol.*, 41, 81–90, 1964.

163. Holmes, W.N. and Phillips, J.G., The avian salt gland, *Biol. Rev.*, 60, 213–256, 1985.

164. Huber, G.C., On the morphology of the renal tubules of vertebrates, *Anat. Rec.*, 13, 305–339, 1917.

165. Ichikawa, I. and Brenner, B.M., Evidence for glomerular actions of ADH and dibutyryl cyclic AMP in the rat, *Am. J. Physiol.*, 233, F102–F117, 1977.

166. Iturriza, F.C., Venosa, R.A., Pujol, M.G., and Quintas, N.B., Alpha-melanocyte-stimulating hormone stimulates sodium excretion in the salt gland of the duck, *Gen. Comp. Endocrinol.*, 87, 369–374, 1991.

167. Jorgensen, C.B., Role of urinary and cloacal bladders in chelonian water economy: historical and comparative perspectives, *Biol. Rev. Cambridge Philos. Soc.*, 73, 347–366, 1998.

168. Junqueira, L.C.U., Malnic, G., and Monge, C., Reabsorptive function of the ophidian cloaca and large intestine, *Physiol. Zool.*, 39, 151–159, 1966.

169. Kattan, G.H. and Lillywhite, H.B., Humidity acclimation and skin permeability in the lizard *Anolis carolinensis*, *Physiol. Zool.*, 62, 593–606, 1989.

170. Khalil, F. and Haggag, G., Nitrogenous excretion in crocodiles, *J. Exp. Biol.*, 35, 552–555, 1958.

171. Kim, J.W., Im, W.-B., Choi, H.H., Ishii, S., and Kwon, H.B., Seasonal fluctuations in pituitary gland and plasma levels of gonadotropic hormones in *Rana*, *Gen. Comp. Endocrinol.*, 109, 13–23, 1998.

172. Kim, S.Z., Kang, S.Y., Lee, S.J., and Cho, K.W., Localization of receptors for natriuretic peptide and endothelin in the duct of the epididymis of the freshwater turtle, *Gen. Comp. Endocrinol.*, 118, 26–38, 2000.

173. Kim, Y.K. and Dantzler, W.H., Relation of membrane potential to basolateral TEA transport in isolated snake proximal renal tubules, *Am. J. Physiol.*, 268, R1539–R1545, 1995.

174. King, P.A. and Goldstein, L., Renal excretion of nitrogenous products in vertebrates, *Renal Physiol.*, 8, 261–278, 1985.

175. Kobayashi, D., Mautz, W.J., and Nagy, K.A., Evaporative water loss: humidity acclimation in *Anolis carolinensis* lizards, *Copeia*, 701–704, 1983.

176. Komadina, S. and Solomon, S., Comparison of renal function of bull and water snakes (*Pituophis melanoleucus* and *Natrix sipedon*), *Comp. Biochem. Physiol.*, 32, 333–343, 1970.

177. Ladyman, M., Bradshaw, S.D., and Bradshaw, F.J., Physiological and hormonal control of thermal depression in the Tiger snake, *Notechis scutatus*, *J. Comp. Physiol. B*, 176, 547–557, 2006.

178. Landmann, L., Lamellar granules in mammalian, avian and reptilian epidermis, *J. Ultrastructure Res.*, 72, 245–263, 1980.

179. Lange, R. and Staaland, H., On the mechanism of water absorption in the rat colon, *Comp. Biochem. Physiol. A*, 40, 823–831, 1971.

180. Laverty, G., Elbrond, V.S., Arnason, S.S., and Skadhauge, E., Endocrine regulation of ion transport in the avian lower intestine, *Gen. Comp. Endocrinol.*, 147, 70–77, 2006.

181. Lavrova, E.A., Membrane coating granules: the fate of the discharged lamellae, *J. Ultrastructure Res.*, 55, 79–86, 1976.

182. Lazari, M.F.M., Alponti, R.F., Freitas, T.A., Breno, M.C., da Conceicao, I.M., and Silveira, P.F., Absence of oxytocin in the central nervous system of the snake *Bothrops jararaca*, *J. Comp. Physiol. B*, 176, 821–830, 2006.

183. Lebrie, S.J., Endocrines and water and electrolyte balance in reptiles, *Fed. Proc.*, 31, 1599–1608, 1972.

184. Lebrie, S.J. and Elizondo, R.S., Saline loading and aldosterone in water snakes *Natrix cyclopion*, *Am. J. Physiol.*, 217, 426–430, 1969.

185. Lebrie, S.J. and Sutherland, I.D.W., Renal function in water snakes, *Am. J. Physiol.*, 203, 995–1000, 1962.

186. Lefevre, M.E., Effects of aldosterone on the isolated substrate-depleted turtle bladder, *Am. J. Physiol.*, 225, 1252–1256, 1973.

187. Lemieux, G., Berkofsky, J., Quenneville, A., and Lemieux, C., Net tubular secretion of bicarbonate by the alligator kidney: antimammalian response to acetazolamide, *Kidney Int.*, 28, 760–766, 1985.

188. Lemieux, G., Craan, A.G., Quenneville, A., Lemieux, C., Berkofsky, J., and Lewis, V.S., Metabolic machinery of the alligator kidney, *Am. J. Physiol.*, 247, F686–F693, 1984.

189. Lemire, M., Contribution à l'étude des structures nasales des Sauriens: structure et fonction de la glande "à sels" des lézards déserticoles, Ph.D. thesis, Université de Pierre et Marie Curie, Paris, 1983 (available on request).

190. Lemire, M., Grenot, C.J., and Vernet, R., Electrolyte excretion by the nasal gland of an herbivorous Saharan lizard, *Uromastix acanthinurus* (Agamidae): effects of single NaCl and KCl loads, *J. Arid Environ.*, 3, 325–330, 1980.

191. Lemire, M., Grenot, C.J., and Vernet, R., Water and electrolyte balance of free-living Saharan lizards, *Uromastix acanthinurus*, *J. Comp. Physiol. B*, 146, 81–93, 1982.

192. Lillywhite, H.B., Water relations of tetrapod integument, *J. Exp. Biol.*, 209, 202–226, 2006.

193. Lillywhite, H.B. and Maderson, P.F.A., The structure and permeability of the integument, *Am. Zool.*, 28, 945–962, 1988.

194. Lorenzo, A., Medina, V., Badia, P., and Gomez, T., Effects of vasopressin on electrolyte transport in the lizard intestine, *J. Comp. Physiol. B*, 159, 745–751, 1990.

195. Lowy, R.J., Rowland, R.M., and Ernst, S.A., Control of ion transport in avian salt gland primary cultures: role of novel secretagogues and second messengers, *J. Cell Biol.*, 101, 478, 1985.

196. Lowy, R.J., Schreiber, J.H., and Ernst, S.A., Vasoactive intestinal peptide stimulates ion transport in avian salt gland, *Am. J. Physiol.*, 253, R801–R808, 1987.

197. Macchi, I.A. and Phillips, J.G., *In vitro* effect of adrenocorticotropin or cortical secretion in the turtle, snake, and bullfrog, *Gen. Comp. Endocrinol.*, 6, 170–182, 1966.

198. Machen, T.E. and Diamond, J.M., An estimate of the salt concentration in the lateral intercellular spaces of rabbit gall-bladder during maximal fluid transport, *J. Membr. Biol.*, 1, 194–213, 1969.

199. Mainoya, J.R., Bern, H.A., and Regan, J.W., Influence of ovine prolactin on transport of fluid and sodium chloride by the mammalian intestine and gall bladder, *J. Endocrinol.*, 63, 311–317, 1974.

200. Marshall, A.T., Intracellular and luminal ion concentrations in sea turtle salt glands by x-ray microanalysis, *J. Comp. Physiol. B*, 159, 609–616, 1989.

201. Marshall, A.T., Hyatt, A.D., Phillips, J.G., and Condron, R.J., Isosmotic secretion in the avian nasal salt gland: x-ray microanalysis of luminal and intracelllular ions distributions, *J. Comp. Physiol. B*, 156, 213–227, 1985.

202. Marshall, A.T. and Saddler, S.R., The duct system of the lachrymal salt gland of the green sea turtle, *Chelonia mydas*, *Cell Tiss. Res.*, 257, 399–404, 1989.

203. Mautz, W.J., Correlation of both respiratory and cutaneous water loss in lizards with habitat aridity, *J. Comp. Physiol.*, 149, 25–30, 1982.

204. Mautz, W.J., Patterns of evaporative water loss. In *Biology of the Reptilia*, Vol. 12, Gans, C. and Pough, F.H., Eds., Academic Press, New York, 1982, pp. 443–481.

205. McVicar, A.J. and Rankin, J.C., Dynamics of glomerular filtration in the river lamprey, *Lampetra fluviatilis* L., *Am. J. Physiol.*, 249, F132–F139, 1985.

206. Medica, P.A., Bury, R.B., and Luckenbach, R., Drinking and construction of water catchments by the desert tortoise *Gopherus agassizii*, *Herpetologica*, 36, 301–304, 1980.

207. Minnich, J.E., Water and electrolyte balance of the desert iguana, *Dipsosaurus dorsalis*, in its natural habitat, *Comp. Biochem. Physiol.*, 35, 921–933, 1970.

208. Minnich, J.E., Excretion of urate salts by reptiles, *Comp. Biochem. Physiol. A*, 41, 535–549, 1972.

209. Minnich, J.E., Adaptations in the reptilian excretory system for excreting insoluble urates, *Isr. J. Med. Sci.*, 12, 854–861, 1976.

210. Minnich, J.E., Water procurement and conservation by desert reptiles in their natural environment, *Isr. J. Med. Sci.*, 12, 740–758, 1976.

211. Minnich, J.E., Adaptive responses in the water and electrolyte budgets of native and captive desert tortoises, *Gopherus agassizii*, to chronic draught, *Desert Tortoise Council Symp. Proc.*, 102–129, 1977.

212. Minnich, J.E., Reptiles. In *Comparative Physiology of Osmoregulation in Animals*, Maloiy, G.M.O., Ed., Academic Press, London, 1979, pp. 393–641.

213. Minnich, J.E., The use of water. In *Biology of the Reptilia*, Gans, C. and Pough, F.H., Eds., Academic Press, New York, 1982, pp. 325–395.

214. Minnich, J.E. and Piehl, P.A., Spherical precipitates in the urine of reptiles, *Comp. Biochem. Physiol. A*, 41, 551–554, 1972.

215. Minnich, J.E. and Ziegler, M.R., Comparison of field water budgets in the tortoises *Gopherus agassizi* and *Gopherus polyphemus*, *Am. Zool.*, 16, 2191976.

216. Molnar, R.E., Fossil reptiles in Australia. In *Vertebrate Palaeontology of Australia*, Vickers-Rich, P. et al., Eds., Pioneer Design Studio/Monash University Publications Committee, Melbourne, Australia, 1991, pp. 605–701.

217. Morel, F., Regulation of kidney functions by hormones: a new approach, *Recent Prog. Hormone Res.*, 39, 271–304, 1983.

218. Moyle, V., Nirogenous excretion in chelonian reptiles, *Biochem. J.*, 44, 581–584, 1949.

219. Mukherjee, S.K. and Dantzler, W.H., Effects of SITS on urate transport by isolated, perfused snake renal tubules, *Pflügers Arch.*, 403, 35–40, 1985.

220. Munsick, R.A., Chromatographic and pharmacologic characterization of the neurohypophysial hormones of an amphibian and a reptile, *Endocrinology*, 78, 591–599, 1966.

221. Murrish, D.E. and Schmidt-Nielsen, K., Water transport in the cloaca of lizards: active or passive, *Science*, 170, 324–326, 1970.

222. Nagy, K.A., Water and electrolyte budgets of a free-living desert lizard, *Sauromalus obesus*, *J. Comp. Physiol.*, 79, 39–62, 1972.

223. Nagy, K.A., Seasonal patterns of water and energy balance in desert vertebrates, *J. Arid Environ.*, 14, 201–210, 1988.

224. Nagy, K.A. and Bradshaw, S.D., Energetics, osmoregulation and food consumption by free-living desert lizards, *Ctenophorus* (=*Amphibolurus*) *nuchalis*, *Amphibia–Reptilia*, 16, 25–35, 1995.

225. Nagy, K.A. and Medica, P.A., Physiological ecology of desert tortoises in southern Nevada, *Herpetologica*, 42, 73–92, 2007.

226. Navar, L.G., Inscho, E.W., Majid, D.S.A., Imig, J.D., Harrison-Bernard, L.M., and Mitchell, K.D., Paracrine regulation of the renal microcirculation, *Physiol. Rev.*, 76, 425–536, 1996.

227. Nishimura, H., Comparative endocrinology of renin and angiotensin. In *The Renin–Angiotensin System*, Johnson, R.A. and Anderson, R.R., Eds., Plenum Press, New York, 1980, pp. 29–77.

228. Nishimura, H., Endocrine control of renal handling of solutes and water in vertebrates, *Renal Physiol.*, 8, 279–300, 1985.

229. O'Shea, J.E., Bradshaw, S.D., and Stewart, T., Renal vasculature and excretory system of the agamid lizard, *Ctenophorus ornatus*, *J. Morphol.*, 217, 287–299, 1993.

230. Pang, P.K.T. and Sawyer, W.H., Renal and vascular responses of the bullfrog (*Rana catesbeiana*) to mesotocin, *Am. J. Physiol.*, 235, F151–F155, 1978.

231. Parvez, S., Ventura, M.A., and Parvez, H., Aldosterone antagonism. In *Antihormones*, Agarwal, M.K., Ed., Elsevier, Amsterdam, 1979, pp. 111–135.

232. Patterson, R., Growth and shell relationships in the desert tortoise. In *Symposium of Desert Tortoise Council*, Las Vegas, NV, Trotter, M., Ed., Desert Tortoise Council, San Diego, CA, 1977, pp. 158–166.

233. Peaker, M., Avian salt glands, *Philos. Trans. R. Soc. Lond. B*, 262, 289–300, 1971.

234. Peaker, M., Excretion of potassium from the orbital region in *Testudo caronaria*: a salt gland in a terrestrial tortoise?, *J. Zool. Lond.*, 184, 421–422, 1978.

235. Peaker, M. and Linzell, J.L., *Salt Glands in Birds and Reptiles*, Cambridge University Press, Cambridge, U.K., 1975, pp. 1–307.

236. Peek, W.D. and McMillan, D.B., Ultrastructure of the renal corpuscle of the garter snake *Thamnophis sirtalis*, *Am. J. Anat.*, 155, 83–102, 1979.

237. Peek, W.D. and McMillan, D.B., Ultrastructure of the tubular nephron of the garter snake *Thamnophis sirtalis*, *Am. J. Anat.*, 154, 103–128, 1979.

238. Perrault, C., Description anatomique d'une grande Tortue des Indes, *Mém. Acad. R. Sci. Paris*, 3, 395–422, 1676.

239. Perschmann, C. Über die Bedeutum der Nierenpfortader insbesondere für die Ausschiedung von Harnstoff und Harnsäure bei *Testudo hermanni* Gml. und *Lacerta viridis* Laur. sowie über die Funktion der Harnblase bei *Lacerta viridis* Laur, *Zool. Beitr.*, 2, 447–480, 1956.

240. Persson, B.-E., Dynamics of glomerular ultrafiltration in *Amphiuma means*, *Pflügers Arch.*, 391, 135–140, 1981.

241. Peterson, C.C., Physiological Ecology of Two Mojave Desert Populations of the Desert Tortoise *Xerobates agassizii:* Effects of Seasonal Rainfall Pattern and Drought, Ph.D. thesis, University of California, Los Angeles, 1993 (available on request).

242. Peterson, C.C., Anhomeostasis: water and solute relations in two populations of the desert tortoise (*Gopherus agassizii*) during chronic drought, *Physiol. Zool.*, 69, 1324–1358, 1995.

243. Porter, P., Physico-chemical factors involved in urate calculus formation. I. Solubility, *Res. Vet. Sci.*, 4, 580–591, 1963.

244. Prange, H.D. and Greenwald, L., Concentrations of urine and salt gland secretions in dehydrated and normally hydrated sea turtles, *Fed. Proc.*, 38, 970, 1979.

245. Rage, J.-C., Phylogénie et systématique des lépidosauriens. Où en sommes-nous?, *Bull. Soc. Herpétol. France*, 62, 19–36, 1992.

246. Randle, H.W. and Dantzler, W.H., Effects of K⁺ and Na⁺ on urate transport by isolated perfused snake renal tubules, *Am. J. Physiol.*, 255, 1206–1214, 1973.

247. Regaud, C. and Policard, A., Sur l'existence de diverticules du tube urinpare sans relations avec les corpuscles de Malpighi, chez les serpents, et sur l'indépendance relative des fonctions glomérulaires et glandulaires dur rein en general, *C. R. Soc. Biol. (Paris)*, 55, 1028–1029, 1903.

248. Reinhart, G.A. and Zehr, J.E., Atrial natriuretic factor in the freshwater turtle *Pseudemys scripta*: a partial characterization, *Gen. Comp. Endocrinol.*, 96, 259–269, 1994.

249. Renkin, E.M. and Gilmore, J.P., Glomerular filtration. In *Handbook of Physiology*. Vol. 8. *Renal Physiology*, Orloff, J. and Berliner, R.W., Eds., American Physiological Society, Washington, D.C., 1973, pp. 185–248.

250. Rice, G.E., Plasma arginine vascotocin concentrations in the lizard *Varanus gouldii* (Gray) following water loading, salt loading, and dehydration, *Gen. Comp. Endocrinol.*, 47, 1–6, 1982.

251. Rice, G.E., Bradshaw, S.D., and Prendergast, F.J., The effects of bilateral adrenalectomy on renal function in the lizard *Varanus gouldii* (Gray), *Gen. Comp. Endocrinol.*, 47, 182–189, 1982.

252. Roberts, J.B. and Lillywhite, H.B., Lipids and the permeability of epidermis from snakes, *J. Exp. Zool.*, 228, 1–9, 1983.

253. Roberts, J.B. and Lillywhite, H.B., Lipid barrier to water exchange in reptile epidermis, *Science*, 207, 1077–1079, 2007.

254. Roberts, J.S. and Schmidt-Nielsen, B., Renal ultrastructure and excretion of salt and water by three terrestrial lizards, *Am. J. Physiol.*, 211, 476–486, 1966.

255. Robinson, G.D. and Dunson, W.A., Water and sodium balance in the estuarine diamondback terrapin (*Malaclemys*), *J. Comp. Physiol.*, 105, 129–152, 1976.

256. Rosenbloom, A.A. and Fisher, D., Radioimmunoassay of arginine vasotocin, *Endocrinology*, 95, 1726–1732, 1982.

257. Saint-Girons, H. and Bradshaw, S.D., Données histophysiologiques sur l'hypophyse de *Tiliqua rugosa* (Reptilia, Scincidae) et notamment la signification fonctionale des cellules gamma, *Arch. Anat. Micro. Morphol. Exp.*, 70, 129–140, 1981.

258. Saint-Girons, H. and Bradshaw, S.D., Aspects of variation in histology and cytology of the external nasal glands of Australian lizards, *J. R. Soc. West. Austr.*, 69, 117–121, 1987.

259. Saint-Girons, H., Lemire, M., and Bradshaw, S.D., Structure de la glande nasale externe de *Tiliqua rugosa* (Reptilia, Scincidae) et rapports avec sa fonction, *Zoomorphology*, 88, 277–288, 1977.

260. Saint-Girons, H., Rice, G.E., and Bradshaw, S.D., Histologie comparée et ultrastructure de la glande nasale externe de quelques Varanidae (Reptilia, Lacertilia), *Ann. Sci. Nat. Zool. Paris*, 3(13), 15–31, 1981.

261. Sandor, T., Corticosteroids in Amphibia, Reptilia and Aves. In *Steroids in Non-Mammalian Vertebrates*, Idler, D.R., Ed., Academic Press, New York, 1972, pp. 253–327.

262. Sandor, T. and Mehdi, A.Z., Corticosteroids and their role in the extrarenal electrolyte secreting organs of nonmammalian vertebrates. In *Steroids and Their Mechanism of Action in Non-Mammalian Vertebrates*, Delrio, G. and Brachet, J., Eds., Raven Press, New York, 1980, pp. 33–49.

263. Sawyer, W.H. and Pang, P.K.T., Responses by vertebrates to neurohypophysial principles. In *Hormones and Evolution*, Barrington, E.J., Ed., Academic Press, London, 1979, pp. 493–523.

264. Sawyer, W.H. and Sawyer, M.K., Adaptive responses to neurohypophysial functions in vertebrates, *Physiol. Zool.*, 25, 84–98, 1952.

265. Schilb, T.P., Effect of a cholinergic agent on sodium across isolated turtle bladders, *Am. J. Physiol.*, 216, 514–520, 1969.

266. Schmidt-Nielsen, B. and Davis, L.E., Fluid transport and tubular intercellular spaces in reptilian kidneys, *Science*, 159, 1105–1108, 1968.

267. Schmidt-Nielsen, B. and Schmidt, D., Renal function of *Sphenodon punctatum*, *Comp. Biochem. Physiol. A*, 44, 121–129, 1973.

268. Schmidt-Nielsen, B. and Skadhauge, E., Function of the excretory system of the crocodile (*Crocodylus acutus*), *Am. J. Physiol.*, 212, 973–980, 1967.

269. Schmidt-Nielsen, K., Borut, A., Lee, P., and Crawford, Jr., E.C., Nasal salt excretion and the possible function of the cloaca in water conservation, *Science*, 142, 1300–1301, 1963.

270. Schmidt-Nielsen, K. and Fänge, R., Salt glands in marine reptiles, *Nature*, 182, 783–784, 1958.

271. Schnermann, J., The juxtaglomerular apparatus: from anatomical peculiarity to physiological relevance, *J. Am. Soc. Nephrol.*, 14, 1681–1694, 2003.

272. Schultz, S.G., The role of paracellular pathways in isotonic fluid transport, *Yale J. Biol. Med.*, 50, 99–113, 1977.

273. Seshadri, C., Urinary excretion in the Indian house lizard, *Hemidactylus flaviviridis* (Rüppel), *J. Zool. Soc. India*, 8, 63–78, 1956.

274. Seshadri, C., Water conservation in *Uromastix hardwickii* (Gray) with a note on the presence of Müllerian ducts in the male, *J. Zool. Soc. India*, 9, 103–113, 1957.

275. Shoemaker, V.H. and Nagy, K.A., Osmoregulation in the Galapagos marine iguana, *Amblyrhynchus cristatus*, *Physiol. Zool.*, 57, 291–300, 1984.

276. Shoemaker, V.H., Nagy, K.A., and Bradshaw, S.D., Studies on the control of electrolyte excretion by the nasal gland of the lizard *Dipsosaurus dorsalis*, *Comp. Biochem. Physiol. A*, 42, 749–757, 1972.

277. Shuttleworth, T.J. and Hildebrandt, J.-P., Vertebrate salt glands: short- and long-term regulation of function, *J. Exp. Zool.*, 283, 689–701, 1999.

278. Shuttleworth, T.J. and Thompson, J.L., Secretory activity in the salt gland of birds and turtles: stimulation via cyclic AMP, *Am. J. Physiol.*, 21, R428–R432, 1987.

279. Shuttleworth, T.J., Thompson, J.L., and Dantzler, W.H., Potassium secretion by nasal salt glands of desert lizard *Sauromalus obesus*, *Am. J. Physiol.*, 253, R83–R90, 1987.

280. Shuttleworth, T.J. and Thorndyke, M.C., An endogenous peptide stimulates secretory activity in the elasmobranch rectal gland, *Science*, 225, 319–321, 1984.

281. Silveira, P.F., Koike, T.I., Schiripa, L.N., Reichl, A.P., and Magnoli, F.C., Plasma arginine-vasotocin and hydroosmotic status of the terrestrial pit viper *Bothrops jararaca*, *Gen. Comp. Endocrinol.*, 109, 336–346, 1998.

282. Silveira, P.F., Schiripa, L.N., Carmona, E., and Picarelli, Z.P., Circulating vasotocin in the snake *Bothrops jararaca*, *Comp. Biochem. Physiol. A*, 103, 59–64, 1992.

283. Skadhauge, E., Hydromineral regulation in lower vertebrates. In *Comparative Endocrinology*, Gaillard, P.J. and Boer, H.H., Eds., Elsevier, Amsterdam, 1978, pp. 197–207.

284. Skadhauge, E., *Osmoregulation in Birds*, Springer-Verlag, Berlin, 1981, pp. 1–203.

285. Skadhauge, E., *In vivo* perfusion studies of the cloacal water and electrolyte resorption in the fowl (*Gallus domesticus*), *Comp. Biochem. Physiol. A*, 23, 483–501, 1967.

286. Skadhauge, E., The mechanism of salt and water absorption in the intestine of the eel (*Anguilla anguilla*) adapted to waters of various salinities, *J. Physiol.*, 204, 135–158, 1969.

287. Skadhauge, E., Excretion in lower vertebrates: function of gut, cloaca, and bladder in modifying the composition of urine, *Fed. Proc.*, 36, 2487–2492, 1977.

288. Skadhauge, E. and Duvdevani, I., Cloacal absorption of NaCl and water in the lizard *Agama stellio*, *Comp. Biochem. Physiol. A*, 56, 275–280, 1977.

289. Sokabe, H. and Ogawa, M., Comparative studies of the juxtaglomerular apparatus, *Int. Rev. Cytol.*, 37, 271–327, 1974.

290. Sougrat, R., Morand, M., Gondran, C., Barré, P., Gobin, R., Bonté, F., Dumas, M., and Verbavatz, J.-M., Functional expression of AQP3 in human skin epidermis and reconstructed epidermis, *J. Invest. Dermatol.*, 118, 678–685, 2002.

291. Stegbauer, S.A., The Effects of Hydration State and Hormones on the Ultrastructure of the Cloaca of Arid-Adapted Lizards, M.S. thesis, University of Wisconsin, Madison, 1979 (available on request).

292. Steinmetz, P.R., Characteristics of hydrogen ion transport in urinary bladder of water turtle, *J. Clin. Invest.*, 46, 1531–1540, 1967.

293. Stokes, G.D. and Dunson, W.A., Permeability and channel structure of reptilian skin, *Am. J. Physiol.*, 242, F681–F689, 1982.

294. Stolte, H., Schmidt-Nielsen, B., and Davis, L., Single nephron function in the kidney of the lizard, *Sceloporus cyanogenys*, *Zool. Jb. Physiol. Bd.*, 81, 219–244, 1977.

295. Taplin, L.E., Osmoregulation in crocodiles, *Biol. Rev.*, 63, 333–377, 1988.

296. Taplin, L.E. and Grigg, G.C., Salt glands in the tongue of the estuarine crocodile *Crocodylus porosus*, *Science*, 212, 1045–1047, 1981.

297. Taylor, G.C., Franklin, C.E., and Grigg, G.C., Salt loading stimulates secretion by the lingual salt glands in *Crocodylus porosus*, *J. Exp. Zool.*, 272, 490–495, 1995.

298. Templeton, J.R., Nasal salt excretion in terrestrial lizards, *Comp. Biochem. Physiol.*, 11, 223–229, 1964.

299. Templeton, J.R., Salt and water balance in desert lizards, *Symp. Zool. Soc. Lond.*, 31, 61–77, 1972.

300. Templeton, J.R., Murrish, D.E., Randall, E.M., and Mugaas, J.N., Salt and water balance in the desert iguana, *Dipsosaurus dorsalis*. I. The effect of dehydration, rehydration, and full hydration, *Z. Vergl. Physiol.*, 76, 245–254, 1972.

301. Templeton, J.R., Murrish, D.E., Randall, E.M., and Mugaas, J.N., Salt and water balance in the desert iguana, *Dipsosaurus dorsalis*. II. The effect of aldosterone and adrenalectomy, *Z. Vergl. Physiol.*, 76, 255–269, 1972.

302. Templeton, J.R., Murrish, D.E., Randall, E.R., and Mugaas, J.N., The effect of aldosterone and adrenalectomy on nasal salt excretion by desert iguana *Dipsosaurus dorsalis*, *Am. Zool.*, 8, 818–819, 1968.

303. Thomas, D.H. and Skadhauge, E., Water and electrolyte transport by avian ceca, *J. Exp. Zool.*, Suppl. 3, 95–102, 1989.

304. Thomas, D.H. and Skadhauge, E., Chronic aldosterone therapy and the control of transepithelial transport of ions and water by the colon and coprodeum of the domestic fowl (*Gallus domesticus*) *in vivo*, *J. Endocrinol.*, 83, 239–250, 1979.

305. Thomas, D.H., Skadhauge, E., and Read, M.W., Acute effects of aldosterone on water and electrolyte transport in the colon and coprodeum of the domestic fowl (*Gallus domesticus*) *in vivo*, *J. Endocrinol.*, 83, 229–237, 1979.

306. Toop, T. and Donald, J.A., Comparative aspects of natriuretic peptide physiology in non-mammalian vertebrates: a review, *J. Comp. Physiol. B*, 174, 189–204, 2004.

307. Tu, M.C., Lillywhite, H.B., Menon, J.G., and Menon, G.K., Postnatal ecdysis establishes the permeability barrier in snake skin: new insights into barrier lipid structures, *J. Exp. Biol.*, 205, 3019–3030, 2002.

308. Turner, F.B., Hayden, P., Burge, B.L., and Roberson, J.B., Egg production by the desert tortoise (*Gopherus agassizii*) in California, *Herpetologica*, 42, 93–104, 1986.

309. Uva, B., Vallarino, M., Mandich, A., and Isola, G., Plasma aldosterone levels in the female tortoise *Testudo hermanni*, *Gen. Comp. Endocrinol.*, 46, 116–123, 1982.

310. van Lannep, E.W. and Komnick, H., Fine structure of the nasal gland in the desert lizard *Uromastix acanthinurus*, *Cytobiology*, 2, 47–67, 1970.

311. van Lannep, E.W. and Young, J.A., Salt glands. In *Transport Organs*, Giebisch, G., Ed., Springer-Verlag, Berlin, 1979, pp. 675–692.

312. Vinson, G.P., Whitehouse, B.J., Goddard, C., and Sibley, C.P., Comparative and evolutionary aspects of aldosterone secretion and zona glomerulosa function, *J. Endocrinol.*, 81, 5P–24P, 1979.

313. Wallace, D.P., Rome, L.A., Sullivan, L.P., and Grantham, J.J., cAMP-dependent fluid secretion in rat inner medullary collecting ducts, *Am. J. Physiol.*, 280, F1019–F1029, 2001.

314. Walsh, P.J. and Wright, P., *Nitrogen Metabolism and Excretion*, CRC Press, Boca Raton, FL, 1995, pp. 1–337.

315. Weinstein, S.W., Proximal tubular energy metabolism, sodium transport, and permeability in the rat, *Am. J. Physiol.*, 219, 978–981, 1970.

316. Wertz, P.W., Lipids of keratinizing tissues. In *Biology of the Integument*, Vol. 2, Bereiter-Hahn, J., Matoltsy, A.G., and Richards, N.W., Eds., Springer-Verlag, Berlin, 1986, pp. 815–823.

317. Wideman, Jr., R.F., Nishimura, H., Bottje, W.G., and Glahn, R.P., Reduced renal arterial perfusion pressure stimulates renin release from domestic fowl kidneys, *Gen. Comp. Endocrinol.*, 89, 405–414, 1993.

318. Will, P.C., Cortright, R.N., Groseclose, R.G., and Hopfer, U., Amiloride-sensitive salt and fluid absorption in small intestine of salt-depleted rats, *Am. J. Physiol.*, 248, G124–G132, 1985.

319. Yokota, S.D., Ophidian kidney preparation for the measurement of glomerular dynamics in real time, *Pflügers Arch.*, 415, 501–503, 1990.

320. Yokota, S.D., Benyajati, S., and Dantzler, W.H., Comparative aspects of glomerular filtration in vertebrates, *Renal Physiol.*, 8, 193–221, 1985.

321. Yokota, S.D., Benyajati, S., and Dantzler, W.H., Renal function in sea snakes. I. Glomerular filtration rate and water handling, *Am. J. Physiol.*, 249, R228–R236, 1985.

322. Yokota, S.D. and Dantzler, W.H., Measurements of blood flow to individual glomeruli in the ophidian kidney, *Am. J. Physiol.*, 258, R1313–R1319, 1990.

11 Osmotic and Ionic Regulation in Birds

Eldon J. Braun

CONTENTS

I. INTRODUCTION

With respect to osmoregulation, birds differ in a number of aspects from the other group of warm-blooded vertebrates—mammals. These differences are highlighted along with the organs and organ systems that function in osmoregulation in birds. These systems are the kidneys, lower gastrointestinal tract, and the nasal salt glands, and each is discussed in turn. As a class of vertebrates, birds have evolved successfully to inhabit almost all environments or biomes that occur on Earth. This wide range of habitats has necessitated the evolution of a variety of osmoregulatory strategies. Often, birds are compared rather negatively to mammals with respect to osmoregulatory capabilities, particularly the ability to conserve body water by producing urines that are hyperosmotic to the plasmas from which they are derived. This can be highlighted by quoting a statement published in 1937 by Marshall and Smith:[46] "It is obvious that the bird's kidney shows glomerular degeneration, as indicated by the very small size and poor vascularization of the glomeruli and by the replacement of the central portion of the tuft by syncytial tissue. It is improbable that increased number of glomeruli can offset this reduction in filtering surface." This review argues that this statement stemmed from a lack of understanding of avian renal function and, more importantly, compares birds and other vertebrate classes with respect to their osmoregulatory capacities.

TABLE 11.1
Presence of Osmoregulatory Organs among Vertebrates

Organ	Fish	Amphibians	Reptiles	Birds	Mammals
Kidney	X	X	X	X	X
Intestine	X	X	X	X	
Bladder	X	X	X		
Gills	X	X			
Salt glands			X	X	
Skin		X			

II. OSMOREGULATORY ORGANS OF VERTEBRATES

From a morphological perspective, inspection of Table 11.1 indicates that mammals are the only class of vertebrates that rely solely on the kidney to regulate and maintain the homeostasis of the extracellular fluid. Moreover, this class of animals possesses a urinary bladder that is relatively impervious to the movement of fluid and one that has a different embryological origin than those of the other vertebrate classes. All nonmammalian vertebrates utilize more than one organ or organ system to regulate and maintain the homeostasis of the extracellular fluid (Table 11.1). Birds fit nicely into this latter category, because as a group they use the kidneys, lower gastrointestinal tract, and extrarenal salt glands to regulate the whole-body fluid balance. The function of the lower gastrointestinal tract is particularly important for birds in osmoregulation.

III. THE ROLE OF THE AVIAN KIDNEY IN OSMOREGULATION

A. MORPHOLOGY

The gross morphology of the avian kidney differs markedly from the typical bean shape of the unipapillate kidney of small mammals (Figure 11.1). The paired avian kidneys fit tightly into the fused lumber and sacral vertebrae (the synsacrum), and the tissue of each kidney is divided into three divisions: anterior, middle, and posterior, with the posterior divisions being the largest.[35]

1. Blood Supply

The kidney is supplied with two afferent blood supplies: the typical high-pressure arterial supply and an afferent venous supply by way of a functional renal portal system. The arterial supply is from two renal arteries that branch from the aorta. One branch goes to the anterior divisions, where the middle and posterior divisions are fed by common branches from the ischiatic arteries that bifurcate to send one branch to the middle divisions; the other branch feeds the posterior divisions (Figure 11.1). The renal portal system is a complex array of vessels delivering venous blood to the peritubular surfaces of nephrons. The persistence of a functional renal portal system in birds is probably related to the phenomenon of glomerular intermittency observed in other nonmammalian vertebrates and which also occurs in birds (see discussion on regulation of the glomerular filtration rate, below). Perfusion of the peritubular surfaces of the renal tubules with portal blood is controlled by a smooth muscle valve located in the common iliac vein that is under the control of the autonomic nervous system.[16] When the valve is closed, portal blood perfuses the peritubular surfaces of the small nephrons in the cortical regions of the kidney (portal blood flow does not enter the medullary cones); when it is open, the blood passes directly to the inferior vena cava to enter the central circulation. The physiological factors that regulate the function of the valve are uncertain but may be related to the systemic blood pressure and the hydration state of the bird. When the blood

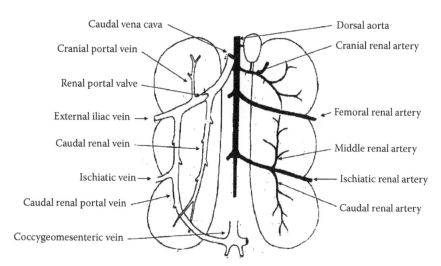

Caudal vena cava — Dorsal aorta

Cranial portal vein — Cranial renal artery

Renal portal valve — Femoral renal artery

External iliac vein →

Caudal renal vein — Middle renal artery

Ischiatic vein → Ischiatic renal artery

Caudal renal portal vein — Caudal renal artery

Coccygeomesenteric vein —

FIGURE 11.1 Illustration of avian kidneys in a ventral view. As depicted, each kidney is composed of three divisions: an anterior, a middle, and a posterior division. The arterial vasculature is shown on the right, and, for clarity, the venous vasculature is shown only on the left. Afferent blood is delivered to the kidney from three renal arteries; the femoral artery passes through the kidney without delivering blood to the renal tissue. Blood is also delivered to the peritubule surface of the nephrons by an afferent venous supply, the renal portal system. The perfusion of the renal tissue by afferent venous blood is controlled by the closing of the renal portal valve. If the valve is open, this blood bypasses the renal tissue and enters the central circulation by way of the caudal vena cava. (Adapted from Hodges, R.D., *The Histology of the Fowl*, Academic Press, London, 1974.)

pressure falls below the autoregulatory range and glomerular filtration ceases, the valve closes to sustain secretion by the renal tubules.[65] The perfusion of the peritubular surfaces of some of the nephrons facilitates their secretory function when no filtration occurs due to glomerular intermittency, such as during water restriction. This secretory function serves to continue the elimination of uric acid, the end product of nitrogen metabolism of birds.

2. Nephron Population and Their Structure

As is the case for the mammalian kidney, the internal organization of the avian kidney shows areas of zonation—cortical and medullary zones—but not as marked as occurs in the unipapillate kidneys of mammals. The internal organization of the avian kidney more resembles the structure of discrete multireniculate mammalian kidneys (Figure 11.2).[5] The cortical area of the avian kidney contains large numbers of short nephrons without loops of Henle that, as a first approximation, resemble the nephrons of reptilian kidneys and therefore have been referred to as *reptilian-type nephrons* or *loopless nephrons*. In the deeper aspects of the cortex (see Figure 11.2) are more complex nephrons with highly convoluted proximal tubules and loops of Henle. These nephrons resemble those found in the outer medulla of mammalian kidneys and therefore have been referred to as *mammalian-type nephrons* or *looped nephrons*. The morphology of the cells of the nephrons is discussed later.

B. Glomerular Filtration and Its Control

1. Morphology of the Avian Renal Corpuscle

Before presenting data on the rates of glomerular filtration, some morphological features of the avian renal corpuscle are presented. Whereas the capillaries within the renal corpuscle of mammals consist of a highly anastomotic network of small capillaries (4-µm luminal diameter), those of the

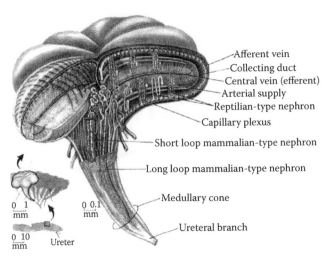

FIGURE 11.2 Illustration of the detailed internal organization of the avian kidney. The whole kidney is shown at the lower left with two successive enlargements indicating more detail of the internal organization. Near the surface of the kidney are small loopless nephrons arranged in a radiating pattern around a central vein. In the deeper regions of the cortex are larger nephrons with highly convoluted proximal tubules, loops of Henle, and distal tubules. A gradual transition occurs from the loopless to the looped nephrons. The loops of Henle are in parallel with collecting ducts and vasa recta, an arrangement that facilitates the functioning of a countercurrent multiple system. (From Braun, E.J. and Dantzler, W.H., *Am. J. Physiol.*, 222, 617–629, 1972. With permission.)

avian renal corpuscle are much simpler. The network of the small loopless nephrons are formed by a single, unbranched capillary (Figure 11.3), and those of the larger looped nephrons may have one or two bifurcations with luminal diameters of 8 μm. Quite possibly the underlying reason for the less complex networks of the avian glomerular capillaries is related to the nature of the avian red blood cells. These cells are nucleated, and the nuclei are attached to the envelope forming the cell. This structure leads to a very ridged, nondeformable cell, whereas the mammalian cell is deformable and can pass through capillaries whose diameter is half that of the cell. This simple morphology of the avian glomerular capillaries led Marshall and Smith to suggest that avian renal corpuscle was tending to evolutionary degeneration.[46]

The avian glomerular filtration barrier appears to be less restrictive than that of mammals, as avian ureteral urine contains large amount of protein (5 mg/mL). Morphological data show that the slit pore size in the kidney of the domestic chicken is 40 to 80% larger than the values reported for mammals (Table 11.2). Although the avian filtration barrier possesses a glycocalyx, selective staining for the anionic charge is not strong or well developed.[20] The significance of the morphology of the avian glomerular filtration barrier is discussed in greater detail below.

2. Glomerular Filtration

From a physical aspect, the primary factors that control the movement of fluid across the filtration barrier of the avian kidney are the same as those for all other vertebrates: hydrostatic pressure and colloid osmotic pressure of the blood. The whole-kidney glomerular filtration rate (GFR) is equal to the sum of the single nephron glomerular filtration rates (SNGFRs). Because of the simplicity of the avian glomerular capillaries noted above, the SNGFRs of birds tend to be lower than similar measurements for mammals; however, because avian kidneys have more nephrons than those of mammals,[18] the whole-kidney GFRs of birds and mammals are the same. The SNGFRs have been quantified for only two avian species: Gamble's quail and the European starling.[5,7] Using the sodium

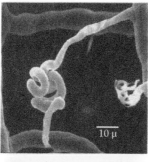

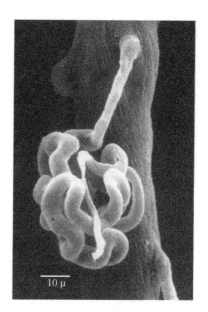

FIGURE 11.3 Scanning electron microscope images of negative casts of glomerular capillaries within avian and mammalian renal tissue. At the upper left is a cast of capillary from a loopless nephron, the illustration on the right is a cast of a looped nephron, and the cast at the lower left is of laboratory rat glomerulus for comparison. The casts of the avian glomerular capillaries appear to show tufts consisting of one unbranched capillary, whereas that of the rat indicates several through-fare channels forming branching networks. Note the scale bars.

ferrocyanide precipitation technique, the mean SNGFRs of the small loopless nephrons were found to be 7 and 14.7 nL/min, respectively, for the larger looped nephrons. By comparison, the SNGFRs of the nephrons within the white laboratory rat kidney are about 32 nL/min.[27] In addition to the SNGFRs measured using the ferrocyanide technique, two studies have employed *in vivo* micropuncture techniques to quantify the SNGFRs.[40,55] The limitation of *in vivo* micropuncture is that only the very smallest of the loopless nephrons on the surface of the kidney are accessible for sampling. This approach yielded SNGFRs of only 0.25 to 0.5 nL/min, values much lower than those determined by the sodium ferrocyanide technique, which permits measurement of the SNGFRs of all nephrons at one point in time.

TABLE 11.2
Glomerular Filtration Barrier Pore Size for Domestic Fowl and Laboratory Rat Kidneys

	Pore Sizes (nm)	
Animal	Endothelium	Epithelium
Domestic fowl	107–147	35
Rat	50–100	10

Source: Data from Casotti, G. and Braun, E.J., *J. Morphol.*, 228, 327–334, 1996.

TABLE 11.3
Glomerular Filtration Rates of Birds

Species	Glomerular Filtration Rate (mL/kg/min)
Domestic fowl	2.5
Desert quail	1.8
Mallard duck	2.5
Canadian goose	1.2
Glaucous-winged gull	1.9
European starling	2.8
Budgerigar	4.4
Turkey	1.3

Source: Data from Braun, E.J., *Comp. Biochem. Physiol. A*, 71, 511–517, 1982

Whole-kidney GFRs of avian kidneys tend to more labile than those of mammals, as birds tend to employ glomerular as well as tubular antidiuresis to regulate water excretion.[59] In mammals, GFRs are more stable, as water excretion is regulated mainly by tubular antidiuresis. Some representative values of avian GFRs are presented in Table 11.3. Arginine vasotocin (AVT), the antidiuretic hormone of birds, appears to play a dual antidiuretic role in conserving body water. At low circulating plasma levels (<5 μU/mL), data indicate an action of the hormone on the collecting ducts to promote the reabsorption of water.[59] At somewhat higher plasma concentrations of the hormone (~16 μU/mL), the whole-kidney GFR is reduced by about 30%. Earlier work suggested that AVT reduces the GFR through a constriction of the afferent arterials of primarily the loopless nephrons,[6] as the looped nephrons continue to filter at normal levels (~15 nL/min).

C. PROXIMAL TUBULE FUNCTION

In general, the proximal tubule of birds functions in a manner similar to that of other vertebrates in that in reabsorbs a large fraction of the glomerular filtrate. The transport capabilities of the avian proximal renal tubule have been studied directly using *in vivo* micropuncture, isolated tubule segments perfused *in vitro*, and primary cell cultures, as well as being inferred from whole-animal studies. Whole-animal studies indicate that the proximal tubules as an aggregate reabsorb about 65% of the glomerular filtrate; however, data from *in vivo* micropuncture studies indicated that only 24% of the filtered load is reabsorbed by the proximal tubules of the smallest nephrons on the surface of the kidney.[40] This large discrepancy is due to the fact that micropuncture can sample only the early segment of the very smallest, loopless nephrons on the surface of the kidney. The micropuncture data do indicate that most ions are reabsorbed isosmotically by the surface nephrons of the kidney. Moreover, data on fluid absorption by isolated perfused proximal tubule segments of short-looped mammalian-type nephrons (2 nL/mm/min) would indicate that 64% of the filtrate is reabsorbed by these tubules, a value that can be predicted based on the SNGFR of these nephrons (7 nL/min) and their average length (3.4 mm). The assumption has been made that solutes filtered are reabsorbed in an isosmotic manner by the proximal tubule.[40]

1. Sodium–Glucose Transport by Proximal Tubules

The plasma glucose concentration of birds is markedly higher than that of mammals when these two groups are compared on a body-mass-specific basis (Figure 11.4).[61] Moreover, the plasma glucose levels of birds appear to be independent of body mass, whereas for mammals the relationship

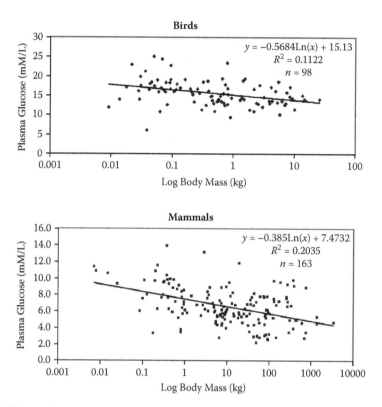

FIGURE 11.4 Plasma glucose concentrations of birds and mammals plotted against the log of body mass. The relationship for birds has a p value of 0.01, but that for mammals has a more significant relationship ($p = 0.001$). Note the much higher plasma glucose concentrations for avian plasma than for mammals. (Data were extracted from the International Species Information System (ISIS) database; actual data are available on request.)

has a slight but significant negative slope.[3,64] In spite of the relatively high plasma glucose concentrations, little or no glucose appears in the ureteral urine of birds.[49] This is remarkable given that whole-kidney GFRs of birds are high, and, coupled with the elevated plasma glucose concentrations, the filtered load of glucose by the avian kidney is also very high. This suggests that the sodium–glucose cotransporters (SGLTs) of the avian proximal tubule must have an enhanced capacity to reabsorb glucose or that they may be present at a greater density. The SGLTs have not been characterized directly; however, their function can be inferred from whole-animal studies and from work on primary cell cultures of the proximal tubule. Glucose loading studies on White Leghorn domestic fowl revealed that the tubular maximum for glucose transport is 4 to 5 times higher than that of the human kidney.[49] Work with primary cell cultures of the proximal tubule indicates that glucose in the medium applied to the mucosal side of the preparation stimulates short-circuit current.[60] Additional evidence for the presence of glucose transporters comes from the application of inhibitors such as phloridzin (competitor) and amiloride, both of which lowered or blocked the short-circuit current.[60] Studies on a similar preparation by Dudas et al.[28] demonstrated that applying glucose to the mucosal surface modulates the short-circuit current. Thus, indirect evidence indicates the obvious presence of SGLTs in the avian proximal tubules; however, given what appears to be the high capacity of the transporters, further molecular characterization of the transporters is warranted. Using the sequences available for the mammalian SGLTs, primers should be designed to isolate the transport proteins and insert them into amphibian oocytes to quantify the V_{max} and K_m. The prediction is that at least the V_{max} is much higher for the avian glucose transporters, compared to those in mammals.

TABLE 11.4
Aqueous Solubility of Uric Acid,
the Salts of Uric Acid, and Urea

Compound	Solubility (mmol/L)
Uric acid	0.381
Ammonium urate	3.21
Sodium urate	8.32
Potassium urate	14.75
Urea	16,650

2. Uric Acid Transport by Proximal Tubule

With respect to nitrogen excretion, birds are uricotelic; that is, they excrete the largest percent of excess nitrogen as uric acid. This compound is actually excreted in its anionic form: urate and its salts. Given the pH of avian plasma (7.4), the pK_a of uric acid, and using the Henderson–Hasselbach equation, it can be calculated that 98% of the uric acid exists in the anionic form in the plasma, leading to the formation of urate salts that have a high aqueous solubility (Table 11.4). This prevents uric acid from precipitating from solution and blocking flow within the renal tubules. The aqueous solubility of the uric acid salts is an order of magnitude higher than the protonated form, which is only 0.386 mmol/L.[13] On average, birds excrete 75% of the nitrogen as urate salts, with the second largest percent (20%) excreted as ammonium salts; however, it has been reported that some birds (nectivores) with a high throughput of fluid excrete a much larger fraction (50%) of the nitrogen as ammonia.[39,53]

The characteristics of urate transport within the proximal renal tubule have been studied employing several preparations. Studies with isolated, perfused tubule segments and primary cell cultures have yielded similar but not definitive data. As it is well accepted that uric acid is avidly secreted by the cells of the proximal tubule, urate transport proteins have been assumed to be located in the basolateral membranes of the cells. Data obtained from isolated segments of proximal tubules perfused *in vitro* indicate that urate accumulates in cells and exits the apical cell membrane on an as-yet undefined carrier[14] (Figure 11.5). The uptake by the basolateral membrane can be competitively blocked by *para*-aminohippuric acid (PAH), suggesting that these two compounds are recognized by the same transport protein, a member of the organic anion transport (OAT) family of proteins (Figure 11.5). Studies using primary cell cultures also show that urate is actively taken up by the basolateral cell membrane and accumulates inside the cells.[20] The net secretory flux is blocked by probenecid and PAH, supporting the perfused tubule studies suggesting that an OAT protein is involved. That an OAT protein is involved was also supported by reverse-transcription polymerase chain reaction (RT–PCR) studies, the results of which showed the expression of genes associated with the OAT family of proteins.[28] What remains to be resolved is how urate actually crosses the interior of cells and the identity of the facilitated exit step across the apical cell membranes.

3. Physical Form of Uric Acid in Avian Urine

The existence of uric acid as salt in the plasma prevents its precipitation from solution; however, within the renal tubules the conditions are altered compared to the plasma. Urate, being a relatively small molecule (168 MW), passes freely through the glomerular filtration barrier. In addition, urate is avidly secreted by the cells of the proximal tubule and this, coupled with the reabsorption of filtered water, raises the intraluminal concentration of urate beyond the solubility limits of uric acid and its salts. This could present a problem if uric acid and its salts precipitated from solution; for

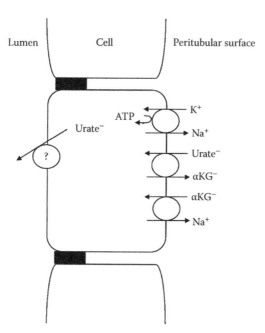

FIGURE 11.5 Scheme depicting the transcellular movement of uric acid across avian proximal tubule cells. Urate on the peritubule surface binds to a transport protein belonging to the organic anion transporter family of proteins (OAT) and enters the cell in exchange for αKG (alpha ketogluteric acid). The aKG reenters the cell in exchange for sodium ions (Na+). The αKG originates from intracellular metabolism. Urate enters the lumen across the apical cells membrane on an as-yet undefined transporter. (Modified from Brokl, O.H. et al., *Am. J. Physiol.*, 266,1085–1094, 1994.)

example, a single crystal of sodium urate is about 8 by 15 μm in size. This potentially could cause a blockage, given that the proximal tubule luminal diameter is about 8 to 10 μm. The formation of crystals is prevented by urate binding to a protein (serum albumin) within the tubule lumen. It was somewhat surprising to learn that the urate binding protein is serum albumin, which apparently enters the tubule by passing through the glomerular filtration barrier.[12] Data indicate that the glomerular filtration barrier of the avian kidney may be somewhat less restrictive, as the effective pore size is larger and there appears to less of a polyanionic charge to the barrier.[20] Uric acid and the matrix protein (albumen) of avian urine take the form of small spherical structures that range in diameter from 0.5 to 14 μm[10] (Figure 11.6). Evidence indicates that the association between uric acid and albumin and the formation of the spheres begins early in the proximal tubule.[19,23] This is probably necessitated by the high concentration of uric acid in the early proximal tubule that occurs as a result of events discussed above.

Empirical observation of avian urine on microscope slides indicates that when the spheres come into close contact with near neighbors they tend to coalesce into larger spheres (i.e., the soap bubble phenomenon). This is prevented *in vivo* by the presence of relatively large amounts of protein (5 mg/mL) in avian ureteral urine that maintains the spheres in a colloidal suspension.[10] The urine of mammals typically contains little or no protein. The potential loss of this protein (and therefore energy) in birds is prevented by the refluxing of the urine into the lower gastrointestinal tract, where the protein is degraded by bacteria and the products are absorbed by the colonic epithelium.[19] Comparison of sodium dodecyl sulfate–polyacrylamide gel electrophoresis (SDS-PAGE) gels run on ureteral urine and excreted cloacal fluid show the complete absence of protein in the excreted fluid.[34] Not only is the protein recovered by the lower gastrointestinal tract, but 60% of the uric acid that enters the colon is also broken down.[1] This is supported by data indicating that radiolabeled glutamic acid (a key component of the uric acid molecule) infused into the cloaca appears in the

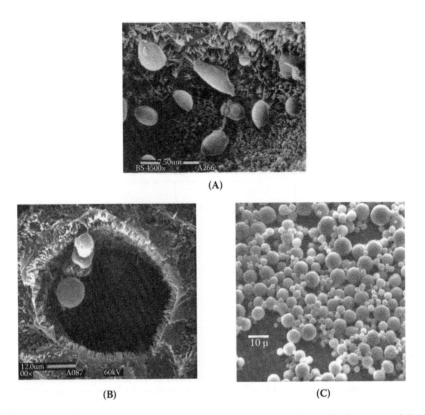

FIGURE 11.6 Spherical structures containing chemically bound uric acid. (A) Early stages of the formation the spheres in the early proximal tubule of an avian kidney; proximal tubule is identified by the presence of the microvillus brush border. (B) Completely formed spheres in the proximal tubule. (C) Scanning electron micrograph of dried avian ureteral; note that no crystals of uric acid are present. By chemical analysis, the spheres were found to be 65 to 70% uric acid.

vasculature leading from the colon.[37] What these data indicate is that a significant amount of the nitrogen sequestered in the uric acid molecule is recycled for metabolic use. This is probably very important for those birds whose dietary intake of nitrogen is low (nectarivorous and frugivorous species). Thus, uric acid plays an integral role in the fluid and electrolyte homeostasis of birds. This topic is addressed further after the urine-concentrating mechanism of the avian kidney is presented.

D. Concentration and Dilution of Urine by the Avian Kidney: Loop of Henle Function

Among the vertebrate classes, only birds and mammals have kidneys that conserve body water by producing urine in which the solutes are in a greater concentration then they were in the plasma from which they were derived (i.e., hyperosmotic urine). In contrast to some mammalian kidneys, avian kidneys in general do not concentrate urine well (Table 11.5). Whereas avian kidneys typically produce urines that result in maximum urine-to-plasma osmolar ratios (U/Posm) of 2 to 2.5, the kidneys of small mammals (desert rodents) can produce urines that result in U/Posms of 20 to 25.[8,43] Why does such a large discrepancy in maximum urine osmolalities exist when small birds (10 g body mass or less) inhabit the same desert environments as excellent urine-concentrating rodents? This question can be answered at several levels: morphological, functional, and behavioral.

TABLE 11.5

Urine-to-Plasma Osmolality Ratios for Selected Birds and Mammals

	Birds			Mammals		
	U/P Ratio[a]				U/P	
Species	Ureteral Urine	Voided Fluid	Ref.	Species	Ratio	Ref.
Domestic fowl	2.0	—	Krag and Skadhauge[38]	Long-nosed bat	1.1	Braun[13]
Ring-necked pheasant	1.5	—	Goldstein and Braun[29]	Nutria	2.5	Braun[13]
Kookaburra	—	2.7	Skadhauge[58]	Mountain beaver	2.7	Braun[13]
Singing honey eater	—	2.4	Skadhauge[58]	Hereford cow	3.9	Braun[13]
Red wattlebird	—	2.4	Skadhauge[58]	Blue whale	4.5	Braun[13]
Bobwhite quail	—	1.6	MacNabb[48]	Bottlenose dolphin	6.1	Braun[13]
California quail	—	1.7	MacNabb[48]	Weddell seal	6.8	Braun[13]
Gamble's quail	—	2.5	MacNabb[48]	Dog	8.7	Braun[13]
Senegal dove	—	1.7	Skadhauge[58]	Cat	10.8	Braun[13]
English sparrow	1.7	—	Casotti and Braun[22]	Cottontail	11.0	Braun[13]
Song sparrow	2.2	—	Casotti and Braun[22]	Marriam's K-rat	21.2	Braun[13]
White-crowned sparrow	2.1	—	Casotti and Braun[22]	House mouse	23.3	Braun[13]
White-winged dove	1.8	—	Krag and Skadhauge[38]	Desert pocket mouse	28.6	Braun[13]
Emu	—	1.4	Skadhauge[58]	Australian hopping	31.2	Braun[13]
Galah	—	2.5	Krag and Skadhauge[38]	mouse		
Glaucous-winged gull	1.9	—	Krag and Skadhauge[38]			
Savannah sparrow	—	1.6	Goldstein et al.[30]			

[a] Calculation based on urine collected from ureters or cloacal voided fluid.

About 10 to 15% of the nephrons in avian kidneys have loops of Henle. Organized in parallel with collecting ducts and vas recta they allow the avian kidney to produce urine hyperosmotic to plasma (albeit to a lesser degree than mammals) through the operation of a counter-current multiplier system. A number of striking differences in the gross morphology, microanatomy, and physiology can be identified when the kidneys of avians and mammals are compared. Internally, the organization of the avian kidney is somewhat similar to that of the large mammals (whales), where the tissue is divided into small functional units or rencules (Figure 11.2).[52] Within the avian kidney, the cortical tissue coalesces into tapering units or medullary cones. It is within these structures that elements are situated that facilitate the functioning of a counter-current multiplier system (i.e., a parallel arrangement of loops of Henle, vasa recta, and collecting ducts). The organization of the avian medullary cone has been suggested to be similar to that of the outer medulla of developing kidney of mammals, specifically that of the neonatal laboratory rat.[44] In the outer medulla of neonatal rat are short-looped nephrons in which the epithelium thickens before the hair-pin turn is formed. Within the avian medullary cone, the epithelium of all the nephrons thickens prior to the hair-pin turn. This thick descending limb makes an important contribution to the recycling of solute within the medullary cones.

In the avian kidney, a sharp change in the cellular morphology of the renal tubule occurs as the transition takes place between the straight segment of the proximal tubule to the thin descending limb of Henle's loop.[11] The thin descending segment consists of low-profile, mitochondria-poor cells. The transition from the low-profile cells to the mitochondria-rich, more cuboidal cells of the thick limb of Henle's loop always occurs prior to the hair-pin turn of the loop.[11] This is in contrast to the same cellular transition that occurs in the longer loops of Henle of most mammalian kidneys, where the epithelium at the hairpin turn has a very low profile; that is, the thin descending limb extends around the loop bend. A computer simulation of the avian counter-current multiplier

TABLE 11.6
Transport Characteristics of Thick Limbs of Henle's Loop

Animal	Water Flux (nL/m/min)	Sodium Flux (pEq/mm/min)	Chloride Flux (pEq/mm/min)
Coturnix quail	-0.01 ± 0.04	242 ± 49	272 ± 33
Rabbit	-0.02 ± 0.01	79 ± 8.4	59 ± 10
Mouse	-0.01 ± 0.04	—	94 ± 8

Source: Adapted from Nishimura, H. and Fan, Z., *Comp. Biochem. Physiol.*, 136, 479–98, 2003.

system[42] incorporating the transport properties[50] of the loop of Henle cells suggested that the transport of sodium chloride by the pre-bend cells is an important element in generating and maintaining the interstitial osmotic gradient.

Another difference from mammal kidneys is that the composition of the interstitial osmolality in the avian medullary cone is a marked departure from that observed for the medulla of most mammalian kidneys. Whereas in the kidneys of mammals sodium chloride and urea contribute solute particles to the interstitial osmolality, in birds the interstitial osmolality is made up almost entirely by sodium chloride. This is a consequence of uric acid being the endproduct of nitrogen metabolism in birds as opposed to urea. With some exceptions, urea as a solute is an integral component of the medullary interstitial osmotic gradient of mammalian kidneys.

1. Operation of the Counter-Current Multiplier System of the Avian Kidney

The permeability and transport characteristics of the avian loop of Henle suggest that sodium chloride is cycled between the ascending and descending limbs of Henle's loop. The descending limb is highly permeable to sodium chloride but, in contrast to mammalian kidneys, has very low osmotic and diffusional permeability to water (Table 11.6).[51] The thick limb of the Henle's loop, including the portion prior to the hair-pin turn, serves as the diluting segment of the nephron. This segment has very low water flux ($J_v = -0.01 \pm 0.04$ nL/m/min), which is similar or even lower than values for segments from mammal kidneys,[51] coupled with very high sodium and chloride fluxes (Table 11.6).

The flux of water in the proximal tubules, distal tubules, loop of Henle segments, and the collecting ducts of the avian nephron is facilitated by the presence of water channels or aquaporins (AQPs), although data have been derived on tissues from only two species. The distribution and physiological response of the avian AQPs appears to be somewhat different from that of mammals (Table 11.7). In the nephrons of mammals, AQP1 is the constitutively present channel that facilitates the transmembrane movement of water in the proximal tubule and descending limb of Henle's loop, as well the vasa recta. Three different water channels can be found in the distal nephron of mammals. These are AQPs 3 and 4 in the basolateral membranes of the late distal tubules and collecting ducts and AQP2 in the apical membranes of the distal nephron. It is the latter AQP that is responsive to antidiuretic hormone and is inserted into the apical cell membrane on demand to facilitate water movement across the epithelium of the distal nephron.

The nephrons in the avian kidney appear to have similar AQPs, at least in terms of number or terminology; however, their distribution and function in the nephron segments differ to a degree. Several laboratories have applied molecular techniques to study the AQPs within the avian kidney. The distribution and function of the AQPs in the avian kidney may be somewhat complicated by the nature of the nephrons present. As pointed out above, the population of nephrons in the avian kidney is made up of small nephrons lacking a loop of Henle (80 to 85% of total) and larger nephrons with loops of Henle. In the loopless nephrons (or reptilian-type nephrons), immunostaining

---

TABLE 11.7

Presence of Aquaporins in Nephrons of Birds and Mammals

Nephron Segment	Birds	Mammals
Cortical nephron		
Proximal tubule (PT)	AQP1	AQP1 (APM, BPM)
Distal tubule (DT)	AQP1	
Medullary nephron		
PT straight	AQP1	AQP7
Descending limb of Henle's loop (DLLH)	AQP1	AQP1 (APM, BPM)
Ascending limb of Henle's loop (ALLH)	AQP1	—
Collecting duct (CD)	AQP2 (APM)	AQP2 (APM, PCs)
	AQP4 (BPM)	AQP3 (BPM, PCs)
		AQP4 (BPM, PCs)

Note: APM, apical plasma membrane; BPM, basolateral plasma membrane; PCs, principal cells.

---

for AQP1 appears in both cell membranes of the proximal tubule, and the staining is very intense in the distal tubules of these nephrons.[24] The presence of AQP1 in the distal tubules of the looped nephrons may allow for the movement of water across this epithelium. The data in this paper do not show information for the distribution of AQP1 in any of segments from looped nephrons.

The movement of water across the epithelium of the distal nephron (late distal tubule and collecting duct) in the avian kidney is facilitated by the presence of AQPs 2 and 4.[47] Using RT-PCR-based cloning techniques, a homolog of AQP2 has been identified in the collecting ducts of the Japanese quail (*Coturnix coturnix*) kidney. The expression of this AQP in *Xenopus* oocytes significantly increased the osmotic water permeability of these cells, thus substantiating the presence of a functional water channel. Two different AQP4s have been cloned from the renal tissue of the Japanese quail.[51] In concert with the AQP2 they could facilitate the movement of water across the collecting ducts; however, when these AQPs were expressed in *Xenopus* oocytes, they did not enhance the movement of water.[51] Thus, either these AQPs do not function as water channels or the insertion of them into *Xenopus* oocytes does not provide the proper environment for their function. At this time, the other basolateral membrane AQP (AQP3) present in other animal renal tissue has not been identified in avian renal tissue.

IV. INTEGRATION OF RENAL AND LOWER GASTROINTESTINAL FUNCTION

Contrary to morphology of mammals, the urinary system of birds does not include a bladder for temporary storage of renal output. Instead, the ureters from the avian kidney convey urine to the terminal portion of the gastrointestinal tract, the cloaca. The urine does not remain in the cloaca to be excreted (except when evacuation of the lower gastrointestinal tract occurs) but is moved by peristaltic action of the cloacal musculature into the colon (rectum). Within the lower gastrointestinal tract, the composition of the urine can be significantly modified. Parenthetically, it has been reported, and I have observed, that one bird, the ostrich, will urinate and defecate separately.[26]

The reverse peristaltic action of the cloaca and lower gastrointestinal tract appears to be controlled locally and not by central (hypothalamic) osmoreceptors. This was clearly demonstrated by experiments carried by Brummermann.[15] *In vivo* surfusion of the hypothalamic area with hyperosmotic saline or mannitol solutions administered via a carotid artery did not influence the rate of peristaltic activity of the lower gastrointestinal tract; however, suffusion of the cloaca with varying saline concentrations or mannitol solutions via the vas deferens of male domestic fowl produced results suggesting that the sodium chloride content of the solutions was sensed (detected) within the cloaca. Hyperosmotic solutions of mannitol had no affect on the motility. It is probable that a type of vanilloid receptor (osmoreceptor) is present in the cloaca to detect the osmotic potential of the fluid within the cloaca, but this has yet to be demonstrated. The motility moving sodium chloride solutions from the cloaca into the colon continued as long as the osmolality (sodium chloride concentration) of the solutions did not exceed the plasma osmolality by 100 mOsm/kg H_2O. This agreed well with previously published data suggesting that the epithelium of the colon could recover water from the lumen as long as the differential between the luminal contents (higher) and plasma did not exceed 100 mOsm.[4]

The data suggesting that urines hyperosmotic to plasma by more that 100 mOsm are not refluxed into the colon are in agreement with the maximum U/Posm produced by avian kidneys. If the urine produced by the kidneys and entering the colon had a significantly greater osmolality, fluid would be drawn in the serosal-to-mucosal direction, causing water loss to the animal. Indeed, unstressed birds produce urine that is only slightly hyperosmotic to plasma which facilitates recovery of fluid by the colonic epithelium. In the literature is one suggestion of a species producing urine resulting in U/Posm values of 4 to 6.[17] These are data generated under laboratory conditions for the salt marsh Savannah sparrow. Experiments on the same species carried out in the field indicated that these birds under natural conditions are not significantly different from a variety of other species in terms of the concentration of ureteral urine produced.[30] The mean U/Posm of the Savanna sparrows in the more recent study was 1.2 mOsm/kg H_2O, again suggesting that birds in general do not produce significantly concentrated urines. In the original study, sparrows were acclimated to drinking increasing concentrations of seawater over a period of weeks, with the endpoint being that several birds drank and survived on full-strength sea water. The fluid output collected from the birds was excreted cloacal fluid, not ureteral urine, and therefore was not representative of renal function but was due to the integrative action of the kidneys and lower gastrointestinal tract. It is quite possible that some of the ingested seawater was not absorbed but passed through the gastrointestinal system to be excreted with the renal output, yielding a fluid that produced the high U/P osmolar ratio. It is essential when studying the renal function of birds that ureteral urine be collected, as excreted fluid represents the integrated product of the kidneys and lower gastrointestinal tract (for birds without functional salt glands).

As mentioned above, the renal system of birds must function in concert with the lower gastrointestinal tract in the maintenance of fluid and electrolyte homeostasis. This includes all birds, with or without functional salt glands. As suggested by a number of studies, this functional arrangement is particularly important in marine birds, where salts glands are a major component in the axis that regulates fluid balance (see Hughes[32] for an excellent review).

In all birds, ureteral urine is refluxed from the cloaca (urodeum) into the colon and digestive ceca, where large populations of bacteria reside that are capable of digesting not only protein but also uric acid. These bacterial populations have been well characterized in several studies[2,19] where segments of these populations have been shown to reproduce and grow vigorously on culture media containing only uric acid as a metabolic substrate, thus substantiating the degradation of uric acid within the lower gastrointestinal tract.[2]

A. STRUCTURE OF DIGESTIVE CECA

Not only does the colon function in osmoregulation, but many birds also have ceca as outpocketings from the colon. These ceca have been shown to augment the absorptive capacity of the colon.[66] The gross structure of the digestive ceca varies greatly from one taxonomic group to another, from being very large, highly coiled saculated structures in galleneous species to being nonexistent in many passerine birds.[25,47] It appears that the presence or absence of ceca is not correlated with specific taxonomic groups but is more associated with the ecology and diet of the species.[25] It is

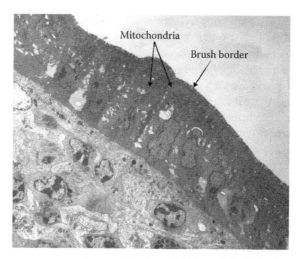

FIGURE 11.7 Histological section of the cecal wall of the English sparrow cecum. The epithelial cells lining the channels within the cecum possess a very dense brush border. On the interior of the cells, just below the brush border, is a very dense aggregation of mitochondria.

clear, however, that urine is refluxed into the ceca, where its composition is modified. Moreover, urine is refluxed in all avian species, including hummingbirds (lacking ceca), in which strands of colloidal urate have been observed beyond the colon into the ileum.[21] It is of interest to note that the small ceca of passerine birds have been considered to be rudimentary or vestigial and therefore having little or no function. Recent morphological studies on the small, vestigial ceca of the English sparrow demonstrated that their internal structure resembles that of a bottle brush (i.e., a central channel with numerous side branches).[54] The lumen of these channels is lined with epithelial cells that possess a very dense brush border. Moreover, within the cells just below the brush border is a very dense aggregation of mitochondria (Figure 11.7). Although functional studies on these ceca have yet to be done, the morphological features suggest that this tissue is not vestigial or nonfunctional.

The epithelial tissue of the avian lower gastrointestinal tract is very sensitive to the homeostatic sodium balance of the animal, and this sensitivity appears to be regulated by aldosterone.[41] These tissues express a diverse array of transport proteins that function to maintain not only osmotic balance but also nutritional balance.[41] Sodium–glucose cotransporters (SGLT1) have been identified in these tissues that recover glucose that may escape absorption by the kidney. In addition, aquaporins (AQP1) are present to facilitate the absorption of fluid.[24] Thus, studies have demonstrated that the renal system and the lower gastrointestinal tract of birds are intractably linked in the regulation of homeostatic fluid and ion balance.

B. Comparison of Urine-to-Plasma Osmolar Ratios of Birds and Mammals

Comparisons of birds and mammals with respect to the conservation of water by producing urines hyperosmotic to plasma are not valid, as the osmoregulatory systems are markedly different between the two groups of vertebrates. Two major features highlight the differences: (1) the lack of a urinary bladder in birds, with the result that ureteral urine enters the lower gastrointestinal tract, and (2) the end product of nitrogen metabolism being uric acid and not urea. Because of the anatomical involvement of the lower gastrointestinal tract in osmoregulation, the avian kidney does not and should not conserve water by producing urines that are significantly hyperosmotic to plasma

(U/Posm greater than 2). A significantly hyperosmotic fluid (urine) entering the lower gastrointestinal tract would draw fluid in a mucosal-to-serosal direction, leading to a negative fluid balance.

Uric acid being the endproduct of nitrogen metabolism in birds is the other reason why U/P osmolar ratios cannot be directly compared between birds and mammals. As it is sparingly soluble in aqueous solutions, uric acid does not contribute to the pool of osmotically active solutes in urine as does urea in mammals, where it can account for approximately 50% of the osmolality of the urine.

Numerous attempts have been made to formulate indices that would correlate urine-concentrating ability with morphological features of the avian kidney, as has been elucidated for the kidneys of mammals, such as relative medullary thickness (RMT).[36] The two measures (U/Posm and RMT) are not relevant when discussing osmoregulation by birds. Because the corticomedullary organization of the avian kidney is not as discrete as occurs in the unipapillate kidney of mammals, it not possible to section the avian kidney to show marked boundaries between cortical and medullary tissue (see Figure 11.2). What is more relevant is to appreciate the unique suite of characteristics that have evolved to allow the wide-ranging habitats of birds. These can be summarized as follows: the evolution of uric acid as an endproduct of nitrogen metabolism, the packaging of uric acid in the urine, the presence of a large amount of protein in the urine, and the modification of the renal output by the lower gastrointestinal tract.

C. Behavioral Ecology of Small Desert Birds and Mammals

From a behavioral perspective, the small desert rodents tend to be active at night (nocturnal) when the ambient temperatures have begun to subside; moreover, they escape the heat of the day by retreating to burrows where the ambient temperature is significantly lower than the ground surface temperature.[63] Small birds employ a different suite of behavioral patterns to escape from the peak daytime temperatures,[67] as they will seek out small crevices in the trunks of trees out of the direct flow pattern of the prevailing air currents to reduce exposure and evaporative water loss.[67] Thus, comparing small desert-dwelling birds and mammals, it would appear that the small mammals are better off; however, both groups of animals survive and thrive in the dry deserts of the southwestern United States.

V. SALT GLAND FUNCTION

Because of the anatomical involvement of the lower gastrointestinal tract in osmoregulation, the avian kidney does not and should not conserve water by producing urines that are significantly hyperosmotic to plasma (U/Posm greater than 2). This leaves birds whose habitats are entirely marine with the concern of producing osmotically free water to meet metabolic needs. Evolution solved this dilemma through the development of structures capable of eliminating salt (primarily sodium chloride) in excess of water and thereby producing osmotically free water. These structures are the nasal salt glands. Apparently all birds possess salt glands, but the degree to which they are developed and functional is determined by the amount and salinity of water and food consumed.[45,62] The fact that all birds possess some form of the salt gland may be a statement about the evolutionary origin of extant species. This suggests that the diet of the ancestor of modern birds may have had a high salt content that forced the evolution of salt glands. In a 1936 publication, Technau examined 106 species from 44 orders for the presence of salt glands. All birds had some form of the gland and its development was dependent on the natural consumption of waters with greater osmolalities than that of plasma.[62]

The majority of the studies on the function of avian salt glands have been carried out using domestic ducks as model systems.[32] Although considerable data have been generated from these studies, the morphological development of their salt glands is dependent on the salinity of the water consumed. Data generated from studies on birds confined to marine habitats (gulls, penguins) highlight the remarkable capacity of the salt glands to generate osmotically free water.[57]

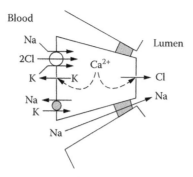

FIGURE 11.8 Mechanism of salt secretion by the avian salt gland. The process is driven by the Na,K-ATPase on the basolateral side of the cell facilitating the entry of chloride on the basolateral side through a Na–K–2Cl mechanism. Chloride exits across the apical membrane through channels with sodium moving through the paracellular route. (From Shuttleworth, T.J. and Hildebrandt, J.-P., *J. Exp. Zool.*, 283, 689–701, 1999. With permission.)

The primary ions secreted by salt glands are sodium and chloride in a hypertonic solution, with minor contributions from other ions.[57] It is generally accepted by those who study salt glands that the main process is secondary active secretion of chloride energized by the basolateral localized electrogenic Na,K-ATPase.[57] In this model of secretion, chloride exits the apical membrane via chloride channels, with sodium crossing the membrane through paracellular routes (Figure 11.8). It appears that control of salt gland function involves central osmoreceptors as first suggested by Schmidt-Nielsen,[56] but with input from volume receptors. Information is transmitted to the glands by way of parasympathetic cholinergic fibers, for which the receptors are of the muscarinic type. Shuttleworth[57] has provided an excellent review of vertebrate salt gland function. As stated by Shuttleworth, the avian salt gland is one of the most potent ion-transporting organs in the animal kingdom. These glands are capable of secreting solutions where the sodium chloride concentration is ten times the plasma concentration. Thus, they are capable of generating sufficient osmotically free water to allow birds to inhabit marine environments where freshwater is at a premium or is not at all available.[33]

VI. SUMMARY

As stated in the opening, from an osmoregulatory perspective birds as a class of vertebrates have evolved successfully to inhabit all environments on Earth. Evolution has provided birds with a suite of organs that function in concert to maintain the consistency of their internal environment and facilitate cellular function which has led to a group of animals with long life spans (compared to mammals of similar body mass). The evolution of specialized organs and organ systems has been complemented by biochemical specializations, probably the most important one being the capability of excreting nitrogen as uric acid. Excreting nitrogen in this form, however, has necessitated the development of mechanisms to prevent this sparingly soluble compound from precipitating from aqueous solutions and mechanisms to recover the glycoproteins that are necessary to facilitate the movement of uric acid through the excretory system. This composite of features of birds makes the study of osmoregulation by birds fascinating and intellectually rewarding.

REFERENCES

1. Anderson, G.L. and Braun, E.J., Post-renal modification of urine in birds, *Am. J. Physiol.*, 248, R93–R98, 1985.
2. Barnes, E.M., The avian intestinal flora with particular reference to the possible ecological significance of the cecal anaerobic bacteria, *Am. J. Clin. Nutr.*, 25, 1475–1479, 1972.

3. Beuchat, C.A. and Chong, C.R., Hyperglycemia in hummingbirds and its consequences for hemoglobin glycation, *Comp. Biochem. Physiol. A, Mol. Integr. Physiol.*, 120, 409–416, 1998.

4. Bindslev, N. and Skadhauge, E., Sodium chloride absorption and solute linked water flow across the epithelium of the coprodeum and large intestine in the normal and dehydrated fowl (*Gallus domesticus*): *in vivo* perfusion studies, *Am. J. Physiol.*, 216, 753–768, 1971.

5. Braun, E.J. and Dantzler, W.H., Function of mammalian-type and reptilian-type nephrons in kidney of desert quail, *Am. J. Physiol.*, 222, 617–629, 1972.

6. Braun, E.J. and Dantzler, W.H., Effects of ADH on single-nephron glomerular filtration rates in the avian kidney, *Am. J. Physiol.*, 226, 1–8, 1974.

7. Braun, E.J., Renal response of the starling (*Sturnus vulgaris*) to an intravenous salt load, *Am. J. Physiol.*, 234, 270–278, 1978.

8. Braun, E.J., Lindstedt, S.L, Nagel, R.B., and Altschuler, E.M., A model for the study of the countercurrent multiplication process within the mammalian kidney, *Proc. VIIth Int. Cong. Nephrol.*, Q-4, 1978.

9. Braun, E.J., Renal function, *Comp. Biochem. Physiol. A*, 71, 511–517, 1982.

10. Braun, E.J., Reimer, P.R., and Pacelli, M.M., The nature of uric acid in avian urine, *Physiologist*, 30, 120, 1987.

11. Braun, E.J. and. Reimer, P.R., The structure of the avian loop of Henle as related to the countercurrent multiplier system, *Am. J. Physiol.*, 255, F500–F512, 1988.

12. Braun, E.J., Boykin, S.L.B., and Schaeffer, R.C., The porosity of the avian glomerular filtration barrier, *Proc. Acta XXI Int. Ornithol. Congress* (special volume of *J. Ornithol.*), 1994.

13. Braun, E.J., Comparative renal function in reptiles, birds, and mammals, *Sem. Avian Exotic Pet Med.*, 7, 62–71, 1998.

14. Brokl, O.H., Braun, E.J., and Dantzler, W.H., Organic anion, organic cation, and fluid transport by isolated perfused and nonperfused avian renal proximal tubules, *Am. J. Physiol.*, 266, 1085–1094, 1994.

15. Brummermann, M. and Braun, E.J., Effect of water deprivation on colonic motility of white leghorn roosters, *Am. J. Physiol.*, 268, R690–R698, 1995.

16. Burrows, M.E. Braun, E.J., and Duckles, S.P., Avian renal portal valve: a reexamination of its innervation, *Am. J. Physiol.*, 245(4), 628–634, 1983.

17. Cade, T.J. and Bartholomew, G.A., Sea water and salt utilization by Savannah sparrows, *Physiol. Zool.*, 32, 230–238, 1959.

18. Calder, W.A. and Braun, E.J., Scaling of osmotic regulation in mammals and birds, *Am. J. Physiol.*, 244, 601–606, 1983.

19. Campbell, C.E. and Braun, E.J., Cecal degradation of uric acid in Gambel quail, *Am. J. Physiol.*, 251, R59–R62, 1986.

20. Casotti, G. and Braun, E.J., Functional morphology of the avian glomerular filtration barrier, *J. Morphol.*, 228, 327–334, 1996.

21. Casotti, G., Beuchat, C.A., and Braun, E.J., Morphology of the kidney in a nectarivorous bird, the Anna's hummingbird *Calypte anna*, *J. Zool. Lond.*, 244, 175–184, 1998.

22. Casotti, G. and Braun, E.J., Renal anatomy in sparrows from different environments, *J. Morphol.*, 243, 283–291, 2000.

23. Casotti, G. and Braun, E.J., Protein location and elemental composition of urine spheres in different avian species, *J. Exp. Zool. A, Comp. Exp. Biol.*, 301, 579–587, 2004.

24. Casotti, G., Waldron, T., Misquith, G., Powers, D., and Slusher, L., Expression and localization of an aquaporin-1 homologue in the avian kidney and lower intestinal tract, *Comp. Biochem. Physiol. A*, 147, 355–362, 2007.

25. Clench, M. H., The avian cecum: update and motility review, *J. Exp. Zool.*, 283, 441–447, 1999.

26. Duke, G. E., Mechanisms of excreta formation and elimination in turkeys and ostriches, *J. Exp. Zool.*, 283, 478–479, 1999.

27. Deng, A., Wead, L.M., and Blantz, R. C., Temporal adaptation of tubuloglomerular feedback: effects of COX-2, *Kidney Int.*, 66, 2348–2353, 2004.

28. Dudas, P.L., Pelis, R.M., Braun, E.J., and Renfro, J.L., Transepithelial urate transport by avian renal proximal tubule epithelium in primary culture, *J. Exp. Biol.*, 208, 4305–4315, 2005.

29. Goldstein, D.L. and Braun, E.J., Structure and concentrating ability of the avian kidney, *Am. J. Physiol.*, 256, R501–R509, 1989.

30. Goldstein, D.L., Williams, J.B., and Braun, E.J., Osmoregulation in the field by salt-marsh Savannah sparrows *Passerculus sandwichensis beldingi*, *Physiol. Zool.*, 63, 669–682, 1990.
31. Hodges, R.D., *The Histology of the Fowl*, Academic Press, London, 1974.
32. Hughes, M. R., Regulation of salt gland gut and kidney interactions, *Comp. Biochem. Physiol. A*, 136, 507–524, 2003.
33. Janes, D.N., Osmoregulation by Adélie penguin chicks on the Antarctic peninsula, *The Auk*, 114, 488–495, 1997.
34. Janes, D.N. and Braun, E.J., Urinary protein excretion in red jungle fowl (*Gallus gallus*), *Comp. Biochem. Physiol. A*, 118, 1273–1275, 1997.
35. Johnson, O. W., Some morphological features of avian kidneys, *The Auk*, 85, 216–228, 1968.
36. Johnson, O.W. and Mugas, J.N., Quantitative and organizational features of the avian renal medulla, *Condor*, 72, 288–292, 1970.
37. Karasawa, Y. and Koh, K., Incorporation of intraportal ammonia-N into blood and tissue nitrogenous compounds in chickens fed a low and high protein diet plus urea, *Comp. Biochem. Physiol.*, 87, 799–802, 1987.
38. Krag, B. and Skadhauge, E., Renal salt and water excretion in the budgerygah (*Melopsittacus undulates*), *Comp. Biochem. Physiol. A*, 41, 667–683, 1972.
39. Korine, C., Vatnick, I., Tets, I.G., and Pinshow, B., The influence of ambient temperature and the energy and protein content of food on nitrogenous excretion in the Egyptian fruit bat (*Rousettus aegyptiacus*), *Physiol. Biochem. Zool.*, 79, 957–964, 2006.
40. Laverty G. and Dantzler, W.H., Micropuncture of superficial nephrons in avian (*Sturnus vulgaris*) kidney, *Am. J. Physiol.*, 243, 561–569, 1982.
41. Laverty, G. and Skadhauge, E., Physiological roles and regulation of transport activities in the avian lower intestine, *J. Exp. Zool.*, 283, 480–494, 1999.
42. Layton, H.E., Davis, J.M., Casotti G., and Braun, E.J., Mathematical model of an avian urine concentrating mechanism, *Am. J. Physiol. Ren. Physiol.*, 279, F1139–F1160, 2000.
43. Lindstedt, S.L. and Braun, E.J., A model for studying the regulation of individual nephron function, *Fed. Proc.*, 37, 816,1978.
44. Liu, W., Morimoto, T., Kondo, Y., Linuma, K., Uchida, S., and Imai, M., "Avian type" renal medullary tubule organization causes immaturity of urine concentrating ability in neonates, *Kidney Int.*, 60, 680–693, 2001.
45. Marples, B.J., The structure and development of the nasal glands of birds, *Proc. Zool. Soc. Lond.*, 4, 829–844, 1932.
46. Marshall, E.K. and Smith, H.W., The glomerular development of the vertebrate kidney in relation to habitat, *Biol. Bull.*, 59, 135–150, 1930.
47. McLelland, J., The anatomy of the avian cecum, *J. Exp. Zool.*, 3(Suppl.), 2–9, 1989.
48. MacNabb, F.M.A., A comparative study of water balance in three species of quail. II. Utilization of saline drinking solutions, *Comp. Biochem. Physiol.*, 28, 1059–1074, 1969.
49. Morgan, C. and Braun, E.J., Glucose handling by the kidney of the domestic fowl, *FASEB J.*, 15, A854, 2001.
50. Nishimura, H., Koseki, C., Imai, M., and Braun E.J., Sodium chloride and water transport in the thin descending limb of Henle of the quail, *Am. J. Physiol.*, 257, F994–F1002, 1989.
51. Nishimura, H. and Fan, Z., Regulation of water movement across vertebrate renal tubules, *Comp. Biochem. Physiol.*, 136, 479–98, 2003.
52. Oliver, J., *Nephrons and Kidneys: A Quantitative Study of Developmental and Evolutionary Mammalian Renal Architectonics*, Harper and Row, New York, 1968.
53. Preest, M.R. and Beuchat, C.A., Ammonia excretion in hummingbirds, *Nature*, 386, 562, 1997.
54. Reyes, L. and Braun, E.J., The functional morphology of the English sparrow cecum, *Comp. Biochem. Physiol.*, 141, 292–297, 2005.
55. Roberts, J.R. and Dantzler, W.H., Micropuncture study of the avian kidney: infusion of mannitol or sodium chloride, *Am. J. Physiol.*, 258, 869–875, 1990.
56. Schmidt-Nielsen, K., The salt-secreting glands of marine birds, *Circulation*, 21, 955–967, 1960.
57. Shuttleworth, T.J. and Hildebrandt, J.-P., Vertebrate salt glands: short- and long-term regulation of function, *J. Exp. Zool.*, 283, 689–701, 1999.

58. Skadhauge, E., Renal concentrating ability in selected Western Australian birds, *J. Exp. Biol.*, 61, 269–276, 1974.
59. Stallone, J.N. and Braun E.J., Contributions of glomerular and tubular mechanisms to antidiuresis in conscious domestic fowl, *Am. J. Physiol.*, 249, 842–850, 1985.
60. Sutterlin, G.G. and Laverty, G., Characterization of a primary cell culture model of the avian renal proximal tubule, *Am. J. Physiol.*, 275, R220–R226, 1998.
61. Sweazea, K.L. and Braun, E.J., Glucose transporter expression in English sparrows (*Passer domesticus*), *Comp. Biochem. Physiol. B*, 144, 263–270, 2006.
62. Technau, G., Die nasendrüse der vögel, *J. Ornthol.*, 4, 511–617, 1936.
63. Tracy, R.L. and Walsberg, G.E., Kangaroo rats revisited: re-evaluating a classic case of desert survival, *Oecologia*, 133, 449–457, 2002.
64. Umminger, B.L., Body size and whole blood sugar concentrations in mammals, *Comp. Biochem. Physiol. A*, 52, 455–458, 1975.
65. Wideman, Jr., R.F. and Gregg, M., Model for evaluating avian renal hemodynamics and glomerular filtration rate autoregulation, *Am. J. Physiol.*, 254, R925–R932, 1988.
66. Williams, J.B. and Braun, E.J. Renal compensation for cecal loss in Gambel's quail (*Callipepla gambelii*), *Comp. Biochem. Physiol.*, 113, 333–341, 1996.
67. Wolf, B.O. and Walsberg, G.E., Thermal effects of radiation and wind on a small bird and implications for microsite selection, *Ecology*, 77, 2228–2236, 1996.
68. Yokota, S.D., Benyajati, S., and Dantzler W.H., Comparative aspects of glomerular filtration in vertebrates, *Ren. Physiol.*, 8, 193–221, 1985.

12 Osmotic and Ionic Regulation in Mammals

Rolf K.H. Kinne and Mark L. Zeidel

CONTENTS

I. INTRODUCTION

In this chapter, we first define in humans the various fluid spaces and their composition. We follow with a description of the neural and hormonal reflexes and the behavioral responses elicited by moderate challenges to the steady state. We then define the mechanisms of these overall responses at the level of effector and target molecules in cells. Finally, we build on these mechanisms of adaptation to moderate challenges to define the mechanisms of adaptation to more extreme habitats in specific organs and in other mammals. Although the scope is broad, we hope that by presenting these adaptive processes we will reveal the amazing precision of the body in maintaining water and electrolyte balance.

II. BODY FLUID COMPARTMENTS AND COMPOSITION

A. BODY WATER AND ITS SUBDIVISIONS IN THE HUMAN BODY

By far the most abundant body constituent, water plays a crucial role in all metabolic processes and in the exchange of solutes between the various tissues and the environment. Water represents a very constant 73% of *lean body mass* (defined as the body weight minus the weight of the fat) in the average human individual. Body water varies, however, in the range of 55 to 60% with the total weight. It is usually lower in the obese person than in a thin person, and in women body water is usually a smaller fraction of the body weight than in men. Thus, for a man with a weight of 70 kg the total amount of water is about 45 L. This water is present in several compartments of the body. The *extracellular fluid* comprises a little more than one third of the total body water (about 17 L); it is made up of the *blood plasma* (3 L) which is separated from the *interstitial fluid* (12 L). In addition are the so-called *transcellular fluids* (2 L), which include the cerebrospinal fluid, gastric juices, and intestinal fluids, as well as the urine; these fluids do not exchange directly with the extracellular fluid but are separated from it by epithelial cell layers. The remainder of the fluid is the *intracellular fluid* (tissue cells, about 25.5 L; blood cells, 2.5 L), which is separated from the extracellular fluid by the plasma membrane of the cells. Most of the intracellular fluid resides in the skeletal muscles, followed by the skin, fat, bones, and liver. The sizes of the various fluid spaces have been determined experimentally using indicator dilution methods, which measure the distribution of labeled compounds in various body spaces (see physiology texts for details).

B. BODY WATER BALANCE

Normally, the daily rate of water intake and generation matches the daily water loss of 2.5 L/day. Water intake includes drinking (1.3 L) and eating (dietary water, 0.9 L), as well as water that is generated from metabolism (metabolic water, 0.3 L) and by the oxidation of carbohydrates (0.6 mL/g), fat (1.1 mL/g), and protein (0.5 mL/g). Water is lost by the excretion of urine (1.5 L), via evaporation through the skin and the respiratory tract (insensible losses, 0.9 L), and in the feces (0.1 L). Changes in any of these routes of water intake and generation or loss can drastically alter the amount of body water. In addition, mammals living under extreme conditions use several strategies to regulate intake/generation and loss so they maintain constant body water content. Indeed, given the enormous range of conditions in which mammals live, from arid to wet environments or in seawater, their ability to maintain total body water remarkably constant defines the concept of homeostasis. As is detailed below—for example, in arid environments—metabolically generated water may be the only way to achieve a net water gain, whereas reduction of water loss by the kidney compensates for the low net gain of water. Thus, although the daily turnover of water of 2.5 L represents a volume equal to about 50% of the blood volume (plasma and blood cells = 5.5 L), this critical volume is kept remarkably constant.

Maintenance of the total body water balance requires massive movements of fluid between body compartments; for example, in the human kidney, 180 L of primary urine are generated per day but only 1.5 L leave the body as urine, requiring the reabsorption of nearly all of the glomerular filtrate in the renal tubules. In the gastrointestinal tract, 7 L are secreted daily into the lumen of the intestine, of which nearly all is reabsorbed. In ruminants, intestinal secretions can reach volumes of 100 L (see below), only a small amount of which is excreted in the feces.

C. Composition of Body Fluids

Because water is freely permeable across most (but not all) biological membranes, the osmolality of the various fluid spaces (except for the transcellular space) is very similar and tightly maintained at 330 mOsmol/kg. In some transcellular spaces (most notably the urine), the presence of membranes that block water flux in some of the epithelia permits these particular fluids to change in osmolality while the other body fluids remain constant.

Although osmolality remains constant in all but the transcellular space, the nature of the osmotically active solutes, the osmolytes, in the various body fluid compartments can vary widely. In the extracellular space (again with the exception of the transcellular spaces), sodium and chloride are the major osmolytes. In the intracellular space, potassium is the main inorganic electrolyte contributing to osmolality, while numerous organic metabolic substrates and intermediates form the balance. Most of these organic molecules are present in millimolar concentrations. Depending on the cell, chloride contributes to a small extent to the osmolality of the intracellular fluid. Although intracellular proteins are present in high concentration (representing about 30% of wet weight in kidney, for example) and contribute to the charge equilibrium (see Donnan equilibrium in physiology texts), their high molecular weight results in a relatively low concentration when compared with other solutes. Thus, potassium and chloride are the main osmolytes that can be exchanged rapidly with the extracellular space without sacrificing the intracellular metabolism.

Some cells that are permanently or intermittently submitted to changes in the osmolality of the extracellular medium during the normal functioning of the body, such as kidney cells in the inner medulla, contain in addition high amounts of low-molecular-weight organic osmolytes that are used instead of potassium to regulate intracellular osmolality and thus cell volume. As an example, inner medullary collecting duct cells exposed in cell culture to a hypertonic medium of 600 mOsmol contain about 40 mOsmol sorbitol, 75 mOsmol myo-inositol, 20 mOsmol taurine, and 75 mOsmol glycerophosphorylcholin.[91] Thus, about half of the intracellular osmolality is contributed by the organic osmolytes. These organic osmolytes are used not only in the human kidney but also in a variety of mammals such as cat, dog, mouse, rat, and rabbit.[91] The mechanisms involved in the regulation of the intracellular concentration of the organic osmolytes have been reviewed previously.[262]

D. Acute Disturbances of Volume and Osmolality of Body Fluids

Constancy of volume or osmolality of body fluids may be threatened by changes in intake or loss of free water, of osmolytes, or a combination of both. Thus, excessive freshwater uptake in drinking leads to hypotonic overhydration characterized by an increase in *extracellular fluid* (ECF) volume and a decrease in osmolality of the ECF. Uptake of a large amount of fluid with a high concentration of osmolytes, which can occur during feeding of mammals living in seawater (open seawater osmolality can reach more than 1000 mOsm/kg), could lead to hypertonic overhydration, in which an increase in osmolality of the ECF is the main disturbance. Uptake of isotonic fluid—a situation induced, for example, by intravenous infusion of isotonic saline in patients or by feeding of bats on fruits—just increases the volume of the extracellular fluid. Loss of water with a low osmolyte content such as during sweating, diuresis, or diarrhea leads to dehydration, which is characterized by a decreased volume of the extracellular fluid and by an increase of the osmolality of the ECF. Loss of blood and early forms of diarrhea where isotonic fluid is lost lead to a decrease of ECF volume but the osmolality remains constant. It thus becomes clear that two main parameters can change when the fluid osmolyte balance of the extracellular fluid is challenged: the *volume* and the *osmolality*, which can deviate in different directions and therefore have to be sensed independently and regulated separately, yet in a coordinated fashion.

Acute changes in *intracellular fluid* (ICF) are most commonly the consequence of changes in the osmolality of the extracellular fluid. Increased extracellular osmolality leads to rapid cell shrinkage, and decreased osmolality leads to cell swelling, which in extreme cases can lead to profound functional impairment of organ functions, in particular when they are encased in a nonexpendable space, as the brain is encased in the rigid skull. Here, during water intoxication accompanied by excessive cell swelling, loss of consciousness and convulsions can occur, in part due to the damage inflicted on neurons by increased intracranial pressure. Even under isosmotic conditions of the ECF, sudden bursts in metabolism (e.g., in the liver during ingestion of a glucose-rich meal or a hormonally induced increase in glycogenolysis) can lead to cell swelling requiring osmoregulation of the intracellular fluid.[101]

III. SENSING OF CHALLENGES

A. THE BODY-FLUID-RELATED NEURONAL NETWORK

The ability of animals to survive in an ever-changing environment requires that the major components of the body fluids be in a dynamic steady state. Even under benign environments terrestrial animals continuously lose water and sodium. To correct water and sodium imbalances and maintain body fluid homeostasis, *reflex mechanisms* and *behavioral responses* are necessary. The reflex mechanisms are comprised of the neuronal responses and endocrine responses; the behavioral responses include seeking out and drinking water and the ingestion of salt-rich food or fluids.[109] After a challenge, the neuronal responses occur in a matter of seconds, endocrine responses can be observed a few minutes, and behavioral responses are realized in tens of minutes or hours. The major difference in the nature of the responses as well as their different latency require an integration of the stimuli and the signaling mechanism. The central nervous system constantly monitors body fluid status and integrates reflex and behavioral responses. Convincing evidence suggests that autonomic responses, endocrine reflexes, and ingestive behaviors are activated by similar afferent signals in the brain and that these signals are processed in the same areas of the central nervous system, the so-called body-fluid-related neuronal network.[109] This integration is the final point in a series of events proceeding from sensors for extracellular volume or osmolality to the generation of signals, their translation into intracellular signaling, and effects on the function of the target cells. In the following sections, these points are considered consecutively and separately, from the organ to the molecular level, but their intricate interconnections should always be kept in mind.

B. OSMOSENSORS

The term *osmoreceptors* was coined by Verney[251] to describe the cells in the brain that shrink in response to an increase in extracellular osmolality and, as a result, effect antidiuretic hormone (ADH) release.[252] Other authors have argued that sodium-sensing cells are also involved in stimulating ADH release.[164] These salt-sensitive receptors are thought to be important when the osmolality of the extracellular fluid is increased by a result of water loss. In humans, the osmo- and sodium-sensing cells are located in the lamina terminalis, the anterior wall of the third ventricle adjacent to the anterior hypothalamus. There, the so-called *circumventricular organs* are located which lack the blood–brain barrier[166,258] and, because of their direct exposure to the blood, form the neuroendocrine interface between the brain and the nontranscellular body fluids.

Osmo- and sodium receptors are, however, found not only in the brain but also in the periphery. Thus, Haberich et al.[95] drew attention to the importance of sensors located in the liver for the maintenance of body fluid homeostasis. These sensors and probably some in the portal vein or mesenteric veins transmit signals via alteration of the vagal afferent nerve activity to the brain that induce increased urine output. These receptors might be advantageous in providing early warnings to buffer large changes in both systemic and brain osmolality due to ingested fluid.

At the molecular level, significant progress has been made since Homer Smith's review in 1957 which he entitled "Salt and Volume Receptors: An Exercise in Physiologic Apologetics."[231] Recently, transient receptor potential vanilloid 4 (TRPV4), an ion channel in the cell membrane, has been shown in a transgenic approach to be necessary for the maintenance of osmotic equilibrium.[147] In TRPV4 gene null mice, the increase in antidiuretic hormone (ADH) synthesis in response to osmotic stimulation is attenuated, whereas renal water reabsorption capacity is not impaired. Hypertonic stress leads to diminished expression of c-Fos-positive cells in the circumventricular organ OVLT (organum vasculosum of the lamina terminalis), indicative of a reduced osmotic stimulation.[147] TRPV4 very probably functions as an ion channel in the transduction of osmotic and mechanical stimuli *in vivo*. The channel conducts, when activated, more cations, particularly calcium, into the cell. The mechanism of activation is unclear; either an osmotic stimulus transduction apparatus is first activated (activation of a kinase?) followed by a gating of the channel or the gating of the channel initiates further osmotic responses.[148]

Another channel recently reported to be a potential candidate for an osmotic sensor molecule is the TRPM7 channel. In cell-attached patch-clamp experiments and in excised membrane patches, the channel open probability of this channel was drastically increased by mechanical stress.[178] The PM7 channel is a member of the TRP channel family, which are sensors for a variety of stimuli.[46] The TRPM (mammalian melastatin-related transient receptor potential) subfamily has eight members. The TRP family was originally described in *Drosophila* and named after its role in phototransduction. Mammalian TRP subunit proteins are encoded by at least 28 genes.[69] The proteins consist of six putative transmembrane domains and intracellular N- and C-termini. They form channels for monovalent or divalent cations with variable selectivity. Some of them are involved in Mg homeostasis of the cells; others modulate the membrane potential. In addition to cell swelling, they are activated by temperature, lipid compounds, and other endogenous or exogenous ligands (for review, see Kraft and Harteneck[128]). Other mechanosensitive channels identified recently include the epithelial sodium channels, voltage-gated sodium channels,[169] and two pore potassium channels (see Hamill[97] for review). Although the exact biophysical mechanisms involved in sensing mechanical stress remain unclear, it appears likely that the mechanism involves the concerted gating of multiple ion channels.

The molecular basis for specific sodium sensors is also rather obscure. In theory, sodium channels could fulfill that role because they possess highly selective sodium binding sites and intramolecular flexibility. The use of a substrate binding site as a sensing mechanism has recently been demonstrated for the sodium–D-glucose cotransporter SGLT whose isoform SGLT3 has predominant (or exclusive) glucose-sensing functions.[263]

In the human, the impulses of the sodium and osmotic sensors in the brain are transmitted from the supraoptic nuclei through the pituitary stalk into the posterior pituitary gland, where they promote the release of ADH. (For details on the mechanisms of release of ADH and its control, the reader is referred to endocrinology textbooks.) It is clear that, in the release process, intracellular calcium and the actin cytoskeleton are involved—the same elements that control volume regulation at the single-cell level.[242] The ADH concentration in the plasma shows a marked increase when plasma osmolality is changed by less than 1%, and the concentration increases in a linear fashion with further increases in plasma osmolality. In contrast, isotonic volume depletion in the range of up to 10% exerts a much smaller effect on plasma ADH concentration but ADH concentration increases almost exponentially when the changes become larger than 10%.[93]

C. Volume Receptors

The blood volume of 5.5 L is one of the most guarded values in the human body. Because it can change without a change in osmolality or sodium concentration, volume sensors must be active. The "*fullness of the blood stream*," as it was termed by Peters,[191] can be sensed by measuring the wall stretch in a distensible system or by determining the pressure in a rigid system.[191] The vascular systems of the body are a combination of both a very high-capacitance venous low-pressure system and a low-capacitance arterial high-pressure system. The former low-pressure system is comprised of the pulmonary circulation, the right heart, and the systemic venous volume and contains about

3.2 to 3.4 L; the high-pressure system includes the left ventricle and the arterial system.[84] As discussed in detail and with great sophistication by Gauer and Henry,[84] due to the large differences in size and elastic resistance of the two systems, when 1000 mL blood are infused into the circulation, only 5 to 10 mL will be accommodated in the arterial system; the rest will enter the venous system.

Because survival of vertebrates on land requires maintenance of adequate circulating blood pressure without severe hypertension, volume receptors are focused on monitoring blood volume. We can best consider the pathways regulating blood volume by examining the response of a person to hemorrhage, followed by infusion of saline. When blood is lost, the rate of return of blood to the right and left atria of the heart is reduced. The *atrial receptors* are the first to respond even to a small loss in blood volume, as they are affected most prominently.[84] As we will see below, reduction of atrial stretch shuts off the release of salt- and water-wasting hormones such as atrial natriuretic peptide. If the blood loss exceeds 7 to 10% of the estimated blood volume, diminished filling of the ventricles occurs. Because the strength of myocardial contraction rises as ventricular filling is increased, a fall in filling results in a fall in cardiac output. Such a fall in cardiac output stimulates *arterial baroreceptors* in the high-pressure, high-resistance part (carotid bodies and aortic arch) of the circulatory system. Afferent fibers carry stimulatory impulses from activated baroreceptors to the hypothalamus. The hypothalamus then increases sympathetic nerve activity and releases ADH. Increased sympathetic nerve activity restores or maintains arterial blood pressure by increasing heart rate and contractility and by increasing peripheral vascular resistance. In addition, increased sympathetic nerve activity results in the release of a cascade of hormones and mediators, including circulating catecholamines, angiotensin II, and aldosterone, which together increase cardiac output, increase vasoconstriction, and reduce salt and water excretion by the kidneys.[94]

When the hemorrhage is halted and saline is infused, the increased fluid returning to the *right atrium* triggers *stretch receptors*, which are located below the endocardium and have pressure-sensitive endplates.[49] Their stimulation leads to a decrease in the release of ADH from the brain and an increased water diuresis. The signals are carried via vagal fibers to the hypothalamus, and a significant correlation between signal activity and intrathoracic blood volume has been observed.[140] Entry of the vagus into the medulla is followed by, among others, strong connections to the supraoptic regions that control the release of ADH.[141,231] The signals also reach areas involved in the secretion of adrenocorticotropic hormone (ACTH), which in turn leads to a small decrease in the aldosterone secretion.

In addition to activation of reflex arcs, atrial stretch leads to direct release of at least two peptide hormones: *atrial natriuretic peptide* (ANP) and *urodilatin*, which also increase sodium excretion by the kidney, although by different mechanisms (see below). The combined effect of the atrial-induced reflex arcs and release of ANP and urodilatin is to reduce sympathetic nerve activity and the release of its downstream hormones and mediators and to induce vasodilatation and salt and water excretion by the kidneys, tending to reduce intravascular volume.

Although the body focuses on regulating blood volume, we must also consider the regulation of extracellular volume because of the rapid change between the blood compartment and the other compartments that comprise extracellular volume. In the normal operating range, blood volume is about 5.5 L, and the volume of the whole extracellular space is about 17 L. This distribution can change dramatically. When the extracellular volume rises considerably above normal,[94] most of the volume is going into the interstitial spaces, causing edema. Thus, these spaces become overflow reservoirs for excess fluid in which fluid can be stored without deleterious effects on the cardio-vascular system. Because the fluids in the circulatory system and in the interstitial space are connected osmotically by the capillary walls that are permeable to water and sodium, the osmotic and sodium sensors discussed above control not only the blood volume but also the osmolality and volume of the total extracellular body fluid. The distribution of salt and water between the blood and interstitial spaces is determined by the balance between the relatively higher hydrostatic pressure in the capillaries, which tends to drive fluid into the interstitial space, and the oncotic pressure of proteins such as albumin that, under normal conditions, remain trapped in the circulation, thus pulling fluid from the interstitial space back into circulation.

D. Other Important Receptors in the Periphery

Other receptors are located in the small arteries, the afferent arterioles, leading to the glomeruli in the kidney. Baroreceptors located in the walls are part of the powerful renin–angiotensin system. A decrease in blood pressure (or increased sympathetic nerve traffic to the kidney) stimulates the release of renin, which converts angiotensinogen produced in the liver and secreted into the circulation to angiotensin I. This peptide is in turn converted to angiotensin II by the angiotensin-converting enzyme present in high concentrations in the lung; angiotensin II is the physiologically active compound controlling blood pressure and body fluid homeostasis (for details see below).

Receptors are also located at the very point of entry of salt and water into the body—the intestine. Even before the cardiac ANP was discovered, it was observed in studies with experimental animals and humans (Lenanne et al.[143,144]) that an oral salt load caused the intestine to release a natriuretic factor that stimulates kidney function in postprandial periods of salt absorption. Two peptides, guanylin and uroguanylin, have been identified that are synthesized primarily in the intestinal mucosa but induce natriuresis and diuresis in the kidney.[31,51,70,71,73,98] In genetically modified mice lacking uroguanylin, the natriuresis following enteral salt loading is blunted, whereas the response to intravenous loading is maintained.[154] This clearly demonstrates a role for uroguanylin in an enteric–renal communication axis.

IV. HORMONES AS SIGNALS AND THEIR MODE OF ACTION

A. Water and Salt Transport Systems in the Main Target: The Kidney

To facilitate an understanding of the action of hormones involved in osmoregulation, the main salt and water transport mechanisms and their location in the nephron are briefly summarized here (see Figure 12.1). The nephron begins with the glomerulus where the primary urine is formed. The proximal tubule follows in which about 67% of the water and salt are reabsorbed in an essentially isotonic way. The main transport systems mediating sodium transport from the glomerular filtrate/tubular fluid into the cell via the luminal brush-border membrane include the sodium/hydrogen exchanger NHE3; sodium–symport systems, such as the sodium–D-glucose cotransporter SGLT1 and SGLT2; and the sodium–phosphate cotransporter NaPi2. Sodium exit from the cells is mediated, as in all tubular epithelial cells, by the Na,K-ATPase located in the basolateral membrane. Considerable sodium flux also occurs via the paracellular pathways by solvent drag. Water is translocated across the luminal membrane by the water channel aquaporin 1 (AQP1) and also flows paracellularly.

The next main point of salt reabsorption (about 25% of the filtered load) is the thick ascending limb of Henle's loop, where sodium entry into the cell and its paracellular transport are mediated by the sodium/potassium–chloride cotransporter NKCC2. The luminal sodium/proton exchanger NHE3 is also involved in sodium entry. Again, the basolateral Na,K-ATPase plays the predominant role in removing sodium from the cells. The sodium transport in the thick ascending limb provides the main driving force for the urine concentration in the medullary counter-current system. Because the apical membrane of this segment is specialized to render it nearly impermeable to water, no appreciable water reabsorption occurs at this point, and the absorption of solute without water makes the tubular fluid increasingly dilute as it flows along this segment.

The distal convoluted tubule follows, where about 5% of the filtered load of sodium is reabsorbed by yet another sodium transporter, the luminal thiazide-sensitive sodium chloride cotransporter (NCC). Significant water reabsorption (about 13% of the filtered load) also occurs. The luminal aquaporin 2 (AQP2) plays a major role. In the subsequent connecting tubule, the epithelial sodium channel ENaC comes into play—again, about 5% of the filtered sodium is reabsorbed. Water uses AQP2 in the luminal membrane, and about 8% of the filtered water is recovered. In the collecting duct, about 3% of the filtered sodium but about 11% of the filtered water are removed from the tubular fluid, leading to hyperosmotic urine. Again, ENaC, Na,K-ATPase, and AQP2 are

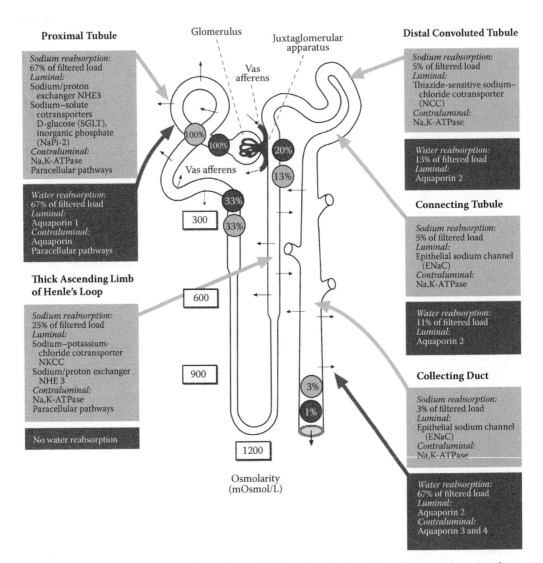

FIGURE 12.1 Scheme of a mammalian nephron with the main sites for sodium (light gray boxes) and water (dark gray boxes) reabsorption and the molecules involved in the transport. The numbers in the circles give the amount of the filtered load remaining after passage of the tubular fluid through this segment. The amount of sodium is given in light gray circles, the amount of water in dark gray circles.

the major players. It should be noted that the scheme presented in Figure 12.1 is quite generalized and that minor shifts in the presence of transporters are seen between species.[6,17,250]

B. ANTIDIURETIC HORMONE

1. Chemistry and Receptors

Antidiuretic hormone is released in its active form as an octapeptide with an intramolecular disulfide bridge. The peptide sequence varies slightly between species. In most mammals, including humans, rats, and dogs, vasopressin contains an arginine in position 8 (arginine–vasopressin), whereas pig vasopressin contains a lysine in position 8 (lysine–vasopressin). The release of ADH, which occurs after a hypovolemic or hyperosmotic stimulus, is followed by its binding to a variety of receptors: V_{1a}, V_{1b}, and V_2. Most important of these receptors for osmoregulation are the renal V_2 receptors.[192]

These receptors are present in the principal collecting duct cells, which are the site of regulation of water excretion by the kidney. Membrane fractionation, perfusion experiments, and immunohistochemical studies have revealed an almost exclusive basal–lateral localization of the receptors.[221] Investigations on several animal models have shown that the effects of V_2 receptor agonists—antidiuresis and increased urine osmolality—are quite similar in all mammalian species investigated, although marked differences between species exist with regard to the relative affinity to various agonists and antagonists.

2. Intracellular Events Elicited by Antidiuretic Hormone

The V_2 receptor is coupled to adenylate cyclase by a heterotrimeric G-protein, G_s. G_s is a guanosine triphosphate (GTP)-binding protein that consists of three subunits: alpha, beta, and gamma. The binding of vasopressin to its receptors causes the alpha subunit to release guanosine diphosphate (GDP) and bind GTP and to dissociate from the trimeric complex. The alpha GTP complex, in turn, activates adenylate cyclase which leads to an increase in the cellular level of cAMP. cAMP increases the activity of protein kinase A (PKA). Protein kinase A then phosphorylates the water channel AQP2 at serine 256.[78,117,175,247,269] This phosphorylation is required for the translocation of AQP2 from intracellular storage vesicles to the apical plasma membrane of the principal collecting duct cells. Like that of the thick ascending limb of Henle, the collecting duct apical membrane exhibits low water permeability in the absence of AQP2. Interestingly, AQP2 forms heterotetramers, and at least three out of the four monomers must be phosphorylated for plasma membrane localization.[113,114,211] The phosphorylation affects only the trafficking of AQP2 to the apical membrane and has no effect on the conductance of the channel for water.[139]

This increased incorporation into the plasma membrane leads to an augmented water flow across the cells that is also facilitated by the basolaterally located AQP3 and AQP4.[107,119] The binding of vasopressin to the V_2 receptor also leads to a rapid increase in intracellular calcium followed by sustained oscillations.[38,61,159,233,267] These phenomena are also important for AQP2 translocation and probably involve calcium released from intracellular stores and a store-operated influx of calcium into the cells. Within AQP2, not only serine 256 but also other regions are important for the proper targeting of the protein to the apical membrane. The proximal region of the C-terminus is essential for localization in the apical membrane, and the C-terminus and N-terminus are essential for trafficking to intracellular storage vesicles and shuttling protein within the cell. The sixth transmembrane domain is also important.[202]

Other cellular elements required for these events are an intact cytoskeleton and a variety of proteins that bind to aquaporin as well as actin and thus determine the direction and speed of trafficking. According to recent models, AQP 2 binds on the vesicle to Sp1, G-actin, and at least 11 other proteins. Sp1 is a specific GTPase-activating protein (GAP) that activates a Ras-related small GTPase, which, besides other physiological roles, is involved in signaling events for the cytoskeleton.[99,176,177] Thus, the intracellular submembraneous actin network can be dissociated locally, allowing the approach of the storage vesicles to the luminal plasma membrane. The other proteins associated with AQP2 could form a force generator complex, providing the machinery to drive the storage vesicles toward the membrane. This model would fulfill the temporal and spatial requirements for targeting the channel to the appropriate membrane.

Compartmentalization of the cAMP signaling is also achieved by protein kinase A anchoring proteins (AKAP18 delta) and a specific phosphodiesterase (PDE4D), which are associated with the aquaporin-studded vesicles and promote a locally defined change in water permeability at the apical membrane and its reversal.[234] The regulation of transport by the targeting of transporters is not unique and is a very important phenomenon in cell biology.[124] The fusion with the plasma membrane appears to follow those events found for the docking and fusion of synaptic vesicles with the presynaptic membrane. Retrieval of AQP2-containing vesicles occurs via constitutive pathways involving clathrin-coated vesicles.[25,26,104,166,222,236,239] Katsura et al.[116] have shown that

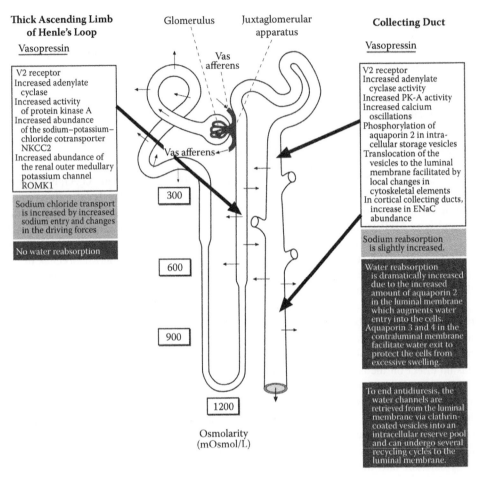

FIGURE 12.2 Sites of action of vasopressin (antidiuretic hormone) on various segments of the mammalian nephron. The resulting effect on sodium transport is given in light gray boxes, the effect on water transport in dark gray boxes. White boxes contain target proteins stimulated by vasopressin.

that this AVP-regulated recycling of AQP2 can occur at least six times with the same repertoire of AQP2 molecules.

In addition to its direct regulation of water transport, AVP also increases sodium transport in the collecting duct via the epithelial sodium channel[11,59,62,63] and urea transport.[210] It also has been shown that in chronic experiments vasopressin increases the abundance of NKCC2 and the renal outer medullary potassium (ROMK) channel in the thick ascending limb of Henle's loop. This leads to an increased sodium reabsorption that could be necessary to maintain the corticomedullary osmotic gradient in view of the increased water reabsorption.[62,63]

The net results of the action of vasopressin on the transport activities in the kidney are summarized in Figure 12.2. The main effect is that water is retained, which lowers the osmolality of the ECF and, in the setting of salt retention, leads to increased ECF volume.

C. THE CARDIAC NATRIURETIC PEPTIDES

1. Chemistry and Receptors

In 1981, de Bold and Sonnenberg showed for the first time that rat atrial extracts contain a potent diuretic and natriuretic factor.[55] This observation led to the isolation of the atrial natriuretic peptide (ANP).[115] ANP is a 28-amino-acid polypeptide with a conserved central core between disulfide-linked cysteines.[9] ANP is primarily found in cardiac atria, where it is stored in secretory granules as a 136-amino-acid prohormone. Upon its release, induced by dilation of the atria and stretch, the prohormone is processed during the release process[58] by a serine protease (corin) into the active peptide. The important role of corin in this event has been recently demonstrated in mice deficient of the corin gene which lack circulating ANP and consequently become hypertensive.[39]

The mechanism by which stretch causes the release of ANP can be summarized as follows: Increases in atrial volume stimulate stretch-activated ion channels in the atrial myocytes which are most likely linked to a G_0 regulatory protein. The formation of prostaglandin $PGF_{2\alpha}$ from arachidonic acid is then stimulated. The prostaglandins increase the release of ANP, and the N-terminal peptides are removed from proANP by the action of corin. Calmodulin/calcium mechanisms also seem to be involved. The stretch-induced ANP release is greatly stimulated by endothelin, which is released from endothelial cells (endocardial cells) after mechanical stretch.[58] The chemical nature of ANP is extremely well preserved in all mammalian species. It has been hypothesized that it evolved as a physiological antagonist to the prevailing salt-retaining homeostatic mechanisms, such as aldosterone, developed when mammals migrated from the salt-rich environment of the seas to the salt-poor environment of the land.[158]

The two main classes of ANP receptors are guanylyl cyclase (GC) receptors and clearance receptors. GC receptors mediate all the known cardiovascular and renal effects; they also are involved in osteogenic and lipolytic effects[138] not covered in this chapter (see Maack[157] for further details). The clearance receptors play an important role in the removal of ANP from the circulation as well as in modulating the local tissue concentration. The GC receptors are unique receptors composed of homodimers that contain in a single molecule an extracellular recognition site for the hormone, a transmembrane domain, and, in the intracellular cytoplasmic domain, the enzyme GC separated from the transmembrane domain by a tyrosine-kinase-like sequence that controls the GC activity.[157,220,255] In the absence of ANP, the GC is inhibited, and the intracellular level of cGMP is very low. Upon binding of ANP to the receptor, conformational changes occur that result in an allosteric desinhibition of the GC activity, and the intracellular cGMP concentration increases.[80,81] The conformational changes involve a unique rotational mechanism. In the dimer, ANP binding causes a twist motion of the two extracellular domains. This motion promotes a rotation of the two domains which now face each other with their ANP-binding domains. This rotation is transduced across the transmembrane helices and reorients the two intracellular domains. This reorientation of the intracellular domains brings the two active sites of the GC into the optimal proximity for catalytic activity, thereby giving rise to the generation of cGMP by the enzyme.[167] After this activation, the affinity of the receptor for ANP decreases markedly and ANP dissociates from the extracellular ligand-binding site. The stimulation of the receptor is terminated, the cyclase is inhibited again, and the cGMP level in the cell returns to its normal low level due to the activity of the phosphodiesterase.

Natriuretic peptide clearance receptors (NPCRs) are much more abundant in many tissues than the GC receptors.[29] They have a single membrane-spanning domain and a very short cytoplasmic domain of 31 amino acids. This short cytoplasmic domain is characteristic of all receptors that transport peptides and proteins into cells. NPCR are apparently not only involved in the removal of ANP from the plasma but also have an additional physiological role, as shown recently in mice in which the NPCR gene had been deleted.[163] After binding to the NPCR, ANP is taken up into the cell by receptor-mediated endocytosis and hydrolyzed in the lysosomes. The free NPCR then returns to the cell surface[47] and can scavenge another ANP from the circulation. A detailed review of other homologs of ANP and their receptors is given in Pandey.[187,188]

2. Cellular Events Elicited by Atrial Natriuretic Peptide

The increase of cGMP elicited by atrial natriuretic peptide (ANP) modulates cellular functions via regulation of specific downstream targets, such as cGMP-gated channels, cGMP-dependent protein kinases, and cGMP-regulated phosphodiesterases. As summarized in Figure 12.3, ANP in the kidney

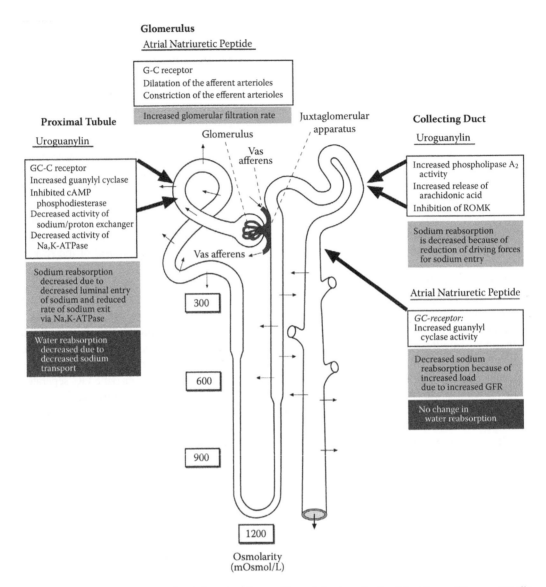

FIGURE 12.3 Sites of action of atrial natriuretic peptide and the intestinal natriuretic peptide uroguanylin on various segments of the mammalian nephron. The resulting effects on sodium transport are given in light gray boxes, the effect on water transport in dark gray boxes. White boxes contain target proteins affected by the peptides.

increases the glomerular filtration rate (GFR), mainly via an increase in glomerular capillary hydrostatic pressure that results from dilation of the afferent arteriole and constriction of the efferent arteriole.[158] The natriuretic effect results from an increase in sodium load to the base of the inner medullary collecting duct, an effect that is essential for a robust natriuretic response and disruption of the load-reabsorption balance in this nephron segment.[9,158] ANP also inhibits sodium reabsorption in the renal tubules, attenuating the increase in salt reabsorption caused by angiotensin in the proximal tubule and reducing sodium-channel-mediated salt transport in the collecting duct. The direct tubular effects occur at lower ANP concentrations; however, full, robust diuresis and natriuresis require a synergistic response that includes both the hemodynamic and tubular effects. ANP is also a powerful antagonist of the renin–angiotensin–aldosterone system, which is discussed below. Urodilatin, a peptide hormone very similar in sequence and structure to ANP, has also been implicated as a potential candidate for a peptide modulating the water and salt handling of the kidney.[105]

D. Intestinal Natriuretic Peptides

1. Chemistry, Receptors, and Cellular Events Elicited by Guanylin Peptides

As reported above, strong experimental evidence indicates that guanylin peptides are involved in body fluid homeostasis and sodium balance. They appear very early in evolution.[130–132] The peptides have 15 to 19 amino acids, and *guanylin* (GN) and *uroguanylin* (UGN) have two intramolecular disulfide bonds connecting cysteines at position 4 and 12 and at position 7 and 15, respectively. Human GN consists of 15 amino acids, and human UGN has 16 amino acids, with major differences in the three N-terminal amino acids. Various mammalian species exhibit differences of 1 to 3 amino acids for the two peptides. The regions close to the cysteines are thereby strongly conserved (for a summary, see Sindic and Schlatter[228]). Both peptides are synthesized as prepropeptides. PreproGN is 115 amino acids long, proGN has 94 amino acids,[56,260] preproUGN has 112 amino acids, and proUGN has 86 amino acids.[160,168] In plasma, GN is present in its inactive proGN form at a concentration of 30 to 40 pM.[122,135,169] In contrast, 50 to 90% of the UGN circulates in its active form at a concentration of 5 to 7 pM.[122,171] The main source of proGN, proUGN, and UGN is the small intestine,[136,172] although mRNA for the peptides has also been found in other tissues (for a summary, see Sindic and Schlatter[228]). The peptides are freely filtered in the glomerulus[121] and appear in the primary urine. In the proximal tubule, proGN is degraded by brush-border membrane-bound peptidases or endocytozed. ProUGN can be activated by the same peptidases without further breakdown. Thus, only UGN appears in the final urine of rats and is the main GN peptide found in opossum urine.[228]

It is interesting that GNpeptides exert both local effects on the intestine and effects in the kidney. The former are exerted from the luminal side by stimulation of a brush-border guanylate cyclase C (GC-C) receptor. The stimulation leads to an increase in intracellular cGMP, which in turn inhibits cAMP phospodiesterase III and thereby also increases intracellular cAMP concentration. The cGMP-activated protein kinase G II and the cAMP-activated protein kinase A increase the secretion of chloride, bicarbonate, and water via activation of the cystic fibrosis transmembrane conductance regulator (CFTR). Furthermore, sodium absorption mediated by the sodium/proton exchanger NHE is inhibited. This acts as a first-defense mechanism against excessive salt and volume uptake (for a summary, see Sindic and Schlatter[228]). As also summarized in Figure 12.3, in the kidney UGN present in the primary urine or generated in the kidney by tubule cells can modify salt and water transport. At the moment, evidence tends to support the latter paracrine mechanism;[228] for example, UGN mRNA increases with salt load in mice and rat kidneys.[74,186]

The intra-nephron distribution of receptors for UGN seems to vary in the kidneys of various species, but GC-C is probably expressed in the proximal tubule and the collecting ducts, the two most important segments in the regulation of water and salt excretion. In proximal tubule cells in culture (OK from opossum kidney, PtK2 cells from kangaroo, and IHKE1 from human kidney), GNP peptides induce an increase of cGMP.[60,72,153,230] cGMP inhibits sodium transport in the proximal tubules.[149] As described above for the action of GNP on the sodium transport of the intestine, the driving forces for sodium movement—the electrical potential across the membrane and the entry and exit steps—are potentially affected (see Figure 12.3).[34,66,230]

In addition, UGN activates another receptor that belongs to the clearance receptor family: NPR-C. Evidence for such an additional receptor pathway has been provided in *in vitro* studies as well by the fact that in GC-C-deficient mice GN peptides still cause natriuresis, kaliuresis, and diuresis.[35,229] These receptors provide the signaling pathway for UGN in the cortical collecting duct, where sodium reabsorption via the epithelial sodium channel ENaC takes place. After binding to the receptor the membrane-bound phospholipase A$_2$ is activated, and arachidonic acid is released from the phospholipids of the membrane. The arachidonic acid in turn inhibits the ROMK channel for potassium located in the same luminal membrane.[229] Inhibition of the channel depolarizes the cells and reduces the driving force for sodium entry; thus, transcellular sodium transport decreases. The cortical collecting ducts (and later parts of the collecting duct) probably also exhibit luminal chymotrypsin activity that destroys the guanylin synthesized and released by the tubular segments located upstream.[77] The actions of the natriuretic peptides on the kidney are summarized in Figure 12.3; the net effect is a loss of sodium (and water), which reduces the increased extracellular fluid volume.

E. The Renin–Angiotensin–Aldosterone System

In one sensor system both blood volume and sodium concentration are the signals that evoke a homeostatic response. It is located in the kidney in the afferent arterioles of the glomerulus which are connected to the juxta-glomerular apparatus (JG). This system senses both blood volume (in the form of pressure exerted on the arterioles) and the sodium chloride concentration in the fluid entering the distal tubule. The juxta-glomerular apparatus consists of specialized myoepithelila cells (the macula densa), which are connected via specific baroreceptors to the afferent arterioles of the glomeruli. In addition, they are exposed to the distal tubular fluid and sense the sodium chloride concentration via the Na–K–2Cl transport system, as furosemide, a specific blocker of the transporter, inhibits glomerulotubular feedback.[24] Glomerulotubular feedback—by which sodium chloride load in the distal tubule is controlled by changes in GFR involving constriction of the afferent arterioles to the glomerulus—constitutes a local phenomenon probably mediated by adenosine; however, when blood pressure drops (or blood volume is decreased), renin is released from the JG into the blood, and the renin–angiotensin–aldosterone system (RAAS) is activated.

1. Renin

a. Chemistry and Action

Renin, a 340-amino-acid aspartylprotease, is produced in the macula densa cells initially as the precursor prorenin. The prorenin accumulates in a pseudocrystal form in the protogranules that bud from the *trans*-Golgi network. These protogranules are modified by the addition of membrane material to immature granules that can be secreted by the constitutive pathway and account for the fact that 80% of the peptides present in the plasma are prorenin. During intracellular maturation, 23 amino acids (presegment) and then 43 amino acids (prosegment) are removed, and the mature granules that contain only renin are subjected to regulated exocytosis (for a review, see Reudelhuber[198]). This secretory pathway is not the normal pathway observed in other peptide-secreting cells; the reason for this difference is unclear but might be related to the specific adaptation of the myoepithelial cells to their function as peptide-secreting cells. Regulation of renin release involves intracellular calcium as well as the adenylate cyclase. In addition to the changes in secretion induced by low pressure and low sodium chloride, renin release is also influenced by neural mechanisms. Increased sympathetic nerve activity increases the release of renin, which mediates the response via activation of the adenylate cyclase system. Several hormones also modify renin release. Circulating catecholamines increase renin release. As part of a feedback mechanism to maintain appropriate activity of the renin–angiotensin system, elevation of angiotensin II decreases renin secretion. ANP also inhibits renin release, both by reducing sympathetic nerve activity (and circulating catecholamines) and by direct effects on the macula densa.

2. Angiotensin II

a. Chemistry and Receptors

In the blood, renin acts on angiotensinogen, which is produced predominantly in the liver. Angiotensinogen is a 452-amino-acid glycoprotein of varying molecular weight (between 61.2 and 65.7 kDa). Its sequence has homology to the serpin family, which is thought to be derived from an ancestral gene coding for a serine protease inhibitor. Other members of this family include α1-antitrypsin and antithrombin III. Thus, during evolution, a former inhibitor has been modified into a precursor for an active hormone.[156,241] Renin cleaves 10 amino acids of angiotensinogen off of the N-terminal to release angiotensin I; angiotensin I is further processed by the removal of eight amino acids by the angiotensin-converting enzyme (ACE) to angiotensin II (ANGII). ACE is located mainly in the lung but has recently also been found in the kidney together with other angiotensin-converting enzymes.[28] Because other tubular sites of angiotensin production have been identified, the possibility of an intracrine renal angiotensin II signaling mechanism exists.[270]

It is agreed that most of the functional responses to ANGII are mediated by the AT1 receptor. The AT1 receptor is a member of the large super family of G-protein-associated receptors that have seven discrete transmembrane alpha-helices. Similar structures are found in rhodopsin and, for example, in the vasopressin receptor. The extracellular hormone binding domain is critically determined in its structure by disulfide bridges connecting extracellular loops of the protein.[18,44,262] The binding of angiotensin II thereby appears to occur at the junction of the extracellular space and the plasma membrane. Intracellularly, the receptors couple to heterotrimeric G-proteins that are regulated by GTP.[32] Stimulation of the AT1 receptor activates phospholipase C, the opening of a dihydropyridine-sensitive voltage-gated calcium channel, and inhibition of the adenylate cyclase.[37,179] Other so-called AT2 receptors also show species-dependent subtypes and differ in their binding specificity and tissue distribution.[18,23,41,92] The AT2 receptors have been discussed as functional agonists of AT1 receptors. Their role in the regulation of functional properties of the proximal tubule remains to be determined as well as the role of the intracellular angiotensin found in proximal tubule cells.[96,146,270]

b. Cellular Events Elicited by Angiotensin II

Angiotensin II has multiple actions on a variety of tissues.[41] It stimulates salt and water uptake in the proximal tubule and the gut, stimulates aldosterone secretion from the zona glomerulosa in the adrenal gland, increases the resistance in the afferent and efferent arterioles of the glomeruli, and increases thirst (i.e., is dipsogenic). More specifically, as summarized in Figure 12.4, in the proximal tubule angiotensin II increases Na,K-ATPase activity as well as bicarbonate transport.[83] It also increases sodium–D-glucose cotransport and thereby fluid reabsorption.[82] Other sodium–solute cotransport systems affected by angiotensin include the luminal sodium/proton exchanger NEH3 and the contra-luminal electrogenic sodium/bicarbonate transporter NBC1, the main transporters involved in bicarbonate reabsorption in the proximal tubule (see Boron[22] for a review). Evidence suggests that in long-term experiments angiotensin promotes the expression of both the Na/H exchanger as well as the sodium–bicarbonate cotransporter.[137,146,245,264] Also, the medullary TALH is a target of angiotensin-augmenting expression of NaKCC 2 as well as of luminal NHE3 and the basolateral electroneutral sodium–bicarbonate cotransporter BSC1.[137] The effect on the latter two systems might be secondary to the change in bicarbonate reabsorption in the proximal tubule as the response was blocked by acid loading of the animals. The effects of angiotensin on the kidney are summarized in Figure 12.4. The net effect of the action of angiotensin on the kidney is the increased absorption of sodium and water to reestablish the vascular volume to normal levels.

3. Aldosterone

a. Chemistry, Receptors, and Prereceptor Specificity

Aldosterone belongs chemically to the group of C21 steroids from which both the glucocorticoids and the mineralocorticoids are derived. The mineralocorticoid aldosterone is chemically distinct from the glucocorticoids in that it carries an aldehyde function at C18, which is introduced into the molecule by the aldosterone synthase cytochrome P450 (CYP11B2 or P450c18).[112] This enzyme is regulated in its expression in the cells of the zona glomerulosa of the adrenal gland by sodium restriction in the diet.[112] When sodium is restricted, mRNA for this enzyme and the enzyme activity

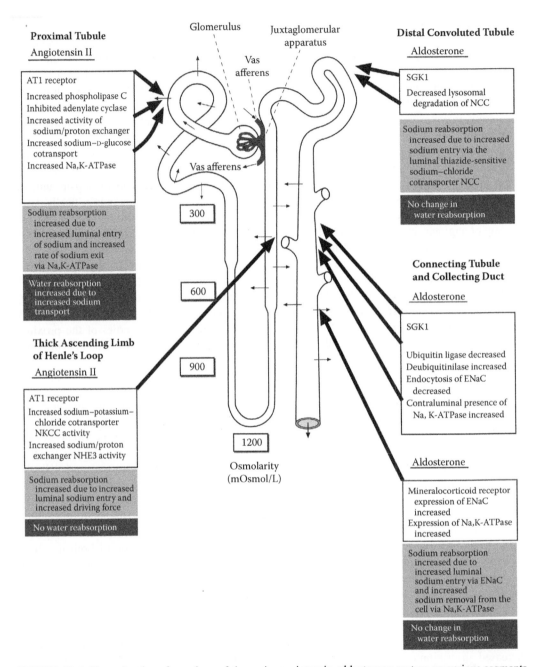

FIGURE 12.4 Sites of action of members of the renin–angiotensin–aldosterone system on various segments of the mammalian nephron. The resulting effects on sodium transport are given in light gray boxes, the effect on water transport in dark gray boxes. White boxes contain target proteins affected by the hormones.

itself increase drastically; at the same time, aldosterone levels in the plasma rise. This response is very specific and not observed for a key enzyme in glucocorticoid synthesis (CYP11B1 or P45011β) and is mediated by an increase in plasma angiotensin II, as the response is completely inhibited by antagonists of angiotensin II receptor AT1.[112] Interestingly, the expression of the AT1 receptor also increases under these conditions, providing a potentiation of the angiotensin II regulation of the enzyme. The stimulation of the aldosterone synthase gene by angiotensin II involves a variety of transcription regulatory genes, which are currently being identified.[205]

At the level of the target cells, aldosterone exerts genomic and nongenomic actions.[256] The former require at least 45 minutes to occur in experimental setups and are blunted in the presence of inhibitors of mRNA and protein synthesis; the latter occur much more quickly and initially are not dependent on transcription and translation. The genomic actions involve a *mineralocorticoid receptor* that was cloned in 1987.[8] The human receptor belongs to the steroid/retinoid/orphan receptor family of nuclear transactivating factors. It is the largest member (984 amino acids) in the family; in particular, the N-terminus is very variable. In contrast, the central DNA binding domain is the area of maximally conserved identity over a stretch of 66 to 68 amino acids. It contains nine invariant cysteines, eight of which complex two zinc ions to form the zinc fingers that interact with high specificity with the DNA. The hormone binding site is located in the C-terminus and shows only about 50% identity with the glucocorticoid receptors in the mineralocorticoid family,[256] whereas the DNA binding domain has a 90% identity.[106]

Functional expression studies of the receptor revealed a surprising result. As expected, the receptor had a high affinity to aldosterone ($K_d = 1.7$ nM), but it did not discriminate well between mineralocorticoids and glucocorticoids. Thus, in view of the simultaneous presence of the glucocorticoid cortisol and the mineralocorticoid aldosterone in the plasma, other means of cortisol exclusion from the receptor in mineralocorticoid target tissues had to be found. This crucial *prereceptor-specificity*-conferring role is provided by an enzyme that inactivates the glucocorticoids but not aldosterone in the aldosterone-sensitive tissues.

This enzyme is the 11-β-hydroxysteroid dehydrogenase (11βHSD2),[2,4] which has a very high affinity for cortisol and cortisone, the main glucocorticoids in humans. The enzyme is present in high concentrations in the epithelial (kidney, colon, salivary glands, sweat glands) cells, which are the target of aldosterone. In evolution, it is at the stage of the first appearance of terrestrial mammals that aldosterone acquires this specific mineralocorticoid function due to the simultaneous expression of the mineralocorticoid receptor and 11βHSD2 in salt-transporting epithelial cells. This development therefore helped in the conquest of land[50] by mammals requiring specific measures for salt preservation. In addition to the need for 11βHSD2 in salt-transporting epithelia, recent results indicate that the receptor itself has some domains that interact specifically with aldosterone, so the profile of responses elicited by aldosterone and glucocorticoids in the cells differs somewhat.[75,204]

b. Early Actions on Target Cells and Transporter Trafficking

The first report on a rapid action of aldosterone on the urinary electrolyte secretion in dogs was published in 1958 by Ganong and Mulrow[79] and was confirmed in rats in 2005 by Rad et al.[197] In renal cells in culture, aldosterone was shown to rapidly increase intracellular calcium and pH, the reason for the change in intracellular pH being the activation of a proton conductance and thereby an increased rate of sodium/proton exchange. The response appears to be mediated by protein kinase C-alpha (PKCα). In some studies, an increase in cAMP[30] was also reported. Similar results have been obtained in colon tissue and cells.[76] It remains to be established how these fast effects are integrated into the regulatory role of aldosterone for sodium, intravascular volume, and potassium homoeostasis and whether or not they are also mediated by the mineralocorticoid receptor.

One of the longer lasting effects of aldosterone on tight epithelia in the kidney is to increase the sodium entry into the cells at the luminal cell side mediated by the epithelial sodium channel ENaC and to accelerate the removal of sodium from the cell at the contraluminal side by the primary active Na,K-ATPase (see above). One of the earliest mediators of these changes appears to be a member of the serine/threonine kinase family, the SGK1 (for serum/glucocorticoid-regulated kinase).[43,151,170] *In vivo* and *in vitro* it has been demonstrated that within 30 minutes aldosterone increases the amount of SGK1 mRNA in cortical collecting duct cells[173] and in A6 toad kidney cells. In the same cells and in rat kidney also, an increase in the level of enzyme protein was observed. The increased transcription and expression of SGK1 involved the mineralocorticoid receptor, as it was inhibited by a specific antagonist (ZK91857).[43]

The increased presence of sodium channels in the luminal membrane of the sodium-transporting cells is at least partly brought about by a reduced rate of endocytosis in the normal exo-/endocytosis trafficking process of the channel. Apical targeting of the channel is controlled by the ubiquitin ligase Nedd4-2; this enzyme is inhibited by phosphorylation through SGK1, so retrieval of the channel from the membrane is slowed down.[189] It appears that the N-terminal of SGK1 is necessary for the stimulatory effect on the ENaC-dependent sodium current.[174] SGK1 also seems to be involved in the targeting of Na,K-ATPase to the membrane when the intracellular pool of the enzyme has been increased by aldosterone.[268] In addition, low sodium intake (increased levels of aldosterone) drastically reduces ubiquitin ligase Nedd4 immunostaining in aldosterone-sensitive distal nephron segments.[16,152] Recently, deubiquitinilase, an enzyme that removes ubiquitin already attached to ENaC, was found to be expressed in mouse distal nephrons. This enzyme is also induced by aldosterone and removes from ENaC signals that target it for rapid removal from the apical plasma membrane.[65] The increased expression of SGK1 in the cortical collecting tubules is a primary event in the early antinatriuretic and kaliuretic responses to physiologic concentrations of aldosterone. Induction of alpha-ENaC mRNAs may play a permissive role in the enhancement of the early responses; these effects may be necessary for a full response but do not by themselves promote early changes in urinary Na$^+$ and K$^+$ excretion.

Another important member of the chain of events leading to the effect of aldosterone on its target tissues is another kinase, WNK4 (without N [lysine] kinase). This kinase is expressed in the distal nephron and other salt-transporting epithelia[65] and has been shown to be a regulator of each of the major Na, Cl, and K flux pathways.[110,261,265] The enzyme inhibits epithelial sodium channel expression when coexpressed with ENaC RNAs in *Xenopus* oocytes.[201] Interestingly, no kinase activity is required; the enzyme seems to interact directly with the carboxy termini of the beta and gamma subunits of ENaC, which contain the signals for targeting the proteins to the apical membrane. Similarly, WNK4 interacts with the carboxyl terminus of thiazide-sensitive NCC. It is not clear whether or not the kinase activity of WNK4 is necessary for inhibition of NCC expression.[261,265,266] In this instance, the reduced expression of the transporter at the cell surface is probably caused by a stimulation of lysosomal degradation, which leads to an increased rate of removal from the membrane.

In addition to sodium channels, the potassium channel located in the distal tubule (ROMK) is regulated in its membrane surface expression by WKN4.[103] Here, WNK4 stimulates clathrin-dependent endocytosis of ROMK1. The stimulation of endocytosis of ROMK1 by WNK4 requires specific proline-rich motifs of the WNK and does require its kinase activity. WNK4 interacts directly with ROMK1 as well as with intersectin, a multimodular endocytic scaffold protein; both interactions are crucial for the stimulation of clathrin-dependent endocytosis of ROMK.[103] Thus, for each sodium transport protein a different mechanism exists for controlling the expression of the active transport proteins on the membrane surface.[200]

Aldosterone levels are increased in response to two physiological stimuli: intravascular volume depletion, as discussed thus far, and high plasma K levels. The restoration of intravascular volume requires increased renal NaCl reabsorption but should not lead to a change in potassium secretion. High potassium levels in the plasma should be compensated for by an increased renal excretion of potassium but ideally not by a change in sodium chloride reabsorption. In the collecting duct, increased reabsorption of sodium (as occurs following aldosterone stimulation) via apical channels leads to an increasingly negative potential (lumen compared with basolateral membrane), which favors potassium secretion via apical membrane ROMK channels. Thus, the question arises as to how the kidney distinguishes between hypovolemia and hyperkalemia to achieve the proper physiological response. The discriminatory element orchestrating the differential responses appears to be WNK4.[110] It is postulated that WNK4 has at least three different states. At the basal state (WNK4.1), in the distal tubule, the kinase inhibits the plasma membrane surface expression of both ENaC, the main element in sodium reabsorption, and ROMK, the potassium channel essential for potassium secretion. Low intravascular volume—associated with increased plasma aldosterone and angiotensin II levels — leads to the appearance of WNK4.2, which alleviates the inhibition of ENaC plasma membrane surface expression but augments the inhibition of ROMK plasma membrane surface expression, thus promoting a net increase in sodium reabsorption without a concomitant potassium loss. Hyperkalemia results in increased aldosterone signaling via SGK1 (see above). SGK1 phosphorylates WNK4 at a specific site (WNK4.3 state), and the inhibitory effect of WNK4 on both ENaC and ROMK plasma membrane surface expression is lost. Potassium secretion is stimulated by a higher density of the potassium channels in the luminal membrane of the distal tubular cells as well as by the increased intraluminal negative transmembrane electrical potential difference, which adds to the driving forces for potassium secretion.[238] That such combinations can indeed exist is evident from investigations on pseudohypoaldosteronism type II, where mutations in WNK4 have been found that feature increased NaCl reabsorption and decreased potassium secretion.[201] Because more recent studies indicate that other isoforms of the WNK kinase family interact with specific transporters and with WNK4 itself, it appears that the scheme presented above represents a simplification that will grow more complex with further study.[89,142,200,238]

c. Induction of Transport Systems by Aldosterone

To mediate a graded response to hypovolemia, both in time and in magnitude, aldosterone not only drives the trafficking of transporters from intracellular pools to the cell surface but also increases the size of the intracellular pool of transporters by stimulating *de novo* synthesis of transporters or components of them. Initial observations on the mechanism of action of aldosterone on sodium transport were made in the toad bladder and in the A6 cell line derived from toad kidney, and, for the first time, the dependence of the action of the hormone on mRNA and protein synthesis was observed (see review by Edelman 1981[64]). Investigators initially concentrated only on Na,K-ATPase, the sodium transport system in which activity could be measured biochemically and antibodies were available. Thus, Geering et al.[85,88] showed that the rate of biosynthesis of the alpha and beta subunits of the enzyme increased about 2.5-fold in toad bladders after 18 hours of exposure to aldosterone. They also observed that this effect was not dependent on increased sodium entry into the cells. The induction of enzyme activity (increase in total cellular pool) seems, however, to be rather low, so changes in surface expression (see above) are probably the main factors increasing the exit step in transepithelial sodium transport.

The main rate-limiting step for transepithelial sodium transport is sodium entry into the cells mediated by ENaC. The channel consists of three subunits, alpha, beta, and gamma. The alpha subunit is thought to play an essential chaperone role in the trafficking of the channel to the cell surface[20] in addition to forming part of the sodium channel pore. The three subunits are affected differently by moderately low and high sodium intake or by differences in the aldosterone level in the plasma. In the kidneys of mice kept on a high-sodium diet, the alpha subunit was undetectable in the late portion of the distal tubule and the medullary collecting duct, whereas the beta and gamma subunits were clearly detectable, but only in the cytoplasm. For the low-sodium diet, two phenomena occurred: (1) induction of the alpha unit, which appeared in the apical membrane and subapical vesicles, and (2) redistribution of the beta and gamma subunits to the same location, suggesting the assembly of active sodium channels and their insertion into the apical cell membrane.[150]

It is interesting to note that this mechanism differs from the effect of vasopressin on the sodium channel (i.e., vasopressin-induced upregulation of channel abundance involves induction of the beta and gamma subunits of the channel).[59,63] In another aldosterone-sensitive segment of the nephron (the late distal convoluted tubule), the thiazide-sensitive sodium chloride cotransporter (NCC) is the main route for sodium entry into the cells. This transporter is also an aldosterone-induced protein. In rats on a low-sodium diet, a large increase in transporter protein was found in plasma membranes isolated from the kidney cortex, where this segment is located, but not in membranes isolated from the renal medulla. Within the cells, the increased expression was predominantly found in the apical membranes.[118]

This finding points to the important role of the late distal convoluted tubule and the connecting tubule to achieve sodium and potassium balance. This is also evident from studies where the alpha-ENaC gene was specifically deleted in the collecting duct of transgenic mice. These animals survived well and were able to maintain sodium and potassium balance even under conditions of salt restriction.[207]

As mentioned above, aldosterone levels also increase during hyperkalemia and the kidney responds with an increased excretion of potassium. The main mediator of this secretion is ROMK. Adrenalectomy downregulates ROMK in the renal cortex of rat kidneys, parallel with the decrease in the alpha and beta subunits of Na,K-ATPase.[253] In the same vein, aldosterone increases ROMK mRNA in rat kidney. Interestingly only some of the known isoforms are affected—namely, ROMK2, 3, and 6.[12]

Despite the wealth of knowledge regarding possible targets for aldosterone and its possible genomic and nongenomic signaling pathways (which are partially summarized in Figure 12.4), the exact sequence of events with regard to temporal and spatial resolution and to permissive vs. determining roles are still not completely resolved (see, for example, Muller et al.[170]). Even the involvement of glucocorticoid receptors in the mineralocorticoid response of some nephron segments has been postulated recently. This might be an expression for redundancy in control systems often used in biology. It seems clear, however, that under extreme conditions such as low-sodium diets the mineralocorticoid receptors in the principal cells of the collecting duct and in the cells of the late connecting tubule are necessary to elicit the desired renal response of sodium retention. Mice without mineralocorticoid receptors in these cells show increased sodium and water excretion under sodium restriction compared to control mice.[206]

Also, the parts of the tubule where the early and late effects of aldosterone are observed may differ within the aldosterone-sensitive distal nephron segments.[207] These include the end of the distal convoluted tubule (late DCT), the connecting tubule (CNT), and the collecting duct (CD), with its subdivision of outer medullary collecting duct (OMCD) and the inner medullary collecting duct (IMCD) (see below).[151] Last, but not least, it has also been noted that mineralocorticoids regulate the activity of mitochondrial enzymes such as citrate synthase, the entry step into the tricarboxylic acid cycle,[120,125] thereby providing more cellular energy for the active reabsorption of sodium by these cells.

V. BEHAVIORAL AND METABOLIC RESPONSES TO CHALLENGES OF WATER AND ELECTROLYTE HOMEOSTASIS

A. THIRST AND SALT APPETITE

Thirst and salt appetite are behavioral responses to losses of water and salt and, in concert with the reflex neural and endocrine responses discussed above, are critical for reestablishing body-fluid homeostasis. Like their counterparts, they are under the control of influences arising from changes in intravascular volume or osmolality in the extracellular fluid, either directly or indirectly via endocrine factors such as angiotensin and aldosterone that are induced in the early attempts of the body to regain homeostasis. Their integration is depicted in Figure 12.5.

Osmoreceptors and sodium receptors controlling thirst and salt appetite are present in both the brain and in the periphery. In the goat, direct injection of hypertonic saline into the carotid artery elicits drinking.[7] These central osmoreceptors are located in the circumventricular organs that lack the blood–brain barrier and thus can sense changes in blood osmolality. Peripheral receptors are located in the splanchnic regions and include sensors in the stomach, the small intestine, and the portal veins.[129,237,243,244]

Defining the effect of hormonal effectors on thirst and salt appetite centers is complicated by the fact that it is difficult to distinguish between direct effects on brain behavioral centers and effects that result from the peripheral action of the hormones during disturbances such as sodium retention

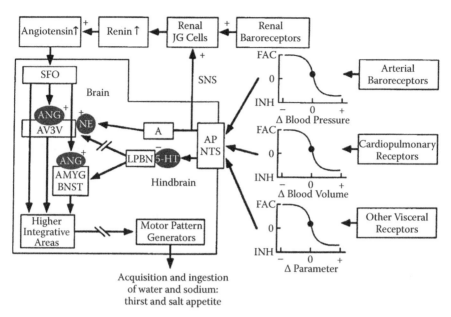

FIGURE 12.5 Diagram depicting the nature of neural and hormonal inputs into the brain and the central neural pathways that mediate sensory integration of signals for the generation of drinking (thirst) and sodium ingestions (salt appetite). Both inhibitory and excitatory inputs from the periphery derive from arterial and cardiopulmonary baroreceptors and probably other visceral receptors. Information carried in afferent nerves projects mainly to the nucleus of the tractus solitarius (NTS) and the area postrema (AP). Angiotensin (ANG) acts in the form of angiotensin II on angiotensin type 1 receptors in the subfornical organ (SFO). Information is then processed within various area of the brain and integrated in higher integrative areas. As messengers angiotensin, norepinephrine (NE), and serotonin (5-HT) are involved. From these areas the behavioral changes are activated. Abbreviations for brain areas involved: AV3V, anteroventral third ventricle; AMYG, nucleus amygdalis; BNST, bed nucleus of the stria terminals; LPBN, lateral parabrachial nucleus. (For a detailed discussion, see Johnson and Thunhorst.[109])

and changes in blood volume or blood pressure. Thus, renin[67,68] and angiotensin II are dipsogenic (i.e., they induce drinking of water) as is central cholinergic activation. Drinking is inhibited by central release of serotonin, by distension of the stomach and the small intestine,[237] and by ANP.

Angiotensin II is also natriorexigenic (i.e., it increases salt appetite). The relative importance of peripheral angiotensin (generated in the bloodstream) vs. central angiotensin (generated in the brain) in this response apparently differs from species to species. Thus, in rats, central angiotensin seems to be more important, whereas in sheep peripheral angiotensin seems to be the more effective stimulus. Aldosterone also enhances sodium intake. An antinatriorexigenic response is elicited by serotonin in the nucleus amygdalis.[156] Furthermore, the natriorexigenic response to angiotensin II is in some instances attenuated by oxytocin.[19]

What becomes clear from reviewing the literature is that there exist close connections between the peripheral control systems and autocrine regulation within the brain. In this context, it is of note that the brain contains aldosterone as well as angiotensin-sensitive neurons and that it itself is capable of producing angiotensinogen and processing it to angiotensin II. The main communication point between the brain interior and hormones in the plasma is the subfornical organ, which is part of the circumventricular organs, which are devoid of the blood–brain barrier that otherwise would impede the passage of the peptide angiotensin II to its neuronal receptor. At these neuronal receptors, angiotensin II is apparently capable of eliciting two different responses, depending on the signaling pathway activated. The inositol triphosphate (IP3) pathway mediates the dipsogenic response, whereas the mitogen-activated protein (MAP) kinase pathway is responsible for the enhanced salt appetite.[52] The points where inhibitory and excitatory signals from the peripheral baroreceptors and other visceral receptors enter the brain network are the area postrema and the nucleus tractus solitarius. Recently, the nucleus solitarius has also been shown to contain neurons that express both mineralocorticoid receptors and HSD2, which makes them selectively responsive to aldosterone.[86] They are located in an area where, again, the blood–brain barrier is lacking, thus exposing the neurons to the bloodborne hormone. Interestingly, they receive signals not only via the aldosterone level but also other signals that remain, for example, after rats have been adrenalectomized.[87] Mineralocorticoids are also involved in regulating the activation of oxytocinergic and vasopressinergic neurons.[203] This indicates that ADH also fine-tunes the behavioral responses.

Noradrenergic inputs and the regulatory effects of 5-hydrotryptamin (serotonin) action on the nucleus amygdalis[155] are processed in higher integrative areas.[109] The current view of these complex interactions is summarized in Figure 12.5. Further studies are necessary to reveal at a cellular and molecular level the important reactions of the body that regulate the intake side of water and electrolyte metabolism.

B. Metabolic Water

1. Marine Mammals

In some habitats behavioral responses cannot be satisfied, such as when no freshwater is available. Under these circumstances, water absent from the diet is generated from the oxidation of nutrients and is critical to osmoregulation. As already stated above, about 1 mL of water is formed per gram of fat oxidized and about 0.6 mL per gram of carbohydrate. Metabolic water is of outmost importance under two apparently very different habitats: the sea and the desert. As a side note, metabolic water is also very important during hibernation.[54]

Mammals living in the sea, such as pinnipeds (seals and sea lions) and cetaceans (whale, dolphins, and porpoises), have only very limited access to free water and therefore depend heavily on water generated metabolically. In addition, although their kidneys show very specific features such as reniculations (i.e., each kidney is made up of hundreds of individual lobes), which allow for a large volume of urine to be processed, and an increased thickness of the medulla, their concentrating ability is only twice as high as that found in humans—between 1200 and 2400 mOsm/kg compared to 1400 in humans.[182] Thus, the kidney can produce urine that has a higher osmolality and salt content than seawater and thereby regain free water, but a much higher potency would be expected from the structure of the kidney and the extreme environmental salinity (see discussion below).[249] The ability of these animals to maintain their osmolality without relying solely on the kidney as the main osmoregulatory organ results from the fact that these mammals do not drink seawater and instead rely on dietary water and on water generated from their extensive fat reserves; for example, 600 mL of water are generated from endogenous fat stores by a seal with a body weight of 120 kg.[145] The need for metabolic water to maintain a positive water balance becomes even more important during periods of fasting, which are a natural component of the life history of pinnipeds.

The duration of (terrestrial) fasting varies among species (for a review, see Riedman[159]); harp seal pups naturally fast for about 6 weeks, but elephant and gray seal pups are able to maintain their water balance during their 1.5 to 3 months of terrestrial fasting.[182] During this period of time, the fasting seal pups rely primarily on the water derived metabolically from the catabolism of their fat stores.[36,181] The use of fat as the primary source for caloric requirements also reduces the need for the kidney to excrete urea and aids in water preservation, as does the reduced glomerular filtration rate.[1,183] Whether the renal responses concerning water resorption are mediated in seals by AVP (and to some extent by aldosterone) as in terrestrial animals is still a matter of debate.[186] During fasting, the availability of sodium is limited, so hyponatremia must be prevented. This is achieved by activation of the complete renin/angiotensin/aldosterone axis that has been shown to be operative in California sea lions[162] and northern elephant seal pups.[185] The hormonal mechanisms by which the northern elephant seals maintain water and electrolyte balance appear to be similar regardless of age, as angiotensin II and aldosterone are also increased with fasting in breeding adult male northern elephant seals.[184] How the body-water homeostasis and the increase in the oxidation of fat are coordinated remains to be established.

2. Small Desert Mammals

Metabolic water is also one of the main sources of free water in some small desert animals such as the kangaroo rat, which feeds on dry grains and never has access to freshwater. In a study of water balance published in 1952,[217] it is clear that metabolic water, despite its relative small amount (about 1.3 mL per day for an animal of 35 g, or about 0.6% of total body water), suffices to compensate for water loss in the urine, by evaporation, and in the feces even at very low ambient atmospheric humidities.[217] The kangaroo rat minimizes water loss by producing an exceptionally concentrated urine (about 5000 mOsm/kg in electrolytes and urea), which is partly due to the specific morphological feature of a thin and very long papilla, a broad outer stripe of the outer medulla, and particular morphological changes in the tubular epithelium (see below). In addition, the concentration of ADH in the plasma appears to be very high.[217] These calculations make it obvious that in addition to generating metabolic water, water losses by the kidneys have to be minimized, as do fluid losses from surfaces of the body in contact with the external environment, such as the skin, mucous membranes of the respiratory tract, and the gastrointestinal tract. These factors are considered in the following sections.

VI. SPECIFIC ORGAN ADAPTATIONS TO OSMOREGULATORY DEMANDS

A. THE KIDNEY

As a group, mammals stand out from other vertebrates in that the kidney is the most important osmoregulatory organ.[214] Homer Smith, one of the pioneers in renal physiology, is quoted as saying that in mammals the composition of the blood and the body fluids is determined "not only by what the mouth takes but by what the kidney retains."[195] Members of other vertebrate classes possess potent extrarenal water and solute regulatory mechanism such as the gills in fish, the skin in amphibians, and salt glands in a variety of species. Also unique is the degree of phenotypic plasticity of the kidney with regard to both structure and function. The mammalian kidney can cope with the demands of animals that never ingest free water as well as those that live in freshwater, where the influx of free water into the animals is constant and high. In the same vein, the kidney can compensate for an excess of salt in animals with a salt-rich diet and conserve salt when it is required.

1. Basic Morphological Functional Architecture

This high flexibility is achieved by a variety of mechanisms. As a basis for discussion, we will first consider the basic morphological features of a mammalian kidney[190] typical of one from animals living in a mesic climate with moderate temperatures and a well-balanced supply of moisture (see Figure 12.6).[10] The nephron begins with the *glomerulus*, where ultrafiltration of the blood plasma and formation of the primary urine take place. The glomeruli are located in the cortex of the kidney; they can be located close to the surface or at the junction between the cortex and the medulla. The latter *juxtamedullary glomeruli* are usually larger and have a higher glomerular filtration rate. The primary urine then passes the pars convoluta of the *proximal tubule*. The length of this part and the transport properties differ between those that originate from a surface glomerulus and those that are connected to a juxtamedullary glomerulus. The latter are shorter and have a lower transport capacity for sodium and chloride than the former.[14,108,246] The pars recta of the proximal tubule is

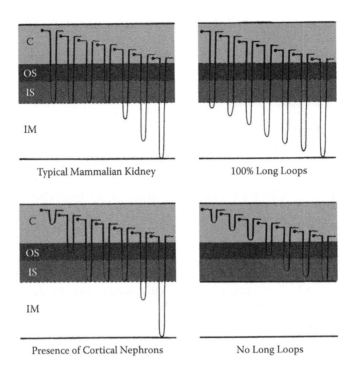

FIGURE 12.6 Different types of nephrons in the mammalian kidney with regard to loop length. C, cortex; OS, outer stripe of the outer medulla; IS, inner stripe of the outer medulla; IM, inner medulla. For further details, see text. (Adapted from Bankir, L. and de Rouffignac, C., *Am. J. Physiol.*, 249, R643–R666, 1985.)

next; it enters the medulla and is part of the outer stripe of the renal medulla. The next segment is the *thin descending limb of Henle's loop*, which normally turns around at the border between the inner stripe of the outer medulla and the inner medulla or continues as a loop into the inner medulla. When it does so, a long loop is formed. A true thin limb appears first in the avian kidney in the folded nephrons of the mammalian type.[53] The occurrence of long loops is apparently a new step in the design of a concentrating kidney and is present only in the mammalian kidney. At the border between the inner medulla and the inner stripe of the outer medulla, the epithelium of the ascending loop changes and the segments turn into the *medullary thick ascending limb of Henle's loop* and the cortical thick ascending limb. The distal *convoluted tubule* located in the renal cortex follows and, via the *connecting tubule*, drains into the *collecting ducts*.

The cortex and in particular the concentric outer and inner stripe of the outer medulla and the inner medulla have specific vascular patterns. Of particular importance are where the nutrient blood flow surrounding the tubular segments originates and the extent to which tubular fluid and intravascular fluid can exchange solutes. The nutrient blood flow of the superficial proximal convoluted tubules is derived from the efferent arterioles of the superficial glomeruli; the blood flow to the outer stripe of the outer medulla is essentially venous, consisting of vasa recta ascending from deeper regions of the medulla. The nutrient supply of the inner stripe comes mainly from the descending branches of the efferent arterioles of the juxtamedullary glomeruli. The blood supply for the inner medulla also comes from descending juxtaglomerular efferent arterioles. The vessels cross the outer medulla within vascular bundles without significant contact to the tubular structures.[10]

This arrangement of tubular structures and blood vessels and their different transport activities and permeabilities provide the spatial and functional requirements for the generation and maintenance of the corticopapillary osmotic gradient in the counter-current system (see textbooks of physiology for details) on which the concentration of the final urine depends. This discussion does not address the role of the pelvic fornices (particularly in the renal handling of urea[126,212]).

2. Adaptations in the Renal Cortex

From the scheme given in Figure 12.6, variations can be derived that can occur during evolution in the adaptation to the environment, species-determined habits, and the available diet. Starting in the cortex, the number of nephrons is the first variable. Several authors have observed a relatively low filtration rate or filtration surface area per gram kidney or gram body weight in desert-adapted animals.[208,259] Thus, mice living in a rather arid environment have significantly fewer glomeruli than mice of a similar size living in a damp environment—the filtration area is almost twice as high in the latter.[57] The same holds for other rodents (see Figure 3 in Bankir and de Rouffignac[10]). This adaptation in GFR is reminiscent of the action of vasopressin in fish, where the number of functionally active glomeruli is varied by constricting the afferent arterioles to reduce water loss.[102] Also, the ratio between superficial and juxtamedullary nephrons varies, and a correlation seems to exist between this ratio and the concentrating ability in mammals.[10] A ratio of 0.8 to 1.5 is found in humans, monkeys, rabbits, and guinea pigs, which generally have a urine osmolality of around 1000 mOsm/kg, which contrasts with a ratio of 2.5 to 5 in jerboa and the pocket mouse, which have a urine osmolality of about 4000 mOsm/kg.

The normal architecture of the mammalian kidney contains a mixture of short loops and long loops of the thin descending limb (see Figure 12.6). Initially, when the importance of the counter-current system for urinary concentration was realized, it was thought that the percentage of long loops can be correlated directly to the concentrating ability of the kidney,[232] but this turned out to be an oversimplification because short loops also contribute to the counter-current system.[40,235,246,254] The ratio of short loops to long loops is about 70:30 in rats,[133,232] 75:25 in the mouse,[135] 85:15 in humans,[190] and 97:3 in the pig.[232] In the rabbit, the ratio is reversed.[111] For mammals living in freshwater, such as the beaver and the hippopotamus, two departures from this basic scheme have been observed; due to the absence of a medulla, they have no long loops and only cortical nephrons.[194,232] Cortical nephrons are also found in humans and occasionally in pigs. On the other extreme, in some kidneys every tubule enters the inner medulla even if only to a small extent, and they all have long loops. This phenomenon is observed for carnivores such as the dog, cat, and fox[13,27,134,232] and might be related to the handling of urea derived from the protein-rich diet.

3. Adaptations in the Renal Medulla

Other variants include the thickness of the outer stripe of the outer medulla and the inner stripe of the outer medulla (see Figure 12.7). In particular, the latter appears to be important because in this area of the kidney the thick ascending limb of Henle's loop is the most abundant tissue. The transport of sodium chloride into the interstitium of the medulla without an associated movement of water provides the driving force for the counter-current system. Bankir and de Rouffignac[10] placed special emphasis on the distinction between these two different zones when comparing medullary thickness with urinary concentrating ability. The thickness of the inner medulla, length of the papilla, and their cross-sectional areas also vary. Beuchat found a significant relationship between the thickness of the inner medulla (corresponding to the length of the thin ascending limbs of the loop of Henle) and concentrating ability but only for species living in mesic environments.[15] The cross-sectional area at the base of the papilla can be related to the volume of the urine that has to be processed. A broad base or several papillae in reniculated kidneys, as found in the marine animals, humans, and bears, might be necessary to process a large volume. The pocket mouse has an extreme low urine flow and a very thin but also very long papilla.[5] Within three species of hedgehogs, the one with the narrowest papilla had the lowest urine output and could withstand a shortage of water much better than the other two.[4]

As pointed out by Bankir and de Rouffignac,[10] both the height of each zone along the cortico-papillary axis (which relates to the length of the particular segments of the loop of Henle) and the volume of each zone (which relates to the number of nephrons present at each level) have to be considered.[10] These authors strongly emphasized the importance of the vascular bed in establishing,

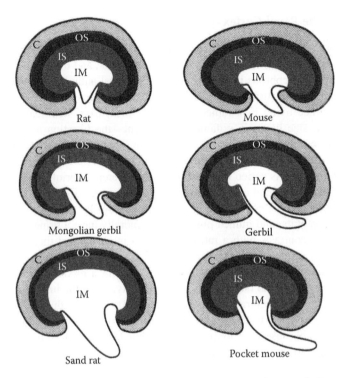

FIGURE 12.7 Kidney cross-sections of various rodents showing the limits and sizes of different kidney zones. C, cortex; OS, outer stripe of the outer medulla; IS, inner stripe of the outer medulla; IM, inner medulla. For further details see text. (Adapted from Bankir, L. and de Rouffignac, C., *Am. J. Physiol.*, 249, R643–R666, 1985.)

maintaining, and dissipating the corticomedullary osmotic gradient, particularly the degree to which the vessels are bundled and are insulated from their surroundings when passing through other medullary zones.

Despite these intricacies, studies continue at the macroscopic level, as initiated by Sperber in 1942.[232] A relation between kidney mass and relative medullary thickness of rodents in relation to habitat, body size, and phylogeny has been described by al Kahtani et al.[3] Such observations appear to be valid if the range of species compared is limited and factors other than the habitat are taken into account. In phylogenetically related phyllostomid bats, a correlation between evolutionary changes in diet and changes in physiological functions in the intestine and in the kidney have been observed.[219] The transition from insectivore to nectavore and frugivore resulted in changes in the levels of intestinal sucrase and maltase activity. More importantly, in the context of this chapter, frugivory and nectarivory bats, which have to cope with an excess of free water and have to conserve electrolytes, have developed kidneys with a large cortex, an undivided outer medulla (probably mostly outer stripe of the outer medulla), and no inner medulla and papilla. A large volume of ultrafiltrate can be generated which can be modified effectively in the thick ascending limbs of Henle's loop to preserve the electrolytes.[219] Interestingly, an almost similar arrangement is found in the freshwater-living mountain beaver, which has a large cortex, broad outer stripe of the outer medulla, and no inner medulla.[193,194]

These examples demonstrate that many morphological factors in the kidney can be modified depending on the environment, habits, and diet of a particular species. It is possible to adapt to the availability of, or exposure to, water and salt in more than one way, and different routes and different combinations have emerged during development.

B. The Respiratory System: The Nasal Counter-Current Heat Exchanger

Inspired air becomes very rapidly saturated with water by the moist surfaces of the early respiratory tract. In the absence of heat stress, the evaporation of water by respiration from mammals is found to be rather uniform when related to oxygen consumption.[216] Assuming that the 21% O_2 in the atmospheric air has been reduced to 16% in the exhaled air and that the exhaled air is saturated with water vapor at body temperature, a value of 0.84 mg water per mL O_2 is obtained. Levels in humans, the albino mouse, and the albino rat come close to this value, but several desert animals such as the kangaroo rat, the pocket mouse, and the hamster show a much lower evaporation of 0.54 mg water per mL O_2. In his review, Schmidt-Nielsen mentions three possible ways in which an animal can reduce evaporation from the respiratory tract: (1) The exhaled air is not saturated with water vapor, (2) oxygen extraction is higher than the assumed 5%, and (3) air is exhaled at a temperature below the body temperature. He dismisses the first possibility because of the abundance of moisture all over the respiratory tract, and he dismisses the second possibility for the kangaroo rat because no evidence has been found for a shift in the affinity of the hemoglobin to lower O_2 tensions or specific mechanisms to cope with the increased CO_2 tension that would result. As we will see later, this does not hold for all mammals. The third possibility is realized in the kangaroo rat, for which the exhaled air can be 14 degrees below the body temperature and even below the temperature of the ambient air. This startling observation can be explained by a heat-exchange system in the nasal passage airways. During inspiration, the inhaled air is humidified, which leads to a loss of heat on the nasal surfaces by water evaporation; thus, the temperature of these surfaces decreases and can even be below the ambient temperature. During exhalation, the air that leaves the lungs is at body temperature and saturated with water. When the air passes along the cooler nasal surfaces, water condenses and is retrieved. The extent to which this can occur is between 56 and 88% for the kangaroo rat, with an average of about 58% for animals living in arid zones and only 27% for terrestrial animals living in temperate zones.[145] A more detailed analysis of the nasal counter-current heat exchange system is given by Schmidt-Nielsen et al.[216]

The counter-current system in the kidney is separated in two channels and functions because the two fluids moving in opposite direction travel in two different neighboring tubes (*spatial separation*). In the heat exchanger in the nasal passages, the flow is also in two opposite directions, but it is separated in time (*temporal separation*).[216] For the exchangers to be effective, equilibrium between the content and the surface of the tubes has to be reached. This is facilitated by a large surface area, low flow rate, and short distance to the wall. Accordingly, the nasal passages of the kangaroo rat have a complex system of narrow turbinates by which the surface area is increased and the distance from the center of the air stream to the surface is less than 0.1 mm.[216]

It is interesting to note that the northern elephant seals are even more efficient than small desert animals in reducing evaporative water loss, as they recover up to 92% of the water initially added to the air. This is also achieved by extended nasal turbinates as well as by reducing the rate of breathing. Fasting elephant seal pubs exhibit about 9 hours of apneustic breathing during the day, with durations of apnea of 4 to 8 minutes. During apnea, the metabolic rate is reduced and the oxygen extraction efficiency is increased. This mechanism, which is closely similar to the diving response, seems to preadapt these animals to face extended periods of water deprivation with behavioral strategies necessary to survive them.[145]

C. The Skin: Sweating and Body Temperature Regulation

The skin represents a large interface between the body and the environment and thus provides a large surface for the exchange of gases, water, and heat. This exchange has to be tightly controlled so the water and electrolyte homeostasis of the body is not put in jeopardy. The skin has to be rather impermeable to salt and water for mammals living in seawater and in freshwater—the permeability is indeed minimized but not completely abolished. The heat exchange is also reduced in marine mammals by the thick insulation provided by subcutaneous fat (blubber) or fur (e.g., in beavers and otters). The question of heat exchange becomes particularly important for animals living in the desert.[215] Small animals have developed some specific behaviors to avoid exposure to the sun and the soil, which can reach up to 70°C in open spaces. They prefer to stay in the shade when they are active during the daytime, or they are nocturnal and escape the heat in burrows that are cooler and more humid.

For large mammals living in the desert, heat dissipation becomes a major challenge. Regulation of the body temperature usually involves the evaporation of water from the body surface, be it by sweating in humans, camels, and horses; panting in dogs; or spreading saliva on the fur as in mice. A human may sweat at a rate of 1 to 1.5 L/hr; thus, the loss of water during a hot desert day can be as high as 15 L. So much water cannot be saved by increasing, for example, the renal concentration efficiency. Even if it could be twice as high, only an additional volume of about 150 mL would be saved per day. Thus, the water and the electrolytes lost by sweating must be constantly replenished; if done so, humans can survive pretty well in the desert. But, animals that have no or very limited access to water must develop other strategies to reduce the expenditure of water for heat regulation. Investigations on the camel have revealed some of these mechanisms. One important factor is that the body temperature of the camel is not kept constant but fluctuates during the day and can reach up to 40°C; thus, heat is stored in the body rather than dissipated. This stored heat is then given off to the cool surroundings during the night, as the body temperature drops to about 34.5°C without the expenditure of water. In addition, raising the body temperature reduces the gradient between the temperature of the ambient air and the body, thus reducing the heat transfer. The fluctuations in body temperature are particularly large in dehydrated camels, for which water conservation is a major concern. The fluctuations are smaller in hydrated camels, as they can use some water for evaporation.[215,218] Such fluctuations are also observed in large East African hooved animals.[240] Fur layers can provide an effective insulating layer between the source of heat and the cooler body; in the camel, for example, removal of the fur increases the water expenditure for heat regulation by 50%.[218]

Developing the means to regulate body temperature is apparently advantageous from an evolutionary point of view. It is hypothesized that the gain of diffuse thermoregulatory eccrine sweat glands by early bipedal hominids allowed them to leave the shady forests and cool woodlands of their ancestors and to expand their range into the hot, open savannah grasslands some 3 to 4 million years ago. Almost no other primate can regulate its body temperature by sweating.[48]

Sweat contains NaCl, the concentration of which can increase up to 100 mM NaCl at higher rates; thus, the electrolyte balance has to be controlled. In a nonacclimated human, up to 30 mg of salt can be lost during heat exposure or during dissipation of heat generated metabolically through exercise.[223] It is also interesting that in humans acclimated to heat the salt content of the sweat is reduced to decrease the loss of salt during body temperature regulation.[161,209] This effect is probably a consequence of the increased level of aldosterone, which increases sodium reabsorption not only in the kidney but also in the sweat gland secretory ducts (as well as in the colon).[21]

D. The Gastrointestinal Tract in Ruminants

1. Salivary Gland, Gastrointestinal, and Hepatoportal Circulation in Ruminants

It was emphasized earlier that considerable inter-organ cycling of water and electrolytes takes place in the gastrointestinal tract. In humans, this amounts to about 7 L but reaches much higher values in ruminants, which continuously produce saliva to buffer the fermentation process in the rumen. In the small intestine, the fluid reabsorbed is almost isotonic; the removal of water and salt occurs in the later parts such as the colon. In this epithelium, aldosterone as well as antidiuretic hormone can exert their actions, and the water content of the feces can be reduced significantly. In small desert animals, this fluid and vitamins produced by the intestinal microflora are recovered by ingestion of the feces.[215]

Ruminants living in arid areas have a particular challenge in that their digestive process appears to obligate large potential water losses that must be recovered. Ruminants originated in the Eocene (about 55 million years ago) and became the dominant herbivores on Earth in the Miocene (23.8 million to 5.3 million years ago) when grasslands appeared in conjunction with drier climates.[248] The considerable anatomical adaptations include, in particular, development of the spacious fore-stomach (the reticulorumen or simply rumen), which serves as a digestion chamber but also has special functions during dehydration and rapid rehydration. Because the digesta in ruminants contribute about one quarter of the animal's body weight, the gastrointestinal tract and the related organs such as the salivary gland and the liver become vitally important for the water and electrolyte homeostasis of the body. The salivary gland, gastrointestinal tract, and hepatoportal circulation can reach enormous values, ranging from 6 to 16 L in sheep and up to 250 L in lactating dairy cows.[227] Desert ruminants such as the ibex (*Capra ibex nubania*) and the bighorn sheep are typical examples of the strategy commonly adopted in dry areas—namely, a combination of a frugal water economy and a capacity to endure severe dehydration and rapid rehydration.[224]

2. Dehydration and Rehydration in Ruminants

During dehydration, the rumen and the gut serve as water reservoirs from which water and electrolytes can be recovered when needed. This is evident from the fact that about 60% of the water lost during dehydration is provided by the rumen and gut.[225] In addition, in ruminants exposed to heat stress or a regimen of dehydration/rehydration cycles, the water content of the rumen increases, supporting the important physiological role of the rumen in osmoregulation.[226] Further-more, during hot hours net absorption and net outflow of fluid from the rumen are higher than the volume of the saliva secreted; thus, a net transfer of fluid from the gastrointestinal tract into the blood occurs. The rumen appears to be used as a water reservoir during the day and as a fermentation vat during the night. The rumen is also the buffer when water is accessible after days of dehydration. The water ingested can comprise between 15 and 40% of the body weight. This poses the big challenge of avoiding water intoxication and retaining the ingested water, as another period of dehydration may follow immediately (for review, see Olsson[180]). Several mechanisms prevent flooding of the body with water with its potentially deadly consequences. First, the rumen acts as a diffusion barrier.[45] The rumen epithelium has highly developed tight junctions in the stratum granulosum, which is the second layer close to the lumen of the rumen.[90] Second, the rate of secretion of the hypotonic saliva is accelerated when the osmolality of the portal blood decreases or when the plasma volume is expanded; thus, the increased rate of water recycling within the salivary, gastrointestinal tract, and hepatoportal circulation prevents a drop in the osmolality of the plasma.[33] In parallel to the changes in salivary secretion, urine flow drops immediately following drinking in a variety of ruminants (for references, see Silanikove[225,227]), thus the ingested water is retained. This reflex of renal vasal constriction in response to portal vein distension[127] is quite in contrast to the reflexes elicited by the low-capacity vessels in humans which lead to a water diuresis.[95] Thirst satiation in ruminants is similar to that of other mammals in that it is a function of mechano- and osmoreceptors that record the volume of the receptacle acting as the first reservoir—the rumen[100] (stomach in other mammals).

VII. CONCLUDING REMARKS

The advent of improved technologies as well as the explosive gain in knowledge due to deciphering the genomes of various mammals have significantly improved our understanding of the basic principles and the molecular mechanisms involved in water and electrolyte homeostasis of mammals. A particular improvement in our understanding has resulted from the ability to generate animals (primarily mice) with singular gene defects in well-defined regions of the body. It has also been of great help that evolution continues and manifests itself as genetic diseases in humans, to the benefit of science but to the disadvantage of the affected.

ACKNOWLEDGMENTS

The authors would like to thank David Evans for the opportunity to revisit so many important papers written by contemporaries, most of them being close friends and acquaintances. We are also grateful for having been able to discover during the preparation of the manuscript so many strategies for osmoregulation we had neither heard nor dreamt of. Special thanks also go to the departmental and secretarial staff, in particular the extremely able and competent secretary of R.K., Christine Riemer, for her outstanding organizational skill and computer proficiency, as well as her endurance and patience. Without her, the writing of this chapter while traveling between two continents would not have been possible. Also, the enthusiastic and untiring support of the staff of the library of the Max Planck Institute, particularly Christiane Berse, Jürgen Block, and Jan-Helge Ralle, is gratefully acknowledged. The outstanding creative artwork was provided by Claudia Pieczka and Melanie Wilkesmann.

REFERENCES

1. Adams, S. H. and Costa, D. P., Water conservation and protein metabolism in northern elephant seal pups during the postweaning fast. *J. Comp. Physiol. B*, 163, 367–373, 1993.
2. Agarwal, A. K., Mune, T., Monder, C., and White, P. C., NAD(+)-dependent isoform of 11 beta-hydroxysteroid dehydrogenase: cloning and characterization of cDNA from sheep kidney. *J. Biol. Chem.*, 269, 25959–25962, 1994.
3. al Kahtani, M. A., Zuleta, C., Caviedes-Vidal, E., and Garland, Jr., T., Kidney mass and relative medullary thickness of rodents in relation to habitat, body size, and phylogeny. *Physiol. Biochem. Zool.*, 77, 346–365, 2004.
4. Albiston, A. L., Obeyesekere, V. R., Smith, R. E., and Krozowski, Z. S., Cloning and tissue distribution of the human 11 beta-hydroxysteroid dehydrogenase type 2 enzyme. *Mol. Cell Endocrinol.*, 105, R11–R17, 1994.
5. Altschuler, E. M., Nagle, R. B., Braun, E. J., Lindstedt, S. L., and Krutzsch, P. H., Morphological study of the desert heteromyid kidney with emphasis on the genus *Perognathus. Anat. Rec.*, 194, 461–468, 1979.
6. Amemiya, M., Loffing, J., Lotscher, M., Kaissling, B., Alpern, R. J., and Moe, O. W., Expression of NHE-3 in the apical membrane of rat renal proximal tubule and thick ascending limb. *Kidney Int.*, 48, 1206–1215, 1995.
7. Andersson, B., The effect of injections of hypertonic NaCl-solutions into different parts of the hypothalamus of goats. *Acta Physiol. Scand.*, 28, 188–201, 1953.
8. Arriza, J. L., Weinberger, C., Cerelli, G., Glaser, T. M., Handelin, B. L., Housman, D. E., and Evans, R. M., Cloning of human mineralocorticoid receptor complementary DNA: structural and functional kinship with the glucocorticoid receptor. *Science*, 237, 268–275, 1987.
9. Atlas, S. A. and Maack, T., Atrial natriuretic factor. In *Handbook of Physiology: Renal Physiology*, Windhager, E. E., Ed., Oxford University Press, New York, 1992, pp. 1577–1673.
10. Bankir, L. and de Rouffignac, C., Urinary concentrating ability: insights from comparative anatomy. *Am. J. Physiol.*, 249, R643–R666, 1985.
11. Bankir, L., Fernandes, S., Bardoux, P., Bouby, N., and Bichet, D. G., Vasopressin-V_2 receptor stimulation reduces sodium excretion in healthy humans. *J. Am. Soc. Nephrol.*, 16, 1920–1928, 2005.
12. Beesley, A. H., Hornby, D., and White, S. J., Regulation of distal nephron K^+ channels (ROMK) mRNA expression by aldosterone in rat kidney. *J. Physiol.*, 509(Pt. 3), 629–634, 1998.
13. Beeuwkes III, R., Efferent vascular patterns and early vascular-tubular relations in the dog kidney. *Am. J. Physiol.*, 221, 1361–1374, 1971.

14. Berry, C. A., Heterogeneity of tubular transport processes in the nephron. *Annu. Rev. Physiol.*, 44, 181–201, 1982.

15. Beuchat, C. A., Structure and concentrating ability of the mammalian kidney: correlations with habitat. *Am. J. Physiol. Regul. Integr. Comp. Physiol.*, 40, R157–R179, 1996.

16. Bhalla, V., Oyster, N. M., Fitch, A. C., Wijngaarden, M. A., Neumann, D., Schlattner, U., Pearce, D., and Hallows, K. R., AMP-activated kinase inhibits the epithelial Na+ channel through functional regulation of the ubiquitin ligase Nedd4-2. *J. Biol. Chem.*, 281, 26159–26169, 2006.

17. Biner, H. L., Arpin-Bott, M. P., Loffing, J., Wang, X., Knepper, M., Hebert, S. C., and Kaissling, B., Human cortical distal nephron: distribution of electrolyte and water transport pathways. *J. Am. Soc. Nephrol.*, 13, 836–847, 2002.

18. Birabeau, M. A., Capponi, A. M., and Vallotton, M. B., Solubilized adrenal angiotensin II receptors: studies on the site of action of sodium and calcium ions, and on the role of disulfide bridges. *Mol. Cell Endocrinol.*, 37, 181–189, 1984.

19. Blackburn, R. E., Demko, A. D., Hoffman, G. E., Stricker, E. M., and Verbalis, J. G., Central oxytocin inhibition of angiotensin-induced salt appetite in rats. *Am J. Physiol.*, 263, R1347–R1353, 1992.

20. Bonny, O., Chraibi, A., Loffing, J., Jaeger, N. F., Grunder, S., Horisberger, J. D., and Rossier, B. C., Functional expression of a pseudohypoaldosteronism type I mutated epithelial Na+ channel lacking the pore-forming region of its alpha subunit. *J. Clin. Invest.*, 104, 967–974, 1999.

21. Bonvalet, J. P., Rossier, B., and Farman, N., Distribution of amiloride-sensitive sodium channel in epithelial tissue. *C. R. Seances Soc. Biol. Fil.*, 189, 169–177, 1995.

22. Boron, W. F., Acid–base transport by the renal proximal tubule. *J. Am. Soc. Nephrol.*, 17, 2368–2382, 2006.

23. Bouscarel, B., Blackmore, P. F., and Exton, J. H., Characterization of the angiotensin II receptor in primary cultures of rat hepatocytes: evidence that a single population is coupled to two different responses. *J. Biol. Chem.*, 263, 14913–14919, 1988.

24. Briggs, J. P. and Schnermann, J., Macula densa control of renin secretion and glomerular vascular tone: evidence for common cellular mechanisms. *Renal Physiol.*, 9, 193–203, 1986.

25. Brown, D. and Orci, L., Vasopressin stimulates formation of coated pits in rat kidney collecting ducts. *Nature*, 302, 253–255, 1983.

26. Brown, D., Weyer, P., and Orci, L., Vasopressin stimulates endocytosis in kidney collecting duct principal cells. *Eur. J. Cell Biol.*, 46, 336–341, 1988.

27. Bulger, R. E. and Dobyan, D. C., Recent advances in renal morphology. *Annu. Rev. Physiol.*, 44, 147–179, 1982.

28. Burns, K. D., The emerging role of angiotensin-converting enzyme-2 in the kidney. *Curr. Opin. Nephrol. Hypertens.*, 16, 116–121, 2007.

29. Butlen, D., Mistaoui, M., and Morel, F., Atrial natriuretic peptide receptors along the rat and rabbit nephrons: [125I] alpha-rat atrial natriuretic peptide binding in microdissected glomeruli and tubules. *Pflügers Arch.*, 408, 356–365, 1987.

30. Carattino, M. D., Edinger, R. S., Grieser, H. J., Wise, R., Neumann, D., Schlattner, U., Johnson, J. P., Kleyman, T. R., and Hallows, K. R., Epithelial sodium channel inhibition by AMP-activated protein kinase in oocytes and polarized renal epithelial cells. *J. Biol. Chem.*, 280, 17608–17616, 2005.

31. Carey, R. M., Evidence for a splanchnic sodium input monitor regulating renal sodium excretion in man: lack of dependence upon aldosterone. *Circ. Res.*, 43, 19–23, 1978.

32. Caron, M. G. and Lefkowitz, R. J., Catecholamine receptors: structure, function, and regulation. *Recent Prog. Horm. Res.*, 48, 277–290, 1993.

33. Carr, D. H. and Titchen, D. A., Post prandial changes in parotid salivary secretion and plasma osmolality and the effects of intravenous of saline solutions. *Q. J. Exp. Physiol. Cogn. Med. Sci.*, 63, 1–21, 1978.

34. Carrithers, S. L., Hill, M. J., Johnson, B. R., O'Hara, S. M., Jackson, B. A., Ott, C. E., Lorenz, J., Mann, E. A., Giannella, R. A., Forte, L. R., and Greenberg, R. N., Renal effects of uroguanylin and guanylin *in vivo. Braz. J. Med. Biol. Res.*, 32, 1337–1344, 1999.

35. Carrithers, S. L., Ott, C. E., Hill, M. J., Johnson, B. R., Cai, W. Y., Chang, J. J., Shah, R. G., Sun, C. M., Mann, E. A., Fonteles, M. C., Forte, L. R., Jackson, B. A., Giannella, R. A., and Greenberg, R. N., Guanylin and uroguanylin induce natriuresis in mice lacking guanylyl cyclase-C receptor. *Kidney Int.*, 65, 40–53, 2004.

36. Castellini, M. A., Costa, D. P., and Huntley, A. C., Fatty acid metabolism in fasting elephant seal pups. *J. Comp. Physiol. B,* 157, 445–449, 1987.

37. Catt, K. J., Sandberg, K., and Balla, T., Angiotensin II receptors and signal transduction mechanisms. In *Cellular and Molecular Biology of the Renin–Angiotensin System,* Raizada, M. K., Phillips, M. I., and Sumners, C., Eds., CRC Press, Boca Raton, FL, 1993, pp. 307–356.

38. Champigneulle, A., Siga, E., Vassent, G., and Imbert-Teboul, M., V_2-like vasopressin receptor mobilizes intracellular Ca^{2+} in rat medullary collecting tubules. *Am. J. Physiol.,* 265, F35–F45, 1993.

39. Chan, J. C., Knudson, O., Wu, F., Morser, J., Dole, W. P., and Wu, Q., Hypertension in mice lacking the proatrial natriuretic peptide convertase corin. *Proc. Natl. Acad. Sci. USA,* 102, 785–790, 2005.

40. Chandhoke, P. S. and Saidel, G. M., Mathematical model of mass transport throughout the kidney: effects of nephron heterogeneity and tubular–vascular organization. *Ann. Biomed. Eng.,* 9, 263–301, 1981.

41. Chang, R. S. and Lotti, V. J., Angiotensin receptor subtypes in rat, rabbit and monkey tissues: relative distribution and species dependency. *Life Sci.,* 49, 1485–1490, 1991.

42. Chen, S., Burgner, J. W., Krahn, J. M., Smith, J. L., and Zalkin, H., Tryptophan fluorescence monitors multiple conformational changes required for glutamine phosphoribosylpyrophosphate amidotransferase interdomain signaling and catalysis. *Biochemistry (Mosc.),* 38, 11659–11669, 1999.

43. Chen, S. Y., Bhargava, A., Mastroberardino, L., Meijer, O. C., Wang, J., Buse, P., Firestone, G. L., Verrey, F., and Pearce, D., Epithelial sodium channel regulated by aldosterone-induced protein sgk. *Proc. Natl. Acad. Sci. USA,* 96, 2514–2519, 1999.

44. Chiu, A. T., Leung, K. H., Smith, R. D., and Timmermans, P. B. M. W. M., Defining angiotensin receptor subtypes. In *Cellular and Molecular Biology of the Renin–Angiotensin System,* Raizada, M. K., Phillips, M. I., and Sumners, C., Eds., CRC Press, Boca Raton, FL, 1993, pp. 245–271.

45. Chosniak, I. and Shkolnik, A., The rumen as a protective osmotic mechanism during rapid rehydration in the black Bedouin goat. In *Osmotic and Volume Recognition,* Alfred Benzon Symposium, Vol. 11, Skadaughe, E. and Jurgensen, C.B., Eds., Munksgard, Copenhagen, 1978, pp. 344–359.

46. Clapham, D. E., TRP channels as cellular sensors. *Nature,* 426, 517–524, 2003.

47. Cohen, D., Koh, G. Y., Nikonova, L. N., Porter, J. G., and Maack, T., Molecular determinants of the clearance function of type C receptors of natriuretic peptides. *J. Biol. Chem.,* 271, 9863–9869, 1996.

48. Cohn, B. A., The vital role of the skin in human natural history. *Int. J. Dermatol.,* 37, 821–824, 1998.

49. Coleridge, J. C. G., Hemingway, A., Holmes, R. L., and Linden, R. J., The location of atrial receptors in the dog: a physiological and histological study. *J. Physiol. (Lond.),* 136, 174–197, 1957.

50. Colombo, L., Dalla, V. L., Fiore, C., Armanini, D., and Belvedere, P., Aldosterone and the conquest of land. *J. Endocrinol. Invest.,* 29, 373–379, 2006.

51. Currie, M. G., Fok, K. F., Kato, J., Moore, R. J., Hamra, F. K., Duffin, K. L., and Smith, C. E., Guanylin: an endogenous activator of intestinal guanylate cyclase. *Proc. Natl. Acad. Sci. USA,* 89, 947–951, 1992.

52. Daniels, D., Yee, D. K., and Fluharty, S. J., Angiotensin II receptor signaling. *Exp. Physiol.,* 92(3), 523–527, 2007.

53. Dantzler, W. H. and Braun, E. J., Comparative nephron function in reptiles, birds, and mammals. *Am J. Physiol.,* 239, R197–R213, 1980.

54. Dark, J., Annual lipid cycles in hibernators: integration of physiology and behavior. *Annu. Rev. Nutr.,* 25, 469–497, 2005.

55. de Bold, A. J., Borenstein, H. B., Veress, A. T., and Sonnenberg, H., A rapid and potent natriuretic response to intravenous injection of atrial myocardial extract in rats. *Life Sci.,* 28, 89–94, 1981.

56. de Sauvage, F. J., Keshav, S., Kuang, W. J., Gillett, N., Henzel, W., and Goeddel, D. V., Precursor structure, expression, and tissue distribution of human guanylin. *Proc. Natl. Acad. Sci. USA,* 89, 9089–9093, 1992.

57. Dewey, G. C., Elias, H., and Appel, K. R., Stereology of the renal corpuscles of desert and swamp deermice. *Nephron,* 3, 352–365, 1966.

58. Dietz, J. R., Mechanisms of atrial natriuretic peptide secretion from the atrium. *Cardiovasc. Res.,* 68, 8–17, 2005.

59. Djelidi, S., Fay, M., Cluzeaud, F., Escoubet, B., Eugene, E., Capurro, C., Bonvalet, J. P., Farman, N., and Blot-Chabaud, M., Transcriptional regulation of sodium transport by vasopressin in renal cells. *J. Biol. Chem.,* 272, 32919–32924, 1997.

60. Dousa, T. P., Cyclic-3',5'-nucleotide phosphodiesterase isozymes in cell biology and pathophysiology of the kidney. *Kidney Int.*, 55, 29–62, 1999.

61. Ecelbarger, C. A., Chou, C. L., Lolait, S. J., Knepper, M. A., and DiGiovanni, S. R., Evidence for dual signaling pathways for V_2 vasopressin receptor in rat inner medullary collecting duct. *Am. J. Physiol.*, 270, F623–F633, 1996.

62. Ecelbarger, C. A., Kim, G. H., Terris, J., Masilamani, S., Mitchell, C., Reyes, I., Verbalis, J. G., and Knepper, M. A., Vasopressin-mediated regulation of epithelial sodium channel abundance in rat kidney. *Am. J. Physiol. Renal Physiol.*, 279, F46–F53, 2000.

63. Ecelbarger, C. A., Kim, G. H., Wade, J. B., and Knepper, M. A., Regulation of the abundance of renal sodium transporters and channels by vasopressin. *Exp. Neurol.*, 171, 227–234, 2001.

64. Edelman, I. S., Receptors and effectors in hormone action on the kidney. *Am. J. Physiol.*, 241, F333–F339, 1981.

65. Fakitsas, P., Adam, G., Daidie, D., van Bemmelen, M. X., Fouladkou, F., Patrignani, A., Wagner, U., Warth, R., Camargo, S. M., Staub, O., and Verrey, F., Early aldosterone-induced gene product regulates the epithelial sodium channel by deubiquitylation. *J. Am. Soc. Nephrol.*, 18, 1084–1092, 2007.

66. Fawcus, K., Gorton, V. J., Lucas, M. L., and McEwan, G. T., Stimulation of three distinct guanylate cyclases induces mucosal surface alkalinisation in rat small intestine *in vitro*. *Comp. Biochem. Physiol. A*, 118, 291–295, 1997.

67. Fitzsimons, J. T., The physiology of thirst and sodium appetite. *Monogr. Physiol. Soc.*, 1–572, 1979.

68. Fitzsimons, J. T. and Moore-Gillon, M. J., Drinking and antidiuresis in response to reductions in venous return in the dog: neural and endocrine mechanisms. *J. Physiol.*, 308, 403–416, 1980.

69. Flockerzi, V., Transient receptor potential (TRP) channels, with contributions by numerous experts. In *Handbook of Experimental Pharmacology*, Flockerzi, V. and Nilius, B., Eds., Springer, Heidelberg, 2007, pp. 1–19.

70. Forte, L. R., A novel role for uroguanylin in the regulation of sodium balance. *J. Clin. Invest.*, 112, 1138–1141, 2003.

71. Forte, L. R. and Currie, M. G., Guanylin: a peptide regulator of epithelial transport. *FASEB J.*, 9, 643–650, 1995.

72. Forte, L. R., Krause, W. J., and Freeman, R. H., Receptors and cGMP signalling mechanism for *E. coli* enterotoxin in opossum kidney. *Am. J. Physiol.*, 255, F1040–F1046, 1988.

73. Forte, L. R., London, R. M., Freeman, R. H., and Krause, W. J., Guanylin peptides: renal actions mediated by cyclic GMP. *Am. J. Physiol. Renal Physiol.*, 278, F180–F191, 2000.

74. Fukae, H., Kinoshita, H., Fujimoto, S., Kita, T., Nakazato, M., and Eto, T., Changes in urinary levels and renal expression of uroguanylin on low or high salt diets in rats. *Nephron*, 92, 373–378, 2002.

75. Fuller, P., The aldosterone receptor: new insights? *Expert. Opin. Invest. Drugs*, 15, 201–203, 2006.

76. Funder, J. W., The nongenomic actions of aldosterone. *Endocr. Rev.*, 26, 313–321, 2005.

77. Furuya, S., Naruse, S., Ando, E., Nokihara, K., and Hayakawa, T., Effect and distribution of intravenously injected ^{125}I-guanylin in rat kidney examined by high-resolution light microscopic radioautography. *Anat. Embryol. (Berl.)*, 196, 185–193, 1997.

78. Fushimi, K., Sasaki, S., and Marumo, F., Phosphorylation of serine 256 is required for cAMP-dependent regulatory exocytosis of the aquaporin-2 water channel. *J. Biol. Chem.*, 272, 14800–14804, 1997.

79. Ganong, W. F. and Mulrow, P. J., Rate of change in sodium and potassium excretion after injection of aldosterone into the aorta and renal artery of the dog. *Am. J. Physiol.*, 195, 337–342, 1958.

80. Garbers, D. L., Guanylyl cyclase-linked receptors. *Pharmacol. Ther.*, 50, 337–345, 1991.

81. Garbers, D. L. and Lowe, D. G., Guanylyl cyclase receptors. *J. Biol. Chem.*, 269, 30741–30744, 1994.

82. Garvin, J. L., Angiotensin stimulates glucose and fluid absorption by rat proximal straight tubules. *J. Am. Soc. Nephrol.*, 1, 272–277, 1990.

83. Garvin, J. L., Angiotensin stimulates bicarbonate transport and Na^+/K^+ ATPase in rat proximal straight tubules. *J. Am. Soc. Nephrol.*, 1, 1146–1152, 1991.

84. Gauer, O. H. and Henry, J. P., Circulatory basis of fluid volume control. *Physiol. Rev.*, 43, 423–481, 1963.

85. Geering, K., Girardet, M., Bron, C., Kraehenbuhl, J. P., and Rossier, B. C., Hormonal regulation of (Na^+,K^+)-ATPase biosynthesis in the toad bladder: effect of aldosterone and 3,5,3'-triiodo-L-thyronine. *J. Biol. Chem.*, 257, 10338–10343, 1982.

86. Geerling, J. C., Engeland, W. C., Kawata, M., and Loewy, A. D., Aldosterone target neurons in the nucleus tractus solitarius drive sodium appetite. *J. Neurosci.*, 26, 411–417, 2006.
87. Geerling, J. C. and Loewy, A. D., Sodium depletion activates the aldosterone-sensitive neurons in the NTS independently of thirst. *Am. J. Physiol. Regul. Integr. Comp. Physiol.*, 292, R1338–R1348, 2007.
88. Girardet, M., Geering, K., Gaeggeler, H. P., and Rossier, B. C., Control of transepithelial Na$^+$ transport and Na,K-ATPase by oxytocin and aldosterone. *Am. J. Physiol.*, 251, F662–F670, 1986.
89. Golbang, A. P., Cope, G., Hamad, A., Murthy, M., Liu, C. H., Cuthbert, A. W., and O'Shaughnessy, K. M., Regulation of the expression of the Na/Cl cotransporter by WNK4 and WNK1: evidence that accelerated dynamin-dependent endocytosis is not involved. *Am. J. Physiol. Renal Physiol.*, 291, F1369–F1376, 2006.
90. Graham, C. and Simmons, N. L., Functional organization of the bovine rumen epithelium. *Am. J. Physiol. Regul. Integr. Comp. Physiol.*, 288, R173–R181, 2005.
91. Grunewaldt, R. W. and Kinne, R. K. H., Osmoregulation in the mammalian kidney: the role of organic osmolytes. *J. Exp. Zool.*, 283, 708–724, 1999.
92. Gunther, S., Characterization of angiotensin II receptor subtypes in rat liver. *J. Biol. Chem.*, 259, 7622–7629, 1984.
93. Guyton, A. C., Ed., *Textbook of Medical Physiology*, Saunders, Philadelphia, PA, 1981, pp. 435–447.
94. Guyton, A. C. and Hall, J. E., Eds., *Textbook of Medical Physiology*, Saunders, Philadelphia, PA, 1996, pp. 1–1074.
95. Haberich, F. J., Osmoreception in the portal circulation. *Fed. Proc.*, 27, 1137–1141, 1968.
96. Hakam, A. C. and Hussain, T., Angiotensin II type 2 receptor agonist directly inhibits proximal tubule sodium pump activity in obese but not in lean Zucker rats. *Hypertension*, 47, 1117–1124, 2006.
97. Hamill, O. P., Twenty odd years of stretch-sensitive channels. *Pflügers Arch.*, 453, 333–351, 2006.
98. Hamra, F. K., Forte, L. R., Eber, S. L., Pidhorodeckyj, N. V., Krause, W. J., Freeman, R. H., Chin, D. T., Tompkins, J. A., Fok, K. F., and Smith, C. E., Uroguanylin: structure and activity of a second endogenous peptide that stimulates intestinal guanylate cyclase. *Proc. Natl. Acad. Sci. USA*, 90, 10464–10468, 1993.
99. Harazaki, M., Kawai, Y., Su, L., Hamazaki, Y., Nakahata, T., Minato, N., and Hattori, M., Specific recruitment of SPA-1 to the immunological synapse: involvement of actin-bundling protein actinin. *Immunol. Lett.*, 92, 221–226, 2004.
100. Harding, R. and Leek, B. F., Rapidly adapting mechanoreceptors in the reticulo-rumen which also respond to chemicals. *J. Physiol.*, 223, 32P–33P, 1972.
101. Haussinger, D., Reinehr, R., and Schliess, F., The hepatocyte integrin system and cell volume sensing. *Acta Physiol. (Oxford)*, 187, 249–255, 2006.
102. Hays, R. M., Water transport in epithelia. In *Urinary Concentrating Mechanisms*, Kinne, R. K. H., Kinne-Saffran, E., and Beyenbach, K. W., Eds., Karger, Basel, 1990, pp. 1–30.
103. He, G., Wang, H. R., Huang, S. K., and Huang, C. L., Intersectin links WNK kinases to endocytosis of ROMK1. *J. Clin. Invest.*, 117, 1078–1087, 2007.
104. Hinshaw, J. E., Dynamin and its role in membrane fission. *Annu. Rev. Cell Dev. Biol.*, 16, 483–519, 2000.
105. Hirsch, J. R., Meyer, M., and Forssmann, W. G., ANP and urodilatin: who is who in the kidney? *Eur. J. Med. Res.*, 11, 447–454, 2006.
106. Hollenberg, S. M., Weinberger, C., Ong, E. S., Cerelli, G., Oro, A., Lebo, R., Thompson, E. B., Rosenfeld, M. G., and Evans, R. M., Primary structure and expression of a functional human gluco-corticoid receptor cDNA. *Nature*, 318, 635–641, 1985.
107. Ishibashi, K., Sasaki, S., Fushimi, K., Yamamoto, T., Kuwahara, M., and Marumo, F., Immunolocalization and effect of dehydration on AQP3, a basolateral water channel of kidney collecting ducts. *Am. J. Physiol.*, 272, F235–F241, 1997.
108. Jamison, R. L., Intrarenal heterogeneity: the case for two functionally dissimilar populations of nephrons in the mammalian kidney. *Am. J. Med.*, 54, 281–289, 1973.
109. Johnson, A. K. and Thunhorst, R. L., The neuroendocrinology of thirst and salt appetite: visceral sensory signals and mechanisms of central integration. *Front. Neuroendocrinol.*, 18, 292–353, 1997.
110. Kahle, K. T., Wilson, F. H., Leng, Q., Lalioti, M. D., O'Connell, A. D., Dong, K., Rapson, A. K., MacGregor, G. G., Giebisch, G., Hebert, S. C., and Lifton, R. P., WNK4 regulates the balance between renal NaCl reabsorption and K$^+$ secretion. *Nat. Genet.*, 35, 372–376, 2003.

111. Kaissling, B. and Kriz, W., Structural analysis of the rabbit kidney. *Adv. Anat. Embryol. Cell Biol.,* 56, 1–123, 1979.

112. Kakiki, M., Morohashi, K., Nomura, M., Omura, T., and Horie, T., Regulation of aldosterone synthase cytochrome P450 (CYP11B2) and 11 beta-hydroxylase cytochrome P450 (CYP11B1) expression in rat adrenal zona glomerulosa cells by low sodium diet and angiotensin II receptor antagonists. *Biol. Pharm. Bull.,* 20, 962–968, 1997.

113. Kamsteeg, E. J., Heijnen, I., van Os, C. H., and Deen, P. M., The subcellular localization of an aquaporin-2 tetramer depends on the stoichiometry of phosphorylated and nonphosphorylated monomers. *J. Cell Biol.,* 151, 919–930, 2000.

114. Kamsteeg, E. J., Wormhoudt, T. A., Rijss, J. P., van Os, C. H., and Deen, P. M., An impaired routing of wild-type aquaporin-2 after tetramerization with an aquaporin-2 mutant explains dominant nephrogenic diabetes insipidus. *EMBO J.,* 18, 2394–2400, 1999.

115. Kangawa, K. and Matsuo, H., Purification and complete amino acid sequence of alpha-human atrial natriuretic polypeptide (alpha-hANP). *Biochem. Biophys. Res. Commun.,* 118, 131–139, 1984.

116. Katsura, T., Ausiello, D. A., and Brown, D., Direct demonstration of aquaporin-2 water channel recycling in stably transfected LLC-PK1 epithelial cells. *Am. J. Physiol.,* 270, F548–F553, 1996.

117. Katsura, T., Gustafson, C. E., Ausiello, D. A., and Brown, D., Protein kinase A phosphorylation is involved in regulated exocytosis of aquaporin-2 in transfected LLC-PK1 cells. *Am. J. Physiol.,* 272, F817–F822, 1997.

118. Kim, G. H., Masilamani, S., Turner, R., Mitchell, C., Wade, J. B., and Knepper, M. A., The thiazide-sensitive NaCl cotransporter is an aldosterone-induced protein. *Proc. Natl. Acad. Sci. USA,* 95, 14552–14557, 1998.

119. Kim, S. W., Gresz, V., Rojek, A., Wang, W., Verkman, A. S., Frokiaer, J., and Nielsen, S., Decreased expression of AQP2 and AQP4 water channels and Na,K-ATPase in kidney collecting duct in AQP3 null mice. *Biol. Cell,* 97, 765–778, 2005.

120. Kinne, R. and Kirsten, R., Effect of aldosterone on activity of mitochondrial and cytoplasmatic enzymes in rat kidney. *Pflügers Arch. Eur. J. Physiol.,* 300, 244–254, 1968.

121. Kinoshita, H., Fujimoto, S., Fukae, H., Yokota, N., Hisanaga, S., Nakazato, M., and Eto, T., Plasma and urine levels of uroguanylin, a new natriuretic peptide, in nephrotic syndrome. *Nephron,* 81, 160–164, 1999.

122. Kinoshita, H., Fujimoto, S., Nakazato, M., Yokota, N., Date, Y., Yamaguchi, H., Hisanaga, S., and Eto, T., Urine and plasma levels of uroguanylin and its molecular forms in renal diseases. *Kidney Int.,* 52, 1028–1034, 1997.

123. Kinoshita, H., Nakazato, M., Yamaguchi, H., Matsukura, S., Fujimoto, S., and Eto, T., Increased plasma guanylin levels in patients with impaired renal function. *Clin. Nephrol.,* 47, 28–32, 1997.

124. Kipp, H. and Arias, I. M., Trafficking of canalicular ABC transporters in hepatocytes. *Annu. Rev. Physiol.,* 64, 595–608, 2002.

125. Kirsten, E., Kirsten, R., Leaf, A., and Sharp, G. W., Increased activity of enzymes of the tricarboxylic acid cycle in response to aldosterone in the toad bladder. *Pflügers Arch. Gesamte Physiol. Menschen Tiere,* 300, 213–225, 1968.

126. Kokko, J. P. and Sands, J. M., Significance of urea transport: the pioneering studies of Bodil Schmidt-Nielsen. *Am. J. Physiol. Renal Physiol.,* 291, F1109–F1112, 2006.

127. Koyama, S., Nishida, K., Terada, N., Shiojima, Y., and Takeuchi, T., Reflex renal vasoconstriction on portal vein distension. *Jpn. J. Physiol.,* 36, 441–450, 1986.

128. Kraft, R. and Harteneck, C., The mammalian melastatin-related transient receptor potential cation channels: an overview. *Pflügers Arch.,* 451, 204–211, 2005.

129. Kraly, F. S., Kim, Y. M., Dunham, L. M., and Tribuzio, R. A., Drinking after intragastric NaCl without increase in systemic plasma osmolality in rats. *Am. J. Physiol.,* 269, R1085–R1092, 1995.

130. Krause, W. J., Freeman, R. H., Eber, S. L., Hamra, F. K., Fok, K. F., Currie, M. G., and Forte, L. R., Distribution of *Escherichia coli* heat-stable enterotoxin/guanylin/uroguanylin receptors in the avian intestinal tract. *Acta Anat. (Basel),* 153, 210–219, 1995.

131. Krause, W. J., Cullingford, G. L., Freeman, R. H., Eber, S. L., Richardson, K. C., Fok, K. F., Currie, M. G., and Forte, L. R., Distribution of heat-stable enterotoxin/guanylin receptors in the intestinal tract of man and other mammals. *J Anat.,* 184(Pt. 2), 407–417, 1994.

132. Krause, W. J., Freeman, R. H., Eber, S. L., Hamra, F. K., Currie, M. G., and Forte, L. R., Guanylyl cyclase receptors and guanylin-like peptides in reptilian intestine. *Gen. Comp Endocrinol.*, 107, 229–239, 1997.

133. Kriz, W., Der architektonische Aufbau der Rattenniere (The architectonic and functional structure of the rat kidney). *Z. Zellforsch. Mikrosk. Anat.,* 82, 495–535, 1967.

134. Kriz, W., Structural organization of the renal medulla: comparative and functional aspects. *Am. J. Physiol.,* 241, R3–R16, 1981.

135. Kriz, W. and Koepsell, H., Structural organization of mouse kidney. *Zeitschrift fur Anatomie und Entwicklungsgeschichte,* 144, 137–163, 1974.

136. Kuhn, M., Raida, M., Adermann, K., Schulz-Knappe, P., Gerzer, R., Heim, J. M., and Forssmann, W. G., The circulating bioactive form of human guanylin is a high molecular weight peptide (10.3 kDa). *FEBS Lett.*, 318, 205–209, 1993.

137. Kwon, T. H., Nielsen, J., Kim, Y. H., Knepper, M. A., Frokiaer, J., and Nielsen, S., Regulation of sodium transporters in the thick ascending limb of rat kidney: response to angiotensin II. *Am. J. Physiol. Renal Physiol.,* 285, F152–F165, 2003.

138. Lafontan, M., Moro, C., Sengenes, C., Galitzky, J., Crampes, F., and Berlan, M., An unsuspected metabolic role for atrial natriuretic peptides: the control of lipolysis, lipid mobilization, and systemic nonesterified fatty acids levels in humans. *Arterioscler. Thromb. Vasc. Biol.,* 25, 2032–2042, 2005.

139. Lande, M. B., Jo, I., Zeidel, M. L., Somers, M., and Harris, Jr., H. W., Phosphorylation of aquaporin-2 does not alter the membrane water permeability of rat papillary water channel-containing vesicles. *J. Biol. Chem.,* 271, 5552–5557, 1996.

140. Langrehr, D. and Kramer, K., Beziehungen der mittleren Impulsfrequenz von Vorhofsrezeptoren zum thorakalen Blutvolumen (Relationships of median impulse frequency of the auricular receptors to the thoracic blood volume). *Pflügers Arch. Gesamte Physiol. Menschen Tiere,* 271, 797–807, 1960.

141. Leaf, A. and Frazier, H. S., Some recent studies on the actions of neurohypophyseal hormones. *Prog. Cardiovasc. Dis.*, 4, 47–64, 1961.

142. Leng, Q., Kahle, K. T., Rinehart, J., MacGregor, G. G., Wilson, F. H., Canessa, C. M., Lifton, R. P., and Hebert, S. C., WNK3, a kinase related to genes mutated in hereditary hypertension with hyperkalaemia, regulates the K^+ channel ROMK1 (Kir1.1). *J. Physiol.,* 571, 275–286, 2006.

143. Lennane, R. J., Carey, R. M., Goodwin, T. J., and Peart, W. S., A comparison of natriuresis after oral and intravenous sodium loading in sodium-depleted man: evidence for a gastrointestinal or portal monitor of sodium intake. *Clin. Sci. Mol. Med.,* 49, 437–440, 1975.

144. Lennane, R. J., Carey, R. M., Goodwin, T. J., and Peart, W. S., A comparison of natriuresis after oral and intravenous sodium loading in sodium-depleted man: evidence for a gastrointestinal or portal monitor of sodium intake. *Clin. Sci. Mol. Med.,* 49, 437–440, 1975.

145. Lennane, R. J., Peart, W. S., Carey, R. M., and Shaw, J., A comparison on natriuresis after oral and intravenous sodium loading in sodium-depleted rabbits: evidence for a gastrointestinal or portal monitor of sodium intake. *Clin. Sci. Mol. Med.,* 49, 433–436, 1975.

146. Lester, C. W. and Costa, D. P., Water conservation in fasting northern elephant seals (*Mirounga angustirostris*). *J. Exp. Biol.,* 209, 4283–4294, 2006.

147. Li, X. C., Navar, L. G., Shao, Y., and Zhuo, J. L., Genetic deletion of AT_{1a} receptors attenuates intracellular accumulation of angiotensin II in the kidney of AT1a receptor-deficient mice. *Am. J. Physiol. Renal Physiol.,* 293, F586–F593, 2007.

148. Liedtke, W., TRPV4 as osmosensor: a transgenic approach. *Pflügers Arch.,* 451, 176–180, 2005.

149. Liedtke, W., TRPV4 plays an evolutionary conserved role in the transduction of osmotic and mechanical stimuli in live animals. *J. Physiol.,* 567, 53–58, 2005.

150. Lima, A. A. M., Monteiro, H. S. A., and Fonteles, M. C., The effects of *Escherichia coli* heat-stable enterotoxin in renal sodium tubular transport. *Pharmacol. Toxicol.,* 70, 163–167, 1992.

151. Loffing, J., Pietri, L., Aregger, F., Bloch-Faure, M., Ziegler, U., Meneton, P., Rossier, B. C., and Kaissling, B., Differential subcellular localization of ENaC subunits in mouse kidney in response to high- and low-Na diets. *Am. J. Physiol. Renal Physiol.,* 279, F252–F258, 2000.

152. Loffing, J., Zecevic, M., Feraille, E., Kaissling, B., Asher, C., Rossier, B. C., Firestone, G. L., Pearce, D., and Verrey, F., Aldosterone induces rapid apical translocation of ENaC in early portion of renal collecting system: possible role of SGK. *Am. J. Physiol. Renal Physiol.,* 280, F675–F682, 2001.

153. London, R. M., Eber, S. L., Visweswariah, S. S., Krause, W. J., and Forte, L. R., Structure and activity of OK-GC: a kidney receptor guanylate cyclase activated by guanylin peptides. *Am. J. Physiol.*, 276, F882–F891, 1999.

154. Lorenz, J. N., Nieman, M., Sabo, J., Sanford, L. P., Hawkins, J. A., Elitsur, N., Gawenis, L. R., Clarke, L. L., and Cohen, M. B., Uroguanylin knockout mice have increased blood pressure and impaired natriuretic response to enteral NaCl load. *J. Clin. Invest.*, 112, 1244–1254, 2003.

155. Luz, C. P., Souza, A., Reis, R., Mineiro, P., Ferreira, H. S., Fregoneze, J. B., and De Castro E Silva, The central amygdala regulates sodium intake in sodium-depleted rats: role of 5-HT$_3$ and 5-HT$_{2C}$ receptors. *Brain Res.*, 1139, 178–194, 2007.

156. Lynch, K. R. and O'Connell, D. P., Molecular, biochemical and functional biology of angiotensinogen. In *Cellular and Molecular Biology of the Renin–Angiotensin System*, Raizada, M. K., Phillips, M. I., and Sumners, C., Eds., CRC Press, Boca Raton, FL, 1993, pp. 131–148.

157. Maack, T., Receptors of atrial natriuretic factor. *Annu. Rev. Physiol.*, 54, 11–27, 1992.

158. Maack, T., Role of atrial natriuretic factor in volume control. *Kidney Int.*, 49, 1732–1737, 1996.

159. Maeda, Y., Han, J. S., Gibson, C. C., and Knepper, M. A., Vasopressin and oxytocin receptors coupled to Ca^{2+} mobilization in rat inner medullary collecting duct. *Am. J. Physiol.*, 265, F15–F25, 1993.

160. Magert, H. J., Reinecke, M., David, I., Raab, H. R., Adermann, K., Zucht, H. D., Hill, O., Hess, R., and Forssmann, W. G., Uroguanylin: gene structure, expression, processing as a peptide hormone, and co-storage with somatostatin in gastrointestinal D-cells. *Regul. Peptides*, 73, 165–176, 1998.

161. Maloiy, G. M. and Edholm, O. G., Thermoregulation, water and electrolyte metabolism of man in the desert. *East Afr. Med. J.*, 52, 97–112, 1975.

162. Malvin, R. L., Ridgway, S. H., and Cornell, L., Renin and aldosterone levels in dolphins and sea lions. *Proc. Soc. Exp. Biol. Med.*, 157, 665–668, 1978.

163. Matsukawa, N., Grzesik, W. J., Takahashi, N., Pandey, K. N., Pang, S., Yamauchi, M., and Smithies, O., The natriuretic peptide clearance receptor locally modulates the physiological effects of the natriuretic peptide system. *Proc. Natl. Acad. Sci. USA*, 96, 7403–7408, 1999.

164. McKinley, M. J., Denton, D. A., and Weisinger, R. S., Sensors for antidiuresis and thirst: osmoreceptors or CSF sodium detectors? *Brain Res.*, 141, 89–103, 1978.

165. McKinley, M. J., McAllen, R. M., Mendelsohn, F. A. O., Allen, A. M., Chai, S. Y., and Oldfield, B. J., Circumventricular organs: neuroendocrine interfaces between the brain and the hemal milieu. *Front. Neuroendocrinol.*, 11, 91–127, 1990.

166. McNiven, M. A., Cao, H., Pitts, K. R., and Yoon, Y., The dynamin family of mechanoenzymes: pinching in new places. *Trends Biochem. Sci.*, 25, 115–120, 2000.

167. Misono, K. S., Ogawa, H., Qiu, Y., and Ogata, C. M., Structural studies of the natriuretic peptide receptor: a novel hormone-induced rotation mechanism for transmembrane signal transduction. *Peptides*, 26, 957–968, 2005.

168. Miyazato, M., Nakazato, M., Yamaguchi, H., Date, Y., Kojima, M., Kangawa, K., Matsuo, H., and Matsukura, S., Cloning and characterization of a cDNA encoding a precursor for human uroguanylin. *Biochem. Biophys. Res. Commun.*, 219, 644–648, 1996.

169. Morris, C. E. and Juranka, P. F., Nav channel mechanosensitivity: activation and inactivation accelerate reversibly with stretch. *Biophys. J.*, 93, 822–833, 2007.

170. Muller, O. G., Parnova, R. G., Centeno, G., Rossier, B. C., Firsov, D., and Horisberger, J. D., Mineralocorticoid effects in the kidney: correlation between alphaENaC, GILZ, and Sgk-1 mRNA expression and urinary excretion of Na$^+$ and K$^+$. *J. Am. Soc. Nephrol.*, 14, 1107–1115, 2003.

171. Nakazato, M., Yamaguchi, H., Kinoshita, H., Kangawa, K., Matsuo, H., Chino, N., and Matsukura, S., Identification of biologically active and inactive human uroguanylins in plasma and urine and their increases in renal insufficiency. *Biochem. Biophys. Res. Commun.*, 220, 586–593, 1996.

172. Nakazato, M., Yamaguchi, H., Shiomi, K. et al., Identification of 10-kDa proguanylin as a major guanylin molecule in human intestine and plasma and its increase in renal insufficiency. *Biochem. Biophys. Res. Commun.*, 205, 1966–1975, 1994.

173. Naray-Fejes-Toth, A. and Fejes-Toth, G., The *sgk*, an aldosterone-induced gene in mineralocorticoid target cells, regulates the epithelial sodium channel. *Kidney Int.*, 57, 1290–1294, 2000.

174. Naray-Fejes-Toth, A., Helms, M. N., Stokes, J. B., and Fejes-Toth, G., Regulation of sodium transport in mammalian collecting duct cells by aldosterone-induced kinase, SGK1: structure/function studies. *Mol. Cell. Endocrinol.*, 217, 197–202, 2004.

175. Nishimoto, G., Zelenina, M., Li, D. et al., Arginine vasopressin stimulates phosphorylation of aquaporin-2 in rat renal tissue. *Am. J. Physiol.*, 276, F254–F259, 1999.

176. Noda, Y., Horikawa, S., Furukawa, T., Hirai, K., Katayama, Y., Asai, T., Kuwahara, M., Katagiri, K., Kinashi, T., Hattori, M., Minato, N., and Sasaki, S., Aquaporin-2 trafficking is regulated by PDZ-domain-containing protein SPA-1. *FEBS Lett.*, 568, 139–145, 2004.

177. Noda, Y., Horikawa, S., Katayama, Y., and Sasaki, S., Water channel aquaporin-2 directly binds to actin. *Biochem. Biophys. Res. Commun.*, 322, 740–745, 2004.

178. Numata, T., Shimizu, T., and Okada, Y., Direct mechano-stress sensitivity of TRPM7 channel. *Cell. Physiol. Biochem.*, 19, 1–8, 2007.

179. Ohnishi, J., Ishido, M., Shibata, T., Inagami, T., Murakami, K., and Miyazaki, H., The rat angiotensin II AT_{1A} receptor couples with three different signal transduction pathways. *Biochem. Biophys. Res. Commun.*, 186, 1094–1101, 1992.

180. Olsson, K., Fluid balance in ruminants: adaptation to external and internal challenges. *Ann. N.Y. Acad. Sci.*, 1040, 156–161, 2005.

181. Ortiz, C. L., Costa, D., and Leboeuf, B. J., Water and energy flux in elephant seal pups fasting under natural conditions. *Physiol. Zool.*, 51, 166–178, 1978.

182. Ortiz, R. M., Osmoregulation in marine mammals. *J. Exp. Biol.*, 204, 1831–1844, 2001.

183. Ortiz, R. M., Adams, S. H., Costa, D. P., and Ortiz, C. L., Plasma vasopressin levels and water conservation in fasting, postweaned northern elephant seal pups (*Mirounga angustirostris*). *Mar. Mammal Sci.*, 12, 99–106, 1996.

184. Ortiz, R. M., Crocker, D. E., Houser, D. S., and Webb, P. M., Angiotensin II and aldosterone increase with fasting in breeding adult male northern elephant seals (*Mirounga angustirostris*). *Physiol. Biochem. Zool.*, 79, 1106–1112, 2006.

185. Ortiz, R. M., Wade, C. E., and Ortiz, C. L., Prolonged fasting increases the response of the renin–angiotensin–aldosterone system, but not vasopressin levels, in postweaned northern elephant seal pups. *Gen. Comp. Endocrinol.*, 119, 217–223, 2000.

186. Ortiz, R. M., Wade, C. E., Ortiz, C. L., and Talamantes, F., Acutely elevated vasopressin increases circulating concentrations of cortisol and aldosterone in fasting northern elephant seal (*Mirounga angustirostris*) pups. *J. Exp. Biol.*, 206, 2795–2802, 2003.

187. Pandey, K. N., Biology of natriuretic peptides and their receptors. *Peptides*, 26, 901–932, 2005.

188. Pandey, K. N., Internalization and trafficking of guanylyl cyclase/natriuretic peptide receptor-A. *Peptides*, 26, 985–1000, 2005.

189. Pearce, D., SGK1 regulation of epithelial sodium transport. *Cell. Physiol. Biochem.*, 13, 13–20, 2003.

190. Peter, K., *Untersuchungen über Bau und Entwicklung der Niere*, Fischer, Jena, Germany, 1909.

191. Peters, J. P., *The Exchange of Fluids in Man*, Charles C Thomas, Springfield, IL, 1935, p. 288.

192. Petersen, M. B., The effect of vasopressin and related compounds at V_{1a} and V_2 receptors in animal models relevant to human disease. *Basic Clin. Pharmacol. Toxicol.*, 99, 96–103, 2006.

193. Pfeiffer, E. W., Comparative anatomical observations of the mammalian renal pelvis and medulla. *J. Anat.*, 102, 321–331, 1968.

194. Pfeiffer, E. W., Nungesser, W. C., Iverson, D. A., and Wallerius, J. F., The renal anatomy of the primitive rodent *Apoldontia rufa* and a consideration of its functional significance. *Anat. Rec.*, 137, 227–236, 1960.

195. Phillips, J. G. and Harvey, S., Vertebrate osmoregulation. *Sci. Prog.*, 67, 335–356, 1981.

196. Potthast, R., Ehler, E., Scheving, L. A., Sindic, A., Schlatter, E., and Kuhn, M., High salt intake increases uroguanylin expression in mouse kidney. *Endocrinology*, 142, 3087–3097, 2001.

197. Rad, A. K., Balment, R. J., and Ashton, N., Rapid natriuretic action of aldosterone in the rat. *J. Appl. Physiol.*, 98, 423–428, 2005.

198. Reudelhuber, T. L., Molecular biology of renin. In *Molecular Nephrology*, Schlöndorff, D. and Bonventre, J. V., Eds., Marcel Dekker, New York, 1995, pp. 71–89.

199. Riedman, M., *The Pinnipeds: Seals, Sea Lions and Walruses*, University of California Press, Berkeley, 1990.

200. Rinehart, J., Kahle, K. T., de Los, H. P. et al., WNK3 kinase is a positive regulator of NKCC2 and NCC, renal cation–Cl⁻ cotransporters required for normal blood pressure homeostasis. *Proc. Natl. Acad. Sci. USA*, 102, 16777–16782, 2005.

201. Ring, A. M., Cheng, S. X., Leng, Q., Kahle, K. T., Rinehart, J., Lalioti, M. D., Volkman, H. M., Wilson, F. H., Hebert, S. C., and Lifton, R. P., WNK4 regulates activity of the epithelial Na^+ channel *in vitro* and *in vivo*. *Proc. Natl. Acad. Sci. USA*, 104, 4020–4024, 2007.

202. Robben, J. H., Knoers, N. V., and Deen, P. M., Cell biological aspects of the vasopressin type-2 receptor and aquaporin 2 water channel in nephrogenic diabetes insipidus. *Am. J. Physiol. Renal Physiol.*, 291, F257–F270, 2006.

203. Roesch, D. M., Blackburn-Munro, R. E., and Verbalis, J. G., Mineralocorticoid treatment attenuates activation of oxytocinergic and vasopressinergic neurons by i.c.v. ANG II. *Am. J. Physiol. Regul. Integr. Comp. Physiol.*, 280, R1853–R1864, 2001.

204. Rogerson, F. M., Yao, Y. Z., Elsass, R. E., Dimopoulos, N., Smith, B. J., and Fuller, P. J., A critical region in the mineralocorticoid receptor for aldosterone binding and activation by cortisol: evidence for a common mechanism governing ligand binding specificity in steroid hormone receptors. *Mol. Endocrinol.*, 21, 817–828, 2007.

205. Romero, D. G., Rilli, S., Plonczynski, M. W., Yanes, L. L., Zhou, M. Y., Gomez-Sanchez, E. P., and Gomez-Sanchez, C. E., Adrenal transcription regulatory genes modulated by angiotensin II and their role in steroidogenesis. *Physiol. Genom.*, 2007.

206. Ronzaud, C., Loffing, J., Bleich, M., Gretz, N., Grone, H. J., Schutz, G., and Berger, S., Impairment of sodium balance in mice deficient in renal principal cell mineralocorticoid receptor. *J. Am. Soc. Nephrol.*, 18, 1679–1687, 2007.

207. Rubera, I., Loffing, J., Palmer, L. G., Frindt, G., Fowler-Jaeger, N., Sauter, D., Carroll, T., McMahon, A., Hummler, E., and Rossier, B. C., Collecting duct-specific gene inactivation of alphaENaC in the mouse kidney does not impair sodium and potassium balance. *J. Clin. Invest.*, 112, 554–565, 2003.

208. Rytand, D. A., The number and size of mammalian glomeruli as related to kidney and to body weight, with methods for their enumeration and measurement. *Am. J. Anat.*, 62, 507–520, 1938.

209. Samueloff, S., Metabolic aspects of desert adaptation (man). *Adv. Metab. Disord.*, 7, 95–138, 1974.

210. Sands, J. M., Molecular mechanisms of urea transport. *J. Membr. Biol.*, 191, 149–163, 2003.

211. Schenk, A. D., Werten, P. J., Scheuring, S., de Groot, B. L., Muller, S. A., Stahlberg, H., Philippsen, A., and Engel, A., The 4.5 Å structure of human AQP2. *J. Mol. Biol.*, 350, 278–289, 2005.

212. Schmidt-Nielsen, B., The renal pelvis. *Kidney Int.*, 31, 621–628, 1987.

213. Schmidt-Nielsen, B. and Pfeiffer, E. W., Urea and urinary concentrating ability in the mountain beaver *Aplodontia rufa. Am. J. Physiol.*, 218, 1370–1375, 1970.

214. Schmidt-Nielsen, K., Osmotic regulation in higher vertebrates. *Harvey Lect.*, 58, 53–93, 1963.

215. Schmidt-Nielsen, K., *Desert Animals: Physiological Problems of Heat and Water*, Clarendon, Oxford, 1964, pp. 1–277.

216. Schmidt-Nielsen, K., The neglected interface: the biology of water as a liquid–gas system. *Q. Rev. Biophys.*, 2, 283–304, 1969.

217. Schmidt-Nielsen, K. and Schmidt-Nielsen, B., Water metabolism of desert mammals 1. *Physiol Rev.*, 32, 135–166, 1952.

218. Schmidt-Nielsen, K., Schmidt-Nielsen, B., Jarnum, S. A., and Houpt, T. R., Body temperature of the camel and its relation to water economy. *Am. J. Physiol.*, 188, 103–112, 1957.

219. Schondube, J. E., Herrera, M., and del Rio, C. M., Diet and the evolution of digestion and renal function in phyllostomid bats. *Zool. Anal. Complex Syst.*, 104, 59–73, 2001.

220. Schulz, S., C-type natriuretic peptide and guanylyl cyclase B receptor. *Peptides*, 26, 1024–1034, 2005.

221. Schwartz, I. L., Shlatz, L. J., Kinne-Saffran, E., and Kinne, R., Target cell polarity and membrane phosphorylation in relation to the mechanism of action of antidiuretic hormone. *Proc. Natl. Acad. Sci. USA*, 71, 2595–2599, 1974.

222. Sever, S., Damke, H., and Schmid, S. L., Dynamin:GTP controls the formation of constricted coated pits, the rate limiting step in clathrin-mediated endocytosis. *J. Cell Biol.*, 150, 1137–1148, 2000.

223. Shibasaki, M., Wilson, T. E., and Crandall, C. G., Neural control and mechanisms of eccrine sweating during heat stress and exercise. *J. Appl. Physiol.*, 100, 1692–1701, 2006.

224. Shkolnik, A., Maltz, E., and Chosniak, I., The role of the ruminant's digestive tract as a reservoir. In *Digestive Physiology and Metabolism in Ruminants*, Ruckebusch, Y. and Thivend, P., Eds., Medical Technical Press, Lancaster, U.K., 1980, pp. 731–742.

225. Silanikove, N., Effects of oral, intraperitoneal and intrajugular rehydrations on water retention, rumen volume, kidney function and thirst satiation in goats. *Comp. Biochem. Physiol. A*, 98, 253–258, 1991.

226. Silanikove, N., Effects of water scarcity and hot environment on appetite and digestion in ruminants: a review. *Livestock Prod. Sci.*, 30, 175–194, 1992.

227. Silanikove, N., The struggle to maintain hydration and osmoregulation in animals experiencing severe dehydration and rapid rehydration: the story of ruminants. *Exp. Physiol.*, 79, 281–300, 1994.

228. Sindic, A. and Schlatter, E., Mechanisms of actions of guanylin peptides in the kidney. *Pflügers Arch.*, 450, 283–291, 2005.

229. Sindic, A., Velic, A., Hirsch, J. R., Kuhn, M., and Schlatter, E., Guanylin peptides regulate K$^+$ conductance via phospholipase A$_2$ and arachidonic acid in principal cells of mouse CCD. *Pflügers Arch.*, 447, S55, 2004.

230. Sindice, A., Basoglu, C., Cerci, A., Hirsch, J. R., Potthast, R., Kuhn, M., Ghanekar, Y., Visweswariah, S. S., and Schlatter, E., Guanylin, uroguanylin, and heat-stable euterotoxin activate guanylate cyclase C and/or a pertussis toxin-sensitive G protein in human proximal tubule cells. *J. Biol. Chem.*, 277, 17758–17764, 2002.

231. Smith, H. W., Salt and water volume receptors: an exercise in physiologic apologetics. *Am. J Med.*, 23, 623–652, 1957.

232. Sperber, I., Studies on the mammalian kidney. *Zool. Bidrag Uppsala*, 22, 249–432, 1942.

233. Star, R. A., Nonoguchi, H., Balaban, R., and Knepper, M. A., Calcium and cyclic adenosine monophosphate as second messengers for vasopressin in the rat inner medullary collecting duct. *J. Clin. Invest.*, 81, 1879–1888, 1988.

234. Stefan, E., Wiesner, B., Baillie, G. S., Mollajew, R., Henn, V., Lorenz, D., Furkert, J., Santamaria, K., Nedvetsky, P., Hundsrucker, C., Beyermann, M., Krause, E., Pohl, P., Gall, I., MacIntyre, A. N., Bachmann, S., Houslay, M. D., Rosenthal, W., and Klussmann, E., Compartmentalization of cAMP-dependent signaling by phosphodiesterase-4D is involved in the regulation of vasopressin-mediated water reabsorption in renal principal cells. *J. Am. Soc. Nephrol.*, 18, 199–212, 2007.

235. Stephenson, J. L. and Mejia, R., Theoretical bounds on the passive concentrating mechanism (abstract). *Proc. Int. Union Physiol. Sci. Sydney*, 419, 1983.

236. Strange, K., Willingham, M. C., Handler, J. S., and Harris, Jr., H. W., Apical membrane endocytosis via coated pits is stimulated by removal of antidiuretic hormone from isolated, perfused rabbit cortical collecting tubule. *J. Membr. Biol.*, 103, 17–28, 1988.

237. Stricker, E. M., Bushey, M. A., Hoffmann, M. L., McGhee, M., Cason, A. M., and Smith, J. C., Inhibition of NaCl appetite when DOCA-treated rats drink saline. *Am. J. Physiol. Regul. Integr. Comp. Physiol.*, 292, R652–R662, 2007.

238. Subramanya, A. R., Yang, C. L., McCormick, J. A., and Ellison, D. H., WNK kinases regulate sodium chloride and potassium transport by the aldosterone-sensitive distal nephron [review]. *Kidney Int.*, 70, 630–634, 2006.

239. Sun, T. X., Van Hoek, A., Huang, Y., Bouley, R., McLaughlin, M., and Brown, D., Aquaporin-2 localization in clathrin-coated pits: inhibition of endocytosis by dominant-negative dynamin. *Am. J. Physiol. Renal Physiol.*, 282, F998–1011, 2002.

240. Taylor, C. R., Strategies of temperature regulation: effect on evaporation in East African ungulates. *Am. J. Physiol.*, 219, 1131–1135, 1970.

241. Tewksbury, D. A., Angiotensinogen: biochemistry and molecular biology. In *Hypertension: Pathophysiology, Diagnosis, and Management*, Laragh, J. H. and Brenner, B. M., Eds., Raven Press, New York, 1990, pp. 1197–1216.

242. Tobin, V. A. and Ludwig, M., The role of the actin cytoskeleton in oxytocin and vasopressin release from rat supraoptic nucleus neurons. *J. Physiol.*, 582(3), 1337–1348, 2007.

243. Tordoff, M. G., Schulkin, J., and Friedman, M. I., Hepatic contribution to satiation of salt appetite in rats. *Am. J. Physiol.*, 251, R1095–R1102, 1986.

244. Tordoff, M. G., Schulkin, J., and Friedman, M. I., Further evidence for hepatic control of salt intake in rats. *Am. J. Physiol.*, 253, R444–R449, 1987.

245. Turban, S., Beutler, K. T., Morris, R. G., Masilamani, S., Fenton, R. A., Knepper, M. A., and Packer, R. K., Long-term regulation of proximal tubule acid–base transporter abundance by angiotensin II. *Kidney Int.*, 70, 660–668, 2006.

246. Valtin, H., Structural and functional heterogeneity of mammalian nephrons. *Am. J. Physiol.*, 233, F491–F501, 1977.

247. van Balkom, B. W., Savelkoul, P. J., Markovich, D., Hofman, E., Nielsen, S., van der, S. P., and Deen, P. M., The role of putative phosphorylation sites in the targeting and shuttling of the aquaporin-2 water channel. *J. Biol. Chem.*, 277, 41473–41479, 2002.

248. van Soest, P. J., *Nutritional Ecology of Ruminants*, O & B Brooks, Corvallis, OR, 1982.

249. Vardy, P. H. and Bryden, M. M., The kidney of *Leptonychotes weddelli* (Pinnipedia, Phocidae) with some observations on the kidneys of two other southern phocid seals. *J. Morphol.*, 167, 13–34, 1981.

250. Verkman, A. S., Roles of aquaporins in kidney revealed by transgenic mice. *Semin. Nephrol.*, 26, 200–208, 2006.

251. Verney, E. B., Croonian Lecture: the antidiuretic hormone and the factors which determine its release. *Proc. R. Soc. Lond. B, Biol. Sci.*, 135, 25–105, 1947.

252. Wade, C. E., Bie, P., Keil, L. C., and Ramsay, D. J., Osmotic control of plasma vasopressin in the dog. *Am. J. Physiol.*, 243, E287–E292, 1982.

253. Wald, H., Garty, H., Palmer, L. G., and Popovtzer, M. M., Differential regulation of ROMK expression in kidney cortex and medulla by aldosterone and potassium. *Am. J. Physiol.*, 275, F239–F245, 1998.

254. Walker, L. A. and Valtin, H., Biological importance of nephron heterogeneity. *Annu. Rev. Physiol.*, 44, 203–219, 1982.

255. Wedel, B. and Garbers, D., The guanylyl cyclase family at Y2K. *Annu. Rev. Physiol.*, 63, 215–233, 2001.

256. Wehling, M., Kasmayr, J., and Theisen, K., Rapid effects of mineralocorticoids on sodium–proton exchanger: genomic or nongenomic pathway? *Am. J. Physiol.*, 260, E719–E726, 1991.

257. Wehner, F., Olsen, H., Tinel, H., Kinne-Saffran, E., and Kinne, R. K. H., Cell volume regulation: osmolytes, osmolyte transport, and signal transduction. *Rev. Physiol. Biochem. Pharmacol.*, 148, 1–80, 2003.

258. Weindl, A., Neuroendocrine aspects of circumventricular organs. In *Frontiers in Neuroendocrinology*, Ganong, W. F. and Martini, L., Eds., Oxford University Press, New York, 1973, pp. 3–33.

259. Weisser, F., Lacy, F. B., Weber, H., and Jamison, R. L., Renal function in chinchilla. *Am. J. Physiol.*, 219, 1706, 1970.

260. Wiegand, R. C., Kato, J., Huang, M. D., Fok, K. F., Kachur, J. F., and Currie, M. G., Human guanylin: cDNA isolation, structure, and activity. *FEBS Lett.*, 311, 150–154, 1992.

261. Wilson, F. H., Kahle, K. T., Sabath, E., Lalioti, M. D., Rapson, A. K., Hoover, R. S., Hebert, S. C., Gamba, G., and Lifton, R. P., Molecular pathogenesis of inherited hypertension with hyperkalemia: the NaCl cotransporter is inhibited by wild-type but not mutant WNK4. *Proc. Natl. Acad. Sci. USA*, 100, 680–684, 2003.

262. Wong, P. C., Hart, S. D., Zaspel, A. M., Chiu, A. T., Ardecky, R. J., Smith, R. D., and Timmermans, P. B., Functional studies of nonpeptide angiotensin II receptor subtype-specific ligands: DuP 753 (AII-1) and PD123177 (AII-2). *J. Pharmacol. Exp. Ther.*, 255, 584–592, 1990.

263. Wright, E. M., Hirayama, B. A., and Loo, D. F., Active sugar transport in health and disease. *J. Intern. Med.*, 261, 32–43, 2007.

264. Xu, L., Dixit, M. P., Nullmeyer, K. D., Xu, H., Kiela, P. R., Lynch, R. M., and Ghishan, F. K., Regulation of Na$^+$/HWNK4 exchanger (NHE3) by angiotensin II in OKP cells. *Biochim. Biophys. Acta*, 1758, 519–526, 2006.

265. Yang, C. L., Angell, J., Mitchell, R., and Ellison, D. H., WNK kinases regulate thiazide-sensitive NaCl cotransport. *J. Clin. Invest.*, 111, 1039–1045, 2003.

266. Yang, C. L., Zhu, X., Wang, Z., Subramanya, A. R., and Ellison, D. H., Mechanisms of WNK1 and WNK4 interaction in the regulation of thiazide-sensitive NaCl cotransport. *J. Clin. Invest.*, 115, 1379–1387, 2005.

267. Yip, K. P., Coupling of vasopressin-induced intracellular Ca^{2+} mobilization and apical exocytosis in perfused rat kidney collecting duct. *J. Physiol.*, 538, 891–899, 2002.

268. Zecevic, M., Heitzmann, D., Camargo, S. M., and Verrey, F., SGK1 increases Na,K-ATP cell-surface expression and function in *Xenopus laevis* oocytes. *Pflügers Arch.*, 448, 29–35, 2004.

269. Zelenina, M., Christensen, B. M., Palmer, J., Nairn, A. C., Nielsen, S., and Aperia, A., Prostaglandin E(2) interaction with AVP: effects on AQP2 phosphorylation and distribution. *Am. J. Physiol. Renal Physiol.*, 278, F388–F394, 2000.

270. Zhuo, J. L., Li, X. C., Garvin, J. L., Navar, L. G., and Carretero, O. A., Intracellular ANG II induces cytosolic Ca^{2+} mobilization by stimulating intracellular AT1 receptors in proximal tubule cells. *Am. J. Physiol. Renal Physiol.*, 290, F1382–F1390, 2006.

Index

A

ABC transporter, 45–46
Acanthamoeba, 75
Acheta domesticus, 261, 264
Acid–base balance
  amphibian urinary bladder, and, 410–411
  gill function
    crustacean, 179
    elasmobranch, 315
    hagfish, 297
  pH effects on osmoregulation, 190
  protozoan cytosol and contractile
    vacuoles, 80
  reptile renal nitrogen metabolism, 464,
    465
  sodium/proton exchange, and, 23
Acidocalcisomes, 71, 74, 87–88
Acid rain, 190
Acipenser
  A. baerii, 335
  A. brevirostrum, 335
  A. naccarii, 335
  A. oxyrhynchus, 335
Actinopterygian fishes, 319, 335–336; *see also*
  Teleost fishes
Active transport, 19–21, 384–385; *see also*
  specific transporter systems
  feasibility analysis, 21–23
  flux ratio analysis, 20–21
  insect Malpighian tubules, 255
  protozoan contractile vacuoles, and, 75
  stoichiometry, and, 22–23
Acylglucosylsterol, 448
Acyrthosiphon pisum, 249
Adrenocorticotrophic hormone (ACTH), 416,
  479, 530
Aedes aegypti, 232, 249, 256–259, 261, 267,
  272
Aedes spp., Malpighian tubules, 249–259
African lungfish, 334–335
Agama stellio, 469
Agnathans, *see* Hagfishes, Lampreys
Aipysurus laevis, 462, 463, 477

Alanine, 111
Albumin, urate binding, 513
Aldose reductase, 45
Aldosterone, 416–417, 485–490, 520, 536,
  539–546
Algal contractile vacuoles, 72–73
Alligator mississippiensis, 464
Alpha (α_1)-adrenergic receptors, 396
Alpha-ketoglutarate, 467
Ambassis interrupta, 336
Amblyrhynchus cristatus, 471, 476
Ambystoma
  A. mexicanum, 386, 413, 414
  A. opacum, 413
  A. punctatum, 398
  A. subsalsum, 375
  A. tigrinum, 373, 375, 377
Amino acids, 39, 76, 98, 111
  crustacean osmoregulation, 167
  mollusc cell volume regulation, 111–115
  uptake from seawater, 114, 148
  ventricular, 113
Ammonia/ammonium, 463
  amphibian excretion, 411
  avian excretion, 512
  crustacean excretion, 176
  crustacean gill function, and, 187
  lungfish excretion, 335
  mollusc uptake and storage, 110
  mollusc volume regulation, and, 115
  reptile kidney function, 463–465
Amoeba proteus, 72, 73, 75, 84
Amphibian, 368–420
  body fluid composition, 368–375
    breeding, and, 373
    external salinity effects, 373–375
    extracellular concentrations, 370
    feeding behavior, and, 371–372
    hibernation, and, 372–373, 385
    intracellular concentrations, 371
    temperature effects, 372
  burrowing behaviors, 374, 377–378
  climate change sensitivity, 419–420
  fossil record, 368

Printed and bound by CPI Group (UK) Ltd, Croydon, CR0 4YY

23/10/2024

01778251-0010